D0138625

ENCYCLOPEDIA OF MICROBIOLOGY

ANNE MACZULAK, PH.D.

FOREWORD BY

ROBERT H. RUSKIN, PH.D.

ENCYCLOPEDIA OF MICROBIOLOGY

Facts On File, Inc.
An imprint of Infobase Learning
132 West 31st Street
New York NY 10001

Library of Congress Cataloging-in-Publication Data
Maczulak, Anne E. (Anne Elizabeth), 1954–
Encyclopedia of microbiology / author, Anne Maczulak; foreword, Robert H. Ruskin.
p. cm.
Includes bibliographical references and index.
ISBN 978-0-8160-7364-1 (alk. paper) 1. Microbiology—Encyclopedias. I. Title.
[DNLM: 1. Microbiology—Encyclopedias—English. QW 13 M177e2011]
QR9.M33 2011
579.03—dc22 2010004551

Facts On File books are available at special discount when purchased in bulk quantities for businesses, associations, institutions, or sales promotions. Please call our Special Sales Department in New York at (212) 967-8800 or (800) 322-8755.

You can find Facts On File on the World Wide Web at http://www.infobaselearning.com

Text design by Cathy Rincon
Composition by A Good Thing, Inc.
Photo research by Elizabeth H. Oakes
Cover printed by Sheridan Books, Inc., Ann Arbor, Mich.
Book printed and bound by Sheridan Books, Inc., Ann Arbor, Mich.
Date printed: April 2011
Printed in the United States of America

10 9 8 7 6 5 4 3 2 1

This book is printed on acid-free paper.

CONTENTS

Appendixes:

FOREWORD

Recently, we have been besieged every day, via every available medium—television, newspapers, radio, the Internet, Twitter, and so on—with reports about the swine flu and H1N1 virus. People are asking themselves questions such as What exactly is a global pandemic? Why are people dying? What can I do to protect myself and my family from this virus? What is a vaccine? How are vaccines made? What is the difference between swine flu and annual flu?

There is a lot of misinformation concerning the current pandemic. What *is* the difference between a bacterial infection and a viral infection? Among young adults in both high school and junior college there seems to be genuine fascination about the general subject of microbiology and the concept of disease. To be sure, much of this interest is due to the current outbreak of swine flu. For some, fear is driving their interest; for others, they have been assigned a paper on some topic pertaining to microbiology; and for still others, they are asking themselves about what types of careers are available in microbiology.

Most people would be astounded to learn of the diverse fields in science supported by microbiology. A partial list would include environmental science, marine science, food science, manufacturing, mining, and, of course, both medical and public health sciences, to name a few. Microbiologists are hired both by public (state and federal institutions) and private companies. The degree and type of education required to work in these fields vary from a high school diploma (with the supervision of a microbiologist) to a Ph.D. In general, however, a four-year university degree in chemistry, biology, or the like, includes at least one full year of microbiology course work.

Encyclopedia of Microbiology is an excellent reference by Dr. Anne Maczulak, whose doctorate is in microbiology and animal nutrition and who is the author of numerous books and professional papers on environmental microbiology and environmental science, including the enjoyable *The Five Second Rule and Other Myths about Germs*. In *Encyclopedia of Microbiology,* Dr. Maczulak addresses many of the questions students considering careers in this field might have. The encyclopedia is also a wonderful resource for the general public who may simply be interested in the world of microorganisms. Arranged as a collection of literate entries, the encyclopedia is enhanced by 13 essays on topics relevant to today's microbiology—for example, global warming and emerging infectious diseases, antibiotics in our meat supply, the microbial hazards of air travel, and bioengineered microorganisms in the environment. The more than 200 entries have also been selected to cover the most recent advances and focus areas in microbiology. Some examples are those on gene therapy, nanobiology, and bioremediation. The encyclopedia includes tables, charts, diagrams, and photos that both highlight and facilitate the understanding of the topics.

The author covers the expansive area of microbiological science well: She provides readers interested in microbiology with a single source that is easy to understand, accessible, and very well written. The major topics in microbiology are organized in alphabetical order, and each entry includes cross-references to related topics and essays, in addition to resources for further reading on the subject.

For those seriously considering microbiology as a career, opportunities exist in teaching, research, and industry, or in public service as a microbiologist at the U.S. Centers for Disease Control or the U.S. Public Health Service. The American Society for Microbiology additionally describes the many subspecialty areas focused on bacteria, fungi, protozoa, algae, and viruses. These are but a few of the professional organizations that deal with microbiology, a science that expands almost daily with our changing environment. *Encyclopedia of Microbiology* is an excellent start to an exploration of this intriguing field of study.

—Robert H. Ruskin, Ph.D.
Director of Laboratory Research, Retired
Water Resources Research Institute
University of the Virgin Islands, St. Thomas

ACKNOWLEDGMENTS

This encyclopedia could not have been written without the guidance I received from my colleagues in microbiology throughout my career. Some of these scientists graciously contributed essays that discuss current issues and problems in microbiology. My thanks go to the following essayists:

- Richard E. Danielson, Ph.D., BioVir Laboratories, Benicia, California
- Carlos Enriquez, Ph.D., Chabot College, Hayward, California
- Wanda C. Manhanke, M.S., St. Louis Children's Hospital, St. Louis
- Kelly A. Reynolds, Ph.D., College of Public Health, University of Arizona, Tucson
- Nokhbeh M. Reza, Centre for Research on Environmental Microbiology, University of Ottawa
- Syed A. Sattar, Ph.D., Centre for Research on Environmental Microbiology, University of Ottawa
- Susan Springthorpe, Centre for Research on Environmental Microbiology, University of Ottawa
- Philip M. Tierno, Jr., Ph.D., Clinical Microbiology and Immunology, New York University

Special thanks are due to Philip M. Tierno, who contributed to the discussions on hygiene and germ transmission. I owe Robert H. Ruskin, Ph.D., a great deal of gratitude for outlining numerous entry topics, proofreading, fact checking, and offering insight on marine microorganisms. I could not have completed this project without his help. I also thank Dana Gonzalez, Ph.D., for input on infectious agents and disinfection. My gratitude also goes to the literary agent Jodie Rhodes, the photo researcher Elizabeth Oakes, and, especially, the executive editor Frank K. Darmstadt, for his encouragement, timely news stories, and belief in this encyclopedia as a valuable resource for students of biology and microbiology.

INTRODUCTION

The tiniest organisms on Earth wield tremendous power over all biota. Plants and animals, arthropods and marine invertebrates, and creatures in the soil and in the oceans all depend on a diverse community of microorganisms. These microscopic beings live on the body and on every surface in our environment. But the degree to which microorganisms affect humanity is staggering even though the microorganisms may live in remote and forbidding habitats or in places halfway around the world. A person cannot make it through a single day without the need for a product or food with a tie to microbial growth. Every one of Earth's functions relates in some way to microbial actions.

In their most obvious roles, microorganisms decompose waste to prevent its overtaking the environment, produce antibiotics, turn milk and vegetables into foods that last longer, and help food digestion in the intestines. But these aspects also reveal the dichotomy of microorganisms. Microorganisms harm the environment on occasion, cause disease and dental caries, and spoil foods. Humans have always struggled with how best to balance the benefits bacteria offer with the threats that bacteria produce. Much less obvious than microorganisms' direct effects on plants and animals are the indirect effects by which they shape the planet. In fact, these hidden activities have barely been explained in science. Scientists nevertheless realize that microbial activities maintain Earth in a condition that supports all life.

Bacteria have the power to cycle essential nutrients such as carbon, nitrogen, and sulfur through plant and animal life and into the atmosphere, and then return these elements to the oceans and continents. Bacteria and fungi together contribute to ways the planet changes each day under our feet by leaching minerals from rock, corroding inorganic matter, and adding or deleting nutrients in water. Microorganisms contribute to the cycling of sediments from Earth's crust to deep in the mantle. These microscopic cells also contribute to the formation of the fossil fuels we use up at an alarming rate today. Bacterial metabolism plays both beneficial and harmful roles in climate change. Microbial metabolic pathways influence weather and even play a part in cloud formation, drought, and warming of the oceans.

The more than 200 entries in *Encyclopedia of Microbiology* present the myriad ways in which microorganisms influence the biosphere. A global theme throughout the encyclopedia begins to become apparent: All microorganisms relate to each other just as all higher organisms relate to all other animate and inanimate things on Earth.

Many of the encyclopedia's entries include biographical sections on scientists who most influenced developments or discoveries in microbiology. Some of these luminaries, such as Louis Pasteur, have been examined in history for more than a century. Other scientists who made critical contributions to microbiology may have faded from the spotlight, but this encyclopedia strives to describe the work accomplished by these equally important figures.

Encyclopedia of Microbiology's second major theme relates to the incredible diversity of microorganisms. Microbiology covers areas such as mycology, the study of fungi, which includes organisms that can grow to an area of hundreds of square acres in soil. The science also ranges to the simplest biological entities of all: viruses. Viruses have evolved to such a simple structure that they can no longer live on their own. The encyclopedia includes a discussion of infective agents that are even more streamlined than viruses, but just as dangerous. These entities, called prions, contain little more than a protein, yet they infect as other pathogens do.

I have also strived to include the most recent technologies relative to microbiology in this book. Entries cover new techniques in microscopy, genetic engineering, gene therapy, and nanotechnology. To cover topical areas that prompt active discussions among scientists, I have included 13 essays. Six of these essays have been contributed by microbiologists currently active and respected in industry and academia. The essays are as follows:

- Antibiotics and Meat
- Where Are Germs Found?
- Realties of Bioterrorism
- Does Immigration Lead to Increased Incidence of Disease?
- Sanitation in Restaurants
- Why AIDS Is Not Going Away
- Will Global Warming Influence Emerging Infectious Diseases?
- Microbes Meeting the Need for New Energy Sources
- The Day Care Dilemma
- Bioengineered Microbes in the Environment
- Do Disinfectants Cause Antibiotic Resistance?
- How Safe Is Air Travel?
- Does Vaccination Improve or Endanger Our Health?

The encyclopedia's appendixes provide resources that students of microbiology will find helpful. These appendixes provide a chronology of milestones in microbiology, the hierarchy of living things in evolution, classification of bacterial and archaeal genera, viruses of animals and plants, and major human diseases caused by microorganisms. Finally, *Encyclopedia of Microbiology* includes a resources section with recommended print resources and Web sites. Each entry also gives a list of pertinent resources for further reading on the topic.

Microbiology has always had controversies. Perhaps the first major conflict concerning microorganisms occurred when scientists argued the theory of spontaneous generation. Microbiology has never lacked controversy since then. The encyclopedia covers these topics either as separate entries or within related subject discussions. Examples of the topics in microbiology that continue to engender debate are bioengineering, gene therapy, vaccination, bioweapons, and nanobiology.

Microbiology has certainly advanced beyond the simple inspection of tiny specks in a microscope. This science now uses very sophisticated technologies. Still, the basic techniques of microbiology relate to the difficult tasks of working with invisible organisms and preventing contamination. *Encyclopedia of Microbiology* describes the important techniques that all microbiologists must master to be successful in their field. Entries cover aseptic techniques, disinfection, sterilization, growth media, staining, microscopy, immunoassays, and recombinant deoxyribonucleic acid (DNA) technology. Other entries include shorter sections on methodology.

Encyclopedia of Microbiology is intended as a valuable reference on the history of microbiology to the technologies on microbiology's horizon. Every topic shows the remarkable influence of microorganisms on this planet and perhaps beyond this planet. This encyclopedia opens the diverse and broad world of microbiology to students and identifies the multitude of specialties within the field that impact health, industry, and the environment.

ENCYCLOPEDIA OF MICROBIOLOGY

aerobe An aerobe is a microorganism that grows in the presence of the gas oxygen. By contrast, an anaerobe requires the absence of oxygen. Most of the aerobes studied in microbiology are bacteria and fungi that grow in the environment and in laboratories in the presence of air. These species are sometimes referred to as true aerobes.

Aerobic microorganisms encompass two specialized subgroups that have specific oxygen needs. The first group contains obligate aerobes, which have an absolute requirement for oxygen in their environment and cannot live without it. The second group, the *microaerophiles,* require minute amounts of oxygen in the environment. At higher oxygen levels, microaerophiles die, so microbiologists must perform special culture techniques to grow microaerophilic species in laboratories.

Aerotolerant and microaerotolerant microorganisms can grow with or without the presence of oxygen at limited levels, but they are usually classified with anaerobes because they often grow best in the absence of oxygen. Aerotolerant microorganisms are indifferent to the presence of oxygen because they use only anaerobic reactions in their metabolism. A microaerotolerant microorganism can survive only if oxygen levels are 5 percent or less. (Air is about 21 percent oxygen.)

True aerobes thrive in the oxygen levels of the troposphere, Earth's lowest layer of the atmosphere. The air's oxygen is in a chemical form called *dioxygen* (O_2) made up of two oxygen molecules connected by a double chemical bond between them; the structure can be written O=O. (Chemists refer to the O_2 molecule as *singlet oxygen.* Molecules in which the outermost pair of electrons rotates in opposite directions are called *singlet* molecules. Oxygen exists also in a more stable triplet form in which the electron pair rotates in the same direction; it is symbolized by the abbreviation $^3O_2$.)

The oxygen that makes up 21 percent of Earth's troposphere shifts between the singlet and triplet forms. Oxygen also accounts for 49 percent of the mass of Earth's crust and almost 90 percent by mass of the planet's oceans. The aerobic bacteria are believed to have evolved after anaerobic bacteria because the early Earth, when organized bacterial cells developed (about 3.8 billion years ago), contained no free oxygen.

The earliest forms of life used simple molecules present in the atmosphere: methane, ammonia, and hydrogen. The earliest atmosphere lacked oxygen; any oxygen occurred only in water molecules, not as free molecules. Energy in the form of a lightning strike, meteor crash, or a similarly violent event powered chemical reactions that combined these simple molecules into more complex molecules, such as amino acids and sugars. Eventually compounds combined to form carbohydrates, lipids, and nucleic acids. Primitive prokaryotes consumed these molecules for energy and as a source of building blocks for making new cells, and all these reactions occurred without free oxygen in the atmosphere. In certain habitats, small building block and energy compounds may have been in short supply, favoring a small number of mutated cells that thrived in the bleak conditions. These mutations allowed the cells to take advantage of the Sun as an energy source and photosynthesis evolved. The earliest photosynthetic prokaryotes probably did not absorb oxygen for their reactions but instead used the inorganic molecules that were more abundant.

Between 3.5 and 3 billion years ago, cells developed the ability to incorporate oxygen into energy-generating pathways. Under the sparse oxygen conditions, an evolutionary advantage bestowed on certain cells the chance to outcompete the anaerobic life around them. These early photosynthetic bacteria, the cyanobacteria, developed and began excreting oxygen as a waste product of their photosynthetic reactions. Over billions of years, the oxygen level in the atmosphere rose, and with it kingdoms of oxygen-metabolizing organisms began to evolve and populate Earth.

In the natural environment, many aerobic bacteria move in the direction of a specific oxygen concentration until they find oxygen levels favorable to them. This movement in response to oxygen levels is called *aerotaxis.* Aerotaxis allows various bacteria to migrate in either of two directions, toward higher oxygen availability or toward lower oxygen availability, depending on the microorganism's needs and the gas conditions around it. True aerobes migrate toward atmospheric levels of oxygen; that is, they prefer being exposed to the air. Microaerotolerant bacteria and microaerophiles seek places where oxygen occurs at levels lower than in air.

AEROBIC HABITATS

Aerobes live in places with access to the air, so they accumulate on surfaces. Aerobes populate the skin, hair, and fur on humans and animals and live on the stalk and leaf surfaces of plants. In soil and in water, aerobes prefer conditions in which their habitat is aerated. Loose topsoil contains high concentrations of aerobes compared with deeper compacted soil containing few pores to allow oxygen to penetrate. Agitated waters such as streams, fast flowing rivers, the ocean surface, and the upper layer of lakes also hold diverse aerobic populations. Still and/or deep waters can contain less aerobes relative to the concentration of anaerobes.

Industrial microbiology uses aerobic microorganisms for manufacturing microbial products because aerobes grow faster than anaerobes and do not require extra effort to maintain oxygen-free conditions. Microbiologists help aerobes grow in industrial bioreactors by agitating aerobic cultures to disperse more oxygen throughout the liquid medium. To do this, bioreactors contain a shaft equipped with paddles (an impeller) or bubblers, which constantly mix the culture liquid. Aerobic bacteria have been used in this way to produce vitamins, industrial enzymes, acids, thickening agents, and foods.

Wastewater treatment also depends on the activity of aerobic bacteria. In one step in wastewater treatment, aerobic bacteria inside a large tank digest organic matter in the water. This aeration tank contains a device that bubbles air through the contents to mix the suspension and to help give the bacteria access to oxygen. Without constant aeration, the aerobic activity would end and the wastewater's solid organic matter would sink and stall the entire treatment process. Laboratory microbiologists similarly mix aerobic cultures in small test tubes as they grow. This is done

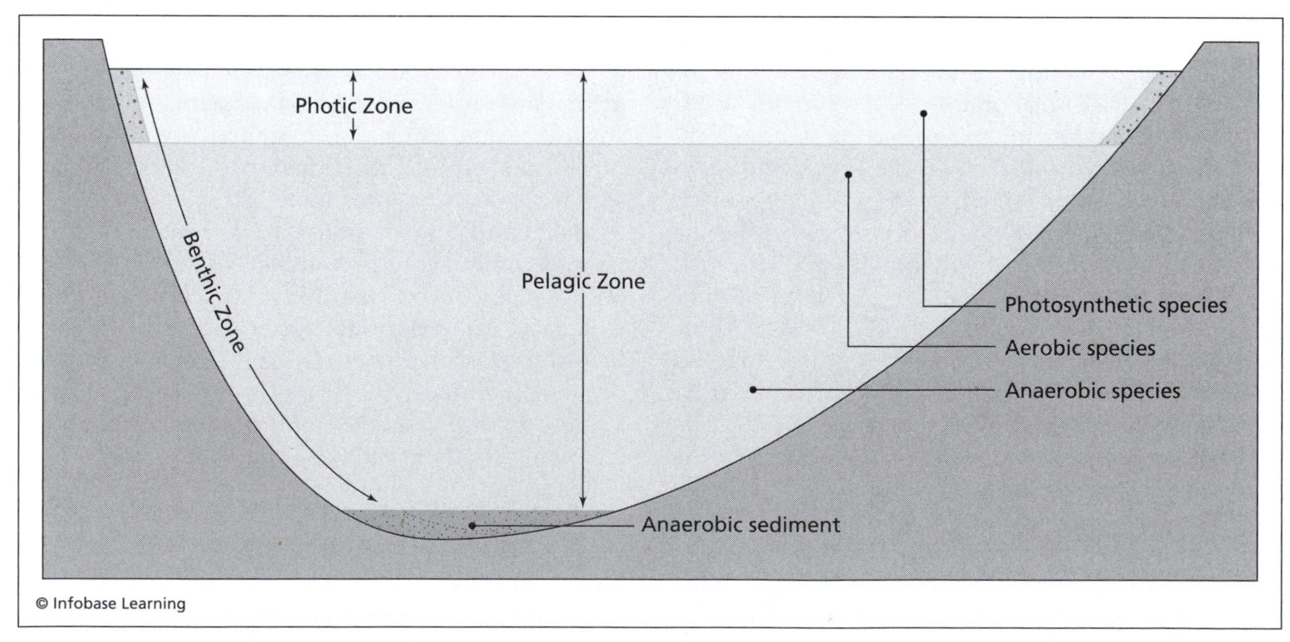

A freshwater lake ecosystem contains a diverse population of microorganisms. Aerobic microorganisms live exclusively in the upper oxygenated layers, in addition to photosynthetic species. Anaerobes occupy deeper oxygen-depleted layers that have high levels of hydrogen sulfide and methane gases.

by putting the capped tubes containing aerobic broth culture on a mechanical rocker that gently mixes the contents by rocking the tubes back and forth.

Aerobic microorganisms do not use Earth's atmosphere as a habitat, and they do not reproduce in air, but bacteria, molds, and protozoa can travel in the air. These so-called airborne microorganisms can move great distances, and aerobes are more suited to this type of transmission than anaerobes. Air transmission of microorganisms becomes especially important if one or more of the microorganisms cause disease, that is, if they are pathogens. Some pathogens travel through the air in aerosols (tiny moisture droplets) that cover distances from several feet to a mile in breezes and in prevailing winds. Mold spores, for example, use airborne transmission as the main way of dispersing and starting new generations.

Survival of aerobes in the air depends on two factors in addition to oxygen. First, humidity affects the length of time aerobes survive in the air. Prolonged exposure to low humidity dries out cells and makes them vulnerable to damage in a process called *desiccation,* which simply means the drying out of matter. The air contains a second group of factors called *open-air factors,* which influence cell survival. Open air factors (OAFs)—an area of study also called *open-air chemistry*—affect the period of time in which microbial cells can survive in the air. Oxygen reacts in the atmosphere to form ozone, a compound containing three oxygen molecules. Ozone in turn reacts with carbon-based compounds in the air that can be damaging to microorganisms. Microbial damage is caused specifically by ions, which are charged—positive or negative—chemicals that result from ozone reactions in the air. OAFs such as ions produced in these reactions damage deoxyribonucleic acid (DNA), proteins, enzymes, and membranes. Since their discovery in the 1970s, OAFs have remained the focus of many researchers investigating ways disease is transmitted and the methods in which bioweapons could potentially be spread.

Many OAFs are probably yet to be discovered, but, in general, scientists know that OAFs are chemical air pollutants that react very readily with other chemicals, often ozone. This observation suggests that oxygen actually participates in reactions that might kill aerobes. Yet how could aerobes be harmed by a molecule that they require for growth? Aerobes illustrate one of the interesting curiosities in biology: Oxygen is a poison to almost all living cells. Yet higher organisms, invertebrates, and aerobes thrive in the presence of oxygen and even make oxygen part of their metabolism. They do this by having evolved a system specially designed to protect the cell contents from oxygen poisoning, thus allowing oxygen to serve as a key component of cellular energy production.

OXYGEN AND MICROBIAL GROWTH

Earth's primitive life depended on metabolism geared to the chemistry found on the land and in the atmosphere. Oxygen has a proclivity to create chemically unstable molecules that would have destroyed simple metabolic pathways billions of years ago. Oxygen participates in a type of reaction called *oxidation* in which one molecule loses electrons to another molecule. The molecule that accepts extra electrons is said to be reduced, and the entire process is reduction-oxidation, or a redox reaction. A molecule that is becoming oxidized (giving up electrons) must transfer those electrons to an intermediary in the redox process. Oxygen plays the role of an electron acceptor in the metabolism of many modern organisms. All of this would work flawlessly in microbial cells and in other organisms' cells were it not for other chemicals produced during oxidation. The microbiologist Moselio Schaechter explained in his 2006 book *Microbe,* "The relationship of living organisms to oxygen is complex, since oxygen and its metabolic derivatives are extraordinarily toxic to cells." In order to evolve in the presence of oxygen, bacteria would be required to develop one or more systems for controlling oxygen toxicity.

Oxygen's danger in living cells arises because the molecule readily accepts free electrons in a step that forms an unstable molecule called a *free radical*:

$$O_2 + 1 \text{ electron} \rightarrow O_2\cdot$$

The free radical shown in the equation ($O_2\cdot$) is a negatively charged superoxide radical. Inside the cell, superoxide radicals combine with additional electrons to form hydrogen peroxide (H_2O_2), which leads to the formation of hydroxyl radicals ($OH\cdot$). Superoxide, hydroxyl radicals, and hydrogen peroxide are all strong oxidizing agents and very reactive chemicals inside cells. They lead to the formation of additional free radicals (signified by the black dot beside the chemical formula). Each of these compounds damages cell constituents such as proteins, enzymes, fats, and membranes.

Aerobic microorganisms protect themselves from toxic and reactive oxygen radicals with the following three different enzyme systems:

1. Superoxide dismutase neutralizes the superoxide radical by the following reaction:

$$2\, O_2^{-}\cdot + 2\, H^+ \rightarrow O_2 + H_2O_2$$

2. Catalase destroys hydrogen peroxide to make water and oxygen gas:

$$2\, H_2O_2 \rightarrow 2\, H_2O + O_2$$

How Different Microorganisms Use Oxygen

Category	Growth in Air	Growth without O_2	Contains O_2-Destroying Enzymes	Energy Metabolism	Summary
aerobe	yes	no	yes	aerobic respiration	requires O_2; cannot ferment
microaerophile	slight	yes	small amounts	aerobic or anaerobic respiration	requires low O_2 levels
strict anaerobe	no	yes	no	anaerobic respiration	killed by O_2; ferments
facultative anaerobe	yes	yes	yes	aerobic or anaerobic respiration, or fermentation	alters metabolism depending on O_2 levels
aerotolerant	yes	yes	yes	fermentation	unaffected by presence or absence of O_2

3. Peroxidase also destroys hydrogen peroxide but without forming O_2:

$$H_2O_2 + 2\,H^+ \rightarrow 2\,H_2O$$

Most aerobes have at least one of these three enzymes to destroy oxygen inside the cell. This makes the normal oxygen levels inside microbial cells very low. Anaerobes do not contain superoxide dismutase, catalase, or peroxidase, explaining why these species cannot live long in the presence of air. The table above summarizes the mechanisms of true aerobes and other types of microorganisms in dealing with oxygen in their environment.

Oxygen's use as an electron acceptor puts it at the core of energy metabolism in aerobic cells. As evolution progressed, species with aerobic metabolism came to dominate species dependent on anaerobic conditions, mainly because of the efficiency of aerobic energy production compared with energy production in anaerobes.

AEROBIC METABOLISM

The millions of species of aerobic microorganisms living on Earth have cyanobacteria to thank for their existence. The early Precambrian atmosphere contained no oxygen, so photosynthesis involved other compounds to accept electrons in energy-producing steps. As a consequence, methane, sulfides, and other compounds were released into the air. But cyanobacteria made a critical shift in this process: They produced oxygen as an end product of their energy metabolism. For million of years, most of the gas dissolved in the oceans until the waters became oxygen saturated. Some of the excess oxygen then bonded with inorganic molecules. The red iron oxides that can be found today in iron ore give evidence of one of these ancient oxidation reactions.

Oxygen at first entered the atmosphere slowly. About 2.2 million years ago, Earth's biological activity reached a critical point in which oxygen release took place faster. Anaerobic species retreated to habitats that remained mostly oxygen free: deep sediments and stagnant waters. Aboveground, meanwhile, aerobic species blossomed. Various theories have tried to explain why the atmospheric oxygen levels began their rapid increase two million years ago. It is thought that the rise of eukaryotic cells containing chloroplasts—the earliest plants—accelerated this increase in the atmosphere's oxygen content.

Aerobic respiration takes place in the cell membrane of aerobic bacteria and in the mitochondria of eukaryotes. In either case respiration uses oxygen to receive electrons from a series of energy-producing reactions. Oxygen is said to be the final electron acceptor in respiration or an electron sink. (Anaerobic respiration is similar, except molecules other than oxygen, such as carbon dioxide, act as electron acceptors.)

Aerobic respiration consists of two components. The first is the Krebs cycle, in which a circular series of steps releases energy in the form of a compound called *acetyl coenzyme A,* abbreviated acetyl CoA. When an aerobic cell oxidizes glucose, the cell produces two molecules of acetyl CoA, and each enters the Krebs cycle. Two cycles run for each one molecule of glucose taken in by an aerobic cell. The cycle then links to the second component of aerobic respiration, a chain of reactions that transfer electrons

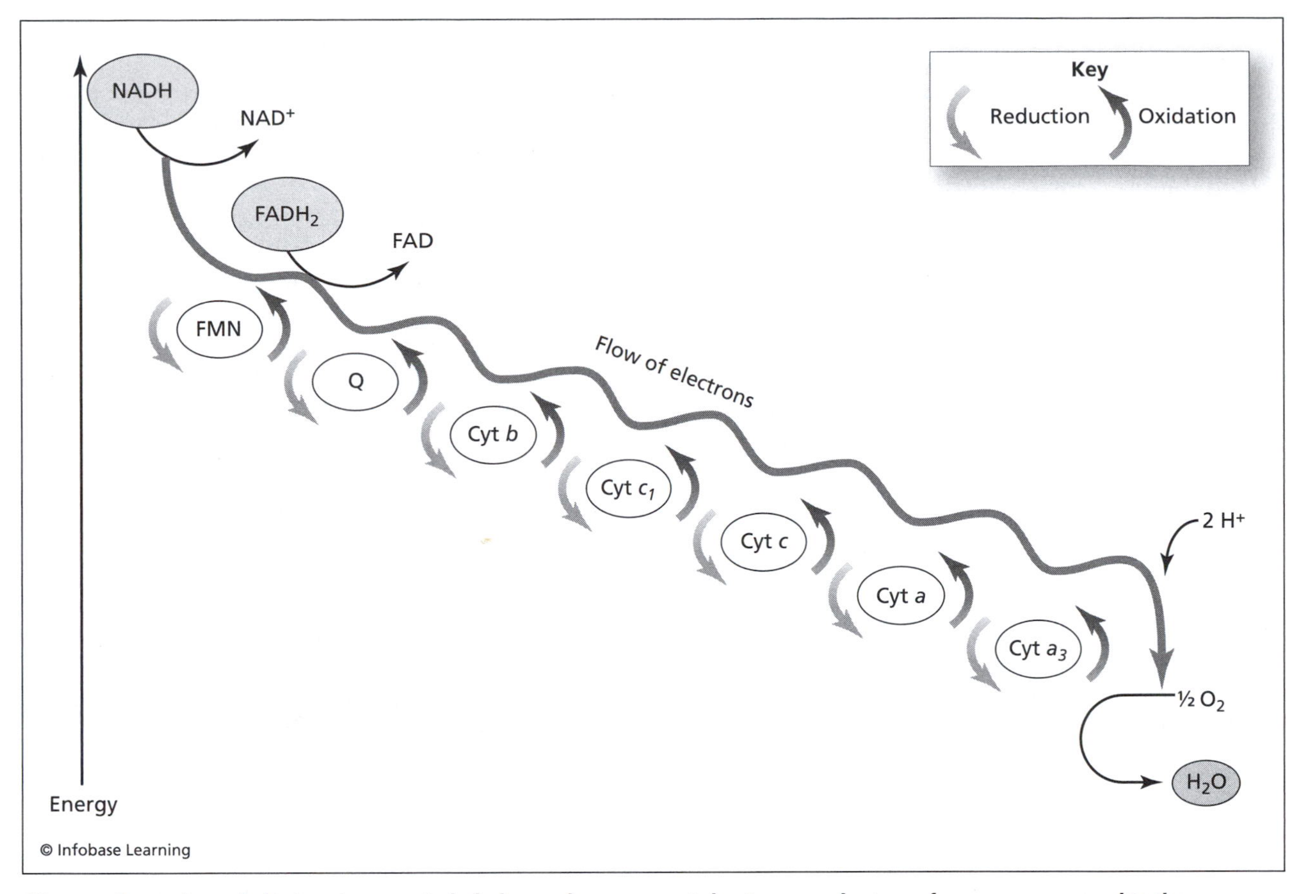

The membrane-bound electron transport chain in aerobes uses proteins to carry electrons from one compound to the next, with each of the redox reactions releasing energy to make ATP. NADH and NAD+ are forms of nicotinamide adenine dinucleotide; FADH2 and FMN are flavin proteins; Q is the protein ubiquinone; and "cyt" refers to cytochrome proteins. The chain's final products are water and the energy stored in ATP.

from one compound to the next until the electrons reach an oxygen molecule. This series of steps is called an *electron transfer system* or *electron transport chain*. Large compounds called flavoproteins, cytochromes, and coenzyme Q (a quinone) transfer the electrons from the beginning of the transport chain to the end oxygen molecule.

Cells store energy made in electron transport in the form of adenosine triphosphate (ATP). Aerobes use a process called *oxidative phosphorylation* to transfer the energy produced by the electron transport chain and store it in ATP's phosphate bonds. This entire aerobic respiration has an important advantage over anaerobic metabolism. It produces

Energy Production as ATP in Aerobic Respiration

Metabolic Step	Method of Making ATP	ATP Yield
glycolysis	glucose oxidized to pyruvic acid	2
	2 electron carriers made per glucose	6 (from electron transport chain)
formation of acetyl CoA	2 electron carriers made	6 (from electron transport chain)
Krebs cycle	completes the degradation of glucose to carbon dioxide and by-products	2
electron transport chain	8 electron carriers produced from the Krebs cycle	22
Total ATP Produced		**38**

38 ATP molecules from each glucose molecule used up in respiration, illustrated in the table on page 5. Anaerobic fermentation, by contrast, produces only 2 ATP molecules per glucose.

Eukaryotic aerobes produce only 36 total ATPs in respiration compared with the prokaryotes' 38 ATPs. Eukaryotes must shuttle electrons across their mitochondrial membranes during respiration, and this process takes energy. (Prokaryotes are not burdened by this membrane shuttle because they do not have membrane-bound organelles such as mitochondria.) Eukaryotes spend two ATPs to power the shuttle system. Once a eukaryotic cell transfers electrons from the cytoplasm to inside the mitochondria, the cell then begins its electron transport chain.

Aerobes are not more critical or less critical to life on Earth than anaerobes. Both have been part of the evolution of multicellular organisms, and both contribute to Earth's present ecosystems. In most microbiology laboratories, however, aerobic species dominate studies on the inner workings of microorganisms because of the ease with which aerobes can be grown.

In 2005, the Massachusetts Institute of Technology microbiologist Edward DeLong explained to the *Los Angeles Times* the importance of microbial life on Earth: "Microbes are the master chemists of the planet. Even though they are small, they are the engines that drive the conversion of matter and energy that produce our atmosphere and influence our climate." Humans and other life could not exist without the combined activities of aerobic microorganisms.

MICROAEROPHILES

Microaerophiles cannot live in atmospheric oxygen levels but do require small amounts of oxygen in their environment. For this reason, microaerophiles are said to live in microaerobic environments, defined by oxygen levels lower than 10 percent.

The bacteria *Campylobacter jejuni* and *Magnetospirillum magnetotacticum* (also known as *Aquaspirillum magnetotacticum*) offer two examples of microaerophiles that live distinctive lifestyles. *C. jejuni* is a food-borne pathogen that causes gastrointestinal illness; contaminated meats and dairy products are its most likely source—it prefers an atmosphere of 5 percent oxygen and 10 percent carbon dioxide (CO_2). For this reason the food industry avoids packaging meats at these levels of oxygen and carbon dioxide. *M. magnetotacticum* possesses the unique ability to use both aerotaxis and magnetotaxis. In other words, this species moves toward favorable oxygen levels but it also moves in response to Earth's magnetic poles. *M. magnetotacticum* contains tiny magnetosomes, similar to magnets, inside each cell. In aqueous environments the magnetosomes make the cells travel downward until they reach mud or sediment. As the cells burrow into the bottom of ponds or lakes, they find suitable environments low in oxygen. Microaerophiles such as these two microorganisms express in their own ways the special capabilities and the strict requirements of microaerophilic life.

True aerobes do not demand the fastidious conditions that microaerophiles need, so they will remain the primary microorganisms for studies and for industries such as biotechnology, industrial raw materials, and wastewater treatment.

See also ANAEROBE; METABOLIC PATHWAYS; METABOLISM; TRANSMISSION.

Further Reading

Dyer, Betsey Dexter. *A Field Guide to Bacteria.* Ithaca, N.Y.: Cornell University Press, 2003.

Garrity, George M. *Bergey's Manual of Systematic Bacteriology.* Vol. 1, *The Archaea and the Deeply Branching Phototrophic Bacteria,* 2nd ed. New York: Springer-Verlag, 2001.

———. *Bergey's Manual of Systematic Bacteriology.* Vol. 2, *The Proteobacteria (Part C) The Alpha-, Beta-, Delta-, and Epsilonproteobacteria,* 2nd ed. New York: Springer-Verlag, 2005.

Hotz, Robert Lee. "Lofty Search for Life in New York." *Los Angeles Times,* 17 April 2005. Available online. URL: http://articles.latimes.com/2005/apr/17/nation/na-venter17. Accessed March 9, 2009.

Needham, Cynthia, Mahlon Hoagland, Kenneth McPherson, and Bert Dodson. *Intimate Strangers: Unseen Life on Earth.* Washington, D.C.: American Society for Microbiology Press, 2000.

Panno, Joseph. *The Cell: Evolution of the First Organism.* Rev. ed. New York: Facts On File, 2010.

Prescott, Lansing M., John P. Harley, and Donald A. Klein. *Microbiology,* 6th ed. New York: McGraw-Hill, 2005.

Schaechter, Moselio, John L. Ingraham, and Frederick C. Neidhardt. *Microbe.* Washington, D.C.: American Society for Microbiology Press, 2006.

Tortora, Gerard J., Berdell R. Funke, and Christine Case. *Microbiology: An Introduction,* 8th ed. San Francisco: Benjamin Cummings, 2004.

aeromicrobiology Aeromicrobiology covers a specialty concerned with the types and numbers of microorganisms in the air. *Airborne microorganisms* is a term that refers to any microorganism that can be transmitted through the air. Aeromicrobiology focuses on the types of microorganisms in the air and the modes and patterns of their transmission through air. Bacteria, fungal spores, and viruses constitute the main microorganisms that become airborne under certain conditions in the environment.

Airborne transport of microorganisms uses two methods: aerosols and solid particles. Aerosols are fluid droplets that can be light enough to travel long distances through the air, usually aided by a breeze or other source of power. The term *bioaerosols* refers to aerosols that contain a living microorganism. Aeromicrobiology encompasses a specialty in science that focuses on the physics of aerosol and bioaerosol movement over small and large distances as well as the patterns that these droplets use when settling out of the air.

Small solid particles act as the second method by which microorganisms can travel in air. Examples of particles that carry microorganisms are the following: dust, pollen, soil particles, emissions, soot, smoke, feathers, hair, dander, fibers, or any other tiny particles that can remain in the air for periods of ranging from several seconds to several hours. Agriculture contributes a portion of these so-called particulate pollutants from farmyard wastes, fertilizers, and soil that run off cultivated fields with rain. Industrial emissions also contribute soot and smoke; construction sites or demolitions put large amounts of dirt, dust, and fibers into the air, and vehicle emissions also release small particles.

Microbiologists have discovered that raindrops and snowflakes carry bacteria. The Louisiana State University biologist Brent Christner has reexamined the idea that bacteria could influence precipitation, proposed more than 25 years ago by the microbiologist David Sands of Montana State University. Christner pointed out in 2008: "Transport through the atmosphere is a very efficient dissemination strategy. . . . We have found biological ice nuclei in precipitation samples from Antarctica to Louisiana—they're ubiquitous." The microorganism acts as a nucleus for the condensation of moisture or for ice crystal formation. Microbiologists and climate experts pursuing these studies have suggested that microorganisms in the air might help induce precipitation in times of drought. No one has yet devised a method, however, for putting airborne microorganisms to use in causing rain or snow.

Aeromicrobiology focuses on the following three main aspects: (1) the types of airborne particles and how they move, (2) airborne transmission of plant and animal diseases, and (3) indoor and outdoor air quality. Each of these areas depends on methods for identifying the microorganisms in the air and ways they move from place to place as influenced by winds, air turbulence, and humidity. In general, winds increase the distances of airborne transmission, high turbulence increases overall numbers of airborne particles and their settling rate from the air, and humidity keeps moisture-containing particles in the air longer.

Some industries require a thorough knowledge of aeromicrobiology because airborne microorganisms have the potential to contaminate their products. For example, makers of packaged food products (soups, juices, and powdered mixes) and manufacturers of products to be used on the body (shampoos, skin lotions, and cosmetics) monitor the numbers and types of bacteria, fungi, and viruses that may be in a manufacturing plant's air. Drug manufacturers that produce sterile medicines must also prevent airborne contamination of drugs that will be ingested by patients or injected into the body.

Microbiologists familiar with aeromicrobiology play a valued role in these industries because airborne contaminants can cost millions of dollars to be lost to ruined products. Contaminated products that manufacturers do not catch could additionally be a serious health threat to people using such products. The microbiologist Gary Andersen of the Lawrence Berkeley National Laboratory in California explained in the laboratory's newsletter, "We found that there are a lot of airborne bacteria, including pathogens, which we did not know are out there." Studies of outdoor and indoor microbiology have become an important part of industry and medicine.

In medicine, surgical procedures have developed as a result of the understanding that airborne microorganisms cause contamination and infection. Operating room employees take special measures to protect open surgical incisions from airborne particles and contamination of surgical instruments. The main methods used in medical care for preventing hazards from airborne microorganisms and bioaerosols are the following:

- wearing masks and body coverings
- high-efficiency, small-pore air filtration systems
- use of antiseptics before and after surgical incisions
- use of disinfectants in medical care rooms
- sterilization of equipment and protective covering until use

Because airborne contamination is a continual concern in microbiology, prevention methods must be carried out rigorously.

TYPES OF PARTICLES IN AIR

Most airborne microorganisms stay in the lower layers of the troposphere, where air turbulence stirs the movement of fine particles. Bacteria, viruses, and fungi are the most common airborne microorganisms. Bacteria and viruses travel aboard any particle that becomes airborne. Some bacterial species, however, are airborne on their own. For example, *Actinomycetes* bacteria resemble fungal spores in the way they travel on a breeze without depending on particles or aerosols. *Actinomycetes* and fungal spores both possess physical characteristics that help them withstand dry conditions and increase the time they can remain in the air.

Protozoa and algae do not commonly travel by air. The few protozoa known to travel in the air do so exclusively in moisture droplets.

Of the approximate 100,000 species of fungi known to exist, a large portion use the air in part of their reproductive cycle. Most molds, for example, produce a structure called a *sporangium,* which is a baglike structure filled with of thousands of tiny spores. When the sporangium becomes full, it bursts open and releases its spores in a step called *launching.* The force of launching disperses spores over wide areas; spores travel distances from a few inches to several miles. Fungal spores range in size from 1 to 100 micrometers (µm), and this small size combined with the spore's outer coat seem perfectly designed for air travel. Molds consequently have evolved three characteristics that help them spread to new places for reproducing: small size, spore coat structure, and launching.

As mentioned, aerosols aid airborne transmission of microorganisms. Some microorganisms need the moisture inside aerosols to stay alive as they travel through the air; protozoa and many viruses are examples. Bioaerosols can stay in the air for several seconds to several days and, as fungal spores do, they travel long distances. In medicine, the types of microorganisms within bioaerosols are an important factor in disease transmission. This is because airborne transmission is one of the main recognized modes of disease transmission. In fact, health professionals often use the term *bioaerosol* specifically for pathogen-containing aerosols. Disease-transmitting bioaerosols may be composed of various materials known to carry infectious microorganisms: contaminated water, blood, saliva, fecal wastes, or mucus.

The size of a bioaerosol determines the type of microorganism it can hold. The following three categories, called *modes,* describe bioaerosol sizes: (1) Nuclei mode bioaerosols are those less than 0.1 µm in diameter and transmit viruses; (2) accumulation mode bioaerosols range from 0.1 to 2 µm and can carry viruses, bacteria, or small fungal spores; and (3) coarse mode bioaerosols are larger than 2 µm so are big enough to carry protozoa in addition to all the other types of microorganisms. The patterns by which bioaerosols disperse impact health and disease transmission. The microbiologist Andersen told *Scientific American* in 2006: "It's important to do a microbial census to see what's in the air we breathe. I believe it's going to change as the climate changes. We may see very different populations of microbes in the air and that may have some health implications." Changes in climate will affect humidity, rain patterns, winds, and storms, which will all change the way various sized bioaerosols carry disease from place to place.

Some microorganisms excrete toxins into the air. These toxins, then, damage cells or tissue when they are inhaled at a dose of as small as 1 µm or less. The toxin produced by the bacterial species *Clostridium botulinum,* type A botulinum toxin, is lethal at an inhaled dose of 0.3 µm. For this reason, experts in extreme pathogens have identified the botulinum toxin as a potential biological weapon, or bioweapon. The botulinum toxin has the ability to be transmitted through the air and only a minuscule dose will cause sickness or death.

Fungi release toxins called *mycotoxins* that enter the air when contaminated soils are stirred and breezes pick up small dirt particles. Mycotoxins also present a breathing hazard inside water-damaged buildings that are contaminated with extensive mold growth. The molds *Claviceps, Aspergillus, Penicillium,* and *Stachybotrys* produce the most familiar airborne mycotoxins in microbial and health studies, but many other molds also produce similar toxins. *Stachybotrys* becomes especially dangerous when it grows into large masses inside the walls of flood-damaged buildings. Masses of fungal colonies trap enormous numbers of *Stachybotrys* spores, which then excrete a large dose of toxin into the building's air. This is one of several factors that contribute to a situation called *sick building syndrome,* in which contaminated buildings cause health problems to humans and pets.

AIRBORNE DISEASE TRANSMISSION

Humans inhale into their lungs particles of less than 10 µm in diameter. Gary Andersen's team has found 10,000 types of bacteria in grit floating in city air—the study did not measure the additional numbers of fungi and viruses—so every day people inhale an enormous amount and variety of microorganisms. A small, unknown percentage of these microorganisms cause disease.

Humidity, airflow patterns, and turbulence in the atmosphere affect the spread of diseases caused by airborne transmission of pathogens, or disease-causing microorganisms. The following table summarizes the main airborne pathogens of plants, animals, and people. The influenza virus—the cause of flu—infects human populations each year. Much of the flu's spread depends on how flu-containing bioaerosols transmit from person to person. In flu season, for instance, the disease transmits easily between people crowded together indoors in cold, rainy weather. But health experts have suspected that more general conditions in the atmosphere, such as humidity and winds, keep the virus active and allow it to spread over a much wider range. As a result, entire communities suffer flu epidemics rather than a small number of isolated cases.

Humans spread airborne germs by sneezing, coughing, talking, laughing, or breathing. A single sneeze expels about 20,000 mucus droplets that contain bacteria and viruses. Coughing and sneezing expel these bioaerosols with more force than talking or laughing, so the bioaerosols travel farther. Bioaerosols launched by coughing or sneezing are large—5–10 µm or more—and heavy, and as a result they usually travel no farther than about three feet (1 m). The smallest bioaerosols evaporate while they are in the air and remain airborne for longer periods than large bioaerosols, and as a consequence these small particles cover greater distances. A susceptible person becomes infected with pathogens after inhaling either large, moist droplets expelled from someone nearby or tiny, dry particles that have traveled several feet.

The World Health Organization (WHO) stated in its 2007 report, "Global Surveillance, Prevention and Control of Chronic Respiratory Diseases," that "preventable chronic respiratory diseases . . . constitute a serious public health problem in all countries throughout the world, particularly in low and middle income countries and in deprived populations." The WHO report further stated that chronic respiratory diseases account for four million deaths annually, and of these cases, airborne transmission makes up the main mode of disease transmission. The most prevalent respiratory infections seen worldwide are the following: influenza, measles, tuberculosis, respiratory syncytial virus (RSV), severe acute respiratory syndrome (SARS), and *Streptococcus pneumoniae* infection. Bacteria cause tuberculosis and *Streptococcus* infection; viruses cause the other illnesses listed.

The health community recommends several preventative measures against airborne diseases. Vaccination programs, such as yearly flu shots, provide protection against disease, and face masks, indoor air filters, and indoor ventilation also help prevent airborne transmission.

The following table shows that transmission can spread disease even though individuals may not be in close proximity. Diseases in fact spread when individuals are quite far apart. Most diseases in trees, for example, occur because pathogens become airborne and travel over long distances. A large portion of pathogens that infect tree populations depend on airborne transmission.

The Science of Aerosols

Formulas derived from physics predict distances that aerosols migrate through air, how they disperse, and how they land on surfaces. Mathematical equations also estimate the survival of bioaerosols in the atmosphere, and these equations can be used to calculate the time a particle will spend airborne and distance it will probably travel. This aspect of biology relies on two laws of physics: Newton's laws that describe the drag force put on moving objects, and the Stokes' law, which relates drag to particle size, speed, and the air's ambient (the immediate surroundings) conditions.

From these calculations, physicists categorize the types of airborne transport that an aerosol may take over a certain period. (It does not matter whether the aerosol contains moisture or is dry, or whether it contains microorganisms.) The first category is called *short-term airborne transport* or *submicroscale transport*. Aerosols in this mode of travel stay airborne for less than 10 minutes and cover distances of less than 328 feet (100 m) from their origin. Microscale transport composes the second category, in which aerosols stay in the air from 10 minutes to an hour and may travel almost two thirds of a mile (1 km). Third, mesoscale and macroscale transport denote long-distance transmission in which aerosols remain airborne for days at a time and travel at least 62 miles (100 km). Mesoscale transport lasts for days and extends up to about 62 miles, while macroscale transport may last longer than mesoscale and extend beyond mesoscale distances.

The size of an aerosol and the energy used to launch it determine the distance it travels. Physicists calculate a value that converts all of these relationships into a single number, called a *Reynolds number*. This value was named after the Irish engineer Osbert Reynolds, who, in the 19th century, developed relationships for particle movement through the air. The calculations begin with the equation that follows. In it, the minimal energy needed to launch, or accelerate, a particle is $1/2\ mv^2$, where m is mass and v is velocity (speed at the time of launch). The energy to accelerate an object with density p and radius r becomes:

$$E = 2/3\ \pi\ pr^3v^2$$

Once v is solved, it may be entered into a second equation that relates distance d to velocity v (T is a time constant):

$$d = v\mathrm{T}$$

Distance traveled is proportional to the square root of launching energy and radius. In other words, no matter the energy with which they are launched, large aerosols travel farther than small aerosols.

Major Airborne Pathogens

Pathogen	Disease	Host
	BACTERIA	
Bacillus anthracis	pertussis (whooping cough)	humans
Brucella species	brucellosis	hoofed animals, dogs
Klebsiella pneumoniae	pneumonia	humans, primates
Legionella species	legionellosis	humans
Mycobacterium tuberculosis	tuberculosis	humans, primates
Neisseria meningitides	meningitis	humans
	FUNGI	
Aspergillus species	aspergillosis	humans, domesticated animals
Ceratocystis ulmi	Dutch elm disease	Dutch elms
Coccidioides immitis	coccidiosis	humans, primates, birds
Cryptococcus species	cryptococcosis	birds, cats, humans
Histoplasma capsulatum	histoplasmosis	humans
Puccinia species	rusts	wheat, rye, junipers, oat
Phytophthora infestans	potato blight	potatoes
	VIRUSES	
Aphthovirus	hoof and mouth disease	cattle, sheep, swine
Arbovirus	eastern equine encephalomyelitis	horses
Hantavirus	pulmonary syndrome	humans
Herpes	chicken pox	humans
Influenza	influenza	humans, birds, horses, swine, marine mammals
Morbillivirus	distemper/measles	cats, dogs, humans
Picornavirus	common cold	humans
	PROTOZOA	
Pneumocystis carinii	pneumocystosis	humans

Diffusion, the dispersal or scattering of aerosols, provides another characteristic of airborne transmission. Turbulence in the air puts drag on any particle's movement, so Reynolds took all the factors of size, speed, and drag into consideration to develop an equation that puts a value on turbulence:

Reynolds number = velocity × dimension/viscosity

Reynolds showed that bioaerosols disperse over a larger area when they travel in swift breezes, but dispersal also decreases as the air's viscosity increases. Reynolds numbers above 2,000 indicate turbulent conditions, and the higher the number, the greater the turbulence. A high Reynolds number means that aerosols will disperse over a large area from their launching point.

AIR SAMPLING

The concentration of microorganisms in the air can be determined by taking samples of the air. Air sampler machines help assess airborne particles in buildings that make food products, drugs, or products for the body. Air samplers are portable battery-powered devices that draw in as much as 265 gallons (1,000 l) of air. All of the bioaerosols entering the sampler land on a solid agar surface, such as a petri dish filled with agar medium. After the sampling period, a microbiologist removes the agar plate and incubates it to grow the microorganisms that have been captured by the air sampler.

Air sampling did not develop quickly into the handy methods used today. The U.S. Public Health Service used a sampling technique, in 1915, in which air flowed through a bed of sand. Their scientists then washed the sand to remove microorganisms and inoculated a small volume of the washings to agar plates. Microbiologists seeking less laborious techniques adopted the settle plate method for sampling air. In the settle plate method, agar plates left open and horizontal for one to several minutes collect particles that slowly fall from the air. After incubation, the number of colonies that grow from microorganisms that have settled by gravity onto the agar provides an air microbial count. The settle plate method offered the advantages of easy use and low cost. Eventually microbiology would require a more accurate sampling method for assessing the number of microorganisms in the air rather than counting only cells that settled out of the air. Microbiologists needed a device to measure microbial concentration per volume of air to assess air quality better.

Mechanical air samplers, based on models developed almost 150 years ago, would be remodeled to fill the need for accurate air measurements. An air sampler invented in Germany, in 1861, differentiated particles by size. The Hess sampling tube forced air through a vertical chamber to capture a known amount of air. Various inventors attempted to improve on the Hess sampler until 1958, when the bacteriologist A. A. Andersen built a device that controlled the sampling time and the sample volume. Andersen's sampler used a series of horizontal agar plates to capture large particles at the upper levels and increasingly smaller particles at lower levels as air flowed downward. The sampler came to be known as the Andersen cascade impactor because particles drawn into the collector impacted the agar plates as they cascaded down through the chamber. The number of colonies captured per cubic meter of air is called an Andersen value. Many laboratories today use Andersen air samplers to produce a reliable and accurate assessment of the numbers and types of microorganisms in their airspace.

INDOOR AIR MICROBIOLOGY

Microbiologists use air samplers to measure indoor air quality as well as outdoor air quality. One reason for monitoring indoor air is to assess sick building syndrome. Just a few decades ago, few scientists believed sick building syndrome existed, but as the field of aeromicrobiology grew, scientists learned that many harmful chemicals and much biological matter rose to high levels indoors. Climate-controlled buildings with little outside ventilation caused particular problems unless the ventilated air had been filtered to clean out most particles.

Many poorly ventilated buildings contain molds, bacteria, and viruses in the air in higher concentrations than are found outdoors. Indoor air turbulence, ventilation, temperature, and humidity affect indoor airborne microorganisms as they affect the same microorganisms outdoors. Weather and the time of year also influence indoor microorganisms. The indoors possesses certain characteristics that are not present outdoors, and these factors have been implicated in contributing to sick building syndrome. Five main factors that influence indoor air quality are the following:

- Modern buildings are sealed to maintain temperature/humidity.
- Indoor circulation systems disperse microorganisms.
- More people congregate indoors and closer together.
- Indoors, people are often close to pets and pet dander.
- People who have illnesses usually confine themselves to the indoors.

Three means of improving indoor air quality are good ventilation, effective filter systems, and chemicals that clean the air. Ventilation controls the concentration of indoor bioaerosols and may be done by something as simple as opening a window. Open windows provide a passive method of ventilation to help rid the indoors of pathogen-containing bioaerosols, especially if a member of a household or a coworker is sick. Open windows may not, however, decrease the overall concentration of airborne microorganisms because more bioaerosols enter on the breeze as others flow out.

Active ventilation uses a powered device to force air through a filter that retains bioaerosols. Buildings that require extremely clean conditions, such as medical clinics, may use high-efficiency particulate air (HEPA) filters to clear the air. HEPA filters contain pores as small as 1 µm diameter. Bacteria, fungal spores, dust particles, pollen, and most viruses cannot fit through HEPA filter pores and so do not return to the indoors with circulated air.

Some chemical products reduce the amount of microorganisms in indoor air. These products have been tested in laboratories, where they have been shown to reduce microbial numbers by at least three logarithms (99.9 percent). In real-life household conditions, nevertheless, microbiologists cannot be sure that these products effectively overcome all the factors that contribute to bioaerosol levels.

OUTDOOR AIR MICROBIOLOGY

Airflow, air turbulence, mixing, updrafts, temperature, and humidity affect the transmission of bioaerosols and also affect the amount of time pathogens remain infectious outdoors. In general, moist, warm conditions allow pathogens to stay infectious longer than very dry conditions. There is also seasonal variation; Andersen values tend to be higher in summer and fall than in winter and spring. Year-round, the majority of airborne microorgan-

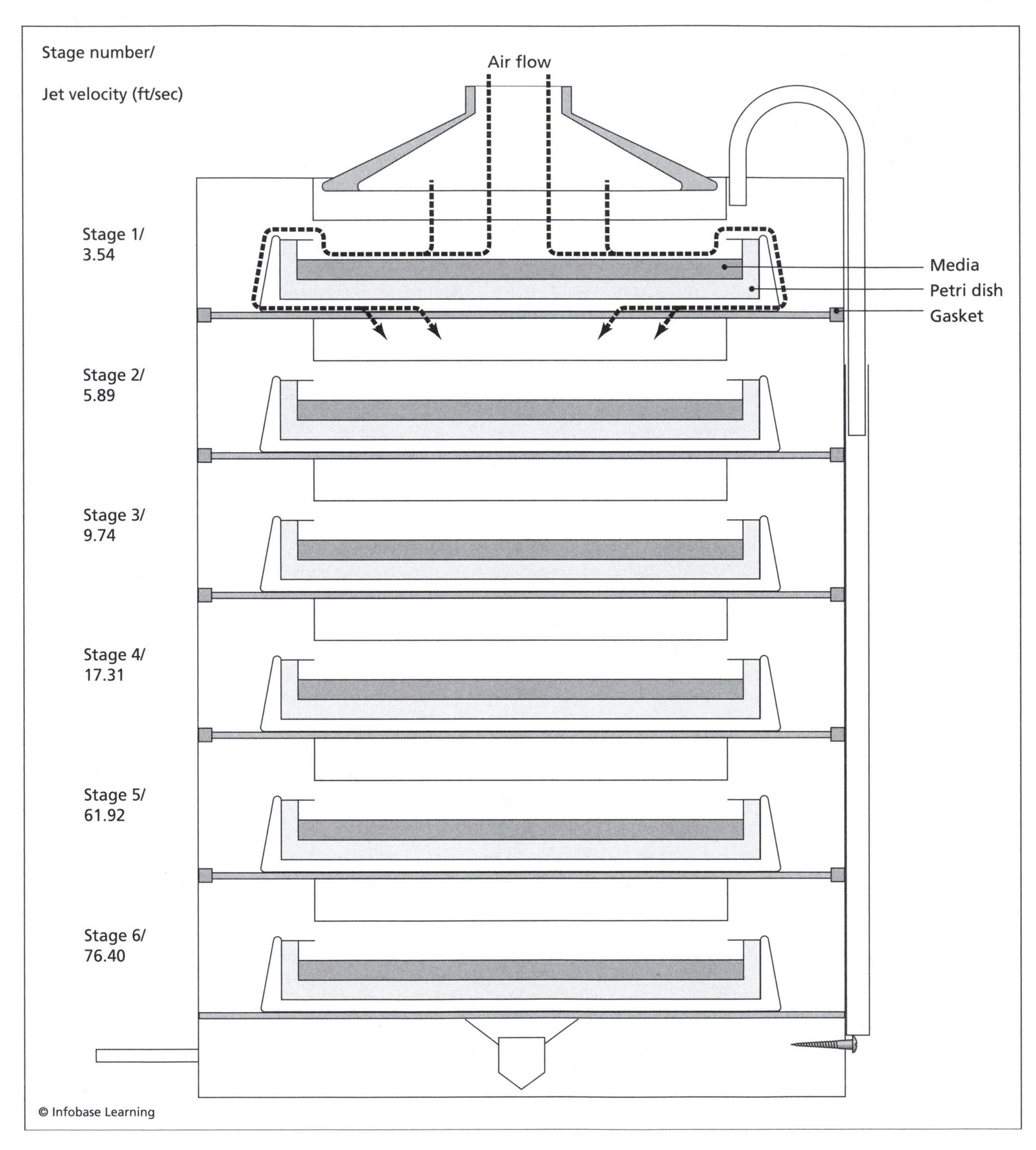

The Andersen air sampler measures the amount of different sized particles in air. As air enters the sampler, large particles (large bacteria, mold spores, dusts, and fibers) land on the plates in stages 1 and 2. Smaller particles (aerosols, soot, and smoke) flow to stages 3 through 6. A microbiologist removes the agar plates at each stage and incubates them to determine the number of microorganisms.

isms are in the atmosphere within 0.062 mile (0.1 km) of Earth's surface.

Human activities that change the concentration of outdoor aerosols include construction projects, tilling of agricultural fields, harvesting, road building, and vehicle traffic. Natural activities that play a part are violent storms, floods, volcanoes, and wind and breezes. Wind alone is known to carry plant pathogens many hundreds of miles. Tree disease may move from an infected area to uninfected areas in sequence until the disease has moved several hundreds of miles along the path of prevailing winds. Aeromicrobiology experts are beginning to understand that airborne transmission may extend much farther than previously thought. For instance, environmental microbiologists have discovered that dust particles from the Sahara can be detected in air samples taken in North America.

Aeromicrobiology will be a continuing area of exploration. Much less information has been gathered on the types and amounts of microorganisms in the air compared with those found in water and soil. Future studies in aeromicrobiology will help answer questions on human, animal, and plant disease transmission; weather; and the ecology of microorganisms on Earth.

See also *ASPERGILLUS;* CLEAN ROOM; *CLOSTRIDIUM;* FUNGUS; INDUSTRIAL MICROBIOLOGY; PASTEUR, LOUIS; *STACHYBOTRYS;* TRANSMISSION.

Further Reading

Biello, David. "Microbe Census Reveals Air Crawling with Bacteria." *Scientific American,* 19 December 2006. Available online. URL: www.sciam.com/article.cfm?id=microbe-census-reveals-ai. Accessed March 9, 2009.

Dowd, Scot E., and Raina M. Maier. "Aeromicrobiology." In *Environmental Microbiology,* edited by Raina M. Maier, Ian L. Pepper, and Charles P. Gerba. San Diego: Academic Press, 2000.

Jjemba, Patrick, K. *Environmental Microbiology: Principles and Applications.* Enfield, N.H.: Science Publishers, 2004.

Krotz, Dan. "Study Finds Air Rich with Bacteria." *Berkeley Lab Research News,* 18 December 2006. Available online. URL: www.lbl.gov/Science-Articles/Archive/ESD-air-bacteria.html#. Accessed March 9, 2009.

Science*Daily*. "Evidence of 'Rain-Making' Bacteria Discovered in Atmosphere and Snow." 29 February 2008. Available online URL: www.sciencedaily.com/releases/2008/02/080228174801.htm. Accessed March 9, 2009.

Shelton, Brian G., Kimberly H. Kirkland, W. Dana Flanders, and George K. Morris. "Profiles of Airborne Fungi in Buildings and Outdoor Environments in the United States." *Applied and Environmental Microbiology* 68 (2002): 1,743–1,753.

World Health Organization. "Global Surveillance, Prevention and Control of Chronic Respiratory Diseases." Geneva, Switzerland, 2007. Available online. URL: www.who.int/gard/publications/GARD%20Book%202007.pdf. Accessed June 4, 2010.

agar Agar is a gellike substance composed of a polysaccharide made by marine algae. Microbiology uses agar as a solid surface for growing bacteria and fungi because of its properties of melting and solidifying, holding moisture, and providing an inert surface for microbial growth. The gellike characteristic of agar results from a polysaccharide called *galactan,* which is found in red algae. The red algae *Gelidium* and *Gracilaria* serve as the main commercial sources of agar supplied to microbiology laboratories today, though they may be labeled simply as "seaweed" on the agar's container.

Agar offers a useful attribute in that few microorganisms can degrade agar's constituents. Red algae in the ocean produce a large and long molecule called a *polymer,* which contains a string of repeating galactose sugar units. Most bacteria evolved without the need for breaking down agar's polymer, so bacteria do a poor job splitting the galactose-galactose linkages in agar formulations. This allows microbiologists to use agar as an inert support substance for studying microbial characteristics.

Agar-based media enable microbiologists to grow a very wide variety of microorganisms in the laboratory for studying microbial behavior, enzyme activities, and morphology (cell or colony appearance). Agar's main attribute is its capacity to form a solid or a semisolid surface. Solid agar enables microbiologists to perform the following studies: colony morphology of bacteria, yeasts, and fungi; enzyme activity assays; antibiotic activity testing; and nutrient requirement tests. Semisolid agar helps in studying motility and chemotaxis because motile species move through the semisolid material while nonmotile species will not move. Studies on nutrients, growth factors, and antibiotics have been aided by the ability of agar to permit certain substances to diffuse into it. Agar diffusion studies wherein an antibiotic diffuses into an agar medium previously inoculated with bacteria have been the main test method for assessing antibiotic activity in a test called the *Kirby-Bauer test.*

Purified agar granules disperse into water when the mixture is heated to over 212°F (100°C). Then as the mixture cools to about 104–115°F (40–45°C) it solidifies into a translucent gel. Once gelled, it remains stable for several weeks and up to 150°F (65°C). Solid growth media contain from 3.5 to 6 percent agar; semisolid formulas usually contain 0.5–1.5 percent agar.

AGAR'S HISTORY IN MICROBIOLOGY

While agar is the backbone of microbiology, its discovery was a stroke of luck. In the 1800s, microbiology was coming into its own as a specialized branch of biology. The properties of bacteria in nature and in medicine fascinated the pioneers of the new science of microscopic particles, but these scientists lacked a reliable way to grow microorganisms in a laboratory. The German physician Robert Koch (1843–1910) experienced just such frustration, so, in the late 1800s, Koch teamed with others in his laboratory to devise a better way to maintain the bacteria they had isolated from patients. One of Koch's colleagues, the physician and bacteriologist Walther Hesse (1846–1910), had been trying and rejecting growth medium formulas. Hesse's wife, Angelina (1850–1934), meanwhile had devised a mixture from an East Indian gelatin called *agar-agar* she had found at a local market. She soon began using the substance to keep jellies and puddings solid in warm weather. Angelina, perhaps to quiet her husband's complaints from his laboratory, suggested to him that he and Koch try the gelatin in their experiments. It might, she offered, hold nutrients for bacteria as well as it held fruit juices in homemade preserves. Walther Hesse took his wife's advice and used it to develop a stable solid medium. The Hesses' grandson, Wolfgang Hesse, recounted in 1992: "[Angelina] had learned about this material as a youngster in New York from a Dutch neighbor who had emigrated from Java. The practical application of this kitchen secret was to bring major recognition to the Hesses, more today than during their lifetime." The discovery of agar had truly been a serendipitous advance in science.

As Wolfgang Hesse implied, the Hesses asked for and received little notice for their breakthrough in culture technique. By the late 1880s, Koch had been publishing articles on the best methods for growing bacteria, and he made only a fleeting mention of "gelatin" and no mention at all of either Hesse. In his 1881 article, "Methods for the Study of Pathogenic Organisms," Koch described a medium "which would be firm and rigid. The most useful way to attain this end is to add gelatin to the nutrient liquid. . . . I have determined that the best concentration of gelatin for these purposes is 2.5 to 3 percent." With that, agar became a staple in every succeeding study in microbiology.

During the same period, the German microbiologist Richard J. Petri—better known as R. J. Petri—designed a small dish that could be stacked on shelves inside an incubator. Petri wrote of his invention in 1887: "In order to perform the gelatin plate technique of Koch, it is necessary to have a special horizontal pouring apparatus. . . . I have been using flar double dishes of 10–11 cm [3.9–4.3 inches] in diameter and 1–1.5 cm [0.4–0.6 inch] high." Koch's need for pure bacterial cultures, the Hesses' culinary contributions, and Petri's simple innovation forever changed the course of microbiology. Few advances in agar use were made for the next century until the emergence of disposable equipment in modern microbiology. Solidified agar forms a firm material that adheres to laboratory dishes and tubes, and the time and effort of cleaning used agar out of plates at the end of each study grew to be a problem. Microbiology laboratories now use disposable plastic petri dishes, tubes, and pipettes to replace glass. This disposable plasticware has been an overlooked advantage for lowering costs and speeding up laboratory work involving agar.

TYPES OF AGAR

Nutrients are added to molten agar to make hundreds of varieties of microbiology media. When microbiologists refer to *agar* media, they mean the formulas containing water, agar, and all the nutrients needed by microorganisms for growth. Nutrient agars grow a diverse variety of bacteria or fungi. Specialized agars consist of formulas for growing only a certain type of microorganism. Selective agars and differential agars represent the two most commonly used specialized agars.

Selective agars favor the growth of one type of microorganism over others. Sabouraud agar, for example, enables fungi to grow but inhibits the growth of most bacteria. Differential agars give more specific growth responses and distinguish one type of bacteria from other types of bacteria. Differential agars do this by containing ingredients that enhance the growth of some bacteria while inhibiting or having no effect on others. For example, MacConkey agar is a differential medium containing a red dye and the sugar lactose. It differentiates between lactose-fermenting bacteria, whose colonies turn red during incubation, and nonfermenting bacteria, which are colorless.

Most selective and differential formulas take the form of agar media rather than liquid media, referred to as *broths*. This is because the purpose of selection and differentiation is to isolate one type of microorganism from all others. By inoculating a solid agar surface in a technique called *streaking*, a microbiologist can produce isolated colonies on the agar. These single, distinct colonies become visible during incubation. They are usually pure, meaning they contain the exact clones of a single parent cell. Pure cultures are necessary in many areas of microbiology, of which the most important are the following: identification techniques, selection of an antibiotic treatment, or carrying out of genetic engineering. None of these activities could take place without a pure colony of a single microorganism as their starting point.

An overlay is an agar formula composed of two layers. A microbiologist prepares an overlay in a petri dish by pouring molten agar, already inoculated with bacteria, onto a layer of solidified agar. The upper layer is called the *overlay* and usually contains no more than 0.75 percent agar in water. This low level of agar produces a fairly soft layer even after it cools. Inside the overlay, bacteria disperse into the agar. During incubation, the bacteria grow and colonies in the agar become visible. If the microbiologist uses a very concentrated inoculum, the overlay becomes cloudy with millions of bacteria.

Overlay cultures serve two main purposes. The first purpose relates to the testing of antibacterial compounds, such as antibiotics and biocides. To conduct the test, a microbiologist puts a drop of antibacterial compound into the cooling overlay and then incubates the plate. After incubation, an area around the drop appears clear because the antibacterial compound killed all the bacteria within a certain distance of the drop. The rest of the overlay remains cloudy with healthy, growing bacteria. Virologists (microbiologists who specialize in growing viruses) take advantage of the second purpose of overlays. Viruses that attack bacteria cause zones of clearing in overlay agar in a similar manner to the way antibacterial compounds work. In virus studies, these clearing zones are called *plaques*. The plaque technique helps virologists determine the activity and concentration of specialized viruses called *bacteriophages*, which attack bacteria and no other types of cells.

When agar solidifies on a slant, it provides more surface area for growth of tube cultures. ***(Chiang Mai University, Division of Clinical Microbiology)***

TECHNIQUES USING AGAR MEDIA

Molten agar poured into either petri dishes or test tubes takes the shape of the container once the agar solidifies, and these different forms of agar serve different uses in microbiology. In petri dishes, molten agar spreads out into a flat, smooth surface that fills the dish. Once the agar has solidified, it takes the shape of the petri dish (also called a *petri plate*), and together they are known as an *agar plate*. A pour plate is any petri dish that has had agar medium poured into it and allowed to solidify. The resulting smooth gellike solid surface is, then, ready for inoculation. (The terms *pour plate* and *agar plate* have equivalent meanings.)

Molten agar can also be poured into tubes instead of petri dishes. Tubes that are positioned at an angle when the agar solidifies allow the agar also to harden at an angle. This type of agar surface is called a *slant*. Slanted agars provide a larger surface area for bacteria or fungi that grow better in tubes than on plates. Molten agar poured into upright tubes, by contrast, constitutes an agar deep, also called an agar *butt*. Deep agar tubes are useful for studying bacterial motility in semisolid medium.

Microbiologists inoculate agar plates by one of two methods: streaking or spreading. Streaking is done with a small wire loop that holds a drop of broth culture, and by dragging or streaking, the loop is drawn gently over the agar surface. By doing this, a single line of colonies forms when the plate incubates. The colonies at the outermost end of the streak are often isolated from the others, and these isolated colonies are considered pure colonies. Spread plates differ from streak plates by holding an inoculum made as a wide swath across the agar surface. After incubation, a spread plate contains a uniform layer of adjacent colonies called a *lawn*.

The properties of agar have made it superior to other gellike substances for procedures in microbiology. Agar has only two disadvantages: The supply of agar worldwide varies, depending on seaweed supply, and, because agar is a natural material, it varies in composition from one batch to the next. Some compounds serve as agar substitutes: silica gel, carrageenan, gellan gum, and pectins. All of these except silica gel contain long polysaccharides made by microorganisms. Because these materials have natural sources, they vary from batch to batch on occasion in the same manner as agar. Silica gel requires an extra step when being used in microbiology because it contains silicon dioxide (SiO_2) units and contains no carbon. To use silica gel for growing bacteria, a microbiologist must add a carbon source. Despite the greater availability of silica gel, carrageenan, gellan gum, and pectins as agar substitutes, microbiology laboratories prefer the properties of agar over all other substances.

IMPORTANT AGARS IN MICROBIOLOGY

Hundreds of agar-containing formulas can be made for the thousands of species known to microbiology. Different specialties use different agar formulas. For instance, the agar media used in a hospital's clinical microbiology laboratory differ from those needed by marine microbiologists. Some agar media are prevalent across all fields of microbiology. The common ones are highlighted in the table. Almost all of the examples given have variations and can be altered into selective or differential agar.

The invention of new agar-based formulas for growing unique bacteria and fungi has grown into a specialty of its own in microbiology. Hundreds of agar formulas exist that provide much more specialized conditions than the formulas shown in the table, which are listed in their approximate order of popularity. Though Walther Hesse and Robert Koch probably appreciated their discovery, agar has become an overlooked part of microbiology. Without agar, microbiology would not have advanced as quickly as it has in the areas of medicine, environmental studies, and biotechnology.

See also BACTERIOPHAGE; CLINICAL ISOLATE; COLONY; CULTURE; DIFFUSION; MEDIA; PLAQUE.

Further Reading

Cappuccino, James G., and Natalie Sherman. *Microbiology: A Laboratory Manual,* 8th ed. San Francisco: Benjamin Cummings, Pearson Education, 2008.

Difco and BBL Manual. Available online. URL: www.bd.com/ds/technicalCenter/inserts/difcoBblManual.asp. Accessed March 9, 2009.

Gerhardt, Philipp, ed. *Manual of Methods for General Bacteriology.* Washington, D.C.: American Society for Microbiology Press, 1981.

Hesse, Wolfgang. "Walther and Angelina Hesse—Early Contributors to Bacteriology." *American Society for Microbiology News* 58 (1992): 425–428.

Koch, Robert. "*Zur Untersuchung von Pathogenen Organismen*" (Methods for the Study of Pathogenic Organisms) *Mittheilungen aus dem Kaiserlichen Gesundheitsamte* 1 (1881):1–48. In *Milestones in Microbiology,* translated by Thomas Brock. Washington, D.C.: American Society for Microbiology Press, 1961.

Petri, R. J. "Eine kleine Modification des Koch'schen Plattenverfahrens" (A Minor Modification of the Plating Technique of Koch). Centralblatt für Bacteriologie und Parasitenkunde 1 (1887): 279–280. In *Milestones in Microbiology,* translated and edited by Thomas Brock. Washington, D.C.: American Society for Microbiology Press, 1961.

Prescott, Lansing M., John P. Harley, and Donald A. Klein. *Microbiology,* 6th ed. New York: McGraw-Hill, 2005.

algae (singular: alga) Algae make up a diverse group of eukaryotes that perform photosynthesis. They are referred to as a *heterogeneous* collection of organisms because they include unicellular, multicellular, and filamentous forms. Algae include not only microscopic plankton cells but also seaweeds and kelps that grow to about 165 feet (50 m) in length. Though many resemble plants, they lack a vascular system like that found in higher plants.

Algae have been found in all climates and in places as diverse as tropical settings, polar regions, and even deserts. Algae grow in the following aquatic environments: marine waters (such as oceans and bays), freshwater lakes and streams, and brackish waters, meaning waters with high levels of salt but not as salty as the ocean. Algae have also been isolated from rice paddies, hot springs, and the caustic waters of hazardous waste sites. If soil, tree trunks, or leaves on plants hold sufficient moisture, algae will grow in those places as well.

The algae that float freely in water are called *planktonic algae,* while species that attach to submerged surfaces are known as *benthic* or *sessile algae.* These aquatic algae belong to a group of organisms called *phytoplankton* that lives in the world's oceans and other bodies of water and serves as a food source for marine organisms. Phytoplankton makes up the lowest level in aquatic food chains and many other forms of life depend on it.

Phycology, or algology, is the study of algae, a specialty of microbiology that spans simple single-cell species to complex organisms; a person who studies and grows algae is a phycologist.

Algae play a variety of roles in the environment and in industry. In the environment, all of Earth's collective mass of algae helps pull carbon dioxide out of the atmosphere, which is a benefit in reducing greenhouse gases. But algae also present a hazard in the environment under certain conditions. For example, environmental algae begin to grow very rapidly when nutrients enter their normally nutrient-sparse water habitat. A sudden influx of nitrogen- or phosphorus-containing compounds causes this growth response, which results in an *algal bloom.* Algal blooms threaten normal aquatic ecosystems and can kill marine life through toxins secreted by the algae.

Outside marine environments, algae can present hazards due to their toxin production or they may be harmless nuisances in pools, ponds, and fish tanks.

Common Examples of Agar Media Used in Microbiology

Name	Description	Types of Microorganisms It Grows
nutrient agar	carbon and nitrogen sources to meet general growth requirements	wide variety
blood agar	carbon, nitrogen, amino acid, and vitamins supplied with addition of 5–10% blood	microorganisms with specific nutrient requirements or difficult to maintain in a laboratory (fastidious)
chocolate agar	heated blood (gives agar a chocolate color) makes growth factors available	fastidious gram-positive cocci
MacConkey agar	addition of carbohydrates differentiates gram-negative species	gram-negative bacteria
LB (Luria Bertani) agar	rich formula with amino acids, nucleotide precursors, vitamins, and other growth factors	*Escherichia. coli* in molecular biology experiments
tryptic soy agar	carbon and nitrogen sources to meet general growth requirements	various microorganisms including fastidious bacteria
HPC (heterotrophic plate count) agar	sufficient nitrogen and carbon for general nutrient requirements	heterotrophic (use variety of carbon sources) bacteria in water
standard methods agar	meets general growth requirements for counting variety of bacteria	aerobic bacteria from food and water
XLD (xylose, lysine, deoxychocolate) agar	carbon supplied in form of fermentable carbohydrate plus nitrogen source and a dye	enteric bacteria differentiated by growth response and colony color
R2A agar	low in nutrients to mimic conditions in water	bacteria in drinking water
Sabouraud agar	general nutrient agar with low pH	yeast, mold, and bacteria that grow in acidic conditions
potato dextrose agar	contains liquid soak from potatoes	yeast and mold from foods
BHI (brain heart infusion) agar	contains liquid soak from calf brain and heart	fastidious streptococci, pneumococci, and meningococci; susceptibility testing
endo agar	lactose and dye added to differentiate lactose from non-lactose fermenters	coliform bacteria
phenol red agar	carbohydrate and dye added to differentiate fermentations	gram-negative enteric and some gram-positive bacteria
EMB (eosin methylene blue) agar	eosin and methylene blue dyes added to differentiate lactose from nonlactose fermenters	gram-negative enteric bacteria

ALGAL STRUCTURE AND METABOLISM

A thin cell wall surrounds algal cells, and, inside the cell, a membrane encloses each organelle. Organelles are distinct bodies within eukaryotic cells that carry out specific cell activities. In algae, an organelle called a *chloroplast* contains flat sacs called *thylakoids*. Many thylakoids stack into layers called *grana* to fill the chloroplast, and these stacks serve the cell by storing chlorophyll pigments, which play a major part of photosynthesis. The chloroplasts also contain protein-based pyrenoids, structures that store the carbohydrates produced during photosynthesis.

As other eukaryotes do, algae contain a nucleus, mitochondria, endoplasmic reticulum, Golgi bodies, and a plasma membrane. Most algae also have a contractile vacuole that takes in and releases water to maintain the cell's water balance. Other vacuoles help digest food and store nutrients. Motile algae possess from one to three taillike flagella on their outer surface.

The vegetative (nonreproducing part of a life cycle) body of an alga is called a *thallus*. The thallus participates in one of the three following types of asexual reproduction: (1) fragmentation, in which the thallus breaks into several pieces that each grow into a new cell; (2) spore formation, whereby algae disperse spores that develop into cells or split into two new cells by binary fission; and (3) sexual reproduction, in which algal cells form haploid female or male structures (gametes) equivalent to egg or sperm cells in higher organisms. Gametes combine to produce a diploid zygote, which undergoes the process called *meiosis* in which diploid deoxyribonucleic acid (DNA) divides into haploid copies. At this point, the reproductive process returns to the first step and recommences.

Water molds of the phylum *Oomycota* produce asexually in a way similar to that used by other algae, but water molds differ from other algae by forming asexually produced spores called *oomycotes.* Oomycetes fill sacs called *sporangia,* and when they escape the sporangia the oomycetes use two flagella that enable them to move through moist habitats. These flagellated spores of water molds are called *zoospores.*

DIVISIONS OF ALGAE

The diversity found among algae is explained by their evolution. Molecular identification methods have helped explain the phylogeny, or evolutionary relationships, of algae and their evolution from single-celled organisms to more complex organisms. Molecular methods involve comparing the ribonucleic acid (RNA) from different algae, specifically a subunit of ribosomal RNA (rRNA) called *18S rRNA.* By studying the nucleic acids contained in 18S rRNA, phycologists have learned that different algae evolved independently of each other, yet the modern forms share many characteristics. Algae provide an example of a polyphyletic organism: Their phyla have different origins.

Algal phyla are also referred to as *divisions;* they are, from most primitive to most recent on the evolution scale, the following: Euglenophyta, Pyrrhophyta, Phaeophyta, Chrysophyta, Rhodophyta, Charophyta, and Chlorophyta. These phyla belong to two different kingdoms. For example, Phaeophyta and Rhodophyta reside in kingdom Plantae, but the other phyla belong to kingdom Protista. In addition to rRNA, cell structure and pigments can be used to group algae, yet sometimes these groupings appear to hold little logic. For instance, diatoms belong with the Chrysophyta even though diatom rRNA suggests these microorganisms evolved with members of Phaeophyta. The Cyanophyta had been called blue-green algae for many years, but these organisms actually belong with aerobic photosynthetic bacteria. A widely used classification system for algae is shown in the table on page 19.

Green Algae

Green algae contain diverse structures and live in a variety of habitats, including inside other organisms. Biologists believe green algae most closely relate to the plants living on land. Because of their place in phylogeny, green algae have been proposed as precursors to the evolution of Earth's green plants. Biologists have considered the evolutionary role of algae in terms of a structure called a *holdfast* that is present in some aquatic green algae. Holdfasts anchor algae to surfaces for part of their life cycle yet do not draw in any nutrients as do root systems in terrestrial habitats. Some green algae growing underwater and held in place by holdfasts resemble lawn grass. The grass-green pond scum that grows into long filaments and covers the water's surface also depends on holdfasts to keep the algae attached to the earth.

Chlamydomonas is a genus of flagellated green algae prevalent in freshwater and moist soils. Because of its ability to use sexual or asexual reproduction, *Chlamydomonas* resides at a key point in the evolution of different types of algae. The first evolutionary path contains *Chlorella,* a nonmotile alga that reproduces asexually. *Chlorella*'s reproduction involves the formation of a protoplast (a cell lacking a cell wall), which then divides to form from two to 16 daughter cells that are identical to the parent cell. In *Chlorella* metabolism, the cells ferment the sugar glucose to lactate when growing in anaerobic conditions and use aerobic photosynthesis to make energy when oxygen is available (*Chlorella*'s preferred metabolism).

A second important path in evolution contains *Volvox.* This organism forms a hollow sphere of a single layer of from 500 to 20,000 individual cells. Flagella on each cell beat in a coordinated way to rotate the entire colony and propel it through its watery habitat. When using sexual reproduction, *Volvox* employs male and female gametes each containing half of the chromosome. *Volvox* represents a critical step in evolution in which single cells combined and worked together in coordination as a multicellular organism. *Volvox* thus offers a glimpse at two important phases in the evolution of higher organisms: sexual reproduction in which male and female gametes combine their genes, and independent motility created by cells working together in a coordinated manner.

Red Algae

Red algae consist of unicellular forms or a variety of multicellular forms that grow to three or more feet (1 m) in length. All red algae store energy in a starch molecule called *floridean,* made up of glucose units in α-1, 4 and α-1, 6 linkages. The same α-1, 4 linkages in nature connect glucose units in vegetable starch and animal glycogen. Bacteria use α-1, 6 linkages when storing their sugars.

Red algae contain the red pigment phycoerythrin and the blue pigment phycocyanin, and each of these pigments gives red algae the unique ability to live at depths where other algae cannot. Light from the violet to blue section of the spectrum is the only light that penetrates ocean waters to depths of more than 650 feet (200 m) or more than 800 feet (250 m) in clear waters. At these depths red algae's pigments absorb light in the violet-blue spectrum and transfer the light energy to chlorophyll *a* in the cell's chloroplast. The algae's distinctive red color turns to blue-brown or brown-green when the pigment phycoerythrin breaks down during exposure to bright light. Divers would expect to find the reddest algae in deepwater habitats and less-red algae nearer the surface.

Red algae supply agar, the gelatinous polymer used for preparing solid and semisolid growth media used in microbiology laboratories. In living algae, the agar combines with three additional polymers to form a matrix that strengthens the cells. Outside laboratories, beachgoers sometimes find slippery and rubbery red seaweeds. These characteristics are produced by long polymer compounds that make red seaweeds indestructible in ocean currents. In addition to red seaweeds of the phylum Rhodophyta, the red algae group also contains the green seaweeds Chlorophyta and the brown seaweeds Phaeophyta. The following lists the familiar types of seaweeds and their characteristics.

Brown Algae

As with red algae, brown algae contain no chlorophyll *b*, but they use both chlorophylls *a* and *c* in their photosynthesis. The pigment fucoxanthin gives these algae their characteristic brown and olive colors. Brown algae store food energy in the form of complex carbohydrates, such as the

Some Characteristics of Algae Divisions

Division	Common Name	Cellular Form	Pigments		Cell Wall	Main Habitat
			CHLOROPHYLLS	CAROTENOIDS		
Charophyta	stoneworts	multicellular	*a, b*	α-c, β-c, X	cellulose	f, b
Chlorophyta	green algae	multi- and unicellular	*a, b*	β-c, X	cellulose	f, b, s, t
Chrysophyta	brownish, gold-brown, or yellow-green algae	uni-	a, c_1, c_2	α-c, β-c, ε-c	cellulose or none	f, b, s, t
Bacillariophyceae (a family within Chrysophyta)	diatoms	uni-	*a, c*	β-c, X	silica, calcium carbonate ($CaCO_3$), chitin	f, b, s, t
Euglenophyta	Euglena	uni-	*a, b*	β-c, X	none	f, b, t
Phaeophyta	brown algae	multi-	*a, c*	β-c, X	cellulose, alginic acid	b, s
Pyrrhophyta	dinoflagellates	uni-	a, c_1, c_2	β-c, xanthins	cellulose or none	f, b, s
Rhodophyta	red algae	multi-	*a*	X	cellulose	b, s

Notes: α-c = *alpha*-carotene; β-c = *beta*-carotene; ε-c = *epsilon*-carotene; X = xanthophylls; f = freshwater; b = brackish water; s = salt water, t = terrestrial

polysaccharide laminarin, rather than use simple starches. The food industry harvests brown algae to recover the alginic acid found in their cell walls. Food chemists then convert alginic acid to a compound called *algin,* which is a thickener for foods, cosmetics, and pharmaceuticals. For example, several face moisturizers sold today contain algin to give the product a creamy consistency and hold moisture in the skin.

Brown algae form large kelp beds or kelp forests in which robust holdfasts enable the kelp to withstand dislodging by strong currents or the pounding of waves. Large kelp consists of a stalk called a *stipe* leading upward from the holdfast to a round bladder from which many blades or stalks originate. The blades then grow upward until they reach the water's surface. In many of the world's oceans, kelp beds provide resting places for marine animals such as sea lions as well as feeding sites for small fish and large marine predators. On the U.S. West Coast, lush kelp forests serve as marine habitats off California; they also occur off Victoria Island of British Columbia, and along the Aleutian Islands of Alaska. The respected underwater photographer Chuck Davis wrote an eloquent description of kelp forests in his 1991 book *California Reefs*: "Unlike terrestrial plants that transport nourishment upward from root systems, the giant kelp plant doesn't have roots in the truest sense of the word. It must use its entire surface area to absorb nutrients from sea water. It also relies heavily on its surface canopy to capture solar energy. The products of photosynthesis are then conducted downward to low-light areas. . . . By this means, the plant forms extensive undersea forests that thrive in the shadow of its surface umbrella." Brown algae have created very distinct and specialized marine habitats throughout the world.

East coast kelp forests grow smaller than those in the West and are limited to an area from Nova Scotia to Cape Cod. As in West Coast kelp forests, eastern kelp serves as important habitat for seals, sea lions, otters, and marine grazers such as urchins and the starfish that prey on them. The world's most famous kelp bed grows in the Sargasso Sea in the North Atlantic Ocean 100 miles (161 km) east of the North American coast. The Sargasso Sea extends from Cape Hatteras, North Carolina, to Cuba and lies in the middle of several currents—the Gulf Stream and the Canaries Current are the main currents—that circulate counterclockwise around this 200-mile (322-km) region; see the color insert on page C-1. According to the season, the Sargasso Sea can stretch many hundreds of miles beyond than its normal area.

The entire area of the Sargasso Sea, called the *sargassum* and infamously known as the Bermuda Triangle, contains a mass of brown algae that remains in the sea's vortex in the center of the swirling currents. The algae mostly contain members of the genus *Sargassum,* primarily *S. natans.* This kelp mat provides a unique habitat of high-salt waters and distinct weather, which both supply breeding grounds for species more commonly found along coasts, including shrimp, crabs, worms, coastal fish, and eels. The Sargasso Sea comprises a floating ecosystem seen nowhere else on the globe. More than 50 fish species have contact with the Sargasso Sea, and the area provides habitat or feeding grounds to an enormous collection of invertebrates.

Dinoflagellates

Free-floating unicellular algae belonging to Pyrrhophyta are one of many types of organisms categorized as plankton. Most dinoflagellates live in salt waters, where they serve as the base for marine food chains. Other than this role, dinoflagellates offer few known benefits to humans except their attractive bioluminescence (also called *phosphorescence*), which glows in ocean waters at night.

Species belonging to *Pyrodinium, Noctiluca,* and *Gonyaulax* are the most common producers of bioluminescence. Some dinoflagellates and diatoms produce neurotoxins that are lethal to fish and other marine animals and harmful to people who eat contaminated seafood. The genus *Alexandrium* and the genera *Gymnodinium* and *Fibrocapsa* cause paralytic shellfish poisoning and neurotoxic shellfish poisoning, respectively. The dinoflagellate *Pfiesteria piscicida* has become a serious health threat that has proliferated in waters along the U.S. East Coast. This neurotoxin producer has killed billions of fish and caused severe symptoms in people who have contact with contaminated waters.

In the 1990s, *P. piscicida* caused massive numbers of fish deaths (called *fishkills*) from Chesapeake Bay to North Carolina. The North Carolina marine biologist JoAnn Burkholder discovered *Pfiesteria* in 1988 and named it after the dinoflagellate expert Lois Pfiester. Burkholder discovered that workers handling fish from poisoned waters contracted neurotoxic shellfish poisoning characterized by memory loss, headaches, skin rashes, upper respiratory irritations, and gastrointestinal ailments. She eventually traced the symptoms to two *Pfiesteria* toxins, one that stuns an infected fish and another that causes cellular damage. In 2008, the river ecologist Dean Naujoks gave a keynote speech at a meeting to recognize Burkholder's accomplishments: "Her research linking *Pfiesteria* toxins to massive fish kills, nutrient pollution and human illness led to hundreds of millions of dollars in water quality improvements and addi-

Major Types of Algal Seaweed

Type	Seaweed Characteristics	Habitat Characteristics
red	includes corallines that contain calcium carbonate used for making prostheses and for making foods, such as Japanese nori; carrageenans that give foods texture	marine intertidal and subtidal zones
brown	large filamentous forms are common, such as kelp, but no unicellular forms; harvested for alginic acid	prefers cold marine waters
green	chlorophylls *a* and *b* present in the same proportion as in higher plants; varies from unicellular to multicellular aquatic plants	fresh and marine waters

tional scientific research for the Neuse River [in North Carolina] and other coastal estuaries." With increased knowledge about this harmful microorganism, workers along infected waters have been better prepared to prevent infection.

Red tides and toxic blooms like that caused by *P. piscicida* have been increasing worldwide. Nutrient influx from agricultural runoff has been blamed as a major cause. Solving the dynamics of algae blooms is often difficult because of the complicated life cycles of the species involved, especially dinoflagellates. *P. piscicida* is a particular challenge for marine biologists to study because its life cycle consists of at least 24 different stages. One of the stages has become notorious because during this period *P. piscicida* cells use chemotaxis to detect fish as they swim nearby. The pathogen then becomes a predator and swims after the fish until it draws near enough to excrete its toxin.

Diatoms

Diatoms belong to the classification of golden-brown/yellow-green algae that use chlorophylls for energy production and contain the brown pigment fucoxanthin. Diatoms also make up a portion of heterogeneous organisms called *phytoplankton,* which are plankton of plant origin. Diatoms reproduce through sexual or asexual means and include at least 10,000 species with their own unique structures. Because of the large size of this group, many biologists prefer to classify diatoms as a distinct division of algae.

Diatoms possess some of the most beautiful and intricate structures in nature. Each diatom species possesses a characteristic shape; the body is called a *frustule,* and it is composed of two sections or halves called *thecae.* If the sections differ in size, the larger piece is called the *epitheca* and the smaller of the two is the *hypotheca.* Thecae fit together by overlapping and then bind with a material composed of silica. Diatoms are, in fact, unique in the biological world because they require silicon in their cell wall—as crystallized silica, $Si(OH)_4$—and some species require silicon for gene expression as well. Diatoms' exquisite frustules seem to disprove one maxim of biology, that there are no straight lines or 90° angles in the natural world.

Diatoms provide the microbial world with another unique feature: decreasing cell size with each new generation. In asexual reproduction, diatoms construct new theca inside the parent before the cell divides. Each successive generation produces smaller and smaller cells, in contrast to binary fission in bacteria, which produces two daughter cells that are replicas of the parent cell. Diatoms must find a way to return to their original size. When diatom cell size has diminished by about 30 percent, diatoms begin to reproduce sexually to form a resting cell called an *auxospore.* After the resting phase, a protoplast emerges from the auxospore and quickly expands to normal size before the cell builds a new rigid outer wall. Diatoms in this way provide a rare example in nature in which a protoplast plays an active role in a microbial life cycle. In bacteria, protoplasts form only when harsh environments damage cell walls, but the protoplasts never become part of bacteria's life cycle.

Diatoms are divided into the three following categories: (1) Centric diatoms, which inhabit marine waters and may be composed of either chains of interlocking frustules or free-floating planktonic cells; (2) pennate diatoms, which live in marine water or freshwater as well as the moisture on rocks or in soils; and (3) diatoms of the *Triceratium* species, which does not fit into either of the previous groups. Diatoms that stick to surfaces do so by secreting a mucuslike substance called *mucilage.* Mucilage forms weak bonds between diatoms and various surfaces so that the diatoms can glide across submerged surfaces rather than live anchored.

Diatomaceous earth contains a collection of fossilized frustules and has some value in industry. Diatomaceous earth has been used in toothpaste and in polishes because it is abrasive, and diatomaceous earth also serves as a low-cost material in large filters, such as swimming pool filters.

Diatoms produce a lethal neurotoxin called *domoic acid,* first discovered in mussels infected with the diatom *Pseudonitzschia* or the red alga *Chondria armata.* The neurotoxin can cause illness in people within 30 minutes to 24 hours after eating infected seafood. In severe cases, victims suffer permanent short-term memory loss in a condition called *amnesic shellfish poisoning* (ASP).

Marine and coastal animal populations have also suffered from domoic acid poisoning. In 1991, for example, pelicans fishing along the California coast began dying from a poisoning identified as domoic acid after eating anchovies. This incident provided the first solid evidence that domoic acid infection was not confined to the marine shellfish mussels, oysters, and razor clams; the poison could also be found in the nonmuscle tissue of anchovies, sardines, crab, and lobster. Since then, much of the research in domoic acid poisoning has been on seals and sea lions. The biologist Joe Cordaro of the National Marine Fisheries Service told the University of California–Santa Barbara *Daily Nexus* in 2003, "I've been recording numbers since 1998, and it [poisoning] seems to be happening on a yearly basis. Last year, over 1,000 animals came in [to the local Marine Mammal Care Center]." Domoic acid poisoning continues to threaten marine mammal health along the California coast.

Other algae that threaten the health of people or marine life consist of several species of green algae *(Chlorophyta),* gold-brown algae *(Chrysophyta),* and certain dinoflagellates; all have caused severe skin irritations and allergies in humans. Fishermen who handle infected catch have the highest risk of skin irritations, but marine biologists have not yet determined whether the symptoms arise from a toxin or an allergen or perhaps another compound altogether.

Water Molds

Deoxyribonucleic acid (DNA) analysis relates the group of organisms known as *water molds* to diatoms and dinoflagellates. Many terrestrial species of water molds cause plant diseases and create increased costs for the agriculture industry. The blight mold *Phytophthora infestans,* a type of water mold, affected history when it caused Ireland's potato famine in the 1800s. Other water molds cause infections in healthy vegetation: White rusts infect chrysanthemums, and *Phytophthora* species, which are pathogens in Australian eucalyptus trees, cause sudden oak disease in the western United States.

Euglena

The genus *Euglena* belongs to the Euglenophyta division. It has been studied under microscopes by generations of students learning about cell structure, organelles, and cell motility. *Euglena* species possess elongated single cells that range in length from 15 to 400 μm and contain one to three flagella. The flagella are all positioned on one side of the cell, in what is termed a *paraflagellar arrangement.* Only one flagellum at a time protrudes from an anterior pocket of a *Euglena* cell. This pocket called the *cytostome* or *reservoir* is a distinguishing feature of *Euglena.*

Euglena possesses two modes of motility. First, the single flagellum propels *Euglena* through the water. An eyespot located beside the reservoir and containing the light-sensing pigment β-carotene allows the cell to move toward light. In the dark, *Euglena* cannot use photosynthesis to make energy, so it depends on the cytostome to ingest organic matter and so provide the cell with nutrients. Each cell also contains an inner protein membrane called a *pellicle* that provides support yet allows the cell to retain some flexibility. This flexibility benefits *Euglena* by giving it its second mode of propulsion through water. This method involves the repeated swelling and constricting of the cell. By doing this, *Euglena* can crawl along submerged surfaces rather than swim.

Euglenophyta belongs to the Entamoebae group of eukaryotes. They all have similar rRNA composition, and, as do all eukaryotes, they contain disk-shaped plates inside their mitochondria called *cristae.* Cristae may serve the purpose of providing an increased membrane area for carrying out energy metabolism.

Euglena is an important representative of the evolution of higher organisms for three reasons: (1) They respond to a stimulus; in this case light acts as the stimulus; (2) they maintain a distinct cell structure; and (3) they are equipped to use alternative means of locomotion. *Euglena* may be thought of as a cross between a plant and an animal because it performs plantlike photosynthesis but also moves in the direction of a stimulus as animals do. Also similarly to an animal lifestyle, *Euglena* takes in nutrients even when there is no sunlight, but, unlike

These phytoplankton, called diatoms, were found living between ice crystals in the sea ice of McMurdo Sound, Antarctica. ***(NOSS/Department of Commerce)***

higher life-forms, *Euglena* shuns sexual reproduction and instead propagates by binary fission. Each cell splits in two along its long axis with one half getting the active flagellum and the other half left with the task of constructing its own new flagellum.

ALGAE IN NATURE AND ALGAL BLOOMS

Algae have their own specific aquatic habitats. Their distribution in waters depends on their pigments and the amount of light available for photosynthesis. Red algae live in depths from 656 to 820 feet (200–250 m). Brown algae holdfasts, by contrast, connect to the sea bottom at depths of up to 330 feet (100 m) and allow the algae to extend up to the water's surface. Unicellular green algae, dinoflagellates, and diatoms must live where short wavelength light penetrates the water; their habitat ranges from the surface to a depth of no more than 33 feet (10 m), depending on the clarity of the water. When photosynthesizing, *Euglena* cells live near the water's surface and often form green blooms of many billions of cells that cover still ponds bathed in sunlight. If places on land hold high amounts of moisture, green and brown algae, diatoms, and the Euglenophyta will also live there.

Earth's algae play a crucial part in biogeochemical cycling of some of the earth's essential nutrients: nitrogen, phosphorus, sulfur, and carbon. Algae play a role in almost every food chain due to their capture (called *fixing*) of atmospheric carbon dioxide to begin the carbon cycle. During photosynthesis, algae use energy to convert carbon dioxide into carbohydrates for storage. In this process, the cells release oxygen into the atmosphere. Planktonic algae occupying the upper layers of the oceans may contribute up to 80 percent of Earth's atmospheric oxygen.

Pollution destroys algae's ability to produce oxygen, and with enough pollution an algal bloom grows. To begin, rain washes large amounts of organic matter into tributaries. Organic compounds from agricultural fertilizers, farm and feedlot wastes, and municipal runoff then cause a surge in the growth of planktonic algae. Although this influx of nitrogen and phosphorus might be thought of as a benefit to Earth by reducing carbon dioxide levels and increasing oxygen, the sudden bloom of algal growth indicates that waters have been polluted with organic wastes, including sewage. As billions of algal cells die, bacteria grow to large numbers, as they decompose the bloom. The large-scale bacterial activity consumes enormous amounts of oxygen in a short period. A dead area then forms in the pond, lake, or sea where this oxygen depletion has occurred, a process known as *eutrophication.* Not only do algal blooms create eutrophicated waters, but the large numbers of algae may produce neurotoxins that put marine and human health at risk.

Algal blooms cause an indirect effect on aquatic life when the algae grow into thick mats over the water's surface and block sunlight from reaching aquatic plants. After aquatic plants and grasses die, the numbers of mollusks, finfish, and diving waterfowl that feed on them begin to decline, as do those of large predators that seek these animals. An algal bloom can destroy an entire land-water ecosystem this way. Indirectly, blooms also influence where people will choose to vacation for fishing, snorkeling, diving, and bird-watching.

ROLES OF ALGAE IN INDUSTRY

Algae serve humans in many ways from inexpensive diatomaceous earth in cleansers and filters to foods. Students observe algal cells to learn about cell structure and function, and certain algae produce agar for use in microbiology studies. Seaweed has been part of Asian diets for centuries, and it has increasingly become part of diets in other parts of the world. Seaweeds sold as food are the following: nori, kelp, alaria, dulse, and digitata. These items are rich in protein and important sources of amino acids that are usually low in other common foods. Seaweeds also provide β-carotene and B-complex vitamins plus a variety of essential macro- and micronutrients: calcium, chloride, potassium, sodium, phosphorus, magnesium, iron, zinc, copper, manganese, iodine, selenium, molybdenum, and chromium.

The food industry uses alginate and carrageenans extracted from seaweeds to add consistency to packaged foods. Alginate thickens puddings, yogurts, sauces, gravies, and syrups, and personal care products such as lotions and toothpaste. The paper and textile industry uses alginate to absorb water generated during production steps, and the drug industry uses it as an inert (inactive) ingredient in drugs. For example, the inert portions of tablet and capsulated drugs contain alginates from red and brown algae. Alginate and carrageenans are safe to consume when used in foods and drugs.

Kelp has been used as protection against radioactive iodine 131 released in small amounts each day from nuclear reactors. It serves this medical purpose because kelp contains a high concentration of iodine 127, which when ingested may block the body's absorption of iodine 131. The University of Delaware oceanographer Geaorge Luther said in 2008 to the campus newspaper *UDaily,* "Brown kelp has 1,000 times more iodine than what is in the sea, and it is always taking it on." Since the Middle Ages, eating kelp has been used as a cure for goiter, an enlargement of the thyroid gland due to lack of iodine in the diet.

Seaweed has also gained acceptance in Western countries as a fertilizer for home gardens and in agriculture. Coastal populations outside the United States have long used dried seaweed as a fertilizer and for soil conditioning because it is a good source of nutrients for plants.

A new industry based on the harnessing of algae as an energy source for human use has been growing. This invention, called a *biocell,* generates energy from biological activities such as photosynthesis rather than chemical reactions. In 2007, the *San Francisco Chronicle* reporter David R. Baker wrote of one of the new enterprises pursuing algae energy: "The algae beneath Harrison Dillon's microscope could one day fuel your car. Dillon's Menlo Park [California] company, Solazyme, has tweaked the algae's genes to turn the microscopic plant into an oil-producing machine. If everything works . . . vats of algae could create substitutes for diesel and crude oil." So far, algae biocells have been made to generate small electrical outputs, enough to power a hand calculator, for instance. Larger outputs of energy from algae may be part of future energy production.

Culturing Algae

Algae's nutrient requirements resemble those of photosynthetic land plants; they require oxygen, carbon dioxide, water, minerals, and a light source. As are bacterial cultures grown in a laboratory, algae can be grown on agar plates or in broth. Algal growth is unique, however, because many species form filaments that create a large mass on solid or liquid media.

Algae require up to three weeks of incubation at room temperature with periods of light alternated with dark periods. Generation time among the algae, the time needed for a cell to divide into two new cells, varies by species and ranges from six hours to more than 80 hours. This long growth period increases the risk of contamination in algal cultures, so algologists (microbiologists who specialize in growing algae) must use good aseptic techniques and take extra steps to prevent contamination. Three precautions help prevent contamination in algae cultures. First, an algologist supplements the growth medium with antibiotics to prevent bacteria or molds from entering the culture. Second, algae cultures may be exposed to low doses of ultraviolet light, which does not harm the algae but does kill many contaminants. Third, enrichment media have been designed that favor the growth of algae over other microorganisms. For example, seawater medium contains a blend of salts that mimic the composition of marine water. Marine algae grow well in it, while most potential contaminants cannot tolerate the salty conditions.

The diverse world of algae holds many benefits for humans and contains clues on the evolution of life on Earth. Algae have played an important part as a teaching tool for learning about eukaryotic cell structure, and science continues to find valuable uses for algae in industry.

See also ASEPTIC TECHNIQUE; BLOOM; CYANOBACTERIA; EUKARYOTE; PLANKTON; PROTOPLAST.

Further Reading

Anderson, D. M. "The Growing Problem of Harmful Algae." *Oceanus Magazine,* July/August 2004. Available online. URL: www.whoi.edu/oceanus/viewArticle.do?id=2483. Accessed March 10, 2009.

Baker, David R. "Green Valley." *San Francisco Chronicle,* 4 March 2007.

Bryant, Tracey. "Iodine Helps Kelp Fight Free Radicals and May Aid Humans, Too." *University of Delaware UDaily,* 17 June 2008. Available online. URL: www.udel.edu/PR/UDaily/2008/jun/iodine061708.html. Accessed March 10, 2009.

Davis, Chuck. *California Reefs.* San Francisco: Chronicle Books, 1991.

Graham, Linda E., and Lee W. Wilcox. *Algae.* Upper Saddle River, N.J.: Benjamin Cummings, 1999.

Microscopy-UK. "Algae." Available online. URL: www.microscopy-uk.org.uk/index.html?http://www.microscopy-uk.org.uk/pond/algae.html. Accessed March 24, 2009.

Naujoks, Dean. "Speech Honoring Dr. JoAnn Burkholder for Receiving the River Network Jim Compton Lifetime Achievement Award, Cleveland, OH, May 4, 2008." Available online. URL: http://www.neuseriver.org/images/River_Hero_Award_II_2008_Burkholder_5-2008_1_.pdf. Accessed March 10, 2009.

Phycological Society of America. Available online. URL: http://www.psaalgae.org. Accessed March 10, 2009.

Sherr, Barry F., E. B. Sherr, David A. Caron, Dominique Vaulot, and Alexandra Z. Worden. "Ocean Protists." *Oceanography* 20 (2007): 130–134.

Thomas-Anderson, Melissa. "Domoic Acid Sickens Sea Mammals." *University of California-Santa Barbara Daily Nexus,* 21 May 2003. Available online. URL: www.dailynexus.com/article.php?a=5335. Accessed March 10, 2009.

Van den Hoek, Christiaan, David Mann, and Hans M. Jahns. *Algae: An Introduction to Phycology.* Cambridge: Cambridge University Press, 1996.

amoeba (plural: amoebae) An amoeba is a type of protozoa characterized by the ability to change shapes. Amoebae move by extending part of their cell forward like a foot and then filling the extension with cell contents. Each step in this type of locomotion called *pseudopodia* moves the cell through its watery environment. The lobes or extensions are called *pseudopods.*

Amoeba cells grow larger than bacteria, but the continually changing shape makes size measurements difficult. Amoeba can range from about 40 micrometers (μm) to more than 600 μm in diameter. Their ever-changing structure makes amoebae equally difficult to identify to species or even genus level under a microscope. As bacteria do, amoebae reproduce asexually by splitting in two in a process called *binary fission.*

As all other eukaryotic cells do, amoebae have distinct membrane-enclosed organelles such as a nucleus (some species have many nuclei), mitochondria, water-containing contractile vacuoles, and food vesicles in their cytoplasm. Amoebae are also divided into two general groups based on morphology (cell structure): naked amoebae and shelled amoebae. Naked amoebae possess a flexible membrane that holds in the cell contents. Shelled amoebae contain an outer structure called a *shell* or a *test,* which provides the cell with a characteristic shape. The shell is rather fragile, however, and does not provide the physical protection and strength seen in other protozoa that make much stronger forms called cysts. Many shelled amoebae travel by pseudopodia by extending a pseudopod out of their shell and then moving in that direction.

Amoebae possess methods of mobility without the use of pseudopods. For example, amoebae move by changing the consistency of their cytoplasm. Cytoplasm is the aqueous material inside all prokaryotic and eukaryotic cells. In eukaryotes, cytoplasm occurs inside the membrane but outside the nucleus and other organelles. Amoebae change their cytoplasm depending on their need to move. When they prepare to become motile, their cytoplasm becomes soft and fluid, and it is referred to as *plasmasol.* Plasmasol gives the cell flexibility. When the cell stops moving, the cytoplasm becomes slightly firmer and is called *plasmagel.* Microbiologists have not yet fully explained how plasmasol turns into plasmagel and back. It is known, however, that high pressure on the outside of the cell hastens plasmagel's breakdown to fluid and the compound hyaline gives the gel its firmness, especially inside pseudopods.

Because they are found only in aqueous environments, amoebae are good at floating with the currents as another means of motility. To do this efficiently, amoeba cells extend several pseudopods in different directions at once to create a starlike shape. Much as a sailboat unfurls its sails, these pseudopods help the cell float and ride on the current.

Under the microscope or in a video, amoebae appear rather sluggish when they move through water. Amoebae in nature, in fact, move quickly to catch their food. Amoebae eat by engulfing their prey, which may include small bits of organic matter, bacteria, algae, or other protozoa. They have no trouble catching nonmotile algae and bacteria by extending pseudopods around them and engulfing them. But ciliated protozoa and flagellated bacteria contain many short cilia or long flagella, respectively, which make them fast swimmers. Amoebae must move at least as fast to catch these cells.

Amoebae extend their cell bodies entirely around their food or prey before actually touching it. Once completely surrounding the item, the amoeba cell gathers it into its cytoplasm in a process called *pinocytosis*. Pinocytosis is similar to phagocytosis that occurs in the mammalian bloodstream in which white blood cells envelop foreign materials in the blood. The amoeba draws the food into a vacuole into which it releases digestive enzymes. The enzymes degrade the food into small pieces that the cell can then absorb into its cytoplasm. Amoebae meanwhile excrete wastes and carbon dioxide through their cell membrane and take in oxygen through the membrane. They carry out oxygen and carbon dioxide exchange by simple diffusion. Simple diffusion is a process that does not cost the cell any energy to perform because materials cross the membrane without the need for active transport systems. By contrast, waste excretion requires energy.

Amoeboids is a term for microbial species that are not true amoebae but move as amoebae do by using pseudopods. This type of motility is called *amoeboid movement*. Examples of organisms that use amoeboid movement are slime molds, water molds, members of the protozoal subphylum Sarcodina, and some algae (Chlorarachniophytes).

AMOEBA HABITATS

Amoebae's fragile cell structure forces them to inhabit moist environments. When amoebae live in soils rather than water, they must inhabit only soils with high moisture content. The digestive tract of humans and other animals provides an aqueous environment where amoebae exist and can cause potential health problems. The vast majority of amoebae, however, live as harmless inhabitants of the environment in ponds, lakes, and other natural bodies of water.

Aquatic amoebae can live in either freshwater or saltwater habitats. Freshwater habitats include ponds, ditches, and slow, moving waters high in organic matter. Some amoebae live in marine habitats, but most saltwater amoebae prefer brackish (salty, but not as salty as the ocean) waters rather than ocean waters.

Amoebae have certain requirements for the type of water in which they live due to their need for isotonic conditions. *Isotonic conditions* describes a situation in which the pressure outside a cell equals the pressure inside the cell. Unlike bacteria, amoebae do not have a rigid cell wall to protect them from too much or too little pressure in their surroundings. Most cells rely on the process of osmosis to regulate pressure; osmosis is the transfer of water into or out of a cell to keep the pressure of the contents equal to the pressure in the environment. But because amoebae do not have this ability and also lack a cell wall, both hypertonic and hypotonic conditions can injure or kill amoebae. In hypertonic conditions, the concentration of matter outside the cell is greater than that inside. When this occurs, water rushes out of the cell, and it shrivels. In hypotonic conditions, the concentration of matter outside the cell is less than that inside. Water then rushes into the cell, and it eventually bursts. Amoebae, in nature, seek habitats that provide the correct osmotic conditions for keeping them alive. Microbiologists who study amoebae in a laboratory specialty called *protozoology* ensure that the osmotic pressure is safe for the cells being studied. For this purpose, laboratory media usually contain 7.5 grams of salt (NaCl) along with other nutrients per liter of water to provide osmotic conditions favorable to amoeba cells.

NONPATHOGENIC AMOEBAE

Amoebae belong to the order of protozoa called Amoebida. Most species within this order live in the environment where they digest organic matter and do not affect human or veterinary health. Common nonpathogenic Amoebida genera of freshwater habitats are *Amoeba, Echinamoeba, Mayorella, Rosculus, Saccamoeba, Thecamoeba, Trichamoeba, Vannella,* and *Vexillifera*. Of these, *Vannella* also grows well in marine environments, usually by living on seaweed. The genera *Flabellula* and *Platyamoeba* have also been found in marine waters. *Mayorella* species live in digestion tanks at sewage treatment plants in addition to their natural freshwater habitats. Of the Amoebida, some genera are inhabitants of animal intestines. The most well-known of these genera are the *Entamoeba,* which, other than the parasite *E. histolytica,* live as harmless inhabitants in the digestive tract of humans and other animals.

Little information has been gathered on the role amoebae play in the digestive tract. Because these microorganisms actively envelop and digest organic matter by pinocytosis, they may contribute to food digestion. When amoebae die and their cells lyse (break apart), the host animal absorbs the digested matter and uses it as nutrients in its own metabolism. If this is the main benefit of amoebae in digestion, then it is similar to the many other types of protozoa known to inhabit the intestines. On occasion, however, a more dangerous species finds its way into the digestive tract via food or water. When it causes disease or death, it is classified as a pathogen.

PATHOGENIC AMOEBAE

One species of *Entamoeba* causes a devastating form of diarrhea that afflicts millions throughout the world each year. The illness is amoebic dysentery, or amoebiasis, caused by *E. histolytica*. Dysentery is any severe diarrhea containing blood and mucus. Amoebic dysentery spreads in food or water contaminated with fecal matter. In severe cases, dehydration caused by the diarrhea can lead to death. In amoebic dysentery, *E. histolytica* cells invade the epithelial cells lining the intestinal tract. The damage can be so great as to cause abscesses within the intestinal wall and allow the infection to invade the liver and other organs. Worldwide, about 10 percent of people carry *E. histolytica* without developing symptoms but they can spread the disease to others.

E. histolytica and other *Entamoeba* species form a rugged cyst that may help them withstand harsh conditions in the environment. Undoubtedly this cyst helps *E. histolytica* remain dangerous as it spreads through a population by way of food or water. Normal chlorine levels used by water treatment plants for disinfecting drinking water do not kill *E. histolytica* cysts, but the cysts can be damaged by very cold (less than -40°F [–5°C]) or very hot (greater than 104°F [40°C]) temperatures or by drying.

E. histolytica thrives in the anaerobic conditions found in the digestive tract. Inside the intestines, the cysts release trophozoites of 10–60 μm in diameter.

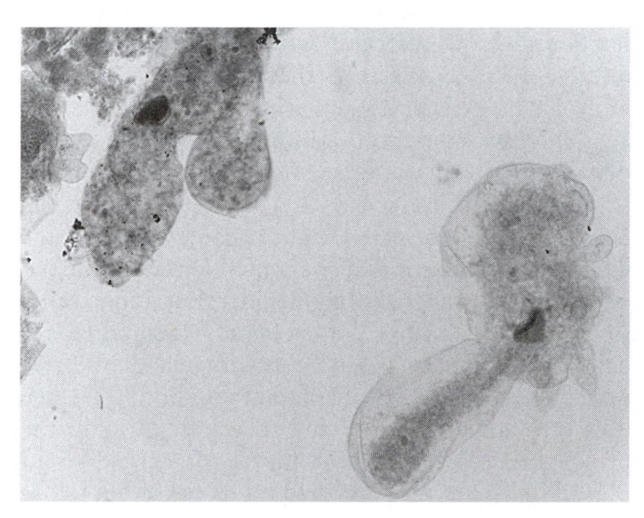

An amoeba cell contains organelles such as the nucleus *(n)*, contractile vacuole *(cv)*, phagocytic vacuoles *(wv)*, and a vacuole containing food *(fv)*. ***(Edmund B. Wilson.* The Cell in Development and Inheritance. *The Macmillan Company, 1900)***

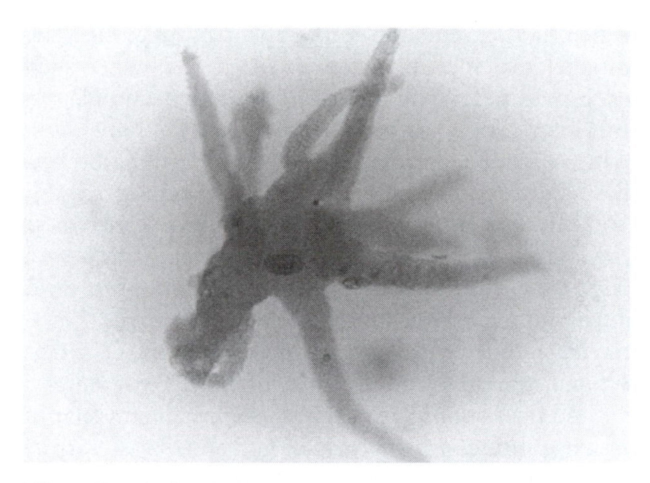

When disturbed by irritations in its environment, *Amoeba proteus* assumes a star shape that helps it move faster in a flowing current. ***(David Byres)***

Trophozoites act as the feeding form of *E. histolytica* because in this form the pathogen eats through the intestinal lining. Inside the body, trophozoites divide and form new cysts, which then spread the disease to other people, when excreted in fecal matter. *Entamoeba* is defined as an *obligate parasite* because it depends on getting nourishment from a host animal for its survival.

Amoebic dysentery afflicts humans, primates, and domestic dogs and cats. In humans, the infection moves from the intestines to the liver, leading to the liver disease hepatitis. Doctors diagnose *E. histolytica* infection by looking for trophozoites in a stool sample. In humans, antibodies against the trophozoites usually appear in the blood about seven days after infection, but these antibodies may be only marginally able to fight off the pathogen in most people.

Doctors treat amoebic dysentery by giving the antibiotic metronidazole followed by iodoquinol. In addition, people who travel outside the United States should avoid ingesting unboiled water, ice made from unboiled water, and fresh fruits and vegetables. Doctors additionally advise travelers to tropical areas to take extra precautions against this illness known as *travelers' dysentery*. India and Mexico have been cited as particularly high-risk regions for contracting *travelers' dysentery* when visitors do not take the necessary precautions.

For animals, most veterinarians follow a similar course for dogs to that used for people, that is, treatment with metronidazole. *Entamoeba invadens* looks almost identical to *E. histolytica,* but although it infects reptiles, it has never been shown to infect mammals.

Amoebic dysentery has declined in the United States over several decades as safer drinking water sources and better sanitation developed. But it is still a health threat in the world's tropical regions. The disease also increases during times of disaster such as hurricanes, typhoons, floods, and tsunamis. In these situations, water treatment plants become overwhelmed with contaminated water and pathogens have a greater chance to enter clean water supplies.

Microorganisms other than amoebae cause their own forms of dysentery: the bacteria *Shigella* and *Escherichia coli,* the protozoan *Balantidium coli,* and a number of viruses, especially rotavirus, coronavirus, and viruses of the intestinal tract. *E. histolytica* is the only known cause of amoebic dysentery.

Acanthamoeba is a freshwater inhabitant that in humans can cause eye infections (amoebic keratitis) as well as meningoencephalitis, an inflammation of the brain and the connective tissue (the meninges) that protects the brain and the spinal cord. Infections have been attributed to swimming in contaminated freshwaters. Amoebic keratitis (AK) has become an increasing problem in medicine, particularly among contact lens wearers. Andrea Boggild reported in a 2009 issue of *Journal of Clinical Microbiology,* "The overwhelming majority of cases of AK occur in immunocompetent contact lens wearers, and outbreaks have been linked to contact lens solutions contaminated with acanthamoebae or to those that fail to effectively decontaminate lenses." AK outbreaks of more than 100 people at a time have led to recalls of lenses and cleaning solutions. The public health physician Basilio Valladares of the University of La Laguna in the Canary Islands said in a 2008 news release, "When people rinse their contact lens cases in tap water, they become contaminated with amoebae that feed on bacteria. They are then transferred to the lenses and can live between the contact lens and the eye. This is particularly worrying because commercial contact lens solutions do not kill the amoebae." Scientists studying contact lens cases at this university found almost 70 percent of the cases contaminated with *Acanthamoeba,* and 30 percent of those organisms were highly pathogenic varieties.

Most *Paramoeba* species live as harmless free-floating amoeba of salt water. But *P. perniciosa* infects blue crabs common in the Chesapeake Bay and other places along the U.S. East Coast. Other *Paramoeba* species have infected and killed large populations of sea urchins in the North Atlantic along Nova Scotia.

Amoebae require certain conditions in water habitats, most notably proper isotonic conditions. Much remains unknown on the life cycle, nutrient requirements, and disease aspects of amoebae. Because these microorganisms inhabit a wide range of habitats, they probably carry out many activities in the environment that have not yet been discovered.

See also BINARY FISSION; EUKARYOTE; HEPATITIS; MOTILITY; OSMOTIC PRESSURE; PATHOGEN; PROTOZOA.

Further Reading

Boggild, Andrea K., Donald S. Martin, Theresa YuLing Lee, Billy Yu, and Donald E. Low. "Laboratory Diagnosis of Amoebic Keratitis: Comparison of Four Diagnostic Methods for Different Types of Clinical Specimens." *Journal of Clinical Microbiology* 47 (2009): 1,314–1,318. Available online. URL: http://jcm.asm.org/cgi/reprint/47/5/1314. Accessed November 4, 2009.

Medical News Today. "Contact Lenses Are Home to Pathogenic Amoebae." News release, October 21, 2008. Available online. URL: www.medicalnewstoday.com/articles/126235.php. Accessed March 10, 2009.

Patterson, David J. "Amoebae: Protists Which Move and Feed Using Pseudopodia." Tree of Life Web Project. Available online. URL: http://tolweb.org/notes/?note_id=51. Accessed March 10, 2009.

University of Edinburgh. Genes and Development Group. Available online. URL: www.bms.ed.ac.uk/research/others/smaciver/amoebae.htm. Accessed March 10, 2009.

anaerobe An anaerobe is any microorganism that cannot grow well in the presence of free oxygen (O_2). Put another way, anaerobic microorganisms thrive where oxygen is absent and where other forms of life that rely on respiration cannot exist.

Anaerobes can be divided into the following four major groups: (1) obligate or strict anaerobes that cannot tolerate oxygen and die when exposed to it; (2) facultative anaerobes that do not require oxygen and can live in its presence or its absence but grow slightly better when oxygen is available; (3) aerotolerant anaerobes, which are unaffected by oxygen levels and grow the same with or without oxygen; and (4) microaerotolerant anaerobes that can tolerant oxygen levels at no higher than 5 percent.

The atmosphere of primitive Earth before life began contained no free oxygen. Life that formed more than 3.8 billion years ago did so when elements combined and formed simple organic compounds. These reactions required energy, which was supplied by lightning, ultraviolet light, heat from volcanoes, or heat generated from asteroids crashing to Earth. The earliest life developed in an atmosphere of methane (CH_4), hydrogen (H_2), and ammonia (NH_3). The only oxygen was bound to water vapor in the atmosphere. When the first bacteria emerged about 3.8 billion years ago, they had no recourse but to use the oxygenless conditions.

Anaerobic bacteria evolved in an environment of intense heat from volcanoes and steam fissures in the earth. Eruptions from the earth disgorged toxic compounds. Ultraviolet light pounded onto a planet that contained no or very little greenhouse effect to hold in warmth. Early in the evolution of prokaryotes, the path of evolution leading to higher life-forms split to form two paths: one leading to bacteria and one leading to microorganisms called *archaea*. Archaea now compose a separate kingdom, of which many members still thrive in extreme environments similar to those of early Earth: intense heat, high chemical concentrations, toxic compounds, strong acids, and gases other than oxygen. Many anaerobic bacteria still withstand these extreme conditions, but other anaerobes evolved as Earth changed into today's more temperate conditions.

Anaerobes remained on Earth even as photosynthetic microorganisms and plants began to arise and filled the air with oxygen. By about 2.5 billion years ago, the atmosphere contained 1 percent oxygen, far less than the 21 percent today. Over millennia, however, oxygen levels increased. Rather than adapt to the rising levels of free oxygen, anaerobes found habitats protected from the atmosphere. In certain habitats, aerobic bacteria had already consumed all the oxygen, so anaerobic microorganisms flourished. Many of today's known anaerobes continue to depend on the oxygen-consuming activity of aerobes before the anaerobes can take over a habitat and multiply.

The French bacteriologist Louis Pasteur (1822–95) spent a good portion of his scientific life studying the anaerobic reactions in alcohol-producing fermentation. By doing so, he laid the foundation of modern microbiology, and his experiments on fermentation also built a store of knowledge on anaerobes. Pasteur published "Animal Infusoria Living in the Absence of Free Oxygen, and the Fermentations They Bring About," an 1861 article describing fermentation that produced the simple organic compound butyric acid. In his article, Pasteur described microscopic cylindrical rods gliding through the fermentation liquid. He called this form of life an infusorium and defined it as "the first example of an animal living in the absence of free oxygen." Because of Pasteur's growing status in science, most of his peers accepted his theory even though it was contrary to their long-held beliefs about life-forms. Pasteur's article is now recognized as the first known report of anaerobic activity among microorganisms.

Anaerobic fermentations producing alcohols became a common area of study in early microbiology. Scientists found fermentation experiments easy to conduct, and the reactions produced an array of interesting end products from various sugars. The many biochemical pathways in fermentation soon fascinated both biologists and chemists. Sir Arthur Harden (1865–1940) was awarded the 1929 Nobel Prize in chemistry for describing the details of alcohol fermentations that had been employed in wine making since early civilization.

Both anaerobic and aerobic bacteria contribute to Earth's biogeochemical cycles. These cycles are essential pathways in which nutrients and energy are reused by living plants and animals. Anaerobes have been responsible for producing Earth's deposits of fossil fuels—coal, oil, natural gas—from decomposing the planet's organic waste matter millions of years ago, a process anaerobes continue to perform today. Methane that has built up in pockets between underground rock formations was created by anaerobes; today this gas presents a hazard to miners working in deep underground shafts. Anaerobes also aid in digesting foods in animal digestive tracts because of their ability to decompose organic matter in oxygenless environments.

Anaerobic bacteria are important in clinical microbiology because these microorganisms cause several human and animal infections. Clinical microbiologists use special methods to grow and identify unknown anaerobic bacteria isolated from patients. Industrial microbiology prefers aerobic to anaerobic species for manufacturing biological products because anaerobes do not grow as fast as microorganisms that use oxygen. Anaerobes also play a useful role in wastewater treatment as a tool in digesting sludge high in organic matter.

The table presents the main anaerobic bacteria studied in microbiology and their growth preferences regarding oxygen.

ANAEROBIC HABITATS

In nature anaerobes grow in deep sediments beneath Earth's crust. Microbiologists recover anaerobic bacteria from places where oxygen has been used up by aerobes or where oxygen has been displaced under pressure. Deep ocean and freshwaters contain anaerobes, as do stagnant ponds, still lakes, bogs, swamps, and slow-moving rivers. In these places, anaerobes congregate in the deepest regions and in sediments.

Anaerobes also survive in places that at times can contain large levels of oxygen: soils, the skin surface, inside the mouth, and within biofilms. The reason anaerobes survive in these places is that they find oxygen-free places, called microenvironments. A microenvironment refers to a location that holds unique, highly specialized characteristics. In soil, anaerobes find small pockets between soil particles that contain no oxygen, especially in water-saturated soils. Oral anaerobes find microenvironments between the gum and teeth. Often these places have been made anaerobic by aerobes that first use up all the oxygen, making the conditions favorable for anaerobes to grow.

Anaerobic Bacteria Genera

Obligate Anaerobes	Facultative Anaerobes	Aerotolerant Anaerobes	Microaerotolerant Anaerobes
Bacteroides	*Actinomyces*	*Bifidobacterium*	*Campylobacter*
Clostridium	*Escherichia*	*Propionibacterium*	*Clostridium*
Desulfovibrio	*Lactobacillus*		*Helicobacter*
Eubacterium	*Propionibacterium*		
Fusobacterium	*Salmonella*		
Peptostreptococcus			
Porphyromonas			
Selenomonas			
Succinovibrio			
Veillonella			

Note: See Appendix V for microbial classifications

When microbiologists discover new bacteria in the environment or in medicine, they determine the oxygen needs of these bacteria in a laboratory. Oxygen is toxic to most bacteria, whether the bacteria are aerobic or anaerobic. Aerobes, facultative anaerobes, and aerotolerant bacteria fight oxygen's toxic effects with enzymes they possess for just such purposes. Anaerobes do not contain these enzymes because in their normal habitats they are not exposed to oxygen. When microbiologists grow anaerobes in a laboratory, they carry out extra steps to assure the anaerobes do not become exposed to the air. *Fastidious anaerobes* is a term used for microorganisms with the strictest requirements for an oxygenless environment. Although some anaerobes can withstand small exposures to oxygen, fastidious anaerobes cannot and die if even a tiny amount of oxygen pervades the medium in which they grow.

To test a culture's ability to grow in the presence of oxygen, a microbiologist inoculates a tube of broth medium and then incubates it in an upright position without shaking or mixing it. Anaerobes grow only in the region of the tube where oxygen levels allow their survival. Aerobes, by contrast, cluster toward the upper surface of the broth, where oxygen is highest.

Microbiologists prepare anaerobic media by flushing out all air and replacing it with an inert gas, such as nitrogen, or a nitrogen–carbon dioxide mixture. A microbiologist can test a medium's oxygen levels by measuring a value called the reduction-oxidation (redox) potential (Eh). Commercially prepared redox dyes added to media give an indication of the oxidized state (free oxygen present) or reduced state (free oxygen absent). The inert gas can be bubbled through the medium until the redox dye changes colors. Reasurian serves this purpose as follows:

oxidized reasurian dye (dark pink) $\rightarrow$ reduced resorufin dye (colorless)

The standard redox potential (Eh_0) of the preceding reaction is -51 millivolts (mV) at pH 7; when the redox potential drops to -51 mV in the medium, the formula turns from pink to colorless. The Eh_0 of most dyes increases 30–60 mV for each unit of pH decrease.

The pH of the medium affects the redox potential of various dyes because pH is an indication of the concentration of electrons available for reducing oxidized compounds. The standard Eh_0 values published in tables represent conditions at pH 7.0. Low Eh_0 values of -100 to -360 mV indicate a condition called a *reducing environment* in which electrons are readily available to reduce any oxidized compounds.

Facultative anaerobes can lower the Eh of natural environments themselves by using up all the oxygen. In oral bacterial communities, these bacteria lower the Eh to at least -100 mV, where obligate anaerobes grow. Fastidious anaerobes often cannot live at Eh values above -300 mV.

Depending on the environment, anaerobes can be dangerous pathogens or useful microorganisms. Pathogenic anaerobes cause skin infections in wounds that have injured tissue containing tiny airless pockets. For example, *Clostridium perfringens* bacteria infect wounds, followed by the development of gas gangrene. *C. perfringens* and other members of *Clostridium* also cause serious food-borne illnesses. *Clostridium botulinum* illness results from eating contaminated packaged foods held inside a container that has been purged of air during manufacture. *C. perfringens* illnesses may be caused by contaminated food or water.

Animals and some insects harbor large populations of anaerobic bacteria and protozoa in their digestive tracts. Ruminant animals (or ruminants, animals with a compartmentalized stomach containing a large organ called the *rumen*), such as cattle, sheep, and goats, depend on anaerobic microorganisms to digest dietary fiber. Nonruminant animals such as humans, pigs, dogs, and cats (animals with a single-compartment stomach) also harbor have anaerobic populations in their lower digestive tract that contribute to fiber digestion. Intestinal anaerobes also make vitamins and amino acids that the host animal absorbs and uses for its own metabolism. Ruminants receive an added benefit from anaerobic metabolism of dietary fiber due to the large amounts of organic compounds called *volatile fatty acids* (VFAs) produced by the anaerobic breakdown of sugars. Ruminant animals depend in great part on the VFAs as their energy source. Insects behave as ruminants do in the way they rely on anaerobic populations to help with digestion. Termites and cockroaches are two examples of insects that could not live without their population of anaerobes. Termites in particular use the ability of anaerobes to break down woody fibers in the termite digestive tract.

ANAEROBIC METABOLISM

All living cells must possess ways to break down nutrients for building new cells and for providing energy to run the cell's activities and maintenance. The biochemical pathway called *glycolysis* serves anaerobic as well as aerobic mammalian cells as the main process for converting six-carbon glucose to three-carbon pyruvic acid. Aerobic bacteria and mammalian cells shunt pyruvic acid into aerobic respiration via the Krebs cycle. This respiration depends on oxygen as a crucial component in energy generation. Oxygen acts as a receptor for electrons that pass through an energy-generating series of reactions called the *electron transport chain*.

Anaerobes do not use oxygen for accepting electrons. They instead use nitrate (NO_3^-), sulfate (SO_4^{2-}), or carbonate (SO_3^{2-}). By using these molecules to accept electrons from the energy-producing steps in the cell, anaerobes produce the gases nitrogen (N_2), hydrogen sulfide (H_2S), hydrogen (H_2), carbon dioxide (CO_2), and methane (CH_4). Some anaerobes produce only nitrite (NO_2^-) or nitrous oxide (N_2O) from nitrate, and carbon dioxide (CO_2) from carbonate. Gas production is representative of anaerobic digestion. Bubbles that drift from the muddy bottom of a still pond are probably carrying methane, carbon dioxide, hydrogen, and hydrogen sulfide produced by anaerobic bacteria in the pond's sediments.

After glycolysis, an anaerobe uses either anaerobic respiration or fermentation to produce energy for cellular activities. Anaerobic respiration resembles aerobic respiration, but in anaerobic respiration, the cell uses only part of the Krebs cycle. As a result, anaerobic species generate much less energy from their metabolic pathways than aerobic species.

Anaerobes also carry out fermentation to obtain energy from sugars, amino acids, organic acids, and nucleic acids. The microorganisms use either of two types of fermentation: lactic acid fermentation, in which only lactic acid is produced, or alcohol fermentation, which results in a variety of end products. People have for centuries used the end products from anaerobic fermentation to preserve foods (fruit juices preserved as wine) or to develop new foods (sauerkraut made from cabbage).

Fermentation has been an effective way to preserve foods that would eventually spoil at room temperature or colder storage temperatures. The acids produced by certain anaerobic microorganisms chemically change components in food. The acid also drops the pH to a low level, which wards off the growth of other microorganisms. In this way, acid fermentation acts in both food production and food preservation. Lactic acid bacteria have played perhaps the biggest role in this type of food preservation by acting on raw dairy products and converting them to more stable foods. This group of bacteria is so named because of the end product of their anaerobic fermentations. Lactic acid reacts with proteins in milk to curdle them. Depending on the addition of other ingredients and the fermentation conditions, the final products are a wide variety of cheeses, yogurt, preserved vegetables, and drinks. The table on page 33 lists the main fermentation end products in use today.

Anaerobic bacteria also carry out an important step in wastewater treatment. Wastewater treatment plants contain at least one covered digestion (or digester) tank that receives organic sludge left over from other parts of the treatment process. This sludge contains materials that are difficult to digest in regular aerobic metabolism. Inside the digester tank, the sludge combines with a diverse population of anaerobic bacteria, called a *mixed population*. The various bacteria in the mixed population work in complementary ways to decompose the sludge and reduce its total volume. Their complementary actions take the form of two steps. First, some anaerobes partially digest the sludge to products called *intermediary compounds*. In a second step, other bacteria use the intermediary compounds as food to complete the digestion of the wastewater sludge.

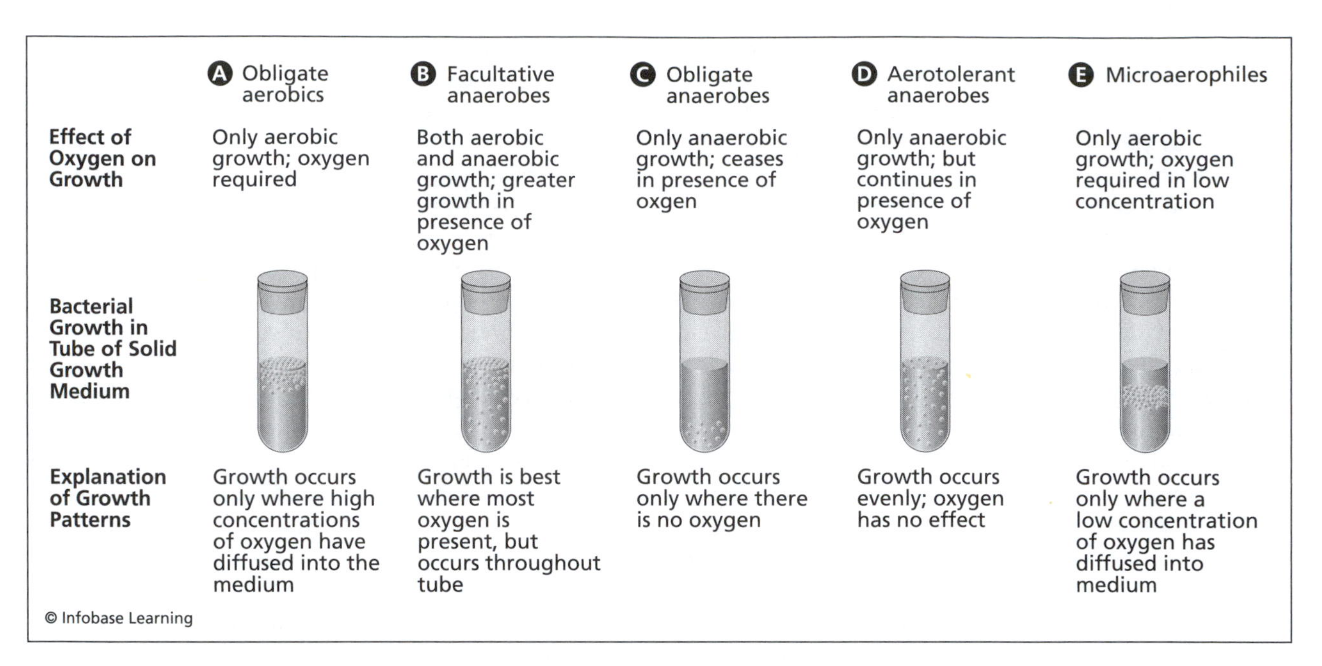

Oxygen-requiring bacteria grow in the aerated upper layers of nonagitated broth cultures. Other bacteria grow in regions that suit their requirements for reduced oxygen levels.

Because the anaerobes work slowly, the sludge digestion step at wastewater treatment plants usually occurs separately from other parts of the treatment process, which must keep up with a steady inflow of wastewater. In digesters, as in most other anaerobic metabolism, methane gas represents the primary end product. Many wastewater treatment plants take advantage of the methane produced in the digester tank by routing it to a burner for energy production. Methane produced by anaerobic fermentation has the same capacity to generate energy as natural gas. Wastewater treatment plants that use methane this way are called waste-to-energy plants.

Cows produce large amounts of methane from the digestion in their rumen. Each cow eliminates a portion of this methane by belching and eliminates the rest with manure. Innovative farm owners have proposed capturing the methane that manure emits to serve as an energy source, an operation known as *cow power.* In 2008, the correspondent Ralph Dannheisser wrote, "'You can't make a silk purse out of a sow's ear,' an old saying has it. But a Vermont utility is accomplishing an equally remarkable transformation, turning cow manure into electric power—and in the process both helping to reduce pollution and giving an economic boost to hard-pressed dairy farmers." Methane is a greenhouse gas that has contributed to global warming.

FACULTATIVE ANAEROBES

Facultative anaerobes are microorganisms that would normally grow without oxygen but can continue growth if they are exposed to oxygen. The enteric bacteria and many of the lactic acid bacteria are examples of facultative anaerobes. Facultative bacteria concern microbiologists for two reasons: They can be pathogens and they are food spoilage bacteria.

Enteric bacteria are normal inhabitants of the animal digestive tract, where they help break down food and supply the body with nutrients. They leave the body and enter the environment when they are shed in human and animal wastes. If enteric species were obligate anaerobes, many would soon die when exposed to the air. But because enteric bacteria are facultative, they can travel through the environment in soil, water, and food, even in the open air. People who ingest a large enough dose (the actual number of cells ingested) of enteric bacteria risk contracting a food-borne illness called gastroenteritis. The most famous of all facultative anaerobes is *E. coli*, but *Salmonella, Shigella, Klebsiella, Serratia,* and *Proteus* also cause illness in humans; *Erwinia* is a facultative enteric genus pathogenic to plants. *E. coli, Klebsiella,* and *Proteus* tend to be opportunistic pathogens. That is, they are normally harmless bacteria that cause infection only when presented with an opportunity. Two additional groups of pathogens are the *Vibrio* bacteria, waterborne pathogens, and *Haemophilus,* the cause of respiratory and other infections in humans and animals.

Lactic acid bacteria can be beneficial in certain situations and a pest at other times. They are one of the most widely used groups of microorganisms for making food products. In other instances, lactic acid fermentations overrun the food-making process and cause spoilage. Large amounts of lactic acid can spoil beer, meat, and milk.

Most facultative anaerobes grow faster when exposed to oxygen than when grown under anaerobic conditions. This seems the opposite of what would be expected, considering they are classified as anaerobes and not aerobes. But as discussed, aerobic respiration is more efficient than anaerobic metabolism, and all microorganisms pick the most efficient way to keep their cells going.

ANAEROBIC CULTURE TECHNIQUES

Growing microorganisms without exposing them to oxygen requires specialized skills beyond the normal skills of microbiology. Anaerobic culture techniques require tubes and jars that maintain an airtight seal. In addition, microbiologists must fill the extra space inside each culture's container with an oxygen-free gas. This area is called the headspace. Nitrogen–carbon dioxide–hydrogen mixtures are commonly used for filling the headspace in anaerobic culture tubes and lowering the medium's Eh. Microbiologists use four different techniques to ensure their cultures remain oxygen-free: (1) the Hungate culture technique, (2) the serum tube culture technique, (3) the use of anaerobic jars, and (4) the use of anaerobic chambers.

The Hungate culture technique was developed, in the early 1960s, by the microbiologist Robert Hungate (1906–2004) at the University of California–Davis. The technique begins with test tubes containing sterile medium formulated to provide all of an anaerobe's essential nutrients. A microbiologist then inoculates the medium and immediately plugs the tube with a rubber stopper to prevent air from entering during incubation. Skilled microbiologists complete this step in one rapid action. The medium and the bacteria must also be protected from the air during routine inoculations and transfers that are part of normal culture methods. To do this, the Hungate technique calls for a gentle stream of oxygenless gas to pass over the opening any time a microbiologist removes the rubber stopper. The gas displaces any oxygen that might drift into the tube and creates an anaerobic headspace once the tube is again stoppered.

The serum tube technique is actually a modification of the Hungate technique. Rather than risk the chance that oxygen will enter open vessels, a microbiologist uses a syringe preflushed with inert gas to transfer anaerobes. The microbiologist then inoculates sealed serum tubes containing sterile medium by injecting the inoculum through a rubber diaphragm in the center of the tube's seal. Like Hungate tubes, serum tubes are prepared in oxygen-free conditions and sealed under an anaerobic gas mixture.

The tube methods described previously require some expertise. Even skilled microbiologists who work mainly with aerobes must practice anaerobic techniques before beginning to grow anaerobes in a laboratory. Anaerobic jars partially solve this problem. In this case, the microbiologist inoculates agar plate cultures with anaerobes and immediately puts them into an anaerobic jar. Anaerobic jars are small

Industrial Uses for Fermentation End Products

Fermentation Product	Starting Material	Industrial Product	Example Microorganism
ethanol	malt extract	beer	*Saccharomyces* (y)
	fruit juices	wine	
acetic acid	ethanol	vinegar	*Acetobacter* (b)
lactic acid	milk	cheeses, yogurt	*Lactobacillus* (b), *Streptococcus* (b)
	grain and sugar	breads	*Lactobacillus*
propionic acid and carbon dioxide	lactic acid	Swiss cheese	*Propionibacterium* (b)
acetone	molasses	industrial solvents	*Clostridium* (b)
gycerol	molasses	industrial, medical lubricant	*Saccharomyces*
citric acid	molasses	natural flavoring	*Aspergillus* (f)
sorbose	sorbitol	vitamin C	*Gluconobacter* (b)

Note: (b) = bacteria, (f) = fungi, (y) = yeast

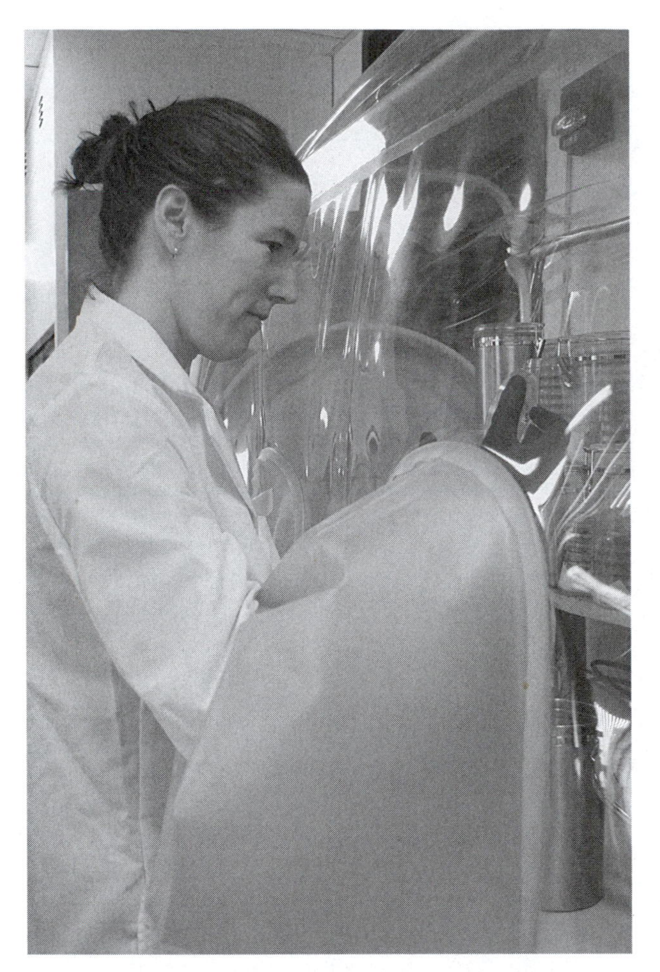

Anaerobic chambers allow microbiologists to work with cultures in an oxygen-free atmosphere. ***(Geobacter Project, University of Massachusetts, Amherst)***

containers, about the size of a shoebox. The microbiologist places a hydrogen generator material into the jar along with the inoculated plates; the material releases hydrogen gas or carbon dioxide gas when water is added. After the jar has been closed with a tight seal, the gas generator removes any oxygen from the inside of the jar and emits an anaerobic gas. Anaerobic jars typically hold a stack of one or two dozen plates, and the entire jar fits in an incubator. Smaller bags, called Bio-Bags, which hold no more than two plates, are also available for anaerobic cultures using the same principles as jars.

Anaerobic chambers, or small glove boxes, allow microbiologists to carry out all the same inoculations, cultures, and transfers as in aerobic microbiology. The difference is that the operator conducts all the steps inside an oxygen-free clear plastic chamber equipped with sleeves and gloves. The microbiologist sits outside the chamber and wears the gloves to carry out manipulations inside the chamber. Anaerobic chambers are usually filled with gas mixtures of 80–90 percent nitrogen, 5–10 percent hydrogen, and 5–10 percent carbon dioxide. Many hold small incubators so all inoculations and incubations can take place inside the chamber.

Fastidious anaerobes are obligate anaerobes that are extremely sensitive to any exposure to air. These species often require use of the Hungate or the serum tube methods in order to grow. Clinical anaerobes recovered from medical patients usually grow well in anaerobic jars or chambers, and most clinical microbiology laboratories have either jars or chambers or both.

Anaerobic media require more effort to prepare than aerobic media because they also must be completely free of oxygen. Anaerobic media are usually boiled during preparation to drive off excess oxygen and the containers quickly sealed with an anaerobic headspace. Fastidious anaerobes require an oxygen-free system with a negative Eh. Microbiologists create these conditions in anaerobic media by adding reducing agents such as cysteine, sodium thioglycolate, or other compounds containing a sulfhydryl group (-SH). The addition of resazurin acts as a redox indicator. Even with all these precautions, some anaerobes remain very difficult to maintain in laboratory cultures. In the world of microbiology, many more studies have been done on aerobic bacteria than have been completed on anaerobic bacteria.

See also ARCHAEA; FERMENTATION; LACTIC ACID BACTERIA; METABOLISM; ORAL FLORA; RUMEN MICROBIOLOGY.

Further Reading

Dannheisser, Ralph. "'Cow Power' Program Converts Animal Waste into Electricity." 21 February 2008. Available online. URL: www.america.gov/st/env-english/2008/February/20080221103802ndyblehs0.5918238.html. Accessed March 10, 2009.

Engelkirk, Paul G., Janet Duben-Engelkirk, and V. R. Dowell. *Principles and Practice of Clinical Anaerobic Bacteriology.* Belmont, Calif.: Star, 1992.

Hungate, Robert E. "A Roll Tube Method for Cultivation of Strict Anaerobes." In *Methods in Microbiology,* edited by J. R. Norris, and D. W. Ribbon. New York: Academic Press, 1969.

———. *The Rumen and Its Microbes.* New York: Academic Press, 1966.

Sutter, Vera L., Diane M. Citron, Martha A. C. Edelstein, and Sydney M. Finegold. *Wadsworth Anaerobic Bacteriology Manual,* 4th ed. Belmont, Calif.: Star, 1985.

anthrax Anthrax is a bacterial disease that attacks humans and animals and is caused by infectious endospores of *Bacillus anthracis*. An endospore is a thick-walled, dormant cell able to withstand harsh conditions, such as heat, drying, and exposure to chemicals. Anthrax and other *Bacillus* spores have been shown to remain alive for centuries. Anthrax possesses all of the characteristics of an acute disease: It worsens quickly after infection; it is severe; and illness lasts for a short period. In humans, anthrax is contracted in one of three ways: through skin contact with the endospores, by inhaling endospores into the lungs, or by ingesting them. Skin contact and inhalation are the most common modes of transmission; infection caused by ingesting endospores is very rare.

Although anthrax is an ancient disease, *B. anthracis* has developed into a feared pathogen in recent years because of its potential as a biological weapon. The specific cause of anthrax has been known for a century even though it has always been a fairly rare disease. In the United States, the Centers for Disease Control and Prevention (CDC) estimates anthrax incidence of about one case in 300 million people per year. The World Health Organization (WHO) proposes that anthrax is similarly rare worldwide, but incidence increases in sub-Saharan Africa, parts of Asia, southern Europe, and Australia.

Those regions in Africa, Asia, Europe, and Australia have higher incidences of anthrax today because of the prevalence of animal-based work in certain areas. Throughout history, anthrax has been associated with jobs in which people work with contaminated textiles, animal hides, animal hair, wool, or contaminated feces. Direct contact with animals, their products, or soil containing *B. anthracis* endospores is thought to account for at least 95 percent of the world's anthrax cases. People who handle these animal products or work with contaminated soils have an increased risk of infection due to disease transmission called *contact transmission*. In contact transmission, a person becomes infected by touching an object carrying the pathogen.

Sporadic cases of anthrax worldwide occur on an average of once a year, usually on farms. Public health departments manage most of these cases quickly to prevent spread of the endospores. During a 2008 outbreak of anthrax in cattle in which 13 animals died on a farm in Sweden, the infectious disease expert Bengt Larsson said to a Swedish paper, "This disease is not to be taken lightly. The disease is classified under epizootic [infecting many animals in defined area] legislation which shows just how serious it is and the department of agriculture can decide on the appropriate measures." The Swedish farm followed the standard actions for managing further outbreak: The farm's owner put all the cattle under quarantine and workers who had had contact with the animals were treated with antibiotics.

THE HISTORY OF ANTHRAX

Humans and their meat-producing and wool-bearing animals have had a tumultuous history with *B. anthracis*. Anthrax has been blamed for the biblical fifth and sixth plaques in Egypt occurring in 1500 B.C.E. in which a "plague of boils" had been described as infecting the pharaoh's cattle. Virgil, who lived from about 70 to 19 B.C.E., wrote in *Georgics* of an epidemic that had the symptoms of anthrax and spread from animals to humans.

Epidemics in cattle in Russia and Europe from the 1600s to 1700s have been attributed to anthrax. In the 1600s, the black bane spread throughout Europe, killing thousands of cattle and probably spreading to hundreds of people. In 1769, the French physician Jean Fournier named the disease *charbon malin* (black malignancy) for the black lesions it caused on the skin. Fournier spent much of his career writing about the connection of animal sicknesses and human epidemics. He was perhaps the first person to propose a link between the malady and people who handled untreated hides and wool. Later in the 18th century, anthrax became known as *wool sorter's disease* or *rag picker's disease* because of its connection with those activities.

The science of microbiology developed throughout the 1800s. With better microscopes, microbiologists inspected details of the rod-shaped cells isolated from lesions in animals. Others tried to find the relationship between disease and these small "filiform bodies," as described by the French physician Pierre Françoise Olive Rayer and the biologist Casimir-Joseph Davaine.

Proof of the connection that many scientists believed existed between the rod-shaped cells and the disease continued to evade those who researched anthrax. The German physician Robert Koch (1843–1910) introduced a plan for proving the microbial cause of not only anthrax, but other infectious diseases as well. Koch's theory involved infecting healthy animals with blood from infected animals and observing the outcome. By the late 1800s, Koch developed a written procedure for showing whether and how any particular microorganism causes disease. The steps of this procedure became known as Koch's postulates.

In 1877, Koch tested the postulates in an experiment on anthrax and definitively showed that the *Bacillus* microorganism caused the dreaded disease. By 1881, the biologist William S. Greenfield had developed a vaccine against anthrax for livestock. Most of the credit for the anthrax vaccine fell to another microbiologist, famous at the time for his discoveries in disease transmission, fermentation, and metabolism. The French biologist Louis Pasteur (1822–95) followed Greenfield's lead the same year, 1881, by heating anthrax microorganisms and using them to make a vaccine for sheep. Pasteur's vaccine staved off a serious outbreak in sheep taking place at the time in France and launched a career that included a good share of fame. Remarkably, the French veterinarian Jean Joseph Henri Toussaint (1847–90) developed his own version of an anthrax vaccine, also in 1881, but Pasteur's fame overshadowed Toussaint, as it had Greenfield.

New anthrax vaccination programs helped public health officials break the chain of transmission from livestock to humans and so reduce the incidence of anthrax in the human population. Another factor helped combat anthrax. In the late 1930s, economies began shifting away from agriculture to manufacturing. In the United States, fewer people earned a living by handling animal skins and hides; thus anthrax became less a threat to the general public.

Anthrax's history did not fade, however, with effective vaccines. In 1915 during World War I, German agents infected horses, mules, and cattle bound for shipment to Allied countries. The Germans sent infected reindeer to Norway. These attempts at spreading disease to their enemies are believed to be the first time a biological weapon, or bioweapon, had been used in wartime. (A bioweapon is any weapon in which the lethal action is produced by a biological substance.)

Although the postwar Geneva Convention, of 1925, banned all bioweapons, secret work on anthrax weapons continued. In the 1930s, the Japanese imperial army built weapons containing anthrax in preparation for World War II. During the war, the United States, Britain, and Adolf Hitler's Germany built similar anthrax weapons, but they were never used by either side. The threat understandably put health officials in several countries on edge. Occasional outbreaks in the 1950s and 1960s sent CDC scientists rushing to see whether the cause was natural or an anthrax-laced weapon. These outbreaks never spread far, but they took a small number of lives. In 1957, for example, nine employees of a textile mill in New Hampshire contracted anthrax and died.

The CDC continued tracing infrequent anthrax cases over the following decades. As anthrax incidence declined, the disease seemed no longer to pose a serious threat to the U.S. population. That peace of mind ended in 2001 with the resurgence of terrorist acts, this time on American soil.

In 2001, one week after the September 11 terrorist attacks on New York City and Washington, D.C., a letter containing anthrax endospores arrived at a television station's New York office. Other random incidents happened throughout the United States in the next few months. In Florida, one man died after inhaling anthrax endospores at his workplace. The cause of these anthrax cases has never been found. (One suspicious outbreak of anthrax occurred in the Soviet Union in 1979 in which 64 people died. Many nations' leaders believed the outbreak originated in a laboratory researching bioweapons, though this was never proved.)

President Bill Clinton initiated the Anthrax Vaccine Immunization Program (AVIP), in 1997, for the purpose of protecting all U.S. active military personnel assigned to combat zones. The program was suspended for a period for additional studies on the safety of the vaccine itself, required by the U.S. Food and Drug Administration (FDA). In 2006, Deputy Secretary of Defense Andrew England issued a memo to the military branches saying, "Based on the continued heightened threat to some U.S. personnel of attack with anthrax spores, the Department of Defense will resume a mandatory Anthrax Vaccine Immunization Program, consistent with the Food and Drug Administration guidelines and the best practice of medicine, for designated military personnel." Nonmilitary civilians and defense contractors who work in these high-risk geographic areas also receive the vaccine.

THE DISEASE IN HUMANS

Anthrax is called a zoonotic disease, or zoonosis, because it occurs in wild or domestic animals but can be transmitted to humans. It is not transmitted from person to person. Anthrax endospores follow three modes of entry into the body, each mode associated with a specific type of anthrax disease: cutaneous anthrax, inhalation (also called pulmonary or respiratory) anthrax, or gastrointestinal anthrax. Cutaneous anthrax accounts for 95 percent of the world's anthrax cases, but all three types are can be fatal if not treated.

Cutaneous anthrax arises from contact transmission. *B. anthracis* endospores usually enter through a break in the skin caused by a cut, wound, or burn. In cutaneous anthrax, *B. anthracis* cells do not usually enter the bloodstream but rather stay near the site of infection, where they cause a swelling in the skin called a *papule*. The infection progresses from a papule to a larger skin ulcer covered by a black scar. (The name *anthrax* derives from the Greek word for coal.) Mild fever and loss of energy accompany most cutaneous anthrax cases, which are successfully treated with antibiotics. In rare instances, the cutaneous infection enters the bloodstream. When this happens, the disease is likely to be fatal.

Inhalation anthrax is caused by breathing in *B. anthracis* endospores. The endospores enter the lungs, and from there the bacteria infect the bloodstream and multiply. *Sepsis* is the term used for a condition in which microorganisms infect and multiply in the bloodstream. *B. anthracis* sepsis infects organs and causes the patient to suffer septic shock within two or three days. Death follows within the next 24–36 hours; the mortality rate for this form of

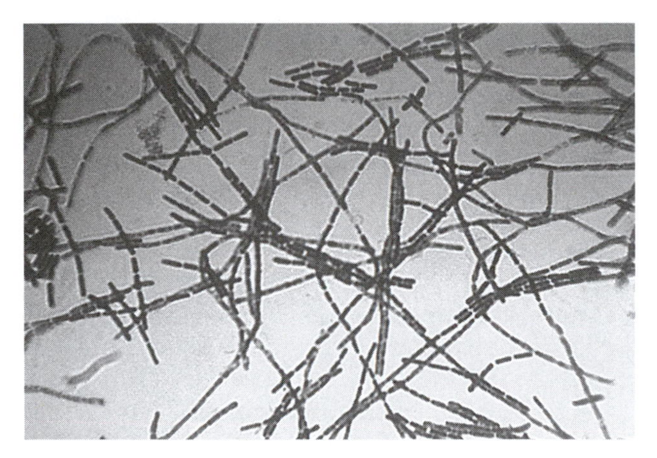

Bacillus anthracis*, the anthrax pathogen, develops straight rods in the vegetative (growing) form. The lighter, thicker links in the bacillus chain are endospores.** ***(Y. Tambe, CDC)

anthrax approaches 100 percent. In the early stages of inhalation anthrax, a few days after infection, patients usually begin coughing and have a mild fever and chest pain. Antibiotics can arrest the infection at this point and decrease the mortality rate to about 1 percent.

Gastrointestinal anthrax results from eating undercooked foods contaminated with endospores. Its symptoms are nausea, abdominal pain, and bloody diarrhea. In addition, ulcers form inside the mouth, in the throat, and in the intestines. Public health departments see very few cases of gastrointestinal anthrax worldwide each year. The mortality rate for this form of the disease is about 50 percent.

Inside the body, *B. anthracis* cells revert from their spore form to normal *Bacillus* cells, called *vegetative cells*. Vegetative cells then produce a toxin, which causes the symptoms of all three types of anthrax. The toxin contains three proteins that fit together in a specific way in order for the toxin to be active. Separate, these components do not harm the body. The complete active toxin works by entering the body's cells and blocking the activity of enzymes essential for normal cell development.

ANTHRAX VACCINATION

Although the general public in the United States does not receive a vaccine for anthrax, livestock owners have for many years given their herds a vaccine made of live, attenuated (inactivated) *B. anthracis*. But vaccines made from live cells have been thought to be too risky for human use.

Anthrax vaccine can be made from one of the three *B. anthracis* toxin proteins. Only 0.007 percent of people receiving this attenuated vaccine have been estimated to suffer side effects, but this protein vaccine has been suspected of being less effective for immunizing humans than the animal vaccine. The protein-based vaccine furthermore requires a regimen of six injections over 18 months, followed by annual boosters. This type of vaccination does not appeal to most people.

Since 1970, the CDC has followed studies on a more effective vaccine made from live endospores, similar to the livestock vaccine. This anthrax vaccine has not been used in the general population because U.S. health officials hesitate to use new vaccines that are largely untested.

The United States currently holds a stockpile of 10 million doses of anthrax vaccine, called the Strategic National Stockpile (SNS), which is stored in case of a national emergency caused by an anthrax bioweapon attack. The product is named Anthrax Vaccine Adsorbed (AVA), made by a company in Michigan, and is the only anthrax vaccine licensed for use in the United States by the U.S. Food and Drug Administration (FDA). The manufacturer produces the AVA vaccine from an avirulent (not capable of causing disease) strain of *B. anthracis* known as *V770-NP1-R*. The vaccine contains the filtrate—the liquid that passes through a filter—from a culture of V770-NP1-R grown in broth medium. Because the filter removes the cells from the filtrate, the AVA vaccine contains no live *B. anthracis* cells.

The CDC has said about the SNS, "The SNS is a national repository of antibiotics, chemical antidotes, antitoxins, life-support medications, IV [intravenous] administration, airway maintenance supplies, and medical/surgical items." The Department of Homeland Security furthermore assures communities that the SNS is capable of delivering needed supplies to any site within 12 hours.

ANTHRAX AS A BIOWEAPON

Inhalation anthrax is the form that concerns scientists as a potential bioweapon. This is because the endospores can be dried into a fine powder and applied to common items such as mail, public transportation, or drinking water systems. The fine particles additionally disperse into the air and can travel airborne for great distances, perhaps up to miles. The dispersal of anthrax spores through the air is called an *aerosol route of exposure*. Aerosols are tiny particles or moisture droplets that travel in the air. A bioaerosol, such as one containing *B. anthracis* endospores, is an aerosol that contains a biological component.

B. anthracis has been studied in the United States since 1943 during the height of World War II. At least three government laboratories have worked on this and other potential bioweapon microorganisms: Fort Detrick, Maryland; Horn Island, Mississippi;

and Granite Peak, Utah. During the Korean War (1950–53), the U.S. government built a plant in Pine Bluff, Arkansas, to be devoted to bioweapon production. Opinions differ among leaders and within the public on the true threat of bioweapons in the United States and abroad. Regardless of the degree of the potential danger, the CDC and other U.S. government agencies provide the public with information about the anthrax pathogen, the disease, and its causes.

Among the information published by the CDC are tips on what U.S. residents can do about anthrax. Vaccinations are not available for the general public. Only members of the military or people working with *B. anthracis* in microbiology laboratories receive the vaccine. The CDC advises people to prevent anthrax infection by being aware of the common carriers of anthrax: animal products and soil. The CDC advises people to be familiar with the anthrax symptoms and to seek medical help immediately if they think they have been exposed to the *B. anthracis* endospores. In addition, the patient or doctor should call local law enforcement immediately. During the anthrax scare of 2001, federal and local law enforcement agencies warned the public to be wary of delivered packages or envelopes containing a tan, off-white, or white powder. The CDC gives the following six additional tips for preventing infection from *B. anthracis* or other dangerous microorganisms:

1. Know the symptoms of anthrax.
2. Wear gloves and face masks if handling large amounts of mail.
3. Avoid touching hands to eyes, mouth, or nose.
4. Wash hands frequently with warm, soapy water.
5. Bandage all cuts and open wounds until healed over with a scab.
6. Clean the skin with rubbing alcohol if it has contact with suspicious items.

These guidelines work for almost all infectious microorganisms in addition to anthrax. In fact, such measures of good hygiene help prevent infections from intentional release of bioweapons as well as accidental releases.

Anthrax experts list the following as activities in which people should take extra precaution to prevent infection:

- conducting studies on soil
- tilling, plowing, or cultivating agricultural soils
- digging and planting in garden soils
- working with livestock and livestock wastes
- working with untreated livestock hair, wool, furs, or hides
- digging at archaeological sites
- working at excavation or construction sites with earthmovers
- working in laboratories or clinics that contain *B. anthracis*
- serving in the military
- sorting, handling, or delivering mail
- living in regions known to have had anthrax cases

In normal conditions in which no bioweapon threat is known to exist, anthrax exposure remains highest among people carrying out nonmilitary and nonmail delivery jobs in the preceding list.

ANTHRAX INFECTIONS IN DOMESTICATED ANIMALS

Cattle, sheep, goats, horses, and wild herbivores can ingest *B. anthracis* endospores when grazing on land containing the microorganism. In rare cases, birds have contracted anthrax disease. *Bacillus* is a common soil microorganism, and its endospore has the ability to withstand extremes in moisture, drying, heat, cold, and toxic chemicals. The endospores remain alive whether they are in the ground or clinging to grasses or hay grown on infected soils.

In veterinary medicine, anthrax goes by the names of *splenic fever, Siberian ulcer, Charbon,* or *Milzbrand,* and it has been reported in animals on every continent except Antarctica. Veterinary anthrax tends to appear where the soil is either neutral or alkaline, rather than acidic.

The *B. anthracis* toxin causes the same symptoms in animals as in humans. The infection's incubation period is about three to seven days. Cattle, sheep, and goats suffer a variety of escalating symptoms consisting of the following: depression, stupor, staggering and trembling, convulsions, cardiac arrest, collapse, and then death. Sometimes the animal dies so quickly that livestock owners barely notice most symptoms. Acute anthrax is a disease that has a fast onset and runs a short course before ending, either in death or in recovery. In some anthrax cases, the symptoms arrive even more rapidly and produce a sudden violent outcome. This condition is called *peracute anthrax*. In cases of acute or peracute anthrax, livestock owners often realize their animals have been infected only after they find dead animals in the field.

Horses undergo slightly different symptoms than other herbivores: fever, chills, colic, anorexia (loss of appetite), lethargy, weakness, bloody diarrhea, and swelling at various parts of the body. In wild herbivores, such as deer, symptoms resemble those in cattle.

Dogs, cats, pigs, and wild carnivores also endure many of these typical anthrax symptoms, usually from ingesting endospores. Acute septicemia, which is a contamination of the blood with bacteria, occurs. Sometimes the throat swells until the animal suffocates. These animals also may contract a chronic (long-lasting) form of the disease in which damage occurs to the lymph system and the intestinal tract.

Vaccination programs in the livestock industry have helped reduce the incidence of anthrax. Veterinarians today prescribe the Sterne vaccine to immunize livestock. This vaccine contains live *B. anthracis* given two to four weeks before the season in which outbreaks are expected. At least one week after vaccination, animals receive antibiotics to eliminate any live *B. anthracis* cells remaining in their bloodstream.

Livestock owners should report to local agriculture agencies any anthrax outbreaks in their animals. The local agricultural official or veterinarian then orders the quarantine of infected animals and those suspected of being infected. Stalls, pens, milking parlors, barns, and equipment used on or around animals must be disinfected. Animal carcasses must be removed from the farm quickly and safely disposed. It is important to control insects and rodents around farms to decrease the spread of anthrax from sick to well animals, and workers should also follow good hygiene. Some livestock owners take an extra step by removing soil they suspect contains *B. anthracis* endospores. These programs have helped keep the cases of anthrax to a minimum in the United States each year. Considering that this microorganism can hide in the ground for decades or much longer, humans will never have the luxury of giving up their diligence to prevent anthrax infection.

See also BACILLUS; BIOWEAPON; KOCH'S POSTULATES; SPORE.

Further Reading

Brock, Thomas D. *Robert Koch: A Life in Medicine and Bacteriology.* Washington, D.C.: American Society for Microbiology Press, 1998.

Centers for Disease Control and Prevention. "Strategic National Stockpile (SNS)." Available online. URL: http://emergency.cdc.gov/stockpile. Accessed March 11, 2009.

England, Andrew. "Memorandum for Secretaries of the Military Departments." 12 October 2006. Available online. URL: www.anthrax.mil/documents/972OSD15400-06.pdf. Accessed March 30, 2009.

"Focus on anthrax: A special online focus." *Nature.* Available online. URL: www.nature.com/nature/anthrax/index.html. Accessed on June 9, 2010.

Guillemin, Jeanne. *Anthrax: The Investigation of a Deadly Outbreak.* Berkeley: University of California Press, 1999.

Holmes, Chris. *Spores, Plagues and History: The Story of Anthrax.* Dallas: Durban House, 2003.

Kahn, Cynthia M., and Scott Line, eds. *The Merck Veterinary Manual,* 9th ed. Whitehouse Station, N.J.: Merck, 2005.

The Merck Veterinary Manual. 2006. Available online. URL: www.merckvetmanual.com/mvm/index.jsp. Accessed March 25, 2008.

"Swedish Farm Hit by Anthrax Outbreak." The Local: Sweden's News in English. 13 December 2008. Available online. URL: www.thelocal.se/16332/20081213. Accessed March 11, 2009.

antibiotic An antibiotic is any substance that in low concentrations inhibits or kills susceptible microorganisms. The earliest antibiotic discoveries all fulfilled this definition and were also substances made by one microorganism to kill other microorganisms. Today many drug companies, such as Merck and Pfizer, make many antibiotics synthetically without the need for a microbial source.

Microorganisms in nature secrete antibiotics for two purposes: to protect the cell from attack from other species and to eliminate competition from other microorganisms for nutrients and habitat.

Antibiotics in use today in human medicine, veterinary medicine, and for the control of plant diseases are natural, semisynthetic, or entirely synthetic. Semisynthetic antibiotics are compounds made by a microorganism, then chemically modified in a laboratory; synthetic versions originate entirely in a laboratory. All antibiotics are chosen for treatment because they harm the intended pathogen but do not cause harm to the patient or plant, an ability called *selective toxicity*. Not all microorganisms produce antibiotics, but thousands of natural antibiotics, nevertheless, occur in nature. Human medicine uses a relatively small number of these antibiotics, chosen because they exhibit the best selective toxicity.

The U.S. Food and Drug Administration (FDA) classifies antibiotics as drugs and regulates the testing and sale of these substances. As drugs, antibiotics are prescribed for one of two reasons: as therapeutic agents or as prophylactic agents. Therapeutic antibiotics treat an existing infection; prophylactic antibiotics guard against new infection or recurrence of infection. Because a single antibiotic does not kill every different type of microorganism, physicians prescribe an antibiotic for a patient on the basis of its ability to kill a specific pathogen. An antibiotic effective on a small number of microorganisms is called a narrow-spectrum antibiotic. Such an antibiotic would be useful when a patient suffers from a known infectious disease. But when an infection involves more than one microorganism or unknown microorganisms, doctors choose an antibiotic that attacks a wide variety of species, called a broad-spectrum antibiotic.

Unfortunately, antibiotics were not chosen with care in the early years of their use. They were prescribed seemingly for any and all ailments. As a consequence, many microorganisms have now evolved to possess resistance to antibiotics. Careful prescribing of antibiotics today must take into account both the pathogen causing an infection and any possible antibiotic resistance that the pathogen might possess.

THE DISCOVERY OF WONDER DRUGS

In the early 1900s, the German physician Paul Ehrlich (1854–1915) discovered the first known antibiotic. As a medical student, Ehrlich sought a "magic bullet" that would kill pathogens but not harm patients. Ehrlich had in his studies come to appreciate the complex relationship between microorganisms and compounds having very specific structures. Ehrlich proposed that if a compound could be made that targeted a specific disease-causing agent, then a drug could be invented to kill the microorganism and only that microorganism. This magic bullet had been a wish of many physicians for more than a century, but little success was achieved.

In a 1908 presentation to a medical audience, Ehrlich described his logic behind one of many arsenic-containing compounds he tested against infections in animals: "After the structure had been determined, it was now possible to produce a large series of related compounds, all of which were organic compounds of arsenic acid. Various groups could be introduced on the amino group, or it could be combined with various acidic groups, or it could be coupled with aldehydes." Ehrlich summarized the science of antibiotic discovery and synthesis that would continue to the present day, that is, identifying the portion of a compound responsible for the antibiotic's activity and trying substitute structures on the antibiotic molecule with the goal of enhancing the drug's activity.

It may be fortunate that Ehrlich possessed the naïveté to believe he could find this elusive medical breakthrough through what had been a hit-or-miss approach to drug discovery. In his laboratory, Ehrlich synthesized and tested hundreds of compounds for use against sleeping sickness and syphilis and discovered hundreds of failures. But in 1910 he made an arsenic-containing substance, called arsphenamine, or Compound 606; it was the 606th compound he had prepared. Ehrlich showed that Compound 606 inhibited *Treponema* bacteria that cause syphilis. A drug company adopted the invention and began to sell it under the trade name *Salvarsan*. Salvarsan became the first commercially sold antibiotic and the first effective synthetic one. (Ehrlich's studies also produced early clues to the emerging phenomenon of antibiotic resistance.) But Ehrlich considered the search for a magic bullet only a partial success because though Salvarsan worked, it seemed too general as a treatment. Ehrlich encouraged his colleagues that the search for antibiotics was in its infancy. "I am well aware that animal experiments allow initially no conclusion for the therapy of man," he said. "If this substance [an antibiotic against trypanosome protozoal parasites] does not prove to be suitable in human pathology, this does not mean that we should throw in the sponge and give up all hope. Rome was not built in a day! Therefore, we must continue to stride forward on the path which has now been clearly revealed before us." Ehrlich had presaged the long and sometimes frustrating search for new antibiotics.

Unknown to Ehrlich, a medical student in France in 1896 had already discovered a natural antibiotic. Before graduating, Ernest Duchesne (1874–1912) wrote a thesis on his observations in which many bacteria were killed in the presence of a certain mold. This finding would become a major breakthrough in the history of medicine, yet it seems Duchesne failed to see the meaning hidden in his notes. He did not mention his findings to fellow scientists and the discovery was unnoticed.

Thirty-two years later at St. Mary's Hospital in London, a mold contaminated bacterial left unattended in the laboratory of Alexander Fleming (1881–1955). Fleming took note of the unusual growth in the contaminated petri dishes he was about to discard. He noticed also that bacteria had not grown in any areas of the plate where the mold had grown. The mold turned out to be *Penicillium,* the very same that Duchesne had studied. In coming months, Fleming extracted a substance from *Penicillium* that would kill bacteria even without mold present. Unlike Duchesne, Fleming possessed the gift of talking up his discoveries. But he stirred little interest in the mold extract among his peers so turned to other topics for study.

In 1938, an Oxford University pathologist with an interest in antibacterial compounds, Howard Florey (1898–1968), teamed with his assistant Ernst Chain (1906–79) to reexamine Fleming's work. Florey and Chain showed that the *Penicillium*'s secretion was active not only in petri dishes but also in mice that had been infected with lethal doses of staphylococci or streptococci. Doses of the crude preparation from mold cultures cured infections in animals, and then in humans. Furthermore, the drug seemed safe for the patients. One year after they had begun their experiments, Florey and Chain introduced this new compound, penicillin, to the medical world. The microbiologist Selman Waksman (1888–1973) soon coined the term *antibiotic,* and his laboratory began to screen tens of thousands of compounds in search of another like penicillin. In 1944, Waksman's team discovered streptomycin, an antibiotic made by *Streptomyces* bacteria. The dreams that had pestered Ehrlich decades earlier had been realized: A magic bullet that would kill pathogens but not harm the patient had become a reality.

Between streptomycin's discovery and the early 1980s, drug companies launched an intense search for new antibiotics and discovered at least 100 promising leads. The tedious process consisted of screening hundreds upon hundreds of bacteria against mold extracts. Often luck played a big part in the search for the next *wonder drug.* An employee of New Jersey's Pfizer Company, one day in 1950, found actinomycete bacteria growing outside his laboratory. The substance he and fellow scientists extracted from the bacteria became Terramycin. The Pfizer scientist Lloyd Conover followed the work, in 1952, by modifying the natural compound in his laboratory to develop tetracycline, which became one of the most prescribed antibiotics for the next several decades. Years later, Conover modestly described his breakthrough: "I was lucky. I was at the right place at the right time, with an opportunity to pursue my own scientific hunches in an unexplored area." Conover underplayed his contributions to medicine, and antibiotic discovery has been the story of many chance occurrences and some luck.

Since the 1950s, few new drugs had made so great an impact or had been prescribed so liberally as the expanding list of wonder drugs. Within 20 years of the first sale of penicillin, a new discovery emerged: antibiotic resistance. Microorganisms had begun to acquire defensive tactics against antibiotics. Resistant pathogens became more difficult to kill with antibiotics that had once worked well against them. Resistance began to alter the course of tuberculosis, sexually transmitted diseases, acquired immunodeficiency syndrome (AIDS), and other bacterial and fungal infections. Antibiotic resistance grew into a health concern, reaching crisis levels throughout the world.

TESTING ANTIBIOTIC ACTIVITY

For many years, antibiotic testing remained almost as tedious as in Waksman's time. Tests were designed to find a handful of antibiotics among thousands of compounds that would kill a specific pathogen. Then, a microbiologist would compare the new finds to determine which antibiotic had the best activity against a single pathogen or a variety of pathogens. As recently as 2006, the *Microbe* magazine reporter Jeffrey L. Fox wrote, "The race for new antibiotics is a fragmented event, and researchers who persist in it tend to find themselves headed in disparate directions and moving at less-than-blistering pace." Drug companies have developed more sophisticated methods of synthesis since Ehrlich's time, but much antibiotic testing continues to rely on two time-honored methods: the Kirby-Bauer test and the minimum inhibitory concentration (MIC) test.

The Kirby-Bauer test is an example of a disk diffusion test. In this method, small (6 mm in diameter) filter paper disks are soaked with an antibiotic, dried, and then placed on an agar plate containing a single sheet of pathogen culture. As the plates incubate, the bacteria grow into a visible sheet covering the agar surface. This continuous sheet of bacterial growth is called a *lawn.* Clear areas often form around different disks that contain various antibiotics. These clearings, called *zones of inhibition,* result from the antibiotic's ability to prevent bacteria from growing in the vicinity of the disc. Ineffective antibiotics produce no zone of inhibition, and bacteria grow right up to the edges of the disk. Additionally, the zone diameters can be measured with a ruler to find the best antibiotic (the largest zone). By putting varying levels of antibiotic into disks, a microbiologist can estimate an MIC, that is, the minimal concentration of antibiotic needed to kill bacteria.

Similar tests have now been developed for use on a single paper strip containing a gradient of antibiotic levels. After incubation on a lawn, the MIC is determined on the basis of the growth surrounding the strip (called the Etest). Automated test tube methods are also available for testing antibiotics against a large number of microorganisms. Recently, computer programs have been designed to construct unique virtual structures that would work in killing bacteria.

ANTIBIOTIC CLASSES AND THEIR SOURCES

By testing the activity of antibiotics, they can be grouped by the type of microorganisms they kill or by their source. Antibiotics are either antibacterial or antifungal; they attack bacterial or fungal pathogens, respectively. (Antibiotics are not active against viruses, which must be treated with antiviral drugs, or protozoa, which are treated with antiprotozoal drugs.)

Broad-spectrum antibiotics work against a wide range of gram-positive and gram-negative bacteria. (A gram reaction is determined by the manner in which various bacteria accept a stain in the Gram stain procedure.) Narrow-spectrum antibiotics are effective against a smaller range of microorganisms. Often narrow-spectrum antibiotics attack gram-positives or gram-negatives, but not both. Some are narrower still and attack only one type of bacteria, such as antibiotics that target the tuberculosis pathogen.

Antibiotics terminology varies regardless of the type of microorganisms inhibited. They may be microbicidal or microbistatic. Microbicidal antibiotics kill the target pathogen; microbistatic antibiotics merely inhibit the growth of a pathogen but might not kill it.

Physicians select these drugs on the basis of the pathogens they kill. The source of an antibiotic is less important to them. Nevertheless, antibiotics have customarily been grouped according to source for many years. Examples of these classifications still in use are shown in the table on page 43.

Streptomyces bacteria are a source of many antibiotics. Several species are known to make specific antibiotics, but there are more than 500 *Streptomyces* species and many have not yet been studied for antibiotic production. Since Fleming's day, more bacteria than molds have been employed in antibiotic production for human and veterinary medicine. In the table on page 43, only *Penicillium* and *Cephalosporium* are molds.

Chemists working on the synthesis of new antibiotics classify these compounds by structure. This method of categorization makes sense because antibiotic structure usually relates to antibiotic activity. For example, antibiotics of various structures may selectively target membranes, proteins, or genetic material inside a cell. The major groups by structure are the tetracyclines, penicillins, cephalosporins, sulfonamides, aminoglycosides, quinolones, and macrolides.

Tetracyclines

Tetracyclines are made by *Streptomyces* or produced as semisynthetic or partial synthetic antibiotics. The semisynthetic versions contain minor changes in the structure of the natural compound to produce new activity. Tetracyclines are composed of four rings containing hydroxyl, oxygen, or chloride. Chlortetracycline, doxycycline, oxytetracycline, and tetracycline are the most commonly used. These are broad-spectrum antibiotics especially active against *Brucella* (cause of bovine aborted fetuses), *Chlamydia* (sexually transmitted disease), *Mycoplasma, Rickettsia,* and *Vibrio* (water contaminant).

Penicillins

The penicillins have a nitrogen-containing ring structure called a β-lactam ring, which is thought to be essential for killing bacteria. Most penicillins used today are synthetic: ampicillin, carbenicillin, methicillin, nafcillin, oxacillin, penicillin V, and ticarcillin. Only penicillin G is a form made entirely from mold cultures. Though penicillin was the first commercial antibiotic to receive widespread use, its mode of action is still not well known. An antibiotic's mode of action refers to the mechanism it uses to kill bacteria. The best understood mode of action in penicillins is the interference in cell wall peptidoglycan synthesis by the β-lactam part of the antibiotic. Many bacteria have become resistant to this mode of action, so synthetic penicillins now contain altered rings to outsmart the bacteria. Over time, however, new generations of microorganisms develop resistance to most of the new penicillins.

Cephalosporins

The cephalosporins (also cefalosporins) are either naturally produced or semisynthetic. They are similar to penicillins because they have a β-lactam ring. The original cephalosporin was discovered in *Cephalosporium* mold in 1948. Three generations of cephalosporins have since been invented because bacteria developed resistance to earlier versions. First-generation antibiotics are those discovered in mold cultures. Second- and third-generation antibiotics are synthetic forms designed for better activity against pathogens. Some common cephalosporins in use today are cephalothin, cefoxitin, cefoperazone, ceftriaxone, cephalexine, and cefixime.

Sulfonamides

Sulfonamides (sulphonamides) are also known as sulfa drugs. They all contain sulfur and inhibit a wide range of bacteria and some protozoa. Sulfa drugs work by interfering with B vitamin synthesis. But because their activity is inhibitory rather than cidal, they are used in combination with other antibiotics. Sulfonamides have become ineffective against many infections because of resistance. Some of the sulfonamides are sulfamethoxazole, sulfisoxazole, sulfacetamide, sulfadiazine, sulfathiazole, sulfadimidine, and sulfamethizole.

One sulfa drug nicknamed TMP-SMZ is a mixture containing sulfamethoxazole (SMZ) and trimethoprim (TMP). Microbiologists noticed that both antibiotics work better together than either one does separately. This cooperative situation between two antibiotics is called synergism.

Aminoglycosides

Aminoglycosides all have in common a cyclohexane ring and a sugar containing an amino group. They interfere with protein synthesis in bacteria in two ways: by binding with ribosomes and by blocking messenger ribonucleic acid (mRNA), both substances essential for constructing proteins out of individual amino acids. Many aminoglycosides have been used effectively in medicine for years, especially on infections caused by gram-negative bacteria.

Common antibiotics in this group are gentamicin, kanamycin, lividomycin, neomycin, paramomycin, ribostamycin, streptomycin, and tobramycin. Aminoglycosides have two disadvantages in medicine. First, resistance to this group has grown to dangerous levels. So numerous are the species resistant to streptomycin, for instance, physicians no longer prescribe it. Second, aminoglycosides cause the following side effects: renal failure, loss of balance, nausea, deafness, and allergic responses.

Quinolones

All quinolone antibiotics contain a 4-quinolone ring, which includes nitrogen and a carboxyl group. Early versions of quinolones (cinoxacin, nalidixic acid, oxolinic acid, and pipemidic acid) had limited activity. New versions were developed with fluorine as part of the molecule's ring. These new fluoroquinolones are ciprofloxacin, danofloxacin, norfloxacin, ofloxacin, pefloxacin, and enoxacin. Quinolones and fluoroquinolones are effective for treating urinary tract infections, enteric bacteria infections, respiratory tract infections, and sexually transmitted diseases.

Macrolides

Macrolides are complex structures part of a group known as *MLS antibiotics*. The MLS antibiotics—macrolide, lincosamide, and streptogramin B—have

Commonly Used Antibiotics

Drug	Effect	Spectrum	Source
ampicillin	cidal	broad	semisynthetic
amphotericin B	static	narrow (fungi)	*Streptomyces*
bacitracin	cidal	narrow (gram +)	*Bacillus*
carbenicillin	cidal	broad	semisynthetic
cephalosporins	cidal	broad	*Cephalosporium*
chloramphenicol	static	broad	*Streptomyces*/synthetic
ciprofloxacin	cidal	broad	synthetic
clindamycin	static	narrow (gram + anaerobes)	synthetic
dapsone	static	narrow (mycobacteria)	synthetic
erythromycin	static	broad	*Streptomyces*
gentamicin	cidal	narrow (gram -)	*Micromonospora*
griseofulvin	static	narrow (fungi)	*Penicillium*
isoniazid	cidal	narrow (mycobacteria)	synthetic
kanamycin	cidal	broad	*Streptomyces*
methicillin	cidal	narrow (gram +)	semisynthetic
neomycin	static	broad	*Streptomyces*
oxacillin	cidal	narrow (gram +)	semisynthetic
penicillin	cidal	narrow (gram +)	*Penicillium*
polymyxin B	cidal	narrow (gram -)	*Bacillus*
quinolones	cidal	narrow (gram -)	*Streptomyces*
rifampin	static	broad	*Streptomyces*
streptogramins	cidal	broad	synthetic
streptomycin	cidal	broad	*Streptomyces*
sulfonamides	static	broad	synthetic
tetracyclines	static	broad	*Streptomyces*
vancomycin	cidal	narrow (gram +)	*Streptomyces*

Note: gram + = gram-positive; gram - = gram-negative

different structures, but they all act by disrupting ribosomes, and they all have broad-spectrum activity. Erythromycin is the most prevalent of the natural macrolides, and it is often used for patients who are allergic to penicillin. Additional macrolides are angolamycin, carbomycin, cirramycin, clindamycin, lankamycin, leucomycin, methymycin, niddamycin, oleandomycin, relomycin, and spiramycin.

Other Antibiotics

Vancomycin is a massive ($C_{66}H_{75}Cl_2N_9O_{24}$) molecule made by actinomycete bacteria. Its structure does not belong in any of the categories discussed previously. As penicillin does, it blocks peptidoglycan synthesis. Vancomycin became widespread, in the 1970s, as an alternate treatment for penicillin-resistant staphylococci and enterococci. But vancomycin-resistant bacteria now cause infections in hospitals and outpatient clinics. The major resistant microorganism is called *VRE* for vancomycin-resistant *Enterococcus.*

Chloramphenicol has a simpler structure than other antibiotics. It works by interfering with ribosome function. Chloramphenicol is normally bacteriostatic and must be used at a high dose to kill bacteria. But at high doses it causes side effects such as allergies and nerve damage. Chloramphenicol should be used only when no other antibiotics can stop an infection.

ANTIBIOTIC RESISTANCE

Antibiotic resistance is the ability of a microorganism to repel the effects of an antibiotic. Almost all of the antibiotics mentioned here have less effectiveness today because of resistance. Resistance develops through random mutations in DNA from one generation to the next. Once it has become part of a cell's makeup, it protects the cell in one of three ways: (1) The cell may develop the ability to prevent antibiotics from penetrating its cell wall; (2) the cell allows an antibiotic to enter but then pumps it back out before it causes damage; or (3) the resistant cell makes enzymes that destroy the antibiotic.

Genes for resistance are found both on the chromosome—the main store of DNA in the cell—and on small pieces of DNA outside the chromosome called plasmids. Bacteria share plasmids between them, an arrangement that spreads resistance from one cell to another and on occasion from one species to another species. Because bacteria grow rapidly, resistance can spread through a population of microorganisms very quickly.

Resistant pathogens threaten the well-being of patients in hospitals and individuals in nursing homes, day care centers, outpatient clinics, and athletic clubs. Some diseases once thought to be under control are returning to human, animal, and plant populations because of resistant pathogens. *Reemergence* refers to the return of a disease that was once nearly eradi-

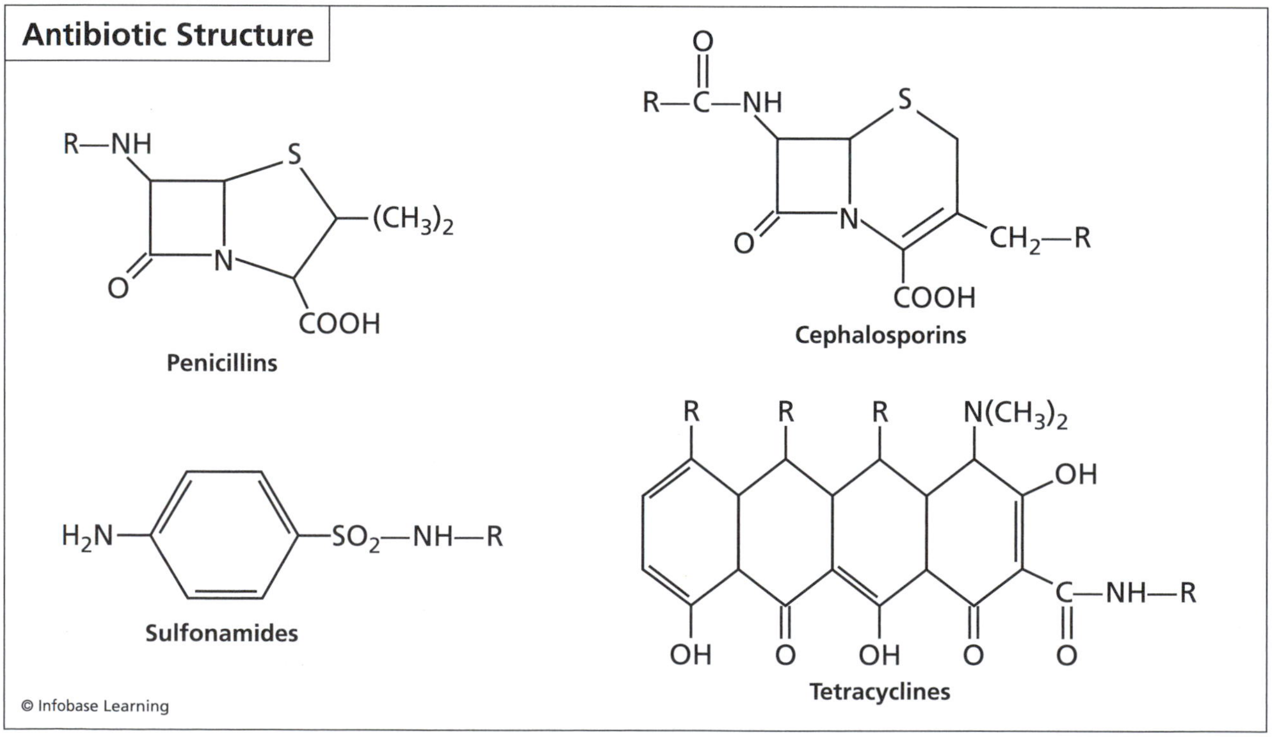

Antibiotic Structure

Penicillins

Cephalosporins

Sulfonamides

Tetracyclines

Antibiotic structure determines the antibiotic's ability to penetrate bacterial membranes. Most antibiotic classes contain at least one ring structure.

cated. Tuberculosis, for example, has reemerged to become a health threat in many parts of the world. The rise of antibiotic-resistant *Mycobacterium* strains is at the root of the return of tuberculosis.

Antibiotic companies try to outwit bacteria by making new synthetic antibiotics with a slightly different structure from the previous one. First-generation penicillin G is an example. This drug had been overprescribed and used incorrectly for many years, leading to penicillin-resistant bacteria. Chemists in the drug industry then invented second-generation methicillin and ampicillin. These antibiotics worked for a few years, until bacteria became resistant to them also. One example microorganism is methicillin-resistant *Staphylococcus aureus* (MRSA). MRSA has made methicillin almost useless for treating staph infections (any infection caused by staphylococci). Some antibiotics are now in third or fourth generations, and it is not difficult to predict a new wave of resistant microorganisms will soon evade these antibiotics as well. *Second-generation, third-generation,* and so forth, refer to each newly modified structure of an antibiotic.

For many years, meat and milk producers gave antibiotics to their animals to protect them from infection and increase growth, perhaps also to produce overall healthier animals. Opponents of this practice have argued that the widespread antibiotic use creates a number of health threats to humans and the environment. Specifically, antibiotic in meat may reduce the effectiveness of the antibiotic should a doctor need to prescribe it for a patient. Also, antibiotics excreted into the environment might have damaging effects on wildlife. Some scientists have countered that these effects are negligible, and antibiotics help support world food production. The veterinary researcher H. Scott Hurd estimated the incidence of *Campylobacter* illness that could be expected from antibiotic-resistant bacteria isolated from meat animals. Hurd wrote in *Microbe* in 2006, "We estimated that the greatest probability of compromised human treatment (not death) in the U.S. due to macrolide-resistant campylobacteriosis for all meat commodities combined was less than 1 in 10 million per year. For pork and beef, the probabilities were 1 in 53 million and 1 in 236 million, respectively. For poultry it was about 1 in 14 million." These odds should be encouraging to people who worry about the effects of antibiotics in their food. Many people are nonetheless very worried.

Some consumer groups have taken a strict watchdog role regarding antibiotics in food. For example, *Natural News,* which advocates healthy organic foods, reported in 2008, "Tyson Foods, the world's largest meat processor, and the second largest chicken producer in the United States, has admitted that it injects its chickens with antibiotics before they hatch, but labels them as raised without antibiotics anyway." The U.S. Department of Agriculture has ordered the company to remove the "Raised without Antibiotics" label.

Some countries, especially in Europe, have now banned antibiotics for beef, swine, and poultry production, and the World Health Organization (WHO) has warned of the increasing threat of drug-resistant microorganisms due to indiscriminate use of antibiotics on farms. The WHO has stated, "Studies in several countries, including the United Kingdom (UK) and USA, have demonstrated the association between the use of antimicrobials in food animals and antimicrobial resistance." Strong feelings arise on both sides of this debate, which probably will continue.

SOLUTIONS TO THE RESISTANCE PROBLEM

Antibiotic manufacturers continue searching for compounds that will confound the defensive mechanisms of bacteria. Ways to do this are to synthesize new antibiotics with structures never before seen in nature, as already mentioned, or to find new natural sources of antibiotics.

Many microbiologists believe there are thousands of natural antibiotics not yet discovered. In the past decade, marine seaweeds, algae, sponges, and corals have been investigated as new sources of antibiotics for treating animal and plant diseases. For example, a variety of antibiotic called the *cribrostatins* made by the blue marine sponge *Cribrochalina* has been shown to attack penicillin-resistant *Streptococcus* and other resistant gram-positive bacteria. The biologist Julia Kubanek of Georgia Institute of Technology remarked to *Science Daily* in 2003, "Seaweeds live in constant contact with potentially dangerous microbes, and they have apparently evolved a chemical defense to help resist disease." New antibiotics from environments quite different from the places where humans and other terrestrial animals live may soon provide the next generation of antimicrobial drugs.

The Advantages and Disadvantages of Synthetic Antibiotics

In the 1950s, the Eli Lilly Company experimented with the structure of penicillin to make it more effective. Ampicillin and amoxicillin were two of their chemists' first synthetic antibiotics. Making synthetic chemicals as complex as an antibiotic was a daunting task at the time. Before long, chemists realized that only slight changes to an antibiotic's structure could improve it. They decided on semisynthetic antibiotics. Semisynthetic and completely synthetic antibiotics give at least four benefits compared with the natural form, as follows:

1. They have higher activity at lower concentrations.

2. They provide a broader spectrum of activity.

3. Most deliver improved absorption, distribution, metabolism, and excretion by the body.

4. They give fewer side effects.

Synthesis also streamlines the challenge of finding new antibiotics. For many years after Florey and Chain's work, companies manually screened thousands of extracts against hundreds of pathogen cultures. Not only was the process slow: it was inefficient. One company was said to have screened 400,000 cultures from which it found only three useful antibiotics. Automated methods sped up the work but did not make it more efficient. Of the hundreds of thousands of substances screened by drug companies, the current list of antibiotic-producing organisms is still quite small. Synthesis of new compounds offers a brighter potential for finding new drugs at a faster pace.

Despite encouraging results from synthetic and semisynthetic antibiotics, they sometimes cause harmful side effects. Several broad-spectrum antibiotics in use today cause allergic reactions, gastrointestinal ailments, kidney damage, or nerve damage.

A new approach called *combinatorial biosynthesis* is under way to solve the problem of side effects. In this method, the DNA of microorganisms is altered so the cells produce effective antibiotics that do not cause serious side effects. *Streptomyces* has already been engineered this way to produce a new macrolide related to erythromycin. Combinatorial biosynthesis also is a source of an antibiotic called a *bacteriocin*. Bacteriocins are proteins produced by bacteria to kill other similar bacteria. The microorganism used in wine making, *Saccharomyces*, is a yeast that makes pediocin, which is similar to bacteriocins. Pediocin destroys the bacteria that contaminate wines during fermentation.

ANTIBIOTIC MODE OF ACTION

Mode of action is also referred to as *mechanism of action* because it is the mechanism by which an antibiotic kills a bacterial cell. Antibiotics against bacteria work by the following five different modes of action:

1. inhibitors of cell wall synthesis

2. inhibitors of protein synthesis

3. inhibitors of nucleic acid synthesis

4. antibiotics that injure membranes

5. inhibitors of the synthesis of essential metabolites

Antibiotics against fungi have three similar modes of action:

1. disruptors of membrane sterols

2. disruptors of cell walls

3. inhibitors of nucleic acid synthesis

Antibiotic Groups by Mode of Action

Mode of Action	Structure Type
ANTIBACTERIAL ANTIBIOTICS (EXAMPLES ARE IN PARENTHESES)	
inhibitors of cell wall synthesis	β-Lactam; polypeptide; antimycobacterium
inhibitors of protein synthesis	aminoglycosides; tetracyclines; macrolides
antibiotics that injure membranes	polymyxins
inhibitors of nucleic acid synthesis	rifamycins (rifampin); quinolones
inhibitors of synthesis of metabolites	sulfonamides
ANTIFUNGAL ANTIBIOTICS	
disruptors of membrane sterols	polyenes (amphotericin B); azoles (clotrimazole, miconazole, ketoconazole); allylamines (naftifine)
disruptors of cell walls	echinocandins (caspofungin)
inhibitors of nucleic acid synthesis	flucytocine

There exist a few antibiotics, such as griseofulvin, that fall into none of the categories described. Griseofulvin interferes with mitosis in fungi.

Since most antibiotic structures are also related to their mode of action, they can be grouped in yet another way, by structure. The main groups are shown in the table.

Sometimes mode of action is helped by a second antibiotic in an interaction called *synergism*. In synergism each antibiotic can be used at a lower dose than if each were given alone. But synergism does not work with any random antibiotic pairing. Antagonism can occur in which one antibiotic blocks the action of the other. For example, tetracycline acts as an antagonist to penicillin's action on cell wall peptidoglycan synthesis.

SPECTRUM OF ACTIVITY

An antibiotic's spectrum of activity is the range of different types of microorganisms that it kills or inhibits. Most antibiotics work against either bacteria or fungi, but seldom do they affect both. Some antibiotics work on either gram-positive or gram-negative bacteria or on a specific type of microorganism. These narrow-spectrum drugs are useful when a physician or veterinarian is certain of the pathogen causing an infection. In some cases, an infection involves a number of pathogens, some unidentified, so a broad-spectrum antibiotic is the best choice.

Broad-spectrum antibiotics kill susceptible pathogens, but they may also attack the body's normal flora, which protect it against invasion from opportunistic pathogens. In this event, a superinfection may occur. In a superinfection, most of the normal flora disappears except one. This one species begins to grow quickly to large numbers because no other bacteria compete against it. It may then cause its own infection. An example of a superinfection is a *Candida* yeast infection that occurs during antibiotic treatment of a bacterial infection. The antibiotic eliminates the native bacteria on the skin but has no effect on eukaryotic yeast cells. The yeast, no longer held in check by skin bacteria, then causes candidiasis.

ANTIBIOTICS AGAINST PLANT PATHOGENS

In the 1950s, plant pathologists wondered about the potential of the new wonder drugs for plant diseases. After screening hundreds of compounds, they found a small number that killed plant pathogens.

Only 0.1 percent of the millions of pounds of antibiotics produced yearly in the United States is used on plants. Oxytetracycline and streptomycin are the only two used in the United States. They are used to combat fire blight caused by *Erwinia amylovora* in apple and pear trees, as well as a lethal yellowing of coconut palms from bacteria similar to *Mycoplasma*. Oxytetracycline is the only antibiotic permitted by the USDA for injection into plants. By contrast, streptomycin may only be sprayed onto leaves and other outer portions of the plant.

Resistance among plant pathogens has become as great a concern in agriculture as it is in medicine. The USDA monitors antibiotic use and tries to control the spread of resistance by reviewing how each antibiotic is used and to which plants it may be applied. Every plant antibiotic is sold with information on the intended plants, application instructions, period of time the antibiotic may be applied, and safety precautions for handling it. Nevertheless, antibiotic resistance in plant pathogens is growing.

See also ANTIMICROBIAL AGENT; BACTERIOCIN; DIFFUSION; MINIMUM INHIBITORY CONCENTRATION; MODE OF ACTION; *MYCOPLASMA*; OPPORTUNISTIC PATHOGEN; PENICILLIN; PLASMID; RESISTANCE; *RICKETTSIA*; SPECTRUM OF ACTIVITY; SYMBIOSIS.

Further Reading

Chain, E., H. W. Florey, A. D. Gardner, N. G. Heatley, M. A. Jennings, J. Orr-Ewing, and A. G. Sanders. "Penicillin as a Chemotherapeutic Agent." *Lancet* 1 (1940): 226–228.

Ehrlich, Paul. "Ueber moderne Chemotherapie" (Modern Chemotherapy). *Beiträge zur experimentellen Pathologie und Chemotherapie* (1909): 167–202. In *Milestones in Microbiology*, translated and edited by Thomas Brock. Washington, D.C.: American Society for Microbiology Press, 1961.

Fox, Jeffrey L. "Race to Antibiotics Is Slow, Fragmented." *Microbe*, March 2006.

Gutierrez, David. "Tyson Foods Injects Chickens with Antibiotics before They Hatch to Claim 'Raised without Antibiotics.'" Natural News, 9 November 2008. Available online. URL: www.naturalnews.com/024756.html. Accessed March 12, 2009.

Hurd, H. Scott. "Assessing Risk to Human Health from Antibiotic Use in Food Animals." *Microbe*, March 2006.

McManus, Patricia S., and Virginia O. Stockwell. "Antibiotic Use for Plant Disease Management in the United States." Plant Health Progress. Available online. URL: http://www.apsnet.org/education/feature/antibiotic/Top.htm. Accessed March 12, 2009.

Prescott, Lansing M., John P. Harley, and Donald A. Klein. "Antimicrobial Chemotherapy." In *Microbiology*, 6th ed. New York: McGraw-Hill, 2005.

Reiner, Roland. *Antibiotics: An Introduction*. Stuttgart, Germany: Georg Thieme Verlag, 1982.

Rx List. Available online. URL: www.rxlist.com/script/main/hp.asp. Accessed March 12, 2009.

Science*Daily*. "Seaweed Surprise: Marine Plant Uses Chemical Warfare to Fight Microbes." 30 May

Antibiotics and Meat

by Wanda C. Manhanke, M.S., *St. Louis Children's Hospital, St. Louis, Missouri*

When my son got an infection in an insect bite this past summer, I was not surprised to learn that the infecting bacterium was a *Staphylococcus aureus.* I was surprised, however, when I saw the antibiotic profile of the offending agent. There were few antibiotics that could be used to treat the offender. The organism was amazingly resistant! Penicillin, oxacillin, and their derivatives were ineffective, as were erythromycin and clindamycin.

Ours is not a household that incorporates a lot of antimicrobial therapy. I am negative about the use of hand washes that guarantee the destruction of microbial populations, I do not wipe my counters down with disinfecting wipes, and triple antibiotic ointment does not have shelf space in my medicine cabinet. And yet, an organism isolated from one of the household members had an amazingly resistant profile.

If this were an isolated incident, it would, perhaps, be remarkable. And, if the only organisms experiencing increased patterns of resistance were the *Staphylococcus,* that, too would be remarkable. However, over the last 20 years, one trend that has been constant is the acquisition of antimicrobial resistance by bacteria. Drugs that could easily have treated a community acquired staphylococcal infection a decade ago are, today, ineffective. Organisms that once were susceptible to and easily treated with penicillin are now resistant and can be associated with poor clinical outcomes and sometimes even death.

Bacteria become resistant to antimicrobials in one of two ways: A spontaneous mutation can occur in their chromosomal genes, or new genes or sets of genes can be acquired from another species or from the environment. New genes or sets of genes can be acquired through the processes of conjugation, transformation, and transduction.

Conjugation is the most common mechanism by which resistant genes are acquired. It is a genetic exchange mechanism between bacteria that requires cell-to-cell contact. The bacterial pilus functions to establish contact with another cell and acts as a tube through which the DNA is passed during the conjugative process. Genes that encode for antimicrobial susceptibility can be found on extrachromosomal pieces of DNA referred to as plasmids. The plasmid is a piece of DNA that acts independently of the chromosome. In the process of conjugation, clinically significant organisms that have contact with innocuous environmental organisms can acquire these pieces of DNA and incorporate them into their DNA, thus creating clinically significant organisms with increased resistance factors.

A population of microbes can contain a few cells that are drug resistant as a result of a mutation or the acquisition of drug resistance. Having the characteristic of resistance to a given antibiotic gives the organism no particular advantage if the drug is not present in the habitat. The number of resistant forms remains low. If the population is somehow exposed to the drug, for example, during a course of antimicrobial therapy, the resistant organisms are the ones to survive and the offspring of subsequent populations will be resistant. In biological terms, there is a natural selection for the resistant organisms. As these organisms grow and divide, the entire population will eventually become resistant.

The likelihood of an organism's encountering a drug in its habitat is not small. One need only survey the many common household items that have as their aim the complete destruction of the microbial world to realize the prevalence of disinfecting and antimicrobial agents in our environment.

Increased use and misuse of antibiotics in human disease treatment and in agriculture can also contribute to the presence of antimicrobial agents in the environment. One such practice that plays a role in ensuring that populations of microorganisms will encounter an antimicrobial in their habitat is that of incorporating antibiotics into the feed of livestock animals.

Penicillin gained widespread use in the treatment of human diseases during the 1930s, and it paved the way for similar discoveries in veterinary science and medicine. Selman Waksman's discovery of streptomycin while working at a New Jersey Agricultural Experiment Station opened the door for the use of antimicrobials in livestock production. Streptomycin, alone, helped to wipe out bovine tuberculosis and mastitis in dairy cattle.

One observation made as antibiotics began to be used in agricultural herds was that not only was the health of the herd improved, the animals grew faster on the same amount of feed. Antibiotics began to be added to the feed of all animals in the herd, whether they were ill or not. By the end of the 1940s, vitamins, proteins, other nutrients, and antibiotics were all available in manufactured animal feeds. The use of antibiotics changed the way livestock animals could be produced. Poultry, swine, and dairy and cattle feedlots showed dramatic increase in size, as the constraints imposed by the possibility of infectious diseases were eliminated through the use of antibiotic-supplemented feeds.

Currently, in the United States, on an annual basis, 25 million pounds of valuable antibiotics, in some cases, the same antibiotics used in human disease treatment, are fed to agricultural animals for nontherapeutic purposes such as as growth promotion. This represents an increase from 16 million pounds, in the mid-1980s, and is roughly 70 percent of total U.S. antibiotic production. This is a practice that, in the words of Charles Benbrook of the Union of Concerned Scientists, "is sobering."

Nontherapeutic use of antibiotics has been especially common in poultry production. The discovery that low doses of antibiotics make chickens grow faster was made in 1950. By the 1970s, 100 percent of all commercially raised poultry in the United States was raised with antibiotics. The use of fluoroquinolone drugs was approved for poultry production, in 1996. The use of these drugs for agricultural purposes presents a model of how resistance can develop. The approval of their use as a feed additive coincides with the sharpest rise seen in fluoroquinolone resistance. This is the most recently approved class of antibiotics, and the expectation by the FDA was that these particular antibiotics would remain effective for a long time.

The overuse of antibiotics in livestock production creates a serious threat to human health. When the antibiotics used in livestock production are the same as those used in human medicine, there is increased risk that resistance developing in organisms infecting humans will pose a public health threat worldwide. At a conference held by the World Health Organization (WHO), it was concluded that the major transmission pathway for resistant bacteria is from food animals to humans.

Antibiotic-resistant bacteria can be transmitted to humans in several ways. Examples include the consumption of meat contaminated with antibiotic residues or resistant bacteria during slaughter, as is seen with the *E. coli* 0157 bacterium; direct spread of antibiotic-resistant organisms by farmers and farmworkers to family and community; or contamination of local waterways and groundwater by bacteria found in animal excrement. All three examples illustrate threats to human disease management, and the latter two examples illustrate ideal opportunities for clinically significant organisms to interact with organisms carrying resistant properties with the possibility of becoming more resistant themselves.

It is a dilemma. On the one hand are the meat producers who are charged with creating a safe and plentiful food supply to meet the needs and demands of consumers and earn a profit while doing so. They believe they require the use of antibiotics as a means of achieving that end. On the other hand is the medical community, who watch as previously valuable antibiotics become ineffective and previously easily treated infections become life-threatening.

Antibiotic resistance is inevitable. And increased bacterial resistance cannot be solely linked to subtherapeutic use in the animal industry. Misuse can and does occur within the medical community. Practices such as prescribing broad-spectrum antibiotics (capable of wiping out all types of bacteria) and overprescribing antibiotics can also contribute to the development of resistance.

Prudent use of and restriction of antibiotics to treatment of human and animal diseases are essential. The Preservation of Antibiotics for Medical Treatment Act 2007, endorsed by the American Medical Association, the American Academy of Pediatrics, the Infectious Disease Society of America, the American Public Health Association, and others, makes several proposals for the protection of human disease—fighting antibiotics. Among these proposals is an amendment to the Federal Food, Drug and Cosmetic Act to withdraw approval from feed-additive use of seven specific classes of antibiotics: penicillins, tetracyclines, macrolides, lincosamides, streptogramins, aminoglycosides, and sulfonamides. Each of these classes contains antibiotics used in human medicine. The bill would ban only the feed-additive uses of the drugs for "nontherapeutic" purposes, that is, in the absence of any clinical sign of disease in the animal for growth production, feed efficiency, or weight gain. Sick animals would still receive treatment and drugs would be available for the purpose of legitimate prophylaxis. Meat producers would be left with the option of using antibiotics not used in human medicine. The Senate version of the bill also authorizes funds to help farmers defray the costs incurred in the phasing out of the use of medically important antibiotics.

As with most production costs, the consumer eventually pays. The National Academy of Sciences estimates that a total ban on nontherapeutic antibiotic use would raise meat prices five dollars to $10 per person annually. This is a small price to pay to protect our antibiotic arsenal and to ensure that bacterial infections remain treatable.

See also ANTIBIOTIC; PENICILLIN; RESISTANCE.

Further Reading

Bon Appetit Management Company. "Poultry Raised without Routine Use of Antibiotics." Available online. URL: www.circleofresponsibility.com/page/19/poultry.htm. Accessed November 8, 2009.

Living History Farm. "Antibiotics and Feed Additives." Available online. URL: www.livinghistoryfarm.org/farminginthe40s/crops_09.html. Accessed November 8, 2009.

Todar, Kenneth. "Bacterial Resistance to Antibiotics." In *The Microbial World.* 2008. Available online. URL: http://textbookofbacteriology.net/resantimicrobial.html. Accessed November 8, 2009.

Union of Concerned Scientists. "The Preservation of Antibiotics for Medical Treatment Act of 2007." June 8, 2007. Available online. URL: http://ucsusa.wsm.ga3.org/food_and_environment/antibiotics_and_food/the-preservation-of-antibiotics-for-medical-treatment-act.html. Accessed June 12, 2009.

2003. Available online. URL: www.sciencedaily.com/releases/2003/05/030530082615.htm. Accessed March 12, 2009.
Tortora, Gerard J., Berdell R. Funke, and Christine L. Case. "Antimicrobial Drugs." In *Microbiology: An Introduction*. San Francisco: Benjamin Cummings, Pearson Education, 2004.
World Health Organization. "Use of Antimicrobials outside Human Medicine and Resultant Antimicrobial Resistance in Humans." Fact sheet, January 2002. Available online. URL: www.who.int/mediacentre/factsheets/fs268/en. Accessed March 12, 2009.
Zaidan, Abe. "Inventors to Be Inducted into Hall of Fame: Chemist Says Luck Played Role in Wonder Drug." *Cleveland Plain Dealer*, 24 April 1992.

antimicrobial agent An antimicrobial agent is any chemical, biological substance, or gas that kills or inhibits the growth of microorganisms. Antimicrobial agents belong to one of the two following broad groups: drugs or chemical biocides. Antimicrobial drugs consist of natural antibiotics, synthetic antibiotics, bacteriocins, and other types of drugs that inhibit bacteria, fungi, protozoa, viruses, malaria, or tuberculosis. Synthetic compounds today compose most of the latter group.

Antimicrobial chemotherapy is the use of an antimicrobial agent to treat disease caused by a pathogen, also termed *infectious disease*. Chemical biocides, by comparison, kill or inhibit microorganisms on inanimate objects or in water. A *biocide* is a substance that kills any living thing, but microbiology reserves this term for chemical compounds rather than antibiotics or other drugs. The main chemical biocides are disinfectants, sanitizers, sterilants, preservatives, and antiseptics. These substances contain formulas that act on bacteria, fungi, algae, protozoa, viruses, malaria, or tuberculosis. Many chemical biocides work against more than one of these types of microorganisms at the same time.

ANTIMICROBIAL DRUGS

A drug is any substance that works in or on the body to treat a disease condition. The U.S. Food and Drug Administration (FDA) controls the manner in which all drugs may be tested, sold, and used in the United States. Drugs such as antibiotics have impacted human health since their first use in the 1940s. Within the past several decades, however, a growing concern over the resistance of many bacteria and viruses to antibiotics has changed the way these drugs have become viewed in medicine and by the public.

Microorganisms destroyed by any antimicrobial drug are said to be *susceptible* to the drug. Conversely, resistant microorganisms are those that have developed a means to ward off the effects of an antimicrobial agent.

Any antimicrobial drug must have certain characteristics in order to restore a person or an animal to health. First, it must be effective against the pathogen causing an infection. Broad-spectrum antimicrobial agents kill or inhibit a wide variety of microorganisms; limited-spectrum agents kill or inhibit only one or a few types of microorganisms. Limited-spectrum antimicrobial agents are also called narrow-spectrum. Physicians and veterinarians must select the correct agent to act on a specific microorganism or group of microorganisms. Second, the antimicrobial agent must destroy pathogens at the same time it causes no harm to the patient. This quality is called *selective toxicity*. The most selective and safe drugs are those with a mode of action against only the pathogen's activities, activities that mammalian cells do not possess. *Mode of action* refers to the manner in which an antimicrobial agent works to damage prokaryotic or eukaryotic cells. One example of selective toxicity occurs in antibiotics that destroy bacterial cell walls. These antibiotics are safe for most humans and animals because mammalian cells do not have cell walls. The less selective toxicity a drug possesses, the more chance there is of side effects occurring in the patient. In truth, almost all antimicrobial drugs have some side effects in patients.

Other desirable characteristics of antimicrobial agents, in addition to effectiveness and safety, are as follows.

- quick-acting
- effective in low doses
- able to exit the body rapidly
- easy to administer to the patient

Today's antimicrobial drugs may be categorized in more than one way. They belong to various groups according to structure, source, the type of microorganisms they destroy, or their mode of action. The general groups of antimicrobial drugs are the following:

- antibacterial—effective against bacteria
- antifungal—effective against fungi, including yeasts
- antiviral—effective against viruses
- antiprotozoal—effective against protozoa

In each of the groups, the drug is either static or cidal. Static drugs collectively make up a group called microbistats. Microbistats inhibit the growth of microorganisms but do not necessarily kill them. Cidal drugs are referred to as microbicides. Microbicides kill microorganisms. A number of drugs are microbistatic at low doses but become more effective, thus microbicidal, at higher doses.

CHEMICAL BIOCIDES

Chemical biocides are substances that kill microorganisms on nonliving surfaces, in drinking water, or in products used by consumers such as foods, cosmetics, and paints. *Germicide* is another term for a chemical that kills *germs,* an informal term for harmful microorganisms. (Some chemical formulas merely inhibit microorganisms; they do not kill them. By convention, however, these inhibitory chemicals tend to be grouped along with true biocides.)

The U.S. Environmental Protection Agency (EPA) oversees laboratories that test chemical biocides, and EPA scientists categorize biocides in two different ways. The first method of classification uses the type of microorganisms that the chemical destroys. Therefore, chemical biocides may be bactericides, fungicides, algicides, or sporicides. These products kill bacteria, fungi, algae, and bacterial spores, respectively. Chemicals that kill more than one type of microorganism are usually called biocides or antimicrobial products. The second method of categorizing biocides uses the chemical's level of effectiveness. In the world of biocides, effectiveness is called *efficacy.* Biocides fall into the following three main efficacy categories:

- Sterilants, also called *sporicides,* kill all microorganisms of every type, including bacterial spores.
- Disinfectants kill all microorganisms other than bacterial spores.
- Sanitizers reduce the numbers of bacteria to safe levels.

Disinfectants mimic antibiotics through their ability to be either broad-spectrum or limited- (narrow-) spectrum. Broad-spectrum disinfectants kill a variety of gram-negative and gram-positive bacteria. Some broad-spectrum disinfectants kill viruses and fungi in addition to bacteria. Limited-spectrum disinfectants kill either gram-negative or gram-positive bacteria but not both.

The EPA recognizes additional specialized biocide categories. Formulas may be intended to kill only one type of microorganism, as follows: fungi (fungicides), viruses (virucides), algae (algicides), or the unique *Mycobacterium* bacteria, which cause tuberculosis (tuberculocides).

Health care clinics, veterinary clinics, food production plants and restaurants, day care centers, schools, nursing homes, and public restrooms should receive regular cleaning with one of these three types of biocides. Many people also use them at home and in offices, cars, and kennels. Biocides alone do not, however, assure that a family, patients, or other members of the public will avoid infection. The Centers for Disease Control and Prevention has highlighted seven important steps in preventing infection, as follows:

1. frequent and proper hand washing
2. careful handling and preparing of foods
3. immunization
4. proper care and handling of pets
5. avoiding contact with wild animals
6. appropriate use of antibiotics; avoiding antibiotic overuse
7. routine cleaning and disinfecting of surfaces with biocides

Drinking water treatment plants use a special type of disinfection to kill pathogens in water before it is distributed to the community. Treatment plants mainly use chlorine compounds, but other types of disinfection (ozone, ultraviolet radiation, and filtration) also kill pathogens. Operators of recreational waters also disinfect water to prevent the spread of pathogens. Some places where disinfects help keep swimmers safe from infection are swimming pools, wave pools, water rides, hot tubs, whirlpools, and spas.

Antimicrobial agents cannot remove all risks of infection, and questions have arisen in the past few decades on microbial resistance to them, side effects, and limited efficacy. But, in general, antimicrobial agents reduce illness by halting the spread of infection through the air and in water, food, and nonfood products. The University of Arizona microbiologist Charles Gerba has led many studies on the effects of disinfectants on human health. "When you don't use a disinfectant product," Gerba explained, "you just spread germs around and give them a free ride. . . . Disinfectants do reduce illness." Used in the appropriate situations, antimicrobial agents can enhance the well-being of humans and domesticated animals.

See also ANTIBIOTIC; ANTISEPTIC; BACTERIOCIN; BIOCIDE; DISINFECTION; MODE OF ACTION; PRESERVATION; RESISTANCE; SPECTRUM OF ACTIVITY.

Further Reading

Block, Seymour S., ed. *Disinfection, Sterilization, and Preservation,* 5th ed. Philadelphia: Lippincott Williams & Wilkins, 2000.

Centers for Disease Control and Prevention. Available online. URL: www.cdc.gov. Accessed March 12, 2009.

Mollenkamp, Becky. "Germ Warfare: Cleaners and Disinfectants." CleanLink: Sanitary Maintenance, August 2004. Available online. URL: www.cleanlink.com/sm/article.asp?id=1347. Accessed March 12, 2009.

National Institutes of Health. U.S. National Library of Medicine. "Antibiotics." Available online. URL: www.nlm.nih.gov/medlineplus/antibiotics.html. Accessed March 12, 2009.

The Soap and Detergent Association homepage. Available online. URL: www.cleaning101.com. Accessed March 12, 2009.

Tortora, Gerard J. "Antimicrobial Drugs." In *Microbiology: An Introduction.* San Francisco: Benjamin Cummings, 2004.

———, Berdell R. Funke, and Christine L. Case. "The Control of Microbial Growth." In *Microbiology: An Introduction.* San Francisco: Benjamin Cummings, 2004.

U.S. Environmental Protection Agency. "Antimicrobial Pesticide Products." Available online. URL: www.epa.gov/pesticides/factsheets/antimic.htm#types. Accessed March 1, 2009.

antiseptic An antiseptic is any chemical that reduces the number of pathogens on human or animal skin. Antiseptic products create asepsis, which is the absence of pathogens. Because they act on the outside of the body, they are called topical products, and they also belong to a broader category of chemical and biological substances called antimicrobial agents.

ANTISEPTICS IN MEDICAL HISTORY

In the mid-18th century, the Scottish physician John Pringle (1707–82) pondered the effect that strong chemicals seemed to have on halting the spoilage of food. Pringle and other scientists of his time had been inspired by the article "Natural History" published by the English author and philosopher Sir Francis Bacon (1561–1626) a year before his death. In it, Bacon discussed the relationship between the use of chemicals and the eradication of *putrefaction,* a general term for the decomposition of animal matter. He proposed that certain chemicals blocked the reactions leading to food spoilage and other chemicals could similarly halt gangrene infections in the skin. Bacon tested his theories by studying various chemical treatments—strong acids, salting, sugaring, and protection of items from exposure to air. Many of these ideas were not new; curing foods with salt, smoke, or acid had been handed down through generations ever since its use in ancient societies.

Fifty years after Bacon's publication, the Dutch tradesman Antoni van Leeuwenhoek (1632–1723) found that he could slow the activity of protozoa by dousing them with sulfuric acid, salts, sugars, and even wine. By arresting the protozoa's movements, van Leeuwenhoek made them easier to observe under his rudimentary microscope. But Pringle's peers in the medical community remained skeptical of the use of harsh chemicals in treating patients. It might be useful to kill those tiny "animalcules" that van Leeuwenhoek had observed in his laboratory, some argued, but chemicals had no place for use on human skin. Pringle nevertheless saw value in selecting certain chemicals for treating surface (skin) injuries, particularly on the battlefield. He set up experiments to test a variety of compounds in a range of concentrations against microorganisms that caused putrefaction. He discovered camphor, acids, and bases were the most effective in cleaning wounds.

In 1752, John Pringle collected his observations and discoveries on injury treatment in *Observations of Disease of the Army,* a book that provided the most useful information for the period on sanitation and stopping infection through the use of antiseptics, plus additional theories on infectious disease, throughout six editions. In the book, Pringle may have been the first person to coin the term *antiseptics* when he wrote, "Were putrefaction the only change made in the body by contagion, it would be easy to cure such fevers, at any period, by the use of acids, or other 'antiseptics.'" (Pringle also introduced here the idea of contagion, or the spread of disease from person to person, and he introduced the term *influenza.*)

Over the next 200 years, microbiological science as well as chemistry blossomed with the development of methods for synthesizing new chemicals and sensitive instruments to study the purity and the structure of these chemicals. Although microbiologists were expanding their knowledge on the basics of microorganisms, they were slow in fully realizing the relationship between infection and the antiseptics proposed by Pringle. Antiseptics for preventing infection were rather limited in the 18th century. Vinegar, which had first been used by ancient Roman legions, still served as the antiseptic of choice by doctors treating navy seamen during the American Revolution. Surgeons eventually began exploring other chemicals in intervals free of the heat of battle. Mercury-containing ointments as well as hypochlorite (bleach), phenol, and carbolic acid worked well but may have caused far more harm to the patient than to the infectious microbes. Well into the 1800s, surgeons on both sides of the Atlantic continued to take opposing viewpoints on the use of strong chemicals on human skin. Some, however, began to accept the idea of chemical antiseptics. One surgeon in England not only accepted antiseptics but believed that his patients' lives depended on them.

In the mid-1800s, the English surgeon Joseph Lister (1827–1912) considered the pros and cons of putting chemicals on open wounds. His contemporary, the microbiologist Louis Pasteur (1822–95), had already completed a series of experiments showing that germs related directly to putrefaction. Lister used Pasteur's evidence and his own convictions to declare that surgical wounds demanded aseptic conditions. He began cleaning his surgery room more thoroughly than other surgeons had been tending theirs at the time. Lister went so far as to sterilize all his surgical instruments, a process that surgeons had not previously considered.

Spontaneous generation was the prevailing scientific theory at the time. As Lister tried to convince other surgeons to create sanitary conditions in preparation for surgery, many of his peers could not see his logic because they believed germs arose spontaneously from forces in nature and the air. Surgeons of that period viewed sterilization of their instruments as an unnecessary luxury. They responded to Lister's ideas more often with disregard, even laughter, than with respect. In an 1867 presentation in Dublin to the British Medical Association, Lister made several pointed comments to his detractors. In referring to hospital care, Lister said, "Previously to its [antiseptic] introduction, the two large wards in which most of my cases of accident and of operation are treated were amongst the unhealthiest in the whole surgical division of the Glasgow Royal Infirmary. . . . I have felt ashamed, when recording the results of my practice, to have so often to allude to hospital gangrene or pyaemia [blood infection]." Although Lister had slowly won over some physicians, he emphasized, "The point of the fact [affect of antiseptic use] can hardly be exaggerated." Lister's most powerful response to those in the medical community who remained unmoved was to lead by example.

Lister sterilized instruments, washed the floors and walls of his surgery with phenol solution, and also sprayed phenol over wound dressings and around each surgical incision. With this cleaning regimen, an increasing percentage of Lister's patients recovered from surgery without infection. After Lister's speech in Dublin, and his article "On a New Method of Treating Compound Fracture, Abscess, etc., with Observations on the Conditions of Suppuration" in the same year, an increasing number of surgeons began tinkering with antiseptics such as alcohol, iodine, and hydrogen peroxide. As did Lister, they scrubbed surgery rooms with phenol solutions to remove all dirt and any unseen pathogens.

Alcohol, iodine, and hydrogen peroxide have remained standard antiseptics in medical care. As a final validation of Lister's work, medical professionals today use these substances as Lister had described almost 150 years ago: for presurgery preparation and for treatment of open wounds and surgical incisions. Mercury compounds had also been included in the list of useful antiseptics for decades before scientists began to understand their dangers in the body. Today mercury compounds are rarely used as antiseptics.

From the 1970s to mid-1990s, the U.S. Food and Drug Administration (FDA) and the U.S. Environmental Protection Agency (EPA) met to decide how they would split responsibilities in regulating chemicals that kill microorganisms. The FDA has been regulating antiseptics since 1972 and continues to do so today. The EPA meanwhile controls the use of substances that remove germs from inanimate objects, such as disinfectants.

ANTISEPTIC CHEMICALS

The most popular antiseptics in use in medicine and veterinary medicine are alcohols, quaternary ammonium compounds (quats), various iodine formulas, and hydrogen peroxide, each representing one of four different chemical categories of antiseptics. The table on page 54 presents the main chemicals used for removing germs from wounds, skin breaks, burns, and surgical incisions. Some have very specialized uses such as the iodine compounds used by diary operators to clean cows' teats before milking.

All of the example antiseptics in the table, except acridines, inhibit microorganisms by destroying the normal structure and function of their cell membrane. Acridines interfere with the cell synthesis of nucleic acids and nucleic acid function. At least one of three common antiseptics is found in almost every household medicine cabinet today, as they were 100 years ago: rubbing alcohol, tincture of iodine, and hydrogen peroxide.

Alcohol has been used since antiquity to control odors, cleanse the skin, and rinse dirt from foods. The ancients preferred wine to carry out these duties, but its use may have been split about evenly between external and internal use! As the science of microbiology bloomed, microbiologists noticed that wine's activity was not strong enough to kill large numbers of germs. Though microbiologists understood Lister's theories, alcohol did not seem to them to be their answer. But in the 1880s, the German physician Robert Koch (1843–1910) changed their minds by demonstrating ethyl alcohol's capacity to kill pure cultures of bacteria. Microbiologists reexamined ethyl alcohol (ethanol) and other alcohols. By the dawn

Some Common Antiseptics Used in Human and Veterinary Medicine

Chemical Group	Example Antiseptics
acridines	acriflavine, aminoacridine
alcohols	isopropyl rubbing alcohol
chlorines	chlorhexidine, chloramine
iodine formulas	tincture of iodine, iodophor
peroxides	hydrogen peroxide, benzoyl peroxide, peracetic acid
phenols (carbolic acid)	phenol, bisphenol, triclosan
quats	benzalkonium chloride
salicylic compounds	salicyclic acid, salicylanide

of the 20th century, isopropyl alcohol (isopropanol) and ethyl alcohol had become accepted as the best forms for killing a variety of microorganisms. By the 1920s, researchers had shown that alcohol worked well in removing microorganisms from the hands and other parts of the body. Most of those studies had taken place in laboratories, however, and doctors avoided the alcohols because they dried and cracked their patients' skin. The damage to the skin, in turn, would increase risks of infection. Although alcohol had proven to be a reasonable choice as an antiseptic, its effect of drying out the skin led to only limited use.

Despite their drawbacks, alcohols are effective antiseptics because they work quickly on bacteria by denaturing membrane proteins and fats. Alcohols destroy the structure of proteins and enzymes that these molecules need to function, and they also dissolve membrane lipids. Alcohol antiseptic activity is best when the compound has no more than 10 carbons in its chain structure. Alcohols composed of more than 10 carbons become less soluble in water as chain length increases, making them less able to permeate membranes and disrupt cell activity. In 1903, Charles Harrington and H. Walker discovered another property affecting activity in addition to chain length: alcohol's antimicrobial activity depends on its percent solution in water. Alcohol, when slightly diluted with water, may permeate cells faster than undiluted alcohol. In addition, the small amount of water delays alcohol's rapid evaporation and so allows it to stay on the skin longer than if undiluted. Most current alcohol solutions sold in stores are from 60 percent to 90 percent water. For example, isopropyl rubbing alcohol (isopropanol), a 70 percent formula, is a common household antiseptic. Ethanol is most effective at 60–70 percent. Newer alcohol-based antiseptics have also been formulated to prevent drying the skin. Alcohol hand washes, for example, contain a small amount of glycerol to help preserve the skin's moisture. Examples of alcohols with antimicrobial activity are listed in the table on page 55.

As mentioned, alcohols work fast, and they also help other antiseptics act faster. For example, when either chlorhexidine gluconate or iodine compounds called *iodophors* are used in a mixture with alcohol, their antimicrobial action proceeds faster than when they are used alone.

Iodine's history in medicine dates to the 1800s. Surgeons manning tents on battlegrounds during the Civil War relied on a 5 percent solution of iodine in diluted alcohol, a mixture known as *tincture of iodine*. But this form of iodine stained and had a bad odor, and the chemical did not easily dissolve in water. Physicians began to replace tincture of iodine in surgeries with iodophors. Iodophors used today are compounds made of water-soluble polymers (long-chain compounds) that act to hold iodine in a homogeneous mixture. Iodophors also spread easily on the skin compared with the older tincture mixes. Povidoneiodine (PVP-I) is an iodophor in which iodine combines with a polyvinyl compound, which helps distribute the iodine molecules over the skin. In either tincture or iodophor form, iodine kills bacteria, fungi, and viruses. It acts on these agents by penetrating the cell wall and membrane and then destroying amino acids and unsaturated fats.

Hydrogen peroxide was first used in 1858 by the English physician Benjamin W. Richardson (1828–96) for reducing odors emanating from spoiled substances. He had joined his colleagues at the Medical Society of London in agreeing that odor and infection were linked, yet the medical profession would wait for the emergence of Joseph Lister's publications to draw a sound connection between microorganisms and infection. During most of Richardson's career, any chemical that killed odor had been considered a worthwhile aid in surgery and medical treatment. For that reason, medical offices and homes had on hand a 3 percent solution of hydrogen peroxide, not for killing germs but for eliminating odors.

Hydrogen peroxide acts in a similar manner to bleach because it is a strong oxidizing agent. Oxidizing agents produce a series of reactions inside cells that create molecules called *free radicals*. Free radicals are chemically unstable molecules that react readily with other compounds inside cells. The reactions destroy enzymes, denature proteins, and disrupt membranes. Hydrogen peroxide kills bacteria, yeasts, viruses, and even the difficult-to-kill bacterial spores of *Bacillus* and *Clostridium*.

Hydrogen peroxide's antiseptic activity is effective but short-lived because it is degraded by catalase enzyme. Mammalian cells and aerobic microorganisms produce catalase to destroy the small amounts of hydrogen peroxide released naturally in the reactions inside cells, thereby protecting the cells from free radical damage. In protecting cells, catalase acts as an antagonist to hydrogen peroxide, meaning it takes away some of the chemical's effectiveness. Despite interference from catalase, hydrogen peroxide can be a useful antiseptic. Current solutions usually contain 3.5 percent hydrogen peroxide plus stabilizers to preserve the chemical.

USES OF ANTISEPTICS

An antiseptic removes native flora and transient microorganisms from the skin. Native flora are the microorganisms that normally live on the skin without causing harm to a healthy host. Transient microorganisms are on the skin for short periods, and, unlike native flora, they are not part of the body's normal microbial community. Any break in the skin's continuous protective layer gives both native and transient species a chance to start an infection, so an antiseptic must be effective against all types of microorganisms.

Medical providers use antiseptics before making surgical incisions, giving injections, drawing blood samples, and treating trauma to the skin in the form of cuts, scrapes, or punctures. The Association for Professionals in Infection Control and Epidemiology (APIC) recommends that health care providers look for the following five characteristics in a good antiseptic:

1. fast action—acts quickly and is effective on the first application
2. persistence—stays on the skin to prevent regrowth of microorganisms
3. breadth of spectrum—activity against a wide range of microorganisms
4. efficacy—kills microorganisms
5. safety—nonirritating and nontoxic

Some antiseptics lack one or more of the attributes listed here that make antiseptics superior choices for presurgery preparations. As mentioned, hydrogen peroxide kills a wide range of microorganisms but it degrades quickly. Said another way, it has a broad spectrum of activity but lacks persistence on the skin. Triclosan and the quats work better against gram-positive bacteria than gram-negative, and so each has a narrow spectrum of activity. Safety considerations may affect the choice of antiseptics. Alcohols cause minor skin irritation and chlorhexidine can damage mucous membranes, so these chemicals would not be good choices for use on patients with serious skin trauma. The characteristics indicated make certain antiseptics better suited for some roles than for others.

Antiseptics can be classified by chemical type, as shown earlier, but classification has been a frustrating exercise over many years because most antiseptics share some characteristics. In 1913, the sanitation experts Dakin and Dunham expressed their frustrations in their *A Handbook of Antiseptics*: "For various reasons it is quite impossible to formulate a perfectly logical classification of antiseptics. In the first place, almost every soluble substance, provided it

Alcohols with Antimicrobial Activity

Alcohol	Structure	Properties
ethanol (ethyl alcohol)	CH_3CH_2OH	active against bacteria, including the tuberculosis *Bacillus;* viruses; and fairly effective against fungi
isopropanol (isopropyl alcohol, rubbing alcohol)	$CH_3CH_2CH_2OH$	similar to ethanol
benzyl alcohol	aromatic (contains a ring structure); $C_6H_5CH_2OH$	active against bacteria

Types of Currently Used Antiseptic Products

Product	Target Population	Where It Is Used
	HEALTH CARE ANTISEPTICS	
hand wash/hand sanitizer	health care professionals, patients	hospitals, clinics, nursing homes
preoperative skin preparation	surgeons, nurses, presurgery	hospitals
surgical hand scrub	surgeons, presurgery	hospitals
	GENERAL CONSUMER ANTISEPTICS	
hand wash/hand sanitizer	general population	homes, workplace, day care centers
body wash	general population	homes
	FOOD HANDLER ANTISEPTICS	
hand wash/hand sanitizer	commercial food handlers	restaurants, food carts, food processing operations

can be obtained in sufficient concentration, is capable of exerting some antiseptic action." The authors surmised that, before long, a scientist would be required to classify almost every known chemical as an antiseptic, a process they correctly viewed as "obviously useless and unnecessary." Today the FDA classifies antiseptics according to their intended use rather than chemical composition, summarized in the table.

Health care professionals take into account the two ways of classifying antiseptics before selecting one for use on patients: the advantages and disadvantages of the antiseptic's chemical ingredients and the intended use of the antiseptic product.

ANTISEPTICS IN VETERINARY MEDICINE

Antiseptics are used in veterinary medicine for the same purpose as they are in human medicine. The most common antiseptics used in veterinary medicine are alcohols, chlorhexidine, hydrogen peroxide, peracetic acid, acetic acid at 1 percent, iodine solutions, phenol, and soda lye at 2 percent.

Veterinarians avoid chlorine antiseptics such as hypochlorite solutions (bleach) because of their strong odor and because they have an irritating effect on the skin and when inhaled. For many years, veterinarians relied on mercury- or silver-containing compounds. (These compounds were occasionally also used on humans.) Mercuric (mercury-containing) compounds are toxic and cause environmental damage to ecosystems so they are no longer used in medicine. Aqueous silver solutions can irritate the skin, but a 0.5 percent solution sometimes still serves as an application to surgical dressings.

Pine tar belongs in the phenol chemical category of antiseptics. It has been used for many years on bandages for dressing wounds of the hoof and the horn. Coal tar creosote is a preservative oil used on rare occasions to treat inflammations of the skin, hoof, or horn. Coal tar creosote causes cancer, however, so veterinarians avoid using it.

METHODS OF TESTING ANTISEPTICS

Hand scrubs, hand washes, and preoperative skin preparations are tested by a variety of methods on human volunteers. APIC publishes the methods that the FDA accepts for showing how well an antiseptic works and the manner in which it will be used, according to the table. Scientists test surgical hand scrubs—intended for scrubbing up prior to surgery—by putting a small volume of antiseptic into a surgical glove worn by a volunteer. After massaging the hand to dislodge any bacteria on the skin inside the glove, the tester removes a sample of the liquid from the glove and then uses aseptic techniques and regular culture methods to determine the number of bacteria that the antiseptic has killed. (This is done by comparing a sample from a similar gloved hand treated with water instead of antiseptic.) Similar methods have been adapted for testing antiseptic activity on other parts of the body, using specialized sampling cylinders instead of gloves.

Laboratory testing shows that antiseptics do not remove every single bacterial cell from the skin. Antiseptics do, however, lower to safe levels the amount of microorganisms on the skin. Health professionals have struggled with the difficult question of what constitutes a safe amount of bacteria. Joseph Lister showed that the safe level of bacteria on a patient is any level that cannot cause infection. The FDA accepts today's antiseptics on the basis of their ability to kill enough microorganisms to make the skin safe for incision, needle puncture, or other breaks in the body's protective barrier. Decades of antiseptic use in preventing infection have provided evidence of antiseptic safety.

Antiseptics entered use in medicine more slowly than perhaps they should have; acceptance took time to overcome long-held beliefs in spontaneous generation. With the widespread use of antiseptics in health care today, however, serious infections and deaths can be controlled to a degree not seen a century ago.

See also ANTIBIOTIC; ANTIMICROBIAL AGENT; DISINFECTION; GERM THEORY; LISTER, JOSEPH; LOGARITHMIC GROWTH; SANITIZATION; SPECTRUM OF ACTIVITY.

Further Reading

Ascenzi, Joseph M., ed. *Handbook of Disinfectants and Antiseptics.* New York: Marcel Dekker, 1996.

Association for Professionals in Infection Control and Epidemiology. Available online. URL: www.apic.org. Accessed March 13, 2009.

Block, Seymour S. *Disinfection, Sterilization, and Preservation,* 5th ed. Philadelphia: Lippincott Williams & Wilkins, 2000.

Brewer, Timothy F., Richard P. Wenzel, and Jean-Paul Butzler. *A Guide to Infection Control in the Hospital,* 2nd ed. Hamilton, Canada: B. C. Decker, 2002.

Chinnes, Libby F., Anne M. Dillon, and Loretta L. Fauerbach. *Homecare Handbook of Infection Control,* 2nd ed. Washington, D.C.: APIC and Missouri Alliance for Home Care, 2002.

Dakin, Henry D., and Edward K. Dunham. *A Handbook of Antiseptics.* New York: MacMillan, 1918. Available online at URL: http://books.google.com/books?id=9-APAAAAYAAJ&dq=%22Handbook+of+Antiseptics%22+Dakin+Dunham&printsec=frontcover&source=bl&ots=itOOnbcUjJ&sig=6H10aKnP6VV35b0aglwxefjZZ7g&hl=en&ei=oZ-6Sa2OAoH0sAOWhoEs&sa=X&oi=book_result&resnum=1&ct=result#PPA2,M1. Accessed March 12, 2009.

Lister, Joseph. "On a New Method of Treating Compound Fracture, Abscess, and So Forth; with Observations on the Conditions of Suppuration." *Lancet* 1 (1867): 326–357.

———. "On the Antiseptic Principle in the Practice of Surgery." *British Medical Journal* 2 (1867): 246–248. Available online. URL: www.pubmedcentral.nih.gov/articlerender.fcgi?artid=2310614. Accessed March 13, 2009.

Marples, Mary J. *The Ecology of the Human Skin.* Springfield, Ill.: Charles C Thomas, 1965.

Pringle, John. *Observations on the Diseases of the Army in Camp and Garrison.* London: Millar, Wilson and Payne, 1752.

Archaea (Archea) *Archaea* is the name for one of three domains in biology. The microorganisms in this domain were at one time classified as bacteria because they resembled most other bacteria under a microscope. As microbiological techniques improved during the 20th century, microbiologists suspected that archaea (also referred to as archaeans) behaved in ways unique among microorganisms. In the 1970s, the American microbiologist Carl Woese of the University of Illinois redefined the microbial world by classifying microorganisms not by their outward appearance and internal structures, but by their genetic makeup. Specifically, Woese devised a classification scheme based on the composition of ribosomal ribonucleic acid (rRNA), a component of the protein synthesis systems in all microorganisms. Woese found that archaeans possessed some features that related them to bacteria, but other features that connected them more closely to eukaryotic cells than to bacteria.

In classification schemes that divide living cells into two categories, prokaryotic and eukaryotic, the archaea fall in with prokaryotes. That is, they lack membrane-enclosed structures (organelles) inside their cells, and their deoxyribonucleic acid (DNA) floats freely in the cytoplasm. Eukaryotes, by contrast, are more organized cells with membrane-enclosed organelles. The rRNA studies showed archaeans to be neither true bacteria nor true eukaryotes. The new three-domain scheme of living livings, called *biota,* departed from the previous hierarchy of life in which all biota had been divided into five kingdoms with archaea joining bacteria in kingdom Monera.

Carl Woese and his fellow geneticist George Fox proposed their new hierarchy of life, based on the rRNA studies. They wrote in 1977, "We are for the first time beginning to see the overall phylogenetic structure of the living world. It is not structured in a bipartite way along the lines of the organizationally dissimilar prokaryote and eukaryote. Rather, it is (at least) tripartite, comprising (i) the typical bacteria, (ii) the line of descent manifested in eukaryotic cytoplasms, and (iii) a little explored grouping, represented so far only by methanogenic bacteria." The "little explored grouping" Woese referred to were the archaea.

The new classification scheme prompted biologists to reexamine members of each kingdom. Science often involves the investigation of ideas, followed by debate over new findings, and then reinvestigation. Few new ideas enter the mainstream of science without this scientific debate. Microbiologists turned their attention to the archaea; perhaps many did so to support Woese, and probably many others intended to prove him wrong.

In-depth studies of the archaea, after Woese's proposal in 1977, were aided by three disciplines: (1) electron microscopes, (2) methods for sampling genetic material directly from nature—pioneered in the 1980s by Norman Pace of the University of Colorado, and (3) frequent advances in DNA and RNA analysis. Microbiology gathered information on archaeans that had not been pursued with as much fervor in the past. What microbiologists discovered was an incredibly diverse group of microorganisms in both their genetic makeup (genotype) and their lifestyle (phenotype).

Microscopically, archaea resemble bacterial rods and cocci, but the archaean cell membranes and flagella contain distinctive components unlike anything found in other microorganisms. Members of domain Archaea also live in places that few other organisms on Earth could withstand. The Archaea are now known as the microorganisms most likely to be found in harsh environments characterized by extreme heat, acids, bases, high salt concentrations, and caustic chemicals, conditions that most other biota find uninhabitable. Microorganisms able to live in such conditions are called *extremophiles,* and the archaea make up the majority of known extremophiles. The archaea, in fact, inhabit places that resemble conditions on Earth when life was just beginning to emerge.

Early studies on archaea suggested that these microorganisms required anaerobic (lacking oxygen) conditions and lived only in extreme environments characterized by very hot temperatures. Biologists made the logical assumption that archaea were ancestors of present-day bacteria. Genetic studies similar to those performed by Carl Woese have now shown that the archaea and the bacteria split very early in evolution. About three billion years ago, the archaea and the bacteria evolved on separate, parallel paths. (The first primitive bacteria appeared 3.5 billion years ago.)

The archaea that have been identified and studied today cover much more diversity than biologists suspected, in the 1970s–1980s. Not all archaea are anaerobes or extremophiles. Species of archaea have been shown to live in aerobic conditions or to be facultative anaerobes, meaning they prefer oxygenless conditions but can survive if exposed to oxygen. Other archaea are obligate anaerobes, which are cells that cannot live in the presence of oxygen. Although archaea certainly inhabit extreme environments, microbiologists discovered, in the early 1990s, that some members of the domain prefer less stressful environments. For example, methane-producing archaea living in the digestive tract of animals and humans flourish at 98°F (37°C). Microorganisms such as these that grow best at moderate temperatures are called *mesophiles.*

Perhaps the most important revelation of genetic testing was evidence that proved bacteria did not evolve from archaea. Archaea and bacteria share some common genes, but most biologists now realize that these two microorganisms are unrelated and occupy diverse niches in biology. (Some texts continue to refer to archaea as bacteria and often use the misleading terms *archaebacteria* for archaea and *eubacteria* for bacteria, both in the prokaryotic kingdom.)

CHARACTERISTICS OF ARCHAEA

Archaea live in places that are difficult for microbiologists to sample. For this reason, few data accumulated on archaea while microbiology made constant discoveries on bacteria, viruses, molds, and so forth.

Archaea, as do bacteria, depend on a cell wall to shield them from the outside world, and their cell membrane is vital for energy-producing reactions. The archaea's inner structures also resemble those in bacteria. But archaeans differ from bacteria in the four following ways: (1) cell membrane composition, (2) cell wall structure, (3) structure and development of flagella, and (4) gene expression, which is the conversion of information carried in genes to new proteins. Archaean gene expression operates more like that in eukaryotes than in bacteria.

The lipids (fatlike compounds) in archaean cell membranes differ in structure from bacterial membrane lipids. Bacterial lipids mostly contain long-chain hydrocarbons called *fatty acids,* which connect to a glycerol molecule. This structure allows bacterial membrane lipids to orient in a lipid bilayer: Polar hydrophilic (attracted to water) ends point toward the aqueous outside of the cell or toward the cytoplasm inside the cell. The nonpolar hydrophobic (repelled by water) tails point into the fatty middle of the membrane. Archaean lipids, by contrast, contain two polar ends so that the lipids span the entire cell membrane. This arrangement enables archaea to respond to their environment differently than bacteria or eukaryotes, which also use a lipid bilayer. Archaean membranes also contain a large 30-carbon compound called *squalene.* Squalene does not decompose at high temperatures, so it probably gives archaea their ability to withstand very high temperatures that destroy other biological membranes.

The single-layered membrane of archaea provides more strength than bilayered membranes. A mixture of lipids containing as many as 40 carbons, especially structures called tetraether lipids, provides this durability. Tetraether lipids play a part in enabling archaean extremophiles to thrive in the temperatures above 750°F (400°C) found in volcanoes and near steam vents on the ocean floor.

Some species in domain Archaea possess peculiarities in their cell walls as well as their membranes. The stout cell wall provides protection for almost all bacteria, but some archaea depend instead on a single S-layer, a structure made of repeating proteins and glycoproteins (proteins with sugars attached to them). Archaea with S-layers tend to be more sensitive to foams and other substances in the environment that act as detergents. Other archaean species contain a cell wall similar to bacterial walls, but they differ in two other ways. First, these archaea contain different polysaccharides (long sugar chains) than found in bacteria. Second, the archaean cell wall does not contain peptidoglycan, as in bacteria, but rather a compound called pseudomurein. Archaea with pseudomurein react gram-positive in the identification method called the Gram stain procedure. Both peptidoglycan and pseudomurein are composed of linkages and bridges connecting their long-chain units called polymers. In this way, both compounds provide the cell with strength.

The presence of pseudomurein in place of peptidoglycan in the archaean cell wall explains why archaea resist many antibiotics that kill bacteria. Many antibiotics, such as penicillin and vancomycin, act by interfering with normal peptidoglycan synthesis. This, then, leaves a bacterial cell vulnerable to damage and death. Archaea with pseudomurein have been thought of as antibiotic resistant, but more accurately, these archaea simply ignore any antibiotic that acts by destroying peptidoglycan. Domain Archaea holds no known pathogens to animals or plants.

Archaea build their taillike structures called *flagella* in the opposite manner from that bacteria use to construct their flagella. Archaea assemble their flagella from the base, rather than at the tip; bacteria do the opposite. Once the flagella have been made, they act the same way in both archaea and bacteria for propelling the cells in their environment.

Finally, archaean genetics share some features with bacteria genetics and have other features similar to those in eukaryotes. Archaean DNA is a circular structure, as in bacteria, but the total set of archaean genes (the genome) numbers about half of that in bacteria. Archaea additionally do nor possess plasmids, small pieces of DNA that float in the cytoplasm separately from the main DNA molecule. About 50 percent of archaean genes are like those in bacteria and eukaryotes, meaning, of course, that about half of the genes are unique to archaea. While archaean RNA resembles bacterial RNA, the enzymes and proteins archaeans use for replicating DNA resemble the enzymes and proteins of eukaryotes.

IMPORTANT GROUPS OF ARCHAEA

Domain Archaea contains two phyla, Crenarcheota and Euryarchaeota. Noteworthy archaeans from these two groups are described in the table on page 62. Crenarchaeota contain extremophiles called *thermoacidophiles*. These microorganisms grow in temperatures of 160–235°F (70–113°C) and at acidic pH as low as 2.0, so they offer an example of the special skill archaeans have for living in uncomfortable places. The phylum is diverse: It contains aerobes, anaerobes, and a variety of metabolisms. But the thermoacidophiles tend to have in common a need for sulfur or sulfur-containing compounds. The genera *Desulfurococcus* and *Sulfolobus* are examples of archeans with all these characteristics. These two genera live in the sulfur cauldrons dotting Yellowstone National Park, boiling habitats rich in sulfur deposits. Some of the Crenarcheota grow at even higher temperatures and so can be called *hyperthermophiles*.

A more mysterious group of Crenarcheota lives in the ocean. They are not thermophiles or acidophiles, and they probably do not use sulfur in their metabolism since ocean water contains low levels of sulfur. These Crenarcheota have not been studied in laboratories, but microbiologists know they exist in the ocean in large numbers, and so they may play an important role in ocean ecology.

The phylum Euryarcheota contains more diversity than Crenarcheota and includes methanogens and halophiles. Methanogens are microorganisms that produce methane gas (CH_4); halophiles live in very salty environments.

Methanogens are obligate anaerobes that carry out methanogenesis, which is the production of methane. Obligate anaerobes cannot live in the presence of oxygen. With these special characteristics, it is not surprising that methanogens live in unique habitats. They thrive in organic-rich habitats such as mud at the bottom of still ponds, the stomach of ruminant animals, and even the digestive tract of some insects.

The gases produced by methanogens of ruminant animals have become an interest of scientists developing new sources of energy—cow gas as a renewable energy source! Researchers experiment with ways to capture this large source of methane gas produced daily by millions of cattle. By using the methane produced in ruminant food digestion, scientists hope to fulfill two objectives. First, methane is a greenhouse gas, so diverting it for human use rather than releasing it to the atmosphere may help keep global warming in check. Second, ruminant methane as an energy source will conserve fossil fuels, thereby also providing benefits to the environment. In a 2008 scientific journal, the researchers Michael E. Webber and Amanda D. Cuellar of the University of Texas wrote, "In light of the criticism that has been leveled against biofuels, biogas [methane] production from manure has the less controversial benefit of reusing an existing waste source and has the potential to improve the environment." Biofuels are any energy-containing fuels made from biological sources rather than coal or oil.

Ironically, the methanogenesis performed by archaea billions of years ago also produced natural gas, a fossil fuel. Natural gas used for heating and cooking is obtained from vast underground reserves that the petroleum industry recovers and distributes throughout the world. This methane originated in Earth from the gradual settling of organic compounds. Over millions of years, the enormous pressures in the deep earth turned the organic compounds into carbon-rich sediments. As ancient archaea decomposed the organic matter, they released methane gas, which now fills large underground pockets that usually lie near coal.

Halophile archaea live a very different lifestyle from methanogens. Halophiles withstand high salt (NaCl) concentrations. These microorganisms can live in the oceans and saltwater bays and estuaries, but many Euryarcheota grow in much higher-salt environments. The microorganisms that prefer places nearly saturated with salt are called *extreme halophiles*. Among the archaea, these microorganisms are often referred to as *haloarchaea*. Haloarchaea require salt concentrations that are typical of brine, which is salt-saturated water with NaCl concentrations up to 4 molar (M). If the salt concentration falls to 1 M, which is still very salty water, haloarchaea begin to die.

Unlike methanogens in phylum Euryarcheota, haloarchaea are strict aerobes; oxygen must be available for their survival. If oxygen levels fall in halo-

Examples of Archaea Diversity

Domain Archaea	Characteristics	Habitats	Example Genera
	PHYLUM CRENARCHEOTA		
sulfur metabolizers	gram-negative, thermophilic, acidophilic, usually strict anaerobes, but some aerobes and facultative	hot sulfur springs	*Desulfurococcus* *Sulfolobus* *Pyrodictium* *Thermococcus*
	PHYLUM EURYARCHEOTA		
methanogens	obligate anaerobes, produce methane, some produce carbon dioxide and hydrogen	digestive tracts, sediments	*Methanococcus* *Methanosarcina* *Methanobacterium*
extreme halophiles	aerobic, require NaCl greater than 1.5 molar; some contain bacteriorhodopsin, slightly thermophilic	brine, salt lakes, salt mines	*Halobacterium* *Halococcus*
sulfate reducers	obligate anaerobes, use sulfate and thiosulfate, produce hydrogen sulfide, hyperthermophilic	hydrothermal vents at ocean bottom	*Archeoglobus*
cell-wall-lacking	facultative	coal mines, active	*Thermoplasma*
pleomorphic archaeans	anaerobes, no cell wall, irregular shapes, unique cell membrane high in mannose	volcanoes	*Ferroplasma*

archaea habitats, the cells activate an emergency energy-producing system similar to photosynthesis. In this process, the haloarchaea create special regions in their membranes called *purple membranes.* These regions consist of a rudimentary type of photosynthesis that needs only light and the reddish protein bacteriorhodopsin to perform energy-producing reactions. In low-oxygen salt water, haloarchaea seek what little oxygen they can find by moving toward the water's surface. When this happens, billions upon billions of haloarchaea cells turn the surface waters red. Once haloarchaea find favorable locations containing adequate levels of both salt and oxygen, they revert to their normal energy process, called *aerobic respiration.*

Methanogens and halophiles represent two of the better known archaea, but this group of bacteria has such large diversity that archaea continue to present science with a rich store of questions yet to be answered. The table on page 62 summarizes the main groups of archaea, categorized by metabolism or by cell structure.

EXTREMOPHILES

Any study of extremophiles could not be done without also studying archaea. Some texts equate extremophiles with archaea, but studies since the 1990s have uncovered a growing list of archaea that do not live in extreme environments. Nevertheless, the lifestyles of extremophiles and extreme archaea can be considered equivalent.

Extremophiles live in habitats that are hostile to most other forms of life. Archaea researchers at the University of California–Berkeley have summarized the life of archaea on their Web site: "Archaeans include habitats of some of the most extreme environments on the planet. Some live near rift vents in the deep sea at temperatures well over 100 degrees Centigrade [212°]. Others live in hot springs [such as those in Yellowstone National Park], or in extremely alkali or acid waters. They have been found thriving inside the digestive system of cows, termites, and marine life where they produce methane. They live in the anoxic [without oxygen] muds of marshes and at the bottom of the ocean, and even thrive in petroleum deposits deep underground. . . . Some archaeans can survive the dessicating effects of extremely saline waters . . . they are also quite abundant in the plankton of the open sea." Despite the bounty of activities that archaea perform, scientists have yet to mine all of the potential benefits from these microorganisms.

Some of the known extremophile lifestyles among archaea are the following:

- thermophiles—growth at 120–150°F (50–65°C)
- hyperthermophiles—growth at 150–230°F (65–110°C) or higher
- halophiles—growth at salt concentrations above 0.2 M
- extreme halophiles—growth at salt concen trations greater than 1.5 M
- acidophiles—growth at below pH 4
- thermoacidophiles—growth at high temperature and acidic conditions

Studies on archaea in extreme environments may take a new turn in the near-future. As microbiologists unearth new information on how archaea live, they may use these microorganisms as models of life beyond Earth. The U.S. Geological Survey researcher Francis Chappelle was quoted in *Scientific American,* in 2002, on the subject of life on Mars. "The water deep within these volcanic rocks," Chappelle said of formations deep below a mountain range in Idaho, "has been isolated from the surface for thousands of years. It is devoid of measurable organic matter, but contains significant amounts of hydrogen." Chappelle's research team found that the microorganisms living in this remote habitat were archaea. The archaea seem suited to a remote environment without sunlight and with only hydrogen as a ready energy source. By learning all they can about archaea, biologists hope to strategize ways to search for life on other planets.

Carl Woese pondered the next steps in studying archaea in the larger realm of biology. In 2004, he wrote in *Microbiology and Molecular Biology Reviews,* "I think the 20th century molecular era will come to be seen as a necessary and unavoidable transition stage in the overall course of biology . . . [but] knowing the parts of isolated entities is not enough." Woese has advocated an "'eyes-up' view of the living world, one whose primary focus is on evolution, emergence, and biology's innate complexity." Such an understanding of life on Earth and in other locations of the solar system could not be accomplished without studying the archaea.

See also BIOREMEDIATION; DOMAIN; EXTREMOPHILE; METHANOGEN; POLYMERASE CHAIN REACTION; RUMEN MICROBIOLOGY.

Further Reading

Blum, Paul. *Archaea: Ancient Microbes, Extreme Environments, and the Origin of Life.* San Diego: Academic Press, 2001.

Cavicchioli, Ricardo. *Archaea: Cellular and Molecular Biology.* Washington, D.C.: American Society for Microbiology Press, 2007.

Dowd, Scot E., Daved C. Herman, and Raina M. Maier. "Aquatic and Extreme Environments." In *Environmental Microbiology,* edited by Raina M. Maier, Ian L. Pepper, and Charles P. Gerba. San Diego: Academic Press, 2000.

Dyer, Betsey D. *A Field Guide to the Bacteria.* Ithaca, N.Y.: Cornell University Press, 2003.

Graham, Sarah. "E.T. Life Forms Might Resemble Newly Discovered Microbial Community." *Scientific American,* 17 January 2002. Available online. URL: www.sciam.com/article.cfm?id=et-life-forms-might-resem. Accessed March 14, 2009.

Science*Daily.* "Cow Power Could Generate Electricity for Millions." 25 July 2008. Available online. URL: www.sciencedaily.com/releases/2008/07/080724064840.htm. Accessed March 14, 2009.

University of California. "Introduction to the Archaea, Life's Extremists. . . ." Available online. URL: www.ucmp.berkeley.edu/archaea/archaea.html. Accessed March 14, 2009.

Virtual Fossil Museum. "Precambrian Paleobiology." Available online. URL: www.fossilmuseum.net. Accessed March 12, 2009.

Woese, Carl R. "A New Biology for A New Century." *Microbiology and Molecular Biology Reviews* 68, no. 2 (2004): 173–186. Available online. URL: www.ncbi.nlm.nih.gov/pubmed/15187180?dopt=Abstract. Accessed March 14, 2009.

———, and George E. Fox. "Phylogenetic Structure of the Prokaryotic Domain: The Primary Kingdoms." *Proceedings of the National Academy of Sciences* 74, no. 11 (1977): 5,088–5,090. Available online. URL: www.pnas.org/content/74/11/5088.full.pdf+html. Accessed March 14, 2009.

aseptic technique *Aseptic technique* is a collective term for all of the procedures taken in microbiology laboratories and in medicine to prevent contamination, which is the presence of unwanted microorganisms. Proper aseptic technique depends on the use of sterile equipment and sterile growth media, explaining why aseptic technique is sometimes referred to as *sterile technique.* But aseptic conditions and sterile conditions bear a discreet difference: Asepsis equals the absence of unwanted microorganisms, while sterility is the absence of all microorganisms.

Sterilization plays a critical role in aseptic technique for three reasons. First, sterilization ensures that a microbial culture contains only the desired microorganism and no other. The absence of all unwanted microorganisms enables microbiologists to study specific characteristics of species without interference from a contaminant. Second, surgeons require sterile conditions and aseptic techniques to prevent wounds or surgical incisions from becoming infected. Third, aseptic techniques help manufacturers reduce the chance of contaminating products with pathogens during its production. For example, products intended for use in the eyes require production methods using aseptic processes.

Aseptic techniques require that any surface or liquid that has contact with a pure culture or with an open wound should, first, be sterilized. In microbiology laboratories, technicians typically sterilize the following equipment before use: petri dishes, flasks, bottles, tubes, caps and lids, inoculating loops, forceps, syringes, and filters. All growth media, dilution solutions, and additives to the media must also be sterile before they are used for pure cultures. In medicine, technicians sterilize instruments, bandages, dressings, and bedding, among other items used in hospital and outpatient clinics, before a physician uses any of these items to treat patients.

The following operations use aseptic techniques in all or part of their activities: clinical laboratories, medical and veterinary clinics, dental offices, food processing plants, drug manufacturing plants, medical instrument supply companies, personal care product manufacturing plants, and semiconductor production rooms. In these industries, microbiologists are said to achieve *clean* (aseptic) conditions when they perform all of the steps needed to assure that no contamination has taken place.

Physicians and veterinarians require aseptic techniques to diagnose disease and treat their patients accurately. Every medical microbiologist must keep pure cultures free of contamination in order to identify a microorganism that could be a potential pathogen. Doctors use aseptic techniques to prevent spreading infection from one patient to the next. Also in medical care, the solutions and instruments that touch patients must be handled in aseptic fashion to prevent transmitting an infectious microorganism to the patient.

Although health care facilities use routine procedures for keeping unwanted microorganisms away from patients, this was not always the case. The idea of aseptic technique developed as biologists learned more about disease and its relationship to individual microorganisms.

THE DEVELOPMENT OF ASEPTIC TECHNIQUE

In 1861, the French microbiologist Louis Pasteur (1822–95) demonstrated the principles of contaminant-free conditions in a simple experiment. In the first step, he boiled broth inside a flask to kill any microorganisms present in the mixture. Then he left the flask open to the air. Before long, microbial growth appeared in the liquid, caused by microorganisms from the air that had drifted into the flask and contaminated the broth. Pasteur had also constructed a second flask similar in every way to the first except its neck had been bent into an S shape to prevent airborne germs from entering. No growth occurred in this flask. Pasteur's experiment demonstrated how microorganisms in the environment could easily contaminate sterile items and even people unless the items were carefully protected. The S-shaped flasks used in Pasteur's experiments, dating to the 1860s, remain sterile to this day and are on display at the Pasteur Institute in Paris.

At the same time, Pasteur set up his experiments, the English surgeon Joseph Lister (1827–1912) began stressing the need for asepsis during surgery. Lister had paid rapt attention to Pasteur's theories, as well as the ideas of the Hungarian physician Ignaz Semmelweis (1818–65), who had proposed two decades earlier that hand washing helped prevent infection. Lister concluded that a strong connection existed between cleanliness and the prevention of infection. But Lister's ideas would probably not have gained acceptance at all without the growing stature of Pasteur's name in science and his experiments proving the germ theory. The combined work of Semmelweis, Pasteur, and Lister refocused the medical world from spontaneous generation to the methods used today to prevent contamination and infection.

Modern medical clinics and surgery facilities require basic aseptic techniques for the purpose of stopping the transmission of pathogens. The principles of medical asepsis in these settings are followed by using proper hand washing methods, good personal hygiene, housekeeping to reduce dust and dirt, clean and sterilized instruments and other items that touch patients, and the proper use of antiseptics and disinfectants.

METHODS FOR ASSURING ASEPTIC CONDITIONS

Today methods for reducing the spread of contaminants have changed little since the days of Pasteur, Lister, and other pioneers in microbiology. Microbiologists use three means of achieving aseptic conditions. The first, as mentioned, is the use of sterilized equipment. Solid items and liquids can be sterilized by a variety of techniques; the main methods are steam heating in an autoclave, boiling, liquid filtration, gas exposure, and irradiation. A second technique used in microbiology laboratories is called *flaming*: the action of heating of an instrument for a few seconds to kill any contaminants it carries. For example, microbiologists flame metal inoculating loops by immersing the loop in the flame of a Bunsen burner—a laboratory appliance that provides a steady flame by igniting natural gas or propane—or in a small incinerator. All cells on the loop are dead when the metal begins to glow red. Bunsen burner flaming also works well for decontaminating metal forceps and the rims of

Robert A. Thom's painting of Louis Pasteur shows the microbiologist with the "swan-necked" flask he used to disprove the theory of spontaneous generation. Mme Pasteur watches from the background. ***(University of Michigan Health System)***

open culture tubes and flasks. Third, microbiologists reduce the chance of contamination by disinfecting their work area before and after each use. Alcohol, commercial disinfectants, and ultraviolet irradiation each disinfect surfaces when used properly. Work areas that should be decontaminated are benchtops, walls, floors, and the inner surfaces of incubators.

Most microbiology laboratories today have on hand a supply of sterilized plastic disposable equipment. These presterilized items have streamlined microbiology by eliminating the need for a technician to sterilize equipment. Items that can be purchased presterilized include inoculating loops, petri dishes, pipettes, culture bottles, flasks, and tubes. After using disposable equipment in culture methods, microbiologists simply discard the contaminated items.

ASEPTIC TECHNIQUES IN MICROBIOLOGY LABORATORIES

Microbiologists work with pure cultures by using certain aseptic methods common to all microbiology laboratories. For instance, bacterial and fungal cultures require periodic transfer from used-up medium to fresh growth medium in order to give the cells a new supply of nutrients. To begin, the microbiologist removes the culture tube's cap or plug and briefly heats the rim using a Bunsen burner. This step serves two purposes: It removes any contaminants near the culture tube's rim, and it creates a momentary updraft to draw airborne particles away from the tube's mouth. The next step is called a *transfer*. To transfer a culture, a microbiologist immerses a sterile loop (or a pipette) in the culture and carries a small volume to fresh sterile medium. A similar process carries cells growing on agar to a fresh agar surface. Before reclosing the newly inoculated tube, the microbiologist again briefly heats its rim in the burner's flame.

Microbiology laboratories follow procedures based squarely on Pasteur's experiments with S-flasks. Some of the simple yet necessary actions that keep out contamination are the following:

- covering all petri dishes
- closing all vessels with caps, plugs, rubber stoppers, or gauze
- checking all uninoculated media for contami nation before use and discarding all contami nated items
- avoiding breathing on cultures or excessive talking when handling open vessels

Specialists in anaerobic microbiology handle their cultures in a different manner for maintaining aseptic conditions than they use for aerobic cultures. Anaerobic species require an oxygen-free environment, so microbiologists often transfer them inside sterile syringes that have been flushed of oxygen with another gas, such as nitrogen. As with aerobic cultures, anaerobic microorganisms must be grown on sterilized media and exposed only to sterilized vessels and equipment.

In terms of maintaining aseptic conditions, anaerobic cultures offer an advantage because they must be kept closed at all times to exclude oxygen. This also helps prevent contamination. Aerobic cultures do not require airtight conditions, so microbiologists often use a specially designed work area called a *laminar flow hood*. A laminar flow hood comprises an open chamber set atop an ordinary laboratory bench. Small hoods accommodate one microbiologist's work activities, and larger units allow two microbiologists to work side by side.

Aseptic techniques are easier to maintain inside laminar flow hoods because these units are enclosed on all sides except where the microbiologist is stationed. Each hood contains specialized vents in the work surface and in the ceiling that allow air to flow across the work area in flat sheets, called *laminar flow*. Laminar flow of air minimizes the chance that an open vessel will receive an unwanted particle. Once air has passed through the hood, it exits through a high-efficiency particulate air filter (HEPA). A HEPA provides controlled air filtration by passing air through very small pores. While air passes through the filter, small particles become trapped on the filter. A portion of HEPA-filtered air leaves the hood through an exhaust, and a portion recirculates into the hood.

Microbiologists conduct most routine work with bacteria either in a laminar flow hood or at an open benchtop. Mold cultures, however, should always be used inside a laminar flow hood because mold spores travel easily through the air. Mold spores release into the air, then create a higher risk of contaminating all other aseptic activities in the laboratory. Finally, microbiologists who work with viruses must also use a laminar flow hood. Viruses cannot propagate on their own and must be raised in living tissue, a process called *tissue culture*. Tissue culture is highly susceptible to contamination, so it must take place under strict aseptic conditions in a laminar flow hood.

Continuous cultures in large vessels called *bioreactors* contain a constant inflow of fresh medium and a constant outflow of used medium, which contains cells, cell debris, and end products. Because bioreactors run constantly for days at a time, the chance of contamination increases. Proper sterilization of bioreactor contents and aseptic technique become essential for maintaining proper growth inside the vessel.

Bioreactors sterilize the liquid medium within the inner chamber. Operators maintain asepsis when setting up the bioreactor culture by using only sterilized tubing and connections and by decontaminating all the bioreactor's ports before opening them. Technicians swab each port with alcohol or other disinfectant before and after each use. All additions to the culture and all samples taken from the culture must be performed with sterile solutions, pipettes, and any other equipment that has in contact with the bioreactor's contents.

ASEPTIC MANUFACTURING

Aseptic manufacturing is the creation of products or drugs in contaminant-free conditions. In manufacturing, *aseptic filling* refers to the preparation and packaging of products so that contaminants are not transmitted to consumers. To do this, manufacturers make use of disinfectants, sterilization, sterility testing, and strict monitoring of the manufacturing area. *Environmental monitoring* refers to any process in which a person checks for the presence of contaminants within a certain area, which includes the air, water supply, and hard surfaces.

Drug manufacturing plants that produce injectable products such as vaccines have designated clean areas where aseptic conditions exist and where aseptic filling takes place. Operations within clean areas may include sterilization, sanitization, rinsing, filling, and capping. Plant operators also keep clean areas at an air pressure slightly higher than the atmosphere's normal pressure. This pushes airborne particles away from the work area rather than drawing them toward the work area. The pressurized air additionally passes through a HEPA filter before entering the room. In aseptic filling, all bottles, jars, cans, or other product containers and their closures must be sterilized or cleaned very well before the food, drug, or body product is put into them.

Manufacturers of sterile injectable drugs use rigorous steps to ensure microorganisms do not enter their products. Rather than relying on aseptic conditions, these drug manufacturers must employ sterile conditions for the production and packaging of the drug. All of these activities take place inside an area called a *clean room*. Clean room workers don special protective clothing and use only sterile equipment when working in this area.

Sterile drug manufacturers test the success of their aseptic filling by conducting sterile media fill tests. In this method, workers package sterile broth medium in place of the actual product. After the broth moves through the production equipment, the workers fill it into the same sterile containers as used for packaging the product, and then put the containers in incubators set at various temperatures. One incubator is usually set at room temperature, and at least one other incubator is set at a temperature preferred by human pathogens (99°F [37°C]). If all the containers complete the incubation period free of growth, that means the production maintained aseptic conditions. This result signals the workers that the equipment is safe for making the drug.

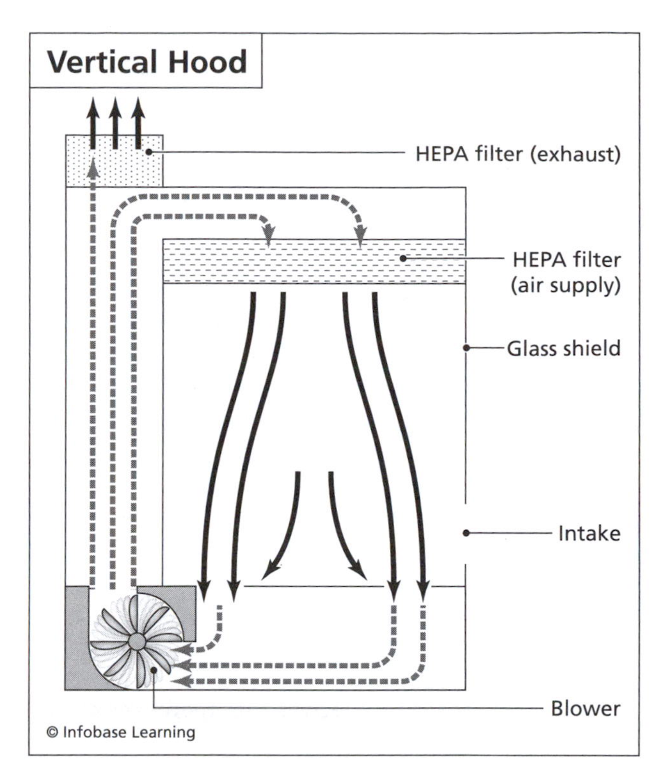

Laminar flow hoods create airflows in either horizontal or vertical direction. To use the vertical airflow hood pictured here, a microbiologist sits outside the glass shield and carries out activities inside the hood. Parallel sheets of downward-flowing HEPA-filtered air prevent microorganisms from exiting the hood and protect the worker.

ASSURING ASEPTIC CONDITIONS

Expertise in aseptic techniques does not come easily; microbiologists attain this skill through practice. Deft and quick manipulations are part of the technique, and a steady hand is a plus. In maintaining aseptic conditions, a microbiologist must learn to see, in a sense, microscopic particles even though they cannot truly be seen with the unaided eye. A key principle helpful in all aseptic techniques proposes that if a nonsterile item has contact with a sterile item, a microbiologist should assume that aseptic conditions have been destroyed.

Microbiologists check all prepared media for signs of contamination before they use them in aseptic techniques. Although gross contamination is easy to see—a moldy orange in the refrigerator provides the telltale sign of mold contamination—microorganisms are invisible. Microbiologists investigate the presence of unwanted microorganisms using a step called *preincubation*. In preincubation, microbiologists incubate all uninoculated media before using them. Any media showing evidence of contamination should be discarded. Signs of contamination are the following: cloudiness, obvious microbial growth, altered color, strong or unusual odors, or evidence of gas production (bubbles).

Workers in large production plants and in large or small microbiology laboratories carry out periodic environmental monitoring. Environmental monitoring programs support aseptic techniques by checking for the presence of unusually high amounts of microorganisms on surfaces or in the air. Environmental monitoring also identifies particular trouble spots where contamination appears to occur more than usual. For example, a monitoring program might show that the air and the equipment have been properly cleaned and maintained under aseptic conditions, but the water entering the facility contains a high concentration of bacteria.

Environmental monitoring uses three common methods for sampling hard surfaces for the presence of microorganisms, as follows:

- swabbing, the use of cotton swabs
- sampling with agar paddles, contact plates, or petri film
- rinsing

Each of these methods can be used on equipment, piping, work areas, floors, walls, ceilings, sinks, and faucets.

Swabbing consists of wiping an area with a sterile cotton swab, then breaking off the swab tip in a tube of sterile growth medium, followed by incubation of the tube. Swabs are useful for sampling uneven or very small surfaces. Microbiologists also use rectangular agar paddles, round contact plates, or square petri film to sample larger, flat surfaces. Each device collects a sample from a surface by pressing the agar against the surface, and then incubating. Paddles, contact plates, and petri film all contain sterile agar that picks up any microorganisms present on the sampled surface. Each also contains a visible grid so that the microbiologist can determine the number of microorganisms in a square inch (6.45 cm^2) of sampled surface, called an area count. This result can be determined simply by counting the number of microbial colonies that have grown within the grid during incubation. (Swabbing also lends itself to a similar area count by laying a template of known square inches on the surface to be sampled, then swabbing inside the entire defined area.)

Rinsing offers more thorough sampling than either agar or swabs for recovering microorganisms, especially when equipment to be sampled is large or complicated and contains few flat surfaces but many nooks and crannies. In this method, workers remove a piece of equipment from the production line, and then rinse it in sterile water. A microbiologist tests a portion of the rinse liquid for the presence of microorganisms by transferring a small volume to growth medium and incubating it.

Air sampling employs either settle plates or mechanical air samplers. Settle plates are agar plates left open to allow airborne microorganisms to fall by gravity onto the agar during a set sampling period, from one to 24 hours. Mechanical air samplers draw in a set volume of air over a specific period and deposit the air's microorganisms on an agar plate inside the device.

Water sampling uses specialized techniques that are part of water quality testing and the wastewater treatment industry. Water samples may be taken directly from a building's main intake port, faucets inside a building, or distribution lines. The microorganisms found in water usually differ from the varieties found in air and on surfaces.

Aseptic techniques provide the foundation of microbiology studies as well as safe practices used in medi-

Where Are Germs Found?

by Anne Maczulak, Ph.D.

Microorganisms are everywhere. Bacteria cover every surface in nature, on plant and animal bodies, and on areas inside homes and buildings, cars and airplanes, and hospitals and day care centers. Mold spores exist on almost as many surfaces as bacteria, and viruses can remain infectious from a day to a week on an object such as a doorknob. Of all microorganisms, algae and protozoa live in more specialized habitats that require moisture.

Microorganisms as a single group of living things have been called ubiquitous because of their wide range of habitats on Earth. But the bad microorganisms called *germs* are not found everywhere. Germs cluster in places that favor their ability to propagate infection: the digestive tract, skin, mouth, respiratory tract, and any inanimate object in the route of germ transmission. People can protect against infections and colds and flu by remembering that microorganisms are everywhere, but they can reduce infections and disease by paying attention to certain places that microbiologists call germ *hot spots.*

Microbiologists know where germs can be found in a school, day care center, or house even though germs are invisible. For nonmicrobiologists, finding microscopic objects in the surroundings may seem impossible. A few general rules help give clues to the germ hotspots in places where people live and work.

Most people have learned from news feature stories that bacteria cover surfaces of the home, school, day care center, and nursing home as well as public restrooms, but they may forget or fail to realize other places where germs can be found in high numbers. Germs populate business offices, grocery stores, hospitals, mass transportation vehicles, medical offices, and retail stores. Other than sink and shower drains, which have very high numbers of germs, hotspots almost always occur in these places on objects touched frequently and repeatedly by many different people. Water or moisture also enhances a germ's ability to cling to a surface and remain infectious. For example, a dry desktop in a library would probably keep a flu virus active for less time than a kitchen counter that receives periodic drops of water and wipes with a moist sponge. But any inanimate surface that has just been sneezed on by a sick person also qualifies as a potential germ hotspot. Cold and flu viruses stay active under these circumstances for three to four days, and bacteria may be alive for a week or longer. To assess a potential hotspot, consider three factors:

- Is the surface frequently touched or handled by many different people?
- Does the surface receive periodic exposure to water (tap water, juices from foods, condensation, or sneezing or coughing)?
- Are nutrients available on this surface (for bacterial growth, not necessarily for viruses)?

cal care and some types of manufacturing to prevent infection by unwanted microorganisms. These methods developed over a long period in the history of microbiology and are the result of elegant experiments by a number of important names in science. The aseptic techniques practiced today are a direct result of science's acceptance of the germ theory in disease transmission.

See also CLEAN ROOM; CULTURE; DISINFECTATION; FILTRATION; GERM THEORY; GRAM STAIN; LISTER, JOSEPH; PASTEUR, LOUIS; SAMPLE; STERILIZATION.

Further Reading

Agustin, Sofronio, and Holly Williams. "Aseptic Technique." March 12, 2007. Available online. URL: www.microbelibrary.org/asmonly/details.asp?id=2563&Lang=. Accessed March 14, 2009.

Gerhardt, Phillip, ed. *Manual of Methods for General Bacteriology.* Washington, D.C.: American Society for Microbiology Press, 1981.

Micro eGuide. "Laboratory Safety: Aseptic Techniques." Available online. URL: www.microeguide.com/safe_at_frame.htm. Accessed March 14, 2009.

Science Communication Network. "Aseptic Techniques." Available online. URL: www.ems.org.eg/esic_home/data/giued_part1/Aseptic_Techniques.pdf. Accessed March 14, 2009.

Aspergillus *Aspergillus* is a genus of fungi, widespread in nature, that includes almost 200 species, several of which cause disease in humans and other animals. Species of *Aspergillus* also play important roles in industrial microbiology and food production. Despite the value of *Aspergillus* in industry, this microorganism can also cause trouble in food and other product manufacturing when it contaminates formulations.

Aspergillus is a ubiquitous microorganism found in natural nonextreme environments. The fungus exists in any place supplied with organic matter, some moisture, and a moderate temperature range of 50–120°F (10–50°C).

The outdoors and the indoors both provide favorable conditions for *Aspergillus*. Although it lives as a normal inhabitant of soil, it also readily enters the indoors when breezes blow fungal spores inside. *Aspergillus* grows in cool, moist places inside build-

By answering yes to these questions, anyone can predict where germs are found in the home. Ironically, people now clean their bathrooms better than their kitchens, knowing that germs lurk on bathroom surfaces. Kitchens receive enormous influxes of germs, however, from foot traffic from the outdoors, groceries, raw meats, pets, and various other items that usually enter a house first through the kitchen.

The home's germ hotspots are sink faucet handles, appliance handles, microwave and remote controls, sinks, countertops, shower curtains, floors, and trash cans. In an office environment, keyboards, computer mice, phones, soda machine buttons, copier buttons, and automated teller machine (ATM) buttons carry germs. These items also can be found in schools, which additionally have shared desks, shared lockers, showers, locker rooms, gym equipment, and cafeteria surfaces. In day care facilities, infants spread fecal bacteria and viruses to the floor, blankets, toys, tables, and chairs.

Inanimate hard surfaces, such as most of the items listed in the preceding paragraph, can be disinfected to remove germs. As long as the user follows the directions on applying the product and for how long, most germs can be eliminated. But they eventually return, so good personal hygiene (hand washing, covering the mouth and nose when sneezing or coughing, and avoidance of touching hands to the face) and sensible use of a disinfectant or sanitizer break the transmission of germs.

Nonhard surfaces can be much more difficult to manage when controlling the spread of germs. Clothes, bedding, upholstered furniture, carpets, and curtains, plus food, pets, and other people, can carry germs. Disinfection (or sanitization alone) cannot compensate for personal hygiene practices that lower a person's risk for infection. In healthy people with strong immune systems, germs should be an infrequent pest rather than a real danger. Only high-risk health conditions caused by age, pregnancy, chronic disease, or a weakened immune system increase the need for extra vigilance against germs. Despite all of these warnings, it is comforting to remember that the vast majority of microorganisms on Earth cause no harm to humans, and many provide benefits without which life could not continue.

See also BACTERIA; COMMON COLD; INFLUENZA; TRANSMISSION.

Further Reading

Bakalar, Nicholas. *Where the Germs Are: A Scientific Safari.* Hoboken, N.J.: John Wiley & Sons, 2003.

Brown, John C. *Don't Touch That Doorknob! How Germs Can Zap You and How You Can Zap Back.* New York: Warner Books, 2001.

Maczulak, Anne E. *The Five-Second Rule and Other Myths about Germs.* Philadelphia: Thunder's Mouth Press, 2007.

Tierno, Philip M. *The Secret Life of Germs.* New York: Simon & Schuster, 2001.

ings and can affect human health at high concentrations by causing allergies or other, more serious ailments. In the controlled conditions of laboratories, *Aspergillus* has become a common focus of study.

CHARACTERISTICS OF *ASPERGILLUS*

Aspergillus is one of several fungi more commonly known as *molds*. Molds are multicellular fungi that grow into visible fuzzy or fluffy colonies. This fuzziness results from the growth of mycelia, which are bundles of long branching filaments that intertwine as they grow. *Aspergillus* also produces spores, which act as the means for dispersal and reproduction. Spores travel through the air; this mode of transmission is called *airborne transmission*. Eventually, the spores settle out of the air onto a surface, where, if organic matter, nutrients, and moisture are present, they germinate within two to three days. Spore germination entails the process of growth from the spore form of a microorganism. Fungi such as *Aspergillus* germinate from fungal spores; endospore-forming bacteria also germinate. Although the term *spore* has been used for both fungal structures and bacterial endospores, such as those of *Bacillus* or *Clostridium*, these two different types of spores are unalike.

Upon germination, *Aspergillus* spores multiply and form colonies that appear fuzzy and black or greenish. The mold that grows on spoiled fruit in a refrigerator provides a familiar example of *Aspergillus* growth. *Aspergillus* also grows on nonfood items such as leather, clothing, bathroom tiles, and carpets. The mold's favorite conditions are cool, dark places with high humidity and little airflow. Closets, cellars, and attics serve as common indoor habitats for *Aspergillus*.

Fully germinated *Aspergillus* contains structures typical of molds: mycelia and hyphae. Mycelia are long, stringy filaments that grow into the familiar tangles or fuzz that people associate with most molds. Each mycelium is composed of bundles of individual hyphae. A single hypha is a stringlike growth of about 2.5–8.0 micrometers (μm) in diameter.

Aspergillus hyphae contain characteristics that are not found in all molds. For example, their hyphae are called *septate*, meaning structures that contain distinct sections separated from one another by cell walls. *Aspergillus* hyphae also tend to grow in an upward direction, which gives colonies their fluffy appearance. Hyphae may also grow an appendage called a *conidiophore*, which extends outward or upward from the main hypha structure and supports a round end or head. This most distal part of the conidiophore holds hundreds of small (2–5 μm in diameter) conidia, an asexual reproductive component. The conidia are usually recognized and referred to as mold spores.

Mycologists—scientists who study fungi—identify *Aspergillus* and other molds by examining spores in a microscope. Both conidiophores and spores have characteristic features that help mycologists identify many molds to the genus level and sometimes to the species level. *Aspergillus* is a monomorphic fungus, meaning it exists in only one physical form. This makes *Aspergillus* easier to identify under a microscope than other molds that can take several different physical forms (pleomorphic molds).

Aspergillus spores travel from several feet to a mile (1.6 km) or more in the air, depending on humidity, airflow, and air turbulence. As do other mold spores, *Aspergillus* spores possess a shape that enables them to stay airborne for long periods.

The dual personality of *Aspergillus* makes it both a health threat and a source of useful compounds. The table on page 70 lists the most familiar *Aspergillus* species and illustrates how they can be either harmful or helpful.

ASPERGILLUS AS A PATHOGEN

People and animals inhale airborne *Aspergillus* spores into their respiratory passages. This route often begins an infection or an allergic reaction in the respiratory tract. In rarer cases, inhaled spores lead to infections of organs in addition to the lungs or other parts of the respiratory tract.

Members of the *Aspergillus* genus cause a group of diseases collectively known as *aspergillosis*. *Aspergillus* mold often acts as an opportunistic pathogen, which is any microorganism not normally dangerous to the body but able to cause infection under favorable conditions. For example, people who have weakened immunity are more vulnerable to opportunistic infections than healthy persons with strong immune systems. Certain members of the population compose a subpopulation that is at higher-than-normal risk of contracting opportunistic infections, such as the following groups: the elderly, young children and infants, pregnant women, people with a long-term (chronic) existing disease, people with acquired immunodeficiency syndrome (AIDS), chemotherapy patients, or organ transplant recipients.

Aspergillosis has been further differentiated into the four following varieties:

- noninvasive bronchopulmonary aspergillosis (ABPA)—predominantly an allergic reaction within the bronchia of the lungs, especially in asthma sufferers
- invasive pulmonary aspergillosis (IPA)—attacks one or more of the body's internal organs after spreading from the lungs

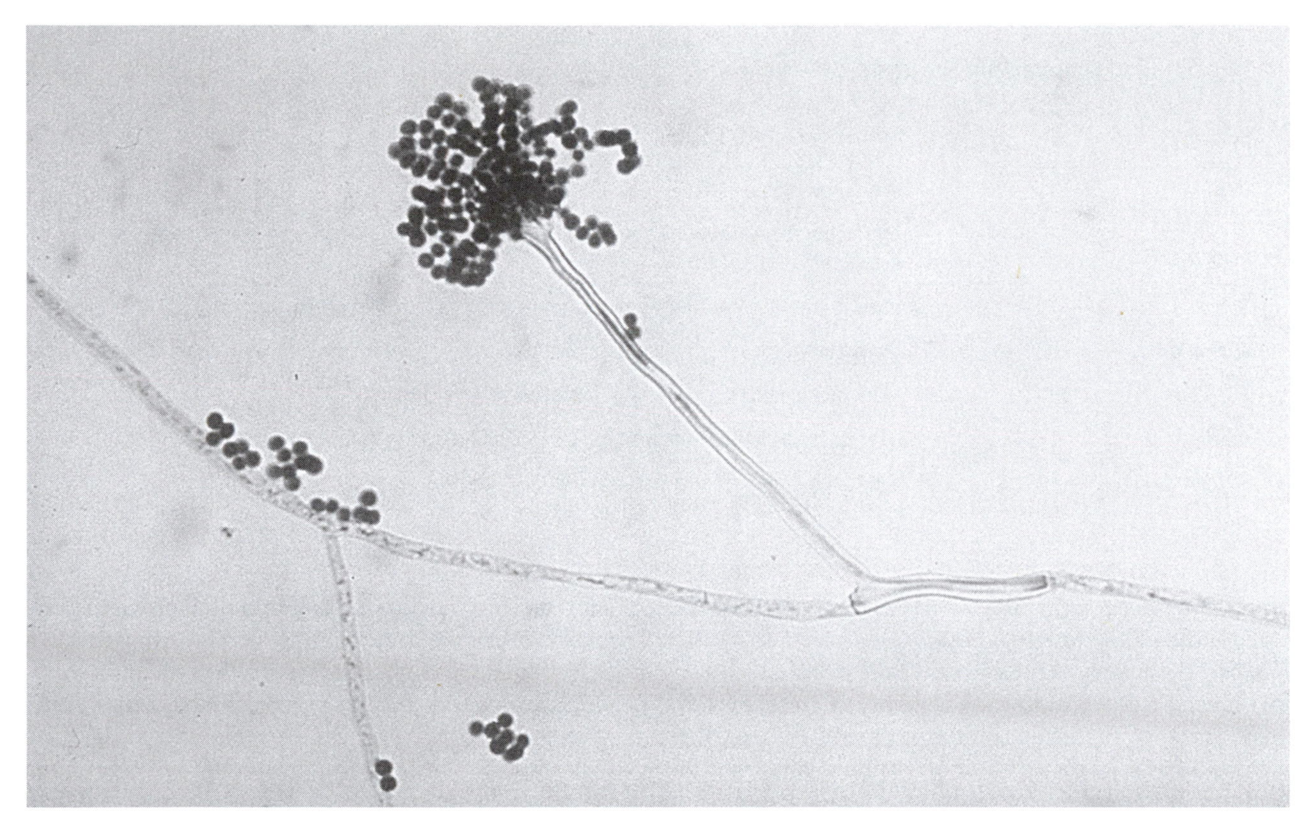

This photo shows *Aspergillus* vegetative hyphae holding small clusters of conidia and an aerial hypa, or conidiophore, extending upward and holding numerous conidia, which will release and form new growth. ***(CDC, Public Health Image Library)***

- chronic pulmonary aspergillosis—masses of mold cells grow and cover much of the inner surface of the bronchia
- skin aspergillosis—infections of the skin

Opportunistic *Aspergillus* infections most commonly cause irritation to the sinuses and upper respiratory tract. Deeper penetration of the respiratory tract results in ABPA, in which both asthma and nonasthma sufferers develop allergy to the mold spores. Invasive infections of IPA also start in the lungs. When either *A. fumigatus* or *A. flavus* infects the lungs, the mold typically does not invade the bloodstream but instead grows directly on lung tissue. A mass of mold growth called an *aspergilloma* may cover the bronchial sacs and tubes of the lungs. This condition makes breathing very difficult by reducing the infected person's lung capacity, the volume of air a person can inhale with a single, deep breath. If *Aspergillus* escapes the lungs and invades the bloodstream to cause IPA, the microorganism infects other tissue or organs and can lead to death.

A. fumigatus and *A. flavus* have also been implicated in an ailment called called *allergic aspergillosis,* which develops within one to a few minutes after a person inhales spores. In these cases, a person may experience difficult breathing or more serious reactions such as asthma. Allergic aspergillosis has been linked to sick building syndrome (SBS), which is a situation in which building occupants develop health problems related to indoor air of poor quality. This circumstance is also called *building-related illness.* Chemicals in addition to microorganisms have been blamed for SBA, but finding the exact cause presents difficulties: Hundreds of chemicals, particles, pollen, molds, and other substances in the air may contribute to SBS. For this reason, experts on indoor air quality (IAQ) have not reached agreement on the extent to which *Aspergillus* or other molds contribute to SBS. A 2008 update on SBS, in *BusinessWeek,* illustrated the reason why this illness has challenged physicians and microbiologists: "There are a number of indoor air pollutants that contribute to poor IAQ and the spread of airborne disease. These include biological contaminants such as molds and bacteria, and combustion pollutants like carbon monoxide and toxic particles. Even the building itself is a factor, since toxic substances emitted from building materials and furnishings degrade IAQ." The following substances have been studied as causes of SBA: paint, varnish, carpet, flooring, insulation, adhesives, and particleboard. Because of its prevalence almost everywhere, *Aspergillus* continues to receive blame today for at

Important *Aspergillus* Species

Species	Main Characteristics
A. niger	produces industrial enzymes and organic acids
A. flavus	opportunistic pathogen in humans and domesticated animals; causes food spoilage, produces aflatoxins
A. oryzae	used in food production; fermentation products from soybeans and corn
A. fumigatus	opportunistic pathogen in humans and domesticated animals; produces aflatoxins
A. nidulans	used in genetic engineering to increase cell growth rates
A. glaucus	used in food production; fermentation products from meat and fish
A. terreus	produces a cholesterol-lowering drug

least some aspects of SBS along with another unrelated mold called *Stachybotrys*.

Spores attach to the skin as readily as they attach to objects inside buildings. They have been recovered from the skin surface of humans, particularly on the external parts of the ear and on the hair, and on pets and on livestock. *Aspergillus* normally causes no harm on the skin and only poses a threat of infection if the skin receives cuts, burns, or similar wounds. *Aspergillus* species, especially *A. fumigatus*, can be common in decaying vegetation, so people who tend compost piles, gardeners, and farmers may have a higher risk of exposure to the spores.

AFLATOXIN

A. flavus and *A. parasiticus* produce a toxin that can cause a mild case of aspergillosis to turn serious. The agent called *aflatoxin* is one of many different mycotoxins, which are any toxins produced by a fungus. Aflatoxins injure humans, and domesticated animals and retard the growth of some plants. In animals, aflatoxins mainly damage the immune system and the liver. Liver exposure to a high dose of aflatoxin in animals has been associated with liver cell death, cirrhosis (liver scarring and loss of function), and liver cancer. *Aspergillus* also produces the toxins asteltoxin, gliotoxin, patulin, and the ochratoxins; less information has been gathered on these mycotoxins compared with aflatoxin.

Aflatoxin poisoning occurs rarely in humans. Damage to the liver and to the digestive tract appear to be two of the main outcomes in humans, and the poisoning is not likely to be fatal.

Doctors usually treat aspergillosis indirectly by focusing on any underlying disease. This is because fungal infections such as those from *Aspergillus* are known to be very difficult to cure. Once the treatment has rid a patient of any underlying disease, the person's immune system may regain strength to a degree that it can fight the *Aspergillus* infection. Individuals who have cancer, severely damaged immunity, or debilitating diseases that accompany aging have a much more difficult time combating *Aspergillus* infection. The antifungal drug itraconazole may help if the infection has not spread.

In veterinary medicine, aflatoxins ingested with moldy hays and grains cause a variety of symptoms: ataxia (loss of muscle coordination), tremors, spasms, and convulsions followed by collapse. Veterinary aspergillosis is also known as *mycotic pneumonia* or *pneumomycosis*. Calves can contract pneumonia from inhaling spores or suffer gastroenteritis from ingesting the toxin. Pregnant cattle, sheep, and horses have suffered abortions and other reproductive failures. Fescue foot toxicity occurs as a specific response to aflatoxin in cattle grazing pastures of tall fescue grass contaminated with spores. Cattle with fescue foot toxicity exhibit lameness and gangrene of the extremities. Aspergillosis also endangers birds; the disease comes on rapidly and causes fatal pneumonia in adult poultry and in brooder chicks (also called *brooder pneumonia*).

Few veterinary treatments exist for *Aspergillus* infections. The best way livestock and poultry breeders combat infection is by preventing it. They check for moldy grains and hays, reduce dust in barns, and thoroughly clean out pens on a regular schedule. Farm fields known to carry high levels of spores may be left idle for a season or mowed and used for purposes other than grazing.

INDUSTRIAL USES OF *ASPERGILLUS*

In industry, microbiologists grow large volumes of *A. niger* and other *Aspergillus* species for enzyme production. *Aspergillus* has long been known as a good source of a variety of enzymes that break down food constituents. The main examples of *Aspergillus* enzymes that have commercial value are amylase, which digests starch, and lipase, which breaks down fats. Both enzymes are ingredients in laundry detergents for dissolving stains; lipase also contributes in leather tanning.

Aspergillus serves as a source of additional and varied industrial enzymes. Oxidase enzyme, for example, acts as a bleaching agent for making paper and producing fabrics. Makers of fruit juices use *A. niger*'s pectinase enzyme to break down a fibrous molecule called *pectin,* found in apples and pears. Pectin affects the consistency of fruit products, so manufacturers use pectinase to make their products more palatable and easier to digest. Finally, protease enzymes made by *A. oryzae* work as meat tenderizers and leather softeners. Proteases are chosen for these uses because they digest proteins.

Statins are drugs that lower cholesterol levels in the blood. *A. terreus* has become an important natural source of lovastatin, and in 1987 it became the first statin approved by the U.S. Food and Drug Administration (FDA) for treating high cholesterol level. The human body makes cholesterol naturally, but some people either make too much or cannot break down excess cholesterol. Lovastatin works by inhibiting an enzyme the body needs to synthesize cholesterol.

FOOD ITEMS PRODUCED BY *ASPERGILLUS*

Aspergillus is a key microorganism in the food industry, where it has long been used for carrying out fermentations in certain foods. Soy sauce is a food made from steamed soybeans combined with roasted wheat. In the initial step, soy sauce makers inoculate soybean-wheat mixtures with *A. oryzae, A. soyae,* or *A. japonicus*. The fungal enzymes soon begin degrading the soybean and wheat constituents. The producer then adds salt and allows the partially degraded mixture to undergo fermentation with yeast. The fermentation results in the dark brown liquid known as soy sauce.

In Asian diets, *A. oryzae* has been used for making miso (a fermented paste used as a spread or ingredient in dishes) from soybeans. In the Philippines, the same *Aspergillus* produces tao-si (a type of soy sauce), also from soybeans. In Ghana, kenkey (also komi or dokonu) is ground corn fermented by *Aspergillus*. The table below lists other diverse uses of *Aspergillus* in food production.

In addition to the major foods shown in the table, the food industry has found uses for *Aspergillus* as a source of food product ingredients. The following *Aspergillus*-produced ingredients are common in packaged foods: citric acid (a preservative), riboflavin (a vitamin), and glucose oxidase (a preservative).

FOOD CONTAMINATION

Though *Aspergillus* has been used for centuries to make foods, in the wrong foods and under the wrong conditions, it is a harmful food-borne pathogen. *Aspergillus* species grow well in any place that contains organic matter. Fruits, vegetables, grains, and peanuts have been special trouble spots for *Aspergillus* contamination, so food producers must check these foods carefully for molds and dirt before and during

Food Products from *Aspergillus*

Raw Ingredient	Food Product	Role of *Aspergillus*
raw fruit juices	clarified fruit juices	amylase digests starches
beer fermentation mixture	finished beer	cellulase enzyme removes cloudiness
essential oils, fruits, herbs	flavorings	cellulase helps extract flavor compounds
raw milk	concentrated milk products	lactase enzyme removes lactose
cornstarch	corn syrup	glucoamylase enzyme converts most of the starch to sugar
dairy products	cheeses	lactase digests fats and adds flavor
dough mixtures	baked products	protease helps dough consistency and mixing

processing. In fatty foods, *Aspergillus* and other molds produce enzymes that convert fats into compounds called *ketones* that produce bad tastes and odors. Such fatty foods as butter, margarine, and cream contain enough moisture to allow *Aspergillus* to grow, and spoilage soon follows. This type of spoilage wherein fatty compounds degrade into undesirable compounds is called *rancidity*. Food microbiologists control most mold contaminations in food processing plants by properly heating the foods, by lowering the moisture content of the foods, or by doing both.

Toxin poisoning from *Aspergillus* presents a more serious type of contamination than food spoilage. Usually people identify spoiled foods by noticing altered colors or odors. These signals offer a valuable safety warning. But aflatoxins produced in food are invisible. Only careful protection of foods from mold during storage and processing helps prevent aflatoxin contamination.

Aflatoxin poisoning is called *aflatoxicosis*. Humans and other animals that contract aflatoxicosis from foods or feeds, respectively, suffer from damage to the liver and possible liver cancers. This infection is rare in humans, partially because the poisoning often is misdiagnosed. In animals, aflatoxicosis takes two different forms. The first is acute (rapid, severe, and short duration) aflatoxicosis, which leads to death, mainly in livestock. The second form is chronic (long-lasting) aflatoxicosis, which does not kill but causes liver damage and poor growth.

Food and feed suppliers prevent contamination of products by keeping the moisture content low in peanuts, cottonseed, soybeans, corn, cereal grains, and the tree nuts (Brazil, pecans, pistachios, and walnuts). They also try to store these items well below 70°F (21°C) to slow mold growth.

Food analysis laboratories test these foods and feeds for aflatoxin by breaking the food into its components—proteins, fats, fibers, carbohydrates—and chemically analyzing them. Some analyses provide very sensitive detection for the presence or absence of the toxin as well as its concentration. Concentrations as low as micrograms per liter (μg/l), also called *parts per billion,* have been measured by chemical methods.

It should not be surprising that a microorganism as ubiquitous as *Aspergillus* has been studied in detail. This mold has been used as far back in history as ancient civilizations for food production and other uses. *Aspergillus* has also been a nuisance as a food contaminant or a more serious threat to health when it produces aflatoxin. *Aspergillus* symbolizes the major characteristics of many molds: commonplace in the environment; either harmful of harmless, depending on circumstances; and a microorganism that is very hard to avoid or kill.

See also FOOD-BORNE ILLNESS; FOOD MICROBIOLOGY; FUNGUS; INDUSTRIAL MICROBIOLOGY; OPPORTUNISTIC PATHOGEN.

Further Reading

Banwart, George J. *Basic Food Microbiology,* 2nd ed. New York: Chapman & Hall, 1989.

Center for Food Safety and Nutrition, U.S. Food and Drug Administration. "Aflatoxins." Available online. URL: http://www.cfsan.fda.gov/~mow/chap41.html. Accessed March 15, 2009.

Larone, Davise H. *Medically Important Fungi: A Guide to Identification,* 4th ed. Washington, D.C.: American Society for Microbiology Press, 2002.

Merck Veterinary Manual, 9th ed. Whitehouse Station, N.J.: Merck, 2005.

"Sick Building Syndrome: Healing Health Facilities." *BusinessWeek*. August 13, 2008. Available online. URL: www.businessweek.com/innovate/content/aug2008/id20080813_845797.htm. Accessed March 15, 2009.

Bacillus *Bacillus* is a genus of gram-positive, endospore-forming, and rod-shaped bacteria that normally inhabit soil and are also found in water. The general term *bacillus* (plural: bacilli) describes any rod-shaped bacteria.

Bacillus belongs to family Bacillaceae, which is the largest family in the order Bacillales. Bacillales makes up one of two orders in class Bacilli. (The other order in the class is Lactobacillales.) *Bacillus* cells measure 0.5–2.0 micrometers (µm) in width and 1.5–6 µm in length, though a few species have been found to grow to larger sizes. Cell shape consists of straight rods, rather than curved, and the cells possess flagella distributed evenly over their surface to provide motility. This type of arrangement is referred to as *peritrichous flagella*. *Bacillus* species exist as either aerobes or facultative anaerobes. (Facultative species can live with or without oxygen.) They use chemoheterotrophic metabolism, meaning they grow on a variety of organic compounds for energy and carbon.

Bacillus deoxyribonucleic acid (DNA) contains two subunits (called *bases*), guanine (G) and cytosine (C), that make up 32–69 percent of the genus's total DNA. The remainder of the DNA contains the bases adenine and thymine. For this reason, microbiologists group *Bacillus* with other genera of gram-positive bacteria called *low G + C* bacteria. (High G + C bacteria contain, by comparison, higher ratios of G and C to the other two bases.)

The *Bacillus* species look and behave very similarly, and so, for many years, microbiologists struggled to find accurate ways to distinguish individual species from others. In the late 1970s, the University of Illinois microbiologist Carl R. Woese developed a new method for differentiating microorganisms. This method involved analyzing a cellular constituent called *ribosomal ribonucleic acid* (rRNA). This breakthrough changed many of the classifications that had been used in bacteriology for decades. It also enabled microbiologists to make clear distinctions between *Bacillus* species that appeared identical. Although bacillis all look alike, rRNA analysis began to reveal wide diversity in the genus. For example, the following three species appear almost identical under a microscope, yet they perform very diverse functions in the environment:

- *Bacillus thuringiensis* produces a natural insecticide.
- *Bacillus cereus* can contaminate foods.
- *Bacillus anthracis* causes anthrax disease.

Common among all of the *Bacillus* is the ability to convert into an endospore form. An endospore is a dormant, thick-walled form of a cell that enables the bacteria to withstand extremes in heating, drying, freezing, irradiation, and chemical exposure. This spore-forming capability has made *Bacillus* a feared pathogen and a troublesome contaminant, but it also gives this microorganism qualities of commercial value.

IMPORTANT *BACILLUS* SPECIES

Activities carried out by various *Bacillus* species fall into the following four general categories: degradation of organic compounds in nature, commercial sources of antibiotics and bacteriocins, sources of industrial enzymes, and pathogens in humans and animals. Within each of these categories, individual species often have additional distinctive capabilities. Some of the important *Bacillus* species are summarized as follows.

Bacillus thuringiensis

B. thuringiensis (Bt) makes a solid crystal protein during its endospore formation. This protein acts as a poison in more than 100 different species of caterpillars, moths, grubs, and beetles when ingested by the insect. (Fortunately, the Bt crystal does not harm bees that are critical for the pollination of commercial and garden plants.) The protein destroys the insect's digestive processes and so protects any plant upon which the insect preys. Gardening supply stores sell liquid or freeze-dried mixtures of Bt endospores or the toxic protein alone. Growers then spray the Bt product onto their plants to protect against insect infestation.

Colorado State University Extension Service has explained on its Web site the characteristics of the Bt toxin: "Unlike typical nerve-poison insecticides, Bt acts by producing proteins (delta-endotoxin, the 'toxic crystal') that react with the cells of the gut lining of susceptible insects. These Bt proteins paralyze the digestive system, and the infected insect stops feeding within hours. Bt-affected insects generally die from starvation, which can take several days." Even dead *Bacillus* cells carry the toxic protein, and when the cells lyse (break apart) in the digestive tract, they still act as an effective insecticide.

Farmers have used Bt for many years as a natural method for protecting crops. Although scientists had already learned that the Bt toxin works inside insect guts, Bt still held a few secrets on the details of life inside that tiny environment. In 2006, the microbiologist Jo Handelsman and her graduate student Nichole Broderick of the University of Wisconsin set up an experiment to show the degree to which native bacteria inside insects might combat Bt; natural microbial populations are known to prevent outsiders from entering a habitat. Could certain insects contain bacteria that combatted Bt? In other words, could a normally Bt-susceptible insect become resistant with the right bacteria in its gut?

In an experiment using gypsy moths, Broderick found a relationship between Bt and other bacteria, but not the result she expected. "Initially, I was testing the hypothesis that the gut bacteria were actually protecting the moth," she said. Broderick cleared the insects' digestive tracts of their normal bacteria, fed the insects Bt toxin, and reported, "I found that once they [moths] did not have a gut community (of bacteria) I could no longer kill them with Bt." This finding suggested that the Bt toxin works in an unusual partnership with the insect's normal bacteria.

While microbiologists pursue studies on the Bt mechanism inside insects, molecular biologists have begun using the Bt gene in ingenious ways. The gene for the Bt toxin resides on a plasmid in the *Bacillus* cell's watery contents called *cytoplasm*. A plasmid is a circular piece of DNA that many bacterial species possess and that lies separate from the main DNA. Molecular biologists have isolated the Bt gene and put it into other types of bacteria and even into plant cells by a process called *genetic engineering*. For instance, the Bt gene inserted into the DNA of potato plants enables each plant leaf to produce its own Bt insecticide. This method saves income that potato growers would lose to crops destroyed by the Colorado potato beetle and other pests. The insecticide-producing plants also reduce the need for chemical insecticide sprays. Organic farmers and others who oppose genetic engineering either avoid the use of bioengineered bacteria or the spray a mixture of Bt cells directly onto their plants to kill insects.

Bacillus cereus

B. cereus acts similarly to *B. thuringiensis* but does not produce an insecticide. Instead, *B. cereus* acts as a food-borne pathogen. The species also causes rare cases of meningitis in humans and spontaneous abortions in herd animals.

The facultative anaerobic *B. cereus* is a food-borne pathogen found in a variety of foods: meat, milk, fish, cheeses, vegetables, and rice products. The pathogen causes symptoms by producing either of two proteins: a large-molecular-weight protein or a smaller, heat-stable protein. Both types of protein induce characteristic abdominal cramps and nausea. The large protein also causes diarrhea; the smaller protein causes vomiting. (The small protein is sometimes referred to as a *peptide*, which contains fewer amino acids than a typical protein.) Because one form of the *B. cereus* toxin resists heat, cooking contaminated foods may not destroy it completely.

Bacillus anthracis

B. anthracis causes the lethal anthrax disease in farm animals and humans. *B. anthracis* is unusual among *Bacillus* because its cells are nonmotile, meaning they cannot propel themselves under their own power. The large cells (1.5 × 6 μm) possess square ends, making them look rectangular, and they usually form end-to-end chains. This facultative anaerobe produces protein toxins, called *exotoxins*, that it releases into the cell's surroundings. The toxin causes three forms of anthrax disease: cutaneous, pulmonary, and gastrointestinal. Cutaneous anthrax is associated with localized skin infections, pulmonary anthrax results from inhaling *B. anthracis* endospores, and gastrointestinal anthrax arises from ingesting the endospores. Each type of disease becomes a health threat only when cells grow out of the dormant endospore in a process called *germination.*

B. anthracis, as can many *Bacillus* species, can remain in the endospore form in the soil for centuries. When endospores germinate, they revert into a regular reproducing cell form, called *vegetative cells.* Vegetative cells then produce the toxin that causes the disease's symptoms. Anthrax is most often contracted by people who are around farm animals or who frequently handle hides and pelts. This is very rare in the United States; the Centers for Disease Control and Prevention (CDC) report on the Web site the incidence of anthrax since the year 1900 is less than two cases a year.

The U.S. Food and Drug Administration (FDA) and the CDC have identified *B. anthracis* as a possible biological weapon. Although farm animals receive anthrax vaccine on a routine basis to protect livestock investments, only high-risk groups in the general U.S. population have access to a vaccine. The CDC considers the following four groups as high-risk groups in regard to anthrax disease: laboratory workers who handle *B. anthracis,* people who frequently handle hides and furs, people working with farm animals in high-incidence areas, and military personnel.

Bacillus subtilis

Aerobic *B. subtilis* cells take a very slender (0.8 µm wide), elongated shape. This species produces a variety of extracellular enzymes that are useful in commercial products. *B. subtilis* enzymes digest starches, proteins, and gelatins.

B. subtilis provides an example of a species that conducts quorum sensing. In this process, cells monitor their own population density using signal molecules they release into the environment. When the signal molecules reach a certain concentration, they induce a response in the bacteria that produced them. In the case of *B. subtilis,* nutrient-poor conditions make the bacteria respond in one of two ways: endospore formation or cooperative growth. Cooperative growth comprises activities within a colony that enable the cells to take in as much nutrient as possible for as many cells as possible. This capability becomes critical in colonies grown on nutrient-poor agar in a laboratory. Cooperative-growth colonies of *B. subtilis* produce distinctive shapes, such as snowflakelike shapes, that have not been observed by microbiologists at any other time.

Bacillus stearothermophilus

This species is a thermophile, a microorganism that grows in a temperature range of 130–150°F (55–65°C). Microbiologists take advantage of this capability of *B. stearothermophilus* by using it to test for sterile conditions. In this role, the microorganism is called a *sterility indicator.* When sterilizing a large volume of liquid or a densely packed volume of dry materials in an autoclave, a microbiologist adds a vial or a paper strip containing *B. stearothermophilus* endospores to an autoclave along with items to be sterilized. An autoclave is a chamber that kills all microorganisms by applying steam heat under pressure. After the sterilization cycle has completed, the microbiologist inoculates the vial's contents or the strip to growth medium, and then incubates the inoculated medium. After incubation, a lack of growth indicates that the heat-resistant *B. stearothermophilus* has been killed in the sterilization process. This result tells the microbiologist that the sterilization procedure also killed all other microorganisms, because normal contaminants cannot withstand the high temperatures that *B. stearothermophilus* tolerates.

The ability to withstand extreme heat makes this species an extremophile. Other *Bacillus* extremophiles are *B. psychrophilus* (grows at low temperatures) and *B. alcalophilus* (grows at high pH).

Bacillus sphaericus

Bacillus sphaericus provides an example of the hardiness of the bacterial endospore. The very slender (0.5–1.0 µm wide) motile cells of this species form endospores, typical constituents of *Bacillus:* That is, they contain an inner membrane, a middle cortex layer, and a protective spore coat. This species also served to demonstrate the incredible durability of *Bacillus* endospores.

In 1993, the American microbiologist Razl Cano reported a *B. sphaericus*–like microorganism inside a primitive bee that had become trapped in amber 25–40 million years ago. *Discover* magazine described Cano's finding but also pointed out that he had faced some skepticism. The *B. sphaericus* endospores, wrote the reporter Lori Oliwenstein, go into "a state of suspended animation. In times of stress a number of microbes knit themselves a strong, protective protein coat called a spore and slow all their cellular processes until they are effectively (but not actually) dead. Once they sense the presence of sufficient nutrients—a sort of bacterial all's-well signal—they resurrect themselves." But the article also noted, "His [Cano's] critics, however, are not quite ready to raise a glass to him. They say it's impossible for any living creature to have survived for so long. Instead of an ancient microbe, they argue, Cano has simply found a modern contaminant." But *Bacillus* endospores have been recovered from 2,000-year-old tombs, 10,000-year-old fossils, and objects dated much older than the remarkable amber specimen found by Cano. Cano conducted DNA analysis on the

Major Enzymes Produced by *Bacillus*

Enzyme	Its Substrate	End Product
amylase	starch	sugars
protease	proteins	amino acids, peptides
lipase	fats	glycerol and fatty acids
glucanase	glucan polymers	short-chain saccharides
pullulanase	maltodextrans	sugars

amber's *Bacillus* and compared it to modern *Bacillus* DNA and concluded that the two varieties were not related enough to suggest they are contemporaries.

B. sphaericus vegetative cells also produce a toxin that kills *Culex* mosquito larvae feeding in water. Some state health departments spray *B. sphaericus* mixtures on still waters, such as stagnant ponds and pooled rainwater, during the spring and summer to control mosquito populations.

Bacillus polymyxa

This facultative anaerobic, motile species produces a number of enzymes with commercial uses. *Bacillus polymyxa* breaks down starches, the protein casein, gelatin, and pectins. In addition, this species produces polymyxin antibiotics, which are effective against many gram-negative pathogens. These large compounds kill other bacteria by infiltrating the cell membrane and causing cell constituents to leak out. A well-known polymyxin used in human and veterinary medicine is polymyxin B.

Bacillus megaterium

B. megaterium produces a very large cell, 1.5–3.0 μm wide. Because of its size, *B. megaterium* serves as a common teaching tool for studying endospore formation and the life cycle of *Bacillus*. Biotechnology also favors *B. megaterium* in cloning experiments and in plasmid production.

COMMERCIAL USES OF *BACILLUS*

Bacillus has two attributes that make it attractive in industrial microbiology: the rugged endospore and production of a wide variety of enzymes. Because endospores resist damage by heat, cold, and chemicals, manufacturers can include them in product formulas with confidence that the bacteria will remain alive. For example, *Bacillus* makes up the main ingredient in septic tank additives and drain openers as well as the insecticide products already mentioned. The endospore gives these formulas a long shelf life, and *Bacillus*'s enzymes deliver the product's desired effect.

Various *Bacillus* species produce the enzymes listed in the table. Some species, such as *B. subtilis,* excrete more than one.

Makers of cleaners, detergents, stain removers, and food additives also use *Bacillus* enzymes in their products. Other industries take advantage of *Bacillus* activities for certain manufacturing steps. Brewers use ß-glucanase produced by *B. subtilis* to clarify beer, and the food industry uses *Bacillus* enzymes to make high-fructose corn syrup, a sweetener added to a multitude of processed foods.

See also ANTHRAX; BACTERIA; SPORE.

Further Reading

Biello, David. "Bt Pesticide No Longer Kills on Its Own, Overturning Orthodoxy." *Scientific American,* 25 September 2006. Available online. URL: www.sciam.com/article.cfm?id=bt-pesticide-no-killer-on. Accessed March 15, 2009.

Cano, Raúl, Heridrik N. Poinar, Norman J. Pieniazek, Aftim Acra, and George O. Poinar. "Amplification and Sequencing of DNA from a 125–130-Million-Year-Old Weevil." *Nature* 363 (1993): 536–538. Available online. URL: www.nature.com/nature/journal/v363/n6429/abs/363536a0.html. Accessed March 29, 2009.

Centers for Disease Control and Prevention. "Anthrax." Available online. URL: www.bt.cdc.gov/agent/anthrax. Accessed March 15, 2009.

Colorado State University Extension Service. "*Bacillus thuringiensis.*" Available online URL: www.ext.colostate.edu/pubs/Insect/05556.html. Accessed March 15, 2009.

Oliwenstein, Lori. "They Came from the Oligocene, He Said." *Discover,* 1 January 1996. Available online. URL: http://discovermagazine.com/1996/jan/theycamefromtheo651. Accessed March 12, 2009.

Todar, Kenneth. "The Genus *Bacillus.*" Todar's Online Textbook of Bacteriology. Available online. URL: www.textbookofbacteriology.net/Bacillus.html. Accessed March 16, 2009.

bacteria (singular: bacterium) Bacteria are single-celled organisms with a cell wall characterized by the large compound peptidoglycan. They are in the kingdom of prokaryotes, so they lack a true nucleus and organelles surrounded by membrane. Bacteriology encompasses all aspects of bacteria. Specialties in bacteriology include, but are not limited to, the following: pathogens in humans, animals, and plants; clinical isolates; intestinal flora; rumen flora; genetic engineering; serotyping; morphology; environmental studies; food preservation; food production; industrial products; and enzymology.

Bacteria comprise a diverse group of microorganisms that display a wide range of physiologies and live in a variety of habitats. Their diversity allows them to participate in almost every biological activity on Earth. Higher organisms could not exist for long without their native bacteria, that is, the bacteria that normally reside in or on the body. The planet also relies almost entirely on bacteria to cycle nutrients from sediments through animal and plant life, then through the atmosphere and back to the earth. Bacterial numbers reach enormous levels in many habitats on Earth. More bacterial cells live on or in the human body than there are human cells.

The classification of bacteria within the world of living things has not come easily. New classifications of species arose as new methods of identification developed in microbiology. From the 1800s to the 1960s, advances in microscopy enabled microbiologists to see ever-finer structures in and on bacterial cells. During this time, microscopic features (morphology) seemed the best way to group bacteria. But as the science of biochemistry grew, biochemical reactions carried out by bacteria replaced or supplemented groupings based on morphology. In the 1980s, molecular biologists discovered methods for determining the subunit, or base, sequences of deoxyribonucleic acid (DNA). This technique allowed biologists to study how the world's organisms are related and the closeness of those relationships, called relatedness.

Microbiologists began rearranging bacteria classifications on the basis of common ancestries between species, as told by their DNA composition. One technique, DNA hybridization, found genes common to different bacteria that had heretofore been thought of as unrelated. In the process, bacteria that evolved along similar paths were soon distinguished from others that were not as closely related.

Throughout the 1980s and 1990s, laboratories sequenced hundreds of bacterial genes. During that period, the bacteriologists Carl Woese at the University of Illinois and Mitch Sogin at the Marine Biological Laboratory in Wood's Hole, Massachusetts, developed another sequencing method. They determined the nucleic acid sequences of 16S subunits of ribosomal ribonucleic acid (rRNA), the cell structures involved in protein assembly. Although Woese had been pursuing the genetic makeup of cells for much of his career, Sogin took a different route into a specialty that would have a tremendous impact on biology and the study of evolution. Sogin told PBS in 2002, "During the third year of my undergraduate career I reached the realization that I didn't want to be a physician. Molecular biology was an emerging field and I had the opportunity to work with microbiologists and physicists who were joining forces to explore questions in evolutionary biology. I was simply in the right place at the right time." Sogin may have described his contributions in modest terms, but the bacterial family trees established by him, Woese, and their colleagues remain in use today.

BACTERIA ON EARTH

Living things, called *biota,* on Earth can be divided into prokaryotes and eukaryotes. The eukaryotes range from single cells to multicellular plants and animals. All the eukaryotic cells possess membrane-enclosed organelles. Prokaryotes contain two domains: bacteria and the archaea. The 16S rRNA sequencing studies, of the 1990s, have shown that domain Bacteria and domain Archaea evolved separately from each other, very early in the evolution of life. For this reason, most texts divide the world of living things into three domains: Bacteria, Archaea, and Eukarya. Older texts sometimes use the term *eubacteria* to describe the "true bacteria" and the term *archaebacteria* to signify the archaea. In fact, archaea are not bacteria, and the term *archaebacteria* can be misleading.

The actual number of bacterial species on Earth is unknown and may never be known. Determining the number would rely on sampling every type of environment and accounting for every mutation. New species of bacteria are discovered almost daily, but only about 5,000 species have been completely characterized. Microbiologists have discovered additional species that they cannot yet identify or have not devised a way to keep alive in a laboratory. These bacteria are called *VNC* (or VBNC) for "viable but noncultivable."

A second way of thinking about the number of bacterial species on Earth proposes that the numbers of calculated species could be an overestimate. Because bacteria freely exchange genetic material between cells, microbiologists could argue that trying to place each into a genus and species is meaningless; bacteria are all related to each other to some degree.

Many microbiologists estimate that less than 1 percent of all bacteria on Earth have been cultured and characterized. In 1998, the University of Georgia microbiologist William B. Whitman led a team of scientists in estimating the number of bacteria by counting samples from a wide array of habitats as well as carbon content measurements in those places. "By combining direct measurements of the number of prokaryotic cells in various habitats," Whitman said, "we found the total number of cells was much larger than we expected." They found the greatest number of bacteria located in the subsurface of the earth or deep soils and in the ocean. Whitman's estimate equaled five million trillion trillion (5×10^{30}) bacteria!

Whitman's studies also produced a calculation for the mass of bacteria on Earth based on carbon content. Whitman estimated that bacteria total $3.5–5.5 \times 10^{17}$ grams of carbon, or about the same amount as all of Earth's plant life.

Determining the number of bacteria on Earth has been understandably a difficult process. By using the 16S rRNA classification scheme, microbiologists can divide bacteria into groups even if a new discovery has not been identified or named. Microbiology often uses general groupings of bacteria based on a combination of physical and genetic features. For instance, prevalent groups of bacteria often go by nicknames: *spore formers, alphas, spirochetes, lactic acid bacteria,* or *nitrogen fixers.* Nicknames perhaps serve to highlight the extraordinary diversity of bacteria and their many roles. For official scientific naming, microbiologists use *Bergey's Manual of Determinative Bacteriology* to help them classify new bacteria. These five volumes serve as the main reference in systematics of bacteria, which is the science of classifying and naming organisms.

HISTORY OF BACTERIOLOGY

Bacteria's history with humans extends to the earliest documented civilization. Species recovered from remnants of ancient civilizations include *Bacillus* endospores and the *Mycobacterium* that causes tuberculosis. The actual study of these tiny creatures unfolded over centuries, beginning with the explorations of the stars and the seas. Humanity's place in the universe became a question for scholars in the 15th century, when a few explorers peered into the smallest universe rather than outward across the oceans or the stars. The Italian Girolamo Fracastoro (1478–1553) was one such visionary, who proposed that infection transmitted as tiny particles on clothes, bodies, or commonly touched inanimate objects. Fracastoro had essentially defined the core idea of disease transmission, but finding these microscopic specks proved difficult without the technologies that would only emerge in the next century. In 1597, the glassmaker Zacharias Janssen (1585–1632) and his father, Hans, developed a useful instrument for looking at small things by arranging lenses in sequence. By creating this method for magnification, the Janssens had invented the first compound microscope. Antoni van Leeuwenhoek (1632–1723), a tradesman living in Holland, applied his own collection of lenses 75 years later to see tiny "animalcules" in a drop of water.

Van Leeuwenhoek's observations of microscopic life, in 1677, have been credited as the first detailed studies of bacteria at the microscopic level. Viewing a drop of water into which he had ground pepper granules, van Leeuwenhoek wrote, "I found a great plenty of them in one drop of water, which were no less than 8 or 10,000, and they looked to my eye, through the Microscope, as common sand doth to the naked eye." Van Leeuwenhoek made painstaking notes of his discoveries, which modern microbiologists now recognize as very sophisticated studies.

Simple microscopes encouraged scientists to debate, for the next 90 years, about the theory of spontaneous generation, in which living organisms were believed to arise from nonliving matter. The so-called golden age of microbiology, from the mid-1800s to about 1915, encompassed some of microbiology's most important breakthroughs: the germ theory, the principles of infection and disease, and immunity. Hans Christian Gram's (1853–1938) contribution in finding a biological stain for making better observations of bacteria began the science of classifying bacteria according to cellular features. When the structure and function of DNA became revealed in the 1940s and 1950s, microbiologists delved into the bacterial chromosome, the entire collection of a cell's genetic matter. A significant advance occurred in the 1980s, when university researchers transferred pieces of bacterial DNA, or genes, from one species to another, unrelated species. Molecular biology was born. From it, the field of biotechnology grew, as scientists manipulated an increasing variety of bacterial genes to make new products. Certain aspects of genetic engineering and medical therapies would not be available today if not for bacteria.

THE BACTERIAL CELL

The diverse bacteria on Earth range in size from nanobacteria of less than 0.2 micrometer (μm) diameter to the immense marine *Thiomargarita namibiensis,* measuring 0.75 mm in diameter, about the size of the period at the end of this sentence. Most bacteria fall in a range of 0.2–2 μm in width and 4–8 μm in length. In the world of microscopic particles, bacteria are 10–100 times greater in size than viruses and about one tenth the size of a human red blood cell.

Bacterial cells reproduce by binary fission, in which a single parent cell splits asexually to form two identical daughter cells. Each daughter cell assumes the size and shape characteristic of the genus. Bacteria fall into five categories based on shape, as follows:

- cocci—round cells (singular: coccus)
- bacilli—rod-shaped cells (singular: bacillus)
- vibrios—curved rods

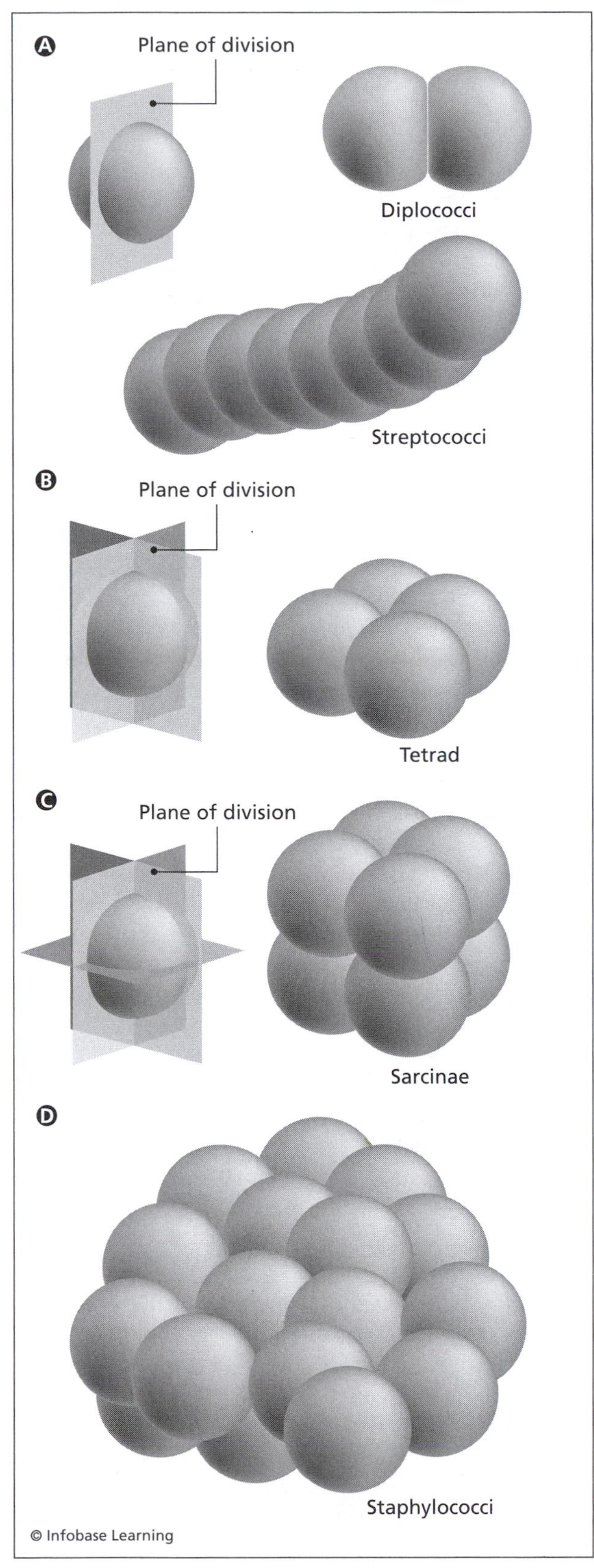

Bacterial species divide from one cell to many in characteristic configurations. Different cell morphologies and orientations of cells aid in genus identification.

- spirochetes—helical or corkscrew shaped
- pleomorphic—many different shapes within a species

Bacteria that are always round, rod-shaped, curved, or helical are referred to as *monomorphic,* because they are genetically programmed to produce only one shape. A few species produce unique shapes, such as the star-shaped bacteria of the genus *Stella*. Bacteria that grow different shapes are called *pleomorphic.*

Cocci and bacilli consist of various arrangements as cells multiply and produce greater numbers of identical cells. Dividing cocci, first, form a pair of joined cells, a diplococcus. Further division leads to one of two arrangements. Some cocci form elongated chains, like a string of pearls, called streptococci. Others form a tetrad of four cells from the diplococcus by dividing in two planes. A second division in three planes forms cubes of eight cells called *sarcinae*. Continued division leads to large numbers of cells in grapelike clusters characteristic of staphylococci.

Bacilli create similar groups of cells as they divide, although chains are more common in bacilli than clusters. Thus, a bacillus forms a diplobacillus, which with continued division forms a chain of streptobacilli. Short, squat bacilli having a rounded appearance are called *coccobacilli.*

Bacteriologists further define bacteria by structures on the outside of cells. Cells with whiplike tails called *flagella* are motile, meaning they have the ability to move on their own. Motility is one characteristic used for identifying bacteria, as is cell morphology, the study of microbial structure. Cell morphology has been helped by advances in microscopy, and it has developed into a separate specialty within bacteriology.

The cell interior contains structures that are common to most bacteria but different from eukaryotic cells. Bacteria are like eukaryotes, in that cytoplasm fills most of the cell contents. Cytoplasm is fairly homogeneous, watery—70–85 percent water by weight—and shapeless. (The outer cell membrane holds cytoplasm together.) All of the cell's reproductive activities take place within the cytoplasm. Reproduction requires a chromosome, the bacteria's main depository of DNA. The chromosome contains each species's entire genetic code, and without it, life could not continue. Bacterial DNA exists in two places in the cell: the nucleoid, a region in the cytoplasm dense in DNA but not surrounded by a membrane, and, in some species, a separate piece of DNA called a *plasmid,* located in the cytoplasm separate from the nucleoid. Bacteria employ plasmids for transferring genes between cells and plasmids often contain genes for antibiotic resistance.

Bacterial cells also contain ribosomes, vacuoles, and inclusion bodies. The ribosomes function in protein synthesis by reading DNA's genetic code. Vacuoles are open spaces in the cytoplasm and shaped by a protein lining. Vacuoles contain gas and serve two main functions: storage of atmospheric gases and creation of cell buoyancy by expanding or collapsing. The role and composition of inclusion bodies vary among species. Some bacteria use inclusion bodies to store nutrients, while others use them for regulating osmotic pressure, which is the pressure of the cell interior relative to the exterior. Vacuoles filled with air are called *gas vacuoles* and may behave similarly to inclusion bodies.

On the outer surface, in addition to flagella, gram-negative bacteria may contain short fimbrae. These appendages measure 5–10 µm long and cover most of the cell. They contribute to motility and allow cells to attach to surfaces. Even smaller outshoots called *pili,* about 2–3 µm long, number no more than 10 per cell and are used for exchanging genetic material between cells.

The bacterial cell wall provides strength and protection from the outside environment and gives each species its characteristic shape. Almost all bacteria are divided into one of two groups based on cell wall structure. These groups are the gram-positive species and the gram-negative species. The distinction between positive and negative results from the cell wall's capacity to absorb the Gram stain. Some species (gram-positive) turn purple when exposed to one of the chemicals used in the Gram stain method. Other species (gram-negative) do not hold this stain and cannot turn purple. Though biologists have invented sophisticated ways to classify bacteria, such as DNA and rRNA sequencing, the Gram stain remains a fundamental technique in every microbiology laboratory.

CLASSIFICATIONS OF BACTERIA

Each domain within the prokaryotic kingdom is divided into phyla, each phylum into classes, each class into orders, and each order into families. Families contain genera, and almost all bacterial genera have more than one species. In bacteriology, bacteria are known by their genus and species name. For example, *Pseudomonas aeruginosa* is a species of the genus *Pseudomonas.*

Domain Bacteria contains 23 phyla (see Appendix V). Over decades of study, bacteriologists have learned much more about some phyla than others. Microbiology has made the biggest strides in characterizing species that have important associations with humans, domestic animals, and plants, or species that carry out key reactions in the environment. The table on page 81 summarizes the main groups studied in bacteriology.

Proteobacteria

Proteobacteria is the largest group of bacteria that have been characterized by bacteriologists. The group is divided into five classes, all having similar base sequences in their RNA: alpha, beta, delta, epsilon, and gamma. This group's physiologies and morphologies are very varied; no single type of metabolism or cell feature represents all species of proteobacteria except that they are all gram-negative.

The alpha proteobacteria include members that perform nitrogen fixation in plants *(Rhizobium),* use methane as a carbon-energy source *(Methylobacterium),* or use chemolithotrophic metabolism *(Nitrobacter),*) which employs inorganic compounds for energy and carbon dioxide as a carbon source. The alpha proteobacteria can live on very low levels of nutrients, and they contribute to Earth's cycling of nitrogen from the atmosphere to the land and biota. A general group called purple nonsulfur bacteria live in freshwater and marine waters, and their need for oxygen varies among species. They gather energy from light and use organic compounds as both electron and carbon sources.

Many beta proteobacteria act as nitrifying bacteria: That is, they convert ammonia or nitrites to nitrates in a chemical step called *oxidation.* Beta proteobacteria tend to use the end products made by anaerobic species, such as hydrogen gas, ammonia, or methane. Many beta proteobacteria live in water environments and in sewage.

The delta proteobacteria contain predators that feast on other bacteria and members that contribute to Earth's sulfur cycle. Anaerobic sulfate- and sulfur-reducing delta proteobacteria live in watery habitats, especially mud, and often grow well in polluted streams and lakes.

Epsilon proteobacteria make up a small group containing human and animal pathogens as well as species that form large mats on water surfaces rich in hydrogen sulfide. These mat communities have been discovered living in harsh oil- or sulfur-polluted places, so they are thought of as extremophiles. Microbiologists who seek new bacteria for cleaning up pollution (a process called *bioremediation*) have looked to epsilon proteobacteria because of their ability to degrade pollutants.

The gamma proteobacteria are the largest and most diverse class in the phylum. *Bergey's Manual* divides them into 14 orders and 25 families. Many gamma proteobacteria are facultative anaerobes, meaning they can live with or without oxygen;

Important Groups of Bacteria

Group Common Name	Phylum	Classes	Important Genera	Important Features
bacteroids	XX: Bacteroidetes	Bacteroides Flavobacteria Sphingobacteria	*Bacteroides* *Prevotella* *Flavobacterium*	degrade complex polysaccharides, rumen inhabitant, sewage treatment
chlamydia	XVI: Chlamydiae	Chlamydiae	*Chlamydia*	sexually transmitted disease in humans
fusobacteria	XXI: Fusobacteria	Fusobacteria	*Fusobacterium* *Leptotrichia*	in habitant of human gastrointestinal (GI) tract
high G + C gram-positives	XIV: Actinobacteria	Actinobacteria	*Actinomyces* *Corynebacterium* *Frankia* *Gardnerella* *Mycobacterium* *Nocardia* *Propionibacterium* *Streptomyces*	aerobic and anaerobic fermentations, soil inhabitants, antibiotic production
low G + C gram-positives	XIII: Firmicutes	Clostridia Millicutes Bacilli	*Clostridium* *Sarcina* *Mycoplasma* *Bacillus* *Listeria* *Lactococcus* *Leuconostoc* *Staphylococcus* *Enterococcus* *Lactobacillus* *Streptococcus*	inhabit mammals, plants, and insects; endospore formation; acid fermentation
photosynthetic bacteria (green bacteria and cyanobacteria)	VI: Chloroflexi X: Cyanobacteria XI: Chlorobi	Chloroflexi Cyanobacteria Chlorobia	*Anabaena* *Gloeocapsa* *Chlorobium* *Chloroflexus*	aerobic and anaerobic photosynthesis
proteobacteria (purple bacteria)	XII: Proteobacteria	alpha, beta,gamma, delta, or epsilon proteobacteria	Beggiatoa Campylobacter Escherichia Rhizobium Salmonella Shigella Vibrio	human and plant pathogens, nitrogen fixation, nitrification, sulfur oxidation and reduction, sewage treatment, fermentations, photosynthesis
spirochetes	XVII: Spirochaetes	Spirochaetes	*Borrelia* *Leptospira* *Treponema*	inhabitant of mammal GI tract, mollusk and insect digestive tract

Note: See Appendix V for microbial classifications

others are aerobes. Gamma proteobacteria also use varied means of generating energy. Gamma proteobacteria contain photosynthetic microorganisms collectively nicknamed the *purple sulfur bacteria.* The purple sulfur bacteria require strict anaerobic conditions and live in sulfide-rich zones in still lakes, swamps, bogs, and lagoons.

Photosynthetic Bacteria

Photosynthetic bacteria belong to three different phyla: Chloroflexi, Chlorobi, and Cyanobacteria. All are gram-negative and contain the pigment chlorophyll. Most of these bacteria live in deep regions of lakes and ponds where much of the light penetration has been blocked by plants living in shallow layers. This may explain why these bacteria use a region of the light spectrum not used by plants. Some of the Chloroflexi even grow in complete dark. The photosynthetic bacteria—some proteobacteria are also photosynthetic, but not included here—use water as an electron source in energy metabolism and generate oxygen.

Chloroflexi contains motile species that grow into orange-red mats in natural hot springs. Nevertheless, many members of Chloroflexi are nicknamed the *green nonsulfur bacteria.* They grow with or without oxygen and use a variety of carbon sources. This phylum additionally contains nonphotosynthetic bacteria (*Herpetosiphon*).

Chlorobi are known as the *green sulfur bacteria.* They are strict anaerobes that use hydrogen sulfide, sulfur, or hydrogen as electron donors and carbon dioxide as a carbon source. Chlorobi have unique vesicles called chlorosomes, which contain chlorophylls *a, c, d,* and *e* and serve as the location for photosynthesis.

The cyanobacteria make up another large and diverse group. Cyanobacteria have a characteristic blue-green color caused by the relative levels of pigments in their cells, mainly phycocyanin. These bacteria do not grow well as pure cultures in laboratories, so bacteriologists have had a difficult time classifying them. For many years, biologists classified cyanobacteria as blue-green algae. Cyanobacteria photosynthesize and fix nitrogen, and, unlike Chlorobi and Chlorofexi, cyanobacteria perform oxygenic photosynthesis, meaning they produce oxygen.

Low G + C Gram-Positive Bacteria

Gram-positive bacteria fall into two groups: low G + C and high G + C. Low G + C bacteria generally have less than 50 percent of their DNA as guanine and cytosine. High G + C DNA bacteria have more than 50 percent guanine and cytosine.

Three important types of bacteria belong to the low G + C gram-positives: the mycoplasmas, the gram-positive rods and cocci, and endospore-forming bacilli. Members of the genus *Mycoplasma* are unique because they lack a cell wall because of their inability to make peptidoglycan. Without a cell wall, *Mycoplasma* cells are fragile and readily lyse (break apart) in liquids containing a high concentration of dissolved molecules. *Mycoplasma* species cause diseases in humans and livestock and are best known as the cause of respiratory diseases in humans, especially tuberculosis.

The most familiar genera in the low G + C group possess similar cell wall structures and are among the most studied bacteria in microbiology. The low G + C bacteria contain the following genera: *Staphylococcus, Streptococcus, Lactobacillus, Micrococcus, Lactococcus, Leuconostoc,* and *Streptomyces.*

The genera *Clostridium* and *Bacillus* are alike because they form a strong, resistant, and protective structure called an *endospore.* Both genera live primarily in soil and contain both rods and cocci. They differ from each other, however, in their metabolism; *Clostridium* is an anaerobe, *Bacillus* is an aerobe.

High G + C Gram-Positive Bacteria

The major gram-positive bacteria containing a high percentage of G + C in their chromosome are the actinomycetes. This is a general group of aerobic species that form branching filaments as they grow. Viewed in a microscope, actinomycetes can resemble filamentous fungi more than they resemble bacteria. A familiar genus within the actinomycetes is *Streptomyces.* This soil inhabitant includes about 500 species. In the environment, *Streptomyces* contributes to nutrient cycling, decomposition of organic matter, and antibiotic production, and it causes some plant and animal diseases.

A few other bacteria in this group are so unrelated to actinomycetes, the classification seems to make little sense. One important example is the *Mycobacterium* genus. These bacteria grow very slowly—taking up to a month in laboratory cultures—and their cell walls contain an unusually high content of fatty substances called lipids. *M. bovis* plays an important role in this group because it causes tuberculosis in humans and most other warm-blooded vertebrates. Examples of animals susceptible to *M. bovis* infection are the following: cattle, sheep, goats, deer, elk, pigs, dogs, cats, monkeys, large apes, and exotic hoofed animals in zoos.

Bacteroids

Bacteroids is a general term for gram-negative bacteria belonging to phylum Bacteroidetes. They live as either strict anaerobes or anaerobes that tolerate only low oxygen levels (aerotolerant). They inhabit the human mouth and intestinal tract and the stomach of ruminant animals. In the intestines, members of *Bacteroides* degrade complex carbohydrates such as starch, pectin, and cellulose. Microbiology historians believe van Leeuwenhoek may have been

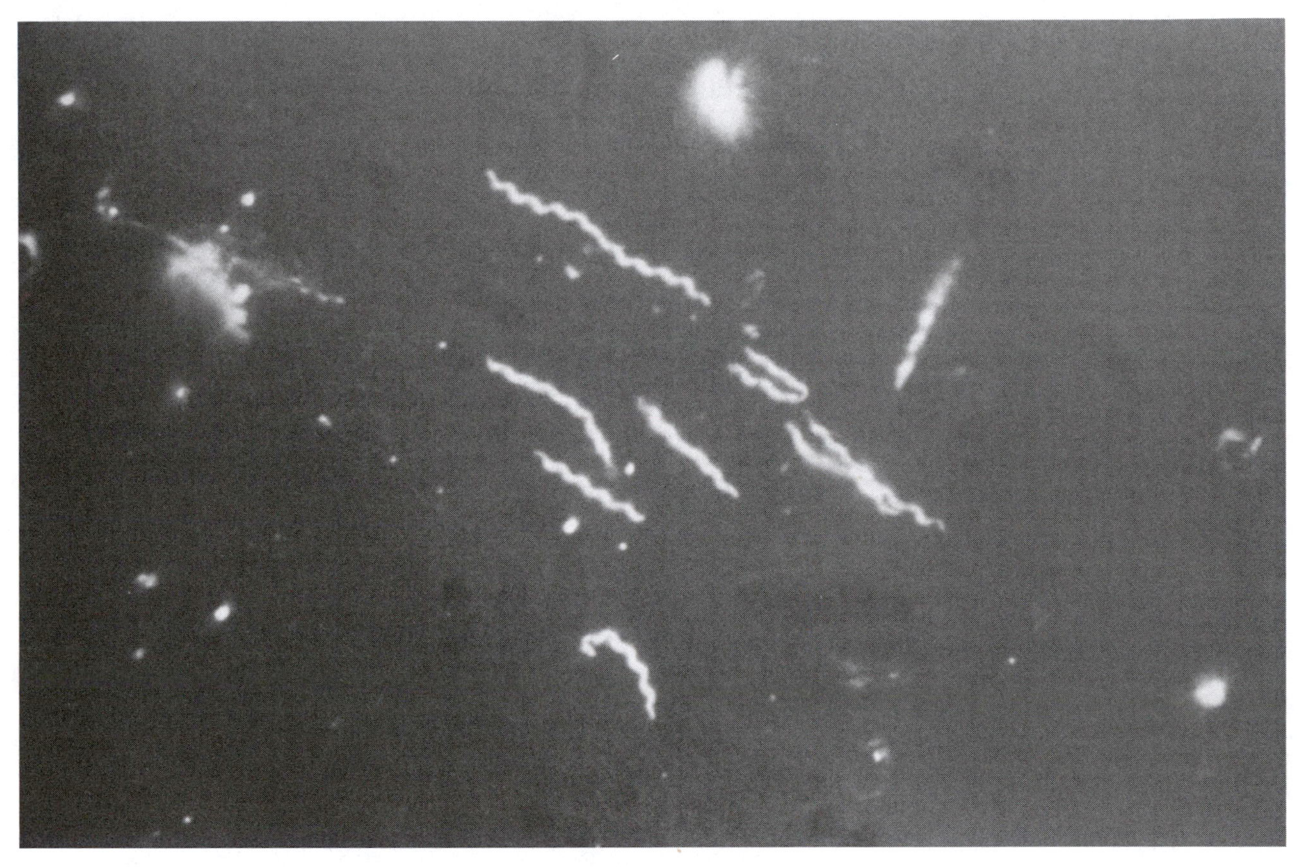

Borrelia burgdorferi* is a spirochete, which is a corkscrew, or spiral-shaped, bacterium. These cells of 10–25 μm in length cause Lyme disease.** ***(CDC, Public Health Image Library)

the first person to observe bacteroides in a microscope because some of his studies focused on matter scraped from between his teeth.

A second notable group within Bacteroidetes belongs to the gliding bacteria. Certain bacteria placed in this group are also part of other classifications. For example, *Myxococcus* also belongs with the delta proteobacteria. Gliding bacteria are motile yet do not possess flagella. They glide across solid surfaces using a mechanism not well understood but believed to include a screwlike twisting motion in some species and flexing or twitching in other species. These bacteria travel at rates ranging from 2 to 600 μm per minute, often in the direction of nutrients. Gliders excrete enzymes particularly active in breaking down paper and other materials containing cellulose. They have been used for many years for retting woody plants, meaning they digest the plants' strong fibers to make them softer. The textile industry also uses gliding bacteria for recovering natural fibers from tough plant matter. As an example, fabric mills use the enzymes from these bacteria to recover textile fibers made from hemp plants. Notable genera in this group are the following: *Desulfonema, Flavobacterium, Flexibacter, Heliobacterium, Myxococcus,* and *Oscillatoria.*

Spirochetes

The spirochetes are so named because of their spiral or corkscrew shape. These gram-negative bacteria grow in a range of oxygen environments and on diverse nutrients. Their unique motility results from filaments that spiral around the outer cell surface; the movement of spirochetes through water is not unlike a screw's being driven into a piece of wood.

Spirochetes live in the human mouth and in the intestinal tract of various animals. The important spirochete genera appear in the following list:

- *Treponema*—includes *T. palladium,* the cause of syphilis
- *Leptospira*—water and soil contaminant carried by domestic dogs and cats; the cause of leptospirosis in animals
- *Borrelia*—cause of relapsing fever and Lyme disease, carried by insects

Chlamydia

Since *Chlamydia* cell walls (gram-negative) do not contain peptidoglycan, their cells are weaker than those of most other bacteria. For this reason,

Chlamydia must live inside other cells. In humans, *Chlamydia* infection gives rise to the following two sexually transmitted diseases: nongonococcal urethritis and lymphogranuloma venereum. Transmission of these bacteria also occurs through the air to cause respiratory infections.

Fusobacteria

The fusobacteria are, like the bacteroids, common gram-negative anaerobes of the mouth and intestinal tract, where they help digest food. Their cell shape is fusiform, meaning they are shaped like spindles, rods that are tapered at each end.

THE ROLES OF BACTERIA

Bacteria play roles in nature, in human and animal health, in association with plants and insects, and as tiny factories for manufacturing industrial materials. In general, their roles may be classified into three major areas: environmental, industrial, and medical.

Environmental microbiology covers bacterial metabolism in the earth, in natural waters, and in the air. Bacteria and fungi in soil and water both degrade organic wastes and recycle the elements carbon, nitrogen, phosphorus, and sulfur, among others in the earth's biogeochemical cycles. In the process of degrading the earth's organic matter, some bacteria produce the gases methane, carbon dioxide, hydrogen, oxygen, and nitrous oxide, which convert to nitrogen gas. Bacteria, therefore, are major contributors to the composition of the atmosphere, and they carry out these reactions even when they are inside plants or animals. Many live in commensal relationships with higher organisms and affect the health of their hosts. In such a relationship, the bacteria help their hosts by digesting nutrients and secreting compounds made during their normal metabolism, which the host then absorbs and uses in its own metabolism. The most important of these secretions are acids, alcohols, antibiotics, proteins and amino acids, enzymes, and vitamins.

Because some bacteria produce large amounts of secretions useful to humans, industrial microbiology developed to take advantage of this activity. Industrial microbiologists harness bacteria in laboratories to produce compounds useful in commercial products. In addition to antibiotics and the other secretions listed, bacteria produce biological polymer compounds, biopolymers, which are long polysaccharides. These compounds act as stabilizers in liquid formulas such as paint, lubricants, absorbents, and additives to drugs and foods.

Food production also relies on specific bacteria. For example, *Lactobacillus* species make a number of dairy products: buttermilk, yogurt, and cheeses. Food microbiologists must also constantly seek the best ways to preserve foods against the many bacteria that cause spoilage or food-borne illnesses.

The field of biotechnology has taken advantage of bacteria's ability to accept new genes into their chromosomes. By inserting genes into bacteria, bioengineers create unique new products, especially for medicine. Biotechnology is now a specialized branch of industrial microbiology that focuses on genetically modified bacteria and fungi.

Medical microbiology not only depends on bacterial products such as antibiotics, but it also involves ways to combat pathogens. Pathogenic bacteria are the bacteria that cause disease. Medical microbiology covers all aspects of pathogen virulence, transmittance, infection, and disease. Clinical microbiology focuses on the bacteria and other microorganisms isolated from patients who are infected with one or more pathogens. This field includes methods for identifying pathogens and selecting drugs to kill infectious bacteria in the body.

Almost every subject in microbiology is connected in some way to bacteria. They have been called the "keepers of the biosphere" because of their vast contributions to the geology and biology on Earth. Evolution of life itself on Earth would not have begun without the formation of prokaryotic cells. It would be impossible to study biology or many of the Earth sciences without an understanding of the world's bacteria.

See also BINARY FISSION; BIOGEOCHEMICAL CYCLES; CELL WALL; CULTURE; CYANOBACTERIA; ENTERIC FLORA; HYBRIDIZATION; IDENTIFICATION; METABOLISM; MORPHOLOGY; MOTILITY; ORGANELLE; PEPTIDOGLYCAN; PLASMID; PROTEOBACTERIA; PURPLE BACTERIA; SPORE; SYSTEMATICS; TAXONOMY.

Further Reading

De la Maza, Luis M., Marie T. Pezzlo, Janet T. Shigei, and Ellena M. Peterson. *Color Atlas of Medical Bacteriology.* Washington, D.C.: American Society for Microbiology Press, 2004.

Dyer, Betsey D. *A Field Guide to Bacteria.* Ithaca, N.Y.: Cornell University Press, 2003.

Garrity, George M., ed. *Bergey's Manual of Systematic Bacteriology. Vol. 1, The Archaea and the Deeply Branching Phototrophic Bacteria,* 2nd ed. New York: Springer-Verlag, 2001.

———. *Bergey's Manual of Systematic Bacteriology. Vol. 2, The Proteobacteria (Part C) The Alpha-, Beta-, Delta-, and Epsilonproteobacteria,* 2nd ed. New York: Springer-Verlag, 2005.

Gest, Howard. *Microbes: An Invisible Universe.* Washington, D.C.: American Society for Microbiology Press, 2003.

Karlen, Arno. *Biography of a Germ.* New York: Anchor Books, 2000.

Needham, Cynthia, Mahlon Hoagland, Kenneth McPherson, and Bert Dodson. *Intimate Strangers: Unseen Life on Earth.* Washington, D.C.: American Society for Microbiology Press, 2000.

Schaechter, Moselio, John L. Ingraham, and Frederick C. Neidhardt. *Microbe.* Washington, D.C.: American Society for Microbiology Press, 2006.

Sea Studios Foundation. "The Shape of Life." PBS interview with Mitchell Sogin. Available online. URL: www.pbs.org/kcet/shapeoflife/explorations/bio_sogin.html. Accessed March 16, 2009.

Sherman, Irwin W. *Twelve Diseases That Changed Our World.* Washington, D.C.: American Society for Microbiology Press, 2007.

Todar, Kenneth. Todar's Online Textbook of Bacteriology. Available online. URL: www.textbookofbacteriology.net. Accessed March 16, 2009.

University of Georgia. "First-Ever Estimate of Total Bacteria on Earth." *San Diego Earth Times,* September 1998. Available online. URL: www.sdearthtimes.com/et0998/et0998s8.html. Accessed March 14, 2009.

bacteriocin A bacteriocin is a protein or a smaller chain of amino acids called a *peptide* made by bacteria to inhibit the growth of other similar bacteria. They differ from antibiotics in two ways: Most antibiotics are not proteins, and, in general, bacteriocins target cells of the same species while antibiotics target cells of different species. Bacteriocins resemble antibiotics in at least one way, however, because they are substances made by one microorganism to kill other microorganisms.

The food industry has used bacteriocins as preservatives in food products such as fermented dairy products and packaged meats. Biotechnology has also become interested in various bacteriocins as alternatives to antibiotics. This new medical role of bacteriocins may become increasingly important in treating diseases caused by pathogens that have developed resistance to most of today's antibiotics.

THE ROLE OF BACTERIOCINS

Bacteria communicate with each other in their environment through the substances they excrete. Bacteriocins and antibiotics participate in this communication by giving bacteria the ability to control the types and amounts of other microorganisms in their vicinity. Bacteriocins also offer an advantage to their producers in habitats where nutrients are scarce. For example, the normal bacteria of the skin consist of a variety of species that have adapted to conditions that may be very dry, moist, oily, or poorly aerated. The species that occupy a niche in the skin's microbial community are often similar and in direct competition with each other for water, nitrogen, salts, and carbon. Bacteria must develop ways to outcompete similar bacteria that occupy the same niche and fight for the same space. Bacteriocins represent one way that bacteria gain advantage over their closely related competitors.

Bacteriocins do not kill the cells that produce them. To protect against self-destruction, producers make other proteins to counteract the effect of their own bacteriocins. Genes that control the synthesis of bacteriocin are said to be *synthetic genes,* and genes that control a protective protein are called *immunity genes.* If a bacteriocin producer were to lose its immunity genes, it would fall victim to the very bacteriocin it makes for killing other cells.

BACTERIOCIN PRODUCTION AND ACTIVITY

Bacteriocins are either cidal or static; they kill or inhibit growth, respectively. Six different modes of action have been uncovered for bacteriocins of either cidal or static effect. These actions take place in the cell wall, cell membrane, or chromosome or on ribosomes, summarized as follows:

1. formation of pores in the cytoplasmic membrane, causing increased permeability and disrupted energy metabolism
2. inhibition of DNA gyrase, which controls the normal spiral twisting of DNA
3. destruction of DNA through deoxyribonuclease (DNase) enzyme activity
4. damage to cell walls by inhibition of peptidoglycan synthesis
5. breakdown of peptidoglycan
6. stopping replication by interfering with ribosomes

Bacteriocin genes can be located either on the main bacterial DNA or on plasmids. Some bacteria may share the bacteriocin genes by transferring plasmids from cell to cell in a process called *gene transfer.* The cell regulates bacteriocin synthesis the same way it regulates the production of other cell constituents, by controlling gene expression. Gene expression is the conversion of information contained in specific genes into specific, functioning proteins.

After making a bacteriocin, the cell excretes it by either of two methods: increased membrane permeability (ability to let substances move in or out) to allow the bacteriocin to escape into the sur-

Bacteriocins

Bacteriocin	Producer	Target microorganism
boticins	*Clostridium botulinum*	*Clostridium botulinum* strains
butyricins	*Clostridium butyricum*	*Clostridium butyricum* strains
colicins	*Escherichia coli* and other enteric bacteria	strains in the same family
enterocin P	*Enterococcus faecium*	*Listeria, Clostridium, Staphylococcus aureus*
epidermin	*Staphylococcus epidermidis*	gram-positive bacteria
killer factor	*Saccharomyces cerevisiae*	yeasts
klebicins	*Klebsiella pneumoniae*	*Klebsiella*
lacticin 3147	*Lactobacillus lactis*	*Listeria monocytogenes*
lactocin 27	*Lactobacillus helveticus*	*Lactobacillus helveticus* strains
lactococcin G	*Lactococcus lactis*	*Lactococcus lactis* strains
megacins	*Bacillus megaterium*	*Bacillus megaterium* strains
monocins	*Listeria monocytogenes*	*Listeria*
perfringocins	*Clostridium perfringens*	*Clostridium perfringens*
pesticin I	*Yersinia pestis*	*Y. pseudotuberculosis and E. coli*
pyocin	*Pseudomonas aeruginosa*	*Pseudomonas*
sakacin A	*Lactobacillus saki*	*Lactobacillus*
staphylococcin 1580	*Staphylococcus*	*Staphylococcus aureus*
ulceracin 378	*Corynebacterium diphtheriae*	*Corynebacterium*

roundings or energy-requiring transport through the membrane. In some species, the cells lyse and die when their bacteriocin passes through the permeable cell membrane. These species do not become extinct, however, because neighboring cells of the same species repress their own synthetic genes and so activate immunity. Meanwhile, the bacteriocin kills all the other susceptible bacteria that have contact with it. By this ingenious process, bacteriocin-producing species survive and competitors disappear.

Bacteria readily absorb proteins and peptides because these compounds are good nitrogen sources. Since bacteriocins also consist of protein or peptide, they gain easy entry into susceptible cells. Some bacteriocins bind to receptor sites on the outside of target cells, which then draw the bacteriocin into its cell cytoplasm. Other bacteriocins bind to the phospholipids within cell membranes and thus work their way into the cell. (Phospholipids are fatty compounds containing phosphorus in the form of phosphate [PO_4^{3-}].) In this case, the bacteriocins are termed *cationic* and are attracted to the oppositely charged anionic phospholipids.

IMPORTANT BACTERIOCINS

Gram-positive and gram-negative species produce bacteriocins, and certain archaea also secrete bacteriocinlike proteins. The archaeal proteins behave as bacterial bacteriocins but have different amino acid sequences.

Some of the most studied bacteriocins are the colicins, microcins, lantibiotics, agrocins, and pediocin. The colicins and microcins are both groups produced by Enterobacteriaceae enteric bacteria. *Escherichia coli, Serratia marcescens,* and *Shigella boydii* are the primary producers of colicins. The colicin group contains a varied collection of high-molecular-weight proteins divided into subgroups. The following list names each subgroup according to its mode of action:

- pore formation—colicins A, B, E, I, K, N, U, and Y
- membrane disrupting—colicin V
- inhibitors of protein synthesis—colicin D
- inhibitors of cell wall synthesis—colicin M

Microcins are low-molecular-weight proteins that belong to subgroups based on their mode of action. Type A microcins disrupt metabolic pathways, type B microcins inhibit DNA replication, and type C microcins block protein synthesis.

The lantibiotics nisin and subtilin, agrocins, and pediocin serve commercial uses, mainly as food preservatives. Lantibiotics are circular proteins produced by gram-positive bacteria such as *Bacillus* and *Lactococcus* to attack other gram-positive bacteria. Nisin made by *Lactococcus lactis* is a lantibiotic used in the food industry as a preservative. In low-acid foods that are heated, nisin inhibits *Clostridium* spores from turning into their lethal form in a process called germination. Nisin mainly preserves canned vegetables and dairy products. Subtilin made by *Bacillus subtilis* is also an effective preservative against contamination by gram-positive bacteria and fungi. Pediocin, produced by *Pediococcus,* serves a role in food production as an inhibitor of *Listeria,* a food-borne pathogen found mainly in cheeses and other dairy products. In addition to being preservatives, pediocins enhance the flavor of cheese during their production. They do this by lysing the cheese-producing bacteria. This, in turn, releases enzymes that contribute to the distinct cheese flavor. In wine making, microbiologists have inserted the pediocin gene into the yeast *Saccharomyces cerevisiae. Saccharomyces* is the normal yeast that ferments fruit juice into wine. With the bacteriocin gene, it also inhibits the growth of contaminants at the same time it carries out its fermentation.

In 2009, a team of U.S. Department of Agriculture scientists reported finding a new bacteriocin in soft cheeses that may be produced in the cheese by bacteria of the digestive tract called *enterococci.* The researchers led by John A. Renye reported, "In addition to their role in flavor development, enterococci are also considered desirable due to their ability to inhibit the growth of several food-borne pathogens including: *Staphylococcus* species, *Clostridium* species, *Bacillus* species, and *Listeria* species." Renye observed that the inhibition "stems from their [enterococci's] production of bacteriocins." This bacteriocin appears to work in much the same way as pediocin.

Agrocins produced by *Agrobacterium* are non-protein agents used against plant pathogens. For instance, growers apply agrocin 84 to stone-fruit trees and vines to control crown gall disease. Agrocin 84 also kills a related species, *A. tumefaciens.* Before microbiologists discovered this bacteriocin, *A. tumefaciens* caused severe economic losses in the fruit industries of Australia, the United States, and Europe.

Bacteriocins familiar to industrial microbiologists are summarized in the following table. This is a relatively short list of the enormous number of bacteriocins found in nature. Probably many more have yet to be discovered.

Industrial microbiologists continue looking for unique uses for bacteriocins as preservatives. Physicians have not adapted bacteriocins as readily for treating infectious diseases or as substitutes for antibiotics. Many microbiologists have the same concerns about bacteriocins as they do about antibiotics: the development of resistant microorganisms. They have already shown in laboratory experiments that various bacteria can develop immunity to bacteriocins made by other types of bacteria. Bacteriocin resistance is similar to antibiotic resistance. Resistant bacteria combat bacteriocins by either excreting compounds that destroy the bacteriocin before it enters the cell or using pumps that expel the bacteriocin as soon as it passes through the cell wall.

Bacteriocins have received less attention in bacteriology than antibiotics, yet these compounds hold much the same promise as antibiotics held when they were introduced a half-century ago. Bacteriocins may serve as alternatives to antibiotics and to chemical food preservatives, especially when resistant bacteria have been detected. But an increased use of bacteriocins might possibly generate resistance to these compounds in the same way antibiotic resistance developed.

See also ANTIBIOTIC; FOOD MICROBIOLOGY; GENE TRANSFER; PRESERVATION.

Further Reading

Renye, John A., George A. Somkuti, Moushoumi Paul, and Diane L. Van Hekken. "Characterization of Antilisterial Bacteriocins Produced by *Enterococcus faecium* and *Enterococcus durans* Isolates from Hispanic-style Cheeses." *Journal of Industrial Microbiology and Biotechnology* 36 (2009): 261–268.

Riley, Margaret A., and John E. Wertz. "Bacteriocins: Evolution, Ecology, and Application." *Annual Review of Microbiology* 56 (2002): 117–137.

bacteriophage A bacteriophage, or simply *phage,* is any virus that infects bacteria. Any virus name that ends with *-phage* is a bacteriophage. For example, phages that infect cyanobacteria are called cyanophages.

Phages and all other viruses exist as obligate parasites, meaning they must infect a host cell for their survival. As parasites, the viral uses that cause infection usually harm the host. Bacteriophages differ from viruses that infect animal or plant cells because a phage must have a mechanism that allows it to traverse bacteria's unique peptidoglycan cell wall.

Microbiologists, in the late 1800s, suspected that some sort of tiny life-form attacked bacteria. But these things were invisible in microscopes and passed through filters that caught bacteria. Scientists in laboratories were forced to rely only on their intuition because they had yet to find solid evidence of these mysterious particles. In 1915, the British bacteriologist Frederick Twort (1877–1950) isolated particles from a culture of bacteria that he suspected were not bacteria.

The American Society for Microbiology historian Barnard Dixon wrote, in 2001, "Frederick Twort was a brilliant but eccentric man, who for much of his career, engaged in splenetic and often unreasonable conflicts with Britain's Medical Research Council." Twort's discovery of phages, in 1915, came about by growing sheets of bacteria called *lawns* on agar plates in his laboratory. After incubating the cultures, he occasionally noticed small clear spots in the lawn where bacteria had not grown. In one simple experiment, Twort had accomplished two things: He had discovered bacteriophages, and he had invented a laboratory technique that remains important in virus studies today. But because of Twort's isolation within the scientific community, his observations were almost unnoticed. Two years later, the French Canadian microbiologist Felix d'Herelle (1873–1949), at the Pasteur Institute in Paris, made the connection between the clear zones in bacterial lawns, called plaques, and viruses. D'Herelle wrote, "From the feces of several patients convalescing from infection with the dysentery bacillus, as well as from the urine of another patient, I have isolated an invisible microbe endowed with an antagonistic property against the bacillus of Shiga [a cause of dysentery]." He continued, "The antagonistic microbe can never be cultivated in media in the absence of the dysentery bacillus. . . . This indicates that the anti-dysentery microbe is an obligate bacteriophage." This statement is thought to be the first time the term *bacteriophage* had ever been used.

At the beginning of the 1920s, physicians pursued the idea of developing phage-based therapies for bacterial diseases. In 1921, bacteriophages had been put to use for the first time to treat a skin disease caused by staphylococci. The technique became known as *phage therapy.* Researchers in the Soviet Union and Eastern Europe continued working on phage therapy, but scientists in the United States became more focused on other types of treatments. Some doctors tried phages in their patients, but when antibiotics came into use in the 1940s, phage therapy all but disappeared in the United States. As antibiotic resistance has grown into a major worldwide health concern today, phage therapy might again gain attention.

For phage therapy to work effectively, scientists will need to overcome the immune response that humans and animals produce when a phage enters the bloodstream. A healthy immune system quickly destroys bacteriophages when they are injected into the bloodstream to fight disease.

CHARACTERISTICS OF BACTERIOPHAGES

Viruses have evolved into the ultimate parasite; their bodies have been stripped down to little more than genetic material wrapped inside a protein coat. Viruses lack all other functions for existence as a living thing. Viruses are said to be *metabolically inert,* meaning they are not truly living because they cannot reproduce on their own. Bacteriophages reproduce by first infecting a bacterial cell, and then taking over a portion of the cell's replication system that makes more phage particles.

Bacteriophages occur in any habitat where their host cells live. Their concentration is enormous (up to 10 million per milliliter) in the ocean, but their role there has never been fully explained. Electron microscopy shows that seawater serves as a bounty of bacteriophages for study, but scientists have found it very difficult to keep the phages active in a laboratory. Virologists—scientists specializing in viruses—have conducted most of their studies on bacteriophages that have been more cooperative in laboratory experiments. The main phages that they study infect bacteria such as *Escherichia coli, Bacillus,* and *Pseudomonas.*

Bacteriophages are classified by two features: their morphology, which is their physical shape, and the type of nucleic acid they contain. Bacteriophages take in a variety of shapes that are visible only by electron microscopy. These shapes can be quite distinctive and so act as an identification aid. Nucleic acid analysis provides a more accurate identification system. Viruses belong to either of two main groups: those that contain deoxyribonucleic acid (DNA) or those that contain ribonucleic acid (RNA).

Bacteriophages contain one of four forms of nucleic acid: double-strand DNA (dsDNA), single-strand DNA (ssDNA), double-strand RNA (dsRNA), or single-strand RNA (ssRNA). DNA-containing bacteriophages are thought to be more common and are called simply *DNA phages.* The DNA phages may be further distinguished by the fact that some contain DNA in a linear structure and others carry circular DNA.

Bacteriophage morphology means the shape of the virion, the main body of a virus. Some have a head, called a *capsid,* atop a short tail portion. Capsids are often polyhedral (made up of many surfaces), and when they attach tail-first to the outside of bacteria, they resemble a lunar landing craft. Other bacteriophages are shaped like long thin string beans. Bacteriophages tend to be symmetrical and 23–32 nm in diameter. Several types of bacteriophages have an outer covering called a *lipid envelope* made of fatlike compounds.

Bacteriophages infect bacteria by injecting their nucleic acid into a bacterial cell. Bacteriophages consisting of a capsid and a tail require both pieces in order to do this. Tailless bacteriophages, however, have their own means of infecting bacteria, similar to virus infection of plants and animals.

BACTERIOPHAGE INFECTION

Bacteriophages infect cells, reproduce inside them, and then burst out of each infected cell in a process called *lysis.* Once they burst free, they find other bacteria to infect. This entire process from infection to the next infection is called a *lytic cycle.* The T2 phage that infects *E. coli* releases about 100 new bacteriophages when a single *E. coli* cell lyses.

Bacteriophages infect only specific bacteria, usually no more than one or two species. They follow the same steps in infecting bacteria as other viruses use in plants and animals; the five steps are landing, attachment, tail contraction, penetration, and DNA (or RNA) injection. Bacteriophages are used as models in virus research, and a large portion of the information on all virus infections has been obtained from the bacteriophage lytic cycle. Details of the five steps of the lytic cycle are summarized in the following. DNA phages and RNA phages both carry out this sequence.

- landing—Fibrous arms extend from the virion and connect with specific sites on the outside of bacteria, such as cell wall components, flagella, or pili.
- attachment or adsorption—The tail binds with the receptor site.
- tail contraction—The tail reconfigures so that an inner tube binds to the receptor.
- penetration—Phage enzymes help the tube burrow through the cell wall and make a pore through the cell membrane.
- injection—The phage injects its DNA or RNA into the cell's cytoplasm.

Bacteriophages lacking a tail connect with the bacterial cell using alternate structures. For instance, certain sites on the outside of the capsid fit with complementary structures on the cell surface. After the capsid connects with the cell, the phage injects its DNA or RNA into the cytoplasm.

Once bacteriophage DNA is inside the cell cytoplasm, the host's own DNA replication and protein synthesis stop. The virus also controls the cell's RNA and takes charge of its enzymes. From that point on, the cell becomes a manufacturing plant for making new virus particles. The usual time span from injection to production of phage DNA is about five minutes.

Viral genes, now part of the bacterial DNA, contain instructions on how the cell is to build new viral proteins and lipids. When the capsid is almost complete, the new viral DNA is inserted inside. This is called DNA packaging. About 15 minutes after a phage infects a cell, DNA packaging is complete and the 100 or so new bacteriophages burst forth.

The bacteriophages that lyse and, therefore, destroy their own host cells, are forced to find new cells to infect. These are termed *virulent bacteriophages* or *lytic phages.* It would seem to be a great advantage if the bacteriophage were to let its host cell live. In this way, the cell could produce many batches of bacteriophages, while the phage always has an available host. This relationship, in fact, does exist between some phages and their hosts in a process called *lysogeny.* In this relationship, the bacteriophages are called *lysogenic phages.*

LYSOGENY

Lysogeny is a situation in which the bacteriophage does not take control of the bacterial cell's reproduction. Instead, it inserts its genes into bacterial DNA but allows the bacteria to live. With each cell division, the bacteria produce more viral DNA, which contains all the information needed for constructing new bacteriophages. This collection of viral genes hidden within the bacterial DNA is called a *prophage.* The prophage is often thought of as a latent form of the virus, even though the virion has not yet been built. After multiple cell divisions, the prophage gives instructions for virion production in the induction step. Finally, the prophage genes instruct the cell to lyse and new bacteriophages burst free by the hundreds.

Bacterial cells taken over by a phage in this process are called *lysogenic cells.* Some bacteriophages force lysogenic cells to take on new characteristics in a process known as *bacteriophage conversion.* A conversion might involve a change in the bacterial cell wall or modification of cellular enzymes. Conversion serves a useful purpose for lysogenic phages because the altered cell becomes impossible for other bacteriophages to infect. The prophages hidden inside the cell, therefore, ensure their own survival. Lysogeny is thought to have played a role in the way viruses have evolved along with bacteria and with higher plants and animals.

EXAMPLES OF BACTERIOPHAGES

Bacteriologists and virologists conduct phage research to learn more about DNA replication, mutations, gene expression, and the manner in which diseases progress. Some of the common bacteriophages used in laboratories have unique characteristics also useful for specific studies. These are shown in the table on below.

PHAGE THERAPY

Phage therapy is the use of bacteriophages to stop bacterial infection. Its main advantage resides in its effectiveness against bacteria that have already become resistant to antibiotics. Bacteriophages also provide a very specific defense against pathogens, unlike broad-spectrum antibiotics that kill many bacteria, even harmless species.

Microbiologist d'Herelle first demonstrated the medical worth of bacteriophages in treating soldiers suffering from dysentery during World War I. He went on to use phage therapy on patients who had cholera, typhoid fever, bubonic plague, and other infections. Today phage therapy has limited use in the United States but is prevalent in Poland, Russia, and the Republic of Georgia. These countries have used phage therapy successfully in treating burn patients against infections caused by *Pseudomonas aeruginosa,* a prevalent contaminant of burn injuries. Burn treatment involves the use of a liquid mixture containing a variety of bacteriophages. Doctors spray the mixture onto open burn wounds, where the phages attack *Pseudomonas* at the injured site as well as bacteria on the skin that could contaminate the site. Veterinarians also use bacteriophages to cure diarrhea in calves and infections in livestock and poultry.

Specific bacteriophages known to attack particular pathogens may also be useful in diagnosing disease. In this method, a microbiologist grows lawns of unidentified bacteria isolated from a patient with an undiagnosed illness. After adding specific phages to the lawns, only the lawn containing bacteria susceptible to a known phage will contain plaques. In an example, a microbiologist selects four phages, each one directed against streptococci, staphylococci, micrococci, or enterococci. After four lawns have been incubated, each seeded with a different phage, only the one with antistaphylococcus phage contains plaques. In this way, a clinical microbiologist pinpoints the pathogen as a member of the genus *Staphylococcus.*

Bacteriophage science has made steady progress in varied areas of microbiology: studies of virus infection, the mechanisms of DNA replication, disease treatment, and disease diagnosis. But the full potential of phages in microbiology and medicine may have yet to be discovered.

See also CLINICAL ISOLATE; PLAQUE; VIRUS.

Bacteriophages and Their Hosts

Bacteriophage	Host Bacteria	Characteristics
7-7-1	*Rhizobium*	attaches only to flagella
λ (Lambda)	*Escherichia coli*	widely used in studies on the lytic cycle
Mu	various enteric species	causes mutations in host DNA
CTX Φ	*Vibrio cholerae*	carries gene for the cholera toxin
MV-L2	*Acholeplasma*	new virions exit cell membrane and leave the cell alive
N4	*Escherichia coli*	carries gene for rifampin antibiotic resistance
Φ29	*Bacillus subtilis*	studied for its mechanism of virion building
Φ6	*Pseudomonas*	only known phage with dsRNA
ΦX174	various enteric species	inside cell, the cell covers it with a protective protein
T4	*Escherichia coli* and other enteric species	study model for phage structure and genome

Further Reading

D'Herelle, Felix. "Sur un microbe invisible antagoniste des bacilles dysentériques" (An Invisible Microbe That Is Antagonistic to the Dysentery Bacillus). *Comptes rendus Acad. Sciences* 165 (1917): 373–375. In *Milestones in Microbiology,* translated and edited by Thomas Brock. Washington, D.C.: American Society for Microbiology Press, 1961.

Dixon, Bernard. "Progeny of the Phage School." *ASM News,* September 2001. Available online. URL: http://newsarchive.asm.org/sep01/animalcule.asp. Accessed March 12, 2009.

Kutter, Elizabeth, and Alexander Sulakvelidze. *Bacteriophages: Biology and Applications.* Boca Raton, Fla.: CRC Press, 2005.

Phage Forum. Available online. URL: www.phages.org. Accessed March 16, 2009.

Prescott, Lansing M., John P. Harley, and Donald A. Klein. "Viruses: The Bacteriophages." In *Microbiology,* 6th ed. New York: McGraw-Hill, 2005.

binary fission Binary fission is an asexual replication method in which one cell divides to produce two identical, or almost identical, daughter cells. Bacteria rely on this method of reproduction; many protozoa and algae can use either asexual binary fission or sexual reproduction.

The binary fission process replicates the genetic information within each cell to give identical copies to both daughter cells. This genetic material is deoxyribonucleic acid (DNA), which replicates to form two identical DNA copies. Once the cell has produced the two DNA copies, it undergoes a series of steps that divide the cell into two halves so that each owns one of the copies.

Two newly forming cells split apart in a manner that protects their cellular contents and at the same time gives the two new cells a shape identical to that of the parent cell. As the parent cell splits into two new cells, new cell wall material surrounds each daughter cell. In bacteria, for example, *Staphylococcus* cells always divide to form round cells; *Bacillus* always make straight rods. Some bacterial species form daughter cells that, with each division, remain attached to one another in characteristic formations. That is why *Staphylococcus* cells tend to form grape-like clusters when they replicate and *Streptococcus* cells line up into long chains. The table describes the steps in binary fission.

TYPES OF BINARY FISSION

Binary fission may vary, depending on the type of cell, prokaryote or eukaryote; the shape of the cell; or other factors. Bacterial cell shape gives rise to three different types of fission: transverse, symmetrical, or asymmetrical. Rod-shaped and oval bacteria divide along a transverse plane, meaning the cell splits in half along its long axis, resulting in transverse binary fission. Round bacteria, cocci, often undergo symmetrical binary fission that produces two daughter cells equal in size. Asymmetrical binary fission results in one daughter larger than the other. Bacteria divide according to the type of binary fission characteristic of their species. For instance, *Bacillus* divides symmetrically, while *Caulobacter* divides asymmetrically.

Symmetric bacteria tend to divide in half at the midpoint along the long axis of rods and at the point of widest diameter in cocci. Cocci then follow one of two additional growth patterns: They either form a new cell perpendicular to the last cell to form clumps or form new cells parallel to the previous division to form chains.

Protozoa have unique types of binary fission. Protozoa covered by small hairs called *cilia* (ciliated protozoa) divide by homothetogenic binary fission. This means the daughter cells do not mirror each other. By contrast, protozoa containing flagella (flagellated protozoa) use symmetrogenic binary fission, meaning the daughter cells are mirror images. In either case, oblong protozoa undergo transverse binary fission.

Bacteria grow not by expanding into a larger size, but rather by dividing again and again until their numbers have grown to the billions or more. The rate at which cells double in number with each division is called the *doubling rate,* or *generation time.* The numbers of bacterial cells can grow to numbers on the order of 10^{10} within a few hours. For this reason, microbiologists measure bacterial numbers and growth on a logarithmic scale. In logarithmic growth, each generation contains double the number of cells in the preceding generation as a result of binary fission.

EUKARYOTES COMPARED WITH PROKARYOTES

Protozoa, algae, and some yeasts undergo more complex binary fissions than bacteria. Eukaryotic binary fission includes the following three main differences from the bacterial process. First, membrane-bound organelles in eukaryotic cells must divide so that each daughter cell receives a full complement. Bacteria do not confront this challenge because they lack membrane-bound organelles. Second, eukaryotic cells perform mitosis inside the nucleus. Mitosis is a process of chromosome duplication that precedes cell division. Once the chromosome has duplicated by mitosis, two daughter eukaryotic cells form. Bacteria do not have an organized nucleus, so binary fission requires only chromosome duplication. Third, as the parent eukaryotic cell turns into two new cells, it must rebuild a part

Binary Fission

Step	Event	Description
1	resting	nonreplicating parent cell controls maintenance activities
2	chromosome replication	chromosome begins replicating at a site called the *origin of replication,* producing two origins
3	replication	entire chromosome replicates, and the two origins migrate to opposite ends of the cell
4	division	cell wall of the parent cell begins pinching in the middle
5	fission	center cell wall forms, creating two new daughter cells

of the cell membrane. Prokaryotes also rebuild their membranes, and they also must build new sections of the cell wall for each daughter cell.

When each half of a dividing cell has its copy of the chromosome, a series of steps takes place to complete the separation and make a completely new pair of daughter cells. In bacteria, a cross-wall within the cell wall builds between the two daughter cells being formed. The cytoplasm inside prokaryotic and eukaryotic cells also divides more or less in half during binary fission in a process called *cytokinesis.* Eukaryotes contain two main proteins, actin and tubulin, that during cytokinesis aid the migration of cell materials toward opposite sides of a cell about to divide. Prokaryotes also contain two proteins, FtsZ and FtsA, that act in a similar manner in cooperation with other proteins (FtsN, FtsQ, DivIB, and ZipA). The microbiologist William Margolin wrote in *Microbe* magazine, in 2008, "Although microtubules [made from tubulin] and actin filaments in eukaryotes generally do not interact directly, FtsZ and FtsA do so, together ensuring the integrity of the cell division machine." Most bacteria, such as *Escherichia coli,* rely on these two proteins, but some genera (*Caulobacter,* for example) use a different protein system.

As cellular components migrate to opposite ends of the cell in bacteria, FtsZ creates a protein ring, called a *septal ring* or a *Z ring.* This ring binds to the cell membrane and marks the plane where fission will take place. Margolin described the final process in bacterial cell division as follows "Once this machine fully assembles, the ring contracts and the cell division septum, cleavage furrow, or a combination, depending on the species, forms behind it. Although the mechanism for this process is not known, it might depend on FtsZ pulling while it is tethered to the membrane, the growing division septum in the periplasm [cell contents] pushing on the collapsing ring, or a combination of those two forces. Because some bacteria divide without forming an obvious septum or cell wall, the former mechanism is potentially more universal." But bacteria do rebuild the cell wall during division, so that the daughter cells retain their defense against harmful substances in the environment.

New bacterial and protozoal cells and most algae reach their normal size and shape very quickly. Diatoms (a type of algae) are an exception to this rule, because each binary fission results in daughter cells smaller by about half than the parent cell. Each new generation of diatoms gets smaller and smaller until they stop dividing and form a resting diatom called an *auxospore.* Eventually, the auxospores enter a process called *germination.* The auxospore cytoplasm expands as it takes in nutrients and water. This continues until the diatom regains normal size. The diatom then builds an entirely new cell wall. Diatoms differ in this way from bacteria, because reproducing bacteria use a portion of the parent's cell wall for the cell walls of the two new daughter cells. Diatoms must expend much more energy than bacteria to construct an entire cell wall.

All bacteria reproduce by binary fission, but not all eukaryotes use it. Algae that have the ability to reproduce asexually use binary fission. Examples are green algae (Chlorophyta), golden-brown algae (Chrysophyta), diatoms, and dinoflagellates. Dinoflagellates possess thick protective outer plates called *theca.* As dinoflagellates divide, the daughter cells each share the theca. Soon after dividing, each new cell must make additional thecal plates to complete its protective coat. Most yeasts, which are eukaryotic fungi, use budding for reproduction, but some species, such as *Shizosaccharomyces,* also use binary fission. As mentioned, flagellated protozoa mainly use binary fission for their reproduction. Protozoa of the classes Zoomastigophorea and Phytomastigophorea use binary fission.

Biologists study the processes of binary fission to learn more about how healthy and diseased cells divide in humans and other organisms. In addition, molecular biologists rely on binary fission within cell cultures to produce clones, which are new cells that all possess the exact same genetic makeup as the original parent cell. Understanding binary fission is key to the advances in medicine, drug therapies, molecular biology, and biotechnology.

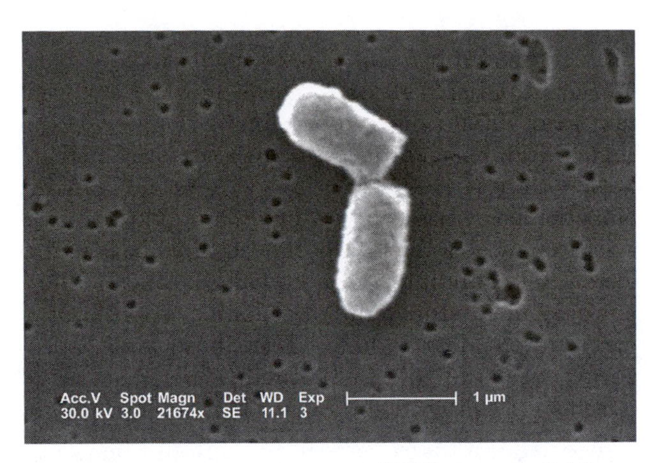

This scanning electron micrograph (SEM) shows a dividing *E. coli* cell; magnification is 21,674 ×. *(CDC, Public Health Image Library)*

See also GENERATION TIME; LOGARITHMIC GROWTH; ORGANELLE; RECOMBINANT DNA TECHNOLOGY.

Further Reading

Campbell, Neil A., and Jane B. Reece. *Biology,* 7th ed. San Francisco: Benjamin Cummings, 2005.

Estrella Mountain Community College. "Online Biology Book." Available online. URL: www.emc.maricopa.edu/faculty/farabee/BIOBK/BioBookTOC.html. Accessed March 16, 2009.

Margolin, William. "What Does It Take to Divide a Bacterial Cell?" *Microbe,* July 2008. Available online. URL: www.asm.org/ASM/files/ccLibraryFiles/Filename/000000004073/znw00708000329.pdf. Accessed March 22, 2009.

biocide A biocide is any substance that kills a living thing. In microbiology, biocides may also be referred to as *microbicides.* Microbicides encompass all chemicals that kill microorganisms on inanimate objects or in water. Antiseptics and antibiotics also kill microorganisms, but these products are not considered biocides because they act on living tissue rather than inanimate objects. All antimicrobial chemicals have also gone by the generic name *germicides.*

Biocides can be named and categorized according to the type of microorganisms they kill, as shown in the table. Individual types of biocides end in the suffix *-cide* to indicate they kill microorganisms. A chemical that inhibits microorganisms but does not necessarily kill them ends in the suffix *-stat.* The table lists widely used microbicides and microbistats.

Very few chemicals kill the endospores made by *Bacillus* and *Clostridium* bacteria. A chemical that kills these hardy cellular forms is said to *sterilize* meaning it eliminates all microbial life. Sterilization is defined as the process of killing every type of microorganism, and chemical sterilants are equivalent to sporicides.

HISTORY OF BIOCIDE USE

Chemical biocides have been studied in microbiology only in the past 200 years. Interest in these chemicals probably grew as the chemical industry itself grew during this period. But people have used chemicals in various forms to kill microorganisms since early in history. The Greek poet Homer may have been the earliest to observe (and to document in writing) housewives' use of "disinfectant sulfurs" to rid the home of decay. About 300 B.C.E., Alexander the Great ordered his men to pour oil on newly built bridges as a wood preservative. Roman societies also experimented with chemicals as wood preservatives, and, in the public baths, they used various herbal oils to cleanse the bath waters.

Through the centuries, individual societies invented effective methods for eliminating the malodors wafting from spoiled items. Wine (alcohols) and vinegar (acids) served this role for generations. Medieval Europe's great plagues began in the 1300s C.E. and continued in waves through the 1600s, and each epidemic forced town officials to confront the mysterious phenomenon of germs. In the 1400s, the Magistry of Health in Venice mandated that all ships docking in port were to be fumigated; they probably used sulfur compounds for this task. Citizens spread sulfur dioxides, vinegars, and even perfumes around homes and buildings to combat the pestilence, though microorganisms had yet to be discovered or understood. In a sense, people living through the plagues fought a blind battle against germs.

The use of chemical biocides came about through the work of three Scottish physicians. The physician John Pringle (1707–82) published his ideas on reducing battlefield infection. His paper "Experiments on Septic and Antiseptic Substances," published in 1750, explored the potential for chemical biocides. During the same period, the navy surgeon James Lind (1716–94) studied ways to reduce the spread of disease aboard ships. His article "Two Papers on Fevers and Infection," published in 1763, proposed a number of preventive measures for infection, including mercury compounds. The surgeon Joseph Lister (1827–1912) pursued Pringle's and Lind's theories and built upon the discoveries of his contemporary in France, the microbiologist Louis Pasteur (1822–95). Lister tried different chemical biocides as disinfectants for preparing the room and instruments before a surgery. Many physicians hesitated to use chemicals on patients, assuming these agents would cause more harm than good.

The first breakthrough in understanding the value of chemical biocides occurred not from human medicine but from botany. In 1807, in France Isaac-Benedict Prevost (1755–1819) showed that copper salts killed a fungus that caused the disease bunt in wheat. He demonstrated the value of a chemical biocide by soaking wheat seeds in copper sulfate solution before planting. The treated seeds produced crops free of fungal disease. Prevost's peers continued to support spontaneous generation, however, and his work met with ridicule. This conservative thinking resulted in tragic consequences as potato blights swept Ireland, in 1845 and 1846, causing thousands to starve. Prevost's method may well have lessened the severity of the epidemics, but no one put it to use to stop the disaster in Ireland. Innovative gardeners first began taking advantage of sulfur compounds for mildews growing on fruiting trees and vines. Less than a decade after Ireland's potato famine, gardening articles touted the use of sulfur chemicals for killing garden pests.

Since the early 1800s, chlorine had already been a treatment for all sorts of bad odors from decaying matter. In the first quarter-century, French morgues applied calcium hypochlorite to floors, walls, and tables, and it soon became an effective antiodor agent in jails, sewers, ships, and stables. Some surgeons even tried weak hypochlorite solutions on wound dressings. By 1827, officials in charge of England's drinking water supply added chlorine to municipal water tanks to rid drinking water of pathogens.

In the 1900s, iodine, mercury compounds, alcohols, and hydrogen peroxide became accepted chemicals for various applications in infection control, odor fighting, and disinfection. In 1935, chemists developed in the laboratory a new class of biocide. These quaternary ammonium compounds possessed antimicrobial activity against many bacteria and seemed safe to use in hospitals and homes. The quaternary ammonium compounds, or quats, now make up one of at least 10 different classes of chemicals in use today for industrial cleaning, hospital and medical clinic disinfection, and household use.

TYPES OF BIOCIDES

Biocides may be grouped by the type of microorganisms they kill, as shown in the table. Biocides may also belong to categories based on their intended use

Types of Biocides

Type	Target Microorganism	Effect
	MICROBIOCIDES	
bactericide	all bacteria except spores	kills
fungicide	all fungi, including yeast	kills
virucide	viruses	kills
algicide	algae	kills
germicide	all pathogens except bacterial spores	kills/inhibits
tuberculocide	tuberculosis pathogens	kills
sporicide	all microorganisms	kills
sterilant	same as sporicide	kills
disinfectant	bacteria and fungi and/or viruses	kills
sanitizer	all bacteria except spores	kills a percentage
preservative	all microorganisms	kills/inhibits
	MICROBIOSTATS	
bacteriostat	all bacteria except spores	inhibits
fungistat	all fungi, including yeast	inhibits
mildewstat	all mildews (a type of fungus)	inhibits

Chemical Biocide Categories and Applications

Biocide Category	Example Compounds	Uses
alcohols	ethyl alcohol, isopropyl alcohol	decontamination of medical surfaces and instruments
aldehydes	formaldehyde, formalin, glutaraldehyde	medical devices
antimicrobial polymers	biguanide polymers, ionenebromides	microbial control in freshwater systems and swimming pools, preservatives
copper and zinc compounds	copper 8-quinolinolate, copper arsenates, zinc naphthenate	fungicides, wood preservatives, fabric treatments
halogens	sodium hypochlorite (household bleach), calcium hypochlorite, chloramine, dichloramine, chlorine dioxide, potassium iodide, iodophors	disinfection of hospital and household surfaces, water disinfection, emergency water treatment
nitrogen compounds	formaldehyde-releasing compounds, nitriles, anilides, isothioazalones	surface, leather, and fabric protectants, preservatives
organotin compounds	tin oxides, bis (tributyltin) oxide (TBTO), tributyltin fluoride (TBTF)	wood and paint preservatives, antifouling agents for boats
peroxygen compounds	hydrogen peroxide, peracetic acid	decontamination of surfaces, swimming pool disinfection, odor control, food production equipment disinfection
phenolic compounds	chlorophenol, bisphenols, nitrophenols, cresols, p-hydroxybenzoic acid	disinfection of hard surfaces, institutional cleaning products
quarternary ammonium compounds (quats)	benzalkonium chloride, benzethonium chloride, cetylpyridinium chloride, ammonium saccharinates	disinfection of hard surfaces, institutional cleaning products
surface-active compounds	sulfonic acids, dodecyl glycines	dairy/food industry sanitizers, disinfection of medical setting walls, floors, and instruments

or their chemical composition. The main intended uses of biocides are the following: disinfectants and sanitizers, food and cosmetic preservatives, paint preservatives, wood preservatives, antifouling agents, and nonfood biocides (for swimming pools, algae treatment of ponds, and water treatment). Intended use and chemical composition are often related as shown by the table. These relationships derive from the safety of the chemical in the setting where it will be used, as well as its effectiveness against particular microorganisms. For example, phenol compounds are strong and caustic chemicals best used for disinfecting floors but not near foods. Chlorine, on the other hand, is a strong disinfectant that dissipates quickly, so it is safe for use near foods and in drinking water.

A small number of biocides, such as mercury, do not fit into any of the categories listed in the table. Mercury's use as a disinfectant may date as far back as that of sulfur. Medical literature of 12th-century Europe first mentioned the use of mercury, but it may have been used much earlier in Egypt and China. Mercury compounds appeared in coatings and paints until the 1970s, when their toxic effects on health and the environment became evident. The U.S. Environmental Protection Agency (EPA) now bans mercury preservatives in paints and pesticides. Physicians have long since stopped using mercury as a biocide or an antiseptic.

The health care profession has similarly phased out aldehydes because of safety concerns. Though they are very effective biocides, formaldehyde and glutaraldehyde produce fumes that irritate the skin and eyes and can be a significant health hazard when they enter the body. They are toxic to living tissues, and formaldehyde may cause cancer. For this reason, aldehydes are now used in only a small, specialized set of applications. Today's medical device industry uses them for disinfecting items such as endoscopes, respiratory instruments, anesthesia equipment, and kidney dialysis machines. Disinfected instruments must therefore be thoroughly rinsed until all the chemical washes off. After this rinsing, the instruments are safe to use for patients.

Formaldehyde has usually been used as a solution called *formalin.* A 100 percent formalin solution contains 40 percent formaldehyde gas dissolved in water. For example, 20 percent formaldehyde in water equals a 50 percent formalin solution. Glutaraldehyde is chemically similar to formaldehyde but is several times stronger for killing spores and other difficult-to-kill microorganisms. Because of the health problems that formaldehyde and glutaraldehyde cause, few people in microbiology or health care rely on these chemicals.

Chlorine

Chlorine is the most effective of all biocides. Chlorine and iodine belong to the halogen biocides; halogens occupy a defined section of the periodic table of elements. The element chlorine is a gas, but chlorine is not found free in nature in this form. Rather, natural chlorine occurs bound to sodium, magnesium, potassium, or calcium as a salt. Sodium chloride (table salt) provides a familiar example of a chlorine-containing salt.

Chlorine acts as a powerful oxidizing agent; that means it creates unstable free radical molecules that are lethal to microorganisms, including bacterial spores. The best-known chlorine biocide is sodium hypochlorite, NaOCl, more commonly known as *household bleach.*

The United States uses chlorine compounds in drinking water, wastewater, and swimming pool disinfection. The EPA requires that drinking water disinfected with chlorine contain no more than 2 parts per million (ppm) of free available chlorine (chlorine not combined with other compounds). Two additional industries that rely on chlorine disinfectants are hospitals and medical clinics, for killing infectious pathogens, and food producers, for controlling contamination in food.

In a 1998 *New York Daily News* article on household germs, the reporter Susan Ferraro wrote, "A bleach solution is a fast, safe and effective means of killing germs, experts say. The recipe: 3/4 cup bleach in a gallon [190 ml per 4.0 l] of water (or three tablespoons in a quart [0.95 l]). The solution will kill 99.9 percent of home germs." No biocide has been proven to be as effective as household bleach in killing microorganisms.

CHARACTERISTICS OF EFFECTIVE BIOCIDES

Biocides must be effective against the microorganisms they are intended to kill. But not every biocide is effective against every type of microorganism. Any biocide should also be effective in killing dangerous microorganisms yet safe for people who become exposed to it. When health care, sanitation, and food production professionals choose a biocide, they select one with as many favorable characteristics as possible related to safety and effectiveness. The following list provides the characteristics of an effective biocide:

- selective toxicity—The biocide kills or injures microbial species but does not harm the per son using it.
- quick-acting—The biocide kills the maximal number of microbial cells within seconds or a few minutes.
- broad-spectrum—The biocide kills a wide variety of microorganisms.
- resistance to interference from organic materials—High concentrations of microbial cells or nonliving organic matter may reduce biocide effectiveness; effective biocides retain their full activity in the presence of organic matter.
- resistance to physical/chemical inactivation—The biocide is not destroyed or neutralized by variations in temperature, pH, salts, or other chemicals.

Few biocides achieve all of the listed characteristics. Sodium hypochlorite may be an effective and quick-acting biocide against all types of microorganisms, but it dissipates quickly and can be inactivated by high pH or organic matter. Users of sodium hypochlorite cannot lower pH to increase effectiveness because at low pH, the solution emits deadly chlorine gas. In summary, even one of the best biocides does not achieve all of the criteria for a perfect biocide.

PRINCIPLES OF BIOCIDAL ACTION

Biocides require optimal conditions in order to be at their most effective in killing dangerous microorganisms. Some formulas work much better as disinfectants when the user first removes dirt and organic matter from a surface to be disinfected. Biocides also require a minimal amount of time to act on a microbial cell and be used at an effective concentration. The time a biocide needs to kill microorganisms is called *contact time*. The table below presents other important factors that influence biocide activity.

BIOPESTICIDES

A biopesticide is an insect-killing biocide derived from a biological source; it is not a manufactured chemical. Many bacteria, fungi, and viruses, or their secretions, can serve as biopesticides because they are known to kill various susceptible insects. The most widely used biopesticide species is *Bacillus thuringiensis*, a bacteria that produces a crystallike compound lethal to insects. DNA from *B. thuringiensis* has been inserted into other bacteria and plants by using genetic engineering. As a result, scientists have created new organisms with the ability to kill insects. Agriculture uses *B. thuringiensis* to ward off pests that attack fruit trees, vegetable plants, field crops, and ornamental plants. A similar species, *B. popilliae*, protects plants against Japanese beetle larvae.

Fungi have been used in a similar fashion to bacteria for producing biopesticides. The following fungal genera have found uses in agriculture:

- *Beauvaria* controls the Colorado potato beetle.
- *Metarhizium* protects sugarcane from frog hopper infestation.
- *Verticillium* controls various insects.
- *Entomophthora* attacks aphids.
- *Coelomyces* controls mosquitoes.

Agricultural specialists and home gardeners often prefer biopesticides to chemical pesticides (another form of biocide). Users of biopesticides prefer them because these substances are natural compounds that do not harm the environment. But many people oppose biopesticides for another reason: They are created by genetic engineering. Opponents of genetic engineering argue that new types of laboratory-invented organ-

Factors Affecting Biocide Activity

Factor	Nature of the Interaction with the Biocide	Effect
concentration of microorganisms	more biocide is consumed by a high concentration of microbial cells than by a low concentration	higher microbial concentrations require longer biocide exposure times (contact time)
types of microorganisms	some species are more resistant to damage from biocides (examples: bacterial spores, *Mycobacterium* bacteria) than other species	biocide exposure time or possibly concentration must be increased to kill certain species
biocide concentration	higher biocide concentrations are more effective than low concentrations against microorganisms, in general; above a certain concentration, further increases have no effect; some biocides work better when diluted (example: 70% alcohol)	biocides are most effective used at the concentration indicated in the manufacturer's directions
exposure time	biocides kill more microorganisms, especially resistant species, when they are allowed to be in contact with the cells for longer than the minimal required contact time	biocides are most effective used for the contact time indicated in the manufacturer's directions
temperature	increased temperature usually increases rate of reactions	many biocides are more effective or work faster at raised temperatures
interfering factors	specific biocide activity may be affected by one or more chemical and physical factors: pH, temperature, organic matter, other chemicals	biocides are most effective when used according to manufacturer's directions

isms might threaten the environment. A staunch opponent of genetic engineering, Jeremy Rifkin, said in a 2001 interview on PBS, "Back in 1983, the United States government approved the release of the first genetically modified organism. In this case, it was a bacteria that prevents frost on food crops. My attorneys immediately went into the federal courts to seek an injunction to halt the experiment. The position I took at the time was that we hadn't really examined any of the potential environmental consequences of introducing genetically modified organisms." Society and individuals have been given two product choices for killing pests. First, they may use chemical pesticides, which are very effective and are not produced by genetic engineering but put long-lasting chemicals into the environment. The second choice is use of biopesticides, which degrade in nature, are not strong chemicals, but require genetic engineering.

Biocides make up an important aspect of microbiology, and they constitute a multibillion dollar industry. Microbiologists in the chemical industry conduct testing to find new and effective chemical biocides that can be safely used to kill pathogens. In biotechnology, scientists also seek nonchemical agents to kill microorganisms. Industries such as medicine, household products, and food processing also require the best biocides available to reduce the chance of contamination. The study of biocides will probably always play a role in microbiology.

See also ANTIMICROBIAL AGENT; *BACILLUS;* DISINFECTION; STERILIZATION.

Further Reading

Blair, J. S. G. "Famous Figures: Sir John Pringle." *Journal of the Royal Army Medical Corps* 152 (2006): 273–275. Available online. URL: http://www.ramcjournal.com/2006/dec06/blair.pdf. Accessed March 30, 2009.

Block, Seymour S. *Disinfection, Sterilization, and Preservation,* 4th ed. Philadelphia: Lea & Febiger, 1991.

Ferraro, Susan. "Invisible Critters on Your Counters Think Your Home Is Clean? News Dispatches a Germ Sleuth to Test the Fridge, the Bathroom, Everything—and the Kitchen Sink." *New York Daily News,* 1 November 1998. Available online. URL: www.nydailynews.com/archives/news/1998/11/01/1998-11-01_invisible_critters_on_your_ c.html. Accessed March 16, 2009.

Hugo, William B., and Allan D. Russell. "Types of Antimicrobial Agents." In *Principles and Practice of Disinfection, Preservation and Sterilization.* Oxford, England: Blackwell Science, 1999.

Knight, Derek J., and Mel Cooke. *The Biocides Business: Regulation, Safety and Applications.* Weinheim, Germany: Wiley-VCH, 2002.

PBS. "Harvest of Fear." Interview with Jeremy Rifkin. Available online. URL: www.pbs.org/wgbh/harvest/interviews/rifkin.html. Accessed March 16, 2009.

Russell, Allan, D. "Antifungal Activity of Biocides." In *Principles and Practice of Disinfection, Preservation and Sterilization.* Oxford, England: Blackwell Science, 1999.

biofilm Biofilm is a mixture of microorganisms living as a community attached to a surface. This three-dimensional layer of diverse microorganisms tends to form on surfaces immersed in a liquid, usually a flowing liquid. Biofilms form on virtually any surface, but they are most common in the following places: water distribution pipes, cooling water distribution lines, toilets, drains, showers, manufacturing equipment, oil drilling equipment, ship and boat hulls, and the rocks, sediment, and plants in streams. In medicine, biofilms raise health concerns by growing on prosthetic devices, contact lenses, teeth, catheters, and feeding tubes. In nature, biofilms are very resistant to harsh conditions; they have been found growing in the acidic steam pools in Yellowstone National Park as well as on glaciers in Antarctica.

Biofilm is a good example of symbiosis, wherein organisms live in close association, either in a beneficial relationship or in a relationship in which one member is harmed. The symbiotic relationship found in biofilm benefits all or most of the microorganisms living in the layer. The film layer protects the microorganisms within it so that they can withstand conditions they might not tolerate outside the biofilm.

Biofilm almost always contains a heterogeneous mixture of any or all of the following types of microorganisms: gram-negative bacteria, gram-positive bacteria, archaeans, fungi, algae, and protozoa. Many nonliving things such as dirt, debris, organic and inorganic compounds, and dead microbial cells also become trapped in thick biofilms.

THE STRUCTURE OF BIOFILM

Cells in biofilm communities have five main advantages not available to planktonic cells, meaning cells that float free in liquids. First, the mixture of biofilm cells and their secretions provide each other protection from strong chemicals and high temperatures. Second, the biofilm layer that adheres to a surface enables the community to stay in place in fast-flowing liquids while planktonic cells wash away. Third, biofilm structure provides a food storage mechanism. Fourth, biofilm efficiently pulls nutrients out of nutrient-scarce liquids, such as drinking water, and makes them available to the community. Fifth, a biofilm community forms a cooperative relationship among its members. In this mutualistic relationship, members share food sources, growth factors, enzyme activities, and protective secretions. Biofilms

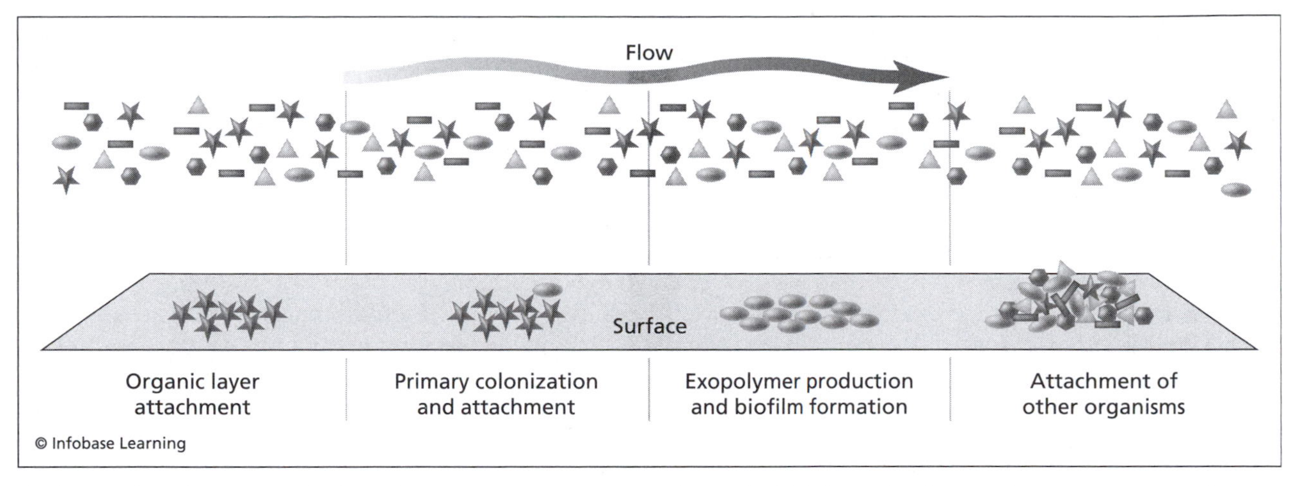

Biofilm begins with a few cells that stick to an inanimate surface. From this initial step, biofilm develops into a community of diverse microorganisms with specialized roles, such as production of polysaccharides that bind the community together. When cell density increases, some cells break away and make new biofilm.

have been described as miniecosystems because of the interrelationships among biofilm species.

As microscopy has become more sophisticated in showing the minute details of the microbial world, the complex of biofilms has been revealed. The biofilm expert J. William Costerton, who invented the term *biofilm,* wrote in his 2007 book *The Biofilm Primer,* "We usually find that well-fed biofilms are unstructured and flat, while less-favored biofilms are highly structured, and (most importantly) biofilms in several natural environments are seen to be composed of tower- and mushroom-shaped microcolonies interspersed between open water channels." These observations led to more detailed studies on how biofilms form a seemingly independent organism composed of many different types of microorganisms.

Biofilm formation starts with the adhesion of organic and inorganic compounds to a surface. Either these compounds diffuse from the liquid, or the liquid's turbulence forces them onto the surface. The accumulation of compounds at the solid-liquid interface is called the conditioning film. A conditioning film provides a higher concentration of nutrients for microorganisms to use as they attach and begin to build up.

In the second step, a few planktonic cells flowing past the conditioning film stick to it. They begin producing a sticky substance, which allows them to remain stationary even though water flows over them. Sometimes the water flow is quite vigorous, but biofilms are known to withstand being washed away even in fast-flowing and agitated waters. This attachment and conversion of a swimming cell to an attached cell take just a few minutes but occurs in two stages. In the reversible attachment step, cells form a weak chemical attraction to the surface. Cells may bump into the surface in this step by chance or by their normal Brownian movement (the random movement of bacteria in water); then van der Waals forces maintain a weak attraction between the cells and the surface. Rapid flows that produce high shear easily dislodge cells in this reversible attachment step. A small number of cells do not rinse away, however. These persistent cells initiate the third step, irreversible attachment, wherein a permanent layer of the cells biologically binds to the surface.

Once the cells attach to the surface, they are no longer planktonic; they are called *sessile* cells. (The word *sessile* is from the Latin for "to sit.") Additional microorganisms attach to the surface to form a monolayer of cells; this monolayer represents the fourth step in biofilm formation. Other cells soon attach and the film thickens. The thickening of biofilm is the fifth step in its formation.

In the final and ongoing step in biofilm formation, the cell numbers increase, and certain bacteria within the colony excrete a gooey exopolymer (a long-chain molecule excreted by a cell) called *extracellular polymeric substances* (EPSs). (Early research in biofilms referred to the EPS as the *glycocalyx.* Glycocalyx is now known to be a more specific polysaccharide capsule made by some but not all bacteria.) EPS contains a complex mixture of large polysaccharides (sugar chains) with a smaller portion (about 15 percent) of proteins. *Pseudomonas aeruginosa* is one of the main producers of EPS, but other species contribute. Only attached cells produce EPS, as if they sense that they belong to a stable community, while free-floating cells do not produce EPS. Microbiologists have shown that EPS also helps the first layer of cells attach to an uncolonized surface.

A mature biofilm is a complex and heterogeneous mixture of alive and dead cells, EPS, nutrients and other compounds pulled from the water, inert substances, and material corroded from the underlying surface. Very thick biofilms contain layers in which microenvironments differ from the outer reaches of the film down to the attachment surface. For example, oxygen concentrations change in the deeper regions of thick biofilms so that aerobic species live near the outer surface (near the flowing liquid), and anaerobic species inhabit deeper layers (near the attachment surface).

Thick, mature biofilms also develop streamers and clusters. Streamers are long filaments that arise from the liquid side of the biofilm and wave in the flowing liquid without detaching from the film—they are like a streamer in the wind. Clusters are packets of cells that break off from the biofilm and follow the downstream flow of liquid. Clusters may be a survival mechanism of sorts for a biofilm that has become very thick and populated by microorganisms, all using the same nutrients arriving with the flow. Biofilm clusters eventually begin another series of steps leading to biofilm formation on a new section of the surface.

The biofilm that stays attached uses factors in addition to EPS to hold the film together and anchored. For example, electrostatic charges between the colonizing bacteria (the first cells to form an attached layer) and the inanimate surface contribute to biofilm development. The surface's composition plays a role in holding biofilms together, too, because its tendency to corrode helps give biofilm a porous surface on which to cling. Biofilms seem to form on almost any type of material: plastics, metal, glass, enamel, ceramic, rubber, wood, and rock. Finally, hydrophobic interactions take place in inner biofilm where little water penetrates. Meanwhile, hydrophilic molecules act to capture nutrients on the biofilm's outer surface.

These stages of biofilm formation are summarized in the table.

Large biofilms develop a structure more complex than a simple sheet of cells that overlie a surface. As they enlarge, biofilms develop unevenly, anywhere from 10 micrometers (μm) to 200 μm thick. A typical biofilm contains thin sections and other regions containing mounds of cell and noncell material. Channels develop through regions of the biofilm so that water flows through, around, and over it all at once. The entire structure protects the living members of the biofilm from harm, especially the cells in the deepest part of the biofilm near the attachment surface. Biofilms resist damage from chemical disinfectants and antibiotics at levels that easily kill planktonic microorganisms. As a result, antimicrobial substances must be at concentrations of up to 1,500-fold higher to kill biofilms than would be needed to kill the same microorganisms in planktonic form.

BIOFILM COMMUNITIES

Biologists think of biofilms as unique miniecosystems because the cells making up biofilm live in cooperation with each other. Some species do not readily stick to surfaces themselves but depend on the ability of others to adhere and produce EPS to hold the system together and protect it. The entire biofilm matrix also serves all its members by storing nutrients that it draws from the water. Some microorganisms growing in water pipes would, in fact, not survive on water's limited nutrient supply without being part of biofilm. In addition, some species

Biofilm Development

Stage	Name	Events
1	conditioning film	soluble and insoluble compounds adhere to the surface
2	reversible attachment	cells attract to the surface and form weak attachments to it
3	irreversible attachment	some cells attach to the surface by producing EPS
4	maturation I	cells multiply and a layer builds across the surface
5	maturation II	more cells from the flowing liquid adhere to the established cell layer and the biofilm thickens
6	maintenance	biofilm grows, pieces break off, and new layers form

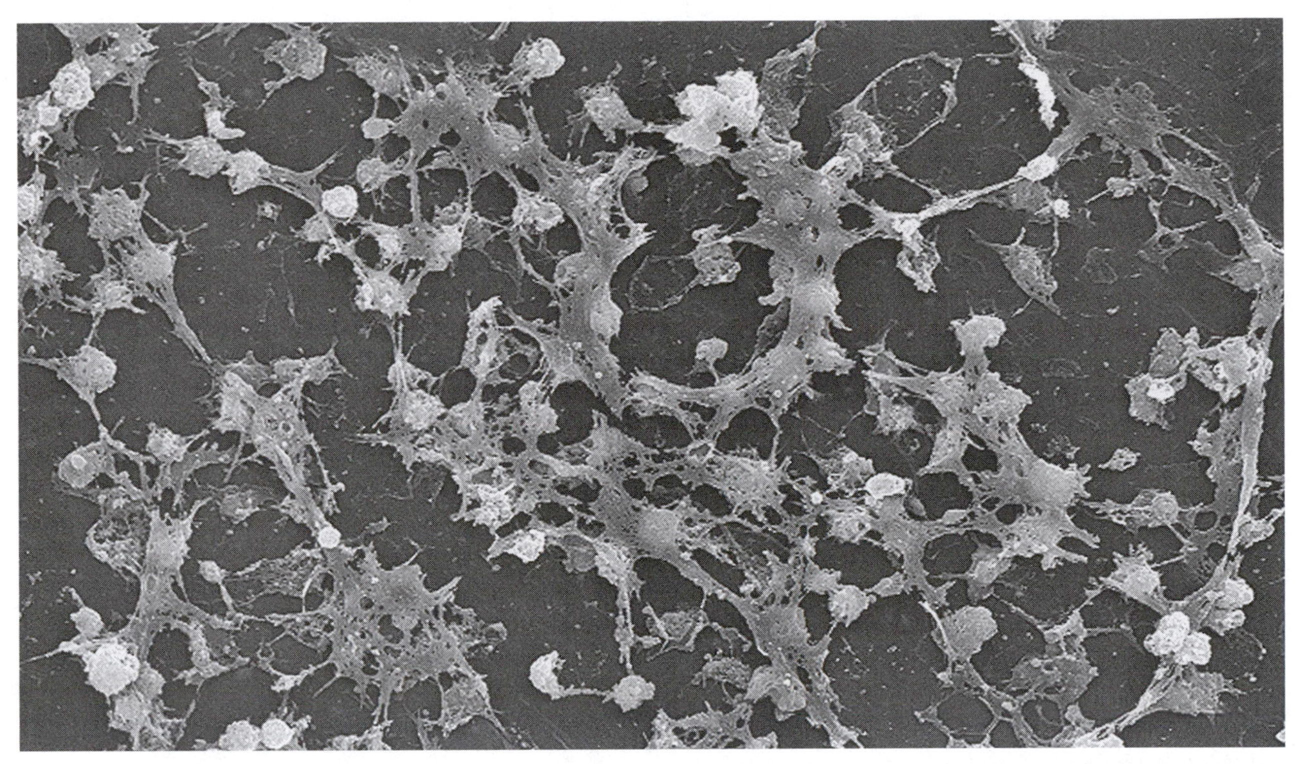

These *Staphylococcus aureus* cocci are growing inside an indwelling catheter and are surrounded by sticky polysaccharide that protects the cells from antibiotics; magnification is 2,363 ×. *(CDC, Public Health Image Library)*

use the end products of other species's metabolism. This serves two purposes: It removes end-product buildup from the area around the producer cells, and it provides a nutrient or energy source for other biofilm members. By removing end-product buildup, species help drive the overall chemical reactions forward.

A major way biofilm members set up cooperation among themselves is through a phenomenon called quorum sensing. Also called autoinduction, quorum sensing is a mechanism by which bacteria monitor their own environment to control the density of their population. Certain bacteria in the biofilm do this by secreting molecules, called *signalers* or *autoinducers,* as they grow. Signal molecules accumulate in the biofilm as its population increases. When the signalers reach a certain concentration, they tell the bacteria that the population has reached its maximal level, or a quorum. A receptor on the cell's surface receives the signal and then starts a process to shut down deoxyribonucleic acid (DNA) replication and cell multiplication.

Quorum sensing prevents populations from becoming too big for the nutrient supply. Pathogens use quorum sensing to help them grow to large numbers, during an infection, for the purpose of fighting off attacks from the body's immune system. Pathogenic biofilms containing *P. aeruginosa* and *Burkholderia cepacia* are thought to use quorum sensing in cooperation to reach a dose likely to hold its own against the body's defenses.

Genetic studies on biofilms reveal the unique abilities that cells develop when they become part of a biofilm. Biofilm species are said to diversify genetically. This means they each acquire traits that serve them and the entire community. These new biofilm traits do not occur in the same species growing as planktonic cells. Some traits microbiologists have recently observed in biofilm species are the following:

- increased ability to disperse and cover a surface
- increased adhesiveness
- increased resistance to antimicrobial substances
- faster growth and biofilm-forming excretions

Molecular biologists have determined that the common biofilm bacterium *P. aeruginosa* alters the way it regulates about 70 of its genes (about 1 percent of its total genetic material) when it becomes part of biofilm. Up to one-half of this species's protein synthesis changes when it joins a biofilm community. As a result, the entire community begins to behave as a unique living thing made up of different types of cells. In other words, it behaves much as a higher multicellular organism does.

HARMFUL BIOFILMS

In medicine, biofilms can contaminate and colonize hip replacements, endotracheal tubes, medical catheters, stents, heart valves, or contact lenses. The film that develops on teeth to form plaque is also a biofilm. Medical researchers now explore the role of biofilms that colonize chronic wounds and prevent them from healing.

Biofilms are known to cause or worsen the following five medical conditions: (1) dental decay, (2) an acute ear infection called otitis media, (3) the respiratory diseases cystic fibrosis and Legionnaires' disease, (4) urinary tract infections, and (5) bacterial endocarditis, an infection of the inner heart. In the latter condition, infection spreads from the heart to other parts of the body, when cells slough off the biofilm and enter the bloodstream.

Medical researchers have investigated ways to prevent biofilm from forming on medical devices to reduce the chance of contamination. Low-energy sound waves have been shown to block cells from attaching to device surfaces. Rodney M. Donlan of the Centers for Disease Control and Prevention (CDC) said, in 2007, "One important advantage of this approach is that the use of antimicrobial agents is not required." Fitting medical devices with a component that continuously emits sound waves at a constant frequency will not be an easy task. But because ebiofilms display strong resistance to chemical biocides, few options have been available for killing biofilms.

Biofilms also affect water quality by creating bad tastes and odors. Biofilms form along the inner surfaces of pipes that distribute drinking water to houses and other buildings. They contribute a steady amount of bacteria and other microorganisms to the water flowing toward taps. Very active biofilms have been known to corrode distribution lines made of cast iron and other metals. This is one reason why many newer water distribution systems use plastic polyvinyl chloride (PVC) piping. Though biofilm still forms on the plastic, chemists originally believed the plastic did not corrode as easily as metal. Scientists now think biofilm may corrode PVC as it does metal.

Since biofilms are highly resistant to chemical disinfectants, the chlorine in water often has little effect on them. To make matters worse, clusters break off periodically in a process called *detachment*. Biofilm detachment causes bacterial numbers in drinking water to fluctuate. After detachment, biofilm rebuilds its lost portions in a process called *regrowth*.

On pipes, drains, or boats and ships, biofilm corrosion weakens metals over time, and, if left

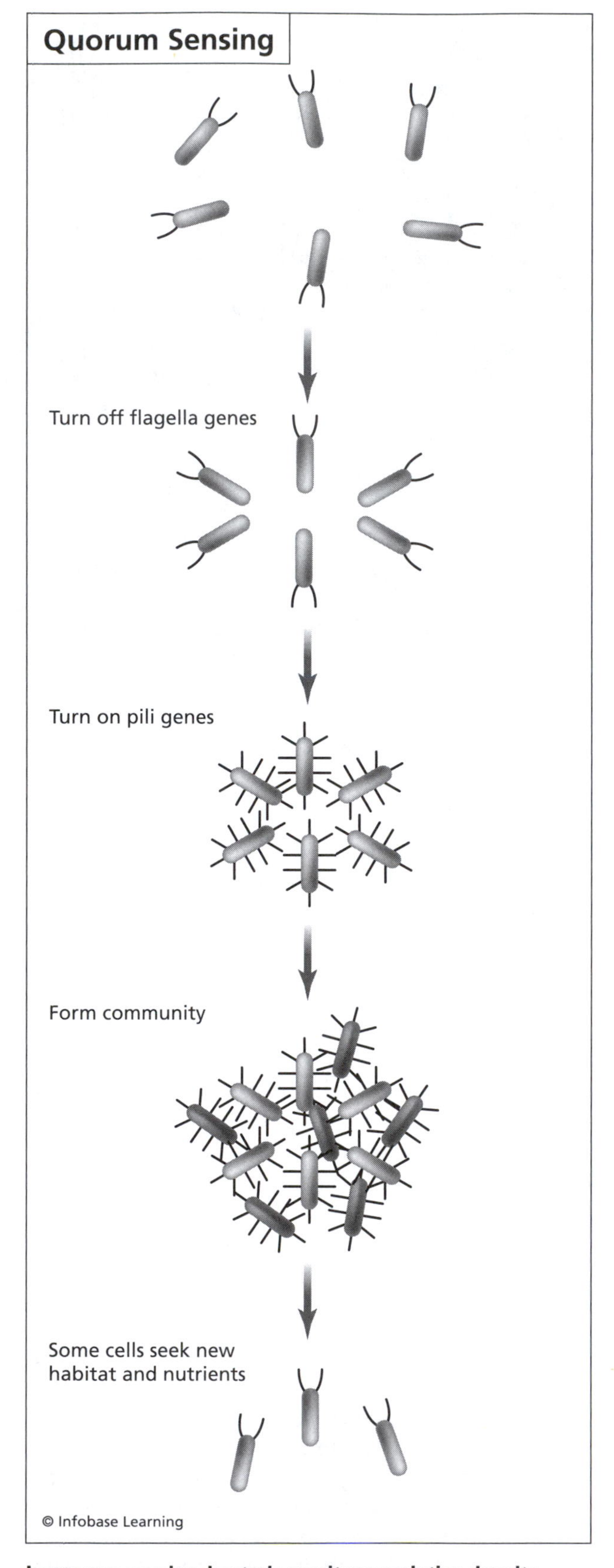

In quorum sensing, bacteria monitor population density by detecting signal chemicals from other bacteria. When signals reach a certain concentration, the bacteria turn on processes to reduce the flagella size and increase the growth of pili, which aid cell-to-cell communication.

unchecked, the corrosion develops into something more serious than a mere nuisance. For example, in some industries biofilms ruin equipment or spoil products. Experts in oil drilling, paper milling, food processing, and oceangoing cargo shipping all must contend with the constant removal of biofilms from their equipment. Damage caused by biofilms costs these industries millions of dollars each year in lost product or in cleaning and repairing of equipment. They are of particular concern in food processing plants, because food ingredients provide a rich conditioning film inside the production line equipment. A single biofilm community in a food processing line will contaminate every batch of food product as it moves through the line.

In addition to sound waves, biofilm researchers study other ways to prevent biofilms from attaching to hard inanimate surfaces. Their first approach is to find materials that repel biofilm attachment. So far, few materials upon which biofilm does not attach have been discovered. The U.S. Navy has experimented with special coatings over the hulls of their ships that will potentially repel bacteria, but this has also been only partially successful. Other methods involve the application of an electric charge that causes the biofilm to detach.

In medical research, microbiologists seek compounds that will inhibit the activities of pathogenic species in biofilms. One technique may be to disrupt the genes associated with quorum sensing. In 2007, the microbiologist Jeffrey L. Fox wrote in *Microbe* magazine, "The term 'sociomicrobiology' alludes to the highly 'structured' communities that can form within biofilms that can prove 'important to their success as pathogens,' says [Peter] Greenberg of the University of Washington." Because persistent biofilm communities may depend on quorum sensing, this aspect of biofilm activity offers a promising way to disrupt biofilms. Harmful biofilms, nonetheless, remain very difficult problems in medicine and in other areas.

BENEFICIAL BIOFILMS

Microbiologists have developed biofilm systems based on the way they work in nature. In the environment, biofilms grow on aquatic plants and on sediments and rocks in streams, rivers, lakes, and wetlands. Here they serve as an important part of food chains: Protozoa and invertebrates graze on biofilm material; they, in turn, serve as food for small fish, then larger fish and terrestrial animals. Biofilms in nature also remove and degrade chemical pollutants before they wash downstream and thus provide a natural form of bioremediation.

Because biofilms efficiently remove substances from the water that flows past them, microbiologists have developed biofilms to clean contaminated water. Biofilms in water treatment plants, wastewater treatment plants, and septic systems break down organic compounds to carbon dioxide and water. In wastewater treatment, biofilms degrade suspended solids in a series of steps that clarify and detoxify the wastewater. Drinking water treatment plants also use biofilms in a similar way. The biofilm develops over particles, making up filters called slow sand filters. As the water slowly trickles through these filters, the biofilm removes organic compounds and metals. Lake, river, and reservoir waters are partially cleaned this way in many communities.

Devices for Growing in Vitro Biofilms

Device	Advantage
annular bioreactor	creates high-shear conditions for studying biofilms in fast-flowing conditions
rotating disk reactor	grows large amounts of biofilm on a round flat disk that can be scraped off
coupon reactor	grows biofilm on coupons made of various materials to study attachment in varying flow speeds
flow cell	compares coupons of different materials at the same time and allows microscopic study of the biofilm
drip flow chamber	uses slow-flow, low-shear conditions like those on medical devices

Environmental scientists have taken this capacity to clean water a step further by building bioreactors that use biofilm to treat polluted water. One such device is a fluidized-bed reactor. This is a bioreactor in which polluted water is slowly pumped upward through a column of biofilm-coated beads. (It is called a *fluidized bed* because the upward flow keeps the beads in suspension, or fluidized, rather than allowing them to sink to the bottom of the bioreactor's chamber.) The biofilm removes organic and inorganic compounds from the water as it flows past. As an option, biofilm may also develop directly on a polluted site, such as an oil spill. Hydrocarbon-degrading species within the biofilm then decompose the oily material coating shorelines and riverbanks.

Confocal microscopy combines light microscopy with scanning electron microscopy. It gives an image superior to regular light microscopy. This technology allows scientists to study the three-dimensional structure of biofilms.

Biologists now believe biofilm communities serve as a study model for the cell-to-cell interactions in higher organisms. These structures may also give clues as to how pathogens and eukaryotic cells give and receive signals in infection, in cancer, and in other diseases. Biofilms have emerged from being a little-understood curiosity in microbiology; they are now known to be of critical importance in medicine, water quality, and environmental science.

See also BIOREACTOR; BIOREMEDIATION.

Further Reading

"A Biofilm Primer." Available online. URL: www.biofilmsonline.com/cgi-bin/biofilmsonline/ed_where_primer.html. Accessed February 17, 2009.

Costerton, J. William. *The Biofilm Primer.* Heidelberg, Germany: Springer-Verlag Berlin, 2007.

Fox, Jeffrey L. "Disrupting Social Lives to Block Bacterial Biofilms." *Microbe,* January 2007.

Lens, Piet N. L. *Biofilms in Medicine, Industry and Environmental Biotechnology.* London: IWA, 2003.

Palmer, Jon, Steve Flint, and John Brooks. "Bacterial Cell Attachment, the Beginning of a Biofilm." *Journal of Industrial Microbiology and Biotechnology* 34 (2007): 577–588.

Sauer, Karin, Alex H. Rickard, and David G. Davies. "Biofilms and Biocomplexity." *Microbe,* February 2007.

biogeochemical cycles Biogeochemical cycles are natural processes that take place on Earth, in which nutrients in various chemical forms move from the nonliving environment to living organisms and back in a circular pattern. These processes are also called *nutrient cycles.* Biogeochemical cycles encompass the atmosphere, water, the earth, and plant, animal, and microbial life. The main biogeochemical cycles on Earth are the following: carbon, nitrogen, phosphorus, sulfur, iron, oxygen, water (the hydrologic cycle), and sediment. The first five of these cycles could not operate without the activity of microorganisms.

Geomicrobiology by definition is the study of the relationship between microorganisms and geologic systems on Earth. Parts of these studies explore the ways in which microorganisms contribute to and benefit from biogeochemical cycling. Geomicrobiologists also examine targeted topics related to the cycling of nutrients or elements through the biosphere, which is the area of Earth where life exists, from the atmosphere to deep sediments and the deep ocean. This discipline, therefore, requires knowledge of the following substances in relation to microbiology: rocks, sediments, and soil; marine water and freshwater; microbial nutrients; and the effect of pollutants on nutrient use.

University courses in geomicrobiology also cover topics that relate living things to Earth's history. For example, the following list provides study areas that are part of this field and the study of biogeochemical cycles:

- development of Earth's geology
- the role of microorganisms in using and con serving Earth's elements
- microbial diversity
- extreme environments
- reduction and oxidation (redox) chemistry
- the coevolution of Earth and the biosphere
- microbial genetics related to activities in the environment

The science of studying relationships between microorganisms and the biosphere has been credited mainly to two scientists who conducted their studies in the late 1800s. The German botanist Martinus Beijerinck studied the relationships between soil bacteria and nutrient use in plants. The Russian microbiologist Sergei Winogradsky conducted his research in France on the physiology of soil microorganisms and the metabolism of nutrients such as nitrogen and sulfur. In 1890, Winogradsky introduced an article of his on bacteria that absorb nitrogen gas from the air by saying, "Besides the organisms which are the subject of the present note, two groups of organisms have been studied which have the ability to oxidize inorganic substances. I have designated them by the names sulfur bacteria and iron bacteria. The first group live in natural waters which contain hydrogen sulfide and do not grow in media lacking this substance. . . . The second group are able to oxidize iron salts, and their life is also closely connected with the presence of these compounds in their nutrient medium." These statements are key to understanding biogeochemical cycles because the microbial conversion of inorganic compounds to organic substances is critical for the continuance of a biogeochemical cycle.

The operation of biogeochemical cycles creates the foundation of two related and rapidly expanding areas in microbiology: environmental microbiology and microbial ecology. Although both disciplines share common subject areas such as the cycling of nutrients, they study the natural world from different viewpoints, indicated in the table.

The environmental microbiologist Raina M. Maier wrote of the value of understanding biogeochemical cycles, in 2000, "These cycles allow sci-

Environmental Microbiology and Microbial Ecology

Subject	Description
environmental microbiology	the study of microbial habitats, microbial activities in those habitats, and the types and amounts of microorganisms in specific places in the environment
microbial ecology	the study of microorganisms' relationships with their environment, and the microorganism-animal-plant relationship that shapes Earth's environments

entists to understand and predict the development of microbial communities and activities in the environment. There are many activities that can be harnessed in a beneficial way, such as for remediation of organic and metal pollutants or for recovery of precious metals such as copper or uranium from low-grade ores. There are also detrimental aspects to the cycles that can cause global environmental problems such as the formation of acid rain and acid mine drainage, metal corrosion processes, and formation of nitrous oxide, which can deplete the Earth's ozone layer." Understanding Earth, environment, and biology cannot proceed without familiarity with the biogeochemical cycles.

THE CARBON CYCLE

Microorganisms decompose once-living organisms: plants, animals, and other microbial life. This step, known as decay, may be thought of as the beginning of a new carbon cycle.

At its most basic, the carbon cycle consists of two equally important conversions of carbon from a gas to a more complex organic compound. When microorganisms decay dead organic matter, these microorganisms are called *decomposers*. Decomposers get carbon for their own use as well as energy by breaking down animal organic matter—proteins, glycogen, fats, and nucleic acids—and plant organic matter—cellulose, hemicellulose, pectin, and lignin. The various species that decompose these substances for their survival are called *heterotrophs*. Heterotrophs (also called *organotrophs*) need an organic carbon source to survive. For example, heterotrophs use sugars, amino acids, or nucleic acids but cannot use the gas carbon dioxide (CO_2).

Heterotrophs oxidize the organic compounds of decomposition in a metabolic pathway called *aerobic respiration*. These microorganisms make energy for growth and maintenance by respiration and produce water and carbon dioxide as end products. The carbon dioxide released into the atmosphere becomes the carbon source for a new group of organisms: photosynthetic organisms.

Plants, algae, and cyanobacteria compose the majority of Earth's photosynthetic life. Organisms that use carbon dioxide as their main carbon source are called *autotrophs*. Autotrophs convert carbon into carbohydrates, proteins, and fats that make up cells.

Aerobic respiration and photosynthesis complement each other for recycling carbon through the atmosphere, soil, and water. As a result, microorganisms and all higher organisms receive the carbon they need for life, whether the carbon occurs in a polar bear eating a seal, an eagle eating a salmon, or a human eating a salad.

The carbon cycle also contains side routes through which carbon flows. The four principal modifications to the carbon cycle are (1) anaerobic fermentation, (2) sedimentation, (3) the methane cycle, and (4) the production of humus.

Anaerobic fermentation involves the breakdown of organic compounds by anaerobic bacteria. Fermentation produces a variety of carbon-containing end products that each take different routes in the larger carbon cycle. Depending on the species of anaerobe, fermentation might produce the gases methane (CH_4), carbon dioxide, or hydrogen. It might also produce alcohols and acids.

The end products of fermentation feed either the sedimentation process or the methane cycle. In sedimentation, decayed organic matter from fermentation sinks into the deep sediments of Earth's crust. Hundreds of thousands of years pass, and as the pressure of Earth's crust bears down on the carbon material, it turns into coal, crude oil, or natural gas. Some of the anaerobic activity on the organic matter serves as the source of natural gas, which is methane.

Other anaerobic microorganisms specialize in producing methane or in consuming methane. If these bacteria do not take part in sedimentation, they participate in the methane cycle. About 75 percent of the methane in the atmosphere originates in biological sources, such as anaerobic microorganisms. Anaerobes produce this gas in a process called mineralization, by which complex organic compounds degrade to carbon dioxide and methane:

$$C_6H_{12}O_6 \rightarrow 3\ CO_2 + 3\ CH_4$$

The anaerobes that execute this reaction are methanogens and belong to domain Archaea. Either the methane rises into the atmosphere as a greenhouse gas, or other microorganisms absorb it for carbon and energy metabolism. Such microorganisms are called *methanotrophs,* and many of them also belong to the archaea.

Some of Earth's carbon takes a fourth route to become part of the large molecule called *humus.* Humus is the end product of organic matter decomposition, in which small breakdown products bind to each other to form the large compound that contains carbon, oxygen, nitrogen, and hydrogen. The enzymes that are part of organic matter decomposition exist in high concentrations in places where active decay goes on, and these enzymes spontaneously bind short breakdown products together to make humus. Humus is very stable in the environment and makes up from 3 percent to 8.5 percent of soil, although these values can extend beyond this range in soils containing high rates of organic decay.

Several different metabolic routes lead to the formation of humus because of the sheer magnitude and diversity of microorganisms in soil. Humus is created from the well-known pathways of carbohydrate, protein, and fat degradation, but it also derives from alcohols and acids produced in fermentation, plant fibers, and even organic pollutants.

THE NITROGEN CYCLE

Nitrogen is the fourth most common element in cells, making up 5 percent, after hydrogen (60 percent), oxygen (25 percent), and carbon (12 percent). Most of the nitrogen resides in proteins, which compose about 50 percent of cells, and the nucleic acids deoxyribonucleic acid (DNA) and ribonucleic acid (RNA), which make up about 15 percent of cell components.

Nitrogen is additionally the main component in Earth's atmosphere, accounting for 78.1 percent; oxygen totals 20.9 percent, and other trace gases make up the remainder. Nitrogen is stored in the biosphere in a diverse variety of storage forms called *reservoirs* or *sinks.* Because of the broad array of nitrogen reservoirs, the nitrogen cycle often serves as an excellent model for studying biogeochemical cycles. The table below summarizes the biosphere's nitrogen reservoirs.

All organisms need the nitrogen cycle to supply the nitrogen needed for making proteins, nucleic acids, and other cell constituents. To begin the cycle, the large reservoir of nitrogen in the atmosphere must be made available for use by living organisms. Because higher plants and animals cannot use nitrogen gas directly, they depend on the activity of soil and water microorganisms to capture nitrogen from the air and convert it to ammonia (NH_3) in a step called *nitrogen fixation.* This energy-demanding process uses the enzyme nitrogenase to convert N_2 to NH_3.

Two types of nitrogen-fixing bacteria capture nitrogen from the air: free-living and symbiotic nitrogen-fixing bacteria. Some bacteria in the soil live within a fraction of an inch (about 2 mm) of

The Earth's Nitrogen Reservoirs

Nitrogen Reservoir	Tons of Nitrogen*	Role in Nitrogen Cycle
	ATMOSPHERE	
N_2 gas	4.3×10^{15}	requires nitrogen fixation by microorganisms
	OCEANS	
biomass (living and nonliving matter)	5.7×10^{8}	actively recycled
organic compounds	3.3×10^{11}	actively recycled
soluble salts (nitrate, nitrite, ammonium)	7.6×10^{11}	actively recycled
dissolved N_2	2.2×10^{13}	minimal contribution
	LAND	
biota (all living things)	2.8×10^{10}	actively recycled
organic matter	1.2×10^{11}	slow recycling
Earth's crust	8.5×10^{14}	no recycling

Notes: Adapted from Maier, Raina R., Ian L. Pepper, and Charles P. Gerba. *Environmental Microbiology* (San Diego: Academic Press, 2000).

* To convert to metric tons, multiply by 0.907

Organisms Involved in Nitrogen Cycle

Nitrogen Cycle	Microorganisms	Type
	MICROBIOLOGY STEP	
Nitrogen Fixation (free-living)	*Acetobacter, Azotobacter, Beijerinckia, Paracoccus, Pseudomonas*	aerobic bacteria
	Azospirillum	microaerophilic bacteria
	Bacillus, Klebsiella	facultative anaerobic bacteria
	Clostridium, Desulfovibrio, Thiobacillus	anaerobic bacteria
	cyanobacteria *Anabaena* and *Nostoc*	aerobic photosynthetic
	cyanobacteria *Chlorobium* and *Chromatium*	anaerobic photosynthetic
Nitrogen Fixation (symbiotic)	*Bradyrhizobium, Rhizobium*	microaerophilic bacteria
	basidiomycetes	mycorrhizal fungi
Nitrification	*Nitrobacter, Nitrosomonas*	aerobic bacteria
Assimilation	microorganisms and plants	aerobic and anaerobic microorganisms, plants
Ammonification	*Alternaria, Mucor, Aspergillus, Penicillium*	fungi
	Bacillus, Pseudomonas	aerobic bacteria
	Clostridium, Desulfovibrio	anaerobic bacteria
	Enterobacter, Klebsiella, Photobacterium, Vibrio	facultative anaerobic bacteria
Denitrification	Alcaligenes, Bacillus, *Flavobacterium, Pseudomonas*	soil bacteria
	Rhodopseudomonas	anaerobic bacteria

plant roots, an area called the *rhizosphere*. The free-living nitrogen fixers consist of aerobes, anaerobes, and photosynthetic cyanobacteria.

Symbiotic nitrogen-fixing bacteria exist inside the roots of agriculturally important plants such as soybeans, peas, beans, peanuts, alfalfa, and clover, collectively known as *legumes*. Legume roots contain small bulges called *root nodules*, where anaerobic nitrogen fixers live in a cooperative relationship with the plant. In this example of symbiosis, the plant root nodule provides low-oxygen conditions and a carbon source for the bacteria, and the bacteria convert nitrogen into a form the plant can use. Inside the nodules, bacterial enzymes incorporate the ammonia into the amino acid asparagine, which then enters the plant's vascular system as cells die and break apart. Ammonia goes directly into amino acid synthesis in a step called *assimilation*, at which point the nitrogen is in a form that can be used by a plant.

Ammonia may also enter a series of steps independent of plant roots and carried out by a separate group of bacteria. This series of steps is called *nitrification*, shown in the following equation:

$$\text{ammonium ion } (NH_4^+) \rightarrow \text{nitrite ion } (NO_2^-) \rightarrow \text{nitrate ion } (NO_3^-)$$

The soil bacterial genus *Nitrosomonas* oxidizes ammonium to nitrite, and *Nitrobacter* oxidizes nitrite to nitrate—these are called *nitrifying bacte-*

ria. Nitrate ions move through soil easily, because they do not bind with soil particles. For this reason, nonlegume plants have evolved to absorb nitrate from the soil as their main nitrogen source.

Decomposer bacteria act on plants and animals that have died and release proteins and nucleic acids. The proteins and other nitrogen compounds break down further to smaller compounds, such as amino acids, until soil bacteria and fungi remove the nitrogen (a step called *deamination*) and convert it to ammonia. This process, called *ammonification*, makes the nitrogen available for reuse by again entering nitrification. Meanwhile, extra nitrates produced by nitrification and not absorbed by plants enter yet another conversion in the cycle, called *denitrification*, shown in the following equation:

$$NO_3^- \rightarrow NO_2^- \rightarrow \text{nitrous oxide } (N_2O) \rightarrow \text{nitrogen gas } (N_2)$$

The gas returns to the atmosphere to complete the nitrogen cycle. In total, the main route of the nitrogen cycle through atmosphere, soil microorganisms, and plant or animal life is the following:

$$\text{nitrogen fixation} \rightarrow \text{nitrification} \rightarrow \text{assimilation} \rightarrow \text{ammonification} \rightarrow \text{denitrification}$$

Human activities can contribute to, but also interfere with, the nitrogen cycle. For example, large amounts of nitrogen fertilizers applied to gardens or agricultural fields, as well as manure from farms, wash away in rains and cause an imbalance in the natural levels of nitrogen in soil or in water. When this happens, microbial growth explodes as a result of the sudden influx of nitrogen, and the resulting high numbers of microorganisms develop a bloom. Blooms harm ecosystems because they disrupt normal nutrient use and oxygen availability in soil and water.

Excess nitrates formed by nitrifying bacteria can also contaminate groundwater sources. Fertilizers and manure that leach into the soil plus leaking septic tanks deliver nitrates to underground sources of drinking water, especially since nitrate is very mobile in soil. In 2005, the physicians Frank R. Greer and Michael Shannon wrote in a pediatrics journal, "Nitrate poisoning resulting in methemoglobinemia [nitrate substitutes for oxygen on the hemoglobin molecule] continues to be a problem in infants in the United States. Most reported cases have been ascribed to the use of contaminated well water for preparation of infant formula." The authors warned that 40,000 infants younger than six months may be receiving water from nitrate-contaminated sources.

Finally, excess nitrous oxide from either biological or industrial sources also presents environmental problems. The nitrous oxide causes the following two hazards: activity as a greenhouse gas that contributes to global warming and destruction of the atmosphere's ozone layer, which protects Earth from excess exposure to ultraviolet light.

The table presents the steps in the nitrogen cycle and the main microorganisms that participate in its reactions.

THE SULFUR CYCLE

The sulfur cycle resembles the nitrogen cycle in two ways. First, chemical conversions of sulfur involve many different oxidation and reduction states, and, second, effects of human activities in the form of acid rain and other air pollution impact the sulfur cycle.

The sulfur cycle contains oxidized sulfate groups (SO_4^{2-}) to reduced hydrogen sulfide gas (H_2S). Sulfur is the 10th most abundant element in Earth's crust, and for this reason bacteria have an ample supply of sulfur compared with more limited supplies of nitrogen. Sulfur appears in cells in amino acids (cysteine and methionine), vitamins, hormones, cofactors needed in energy metabolism, and enzymes. In the sulfur cycle, these organic compounds are called *assimilated sulfur*.

The main pathway of the sulfur cycle is as follows:

$$H_2S \rightarrow \text{elemental sulfur } (S^\circ) \rightarrow SO_4^{2-} \rightarrow \text{assimilated sulfur } (\text{-SH}) \rightarrow H_2S$$

The conversion of hydrogen sulfide to elemental sulfur is called sulfur oxidation. Two very different types of bacteria conduct this process. The first group of sulfur oxidizers contains aerobic chemoautotrophs, which are bacteria that can survive on

Sulfur-Oxidizing Bacteria

Group	Genera	Sulfur Conversion They Perform
aerobic	*Achromatium*, *Beggiatoa*, *Thermothrix*, *Thiobacillus*, *Thiomicrospira*	$H_2S \rightarrow S^\circ$ $S^\circ \rightarrow SO_4^{2-}$ and $SO_3^{2-} \rightarrow SO_4^{2-}$
anaerobic	*Chlorobium*, *Chromatium*, *Ectothiorhodospira*, *Thiopedia*, *Rhodopseudomonas*	$H_2S \rightarrow S^\circ \rightarrow SO_4^{2-}$

Sulfur Cycle Microbiology

Step	Main Microorganisms	Reaction
	MAIN ROUTE	
sulfur oxidation	chemoautotrophic and phototrophic sulfur oxidizers	$H_2S \rightarrow S° \rightarrow SO_4^{2-}$
assimilatory sulfate reduction	*Desulfuromonas*	$SO_4^{2-} \rightarrow$ organic sulfur
mineralization	anaerobic decomposers	organic sulfur $\rightarrow H_2S$
	SHORTCUTS	
dissimilatory sulfate reduction	*Desulfobacter, Desulfococcus, Desulfosarcina, Desulfovibrio*	$SO_4^{2-} \rightarrow H_2S$
dissimilatory sulfite reduction	*Alteromonas, Clostridium, Desulfovibrio, Desulfotomaculum*	$SO_3^{2-} \rightarrow H_2S$
dissimilatory sulfur reduction	*Desulfuromonas,* archaea, cyanobacteria	$S° \rightarrow H_2S$

inorganic chemicals as their energy source and carbon dioxide as the carbon source. These bacteria live in habitats high in hydrogen sulfide at a location where oxygen is also available. Consequently, these microorganisms live at the upper surface of muds, soils, swamps, mining sites, and hot springs. The second group of bacteria are anaerobes and phototrophs, meaning they use sunlight as their main energy source. These bacteria live in shallow waters and sediments. The table describes the main genera and activities in these two groups.

The microorganisms that make up the phototrophic sulfur-oxidizing bacteria also belong to two interesting groups of bacteria: green sulfur bacteria and purple sulfur bacteria. Both groups evolved from primitive cells when Earth's atmosphere contained no oxygen. The green bacteria earned their nickname from their chlorophyll, a pigment that they use in photosynthesis for energy production. The purple sulfur bacteria appear in a range of colors from pink to purple. Because the green sulfur and purple bacteria use photosynthesis, they absorb carbon dioxide and produce sugars. The main reaction carried out by these bacteria in the sulfur cycle is the following:

$$CO_2 + H_2S + \text{light} \rightarrow C_6H_{12}O_6 + S°$$

Once sulfate has been formed in the soil, the sulfur can be converted to form available to higher organisms by reducing the element. A group of specialized sulfate-reducing bacteria carry out this anaerobic process. Higher organisms cannot use the oxidized sulfate form of sulfur for synthesizing cellular compounds, so they rely on sulfate-reducing bacteria to reduce the element to a usable form. Nature provides two processes for reducing sulfate: assimilatory sulfate reduction (ASR) and dissimilatory sulfate reduction (DSR). Higher organisms depend on ASR because this process converts sulfate into forms of sulfur found in amino acids, hormones, and other compounds in the body. DSR, by contrast, serves as an energy-yielding step for certain anaerobic bacteria and converts the sulfate directly to hydrogen sulfide.

Upon decay, dead plant and animal life releases sulfur-containing compounds into the soil; microorganisms break them down into simple inorganic compounds. The complete breakdown of complex organic molecules into simple end products such as carbon dioxide, water, or hydrogen sulfide is called *mineralization,* a process that takes place in some form in all biogeochemical cycles.

DSR might be thought of a shortcut in the sulfur cycle, in which microorganisms derive what they need, but higher organisms receive no direct benefit. The sulfur cycle contains three shortcuts: (1) DSR, (2) reduction of sulfite SO_3^{2-} directly to hydrogen sulfide, and (3) reduction of S° directly to hydrogen sulfide. Even with these shortcuts, the sulfur cycle proceeds in a more simplified manner than either the carbon or the nitrogen cycle. The table describes the main bacteria of the sulfur cycle and their roles.

THE PHOSPHORUS CYCLE

Phosphorus exists almost exclusively in the phosphate form (PO_4^{3-}) in the environment and in biota. This chemical group is required for energy metabolism and energy storage in the body. For example, the energy stored in the phosphate of adenosine triphosphate (ATP) serves as a central compound in aerobic and anaerobic energy production in all cells.

The phosphorus cycle involves changes between soluble and insoluble forms and organic and inorganic compounds, rather than oxidation-reduction reactions of the nitrogen and sulfur cycles. Also, phosphorus does not enter the atmosphere as a gas. The phosphorus cycle is confined mainly to land and the ocean.

The terrestrial phosphorus cycle uses water in moist soils, rivers, and lakes as part of food chains. In the ocean, phosphorus progresses through marine food chains. Long-term phosphate storage occurs in rock and ocean sediments. In both locations, the phosphate cycles very slowly to other chemical forms.

When members of food chains die, decomposer bacteria and fungi release phosphate-containing compounds. Bacteria that excrete the enzyme phosphatase cleave the phosphate groups from compounds, and these molecules dissolve in water as salts. Salts may be absorbed directly by plant roots and enter food chains, or autotrophic bacteria take up the phosphate. Autotrophs are organisms that can use carbon dioxide as their main carbon source. (Multicellular plants are also autotrophs.) Autotrophs carry phosphorus into additional food chains to continue the cycle. When microorganisms capture phosphate groups in this way, the process is called *phosphorus immobilization*. When phosphate-containing compounds reenter the environment when plants, animals, and microorganisms die, the process is mineralization.

IRON AND OTHER METAL CYCLES

Microorganisms perform many transformations on various metals with the oxidation-reduction reaction. The iron cycle is the most studied of microbial metal cycles. The iron cycle, in general, consists of reactions that convert ferrous iron (Fe^{2+}) to ferric iron (Fe^{3+}) and back again. A subgroup of bacteria participate in the iron cycle by including the compound magnetite (Fe_3O_4) as an intermediary. One curiosity of the iron cycle lies in the relationship between aerobes and anaerobes: Aerobic bacteria carry out almost all the oxidation reactions, and anaerobic bacteria carry out all of the reduction reactions.

The iron cycle may take place in the ocean or in terrestrial environments, but it is especially important in the ocean, where iron is scarce. Light helps transform marine iron into forms that are more available for microorganisms and, thus, more readily enter food chains. The chemist Alison Butler of the University of California–Santa Barbara explained in a National Aeronautics and Space Administration (NASA) news release, "We determined that iron bound to oceanic siderophores [iron-binding compounds] react to light. This photochemical reaction helps transform the iron complexes into a form that enables marine organisms to more easily acquire the essential iron." Butler explained that the Sun's energy loosens the connection between iron and oxygen and thus makes the iron more available for bacteria, plankton, and other organisms.

The table shows some of the main bacteria operating the iron cycle.

The aerobic iron-oxidizing bacteria are also referred to as *iron bacteria,* mainly in relation to their effect on corrosion of drinking water distribution pipes. Iron bacteria form a biofilm that adheres to the inner surfaces of iron piping. This bacterial community produces unpleasant odors and tastes in water, but they are not thought to be harmful to humans.

Bacteria also take part in the manganese cycle, which operates in the deep sea, in bogs, and on the surface of rocks. The manganese cycle involves the conversion of the manganous ion (Mn^{2+}) to the maganic

Microbial Iron Cycling

Iron Cycle Microbiology Microorganisms	Reaction
Aerobic Oxidation	
• *Gallionella* at neutral pH	$Fe^{2+} \rightarrow Fe^{3+}$
• *Sulfolobus* at acid pH and thermophilic conditions	
• *Thiobacillus* and *Leptospirillum* at acidic pH	
Anaerobic Reduction	
• *Desulfuromonas*	$Fe^{3+} \rightarrow Fe^{2+}$
• *Ferribacterium*	
• *Geobacter*	
• *Geovibrio*	
• *Pelobacter*	
• *Shewanella*	
• *Aquaspirillum magnetotacticum*	$Fe^{3+} \rightarrow Fe_3O_4$
Anaerobic Oxidation	
• anaerobic purple photosynthetic bacteria	$Fe^{2+} \rightarrow Fe^{3+}$
• archaean *Ferroplasma acidarmanus* at extreme acidic pH	

ion (Mn^{4+}). Example genera that oxidize Mn^{2+} to the manganic ion as part of magnetite (MnO_2) are *Arthrobacter* and *Leptothrix*. *Geobacter* and *Shewanella* reduce magnetite back to the manganous form.

In water environments, aerobic oxidizing reactions of the manganese cycle occur toward the water's surface, and anaerobic reducing reactions take place in deeper, low-oxygen environments.

Bacteria contribute to the cycling of other metals, notably mercury. The mercury cycle has been influenced more by human enterprises than by nature, but bacteria living in sediments under mercury-polluted bodies of water play a role. Anaerobic genera such as *Desulfovibrio* convert free mercury (Hg^{2+}) to methylated forms. Two of the main methylated forms these bacteria make are methyl mercury (CH_3Hg^+) and dimethyl mercury [$(CH_3)_2Hg$]. If sulfide is present in the anaerobic sediments, this element can combine with mercury to form mercury sulfide (HgS).

The study of biogeochemical cycles touches on a vast number of aspects in biology: microbiology, ecology, biochemistry, and nutrition. In addition, the biogeochemical cycles demonstrate the ways in which Earth's biota interacts with nonliving matter in a continual, self-sustaining process.

See also BLOOM; ENVIRONMENTAL MICROBIOLOGY; GREEN BACTERIA; METABOLIC PATHWAYS; MICROBIAL COMMUNITY; MICROBIAL ECOLOGY; PURPLE BACTERIA.

Further Reading

Brock, Thomas D. *Principles of Microbial Ecology.* Englewood Cliffs, N.J.: Prentice-Hall, 1966.

Dyer, Betsey Dexter. *A Field Guide to the Bacteria.* Ithaca, N.Y.: Cornell University Press, 2003.

Espinoza, Leo, Rick Norman, Nathan Slaton, and Mike Daniels. "The Nitrogen and Phosphorus Cycles in Soils." University of Arkansas Agriculture and Natural Resources. Available online. URL: www.uaex.edu/Other_Areas/publications/PDF/FSA-2148.pdf. Accessed March 20, 2009.

Greer, Frank R., and Michael Shannon. "Infant Methemoglobinemia: The Role of Dietary Nitrate in Food and Water." *Pediatrics* 116 (2005): 784–786. Available online. URL: http://aappolicy.aappublications.org/cgi/reprint/pediatrics;116/3/784.pdf. Accessed March 25, 2009.

Maier, Raina M., Ian L. Pepper, and Charles P. Gerba. *Environmental Microbiology.* San Diego: Academic Press, 2000.

National Aeronautics and Space Administration. "Scientists Chart Iron Cycle in Ocean." Earth Observatory News. Available online. URL: http://earthobservatory.nasa.gov/Newsroom/view.php?id=21945. Accessed March 25, 2009.

Winogradsky, Sergei. "Sur les organisms de la nitrification" (On the Nitrifying Organisms). *Comptes rendus de l'Acadimie des Sciences* 110 (1890): 1,013–1,016. In *Milestones in Microbiology,* translated and edited by Thomas Brock. Washington, D.C.: American Society for Microbiology Press, 1961.

biological oxygen demand (BOD) Biological oxygen demand is a measure of dissolved oxygen that microorganisms can use for decomposing organic matter. For this reason, BOD is sometimes referred to as *biochemical oxygen demand.*

The wastewater treatment industry monitors BOD in assessing the quality of treated water released from a treatment plant. BOD gives microbiologists and plant operators an idea of the level of organic matter in water, which may create a threat to aquatic ecosystems when this water enters the environment. Waters high in BOD cause blooms to occur in surface waters in the environment. A bloom is a rapid burst of microbial growth in response to a sudden influx of nutrients. In a bloom, water microorganisms grow so rapidly that they consume all of the water's dissolved oxygen within a localized area. The oxygen-depleted water then asphyxiates fish and other organisms dependent on dissolved oxygen. Waters high in BOD increase the risk of starting a bloom, and consequently BOD has been used as a sign of water pollution.

Oxygen-demanding wastes act as one of the major sources of water pollution, along with inorganic chemicals, organic chemicals, compounds high in nitrates and phosphates, dirt erosion, and radioactive materials. Animal wastes and plant debris also add oxygen-demanding matter to natural waters. Animal feedlots, farms, paper mills, and food processing facilities, each day, contribute tons of substances that raise the BOD levels in water. Examples of specific wastes from human activities and natural events that contribute to BOD are the following: manure, sewage, grass clippings, tree trimmings, leaves, dead plants, food wastes, and drainage from swamps and bogs.

Organisms in addition to microorganisms constantly draw oxygen out of natural waters. Respiration by fish, vertebrates, and invertebrates in water or within sediments requires oxygen. Sediments at the bottom of natural waters have high respiration rates, so scientists have devised a measurement called *sediment oxygen demand* (SOD). SOD works in a similar way to BOD, but it defines conditions in aquatic sediments rather than water. (Some nonbiological reactions called oxidation also consume some of the dissolved oxygen in water and sediment.) Regardless of the source of oxygen-demanding matter, the rule for water quality is as follows: When dissolved oxygen (DO) decreases, BOD increases, and vice versa.

$$\downarrow \text{DO} \rightarrow \uparrow \text{BOD} \quad \text{and} \quad \uparrow \text{DO} \rightarrow \downarrow \text{BOD}$$

A biogeochemical cycle called the *oxygen cycle* replaces the earth's oxygen that aerobic microorganisms and higher organisms need to survive. Photosyn-

Typical DO Values

DO Level (mg/l)	Water Quality
0.0–4.0	poor—some fish and small invertebrates affected or decline
4.1–7.9	fair
8.0–12.0	good
more than 12	needs retest because the water is possibly being artificially aerated

thetic aquatic plants, algae, and bacteria all put oxygen back into aquatic environments and thus maintain the oxygen cycle. Rising BOD levels quickly disrupt normal aquatic ecosystems and severely harm oxygen cycling from the earth to the atmosphere and back.

In aquatic ecosystems, oxygen is at a premium; water normally contains less than 1 percent oxygen. When dissolved oxygen falls below about 5 milligrams per liter (mg/l), many aquatic species become stressed. This stress weakens their health and leads to infection, disease, and death. Meanwhile, other organisms tolerant of low oxygen levels take their place and further deplete the oxygen. As dissolved oxygen falls below 2 mg/l, fish begin dying. Below 1 mg/l, anaerobic bacteria outgrow the struggling aerobic bacteria that remain. As anaerobic species decompose organic matter, they produce gases such as methane, carbon dioxide, hydrogen, and hydrogen sulfide. The release of these gases from aquatic environments indicates that an aerobic ecosystem has turned into an anaerobic one.

The wastewater treatment industry divides BOD into two components: carbonaceous biological oxygen demand (CBOD) and nitrogenous oxygen demand (NOD). CBOD results from the breakdown of organic molecules—cellulose, starch, sugars—into carbon dioxide and water. NOD results from the decomposition of proteins. When microorganisms degrade proteins, they release sugars normally attached to the protein. This reaction leaves behind an ammonia (NH_3) compound, which nitrifying bacteria readily convert to nitrates (NO_3^-). This oxidation step requires more than four times the oxygen needed in converting sugars to carbon dioxide and water. Said another way, the reaction has very high oxygen demand.

WATER QUALITY STANDARDS

The U.S. Environmental Agency publishes information on BOD for natural waters and treated wastewater. The agency dies not set a limit for acceptable BOD levels; each state or local water utility determines the values it targets for its own water. In general, pristine lakes and streams have BOD values of 1–2 mg/l, which indicates very clean water. BOD of 3–5 mg/l indicates moderately clean conditions, and BOD approaching 10 mg/l is an indication of pollution. Very polluted waters have levels of 100 mg/l or sometimes much greater. These levels are shown in the table Typical BOD Values.

Many states and utilities do not target a set BOD level; rather, they monitor BOD in order to achieve a certain DO level. A DO of at least 5–7 mg/l is considered by most municipalities to be clean water. The table (left) summarizes DO levels that have historically indicated the quality of natural waters.

THE BOD TEST

Microbiologists who monitor the operations in wastewater treatment plants measure BOD using a laboratory test developed in the United Kingdom, in the early 1900s. One disadvantage of this test is the long time it requires to give a result. Though BOD tests have been developed to take 5, 10, 20, or as long as 30 days to complete, the wastewater treatment industry almost exclusively uses a five-day test. This is sometimes referred to as the BOD_5. (Five days was not selected on the hard basis of scientific evidence; the Royal Commission on River Pollution in England picked this period with the idea that it would be the average time that pollution takes to flow from its source to the nearest estuary in the United Kingdom.)

The following outline summarizes the steps of the BOD_5 test:

1. Collect 250–300 ml water sample.

2. Measure the sample's DO level by either chemical tests or an electronic probe.

3. Seal the bottle and incubate the sample for five days at 68°F (20°C).

Typical BOD Values

BOD Level (mg/l)	Water Quality
1–2	very good—low amounts of organic matter
3–5	fair—moderately clean with small amounts of organic matter, possibly wastes
6–9	poor—polluted with organic matter, which microorganisms degrade
more than 100	very poor—very polluted with organic waste

4. Measure the DO of the incubated sample.

5. Calculate BOD from the two DO results: BOD = (DO on Day 1) - (DO on Day 5).

During the five-day incubation period, bacteria in the water sample continue degrading any organic matter in the bottle. A large amount of DO in the starting sample results in a large difference between day 1 and day 5 values. This in turn produces a high BOD value. Conversely, small changes between days 1 and 5 indicate waters low in organic matter.

Water bacteria actually take up to 20 days to digest all the organic matter in heavily polluted water. But a test that takes almost three weeks is impractical for wastewater treatment. By the time the results are available, the treated water has long since left the treatment plant. In five days, however, bacteria can degrade 60–70 percent of the organic matter. Microbiologists have learned from years of running the BOD test that five days is a satisfactory period to assess water quality.

Although testing for BOD has drawbacks that are being bypassed by newer test methods, BOD has remained a useful means for assessing general water quality.

See also BIOGEOCHEMICAL CYCLES; BLOOM; WASTEWATER TREATMENT; WATER QUALITY.

Further Reading

Blake, Perry F. *Supplemental Guidance for the Determination of Biological Oxygen Demand and Carbonaceous BOD in Water and Wastewater.* Olympia: Washington State Department of Ecology, 1998.

Clescerl, Lenore S., Arnold E. Greenberg, and Andrew D. Eaton, eds. *Standard Methods for Examination of Water and Wastewater,* 20th ed. Washington, D.C.: American Public Health Association, 1999.

Herman, David C., and Raina M. Maier. "Physiological Methods." In *Environmental Microbiology.* San Diego: Academic Press, 2000.

bioreactor A bioreactor, also called a *fermenter* (or fermentor), is an apparatus used in industrial microbiology to grow large volumes of bacteria or yeasts. *Fermentation* is a term often used for growth inside a bioreactor. This term has a second meaning in microbiology, because fermentation also represents a type of metabolism used in energy production by bacteria or yeasts. In industrial microbiology and in the following discussion, *fermentation* refers to any growth of microorganisms inside a bioreactor. Industrial microbiologists use bioreactors for two purposes: to produce large amounts of microorganisms or to produce large amounts of a product made by microorganisms. In research, microbiologists also use bioreactors to study the growth patterns of microorganisms. Gary Walsh, author of *Biopharmaceuticals: Biochemistry and Biotechnology,* wrote in 2003, "Microbial cell fermentation has a long history of use in the production of various biological products of commercial significance." Walsh has pointed out the following products made in industrial microbiology that today probably come from bioreactors:

- simple organic molecules—acetic acid, ace tone, butanol, ethanol, lactic acid
- amino acids—glutaminc acid, lysine
- enzymes—amylase, cellulase, protease
- antibiotics—bacitracin, penicillin

A bioreactor has two main parts: an inner fermentation vessel and an outer jacket. The fermentation vessel contains liquid medium for growing microorganisms. The jacket is constructed of an inner wall and an outer wall. A space between the jacket's walls can be filled with steam or with hot water, which warms the medium inside the inner vessel. A technician heats the jacket in two different ways, each method done for a different purpose. First, before the medium receives the inoculum, the jacket can be filled with steam, which raises the temperature to 250°F (121°C) under high pressure. High temperature and pressure sterilize the medium inside the fermentation vessel. This assures that all life in the freshly prepared medium has been killed before a microbiologist inoculates the liquid with a desired microorganism. After this sterilization step, the jacket cools to incubation temperatures of 86–99°F (30–37°C) as warm water circulates through it. The jacket then serves its second purpose: It helps maintain a steady warm temperature for helping the microorganisms grow from a small inoculum to millions of cells inside the bioreactor.

Bioreactors range in volume from a few liters to thousands of liters, depending on the microbiologist's objectives. For this reason, the bioreactor size range correlates to three types of activities: bench scale, pilot scale, or industrial scale. In bench scale experiments, microbiologists use small bioreactors holding a few liters and made of either glass or stainless steel. Microbiologists use these units for experiments on microbial cell growth. Such experiments are often called bench scale fermentations. Bioreactors that have capacities of several liters to thousands of liters are made of stainless steel. These large bioreactors hold from 100 to 1,000 l and are called *scale-up* versions, or *pilot scale,* by industrial microbiologists. In scale-up, microbiologists study any special conditions needed for growing large volumes of a microorganism. During scale-up, they adjust several conditions of the fermentation for the purpose of increasing yield. Yield is the amount (usually in grams) of cells or cell product made in a single fermentation. Some of the many conditions in a fermentation that may affect yield are temperature, pH, nutrients, oxygen, special growth factors, and the growth rate of the cells. Bioreactors contain features that enable microbiologists to adjust these and other conditions.

After microbiologists complete the scale-up experiments, they grow the microorganisms in bioreactors that hold thousands of liters. This is *industrial scale* fermentation. Today's industrial fermentations make a variety of products inexpensively compared with synthesis or extracting the materials from sources in nature. Examples of the types of products made in industrial fermentations are enzymes, polymers (long-chain compounds), vitamins, and acids. In addition, the biotechnology industry produces large volumes of bioengineered bacteria and their products. Many new drugs have been produced in bioreactor fermentations.

INOCULUM PREPARATION

All bioreactor fermentations begin with, first, preparation of the medium for growing the microorganisms and second, the inoculum. In industrial fermentations, the medium is usually a general formula that supplies microorganisms with all the nutrients they need for growth. An inoculum is any small volume of a microorganism that, when added to sterile medium, initiates the growth of millions of the same microorganism. Once the microbial cells begin multiplying in the medium, the mixture of medium and cells inside the bioreactor is called a *culture*.

All bioreactor fermentations begin with sterile medium and a pure, or uncontaminated, inoculum. These two factors ensure that only the desired microorganism will be the one growing in the bioreactor and no unwanted microorganisms. Pure cultures are obtained by using aseptic techniques and a series of steps called *isolation*. Aseptic techniques are standard procedures followed in microbiology to prevent contamination, while isolation is the separation of one type of microorganism from all others.

Bioreactor cultures have certain characteristics. With these characteristics, a microbiologist can be confident the fermentation will proceed as it should: That is, the cells will grow, and they will make the desired end product. Three desirable characteristics of microorganisms used in bioreactor fermentations are the following:

- stability—Cells grow well and in the same way generation after generation.
- viability—Cells remain alive and ready to use even if stored for long periods (weeks to months).
- genetic stability—Cells do not spontaneously mutate during storage or during growth.

Industrial microbiologists store microorganisms by freezing, refrigerating, or freeze-drying, a process in which a vacuum takes water out of the culture at the same time it is frozen. A worker revives these stored microorganisms (called the *master culture*) by putting a small amount of the master culture onto fresh medium and incubating it. The new culture is known as a *stock culture*. Next, the job entails making larger and larger volumes of microbial culture in a process called *scaling up the inoculum*. To prepare the inoculum for large-scale fermentations, a microbiologist inoculates a small volume of sterile medium and incubates it. The new culture is then used to inoculate a larger volume. Each inoculum volume is usually 5 to 10 percent of the next volume. Repeating these steps causes the volume of the inoculum to increase progressively, until it its large enough to inoculate a massive industrial-size bioreactor. For example, for inoculating a 3,000 l bioreactor, the inoculum is scaled up until it reaches at least 150 l, 5 percent of the total volume. The intermediate cultures leading from a stock culture to the final inoculum are referred to as the working cultures. A simple example of inoculum scale-up is shown here (v/v = volume inoculum per volume of medium):

10 % v/v   10 % v/v

Primary inoculum → Secondary inoculum → Production

(flask) (bench bioreactor) (large-scale bioreactor)

GROWTH IN BIOREACTORS

The purpose of growing bacteria or yeasts in bioreactors is to make a desired end product. Industrial microbiologists seek to do this in the most efficient way. In other words, they try to maximize the product's yield. Correct nutrients and physical conditions (temperature, oxygen, etc.) help maximize yield. At the same time, the microbiologist tries to minimize the buildup of any factors that inhibit growth, substances such as microbial waste.

Waste products are the normal end products of microbial growth. Cells follow a life cycle of consuming and digesting nutrients, growing and replicating, and then ceasing to grow, then dying. When cells die, they break apart (lysis), and their contents drift into the culture medium. These waste materials build up in a culture over time and inhibit growth. Eventually, enough wastes build up so that even young, growing cells cannot live, and the entire culture dies. Waste buildup in this manner lowers the yield of a desired product.

Cultures grown in bioreactors produce primary metabolites and secondary metabolites. Primary metabolites are compounds made by cells that are directly related to their growth. These compounds may result from the cell's energy-generating reactions or from building cell structures. Primary metabolites are often compounds that industrial microbiology seeks to produce. Examples are amino acids, enzymes, alcohols, and acids.

Secondary metabolites are substances that accumulate in a culture and are not essential for a microorganism's growth. These substances either may be harmless to the cell and have no effect on yield or may interfere with fermentation and alter yield. Some industrial fermentations are set up to produce these compounds rather than primary metabolites. Vitamins, antibiotics from bacteria, and mycotoxins from fungi are examples of secondary metabolites.

Bioreactor cultures make primary metabolites and secondary metabolites at different points in their growth. Primary metabolites appear during periods of active growth. In microbiology, this period is called the *logarithmic phase* and represents the period in which cells multiply at their fastest. Secondary metabolite production correlates with slower cell multiplication. This period is the stationary phase, in which the number of new cells being formed equals the number of dying cells. In more complicated scenarios, microorganisms make secondary metabolites from primary metabolites as they grow.

Fermentation scientists have developed techniques for controlling the growth of bioreactor cultures and thereby have made fermentations more efficient. But a large number of factors affect growth inside a bioreactor, and their relationships can be rather complicated. Fermentation scientists have, therefore, created mathematical equations to predict the effect of growth on cell yield and product yield. The use of these equations to predict yield is called *fermentation modeling*. Modeling helps microbiologists calculate the best conditions for both growth and making a desired product. Sometimes, the best conditions for growth are not the best conditions for product yield, and vice versa.

All of the factors that go into building a fermentation model can be measured by sampling the contents of the bioreactor. Bioreactors have several ports in the outer jacket that extend all the way into the fermentation chamber. Microbiologists use these ports to obtain samples of the culture. They then analyze these samples in a laboratory to learn about the conditions inside the bioreactor. Some laboratories equip their bioreactors with devices that take samples and analyze them automatically. The most common measurements taken on bioreactor cultures, either manually or automatically, are cell concentration, substrate (nutrient) concentration, pH, oxygen levels, and product concentration.

In a single bioreactor culture, the concentration of cells, substrate, or product is

$$(\text{concentration 1} - \text{concentration 2}) \div (\text{time 1} - \text{time 2})$$

Over time, cells and products build up and substrate disappears. These events occur faster at some points in cell growth than at other points. Primary metabolites build up fastest during the logarithmic phase of growth, as mentioned earlier. Eventually, growth in the entire culture slows, and the culture makes no further product.

Fermentation modeling occurs in either batch cultures or continuous cultures. Batch cultures are those in which the entire growth cycle of the microorganisms takes place inside a closed system within the fermentation vessel. Microbial growth starts out slowly, then goes into its logarithmic phase, then its stationary phase, and, finally, the culture begins to die. The entire life cycle is called a microbial *growth curve*. Continuous cultures differ from batch cultures in that fresh medium enters through one port in the bioreactor, and spent medium with wastes, dead cells, and products exits through another port. Microbiologists can control the growth phases by controlling the rate at which the medium flows through the vessel.

Batch cultures are better for making secondary metabolites from most microorganisms, and continuous cultures are best suited for producing primary metabolites.

Modern bioreactors have sensors that automatically measure conditions inside the fermentation vessel. The sensors connect to a computer, which carries out many of the calculations of fermentation modeling. Constant monitoring is helpful for making adjustments in continuous cultures as the microorganisms grow. With automated computer monitoring using sensitive sensors, industrial microbiology produces the best possible conditions

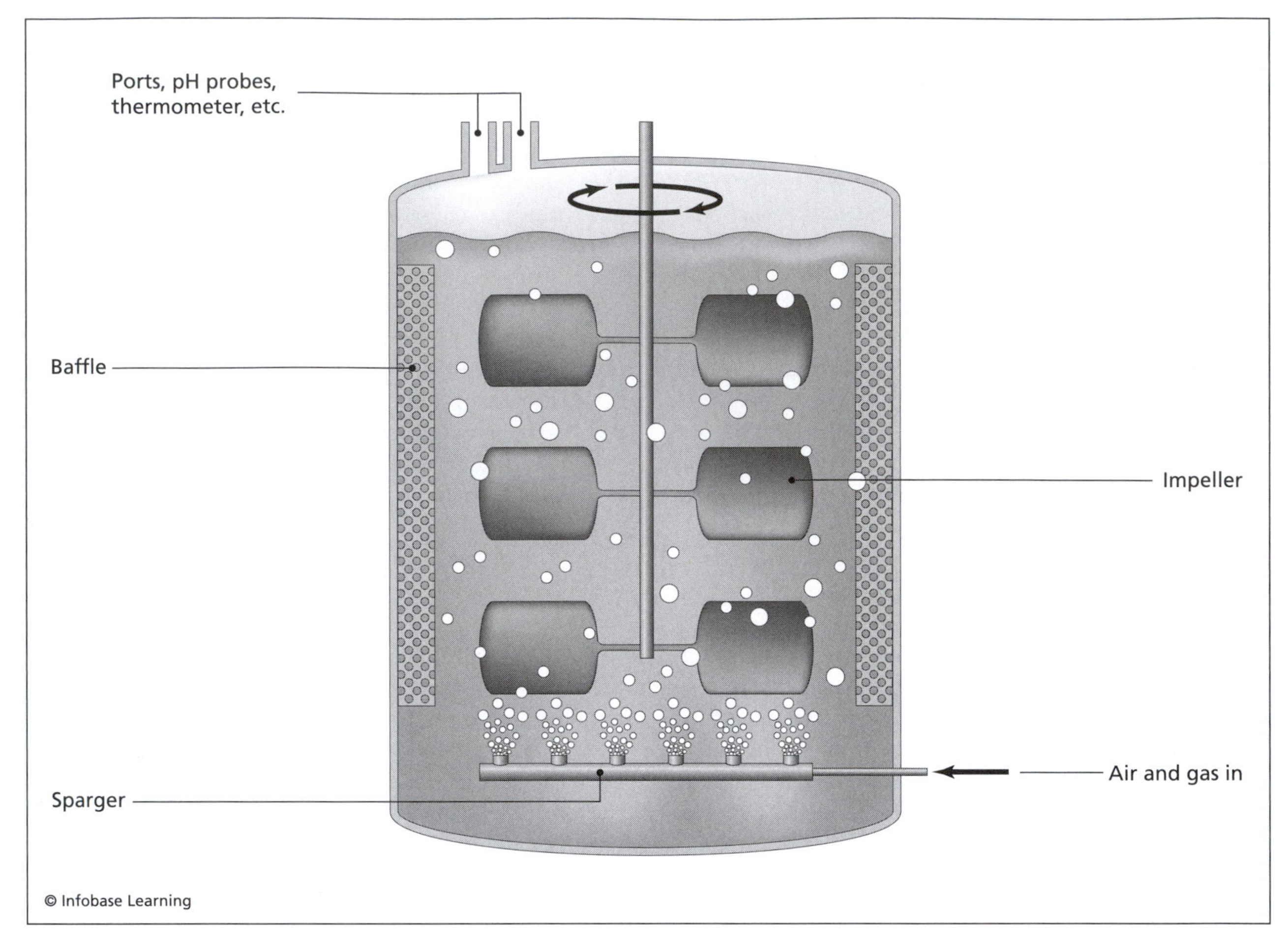

Bioreactors provide optimal growth conditions for microorganisms: nutrients, temperature, pH, and oxygen. Various features maximize mixing and aeration, sampling, and additions. Continuous flow bioreactors create a constant inflow of fresh medium and constant outflow of spent medium.

for increasing the yield of a desired end product. Some of the conditions under constant monitoring and adjustment at most fermentation facilities are temperature, pH, dissolved oxygen, turbidity (cloudiness), oxygen consumption, and carbon dioxide production.

Monitoring produces meaningless results unless the monitoring system works in concert with the adjustment system. This loop in information from the culture and to the culture is called *feedback*. An example of feedback processing is the control of pH. In this example, pH must be held within 0.2 unit above or below 7.0. An electrode takes pH readings in the culture and sends the information to a computer called a *controller*. The controller, then, calculates whether acid or base should be added to the culture; it adds acid if the pH goes to 7.2 and base if the pH goes to 6.8. The controller orders a device to add a small amount of either acid or base to the culture. Some adjustments do not rely on additions of a solution to the culture. For example, dissolved oxygen levels can be controlled simply by changing the mixing speed of the paddles (called impellers) inside the culture vessel.

The newest bioreactors use fiber optics to track the concentrations of the culture's dissolved compounds. Sugars and alcohols each absorb characteristic wavelengths of light, which are measured on an instrument called a *spectrophotometer*. Measurements such as these are called *real-time measurements,* because they describe culture conditions in the present moment. Real-time measurements provide the most valuable means of controlling bioreactors and product yields.

Bioreactors can be set up to grow aerobic (with oxygen) and anaerobic (without oxygen) microorganisms. Aerobic growth is helped by vigorous mixing, a process that aerates the liquid medium. Anaerobic growth is maintained by pumping a gas, such as a nitrogen–carbon dioxide mixture, into the fermentation vessel to keep out any traces of oxygen. Each bioreactor has several ports. Some of these ports are for the monitoring sensors, as mentioned. Others are sampling ports, where technicians remove small volumes of the culture. Culture samples are used mainly for determining cell concentration. Most bioreactors also contain a small port for the addition of antifoam. Antifoam compounds help reduce the bubbling or foaming that commonly occurs in cultures that receive constant mixing. Finally, large bioreactors often have a small window and a lamp inside the fermentation vessel for watching the culture as it grows.

THE DEVELOPMENT OF SPECIALIZED BIOREACTORS

In the early days of industrial microbiology, workers struggled with hundreds of tubes and flasks to manufacture a modest product yield. The first batches of penicillin produced in the United States, for instance, amounted to only a few grams. These cultures could take up to a week to grow. Glass fermentation vessels marked a step forward in supplying larger and continuous amounts of end products. But glass vessels were breakable and difficult to sterilize. Technicians either sterilized them in an autoclave or boiled water inside the vessel for several minutes. Before long, glass bioreactors became impractical for industrial microbiology.

Stainless steel allowed industrial microbiology to develop much larger fermentation volumes more safely than could be accomplished with glass. The metal inner chamber is sterilized in situ, meaning in place, and does not crack or burst, as glass does. Sampling ports were soon added, and after that, the sensors and feedback systems for controlling the growth inside the vessel.

Three different bioreactor designs solve the problem of giving aerobes a constant supply of air. The first is called an *airlift fermenter;* it pumps air in from below. The air circulates through the liquid and then exits from a port at the top. The second version is the jet loop fermenter, in which the entire culture exits at the top of the vessel, flows through tubes to a port at the vessel's bottom, and then reenters. As the culture flows through the circulation tubing, it gathers oxygen. Third, the bubble column fermenter mixes and aerates a culture by bubbling gas from the bottom of the vessel. Bubble column fermentation works well for cultures that are not viscous, or thick, and therefore are easily aerated. Bubble columns also work in anaerobic cultures, but, in this case, oxygen-free gas bubbles through the bioreactor's contents, rather than oxygen. Most bioreactors include baffles along the inner lining of the fermentation vessel. These plates aid in oxygen mixing by adding more turbulence to the contents as the impellers mix the culture.

Certain microorganisms that do not grow well as free-floating cells in liquid medium need a stable surface for attachment. Fixed bed and fluidized bed are two types of bioreactors that provide these microorganisms with an inanimate growth surface. Microorganisms that have attached to a surface are called *immobilized cells.* Fixed bed bioreactors contain dense beads that sit at the bottom of the culture and serve as the growth surface. Fluidized bed bioreactors also contain beads, but they are made of materials that float and circulate within the medium. Beads are usually composed of calcium alginate or calcium pectate, but ceramic, nylon, and even wood particles have been used. Immobilized cells today make high-fructose syrup, aspartic acid, and various biotechnology products.

Continuous flow bioreactors, described previously, were an important innovation in fermentation microbiology. With continuous flow, microbiologists could control almost every aspect of growth and metabolism inside the fermentation vessel. Product yields have been increased and made more efficient than with batch cultures.

INDUSTRIAL FERMENTATIONS

Products made by industrial microbiology serve several other industries, primarily industrial chemicals, agriculture, biofuels, the drug industry, and the food industry (see the table on page 118). About 70 percent of the ingredients now used by the food industry are produced in bioreactors.

Biomanufacturing refers to the production of materials through biological reactions. Most biomanufacturing takes place in bioreactor cultures that make specific end products for the industries discussed earlier. Specific industries that depend on biomanufacturing have perfected its steps by breaking the fermentation process into two main areas: upstream processes and downstream processes. Upstream processes comprise all the steps from the master culture to the bioreactor inoculum. Improvements in upstream processing tend to focus on finding the best microorganisms for making a desired product and the best growth conditions for preparing the inoculum. Downstream processes encompass all the operations connected with harvesting the final product from the bioreactor. The purpose of scale-up studies is to make downstream processes more efficient. Many companies adjust the fine points of their downstream processes in pilot plants, where technicians learn the best ways to convert pilot-scale fermentations to industrial-scale fermentations.

Products Made by Microorganisms in Bioreactor Fermentations

Industry	Product
agriculture	gibberellins, nitrogen-fixing bacteria, pesticides
biofuels	hydrogen, methane, ethanol
drugs	antibiotics, vaccines, steroids, alkaloids, hormones, interferons, nucleotides, monoclonal antibodies, cancer drugs, streptokinase
food processing	vitamins, amino acids, organic acids (acetic, citric, fumaric, gluconic, itaconic, kojic, lactic), enzymes, polysaccharides
food products	dairy, brewing, wine making
industrial chemicals	ethanol, acetone, butanol, 2,3-butanediol, enzymes (lipase, protease, amylase, cellulase, oxidase), biopolymers, biosurfactants

Downstream processes correlate with all the activities that assure a bioreactor culture has produced the correct product and in the most efficient way. The table on page 119 summarizes the main activities that make up downstream processing of a biological product.

Upstream or downstream processes may be ruined by contamination or by equipment failure. Carrying out large bioreactor fermentations, therefore, requires a significant amount of labor. Microbiologists concentrate solely on the problem of contamination. They must find ways to scale up fermentation volumes, yet prevent contaminants from entering at any point from master culture to final product. For this reason, scale-up from the laboratory to the production plant can be an inefficient period in industrial fermentation. Running bioreactors requires time to be set aside for repeated equipment cleaning and sanitization, sterilization, and training. Usually, at least one trial run is conducted in the large industrial bioreactor before production actually starts. This period in biomanufacturing is called downtime.

Downtime can be reduced by using standard techniques in microbiology that help prevent contamination: aseptic techniques, pure cultures, and sterilization. Companies also minimize downtime by learning as much as they can about the favorable growth conditions for a microorganism and for the biosynthesis of the desired product. Optimal conditions for the fermentation can be achieved by adjusting the following: nutrients, precursor compounds for making metabolites, temperature, pH, oxygen supply, and trace growth factors. Paul Kubera, head of the engineering group for a company that manufactures bioreactors, told *Genetic Engineering News,* in 2008, "The technology is improving, and it is now possible to grow culture to higher densities and achieve higher product [concentrations] in smaller reactors." Bioreactor technology will continue to drive the biotechnology industry and be important in drug manufacture and other industries.

Despite the challenges of maintaining and running bioreactors, these instruments serve a critical role in providing biological products that are less expensive and more natural than those produced by chemical manufacturing methods.

See also AEROBE; ASEPTIC TECHNIQUE; CONTINUOUS CULTURE; FERMENTATION; GROWTH CURVE; INDUSTRIAL MICROBIOLOGY; MEDIA; STERILIZATION.

Further Reading

Demain, Arnold L., and Julian Davies. *Manual of Industrial Microbiology and Biotechnology,* 2nd ed. Washington, D.C.: American Society for Microbiology Press, 1999.

Lipp, Elizabeth. "Flexibility Crucial in Bioreactor Systems." *Genetic Engineering News,* August 2008. Available online. URL: www.genengnews.com/articles/chitem.aspx?aid=2563&chid=3. Accessed Match 17, 2009.

McNeil, B., and L. M. Harvey, eds. *Fermentation: A Practical Approach.* Oxford, England: IRL Press, 1990.

Prescott, Lansing M., John P. Harley, and Donald A. Klein. "Industrial Microbiology and Biotechnology." In *Microbiology,* 6th ed. New York: McGraw-Hill, 2005.

Smith, John. E. *Aspects of Microbiology 11: Biotechnology Principles.* Washington, D.C.: American Society for Microbiology Press, 1985.

Tortora, Gerard J., Berdell R. Funke, and Christine L. Case. "Applied and Industrial Microbiology." In *Microbiology: An Introduction,* 8th ed. San Francisco: Benjamin Cummings, 2004.

Walsh, Gary. *Biopharmaceuticals: Biochemistry and Biotechnology,* 2nd ed. Chichester, England: John Wiley & Sons, 2004.

Downstream Processes

Process	Description
fermentation	growing a culture to make a specific end product
cell recovery	centrifugation or filtration to recover whole cells from the culture for the purpose of harvesting an intracellular (inside the cell) product
cell removal	centrifugation or filtration to remove whole cells from the culture fo the purpose of harvesting an extracellular (excreted from the cell) product
cell disruption	breaking apart cells to release a product
initial purification	removal of debris or soluble compounds by either filtration or precipitation
fine purification	biochemical techniques to recover the end product in pure form
analysis	determination of purity, potency, or activity
formulation	addition of any ingredients needed to stabilize or preserve the product
packaging	sterile or aseptic transfer to bottles, tubes, ampoules, etc.
finalization	sealing the package, checking for errors, labeling, and packing for shipment

bioremediation Bioremediation consists of all the procedures employed for using microorganisms to clean up chemical pollution in the environment. Bacteria serve as the main bioremediation microorganism, but fungi also play a part in some pollution cleanups. Bioremediation has been used for contaminated sites on land and in bodies of water, such as lakes, rivers, and wetlands. Today, this method has become increasingly widespread for cleaning the following types of pollution: oil spills, industrial chemical spills, fuels and solvents, pesticides, chlorinated organic compounds, heavy metals, and mining wastes.

Bioremediation has four main advantages over other types of pollution cleanup methods. First, it does not disrupt the land or create noise and dust, as do trucks, bulldozers, dredging equipment, or other large machinery. Second, bioremediation avoids the use of chemicals for neutralizing or binding the pollutants. Third, bioremediation does not disturb wildlife living in the area and, in fact, can help restore the environment. Fourth, bioremediation is easy and inexpensive.

Bioremediation has two potential drawbacks that must be considered along with its advantages. First, microorganisms take a long time to remove hazards from a heavily polluted site, especially in comparison to mechanical methods such as excavation with heavy equipment. Second, bioremediation technology is based on the use of genetically modified organisms (GMOs). GMOs are microorganisms into which a foreign gene has been inserted, giving the microorganism new, beneficial traits.

Although GMOs have been created to clean up highly dangerous compounds in the environment, many people oppose their use because the microorganisms are not normally found in nature. What would happen in nature, they ask, if a GMO escapes into the environment and interiors with nature's processes? This question has been debated by scientists and nonscientists, with perhaps as many people opposing GMOs as those who accept the use of GMOs. The Health and Safety Executive (HSE), an organization in the United Kingdom that monitors workplace safety, has commented on GMOs: "The vast majority of work with GMOs in contained use [measures to prevent GMO escape to the environment] is inherently safe. This is because most work involves the insertion of genes into microorganisms that have been deliberately 'crippled' with disabling mutations so that they will not grow outside of the controlled environment of a laboratory test tube." These disabling genes are sometimes referred to as *suicide genes,* because they ensure a microorganism will die if it escapes into the environment.

Bioremediation includes alternative methods that do not use GMOs. Scientists have developed specialties within the field of bioremediation so that the type of microorganism used to clean up pollution is selected on the basis of its growth characteristics in the environment, its genetic makeup, and the compounds it degrades. Environmental scientists also classify each bioremediation task according to whether pollutants must be fully degraded, partially degraded, or converted into another compound altogether.

TYPES OF BIOREMEDIATION

Four types of bioremediation remove dangerous pollutants from the environment: (1) biodegradation, (2) bioconversion, (3) intrinsic bioremediation, and (4) bioaugmentation. By each of these methods, microorganisms are said to *detoxify* the harmful compounds contaminating soil or water. Remediation specialists, therefore, sometimes refer to bioremediation as the *detoxification* of pollutants.

Biodegradation entails the microbial decomposition of organic compounds. Microorganisms biodegrade organic matter that naturally builds up in the environment—decomposition of organic matter is one of the most important roles of microorganisms on Earth. Examples of the organic matter that microorganisms decompose in nature are decaying plants, broken tree limbs, branches, leaves, dead wildlife, and wastes produced by wildlife. Microorganisms can use the same, or very similar, metabolic pathways (reactions that produce energy for the microorganism) to degrade organic compounds harmful to the environment. Biodegradation is now used to treat organic compounds such as pesticides, solvents, fuels, and munitions.

Three different types of biodegradation occur; they differ in the extent to which a microorganism destroys a compound. The first type of biodegradation involves minor changes to a compound's structure to make it less toxic. For example, a microorganism may remove a chlorine molecule from a chlorinated organic compound, and this small step can reduce the toxicity (capacity to do harm to living things) of this compound. The removal of a chlorine (Cl^-) occurs with substitution by another chemical structure, such as a hydroxyl (-OH) group. The second type of biodegradation is called *fragmentation*. This occurs when microorganisms degrade a toxic compound by breaking it into fragments. Although most fragmentation results in a less toxic molecule than the original molecule, fragmentation may create two new problems. First, the reaction can release new molecules of unknown toxicity, and, second, the fragments can actually be more toxic than the original molecule. Third, microorganisms break pollutants completely in a process called *mineralization*. In mineralization, microorganisms degrade pollutants until nothing remains but carbon dioxide, water, and salts. Mineralization, therefore, renders toxic compounds completely harmless. By contrast, chemical substitutions or fragmentation reactions may merely reduce toxicity but not remove it.

The second mode of bioremediation, bioconversion, resembles the reactions that substitute one chemical group for another in order to reduce a compound's toxicity. This process can be called *neutralization*. Examples of bioconversion reactions are the following: cleaving an aromatic (ring) structure from a more open structure such as a chain, removal of an entire ring structure, or reactions that change the entire chemical structure of a pollutant. The conversion of organic wastes to fuel for energy offers an example of this latter type of bioconversion. Many communities use bioconversion to produce energy from organic wastes such as manure, crop debris, and tree cuttings.

Environmental scientists have developed machinery to carry out biodegradation or bioconversion reactions for treating polluted soils and water. Municipal wastes and excavated polluted soils can be treated in bioreactors containing the microorganisms that will attack the pollutants. In less sophisticated cases, workers put wastes into a large pile and inoculate the waste pile with microorganisms. This mixture of pollutants and bioremediation microorganisms is called a *biopile*. The final two types of bioremediation take place entirely within the soil, so they represent in vivo bioremediation, whereby the process takes place in nature, or literally "in life." These in vivo bioremediation methods are intrinsic bioremediation and bioaugmentation.

Intrinsic bioremediation uses only the microorganisms already living in a polluted area, that is, resident microorganisms or resident flora. Resident bacteria may be capable of breaking down pollutants on their own, but their activity takes a long time. Environmental scientists developed intrinsic bioremediation to speed the cleanup process. Intrinsic bioremediation is also called *natural attenuation,* meaning that people enhance the natural activities in the earth. Scientists do this by adding extra nutrients to a polluted area. The nutrients then help resident bacteria and fungi grow faster and thus degrade pollutants faster than if they were left alone. Nutrients commonly used for intrinsic bioremediation are nitrogen and phosphorus compounds, air to supply oxygen, or methane gas for methanotrophs (bacteria that require methane).

Bioventing offers a specialized method of intrinsic bioremediation in which a pump pushes air down a shaft to underground pollution. The influx of air then helps aerobic bacteria or fungi grow on the toxic compounds in their midst.

Intrinsic bioremediation is currently in use cleaning up pollution at a nuclear weapons facility in South Carolina called the Savannah River Site. Intrinsic bioremediation also helped with the beach cleanup, in 1989, after the oil tanker *Exxon Valdez* ran aground and spilled 33,000 tons of crude oil along the shores of Alaska's Prince William Sound. The Exxon Company investigated intrinsic bioremediation shortly after the spill to help clean up oil-damaged shorelines. In this case, scientists used additives to enhance the intrinsic activities. The Exxon scientist Roger Prince told a group of biologists and reporters in 1993 that the company had spread more than 50,000 pounds [22,680 kg] of material over 74 miles [119 km] of beach to speed the native microorganisms at the site. "Microbes need nitrogen to build more biomass, to grow more rapidly, and eat more of the oil," he told the Bureau of National Affairs. "As they eat the oil it converts the oil into more microbes (at) a conversion rate of 50 percent." At the time, however, few attempts had ever been made at this type of bioremediation on the scale of the *Exxon Valdez* spill.

The *Exxon Valdez* bioremediation actually used a combination of intrinsic activities and bioaugmentation, in which a substance is added to the polluted site to enhance the intrinsic activities. The cleanup crews used a nitrogen-phosphorus mix, spread over the beach, and allowed the natural microorganisms to make use of these nutrients to grow better and degrade more oil, which served as the microbial carbon source. By five years after the spill, hundreds of research articles had been published on the effects, the successes, and the failures of the bioremediation. Despite the fact that scientists could not be certain that the added nutrients would work as hoped, the Alaska project proved the potential of bioremediation.

The scientists who worked on the *Exxon Valdez* cleanup could be said to have augmented the earth's intrinsic activities with adding nutrients. Bioaugmentation has since developed into its own specialty within bioremediation. In addition to nutrients, bioaugmentation can use specially designed microorganisms to aid the metabolism of the soil's natural microorganisms. Biologists develop bioaugmentation bacteria by finding species that already grow on toxic compounds. In the laboratory, they then modify the bacteria by genetic engineering. The result is a new strain of bacteria with improved capabilities. (A strain is a single unique cell line within a certain species, usually possessing a special trait.) Workers then return to the same polluted site and add the new strain to the normal populations already living in the soil. Bioaugmentation by this method has been used for cleaning up fuel spills and oil spills.

Although bioaugmentation cannot by itself restore habitat to its original condition, it complements other cleanup technologies. This technique may play a critical role in the long-term rehabilitation of land polluted by the BP oil rig spill of 2010.

Bioaugmentation offers two advantages. First, it can be used for developing GMOs that contain very specific traits for the particular pollution problem. Second, if a community wishes to avoid the use of GMOs, this bioremediation approach offers the simple and safe use of nutrients rather than GMOs.

Biologists have now developed specialized bacteria to degrade a wide range of pollutants, even radioactive substances. The beginning step involves finding microorganisms in the environment that already use a pollutant in their metabolism. Some of the widely used bioremediation bacteria and fungi and their capabilities are listed in the table. Microorganisms called *extremophiles* have also been applied to pollution cleanup. Extremophiles live in environments where few other organisms survive, and some of them thrive in very acidic runoff from mining operations or in land polluted with heavy metals.

GENETICALLY MODIFIED MICROORGANISMS IN BIOREMEDIATION

The science of making GMOs has entered the mainstream of biology, but many people still worry about the potential dangers of laboratory-created organisms.

Molecular biologists now add safety features to GMOs, such as suicide genes. All bioremediation GMOs begin with the following general process:

1. Discover species growing in the environment and in the presence of a pollutant.
2. In a laboratory, test pure cultures for ability to grow on the pollutant.
3. Select the best grower and determine the gene(s) for the pollutant-degrading enzyme(s).
4. Insert degradation genes into another hardy, fast-growing microorganism.
5. Apply the new GMO to the polluted soil.

Step 4 may include the insertion of a suicide gene. The gene lets a cell die without further replication when a target pollutant has disappeared from the environment. In other words, the GMOs die when their job is complete. Molecular biologists have fine-tuned the suicide system to work on a variety of signals, so the GMOs have decreased opportunity to escape a prescribed area. The most critical advance in suicide genes may be a mechanism by which a cell can detect mutations in its own DNA and immediately shut down all of its operations. If successful, this offers the safest approach to preventing GMOs and unknown mutants of GMOs from escaping into the environment.

GMOs also have potential when being used in mechanical pollution cleanup. In this instance, a bioreactor can be taken to a pollution site and inoculated with GMOs in a liquid medium of nutrients. Polluted soil or water, then, may be pumped into the reactor to allow the GMOs to degrade the toxic chemicals. This method offers the advantage of using GMOs to their fullest capacity, while keeping them contained inside the bioreactor.

THE FUTURE OF BIOREMEDIATION

Bioremediation is a safe and natural way to remove pollution from the environment. Nevertheless, envi-

Microorganisms Used in Bioremediation

Aerobic
Bacillus, Pseudomonas, Arthrobacter, Rhodococcus, Candida (yeast), *Acinetobacter, Phanerochaete chrysosporium* (fungus)
Anaerobic
Wolinella, Dechloromonas, Dechlorosoma
Readily degrades radioactive substances
Deinococcus radiodurans
Readily degrades pentachlorophenols (PCPs)
Sphingomonas chlorophenolica
Readily degrades polychlorinated biphenyls (PCBs)
Desulfitobacterium, Dehalospirillum, Desulfomonile

ronmental scientists and microbiologists confront several obstacles in making bioremediation more effective and more accepted in the public's mind. The first challenge deals with pollutants that most microorganisms cannot metabolize. Pollutants that remain in the environment a long time (decades) because microorganisms do not use them well are called *recalcitrant compounds*. Second, bioavailability of the toxic compound affects the success of bioremediation. Bioavailability of a pollutant is its accessibility to microorganisms. In other words, pollutants that are enclosed in clay that bacteria cannot penetrate are not bioavailable. Pollutants that are not enclosed or otherwise bound to soil are probably more bioavailable. Third, microbiologists must ensure that bioremediation species prefer the pollutant to other compounds in the soil. Microorganisms always select substrates (compounds used in metabolic reactions) that will allow the cell to conserve energy as it grows. For example, bacteria capable of degrading either sugar or starch will use up the sugar first, because it takes less energy than degrading the starch into absorbable pieces, before using it as a carbon-energy source. Bioremediation experts must invent new strains that prefer a toxic compound to other available compounds.

The following list summarizes six microbial factors that affect how well bioremediation works to clean up pollution:

- genetic ability of bioremediation microorganism to degrade the toxic compound
- bioavailability
- pollutant that is not toxic to the microorganisms
- energetically favorable for the microorganism to degrade the pollutant rather than other compounds
- breakdown products safe for the environment
- GMO, if used, safe for the environment and for people using it

Bioremediation has a bright future in pollution cleanup, yet, ironically, this is not a new science because microorganisms have been degrading wastes since life began. Modern bioremediation will advance as microbiologists learn how to design better species for specific tasks and ways to control them in the environment. With these attributes, bioremediation plays a critical role in pollution cleanup.

See also BIOREACTOR; EXTREMOPHILE; GENETIC ENGINEERING.

Further Reading

Alexander, Martin. *Biodegradation and Bioremediation.* San Diego: Academic Press, 1994.

Bragg, James R., Roger C. Prince, E. James Harner, and Ronald M. Atlas. "Effectiveness of Bioremediation for the *Exxon Valdez* Oil Spill." *Nature* 368 (1994): 413–418. Available online. URL: www.nature.com/nature/journal/v368/n6470/abs/368413a0.html. Accessed March 17, 2009.

Bureau of National Affairs. "Bioremediation Effective in Cleanup of *Exxon Valdez* Spill: Company Reports." *Environment Reporter* 23, no. 51 (1993): 3,168–3,169. Available online. URL: www.valdezlink.com/inipol/pages/bna.htm. Accessed March 17, 2009.

Cornell University and Penn State University. "Environmental Inquiry." Available online. URL: http://ei.cornell.edu/biodeg/bioremed. Accessed March 17, 2009.

Health and Safety Executive. "Genetically Modified Organisms (Contained Use)." Available online. URL: www.hse.gov.uk/biosafety/gmo. Accessed March 17, 2009.

U.S. Environmental Protection Agency. Technology Innovation Office. "A Citizen's Guide to Bioremediation." Available online. URL: www.epa.gov/superfund/community/pdfs/suppmaterials/treatmenttech/bioremediation.pdf. Accessed March 17, 2009.

U.S. Geological Survey. "Bioremediation: Nature's Way to a Cleaner Environment." Available online. URL: http://

water.usgs.gov/wid/html/bioremed.html. Accessed March 17, 2009.

biosensor A biosensor is a device that uses a biological reaction for detecting specific substances in the environment. To accomplish this, biosensors contain either whole bacterial cells or a component of bacterial cells, usually enzyme systems. Biosensors detect, within minutes, or even seconds, chemical and biological compounds in the environment. Quick response and sensitivity to very small concentrations of compounds make biosensors an attractive tool in pathogen detection, in foods and water, and as a warning system against biological weapons. See the color insert on page C-1 for a picture of a light-emitting biosensor (lower right).

Medical biosensors have been developed to detect small changes in a person's physiology for the purpose of detecting disease, such as certain blood proteins that may indicate presence of a cancerous tumor. Environmental biosensors detect chemicals or biological agents in the environment. These sensors help scientists to monitor the presence of hazardous chemical pollutants and of biological pollution from wastes and wastewater or, most recently, to detect of bioweapons in the environment. Current biosensor technology involves devices that detect, record, and transmit information.

Scientists developed the first biosensors for practical uses, in the 1980s, at the time when biotechnology was rapidly discovering new ways to use individual cell components or even specific genes. For example, scientists at the Oak Ridge National Laboratory (ORNL), in Tennessee, developed biosensors to detect cancer-causing chemicals in underground sources of drinking water. The ORNL biosensor contained a fiber-optic piece linked to an antibody developed specifically to detect the carcinogen benzene(a)pyrene. As soon as the antibody finds and binds to the target chemical, the light portion of the biosensor causes the emission of fluorescent light. The fiber-optic component detects the fluorescence and sends a signal up the fiber to a monitor. Because the telesensor uses light as its mode of carrying information, it is also referred to as an optical biosensor. This biosensor won for ORNL the 1987 Research and Development 100 Award.

Biosensor technology has since expanded into additional detection systems for a greater number of chemicals and biological substances. The latest medical biosensors are sometimes called *telesensors;*

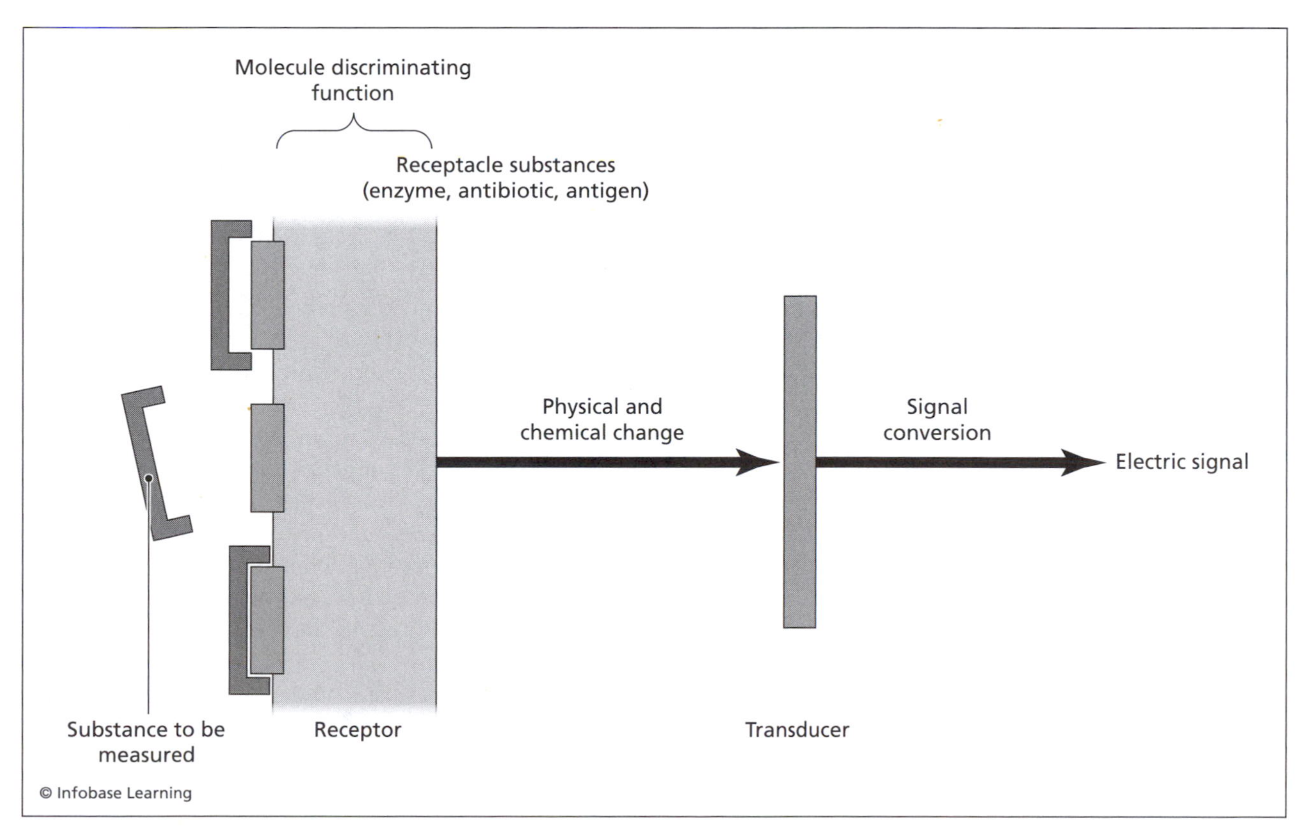

Biosensors use whole microbial cells or parts of cells to detect specific substances in the environment or in the body. Biosensors link specific biological reactions with an electronic or visible light.

they are silicon chips measuring about 2 by 2 millimeters (mm). Telesensors placed on the skin can measure body temperature, pulse rate, and various changes in the body's physiology. As all biosensors do, telesensors offer the valuable advantage of providing this information as it occurs. Biosensors shorten the time that physicians and environmental scientists spend in detecting chemicals in patients or the environment, respectively.

An industry called *bioelectronics* has grown, in the past decade, with the goal of inventing more sensitive and molecule-specific biosensors. In addition to an expanding list of medical uses, the following industries use biosensors: breweries, food production, hazardous site monitoring, wastewater treatment, and water quality testing. Bioelectronics engineers work on improving current biosensors to make them more sensitive, durable, stable, and portable.

COMPONENTS OF BIOSENSORS

Biosensors are of two types: whole cell and cell component. Each type contains two parts; one is called the sensor and the other is the reporter. The sensor detects a specific target substance, and the reporter emits a visible reaction when the sensor detects the target. In many of today's biosensors, both the sensor and the reporter are enzyme systems. In other biosensors, an enzyme serves as the sensor, and an antibody serves as the reporter. Two prominent biosensors that have been built upon bacterial systems are the *lux* biosensor and the streptavidin-biotin biosensor.

Whole cell biosensors use specialized bacteria to detect compounds in their surroundings or slight changes in their environment. These specialized bacteria have two sources: the natural environment or genetic engineering. Microbiologists develop *lux* biosensors by, first, finding bacteria that live in polluted places. The fact that the bacteria can thrive in soil with high levels of toxic compounds suggests that these bacteria have enzymes capable of acting on the pollutants. But simply growing bacteria on a pollutant does not help monitor pollution, unless the bacteria produce a reaction people can see. The second step involves designing bacteria that detect pollution and give a visible signal when they find it. The enzyme luciferase works well in this role and is the main component in *lux* biosensors.

Bacterial species of *Vibrio* and *Photobacterium* genera emit fluorescent light produced by luciferase during their normal metabolism. Microbiologists have removed the luciferase genes from these bacteria and inserted them into the chromosomes of pollutant-sensing bacteria. The result is a whole cell biosensor for pollution that produces a fluorescent and measurable signal. The set of genes that control luciferase production are known as the *lux* operon. An operon is any set of genes that function together to control a single activity. *Lux* operon biosensors now detect environmental and food toxins, pollutants, antibiotics, and pathogens.

A specialized type of whole cell biosensor has been developed in which compounds cause it to stop emitting a signal rather than emitting one. These bacteria normally emit fluorescence as they grow. In the presence of compounds toxic to the bacteria, cells begin to die, and the signal becomes weaker, until it eventually stops. The disappearance of fluorescence indicates the presence of the target compound. These types of whole cell biosensors have been helpful for detecting the presence of toxic chemicals or antibiotics.

Streptavidin-biotin biosensors use a cell component rather than whole cells. In the 1970s, scientists of the pharmaceutical manufacturer Merck and Company discovered a protein in *Streptomyces avidini* bacteria that held a very high affinity for binding the B vitamin biotin. (Biotin found in raw egg whites had long been known to form strong bonds with certain natural proteins, so Merck scientists searched for additional proteins for building a biosensor.) The streptavidin protein made by *S. avidini* worked well in building the basic components of a new biosensor.

The streptavidin-biotin biosensor contains three components:

- a probe containing an electronic component and attached to streptavidin
- a biotin molecule, which binds to streptavidin
- a binder that is linked to the biotin

The binder attaches to the target compound when it finds it in the environment. The biological engineer Marshall Porterfield of Purdue University explained to *Medical News Today,* in 2009, "Biotin and strepavidin are like tiny Lego blocks that are designed to hook together." Scientists use the streptavidin-biotin biosensor in the following stepwise manner:

1. A scientist puts the binder into a sample suspected of containing a certain target compound.
2. The binder attaches to the target compound.
3. The scientist incubates the sample with streptavidin.
4. Streptavidin binds to the biotin to create a complex of target-binder-biotin-streptavidin.

5. The complex activates the probe to create a measurable signal.

Science continues to expand on the type of binders that work in streptavidin-biotin biosensors. The table shows the options currently available in streptavidin-biotin biosensor technology.

THE USES OF BIOSENSORS

Biosensors now serve as important tools in environmental monitoring, health care, various industries, agriculture, and law enforcement. In these applications, biosensors accomplish either one of two objectives: to find a substance that should be present, such as a nutrient in a food product, or to find a substance that should not be present, such as cancer cells in blood. The table summarizes the main uses of biosensors in different disciplines. Many of the specialized biosensor uses contain additional specializations. For example, a biosensor used in environmental science does not simply detect water contamination, which is a broad area. Biosensors used in water quality monitoring are specialized systems for detecting components such as heavy metals, toxic organic chemicals, pesticides, hormones, antibiotics, and pathogens.

Biosensors range in size from a glass slide to the size of a cell phone. Some comprise simple strips that turn color in the presence of the target substance, while others provide a digital readout on a screen and connect to a computerized database.

Despite these improvements, biosensors still cannot cause changes in matter; they only detect characteristics of matter. The next generation of biosensors may be designed for three steps rather than two: (1) sense a target substance, (2) produce a signal indicating the amount of the target substance, and (3) react with the target substance.

NEW BIOSENSORS

The next generation of biosensors will undoubtedly possess improved detection capabilities. For instance, biosensors that now detect organic compounds to levels as small as parts per billion (ppb) may soon detect to parts per trillion (ppt) level or lower. New biosensors will additionally build on their current capabilities of detecting the presence or absence of substances and provide accurate measurements of the substances' concentration. Biosensor technology will follow this advance with units that detect, measure, and then react with a target. For example, the value of biosensors that can find and destroy cancer cells or neutralize toxic chemicals in water cannot be denied.

Nonfluorescent biosensors represent a type of new biosensor that uses an electric pulse rather than

Components and Uses of Streptavidin-Biotin Biosensors

Target Substance	Binder	Probes
antigens	antibodies	bacteriophages
antibodies	antigens	chemiluminescent agents
carbohydrates	lectins	chromophores
cell surface receptors	hormones, toxins	enzymes
cell transport proteins	vitamins, amino acids, sugars	ferritin
cell hydrophobic sites	lipids, fatty acids	fluorescent agents
enzymes	enzyme substrates, coenzymes, cofactors, vitamins, inhibitors	liposomes metals radioactive compounds
lectins	carbohydrate complexes	solid inanimate surfaces
membranes	liposomes	
nucleic acids, genes	deoxyribonucleic acid (DNA), ribonucleic acid (RNA)	

Source: Prescott, Lansing M., John P. Harley, and Donald A. Klein. *Microbiology*, 6th ed. New York: McGraw-Hill, 2005.

Biosensors in Different Disciplines

Biosensors Application	Task
environmental monitoring	drinking water contamination, wastewater treatment, organic and inorganic pollutant monitoring, hazardous chemical detection, and agricultural and domestic waste detection
human and veterinary health	detection of tumors, cancer cells, other cells, proteins, enzymes, biomarker compounds, pathogens, blood components, antibodies, antigens, nucleic acids, and genes
industry	food and dairy contamination, food composition, industrial waste discharge monitoring, testing for production plant contaminants, detection of toxic gas in mines, and measuring fermentation products
plant agriculture	detection of plant nutrients, plant pathogens, diseased cells, growth factors, fragrances, pheromones, and pigments
law enforcement	detection of narcotics, explosives, and bioweapons

light, which is a common signaling system used in current biosensors. In these biosensors, small electrodes carry a signal to the reporter. A transducer in the biosensor then converts the signal to another form that can be seen or measured. For instance, a tiny electrical pulse may be converted to a number on a digital display. These biosensors, therefore, make use of blending biological reactions with electrical activity. Other biosensors in this area of research emit signals in addition to an electrical charge: luminescence (a more general category of light emission than fluorescence), colors, and sound waves.

Biosensors containing the nucleic acids DNA or RNA are called DNA/RNA probes, DNA chips, RNA chips, enzyme chips, or simply biochips. Biochips detect changes in the chromosome of certain microorganisms. Environmental biologists now use biochips to study the ways in which microorganisms respond to changes in their surroundings. A biochip may, for example, detect a microbial secretion that only occurs in the presence of a particular toxin. Many molecular biologists agree that biochips may become one of the best tools for studying gene expression. This technology has uses beyond environmental science, notably the study of diseases in humans, animals, and plants.

The discipline of nanobiology also applies to biosensors that will work at dimensions even smaller than nucleic acids or enzymes. Because biological materials contain a negative charge, they can be bound to minuscule wires called *nanowires,* tube structures with a diameter the size of a molecule. Scientists in nanobiology have employed the streptavidin-biotin system with nanowires as a device to be used in health care. Porterfield has said of this technology, "This is the first time researchers have assembled from the atomic to the biomolecular level all the components you need for a biosensor." Some of the ideas proposed for nano-size biosensors are enzymes associated with nerve communication to study brain and nerve function, enzymes that attach to ethanol to be used as blood-alcohol sensors, detection of stress factors in agricultural plants, and sensors for the acquired immunodeficiency syndrome (AIDS) virus that would bind and neutralize the virus in the bloodstream. Biosensor technologies such as these will develop rapidly along with nanobiology.

See also BIOWEAPON; CHROMOSOME; ENVIRONMENTAL MICROBIOLOGY; GENETIC ENGINEERING; NANOBIOLOGY.

Further Reading

Eggins, Brian R. *Analytical Techniques in the Sciences: Chemical Sensors and Biosensors.* Chichester, England: John Wiley & Sons, 2002. Available online. URL: www3.interscience.wiley.com/cgi-bin/bookhome/114189770. Accessed February 14, 2009.

Marks, Robert S., Christopher R. Lowe, David C. Cullen, Howard H. Weetall, and Isao Karube, eds. *Handbook of Biosensors and Biochips.* Chichester, England: John Wiley & Sons, 2007.

Matrubutham, Udaykumar, and Gary S. Sayler. "Microbial Biosensors Based on Optical Detection." In *Methods in Biotechnology.* Vol. 6, *Enzyme Biosensors: Techniques and Protocols,* edited by A. Mulchandani and K. R. Rogers. Totowa, N.J.: Humana Press, 1998.

Rogers, Kim R., and Marco Mascini. "Biosensors for Analytical Monitoring: General Introduction and Review." Available online. URL: www.epa.gov/heasd/edrb/biochem/intro.htm#environ. Accessed February 15, 2009.

Sayler, Gary S., Udaykumar Matrubutham, Fu-Min Menn, Wade H. Johnson, and Raymond D. Stapleton. "Molecular Probes and Biosensors in Bioremediation and Site Assessment." In *Bioremediation: Principles and Practice.* Vol 1., edited by Subhas K. Sikdar and Robert L. Irvine. Lancaster, Pa.: Technomics, 1997.

Sharpe, Michael. "It's a Bug's Life: Biosensors for Environmental Monitoring." *Journal of Environmental Monitoring* 5, no. 6 (2002): 109–113.

Venere, Emil. "Glucose Precisely Detected by Nano-Tetherball Biosensor." *Medical News Today*, 25 January 2009. Available online. URL: www.medicalnewstoday.com/articles/136508.php. Accessed February 14, 2009.

bioweapon A bioweapon, or biological weapon, is any device used in warfare and containing a deadly microorganism. Specialized microbiology laboratories have, in the past, experimented with weapons containing highly pathogenic bacteria, viruses, or microbial toxins. A pathogen is a disease-causing microorganism; a toxin is a substance that acts as a poison in the body.

Any pathogen or toxin put into a weapon intended to kill humans is termed a *biological agent*. Biological agents can be either microorganisms or substances made by microorganisms. The use of weapons containing biological agents is also referred to as *germ warfare* or *biological warfare*. (*Chemical warfare* is a term also used for weaponry containing lethal nonbiological chemicals. Sometimes the term is also used to describe the use of biologically produced toxins that have been extracted from nature.) *Bioterrorism* refers to the use of bioweapons by one or more individuals who have no affiliation with a national government.

THE HISTORY OF BIOWEAPONS

Many people believe bioweapons are a regrettable invention of the 21st century. On the contrary, they date to the earliest of wars. Ancient societies may have used human or animal waste as biological weapons. These weapons were not lethal, but they probably made enemies pause before storming their adversaries. In the 14th century, the Tartars added lethal power to their weaponry by catapulting the bodies of plague victims over the walls of enemy cities. Historians believe the first of several plague epidemics that ravaged Europe throughout the Middle Ages began in a battle at Kaffa, in 1347, in which such tactics had been used.

In the French and Indian War (1754–67), the British army contaminated blankets with the secretions from smallpox victims in their ranks. They then gave these items to Native Americans, purportedly as gifts, but actually in an attempt to kill off North America's tribes that had sided with the French.

In World War I, Germany worked on ways to infect their enemies with pathogens. One plan involved the contamination of animal feed given to livestock being shipped to the Allied side for feeding the troops. For example, German agents infected sheep bound for Russia with *Bacillus anthracis,* the cause of anthrax disease. Meanwhile, other saboteurs set about inoculating mules and horses of the French cavalry with *Burkholderia mallei.* This bacterial species causes the disease glanders in equines, so Germany probably considered this bioweapon a reasonable way to decimate their enemy's stock. Between 1917 and 1918, the Germans managed to infect mules in Argentina bound for American forces. The outcome included the death of 200 mules. This act did not inflict enough damage on the Americans to impact the war, but it signaled a new era in warfare: the use of deadly pathogens in a planned and controlled way.

World War II became the setting for additional work in laboratories of both the Allies and the Axis countries. Starting in 1932, Japan developed a series of biological weapon experiments by infecting prisoners in Manchuria, China, with a variety of pathogens. In a research facility that became known as Unit 731, prisoners received doses of the following pathogens: *B. anthracis, B. mallei, Salmonella thyphi* (the cause of typhus), *Neisseria meningitides* (meningitis), *Vibrio cholerae* (cholera), *Yersinia pestis* (plague), and smallpox virus. In rural towns, the Japanese are said to have contaminated water supplies and food with *Shigella, Salmonella,* anthrax, *V. cholerae,* and *Y. pestis.* Japan undoubtedly perfected their technique of handling lethal pathogens as it entered the war in 1942. Nonetheless, close to 2,000 Japanese troops died after handling their country's own weapons, as did unknown numbers of civilians working in weapons laboratories.

At the same time, Nazi Germany used the concentration camps as a testing ground for forced infections of *Rickettsia,* the hepatitis virus, and the malaria parasite *(Plasmodium).* The British side, meanwhile, developed bombs that would, upon exploding, release anthrax spores. The British conducted their experiments on Gruinard Island off the Scottish coast. Though these biological weapons were not used, the *B. anthracis* spores did what spores do best: They persisted on the island long after the war ended. Britain finally admitted to years of weaponry research on the island and in 1986 decontaminated the entire island with formaldehyde—itself a toxic chemical.

The United States began its biological weapons production in 1943 at Camp Detrick, Maryland (renamed Fort Detrick in 1956), testing *B. anthracis,* among other pathogens. During the Korean War (1950–53), a second facility, in Pine Bluff, Arkansas, geared up for large-scale pathogen production. Throughout the 1950s, the United States used volunteers for experiments on a variety of microorganisms:

Brucella, Aspergillus, Serratia, and additional *Bacillus* species. From these studies, weapons engineers learned how best to store, transport, and release biological weapons.

During the war in Vietnam (1962–68), China's ally, the Viet Cong, and South Vietnam, which the United States backed, accused each other of using biological weapons on troops and civilians. Fungal toxins called *mycotoxins* were implicated though their use was never proved. The possible use of mycotoxins opened a new era in bioweapons: the use of toxins rather than whole microorganisms. Lethal toxins are usually effective at very low doses, so military planes saw the potential of these compounds as weapons. A toxin extracted from castor beans, called ricin, became the first to be used in a political conflict. In 1978, an assassination attempt on a Bulgarian official (Vladimir Kostov) living in exile in Paris, France, involved a ricin pellet fired from a gun. The attempt was unsuccessful, but 10 days later, the Bulgarian exile Georg Markov died of a ricin-containing pellet shot at him at close range.

The cold war between the United States and the Soviet Union, in the 1960s through the 1970s, saw determined efforts on both sides in bioweapon research. For at least two decades, the separate sides advanced their technologies in putting extremely toxic substances or viruses into weapons. Yet, the work often seemed flawed to even casual observers. Laurie Garrett wrote, in her 1994 book *The Coming Plague,* "In the 1960s when biological warfare research was underway in the United States and the Soviet Union, both the U.S. military and the civilian Public Health Service maintained supplies of special respirators that used ultraviolet light to decontaminate air before it was inhaled. 'Where are those masks now?' Johnson [the U.S. bioweapon expert Karl Johnson] asked. 'Does anybody know?' None of the experts had the slightest idea." Despite advances in technology, the real-life management of bioweapons appeared to be the thornier job.

Throughout the 1990s, scares involving bioweapons occurred in Europe, Japan, and the United States. Plant and microbial toxins, including *B. anthracis,* received most attention. In fall 2001, an unknown person or persons intentionally contaminated pieces of U.S. mail with anthrax spores. The Federal Bureau of Investigation tallied 22 separate incidents of these events in different cities, which resulted in a small number of deaths. The origin of the spores and the person who carried out the attack have never been confirmed. But the scare alerted the public to the hazards of small amounts of biological agents released into a population. This incident may also have been the first case of bioterrorism inside the United States.

The table lists bacteria and viruses that the Centers for Disease Control and Prevention (CDC) has identified as microorganisms with the most potential for use in bioweapons.

In addition to the bacteria and viruses listed in the table, the mold *Aspergillus fumigatus* may be a weapon in spreading lethal aspergillosis.

THE USE OF BIOLOGICAL AGENTS

A biological agent requires many of the same characteristics needed in conventional weapons. Bioweapons must be cost-effective, easy and safe to assemble and effective in reaching their target. In addition, military personnel must be quickly trained to use them. Any successful biological agent would possess the following six additional properties:

- readily disperses in air or water
- invisible, silent, odorless, and tasteless
- ability to be applied as an aerosol or in solution in water
- spreads efficiently from person to person
- effective at small doses
- no or limited vaccines and treatments.

Any bioweapon producer would find it difficult to obtain all of these properties. For instance, biological agents are not always invisible, and some microorganisms do not disperse well in aerosol sprays. Most biological agents would be ineffective if put into a city's water supply. Although public water supplies have been mentioned as easy targets for a bioterrorist, large bodies of water offer a safety mechanism called the *dilution effect.* In this situation, the large volumes of water needed to serve a typical community dilute any biological agent by several million times. This dilution lowers a bioweapon's effectiveness to almost zero. Water treatment plants also employ a battery of disinfectants, filters, digestion steps, and precipitations. Few biological agents can be expected to pass these barriers.

MEANS OF PROTECTION FROM BIOLOGICAL AGENTS

World leaders consult experts to determine the likelihood of attack by any group or individual using a bioweapon. This can probably never be determined with complete certainty. To date, only the armed forces and microbiologists who work with biological agents take precautions against infection.

Microbiologists working at places such as Fort Detrick use protective clothing and handle pathogen cultures only in special biological safety cabinets. Biosafety level-4 laboratories provide enclosed workspaces and extra design features to prevent the escape of microorganisms into the surroundings. Each laboratory contains special airflow patterns and filters that remove virtually every particle from the air. The building where these microbiologists work is further protected by secondary barriers. Secondary barriers are extra steps that are taken to reduce the chance of any dangerous microorganism's contaminating unprotected people. Typical secondary barriers are dedicated entries and exits to laboratories, separate changing rooms for workers to don protective clothing and equipment, airlocks, special seals on doors, controlled air pressures in rooms (usually a negative pressure in rooms containing biological agents), extra exhaust and filtration systems, and dedicated hand washing and shower facilities.

Since fall 2001, the U.S. government has developed programs to protect citizens from potential bioweapon attack. The microbiologist Michael Osterholm of the University of Minnesota explained the bioweapon threats to *Microbe* magazine: "Imagine a bioterrorist lacing a potent toxin into 170,000 pints [80,440 l] of ice cream, ready for being shipped throughout the U.S. Midwest. Within 10 days, some 26,000 consumers who enjoyed that tainted dessert would be ill, with as many half of them dead or near death, while the region and the rest of the country would be subject to panic." This worst-case scenario explains the reasoning behind President George W. Bush's 2004 signing into law Project BioShield.

The BioShield law provides funding for developing better preventive measures, countermeasures, and medical preparedness against infection by biological agents. The Strategic National Stockpile (SNS) supports BioShield by creating storage space for various protective items that would be needed in the event of a bioweapon attack. Some of the items stored and controlled by the SNS plans are the following: antibiotics for anthrax and other agents, antitoxins, vaccines against smallpox and other viruses, medical supplies, medications, and surgical items. The CDC assures that the SNS now stores enough smallpox vaccine for inoculating every person in the United States. Little mention has been made of the supply of respirators that so worried Karl Johnson in the 1960s!

Part of the government's directive toward protecting citizens under the BioShield law is in the area of new tools for detecting biological agents. University laboratories are developing new vaccines for biological agents to which humans have few natural defenses. At the same time, companies investigate products for detecting biological agents in the environment. These so-called early warning systems include deoxyribonucleic acid (DNA) chips and fluorescent kits. DNA chips contain a small piece of DNA from a biological agent. When the chip is exposed to the same agent (bacteria or virus) in nature, its DNA reacts with the complementary strand of the agent's DNA. The chip then gives a visible signal indicating the presence of an infectious agent. These DNA chips are a type of biosensor in which biological components combine with a nonbiological sensing device. Biosensor technology seeks to make devices that detect the presence or absence of a pathogen and give an approximation of the amount of the pathogen in the environment.

Fluorescent detectors work in a similar way to DNA chips because they are designed to give a visible sign when they detect a biological agent: They use recombinant cells that emit fluorescent light in the presence of a pathogen. The cells used in these tools are genetically engineered to contain a gene for producing fluorescence. The light-producing reaction combines with the cell's natural ability to detect a particular biological agent. As DNA chips do, fluorescent devices provide an instant warning of the presence of a bioweapon.

Governments have developed other plans to protect their citizens. Their main protection plan involves large-scale vaccinations. The smallpox preparedness plan offers an example of how an entire population can be protected from a bioweapon through a vaccination program. It would not be practical—and probably would be impossible—to vaccinate every U.S. resident against smallpox in an emergency. The medical community plans to adapt the same measures that were used, from the 1960s to 1980, for eradicating smallpox. This procedure is called *ring containment*. In ring containment, doctors quickly identify all the people in an area who have been exposed to an infectious agent. The community then sets up a vaccination program for anyone who has had contact with infected people. Doctors build an imaginary ring of vaccinated people around the site of the agent's highest infection rates. By containing the spread of infection this way, others in the population who were not infected might safely forgo vaccination and remain safe. This method hastens the vaccination of people who truly need it, yet it also protects unvaccinated people from infection.

The government and health organizations rely on the advice of microbiologists on the best ways to kill a biological agent if it enters a community. The CDC, the U.S. Food and Drug Administration (FDA), and

Bioweapon Bacteria and Viruses and Their Health Threats

Bacteria	Viruses
• *Bacillus anthracis* (anthrax)	• Ebola (hemorrhagic fever)
• *Brucella* (brucellosis)	• Lassa (hemorrhagic fever)
• *Burkholderia mallei* (glanders)	• Marburg (hemorrhagic fever)
• *Chlamydia psittaci* (psittacosis)	• Variola (smallpox)
• *Clostridium botulinum* (botulism toxin)	• Influenza (flu epidemic)
• *Clostridium perfringens* (epsilon toxin)	• Nipah (neurological damage)
• *Coxiella burnetii* (Q fever)	• Sabia (hemorrhagic fever)
• *Escherichia coli* (O157:H7 outbreak)	
• *Francisella tularensis* (tularemia)	
• *Rickettsia prowazekii* (typhus fever)	
• *Salmonella* (typhoid fever, salmonellosis)	
• *Shigella* (shigellosis)	
• *Vibrio cholerae* (cholera)	
• *Yersinia pestis* (bubonic plague)	

the University of Pittsburgh's Center for Biosecurity all provide information on how people can decontaminate buildings, offices, and homes. Professionals trained in decontamination would probably be assigned this duty in a bioweapon emergency.

Two types of decontamination have been advised for suspected bioweapons: surface decontamination and area decontamination. Surface decontamination mimics the procedures microbiologists use to kill unwanted microorganisms, called *contaminants,* in a laboratory. They use strong disinfectants to kill germs on stationary surfaces and sterilization to kill germs on portable items. Items that cannot withstand the high pressure and temperature of autoclave sterilization may need to be disposed of by burying. Area decontamination is for the space inside rooms; it is also called *space decontamination.* In this procedure, microbiologists decontaminate an enclosed space by flooding the area with ethylene oxide gas (ETO). ETO is dangerous to humans if inhaled, but it is also an effective agent against bacteria, molds, yeasts, and viruses. ETO works by permanently damaging the nucleic acids of these microorganisms, causing death. Decontamination experts have also proposed the gases chlorine dioxide and ozone as alternatives to ETO.

The hazards of bioweapons reside mostly in the unknowns of the type of microorganism or toxin being used, its source, and the defensive equipment available to neutralize it. Defenses against bioweapons have been planned by governments mainly speculatively. A population may, furthermore, be forced to react to the hazard after it has already entered their environment. Because the threat of bioweapons has moved to the fore in the past decade, microbiologists have been pursuing new technologies for prevention, protection, and decontamination.

See also ANTHRAX; BIOSENSOR; *CLOSTRIDIUM*; MICROBIOLOGY; PATHOGEN; VIRUS.

Further Reading

Centers for Disease Control and Prevention. "Bioterrorism." Available online. URL: http://emergency.cdc.gov/bioterrorism. Accessed March 17, 2009.

Fleming, Diane O., and Debra L. Hunt, eds. *Biological Safety: Principles and Practices,* 4th ed. Washington, D.C.: American Society for Microbiology Press, 2006.

Fox, Jeffrey L. "BioShield, Other Programs Part of Expanding Federal Antibioterrorist Effort." *Microbe,* July 2005.

Garrett, Laurie. *The Coming Plague.* New York: Farrar, Straus & Giroux, 1994.

Realities of Bioterrorism

In 1997, the Clinton administration had begun implementation of initiatives in preparation for the possibility of attack by weapons of mass destruction (including chemical, radiological, as well as biological agents). Little was made of these efforts. Late in the year 2000, I was asked to speak to an audience about microbial agents of bioterrorism. At the time, this topic was still considered strictly an exercise in speculation and, judging by the audience attendance, considered by many to be an outright science fiction scenario. Little did we know then that within a year, "science fiction" would become harsh reality. By fall 2001, shortly after the September 11 Al Qaeda suicide attacks in New York City and Washington, D.C., the United States was faced with a biological attack by a yet-to-be named person or persons.

What is bioterrorism? Strictly defined, it is the intentional or threatened use of bacteria, fungi, viruses, or toxins from living organisms to produce death or disease in humans, animals, or plants. Bioweapons consist of bacteria, fungi, viruses, or toxins from living organisms that are produced for the purpose of mass destruction.

Although the use of biological weapons has seemingly been recently thrust into our consciousness, the concept of using such agents for the sole purpose of causing harm is almost as old as the concept of war itself. People recognized early that sickness was transmissible. They may not have fully understood the concept of infectiousness, or disease caused by a specific agent as we understand today, but they knew that there was an intrinsic danger associated with sick people and animals or toxic compounds associated with plants. In the sixth century, the Assyrians poisoned the wells of their enemies with ergot toxin, a product from the fungal decomposition of grain. The diseased dead (people and/or animals) may have been placed into a water supply or physically flung over protective walls into a fortified city, during a siege, in order to spread disease and weaken the resolve of an opponent. The early colonization of North America by Europeans was marked not only by the unintentional introduction of infectious agents to the Native American population, but also by intentional exposures. The British purportedly used smallpox against the colonists and knowingly traded blankets that were contaminated with the fluids (exudates) of open smallpox wounds purposefully in order to cause disease. Since the Native Americans had no history of exposure to this agent, they quickly succumbed, and thus the colonists could more easily exert their control over the Native peoples.

Not until the late 1800s, did scientists isolate and describe the specific agents of many diseases (anthrax, cholera, plague). It became clear that disease was not in the realm of providence but rather an interaction between two biological entities. It was also not lost on many that these agents could be isolated, cultivated, and purified for the specific purpose of intentional application to an enemy for the sole purpose of mass destruction. Anthrax was used, in World War I, by the Germans against Allied draft animals (mules and horses). Tick-borne encephalitis was employed, in World War II, by the Japanese in China to weaken resistance of the populace. In the 1950s, the U.S. military experimented with disseminating "harmless" bacteria (including *Serratia*) over the San Francisco Bay area in order to determine the effectiveness of the aerosol delivery of microbes. This experiment was found to have contributed to the death of one individual and may have been linked to illness in about a dozen more. In 1979, in Sverdlovsk, Russia, there was a reported outbreak of a small number of cases of gastrointestinal and cutaneous anthrax. Years later, it was revealed that there were hundreds of fatalities due to the pulmonary form of the disease. The true source was suspected to be a secret Russian bioweapons facility that had experienced an unintentional release of anthrax. The results of a United Nations–sponsored inspection, in the early 1990s, of weapons of mass destruction in Iraq revealed the production of hundreds of thousands of liters of biological weapons, including anthrax, botulinum toxin, ricin, and fungal toxins. A significant amount of this material was loaded into warheads, bombs, and artillery shells but was destroyed after the first Iraqi War (Operation Desert Storm). It was the threat of the existence of these weapons that added to the decision by the United States to invade Iraq a second time (2004). Of course, the major drawback when applying bioweapons on the battlefield is that they are indiscriminate in their effect. That is to say, unless the people applying the agent(s) have a self-protective strategy, they and their allies will also be susceptible to the agent.

Not all biological attacks originate from national military organizations; they have been the result of domestic terrorism. In 1984, it was discovered that a food-borne outbreak of salmonellosis in Dalles, Oregon, was actually intentionally perpetrated. A religious cult had contaminated a salad bar at a restaurant with the plan that if enough local people were sick, they would be incapacitated, they would not be able to vote in a local election, and the cult's political initiatives would prevail. In 1996, a disgruntled labora-

(continues)

Realities of Bioterrorism

(continued)

tory worker intentionally contaminated baked goods intended for coworkers with *Shigella*. Again in 1997, another *Shigella* outbreak occurred in a laboratory, and although there were no arrests, it was suspected to be the work of a disgruntled laboratory worker. The Aum Shinrikyo cult of Japan has been accused of creating and disseminating biological and chemical agents of mass destruction. In 2001, shortly after the Al Qaeda attacks of 9/11, what is believed to be a case of domestic bioterrorism occurred, involving the distribution of highly refined anthrax along the Eastern Seaboard of the United States. Five persons died, and 22 others contracted cutaneous or inhalation anthrax. These cases occurred in Florida; Washington, D.C.; and the New York/New Jersey area.

Although the concept of creating an "ideal" biological weapon is in itself elusive, there are basic characteristics that are sought by the creators of such weapons in order to optimize their effect. One of the first considerations is the pathogenicity of the agent, or the ability of the agent to cause severe disease. Many organisms can infect a host, but fewer actually can establish a state of disease. Many animal pathogens do not survive well outside the host, and, once disseminated, become diluted in the environment; therefore, the ideal agent would be effective at low doses. Once in the environment, the agent should be able to survive under a wide variety of conditions. It should be highly transmissible or contagious, so that secondary infections result from the initial delivery. Thus, it is important that the agent be delivered in aerosol form so that it can penetrate deep within the target host's respiratory system. The agent should be easy and cost-effective to produce and concentrate, and adaptable to weapon delivery systems. In addition, there is some value to a name *(plague or anthrax)* that evokes fear, so that the mere mention of it can be used as a strategic deterrent much in the same way as countries maintain nuclear weapons.

It is also important to remember that such weapons need not only target humans. Targeting livestock and/or food crops can lead to food shortages, low morale, and economic upheaval within the affected country or peoples.

What agents are used as bioweapons? Not surprisingly, the names of these agents are the same ones that have naturally plagued humankind for centuries. Bacterial agents include *Yersinia pestis* (bubonic plague), *Bacillus anthracis* (anthrax), *Clostridium botulinum* (botulism toxin), *Francisella tularensis* (tularemia), and *Brucella* sp. (brucella). There are a variety of viruses that could be used, including smallpox, Marburg, Ebola, and any one of a number of viruses that cause viral hemorrhagic fevers and encephalitis (viruses that infect the brain and neural system). Livestock may be targeted with viral encephalitis diseases such as Eastern and Western equine encephalitis, and avians (chickens) with infectious bursal virus disease and New Castle and influenza viruses. Plant pathogens include a variety of naturally occurring rusts and smuts (fungal diseases) that can destroy or limit crop production.

Even though these agents are considered deadly, they still must infect and then cause disease in the human host. Different agents may have different dose-response patterns, and it may be possible for a single infectious unit to cause disease. Immunity within a population may be such that while most are affected, some will have a higher level of resistance. Further, those populations that live in geographic regions where these agents are indigenous may have developed subpopulations that present greater resistance to disease from the agent. This was demonstrated when Western aid workers attempting to help African populations afflicted with the Ebola virus quickly succumbed to the disease. There are attempts being made to develop vaccines against most of the agents of concern, but vaccine development can be a slow process, and not all vaccines are protective for all persons.

The different agents have varying incubation times, that is, the time delay between infection and onset of disease. Disease from toxins typically occurs within hours of exposure. Demonstrable disease from some bacterial infections may take days (anthrax, plague, or tularemia), whereas viral diseases may take weeks. The longer the incubation period, the more difficult it will be for investigators to pinpoint the initial source. In addition, during the asymptomatic period of infection, the carrier host may be unknowingly communicating the disease to others.

A defense against bioterrorism needs adequate surveillance, reporting, and analysis systems. Since many of these diseases occur naturally, one of the more difficult tasks facing health care providers is discerning the natural occurrences of these types of diseases from an intentional occurrence. Although relatively rare, in the United States there are annual cases of plague (15–20 per year), tularemia (100–200 per year), and brucella (100–200 per year). These are reportable diseases, meaning they must be reported to the Centers for Disease Control and Prevention (CDC), and their occurrence should raise immediate awareness within the health community. Since the anthrax attack of 2001, there has been more education by public health and the Department of Homeland Security of front-line medical staff to make them aware of the symptoms of these relatively rare diseases. An excellent example of the detection of a viral outbreak was the initial occurrence of West Nile virus in New York. It took about 14 days before the public health community realized that the occurrence of a nonnative encephalitis-causing virus was on the rise.

However, for agents such as *Salmonella,* there are hundreds of thousands of cases per year. *Salmonella* may not necessarily be deadly, but it can be effective as a terror weapon by being associated with a food or water source. As previously mentioned, there was an intentional *Salmonella* contamination by a religious cult of a salad bar at a restaurant that resulted in a food-borne outbreak. However, originally when the outbreak occurred, it was believed to be a simple case of food-borne transmission that affected a single restaurant. The true nature of the source of the salmonellae was only discovered years later, when government agents investigated the cult's financial records for evasion of taxes.

Surveillance for human diseases is conducted by the CDC and the national Laboratory Response Network (LRN). LRN is made up of the federal, state, and local public health laboratories, whose mission it is to respond to bioterrorism, chemical terrorism, and other emergencies. The formation of the LRN has led to an enhancement of laboratory capacity and the application of newer technologies in the identification of microorganisms. Other ongoing monitoring efforts include the CDC/USDA/USFDA FoodNet active surveillance network. These laboratories constantly share information about the types of microbes that are isolated from contaminated foods. Primarily tasked with tracking food-borne disease, these laboratories are essential in alerting authorities to outbreaks that may also be the result of bioterrorism. However, laboratory capacity remains a critical concern in the case of an outbreak. For example, during the 2001 anthrax attack, it has been estimated that more than 250,000 environmental samples were generated throughout the nation, even though the contaminated sites were geographically specific and relatively confined. This sample load overwhelmed the public health laboratories. Of course, while some of these samples were made of truly suspicious material, the vast majority were taken out of panic and of highly doubtful linkage to a biological weapon. Fortunately, the number of actual cases of anthrax was relatively small, but when there are many more casualties, there will be many more clinical samples. These will probably be given priority over environmental sample analysis. Are there significant resources to process both the clinical and environmental samples necessary to address both treatment of patients and prevention of further spread of the disease agent(s)?

As of this writing, there is not a reliable immediate detection technology of the microbiological agents in environmental samples, especially viruses. After the anthrax attacks of 2001, a study by the CDC revealed that the rapid "dip-stick" type tests that first responders relied upon were inaccurate and should not be used. So far, there has not been much improvement in the technology that allows a first responder accurately to gauge the content of a biological weapon release.

Before a diagnosis is made, there is also the danger to the laboratory staff processing the clinical and environmental samples. Occupational laboratory-acquired diseases are not rare occurrences. Therefore, laboratorians must be mindful of the source of the samples, including whether they are of a suspicious nature, and take all precautions for self-protection. For the agents listed previously, laboratory diagnosis is not an automated process and requires skilled and trained microbiologists. If there is a shortage of trained public health microbiologists and if those working in the labs succumb to a disease agent, there will be few capable persons to replace them.

The laboratory resources are only part of the response to a bioterrorism attack. In the instance of a biological attack, agencies such as the Department of Defense, Department of Homeland Security, FBI, FEMA, Department of Transportation, and even the EPA are immediately activated. On the local level, these events will be met by the first responders such as firefighters and police. There has been extensive training for these types of events since the attacks of 2001, and most local emergency services have hazardous materials (HazMat) units that are now better prepared for biological agents. However, until the first responders recognize the severity of the attack, they will be most at risk because they will be the first to enter contaminated areas to conduct initial investigations. The sick and injured will then be taken to emergency rooms, where the risk of exposure to the agent will be translated to the nurses and doctors. Hospitals have been tasked to develop plans and train their personnel for such events. Given the budgetary constraints put on most medical facilities, the level of preparedness is not the same facility to facility.

There are plans on how to treat contaminated sites with disinfectants and sterilants. Areas are isolated, sealed, and treated. This is dangerous work, since the

(continues)

Realities of Bioterrorism

(continued)

disinfectants and sterilants are highly toxic in pure form and must be applied by trained professionals so that only the affected areas are treated. Disinfectants/sterilants may include a combination of chlorine dioxide, ozone, and ethylene oxide. In the case of foodborne contamination, lots of affected foods can be identified, removed from distribution, and destroyed. The biological contamination of drinking water could affect a large number of people. However, introducing many of the agents listed earlier into water will subject them to dilution and the deleterious effect of the immediate environment and subsequent water treatment. If a drinking water distribution system were contaminated, a water utility would need to respond by increasing the level of disinfectant and using extensive flushing. Although this will act to reduce the level of the organism, it is also likely that the organism will persist in small concentrations as part of the biofilm of the distribution system. Pathogens are difficult to detect in environmental matrices under the best circumstances; the organisms would be heavily diluted, and it would be an overwhelming and impractical task to guarantee 100 percent eradication. Typically, boiling water for two minutes will destroy viruses, protozoa, and the vegetative forms of bacteria. Boiling water will have little effect on bacterial spores (anthrax). Still, with all the opportunistic pathogens that naturally occur in source water, the water treatment technologies currently employed provide outstanding protection from waterborne microbial disease.

What can you do as an individual to prepare for such an event? The CDC and the American Red Cross suggest that you follow the guidelines for general preparation for natural disasters. There are many similarities between a natural disaster and a biological attack. As was learned in New Orleans in the aftermath of hurricane Katrina, there can be significant biological contamination in the immediate environment from raw sewage mixing with surface water. Precautions in regard to having available potable drinking water and a safe food supply apply in both instances. A biological attack may necessitate a "shelter-in-place" warning, cutting people off from the outside for a prolonged period. Having a plan for food, water, and communication among family members is key in any disaster situation.

All of the preceding discussion is predicated on bioweapon agents that are of a known pathogen. What if an agent is engineered or created such that it is more infectious, is more pathogenic, can elude detection, and, therefore, can delay proper treatment? Although it does require expertise in molecular biology, the openly published research journals are full of the application of techniques of moving the genetic information from one organism to another. In fact, shortly after the 2001 anthrax attacks, there was much debate in regard to the type of detail that could be published by scientists in genetic engineering.

Is this the stuff of science fiction? A former Russian bioweapons expert, Serguei Popov has reported on the efforts of the Soviet Union's creation of such biological agents. For example, the opportunistic pathogen *Legionella pneumophila* typically requires a dose of tens of thousands to cause disease. Popov has described the altering of the organism to a form that requires only a few cells to cause disease. Further, the disease symptoms are not those typical for Legionnaires' disease, thus delaying the treatment of the actual cause. Delayed treatment of legionnaires' disease often leads to death. In addition, Popov reported the creation of an organism dubbed the "horror autotoxicus," which is an altered *Yersina pestis* (bubonic plague) bacterium. The symptoms of plague are displayed, and the exposed person can be treated for the bacteria using standard antibiotic treatment. However, this organism triggers a severe autoimmune response such that the host's own immune system attacks and destroys host tissue. It was estimated that death would occur in as little as three days.

This is the nightmare scenario that worries authorities most, hereto unknown pathogens that target an immunologically defenseless population. Therefore, the question that we must ask is not whether we are ready for a bioterrorism attack, but whether we can be.

See also AEROMICROBIOLOGY; ANTHRAX; BIOWEAPON; TOXIN; TRANSMISSION; VIRUS.

Further Reading

Henderson, Donald A., Thomas V. Inglesby, and, Tara O'Toole, eds. *Bioterrorism—Guidelines for Medical and Public Health Management.* Chicago: JAMA/AMA Press, 2002.

Manning, Frederick J., and Lewis Goldfrank, eds. *Preparing for Terrorism: Tools for Evaluating the Metropolitan Medical Response System Program.* Washington, D.C.: National Academy Press, 2002.

Weinstein, Raymond S., and Kenneth Alibek. *Biological and Chemical Terrorism: A Guide for Healthcare Providers and First Responders.* New York: Thieme Medical, 2003.

Young, H. Court. *Understanding Water and Terrorism.* Denver: BurgYoung, 2003.

—Richard E. Danielson, Ph.D.,
BioVir Laboratories, Benicia, California

Hawley, Robert J., and Edward M. Eitzen. "Biological Weapons—a Primer for Microbiologists." *Annual Review of Microbiology* 55 (2001): 235–253. Available online. URL: www.ncbi.nlm.nih.gov/pubmed/11544355. Accessed March 17, 2009.

University of Pittsburgh Medical Center. Center for Biosecurity. Available online. URL: www.upmc-biosecurity.org. Accessed March 17, 2009.

bloom A bloom is an abundant growth of microorganisms at or near the surface of a body of water. Lakes, reservoirs, ponds, and the ocean represent the most common places for blooms to grow. Blooms usually contain the following types of microorganisms: algae, cyanobacteria, purple sulfur bacteria, and diatoms. Blooms play an important role in environmental health, because many blooms are toxic and harmful to human and aquatic life and aquatic ecosystems. See the color insert on page C-2 (top) for a picture of a *Euglena* bloom.

Microbial blooms form for two different reasons. First, they develop when large amounts of nutrients flow into surface waters, causing a sudden burst of microbial growth in response to the bounty of nutrients entering their environment. This nutrient inflow can have natural sources, such as rain runoff, or can be due to industrial dumping of waste high in nutrients that microorganisms use. Second, blooms form seasonally as a result of changes in water temperature, pH, and physical factors, but not of nutrient inflow.

Blooms made of algae are called *algal blooms* and represent the most familiar blooms that plague coastlines. Nutrient-formed algal blooms have become a growing environmental concern, because they are increasing throughout the world as a result of water pollution. These blooms are often nicknamed *harmful algal blooms* (HABs). HABs occur in salt water or freshwater, in the ocean, lakes, and ponds.

HABs form when nutrient-rich fertilizers and manure wash downstream from feedlots and farms. Nutrients most associated with causing blooms are nitrogen and, to a lesser degree, phosphorus. Organic nitrogen compounds (containing ammonium- or nitrate-nitrogen) and organic phosphorus compounds (containing phosphates) rush into waters and cause a sudden rise in the levels of these two nutrients. When the nitrogen levels become greater than 0.1 mg/L and phosphorus exceeds 0.01 mg/L, a bloom will probably develop. Water temperatures of 60–85°F (15–30°C) and a pH of 6–9 hasten microbial growth, especially of algae. The result is a condition in the water called *eutrophication,* a complete depletion of nutrients and oxygen from water.

High nutrient levels and favorable environmental conditions lead to blooms, and many blooms are followed by eutrophication. Eutrophication consists of a series of four steps in which the sudden influx of organic matter depletes dissolved oxygen from a body of water. The following example describes the events in an algal bloom. In the first step, algae grow to enormous numbers of cells quickly in response to incoming nutrients; that is, they form a bloom. The algae devour nutrients so fast that they grow more cells than the waters can support. The second step then commences, wherein the algae begin to die because the depleted nutrients can no longer sustain the dense population of algae. Third, as algal cells die, bacteria in the water begin to digest them for their own nutrient needs. As a consequence, the next step is the development of a second bloom, consisting of bacteria. The fifth and final step involves large numbers of aerobic bacteria that rapidly use up all the dissolved oxygen, thus eutrophication. These areas of low or no oxygen in a body of water are called *dead zones*.

Eutrophication has a devastating effect on water ecosystems. Fish and crustaceans suffocate in waters depleted of oxygen, the primary reason why these waters transform into dead zones. Other animals dependent on these food sources begin to starve. Waterfowl, hawks and eagles, amphibians, reptiles, muskrat, raccoons, foxes, bears, otters, and marine mammals are examples of the animal life affected. Many of these animals leave the habitat in search of new food sources. Meanwhile, within the water, aerobic microorganisms no longer survive. Anaerobic species take over and emit their normal end products: the gases hydrogen, hydrogen sulfide, and methane. Anaerobic waters cannot sustain a diverse aerobic ecosystem of fish, mammals, and other living things. The overall result of eutrophication becomes a loss of biodiversity.

Eutrophied waters appear along almost all coastlines worldwide that have high human populations and heavy pollution. Blooms and eutrophication occur along the U.S. East Coast (Long Island Sound, the Chesapeake Bay, and the New Jersey shore), the Baltic Sea, and the Mediterranean Sea. A persistent bloom occurs each year in the Gulf of Mexico at the Mississippi River delta. At its peak, this dead zone grows to a size greater than the area of New Jersey.

ALGAL BLOOMS

Algal blooms are perhaps the most familiar signs of water pollution. Algal blooms are notorious for causing beach closing, skin irritation, and more severe illnesses to people exposed to the bloom, and poisoning to aquatic species, mostly shellfish. The health hazards of algal blooms are caused by a variety of neurotoxins (poisons that attack the nervous system). Toxins produced by different algae are collectively known as phycotoxins.

One familiar algal bloom is known as a *red tide*. Red tides consist of blooms of algae that contain a reddish pigment, which turns the water red. Two dinoflagellate algae named *Alexandrium* and *Gymnodinium* are prominent causes of these blooms, but other dinoflagellate species take part, as well, to lesser extents. Dinoflagellates are hardy marine algae in which each cell contains two flagella. *Alexandrium* produces a powerful neurotoxin called *saxitoxin*. Shellfish that consume the toxin producers do not become sick, but they accumulate toxin in their tissues. Humans and wildlife that consume *Alexandrium*-contaminated shellfish fall ill with a condition called *paralytic shellfish poisoning* (PSP). PSP in humans results in numbness of the mouth or the entire face and the extremities.

Gymnodinium produces brevetoxin, which causes an illness similar to PSP called *neurotoxic shellfish poisoning* (NSP). NSP causes a number of neurological symptoms, including numbness, paresthesia (burning, prickly feeling), dizziness, and ataxia (lack of muscular coordination), in addition to gastrointestinal disorders. Symptoms eventually disappear, and usually neither PSP nor NSP causes death.

A second, more dangerous algal bloom is called a *ciguatera bloom*. In this circumstance, the dinoflagellate *Gambierdiscus* invades shellfish near coral reefs. Fish that hunt reefs for their food—sea bass, barracuda, snapper, and grouper—then eat the infected shellfish and begin accumulating the toxin in their tissues. *Gambierdiscus*'s toxin, called *ciguatoxin*, furthermore does not decompose by cooking the seafood. People who have ciguatera poisoning experience severe diarrhea, nausea, vomiting, respiratory paralysis, and tingling in the lips, tongue, and throat.

The Chesapeake Bay and its tributaries in Maryland and Virginia have been home to yet another dinoflagellate that forms harmful blooms. This organism, *Pfiesteria piscicida*, was discovered in the 1990s by the aquatic botanist JoAnn Burkholder at North Carolina State University. In 1997, *Pfiesteria* and similar dinoflagellates caused massive deaths among fish species (called *fish kills*) in the Chesapeake area. As researchers along the Chesapeake studied *Pfiesteria*, they discovered alarming behavior. *Pfiesteria* detects the presence of fish in nearby waters and then swims after the fish and attacks them. (It has been dubbed an *ambush predator*.) Fish that have been victimized by *Pfiesteria* have lesions on their skin, and people who have contact with infected fish also suffer the toxin's effects. Workers who harvest fish in infected waters and even laboratory workers doing studies on *Pfiesteria* have become sick. Their symptoms consisted of respiratory irritation, skin rashes, gastrointestinal illness, headaches, memory loss, disorientation, and emotional irregularities. Some symptoms resolve, but others permanently remain. A proposed *Pfiesteria* toxin has been blamed for effects on nerve function by damaging the normal sodium channels in nerve cells.

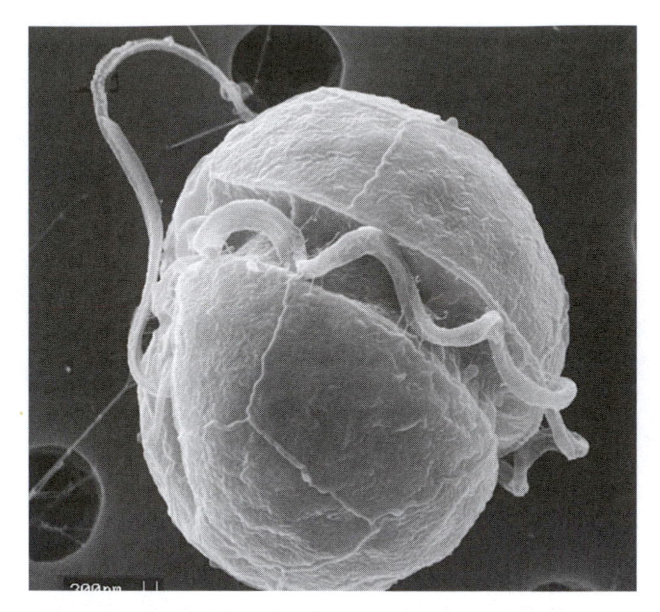

Pfiesteria piscicida* is a dinoflagellate alga that forms dangerous blooms, particularly in estuaries from the Chesapeake Bay to the Carolinas.** ***(Virginia Institute of Marine Science)

Pfiesteria blooms trouble investigators because of unknowns associated with the organism's life cycle, consisting of at least 20 different stages. Marine biologists have recently begun to uncover details about the *Pfiesteria* toxin and the mechanism the organism uses to stalk fish. Because of *Pfiesteria*'s bizarre and complex life cycle, research has been difficult. Although many people who work Chesapeake waters believe the organism poses a serious threat, others have questioned whether any organism can live the complicated lifecycle of *Pfiesteria*. R. Wayne Litaker of the National Oceanic and Atmospheric Administration (NOAA) said to the *Baltimore Sun* in 2002, "Toxic Pfiesteria life cycle stages that don't exist can't be toxic." But in 2007, the chemist Peter Moeller, also from NOAA, completed a decade-long quest to identify the *Pfiesteria* toxin, concluding the substance that stuns fish is a copper-containing compound. Moeller's work also relieved Burkholder, whose work had been criticized from the start. "It's a happy day for our lab," she said after learning of Moeller's breakthrough. "We are vindicated by Dr. Moeller. He quietly persisted and this is no easy toxin group to identify. It is an excellent piece of work." *Pfiesteria* and other algal blooms continue to emerge along Atlantic coasts from North Carolina to Norway. The Pacific Ocean experiences blooms to a lesser extent.

CYANOBACTERIAL BLOOMS

Cyanobacteria (also known as blue-green algae) contribute to blooms and eutrophication mainly in lakes. Cyanobacteria grow in almost all surface freshwaters and in soil. In addition, these bacteria often outcompete many other microorganisms for nutrients. When sudden high levels of nitrogen- and phosphorus-containing organic compounds enter their aquatic environment, cyanobacteria waste little time turning into massive blooms. Some common bloom-forming cyanobacteria that also produce toxins are *Anacystis, Anabaena, Microcystis,* and *Nodularia.* The toxin causes illness in wildlife that drink tainted waters and may lead to animals' death.

Microbial Mats

Microbial mats are complex communities of microorganisms that grow on waters possessing conditions like those found in hot springs, saline (salty) lakes, and intertidal zones. Microbial mats have also been found at the extreme depths of the ocean, where hot vents emit bursts of noxious steam. Cyanobacteria often grow better than other microorganisms in these places, so they tend to predominate in the microbial mats found there.

Microbial mats on the water's surface resemble blooms, because they contain massive microbial communities. Yet mats, which range from about one-eighth inch (0.3 cm) to a half-inch (1.3 cm) or thicker, house a more complicated structure than blooms. Mat structure is striated, meaning that it has distinct layers. This layered structure is called *laminar growth.* Cyanobacteria live at the upper levels of microbial mats, where they take in sunlight and carry out photosynthesis. During daylight hours, microbial mats contain high amounts of oxygen emitted from the cyanobacteria's photosynthesis. As night falls, respiring species quickly consume all of this oxygen. As a result, microbial mats contain anaerobic activities at night.

Microbiologists studying cyanobacteria mats have found a layer of oxidized iron beneath the cyanobacteria but above the other layers. The role of this layer is unknown, but it seems to separate the aerobic activities from anaerobic activities. Beneath the iron layer, purple sulfur bacteria carry out a more rudimentary form of photosynthesis than is found in the cyanobacteria above them. These bacteria give portions of microbial mats their distinctive purple color. The purple sulfur bacteria precipitate the iron layer to make iron sulfide, which is black. Thus, in addition to the green of cyanobacteria and the purple of the sulfur bacteria, microbial mats usually have black regions.

Biologists have studied fossilized 3.5-million-year-old microbial mats called *stromatolites* to learn about ancient microbial communities. It is likely that purple and green sulfur bacteria formed prehistoric microbial mats in the environment scarce in oxygen.

DIATOM BLOOMS

Diatoms belong to a group of unique algae with intricate shell-like bodies that make them appear nonliving. Within the past 15 years, biologists have learned that many diatoms form blooms the same way as other microorganisms. In addition, diatoms produce their own poison called *domoic acid.* One main producer of the toxin is the genus *Pseudonitzchia.* Domoic acid works its way up food chains from shellfish and crabs to the birds and fish that feed on them. Humans and marine mammals become sick from eating contaminated crustaceans and fish. In people, the illness is called *amnesic shellfish poisoning,* because it causes short-term memory loss, among other symptoms.

Long-term domoic acid exposure has poisoned hundreds of marine mammals, such as dolphins and sea lions, and birds along the U.S. West Coast, in recent years. These animals get the toxin mainly from eating infected mussels, sardines, and anchovies. Domoic acid poisons marine mammal nervous systems; its symptoms are disorientation and seizures leading to death. Shorebirds experience similar symptoms and can enter a frenzied intoxicated state. Interestingly, a 1961 outbreak of domoic acid poisoning in gulls and shearwaters in Northern California became the inspiration for the 1963 Alfred Hitchcock horror movie *The Birds.*

OTHER BACTERIAL BLOOMS

Purple sulfur bacteria are anaerobes so nicknamed because these purple-pigmented species oxidize hydrogen sulfide gas and turn it into sulfur, which they then deposit as solid sulfur granules. They grow typically in anaerobic lakes rich in sulfur dioxide and form blooms in bogs and lagoons that are depleted of oxygen. In other words, these bacteria take part in the final anaerobic stage of eutrophication.

In the 1980s, microbiologists began studying a bloom in shallow (29.5 feet [9 m]) Lake Cisó in northern Spain. This bloom mostly contained purple sulfur bacteria of the *Chromatium* genus, plus a small amount of green sulfur bacteria. They found no cyanobacteria in any of their samples. Biologists have since learned that Lake Cisó turns completely anaerobic in the winter, as a result of the sulfur bacteria's activities, but develops more layers in the summer. Other lakes have since been found in which phototrophic bacteria (bacteria that use light as their energy source) dominate the bloom. In these lakes, the bloom forms as a result of oxygen levels rather than organic nutrient levels.

Blooms have increased worldwide as water pollution has increased. Blooms create a health threat to the environment and to humans and animal life, yet many questions remain on the mechanisms of bloom growth. Blooms will remain an important aspect of study in water microbiology and ecology.

See also ALGAE; CYANOBACTERIA; DIATOM; GREEN BACTERIA; PURPLE BACTERIA.

Further Reading

Centers for Disease Control and Prevention. "Harmful Algal Blooms (HABs)." Available online. URL: www.cdc.gov/hab. Accessed March 17, 2009.

Dewar, Heather. "Scientists Challenge Theory on Toxicity of *Pfiesteria;* Study Claims Organism Not as Noxious as Believed." *Baltimore Sun,* 21 June 2002. Available online. URL: http://pqasb.pqarchiver.com/baltsun/access/127598121.html?dids=127598121:127598121&FMT=ABS&FMTS=ABS:FT&type=current&date=Jun+21%2C+2002&author=Heather+Dewar&pub=The+Sun&desc=Scientists+challenge+theory+on+toxicity+of+Pfiesteria+%3B+Study+claims+organism+not+as+noxious+as+believed. Accessed March 17, 2009.

Hartley, Megan. "Scientists Discover the Deadly Toxin Associated with *Pfiesteria.*" Southern Maryland Online, 11 January 2007. Available online. URL: http://somd.com/news/headlines/2007/5163.shtml. Accessed March 17, 2009.

Spivey, Angela. "*Pfiesteria* Stands Charged with Killing Fish and Endangering People as Well. But Could This Be a Case of Mistaken Identity?" *Endeavors Magazine,* Fall 2002. Available online. URL: http://research.unc.edu/endeavors/fall2002/pfiesteria.html. Accessed March 17, 2009.

Woods Hole Oceanographic Institution. "Harmful Algae." Available online. URL: www.whoi.edu/redtide/page.do?pid=9257. Accessed March 17, 2009.

Candida albicans *Candida* is a genus of fungi, specifically a yeast, containing at least 30 different species, the most prominent of which is *Candida albicans*. *C. albicans* possesses large cells that may be round, ovoid, cylindrical, or elongate and 2 to 4 micrometers (μm) in diameter. *Candida* cells reproduce by an asexual process called *budding,* in which a small new cell grows off a part of the parent cell. *Candida* is a normal skin inhabitant; the cells adhere to skin by using tiny appendagers called *fimbriae*.

C. albicans serves as a model for many studies on yeast morphology and physiology. Although *Candida* reproduces by budding and spends most of its life as a single cell—two characteristics of yeasts—this microorganism belongs to a group of fungi called dimorphic imperfect fungi. Dimorphic microorganisms contain species that have two different forms of growth. The term *imperfect fungi* refers to genera that have no known sexual reproduction, even though this type of reproduction is expected because the organisms are eukaryotic.

Candida species live as either commensal organisms or opportunistic pathogens, on human or animal skin. As commensal organisms, they benefit by getting nutrients from the skin surface, but the host does not gain an appreciable benefit. Opportunistic pathogens include any normally harmless microorganism that, under certain conditions, can cause an infection. For example, the microorganisms on the skin cause no harm, but if a person gets a cut and does not treat it, one or more normally benign skin species can infect the wound. When *Candida* causes these opportunistic infections, the disease is candidiasis.

Candidiasis is most often associated with *C. albicans,* but three other species, *C. tropicalis, C. krusei,* and *C. paralopsilosis,* also have been involved in increasing incidence of infection.

Skin bacteria keep *Candida* in check by competing for space and nutrients. Some bacteria also excrete substances that inhibit fungal growth. People who are on long-term courses of antibiotics for bacterial infection also lose some of their harmless skin bacteria. When this happens, *Candida* grows to a larger population than normal. This overgrowth of fungi on the body is called an *opportunistic mycosis*.

Candidiasis, often called yeast infection, usually affects the skin of the female genital region or the mouth. *Candida* infects mucous membranes in these places and can become visible as whitish mucoid growth. Cutaneous (skin) infections tend to be in moist areas such as the groin and the underarms and are often referred to as "yeast infections." More serious systemic infections occur when the microorganism enters the bloodstream, a condition called *candidemia*. The body's mucous membranes usually serve as the main entry route into the bloodstream in these cases.

The Centers for Disease Control and Prevention (CDC) divides *Candida* infections into three different categories according to the site where the infection occurs. These three forms of candidiasis are described in the table on page 140.

In addition to antibiotic therapy, immunocompromised (weakened immune system) conditions increase the risk of candidiasis. The following groups of people may be more susceptible to *Candida* infection, because of weakened immune systems: acquired immune deficiency syndrome (AIDS) patients, chemotherapy patients, persons with diabetes mellitus, and organ transplantation recipients. The University

Forms of Candidiasis

Name	Site of Infection	Main Symptoms	Transmission or Cause
genital-vulvovaginal	genital areas of men and women	itching, rash	antibiotic use, corticosteroid use, pregnancy, sexual transmission (the cause of balanitis in men)
invasive	blood and organs	fever, chills, symptoms associated with affected organs, death	opportunistic infection with injury in the digestive tract
oropharyngeal	mouth and throat	white patches, soreness, difficulty swallowing	opportunistic infection

of Maryland Medical Center Web site has stated, "90 percent of all people with HIV [human immunodeficiency virus]/AIDS develop *Candida* infections." Newborns, because they have not yet developed a fully functioning immune system, are also susceptible to an oral *Candida* infection called *thrush*. Thrush is characterized by the formation of white patches on the tongue and other parts of the mouth.

In veterinary cases, *Candida* infects the skin, digestive tract, mammary glands in dairy cows, and internal organs in more severe infections. Calves, piglets, and birds have the highest risk of serious infection in veterinary medicine.

CANDIDA ALBICANS

Oral infections caused by fungi have been recognized by physicians since the 1800s. A few doctors attempted to isolate the causative agent from oral lesions, but not until 1843 did the French physician Charles-Philippe Robin (1821–85) give a name to the new parasite he recovered from his patients, *Odium albicans,* using the Latin term *albicans* for "white." Within the next century, more than 100 variations of the name appeared in microbiology; *Monilia albicans* became the choice for a while. In 1923, mycologists adopted *Candida albicans* (*Candida* is Latin for "glowing white") as the official name for the organism.

In 2006, the microbiologists Chantal Fradin in France and Barnhard Hube in Germany wrote in *Microbe* magazine, "Among the approximately 150 fungal species known to cause infections in humans, *C. albicans* is one of the rare fungi that are part of the normal human microflora, and is carried by about half the population." As the preceding table indicates, transmission of *Candida* from person to person is less of a health threat than opportunistic infections caused by person's own *Candida* population. This mode of infection is often called *self-infection*.

C. albicans cells are normally oval when living on human skin in a commensal manner. In an opportunistic infection, however, more than 50 percent of the cells develop filaments. This is called the *hyphal form* of the dimorphic organism, and the filaments, called *pseudohyphae* or simply *hyphae,* may help the microorganism penetrate the body's tissue. Clinical microbiologists in hospital laboratories confirm the presence of *C. albicans* infection by exposing cells to serum for two hours at 98.6°F (37°C). The early phase of filamentous growth produces structures called *germ tubes,* which help in diagnosing *C. albicans* infection. Fradin and Bernhard have stated that about 40 percent of *Candida* cells begin forming hyphae in as little as 30 minutes.

In the bloodstream, the hyphal form helps *C. albicans* defend against the body's immune response. The microorganism does this by deactivating some of its genes but turning on a specialized set of genes that direct the synthesis of proteins necessary to warding off the myriad activities of the human immune system. Although *C. albicans* evades the majority of immune actions, it appears to be somewhat vulnerable to cells in the blood called *neutrophils,* which seek foreign particles in the body and then envelop, ingest, and degrade them in a process called *phagocytosis*. "However, the fungus induces a large number of genes to counteract the neutrophils," Fradin and Hube emphasized, "enabling this pathogen to survive in the bloodstream, even if only briefly, to cause life-threatening systemic infections."

Human and veterinary treatments for infections may be any of the following antifungal drugs: itraconazole, miconazole, clotrimazole, amphotericin B, or nystatin. Antibiotic resistance in *C. albicans* is increasing and might create a crisis similar to that of antibiotic-resistant bacteria.

CANDIDA INFECTIONS IN HOSPITALS

Nosocomial infections are infections associated with hospital stays. The pathologist Vasant Baradkar reported in 2008, "The occurrence of fungal infection is rising worldwide. Data from the Centers of Disease Control reveal that, between 1980 and 1990, *Candida* species emerged as the sixth most common nosocomial pathogen (7.2 percent). The increase in the rate of fungal infections has been attributed mainly due to the use of broad-spectrum antibiotics, intravascular devices [devices implanted in blood vessels], and hyperalimentation [intravenous feeding], as well as the ever-increasing number of critically ill or immunocompromised patients in hospital populations." Broad-spectrum antibiotics are drugs that kill a wide variety of microorganisms. Broad-spectrum antifungal drugs have caused an increase in *C. albicans* infections similar to the development of antibiotic resistance in bacteria. General antibiotic therapies induce resistance in organisms that normally live on the body.

C. albicans has been estimated to cause almost 80 percent of nosocomial infections and about 60 percent of hospital-associated candidemia. In 2003, the Tufts University research physician David R. Snydman wrote in the medical journal *Chest,* "The incidence of candidemia—a common and potentially fatal nosocomial infection—has risen dramatically, and this increase has been accompanied by a shift in the infecting pathogen away from *Candida albicans* to treatment-resistant non-*albicans* species." By 2009, the threats from species in addition to *C. albicans* had grown, but *C. albicans* remains a very serious threat. It is the fourth most frequent cause of nosocomial infections and the main cause of hospital deaths due to bloodstream infection, or in the case of *Candida* species, candidemia.

Candida albicans remains a ubiquitous organism that does not normally cause health risks to humans or animals. This yeast does, however, develop into a mild to serious health threat when the body's defenses have been compromised.

See also FUNGUS; NORMAL FLORA; NOSOCOMIAL INFECTION; YEAST.

Further Reading

Baradkar, Vasant P., M. Mathur, S. D. Kulkarni, and S. Kumar. "Thoracic Epyema Due to *Candida albicans.*" *Indian Journal of Pathology and Microbiology* 51, no. 2 (2008): 286–288. Available online. URL: www.ijpmonline.org/temp/IndianJPatholMicrobiol512286-5782804_160348.pdf. Accessed March 19, 2009.

Calderone, Richard A. Candida and *candidiasis.* Washington, D.C.: American Society for Microbiology Press, 2001.

Carlile, Michael J., Sarah C. Watkinson, and Graham W. Gooday. *The Fungi,* 2nd ed. San Diego: Academic Press, 2001.

Centers for Disease Control and Prevention. Available online. URL: www.cdc.gov. Accessed March 19, 2009.

Fradin, Chantal, and Bernhard Hube. "Transcriptional Profiling of *Candida albicans* in Human Blood." *Microbe,* February 2006.

Snydman, David R. "Shifting Patterns in the Epidemiology of Nosocomial *Candida* Infections." *Chest* 123 (2003): 500S–503S. Available online. URL: www.chestjournal.org/content/123/5_suppl/500S.full.pdf+html. Accessed March 19, 2009.

University of Maryland Medical Center. "Candidiasis." Available online. URL: www.umm.edu/altmed/articles/candidiasis-000030.htm. Accessed March 19, 2009.

cell wall The cell wall is an outer protective covering on most bacteria, archaea, fungi, algae, and plant cells. This structure provides a much stronger and more rigid protection to the cell's contents compared with the cell membrane, which lies inside it. Cell walls carry out the following four main functions in bacterial cells: (1) create the characteristic shapes of bacterial genera, (2) protect the cell's contents from the outside environment, (3) help regulate osmotic pressure (the difference between the pressure inside and outside a cell), and (4) protect against damage from drying or freezing. The cell wall of pathogens also contains substances that help them adhere to tissue and begin an infection.

Different types of microorganisms possess different cell wall compositions. In bacterial studies, these differences offer microbiology its major method of identifying unknown bacteria. Bacteria belong to two major groups based on cell wall composition and structure: gram-positive bacteria and gram-negative bacteria. The cell wall of almost all bacteria contains a long-chain molecule, known as a polymer, called peptidoglycan that gives cells their strength.

The cell walls of archaea contain different structural compounds than found in bacteria. The most important of the archaea's features is a group of large compounds called *squalenes,* which help many archaea survive extremely harsh environmental conditions. In algae, cell wall composition varies among the different algae divisions. Algae have in common multilayered cell walls, but some algae are distinct from the others by possessing compounds found in no other cell walls in nature. Fungi, too, possess cell wall components different from anything else in microbiology. Though fungi resemble other microorganisms in containing polymers, the linkages that connect fungal cell wall polymers differ from those of most other microorganisms.

THE STUDY OF CELL WALLS IN MICROBIOLOGY

One of microbiology's most important advances occurred when the Danish physician Hans Christian Gram (1853–1938) sought a means of distinguishing one bacterial cell from another under a microscope. Gram experimented in his laboratory with various biological stains used for tissue specimens in the hope of finding a formula that would make bacteria easier to see. In 1884, Gram wrote, "The differential staining method of [Robert] Koch and [Paul] Ehrlich for tubercle bacilli gives excellent results either with or without counter-staining, since the bacilli stand out very clearly due to the contrast effect." After trying various stains and dyes, Gram discovered a mixture that readily stained some bacteria but left others unstained. By adding a "counterstain," a second stain that would make the remaining unstained cells visible, Gram worked out a series of steps that became a standard method for staining all bacteria, the procedure now known as the Gram stain. This method continues to be a cornerstone of bacteriology.

Although microbiologists of Christian Gram's day did not know the details of bacterial cell wall structure, the Gram stain had actually provided the first clue to the differences between cells that reacted positively in the technique compared with cells that reacted negatively. Gram-positive bacteria contain cell walls made of a thick, somewhat porous layer of peptidoglycan. These species turn dark blue when stained in the Gram procedure. Gram-negative bacteria contain a more complex, layered cell wall than gram-positive bacteria, and their peptidoglycan layer is thinner than in gram-positive cells. Gram-negative bacteria turn deep pink or red when stained in the Gram procedure. Microbiologists learned over the years after Gram's discovery that these distinctions also affect each group's characteristics in the following ways: pathogenicity (the ability to cause disease), susceptibility to antibiotics, susceptibility to chemicals, and spore formation.

A second major advance took place in cell wall study many years later, with the development of transmission electron microscopy (TEM). The German physics student Ernst Ruska (1906–88) built the first transmission electron microscope, in 1931, in Berlin. TEM magnifies cell features many thousand times, so that microbiologists can observe the fine structures of the inner cell, the cell membrane (also called the *cytoplasmic* or *plasma membrane*), and the cell wall. TEM has revealed the layers of gram-positive and gram-negative cell walls and additionally has helped microbiologists define the differences between prokaryotic and eukaryotic cell walls. Some microbiologists devote entire careers to the study of cell walls.

BACTERIAL CELL WALLS

All bacteria except *Mycoplasma* have cell walls. The majority of bacteria fall into into the gram-positive and gram-negative groups; a minority of species contain cells that may stain either positively or negatively, called gram-variable species; and *Mycobacterium* lack this familiar cell wall and possess a different thick protective outer structure. *Mycobacterium* species can be identified using a stain called acid-fast rather than the Gram stain.

Cell walls allow bacteria to live in places few other living things withstand, and for this reason, the cell wall may be an important factor in bacterial evolution and persistence in almost every place on Earth. The cell wall undoubtedly protected the earliest bacteria when life first evolved 3.8 billion years ago on Earth. Earth at that time offered only a very harsh and caustic environment of high temperatures, gases, and acids. In order for later life to evolve, the primitive cells needed to withstand these assaults yet carry out functions that required a standard pH, moisture, and salt content inside the cell.

Domain Bacteria represents one of two domains of prokaryotes. The other is domain Archaea, which is discussed later. Domain Eukarya contains all the organisms composed of eukaryotic cells, which are cells distinguished by membrane-bound internal structures called *organelles*. Of the three domains, bacteria have become a model for comparing prokaryotic cell structure with eukaryotic cell structure, including comparisons of cell walls. Even biologists who spend careers studying eukaryotic lifestyles begin their education in cell structure by studying the bacterial cell wall.

The Gram-Positive Cell Wall

The cell wall of gram-positive bacteria is a thick (30–80 nanometers [nm]) single layer. More than 50 percent of the wall contains peptidoglycan, which is a heterogeneous molecule made of two repeating units (N-acetylglucosamine and N-acetylmuramic acid) connected by two different types of side chains (a peptide and a tetrapeptide). The molecule's strength results from extensive cross-linking between polymer chains that run parallel to the cell's surface. In growing cells, the new chains develop within the matrix of older chains.

Peptidoglycan connects with two other polymers: teichoic acid and teichuronic acid. Both of these large compounds give cells additional strength and create a negative charge on the cell's surface. This negative charge enables bacteria to bind to surfaces such as soil, teeth, living tissue, and inanimate surfaces. Teichoic acid contains long chains of glycerol phosphate or ribitol phosphate. One type of teichoic acid (ribitol) connects only with the peptidoglycan layer, and a second type (glycerol) associates with peptidoglycan and with the cell's cytoplasmic membrane.

Despite the extensive number of studies conducted on peptidoglycan's features and chemistry, much of this compound remains a mystery in biology. The biochemist Samy O. Meroueh of the University of Notre Dame wrote, in 2006, "The 3D structure of bacterial peptidoglycan, the major constituent of the cell wall, is one of the most important, yet still unsolved, structural problems in biochemistry." The Gram stain procedure has indirectly helped microbiologists surmise details of the pepitidoglycan in the cell wall.

In the Gram stain procedure, pores in the intricate peptidoglycan layer allow a dye called crystal violet to enter and reach the cell membrane. Crystal violet does not stain the gram-positive bacteria's cell wall but actually stains the cell's membrane. After cells have been exposed to the crystal violet, they are next immersed in an iodine solution. The iodine forms a complex with crystal violet that cannot escape from inside the cell wall. The result is a permanently colored gram-positive cell. Gram-negative cells, by contrast, do not have extensive cross-linking between peptidoglycan and other polymers. This structure allows the crystal violet–iodine complex flows into and out of the cell wall without staining the membrane. As a result, gram-negative bacteria do not react with crystal violet.

Underneath the gram-positive cell wall lies an area called the *periplasmic space,* or simply *periplasm.* Some enzymes never leave the periplasm and carry out all their reactions there. The cell also excretes some enzymes from the periplasm that traverse the cell wall and work outside the cell. These are called extracellular enzymes.

The cell's membrane lies inside the periplasm and surrounds the entire cellular contents. The membrane serves as the site where much of the cell's energy generation takes place. The membrane also stores energy needed to power the flagella of flagellated motile bacteria.

The gram-positive outer layers are, in order from outside to inside, the peptidoglycan cell wall, periplasm, and cell membrane. All gram-positive species have this structure in common, but many species and even individual strains (unique members of a species) have features on the cell wall's outer surface that make them unique. The gram-positive outer surface may hold small amounts of polysaccharides, which are long chains of sugars. These polysaccharides give many species a characteristic serotype. A serotype is a unique feature on the outside of a cell that makes the cell different from all others. Clinical microbiologists use serotypes to help in identifying unknown pathogens.

Some gram-positive surfaces also contain proteins that help cells interact with the environment and protect them from attack from other microorganisms. For example, the genus *Staphylococcus* contains an S-layer on its surface. The S-layer is a continuous sheet of repeating proteins or glycoproteins (glucose-protein molecules) in a pattern not unlike a tile floor. This layer helps cells adhere to body surfaces and may provide protection from the body's natural defenses. The genera *Aeromonas, Campylobacter, Clostridium,*

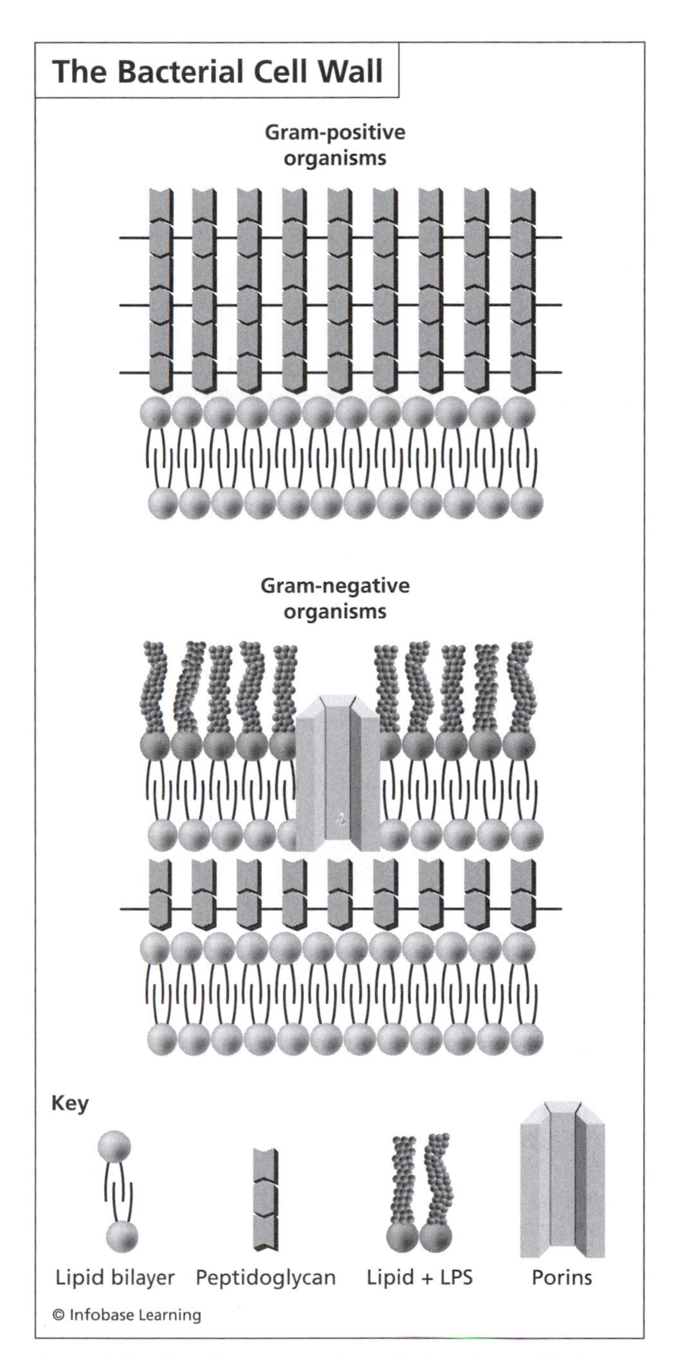

Bacterial cell walls serve as the cell's interface with its environment. The differences between gram-positive and gram-negative cell walls affect bacterial resistance to antibiotics and chemicals and also affect virulence in pathogens.

Lactobacillus, Nitrosomonas, and *Pseudomonas* also contain species that produce S-layers.

The Gram-Negative Cell Wall

Gram-negative cell walls have more complexity than those of gram-positive bacteria, yet they are also thinner than gram-positive walls. The gram-negative peptidoglycan layer is only 2 to 10 nm thick, and it makes up no more than 5 to 10 percent of the cell wall's dry weight (the weight after all the water has been removed).

The gram-negative outer layers are, in order from outside to inside, the outer membrane, periplasm, peptidoglycan cell wall, periplasm, and cell membrane. (Some textbooks describe the entire area between the inner and outer membranes as a single periplasm.) Unlike in gram-positive bacteria, the gram-negative cell wall is often thought of by biologists as the sum of the outer membrane and the peptidoglycan layer. The membrane-peptidoglycan-membrane structure measures only 20 to 25 nm thick, less than the thickness of the gram-positive peptidoglycan cell wall alone.

The gram-negative periplasm and inner cell membrane resemble their counterparts in gram-positive species. Enzymes originate and work in the periplasm, or the cell excretes them from the periplasm to the outside. The gram-negative inner membrane carries out metabolic functions and energy production as in gram-positive bacteria. The cell membrane in both gram-negative and gram-positive species is constructed of a bilayer that follows typical membrane structure found throughout biology. The bilayered membrane contains a phospholipid layer in which the polar (containing a charge) heads point toward the membrane's surfaces and hydrophobic (water-repelling) carbon chains point inward.

All membranes (two in gram-negative cells and one in gram-positive cells) contain proteins that serve different functions. Two types of membrane proteins are distinguished by the tightness of their connection with the membrane: peripheral proteins and integral proteins. First, peripheral proteins lie on or near the outside of the membrane, and cell biologists have found these proteins are easy to remove from the cell surface. Peripheral proteins may serve in helping enzyme reactions or in giving the membrane structural support. The second type of membrane protein is called an *integral protein.* Integral proteins embed in the membrane, rather than attach to the outer surface. Some proteins, called *transmembrane integral proteins,* reach completely across the membrane from the inner to the outer surface. Biologists cannot remove these proteins from cells unless they destroy the membrane with detergentlike chemicals.

Certain transmembrane integral proteins may serve as an aid in transport by forming pores across the membrane. These pores then serve as sites where cellular secretions exit the cell and nutrients enter. For this reason, these specialized proteins are called *porin proteins.* To form a pore, three porin proteins line up to make a channel through the membrane. Small molecules such as monosaccharides (single sugars such as glucose), disaccharides (two sugars linked together), amino acids, and short peptides (amino acid chains) fit through porin channels, and enter the cell. Larger molecules such as vitamin B_{12}, proteins, and synthetic antibiotics cannot fit through porin channels, and the cell must take them in by a different route. Porin channels of different gram-negative species vary in size, so that individual species allow molecules of slightly larger or smaller size ranges to enter the cell. For example, *Escherichia coli*'s porin channels admit molecules up to a molecular weight of 650–850 daltons. Cyanobacteria have much larger channel diameters; they admit molecules up to molecular weights of 70–80 kilodaltons (70,000–80,000 daltons.) (A dalton is a unit for measuring atomic mass; it is equal to one/[Avogadro's number]th of a gram.)

The outer membrane of the gram-negative cell wall is important in giving these bacteria unique features. Lipopolysaccharides, which are large molecules made of carbohydrate and lipid, dot the outermost surface of these cells. Compounds called *phospholipids* and *lipoproteins* make up the membrane itself. All three of these molecules sense the environment in which the bacteria live.

The outer surface constituents of gram-negative cell walls possess two main functions. First, they give the cells a negative charge, which provides some measure of protection from the body's immune response. Second, the long tail of lipopolysaccharides forms stringy molecules that extend into the cell's surroundings. These tails are called O antigens, and they give gram-negative species protection against attack by other microorganisms. O antigens contain sugars that are rare in nature and not easily recognized by predators such as protozoa. The protozoa do not recognize the potential meal when entering into contact with a gram-negative cell, so they leave the bacteria alone and seek other prey. But this feature, which is an advantage in nature, works against gram-negative bacteria when invading the body. Animal immune systems recognize O antigens when gram-negative bacteria invade the body. The immune system then launches a series of steps for the purpose of finding and killing the foreign invaders. The immune system sends cells called *phagocytes* into the bloodstream to engulf, ingest, and destroy the bacteria. Many gram-negative bacteria stay one step ahead of attack by changing their O antigen structure every few generations. The new structure forestalls attack by other microorganisms and by the immune system's cell-eating phagocytes. In this manner, gram-negative pathogens evade detection; this characteristic makes them more virulent, meaning they have enhanced ability to cause disease.

The complete membrane composed of lipopolysaccharides, phospholipids, lipoproteins, and proteins is called the *fluid mosaic model,* fluid because the membrane lacks rigidity. Rather, these compounds make the membrane an ever-changing structure that responds to conditions around it.

If gram-negative cells cannot retain the crystal violet–iodine complex during the Gram stain procedure, why do they turn pink? Christian Gram undoubtedly wanted to see all the bacteria in his microscope, not merely the gram-positive variety. He, therefore, added a final step to the procedure. Since unstained gram-negative cells are almost invisible by light microscopy, the Gram procedure includes a step using a red dye called *safranin.* Safranin stains the cytoplasm and makes gram-negative bacteria appear dark pink under a microscope.

OTHER FEATURES OF THE BACTERIAL CELL WALL

Many bacteria secrete a sticky polymer coat called *glycocalyx*—the word means "sugar coat." The composition of glycocalyx differs depending on the species making it, but it usually contains large amounts of polysaccharides or polypeptides (long amino acid chains). The glycocalyx coating is called a *capsule* when firmly attached to the outside of the cell wall. This strong adhesion forms another layer outside the cell wall.

Some glycocalyx coats contain a loose and disorganized arrangement that microbiologists call a *slime layer.* Capsules and slime layers act as additional defensive tools that enable bacteria to evade attack, or increase their virulence, or do both. Phagocytes have a harder time ingesting encapsulated cells than cells without capsules. Both gram-positive (for example, *Bacillus*) and gram-negative *(Pseudomonas)* bacteria make capsules.

Extracellular polysaccharide (EPS) is a glycocalyx made completely of sugars. EPS enables biofilm to stick to surfaces such as water pipes, teeth, medical implants, and rocks and pebbles submerged in flowing streams. *Pseudomonas,* for example, is a prolific producer of EPS in biofilms. Gram-positive *Streptococcus mutans* causes tooth decay primarily because it forms a firm attachment to teeth by making EPS.

Cell coatings such as EPS and capsules serve bacteria in two other ways: They prevent cells from dehydrating in dry conditions, and they—especially EPS—store a reserve of sugars for times when nutrients become scarce.

Members of the gram-negative family Enterobacteriaceae contain an additional compound in their outer membrane called Braun's lipoprotein. This lipoprotein forms a very strong link between the outer membrane and the peptidoglycan layer. As a result, Braun's lipoprotein anchors the membrane to the layer beneath it and provides the cell with additional strength.

THE CELL WALL IN BINARY FISSION

A new technology called *electron tomography* has been employed in the study of microbial structures such as cell walls. Electron tomography provides observations on cells in magnifications as high as those used in TEM, but it does not require the chemicals that TEM uses; nor does it need stains. Electron tomography, therefore, enables microbiologists to observe structures in a more natural form. This technique has answered some questions on the way a new cell wall develops between two dividing cells during binary fission.

In binary fission, a parent cell splits in two to make a pair of identical daughter cells. To do this, the parent cell must break down part of its cell wall. In binary fission, autolysin enzymes attack the cross-linkages in the peptidoglycan layer and weaken that portion of the parent cell wall. As two cells begin to form, a mechanism in the fission process pinches the rest of the cell—membranes and cytoplasm—in the middle. As a consequence, two daughter cells begin to emerge but remain connected by a bridge of cellular material. The doublet cell begins to build a new rigid wall down the middle. This section of cell wall that crosses through the cytoplasm is called the *septum.* Once the septum connects, two complete cells are born, each with its newly formed cell wall in place.

Bacterial binary fission is perhaps the most efficient process in nature for making two new cells from a single cell. But the cell must spend energy to build the septum and finish building the entire cell wall for the new daughter cells. Binary fission can be considered a vulnerable time for a cell, in which it does not have its full battery of defenses, and it is in an energy-depleting condition. In a general way, this is why many antibiotics work best against growing populations of bacteria with a high proportion of replicating cells.

ANTIBIOTICS AND BACTERIAL CELL WALL FORMATION

Several antibiotics target cell wall synthesis in order to kill bacteria. These antibiotics act in at least two ways. First, they inhibit the enzyme that builds cross-linkages when cells manufacture peptidoglycan. An incomplete peptidoglycan layer leaves the cell vulnerable to osmotic pressures, and the cell lyses. This mechanism is a common mode of action among antibiotics, such as penicillin, ampicillin, cephalosporins, vancomycin, and bacitracin. Second, antibiotics punch holes in the cell wall and perhaps the cell membrane. The cell then dies, as its contents leak into its surroundings.

Bacteria have devised clever ways of building resistance to these antibiotics. Penicillin was the first antibiotic used on a large scale, beginning in the 1940s, but it may also have been the first to begin losing its effectiveness because bacteria developed resistance to it. Microbiologists eventually learned that penicillin-resistant bacteria make the enzyme penicillinase. This enzyme performs the simple task of adding a hydrogen atom to a ring structure in the penicillin molecule, called the β-lactam ring. With this one step, the bacteria turn penicillin into a molecule that no longer disrupts cell wall synthesis.

THE ARCHAEAL CELL WALL

Domain Archaea has several similarities to bacteria. Two obvious similarities relate to the strong cell wall possessed by archaea that provides shape and the characteristic that archaea species stain either gram-positive or gram-negative, depending on cell wall thickness. Cell wall composition across all archaea varies widely, but the domain can be divided into two general groups based on cell wall constituents. The first group contains genera such as *Halobacterium* and *Desulfurococcus* that live in extreme environments. These archaea have an outer wall composed of an S-layer rather than peptidoglycan, and the S-layer attaches directly to the cell membrane. Because these cells lack peptidoglycan, they tend to stain gram-negative. The second group of archaea contains a cell wall of a thick pseudomurein layer or other large polysaccharides. Pseudomurein is similar to peptidoglycan but with slight modifications to the polymer's subunits. These polymers behave as peptidoglycan does to trap the crystal violet–iodine complex during the Gram stain procedure. The result is a gram-positive stained cell. An example of gram-positive archaea is *Methanobacterium formicicum,* an anaerobic species that produces methane gas.

Archaean gram-negative species vary in the proportion of proteins and glycoproteins in their cell walls. As a generalization, methanogens and extremophiles tend to have high amounts of glycoproteins, while other archaea tend to have more protein than glycoprotein. Wall thicknesses among the archaea also vary, from 20 to 40 nm, and some species have double layers.

Perhaps most distinctive of all the archaea's cell wall features is their membrane lipids, especially in the group having an S-layer instead of peptidoglycan. The membrane lipids of archaea might play a protective role similar to that of the peptidoglycan layer in gram-positive bacteria. Squalene and squalenelike compounds make up as much as 70 percent of the membrane. Squalene is long hydrocarbon chain ($C_{30}H_{50}$) that serves as the main membrane lipid, but different species build their membranes from different combinations of squalenelike compounds, some with longer chains and others with shorter chains. This results in very rigid thick-walled cells that stand up to extreme environments that kill bacteria and other microorganisms. For instance, extreme thermophiles live at temperatures of 175–235°F (80–113°C), but many exist in places that are much hotter. These thermophiles depend on membrane lipids with very long hydrocarbons (up to C_{40}) to protect them. Examples of such unusual microorganisms are *Thermoplasma* and *Sulfolobus.*

THE EUKARYOTIC CELL WALL

The eukaryotic cell wall is simpler than the prokaryotic wall and lacks peptidoglycan. As do cell walls in bacteria, the eukaryotic wall supports the cell and protects the cytoplasmic contents. In eukaryotes, however, the cell wall also allows some flexibility, so that cells can change their shape from moment to moment. The flexibility results from differing relative amounts of the polymers cellulose, pectin, chitin, and glucan. Eukaryotes vary from very supple cells to cells that seem to be made of stone.

Algae

Algae are a diverse group of microorganisms with a wide variety of cell wall compositions. Except for *Euglenophyta,* which has no cell wall, most algal walls contain one or more of the following constituents: cellulose, xylan, galactan, mannan, chitin, protein, alginic acid, silica, or calcium carbonate. All except calcium carbonate are large polymers.

The algal cell wall is thin and appears microfibrillar because it contains polymers that are like fibrous strings. The fibrous polymers weave into a mixture of other polymers, called *matrix polymers.* The composition of the matrix varies among species, but usually it contains agar, alginic acid, carrageenan, or fucoidin. Many single-celled algae also possess thecae. A theca is a strong scalelike piece; many of these thecae line up to form a cover on the cell. Algae that have thecae usually lack the more typical microfibrillar structure.

Algal cell walls have another unique quality: Their composition may vary from one generation to the next. For instance, in heteromorphic algae (cells in which their morphology takes many forms), cell walls alternate by generations. For example, the green alga *Derbesia marina* contains either mannan or a cellulose-xylan mixture, and these formulas repeat in alternating generations.

Beneath the cell wall lies a soft cell membrane similar to a bacterial capsule. This membrane controls nutrient transport for the cell and houses the energy-producing machinery for powering flagella in motile algae.

Fungi

Most fungi have cell walls, and the majority of them contain the polysaccharides chitin and cellulose. Chitin is a somewhat flexible molecule that provides structure and strength. The fungal cell wall structure also has an inner microfibrillar support layer, in which proteins and glycoproteins intersperse with polysaccharides. A gellike substance surrounds this layer and behaves in a way similar to bacterial capsule behavior. Fungi that grow hyphae also construct a cell wall around the individual hyphal cells. The hyphal cell wall composition may differ slightly from that of the single-cell form of the fungus (the vegetative form). For instance, in *Mucor,* the hyphal cell wall is high in the compound chitosan, but the vegetative cell wall contains small amounts of chitosan.

In yeasts, chitin is in very low amounts or absent in the cell wall. The strength of each species's wall depends on the amount of branching in its polysaccharides. Yeast cell walls contain glucans, mannans, and a mannan-protein complex that may serve to protect the glucan layer from glucanases. Glucanases are enzymes made by microorganisms to digest glucan, so it can be used as a nutrient. Obviously, yeasts must have a means of protecting themselves from glucanase, or they will be devoured by other microorganisms, and the mannan-protein complex probably provides this defense. Yeast cell walls additionally have unique features called *bud scars,* which remain after cells reproduce and a daughter cell breaks off from the parent cell in a process called *budding.*

THE NEED FOR A CELL WALL

Most cells that normally possess a cell wall cannot survive without it. There are rare times, however, when bacteria in nature lose their cell wall entirely. When microorganisms compete for nutrients and habitat, some species produce enzymes that destroy bacterial cell walls. When the cell wall falls completely away from a cell that has been targeted by these enzymes, only an exposed membrane containing the cytoplasm remains. This structure, called a *protoplast,* is very vulnerable to conditions in the environment, because it does not have its protective cell wall covering it. Protoplasts in an isotonic environment usually swell and then bulge into a round shape. Isotonic conditions are those in which the pressures outside the cell equal the pressures inside it. In the isotonic state, protoplasts can metabolize nutrients, but they do not divide or reproduce. In nonisotonic conditions, protoplasts cannot survive.

An L-form is a cell that is defective or injured so that its cell wall is incomplete. The name *L-form* was coined in 1935 by the bacteriologist Emmy Kleineberger, at the Lister Institute of Preventive Medicine in London. Kleineberger had originally been conducting experiments on pneumonia infection, when she discovered an anomaly in *Streptobacillus* cultures. Strange swollen cells appeared to lack their cell wall. Kleineberger named the structures *L-forms,* presumably in honor of the Lister Institute. L-forms may result when cells have been targeted by enzymes or antibiotics from other microorganisms, or when exposed to extreme temperatures or osmotic pressures. L-form cells can repair themselves in the event the environmental conditions improve. In these cases, they rebuild a complete cell wall and begin to function again as normal cells. L-forms are common in certain gram-positive genera: *Bacillus, Proteus, Streptococcus,* and *Vibrio.* These bacteria actually manage to reproduce through several generations in their L-form.

In the late 1800s, bacteriologists had begun studying odd cellular forms associated with a disease in cattle called *contagious bovine pleuropneumonia.* Details of the cause of the disease came to light, over the next 50 years, as microbiologists tried to uncover the nature of a funguslike organism that was not a fungus, but seemed neither bacterial nor viral. The microorganism ultimately was identified as a new classification of bacteria: *Mycoplasma,* cells that never have a cell wall. Today one of the purposes of L-form studies in laboratories is to discover the traits of *Mycoplasma.* Researchers study how this microorganism causes pneumonia and try to decipher the mode of action by which antibiotics attack cells lacking a cell wall.

The bacterial cell wall is a complex structure that, despite its importance to microbial life, still holds unanswered questions for microbiology. Cell walls provide simple physical protection to cellular contents of the cell membrane and inside the membrane. Biologists continue to study cell walls, especially the bacterial form, to learn more about evolution of life on Earth and the development of higher, more complex organisms.

See also ANTIBIOTIC; ARCHAEA; ELECTRON MICROSCOPY; EUKARYOTE; GRAM STAIN; IDENTIFICATION; MORPHOLOGY; *MYCOPLASMA;* PATHOGENESIS; PEPTIDOGLYCAN; PROKARYOTE; RESISTANCE.

Further Reading

Ehrmann, Michael. *The Periplasm.* Washington, D.C.: American Society for Microbiology Press, 2006.

Gram, Christian. "Ueber die Isolirte Färbung der Schizomyceten in Schnitt- und Trockenpräparaten" (The Differential Staining of Schizomycetes in Tissue Sections and in Dried Preparations). *Fortschritte der Medicin* 2 (1884): 185–189. In *Milestones in Microbiology,* translated and edited by Thomas Brock. Washington, D.C.: American Society for Microbiology Press, 1961.

Meroueh, Samy O., Krisztina Z. Bencze, Dusan Hesek, Mijoon Lee, Jed F. Fisher, Timothy L. Stemmler, and Shahriar Mobashery. "Three-Dimensional Structure of the Bacterial Cell Wall Peptidoglycan." *Proceedings of the National Academy of Sciences* 103, no. 12 (2006): 4,404–4,409. Available online. URL: www.pnas.org/content/103/12/4404.abstract. Accessed March 21, 2009.

Rogers, Howard J. *Aspects of Microbiology 6: Bacterial Cell Structure.* Washington, D.C.: American Society for Microbiology Press, 1983.

Seltmann, G., and O. Holst. *The Bacterial Cell Wall.* Berlin: Springer-Verlag, 2002.

centrifugation Centrifugation is a procedure used for separating particles of varying sizes and densities in a liquid by applying centrifugal force. The piece of equipment that imparts the centrifugal action, sometimes referred to as *spinning,* is called a *centrifuge.*

When a microbiologist centrifuges a sample in a centrifuge, centrifugal force pushes particles in a path that leads them away from the center of the rotation. Centrifugal force (X*g*) represents multiples of gravitational force. In other words, as centrifugal force increases, a rotating object "feels" an increase in the force of gravity. Children on a playground merry-go-round experience centrifugal force as the ride rotates; a washing machine creates centrifugal force during the spin cycle. In microbiology, centrifugation creates this force for one of two purposes: to isolate bacterial cells from a liquid suspension or to separate one type of microbial component from others in a process called *cell fractionation.*

CENTRIFUGES

All centrifuges contain a vertical shaft that holds a rotor equipped with wells, called *sample wells,* for tubes containing a liquid sample to be centrifuged. The centrifuge mechanically rotates the shaft and rotor at a set speed, and, over time, particles of different size (or mass) and density migrate to different levels in the liquid. This separation of particles from a suspension or from other particles is called *sedimentation.*

Centrifuges range in size from small units called *microcentrifuges* that fit on a laboratory bench to large units that process hundreds of liters of liquid. Rotation speeds also vary among different types of centrifuges, from slow-speed centrifugation to extremely fast rotations found in ultracentrifugation. Microbiologists use low- to medium-speed centrifugation to separate large particles such as whole bacterial cells or fungal spores from smaller particles. High-speed centrifugation separates small bacterial cells, cell fragments, and viruses from suspensions. Ultracentrifugation performs the sedimentation of small viruses, cell organelles, and large molecules such as deoxyribonucleic acid (DNA).

Rotor types belong to categories based on the construction of their sample wells. The three rotor types used in centrifugation are fixed-angle, vertical, and swinging-bucket. Fixed-angle rotors hold sample tubes at a set angle relative to the rotor shaft. Vertical rotors also hold tubes in a fixed position throughout centrifugation, but the position is vertical rather than at an angle. The wells in swinging-bucket rotors, by contrast, swing outward during rotation. The table on page 149 gives examples of the rotors recommended for certain procedures in microbiology.

Centrifuge tubes contain special materials that withstand the intense force put upon them during centrifugation. Most centrifuge tubes in use today consist of the plastics polypropylene (PP), polyallomar (PA), polycarbonate (PC), polyethylene terephthalate (PET), polysulfone (PS), or polytetrafluoroethylene (Teflon). Polyethylene usually goes into centrifuge tubes as either low-density (LDPE) or high-density polyethylene (HDPE). Rotor manufacturers recommend tube types on the basis of the maximal force each type of plastic can withstand and of the plastic's compatibility with different chemicals. For instance, polycarbonate and polyethylene terephtalate resist damage from organic solvents better than other plastics, so they are a good choice when centrifuging suspensions containing these liquids. (All centrifuge tube materials are compatible with water and salt solutions.) Thick-walled tubes made of polycarbonate or polypropylene better withstand the forces generated in high-speed ultracentrifugation than do other materials.

CENTRIFUGAL FORCE

During centrifugation, matter migrates outward in a radial direction from the center of rotation. As a rotor rotates or spins, matter inside each centrifuge tube moves toward the bottom of the tube within a centrifugal field. The strength of a centrifugal field, G, at a certain spot within the sample is defined as:

$$G = 4\,\pi^2\,(\text{rpm}^{-1})^2\,r \div 3{,}600$$

In the equation, rpm equals revolutions per minute and r equals the distance in centimeters (cm) from the centrifuge shaft (the axis of rotation) and a specific spot in the sample. The term *r* is more commonly known as the *radius.* Centrifugal force may be increased by increasing either rpm or r. Microbiologists often refer to rpm as the *speed of the centrifugation.*

Relative centrifugal force (RCF) is the force put on matter during centrifugation compared with gravitational force, and it is expressed as a number X*g*. For example, DNA sedimentation may require a RCF of 17,000 X*g* for 10 minutes. Revolutions per minute relate to RCF by the following conversion:

$$RCF = 11.17 \times R_{max} (rpm \div 1{,}000)^2$$

In this equation, R_{max} is the maximal radius in centimeters from the axis of rotation for a given rotor.

Particle features other than size and shape also affect this separation by centrifugation: particle shape, viscosity (thickness) of the liquid, and the density of the liquid. Equations have been developed to take these factors into consideration also when calculating centrifugation speed and time.

TYPES OF CENTRIFUGATION

Batch centrifugation involves sample-filled tubes rotated at a set speed (rpm) for a set period. After the centrifugation ends, a technician recovers the tube's contents by using a pipette or by slowly decanting the liquid phase from the solid pellet. Decanting simply means that the liquid is poured off a solid pellet lodged at the bottom of the tube.

Continuous centrifugation differs from batch centrifugation because it handles larger volumes that continually flow into the centrifuge rotor and out. The inflow begins as a dense suspension of particles, and the outflow exits as a clarified liquid. The solid matter from the inflowing suspension stays inside the rotor on concentric rings.

In either batch or continuous centrifugation, *supernatant* is the term used for the clarified liquid in which most or all solids have been removed.

Differential Centrifugation

This common technique separates solid particles from a suspension; the solids form a pellet at the bottom of the centrifuge tube, and an operator recovers the pellet by decanting the supernatant. Various procedures in microbiology require that a scientist retain the solid pellet (for example, whole cell isolation), but other procedures use centrifugation to recover the clarified supernatant (for example, enzyme purification). The size, shape, and density of the suspension's particles determine how much time is needed and the speed necessary to separate a pellet from a liquid.

Types of Centrifugation Rotors

Purpose of the Centrifugation	Type of Rotor Recommended
cell pellet formation	fixed-angle
differential gradient	vertical, swinging-bucket
density gradient or zonal sedimentation	vertical, swinging-bucket

Density Gradient Centrifugation

In this method, a microbiologist puts a sample onto the surface of a liquid column (inside a centrifuge tube) in which density increases from top to bottom. Sucrose solution, cesium chloride solution, or a mixture containing silica creates a density gradient suitable for separating particles. During centrifugation, a particle of a given density migrates to the part of the gradient having exactly equal density. When the particle reaches this level in the tube and will move no farther. Density gradient centrifugation isolates viruses, cell organelles, or macromolecules from a mixed suspension of other particles.

Isopycnic Centrifugation (also equilibrium density or equal zonal centrifugation)

Isopycnic centrifugation also uses a density gradient. During centrifugation, various particles of different densities form bands at the levels in the gradient having an equal density and move no farther. In this method, several particles or macromolecules (very large molecules) can be separated from each other inside one tube. Isopycnic centrifugation works well for separating bacterial cells, organelles, macromolecules, and nucleoproteins (complexes of nucleic acid and protein).

Rate-Zonal Centrifugation

Rate-zonal centrifugation separates particles of equal density but of differing masses and shapes. A sample applied to the top of the density gradient separates into its constituent particles during centrifugation as a result of the particles' different sedimentation rates. This type of centrifugation is used for ribosomes and various forms of DNA.

Ultracentrifugation

The recovery of macromolecules such as plasmids and DNA segments requires very high-speed centrifugation. (A plasmid is a circular piece of bacterial DNA apart from the cell's main DNA.) Ultracentrifugation achieves speeds of up to 500,000 X*g* in specialized ultracentrifuges, which consist of a refrigerated chamber that holds the rotor and is evacuated to pressures less than atmospheric pressure. Certain ultracentrifuges include optics called a schlieren system, which measures the progress of sedimentation as centrifugation takes place. Ultracentrifugation has now been designed to separate different types of DNA—circular versus linear—as well as different types of viruses.

Dye-buoyant density centrifugation uses ultracentrifuge speeds to separate circular plasmid DNA from other DNA in the chromosome of a single species of bacteria. It is used for separating linear DNA molecules from circular DNA, such as found in plasmids. In this procedure, a technician adds lysed cells to a cesium chloride density gradient containing the dye ethidium bromide. Ethidium bromide molecules only insert between the nucleic acid units of linear (not circular) DNA. This insertion step is known as *intercalating,* and it changes the overall density of the DNA. As a result, less dense linear DNA forms a band separate from denser circular DNA.

Ultracentrifugation also separates plasmids from each other. A scientist does this by using an enzyme to open one circular plasmid but leave the others in their normal form. The opened (nicked) DNA, then, assumes a different density from the circular DNA. As a result, the nicked DNA migrates to a different place in the tube from the circular DNA. Ethidium bromide also fluoresces in ultraviolet light so that the distinct bands of different DNAs become visible.

CELL FRACTIONATION

Microbial cells, such as bacteria, can be divided into their components in a procedure called *cell fractionation.* In cell fractionation, each centrifugation step runs at a higher speed than the last. Fractionation begins with the recovery of a cell pellet from the liquid suspension by differential centrifugation. A microbiologist then lyses the whole cells in the pellet by a technique called *cell disruption.* The microbiologist then saves the cell's contents and centrifuges the mixture by rate-zonal centrifugation. For example, this method could separate organelles from each other and from the cell's chromosome. The microbiologist simply repeats each centrifugation at higher speeds in order to separate organelles farther from each other or to differentiate large molecules. A combination of centrifugations along with laboratory methods called *chromatography* and *electrophoresis* eventually divides even the smallest cellular components into separate distinct fractions. Chromatography is a category of physical and chemical methods that separate molecules from each other. Electrophoresis is the movement of molecules in an electrical field.

Microbiologists use a measurement called a Svedberg unit in cell fractionation. The sedimentation rate of any particle may be determined by using an equation that relates centrifugation time, sedimentation velocity, and r. The final result is the value, S, for a Svedberg unit. Svedberg units help microbiologists differentiate between similar cell components. Ribosomes provide the best example of the use of Svedberg units. Bacterial ribosomes have sedimentation coefficients of about 70S. Each 70S ribosome contains one 30S subunit and one 50S subunit. Eukaryotic cell ribosomes are larger, about 80S, made of one 40S and one 60S subunit. (Svedberg values do not add up mathematically.)

Centrifugation methods have helped microbiologists study the smallest parts of prokaryotic and eukaryotic cells. By doing so, they have made new discoveries related to the size and the composition of large molecules such as DNA, plasmids, and cell organelles. Today's studies on the internal features of microorganisms could not occur without centrifugation.

See also FRACTIONATION; ORGANELLE.

Further Reading

Bates College. "Centrifugation Basics." Available online. URL: http://abacus.bates.edu/~ganderso/biology/resources/centrifugation.html. Accessed March 22, 2009.

Cole-Parmer Technology Library. "Basics of Centrifugation." Available online. URL: www.coleparmer.com/techinfo/techinfo.asp?htmlfile=basic-centrifugation.htm&ID=30. Accessed March 1, 2009.

Lenntech. "Centrifugation and Centrifuges." Available online. URL: www.lenntech.com/Centrifugation.htm. Accessed March 1, 2009.

London South Bank University. "Centrifugation." Available online. URL: www.lsbu.ac.uk/biology/enztech/centrifugation.html. Accessed March 22, 2009.

chemotaxis Chemotaxis is a mechanism in bacteria that enables cells to move toward or away from a chemical. The processes within the cells that carry chemotaxis are collectively called the *chemotactic response.* Chemicals that attract motile bacteria, such as sugars and amino acids, are called *chemoattractants,* and chemicals that repel bacteria are called *chemorepellents.* Some end products of normal metabolism or waste products repel bacteria, perhaps as a means of survival. Because of different reactions to nutrients versus wastes, bacteria use chemotaxis to swim toward higher concentrations of some chemicals and toward lower concentrations of other chemicals. Attractants and repellents in this system are both called *chemoeffectors.*

Chemotaxis helps protect bacteria by enabling them to move away from toxic substances. The University of Wisconsin biochemist Julius Adler, who conducted early studies on chemotaxis, wrote in *Science,* in 1966, "It is clearly an advantage for bacteria to be able to carry out chemotaxis, since by this means they can avoid unfavorable conditions and seek optimal surroundings." Since then, researchers have investigated the modes and the reasons for bacterial chemotaxis in various habitats, in diseases, or in communities such as biofilm. Clearly, bacteria with the ability to move in the direction of nutrients possess an advantage over less motile bacteria or bacteria with a less sensitive capacity for finding nutrients. In either case, chemotaxis requires that a bacterial cell be motile, meaning have the ability to move under its own power.

Taxis refers to any movement of a cell due to a stimulus outside the cell. Chemotaxis allows bacteria to respond specifically to chemicals, but other forms of taxis serve equally important roles in the bacterial world. The following responses are types of taxis: light (phototaxis), oxygen (aerotaxis), temperature (thermotaxis), osmotic pressure (osmotaxis), acids and bases (pH taxis), gravity (geotaxis), and magnetic fields (magnetotaxis). In any type of taxis, the bacterial cell must be capable of three processes. First, it must be able to detect the stimulus. Second, it must have a means of transmitting this information to an organelle responsible for motility. Third, the cell must then respond to the signal.

THE CHEMOTAXIS PROCESS

A cell detects chemical stimuli with proteins called *chemoreceptors* located in or just beneath the cell wall. Bacterial cells may have as many as 30 total attractant and repellent receptors. These chemoreceptor proteins react to certain minimal concentrations of chemoeffectors in the environment in one of two ways. First, they bind directly to the chemical, or, second, they bind to a protein that has attached to the chemical. This binding of receptor to chemical is the first step in initiating movement by the cell. If both attractants and repellents are present at the same time, bacteria compare the signals from each chemoreceptor before responding to either one. Usually, bacteria respond to the chemical present at a concentration that gives a more powerful effect on the cell than an alternate chemical. The correlation between concentration and the cell's response may vary among chemoeffectors and varies from one bacterial species to the next.

A cell's reaction to a moving stimulus that initiates movement is triggered only when the stimulus reaches a minimal level, called the *threshold concentration*. Bacteria do not detect a threshold concentration from a single receptor as if the cells have a sense of taste, smell, or sight. Instead, bacteria sense changes in their environment by comparing a chemical's level at a moment in time, moment 1, to the level that had been had detected at moment 0. In a way, bacteria "remember" the chemical's concentration in order to compare it to the new concentration.

Bacteria contend with two factors that make the concentration of any chemical in their environment change from moment to moment. First, the environment itself changes. For example, bacteria in a slow-moving river might detect an increasing concentration of a pollutant as it flows downriver. If the pollutant is toxic to the bacteria, they will respond by escaping. In a second scenario, a motile bacterial cell swims into a new environment and in so doing detects a change. Perhaps nitrogen compounds increase in concentration as the cell swims forward. This would be very likely in a pond as bacteria travel from nutrient-scarce depths to nutrient-rich places.

The overall change in chemical concentration is called a *gradient,* and a gradient can occur either by staying still in a flowing medium or by moving through a stationary medium. When bacteria move down a concentration gradient, it means they travel from high to low concentration. Moving up a concentration gradient means they travel from low to high concentration.

Chemotaxis's evolution led to bacteria that could react on their own to stimuli and then make an appropriate response to various stimuli. In this manner, chemotaxis in bacteria might represent an important step in the development of cell-to-cell communication and the ability of certain cells in the mammalian immune system to seek foreign bodies in the bloodstream.

Chemotaxis does not induce bacteria to swim in a straight line toward an attractant or away from a repellent. Rather, the process exerts a subtle effect on the normal random tumbling of bacteria, so that overall a cell moves in a direction. In the absence of a chemoeffector, bacteria move in a disorganized fashion through liquids in what is called a *random walk*. In chemotaxis, however, they adjust their movement. Random movement still occurs, but it becomes less frequent and is interrupted by bursts of motility (called a *run*). As a result, a cell migrates not in a straight line, but in a series of runs combined with periods of random movement. Each run in this pattern of fits and starts lasts no more than a few seconds, but the cell eventually arrives at the environment it seeks. When the cell encounters this optimal environment, random tumbling resumes.

Why tumble? It seems bacteria would be more efficient if taking a straight route. In 1998, researchers in physics analyzed the energy needs of *Escherichia coli* cells in a tumble, compared with directional swimming by use of flagella. The physicist Steven Strong surmised that bacteria's alternate modes of motility evolved for a reason: "*E. coli* change direction by entering into 'tumbles,' which have no characteristics which depend on sensory input. It seems likely that, in view of the limited use *E. coli* could make of steering . . . this simple method [tumbling] of direc-

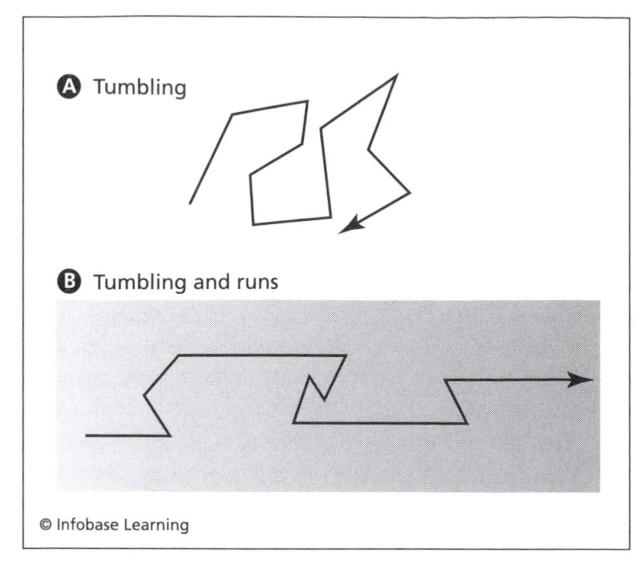

Motile bacteria use (a) random movement and tumbling when no chemical concentration gradient exists. In (b) chemotaxis, the presence of a chemoattractant causes bacteria to reduce the tumbling frequency and lengthen directional runs in response to a concentration gradient.

tion change was evolutionarily preferred because the cost[s] associated with this capability are lower than those that a more developed steering capability would impose." In short, tumbling saves energy, so even when using chemotaxis, bacteria migrate in a specific direction in an energy-conserving manner.

CHEMOTAXIS IN *E. COLI*

Most studies of chemotaxis have been in *E. coli*. Some of the mechanisms *E. coli* uses for chemotaxis illustrate the extraordinary ability of bacteria to sense and react to their environment. At least four different *E. coli* chemoreceptors have been identified; they react to the following compounds: (1) the amino acid serine, (2) the sugar maltose or amino acid aspartate, (3) the sugar galactose or ribose, and (4) dipeptides. *E. coli*'s chemoreceptors are called methyl-accepting chemotaxis proteins (MCPs) because the addition or subtraction of a methyl group (-CH_3) is a hallmark of the binding reaction. These sensitive MCPs on *E. coli*'s surface cause the cell to react to a chemical within 200 milliseconds.

Motility can be created by one of three mechanisms. First and simplest, MCPs bind directly to the chemical, and this binding creates a motility signal inside the cell. A series of reactions in the cell cytoplasm (inner watery component of cells) carry the message from the binding site to the flagellum. In the second mechanism, the chemical binds to a protein-receptor complex. This action causes a signal to be sent to a second protein, which regulates a type of move/do not move message inside the cell. Third, a moving cell that encounters a changing environment activates an MCP system that constantly alters its reactions to the changes by adding or removing phosphate groups (PO_4^{-3}) on a regulator protein.

As in many bacteria, *E. coli*'s motility is derived from its flagella. Bacteria may have a single polar flagellum on one end of the cell or, as in *E. coli*, many flagella on the entire cell surface, called *peritrichous flagella*. When flagella rotate in a counterclockwise direction, cells move in a controlled manner, but clockwise rotation causes a disorganized, tumbling movement. *E. coli* contains more than 30 genes that control receptors, signalers, regulator proteins, and the energy systems that power the flagella. Perhaps the biggest mystery of *E. coli* and other bacteria's chemotaxis arises from the so-called memory of the cell of where it has been.

Bacterial cells moving through a concentration gradient continually compare present conditions to conditions a millisecond in the past. The chemotaxis mechanism would seem to impart memory to bacteria connected to the addition or subtraction of phosphate and methyl groups. Phosphorylation, the addition of phosphate to a molecule, is directly connected to the duration of each run, but methylation has been more closely associated with bacterial memory.

Enzymes that control methylation must activate in response to new conditions in a cell's environment. If the conditions, a new chemical concentration, for instance, become permanent, the methylation enzymes reset themselves in an action called *accommodation*. Accommodation is the gradual decrease of a cell's responsiveness to stimuli, and it results from the addition of methyl groups to chemoreceptors. Humans experience a familiar accommodation when they can no longer smell an odor after being exposed to it for a long, continuous period. The odor remains, but the person's sensory system has adjusted to it through accommodation.

ADAPTATION

Adaptation to the environment led to the evolution of higher organisms that move toward a food source and away from danger. Microbiologists have studied three types of adaptation in microorganisms:

1. Genetic—Mutation and selection cause a cell to develop in a form that is more suited to its environment than other similar cells.

2. Nongenetic—Also called phenotypic adaptation, a cellular system turns on or turns off in response to a stimulus. In microorganisms, the classic example of this adaptation is the induction or repression of a specific enzyme.

3. Behavioral—bacteria respond to environmental factors through chemotaxis, but if the new condition persists, the bacteria adapt, rather than stay in a constant response mode. For example, bacteria may swim toward a higher sugar concentration in their environment, but if the sugar concentration remains high, chemotaxis shuts down, and the bacteria return to random tumbling.

Chemotaxis may provide a second advantage to bacteria in addition to survival: The process helps certain microorganisms form communities. In 2003, researchers from Princeton University in New Jersey and the Curie Institute in Paris demonstrated the concept of bacteria teaming up by using chemotaxis. These researchers showed that bacteria in dense populations sense the presence of many others around them by a process called *quorum sensing*. Individual cells then use chemotaxis to move toward each other to form a group that benefits all the cells. Princeton's Sungsu Park explained, "The bacteria are chasing amino acids released from their own cell bodies during starvation conditions. So by getting close to each other they have a better chance of getting nutrients." This cooperative spirit among bacteria has helped reveal how certain bacterial communities work, communities such as biofilms, mats, and mixed colonies.

Biofilms are complex communities of microorganisms that attach to surfaces submerged in flowing liquids. Biofilms contain many species that share jobs, such as gathering and storing nutrients. Some biofilm members secrete large compounds that provide a protective coat for the entire community. This protective matrix enables biofilms to withstand exposure to chemical disinfectants or extreme temperatures. If, as Park's colleague Peter Wolanin suggested, "The bacteria can actively seek each other out to engage in collective social behavior," then biofilms represent a type of bacterial colony that is stronger than the sum of its individual parts.

Chemotaxis continues to be a key area in the studies of evolution of complex organisms from single-celled bacteria.

STUDYING CHEMOTAXIS

Studies of bacterial motility and its stimuli can be accomplished by using a Dunn chemotaxis chamber. This glass device has the same width and length as a regular microscope slide, but it contains two concentric circular chambers, separated by a circular wall. A microbiologist puts a small volume of culture onto a coverslip, and then inverts the coverslip and places it on top of the two chambers. The wall separating the chambers does not touch the coverslip but rather is within 20 micrometers (µm) of the glass surface. The drop of liquid containing bacteria fills both circular chambers, but cells can move freely between the two chambers because of the 20-µm gap—most bacteria measure only 0.5–5 µm in width. If the two circular chambers contain two different concentrations of a test compound, the microbiologist will be able to view any movement between the chambers by viewing the culture under a microscope.

Dunn chamber experiments usually employ a heated microscope stage to provide the bacteria with their optimal temperature. A device attached to the microscope to produce a time-lapse recording of cell activity also helps in tracking the overall movement of the cells. Depending on the concentration of test compound in each chamber's solution, overall movement may be detected as going toward the inner chamber from the outer chamber, or vice versa. A microbiologist watches in real time as bacteria move in response to a concentration gradient. Specialized computer programs may also be used to collect data from the entire colony of cells and then perform statistical tests to differentiate subtle directional movement from random movement.

Chemotaxis has been a subject that receives moderate attention in microbiology courses, yet this ability of bacterial cells marks a critical step in evolution for two reasons: It demonstrates independent responses of cells to a stimulus, and it suggests a way in which cell-to-cell communication arose for the purpose of forming cell communities.

See also BIOFILM; MOTILITY.

Further Reading

Adler, Julius. "Chemotaxis in Bacteria." *Science* 153, no. 3737 (August 12, 1966): 708–716. Available online. URL: www.sciencemag.org/cgi/content/abstract/153/3737/708. Accessed March 22, 2009.

Bio-Medicine. "Social Mobility: Study Shows Bacteria Seek Each Other Out." Available online. URL: http://news.bio-medicine.org/biology-news-2/Social-mobility-3A-Study-shows-bacteria-seek-each-other-out-4135-1. Accessed March 22, 2009.

Dunn, Graham A. "Using the Dunn Chemotaxis Chamber." April 2006. Available online. URL: www.hawksley.co.uk/downloads/Dunn_Chamber_Hawksley.pdf. Accessed March 22, 2009.

Schultz, Stephen. "Social Mobility: Study Shows Bacteria Seek Each Other Out." *Princeton Weekly Bulletin,* 20 October 2003. Available online. URL: www.princeton.edu/pr/pwb/03/1020/3a.shtml. Accessed March 22, 2009.

Strong, Steven P., Benjamin Freedman, William Bialek, and Roland Koberle. "Adaptation and Optimal Chemotactic Strategy for *E. coli.*" *Physical Review* 57, no. 4 (1998): 4,604–4,617. Available online. URL: www.princeton.

edu/~wbialek/our_papers/strong+al_98b.pdf. Accessed March 22, 2009.

chromosome The chromosome of a prokaryotic or eukaryotic cell comprises all of the genetic material inside the cell. In bacteria and archaea, the chromosome equals the deoxyribonucleic acid (DNA) found within the cytoplasm in a generalized area called the nucleoid, plus, in some bacteria, plasmids, which are strands of DNA outside the nucleoid. In eukaryotic cells such as fungi, algae, and protozoa, the chromosome includes the DNA inside the nucleus, as well as the DNA in energy-producing organelles called *mitochondria*. Unlike bacteria, archaea, or eukaryotes, viruses contain two main types of genetic material that puts viruses into one of two main categories: DNA viruses or RNA (for ribonucleic acid) viruses. The term *chromosome* applies to the DNA in DNA viruses, but this term is usually not used for describing the RNA in RNA viruses.

The chromosome's purpose is to hold all the information of an organism so that it develops, functions, behaves, and reproduces as every other member of its species does. The chromosome ensures that all offspring retain the same characteristics as the parent. For this reason, the chromosome is the focal point of genetics, the study of heredity, or the passing of characteristics from one generation to the next.

THE HISTORICAL IMPORTANCE OF CHROMOSOMES IN MICROBIOLOGY

The British geneticist Frederick Griffith (1879–1941) showed, in 1928, that bacteria could transfer traits from one population to another. The process became known as *transformation*. Oswald Avery (1877–1955) and his team of medical researchers built upon Griffith's work on transformation by demonstrating that the genetic traits of bacteria could be traced directly to an unknown molecule. In 1944, they extracted a material from bacteria in search of this elusive molecule, and by doing so they discovered microbial DNA. But this was not a widely lauded discovery: Theories abounded at the time as to the nature of transformation and whether a single molecule truly existed that could carry information from one population of microorganisms to the next. Avery and his fellow scientists, Colin MacLeod (1909–72) and Maclyn McCarty (1911–2005), reported their discovery in the *Journal of Experimental Medicine*, but even they remained rather tentative about drawing bold conclusions on transformation. Their article stated, "The transformation described represents a change that is chemically induced and specifically directed by a known chemical compound. If the results of the present study on the chemical nature of the transforming principle are confirmed, then nucleic acids must be regarded as possessing biological specificity." Although Avery's peers in science suspected that the chromosome was actually a molecule of some sort, Avery's statements lacked 100 percent conviction and could not sway all of science. Even if the chromosome could contain a bacterial cell's complete genetic makeup, they might have reasoned, surely a human chromosome could not hold all of the information that defined humanity.

With Avery's theory as inspiration, in 1953, the American biologist James Watson (1928–) and the British Francis Crick (1916–2004) proposed a structure for DNA. The design also explained how chromosomes might replicate to form exact copies, a requirement for transformation. In an article of little more than 1,000 words in *Nature*, Watson and Crick wrote with considerable understatement, "We wish to suggest a structure for the salt of deoxyribose nucleic acid (DNA). This structure has novel features which are of considerable biological interest." Watson and Crick gave credit to the chemist Linus Pauling (1901–94) for his proposal, around that same time, that nucleic acids contained three intertwined chains. They also noted the work of the Norwegian physicist Sven Furberg (1920–83), who envisioned a helix-shaped molecule with bases arranged toward the inside of two parallel chains. In truth, Watson and Crick accomplished few laboratory studies of their own, but they had an extraordinary ability to gather the theories of the time and propose a model for DNA's structure. "This structure has two helical chains each coiled round the same axis. We have made the usual chemical assumptions, namely, that each chain consists of phosphate diester groups joining *beta*-D-deoxyribofuranose residues with 3′, 5′ linkages. . . . Both chains follow right-handed helices. . . . The novel feature of the structure is the manner in which the two chains are held together by purine and pyrimidine bases." In a few descriptive paragraphs, Watson and Crick entered history by putting forth the structure of DNA.

By 1956, advanced microscopes enabled scientists to see bacterial nucleoids of dense concentrations of DNA. In the 1960s, a professor of biochemistry at Stanford University, Paul Berg (1926–), took small pieces of DNA and watched them attach to larger DNA molecules. By combining two different types of DNA, Berg demonstrated the basis for recombinant DNA technology. Breakthroughs in DNA replication, the role of RNA in chromosome replication, and special characteristics of heredity occurred quickly in the next decades. In 1983, the geneticist Barbara McClintock (1902–92) received the Nobel Prize in medicine for her work 30 years earlier showing that small segments of DNA could move from one region of the DNA molecule to another region. These segments are now known to be *transposons*, which can cause potential devastation to a cell's activity but have also been shown to contain genes important in infection and disease. For their work on DNA's structure, Watson and Crick shared the Nobel Prize in medicine or physiology with the New Zealander Maurice Wilkins (1916–2004), in 1962; Berg was awarded the Nobel Prize in chemistry in 1980.

In 1943, a hopeful Oswald Avery had said, "If we are right, and of course that is not yet proven, then it means that nucleic acids are not merely structurally important but functionally active substances in determining the biochemical activities and specific characteristics of cells and that by means of a known chemical substance it is possible to induce predictable and hereditary changes in cells. This is something that has long been the dreams of geneticists." Biology now understands that Avery's theories laid the groundwork for studies on chromosomes and heredity.

Today, the fields of genetics, medicine, and biotechnology focus on activities of chromosomes and their replication. The emerging science of gene therapy for curing genetic diseases also requires an intimate knowledge of the chromosome. New technologies enable today's scientists to identify important genes on the chromosomes ranging from bacteria to higher animals.

In 2009, three U.S. researchers won the Nobel Prize in physiology for discovering, in the 1970s and 1980s, how structures called telomeres protect DNA from damage during chromosome replication in eukaryotes. Elizabeth Blackburn (1948–), Carol Greider (1961–), and Jack Szostak (1952–) also uncovered the role that an enzyme called *telomerase* plays in maintaining the telomeres. When the award had been announced, Szostak described their work as "a long-standing puzzle that we were interested in solving. It was only over later years that it emerged, through the work of many people, that this was probably important for aging and cancer." Sophisticated studies such as these begin with an education on the basic workings of microbial chromosomes.

THE STRUCTURE OF BACTERIAL CHROMOSOMES

Chromosomes contain the cell's DNA, and DNA, in turn, serves as the depository of the cell's genes. Genes are short segments of DNA that hold information needed to make products for keeping the cell alive and functioning. Most of microbiology's knowledge of bacterial chromosomes derives from studies on *Escherichia coli,* and *E. coli* chromosome research gave birth to a newer field known as *genomics.* Genomics is the characterization of DNA molecules by determining the order in which the DNA's genes line up, called a *gene sequence.*

Bacterial DNA resides in the nucleoid, which is an irregularly shaped region of the cytoplasm—the nucleoid is also called the *nuclear body* or *nuclear region.* Most bacteria contain only one nucleoid, although some, such as *Vibrio,* contain more than one. Inside the nucleoid, proteins help the DNA coil into a dense structure. In doing this, the proteins bend the large DNA molecule roughly in half, so that genes located far apart on the DNA molecule are drawn closer together. This reshaping of DNA may be important in helping transposons move from place to place on the molecule.

Most bacterial chromosomes contain a continuous circular molecule of DNA, which itself is double-stranded. This double-stranded DNA, designated *dsDNA,* consists of a ladderlike structure with the two strands connected by compounds that form bridgelike connections. The structure described by Watson and Crick contains another characteristic: The strands twist around each other to form a coil. In the nucleoid, DNA coils into a tight disorganized-looking mass referred to as supercoiled DNA. The enzyme DNA gyrase (also called topoisomerase II) carry out most of the folding of the DNA to form the supercoiled structure that greatly compacts the material into the cell's small volume. So dense is the supercoil, a cell measuring about 4 micrometers (μm) in length can contain 1,400 μm of DNA.

Each DNA strand contains alternating deoxyribose sugars and phosphate groups. Each deoxyribose attaches to a nitrogen-containing compound called a *base,* which connects one strand to another while building the rungs of the ladder. Four bases make up these connections between the strands: Adenine always pairs with thymine and cytosine always pairs with guanine. A single base-sugar unit is called a *nucleoside,* and a single base-sugar-phosphate unit is called a *nucleotide. Nucleic acid* refers to the entire DNA molecule made of nucleotides.

As mentioned, bacterial DNA resides in the cell cytoplasm without a membrane to enclose it, but high-powered electron microscopy has shown that the DNA may attach to the cell's outer membrane at places. This attachment may help DNA divide equally when a single parent cell replicates by binary fission to create two daughter cells. Cell replication in bacteria propagates new generations of almost identical progeny cells, so accuracy in DNA replication is critical for maintaining a species.

THE STRUCTURE OF EUKARYOTIC CHROMOSOMES

Eukaryotic DNA resembles bacterial DNA in containing the same bases, base pairing, sugars, and phosphate groups. Eukaryotic DNA differs from prokaryotic DNA, however, in the following ways:

Mitosis and Nonmitosis Events in Eukaryotic Cells

Cellular Event	Cell Cycle Phase	Mitosis Phase	Activity
Mitosis	M	prophase	duplicate chromosome sets move apart
		metaphase	nuclear membrane disappears
		anaphase	chromosomes move toward opposite ends of the dividing cell
		telophase	repackaging of chromosomes in nuclei in two new cells
Nonmitosis		**Nonmitosis Phase**	**Activity**
		interphase	cell growth
	G_1		
	S	interphase	chromosome replication
	G_2	interphase	preparation for mitosis

- packaged inside a membrane-enclosed nucleus
- forms chromatin, an organized rather than loose DNA configuration
- nucleus contains one or more sets of chromosomes
- single linear dsDNA
- DNA molecules associate with proteins called *histones*
- chromatin, fine strands consisting of DNA, RNA, histones, and nonhistone proteins
- chromosome divides in mitosis
- few or no plasmids

Eukaryotic DNA condenses into a densely packed structure, but, unlike prokaryotic DNA, it wraps around histones, which give the large DNA molecule a more organized structure than the prokaryotic chromosome. About 165 base pairs of DNA wind around about nine histones to create a structure called a *nucleosome*. The eukaryotic cell then further twists the nucleosomes into even denser DNA-protein packets.

When eukaryotic cells reproduce, they duplicate then separate the nucleus, so that each daughter cell receives a complete set of chromosomes. Eukaryotes use a process called *mitosis* to complete this division and so produce new nuclei and chromosomes. Mitosis is s critical step in cell biology, because it represents heredity at a molecular level, meaning the molecules that ensure a new generation receives all the characteristics of its parent generation. The table encapsulates the key steps of mitosis.

A eukaryotic cell spends little time in mitosis. Most of its time is spent building structures in the new growing cells and carrying out maintenance activities. This period between mitotic cycles is called the *interphase,* during which time chromosomes disperse into fine chromatin strands in the nucleus, until another replication begins. In preparation for the next mitosis, the chromosomes condense into a more organized form than the thin and dispersed chromatin. Once condensed, the mitotic events begin for the purpose of dividing the chromosome before the next cell division. Overall, eukaryotes perform these same processes, but with a more orderly chromosome orientation.

THE FUNCTION OF CHROMOSOMES

Chromosomes store the information that enables each species to survive. The chromosomes hold all of this information within a genetic code based on an alphabet of four characters: the purine compounds adenine (A) and guanine (G) and the pyrimidine compounds cytosine (C) and thymine (T). Genes contain various arrangements of these four characters known as *bases,* and the multitude of arrangements contain the codes for all the information that defines a species and an individual cell within that species. The term *encode* refers to the conversion of biological information into a code carried by the four bases. An average gene contains about 1,000 units of any of these four bases, and the 1,000 bases can be arranged in $4^{1,000}$ different combinations.

Genes make up the primary unit of information within chromosomes. A gene is a segment of DNA that encodes for a specific function in a cell, and proteins in the form of enzymes carry out most of these cell functions in biology. When the information encoded in genes must be converted into cell proteins, different types of RNA help extract the information from the chromosomes (actually the cell's DNA) and convert this code into a protein or other compound inside the cell. Chromosomes, therefore, play a central role in transferring genetic information from one generation of cells to the next and, as a consequence, from one generation of organisms to the next. The three main events that make up this transfer of genetic information are as follows:

- DNA replication—construction of two identical strands of DNA from one original strand
- transcription—synthesis of RNA under the direction of DNA for the purpose of transferring the genetic code from the four bases to a sequence of amino acids
- translation—synthesis of protein under the direction of RNA (messenger RNA, abbreviated mRNA) from individual amino acids

In all of the processes, the chromosome controls two crucial aspects of genetics: phenotype and genotype. A cell's proteins create phenotype, which is the way an organism looks and functions. For instance, a microbiologist can tell the difference between *Staphylococcus* bacteria and *Candida* yeast on the basis of the phenotypes of each of these organisms—size, shape, staining color, and so on. Put in another context, a mother can tell her identical twins apart based on their phenotypes. The genotype of an organism is its genetic makeup, that is, its complete set of genes that makes the organism what it is. An elephant has a different genotype from a monkey. Genotype determines phenotype, so that one identical twin may be a quarter-inch (0.64 cm) taller than his twin or a split-second faster in a footrace. In these instances, height and speed are phenotypes, and the genes that control height and speed are part of the individual's genotype.

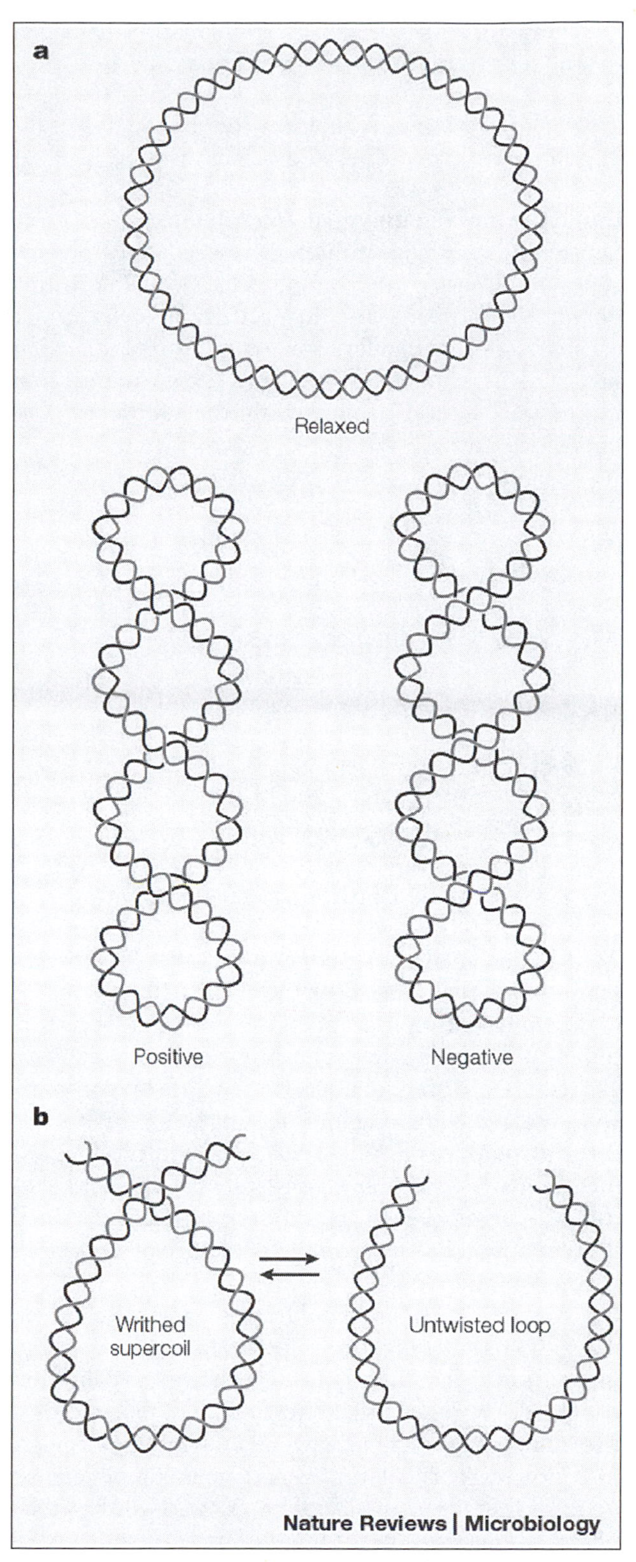

The bacterial chromosome exists as a supercoil with numerous folds. The folds produce localized regions, which are thought to facilitate cellular repair of damaged DNA. ***(Nature Reviews Microbiology 3 [2005]: 157–169)***

The role of chromosomes in biology may be summarized as follows:

1. They store all the genetic information of a cell and a species for future generations.

2. They hold the information for making an individual look as it does and behave as it does.

Genotype and Phenotype in Microbiology

For many years, microbiologists used only phenotypes to identify microorganisms by using stains on microorganisms and inspecting the cells under a microscope. A generation of microbiologists, in fact, did a remarkably accurate job in identifying unknown bacteria using mainly the following phenotypic features.

- cell size
- cell shape
- presence or absence of motility
- stain reaction (usually Gram stain)
- special enzyme activities
- ability to use certain sugars or amino acids
- production of gas from sugar fermentation

Cell size, shape, and the presence of certain features on the surface of the cell, such as flagella, make up the discipline called *cell morphology*. Morphology depends almost entirely on phenotype to characterize different microorganisms. By comparison, the study of enzyme activities, ability to use certain sugars or amino acids, and end products of metabolism, such as gas, constitute biochemical testing. Biochemical testing and morphology combined present a fair (but not complete) picture of a species's chromosome.

Advances in molecular biology have given microbiologists much more powerful tools in studying chromosomes compared with the information provided by biochemical testing and morphology. Molecular techniques now enable microbiologists to characterize the actual genetic makeup of cells by determining the sequence of bases in DNA, the DNA base composition (percentage of guanine plus cytosine in the total DNA), and the sequence of bases in ribosomal RNA (rRNA) and using other specialized techniques in species identification. These molecular techniques also help determine closely related species compared with species that have only a distant relationship. By analyzing the chromosome this way, scientists can strengthen their theories on evolution and the ancestors of present-day species.

The chromosomes hold all the information that scientists need for determining how life developed on Earth, the methods by which organisms evolved, and the characteristics that make up all the known plant and animal species studied today.

See also BINARY FISSION; GENOMICS; MORPHOLOGY; PLASMID.

Further Reading

Avery, Oswald, T., Colin M. MacLeod, and Maclyn McCarty. "Studies on the Chemical Nature of the Substance Inducing Transformation of Pneumococcal Types." *Journal of Experimental Medicine* 149 (1979): 297–326. Available online. URL: http://jem.rupress.org/cgi/reprint/149/2/297. Accessed March 23, 2009.

Campbell, Neil A., and Jane B. Reece. *Biology,* 7th ed. San Francisco: Benjamin Cummings, 2005.

Drlica, Karl, and Monica Riley. *The Bacterial Chromosome.* Washington, D.C.: American Society for Microbiology Press, 1991.

Levs, Josh. "3 Americans Win Nobel for Chromosome Research." October 5, 2009. Available online. URL: http://edition.cnn.com/2009/WORLD/europe/10/05/nobel.medicine. Accessed December 1, 2009.

National Libraries of Medicine. "DNA as the Stuff of Genes: The Discovery of the Transforming Principle." Profiles in Science: The Oswald T. Avery Collection. Available online. URL: http://profiles.nlm.nih.gov/CC. Accessed March 23, 2009.

Snyder, Larry, and Wendy Champness. *Molecular Genetics of Bacteria,* 3rd ed. Washington, D.C.: American Society for Microbiology Press, 2007.

Watson, James D., and Frances H. C. Crick. "A Structure for Deoxyribose Nucleic Acid." *Nature* 171 (1953): 737–738. Available online: URL: http://www.exploratorium.edu/origins/coldspring/ideas/printit.html. Accessed March 23, 2009.

clean room Clean rooms are specially maintained areas designed to prevent contamination of a product. Manufacturing plants owned by pharmaceutical companies use clean rooms for making products free of any contamination from bacteria, mold, dust, fibers, or other particles that may be present on surfaces or in the air. In pharmaceutical manufacturing, drugs intended to be injected into the body must be made in a clean room to assure that a patient will not receive a contaminated dose when a doctor injects the drug into the patient. The principles of clean rooms also apply to medical units that care for burn patients, surgeries, or facilities that care for patients who have severely damaged immune systems.

Workers in clean rooms have several procedures at their disposal for ensuring bacteria, fungi, viruses, protozoa, or inanimate particles do not enter the area. To maintain a clean room under proper germ-free conditions, microbiologists rely on the principles of sterility and sterilization, disinfection, airflow, filtration, and environmental monitoring.

THE LAYOUT OF A CLEAN ROOM

A clean room represents a controlled environment, meaning the conditions inside the room are carefully maintained within certain limits. The primary conditions monitored by microbiologists and kept within specified limits are the following: airflow, air supply, temperature, humidity, airborne particles, and sterility of equipment. Microbiologists monitor each of these conditions and correct any occurrences that stray outside acceptable limits. For example, a drug manufacturer's clean room may require a temperature range of 65–70°F (18–21°C). Automatic systems can monitor temperature continuously and alert workers if any reading falls outside the acceptable range. A worker then resets the room's thermostat and records the date and times that the room fell outside acceptable limits. Clean room operators use the term *out of spec* (*spec* is short for *specifications*) to describe such occurrences outside acceptable limits. Today's clean rooms incorporate designs that reduce the chance of conditions' going out of spec, and, therefore when clean rooms operate as they are supposed to, they are referred to as *within spec*.

Clean rooms consist of three main features: (1) cleanliness, (2) airflow, and (3) filtration. Cleanliness is maintained by disinfecting the floors, walls, ceilings, and all other surfaces and equipment. The junctures from floor to walls and from walls to ceiling consist of a design that eliminates sites where dust can be trapped. Airflow and filtration are part of room design and can be done in several ways to assure that no contamination enters the clean room's working area. Contamination is the presence of any unwanted microorganism or inanimate particles in the clean room, either airborne or on a surface. Microbial contaminants may be bacteria, viruses, molds or mold spores, or aerosols containing any of these microorganisms; inanimate contaminants consist of dust, fibers, hair, dander, or any other tiny bits of clothing, shoes, or the body or from the outdoors. Many inanimate contaminants are not truly inanimate because they carry microorganisms. For example, aerosols are moisture droplets that are so small they can travel long distances (several feet to a mile) in the air. Aerosols often

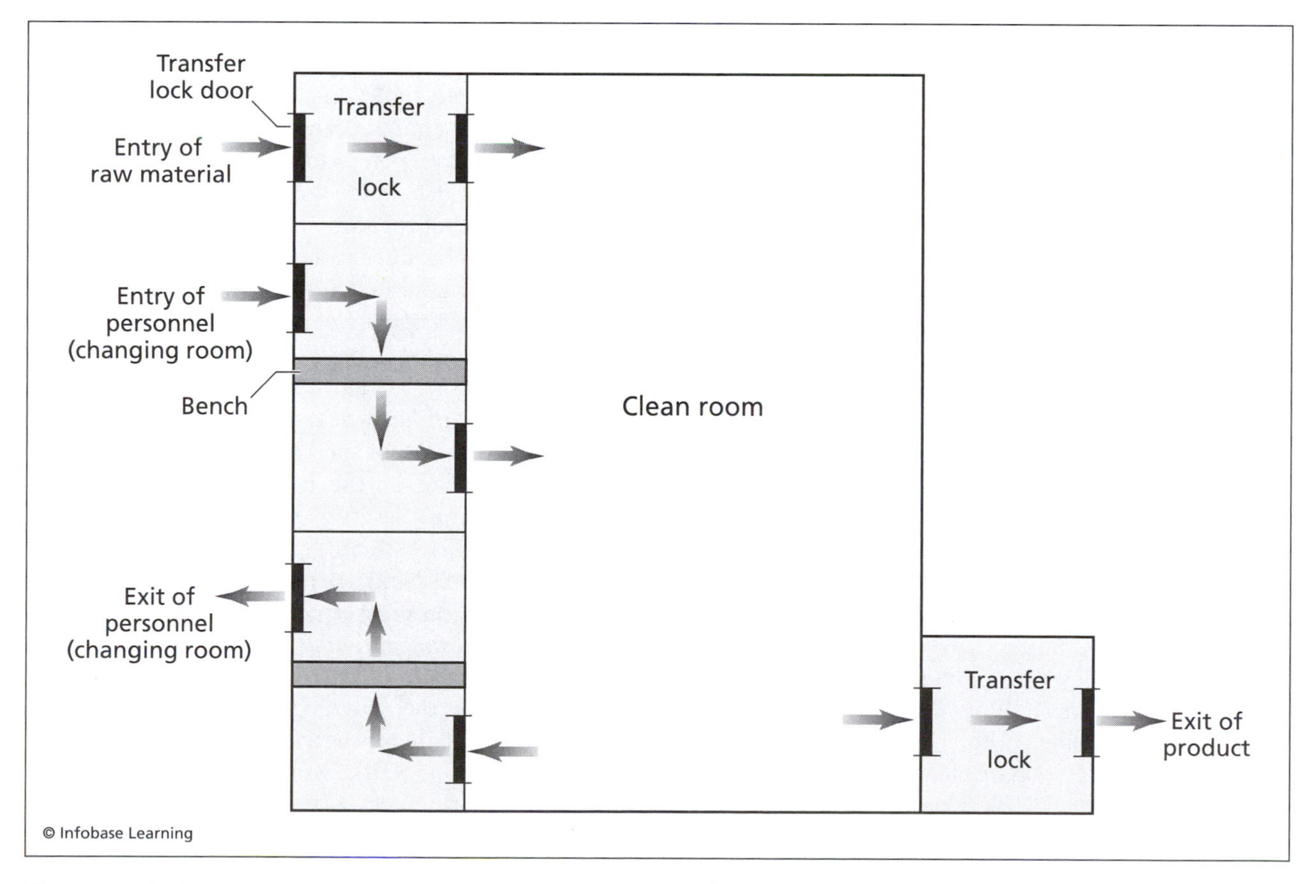

Clean room design ensures that all materials and workers move in a one-way direction so that contaminants cannot enter the sterile-product manufacturing site.

contain microorganisms, and when they do, they are called *bioaerosols*.

Facilities prevent contamination from entering a clean room by adopting two building designs that address cleanliness: (1) preparation (or preparatory) rooms, nicknamed *prep rooms*, and (2) unidirectional workflow. Prep rooms provide space where microbiologists change clothes or otherwise prepare for their work inside the clean room, before entering the clean area. Prep rooms are part of unidirectional workflow, or one-way work flow. Unidirectional workflow assures that people and equipment enter a clean room from one door and exit the clean room at another, separate door. Equipment movement within the clean room may also take a one-way flow so that any accidental contamination affects the fewest possible activities inside the room.

Prep rooms usually consist of a changing room in which workers remove outer garments and put on clean body coverings, including hood, goggles, gloves, and foot coverings. A second prep room contains sterile reagents, solutions, and portable equipment, which may enter the clean area through a separate door or hatch. After a product has been made, it exits the clean area through its own dedicated portal. Likewise, clean room personnel exit the work area through a separate exit door and remove their protective clothing in an outer room.

Controlled airflow also reduces the chance of contamination within a clean room. Air should flow through the work area in a steady smooth stream that carries airborne particles away from sterile activities. Smooth flow also reduces turbulence in the air. Air normally circles and swirls through laboratories, but clean rooms create unidirectional airflow by using air supply ducts and exhaust ducts of equal size and capacity. Unidirectional flow may move either from top to bottom or side to side in the room. By flowing in sheets rather than eddies, the airflow is called *laminar flow*.

Biosafety cabinets use laminar airflow to protect the surroundings from dangerous pathogens. Either top-to-bottom or side-to-side airflow in the cabinet assures that only clean filtered air touches the work area and dirty airflows away from it. Filtration thus plays a critical part in clean room operations.

Clean rooms employ one of two types of filters: high-efficiency particulate air (HEPA) or ultra low-penetration air (ULPA). HEPA filters consist of sheets of pleated filters that remove from the air at least 99.97 percent of particles 0.3 micrometer (μm) diameter or larger. ULPA filters remove even tinier aerosols and viruses that HEPA filters may not stop: 99.9999 percent of particles as small as 0.12 μm. Despite these extraordinary efficiencies for removing contaminants from air, neither filter guarantees 100 percent effectiveness. The air exiting a filter and entering the clean room may not be sterile, even with the best precautions, but a combination of high-efficiency filtration and nonturbulent airflow helps reduce contamination to almost zero.

MAINTAINING CLEAN ROOM CONDITIONS

Clean room personnel follow detailed practices that greatly minimize the chance of contaminating the room's work area and the sterile product. The biotechnology writer Angelo DePalma explained in *Genetic Engineering News*, in 2006, "In biotech, contamination control is 99 percent prevention, 1 percent detection." The main ways to prevent contamination caused by clean room workers are protective garments and good personal hygiene.

Clean room garments prevent hair, dander, dead skin cells, fibers, and dirt from becoming contaminants. No matter how clean normal clothing appears, the mere movement of a person releases more than 500,000 particles of 0.3 μm or larger per minute. Sitting or standing motionless releases hundreds of thousands of particles. Perspiration also releases vapor into the air. Clean room garments have been designed with materials that reduce both of these hazards.

Clean room garments consist of non-fiber-shedding synthetic fabric that breathes, meaning perspiration does not build up inside the suit. The fabric itself should not emit volatile compounds into the air, and it should have pore sizes small enough to retain any particles of 0.2 μm or larger. The hoods and foot covers are made of the same fabric. Gloves are usually made of synthetic nitrile or polyester, which do not emit gases and contain no latex (a cause of allergic reactions). Face masks are composed of materials similar to those in filters; they retain at least 98 percent of microorganisms or particles of 0.1 μm or larger. Goggles consist of hard clear plastic with straps made of materials that do not shed fibers or particles.

Before putting on the protective coverings, personnel do a surgical scrub of their hands and arms and decontaminate their gloves with a 70 percent alcohol solution. Next, personnel about to enter a clean room put on protective clothing in a step called *gowning* or *sterile gowning*. Gowning begins with putting on the pair of decontaminated gloves dedicated solely for the gowning procedure. The entire gowning process then progresses from top to bottom. Workers sit on a bench with a designated "clean" side and a "dirty" side. The worker covers each foot with a foot covering, or bootie, while moving from the dirty to the clean side of the bench. When booting is done properly, the bottom of the foot covering

Clean room procedures minimize the risk of contaminating sterile medical products. ***(Daetwyler Holding Inc.)***

never touches the same floor that had been touched by normal shoes. Finally, personnel remove the gowning gloves and put on a pair of clean room gloves.

Notepads and notebooks used by clean room personnel are composed of specially coated paper that does not shed microscopic paper fibers. As with clothing, paper and other writing implements should not emit volatile compounds. Pens and clipboards and other hard items must be sanitized before being taken into the work area; special pens designed for clean rooms must be used rather than regular pens. Pencils are also banned in clean rooms because of the fibers they shed.

Personal hygiene additionally prevents the introduction of contaminants into clean rooms. For example, clean room workers should bathe frequently and wash hair regularly, and employees who are sick or have skin conditions must not participate in clean room activities. Other hygiene practices advisable for clean rooms are the following:

- no use of cosmetics, colognes, skin medications, or aftershave products
- clothing neat and clean, not frayed, and non-lint-producing
- no wearing of personal eyeglasses without goggles
- smooth and slow movements rather than fast movements
- no scratching, rubbing, or touching the face with hands
- no smoking within a prescribed period before entering the clean room
- no eating, drinking, or gum chewing

ENVIRONMENTAL MONITORING

Clean rooms belong to classifications according to the type of activities that take place in them. Each classification is based on the number of particles allowed in the room's air. The table on page 162 gives clean room classifications published by the U.S. Food and Drug Administration (FDA), which oversees drug manufacture in the United States and U.S. drug companies that manufacture products outside the United States.

Environmental monitoring involves any sampling and testing of air, water, or surfaces to deter-

Clean Room Classifications

Class	Limits (0.5 μm or larger particles per cubic foot)	Metric Class	Limits (0.5 μm or larger per cubic meter)
1	1	M1	10
10	10	M2	100
100	100	M3	1,000
1,000	1,000	M4	10,000
10,000	10,000	M5	100,000
100,000	100,000	M6	1,000,000

mine the presence of contaminants. Drug, nondrug, and food manufacturing companies use environmental monitoring inside their buildings to assure their products are free of dirt and microorganisms. Though the environmental monitoring inside these facilities can be strict, in order to eliminate as much contamination as possible, clean room environmental monitoring is even more detailed, with strict limits on both microorganisms and inanimate particles.

Monitoring for the presence of microbial contaminants occurs in three phases: (1) detection, (2) enumeration, and (3) identification. In order to detect any presence of contaminants, microbiologists take samples from places in the clean room most likely to collect contaminants. Microorganisms and inanimate particles can be collected by a procedure called *sampling*. Sampling in microbiology must overcome two important difficulties: Sampling usually collects only a portion of the material to be studied, and not all microorganisms collected in a sample will grow in a laboratory. Both of these problems complicate the second phase of contaminant monitoring, enumeration. Enumeration is done by various cell- or particle-counting methods for detecting both live and dead microorganisms. Electronic cell counters also offer the advantaged of enabling microbiologists to enumerate microorganisms that do not grow in laboratory conditions.

A technique called a *Hazard Analysis and Critical Control Point (HACCP) Program* has been adapted from the food production industry to assess clean rooms. A HACCP program helps in selecting locations in a room to be sampled on the basis of their likelihood of being contaminated. HACCP procedures involve sampling by wiping surfaces with sterile wipes or swabs or by using special agar plates that collect contaminants directly off a hard surface. Air particles are measured differently, by using air sampling equipment, such as an Andersen sampler. This type of air sampler sorts particles by size by passing the air through sieves containing a range of pore sizes from about 1.0 to more than 7.0 μm in diameter.

After completing the sampling step, the microbiologist counts the number of microorganisms that have been recovered from each area in the clean room; this process is called *enumeration*. Enumeration gives information on the places in the room that have the most contaminants and the places that might be expected to have few or no contaminants. Microbiologists use either culturing techniques or electronic cell counters for enumeration.

The final step in environmental monitoring involves the identification of the main contaminants. Microbiologists use a broad selection of sophisticated identification methods, as well as general, faster techniques to learn about the microorganisms that might invade a clean room. Identification can be the most labor-intensive and time-consuming part of environmental monitoring, but it provides valuable information about where contamination tends to occur in a room and the type of microorganism. Environmental monitoring helps the microbiologist in assessing the following:

- source of the contaminant, such as water or air
- selection of a disinfectant to kill the contaminant
- identification of a test organism that will serve as an indicator of other microorganisms

Indicator organisms become important tools in microbiology when microorganisms in the sampled environment are difficult to collect, culture, or identify.

Clean rooms require a combination of sound laboratory practices that maintain sterility and a monitoring plan that assures that the room remains contaminant-free. With these two components working in concert, clean rooms provide the safest environments for making sterile products.

See also AEROMICROBIOLOGY; CULTURE; DISINFECTION; FILTRATION; HACCP; INDICATOR ORGANISM; SAMPLE; STERILIZATION.

Further Reading

Carlberg, David M. *Cleanroom Microbiology for the Non-Microbiologist,* 2nd ed. Boca Raton, Fla.: CRC Press, 2005.

DePalma, Angelo. "Maintaining Biocontamination Control." *Genetic Engineering News,* 1 June 2006.

Walsh, Gary. *Pharmaceuticals: Biochemistry and Biotechnology,* 2nd ed. Chicester, England: John Wiley & Sons, 2003.

clinical isolate A clinical isolate is a microorganism that is present in a specimen taken from a sick person. The clinical isolates of highest importance are those that cause illness or are suspected of causing illness in the patient. An isolate can be a bacterium, virus, fungus, protozoan, or parasite. These microorganisms are called isolates because of two characteristics. First, the unknown microorganism has been isolated from a patient. Second, as a first step in identifying this microorganism, a microbiologist isolates it from all other microorganisms. When the isolated microorganism has been separated from all other different types of microorganisms, it may also be referred to as a *pure isolate.*

The recovery, isolation, laboratory culturing, and identification of clinical isolates constitute the medical field of clinical microbiology. Many skills work together when a clinical microbiologist conducts all of these steps: specimen sampling, aseptic techniques, culture methods, growth media preparation, identification, and morphology. As a result of all the activities, a microbiologist becomes equipped to identify the isolate with a high level of confidence. This information, then, helps a physician diagnose disease and prescribe an effective treatment. Clinical microbiology, therefore, plays a major role in the medical sciences.

Clinical microbiology accumulates information about microorganisms over time, so that today's clinical microbiologists possess more knowledge for making a correct identification of an isolate than microbiologists had a decade ago. For example, in 1976 at a convention of the American Legion in Philadelphia, more than 100 convention members became sick with a respiratory illness doctors could not immediately identify; more than 30 individuals died of the mysterious illness. Five months after the outbreak, the microbiologist Joseph McDade at the Centers for Disease Control (CDC) identified the disease's cause, but not without months of trials and failures in his laboratory. A 2003 BBC retrospective report on the outbreak explained why the microorganism had been so difficult to identify: "The Legionnaires' Disease bacillus, later named *Legionella pneumophila,* was no ordinary microbe. It could not be grown under typical conditions, being dependent upon ridiculous demands: high levels of the amino acid cysteine and inorganic iron supplements, low sodium concentrations, as well as activated charcoal to absorb free radicals. In addition, it preferred elevated temperatures, which was highly abnormal among pathogens, who preferred near-body temperatures." More than three decades later, clinical microbiologists have much more information on how *Legionella* grows (in water), its preferred nutrients, and its cell and colony morphologies.

Current clinical microbiology depends on the following skills possessed by staffs of clinical microbiology laboratories:

- knowledge of a pathogen's role in disease
- information on the relationship between a pathogen and symptoms
- identification methods for new pathogens
- development of faster and more accurate identification methods
- susceptibility testing of isolates for finding effective drug treatments, such as antibiotics

The skills listed here might well be useless if a seriously sick patient does not receive immediate attention. Therefore, speed becomes a priority throughout clinical microbiology. For this reason, clinical microbiologists have developed standard procedures that speed the process from sampling to identification of a pathogen.

COLLECTION, TRANSPORT, AND PREPARATION OF CLINICAL ISOLATES

Clinical microbiology involves a stepwise approach to handling unknown pathogens in specimens taken from patients. A specimen is any material taken from a patient that is expected to contain a pathogen. The main specimens in human and veterinary medicine are the following: blood, cerebrospinal fluid, feces, mucus, pus, semen, skin, sputum, stomach contents, throat swabs, tissues, urine, vaginal swabs, and wound swabs.

Specimen management consists of three steps: (1) collection or sampling, (2) handling and transport, and (3) preliminary preparation. Each of these steps involves procedures that minimize the chance of contaminating the specimen, preserve it during transport to a laboratory, and maintain any potential pathogen within the specimen. This series of steps must also contain a rigid adherence to *chain-of-custody,* which is a term for the methods by which clinical personnel keep track of the specimen at all times, from the moment it is taken from the patient to the moment a pathogen has been identified.

The main components of chain-of-custody are the following:

- assurance that the specimen remains free of contamination by other specimens or microorganisms
- proper labeling in all phases of clinical microbiology involving the specimen
- prevention of any tampering with the specimen during transport and processing
- identification of a specimen to the proper patient

Chain-of-custody procedures become very important in clinical microbiology because several hospital technicians, nurses, and support staff typically handle specimens from the point of collection to delivery to a clinical microbiology laboratory. In the laboratory, the specimen might also be recorded, stained, or otherwise processed by more than one individual. Clinical microbiology may be summarized as the total activities required for the following tasks: specimen chain-of-custody, collection, transport, preparation, and identification of potential pathogens. Finally, clinical microbiologists must create a relationship of trust and credibility between the laboratory staff and physicians.

Specimen Collection

Each type of specimen for study in a microbiology laboratory has specific requirements for the best methods for collecting and handling the specimen to prevent contamination or degradation. Technicians who collect specimens must have knowledge of aseptic techniques and the principles of sterilization. For example, a technician collecting a urine sample collects urine midstream to ensure that the urine contains microorganisms from the urinary tract and not from the skin outside the body.

Two types of specimens are part of clinical microbiology and disease diagnosis: specimens from the outside of the body and specimens from the inside of the body. Specimens from inside the body, such as blood or cerebrospinal fluid, require invasive sampling, such as a needle that enters the body to withdraw a specimen. Both samplings require aseptic techniques so that no unwanted microorganism contaminates the specimen. Technicians swab the skin with alcohol before taking a blood specimen with a needle and syringe, and they use only sterile needles and syringes.

The protections accorded specimens during collection serve two purposes. First, specific procedures ensure that the specimen does not become contaminated with foreign matter. A contaminated sample will lead to erroneous identification results in the laboratory, which could well lead to an incorrect diagnosis of a disease by the physician. Second, all medical staff must be protected from the specimen to assure they do not receive the pathogen. Hospital staff has contracted diseases such as hepatitis, acquired immunodeficiency syndrome (AIDS), meningitis, *Staphylococcus* infections, and poliomyelitis in isolated cases spanning the past several decades. The Web site Virology-online has described the problem of viral infections as follows: "Needle stick injury [NSI] is a common occupational hazard among health care workers. Most needle stick injuries arise out of unsafe practices and are thus preventable. The greatest proportion of NSIs arises from the action of re-sheathing the needle after taking blood and this practice should be actively discouraged. However, it has been argued that not re-sheathing the needle would increase the risk of other staff such as porters and nurses." The best practices for avoiding contact with a pathogen while working in very close contact with infected specimens remain a critical aspect of health care. Hepatitis B, hepatitis C, and human immunodeficiency virus (HIV) are the viruses that present the greatest risk to hospital staff during specimen collection.

Bacteria, fungi, protozoa, and parasites are very likely to be present in a specimen from a diseased patient, so safety precautions are needed in specimen collection. Health care professionals use the following protective equipment: gloves, masks, protective clothing, lab coats, soaps, alcohol wipes, and disinfectants.

Specimen Transport

Specimen handling and transport must also be done in a manner that eliminates errors and delays. All specimen containers should have a label listing its destination (the laboratory location or a room number) and information needed by the microbiologists who will process the specimen. In clinical microbiology laboratories, labels usually include the following items: patient, hospital, hospital's registration number, patient's location, diagnosis (if available), current drug therapy, attending physician, admission date, and type of specimen (urine, blood, swab, etc.). A technician records all of the label information to retain the patient's and the specimen's pertinent medical information.

Transport procedures focus on preserving the specimen until it arrives at the laboratory. For example, some specimens should be kept on ice; others must be sealed in an anaerobic (oxygen-free) carrier. Some specimens suspected of containing a fastidious pathogen—a microorganism that cannot remain alive for long outside the body—require delivery within 15 minutes.

Specimen Preparation

In the clinical microbiology laboratory, technicians work quickly to prepare the specimen in a way that preserves any pathogens it contains. Technicians usually divide the specimen into two portions: One part goes to preliminary testing, or screening, and one part goes to begin culture methods. Preliminary screens consist of staining a drop of specimen on a glass microscope slide and searching for pathogens in a microscope. The two commonly used staining techniques are the Gram stain and the acid-fast stain. Gram stained microorganisms become visible in specimens using a high-power (magnification 600 ×) light microscope. Acid-fast stained microorganisms require a fluorescent microscope, an instrument that illuminates the specimen in fluorescent light.

Meanwhile, a microbiologist begins culture methods to recover the clinical isolate. These culture methods depend on four different kinds of media, each of which contains ingredients that make microorganisms behave in a certain way. (These media target bacteria, yeasts, and fungi. Viruses, protozoa, and parasites require different specialized techniques.) The following four types of growth media help in identification:

- selective—permits the growth of one type of microorganism while inhibiting others

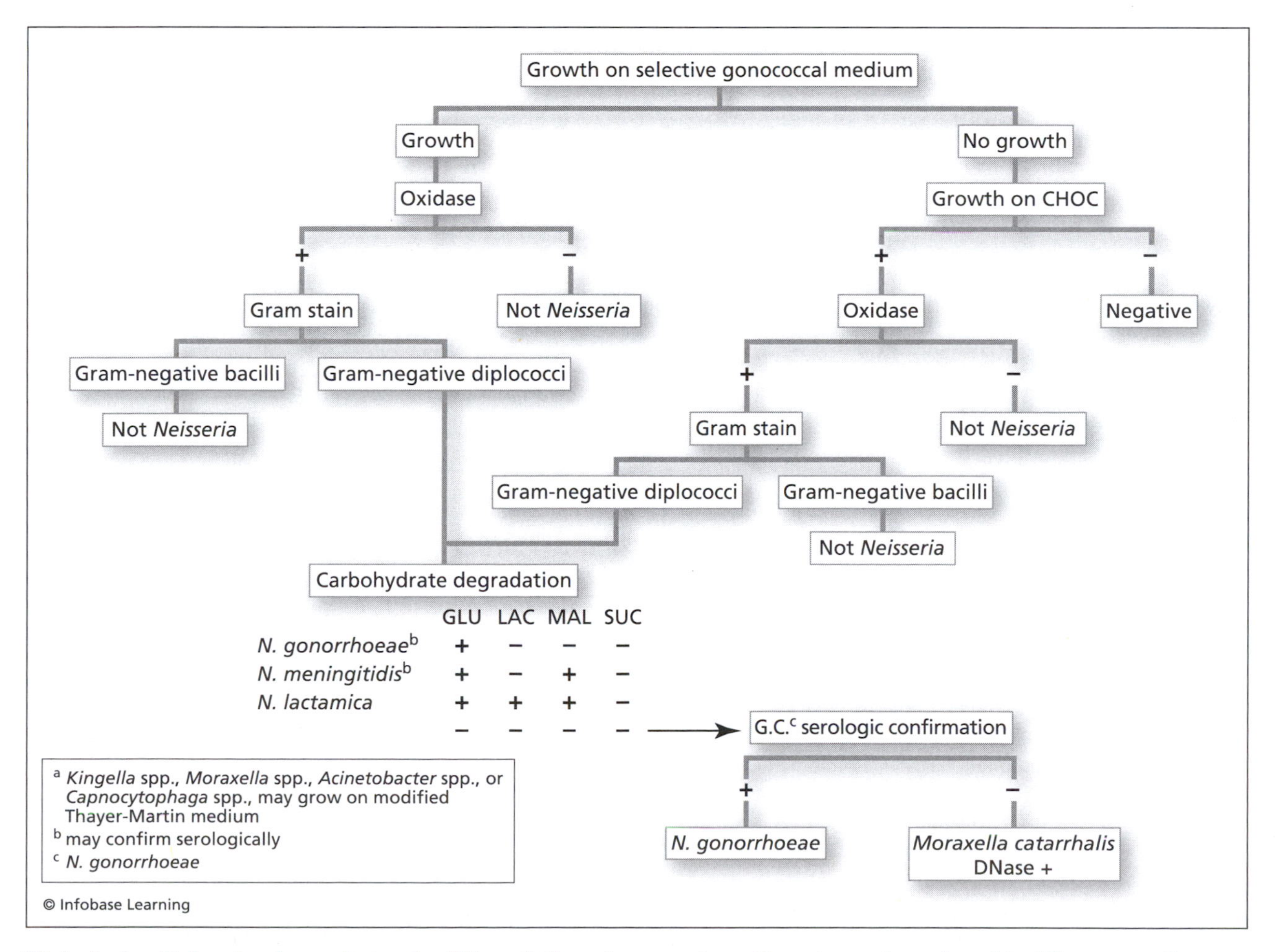

Clinical microbiology involves schemes for differentiating microorganisms. The scheme shown here identifies a sexually transmitted bacterium. Reactions cited as "+" or "-" indicate presence/growth or absence/no growth, respectively. (CHOC = chocolate agar; GLU = glucose; LAC = lactose; MAL = maltose; SUC = succinic acid.)

- differential—differentiates between two or more bacteria by their reaction to chemicals in the medium
- enrichment—contains ingredients that help fastidious (difficult-to-culture) microorganisms grow
- characteristic—tests bacteria for specific characteristics unique to a group of organisms

Microbiologists look for characteristics of the culture on these media, including gas production, motility, enzyme activity, and oxygen requirements. With each piece of evidence, a microbiologist can narrow down the possible identities of an unknown microorganism.

IDENTIFICATION METHODS IN CLINICAL MICROBIOLOGY

Identification of an unknown microorganism is the most critical role of clinical microbiology. By knowing the identity of a clinical isolate, a physician gains a better idea of whether the isolate is a pathogen, the disease associated with this pathogen, a potential therapy, and the long-term effects of the disease on the patient. Obviously, a fast identification of the pathogen can be a lifesaving step. Incubation may take a day or more, however, and this time delay may put a patient at risk. For this reason, clinical microbiology uses a combination of standard incubation methods and rapid methods, which can identify certain common pathogens in an hour or less, compared with a day or more.

Identification tests for bacteria belong to four main categories, shown in the following table.

Molecular methods offer the most sensitive means for identifying bacteria to the genus, species, strain, and even substrain levels. Clinical microbiology may not always need this level of expertise for a treatment to be prescribed for a sick patient. Sometimes, a physician needs only know that an infection has been caused by *Staphylococcus aureus* and that the patient's particular pathogen resists the antibiotic methicillin but can be killed by vancomycin. Molecular methods help, however, in research on clinical pathogens. Molecular methods have also become critical for tracing disease outbreaks.

Microbiologists identify fungi, such as yeasts or molds, by staining the reproductive and vegetative (nonreproductive) structures and observing characteristic features with a low-power (magnification 100 x or less) microscope. Several selective media have also been developed for fungi. Microbiologists learn how to make tentative identifications of fungi

The Major Methods for Bacteria Identification

Test	Description	Advantages	Disadvantages
morphology and staining	distinctive colony shape and color on agar; shape of stained bacteria in a microscope	oldest identification method in microbiology	not sufficient alone; for identification to genus level, must be used in conjunction with other methods
biochemical tests	enzymatic activities on a variety of substrates	identifies bacteria to the species level when used with staining; rapid methods give results in 5 hours	standard methods require about 18 hours of incubation
serology	unknown bacteria tested against a number of antibodies to known pathogens	fast, specific to the strain level, adaptable to computers	requires extra training
phage typing	unknown bacteria tested against a number of viruses, called phages, that are known to attack known pathogens	specific to species or strain level	requires about 18 hours of incubation and requires a stock of phages on hand
molecular methods	pathogen identified by studying its DNA or RNA structure, known as the *base sequence*	very specific for tracking a single strain in a population	technical skills required beyond standard identification methods

by examining colony morphology on selective agar media.

Protozoa and parasites can be also be observed in a low-powered light microscope because they are several times larger than bacteria. Virus identification requires much more intensive techniques, beginning with growing the virus in living tissue by a technique called *tissue culture*. Once a tissue culture is prepared—this process takes several days—the microbiologist observes the virus's effects on the cells growing in culture flasks. More detailed observations of viruses call for electron microscopy.

IMPORTANT PATHOGENS IN CLINICAL MICROBIOLOGY

Clinical microbiologists confront a diversity of pathogens every day; some of the pathogens have recognizable characteristics, while other isolates may never have been isolated before in a laboratory. Over time, most clinical microbiologists learn to expect certain pathogens that tend to predominate in certain specimens. For example, the food-borne pathogen *Salmonella* would be expected to be found in a stool specimen rather than a sputum specimen. Conversely, the tuberculosis bacterium is more likely to be detected in sputum specimens than in stool specimens.

A sick patient in a hospital or an ill person who enters a doctor's office could be harboring any pathogen, including pathogens that have not yet been seen or identified in microbiology. A clinical microbiologist must, therefore, avoid jumping to conclusions after a few steps in the identification scheme. But some pathogens predominate in specimens in localized parts of the world, and this information helps the microbiologist arrive at a correct identification in an efficient manner. The table lists common pathogens, with the caution that this list contains only a small portion of the potential pathogens that infect people.

Many of the bacteria listed in the table can be identified by completing an antigen test on a blood specimen, even though the pathogen is not in the blood. For example, a blood test for the presence of antigens against *Helicobacter pylori* indicates that this bacterial species may be infecting a patient's stomach, a possible cause of gastric ulcers.

SUSCEPTIBILITY TESTING

Once the clinical microbiology laboratory has identified a pathogen, its second crucial role involves susceptibility testing on the organism. Susceptibility testing identifies the specific antibiotics that kill the pathogen, as well as any antibiotics that do not work against the pathogen. Certain susceptibility tests also provide an idea as to the dose of antibiotic that is needed to stop a patient's infection.

Clinical microbiology, therefore, provides physicians with three key pieces of information for curing an infection: (1) the presence or absence of a microbial infection, (2) identification of the pathogen causing an infection, and (3) identification of the antibiotics that kill the pathogen. Without these factors supplied by clinical microbiology, medicine would not be nearly as efficient as it is in diagnosing and treating infectious disease.

In 2008, at the annual meeting of the European Congress of Clinical Microbiology and Infectious Diseases in Barcelona, Spain, the organization's president, Fernando Baquero, stated in his welcome address: "Clinical microbiology and infectious diseases encompass the widest interdisciplinary area of knowledge in medicine. . . . Because of the intrinsic and evolving complexity of man-microbe interactions, infectious diseases cannot be understood

Common Bacterial Pathogens in Clinical Microbiology

Pathogen	Likely Specimen
Staphylococcus aureus, Streptococcus pneumoniae	nasal swab
Mycoplasma, Streptococcus pneumoniae	sputum
Pseudomonas aeruginosa, Staphylococcus aureus	ear swab
Proteus mirabilis, Escherichia coli, Enterococcus	urine
Staphylococcus aureus, Streptococcus, Haemophilus	blood
Neisseria meningitidis	spinal fluid
Staphylococcus aureus, Clostridium	skin
Salmonella, Escherichia coli, Shigella	stool
Treponema, Chlamydia, Neisseria gonorrhoeae	urinary/genital
Pseudomonas aeruginosa, Corynebacterium	tears

without the cooperation of a broad spectrum of scientists. If ever a field of medicine was open to any specialist or expert . . . with the sole condition of a willingness to serve in cooperative efforts . . . that is the field of clinical microbiology and infectious diseases."

Although microorganisms perform myriad services for humans and the earth, the special relationship between a pathogen and a patient can be a life-or-death situation. For this reason, clinical microbiology and knowledge of clinical isolates occupy a central position in all of microbiology.

See also IDENTIFICATION; MORPHOLOGY; SEROLOGY; SPECIMEN COLLECTION; SUSCEPTIBILITY TESTING.

Further Reading

Baquero, Fernando. "Welcome Address, 18th Meeting of the European Congress of Clinical Microbiology and Infectious Diseases, Barcelona, Spain, April 19–22, 2008." Available online. URL: www.akm.ch/eccmid2008. Accessed March 23, 2009.

British Broadcasting Company. "Legionnaires' Disease: A History of Its Discovery." January 16, 2003. Available online. URL: www.bbc.co.uk/dna/h2g2/A882371. Accessed March 23, 2009.

Koneman, Elmer W. *Koneman's Color Atlas and Textbook of Diagnostic Microbiology,* 6th ed. Philadelphia: Lippincott Williams & Wilkins, 2005.

Murray, Patrick, ed. *Manual of Clinical Microbiology,* 8th ed. Washington, D.C.: American Society for Microbiology Press, 2003.

Prescott, Lansing M., John P. Harley, and Donald A. Klein. "Clinical Microbiology." In *Microbiology,* 6th ed. New York: McGraw-Hill, 2005.

Virology-Online. "Needle Stick Injuries." Available online. URL: http://virology-online.com/general/InfectionControl.htm. Accessed March 23, 2009.

Clostridium

Clostridium is the genus name for gram-positive anaerobic bacteria that form endospores. Anaerobic growth is the ability of a microorganism to grow in the absence of oxygen. *Endospores* are very strong forms of a cell that resist drying, heating, freezing, and chemicals. *Clostridium* endospores possess a distinct bowling pin or bottle shape, when viewed in a microscope, a characteristic that distinguishes them from other bacterial endospores, which are usually ovoid in shape. The normal, reproducing cells of *Clostridium,* called the *vegetative form,* are rod-shaped.

The genus *Clostridium* belongs to family Clostridiaceae, order Clostridiales, and class Clostridia of the Firmicutes phylum of bacteria. *Clostridium* is the largest genus in its family. The term *clostridia* can be used as a general name for all of the species in this genus. Four species of clostridia possess activities that affect human and animal health:

- *C. botulinum*—source of lethal toxin and a food-borne pathogen
- *C. perfringens*—common cause of food-borne illness and the cause of gas gangrene
- *C. difficile*—normal inhabitant of the intestines that can cause illness
- *C. tetani*—cause of the neurologic disease tetanus

DISCOVERY OF *CLOSTRIDIUM*

Clostridium species inhabit soils and the intestinal tract of animals, including humans. Because the genus *Clostridium* contains species that live in various habitats, the discovery of different species has followed different paths. The discovery of *C. botulinum* was an important step in the history of this genus, mainly because of the dangers this species has caused and continues to cause in food spoilage.

In the late 1700s, Germany experienced a number of outbreaks of an illness that seemed connected to eating certain sausages. Not until 1817, did the German neurologist Justinus Kerner (1786–1862) detect rod-shaped cells in his investigations into this so-called sausage poisoning. In 1897, the Belgian biology professor Emile van Ermengem (1851–1932) made public his finding of an endospore-forming organism he isolated from spoiled ham and began studying the microorganism in greater detail than Kerner.

Biologists classified van Ermengem's discovery along with other known gram-positive spore formers in the genus *Bacillus.* This classification presented problems, however, because the isolate grew only in anaerobic conditions, but *Bacillus* grew well in oxygen. In 1924, Ida A. Bengtson (1881–1952) published "Studies on Organisms Concerned as Causative Factors in Botulism," an article on the known endospore-forming bacteria of the time. She separated van Ermengem's microorganisms from the *Bacillus* group and assigned them to a new genus, *Clostridium.* By Bengtson's classification scheme, *Clostridium* contained all of the anaerobic endospore-forming rod-shaped bacteria, except the genus *Desulfotomaculum.* (Bengtson was the first woman to be hired by the Hygienic Laboratory of the U.S. Public Health Service, in 1916. In her career, she categorized many additional microorganisms on the basis of toxin production and serology, which is the study of the constituents on the outer surface of cells.)

Microbiologists distinguish *Clostridium* from *Bacillus* by three features: (1) *Clostridium* grows in anaerobic conditions, and *Bacillus* grows in aerobic conditions; (2) *Clostridium* forms bottle-shaped endospores, and *Bacillus* forms oblong endospores; and (3) *Clostridium* does not form the enzyme catalase, while *Bacillus* secretes catalase to destroy toxic byproducts of oxygen metabolism. *Clostridium* can be further distinguished from another bottle-shaped endospore producer, *Desulfotomaculum,* on the basis of the nutrients each genus uses.

Bengston contributed much of the information now known about clostridia and the toxins they produce. In 1920, Bengtson told an audience of public health officials at a meeting in San Francisco, "The value of standard methods for testing and for stating the potency of diphtheria and tetanus antitoxins is universally acknowledged. Before the work at the Hygienic Laboratory was done, establishing United States standards for these products, nothing was known as to the comparative strength of different lots of antitoxin in this country." Bengtson characterized the tetanus toxin made by *C. tetani* and the botulinum toxin from *C. botulinum* and developed a test for determining their potency. This work led to the development of effective treatments for *Clostridium* toxin poisoning.

Though toxin poisoning is associated with eating contaminated food, *Clostridium* is not a resident of food but a normal inhabitant of soil. Tetanus and gas gangrene occur when open wounds become infected with *Clostridium.* Food-borne botulism, by contrast, is from foods contaminated by *Clostridium*-carrying soils before processing and packaging.

CLOSTRIDIUM BOTULINUM

C. botulinum is a dangerous food spoilage and lethal food-borne pathogen in canned and other processed foods. This organism presents a health risk in foods for two reasons: The *C. botulinum* endospores are heat-resistant, and the organism's toxin can be present in foods even though the bacteria have been eliminated.

The toxin-caused food-borne illness, called *botulism,* includes at least seven different forms produced by various versions of the toxin: botulinum toxins A, B, C, D, E, F, and G. Each of these toxins is from a different individual strain of *C. botulinum,* and the cells release them only upon lysis, the physical breaking apart of cells. Each toxin contains a unique set of antigens that prompt the body's immune system to make antibodies against the specific antigens. Botulinum toxins A, B, E, and F cause botulism in humans; toxins C and D affect animals such as cattle, horses, fowl, and some fish. Type G toxins have not been linked to any known outbreaks in human populations, but this toxin has been recovered from tissues during autopsies.

Botulism occurs through infection of a wound or by ingestion, but ingestion is by far the more common route of infection. Food-borne botulism results from ingestion of the bacteria or of their toxin. When toxin rather than the bacteria causes illness, the condition is called *food-borne intoxication.* Canned vegetables, canned soups, sausages and other meat products, and seafood are the main sources of food-borne botulism. The illness may be prevented by cooking all processed foods for a sufficient length of time. Because *Clostridium* produces gas as it grows in closed containers, some spoiled foods may be spotted by the telltale swelling or bulging of the containers. Foods showing signs of gas production should be discarded and never used in preparing any meal.

Van Ermengem found that heating food to 158°F (70°C) for one hour or 176°F (80°C) for 30 minutes, or boiling for five minutes, inactivated the toxin. Any method for destroying the toxin is essential in preventing food-borne illness, because ingestion of only a few nanograms (nm) of toxin-contaminated food can lead to illness or death. The following botulism symptoms usually occur 18 to 36 hours after ingestion of contaminated food: weakness, lethargy, vertigo, constipation, and difficulty in breathing, speaking, or swallowing. Botulism affects infants and adults; infants exhibit slightly more symptoms associated with the nervous system, such as altered crying patterns and irregular head movements.

C. botulinum's toxin is called a *neurotoxin* because it exerts its effect by binding to nerve cells and blocking the release of the compound acetylcholine from nerve endings. Acetylcholine acts in the normal transmission of signals in the body's nervous system, so blocking this compound's activity results in paralysis, because nerves can no longer stimulate muscles. In botulism fatalities, death ensues from severely harmed respiratory activity and asphyxiation. Medical caregivers diagnose food-borne botulism by detecting antibodies to the toxin in blood or feces or by similarly finding the toxin in a food suspected of being the source.

Since several botulism outbreaks in commercially prepared foods in the 1970s, food processors have improved their techniques for preventing contamination in products. Perhaps the most famous outbreak that occurred was due to contaminated soups produced by Bon Vivant Soups of Newark, New Jersey. A batch of vichyssoise soup, which is eaten cold rather than heated, caused a small but deadly outbreak, in 1971. Medical officials, first, were challenged in diagnosing botulism, which had not been seen for many years, and, then, public health departments took on the job of tracking down the source of the food-borne illness and removing unsold portions from markets. As *Time* magazine reported in 1971, "This task is proving complicated. The company processes 4,000,000,000 cans of food a year—mostly soup—under its own name plus thirty-four other labels." The company filed for bankruptcy shortly after the U.S. Food and Drug Administration (FDA) ordered the recall of the soup from store shelves.

The food production industry instituted more precautions against the contamination that bankrupted Bon Vivant Soups, but botulism remained a threat to health. In 2007, an Associated Press release stated, "Each year the CDC [Centers for Disease Control and Prevention] records roughly twenty-five cases of food-borne botulism poisoning. Most involve home-canned foods. CDC epidemiologist Michael Lynch said the last U.S. case of botulism linked to commercially sold canned food was in the 1970s." Botulism cases have become rare in the United States, according to the CDC; most incidences now relate to improper home canning of foods.

The *C. botulinum* toxin causes such powerful effects on the human nervous system that medical research has developed a new use for it. Since 2002, the U.S. Food and Drug Administration (FDA) has approved the toxin for use for removing wrinkles, smoothing skin, correcting involuntary muscle twitches, and correcting some eye muscle disorders. Marketed under the name *Botox,* the toxin paralyzes nerves that control small muscles that cause twitching or skin wrinkling. Botox had originally been approved for use by physicians for treating eye muscle disorders, but the treatment soon expanded to cosmetic uses, mostly on the face. Botox injections, which can be administered only by a licensed physician, smooth facial wrinkles. The effects last about four months before wearing off.

Because a very small dose of toxin can exert a dramatic effect on a person's nerve functions, microbiologists have suggested that the *C. botulinum* toxin should be viewed as a potential bioweapon. The toxin is dangerous when taken into the body by breathing it in or by ingesting contaminated drinking water or food. Without prompt antitoxin treatment, botulism has the potential to cause 100 percent mortality rates, meaning it kills all the people who have been exposed to the toxin. If an antitoxin can be administered within 24 hours of the onset of symptoms, mortality rates fall to about 25 percent.

CLOSTRIDIUM PERFRINGENS

C. perfringens is also found in soil and in the intestines and causes two unrelated illnesses: food-borne illness and gas gangrene. When the microorganism contaminates food, its presence indicates probable fecal contamination. *C. perfringens* is one of the most common causes of food-borne illness outbreaks worldwide, but because medical providers do not monitor it as closely as other food-borne pathogens, *C. perfringens* receives little publicity. Most people are aware of food-borne threats from *Escherichia coli,* which are rare, but do not recognize *C. perfringens* as a much more prevalent cause of food-borne illness.

The *C. perfringens* alpha-toxin causes symptoms in the body similar to those of mild botulism. Painful abdominal cramps and diarrhea begin eight to 22 hours after ingesting a large number of cells. The *C. perfringens* infective dose—the amount of cells needed to cause symptoms—equals several million. Illness lasts about 24 hours but rarely causes death. For this reason, *C. perfringens* is often incorrectly nicknamed the "24-hour flu" or "stomach flu."

C. perfringens contaminates almost any type of food, but it may be slightly more prevalent in undercooked meat, meat products, and gravy. Casseroles that require a long period to heat through completely are also probable sources for this illness. Cafeteria-style meals such as found in schools, colleges, nursing homes, prisons, cruise ships, or supermarket self-serve areas are prone to spreading *C. perfringens* illness for the same reason: either inadequate cooking temperatures or inadequate holding temperatures. As are most food-borne pathogens, *C. perfringens* has been difficult to track for the three following reasons: (1) the illness is not associated with a particular food, (2) symptoms begin quickly and resolve quickly, and (3) people tend not to report minor illnesses to doctors.

Gangrene is a condition in which the body's tissues die—termed tissue *necrosis*—usually as a result of inadequate blood supply. Gangrene has several unrelated causes; one type of gangrene, called gas gangrene, is caused by *C. perfringens* infection of an open wound. Open wounds may be treated with antibiotics or an antitoxin if *C. perfringens* infection is suspected. Most important, however is to clean out the wound in a medical technique called *débridement,* in which dead tissue, foreign matter, and any other dirt are thoroughly cleaned out of the wound. Débridement serves to reduce small oxygen-free pockets in the wound where *C. perfringens* can flourish. A large population of *C. perfringens* growing in anaerobic conditions produces gas from its normal fermentation. The gas quickly builds up under the skin, causing blisters and swelling, pain at the infection site, discoloration of the skin, fever, sweating, and irregular heart rate.

Gas gangrene is uncommon in the United States. Most cases that are found have been caused by *C. perfringens,* but *Staphylococcus* and *Vibrio* bacteria also can cause it, on rare occasions. In addition, wounds allowed to remain dirty, gas gangrene may form in surgical wounds, or the disease may develop on its own in people who have poor circulation, diabetes, or colon cancer. Severe cases in which débridement will not be effective must be treated with surgery and possible amputation.

CLOSTRIDIUM DIFFICILE

C. difficile infections have long been associated with young children in day care, although today more adults, especially the elderly, contract illnesses caused by this microorganism. This may be partly because of growth of the elderly population. People in high-risk situations are more likely than the normal population to contract infection. People living in long-term health facilities or nursing homes or in long-term hospital stays have higher risks of infection due to weakened immune systems. The incidence of *C. difficile* infections in hospitals has increased steadily, since the 1980s. Hospital-related cases doubled between 1993 and 2003, more than doubled between 2000 and 2005, and continue to increase today.

C. difficile inhabits the gastrointestinal tract, and outbreaks are due to fecal contamination of food or other items. Infections arise from eating food contaminated with fecal matter or from self-inoculating with fecal matter on surfaces or inanimate items. Self-inoculation involves receiving a microorganism into the body, usually by touching the hand or fingers to the face—specifically touching the mouth, nasal passages, or eyes. *C. difficile* occurs almost everywhere—in food, water, soil, and numerous surfaces in hospitals. The widespread incidence of the microorganism increases the chance of self-inoculation. Illness symptoms usually consist of watery diarrhea, bloody stools, fever, abdominal cramping, and nausea. These ailments can lead to dehydration and weight loss, and, in severe cases of intestinal inflammation, the infection can be fatal.

Infections caused by this microorganism have been associated with antibiotic treatment for other infections. *C. difficile* has developed resistance to a variety of antibiotics, since at least the 1990s. When a patient has been treated with antibiotics to stop any other type of infection, the antibiotic also kills a large number of normal intestinal bacteria that protect people against many infections. As the antibiotic wipes out these beneficial bacteria, resistant *C. difficile* grows to large numbers in the intestinal tract. It then, releases two toxins, each of which destroys the inner lining of the intestines. As a consequence, the intestinal lining inflames, causing a reduced ability to absorb nutrients and diarrhea. *C. difficile* has become so resistant to a variety of antibiotics and spreads so easily that it has been called a *superbug,* which is any microorganism that has evolved into a very virulent (capable of causing disease) and difficult-to-kill pathogen.

Abraham Sonenshein of the Tufts University School of Medicine said, at the 2007 meeting of the American Society for Microbiology, "The [*C. difficile*] genes responsible for toxin production only seem to be expressed during periods of nutrient deprivation. This is consistent with the view that most disease-causing bacteria express their pathogenicity when they are hungry." Since 2002, a more virulent variety of *C. difficile* has emerged in U.S. and Canadian hospitals. This strain produces a higher level of toxin than typical *C. difficile* and causes a more destructive and deadly disease in humans.

C. difficile infections cause several thousand deaths in the United States each year, yet the illness is preventable in two ways: by limiting the use of antibiotics to only lifesaving situations and by practicing good hygiene. People who prepare meals, hospital patients, and health care providers must wash their hands before handling food or working with patients. Day care workers should also wash hands before preparing any meals, before and after touching children, and after diapering infants. Disposable gloves help reduce the chance of transmitting *C. difficile,* but only if they are changed for each of these tasks.

CLOSTRIDIUM TETANI

The agent that causes the disease tetanus lives in gastrointestinal tracts and soil. *C. tetani* produces the toxin tetanospasmin, which produces the illness's symptoms: stiffness in muscles, spasms of jaw muscles, and progression to violent, painful, and convulsive spasms caused by minor stimuli. The syndrome in which facial muscles become paralyzed is named *lockjaw.* This condition is characterized by tightening of the facial muscles to create a grimace-like expression.

Tetanospasmin enters nerve cells and travels to the spinal cord to create the progressive symptoms of tetanus. Physicians able to act quickly with infected patients administer an antitoxin, but even with treatment, a small percentage of infected people (fewer than 100 persons in the United States) die of tetanus each year. As a preventative, children in the United States receive a tetanus vaccine at about two months of age with four boosters until age six. This vaccination does not give permanent immunity, so adults should receive a booster injection every 10 years.

If a person has sustained a deep wound, tetanus prevention involves a thorough cleaning of the area with soap and water, followed by an antibiotic cream or ointment, then a sterile bandage covering to be changed daily. Very deep wounds in which dirt remains should be treated by a physician.

OTHER CLOSTRIDIA

The *Clostridium* genus contains about 55 different species, most of which provide few commercial benefits to industry. *C. acetylbutylicum* and *C. histolyticum* are two exceptions. *C. acetylbutylicum* produces the industrial solvents acetone and n-butanol from fermentation of the sugar glucose. Many other *Clostridium* species produce similar organic products, but because of the difficulty of growing these anaerobic bacteria, industry tends to use other microorganisms. *C. histolyticum* has been used as a source of the enzyme collagenase, which degrades animal tissue. *Clostridium* species excrete collagenase to eat through tissue and, thus, help the pathogen spread throughout the body. The medical profession uses collagenase for the same reason in the débridement of infected wounds. Débridement involves the removal of foreign material and dead or damaged tissue from a wound.

CLOSTRIDIUM DISEASES IN ANIMALS

Animals incur *Clostridium* infection through contaminated wounds or by ingestion of contaminated foods or soil. As in humans, *C. botulinum* causes a severe and lethal botulism in animals. Scavengers such as vultures or hyena risk infection when feeding on infected carcasses. Botulism causes progressive paralysis, meaning the nerve and muscle functions degrade over time. The paralysis occurs mainly in the respiratory tract and the heart. The disease also causes impaired vision, tremors, and an unusual extension of the neck in some species. For instance, botulism in birds has been described as "limberneck," for the odd involuntary movement of the head. In horses, shaker foal syndrome resembles some of the neurological symptoms in humans and is thought to be caused by the B-type toxin. In addition to nerve and gait disorders, shaker foals experience respiratory failure, loss of appetite, and constipation. Foals less than four weeks old are most susceptible to the infection; death usually occurs 24–72 hours after the onset of symptoms.

C. perfringens and *C. difficile* cause the gastrointestinal disease colitis in horses. Adults and foals show signs of abdominal pain and diarrhea. *C. difficile* is one of the main causes of diarrhea in baby pigs.

Almost all animals, except cats, dogs, and birds, are susceptible to the *C. tetani* toxin if a wound becomes contaminated with soil. Animals receiving deep puncture wounds develop oxygen-free conditions that the tetanus organism prefers, so these types of wounds must receive prompt care. *C. tetani* bacteria in a wound produce large amounts of the tetanus neurotoxin, which then travels in the bloodstream or is absorbed by nerve cells. The toxin eventually reaches the spinal column, where it causes the most damage.

Tetanus in animals causes exaggerated responses to gentle stimuli such as low-level noise or subtle movement. Low sounds or subtle movements may cause an animal to have spasms, unusual or impossible gaits, unusual stance, and lockjaw. Horses become especially susceptible to lockjaw. Sheep, goats, and pigs often fall to the ground when startled and bend backward. On the rare instances of tetanus in cats or dogs, the affected limb becomes paralyzed, followed by paralysis progressing to the anterior limbs.

Other, less common veterinary diseases caused by this genus are the following:

- blackleg in cattle and sheep caused by *C. chauvoei* and found worldwide
- red water disease in cattle caused by *C. novyi* and found in the western United States, South America, Mexico, the Middle East, and Great Britain
- black disease in sheep caused by *C. novyi* and found worldwide
- malignant edema in horses, cattle, sheep, goats, and pigs caused mainly by *C. septicum* worldwide

Despite the rather limited growth conditions of *Clostridium* species, these microorganisms create serious health hazards in a wide variety of forms and in a variety of animal species. *Clostridium* also serves a limited role in industry for producing commercially useful products, but overall this microorganism is more important as a health hazard than as a commercially valuable microorganism.

See also BIOWEAPON; FOOD-BORNE ILLNESS; SPORE.

Further Reading

American Society for Microbiology. "Understanding Why *C. difficile* Causes Disease: It's Hungry." May 24, 2007. Available online. URL: www.asm.org/Media/index.asp?bid=50665. Accessed March 23, 2009.

Associated Press. "Plant Suspected of Botulism Cases Had Production Issues." *Washington Post,* 20 July 2007. Available online. URL: www.washingtonpost.com/wp-dyn/content/article/2007/07/18/AR2007071802407.html. Accessed March 23, 2009.

Bengtson, Ida A. "Standardization of Botulism Antitoxins." Presented at the American Public Health Association meeting, San Francisco, September 16, 1920. Available online. URL: www.pubmedcentral.nih.gov/picrender.fcgi?artid=1353790&blobtype=pdf. Accessed March 23, 2009.

———. "Studies on Organisms Concerned as Causative Factors in Botulism." *Hygienic Laboratory Bulletin* 136 (1924): 1–101.

Hauschild, Andreas H. W., and Karen L. Dodds, eds. Clostridium botulinum: *Ecology and Control in Foods.* Boca Raton, Fla.: CRC Press, 1992. Available online. URL: http://books.google.com/books?id=9xlAfq9GF98C&printsec=frontcover&dq=Clostridium+botulinum:+Ecology+and+Control+in+Food. Accessed March 23, 2009.

Time. "Death in Cans." July 19, 1971. Available online. URL: www.time.com/time/magazine/article/0,9171,905373,00.html?iid=chix-sphere. Accessed March 23, 2009.

coliform The coliform group of bacteria was first described, in 1886, by Theodor Escherich (1857–1911), a German-Austrian pediatrician, who isolated these microorganisms from infants' stool specimens. Escherich identified this general group as any bacteria that possess all of the following characteristics: gram-negative, rod-shaped (called *bacilli*), non-endospore-forming, and capable of fermenting the sugar lactose with the production of acid and gas within 48 hours, when incubated at 95°F (35°C). These four characteristics remain the standard definition for coliform bacteria.

Coliforms are facultative anaerobes, meaning they grow with oxygen but can switch to an alternative metabolism when oxygen is absent. This ability becomes important when certain species of coliforms escape the anaerobic conditions of the intestinal tract and contaminate food or water.

Water microbiology laboratories use coliforms as indicators of possible water contamination from fecal matter. An indicator microorganism is any species or genus whose presence indicates contamination. Although some coliforms live naturally in water and present no known health risks, other coliforms such as *Escherichia coli* normally reside only in animal intestines, so its presence in water provides positive evidence that fecal matter has entered a water system. In water microbiology and in other disciplines within microbiology, coliforms have been defined as any member of the family Enterobacteriaceae having the four main characteristics listed in the first paragraph.

THE COLIFORM BACTERIA

Escherich named his first coliform isolates from infants *Bacterium coli commune* and *Bacterium lactis aerogenes*. Throughout his career, Escherich compiled a store of information on the morphology, cultivation methods, and physiology of numerous intestinal bacteria and built the foundation of coliform studies. The two strains that Escherich had isolated from infants were renamed *Escherichia coli*, in 1919, in honor of his work.

The coliform group contains the following genera: *Escherichia, Citrobacter, Klebsiella, Enterobacter, Edwardsiella,* and *Serratia*. These microorganisms all share similar nutrient needs and energy production; one exception is that the first four in this list all ferment lactose, whereas *Edwardsiella* and *Serratia* do not readily ferment this sugar. These two genera have, therefore, been called *paracolon bacilli* to indicate they are marginally related to the other coliforms.

Water microbiologists monitor coliforms in general, as well as the subgroup called *fecal coliforms*. The fecal coliforms are species of Enterobacteriaceae that are from fecal matter rather than the general environment. Fecal coliforms make up 60–90 percent of total coliforms, and they may be better indicators of fecal contamination in water than general coliforms. Water microbiology distinguishes fecal coliforms from all other coliforms by noting the presence of the following two characteristics:

1. ability to grow at 112°F (44.5°C)

2. production of acid and gas in two types of growth medium: lauryl tryptose broth and EC broth (*EC* is an abbreviation for *E. coli*).

Food microbiology and clinical microbiology laboratories may need to identify coliforms to the genus and species level in order to devise a preservation system or an antibiotic treatment, respectively. Many identification tests differentiate these microorganisms on the basis of their biochemical reactions, cell surface structures, and genetic material. Water microbiologists focus mainly on tests to determine the presence of absence of three coliform classifications: total coliforms, fecal coliforms, and *E. coli*. Clinical microbiologists, by contrast, search for any members of Enterobacteriaceae as possible indicators of a food-borne or waterborne infection.

BIOCHEMICAL CLASSIFICATION OF COLIFORMS

Each microorganism has a unique array of enzymes that enable the cells to use certain nutrients. If certain bacteria cannot use a particular compound for growth, it indicates that the bacteria do not possess the enzymes needed to get energy or carbon from that compound. The enzymes make up what is called a microorganism's *biochemical profile* or simply its *profile*. Simple and commercially available biochemical test kits distinguish several species from each other within 24 hours.

A standard set of tests used for the identification of coliforms are:

- indole test to detect amino acid tryptophan production from indole
- methyl red test to determine the ability to produce acid from a specific formula con-- taining glucose and chains of amino acids called *peptones*
- Voges-Proskauer reaction to determine the ability to form the compounds acetoin and diacetyl from glucose-peptone
- citrate test to determine the ability of bacteria to use citrate as a sole carbon source

These four tests are collectively called the *IMViC tests*—the lowercase *i* is for ease in pronunciation—and they are usually combined with two additional tests: the motility test and the urease test, which detects ammonia and carbon dioxide release from urea. The IMCiV tests alone have 16 different combinations of results. Microbiologists usually focus on the possible IMCiV results summarized in the table to distinguish *E. coli* from other coliforms.

Additional tests narrow down the coliforms, until a laboratory can differentiate each species from biochemical abilities alone. Newer molecular methods using a piece of deoxyribonucleic acid (DNA) can also detect very small numbers of any species having that same DNA sequence. These pieces of DNA, called *DNA probes,* locate just a few cells of a certain species within a large quantity of water or food.

COLIFORMS IN WATER

Water testing laboratories keep a close watch on the coliforms and fecal coliforms found in water samples because their presence may indicate contamination. Because coliforms belong to the Enterobacteriaceae family of bacteria, which is considered to be synonymous with enteric species, the presence of coliforms has been equated with the presence of fecal matter. Some microbiologists argue that coliforms provide a poor assessment of water quality because they can be found in natural waters with no evidence of any fecal contamination. For example, *Citrobacter* commonly occurs in natural waters without any indication that it has originated from contamination. The New York State Department of Health has endorsed monitoring coliforms in water, stating: "It is not practical to test for pathogens in every water sample collected. Instead, the presence of pathogens is determined with indirect evidence by testing for an 'indicator' organism such as coliform bacteria. Coliforms come from the same sources as pathogenic (micro) organisms. Coliforms are relatively easy to identify, are usually present in larger numbers than more dangerous pathogens, and respond to the environment, wastewater treatment, and water treatment similarly to many pathogens. As a result, testing for coliform bacteria can be a reasonable indication of whether other pathogenic bacteria are present." Coliform testing remains a simple and inexpensive way to monitor water quality.

The U.S. Environmental Protection Agency (EPA) controls the standards for drinking water, wastewater, and surface waters, such as rivers and beaches. A standard refers to the allowable limits within which a microorganism, chemical, or particle may be present in water. Though many of the EPA's water standards are expressed as a range of acceptable values, no water should ever exceed the upper limits of these ranges. Over-the-limit test results indicate that the water has been polluted. The EPA calls this upper limit the *Federal Maximum Contaminant Level* (MCL). The MCL for coliforms in drinking water depends on the number of samples a water utility must take from its water source in a month. Most municipal water utilities collect at

IMViC Test Results for Common Coliforms

Coliform	Indole Test	Methyl Red Test	Voges-Proskauer Test	Citrate Test
E. coli	+	+	-	-
Enterobacter aerogenes	-	-	+	+
Klebsiella pneumoniae	-	-	+	+
Serratia marcescens	-	-	+	+

Note: IMViC test results can be abbreviated. For example, *E. coli* produces a "++--" IMViC result

least 40 samples per month. The coliform MCL for these utilities is that no more than 5 percent of total samples can be positive for total coliforms; positive indicates coliforms are present. In water utilities that collect fewer than 40 samples per month, the EPA allows only one positive sample each month. A typical water sample for testing drinking water is 3.4 fluid ounces (100 ml).

Fecal coliforms and *E. coli* have stricter requirements to pass than total coliforms. This is because fecal coliforms and *E. coli* relate more directly to fecal contamination in the water than general coliforms. Water containing fecal microorganisms can cause diarrhea, abdominal cramps, nausea, headaches, and other symptoms if a person ingests the water. This is true not only for drinking water but for recreational waters such as beaches, water parks, swimming pools, and spas. Infants and the elderly as well as persons with weakened immune systems are particularly susceptible to illness from contaminated water, so the EPA has set an MCL of zero for fecal coliforms and *E. coli.*

Water samples that test positive for total coliforms, fecal coliforms, or *E. coli* must be retested to confirm the results. This confirmation is done on the same water sample within 24 hours of the positive result. The EPA requires additional samples if the second test also gives a positive result. Any water supply utility that has numerous and repeated positives for fecal contamination must publish a "boil water alert" to notify the public of a potential health hazard in their water. A boil water alert means that individuals should boil all drinking and cooking water for five minutes at a rolling boil (then allow it to cool) before using it.

LABORATORY TESTING FOR COLIFORMS

Water testing laboratories use three main methods for determining the presence or absence of coliforms in a water sample. The first method, the most probable number (MPN) method, uses the presence or absence of gas production from fermentation to determine the presence of coliforms. The MPN method also estimates the number of coliforms in the original water sample. The second method is the membrane filter method, in which 100 ml of water passes through a thin filter called a *membrane.* The membrane catches the bacteria from the water, and a microbiologist then places it bacteria-side-down on an agar plate. After incubation, the microbiologist counts the bacterial colonies on the agar, equivalent to the number of bacterial cells that were in the filtered water. In this test, the microbiologist often adds a step to confirm whether some colonies are coliforms and others are noncoliforms. A third method of detecting coliforms is the *presence-absence test,* in which the presence of coliforms relates to gas production or some other chemical change in the medium.

The EPA permits other methods to be used in addition to or in place of the three main methods described here. Most of these tests are ready-made commercial kits that contain all the media and sampling equipment a microbiologist needs to detect the presence of coliforms or fecal coliforms. The following list provides the trade names of these products: Colisure, E*Colite, m-ColiBlue24, Readycult Coliforms 100 Presence/Absence Test, Chromocult Coliform Agar with filtration, and Colitag.

Additional biochemical tests target *E. coli* by incorporating the compound 4-methylumbelliferyl-*beta*-D-glucuronide (MUG). *E. coli* contains an enzyme called *glucuronidase,* which cleaves MUG. Various media have been developed to produce a color reaction when glucuronidase, therefore *E. coli,* is present. The three following media are common in microbiology for this purpose:

- *o*-nitrophenyl-*beta*-D-gala ctopyr anoside with MUG (ONPG-MUG)
- violet red bile agar with MUG (VRBA-MUG)
- EC medium with MUG

Many more identification tests have been developed for *E. coli* than for most other bacteria.

COLIFORMS AS INDICATORS OF CONTAMINATION

Coliforms have been useful in predicting potential water contamination, yet they are not perfect as indicator organisms, as noted earlier. The following disadvantages of coliform testing have continued to cause doubt in the minds of microbiologists regarding the value of coliforms, even fecal coliforms, as water pollution indicators. Despite assurances such as those published by the New York Department of Health, the University of Georgia food safety microbiologists Michael Doyle and Marilyn Erickson wrote, in a 2006 opinion article, "Species of Enterobacteriaceae other than *E. coli* are associated with plants and do not indicate fecal contamination, yet they are identified as fecal coliforms by the fecal coliform assay. Hence, *E. coli* is the only valid index for the monitoring of foods containing fresh vegetables."

As suggested earlier, coliform or fecal indicator tests may be confounded by false positive results or false negative results. A false positive result indicates that contamination is present even when there is no contamination. Conversely, a false negative result indicates that the sample contains no contamination when

Total Coliform and Fecal Coliform Standards for Water

Type of Water	Standard (as CFU/100 mL), Not to Exceed
drinking water	1 total coliform
total body contact (swimming)	200 fecal coliforms
partial body contact (boating)	1,000 fecal coliforms
treated wastewater	200 fecal coliforms

Note: CFU = colony-forming unit on agar medium; 1 CFU is equivalent to one cell

it actually does have fecal contaminants in it. The following problems in current testing may lead to false positive or false negative results:

- Coliforms occasionally do not use lactose as normal, and this occurrence causes a false negative test result.
- Large numbers of noncoliform bacteria can suppress coliforms, especially in untreated groundwater, cistern water, and water containing insufficient chlorine disinfectant. The result is a false negative.
- Some coliforms exist as normal members of biofilms that form inside water distribution pipes, and these coliforms do not indicate a health hazard. Their presence in laboratory tests yields a false positive result.
- Coliforms are not a homogeneous group, and finding various species of Enterobacteriaceae does not necessarily mean water is contaminated. Therefore their presence produces a false positives result.
- The genus *Aeromonas* in the family Entero bacteriaceae is a common cause of false positive results in warm weather.
- Disinfectants easily kill coliforms, but other water contaminants resist the same chemical disinfectants yielding a false negative result.

The microbiologists Doyle and Erickson have pointed out another flaw in coliform and fecal coliform testing. "Physicians and public health officials have repeatedly misinterpreted results of the fecal coliform assay when applied to food, beverage, or water samples," they wrote in 2006. "It is not a reliable indicator of either *E. coli* or the presence of fecal contamination. The *E. coli* assay is a more reliable indicator of fecal contamination, although not absolute, and could serve as a replacement for the fecal coliform assay." All of these tests may serve as general indications of water quality until newer and more sensitive DNA probes become more commonplace in testing. Until then, the guidelines in the table have proved useful for assessing water quality in terms of coliform or fecal coliform content.

Years of working with coliforms have allowed water quality microbiologists to make fairly accurate assessments of the potential health hazards of water based on indicator organisms in combination with other more accurate tests. Fecal coliform tests, even with potential drawbacks, serve the same role. As molecular biology techniques become more commonplace in water quality testing, and *E. coli* tests increase in accuracy, coliform testing might become a secondary indicator for water quality rather than a primary indicator.

Coliforms are a widespread group of microorganisms that have been a useful teaching tool in environmental studies, food safety, medicine, and general microbiology. These bacteria are readily available and easy to grow. For these reasons, coliforms will remain a part of basic microbiology teaching and methods.

See also ENTERIC FLORA; *ESCHERICHIA COLI*; INDICATOR ORGANISM; MOST PROBABLE NUMBER; WATER QUALITY.

Further Reading

American Society for Microbiology. "National Primary Drinking Water Regulations: Ground Water." Public Policy Statement, August 9, 2000. Available online. URL: http://www.asm.org/policy/index.asp?bid=3611. Accessed March 26, 2009.

Doyle, Michael P., and Marilyn C. Erickson. "Closing the Door on the Fecal Coliform Assay." *Microbe,* April 2006. Available online. URL: www.asm.org/ASM/files/ccLibraryFiles/Filename/000000002223/znw00406000162.pdf#xml=http://search.asm.org/texis/search/pdfhi.txt?query=coliform+indicator&pr=ASM+Site&prox=page&rorder=500&rprox=500&rdfreq=500&rwfreq=500&rlead=500&rdepth=0&sufs=0&order=r&mode=&opts=&cq=&id=4552fbf28. Accessed March 26, 2009.

Indiana University. "Fecal Coliform Test." Available online. URL: www.indiana.edu/~bradwood/eagles/fecal.htm. Accessed March 26, 2009.

New York State Department of Health. "Coliform Bacteria in Drinking Water Supplies." Available online. URL:

www.health.state.ny.us/environmental/water/drinking/coliform_bacteria.htm. Accessed March 26, 2009.

U.S. Environmental Protection Agency. "Drinking Water Contaminants." Available online. URL: www.epa.gov/safewater/contaminants/index.html#micro. Accessed March 26, 2009.

———. "Evaluation of the Microbiology Standards for Drinking Water." August 1978. Available online. URL: http://yosemite.epa.gov/water/owrccatalog.nsf/1ffc8769fdecb48085256ad3006f39fa/39938a86758964e585256b0600723873!OpenDocument. Accessed March 26, 2009.

colony A colony is a visible mass of cells that has grown from a single cell. Bacteria, yeasts, molds, and algae form colonies on solid agar media. Microbiologists usually try to prepare pure colonies for their studies, meaning the colony contains only one type of microorganism. Clinical microbiology similarly relies on pure colonies as a step in identifying an unknown pathogen. In microbiology, pure colonies serve the following four purposes: (1) identification of microorganisms, (2) determination of the concentration of microbial cells in a suspension, (3) cloning methods, or (4) isolation of pathogens, contaminants, or environmental species. The opposite of a pure colony is a *mixed colony,* which contains more than one microbial species.

Several microbiological techniques produce pure colonies on agar media. The main techniques that microbiologists use to prepare colonies are aseptic techniques, disinfection, and sterilization. Microbiology training also includes the ability to gain clues on the identity of unknown microorganisms by studying the colony. This is because many bacterial, algal, yeast, or mold colonies have distinctive characteristics, which may be seen as distinct color, shape, or size. All of the features that compose the appearance of a colony are called *colony morphology.* Colony morphology and cell morphology are two of the first characteristics checked by a microbiologist to identify a new or unknown species.

Microbiologists grow single, isolated colonies on agar plates by using a method called *streaking.* In this technique, the microbiologist carries a drop of microbial culture in an inoculating loop and then lightly spreads the drop over an agar surface in a continuous line or streak. The concentration of cells decreases as the loop moves over the agar, until only single cells are deposited. After incubation, single cells will have grown into distinct colonies that are isolated from neighboring colonies on the agar surface, meaning the colonies do not touch. The goal of streaking a culture having more than one type of microorganism is to produce a single, distinct, and isolated colony. This single pure colony is called a *colony-forming unit* (CFU), and numbers of colonies on an agar plate are often referred to as the "number of CFUs." By counting the number of CFUs on an agar plate, a microbiologist determines a value called a *plate count.* Plate counts indicate the concentration of microorganisms in liquids such as water and beverages or in solids or semisolid foods.

A single CFU, appearing so simple to the eye, actually contains its own inner metabolism. Generally, cells on the colony's outer edge grow fastest because nutrients and oxygen (in the case of aerobes) are most available. Cells in the center of a colony contend with lower oxygen and nutrient levels and a greater concentration of toxic end products, which all slow their growth rate. During incubation, a colony begins to contain a spectrum of bacterial growth: young, rapidly dividing cells toward the outside and dying and dead cells accumulating on the inside. Special mechanisms might regulate colony growth, as well as communication between cells within each colony.

Although laboratory work relies on pure colonies on an agar plate, colonies do not grow the same way in nature. In nature, microorganisms grow as sheets containing many different types of microorganisms attached to surfaces, called *biofilms,* or as loose cells growing in watery environments, called *planktonic growth.*

Scientists developed the agar plate technique, in microbiology's early years, to help them study specific microorganisms that they recovered from the body or the environment. Growing colonies enabled early scientists such as Louis Pasteur (1822–95), Joseph Lister (1827–1912), Robert Koch (1843–1910), and Hans Christian Gram (1853–1938) to study specific microorganisms related to infections or food spoilage.

TYPES OF COLONIES

Microorganisms grow on solid agar surfaces as either discrete (isolated) colonies or as swarms, which cover an entire agar surface in a sheet. Isolated colonies equal CFUs. Swarming colonies, by contrast, cover the agar surface in a single sheet during incubation. Swarming growth is characteristic of certain bacterial species and thus serves as a good identification aid. For example, some *Proteus* species grow in characteristic swarms.

Colony size and other hallmarks of appearance vary among species and genera. But each cell will always form a colony that looks like all other colonies in its genus or species. Bacterial colonies are either discrete or mycelial. Discrete colonies are single, isolated formations. Mycelial colonies may be less isolated from each other, because they send long thin filaments into the surrounding agar. *Streptomyces* is an example of bacteria that form mycelial colonies. Molds also form large fluffy masses that can cover an entire agar plate surface with mycelia. Mold growth over an entire agar plate makes the task of obtaining single discrete colonies very difficult. Most microbiologists identify molds and other fungi, other than yeasts, less by colony morphology than by examination of the fungal spores under a microscope. Yeast colonies often resemble bacterial colonies by forming discrete and characteristic formations.

Algae forms colonies on agar surfaces with two typical characteristics: The algae tend to grow over the entire surface rather than form discrete colonies, and some algae stick very firmly to the agar. These latter algae must be transferred to fresh sterile media by dislodging a section of agar along with the algae. The microbiologist places the piece of algae upside down on a new agar surface or inoculates a liquid broth with it. Despite these minor difficulties, some algae *(Volvox, Scenedesmus)* have been observed in laboratories in no form other than colonies.

In general, microbiologists look for the following characteristics in growing microbial colonies on agar:

- single, discrete, isolated growth
- no overlap of colony growth
- no overgrowth; heavy growth that covers an entire agar plate surface
- adequate separation of colonies to prevent competition
- all colonies of the same appearance and approximate size
- between 30 and 300 CFUs per plate to allow visual counting

Colonies may be counted by two methods. The first counting method involves manual counting with or without the need for a low-powered microscope. The second counting method uses automated devices that record the number of colonies on a plate and store the count in a computer program.

COLONY COUNTS

A colony count or plate count equals the number of colonies enumerated on an agar plate and is recorded as the number of CFU per plate. Microbiologists usually take, as an average, two or three plates to eliminate any plate-to-plate variability in growth. In addition, plates containing about 30–300 colonies give the most accurate results for two reasons. Fewer than 20–30 CFUs on a plate may not be an accurate assessment of bacterial concentration in a sample. Second, hundreds of colonies that are crowded on a plate might begin to inhibit their neighbors' growth through nutrient depletion or inhibitory secretions. This overgrowth leads to an inaccurate cell concentration assessment. Manually counting hundreds of CFUs on a single agar plate can, furthermore, be tedious and lead to eyestrain.

A colony count tells the microbiologist how many bacterial cells were in the original culture. To make this calculation, the microbiologist extrapolates from the average colony counts back to the culture's concentration in CFU/ml. The following scheme provides an example of the steps in calculating CFU/ml.

1. Dilute bacterial culture 1×10^{-5} (a culture diluted 1/10 five times).
2. Inoculate duplicate agar plates each with 0.1 ml of the 10^{-5} dilution.
3. Incubate the plates.
4. Count number of CFUs on each plate.
5. Plate 1 contains 50 CFUs and plate 2 contains 66 CFUs (example).
6. Calculate the average of the two plates to determine 58 CFUs/plate.

The plate count determined here, 58 CFUs/plate, must be extrapolated back to the original culture by correcting for the dilution; this correction factor is called the *dilution factor.* In this example, the bacterial concentration in the original culture is:

$$58 \times 10 \times 10^5 = 58 \times 10^6 \text{ or } 5.8 \times 10^7 \text{ per ml of culture}$$

In the preceding example, the factor 10 accounts for the 0.1 ml inoculum and the factor 10^5 accounts for the 1×10^{-5} dilution.

Automatic counting equipment can determine the CFU/ml without the need for manual counting as in the example. An instruments called a *spiral plater* mechanically streaks inoculum onto each agar plate in one continuous spiral line. The procedure is known as *spiral plating.* Sections of the entire spiral plate may then be counted manually or automatically, using a device called a laser counter, which scans each plate with a laser beam; each break in the beam is recorded as a colony. Automated colony counters have two disadvantages that manual counting avoids. First, electronic counters cannot distinguish between very small colonies and small particles that are part of the agar. Second, electronic counters that must be preset to detect a certain colony size will not detect smaller colonies.

COLONY MORPHOLOGY

Colony morphology differs by color, size, shape, and edge style. Color and size correlate to the type of microorganism when grown on a specific agar medium. For example, *Escherichia coli* colonies are light tan color on tryptic soy agar but form black colonies on eosin methylene blue (EMB) agar. The nutrients in an agar medium formula may also affect colony size. The presence of nearby colonies, as mentioned, also influences the size of colonies on plates containing dense growth.

Colony morphology is best distinguished by looking at shape. Colony shape has three characteristics:

- form—circular or noncircular shape of the entire colony
- elevation—height of the cell mass
- margin—shape of the colony's edge

A microbiologist holds each agar plate at different angles to assess the height and color of colonies. Some bacterial species form only dense opaque colonies, and others produce clearer colonies. Each characteristic should be studied under standard conditions—same incubation conditions and same age of the culture—because characteristics can change from young to older colonies that have been incubated for a long period. The table summarizes the main characteristics that microbiologists record when examining colony morphology.

Some species produce variations on the common morphologies that help in their identification. For example, some streptococci grown on blood agar form draughtsman colonies, which are raised colonies with steep sides and a flat top. Corynebacteria form daisy head colonies that vary in color outward from the center and have a notched or scallop-shaped edge. Mycoplasmas are known for making fried egg colonies: An opaque, granular center embeds in the agar and then proceeds outward as a smooth surface colony.

Factors in addition to the medium's ingredients and the culture's age influence colony morphology. The main factors that contribute to colony appearance are hardness of the agar surface, moisture on the agar surface, gaseous conditions, and exposure to light. Microbiology laboratories strive to maintain standard procedures to eliminate these factors and, thus, reduce variability in colony morphology.

SPECIAL COLONY TYPES

In the world of biology, few things fit into consistent patterns. This may be especially true in microbiology, where various unique colonies have been discovered. The distinctive features may be due to a specialty of a given microbial species or additional growth factors. Three examples of idiosyncratic colony growth are represented by swarming, motile colonies, and pinpoint colonies.

Swarming colonies and motile colonies are those in which bacterial cells migrate across the agar

Main Characteristics of Bacterial Colony Morphology

Attribute	Terminology	Description
color	pigmentation	colorless, creamy, white, off-white, gray, tan, pink, red, black, etc.
shape	form elevation margin	overall shape circular, irregular, spindle-shaped, etc. height or thickness flat, raised, rounded, etc. clean entire, lobed, undulating, etc.
surface texture	surface	smooth—shiny, glistening, glossy rough—dull, bumpy, granular, matte mucoid—slimy, gummy
density	opacity	opaque (dense) or refractive (transparent or translucent)
colony texture	texture	butyrous—butterlike viscous or mucoid—thick and sticky dry or friable—brittle or powdery

surface during incubation. *Proteus, Serratia,* and *Myxococcus* species swarm and are also referred to as *gliding cells* for the manner in which they move over a solid surface. Swarming occurs in two phases. In the first phase, cells reproduce normally and begin to form a single colony. After several hours of incubation, the first-generation cells, which may be 2–4 μm in length (in the case of *Proteus*), elongate in subsequent generations. In the second phase, these elongated cells at the edge of the colony develop flagella and move outward from the main colony. When these cells replicate, they repeat the pattern of short first-generation cells to long second-generation cells. After several generations have been formed, a series of concentric rings of growth surround the original colony.

Studies of swarming *Proteus* colonies have uncovered behavior called the Dienes phenomenon. In this occurrence, two swarms of two different strains of a *Proteus* species stop when they meet and do not penetrate each other. The sharp line of demarcation between the two swarms is probably due to the secretion of inhibitory substances such as bacteriocins.

Microbiologists can prevent swarming, to some degree, by growing these microorganisms on agar formulated to be extrarigid. Otherwise, swarming serves as a tool in identifying bacteria known to grow in this manner.

Motile cells grow in a different fashion than gliding cells. Most motile species possess flagella that propel them through their environment. On agar, motile cells leave behind a visible trail as they move outward from the original colony. Microbiologists formulate semisolid agar media, which help makes the track through the agar easier to see.

Pinpoint colonies are extremely small colonies that appear on agar as no more than the size of the period at the end of this sentence. Pinpoint colonies may arise from the following three causes: (1) agar formulation or incubation conditions are not optimal, (2) a species normally produces very small colonies on agar, or (3) the pinpoint colonies are contaminants in a culture of more actively growing microorganisms. Microbiologists must determine the causes of pinpoint colonies for the purpose of either eliminating the contaminant or enhancing the growth of the microorganism forming this type of colony.

Smooth-Rough Variation

Smooth-rough variation (S → R variation) in bacterial colonies plays a role in molecular biology for the purpose of following certain genetic traits. For instance, some bacteria form smooth, glossy colonies, when they are first grown and then develop rough or dull colonies, in all subsequent inoculations. This shift from smooth to rough colony appearance has been connected to a number of characteristics within the bacteria's genotype (its genetic makeup), as follows:

- pathogen virulence, usually a reduction
- presence of a mutation
- loss of specific antigens
- changed susceptibility to antibacterial substances
- increased susceptibility to phagocytosis, which is the ingestion of particles by immune system cells
- increased capacity to agglutinate
- altered susceptibility to bacteriocins or bacteriophages

The British geneticist Frederick Griffith (ca. 1879–1941) established, in 1928, the value of studying the variation in bacteria from smooth to rough colonies. Griffith used two strains of the pneumonia-causing bacteria *Streptococcus pneumoniae*, one strain that normally produced a smooth and another that produced a rough colony. Griffith showed that the virulent smooth strain could transfer its virulence to the normally nonvirulent rough strain. Griffith demonstrated that the virulence—the ability of a pathogen to cause disease—was part of the bacteria's genetic makeup. Traits of one type of bacteria could additionally be transferred to other bacteria by growing the two different types together. Griffith called the unknown material a transforming principle, which the biologists Oswald Avery (1877–1955), Colin MacLeod (1909–72), and Maclyn McCarty (1911–2005) showed, 16 years later, to be deoxyribonucleic acid (DNA). The finding laid the foundation for study of the heredity of traits and the development of antibiotic resistance. Griffith based his groundbreaking discovery completely on the appearance of the *S. pneumoniae* colonies.

The ability to grow microbial colonies serves as a cornerstone in almost all microbiology studies. Colonies also can be used as one of many tools for identifying bacteria, although colony morphology cannot be the sole means of an accurate identification. The study of colonies has led to advances in morphology, genetics, nutrient requirements, and medicine.

See also AGAR; BACTERIOCIN; CULTURE; IDENTIFICATION; MORPHOLOGY; SERIAL DILUTION.

Further Reading

American Society for Microbiology. MicrobeLibrary.org. Available online. URL: www.microbelibrary.org/ASMOnly/details.asp?id=2566&Lang=. Accessed March 24, 2009.

Cappuccino, J., and N. Sherman. *Microbiology: A Laboratory Manual,* 8th ed. San Francisco: Benjamin Cummings, 2008.

Gerhardt, Philipp, ed. *Manual of Methods for General Bacteriology.* Washington, D.C.: American Society for Microbiology Press, 1981.

Prescott, Lansing M., John P. Harley, and Donald A. Klein. *Microbiology,* 6th ed. New York: McGraw-Hill, 2005.

common cold Common colds are acute infections of the upper respiratory tract in humans, caused by a variety of viruses. Acute infections have a rapid onset and run a short course, from a few days to no more than two weeks. The common cold is the most prevalent disease in the human population.

The viruses that can cause colds make up an almost endless variety of combinations. Medical literature has suggested that 100–200 different types or subtypes of viruses can cause colds, and these different viruses form hundreds of varying combinations. The human immune system has difficulty protecting the body from every possible mixture of cold viruses that a person might catch, so people do not develop full immunity to colds in their lifetime. Nevertheless, most people acquire a low level of immunity to colds as they age.

Almost every person is familiar with the symptoms and outcome of colds. A normal healthy person will have several dozen colds in a lifetime. Young children (two to seven years old) tend to have six to eight colds a year; adults usually have two to four colds a year. Colds are self-limited illnesses, meaning that a person will recover fully from a cold without treatment. Because of the various combinations of cold viruses, the severity and the duration of colds vary from person to person, and a single individual also experiences cold of different severity in a lifetime.

The treatments available for the common cold act against the symptoms of a cold rather than the infection, although researchers have been testing a variety of treatments on volunteer subjects for the purpose of targeting the virus rather than simply treating the symptoms. Cold symptoms are among the most familiar disease symptoms in all of medicine. On occasion, people confuse colds with influenza (flu). Viruses cause both colds and flu, yet the diseases are caused by different viruses, and the symptoms, though they can be similar, are not the same, as shown in the table below.

The incidence of colds in populations who live in temperate climates increases in colder weather. No

Symptoms of the Common Cold Compared with Influenza

	Common Cold	Influenza
cause	mainly rhinovirus, coronavirus, and parainfluenza virus	influenza A, influenza B
incubation period	12–72 hours	24–72 hours
duration	2 days–2 weeks	acute course lasts 1 week; symptoms persist for several weeks
main symptoms	sneezing, nasal congestion, runny nose, coughing, sore throat	fever, chill, muscle ache, prostration
lesser symptoms	watery eyes, headache, chills, malaise, little or no fever	coughing, sore throat, malaise, dizziness
main mode of transmission	direct contact	inhalation of aerosols
infectious dose	1–30 viruses	depends on strain; often less than 10 viruses
target organ	upper respiratory tract	upper and lower respiratory tract
severity	rarely causes death	causes 20,000–70,000 deaths annually in the United States

study has definitely proven why this occurs. The main theories on cold-weather incidence usually relate to two aspects of transmission. First, people congregate indoors more in cold weather, and that may aid the spread of cold viruses. Second, physiological changes in people during cold weather may contribute to susceptibility to cold viruses.

CAUSES AND TREATMENT OF THE COMMON COLD

The main viruses implicated in causing the common cold are rhinoviruses, coronaviruses, and parainfluenza viruses, but other viruses have been included as possible contributors to cold symptoms: adenoviruses, coxsackieviruses, echoviruses, influenza viruses, or respiratory syncytial viruses. The rhinoviruses have been proposed to be the main cause of at least 50 percent of all colds. Coronavirus, which also causes the sometimes-fatal severe acute respiratory syndrome (SARS), may account for 30–35 percent of colds. Because many viruses typically combine to create a single cold outbreak, the rhinovirus, the coronavirus, or any other single virus is unlikely to act alone in causing infection leading to a common cold.

Treatments against the breadth of viruses that cause colds have been very limited. The rhinovirus alone has more than 100 different serotypes, which are subtypes of a virus characterized by compounds present on the virus's outer surface. A vaccine against so many different viruses and types of individual viruses would be impractical to manufacture, especially since the combination of cold viruses changes from season to season. Vaccine manufacturers furthermore have had limited success in developing vaccines that protect the mucous membranes of the body, where cold viruses attack, compared with other tissues in the body.

Researchers have tried to develop cold treatments by devising molecules to interfere with the attachment between the virus and the cells in the nasal passages. Blocking this step prevents the viruses from invading the body and beginning replicating. These treatments have had little success, however, so most cold therapies continue to treat the symptoms rather than the virus in the hope of making a patient more comfortable as the cold runs its course. The following familiar adage remains the best approach to dealing with colds, "An untreated cold will run its normal course to recovery in a week, whereas the treatment will take seven days." The following treatments are commonly used for cold symptoms:

- antihistamines that lessen the effects of the body's immune reaction to the virus
- decongestants that help with breathing and break up mucus buildup
- nonsteroidal anti-inflammatory medicines to ease the symptoms of the immune response and fight headache

The best cold treatments may actually be in the realm of cold prevention. People can limit the number of colds they catch in a year by understanding how cold viruses move from person to person and how the viruses attack the body. Researchers, meanwhile, have continued to collect data, since the 1970s, on the types of viruses that cause most colds in an effort to find an elusive cure.

In 2009, medical laboratories began reporting that they had determined the genetic codes—the sequence of genes in deoxyribonucleic acid (DNA)—of 99 viruses known to cause nasal infections. The physician Stephen B. Liggett, director of the University of Maryland's cardiopulmonary genomics program, told *HealthDay,* in early 2009, "There has been brilliant work done trying to synthesize compounds against the common cold. But we have not been working with a full knowledge of the genetics of rhinoviruses. Now that we have a full complement of known ones, we see there are subfamilies of rhinoviruses clustering together. The hope is that there could be a drug for each subfamily." Despite this breakthrough, cold viruses remain a moving target to try to control. First, rhinoviruses mutate over time. Though the mutation rates are not rapid, they occur fast enough to stay ahead of new drugs developed against nonmutated viruses. Second, two different strains (related but slightly different versions) of rhinoviruses can exchange genetic material, thereby creating a new virus with new characteristics.

INFECTION BY THE COLD VIRUS

Cold viruses infect the body by associating with the outer nasal passages. Rhinoviruses, in particular, thrive at temperatures slightly lower than normal body temperature, which can be expected to occur in the outer nasal passages. A single cold virus deposited on the nasal mucous membranes is capable of beginning the events leading to a cold. When cold viruses contact the cells lining the nasal passages, the passages themselves transport the viruses deeper into the respiratory system by ciliary action, which is the movement of a particle in a one-way direction by the action of tiny hairlike appendages. The nose's ciliary action pushes viruses to the area of the adenoid lymph gland in the back of the throat. The viruses then attach to the cells in the adenoid area. All of this takes place in a period of 10–15 minutes after the virus first enters the nasal passage.

Cold viruses in the adenoid area of the throat infect a small percentage of all the nasal cells there. Proteins on the outer surface of the cold virus bind to receptor sites on the outside of nasal cells. These receptor sites, called intercellular adhesion molecule-1, allow the permanent binding and invasion of the virus's contents of the interior of the cell. Within 30 minutes of the first virus's entering the nasal passage, viral particles inside the cells begin to replicate and shed new viral particles, which invade other healthy cells.

The body's dramatic reaction to the viruses' proliferation sets in motion a series of events that lead to the familiar and uncomfortable cold symptoms that a person will suffer. The typical response of the body to the cold virus invasion takes the following steps:

1. The body activates immune and nervous system responses.
2. The immune system releases a group of compounds called inflammatory mediators: histamine, kinins, interleukins, and prostaglandins.
3. The mediator compounds cause dilation of blood vessels near the infection site.
4. Mediator-induced leakage of blood vessels and mucus secretions begins to occur.
5. Sneezing and coughing reflexes become activated.
6. The mediators also stimulate pain nerve fibers.

Much of the response to a cold results from histamine, a compound derived from the amino acid histidine and released from immune system cells called *mast cells*. Histamine causes dilation of blood vessels, smooth muscle constriction in the respiratory tract, tissue swelling, mucus production, and itching. Many of these activities are identical to the symptoms in allergies, and, in fact, allergies and colds can appear alike in some individuals.

No drug has been discovered that can attack the cold virus and cure a cold. Cold sufferers take medications to alleviate the worst symptoms and neutralize much of the histamine activity. In general, however, physicians usually recommend bed rest, plenty of fluids, and time to allow a cold to run its normal course in the body. An important aspect of combating colds is the action a person takes to prevent the further spread of a cold. The principal way people contract colds is by inoculating themselves with the cold virus by touching a surface or a person containing the virus and then transferring the virus to the nose or the outer nasal passages.

TRANSMITTING COLDS

The transmission of common colds shows how infectious diseases, in general, can spread through a population. A few simple actions break the transmission of cold viruses from a cold sufferer to a healthy person. Cold sufferers should also understand that they can spread active cold viruses for up to two weeks after the onset of cold symptoms.

Cold viruses can survive up to three hours on a person's skin or on inanimate objects such as stair railings, doorknobs, or telephones. The most likely routes of transmission are by direct touch or by aerosols, which one created by sneezing and coughing and travel through the air. The Centers for Disease Control and Prevention (CDC), the National Institute of Allergy and Infectious Diseases (NIAID) of the National Institutes of Health, and the Mayo Clinic are three of many prominent health organizations that suggest colds spread by either mechanism.

Considering the modes of cold transmission, doctors recommend a number of actions that people can take to stop spreading colds to others or to prevent receiving colds from others. The following preventative measures decrease the chances of transmitting colds:

- Avoid being close to people who have a cold.
- If suffering from a cold, avoid being near healthy people to prevent the spread of colds.
- Keep hands away from eyes, nose, and mouth.
- Cover mouth and nose when sneezing or coughing.
- Wash hands frequently.
- Use a disinfectant on hard surfaces suspected of being contaminated with cold viruses.

The advice listed here resembles the advice that medical professionals give any person on the subject of good hygiene. Hygiene, transmission of infectious agents, and factors of immunity and susceptibility all contribute to the incidence of colds in a population. Stopping the spread of colds has a much larger impact on society and on the economy than many people realize.

COLDS AND THE ECONOMY

The CDC estimates that the U.S. population has one billion colds each year. This cold incidence may lead to as many as 70 million workdays lost each year; each worker who has a cold will lose an average of 8.7 work hours per cold. Even workers who go to work with a cold do not function at their best and cause an undetermined amount of lost efficiency in the workplace. All told, a study published in the *Journal of Occupational and Environmental Medicine,* in 2002, found that the United States economy loses $25 billion through lower worker productivity to colds each year. The article gave the following breakdown, in 2002 dollars, for the way this loss of productivity occurs:

- $8 billion due to worker absenteeism
- $16.6 billion due to on-the-job productivity loss
- $ 230 billion due to caregiver absenteeism

Schools also experience less efficient learning environments affected by colds. The CDC estimated that 22 million schooldays are lost each year in the United States because of colds. Each year in cold weather months, various school districts have high absentee rates, of up to 30 percent, due to colds. In a few serious instances, districts have closed for one to two days to allow students to rest and to keep sick students away from healthy students.

Common colds spread through day care centers in a manner similar to their spread in schools and workplaces. Colds may be more of a concern in day care of very young children, however, because young children tend to touch their faces, share toys, and put objects into their mouths. All of these activities increase the chance of spreading colds. Good cold-prevention hygiene in day care facilities should include the following steps:

- Adults and older children caring for young children should wash hands frequently.
- Surfaces and toys should be cleaned and disinfected daily or more often.
- The facility should use an effective ventilation system.
- Avoid overcrowding during play, nap time, and mealtime.
- Teach children to cover the mouth and nose when sneezing or coughing and to use a tissue to wipe nasal secretions.
- Keep children well nourished and hydrated.

Good common sense has always played a role in cold prevention.

ANIMALS AND COLDS

Animals seen in veterinary practices may have illnesses similar to human colds. The cold viruses that infect domestic animals do not infect humans, however, and the human cold viruses probably do not transmit to other species.

Horses have been diagnosed with colds or flu when presenting a fever, cough, and nasal secretions. Most of these horse "colds" are probably equine influenza. Whether caused by the equine influenza virus or by other viruses unofficially called colds, these infections are highly contagious in horses and must be treated immediately. Horses receive nonsteroidal anti-inflammatory medicines and antibiotics to prevent secondary infections from bacteria.

Veterinarians diagnose colds in dogs and cats that develop various respiratory infections. In dogs, colds may be caused by parainfluenza virus or other viruses associated with more specific symptoms, such as the distemper virus, adenovirus, or kennel cough–causing *Bordetella* bacteria. Cats also contract upper respiratory infections with the same cold symptoms humans endure: sneezing, runny nose, coughing, and watery eyes. These conditions may be either viral or bacterial in nature, so they may be treated with an antiviral drug or an antibiotic, respectively. Dog colds and cat colds are both contagious to members of their species.

The common cold has become a frustrating ailment for individuals and for businesses, yet it does not receive the interest or concern among the public that many other diseases receive. Because of the large amount of lost productive time that colds cause, the common cold is an important area of research in virus studies and in health care.

See also HYGIENE; INFLUENZA; TRANSMISSION; VIRUS.

Further Reading

BBC News. "Echinacea Can 'Prevent a Cold.'" June 25, 2007. Available online. URL: news.bbc.co.uk/2/hi/health/6231190.stm. Accessed February 13, 2009.

Bramley, Thomas, Debra Lerner, and Matthew Sarnes. "Productivity Losses Related to the Common Cold." *Journal of Occupational and Environmental Medicine* 44, no. 9 (2002): 822–829. Available online. URL: www.joem.org/pt/re/joem/abstract.00043764-200209000-00004.htm;jsessionid=JV7ScJHnLRWZyVhWJyfL9Yhv3DyQsY6LSGHGVBw1lpVkLPTMLmRh!1204955331!181195628!8091!-1. Accessed February 13, 2009.

Commoncold, Inc. Available online. URL: www.commoncold.org/index.htm. Accessed February 12, 2009.

Kliff, Sarah. "Can Vitamin C Cure Colds?" *Newsweek,* 15 November 2007. Available online. URL: www.newsweek.com/id/70628. Accessed February 13, 2009.

Mayo Clinic. Available online. URL: www.mayoclinic.com. Accessed February 13, 2009.

U.S. National Library of Medicine. "Genetic Code of Common Cold Cracked." *HealthDay,* 12 February 2009. Available online. URL: www.nlm.nih.gov/medlineplus/news/fullstory_80335.html. Accessed February 13, 2009.

contamination Contamination in microbiology is the presence of any unwanted microorganism or nonliving thing. Contamination from microorganisms or pollen is often referred to as *biocontamination.* Nonbiological entities that microbiologists define as contamination are dust, dirt, fibers, dander, and hair. A vast majority of techniques used in microbiology have been developed for the purpose of preventing contamination.

Contamination prevention is the foundation of sterilization, disinfection, and aseptic techniques. Microbiology could not have progressed as a science without the methods microbiologists have devised to prevent contamination. Many industries also depend on the ability to keep microorganisms out of their products. The following industries use one or more methods for preventing contamination: drug makers, food and beverage processors and producers, personal care product manufacturers, paint manufacturers, and drinking water treatment plants. Although any type of microorganism can contaminate, most contamination prevention focuses on bacteria or fungi.

Despite many obvious reasons why industries do not want dangerous microorganisms in their products, accidental contamination has led to important discoveries. Early civilizations probably discovered new foods from materials that had spoiled. Most spoiled foods emit strong odors and have changes in color and consistency that make them inedible. But, in some instances, the spoilage caused by a contaminant changed the qualities of the food into a different type of food that was palatable and lasted longer when stored. Fermented foods such as beer, wine, sauerkraut, and soy sauce began as accidental contaminations. Without instruments to make possible the observation of microscopic things, ancient civilizations learned about contamination by intuition, though they did not understand the existence of microorganisms.

In the late 1500s to early 1600s, glassmakers began experimenting with the capabilities of lenses. Arranging more than one lens in sequence allowed curious individuals to view microscopic particles. In the 1670s, a low-level official of the town of Delft in the Netherlands studied lenses as a hobby in order to inspect microscopic bits in water, soil, beer, and other materials. With this rudimentary microscope, Antoni van Leeuwenhoek (1632–1723) revealed the microscopic world. Advances in lenses and microscopes enabled later scientists to study and draw conclusions on the presence or absence of microorganisms in food.

The invention of the microscope also allowed scientists to speculate on the activities of tiny organisms in liquids and in moist substances. As early as 1546, the Italian physician Girolamo Fracastoro (1478–1553) had wondered about the presence of invisible agents involved in infection. He called the objects "contagions" but could not make a detailed study of them without a microscope. Fracastoro's musing about the presence of unwanted contagions in infected patients represented the first substantive connection between microorganisms and disease: the germ theory of disease. Equally important, Fracastoro laid the groundwork for understanding the relationship between contaminated wounds and infection.

More than 300 years after the first proposal of *contamination*—this term would not come into use until much later—the English physician Joseph Lister (1827–1912) applied the germ theory and the concept of contamination prevention to his medical procedures. Lister's contemporaries noticed that washing their hands before performing surgery decreased the incidence of infection in their surgery patients. Lister adopted an additional precaution against infection beyond mere hand washing. In 1867, Lister reported in the *British Medical Journal,* "The material which I have employed is carbolic or phenic acid [phenol], a volatile organic compound which appears to exercise a peculiarly destructive influence upon low forms of life, and hence is the most powerful antiseptic with which we are at present acquainted." By using carbolic acid to clean surgical incisions, Lister reduced the incidences of gangrene and other infections that ran rampant in hospitals at the time. Eventually, the medical community accepted Lister's approach to contamination prevention. Medical procedures began to include preoperative and postoperative use of antiseptics.

Another milestone in the history of microbiology occurred in 1929, when the Scottish physician Alexander Fleming (1881–1955) noticed a contamination in the bacterial cultures in his laboratory. "While working with *Staphylococcus* variants a number of culture-plates were set aside on the laboratory bench and examined from time to time," Fleming wrote in a medical journal article. "In the examinations these plates were necessarily exposed to the air and they became contaminated with various micro-organisms. It was noticed that around a large colony of a contaminating mould the staphylococcus colonies became transparent and were obviously undergoing lysis." Lysis is the disintegration of microbial cells; Fleming used this observation to deduce that the contaminating mold had killed the bacteria. Fleming had discovered the antibiotic penicillin, made by the mold *Penicillium,* and the age of antibiotics began.

LABORATORY TECHNIQUES FOR PREVENTING CONTAMINATION

Three main areas in modern microbiology have been developed for preventing contamination: aseptic techniques, sterilization, and disinfection. Aseptic techniques encompass all the procedures that microbiologists use to reduce the chance of a contaminant organism's entering both pure cultures and sterile media. A pure culture is a population of microbial cells that are all identical because they originated from a single parent cell. Research and industrial microbiology depend on pure cultures because a population that has only one type of microorganism allows microbiologists to study the species's characteristics. The presence of a contaminating microorganism alters the growth of any microorganism being studied.

Sterile media represent the starting point for all aseptic procedures. By starting with sterile growth medium, a microbiologist can inoculate the desired microorganism to the medium, confident that no contaminants will alter the growth conditions during incubation. Aseptic techniques and sterilization, therefore, complement each other in contamination prevention.

Aseptic techniques include all methods for keeping pure cultures or sterile media free of unwanted microorganisms. The aseptic handling of cultures and media involves the use of sterilized equipment for transferring culture samples to fresh medium, proper opening and closing of culture vessels, and constant monitoring for the potential presence of a contaminant. Sterilization involves procedures for killing all microorganisms in media or on equipment. Sterilization methods most commonly use heat, gas, or irradiation. Heat sterilization may be either steam heating under high pressure or dry heating in an oven to sterilize media or equipment. Gas sterilization mainly uses ethylene oxide to sterilize equipment. Irradiation comprises exposure of equipment to either ultraviolet light or gamma rays. A successful sterilization renders solid or liquid media, reagents and solutions, and equipment free of all contamination.

Disinfection supports aseptic techniques by killing potential contaminants in a microbiologist's workspace. By disinfecting a laboratory benchtop or other workspace, microbiologists greatly reduce the chance that a microorganism will enter a pure culture or sterile medium. Aseptic techniques usually call for chemical disinfection before and after all handling of cultures and media. Although disinfection cannot assure that contamination will not happen, it works in combination with aseptic techniques and sterilization to make contamination a rare occurrence. The table summarizes laboratory activities that reduce the incidence of contamination. Many of these techniques also work in industrial settings.

INDUSTRY METHODS IN CONTAMINATION PREVENTION

Some industries have a great need to assure that their products will be free of all contamination in order to protect customers from infection. The makers of sterile injectable drugs, foods, beverages, eye care products, and personal care products such as lotions and shampoos must be able to provide products that are safe for consumers. Other industries want to prevent contamination for the purposes of protecting the quality of their product and limiting any decomposition by microbial activity. Manufacturers of paints, fabrics, plastics, leather, and wood products use preservatives in their products to reduce this type of contamination.

Industries start with very basic cleaning methods to reduce the presence of dirt, molds, and bacteria. Some products require much more stringent preventative measures than simple cleaning or even disinfecting. For example, a company that makes shampoo can tolerate a small amount of microorganisms in its product because the preservative in the shampoo will eventually reduce and kill any contaminants. By contrast, a manufacturer of an injectable vaccine must take all possible precautions to ensure that each bottle of vaccine contains no microorganisms. People use shampoos externally on the skin, where many microorganisms already reside; a drug injected into the body would cause a much more critical health problem if it contained a microorganism. The table on page 188 lists the actions taken by various industries to prevent contamination, not only by microorganisms, but also by dusts, fibers, dirt, dander, and pollen. The activities are listed in the table from the least stringent approach, in general, to the most stringent approach in preventing contamination.

All microbiology research and industrial laboratories use some form of quality control to assure that their practices have adequately prevented contamination. Quality control is the active assurance of the quality and integrity of a product. Quality control professionals in industry also take responsibility for the following two tasks: ensuring that no contaminated product is sold to consumers and ensuring that the company takes immediate corrective actions to prevent future contaminations. In the table, HACCP and GMP make extensive use of quality control in all their procedures. HACCP is a program designed to predict where contamination might occur and implement actions to prevent it. GMP is a program designed to assure the FDA that a product has been made to specific standards in order to be safe for consumers to use.

THE ECONOMIC COSTS OF CONTAMINATION

Industries such as food and drug production have long established safeguards against contaminants in their

Laboratory Techniques for Preventing Contamination

Technique	Description
aseptic techniques	the use of sterile pipets, inoculating loops, vessels and closures, reagents, and media for maintaining the purity of the desired microbial culture
disinfection	the use of a chemical biocide before and after all microbiology procedures for the purpose of eliminating unwanted microorganisms from the immediate workspace
irradiation	exposure of workspaces or equipment to be used in aseptic techniques to ultraviolet or gamma irradiation for a period that kills all microorganisms
quality control	monitoring aseptic conditions by checking a pure culture under a microscope and incubating sterilized media before use to assure no growth occurs
sanitization	the use of a chemical biocide before and after all microbiology procedures for the purpose of reducing the number of unwanted microorganisms in the immediate workspace
sterilization	the use of heat, gas, or irradiation to eliminate all unwanted life, including bacterial spores, from media and equipment

products, and these industries continue to find ways to improve the safety of the items sold to consumers. The incidences of contamination remain, nevertheless, fairly common in products made in and outside the United States and can have devastating effects on business. Michael Cox, a manager in the biotechnology industry described to *Genetic Engineering News* the challenge of staying ahead of contaminants in an industry setting: "If you have one bacterium in a reactor, dividing every 20 minutes, versus cells that divide every 24 hours, which one would you bet on?" Cox highlighted the main skill needed in contamination prevention: vigilance against hidden microorganisms that can rapidly grow to very large numbers in seemingly clean environments.

Contaminated products cost industries in two ways. First, the U.S. Food and Drug Administration (FDA) requires that products contaminated with illness-causing microorganisms be recalled and destroyed and all products in the same production lot be held from the market. This costs a company millions of dollars in lost profits. Second, when the public hears of a contamination problem, they may avoid any similar products, even those products are safe. Industry again loses when this occurs. If contamination continues to be seen as an ongoing problem in a certain industry, the stock price of companies in that industry may decline. While recalls and slow sales cost millions of dollars, an entire industry may lose billions of dollars because of contamination. The industry reporter Katherine Glover explained, in a 2009 news release, "The huge food contamination scares in recent years have changed things. They tend to hurt everyone in the affected industries, including companies whose products are not contaminated." The mere perception of a health threat can be almost as damaging to business as a true case of contamination.

In 2009, after an outbreak of illnesses and deaths due to contaminated peanut products, 10 major food industry associations sent a joint letter to the U.S. Congress calling for stricter regulations on food safety. This unprecedented action—industries seldom ask for *more* regulations—alerted the government and the public to the seriousness of contamination and the difficulty many industries have in controlling it. Even so, some businesses try to cut corners in food safety. In late 2009, an *Escherichia coli* outbreak occurred because a large producer of ground beef in New York had decided to streamline testing its product for contamination because it had not found any microorganisms in previous tests. Arnold Gerson, a customer in Massachusetts who had bought some of the tainted beef, said to the *New York Times,* "When you go to a market and pull things off the shelf, you expect things will be safe and O.K. So we've got to be very, very careful." Streamlining or taking shortcuts in food safety testing endangers the health of all consumers.

In 2004, the drug maker Chiron Corporation faced adversity that originated with the bacterial contamination of its Fluvirin influenza vaccine. The FDA had discovered that 4.5 million doses of the 46 million doses Chiron had prepared for the coming flu season contained *Serratia marcescens,* a red-pigmented species of bacteria that often occurs in water. The biotechnology reporter Angelo DePalma wrote of the effects on Chiron, "Chiron shares sank nearly 9 percent on the day it announced it would destroy millions of doses of vaccine and delay shipping uncontaminated product." FDA inspectors worked with Chiron representatives and the manufacturing plant in the United Kingdom where the doses had been made to fix the problems that had caused the contamination. Even though the company's corrective actions solved the problem and averted a health crisis, the FDA stated in its chronology of the event, "Although Chiron's retesting of the unaffected lots of vaccine has been negative for

Industry Methods for Preventing Contamination

Method	Description	Example Industries
cleaning	removal of visible dirt with a soap cleaning solution	restaurants, hotels
sanitization	reduction in the amount of microorganisms to or below safe levels, using a sanitizer product	restaurants, food processing, personal care product manufacturing
use of preservatives	inclusion of a chemical in a product for the purpose of killing any potential contaminants	food processing, personal care product and eye care product manufacturing, paints, wood products
disinfection	elimination of all microorganisms other than bacterial spores	hospitals, nursing homes, day care centers, medical offices
HACCP	identification and mapping of all potential sources of contamination and design of specific prevention measures for those sources	food production, drug manufacturing
sterilization	elimination of all microorganisms, including bacterial spores	medical instrument manufacturers, health care facilities
good manufacturing practices (GMP)	adherence to specific FDA regulations for preventing and eliminating contamination	drug manufacturing, food production, medical device manufacturing
use of clean rooms	use of specialized work areas that are designed to prevent all contamination from entering a product	manufacturers of injectable drugs

Note: HACCP = hazard analysis and critical control point

contamination, FDA has determined that it cannot adequately assure the sterility of these lots to our safety standards." In summary, even experts at the FDA, who understand industry procedures, will lose confidence in an operation or a company that does not take the proper steps to prevent contamination.

See also ASEPTIC TECHNIQUE; DISINFECTION; FLEMING, ALEXANDER; GERM THEORY; LISTER, JOSEPH; SANITIZATION; STERILIZATION.

Further Reading

DePalma, Angelo. "Maintaining Biocontamination Control." *Genetic Engineering News,* 1 June 2006.

Fleming, Alexander. "On the Antibacterial Action of Cultures of a *Penicillium,* with Special Reference to Their Use in the Isolation of *B. influenzae.*" *British Journal of Experimental Pathology* 10 (1929): 226–236.

Glover, Katherine. "Food Industry: Regulate Us Please!" January 28, 2009. Available online. URL: industry.bnet.com/food/1000402/food-industry-regulate-us-please. Accessed February 15, 2009.

Lister, Joseph. "On the Antiseptic Principle in the Practice of Surgery." *British Medical Journal* 2 (1867): 246–248.

Moss, Michael. "*E. coli* Outbreak Traced to Company That Halted Testing of Ground Beef Trimmings." *New York Times,* 12 November 2009. Available online. URL: http://www.nytimes.com/2009/11/13/us/13ecoli.html?_r=1&scp= 3&sq=food+contamination&st=nyt. Accessed December 4, 2009.

U.S. Food and Drug Administration. "2004 Chiron Flu Vaccine Chronology." October 16, 2004. Available online. URL: www.fda.gov/oc/opacom/hottopics/chronology 1016.html. Accessed February 15, 2009.

continuous culture Continuous culture (also *continuous flow culture*) is a method of growing microorganisms in liquid medium so that the cells grow in a continuous logarithmic manner. The culture method accomplishes this with a design that allows a constant inflow of fresh nutrients in sterile liquid medium and a constant outflow of used medium (called *spent medium*) containing live and dead cells, wastes, and end products.

Continuous cultures offer two benefits for research and for industry: the constant logarithmic growth phase and operation of the system for days, weeks, or even months. Noncontinuous cultures, called *batch cultures,* do not offer either of these advantages. Batch cultures in test tubes or flasks contain microorganisms that follow a sequence of events that constitute the growth curve: lag phase, logarithmic phase, static phase, and death phase. For most aerobic bacteria, the growth curve takes place in one to three days; it lasts two days to two weeks (in anaerobic bacteria). By contrast, a microbiologist can control a continuous culture so that the majority of cells pass through a brief lag phase, when initial growth is slow, and then the entire culture remains in logarithmic growth for much longer periods.

Continuous cultures systems are also called *open cultures* or open-system cultures because the conditions inside the vessel can be varied at will by a microbiologist. Industries that employ very large bioreactors to grow several thousands of gallons of culture often refer to continuous culture as *mass culture.*

Continuous cultures have advantages over batch cultures because continuous growth mimics a microorganism's growth in nature. Microbiologists also can control the conditions in the open culture by adjusting two parameters: flow rate and growth factor supply. Flow rate is the speed in which medium enters and exits the bioreactor. Growth factors commonly adjusted in continuous cultures are:

- carbon source
- nitrogen source
- vitamins and cofactors
- carbon dioxide
- oxygen
- pH
- growth temperature
- cell density
- end product buildup

Despite the advantages of continuous cultures in studying the metabolism of a microorganism, these systems require more work than batch cultures. Continuous culture techniques have also fallen victim to the trend in microbiology away from studies on whole, living cells and toward molecular studies on genes. The Molecular biologists Paul A. Hoskisson and Glyn Hobbs wrote in the journal *Microbiology,* in 2005, "The heyday of continuous culture was in the 1960s, when its versatility and reproducibility were used to address fundamental problems in diverse microbial fields such as biochemistry, ecology, genetics and physiology. The advent of molecular genetics in the 1970s and 1980s led to a decline in the popularity of continuous culture as a standard laboratory tool." The authors explained that continuous culture methods have been revived because of the large amount of data that a single study can generate. The heart of microbiology will always be the manner in which living cells use nutrients and secrete products. For this reason, continuous culture remains valuable.

Three key components of the cell growth within a continuous culture are substrate, primary metabolites, and secondary metabolites. A substrate is the material that microbial enzyme systems use for the microorganism's energy and growth. Carbon sources commonly play the role of substrate in continuous cultures. A primary metabolite is any substance produced by a microorganism by growth-associated metabolism. In other words, the microorganism would not be able to grow without the reactions that produce primary metabolites as end products. Ethanol produced from yeast fermentations of fruit juices is an example of a primary metabolite. A secondary metabolite is a substance produced by a microorganism that is not associated with growth; the microorganism can grow whether it does or does not produce this substance. An antibiotic produced by a mold is an example of a secondary metabolite.

Industries such as biotechnology use continuous cultures to produce compounds that are useful for various purposes. For instance, an enzyme company might produce amylase, the enzyme that digests starch, from cultures of *Bacillus subtilis* bacteria, and, then, sell the amylase to breweries for making beer. A different biotechnology company could be in the business of producing a drug such as the blood clot treatment streptokinase made by *Streptococcus* bacteria. Industries that use continuous cultures for making large amounts of similar products or raw materials for other industries regulate all aspects of continuous cultures to optimize the amount of product, its purity, and the time required to make it.

Most large-scale industrial continuous cultures began with studies in a laboratory using a much smaller culture vessel called a *chemostat.* Using a chemostat to control growth of a continuous culture in a laboratory, a microbiologist studies the aspects

A chemostat supports continuous growth. An outer jacket of circulating warm water incubates the culture in the inner vessel. An electronically controlled stir bar outside the culture (white rod) rotates, thus rotating another stir bar within the culture (not visible). The machine at left adjusts flow rate and connects to probes for monitoring/adjusting temperature, pH, and oxygen. Other ports and jars enable sample collections. ***(Dr. Pak-Lam Yu, School of Engineering and Advanced Technology, Massey University)***

of the way a microorganism grows, the nutrients it needs, the products it secretes, and all the conditions that promote growth in increasingly larger systems. To accomplish these tasks, a microbiologist must study the growth kinetics of a microorganism in continuous culture.

GROWTH KINETICS

Growth kinetics refers to the constantly changing conditions in a growing microbial culture. Two critical factors in continuous culture growth kinetics are substrate concentration and specific growth rate. Substrate concentration is the amount of substrate supplied by the fresh medium that constantly enters the system. For example, a glucose concentration of 2 grams/liter represents a substrate concentration that might be used in a given continuous culture. Specific growth rate equals the amount of times the culture's entire volume is replaced in one hour. A specific growth rate of 0.5 per hour, or 0.5 hr^{-1}, means that one-half of the culture volume was replaced with fresh medium in an hour's time.

The science that describes continuous cultures relates to a complex subject called *enzyme kinetics*. Enzyme kinetics uses various mathematical equations to relate reaction rate of an enzyme to the concentration of the enzyme's substrate and the amount of enzyme present. In microbiology, similar equations relate microbial growth rate to the substrate concentration and the microbial cell mass inside the culture vessel. A key concept in continuous culture allows microbiologists to study these relationships and use them to optimize growth and the production of primary or secondary metabolites: growth rate equals flow rate.

The growth rate of cells in a continuous culture relates directly to the flow rate of medium into and out of the culture vessel. (Inflow and outflow are equal in almost all continuous cultures in use today.) This flow rate has been called dilution rate (D) because fresh incoming medium acts to dilute the culture's microbial cells. At fast D, the cells receive ample new substrate to digest, so the cells grow fast and keep up with dilution rate. At slow D, the cells are starved for substrate and quickly digest any new substrate that enters the culture. The microbial culture adjusts its growth rate to a slower rate in accordance with the slow substrate supply. Microorganisms naturally adjust this way for two reasons. First, at fast D, the microbial cells can grow to their maximal rate and at their peak efficiency. Second, at slow D, cells must slow their growth, or else there will be too many cells and not enough substrate to keep them all alive. The kinetics of a continuous culture therefore includes the factors described in the table below.

Microbiologists adjust or monitor all of the factors of growth kinetics with the goal of achieving a steady-state condition. In a steady-state condition, a continuous culture's substrate level and D have been adjusted for the best growth so that there is neither too much substrate nor too little. By adjusting conditions that cause some of the problems listed in the table, microbiologists can restore a culture to a steady state.

Factors in Continuous Culture Kinetics

Factor	Description
dilution rate, D	volume turnover per hour
flow rate	volume of medium entering (or exiting) the system per hour
substrate concentration	amount of substrate as weight per volume in the fresh medium
doubling time (generation time)	rate at which cells replicate by binary fission
cell yield	mass of microbial cells produced per volume of culture
steady state	cell concentration remains proportional to D or flow rate
washout	cell doubling time cannot keep up with a fast flow rate, and the microorganisms eventually disappear from the culture
limiting nutrient	the nutrient present at a low concentration so that growth cannot continue without it, even when all other nutrients are at adequate levels

THE CHEMOSTAT AND THE TURBIDOSTAT

Laboratory-scale continuous cultures rarely reach more than a few liters. These cultures grow in two types of vessels: a chemostat and a turbidostat. A chemostat contains fresh sterile medium that enters the vessel at the same rate at which spent medium flows out. The main controlling factors in running a chemostat are D and substrate concentration. A turbidostat is a vessel containing a device that optically measures the absorbance of light by the culture, which indicates cell density. The table summarizes the main differences between these two pieces of equipment.

In 1889, the Russian microbiologist Sergei Winogradsky (1856–1953) developed a miniature chemostat by putting a drop of various bacterial cultures on a microscope slide, covering it with a thin glass square—today called a coverslip—and, then, replenishing the drop with a nutrient solution several times a day. Winogradsky described the usefulness of this technique in an 1889 technical article: "By imitating conditions in nature where sulfur bacteria occur, I have been able to attain growth in drops of water for weeks or months of *Beggiatoa, Thiothrix,* and other species." This would be one of many methods devised by Winogradsky to mimic natural growing conditions for studies on microbial life.

In 1950, the French biochemist Jacques Monod (1910–76) and a separate scientific team, Aaron Novick and Leo Szilard of the University of Chicago, simultaneously proposed a model for a much larger chemostat for collecting a variety of measurements. Each of these scientists recognized the value of a technique for growing microorganisms in response to different nutrients at varying levels. The chemostat enabled them to study microbial physiology under a range of environmental conditions. Monod's earlier studies leading to the modern chemostat involved the relationships between the growth rate of a microbial population and the amount of nutrients provided to that population.

In a chemostat, the final cell density in the culture depends on the concentration of the limiting nutrient. The characteristics in a chemostat culture are usually described in the following relationship based on Monod's studies, where D is dilution rate, f is flow rate as milliliters per hour (ml/hr), and v is the vessel volume:

$$D = f / V$$

For example, if f = 30 ml/hr and V = 100 ml, D = 0.3 hr^{-1}.

This relationship also shows that as dilution rate increases, flow rate increases in a proportional manner. But at any given flow rate, a high dilution rate requires a low volume in the vessel, and a low dilution rate requires a large-volume vessel. Almost all microbiologists control D by controlling flow rate, rather than changing vessel volume.

Chemostats also have been used to demonstrate that as D increases, the culture's doubling time or generation time decreases. Put simply, the microorganisms must increase their growth rate to keep up with an increased D. If the vessel's flow rate increases to a very fast rate, the microorganisms cannot maintain a growth rate fast enough to keep up with the flow. More cells then exit the chemostat than new cells can grow to replace them; this condition is called washout.

To prevent washout, D must be held within a certain range of rates. Very low D supplies only a limited amount of nutrients for the continuous culture. Cells must use a portion of the incoming nutrients for cell maintenance before the cells can divide and grow. Once this maintenance energy has been supplied, the cells use the remaining energy supplied by the substrate for replication and growth.

A turbidostat, by contrast, relies less on the energetics of D, substrate concentration, and doubling time. A turbidostat regulates cell density by constantly measuring the absorbance of light shined through the liquid and then making continual adjustments to speed or slow the flow rate. Increasing absorbance, a condition described as turbidity, indicates higher cell densities. When cell density becomes too high, the turbidostat speeds the flow rate; when cell density drops too low, the turbidostat slows the flow rate.

Other types of specialized culture equipment mimic the design of the turbidostat, but they use control mechanisms other than turbidity. For example, a pH-stat regulates growth by monitoring and adjusting pH; a CO_2-stat regulates growth by monitoring and adjusting carbon dioxide levels emitted by the culture. Microbiologists work with biological engineers to develop continuous culture vessels that regulate oxygen uptake, micronutrient uptake (nutrients required in very small amounts), and specific primary or secondary metabolites.

Continuous cultures allow microbiologists to study growth at very slow rates or at very low nutrient levels, both conditions likely to be found in nature. Microorganisms in nature seldom have the perfect conditions that laboratory batch cultures receive with ample nutrients and incubation at the microorganism's optimal temperature. Microorganisms in nature often receive a limited supply of nutrients, which keeps the cells in a constant state of growth, steady state, rather than the extremes seen in a batch culture growth curve.

Some studies have been designed around conditions that contain characteristics of both continuous cultures and batch cultures. Three methods are intermediate between continuous and batch growth: the successive transfer method, fed-batch culture, and extended batch culture. These systems are described as follows:

- successive transfer method—A portion of the culture is periodically transferred to sterile medium.
- fed-batch culture—Nutrient solution is added to a culture at periodic intervals.
- extended batch culture—Fed-batch culture in which volumes are periodically removed to extend the time the culture can continue to grow.

Intermediate cultures resemble continuous cultures by the following relationship:

Cells added to the system - cells removed from the system + cells produced during growth - cells consumed through death = cells accumulated in the system

Batch–continuous culture intermediate methods offer two advantages. First, these methods are easier to maintain than continuous cultures for long periods without contamination. Second, the methods replicate certain natural conditions of microbial growth, such as blooms, wherein nutrients enter a microbial habitat intermittently.

INDUSTRIAL BIOPROCESSING

Industrial-scale bioprocessing encompasses all the materials and methods for producing a specific end product that has a monetary value. In addition to microbial mass culture, bioprocessing includes the growth of mammalian cells. Both types of bioprocessing can be divided into two different phases: upstream processing and downstream processing.

Upstream processing begins with the first small-volume inoculation of cells to sterile growth medium. A microbiologist grows one test tube culture of the desired microorganism and then, after the culture has incubated and grown, transfers a portion to larger volume, usually in a 100-ml flask, rather than a 10-ml test tube. After incubation and growth, the culture in the flask becomes known as the starter culture. Starter cultures serve as the first step in scale-up, which is the process of developing large-volume cultures from small-volume laboratory cultures. Microbiologists transfer the microorganism into increasingly larger volumes of medium.

Chemostats and Turbidostats

Feature	Chemostat	Turbidostat
dilution rate, D	remains constant	continually varies
main growth control	limiting nutrient	cell density
main control mechanisms	substrate disappearance, primary or secondary metabolites	light absorbance in the culture (turbidity)
optimal condition	low D	high D

These cultures are called *scale-up cultures,* and they lead to a final culture, which the microbiologist uses to inoculate an industrial bioreactor filled with sterile medium. When the large industrial bioreactors have been inoculated, this phase of bioprocessing is often called *production-scale fermentations* or production-scale cultures.

Downstream processing commences with the production-scale fermentation. Downstream processing encompasses all of the procedures used in harvesting the desired end product: cells, primary metabolites, or secondary metabolites. The goal of downstream processing is to increase cell yield, which is a measure of bioprocessing efficiency.

cell yield = mass of product formed ÷ mass of substrate consumed

Although the detailed steps of all end product harvests differ, most downstream processing includes the following basic activities:

1. recovery of product by centrifugation or filtration
2. secondary high-speed centrifugation or fine-pore filtration to remove small debris
3. purification steps
4. purity testing
5. stability testing
6. packaging

Industrial bioprocessing produces large amounts of final product by using several bioreactors that hold 3,000 gallons (11,356 l) or more of culture. Production facilities employ specially trained technicians, who monitor the cultures for proper growth conditions, accidents, and contamination. A critical part of the responsibilities in production plants relates to the monitoring of the water used in the culture media, called *process water.*

Bioprocessing requires about 3,602 gallons (13,631 l) of water to make one pound (0.45 kg) of final product (30,000 l/kg). Tap water cannot be used because it contains unknown chemicals and microorganisms; sterilization kills the microorganisms but does not destroy the chemicals. Even after sterilization, the microbial debris remains in the sterilized tap water. Water purification, therefore, plays a critical role in bioprocessing. Purified water can be made from tap water that has gone through the following treatments: (1) filtration to remove large particles, (2) passage through a carbon filter to remove some organic material and salts, (3) treatment with resins that remove additional organic chemicals, and (4) ion exchange, which removes charged ions such as sodium (Na^+), calcium (Ca^{2+}), and chlorine Cl^-). Process water purification usually includes one or more additional steps, such as distillation, which is the boiling and condensation of water to separate it from impurities; additional ion exchange; or reverse osmosis. Reverse osmosis is a specialized type of filtration that takes small molecules out of water.

Continuous culture vessels require mixing to keep cells suspended and to distribute substrate. The agitation leads to foam formation on the liquid's surface that interferes with the system by clogging tubes and sampling ports and increasing the chance of contamination. Microbiologists add small amounts of chemical antifoam agents to reduce the problem of foaming.

Bioprocessing facilities follow strict procedures to make a pure product free of contamination. Production-scale bioprocessing requires weeks to shut down, clean, sterilize, and start up again if a contamination occurs. In this way, continuous cultures require extra work that batch cultures seldom demand.

CONTAMINATION PREVENTION IN CONTINUOUS CULTURES

Continuous cultures' long periods of operation increase the chances for contamination. Continuous cultures in chemostats, turbidostats, or massive bioreactors also contain more parts than batch cultures, which usually grow in tubes or flasks.

Contamination prevention begins with sterile medium and a sterilized vessel to hold the culture. Tubing, sampling ports (for sample withdrawal), and injection ports (for addition of nutrients or other growth factors) must be sterilized before use, as well, as all reagents to be added to the culture. Microbiologists use two methods to prevent contamination in small-scale continuous cultures: swabbing ports with an alcohol solution before and after each use and attaching fine-pore filters to injection ports. In some cases, a port that supplies oxygen or other gases must be equipped with a special filter called a high-efficiency particulate air (HEPA) filter to remove impurities from the gas.

Industrial bioprocessing depends on a program called cleaning-decontamination-sanitation (CDS) to reduce contamination. Before a new production begins, technicians apply chemical biocides to two main areas of the production facility: surfaces that do not have contact with the product and surfaces that have direct contact with the product. Examples of such surfaces are as follows:

- noncontact surfaces—walls, floors, work benches, ancillary equipment

- direct contact surfaces—bioreactor internal culture vessel, media and product tubing, sampling and injection ports, filters, product collection vessel

Noncontact surfaces require cleaning and sanitization or disinfection with an effective biocide product. Direct contact surfaces require sterilization. Production facilities follow two methods in their CDS programs. The first method is called *cleaning in place* (CIP), in which surfaces and equipment are decontaminated without disassembly. In CIP procedures, workers pump cleaning solutions and biocides through the system's fixed piping and vessels. Purified water then rinses out the chemicals. Next, workers sterilize the equipment by sending steam through the system. The second method, called *cleaning out of place* (COP), requires complete disassembly of the production line. Workers clean and sterilize the individual parts before reassembling the equipment.

Continuous culture has been a powerful tool for studying the metabolism of microorganisms as it would be likely to occur in nature. Continuous cultures, however, present a complex system to run and to control, and they often have an added disadvantage of the high probability of contamination. But the advantage of being able to control many growth conditions of a culture, including the rate in which microorganisms grow, far outweighs the complexities. Continuous culture methods will remain an important aid in learning about the role of microorganisms in ecology and in the manufacture of valuable biological products.

See also BIOREACTOR; FILTRATION; GROWTH CURVE; HYGIENE; LOGARITHMIC GROWTH; WATER QUALITY.

Further Reading

Hoskisson, Paul A., and Glyn Hobbs. "Continuous Culture—Making a Comeback?" *Microbiology* 151 (2005): 3,153–3,159. Available online. URL: mic.sgmjournals.org/cgi/reprint/151/10/3153. Accessed February 17, 2009.

Monod, Jacques. "La technique de culture continue, théorie et applications." *Annual Report of the Institute Pasteur* 79 (2005): 390–410.

Novick, Aaron, and Leo Szilard. "Description of the Chemostat." *Science* 112, no. 2920 (1950): 715–716.

Panikov, Nocolai S., and Stuart Shapiro. "Archetypes of Modern Continuous Culture Methodologies." *SIM News*, November/December 2008.

Winogradsky, Sergei. "Recherches Physiologiques sur les Sulfurbactéries" (Physiological Studies on the Sulfur Bacteria). *Annales de l'Institut Pasteur* 3 (1889): 49–60. In *Milestones in Microbiology*, edited by Thomas Brock. Washington, D.C.: American Society for Microbiology Press, 1961.

Cryptosporidium *Cryptosporidium* is a single-celled protozoan and a parasite that is pathogenic to humans and other animals. The *Cryptosporidium* life cycle contains many different stages, as the microorganism passes through animals, into the environment, and then into freshwater sources. Once the microorganism has infected a host, it is shed in feces. Fecal contamination of water, therefore, serves as the main way of contracting a *Cryptosporidium* infection. Most medical outbreaks in humans have been related to drinking contaminated water or ingesting fresh vegetables that have been rinsed with contaminated water. The resulting disease is called *cryptosporidiosis*.

Cryptosporidium belongs to a group of protozoa called sporozoa in the Apicomplexa phylum, which all have a life cycle stage that produces a cell form called a *sporozoite*. The *Cryptosporidium* sporozoite is called a merozoite, which is the infectious form of the microorganism. More detailed taxonomy of *Cryptosporidium* has been debated. Various species of *Cryptosporidium* have been classified in the past on the basis of the host that they infect rather than their genotype, which is a microorganism's genetic makeup. Most studies have been made on *C. parvum,* which infects humans and other mammals, especially cattle. Calves have a high susceptibility to *C. parvum* infection, so they have been implicated as a major reservoir for the human infection. A second species that has been found in humans, *C. hominis,* has not been studied as much as *C. parvum* because no evidence exists to suggest it causes illness.

The life cycle of *Cryptosporidium* can take place all within the same host animal or from a reservoir animal, such as cattle, to humans. The complex life cycle contains both sexual and asexual reproduction and six distinct stages. Stage 1 begins with a form called a trophozoite that has infected the small appendages, or microvilli, on the epithelial cells lining the small intestine. Each trophozoite contains four banana-shaped, motile merozoites that exit when the trophozoite ruptures. This stage is called *excystation.* Stage 2 comprises the infection of additional intestinal epithelial cells by the merozoites. Stage 3 takes place when the merozoites reproduce, by either sexual or asexual means, to form a zygote. In stage 4, the zygotes sporulate, meaning they develop the outer wall that characterizes them as either a thin-walled or thick-walled oocyst. The oocysts, then, either exit or reinfect the host in stage 5. A new trophozoite develops in stage 6.

The veterinary researchers Mark Goodwin and Dan Biesel have described the *Cryptosporidium* life cycle stages as shown in the table.

Prevention of *Cryptosporidium* infection in humans depends on the proper treatment of drinking water. Treatment is designed to break the microorganism's life cycle at the oocyst stage, because this form exists in the environment. When farm animals or wildlife shed *Cryptosporidium* oocysts in their wastes, some fecal matter may wash into streams and into drinking water sources such as a river. A portion of this water and other runoff water from rains goes to wastewater treatment plants, which remove the oocysts.

The American Water Works Association (AWWA) acts as an industry representative for municipal water treatment facilities and examines the ongoing successes and problems in controlling *Cryptosporidium* in drinking water. The table on page 197 summarizes the typical levels of *Cryptosporidium* oocysts that could be expected to be found in U.S. surface waters from data compiled by the AWWA. These sources have little exposure to farming wastes; sources identified as pristine have no significant exposure to any type of human activity, including farming.

A *Cryptosporidium* oocyst is 4 to 6 micrometers (µm) in diameter, fairly large for a microorganism. Filtration, therefore, works well in removing many of the oocysts. Untreated wastewater from towns that are not near farms or wildlife habitat usually contains one to 10 oocysts per 100 ml, which wastewater treatment plants remove. The incidence of *Cryptosporidium* in water greatly increases, however, when storms overwhelm wastewater treatment facilities as well as drinking water sources such as reservoirs. Most of the largest cryptosporidiosis outbreaks have resulted from either large rainstorms or faulty operations at a wastewater treatment plant.

CRYPTOSPORIDIUM OUTBREAKS

Cryptosporidium infection occurs mainly in two ways: ingesting contaminated drinking water or ingesting natural surface waters such as streams. Researchers have calculated that the 50 percent infective dose (ID_{50}) in humans is 132 oocysts. In other words, half of the normal healthy population will develop an infection by ingesting this dose of oocysts. (ID_{50} provides a more accurate estimate of infection than calculating the infective dose in 100 percent of a population because it excludes persons that have unusual reactions to *Cryptosporidium,* such as individuals who do not get sick regardless of dose size.)

Cryptosporidium was not recognized as a pathogen in humans until 1976. Although microbiologists had observed oocysts in water as early as 1907, the microorganism generated little interest because no one knew it caused human illness. Not until the 1950s, did researchers relate *Cryptosporidium* to infection in birds, and by the 1960s, scientists had begun accumulating evidence that the microorganism infected mammals. In the 1980s, medical researchers determined that oocysts did, in fact, cause cryptosporidiosis in humans. Still, the disease appeared to be rare and raised no alarms in the medical community.

In 1993, the first major U.S. cryptosporidiosis outbreak occurred in Milwaukee, Wisconsin. In spring that year, pharmacists noticed an increase in requests for treatments for watery diarrhea; grocers noted unusually high purchases of toilet paper. Through March and into April, the illness affected Milwaukee and neighboring communities, and it had a particularly severe effect on people who had damaged immune systems because of the acquired immune deficiency syndrome (AIDS). Residents became leery of their tap water, and public health officials warned people to boil tap water before using it in the hope of killing an unknown pathogen. An entire city seemed to be under siege, yet doctors may have felt helpless in not knowing the cause of the outbreak that had already infected 200,000 people.

On April 8, 1993, the *Milwaukee Journal* correspondent Marilynn Marchione reported,

> Milwaukee physician Thomas A. Taft is a hero today. Taft, 38, an infectious disease specialist, suspected that a protozoan was the culprit that had sickened thousands of people in the Milwaukee area, and he led public health officials to identify it. City Health Commissioner Paul Nannis said Thursday that Taft called the Health Department late Wednesday afternoon to say that a stool sample from an elderly woman Taft was treating at West Allis Memorial Hospital had cultured positive for the protozoan *Cryptosporidium*. City health officials then faxed an alert to all city labs and hospitals to test for the organism, and by 8 p.m., eight samples had tested positive. 'Once people knew to look for this, we found eight cases quickly,' Nannis said. 'We expect to find more today.'

Doctors indeed found more. By the time the *Cryptosporidium* outbreak had been controlled, more than 403,000 people had been made sick—more than 100 would die—by the microorganism.

A team of public health scientists led by William R. MacKenzie then conducted phone interviews of people in the Milwaukee area affected by the outbreak. They also studied the records from two water treatment plants, between March 1 and April 28, the period of the outbreak. The team additionally studied ice cubes made in households from tap water during the outbreak. MacKenzie's results indicated that the municipal water turbidity, or cloudiness, had increased almost 15 percent during the outbreak—a sign of improperly treated water—and ice cubes contained 7–13 *Cryptosporidium* oocysts per 100 ml of ice water.

Despite the clues gathered by MacKenzie and other water quality specialists, the Milwaukee outbreak had been very difficult to diagnose at the time of the outbreak. MacKenzie wrote, in a 1994 *New England Journal of Medicine* article, "Despite communitywide diarrheal illnesses in Milwaukee, the recognition of *Cryptosporidium* infection as the cause of this outbreak was delayed for several reasons. The constellation of gastrointestinal symptoms (e.g., diarrhea, abdominal cramping, and nausea) and the constitutional signs and symptoms (e.g., fatigue, low-grade fever, muscle aches, and headaches) reported by Milwaukee-area residents led many physicians to diagnose viral gastroenteritis or 'intestinal flu' without further investigation." In fact, the symptoms of cryptosporidiosis resemble the symptoms of a large number of other waterborne infections; that is one reason why these waterborne illnesses have been very difficult to diagnose when an outbreak first occurs.

Looking back on Milwaukee's health crisis, investigators learned that a high snowmelt and unusually heavy rainfalls had occurred shortly before the outbreak, causing streams leading into the area to flood. A number of cattle farms near the city also added wastes to the large amount of runoff. One of the two water treatment plants appeared to have used faulty methods for keeping pathogens out of the water, but the heavy inflow of untreated water may have overwhelmed a system that normally would have provided safe drinking water.

Five years after Milwaukee's outbreak, another *Cryptosporidium* outbreak, in Sydney, Australia, led three million residents to boil their drinking water that winter. The Australian outbreak was found to be connected to a contaminated public swimming pool. Sporadic *Cryptosporidium* outbreaks continue to occur. In addition to poorly treated drinking water and untreated surface waters, they have been linked to recreational waters and food.

WATER TREATMENT FOR *CRYTOPSPORIDIUM*

Water treatment plants have a large responsibility in keeping *Cryptosporidium* oocysts out of tap water. A single infected calf can excrete 10 trillion oocysts in a day that survive for weeks in surface waters, so the amount of oocysts entering treatment plants with water can be quite high. The *C. parvum* oocyst presents an additional problem, because it is very resistant to the normal chlorine disinfection used for drinking water treatment.

Filtration systems at water treatment plants that remove particles of 3 μm and larger can remove most

Cryptosporidium Life Cycle Stages

Stage	Name	Event
1	excystation	infective sporozoites are released from mature oocysts
2	merogeny	asexual multiplication to form a meront, which contains four merozoites
3	gametogeny	sexual reproduction to form gametes
4	fertilization	two different types of gametes formed in stages 2 and 3 combine to form a zygote
5	oocyst wall formation	80 percent of zygotes form thick-walled oocysts; 20 percent form thin-walled oocysts
6	sporogeny	sporozoite formation in the oocysts to begin another cycle

***Cryptosporidium* Occurrence in U.S. Surface Waters**

Source	% Samples Positive for *Cryptosporidium*	Range of Oocyst Concentration, per gallon (per l)
stream	73.7	0–907 (0–240)
stream and river system	77.6	0.15–68 (0.04–18)
river	31.8	0.04–286 (0.01–75.7)
river and lake system	87.1	0.26–1,830 (0.07–484)
lake	64.5	0–83 (0–22)
pristine river	32.2	not determined
pristine lake	52.9	not determined

Note: data from at least 20 samples

of the *Cryptosporidium* to make drinking water safe. But no treatment has yet been developed to rid tap water of all oocysts; four to 10 oocysts per gallon (1–3 oocysts/l) may occasionally be present in tap water, although most U.S. water utilities provide water with far lower concentrations of *Cryptosporidium* oocysts—fewer than 0.01 average oocyst per gallon (0.04 oocyst/l). Low and infrequent incidences of *Cryptosporidium* in tap water have not caused health threats in the normal healthy population. Individuals who have higher-than-normal health concerns should, however, take the precautions recommended by the Centers for Disease Control and Prevention (CDC) and other health agencies. The following groups in the population considered high-risk health groups must take extra precautions against infection because of weakened immune systems:

- AIDS patients
- organ transplant patients
- chemotherapy patients
- elderly adults
- persons with chronic debilitating disease

The recommendations for these high-risk groups in the table on page 198 greatly reduce the chance of *Cryptosporidium* oocysts in drinking water or other sources of infection. These actions are listed in the table in order of effectiveness against infection, from the most effective to least effective. *Cryptosporidium* infections can spread to other organs in AIDS patients. For this reason, AIDS patients make up a special group who should always use at least one of the precautions in the table.

Bottled water does not necessarily offer any greater safety against *Cryptosporidium* than tap water. Only bottled water that contains information on the label that the water has been treated in either one of two ways should be considered *Cryptosporidium*-free: reverse osmosis or a filtration unit graded as an "absolute 1 μm-rated filter."

PATHOLOGY OF *CRYPTOSPORIDIUM*

The precautions described here for healthy individuals and persons belonging to high-risk health groups reduce the incidence of *Cryptosporidium* infection. Medical studies conducted in developed countries, including the United States, indicated that 25–35 percent of the population have had a *Cryptosporidium* infection at some time in their life. The incidence rate in the United States is about 1.6 cases per 100,000 people. Incidence rates are slightly higher in children and can be two- to threefold higher in some developing countries.

Cryptosporidiosis symptoms begin two to 10 days after ingesting an infectious dose of oocysts. The pathogen causes severe damage to the intestinal lining, which leads to some of the characteristic symptoms: painful abdominal cramps, watery diarrhea, dehydration, and weight loss. Additional symptoms

Precautions against *Cryptosporidium* Infection

Precaution	Effect
boiling water at a rolling boil for at least one minute	kills oocysts
using water filter with "absolute" pore size of 1 µm or smaller	removes oocysts from water
using reverse osmosis house water treatment that is certified for "cyst reduction"	removes oocysts from water
avoiding drinking or swallowing water from streams, lakes, or recreational waters	reduces chance of ingestion of oocysts
thorough hand washing after exposure to fecal matter: soap and warm water for at least 20 seconds	reduces chance of self-inoculation with oocysts

are nausea, vomiting, and fever, which may be connected to an unknown toxin released by merozoites. The extent of damage to the intestines in cryptosporidiosis leads to a reduced nutrient uptake, which can be a greater health threat to young children and pregnant women.

The symptoms usually last one to two weeks but can continue for up to four weeks in some individuals. Cryptosporidiosis symptoms also recur in many people for 30 days or longer as new merozoites emerge from oocysts that have infected the intestinal lining. This recurrence of the illness due to the cyclical nature of the microorganism is called autoreinfection.

No treatments have been effective in killing *Cryptosporidium* once it has infected a person; most treatments act on reducing the voluminous water loss during the illness. Fat absorption is especially decreased, so that infected persons may develop steatorrhea, which is an elevated level of fats in the stool. While the intestine decreases its fluid and electrolyte uptake, it begins an increased flushing of fluids into the intestines, leading to the severe diarrhea seen in cryptosporidiosis.

Infected persons may shed up to one billion oocysts in a single bowel movement. This high number of fecal oocysts explains why infection can be further spread in recreational waters, by hands dirtied with fecal matter, or in feces-contaminated drinking water or food. The oocysts may continue to be shed up to 50 days after the illness's symptoms have ended.

VETERINARY PATHOLOGY OF *CRYPTOSPORIDIUM*

Cryptosporidium infection in mammals appears to develop in the same way as it does in humans by attacking the microvilli of the small intestine. An exception is *C. bialeyi* infection, in birds, which targets the respiratory system.

Cryptosporidium mostly infects young animals. Calves of two weeks of age are most susceptible to infection and may also serve as the main reservoir for human infection. The dose of oocysts that make calves sick can be as low as 10 oocysts.

Cryptosporidiosis occurs worldwide, and, as in humans, the animal disease is self-limiting, meaning an animal can recover on its own without treatment. Animals that have severe diarrhea and are suspected of having cryptosporidiosis should be isolated from healthy animals whenever possible. The isolation should also last for a few to several weeks after the illness's symptoms have ended.

The table below summarizes the main occurrences of *Cryptosporidium* infection in animals.

People who work with animals may have a higher risk of *Cryptosporidium* infection. The following occupations might require extra precautions against fecal contamination from animals: veterinarians, veterinary students, veterinary technicians, farmworkers, animal handlers, and people who have contact with infected animal products.

***Cryptosporidium* Infection in Animals**

Animal	Description of the Disease
calves	mild to moderate diarrhea, weight loss, emaciation, death
lambs, young goats	diarrhea, secondary infections
foals	diarrhea, death in immune-weakened animals
piglets	poor digestion, diarrhea
turkeys, chickens, quail	infections in all parts of the lungs, death
caged birds, chicks	poor digestion, secondary infections
reptiles	regurgitation, weight loss, chronic debilitation
dogs, cats	diarrhea, poor appetite, weight loss

Cryptosporidium Species

Species	Primary Host	Location of the Infection
C. andersoni	cattle	abomasum (ruminant animal's stomach)
C. baileyi	birds	lungs, cloaca
C. canis	dogs	small intestine
C. felis	cats	small intestine
C. galli	birds	not determined
C. hominis	humans	small intestine
C. meleagridis	birds	small intestine
C. molnari	fish	stomach, small intestine
C. muris	mice	stomach
C. nasorum	tropical fish	stomach, small intestine
C. parvum	humans and other mammals	small intestine
C. suis	pigs, cattle	small and large intestines
C. saurophilum	snakes, lizards	stomach, small intestine
C. serpentis	snakes, lizards	stomach
C. wrairi	guinea pigs	small intestine

Source: Companion Animal Parasite Council

Veterinarians diagnose *Cryptosporidium* infection in animals the same way human infections are diagnosed: serology testing. Serology identifies unique molecules that occur on the outside of microbial cells or cysts. Genotype studies have also been conducted on *C. parvum* with at least four different genotypes discovered. Four different *C. parvum* genotypes infect different animals:

- humans, cattle, and other mammals
- mice and bats
- koalas and kangaroos
- ferrets

The table above summarizes various known and proposed *Cryptosporidium* species and their pathology.

Medical science and microbiology have gained knowledge about *Cryptosporidium* since the 1993 Milwaukee outbreak. However, *Cryptosporidium* remains largely a mysterious microorganism. Drinking water quality professionals have the greatest responsibility in keeping *Cryptosporidium* out of the general population. Despite improving technology in water treatment, *Cryptosporidium* causes periodic outbreaks worldwide in humans and in animals.

See also FILTRATION; PROTOZOA; SEROLOGY; WATER QUALITY.

Further Reading

Centers for Disease Control and Prevention. "Cryptosporidiosis." Available online. URL: www.cdc.gov/crypto/index.html. Accessed February 18, 2009.

Fayer, Ronald, and Lihua Xiao, eds. *Cryptosporidium and Cryptosporidiosis,* 2nd ed. Boca Raton, Fla.: CRC Press, 2007.

MacKenzie, William R., et al. "A Massive Outbreak in Milwaukee of *Cryptosporidium* Infection Transmitted through the Public Water Supply." *New England Journal of Medicine* 331, no. 3 (1994): 161–167. Available online. URL: http://content.nejm.org/cgi/content/full/331/3/161. Accessed February 18, 2009.

Marchione, Marilynn. "Doctor Solves Crucial Part of Illness Puzzle." *Milwaukee Journal,* 8 April 1993.

culture A culture is a population of cells that have grown in broth medium or on agar medium during incubation. Incubation is a set period of time at a specific temperature to allow growth of microbial cells from a few cells to many thousands or millions. Cultures may be composed of bacteria, fungi, algae, protozoa, plant cells, or animal cells. (Viruses are not typically referred to as cultures because they do not replicate on their own.) The term is also used in microbiology as a verb: To *culture* is to conduct all the necessary procedures for growing and maintaining a microorganism in a laboratory.

Cultures are categorized on the basis of the following three main characteristics:

1. type of medium used for growing the microorganisms: broth versus agar
2. type of growth: aerobic versus anaerobic
3. type of microorganism: mold versus bacteria

Cultures usually belong to more than one category at the same time. As the table shows, cultures are categorized in a number of ways. The table on page 207 also provides the terminology used by microbiologists to describe different culture conditions. For example, a supervisor may tell a microbiologist to prepare a bacterial culture, but this is not enough information to enable the microbiologist to fulfill the request. The following statement provides all the information needed to prepare a specific culture: "Grow a *Streptococcus* culture on a trypticase soy agar slant." This request describes the type of microorganism (*Streptococcus* bacteria), the type of medium (agar), the specific type of agar (trypticase soy), and the physical form of the agar (slant tube).

Microorganisms are too small to be seen with the unaided eye. Putting 10 bacterial cells into a tube containing 10 ml of clear broth would not change the appearance of the liquid; the broth remains clear. Some microbiologists might not consider this to be a culture. After incubation, during which cells multiply into the billions, the cells cloud the broth or cover an agar surface with visible colonies. The turbid (cloudy) broth or the agar plate dotted with colonies is then referred to as a culture. See the color insert on page C-2 (lower right).

Eventually, cultures must be transferred to fresh sterile medium. This transfer gives the cultured cells a new supply of nutrients and removes wastes. A *subculture* is a term used for the new culture that arises from the original culture. Microbiologists often store cultures in a refrigerator or freezer and then periodically subculture to new media. The original culture in storage is called a *stock culture,* and the removal of a small volume (10–100 microliters [µl]) of the stock culture to new sterile medium is called a *transfer.*

Today's culture techniques developed from early microbiology 300 years ago. The German physician Rudolf Virchow (1821–1902) perhaps unwittingly described cultures best, in 1858, in support of the cell theory. His simple observation, "*Omnis cellula e cellula*" [every cell originates from another cell] laid the foundation for microbiology. Virchow's statement also summarized the purpose of a culture: that is, to propagate a microorganism without contamination for future study.

Modern culture techniques require proficiency in the following skills: aseptic techniques, sterilization, microscopy, cell morphology, and biological staining. Microbiologists must also have knowledge of media preparation, spectroscopy, biochemical testing, and the handling of biohazards. The newest specialized culture techniques additionally require a background in cellular metabolism, deoxyribonucleic acid (DNA) replication, and taxonomy.

Various Ways to Categorize Cultures

Classification of a Culture	Example
required growth conditions	batch versus continuous
type of medium	broth versus agar
purpose of the medium	enrichment versus isolation
configuration of the agar	plate versus slant
type of inoculation	streak versus lawn
type of biota	mold versus tissue
vessel handling requirements	shake flask versus roll tube
outcome	pure versus contaminated

HOW TODAY'S CULTURE METHODS DEVELOPED

When a microbiologist isolates a new microorganism never before studied, the process traces the history of all the scientists who contributed to the growth of microbiology. Microbiology began with the work of the Dutch clothier Antoni van Leeuwenhoek (1632–1723), who is credited with developing the first microscope for studying microbial life, in 1676. Van Leeuwenhoek had been trying various lens configurations inside a simple tube for inspecting fabric threads. Curious, van Leeuwenhoek began putting other items under his lens, and he soon saw microorganisms in a mixture of pepper and water. His sketches of those "animalcules" were sufficiently detailed and so accurately captured cell morphology that scientists today recognize the microorganisms van Leuwenhoek studied as *Bacillus.*

Culture techniques advanced again in the laboratory of the French microbiologist Louis Pasteur (1822–95). Among Pasteur's many contributions to the blossoming field of microbiology, his fermentation studies and experiments with airborne microorganisms helped define the concept of a germ. The germ theory allowed microbiologists to understand the existence of single-celled structures that carried out complex biological reactions. To do their studies, they needed dependable culture techniques. The German mycologist Julius Oscar Brefeld (1839–1925) proposed, in 1870, that these studies would work best if a scientist could obtain millions of one type of microorganism with no other species present. Brefeld had developed the first notion of *pure cultures,* which is a critical part of today's microbiology. Pasteur meanwhile conducted his research with cultures containing more than one microorganism. These are now called *mixed cultures,* and they have been most useful in studying the activities of microorganisms in the environment.

In 1878, the English surgeon Joseph Lister (1827–1912) invented a clever way to make a pure culture, by isolating a single microorganism from others. But Lister's procedure was tedious to conduct in a laboratory, so, in the late 1800s, the German physician Robert Koch (1843–1910) streamlined it. For his innovation, Koch took advantage of a serendipitous finding by one of his colleagues, Walther Hesse (1846–1910). Hesse's wife, Angelina (1850–1934), had found a gelatinlike material called *agar,* which she used in her kitchen. Hesse and Koch adapted agar for making solid culture media, which became a major step advantage for culturing new microorganisms. In 1887, the German bacteriologist R. J. Petri (1852–1921) designed a new dish for aiding what he called "Koch's gelatin plate technique." Petri's dish saved space in laboratories and inside incubators, because a scientist could stack many dishes in a single column. Petri's dish—now called a *petri dish* or *petri plate*—also provided a suitably sized area for studying the appearance of bacterial or fungal colonies. (A single plate also conveniently fit in one hand.) An often-overlooked modification in modern microbiology relates to disposable plastic petri dishes, tubes, and pipettes. Disposable plasticware has replaced reusable glassware in numerous culture techniques and has become an important time-saving innovation.

The German physician Paul Ehrlich (1852–1915) refined culture techniques, in 1882, with the introduction of biological stains that made microorganisms easier to see under a microscope. Two years later, the Danish scientist Hans Christian Gram (1853–1938) developed a stain that differentiated bacteria into two main groups. The Gram stain remains an essential part of culture techniques in laboratories throughout the world. With the Gram stain and other differential stains, scientists such as Ferdinand Cohn (1828–98) gained the ability to divide bacteria into groups based on their appearance, called *morphology.* Cohn devised such a scheme in 1887.

Armed with the knowledge that different bacteria might prefer different nutrients, microbiologists devised new formulas for growing specific microorganisms on solid agar or in liquid broth. In the early 1900s, the Dutch soil microbiologist and botanist Martinus Beijerinck (1851–1931) developed differential media, an important aid in separating one type of bacteria from other types. Microbiologists now had two powerful tools for culturing microorganisms: stains to help identify microorganisms and media to isolate one type of microorganism from another.

METHODS OF CULTURE PREPARATION

Two of the most common types of cultures used in microbiology today are broth batch cultures and agar plate cultures. Broth cultures consist of liquid medium in tubes or flasks, which supplies microorganisms with all the nutrients they need for multiplication and growth. Broth cultures are used for growing very large volumes of microorganisms and in certain identification techniques, such as enzyme tests. The most probable number method for determining cell concentration and preparations of inoculum also requires broth cultures. A broth culture made specifically for inoculating sterile media is sometimes called an *inoculum culture.* Agar cultures are grown on a solid agar surface, and they work best for isolating single colonies, studying colony morphology, and testing antimicrobial compounds. Specific techniques have been adopted in microbiology to aid the growth of both broth and agar cultures.

Microbiologists use three techniques for obtaining the best growth in broth cultures: shaking, closed cultures, and open cultures. Some broth cultures grow better if placed on a machine that constantly shakes the contents during incubation. Shaking aerates the liquid and provides more oxygen to aerobic microorganisms. Microbiology laboratories usually own two types of shakers. First are mechanical test tube rockers, which provide a continuous mixing motion that helps aerate cultures growing in tubes. Second are platform rotary shakers, which also agitate growing cultures; in this case, the cultures are referred to as *shake flasks* and contain larger volumes than test tubes. Microbiologists control the amount of aeration in tubes on rockers or in shake flask cultures by adjusting the instrument's speed.

A *closed culture* refers to a broth batch culture that receives a single inoculation with no further additions of inoculum or fresh medium; used medium and cannot be removed. In these conditions, cells begin growing slowly, then enter a very rapid period of multiplication, and then eventually slow their growth and begin to die. Such a closed culture offers the advantage of being easy to prepare and monitor. But closed systems have the disadvantage of exposing cells to ever-decreasing levels of nutrients and continually increasing levels of wastes. The entire life cycle of bacteria in closed cultures, called the bacterial *growth curve,* takes place in a batch culture. Many studies in microbiology use growth curves to learn characteristics of species, and for this purpose closed cultures have been helpful.

An open culture, also known as a *continuous culture,* is the opposite of a batch culture. Continuous cultures grow in vessels fitted with an inflow port and a separate outflow port. Sterile nutrient-rich broth medium continually enters the vessel, and nutrient-depleted medium continually exits. The depleted medium contains some live cells, dead cells, cell debris, and waste substances.

Microbiologists use an electronic pump to control the speed at which liquid flows through a continuous culture vessel, and this, in turn, controls the rate at which the microorganisms grow. In continuous cultures, the flow rate is directly related to the cells' growth rate. Microorganisms inside the vessel must grow fast enough to keep up with the flow, or else they wash out the exit port, and the culture disappears. Microbiology uses continuous cultures to study growth characteristics in steady-state conditions, meaning the microbial cells do not need to speed up or slow down their metabolism as they do in batch cultures. Industrial microbiology also uses continuous flow bioreactors to make cells or cell products such as enzymes, vitamins, alcohols, acids, antibiotics, or other products created by genetic engineering.

Agar cultures grow on the gelatinous solid surface provided by agar (1.5–3.0 percent) dissolved in a solution of nutrients in water. (Agar by itself does not supply enough nutrients for microbial growth.) Agar cultures are used for two main purposes: study of a colony morphology and isolation of a pure colony. Colony morphology simply refers to the appearance of a colony growing on agar—white, pink, raised, flat, large, small, and so on. Almost all bacteria and yeasts develop colonies distinctive for their genus. Each colony contains many millions of cells that have descended from one original cell that had been deposited on the agar surface during inoculation. In order to obtain a pure isolate of a microorganism, a microbiologist selects a colony that is separated from all other colonies on the agar surface. The microbiologist then uses a sterile inoculating loop (a wire with a small loop at the end) to transfer the colony to new sterile medium. After incubation, the single pure isolate will have grown into another population of identical cells. (The converse of a pure culture is a contaminated culture, one that contains one or more unwanted microorganisms.)

Agar cultures in tubes behave the same way as agar plates but are used for different purposes. Microbiologists prepare agar tubes by pouring molten agar into glass test tubes, sterilizing them, and then allowing the agar to cool and solidify. Cooling the tubes in an inclined position causes the agar medium to harden at an angle, similarly to tipping a glass of water 45 degrees. By preparing tubes at an angle, called an agar *slant* or *slope,* a microbiologist increases the agar's surface area. The slant configuration makes inoculation easier and gives the microorganisms more space to grow. Square bottles called *serum bottles* laid on their side also create a large agar surface for inoculation and growth.

In contrast to preparing a slant, agar may be poured into upright tubes and then allowed to solidify. The agar is said to create a *butt* in the tube. Microbiologists inoculate this type of agar tube by carrying a small amount of inoculum on a straight wire called an *inoculating needle.* The microbiologist pushes the needle into the agar to a depth of one to two inches (2.5–5.1 cm). Cultures prepared in this manner are called *stab cultures.*

A third technique in microbiology, called a *biphasic culture,* combines agar with broth. A biphasic culture contains a thick layer of agar in a flask, which is then overlaid with a thinner layer of broth medium. Once the broth portion has been inoculated, this type of culture can be incubated with or without shaking to aerate it. The principle of biphasic cultures is to increase the total yield (cells per volume) in the culture by allowing the agar to serve as an added supply of nutrients. As the culture grows and uses up nutrients in the broth, additional nutrients diffuse out of the agar and into the broth. This technique prolongs the period in which cells reproduce rapidly and delays the culture's advances into slower growth and the death phase.

SPECIALIZED CULTURES

A significant amount of microbiology relies on isolated and pure colonies grown on agar. Pure colonies make possible the following tasks: studying colony morphology, isolating and identifying new species, diagnosing disease by isolating the disease-causing microorganism, identifying contaminants in food or other products, and performing antibiotic testing. An enrichment culture provides a way to separate one microorganism from a heterogeneous mixture of other microorganisms before beginning these tasks.

Various formulas that make up the hundreds of different agars used in microbiology induce some microorganisms to express characteristics not seen in any other microorganism. For example, selective media suppress the growth of most microorganisms but allow the growth of a few species. Potato dextrose agar acts as a selective medium because it grows molds but suppresses the growth of most bacteria; it selects for mold growth.

Selective media separate types of microorganisms from other types, but they may not be able to isolate a single particular microorganism, such as a pathogen from a stool sample. An enrichment medium helps do this because it contains an ingredient that suppresses the growth of all microorganisms except the one of interest. Enrichment media thus isolate particular microorganisms rather than a general group. The formula can be either a broth or an agar, and the culture that grows in this medium is called an *enrichment culture.*

Baird-Parker agar with added egg yolk tellurite (EY tellurite enrichment) provides an example of an enrichment medium that allows the growth of only bacteria in the *Staphylococcus* genus. The EY tellurite enrichment contains ingredients that enable several different bacteria to grow, but it also contains glycine and sodium pyruvate that especially stimulate staphylococcal growth. In addition, the EY tellurite formula includes lithium chloride and potassium tellurite, two chemicals that suppress all bacteria except *Staphylococcus.* The staphylococci develop distinct shiny black colonies on the agar. Any other bacteria that manage to grow—*Bacillus* grows a little on EY tellurite and produces brown colonies—do not resemble the staphylococci.

Enrichment cultures have also been useful in bioremediation. Scientists devise enrichment formulas to find bacteria that degrade pollutants in the environment. For example, a soil sample may contain the pesticide heptachlor. In a laboratory, a microbiologist prepares enrichment medium containing heptachlor as the sole carbon source. Only bacteria able to degrade heptachlor as a carbon source will grow on the enrichment medium. Scientists, then, use these special bacteria to clean up land polluted with the pesticide.

INOCULATING A CULTURE

Three styles of inoculating agar plates are prevalent in microbiology, two manual techniques and one automated technique. The common manual methods are streaking and spreading. Streaking is performed by gently dragging an inoculating loop across an agar surface in a continuous line. The streaking pattern on the agar surface yields a heavy concentration of cells at the start of the streak and a very low concentration at the end. Various streak patterns are used by different microbiologists, all with the purpose of distributing the inoculum in a way that produces several isolated, pure colonies after the plate has been incubated. Spreading, by contrast, uses an L-shaped stick rather than a loop for inoculating a sterile agar surface. The stick, which is nicknamed a *hockey stick* because of its shape, draws a film of inoculum across the agar. After incubation, the resulting growth is as a thin layer, called a *lawn,* on the agar surface. Lawn growth produces larger amounts of cells rather than single, isolated colonies.

Spiral streaking, or spiral plating, is an automated plate inoculation method that serves two purposes: It produces isolated colonies, and it aids in calculating the number of cells per volume of broth culture. In spiral plating, an instrument called a *spiral plater* inoculates an agar surface by ejecting a steady stream of inoculum onto an agar plate as the plate rotates on a turntable. The inoculation begins in the middle of the plate and progresses in a single Archimedes spiral toward the outer edge of the plate. As it progresses, the concentration of cells plated on the agar decreases, resulting in several isolated colonies. Cell concentration relates to the distance of the isolated colonies from the center of the spiral. After the spiral plate has incubated, a second instrument, called a *plate counter,* scans the entire agar surface with a laser light beam that counts the colonies—each break in the beam equals one colony. The instrument then calculates the concentration of cells in the original inoculum. The term *laser colony counting* is synonymous with *spiral colony counting* in this technique. Microbiologists may also count spiral plates manually by determining the number of colonies in designated sections of each plate.

Cultures respond to different types of nutrients in the medium and microbiologists use this growth response and other characteristics of cultures to assess purity of the culture and even suggest an approximate identity. Certain species produce distinctive odors, colors, colony morphology on agar, or pellicle formations in broth. For instance, a pellicle is a layer of microbial growth over a broth surface, which is typical of certain species such as *Pseudomonas* bacteria. A pellicle may appear as a film or as a thick mat, which can be viscous, sticky, or slimy. Bacterial and fungal pellicles consist of live cells, cell fragments, and extracellular compounds excreted by living cells. Pellicles may have developed as an aid to survival in the evolution of microorganisms. For example, pellicles help aerobic bacteria migrate upward in an aqueous environment, where oxygen levels are higher than in deep regions. Other characteristics of cultures are:

- unique enzyme activity
- colony swarming
- motility
- colony morphology
- colony color
- cell morphology
- endospore formation
- pellicle formation
- capsules and slime layers
- gas production
- staining characteristics
- hemolysis (lysis of blood cells)
- fungal spore, hyphae, and conidia morphology

Microbiologists gain considerable experience in recognizing certain common microorganisms on the basis of these characteristics.

CELL COUNTING IN BROTH CULTURES

Microbiologists often must determine the cell concentration of bacterial cultures. On agar plates, this is simply done by counting the visible colonies. Broth cultures require different methods for determining the concentration of cells floating free in the culture liquid. Microbiologists use either a direct cell count or a viable cell count. Direct counts determine the concentration of all cells, living or dead. Viable counts determine the concentration of only the living cells.

Direct counts are made by using either a manual counting chamber or an automated flow cytometer. Manual counting chambers have specialized glass slides with a depression called a *well* that holds a very small volume of broth culture. Two commonly used counting chambers are the hemacytometer (or hemocytometer) and the Petroff-Hausser counting chamber. The hemacytometer was designed for counting blood cells but works well in microbiology for counting the number of bacterial cells or fungal spores in a given volume of broth. The Petroff-Hausser chamber was designed specifically for obtaining direct cell counts on bacterial cultures. This chamber is composed of a thick glass slide with a grid etched into the well. A trough boundary surrounds the well and keeps the broth within the grid area. The microbiologist pipettes a small volume of diluted culture into the well and covers it with a thin glass coverslip. Excess liquid runs into the trough. The volume of liquid outlined by the grid may be as small as 0.1 mm^3. Using a microscope, a microbiologist counts the number of cells in the grid to calculate cells per volume of culture.

Counting chambers such as Petroff-Hausser units provide quick results without waiting for a culture to incubate, but this manual method has disadvantages: (1) It does not distinguish between dead and live cells, (2) it relies on a very small volume of culture to determine concentration in a much larger culture volume, (3) the method is difficult for counting large motile cells such as protozoa, and (4) the method is tedious and causes eyestrain. Some of these disadvantages may be bypassed by counting large cells by flow cytometry, which is an automated method that does not require long periods at a microscope. As with manual counting, flow cytometers do not distinguish between dead and live cells.

Viable counts determine only the living cells and do not detect dead cells. For making a viable count, a microbiologist aseptically transfers a small volume (usually 1 ml) of the broth culture to water or buffer (a solution containing molecules that maintain pH in a specific range). The microbiologist then prepares a serial dilution of this suspension by transferring small volumes of successive 1:10 dilutions. The final one or two dilutions, meaning the most diluted samples, serve as inoculum for agar plates. After inoculating and incubating the plates, the microbiologist counts the number of colonies (plate counts), expressed as colony-forming units (CFUs). By multiplying the average CFU/plate by the amount in which the culture had been diluted, the microbiologist arrives at the concentration of the original culture. For example, if an average plate count of a 1:1,000 dilution equals 55, then the cell concentration in the original culture is:

$$55 \times 1{,}000 = 5.5 \times 10^4 \text{ per ml}$$

Viable counts, though labor-intensive, are thought to be the most accurate method for determining a culture's cell concentration.

Three cautions must be maintained when using viable counts to determine cell concentration. First, media that encourage the best growth of microorganisms should be chosen. Second, some microorganisms live in the environment, but microbiologists have yet to find laboratory conditions that stimulate these microorganisms' growth. These cells are called *viable but noncultivable* (VNC or VBNC). Some of the theories as to why VNC microorganisms do not grow in laboratories are as follows:

- unsuitable medium
- stressed cells, such as cells subjected to starvation
- dormant cells, which are alive but not actively metabolizing
- survival mechanism for avoiding injury from antibiotics

Stressed cells, such as bacteria isolated from an environment that has very low nutrient levels, may become VNC in a laboratory when placed into a medium with a rich supply of nutrients. In this situation, stressed cells may have deactivated some of their protective mechanisms that detoxify by-products of metabolism. Though the bacteria do not need these mechanisms in an austere environment, they need them when transferred to the favorable conditions offered by laboratory media and incubation. As a result, the cells experience a type of shock and cannot grow. The microbiologist Todd Steck of the University of North Carolina told an audience at the 2007 American Society for Microbiology national meeting, "VBNC cells are viable yet they do not undergo sufficient division to give rise to visible growth on nonselective growth medium. Classifying cells that don't grow as being viable is controversial. But over 50, mostly gram-negative, bacterial species have been documented to become VBNC. The existence of the VBNC state is thought to be a long-term survival mechanism initiated in response to environmental stresses." Perhaps the greatest concern surrounding VBNC occurs in clinical microbiology facilities that are responsible for isolating and identifying potential pathogens from ill patients quickly and accurately. Pathogens that behave as VBNC microorganisms can present serious health threats.

SYNCHRONOUS BROTH CULTURES

Most microbiology work involves nonsynchronous cultures that contain a mixture of newly formed cells, rapidly dividing cells, slowly dividing cells, and dying cells. A synchronous broth culture differs from nonsynchronous growth because all of its cells pass through the same phase of growth at about the same time. Synchronous cultures help in the production of substances, such as enzymes, that a microorganism secretes only at a very specific point in its cell cycle.

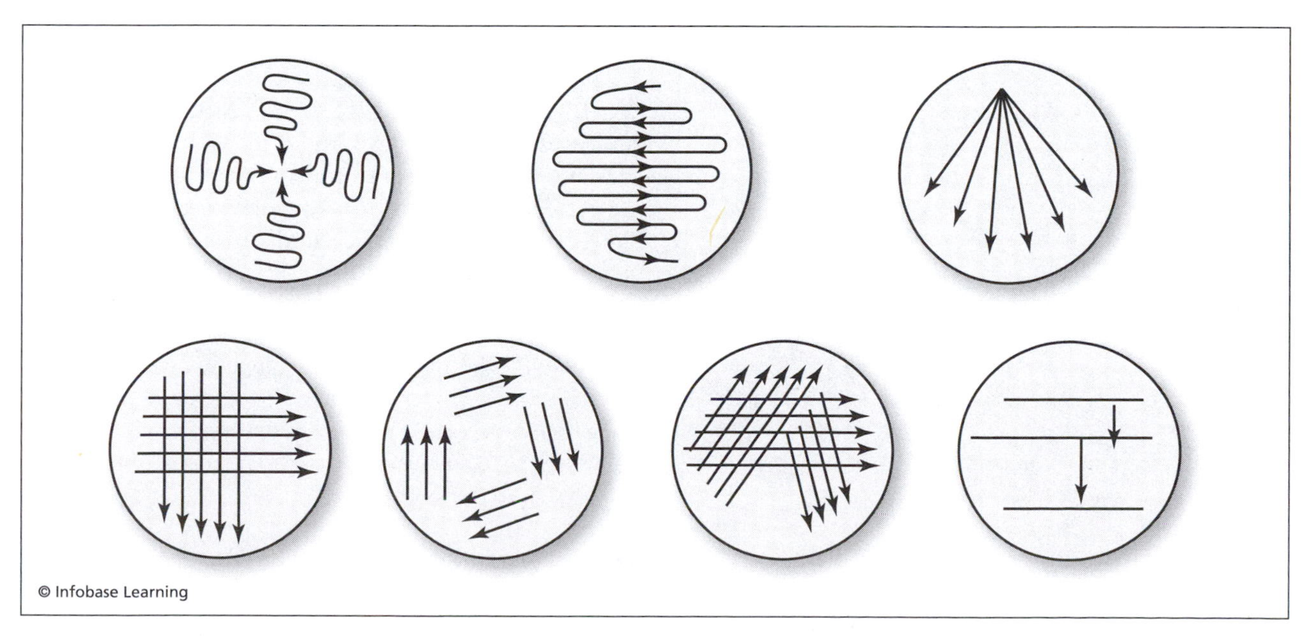

Microbiologists use various streaking techniques for producing single isolated colonies containing only one cell type (called clones) on an agar plate.

To make a synchronous culture, a microbiologist uses one of two methods: altered culture conditions or physical cell separation. By the first method, the microbiologist controls the culture's conditions in a way that makes all the cells grow and multiply together. Some of the effective culture techniques for achieving this are the following:

- addition of a protein inhibitor followed by a sudden switch to noninhibitory medium
- limiting, then adding large amounts of a nutrient
- exposing the culture to sudden temperature changes
- controlling light-dark cycles for growing photo synthetic algae or cyanobacteria

Physical cell separation involves sorting a culture's cells by density or size. In bacteria, the density and size of cells about to divide differ from those of cells in the other phases of growth. An instrument called a *cell sorter* works similarly to a flow cytometer in which cells passing through a narrow channel scatter the light in a laser beam. The amplitude of the scattered light gives an estimate as to each cell's size and other features. A scientist may additionally put in the cell suspension a fluorescent molecule attached to an antibody that seeks out only a particular species. The cell sorter detects each cell that excites the fluorescent reaction when antibody meets its target. The Institute for Systems Biology has explained on its Web site, "Cell sorters differ from cytometers in their ability to separate cells of interest from a complex mixture. Once a cell has been cytometrically characterized, the sorter uses a combination of electronic delays, electrostatic charging, and a static electromagnetic field to separate the chosen cell from the other cells on solution." Cell sorting can be a valuable tool in finding pathogens, food-borne contaminants, and specialized microorganisms of the environment in addition to its role in synchronous cultures.

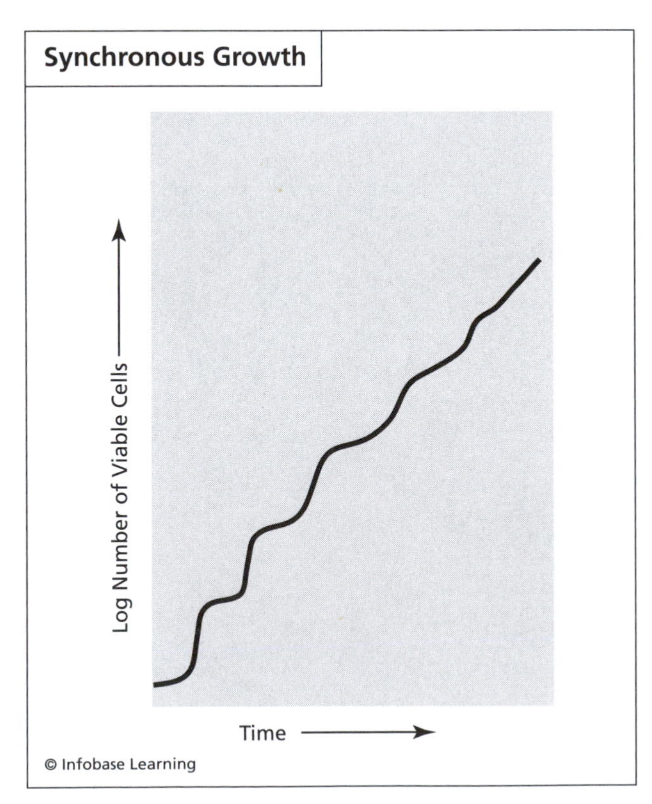

Growth in synchronous cultures is characterized by a population of cells of about equal age because they all divide at about the same time.

CULTURE COLLECTIONS

Culture collections are depositories that store the world's known microorganisms that have been grown in a laboratory. About 20 large culture collections throughout the world are run by nonprofit organizations, businesses, governments, or universities. The table lists the important culture collections used by microbiologists. Some specialize in only certain types of microorganisms, and others strive to store a wide variety of all types of microorganisms.

THE NATURE OF GROWING CULTURES

Growing a microbial culture is not unlike tending to plants in a garden. The biology of microorganisms, as with of any living thing, involves variability. One culture may grow to a dense cell concentration, while the very next culture of the same microorganism shows only moderate growth. A person would not expect every tree in the forest to be of equal height, and, likewise, microbial cultures are influenced by a multitude of known and unknown factors. Culture techniques require experience and intuition to nurture the growth of microorganisms, so the final cultures are similar from one experiment to the next. This may explain why culture techniques depend on many hands-on procedures and relatively few automated methods compared with the physical sciences.

See also AGAR; ASEPTIC TECHNIQUE; BIOREACTOR; COLONY; CONTINUOUS CULTURE; DIRECT COUNT; GRAM STAIN; GROWTH CURVE; MEDIA; MICROSCOPY; SERIAL DILUTION; STAIN.

Further Reading

Cappucino, James G., and Natalie Sherman. *Microbiology: A Laboratory Manual.* Menlo Park, Calif.: Benjamin Cummings, 2004.

Drew, Stephen W. "Liquid Culture." In *Manual of Methods for General Bacteriology,* edited by Phillip Gerhardt.

Important Culture Collections

Acronym	Organization	Specialty
ATCC	American Type Culture Collection, United States	diverse collection of environmental bacteria, fungi, algae, protozoa, viruses including phages, and animal and plant cell lines
BCCM	The Belgium Coordinated Collections of Microorganisms	similar to ATCC
CNCM	National Culture Collection of Microorganisms, France	bacteria, recombinant *E. coli,* fungi including yeasts, phages, human and animal viruses, human and animal cell lines
DSMZ	German Collection of Microorganisms and Cell Cultures	human, animal, and plant cell lines; plant viruses; bacteria; archaea; plasmids; phages; fungi including yeasts
JCM	Japan Collection of Microorganisms	environmental and biotechnology bacteria, archaea, and fungi including yeasts
NCCB	The Netherlands Culture Collection of Bacteria	similar to ATCC, specializing in fungi
NCIMB	National Collections of Industrial, Marine and Food Bacteria, Scotland	bacteria, plasmids, and phages from industrial, food, and marine microbiology
NRRL	Agricultural Research Service Culture Collection, United States	environmental, agricultural, and food microbiology bacteria and fungi including yeasts
SAG	Culture Collection of Algae, University of Göttingen, Germany	algae and cyanobacteria from land and water environments
UKNCC	U.K. National Culture Collection, United Kingdom	similar to ATCC

Washington, D.C.: American Society for Microbiology Press, 1981.

Institute for Systems Biology. "Cell Sorting." Available online. URL: www.systemsbiology.net/Scientists_and_Research/Technology/Data_Generation/Cell_Sorting. Accessed March 26, 2009.

Kreig, Noel R., and Phillip Gerhardt. "Solid Culture." In *Manual of Methods for General Bacteriology,* edited by Phillip Gerhardt. Washington, D.C.: American Society for Microbiology Press, 1981.

Steck, Todd. "Antibiotics May Cause UTI-Causing Bacteria to Become Dormant." Presented at the General Meeting of the American Society for Microbiology, Toronto, May 21–25, 2007. Available online. URL: www.asm.org/Media/index.asp?bid=50155. Accessed March 26, 2009.

cyanobacteria Cyanobacteria comprise a large, diverse group of bacteria that perform photosynthesis. Their photosynthetic ability and a highly organized system of internal membranes make cyanobacteria unusual among the prokaryotes, and, in fact, the cyanobacteria used to be classified as algae because of their resemblance to the eukaryotic cells. Advances in electron microscopy and molecular genetics, in the past several decades, enabled biologists to study cyanobacteria in increasing detail and reclassify them as bacteria. (The term cyanobacteria—not capitalized—is a general name for all the genera in the class Cyanobacteria.

CYANOBACTERIA AND EVOLUTION

Bacteria developed on Earth almost four billion years ago. Until about 2.5 billion years ago, cyanobacteria dominated every habitat that received abundant moisture or periodic rainfalls and flooding. Mats of live microorganisms formed on bodies of water and over time these dense mats trapped sediments. As the sediments grew thicker, the photosynthetic cyanobacteria migrated upward to stay within reach of the Sun's light. Over time, additional sediment layers alternated with living cyanobacteria layers and built up to form mounds. The mounds and layers of trapped sediment and cyanobacteria created fossils known as *stromatolites*. Stromatolites have provided a vast amount of information for study in the specialized field of *micropaleontology*, the study of microscopic fossils or microfossils. The oldest stromatolites are in Western Australia and South Africa and date to 3.5 billion years old. The table on page 210 lists additional places where micropaleontologists have found the largest and oldest collections of these microfossils.

Cyanobacteria played a critical role in the earth's early metabolism, when oxygen was scarce or absent. The early nonphotosynthetic bacteria and even the earliest photosynthetic bacteria did not generate oxygen; oxygen was toxic to bacteria. Rather than using water and oxygen for driving metabolic reactions forward, they used the large amounts of hydrogen sulfide (H_2S) in ancient Earth's atmosphere. Prehistoric bacteria lived in an oxygenless world on the primitive compounds that they found. But an organism that could make its own compounds for growth, rather than rely on ready-made simple compounds, would have a stunning advantage on Earth, especially if it could use the ample energy arriving on Earth in sunlight. The development of photosynthesis represented a pivotal innovation in the evolution of life. Although scientists agree that young Earth's atmosphere contained virtually no oxygen, many questions remain on how cyanobacteria led the way in the fundamental step to using oxygen for life.

The development of a pigment that captured sunlight would have had to happen to begin the evolution of photosynthesis. Perhaps a mutation occurred in which a cell made a pigment by mistake instead of the correct compound. The British biochemist Roger Lewin explained, in 1982, the prevailing theory, in simple terms, on cyanobacteria and the development of life on Earth:

> As we have seen [in studies of microfossils], the first oxygen producers were cyanobacteria. They developed the ability to use the abundant supplies of water in their photosynthetic process; as a consequence, they were more flexible in their habits. Under certain conditions they simply switch off their water-using photosynthetic system and revert to the more primitive hydrogen sulfide mechanism. Such a switch may occur when oxygen falls to low levels in the environment. Presumably, cyanobacteria originated at a time when fluctuations in oxygen availability were part of daily experience in the early world. The high level of oxygen in today's atmosphere is the direct result of the metabolism of the cyanobacteria, and more advanced photosynthesizers that evolved later.

As cyanobacteria produced more and more oxygen, Earth's constitution shifted from one dominated by anaerobic microorganisms to conditions dominated by an evolving array of photosynthetic and oxygen-respiring life-forms. The atmosphere 2.3 billion years ago during the Proterozoic Era became oxygenated.

In this era, the planet's oceans were frozen from the poles to the equator. The relatively rapid growth of cyanobacteria, which could outcompete all other bacteria by virtue of its photosynthesis, began putting increasing amounts of oxygen into the air. The oxygen reacted with the atmosphere's abundant methane gas (CH_4), which helped hold some of the sun's warmth in the atmosphere—the primary role of a greenhouse gas. As oxygen and methane reacted to form carbon dioxide, methane levels declined, and carbon dioxide levels rose. Although carbon dioxide is known today as a major greenhouse gas, methane is actually 22 times more effective than carbon dioxide in retaining the planet's warmth. With the methane disappearing, Earth's temperature started to plunge, to -58°F (-50°C), and a glacial epoch began.

The cyanobacteria probably remained the most abundant life-form for the next 1.8 billion years, during the inevitable rewarming of Earth. About 1.5 billion years ago, fossil records show that primitive eukaryotic cells began to appear. As Lewin wrote, "The old order was inexorably overthrown, and today the diminished ranks of photosynthetic bacteria occupy some of the most extreme and—to us and the rest of the oxygen-dependent organisms—inhospitable environments, such as hot sulphurous pools." Fossil studies also indicate that, in a period between 545 and 500 million years ago, multicellular eukaryotic organisms and the earliest land plants emerged.

Today's photosynthesizing plants owe their capabilities to a process called *endosymbiosis*, which occurred between ancient cyanobacteria and the most primitive of eukaryotic cells. Endosymbiosis is the state in which one microorganism lives inside another microorganism. Structures called chloroplasts inside the cells of photosynthetic algae and plants—possibly also mitochondria—are remnants of cyanobacteria that early cells captured, perhaps as food. This idea had been proposed, in 1905, by the Russian botanist Constantin Mereschkowsky (1855–1921). The origins of Mereschkowsky's insight have been lost to history, but it may be safe to presume that his theory shed a startling new light on the theory of evolution proposed by Charles Darwin (1809–82), a few decades earlier.

In 1999, the German botanists William Martin and Klaus W. Kowallik wrote of the endosymbiosis theory in the *European Journal of Phycology*. In their article, the authors explained that biologists had quickly accepted endosymbiosis, in the first part of the 20th century, but the proposal then hit a period of resistance. "It fell into disfavour shortly after the First

World War, for reasons that are very difficult to summarize briefly, and remained scorned for 50 years." Many biologists returned to the idea that prevailed before the endosymbiosis theory emerged: Internal eukaryotic cell structures developed as a result of the actions of a parent cell and not the invasion or ingestion of a foreign cell. Mereschkowsky's theory on the role of cyanobacteria in the evolution of eukaryotic cells became the almost universally accepted theory, in the 1970s, with an accumulated body of evidence from fossil studies, molecular biology, and advanced tools such as electron microscopy.

Science has several key areas of study to conquer regarding cyanobacteria. Some of the questions that spur scientific debate are the following:

- Are cyanobacteria history's first oxygen producers, or did other bacteria precede them?
- How did cyanobacteria's current photosynthesis pigments evolve?
- In what sequence did heterocysts and akinetes develop and why?
- Are ancient purple sulfur bacteria the precursors to cyanobacteria?

It is not an exaggeration to say that the study of cyanobacteria represents the study of life's evolution on Earth.

CLASSIFICATION OF CYANOBACTERIA

Cyanobacteria have long puzzled biologists because of their diverse structures inside the cell, as well as the external forms that they have in nature. The microorganisms can live as single cells (unicellular), as aggregates of cells that form a ball-like structure, or in a filamentous form. The single cells reproduce by binary fission; aggregates—also called *colonies*—reproduce by multiple fissions, in which all the cells replicate in somewhat coordinated fashion; and long filaments replicate by breaking off pieces, which then continue to reproduce by budding. (Budding is a means of reproduction used by yeast and some bacteria, in which a parent cell gives rise to a smaller daughter cell, which breaks loose and then grows to normal size.) Some filamentous cells develop an akinete form, which is a thick-walled dormant cell that resists harsh environments. Akinetes resist prolonged drying, a characteristic that serve as an important survival tool when the normally aquatic cyanobacteria are deprived of moisture.

Cyanobacteria have been very difficult to grow as pure cultures in laboratories. For this reason, many unknowns about the structures and functions of cyanobacteria exist. The internal cell structures of cyanobacteria distinguish this group from most other bacteria. The pigment cyan gives these microorganisms their bluish green color; cyanobacteria are also referred to as blue-green bacteria or cyanophyta. Some cyanobacteria appear more black or bluish black than green. Cyanobacterial photosynthesis closely resembles that in eukaryotic cells. The photosynthetic activity occurs on internal membranes called thylakoids that are lined with particles called phycobilisomes, which hold the cell's photosynthetic pigments. Chlorophyll, carotenoids, and phycobillin make up the main pigments.

The cell wall of cyanobacteria is gram-negative, but it possesses features seen in gram-positive bacteria. For example, cyanobacteria have a strong protective layer outside the cell membrane.

Cyanobacteria do not possess a taillike appendage called a flagellum, which most motile bacteria use for movement, yet many cyanobacteria can move about in their environment. In some cases, the mechanism for motility in cyanobacteria remains unknown, but some genera such as the marine *Synechococcus* can move by gliding. Gliding is a means of motility in which a cell moves along a solid surface rather than swims through a liquid.

Because the cyanobacteria had been classified for many years as algae, their main characteristics had been studied in relation to other plant life. Therefore, rather than conducting extensive studies on cyanobacteria's cell walls, enzymes, motility, and microbial metabolism, biologists investigated features called *botanical field marks*: types of colony formations, types and frequency of filament branching, and the presence or absence of an outer coating called a sheath.

Biologists have spent considerable effort in reorganizing cyanobacteria to fit logically with other types of prokaryotic cells. This organization places cyanobacteria into phylum X Cyanobacteria, which contains one class, also called Cyanobacteria. Unlike most other bacterial taxonomy, the class Cyanobacteria does not contain orders but rather contains five subsections of cyanobacteria. Each subsection contains at least six genera; the following classification scheme provides an example:

Phylum X: Cyanobacteria

    Class I: Cyanobacteria

        Subsection I

            Genus: *Cyanobium*

Some genera contain more than one species, but other cyanobacteria genera do not contain well-defined species. For this reason, most cyanobacteria studies focus on a specific genus (see Appendix V).

Sites of Ancient Cyanobacteria Stromatolites

Location	Sites
Australia	• North Pole, Western Australia • Shark Bay, Western Australia
Canada	• Yellowknife, Northwest Territories • Thunderbay and Gunflint, Ontario • Waterton Lakes National Park, Alberta
South Africa	• Barberton Mountain Land • Transvaal Dolomites
United States	• Glacier National Park, Montana • Petrified Sea Gardens, New York • Medicine Bow National Forest, Wyoming

TYPES OF CYANOBACTERIA

The five cyanobacteria subsections have been based on five main characteristics of these microorganisms: (1) general shape, (2) reproduction and growth, (3) production of a special cell called a heterocyst that takes in atmospheric nitrogen, (4) the percentage of the nucleic acids guanine plus cytosine (% G + C) in the microorganism's genetic material, and (5) additional distinctive properties. The cyanobacteria subsections are described in the table on page 211.

Microbiologists have learned to classify many cyanobacteria into their proper subsection just by examining the cells under a microscope. These microorganisms look bluish green when alive because of the presence of chlorophylls *a* and *b,* and they appear yellowish or reddish when decomposing. Dead cells often appear dark gray or brown. Microscopic examinations separate the five cyanobacteria subsections into the following proposed names:

I. Chroococcales—rods or cocci as single cells or sometimes in aggregates or in layers

II. Pleurocapsales—single or aggregate cocci

III. Oscillatoriales—long filaments of tiny identical cells

IV. Nostocales—long filaments of irregularly sized cells

V. Stigonaematales—long filaments of unalike cells, forming branches

The table on page 211 illustrates the large amount of diversity among cyanobacteria, yet these bacteria contain additional specialties. Structurally, genera of subsections III and IV produce trichomes with their normal reproducing filaments. Trichomes are branches on a filamentous microorganism, but these branches contain only vegetative cells, that is, cells that do not reproduce. For this reason, trichomes are often called *false branches.*

Cyanobacterial metabolism can be diverse. These organisms range from aerobic to anaerobic growth, and while many use light-requiring photosynthesis, other cyanobacteria can carry out photosynthesis in the dark.

CYANOBACTERIA METABOLISM

Cyanobacteria possess some of the widest diversity of metabolism in all of bacteriology. In general, most cyanobacteria are aerobic, photosynthetic microorganisms that take in carbon dioxide (CO_2) and produce oxygen (O_2). Their main energy-producing metabolic pathway is called the Calvin-Benson cycle, which is a method of storing the carbon from carbon dioxide in reduced organic compounds. (Reduced organic compounds contain many hydrogen molecules attached to the compound's carbon backbone.) Most cyanobacteria also store the energy produced by photosynthesis in a polysaccharide, a long-chain molecule made of sugars, and then use the polysaccharide as an energy source for the cell's maintenance in periods of darkness.

Many cyanobacteria have developed additional capabilities. For instance, specialized genera perform other types of metabolism that are quite different from the normal aerobic photosynthetic mechanism. The table on page 212 provides examples of the major specialties that can occur among members of the cyanobacteria.

Abundant formations of cyanobacteria growing in the ocean have led investigators to seek new and potentially useful compounds from these microorganisms. The oceanographer William Gerwick wrote about his research team's expedition to recover marine cyanobacteria capable of producing compounds with potential anticancer activity. "We launched a program in 1993," Gerwick wrote in *Microbe* magazine in 2008, "surveying marine algae and cyanobacteria in Curaçao and in the southern Caribbean for bioactive natural products. The extract of one shallow-water marine cyanobacterium, *Lyngbya majuscule,* was highly active when tested against a cancerous mammalian cell line, and this finding led us to isolate a lipid we named curacin A." Gerwick eventually found that *Lyngbya* species of subsection III produced nearly 300 distinct substances with potentially useful activities. Synthetic compounds based on curacin A's structure are now being tested in clinical studies on cancer.

Of the almost 800 compounds that Gerwick and others have discovered in marine cyanobacteria, most are made by genera of subsection III. Cyanobacteria researchers focus mainly on secondary metabolites, which are compounds made by microorganisms but are not necessary for normal metabolism, substances such as antibiotics, toxins, and vitamins. Of the compounds discovered so far, marine cyanobacteria contribute the following amounts:

- subsection III, Oscillatoriales—389 compounds, 49 percent of total
- subsection IV, Nostocales—210, 26 percent
- subsection I, Chroococcales—122, 15 percent
- subsection II, Pleurocapsales—48, 6 percent
- subsection V, Stigonematales—29, 4 percent

Gerwick noted another clear advantage of using cyanobacteria as sources of potential new drugs: "Because cyanobacteria colonies grow in such profusion, we collect them by hand in large enough quantities to investigate their chemical, pharmacological, and genetic properties." This method offers a clear contrast to most other studies of environmental microorganisms, which require intensive searches and samplings to take live microorganisms back to a laboratory.

CYANOBACTERIA IN THE ENVIRONMENT

The incredible diversity of cyanobacteria metabolism and cell types enables this group of bacteria to inhabit a very wide range of environments. Cyanobacteria have been found worldwide, and, interestingly, they occupy many places where eukaryotic cells cannot survive. Cyanobacteria have, therefore, been known to dominate certain environments. Betsey Dexter Dyer, author of *A Field Guide to Bacteria,* wrote, in 2003, about the cyanobacteria, "Wherever there is moisture or the potential for sporadic moisture in an area reached by light, there is the possibility of finding cyanobacteria." By using this simple description of cyanobacterial habitats, microbiologists have gone forth into extreme environments to find cyanobacteria. The list of aquatic places where they have found cyanobacteria thriving is truly remarkable, as follows:

- natural freshwater, brackish water, and marine water
- wastewaters
- fountains
- hot and mineral springs
- glacier ice
- salt lakes and salt works
- mudflats and salt marshes
- shoreline rock formations

Cyanobacteria have also been discovered in surprising terrestrial locations. Microbiologists have long known that cyanobacteria could be found in moist soils or on rocks wetted by ocean sprays or streams. But additional forays into the following places have turned up cyanobacteria when looking for

Cyanobacteria Subsections

Subsection	General Shape	Reproduction and Growth	Heterocysts	% G + C	Other Properties
I	single rods or cocci; aggregates	binary fission; budding	no	31–71	nonmotile
II	single rods or usually cocci; aggregates	multiple fission	no	40–46	some motile
III	unbranched filaments	binary fission in a single plane	no	34–67	usually motile
IV	unbranched filaments	binary fission in a single plane	yes	38–47	often motile
V	filaments with either branches or more than one row of cells	binary fission in more than one plane	yes	42–44	produces akinetes

Source: Prescott, Lansing M., John P. Harley, and Donald A. Klein. *Microbiology,* 6th ed. New York: McGraw-Hill, 2005.

Specialized Metabolism within Cyanobacteria

Specialization	Example Genus (Subsection)
anaerobic metabolism of stored polysaccharides	*Oscillatoria* (III)
photosynthesis that may not produce oxygen (anoxygenic)	*Cyanothece* (I)
fermentation that produces lactic acid	*Nostoc* (IV)
use of only light or inorganic chemicals as energy source when using organic compounds as the carbon source	*Calothrix* (IV)
nitrogen fixation (capture of atmospheric nitrogen for use by plants) by heterocysts	*Gloeothece* (I)
nitrogen fixation during periods of darkness	*Anabaena* (IV)

other types of bacteria: pavements, building exteriors, caves, works of art, and deserts and dunes. Places that receive periodic exposure to rain such as buildings or persistently moist places such as caves are suited for cyanobacterial growth. But how do cyanobacteria find a home in a desert or on an object inside a temperature, and humidity-controlled museum? Filamentous cyanobacteria manage to exist in these unlikely places because dead cells in filaments store water and heterocysts help the microorganism remain alive for extended dry periods. In deserts, cyanobacteria usually find small habitats where a moss or lichen has adhered to the underside of a rock. Since these organisms retain moisture, the cyanobacteria carve out an existence by growing in a layer beneath the moss or lichen and on top of the rock's surface.

Life in Aquatic Habitats

Cyanobacteria have become associated with almost all aquatic environments. In or on water, cyanobacteria form huge aggregates large enough to see. These microorganisms can form masses called microbial mats that float on the water's surface and extend from a half-inch (1.3 cm) to several feet below the surface. Cyanobacteria also form a layer called felt on submerged rocks as well as a film that covers the submerged part of plants, sometimes called *fuzz.*

Microbial mats offer perhaps the best-known example of cyanobacteria in aquatic habitats. A microbial mat consists of a layer of many different types of microorganisms that interact with each other and can recycle all of the nutrients they need in a process called biogeochemical cycles. A large mat containing a diversity of microorganisms can recycle nutrients to the point where it is completely self-sufficient.

Microbial mats live on freshwater and marine waters, salt lakes, and hot springs. A fully developed microbial mat is composed of layers that always occur in the same arrangement, in which cyanobacteria dominate other microorganisms at the top exposed to the most sunlight. Sometimes, a thin layer consisting of sand or organic debris lies atop the cyanobacteria, but this layer does not interfere with photosynthesis.

The layers of the microbial mat each carry out a specialized duty in the overall metabolism of the mat community. Distinct aerobic and anaerobic layers, separated by a layer of sediment high in oxidized iron, characterize microbial mats. The origin of the oxidized iron is not completely known, but it probably serves in part to erect a barrier between the aerobic activities of cyanobacteria and the anaerobic activities of purple sulfur bacteria lying beneath the oxidized iron. These roles are described in the table.

Microbial mats possess their own diurnal rhythms, meaning they behave differently in the daytime compared with the nighttime. Motile *Beggiatoa* bacteria, not a member of the cyanobacteria, offers an example of the diurnal activities of microbial mats. *Beggiatoa* lives in mats composed of cyanobacteria and purple sulfur bacteria. But these bacteria avoid light, oxygen, and hydrogen sulfide gas (H_2S). The microorganisms, therefore, live only in a limited region between the oxygen-producing cyanobacteria layer and the H_2S-producing purple sulfur bacteria layer; they cannot venture out of this area without being killed by the high oxygen concentrations above and the high H_2S concentrations below. At the interface between the aerobic and anaerobic layers, *Beggiatoa* uses small amounts of oxygen to oxidize reduced sulfur compounds that drift upward from the deeper anaerobic layer. In this manner, *Beggiatoa* generates energy for its metabolism, maintenance, and growth. At nighttime, microorganisms other than cyanobacteria respire and begin to use up the oxygen in the top layer. *Beggiatoa* glides upward at night in the darkness and low-oxygen conditions. As the sunlight begins returning and cyanobacteria photosynthesis again begins to pump out oxygen, *Beggiatoa* glides downward, until it reaches its safety zone within the microbial mat.

Microbial mats have also contributed to discoveries in paleontology. Some ancient animal fossils are the products of animals that died, fell onto a microbial mat, and were only partially decomposed because the anaerobic layer prevented full decomposition. The mats grew over the bones, which iron-metabolizing bacteria eventually mineralized: That is, they increased the mineral content of the bone. These ancient mats have produced so-called death mask fossils, which appear reddish because of the iron oxide in them.

The aquatic cyanobacteria have been called the world's most important bacteria because of the nutrient recycling they perform and the role they have had in the evolution of life on Earth. Before the early 1970s, aquatic cyanobacteria had been thought to live only in freshwater. By 1980, however, marine biologists from the Woods Hole Oceanographic Institute (WHOI) and Massachusetts Institute of Technology had discovered the cyanobacteria *Synechococcus* and *Prochlorococcus* in the ocean. John Waterbury of WHOI accompanied the expeditions, in 1977, to find marine bacteria. He described the findings, in 2004, in *Oceanus* magazine: "We knew right away that *Synechococcus* was something important by the impressive numbers of them in seawater samples. Since 1977, they have been found everywhere in the world's oceans when the water temperature is warmer than 5°C [41°F] at concentrations from a few cells to more than 500,000 cells per milliliter (about 1/5 of a teaspoon), depending on the season and nutrients. This amazing abundance makes them a source of food for microscopic protozoans, the next organisms up the food chain that ends in fish and humans." It turns out that the marine cyanobacteria are the most abundant organisms on the planet.

Metabolism in a Microbial Mat

Layer	Role
cyanobacteria	aerobic photosynthesis produces organic compounds and oxygen for other microorganisms in the upper layer
oxidized iron	barrier between aerobic and anaerobic functions
purple sulfur bacteria	use sulfide, a compound produced in the layer by sulfur-reducing bacteria
iron sulfide	collects the sulfide that precipitates out of the layer above in the form of iron sulfide (FeS)

Life in Terrestrial Habitats

Cyanobacteria do well in terrestrial habitats that receive sporadic moisture, low light, or extreme conditions. In fact, cyanobacteria can thrive in these places, which eukaryotes such as algae find inhospitable.

On barren land that has not supported living things for a long time—the period after a fire, for example—cyanobacteria often act as the first inhabitant in a process called ecological succession. In ecological succession, new plant and animal communities establish themselves, over time, in an area, and then they are replaced by a series of different, usually larger, and more complex organisms. Lichens and mosses usually follow cyanobacteria in ecological succession (see the color insert on page C-2, lower left). The cyanobacteria also support the small plants that arrive next in the mostly barren environment. They do this by capturing nitrogen from the atmosphere (nitrogen fixation) and so give the plants the nitrogen they need to live.

Lichens may be composed of either algae-fungi or cyanobacteria-fungi, called cyanolichen. In either type of lichen, the photosynthetic organism provides carbon and nitrogen compounds to the fungus, and the fungus provides protection and moisture for the microorganism. Cyanolichens account for only about 8 percent of all the world's lichens, but they provide a simple example of a symbiotic relationship between cyanobacteria and higher organisms. In symbiosis, two organisms live together in a cooperative relationship.

Cyanobacteria create symbiotic relationships with the following plant life: the large green eukaryotic alga named *Codium* (a seaweed); bryophyte plants, consisting of mosses, liverworts, and hornworts; ferns; cycads (tropical nonflowering seed plants); tropical bromeliads that have large water-holding leaves; and gunnera, a tropical wetland plant. The cyanobacteria usually live in the upper soil near the plant stalk and perform nitrogen fixation. In nitrogen fixation, the cyanobacteria take nitrogen out of the atmosphere and convert it to a form, such as ammonia (NH_3), that the plant can use for its growth.

Relationships between Cyanobacteria and Animals

Animal	Relationship with Cyanobacteria
crabs	cyanobacteria form a community with algae and other organisms on the outer shell
elephants	during mating season, male elephant genitalia develop a film of algae and cyanobacteria
flamingos	cyanobacteria pigments in the crustaceans eaten by flamingos colors the birds' plumage
lobsters	cyanobacteria live in large numbers in the lobster's bronchia
sponges	cyanobacteria throughout the sponge might provide a food source
three-toed sloths	cyanobacteria and insects living in the coat provide camouflage coloring for the sloth

Cyanobacteria also have intriguing associations with some animals. Examples of interactions between animals and cyanobacteria on land and in the marine environment are shown in the table above. The reason for many of these relationships remains unknown.

People have made use of cyanobacteria for centuries, mainly as a food. *Nostoc* is part of the diet in parts of Asia and Central and South America. In Western diets, people use *Spirulina* as an antioxidant and a boost to the immune system.

The animal-cyanobacteria relationships are interesting, but they do not tell the extraordinary history of cyanobacteria on Earth and their role in the evolution of higher plants and animals.

See also ALGAE; BIOGEOCHEMICAL CYCLES; METABOLISM; MICROBIAL ECOLOGY; NITROGEN FIXATION; PHOTOSYNTHETIC BACTERIA; SYMBIOSIS; TAXONOMY.

Further Reading

Dyer, Betsey Dexter. *A Field Guide to Bacteria.* Ithaca, N.Y.: Cornell University Press, 2003.

Gerwick, William H., R. Cameron Coates, Niclas Engene, Lena Gerwick, Rashel V. Grindberg, Adam C. Jones, and Carla M. Sorrels. "Giant Marine Cyanobacteria Produce Exciting Potential Pharmaceuticals." *Microbe,* June 2008.

Martin, William, and Klaus V. Kowallik. "Annotated Translation of Mereschkowsky's 1905 Paper *'Über Natur und Ursprung der Chromatophoren im Pflanzenreichi.'*" *European Journal of Phycology* 34 (1999): 287–295. Available online. URL: journals.cambridge.org/action/displayAbstract?fromPage=online&aid=47603. Accessed February 20, 2009.

Thajuddin, N., and G. Subramanian. "Cyanobacterial Biodiversity and Potential Applications in Biotechnology." *Current Science* 89, no. 1 (2005): 47–57. Available online. URL: www.ias.ac.in/currsci/jul102005/47.pdf. Accessed February 20, 2009.

University of California. "Introduction to the Cyanobacteria: Architects of the Earth's Atmosphere." Available online. URL: www.ucmp.berkeley.edu/bacteria/cyanointro.html. Accessed February 19, 2009.

Virtual Fossil Museum. Available online. URL: www.fossilmuseum.net/index.htm. Accessed February 20, 2009.

Waterbury, John. "Little Things Matter a Lot." *Oceanus* 43, no. 2 (2004): 1–5. Available online. URL: www.whoi.edu/oceanus/viewArticle.do?id=3808. Accessed February 20, 2009.

diatom Diatoms are one-celled algae, 3 to 4 micrometers (μm) across, with a hard shell made of silica, the mineral form of silicon dioxide (SiO_2). Biologists categorize diatoms in the diverse category of aquatic living things called *plankton,* which consists of small organisms and microorganisms that serve as the foundation of marine food chains. Plankton and diatoms are among the most abundant photosynthetic organisms on Earth. Diatoms make up the largest portion of marine plankton and can also be referred to as *phytoplankton,* or plankton of plant origin. In their role as plankton, diatoms play a crucial role in Earth's nutrient use, by making energy and recycled nutrients available to more complex organisms.

More than 10,000 species of diatoms exist. Diatoms occupy a single family named Bacillariophyceae that is part of Chrysophyta, or simply golden algae, one of eight different divisions of algae. Characteristically of algae, diatoms carry out photosynthesis and may be responsible for 20–25 percent of all photosynthetically made organic carbon on Earth. Diatoms, therefore, play two roles that are vital for human life: as the foundation of food chains and as producers of atmospheric oxygen.

Biologists divide diatoms into two major types and one minor type. The major divisions contain the centric diatoms and the pennate diatoms. These two types differ in cell shape and in habitat. The genus *Hemidiscus* comprises the third type. The main distinctions among these types of diatoms are shown in the table below. Most diatoms are nonmotile, but when a species is motile, it tends to use the motility method shown in the table.

DIATOM STRUCTURE

Centric diatoms are characterized by having radial symmetry in shape, meaning the shape is usually a symmetrical disk or ball. Centric diatoms possess surface markings that radiate from the center. Pennate diatoms, by contrast, tend to have bilateral symmetry, characterized by an elongated—oval,

Types of Diatoms

Type	Characteristic Shape	Motility	Habitats
centric	radial symmetry	single flagellum	mainly marine
pennate	bilateral symmetry	gliding in some genera	freshwater and marine
Hemidiscus	asymmetrical or three-, four-, or five-fold rotational symmetry (triangle, square, or star, respectively)	nonmotile	freshwater and marine

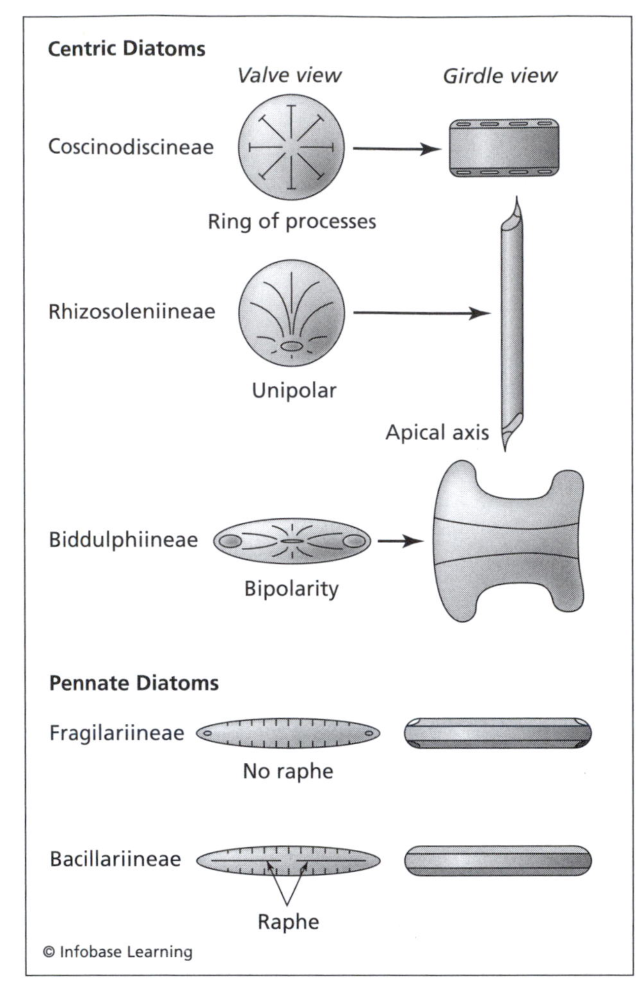

Two orders of diatoms are differentiated by their shape. Centric diatoms orient around a central point called an annulus, or central areola. Pennate diatoms follow a line or a plane. Most diatoms that make up marine plankton are centric diatoms.

spindle, or oblong—shape with surface markings at right angles to the long axis. Some diatoms form long chains of identical cells.

Each diatom species has a characteristic shape, an often intricate ornamentation, which microbiologists use to identify diatoms. The diatom body, called a *frustule,* consists of two sections called *thecae,* or valves. Thecae can have very elaborate architectures unlike anything else found in nature. If the sections differ in size, the larger piece is called the *epitheca* and the smaller of the two is the hypotheca. Thecae fit together by overlapping, and the cell produces a material composed of silica to bind the pieces together. This durable crystallized silica [$Si(OH)_4$] cell wall provides protection for the soft cytoplasm interior.

Cytoplasm in diatoms is referred to as *protoplasm* and has no cell membrane surrounding it. The protoplasm contains the diatom's nucleus, where it holds its deoxyribonucleic acid (DNA), and storage compartments called *fat globules.* Large chloroplasts make up the greatest volume of the diatom interior for carrying out photosynthesis and giving diatoms their color. Different diatoms contain different chloroplast arrangements, from a few large chloroplasts to many smaller chloroplasts. The main pigments in diatom chloroplasts are chlorophyll, beta-carotene, and fucoxanthin.

The almost indestructible diatom shell is a source of fossils in marine environments. When large organisms eat diatoms, they eliminate the indigestible shell portion, which sinks to the ocean bottom. Since the Cretaceous period, 140 million–65 million years ago, diatoms have made large deposits of chalky fossilized shells. These deposits become an abrasive material known as *diatomaceous earth,* or diatomite. Commercial enterprises mine diatomaceous earth and sell it to makers of paints, cleansers, polishes, and toothpaste, and as a material in water filters. Studies on various diatom fossil deposits have been helpful in assessing the condition of the Earth at various time points in its history. That is because diatoms have been able to live under a very wide range of conditions, from extreme heat to extreme cold.

Electron microscopy has allowed taxonomists to classify diatoms in more detail than in the past, on the basis of features of the frustule. Diatoms have been classified, since 1990, into two orders, comprising the centric and the pennate diatoms, and suborders, based on specific shapes within those two groups. Centric diatom suborders are differentiated by the arrangement of striations in the shell when viewing the diatom from above. The pennate diatom suborders are distinguished by the presence or absence of a slit called a *raphe* that runs the length of one theca, interrupted only by a central nodule. The presence of a raphe indicates whether a pennate diatom is motile or nonmotile; all species that have gliding motility also have a raphe. The diatom classifications are described in the table.

DIATOM LIFE CYCLE

Diatoms reproduce asexually by constructing a new theca inside the parent before the parent cell divides. Before division, the nucleus moves to the center of the cell, and the protoplasm swells, pushing apart the two thecae. Mitosis then occurs, wherein the DNA replicates and the cell divides, so that each new cell receives a copy of the DNA.

When a diatom divides, each new diatom takes as its epitheca (the larger of the two thecae) a theca from the parent—one daughter gets the parent's original epitheca, and the other daughter gets the parent's hypotheca. Within 10–20 minutes, most diatoms complete building a new theca to complement the one received from the parent.

Diatom Classifications

Order		Shape
Biddulphiales (also named Centrales)		centric
	SUBORDER	
	Coscinodiscineae	symmetric radial
	Rhizosoleniineae	unipolar radial
	Biddulphiineae	bipolar
		pennate
	SUBORDER	
Bacillariales (also named Pennales)	Fragilariineae	raphe absent
	Bacillariineae	raphe present

Each successive generation of diatoms produces smaller and smaller cells because the new thecae were produced inside a parent diatom. Diatoms must find a way to return to their original size, or they will seemingly vanish. Diatoms, in fact, do reestablish their normal cell size within a new population. When cell size diminishes by about 30 percent, diatoms begin to reproduce sexually to form a resting cell called an auxospore. After the resting phase, a protoplast emerges from the auxospore, and it quickly expands to normal size before the cell builds another rigid outer cell. This is a rare example in nature, in which a protoplast plays an active role in a microbial life cycle. In bacteria, protoplasts form only when harsh environments damage cell walls, but the protoplasts never become part of bacteria's life cycle.

Some diatoms alternative between a reproductive phase and a dormant stage characterized by a thicker than normal cell wall. This resting form is called a *statospore*. Statospores possibly form in conditions of low nutrients or other hardships in the environment. When nutrients and sunlight increase, a cell emerges from the statospore and reproduces as normal.

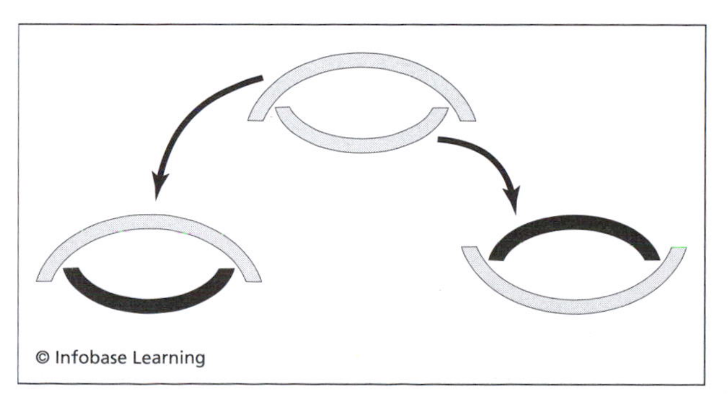

Diatom reproduction involves the contribution of one theca from the parent to each of two daughter cells. The resulting daughter cells will be smaller than the parent.

DIATOMS AND ECOLOGY

Diatoms occur in aquatic environments in either free-floating form called *planktonic diatoms* or attached to surfaces. Planktonic diatoms float near the water's surface as a result of pockets of air trapped in the frustule's elaborate architecture. Only pennate diatoms attach to surfaces, and they do this by secreting mucilage, a sticky mucuslike compound. The mucilage forms weak connections between the diatom and the various submerged surfaces it touches. Because pennate diatoms tend to be submerged, they are called *benthic diatoms* rather than planktonic diatoms.

Both planktonic and benthic diatoms require sunlight for photosynthesis, so they stay in shallow water called the *photic zone*. This zone extends to about 656 feet (200 m) deep. In sunlit aquatic environments, diatoms can live in salt levels ranging from no-salt freshwater to hypersalty conditions found in salt lakes, as well as a range in water temperatures.

Diatoms live in a very large habitat, considering that water covers three fourths of the planet's surface. Diatoms contribute a significant amount of the atmosphere's oxygen, removal of carbon dioxide, and recycling of nutrients. Scientists have recently studied ways to enhance these benefits to the environment. In the late 1990s, scientists studying diatom genes discovered that diatoms have 75 genes involved in silicon metabolism, not surprising since silicon plays a major role in frustule development. But the studies also uncovered 84 genes related to metabolism when either silicon or iron becomes scarce in the water. This finding suggests that iron also has a key role in diatom physiology.

Scientists have pondered innovative ways to halt global warming by affecting the number of diatoms in the ocean. In 2008 a *ScienceDaily* article explained the potential benefit of and concerns about the process: "Considering that 30 percent of the world's oceans are iron-poor, some scientists have suggested fertilizing such areas with iron so diatoms become more numerous and absorb more carbon dioxide from the atmosphere, thus putting the brakes on global warming. If, however, adding iron causes diatoms to change the thickness of their shells then perhaps they won't be as likely to sink and instead would remain in the upper ocean where the carbon they contain might be released back to the atmosphere as they decay or are eaten." The massive amount of diatoms in the ocean certainly suggests that these microorganisms play a critical role in ecology.

DIATOM-CAUSED DISEASES

Despite the benefits offered by diatoms, these microorganisms also produce a marine-associated poison called domoic acid belonging to a class of organic compounds called *tricarboxylic acids*. Domoic acid poisoning in people was first discovered, in 1987, on Prince Edward Island, Canada, among persons who had eaten mussels infected with the diatom *Pseudonitzschia*. Domoic acid poisoning causes symptoms anywhere from 30 minutes to 24 hours after a person has eaten infected seafood. The compound is a nerve toxin, or neurotoxin, a substance that harms the nervous system. Severe poisoning called *amnesic shellfish poisoning* (ASP) has also been linked to eating anchovies and clams.

Marine and coastal animals also suffer from domoic acid poisoning similar to ASP in humans. In 1991, biologists first noticed that pelicans fishing for anchovies along the California coast began dying of domoic acid intake. This observation gave the first solid evidence that domoic acid infection was not confined to marine shellfish, and marine biologists have since expanded the list of potential sources of domoic acid due to infection with diatoms: mussels, oysters, razor clams, krill, and the nonmuscle tissue of anchovies, sardines, crab, and lobster.

In people, ASP causes any of the following symptoms in minor poisoning cases: nausea, vomiting, abdominal cramping, and diarrhea. In severe ASP cases, the symptoms are headaches, disorientation, hallucinations, short-term memory loss, difficult breathing, seizures, coma, and, in extreme cases, death. ASP can have to a mortality rate as high as 4 percent.

Along with the lengthening list of marine life that transmit domoic acid poisoning, marine biologists have found that California sea lions, dolphins, otters, and birds in addition to pelicans fall victim to the neurotoxin from eating infected fish and shellfish. Marine biologists have also wondered whether domoic acid has affected whale populations. In 2007, *Plankton News* reported, "Dr. Spencer Fire, Coordinator for NOAA's [National Oceanic and Atmospheric Administration's] Analytical Response Team, tested 18 samples from whales collected between 1997 and 2007. Specimens tested were strandings from the coast of SC [South Carolina] including pygmy sperm whales, dwarf sperm whales, and a beaked whale. DA [domoic acid] concentrations were found in four of the whales with amounts ranging between 24 and 265 ng/g [nanograms per gram]. This result indicates some toxin exposure, but it is not necessarily considered the cause of mortality." NOAA has since found that whale deaths in the Georges Bank off Cape Cod, Massachusetts, have been due to domoic acid poisoning.

Domoic acid binds to a site on the surface of marine mammal nerve cells called a glutamate receptor. The disabled receptor cannot carry out its normal job of helping ions (charged molecules) travel through the nerve cell membrane, so nervous system function is damaged, and eventually the affected nerve cells die. In advanced domoic acid poisoning of brain tissues death results.

Marine mammals that have been poisoned by domoic acid have a difficult time staying afloat and breathing. Many of the affected marine mammals haul out onto a beach, if possible, to rest and survive. Such beached marine mammals as sea lions have shown head weaving and bobbing, bulging eyes, excessive drooling, disorientation, and poor coordination described as "drunken movements." The most seriously ill animals have seizures. In the worst scenario, dead sea lions or dolphins wash ashore. Marine mammal rehabilitation centers take in sick animals to nourish them and allow time to recover, but marine mammal deaths have, nonetheless, been rising since 1991, when the large-scale outbreaks first began.

Diatoms require a significant amount of future study to determine all the roles they play in the environment and to devise ways to prevent this important photosynthetic microorganism from being a health threat.

See also ALGAE; CELL WALL; MARINE; MICROBIOLOGY.

Further Reading

National Oceanic and Atmospheric Administration. "Domoic Acid." Available online. URL: www.nmfs.noaa.gov/pr/pdfs/health/domoic_acid.pdf. Accessed February 23, 2009.

———. *Plankton News*. "Investigating a Link between Domoic Acid and Whale Mortality." September 2007. Available online. URL: chbr.noaa.gov/pmn/downloads/PlanktonNews_September2007.pdf. Accessed February 23, 2009.

Science*Daily*. "Could Tiny Diatoms Help Offset Global Warming?" January 26, 2008. Available online. URL: www.sciencedaily.com/releases/2008/01/080123150516.htm. Accessed February 23, 2009.

———. "Toxic Algal Blooms May Cause Seizures in California Sea Lions." June 10, 2008. Available online. URL: www.sciencedaily.com/releases/2008/06/080609103232.htm. Accessed February 23, 2009.

diffusion Diffusion refers to the act of a substance's spreading through another substance. Microbiology involves two different and unrelated types of diffusion: diffusion of nutrients through a cell membrane and diffusion of compounds through agar. Membrane diffusion is a process by which a microorganism takes in nutrients and excretes wastes or other substances, such as antibiotics. Agar diffusion techniques, however, relate to a variety of laboratory tests for determining the activity of antibiotics and other biocides on the growth of a microorganism.

MEMBRANE DIFFUSION

Nutrients enter microbial cells by one of two methods: passive transport or active transport. Passive transport allows a substance to diffuse from a place of high concentration across the cell membrane to a place of lower concentration without any expenditure of energy by the cell. Active transport, by contrast, requires that a cell use energy to move a substance across the cell membrane, usually moving the substance from a place of low concentration to a place of high concentration. Active transport relies on the energy-generating systems in the cell membrane, and this energy is usually stored in the form of the compound adenosine triphosphate (ATP). Passive transport using diffusion spares ATP for other uses, so it represents an energy-conserving activity.

Passive transport consists of three different modes of diffusion, shown in the table.

Cell membranes act as selectively permeable (containing pores) membranes that allow some substances into microbial cells but shut out other substances. Microorganisms mainly select nutrients and water for introducing into the cell. They selectively block biocides and toxic compounds from entering the cell interior.

Simple diffusion works best for absorbing small molecules. Glycerol may be one of the largest compounds to use simple diffusion to enter microbial cells. Glycerol contains only three carbons, and each one has a hydrogen and a hydroxyl group (OH) attached. In fact, beyond oxygen, carbon dioxide, water, and glycerol, very few other compounds use this process for crossing the cell membrane. Of the three types of passive transport, simple diffusion works slowest and is the least selective in the compounds it allows entry, providing the compounds are small enough to traverse the cell membrane by this mechanism.

Facilitated diffusion relies on a selective mechanism for transporting molecules across the cell membrane. Proteins in the cell membrane called *transporters* form a pore for certain molecules to use in moving across the membrane. When a nutrient such as sugar must be carried into a cell, facilitated diffusion first begins with the binding of the transporter to the sugar molecule. The sugar makes the transporter change shape in a way that carries the sugar through the membrane and releases it on the opposite side. Once the sugar has detached, the transporter recoils in a way that puts it back in its original shape, ready to bind to another sugar

Modes of Passive Transport

Mode	Description	Example Molecules Transported
simple diffusion	movement of molecules from high to low concentration until they are evenly distributed on both sides of the membrane	oxygen, carbon dioxide
facilitated diffusion	movement of molecules from high to low concentration with the help of a membrane-bound transporter molecule (also called a channel protein or permease)	sugars, amino acids, vitamins
osmosis	movement of a solvent (a liquid that contains dissolved molecules) from low solvent concentration to high concentration	water

molecule. Each different transporter in a membrane selects specific substances to carry; facilitated diffusion is, for this reason, called substrate-specific.

Transporters can be affected by a process called competitive inhibition, in which a molecule that closely resembles the desired nutrient may fit into the transporter and block nutrient uptake. Competitive inhibition has been considered as a way to kill eukaryotic cells such as cancer cells. This treatment would be unlikely to act as an antibacterial agent because relatively few bacteria use facilitated transport.

Escherichia coli uses facilitated diffusion to transport glycerol into its cell. The *E. coli* transport system contains a specialized feature called a major intrinsic protein (MIP) channel, which has been found in all plant and animal cells, as well. An MIP channel consists of a string of amino acids called a polypeptide that carries out the transport by building a pore for specific molecules. In addition to *E. coli,* MIP channels have been discovered in *Lactobacillus lactis* bacteria. Other microorganisms that rely on MIP channels are the yeast *Saccharomyces cerevisiae* and some of the microorganisms of domain Archaea.

Simple and facilitated diffusion differ from osmosis in two main ways. First, these transport methods mainly carry uncharged molecules. Second, both types of diffusion transport molecules from high to low concentration. This process is referred to as moving down a concentration gradient, or using a down gradient. Osmosis performs the important role of moving substances up a concentration gradient.

Cell membranes use osmosis to regulate the concentration of materials inside the cell relative to the outside. Osmosis, therefore, prevents cells from bursting in a very concentrated liquid, and it prevents them from shrinking when in a very concentrated liquid. Microorganisms use osmosis mainly in the movement of water into or out of the cell for the purpose of maintaining the desired pressure inside the cell. Most bacteria live in environments where the concentration of dissolved compounds is less than the concentration of materials inside the bacterial cell, a situation called *hypotonic conditions.* Most bacteria depend on their rigid cell wall to prevent bursting, or lysis, but bacteria that have weak cell walls (some gram-negative species) or no cell wall (mycobacteria) depend on osmosis for survival.

AGAR DIFFUSION

Microbiologists have developed several methods that take advantage of the diffusion of chemicals through agar for the purpose of testing microbial resistance or susceptibility to a material such as an antibiotic. These methods may be divided into two main categories that both require solidified agar plates: diffusion tests and disk diffusion tests. In diffusion tests, a microbiologist prepares wells in the agar, into which a chemical to be tested can be poured. Disk diffusion testing requires a flat agar surface upon which the microbiologist places paper disks that have been saturated with the test chemical. In both the well method and the disk method, the activity of an antibiotic or a chemical biocide against specific bacteria can be measured.

The Kirby-Bauer test and the E test are two main methods that use the principle of an agent, such as an antibiotic, diffusing through agar. The Kirby-Bauer method consists of the following steps:

1. A pure culture of specific bacteria is spread in an even film over the surface of an agar plate, containing Mueller-Hinton agar.

2. Filter paper disks each impregnated with known concentration of different antibiotic solutions are placed apart from each other on the inoculated agar surface.

3. The microbiologist incubates the plates.

4. After incubation, the microbiologist notes which concentrations of antibiotic have created a zone around the disk containing no growth.

The Kirby-Bauer test results provide two important pieces of information. First, the results tell the microbiologist whether an antibiotic has been effective in killing the specific bacteria in the assay. Second, the results may indicate which antibiotic is most effective, when compared with the others in the test.

An antibiotic is considered effective in the Kirby-Bauer test if it causes a zone of no bacterial growth surrounding its disk. This area, called the *zone of inhibition,* shows where the antibiotic has killed the bacteria. All other areas of the plate distant from any antibiotic disk will contain a dense film of live bacteria called a lawn. A microbiologist can measure the diameter of the zone of inhibition with a ruler; in general, the larger the zone, the more effective the antibiotic. Put another way, the Kirby-Bauer test indicates whether a microorganism is susceptible, intermediate, or resistant to a given antibiotic. These results can be described as follows:

- susceptible: The specific bacterial species cannot survive in the presence of the antibiotic.

- intermediate: The bacteria can grow, but poorly, in the presence of the antibiotic.

- resistant: The bacteria can grow in the presence of the antibiotic.

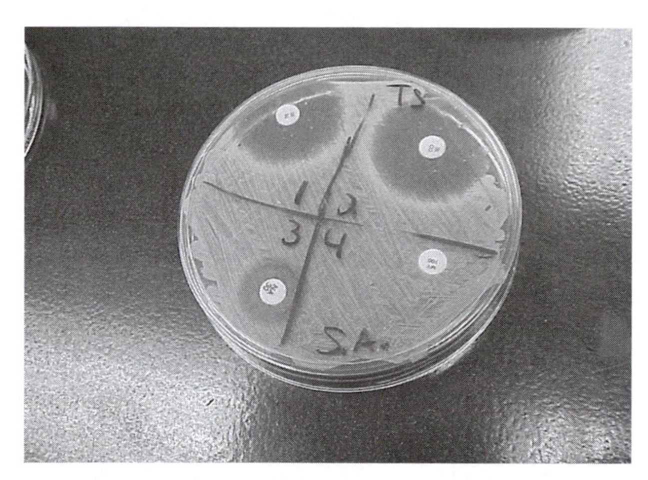

In this Kirby-Bauer test, the larger clear zones of inhibition in the bacterial lawn (quadrants 1 and 2) indicate greater diffusion of the antibiotic out of the paper disk and into the agar. ***(Feu Med Tech AKO)***

The Kirby-Bauer test results require one vital assumption: All antibiotics diffuse through agar in exactly the same manner. In fact, some antibiotics, such as antibiotic A, diffuse better than others and so create a large zone of inhibition. Another antibiotic, antibiotic B, might work very effectively against bacteria, but it does not diffuse well through agar. This scenario would suggest that antibiotic A works better than antibiotic B, when, in reality, antibiotic A simply diffuses better. For this reason, microbiologists use Kirby-Bauer results to find effective antibiotics and eliminate ineffective antibiotics but turn to more specific testing thereafter.

The newer, more detailed E test (for *epsilometer*) uses a plastic strip containing antibiotic rather than a disk. A microbiologist sets up the E test in a similar way to setting up the Kirby-Bauer test. One side of the strip contains an antibiotic that diffuses into agar during incubation. The strip's other side contains a scale with marks designating a concentration gradient for the antibiotic. After incubation, a microbiologist reads the zone of inhibition directly against the scale. This method results in a value called minimum inhibitory concentration (MIC), which is the lowest concentration of antibiotic that affects a certain bacterial species.

Today's clinical microbiology laboratories supplement Kirby-Bauer test results when possible, but this simple and inexpensive method remains a primary tool in clinical microbiology.

See also AGAR; CLINICAL ISOLATE; IDENTIFICATION; OSMOTIC PRESSURE.

Further Reading

Mendoza, Myrna T. "What's New in Susceptibility Testing?" *Philippine Journal of Microbiology and Infectious Diseases* 27, no. 3 (1998): 113–115. Available online. URL: www.psmid.org.ph/vol27/vol27num3topic5.pdf. Accessed February 23, 2009.

Shoeb, Hussein. "Antimicrobial Susceptibility Testing (Kirby-Bauer) Animation." January 30, 2008. Available online. URL: www.microbelibrary.org/asmonly/details.asp?id=2754&Lang. Accessed February 23, 2009.

direct count A direct count, or direct microscopic count, is one of several methods used in the process called microbial enumeration. Enumeration—often simply called *counting*—may be any method that enables a microbiologist to determine the number of microorganisms in a known weight or volume of material. Two major categories of enumeration available to microbiologists are direct counting and indirect counting. These methods are described in detail in the following table.

In a direct count, a microbiologist puts a known volume of a microbial suspension into a defined area of a specialized type of microscope slide. The microbiologist then counts the number of cells in a given field of the slide, marked by a small grid etched into the slide itself, under a microscope. After counting multiple fields, the microbiologist averages the counts from all fields and then multiplies the number by a factor that converts the result into cells per milliliter (ml). Accurate direct counts depend on good techniques in sample preparation and dilution.

Direct counts are suitable for the following microorganisms: bacteria, yeasts, fungal spores, algae, and protozoa. Direct counts help when a microorganism does not grow on standard laboratory media. Live cells that exist in the environment but do not grow in laboratory conditions are known as *VBNC* for "viable but not culturable" cells. Direct count methods provide a way to estimate the numbers of VBNC cells in a suspension.

Indirect counts rely on culture methods whereby a diluted suspension of cells is inoculated to agar medium, incubated, and then the resulting colonies that grew from single cells are counted. Accurate indirect counts depend on good aseptic techniques, culture methods, and dilution.

Direct counting's oft-stated disadvantage lies in its failure to differentiate between live and dead cells. This may be critical when testing the effectiveness of an antibiotic, disinfectant, or any other type of biocide against microorganisms. On the other hand, some industries put less emphasis on live or dead cells because they want all cells out of their final product. Examples of industries that seek to eliminate most or all cells from their products are the following: food production and processing, drug manufacture, personal care products, and drinking water treatment. Within drug manufacturing, the makers of sterile injectable drugs such as vaccines must assure that their product is free of all microorganisms, living or dead.

MICROSCOPIC DIRECT COUNTS

Counting microorganisms under a microscope provides information in addition to counts per milliliter, such as the size and shape of the microorganisms present. A person using this method must keep in mind, however, that the volume of material counted is a fraction of a milliliter, so a slight inaccuracy in the count becomes a large inaccuracy when extrapolating the results. For this reason, direct counts tend not to be used for environment assessments such as water samples or products. Direct counts serve best for monitoring cell numbers in laboratory conditions or for counting cells that cannot be counted any other way.

Microbiologists use three different types of counting chambers for microscopic direct counts: the Petroff-Hausser counting chamber, the Neubauer counting chamber, and the hemocytometer. All of these counting chambers contain a grid called a Neubauer grid, etched into the chamber's glass and visible in a microscope's field. The grid provides a set area that will contain the cell suspension, and a microbiologist counts the cells held in a known volume. Although either prokaryotes or eukaryotes can be used in these chambers (chamber volumes are shown in the table), for increased accuracy and ease of use, microbiologists tend to use the chambers for counting the following:

- Petroff-Hausser—bacteria in a single counting chamber
- Neubauer—two bacterial suspensions in adjacent counting chambers

Direct Counting and Indirect Counting Methods

Type	Description	Advantage	Disadvantage
		Direct Counts	
microscopic	counting cells in a very small volume under a microscope	fast results	counts include dead cells as well as live (viable) cells
electronic	each cell in a suspension forced through a small orifice causes a brief electrical resistance	fast results	expensive
		Indirect Counts	
colony counts	a cell suspension is inoculated to an agar plate and incubated, and the resulting colonies are counted	accurate for counting live cells and eliminating dead cells	24–72 hours for most results; detects only cells that can grow on medium
most probable number (multiple tube method)	number of cells in a suspension is inferred mathematically from the growth in a series of tubes	useful for bacteria that grow only in liquid media or suspensions of more than one type of bacteria (mixed cultures)	reduced accuracy; time-consuming
optical density (light scattering)	cell density is related to amount of light absorbed or transmitted by the liquid suspension	nondestructive to the cells	reduced accuracy

- hemocytometer—bacteria and eukaryotes, especially protozoa

Large microorganisms such as yeasts, protozoa, and algae are better suited for the hemocytometer or electronic counting method than the Petroff-Hauser chamber. Each type of chamber has instructions that describe the area in square millimeters (mm^2) in which cells will be counted and the well's total volume in cubic millimeters (mm^3).

Counting chambers contain a ruled grid that usually covers 9 mm^2. The grid contains 400 small squares and 25 large squares, each of which contains 16 smaller squares. A single large square covers an area of 1/25 mm^2; a single small square covers an area of 1/400 mm^2. A microbiologist might choose to count up to 25 of the large squares if a dilution does not contain many cells but count up to 16 of the small squares if a suspension contains many cells. To complete any direct count using a counting chamber, a person must know the values for the following:

- the number of squares used for counting cells
- the volume of suspension that the square contains, which is based on the depth of the well
- the amount of the original suspension is diluted before being used to fill the chamber
- 1,000 mm^3 = 1 cubic centimeter = 1 ml

The only remaining parameter that must be determined is the average number of cells counted per square. The following procedure provides an example of direct counts using a Neubauer counting chamber.

A microbiologist dilutes a cell suspension 100-fold; this dilution may be expressed as 1×10^2. The counting chamber should be clean and dried before use. A counting chamber can be filled by positioning a glass coverslip atop the counting grid then touching a very small pipet—micropipettes hold volumes in the range of 10–100 microliters (µl)—filled with suspension to the side of the chamber and coverslip. The chamber's well fills automatically by pulling suspension out of the pipette by a process called capillary action.

Once the microbiologist has properly filled the counting chamber, viewing the suspension under a microscope may show that the suspension is fairly dilute; the microbiologist decides to count cells found within the larger squares. If the depth of the well is 0.1 mm and the area is known to be 1/25 mm^2 (equal to 0.04 mm^2), the microbiologist knows that the volume in which cells will be counted is:

$$0.1 \text{ mm} \times 0.04 \text{ mm}^2 = 0.004 \text{ mm}^3$$

The microbiologist counts the number of cells in each of the 25 squares and arrives at a total of 176 cells. The volume of the suspension counted is:

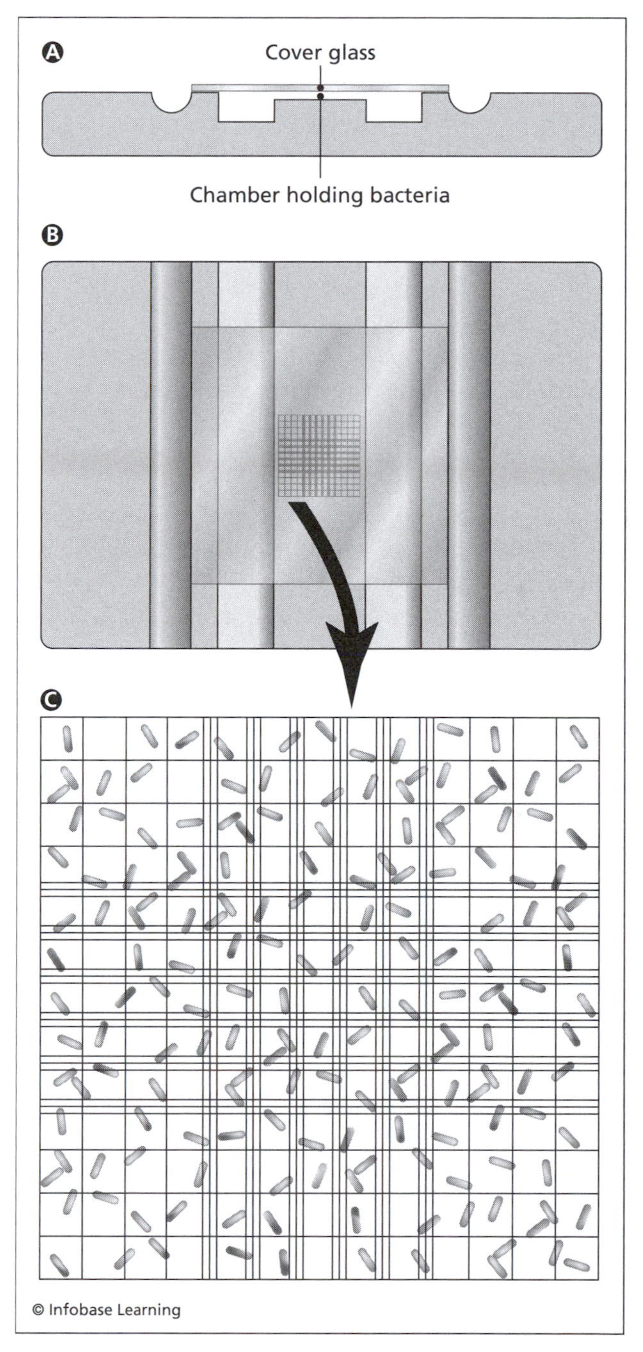

In a Petroff-Hausser counting chamber, a small well (a) holds a volume of bacterial suspension. From above the chamber (b), a microbiologist views a cell-counting grid of a specific area. Under magnification of about 600 × (c), the microbiologist counts the number of cells in either of two ways: (1) low-density suspensions are counted in the central grid of 25 small squares, or (2) high-density suspensions are counted in five of the small squares (usually four outer corners and the center square).

Direct Microscopic Count Chambers

Chamber	Well Volume
Petroff-Hausser	$0.02\ mm^3 = 2 \times 10^{-5}$ ml
Neubauer	$0.1\ mm^3 = 1 \times 10^{-4}$ ml or $0.00025\ mm^3 = 2.5 \times 10^{-8}$ ml
hemocytometer	$0.1\ mm^3 = 1 \times 10^{-4}$ ml

$$25 \text{ squares} \times 0.004 \text{ mm}^3 \text{ volume per square} = 0.1 \text{ mm}^3$$

The suspension counted under the microscope, therefore, has 176 cells per 0.1 mm^3, which can also be expressed as cells per milliliter (ml), as follows:

$$176 \text{ cells per } 0.1 \text{ mm}^3 \times 1{,}000 \text{ mm}^3 \text{ per ml} = 17{,}600 \text{ cells/ml}$$

Because the original suspension had been diluted 100-fold before counting, the number of cells in the original undiluted suspension is calculated as follows:

$$17{,}600 \times 100 = 1{,}760{,}000 \text{ or } 1.76 \times 10^6 \text{ cells per ml}$$

In a second example, a microbiologist performs a direct count on a very dense suspension diluted 100-fold, counting 112 cells in a total of five of the smallest squares. The cell concentration is the original suspension is:

$$0.1 \text{ mm} \times 0.0025 \text{ mm}^2 = 0.00025 \text{ mm}^3 \text{ volume per square}$$

$$5 \text{ squares} \times 0.00025 \text{ mm}^3 = 0.00125 \text{ mm}^3 \text{ total volume counted}$$

$$112 \text{ per } 0.00125 \text{ mm}^3 \times 1{,}000 \text{ mm}^3 \text{ per ml} \times 100 \text{ dilution factor} = 8.96 \times 10^9 \text{ cells/ml}$$

ELECTRONIC DIRECT COUNTS

Electronic instruments for measuring bacterial concentration have replaced many of the manual direct counting methods, especially in industry. In electronic counting, a cell suspension is forced through a narrow chamber connecting two fluid-filled compartments, and electrodes in each compartment are set to measure the ability of the system to conduct electricity. Although the conductance of electricity is high in the fluid, the chamber through which cells flow is so narrow that each cell momentarily impedes the electrical current as it moves through the chamber. The instrument measures the momentary decreases in electrical conductivity as a pulse, which the device then interprets as a count. Increased resistance to the conductivity is equated with increased cell concentration.

The electronic method described here is referred to as *impedance spectroscopy.* The device measures the amount of impedance to the electrical current, and spectroscopy is any technique that measures the transmission of some type of energy (electrical or light) across a material. The most recognized impedance device in microbiology is called a Coulter counter.

The American engineer Wallace H. Coulter (1913–98) developed the theory—to be known as the Coulter principle—upon which electronic cell counters are based. The Coulter principle provided the foundation of how to count and size particles electronically in a flowing liquid. Coulter's diverse experiences in science enabled him to meld two disparate specialties to invent a new method for counting particles. In Coulter's early career in the 1930s, he visited many hospital laboratories to develop medical equipment. After serving in World War II (1939–45), Coulter accepted a navy project for improving the paint used on ships. Coulter's task was to find a way of standardizing the particles contained in paint. In order to standardize particle size, he first had to devise a way of measuring the size of each particle in a sample. Coulter recalled the sight of medical technicians straining at microscopes to count blood cells and realized they needed an automated method for counting and sizing cells. Using his training in electrical engineering, Coulter and his brother, Joseph, built the first instrument for counting blood cells in a flowing suspension. In 2000, the Massachusetts Institute of Technology said of Coulter, "In the basement of his home, Wallace H. Coulter conceived a principle, then invented a device, that revolutionized the analysis of microscopic particles in fluids, most notably blood." The Coulter counter and a related device called a *flow cytometer* actually aided two disciplines at once, microbiology and medicine.

Coulter counters and similar devices designed to measure and record impedance have made direct cell counts much easier than the labor-intensive manual counting with counting chambers. The ease of use of electronic counters has been counterbalanced by some disadvantages of this type of direct counting. As with counting chambers, electronic counters do not work well with filamentous structures such as mold filaments or *Actinomyces* bacteria. Other

Areas of Concern in Electronic Direct Counts

Problem	Cause	Possible Solution
small cells	bacteria of less than 0.4 μm diameter may be difficult to detect	use of electronic counting for estimates only, reserving the method for larger-cell species
turbulence	uneven flow patterns in the suspension moving through the chamber	setting of discriminator dial to exclude background noise (also loses some information on cell concentration)
incomplete fission in cells	aggregates and cells that have not completed reproduction by binary fission are counted as one cell	mild physical treatments that break up many aggregates and some paired cells
coincidence	two cells pass through the chamber in tandem and are detected as one cell	varying of the dilution of the suspension
clogging	cells or debris in the suspension clogs the chamber	filtration to remove small debris

disadvantages of electronic counting can be overcome with advances in engineering or by incorporating a certain margin of error by the microbiologist in any suspension being counted. The table below describes some of the disadvantages of electronic direct counts, listed in order of importance.

Flow cytometry has been used in medicine to count, measure, and analyze the physical characteristics of blood cells, and microbiologists have adapted flow cytometers for studying microorganisms. Flow cytometers use a laser beam to measure particles as they move past a detector, and the electronic component converts the pulse into a readable signal. Microbiologists have devised methods for staining bacteria with chemicals that emit fluorescence when exposed to light. By using different stains that attach to specific bacteria, flow cytometry can be employed to count different types of cells.

Direct counting methods offer advantages and disadvantages to microbiologists. The technique can supplement other means of determining the concentration of cells in a liquid, or it can be used alone. If using direct counts as the only method for cell counting, a person must remember some of the drawbacks to the technique that can affect accuracy.

See also CULTURE; GROWTH CURVE; MICROSCOPY.

Further Reading

Koch, Arthur L. "Growth Measurement." In *Manual of Methods for General Bacteriology,* edited by Philipp Gerhardt. Washington, D.C.: American Society for Microbiology Press, 1981.

Massachusetts Institute of Technology. "Wallace Coulter." Inventor of the Week Archive. August 2000. Available online. URL: http://web.mit.edu/invent/iow/coulter.html. Accessed March 10, 2009.

Rice University. Experimental Biosciences. "Using a Coulter Counter." Available online. URL: www.ruf.rice.edu/~bioslabs/methods/microscopy/cellcounting.html. Accessed March 10, 2009.

disinfection Disinfection is the destruction of all living microorganisms other than bacterial endospores from inanimate surfaces. The term *disinfectant* refers to a chemical that destroys bacteria and fungi or inactivates viruses. The main purpose of disinfectants is to kill pathogens, which are disease-causing bacteria, fungi, or viruses.

Disinfection may be achieved through either chemical means or physical processes. Chemical disinfectants often are referred to as biocides, which are any substances that kill *(-cide)* life *(bio-)*. Physical disinfection may be accomplished through applying heat or irradiation to the inanimate surface.

Disinfectants can be further categorized by the range of microorganisms they kill. For example, a broad-spectrum disinfectant kills both gram-positive and gram-negative bacteria. A limited-spectrum disinfectant, by comparison, kills only one type of bacteria, destroying gram-positives but not gram-negatives, or vice versa. All disinfectants destroy bacteria to some degree, but certain disinfectants have added power against additional types of microorganisms. As examples, antifungal disinfectants kill fungi in addition to bacteria, and antiviral disinfectants kill viruses in addition to bacteria.

Chemical disinfectants are part of a large and general group of compounds called *antimicrobial agents*. An antimicrobial agent is a chemical directed against either microorganisms on inanimate surfaces or microorganisms on or in living tissue. Disinfectants are intended only for use on inanimate surfaces such as counters, sinks, and floors. By contrast, antiseptics or drugs such as antibiotics are antimicrobial agents that work on or in the body; that is, they are intended for use on living tissue. All of these compounds can be referred to simply as *antimicrobials*.

CHEMICAL DISINFECTION

Several classes of chemical biocides kill bacteria, fungi, and viruses. They also work in a variety of ways by damaging membranes, proteins, or nucleic acids (DNA or ribonucleic acid [RNA]). The best disinfectants share certain attributes that make them valuable for use around people and animals. The most important attributes that a chemical disinfectant should possess are the following:

- fast acting
- nonirritating to the skin or eyes or when inhaled
- unaffected by physical extremes in temperature or light
- unaffected by the presence of other chemicals
- active in the presence of high levels of organic matter
- noncorroding to inanimate surfaces
- stable over long periods (several months)
- able to remove small amounts of dirt

A single chemical that can achieve all of these attributes would be rare. Almost any disinfectant can be fully or partially inactivated by substances in nature or by physical activities. Disinfectants have been reduced in activity by the following three main influences:

1. external physical environment—temperature, pH, moisture, light, interfering compounds
2. type of microorganism—bacteria, mold, yeast, virus, algae, protozoa
3. ability of the microorganism to neutralize the disinfectant

Each of the preceding factors should be considered when choosing a disinfectant for a specific job. Some disinfectants lose activity with changes in temperature or pH. For instance, certain chemicals work better at temperatures slightly higher than ambient (the natural temperature of the surroundings). Phenol and alcohols kill molds, yeasts, and bacteria slightly better at warmer temperatures than at cold temperatures; hypochlorite solutions are most effective at pH of about 6.5–8.

Compounds that have contact with the disinfectant contribute to making the chemical more or less effective in killing microorganisms. Organic matter interferes with a disinfectant's antimicrobial activity, and organic substances take many forms: blood, serum, sputum, pus, dead skin cells, food particles, residues of dried milk, fecal matter, and other microorganisms. Because of the interference that organic matter exerts on disinfectants, the chemicals always work better on cleaned surfaces. Cleaning with detergent and water to remove excess organic matter helps in disinfecting a surface. Some metals, such as magnesium, zinc, and copper ions, also interfere with disinfectants.

Partitioning is the separation of materials into different liquid phases such as oil and water. Disinfectants prone to partition into a water phase or an oily phase of a mixture lose their effectiveness against microorganisms. This poses particular problems for chemicals that move into the oil phase of mixtures, because microorganisms prefer to stay in the aqueous phase (the water phase) of almost all mixtures. Chemicals that partition toward oil (lipophilic chemicals) or away from water (hydrophobic chemicals) should be avoided when selecting a disinfectant for oily surfaces.

Perhaps the chief factor in how well a disinfectant works is the microorganism itself. Different types of microorganisms are susceptible to different chemicals, because of their outer structure. The cell wall of bacteria, for instance, provides strong protection for the cell contents against various assaults from the outer environment. Some viruses contain a protective lipid (fatty) layer that resists disinfectants. People who wish to kill a specific microorganism—someone sharing a household with a flu sufferer, for example—would be wise to select a disinfectant that says it kills the influenza virus on its label.

Disinfectants are used in hospitals, outpatient clinics, veterinary medical offices, other medical offices, day care centers, nursing homes, and residences. Industrial disinfectant formulas are used in hospitals and other health care settings, restaurants and

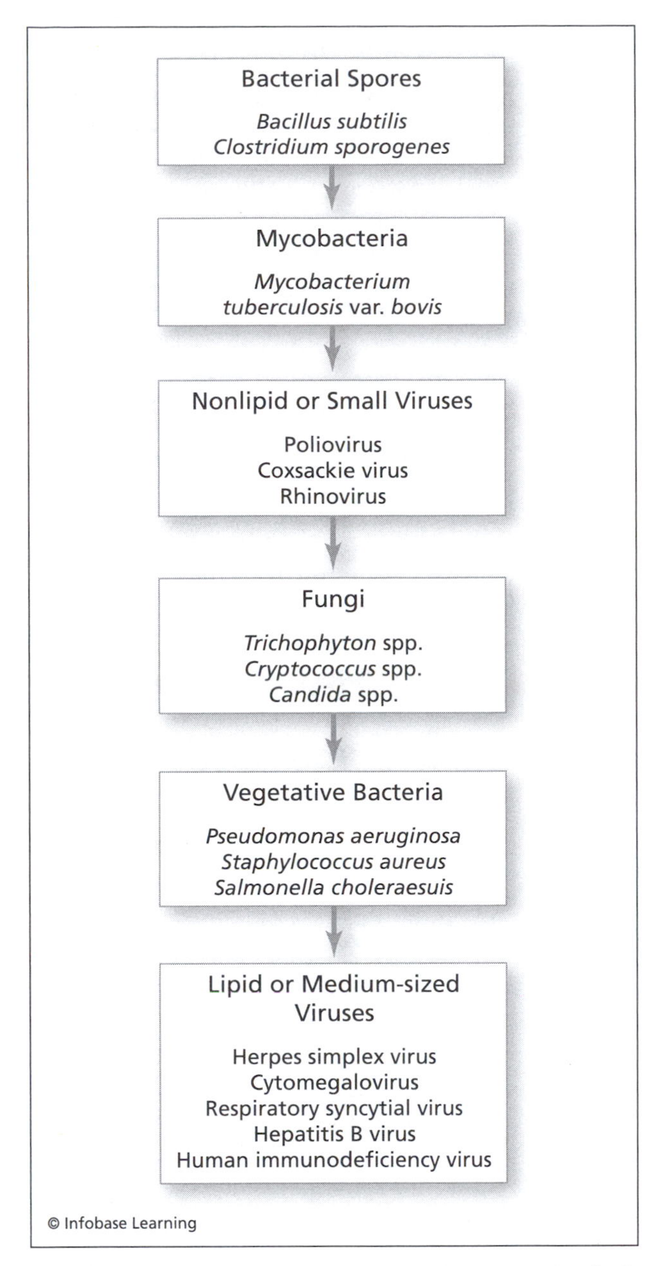

Various microorganisms differ in their resistance to chemical biocides. Bacterial endospores and *Cryptosporidium* cysts resist almost all chemicals. Viruses with lipid coats called envelopes exhibit the least resistance to chemical biocides.

other food preparation areas, cruise ship kitchens, and food manufacturing plants. The disinfectant formulas used in all of these categories contain the following four main ingredients: (1) the chemical that destroys microorganisms, (2) an inert carrier such as water, (3) a cleaning agent, and (4) a fragrance. Ethylenediaminetetraacetic acid (EDTA) might also be included, because this compound helps chemical disinfectants enter bacterial cells.

Today's common disinfectants for hard inanimate surfaces for hospitals, businesses, or households are hypochlorites, quaternary ammonium compounds, phenol-derived compounds, ionic surfactants, nonionic (amphoteric) surfactants, and various nitrogen-containing compounds and polymers (long-chain compounds). Surfactants, also called *surface active agents,* break the surface tension of water and allow chemicals to have intimate contact with nonliving surfaces or living cells. (Surfactants are sometimes simply called *detergents.*) Ionic surfactants are surfactant molecules that hold either a positive charge or a negative charge in water; nonionic surfactants disperse in water because one part of the surfactant molecule is positively charged and another part is negative charged.

Surfactants aid disinfectant activity in several ways in addition to breaking the water-surface or oil-water tension. First, surfactants sometimes provide stability to a formula by keeping the oil and water components in a homogeneous mixture. Second, surfactants loosen dirt, thus helping the chemical disinfectant to work better. Third, as mentioned, surfactants make microbial cells more susceptible to damage from chemicals. Finally, surfactants boost the activity of chemicals that already are fairly effective in killing microorganisms.

TESTING CHEMICAL DISINFECTANTS

Microbiologists test disinfectants to determine the most effective chemicals for destroying bacteria or other microorganisms. Two different methods are suitable for this testing: disk diffusion and the use-dilution test.

The disk diffusion method acts as either a screen for discovering new disinfectant chemicals or an actual method for testing the strength of a chemical in killing disease-causing microorganisms (pathogens). A screen is any procedure that serves as an evaluation of a new chemical disinfectant. (Screening is the process of evaluating any new compound for use in products such as disinfectants or antibiotics.)

In the disk diffusion method, a microbiologist soaks small paper disks (about 4 mm in diameter) in a solution containing the chemical to be evaluated. The microbiologist then places each disk containing a different chemical onto an agar surface that has been inoculated with a thin layer of one species of bacteria. While incubating, the layer of bacteria develops into a visible sheet of growth called a *lawn.* If a disinfectant has been effective in killing the bacteria during incubation, a clear zone forms around the disk where no bacteria grow. This zone, called a *zone of inhibition,* occurs because the chemical slowly diffuses from the disk into the surrounding agar. Susceptible bacteria will not grow in the area of the disinfectant. This disk

Use-Dilution Test Results

Test Result	Interpretation
no growth in 59 or 60 of 60 test tubes	effective disinfectant
weak growth or mixed results	chemical with weak disinfectant activity
growth in 2 or more of 60 test tubes	ineffective as a disinfectant

diffusion method has been a helpful screening tool in microbiology because many different chemicals may be tested against various microorganisms in a short period, usually an incubation period of 18–24 hours.

The use-dilution test offers a more finely tuned test than disk diffusion. The use-dilution test assesses the specific concentration of a disinfectant against bacteria and determines the exact amount of time required to kill the bacteria. (The method has been named *use-dilution* because it evaluates a disinfectant in the exact concentration in which it should be applied to a surface. In other words, this method determines the amount a chemical should be diluted before being used to disinfect a surface.)

Many disinfectants sold for household use do not require dilution; they are effective when used directly from the container. These products are called full strength, undiluted, or neat formulas. Other disinfectants should be diluted in water before they are used. Bleach and strong industrial disinfectants usually require dilution before being used. The use-dilution test accommodates any dilution that a product requires before it is used to kill microorganisms.

Microbiologists cannot test disinfectants against all known bacteria in the world, so three bacteria serve to represent the most common pathogens. *Staphylococcus aureus* represents gram-positive bacteria, *Salmonella choleraesuis* represents gram-negative bacteria, and *Pseudomonas aeruginosa* represents the other species that may be more prevalent in hospitals than in households.

A microbiologist begins testing by putting at least 1 million cells of a bacterial species as a thin layer onto a set of small stainless steel cylinders (about 0.75 inch [19 mm] tall and of about 0.65 inch [16.5 mm] outer diameter). The microbiologist then immerses the set of 60 replicate cylinders in the diluted (or full-strength) disinfectant. At an exact predetermined time, called the *contact time,* the tester transfers the cylinders to individual tubes of sterile broth medium. As the broth tubes containing the cylinders incubate, any surviving bacteria will grow into a visibly cloudy culture in the tube. The effectiveness of a disinfectant can be determined by the number of broth tubes that contain no growth. The U.S. Environmental Protection Agency, which monitors the testing of disinfectant products in the United States, requires that 59 of 60 tubes for a single species and a single disinfect concentration have no evidence of growth in order for the chemical to be called a disinfectant.

Possible results of use-dilution testing are summarized in the table (left).

The success or failure of a chemical passing the use-dilution test depends on the contact time. Some chemicals require 10 minutes of contact time with bacteria to pass the test, while others need only a few minutes, one minute, or even several seconds to pass the test. A disinfectant that works well in killing bacteria in a 10-minute contact time may, for example, give the following results if tested for only five minutes (see table below).

Additional test methods have been developed to determine disinfectant effectiveness. Some of the tests listed in the table below have been based on the use-dilution test, and others have been developed to test disinfectants by a different method.

PHYSICAL DISINFECTION

Humans used heat as a means to kill potentially dangerous microorganisms in food even before science understood the true nature of microbial life. Heating remains an effective method for destroying contaminants in foods and some nonfood products. Intense heat is so effective in killing microorganisms that it can kill bacterial endospores, which are resistant to almost all forms of chemical disinfection. When this happens, the material is said to be *sterilized.* Sterilization is the process of killing all microbial life, including endospores, while disinfection kills all microbial life except endospores.

Example Use-Dilution Test Results at Shortened Contact Time

Test Result	Interpretation
weak growth or mixed results	possible disinfectant activity at less than 10 minutes but more than five minutes contact time
growth in most test tubes	ineffective as a disinfectant at five minutes contact time

Alternate Methods for Evaluating Chemical Disinfectants

Test Method	Summary of the Method
hard surface carrier test	similar to use-dilution but the test surface is on glass rather than stainless steel and the concentration of bacteria is set to a standard level
capacity test	broth culture method to determine the capacity of a disinfectant to destroy increasing numbers ofmicroorganisms
Kelsey-Sykes test	a capacity test designed to mimic actual conditions by including dirt and testing against bacteria most resistant to the disinfectant
phenol coefficient method	compares a disinfectant's activity to that of phenol
Chick-Martin test	similar to the phenol coefficient method but designed to mimic actual conditions by including dirt and testing against bacteria most resistant to the disinfectant
Rideal-Walker test	a phenol coefficient method in which a given dilution of disinfectant is directly compared to phenol activity
suspension test	any test in which the disinfectant is tested in a liquid rather than on a hard surface
5-5-5 suspension test	determines the lowest disinfectant level that kills 10^5 cells by 5 logs (units on a logarithmic scale) in five minutes

Food production and other industries destroy microorganisms on metal equipment by heat washing the equipment with hot water washes or exposing it to steam. In food processing, manufacturers sometimes need only reduce the numbers of bacteria on their equipment to safe levels rather than kill all microorganisms. In this instance, the equipment is said to be sanitized rather than disinfected.

Overall, the three main processes used for killing potentially dangerous microorganisms can be ranked from the most effective to the least effective as follows:

sterilization → disinfection → sanitization.

Ultraviolet (UV) irradiation is another physical disinfection method that uses energy in the form of electromagnetic waves to destroy microorganisms. Light in the ultraviolet portion of the electromagnetic spectrum of wavelengths of 220–300 nanometers (nm) kills bacteria, fungi, and viruses by damaging their DNA or RNA beyond repair and possibly by causing the formation of toxic chemicals inside the cells. Microbiologists use germicidal UV lamps to destroy any contaminants that may be on laboratory benches and equipment. In addition, UV light has been used to eliminate microorganisms from indoor air, to disinfect dental equipment, and to disinfectant drinking water. UV treatment, in the proper circumstances, can sterilize, disinfect, or sanitize surfaces. The following factors affect the efficacy of UV treatment in killing microorganisms on surfaces or in water or air:

- composition of the surface: metal, glass, wood, and so on
- volume of air or water to penetrate
- energy level of the irradiation
- distance of the light source from the surface
- presence of inorganic dirt or organic matter
- interference of items or shadows
- concentration of microorganisms
- angle of the UV light hitting the surface
- humidity
- hours of use of the germicidal UV lamp

UV lamps do not emit a constant output of energy over time and may eventually lose the ability to kill microorganisms. For this reason, UV lamps must receive periodic testing to assure they emit enough energy in the form of UV irradiation to disinfect matter.

HIGH-LEVEL AND LOW-LEVEL DISINFECTION

Health care professionals categorize disinfection according to the types of items that must be treated. For instance, a stethoscope that touches only unbroken skin requires much less stringent treatment than a dental instrument that probes the teeth and gums. Disinfection is, therefore, categorized as high-level, intermediate-level, and low-level treatment. High-level disinfection removes microorganisms from medical instruments called *critical items*. These are medical devices that enter tissue or the vascular system or allow blood to flow through them. An improperly disinfected critical item presents a very high risk of causing infection. Examples of critical items are scalpels, needles, surgical instruments, dental instruments, stents, prostheses, and intravenous tubing and devices. High-level disinfection may be considered equivalent to sterilization because its purpose is to kill all microbial life, including of bacterial endospores.

Intermediate-level disinfection is equivalent to the disinfection that kills all microbial life except bacterial endospores. This treatment is used on medical devices called *semicritical items,* which are items that contact mucous membranes or broken skin (wounds, cuts, burns, etc.). These items carry the risk of transmitting infection, but the risk is considered to be lower than for critical items. Examples of semicritical devices are surgical scissors, ophthalmic instruments, endoscopes, aspirator tubes, bronchoscopes, and urinary catheters.

Low-disinfection treats equipment called *noncritical items* that have contact with intact skin. The following noncritical items have a low risk for transmitting infection: blood pressure cuffs, gloves and face masks, stethoscopes, electrocardiogram electrodes, and X-ray machine surfaces.

Speaking of disinfection in terms of high-level or low-level activity can be misleading because disinfection is an absolute condition. Just as there exists no high-level or low-level sterility, an object is either disinfected or not disinfected. For practical purposes in using this terminology, the following relationships provide a clearer picture of these treatments.

- High-level disinfection is equivalent to sterilization.
- Intermediate-level disinfection is equivalent to broad-spectrum chemical disinfection.
- Low-level disinfection is equivalent to limited-spectrum disinfection.

WATER DISINFECTION

Drinking water, treated wastewaters and sewage, and swimming pool and spa waters must be disinfected to remove infection-causing microorganisms. The four main methods used in water disinfection are (1) chlorine compounds, (2) chlorine dioxide gas, (3) ozone gas, and (4) UV irradiation.

Drinking water receives more than one disinfection step in most water treatment facilities before the water is distributed to homes and businesses. Chlorine gas is an effective chemical for this step, but it is very toxic, so other chlorine compounds have become more common in treating municipal water supplies. The main chlorine disinfectants for this purpose are hypochlorites, chloramines, and chlorine dioxide. Organic matter lessens the effectiveness of all chlorine treatments, and pH also impacts chlorine activity. To overcome the loss of any chlorine that may be inactivated by organic matter, water treatment plants put an extra amount of chlorine into water, called *residual chlorine.* Most municipal water supplies have residual chlorine levels of 0.5 to 2 parts per million (ppm) when they exit the treatment plant. A ppm is equal to 1 mg/ml.

Ozone and UV irradiation may also be used for water disinfection. Ozone is a highly reactive gas that causes damage to microbial membranes and enzymes. Ozone is less affected by pH and temperature than chlorine, and many people prefer it because it does not have the taste and odor associated with chlorinated water. UV treatment similarly produces no taste or odor problems, but UV effectiveness rapidly declines if the water contains high amounts of particles or dissolved organic matter. In addition, UV light cannot penetrate large volumes of water contained in a single tank; water must flow through a narrow channel so that it will be entirely exposed to the UV light.

Wastewaters receive chlorination before leaving treatment plants so that no infectious microorganisms are released into the environment. Disinfection of wastewater presents difficulties because of the high organic content of these waters and constantly changing levels of matter. Wastewater disinfection, therefore, requires higher chlorine residual levels of at least 3 ppm.

Swimming pools, spas, hot tubs, wading pools, and water amusement park rides also require disinfection to prevent the spread of infection. Chlorine compounds are most common for treating these waters plus 0.5–3 ppm of copper sulfate to kill algae. The residual chlorine levels in these types of waters are set by individual states; most fall in the range of 0.5–1 ppm residual chlorine.

Disinfection is a critical area in microbiology for the purpose of stopping the spread of infection, reducing the transmission of disease by way of food or water, and preventing contamination in microbiology laboratory studies. It has made a significant impact on maintaining health for centuries. Because of a wide array of chemical disinfects and physical means of disinfection, this aspect of microbiology is well equipped to confront almost any known or unknown pathogen that may be in the environment.

See also BIOCIDE; DIFFUSION; PATHOGEN; SANITIZATION; STERILIZATION; WATER QUALITY.

Further Reading

Block, Seymour S., ed. *Disinfection, Sterilization, and Preservation,* 5th ed. Philadelphia: Lippincott Williams & Wilkins, 2000.

Cockren, Archibald. *Alchemy Rediscovered and Restored.* New York: Cosimo Classics, 2007.

Hoffman, Peter, Graham Ayliffe, and Christine Bradley. *Disinfection in Healthcare,* 3rd ed. Hoboken, N.J.: Wiley-Blackwell, 2004.

Lenntech Water Treatment and Air Purification Holding B.V. "Conditions for Water Disinfection." Available online. URL: www.lenntech.com/water-disinfection/swimming-pool-disinfection.htm. Accessed March 30, 2009.

Russell, Allan D., William B. Hugo, and Graham A. J. Ayliffe, eds. *Principles and Practice of Disinfection, Preservation and Sterilization,* 3rd ed. Oxford, England: Blackwell Science, 1999.

Rutala, William A., ed. *Disinfection, Sterilization and Antisepsis in Healthcare.* Washington, D.C.: Association for Professionals in Infection Control and Epidemiology, 1998.

DNA fingerprinting DNA (deoxyribonucleic acid) fingerprinting, also called *DNA typing,* comprises the methods in genetic engineering used for identifying bacteria and viruses. This technique has been applied mainly to the following three areas: (1) forensic medicine, which is used for paternity testing; (2) analysis of crime scene evidence; and (3) tracing the origin of food-borne and waterborne pathogens.

In the detection of food-borne bacteria or viruses that cause food-borne outbreaks and in crime scene analysis, the amount of DNA available may be much smaller than the quantity needed for doing DNA fingerprinting. Therefore, technicians usually employ a technique called *polymerase chain reaction* (PCR) to replicate a DNA sample before analyzing the DNA by fingerprinting.

THE PRINCIPLE OF DNA FINGERPRINTING

A single DNA molecule is larger than most other molecules in microbial cells. Each species of microorganisms contains a unique composition of DNA subunits, but unraveling a DNA molecule and determining the entire sequence of subunits would be very time consuming and expensive. DNA fingerprinting uses a shortcut in connecting a specific DNA to its source by studying a small piece of DNA rather than the entire molecule.

The fingerprinting method compares the sequence of nucleotides in a piece of DNA to the nucleotide sequence in a known sample of DNA. A nucleotide is a single unit of DNA or ribonucleic acid (RNA) containing the following three components: (1) a sugar; (2) a chemical group called a *phosphate group,* which is composed of phosphorus and oxygen molecules; and (3) a purine or pyrimidine compound. In DNA, the sugar component is deoxyribose. Purines and pyrimidines are large compounds containing carbon, nitrogen, and a chemical structure that forms a ring. DNA contains the purines adenine and guanine and the pyrimidines thymine and cytosine. Each living organism contains DNA made up of genes that are characterized by distinct sequences of nucleotides, which form the basis of DNA fingerprinting just as a person's fingerprint serves as a unique identifier that is not found in any other person. With the exception of identical twins, each person, animal, plant, or microbial species contains its own unique DNA. DNA fingerprinting is based on the principle of unique DNA in every organism on Earth.

THE DNA FINGERPRINTING METHOD

DNA fingerprinting is the most sensitive method for identifying an individual, an animal, or a microorganism. Although the method is not difficult, it requires precautions against contaminating the DNA sample to be analyzed with DNA from another organism.

The DNA fingerprinting method consists of the following steps:

1. extracting of DNA from an unknown microorganism

2. segmenting of the DNA with specific DNA-splitting enzymes called *restriction endonucleases*

3. combining of the segments with a piece of DNA (called a *DNA probe*) from a known microorganism

4. marking each segment with a molecule (called *tagging*) that can be detected by X-ray analysis

5. arranging the segments by size using a process called *electrophoresis*

6. exposing the segments, separated by size, to X-ray radiation

Methods Used in DNA Fingerprinting

Method	Description
DNA extraction	recovery of intact DNA from cells using salt solutions and buffers (solutions that have been formulated to a specific pH)
polymerase chain reaction (if necessary)	a cyclical heating-cooling process that makes millions of identical copies of small pieces of DNA in quantities that can undergo further analysis
segmentation	cleaving of DNA at specific sites by restriction endonuclease enzymes
tagging	attaching a radioactive molecule to known and unknown DNA so that the DNA segments will be visible when exposed to X-rays
electrophoresis	the migration of different-sized molecules through a gel exposed to an electric current; each different-sized molecule migrates to a different location in the gel
X-ray radiation	exposure to X-rays to make the radioactively-tagged DNA visible in a dried gel
analysis	comparison of the migration pattern of the unknown DNA to DNA from known microorganisms

A microbiologist uses the same process described here on both DNA samples and DNA taken from a known microorganism. After exposing both segmented DNAs to X-rays, the microbiologist compares the pattern of the segments produced by the electrophoresis technique. The table describes the techniques that compose DNA fingerprinting.

TRACING PATHOGENS IN FOOD AND WATER

DNA fingerprinting has been a useful method for tracing and identifying sources of food-borne or waterborne disease. Fingerprinting also helps environmental microbiologists study the types of microorganisms found in soil or natural waters regardless of whether the sample contains pathogens. In environmental microbiology, when scientists extract DNA from a mixed community of microorganisms in soil or water, they are said to be studying *community DNA*. Community DNA has been used to identify the predominant types of DNA and genes belonging to microorganisms in the environment. Public health microbiologists who seek to find the source of a disease outbreak require more specific methods of DNA fingerprinting. They must identify a pathogen to the genus or species level for the purpose of assessing the health threat that the microorganism poses to a human community.

Microbiologists identify the cause of food-borne or waterborne disease outbreaks in a method called *trace-back*. A trace-back is the process by which a microbiologist finds a food or water pathogen and then retraces all of the steps leading from the contamination to its original source. For example, in 2006, spinach that had been harvested and packaged in California caused illness in at least 150 people in 23 states. At the time of this widely distributed outbreak, doctors did not know the microorganism that had caused the illness or its source. A trace-back conducted by public health officials identified a common food among all the people who had been sick: fresh spinach. In addition, clinical microbiologists identified a dangerous variety of *Escherichia coli* called *E. coli* O157:H7. With these two clues in hand, microbiologists conducted a trace-back to the outbreak's origin with the help of DNA fingerprinting.

In September 2006, the *Los Angeles Times* broke the news, "A New Mexico laboratory was able to isolate potentially deadly bacteria in a bag of spinach that had sickened a resident—a step hailed Wednesday as a significant break in the search for the source of a nationwide *E. coli* outbreak." The laboratory made its breakthrough by a combination of hard work in testing hundreds of different spinach batches and use of an essential fingerprinting tool called a *probe*. Pathogens can be identified to the species or subspecies level with the use of DNA or RNA probes. A probe is a segment of either DNA or RNA usually containing between 15 and 40 nucleotides from a known microorganism. Biotechnology companies make probes from known DNA or RNA by cutting the molecule at specific sites with restriction endonucleases. Microbiologists purchase a variety of probes for the purpose of comparing the genetic material of unknown food or water isolates to the distinct sequence in the probe.

After a microbiologist returns to the laboratory with a food to be tested for pathogens—a suspect bag of fresh spinach, for instance—all of the microorganisms in the bag of spinach are recovered and isolated. Simple biochemical testing and growth on specialized agars called *differential* or *selective agar* help narrow down the likely choices of the pathogen's identity. A microbiologist then extracts DNA (or RNA) from the suspected pathogens and commences the DNA fingerprinting technique.

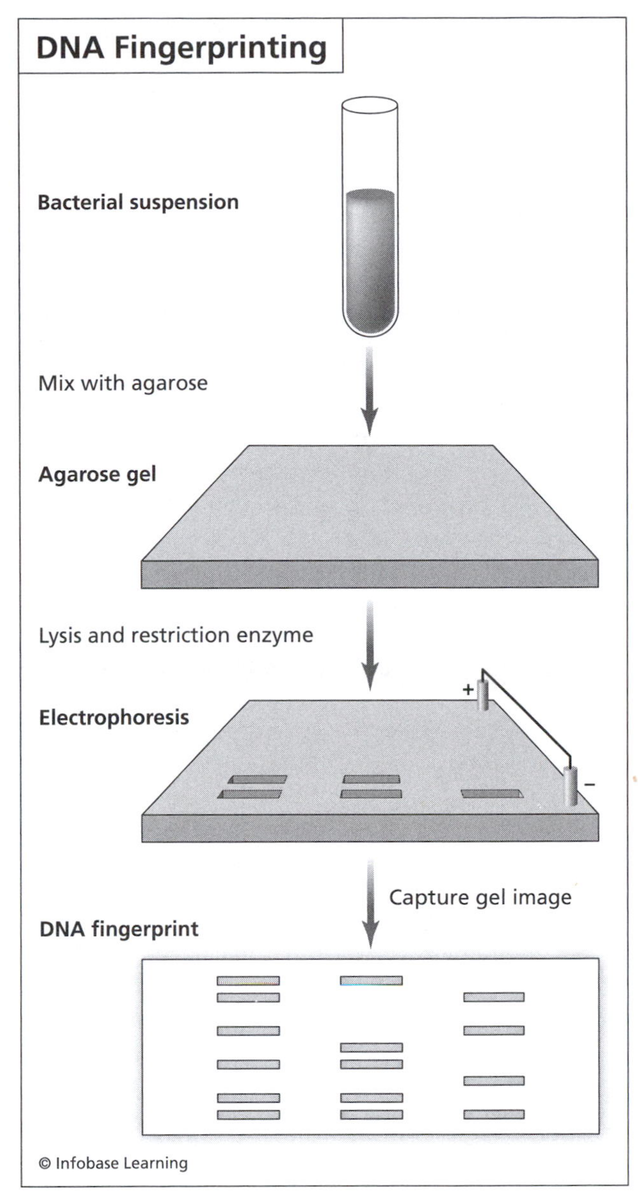

DNA fingerprinting identifies microorganisms to the species level and can even identify particular strains of a species. This is because the DNA fingerprint of every living thing is unique.

DNA fingerprinting's key step takes place when a DNA probe with a specific nucleotide sequence binds with a complementary segment of DNA from the unknown microorganism. DNA is a double-stranded molecule shaped like a spiral ladder in which the purine and pyrimidine bases of one side of the ladder bind to specific bases on the opposite side of the ladder. This connection forms the ladder's rungs. The specific base-to-base binding in DNA is called *complementary binding.* A DNA probe binds only with DNA segments that are exactly complementary to it in base sequence. A DNA probe will make this exact match only with another microorganism from the same species. In this way, a DNA probe from *Salmonella* will bind only with other *Salmonella* DNA. Likewise, a DNA probe from *Staphylococcus* will bind only with *Staphylococcus* DNA and not bind with *Salmonella* DNA.

By testing a DNA probe against many different bacteria from various batches of food suspected of contamination, microbiologists can trace the source of an outbreak to a single strain (a subgroup within a species) of bacteria. Testing for the presence or absence of the dangerous strain in various batches of food can lead public health officials to a particular batch grown at a specific farm.

After an *E. coli* O157:H7 food outbreak in 1993, the Centers for Disease Control and Prevention (CDC) created a DNA fingerprinting program called PulseNet, to be used by a network of laboratories across the United States. The laboratories belong to the CDC or to state or local public health departments. Any laboratory in the nation that has isolated a suspected cause of food-borne or waterborne illness may submit DNA fingerprinting results to a PulseNet laboratory. The authorized laboratory then compares the fingerprinting results to a large database of DNA sequences from known pathogens. PulseNet has set up cooperative connections with similar programs in Canada and with countries in Asia, Latin America, Europe, and the Middle East. The PulseNet section of the CDC's Web site has explained, "PulseNet plays a vital role in surveillance for and the investigation of food-borne illness outbreaks that were previously difficult to detect. Finding similar patterns through PulseNet, scientists can determine whether an outbreak is occurring, even if the affected persons are geographically far apart. Outbreaks and their causes can be identified in a matter of hours rather than days." PulseNet's DNA fingerprinting program, therefore, helps overcome two of the major obstacles in food-borne illness trace-back: the time required to test different probes against an unknown DNA and the ability to link the health condition of people who travel away from an outbreak's source.

Trace-back of waterborne pathogens works the same way as in food-borne illness trace-backs. The source of a waterborne illness can be more difficult to find, however, for two main reasons. First, water systems provide a constantly flowing source of water, rather than discrete batches, as found in most foods. Second, the microorganisms in water undergo a natural rise and fall, depending on season, amount of disinfectant in the water, distance of a tap from a water treatment plant, and presence of thick colonies of microorganisms called biofilms that grow inside water distribution pipes. DNA fingerprinting may, nevertheless, work well in assessing the overall quality of water. For example, DNA fingerprinting has been

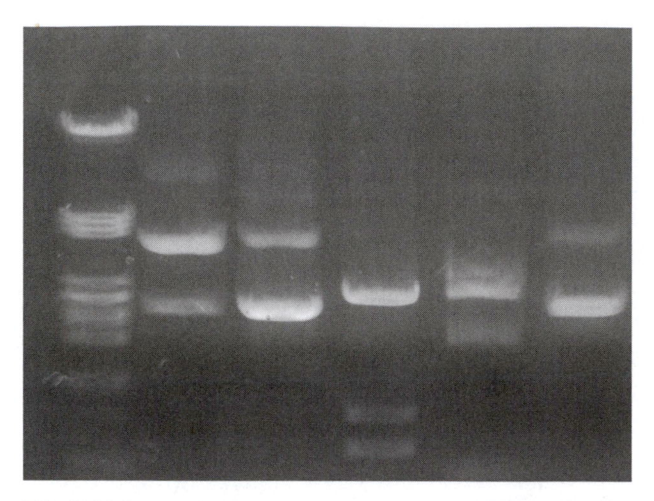

This DNA fingerprint pattern suggests that the DNA fragments in columns 2, 3, and 6, from left, are alike. Column 1 contains various molecular weight controls. ***(Frank LaBlanca)***

used to show the presence of enteric viruses in beach water used for swimming.

DNA FINGERPRINTING IN MEDICINE

DNA fingerprinting has become instrumental in the field of epidemiology, which is the study of disease outbreak and spread through a population. Public health officials use DNA fingerprinting in the following two ways: for tracing the source of a disease outbreak, and for conducting surveillance of the health of a population.

Disease outbreak studies occur in a similar manner to studies of food-borne outbreaks. In medicine, however, the method tends to be used for identifying infectious microorganism that may be causing a single outbreak or an epidemic. Because the fingerprinting can very specifically identify a microorganism, the health community uses it either to distinguish single isolated outbreaks from one strain or to determine whether an illness is occurring in isolated sporadic cases rather than an outbreak.

Medical surveillance studies have helped medical researchers learn about the rise or fall of diseases in populations over time. Epidemiologists use such data to build massive databases of information that correlate disease with other factors of a population such as age, income, rural versus city living, or smoking versus nonsmoking. Epidemiological databases have grown in size to a point where scientists can do very exact statistical studies on disease patterns. For example, the CDC has said that the data collected from thousands of DNA fingerprinting experiments allow public health officials to perform the following tasks:

- find links, if present, between sporadic cases of a disease
- identify related cases and separate unrelated cases of a disease
- identify an outbreak that is too widely dispersed to detect by other means
- track microorganisms that are very common so the minor changes in disease caused by these microorganisms are difficult to detect

As the sensitivity of DNA fingerprinting increases, the medical community will become better able to track disease and spot minor changes in health before an outbreak occurs.

Clinical laboratories in hospitals have also adopted DNA fingerprinting to help identify antibiotic resistance in clinical isolates from patients. By searching for and finding a gene that correlates to resistance, a microbiologist can help a doctor save valuable time in find the correct treatment to fight an infection.

IMPROVEMENTS IN DNA FINGERPRINTING

Scientists must correct the limitations of current DNA fingerprinting techniques to make them an even more powerful tool in medicine and public health.

The DNA fingerprinting technique requires good training and a high level of skill. There is no substitute in learning DNA extraction or electrophoresis for practicing these methods many times. Such sensitive methods also tend to vary in results from person to person, even if they work side by side. DNA extraction and electrophoresis are two of the many skills in microbiology that rely, in part, on the intuition of a scientist—intuition that can only be attained through experience.

Trustworthy identification of a pathogen using DNA fingerprinting is only as good as the probe and the database. Some microbial DNA simply does not match up with any of the probes commercially available. The final step in comparing DNA segments, called bands in electrophoresis, also requires skill. The resulting band patterns—the fingerprint—of a microorganism may not be as obvious as one might hope.

DNA fingerprinting offers more advantages to microbiology than disadvantages. This technique has already helped solved numerous troubling, sometimes deadly, food-borne pathogen outbreaks. But more precision and accuracy can be added to DNA fingerprinting. The future requires larger databases of DNA sequences, a wider array of specific probes,

The Five Kingdoms of Biota Classification

Kingdom	Members
Monera	bacteria and other single-celled microorganisms (prokaryotes, which do not contain membrane-enclosed internal structures)
Protista	eukaryotic microorganisms (cells that contain membrane-enclosed internal structures)
Fungi	yeasts, molds, and mushrooms
Plantae	plants containing internal vascular systems
Animalia	invertebrates, arthropods, reptiles, amphibians, birds, marsupials, and mammals

methods for improving the ease and accuracy of the technique, and time-saving steps that do not prevent from correct identifications of pathogens.

See also AGAR; EPIDEMIC; FOOD MICROBIOLOGY; IDENTIFICATION; POLYMERASE CHAIN REACTION.

Further Reading

Brinton, Kate, and Kim-An Lieberman. "Basics of DNA Fingerprinting." Available online. URL: protist.biology.washington.edu/fingerprint/dnaintro.html. Accessed March 10, 2009.

Engel, Mary. "Lab Definitively Links *E. coli* Outbreak to Contaminated Spinach." *Los Angeles Times,* 21 September 2006. Available online. URL: articles.latimes.com/2006/sep/21/local/me-spinachside21. Accessed March 10, 2009.

PulseNet. Centers for Disease Control and Prevention. Available online. URL: www.cdc.gov/pulsenet. Accessed March 11, 2009.

Secrets of the Dead. New York: Educational Broadcasting Corporation, 2006. Available online. URL: www.pbs.org/wnet/secrets/previous_seasons/lessons/lp_virus.html. Accessed March 11, 2009.

domain A domain is a classification of living things on Earth based on the composition of an organism's genetic material. In the 1800s, the British naturalist Charles Darwin (1809–82) proposed that natural selection led to the similarities and to the differences found among all *biota,* a term for Earth's living things. Scientists who followed Darwin spent considerable effort trying to classify the creatures on Earth in a logical system. These attempts gave birth to two related disciplines: taxonomy, which is the science of classifying organisms, and systematics, which is the science of organizing the classifications into a hierarchy based on the way the organisms evolved. In the process of developing these classification schemes, scientists created the concept of the domain.

Science has identified three different cell types that now represent three different domains among Earth's biota: the Archaea, the Bacteria, and the Eukarya. These three domains have been distinguished from each other by the composition of their ribosomal ribonucleic acid (rRNA). Ribosomal RNA makes up an organelle within cells called the ribosomes, which are small sites where enzymes assemble new proteins.

Prior to 1978, science classified organisms into kingdoms that had been proposed by the American ecologist Robert H. Whittaker (1920–80). After years of study, Whittaker developed a five-kingdom system for classifying organisms, summarized in the table.

Although Whittaker's classification scheme had undeniable logic based on the cell types and structures of the kingdom members, advances in molecular studies enabled scientists to learn about the genetic makeup of organisms in far greater detail, shortly after Whittaker proposed his system, in 1969. In 1978, the American microbiologist Carl R. Woese (1928–) developed classifications using his analysis of rRNA composition and, therefore, the relatedness of each organism with every other organism on Earth. Woese explained, in 1977, "The biologist has customarily structured his world in terms of certain basic dichotomies. Classically, what was not plant was animal. The discovery that bacteria, which initially had been considered plants, resembled both plants and animals less than plants and animals resembled one another led to the reformulation of the issue in terms of yet more basic dichotomy, that of eukaryote versus prokaryote." But scientists soon learned that the eukaryote-prokaryote distinction created more confusion than clarity. These cell types shared many features that suggested they did not belong to distinctly different paths in evolution.

Carl Woese's analysis of the relationships of organisms based on rRNA content led biologists to radically revise the classification systems they had been using again. The living world, Woese proposed "is not structured in a bipartite way along the lines of the organizationally dissimilar prokaryote and eukaryote. Rather, it is (at least) tripartite, comprising (i) the typical bacteria, (ii) the line of descent manifested in eukaryotic cytoplasms, and (iii) a little explored grouping, represented so far only by methanogenic bacteria." The methanogenic, or methane gas–producing, bacteria turned out to be microorganisms distinct from bacteria. This latter group is now recognized as the archaea, bacterialike microorganisms that possess metabolism and structures different from those of either bacteria or eukaryotic cells. Biology now uses the three-kingdom scheme: domain Archaea, domain Bacteria, and domain Eukarya.

The new classification made news in biology. The *New York Times* reported in 1977, "Scientists studying the evolution of primitive organisms today reported the existence of a separate form of life that is hard to find in nature. They described it as a 'third kingdom' of living material, composed of ancestral cells that abhor oxygen, digest carbon dioxide and produce methane." But many of Woese's peers found it difficult to reconcile the new three kingdoms with the previous five kingdoms. In fact, the three domains lie above the kingdom level, and prokaryotes and eukaryotes no longer hold as much meaning as they once did in defining different organisms.

Biologists continue to study the genetic makeup of organisms to determine biotas' common ancestor. This common ancestor is not yet agreed upon in science, but biologists do know that bacterial evolution may have split off from the evolution of archaea and eukaryotes very early in the evolution of life. Genetic studies have suggested that the bacteria and the archaea evolved parallel with each other and may have swapped genes along the way.

Many scientists chose to distinguish the bacteria-like archaea from all other bacteria, so they adopted the term *eubacteria* to separate "true bacteria" from the archaea. But because arachaea are not bacteria at all, the term *eubacteria* has been discarded in favor of the three-domain system; the three-domain system also prevents confusion with the Domain Eukarya. Although other hierarchies of life have been proposed and accepted by scientific circles, the three-domain system has become the most widely accepted to date.

The table describes some of the important distinctions that characterize the three domains.

DOMAIN ARCHAEA

The Archaea contain two phyla. Phylum Crenarchaeota contains only three genera. The larger phylum Euryarchaeota contains seven classes, each of which has at least one genus. Both phyla are home to a very diverse group of microorganisms that represent all types of cell morphologies—rods, cocci, helixes, and other—and gram-negative as well as gram-positive members. Although Woese originally described the archaea as mainly anaerobic, methane-producing microorganisms, microbiologists have since discovered aerobic and facultative anaerobic species.

Many archaea live in the earth's harshest environments. For this reason, they are also extremophiles, microorganisms that thrive in extreme environments. Archaea have been recovered from environments of extremely hot, cold, acidic, salt, or high-pressure conditions. Some archaea have adopted a less extreme lifestyle by living inside the digestive tract of animals. These intestinal microorganisms absorb hydrogen and carbon dioxide and expel methane as part of their normal energy metabolism.

DOMAIN BACTERIA

The Domain Bacteria, known simply as *bacteria,* include some of the most studied species in microbiology. The diversity and breadth of activities among the bacteria exceed those of any other type of organism. Bacteria additionally make up the most numerous group of biota on Earth. For example, a single cubic centimeter of soil can contain 5 billion to 50 billion bacterial cells. Microbiologists have estimated that they have recovered and studied less than 10 percent of all bacterial species present on Earth. In 2002, the microbiologist Vigdis Torsvik and

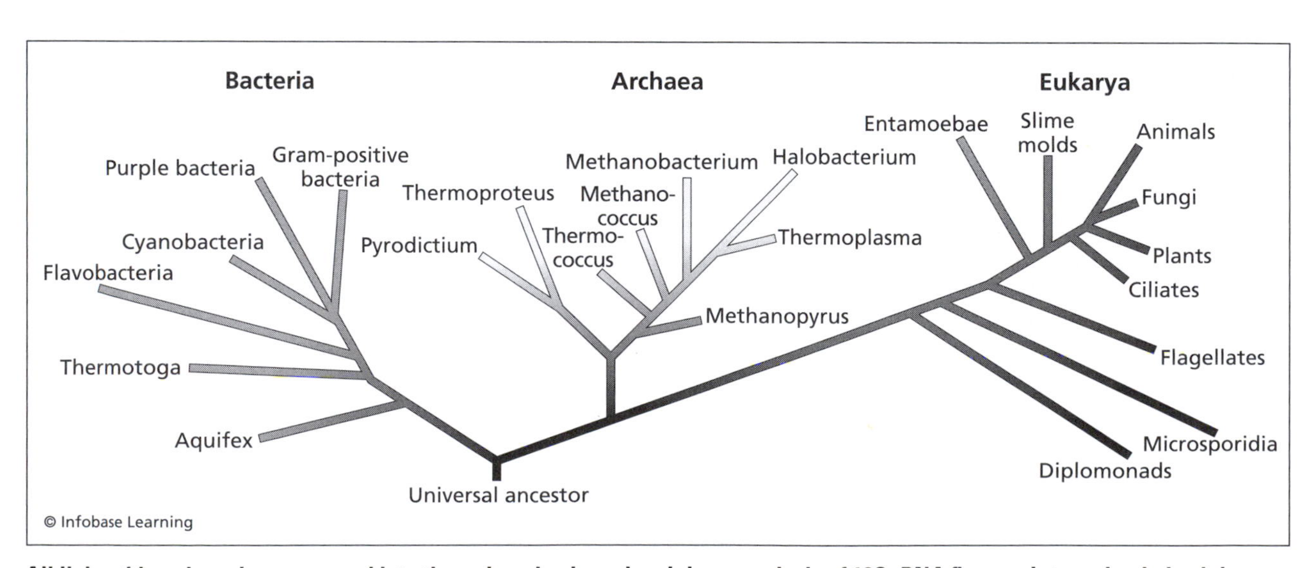

All living things have been grouped into three domains based mainly on analysis of 16S rRNA fingerprints and cel physiology.

The Three Domains of Life

Feature	Archaea	Bacteria	Eukarya
cell type	prokaryotic	prokaryotic	eukaryotic
cell wall	varied composition; no peptidoglycan	peptidoglycan	various carbohydrates
membrane-enclosed nucleus	absent	absent	present
complex internal membrane-enclosed structures	absent	absent	present
membrane lipids	some branched hydrocarbons (organic compounds with a carbon backbone and many hydrogen molecules)	unbranched hydrocarbons	unbranched hydrocarbons
antibiotic sensitivity	no	yes	no
presence of an rRNA loop that binds to ribosomes	absent	present	absent
circular DNA	yes	yes	no
histone proteins attached to the DNA	present	absent	present
habitat	usually extreme; some temperate	extreme or temperate	temperate
methane production	yes	no	no
nitrogen fixation (absorption of nitrogen gas from the air)	yes	yes	no
photosynthesis	absent	present in some species	present
chlorophyll-based photosynthesis	no	present without chloroplasts	present in chloroplasts

colleagues at the University of Bergen in Norway remarked in *Science* magazine, "Diversity estimates for natural bacterial communities have traditionally depended on cultivable species, but results from the use of molecular techniques to measure diversity suggest that reliance on culture has led to a longstanding underestimate of bacterial diversity." These estimates, furthermore, focus on bacteria that inhabit the earth surface, but microorganisms of deeper layers of earth make up what could be a tremendous collection of undiscovered species. Torsvik proposed that the carbon mass of all the subsurface microorganisms probably composes the largest single type of biological mass on Earth.

Bacteria belong to 23 phyla that contain more than 1,000 genera and several thousand species. Although nonscientists often associate bacteria with disease, a very small percentage of bacteria infect animal and plant life to cause harm. The majority of bacteria carry out essential reactions without which all other life on Earth would not survive. The science author Alexandre Meinesz wrote in his 2008 book, *How Life Began,* "Between the first miniscule bacteria that left a trace on Earth and the present day, more than 3.5 billion years have passed. For nearly one-third of the time life has existed in our planet, bacteria reigned alone. There were only microbes of different shapes and functions. How many species? We don't know, we will never know." The domain that contains the world's bacteria will probably be a life-form that will never be completely understood.

DOMAIN EUKARYA

Domain Eukarya contains what students have come to describe as "higher-life forms." The eukarya do contain the largest multiple-celled organisms on Earth, from blue whales to giant sequoia trees, but they also include single-celled yeasts with relatively primitive activities. In microbiology, domain Eukarya includes fungi (molds and yeasts), algae, and protozoa. These microorganisms are characterized by some or all of the following hallmarks of so-called higher life:

- independent movement
- food acquisition
- sexual reproduction
- embryo formation
- use of a variety of nutrients

Most eukarya live as aerobes; anaerobic protozoa that inhabit animal digestive tracts are an exception. Eukarya metabolism is called *heterotrophic* metabolism because of the need or the ability to use a variety of nutrient sources. Finally, the eukarya have evolved with the ability to react to stimuli in the environment and form colonies in which cells work in coordination as a single organism. Bacteria do this, too, but in a much more primitive manner, in the processes of quorum sensing and chemotaxis.

Scientists working in protozoology (study of protozoa), algology (study of algae), mycology (study of fungi), and botany (study of plants) tend to use kingdoms to classify organisms more than microbiologists who study mainly archaea and bacteria. For that reason, biology courses often cover the eukarya in sections that relate to the older classifications: Kingdom Animalia, Kingdom Plantae, and Kingdom Fungi. Kingdom Protista is usually reserved for all unicellular (single-celled) life that Robert Whittaker had originally divided into Monera and Protista.

VIRUSES AND THE THREE DOMAINS

Viruses do not belong to any of the three domains that contain all the life-forms on Earth. Viruses are not defined as cells because they cannot live on their own. In truth, viruses do not live at all; they are obligate parasites, meaning they could not exist without being inside an independently living host cell. Viruses replicate by taking over the reproductive mechanism of the cell they invade.

Biologists have puzzled over the relationship of viruses to the three domains. Two theories have been proposed, rejected, and revisited over the past few decades. First, viruses may have evolved as bits of either cellular deoxyribonucleic acid (DNA) or RNA that escaped from a living cell but retained some functions, such as invasion of cells and replication. A second theory has proposed that viruses are the remnants of cell lines that for some reason degenerated over time. Instead of evolving to adapt extra capabilities, these cells lost functions until they could not survive on their own. Virus structure and general function are nothing like those in the three domains, yet viruses possess an exquisite ability to take over the DNA activities inside cells. In some ways, viruses appear to be the most primitive of beings; in other ways, they seem to possess the most diabolical of lifestyles to force other organisms to live for them. In that second sense, viruses have evolved into very sophisticated particles.

Viruses have undoubtedly evolved with simple cells and perhaps have contributed to cell evolution by carrying genetic material from one organism to another. The relationship of viruses to cells may provide insight into the evolution of the three domains.

See also AEROBE; ANAEROBE; ARCHAEA; BACTERIA; EXTREMOPHILE; SYSTEMATICS; TAXONOMY; VIRUS.

Further Reading

Lyons, Richard D. "Scientists Discover a Form of Life That Predates Higher Organisms: Scientists Discover Distinct Life Form." *New York Times*, 3 November 1977. Available online. URL: http://select.nytimes.com/gst/abstract.html?res=F50F11FC3C5E167493C1A9178AD95F438785F9&scp=1&sq=third+form+of+life&st=p. Accessed March 12, 2009.

Meinesz, Alexandre. *How Life Began—Evolution's Three Geneses*. Translated by Daniel Simberloff. Chicago: University of Chicago Press, 2008.

Torsvik, Vigdis, Lise Øvtreås, and Tron Frede Thingstad. "Prokaryotic Diversity—Magnitude, Dynamics, and Controlling Factors." *Science*, 10 May 2002.

Tortora, Gerard J., Berdell R. Funke, and Christine L. Case. *Microbiology: An Introduction*, 8th ed. San Francisco: Benjamin Cummings, 2004.

Woese, Carl R., and George E. Fox. "Phylogenetic Structure of the Prokaryotic Domain: The Primary Kingdoms." *Proceedings of the National Academy of Sciences* 74, no. 11 (1977): 5,088–5,090.

electron microscopy Electron microscopy is a specialty within microscopy, the science of magnifying objects that cannot be seen with the unaided eye. Electron microscopes use an electron beam to produce an image of structures too small to be seen by the most powerful light microscopes.

Standard light microscopes that exist in every microbiology laboratory enable microbiologists to observe specimens at magnifications of 100 to 2,000 times. These instruments allow them to examine cells of about 0.5 micrometer (μm) in diameter or width, where 1 μm is equal to one millionth (10^{-6}) of a meter or one thousandth (10^{-3}) of a millimeter. By comparison, electron microscopy achieves magnifications up to 1 million times the specimen's actual size (1,000,000 x) and has extended science's power to see into the world of things no larger than 1 nanometer (nm). A nanometer (nm) is one billionth (10^{-9}) of a meter and represents the scale of objects studied in the discipline called *nanobiology.* Put another way, light microscopy aids in the examination of cells, while electron microscopy enables scientists to examine viruses, molecules, and even atoms. These objects measure about 0.2 μm or less in width.

In 2008, the Cornell University professor of engineering physics David Muller explained, "The current generation of electron microscopes can be thought of as expensive black and white cameras where different atoms appear as different shades of gray." But even these capabilities are being improved. Muller described a new, advanced electron microscope at Cornell, saying, "This microscope takes color pictures—where each colored atom represents a uniquely identified chemical species." Microscope capability has long been described in terms of *resolution,* which is the capacity of a microscope to distinguish between two different objects of very small size. Electron microscopy has now achieved resolution at the atomic level.

In microbiology, electron microcopy has been used principally for viewing the physical features of viruses; the organelles inside cells; large molecules called *macromolecules,* such as deoxyribonucleic acid (DNA); and tiny surface appendages. To achieve this extreme power of magnification, electron microscopists prepare samples differently than microbiologists prepare slides for light microscopy. Microscopists also receive training in methods such as slicing microbial cells, and they must understand the physics of controlling electron beams. The table on page 240 lists fundamental differences between electron and light microscopy.

Electron microscopy entails two disadvantages that light microscopy avoids: complexity and high cost. For this reason, light microscopes equip microbiology laboratories the world over, but far fewer companies and universities own electron microscopes.

Two types of electron microscopes are used in industry or at universities. The first type is the transmission electron microscope (TEM), which has a resolution of less than 5 nm and produces images of the interior of microbial or other cells. The second type is the scanning electron microscope (SEM), which focuses on the surface of cells and has a resolution of about 10 nm.

THE DEVELOPMENT OF ELECTRON MICROSCOPY

Microscopy began in 1595, when the Dutch eyeglass maker Zacharias Jannsen (1580–1638) put lenses together to magnify objects up to 20 times their normal size. A century later, Antoni van Leeuwenhoek (1632–1723) had built a more sophisticated assembly of lenses to achieve 300 × magnification of microscopic creatures in water. Discoveries in biology soon accelerated; in part through the aid of microscopes in studies of tiny features of the biological world. By the 20th century, light microscopes magnified to 1,000 × and enabled microbiologists distinguish features as close as 0.2 µm apart. Scientists yearned to learn more of the vast submicroscopic world, but light microscopes had reached their limit of magnification.

In 1931, the German electrical engineer Max Knoll (1897–1969) and the physicist Ernst Ruska (1906–88) conducted studies on the short wavelengths of electron beams compared with light wavelengths. While still a student in Knoll's laboratory, Ruska built the first TEM, in Berlin, that year. Ruska and Knoll developed the instrument entirely on the principles of light microscopy, but instead of directing light to penetrate a sample, they used electron beams.

Ruska then collaborated with other graduate students to improve on the design of cathode ray oscilloscopes for the purpose of investigating the effect of magnetic waves on electric current. These studies led to the development of the polschuh lens, a wire coil device that redirected electron beams in a process similar to the way magnetic fields deflect light beams. This advanced lens remains the basic component in all magnetic high-resolution electron microscopes, to this day. In his acceptance lecture for the 1986 Nobel Prize in physics for contributions in optics and electron microscopy, Ruska recalled, "Through the personality of Max Knoll, there was a companionable relationship in the group, and at our communal afternoon coffee with him the scientific day-to-day problems of each member of the group were openly discussed." Ruska's understated remarks were indeed modest; his discovery opened a new world in biology.

In North America, about the same time (1938) Ruska and Knoll were conducting wave studies, the Canadian James Hillier (1915–2007) built an

Characteristics of Light and Electron Microscopes

Feature	Light Microscope	Electron Microscope
highest practical magnification	1,500 ×	more than 100,000 ×
best resolution	0.2 µm	0.5–10 nm
type of lens	glass	electromagnet
method of changing magnification	adjust objective lens	adjust current to magnetic lens
method of focusing	adjust lens position	adjust current to magnetic lens
light or radiation source	light, usually visible spectrum	electron beam
required conditions	none	vacuum
means of achieving contrast	varying light absorption	electron deflection
specimen preparation	biological staining	embedded thin sections; metal coating; freeze-etching
specimen mount	glass slide	copper grid
cost	$700–$5,000	$100,000s to $1 million
typical specimens viewed	bacteria, fungi, algae, and protozoa whole cells	bacteria, fungi, algae, and protozoa organelles; viruses; macromolecules
typical features viewed	shape; size; stain reaction	shape and size of subcellular features

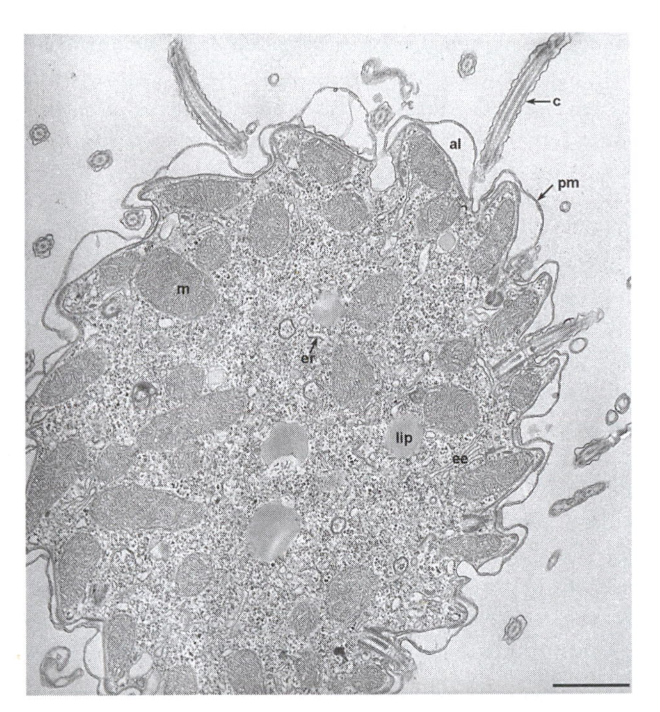

This transmission electron microscopy (TEM) image of *Tetrahymena pyriformis* shows the intracellular structure of this ciliated *(c)* protozoan; magnification is 5,500 ×. "Five Structure, Reconstruction and Possible Functions of Components of the Cortex of *Tetrahymena pyriformis*." *(Richard Allen. Journal of Protozoology 14 (1967): 553–565)*

electron microscope that became the prototype for today's models. It had taken 300 years to perfect the light microscope, whereas less than 40 years the electron microscope had emerged.

With new powers of resolution, biologists developed a new field of study called *descriptive biology*. In this discipline, scientists no longer had to guess about the details of microbial cell structure; they could now view cells at 1,000,000 × with TEM and watch many of the inner workings that had only been described in theory. Biology began to delve into studies at the subcellular level, not just in theory but in actual observation.

Soon electron microscopes revealed cell organelles and viruses. By 1945, the medical community had adopted electron microscopy for tracking the progress of a disease through tissue. Albert Claude (1899–1983), Christian de Duve (1917–), and George Palade (1912–2008) as a team used electron microscopes to study cell metabolism. In 1974, they each were awarded the Nobel Prize in medicine for their detailed studies on organelles.

The SEM's development followed closely behind the introduction of the TEM. The British engineer Charles Oatley (1904–96) developed an SEM, in the 1940s, based on his World War II assignment as a radar operator. The research community may not have applauded the SEM as it had the new TEM images, perhaps because SEM showed only the outer surface of cells, and scientists were still enjoying the newborn fascination with the TEM images.

Not until 1965 did the first SEM, the Stereoscan, enter commercial use. SEM images were every bit as remarkable and exquisite as TEM images, yet Oatley and his students at the University of Cambridge worked in almost complete anonymity as they fine-tuned SEM imaging, improved its detection limits, and enhanced resolution. The scientific world learned of Oatley's invention mainly because of articles published by the German physicist Manfred von Ardenne (1907–97), who had been fascinated by the technique of scanning an object with electrons to produce an image. The physicist K. C. A. Smith of the University of Cambridge said at an electron microscopy conference, in 1997, "The development of the SEM will forever be linked with the names Manfred von Ardenne and Charles Oatley; but the manner of their contributions and the outcome of their work could not have been in more marked contrast. Although in 1935 M. Knoll produced the first scanned image of a surface, it was von Ardenne who, in a relatively short burst of inspired activity just prior to World War II, laid the foundations of both transmission and surface scanning electron microscopy. . . . His involvement with the electron microscope, however, came to an abrupt end when all his apparatus was destroyed in an air raid in 1944." After the war, von Ardenne moved into other areas of science. "It was Charles Oatley who took up the baton in 1947, but in von Ardenne's relatively brief encounter with the subject, he spent the better part of two decades before reaching a successful conclusion with the worldwide acceptance of the SEM as one of the most powerful and productive methods of microscopy yet invented." SEM is now considered as valuable as TEM for inspecting microscopic particles.

TRANSMISSION ELECTRON MICROSCOPY

TEM now provides scientists with images of structures as small as 1 nm inside cells. To produce these images, an electron microscopist must first prepare the sample for the TEM process. The preparation begins by slicing cells into even sheets called *thin sections* through which the electron beam can pass. As the electrons hit various substances in the thin section, the electrons are deflected, creating the basis for forming an electron microscope image. But electrons change course very slightly when passing through biological matter, especially thin sections of 20 to 100 nm in thickness. Specimens must be prepared in a way that increases the electron deflection and thus gives a better image of a cell's interior. These deflected electrons travel to a fluorescent screen upon which the image forms.

The entire process, from a cell suspension to a finished image, demands very specialized skills. The image must additionally be of high quality so that a microbiologist can use it for studying cellular features. A team of electron microscopy specialists starts with a thick piece of tissue or a mass of microbial cells and prepares a delicate, clean, and undamaged thin section. The major steps involved in this process of making a TEM image are the following:

1. fixing
2. embedding
3. preparing thin sections
4. staining
5. shadow casting
6. focusing

Using bacterial cells as an example, a microbiologist centrifuges the cells to remove them from liquid growth medium and then repeats the process with water or buffer in a step called *washing.* A technician then prepares an embedded sample of cells by, first, treating them with glutaraldehyde, a step called prefixation, and, then, with osmium tetraoxide in a postfixation step. The glutaraldehyde reacts with most cellular molecules to make them more stable and less prone to physical damage. The osmium specifically stabilizes lipids (fatty compounds). The glutaraldehyde-osmium steps are together called *double fixation.* The technician then dehydrates the specimen by rinsing it with alcohol and then embeds it in a waxlike substance made up of epoxy resins. (Dehydration helps the electron beam pass through the specimen when inside the electron microscope.)

The embedded specimen can then be more easily cut into thin sections on an instrument called a *microtome.* A microtome houses a fine knife made of glass or diamond and cuts the thin sections as the specimen moves past the blade. The technician views the sectioning process under a dissecting microscope (a low-powered light microscope) attached to the microtome. The microscopist takes the thin cell sections, held rigid in the embedding resin, and attaches each section to a copper mesh platform that will support it during the remaining activities.

The fixing, dehydrating, and embedding steps all have the risks of causing distortion or shrinking in the specimen. As a result, the final image may contain features that appear to be part of the cell but are actually preparation errors. These distortions are called *artifacts* and should be reduced or minimized with careful handling.

In the next step, staining, electron microscopy requires stains that will deflect electrons, rather than produce a visual color, such as the Gram stain method used for light microscopy. TEM staining uses heavy metals such as lead, osmium, molybdenum, or tungsten. The metal stain accumulates in certain organelles and forms dense spots that deflect more electrons than other regions of the thin section.

Electron microscopists use two different types of staining for making a TEM image. The first is called *positive staining;* in it, metals attach to specific organelles to highlight those objects. In the second method, called *negative staining,* the metals attach to areas around or outside the specimen and so illuminate a larger region. Negative staining has been most useful in examinations of viruses, flagella, and macromolecules.

A TEM microscopist might also employ a technique in step 5 to improve the final image. The method called *shadow casting* exposes either a positively or a negatively stained specimen to a shower of metals at a 45° angle. The metal—platinum or gold is common—piles up on one side of an organelle in the same way snowdrifts accumulate against a fence. The clear area on the far side of the drift is called the *shadow.* Shadow casting creates three-dimensional images that help show the size and shape of submicroscopic features.

Step 6, focusing, in electron microscopy involves the control of the electron beam, in contrast to light microscopy, in which glass lenses control the path of light. TEM uses a vacuum chamber in which the specimen rests. The vacuum reduces excess scattering of electrons and so sharpens the black-and-white image. (Electron microscopists must do extra steps to produce color images; this process, called *colorizing the image,* occurs after the TEM image has been made.)

The electron microscope consists of a set of electromagnetic lenses, which the microscopist adjusts for focusing and setting brightness. The lenses also control the direction of the electrons emitted from a source called an *electron gun* at the top of the instrument. A heated filament inside the gun generates the beam, while a large electrical difference from the top of the gun to its exit serves to accelerate the electrons. The electrons travel downward between the lenses to hit the specimen on the platform. A fluorescent screen or a photographic plate captures the black-and-white image created by the deflected electrons. The final image is called a *transmission electron micrograph.* Electron microscopes also have a fitting for 35 mm or digital cameras.

TEM offers undeniable power to view objects that all other instruments cannot match. This attribute is counterbalanced, however, by the fact that TEM shows a very limited area of a specimen. For this reason, microbiologists must be careful in drawing conclusions about intact whole cells based solely on images produced by TEM.

SCANNING ELECTRON MICROSCOPY

SEM images are formed by scanning the outer surface of a microorganism with an electron beam. As a result, SEM images provide a more three-dimensional image than most TEM images.

In this technique, electrons called *primary electrons* exit the electron gun and hit the specimen. Many of the electrons deflect off the specimen's surface as the beam moves back and forth over the terrain. Meanwhile, a detector captures these so-called secondary electrons on a component called a scintillator, which converts each electron's energy into a flash of light. A photomultiplier tube amplifies each signal and sends the information to a cathode ray tube, similar to an older-style television. The microscopist views the image on the screen as it develops. The SEM also contains a separate cathode ray tube for photographing an image at any point during the scan.

The principle behind forming an image from secondary electrons is straightforward: High ridges and low valleys on a specimen's surface cause different deflections to the electron scan. High ridges deflect a large number of electrons, and valleys deflect fewer. More secondary electrons from raised areas, therefore, reach the detector and create lighter areas in the image because they generate more flashes of light. Depressions and valleys remain darker on the image that shows the overall topography of a microscopic particle.

A scanning electron microscopy (SEM) image shows structural differences among a variety of plant pollens; magnification is 500 ×. ***(Dartmouth Electron Microscope Facility, Dartmouth College)***

SEM images, usually at a magnification of 1,000 × to 10,000 ×, distinguish between objects separated by 10 nm or less. SEM has been a valuable tool for viewing the growth of microorganisms on surfaces. For that reason, SEM is helping in studying bacteria attached to the teeth, intestinal lining, and skin, as well as biofilms that grow on inanimate surfaces.

SEM does not provide a view of internal structures of cells as TEM can, but SEM shows morphology of whole cells much better than TEM. As with TEM, SEM studies must be performed with the knowledge that only a tiny portion of an object is shown.

SPECIAL TECHNIQUES IN ELECTRON MICROSCOPY

TEM had for many years taken second place to SEM in its ability to provide a three-dimensional image. A technique called freeze-etching has overcome this drawback by giving shape and dimension to cell organelles viewed by TEM. In this method, a microscopist rapidly freezes the specimen in liquid nitrogen to -385°F (-196°C), then warms it in a vacuum chamber to about -212°F (-100°C). The specimen becomes brittle in this treatment, and it breaks along its weakest points. Many of the outermost layers tend to break away when a knife is applied to the frozen specimen. The result is a topography map of the inside of a cell, where organelles can be seen in the broken cell as rises or depressions.

Freeze-etching in TEM eliminates the use of chemicals needed for fixation and skips the dehydration and embedding steps. The technique, consequently, lessens the risk of producing artifacts. Freeze-etching has been especially valuable in the study of cell membranes. Biological cell membranes tend to develop into bilayer structures, wherein long hydrocarbon chains line up so that their water-attracting (hydrophilic) ends point outward and water-repelling (hydrophobic) ends point into the membrane's inner layer. The inner layer contains hydrophobic fatty substances, while the cell cytoplasm or the cell's exterior is bathed in watery materials. Freeze-etching has enabled biologists to break the bilayer apart so that they can view the once-hidden inner layer of the cell membrane.

Improvements in electron microscopy have continued for the purpose of resolving smaller and smaller objects. Three technologies in electron microscopy have advanced the capabilities of TEM and SEM. These are scanning electron transmission microscopy (STEM), low-temperature electron microscopy (LTEM), and three-dimensional electron microscopy (3DEM).

STEM, as the name implies, combines features of TEM and SEM. STEM had been introduced, in the early 1970s, for studying biological matter, but the technology languished for 20 years, before the field of electron microscopy made it part of routine microscopic studies.

In STEM, electrons pass through a thin section as they do in TEM, but instead of measuring deflection of the electron beam, STEM measures the energy of the deflected electrons after they pass through and emerge from the specimen. Detectors behind the specimen measure the energies of the electrons; those deflected by denser matter lose more energy than those passing through the specimen with little impedance. From 60 to 100 percent of the electrons in the incoming beam emerge for capture.

STEM can be used to construct both dark-field and bright-field images on a screen. Dark-field microscopy creates images when light hits a specimen at an angle so that only reflected light is measured. Bright-field microscopy creates images by passing light through a specimen. In STEM, these principles have been adapted for electron beams rather than light illumination. STEM also makes use of metal stains like those used in TEM to create a clearer image. Overall, the resolution of STEM resembles that of TEM.

LTEM is also called *cryoelectron microscopy* because it requires specimens that have been cooled quickly in a process similar to freeze-etching. Specimens prepared for LTEM are not dehydrated, however, so water remains inside the cell or inside organelles. The rapid freezing to very low temperatures converts the water to a glasslike material rather than the crystals that form in normal freezing. This formation of a glasslike ice by extreme freezing is called *vitrification*. An electron beam hits an LTEM specimen inside an ultracold vacuum chamber. A detector captures and measures the deflected electrons and converts them to an image. Vitrified specimens produce less distortion and fewer artifacts than regular TEM, and this method has enabled microscopists to create images of large proteins, glycoproteins, and detailed features of viruses.

The 3DEM technique also uses vitrified specimens. In this method, the microscopist embeds the specimen in ice. The embedding provides support for molecules, as well as fragile cell organelles such as vacuoles. Both LTEM and 3DEM are also referred to as *three-dimensional electron tomography;* tomography is the study of any type of thin section. The standard resolution of 3DEM is 2 nm, but new technologies will soon make improved resolutions commonplace. As in LTEM, 3DEM greatly reduces distortion and artifacts, compared with the standard TEM method.

The freezing methods in electron microscopy have been especially helpful in preserving structures, from delicate components of cells to macromolecules. Vitrified specimens now provide insight into how atoms are arranged in a molecule as well as the detailed arrangements of larger parts of the cell. The science of gathering information on molecular arrangements of compounds is called *crystallography*. This discipline has become the main method for examining atomic structures in biology.

Electron microscopy represents one of the most important advances in modern cell biology. Before these instruments came into use, microbiology had reached its limits on the information it could obtain on the microbial world. Electron microscopy has enabled microbiologists to study cells at the molecular level, and this specialty will probably contribute to the ongoing progress in studies of the ultrasmall, such as nanobiology.

See also MICROSCOPY; MORPHOLOGY; NANOBIOLOGY; ORGANELLE.

Further Reading

Bozzola, John J., and Lonnie D. Russell. *Electron Microscopy,* 2nd ed. Sudbury, Mass.: Jones & Bartlett, 1998.

Dennis Kunkel Microscopy. Available online. URL: www.denniskunkel.com. Accessed March 29, 2009.

Ruska, Ernst. "The Development of the Electron Microscope and Electron Microscopy." Nobel lecture presented December 8, 1986, in Stockholm, Sweden. Available online. URL: http://nobelprize.org/nobel_prizes/physics/laureates/1986/ruska-lecture.html. Accessed March 12, 2009.

Science*Daily*. "New Electron Microscope Identifies Color-Coded Atoms." February 22, 2008. Available online. URL: www.sciencedaily.com/releases/2008/02/080221153725.htm. Accessed March 28, 2009.

Smith, Kenneth. "Charles Oatley: Pioneer of Scanning Electron Microscopy." Presented at the Annual Conference of the Electron Microscope Group of the Institute of Physics, Cambridge, England, September 2–5, 1997. Available online. URL: www2.eng.cam.ac.uk/~bcb/semhist.htm. Accessed March 26, 2009.

University of Nebraska. "Electron Microscopy." Available online. URL: www.unl.edu/CMRAcfem/em.htm. Accessed March 29, 2009.

Watt, Ian M. *The Principles and Practice of Electron Microscopy,* 2nd ed. Cambridge: Cambridge University Press, 1997.

emerging disease An emerging disease is one that is seen for the first time, that first appears in a new geographical area, or that has an abrupt increase in incidence. Diseases that fit any of these criteria and are caused by the transmission of a microorganism are called *emerging infectious diseases*. The National Institute of Allergy and Infectious Diseases (NIAID) qualifies emerging diseases as diseases that have been characterized, for the first time, within the last two decades.

A reemerging infectious disease is similar; this is a disease that had been previously in a population, disappeared or became dormant, and then reappeared in high incidence.

Emerging and reemerging infectious diseases have always been a part of medical care. Many of the ailments that have become familiar to modern medicine, such as virus infections and sexually transmitted diseases, date to ancient societies. People in industrialized countries may have a sense of confidence that new diseases no longer appear in society, and excellent medical care will manage any unexpected outbreaks. On the contrary, new diseases continue to emerge in current history: *Legionella* infection in 1976, acquired immunodeficiency syndrome (AIDS) in 1980, *Escherichia coli* O157:H7 in 1982, severe acute respiratory syndrome (SARS) in 2002. A deadly influenza H1N1 strain of influenza reemerged in 2008.

Medicine in the 21st century must control diseases that were believed to have been eliminated but are now returning. Sometimes, these diseases reemerge in a more virulent form. Virulence is the ability of a pathogen to cause disease. As a consequence, modern medicine has been confronted with two challenges: the ongoing battle against new, emerging diseases and vigilance against unexpected reemergence of some diseases thought to have been beaten. Many specialties in the medical profession have concentrated on the puzzle of why medical advances cannot prevent emerging or reemerging diseases, and, further,

Factors That Contribute to Infectious Disease Emergence and Reemergence

Factor	How It Influences Disease Transmission
climate change	warmer temperatures enhance breeding of insects that are vectors for disease transmission
evolution	development in microorganisms of resistance to antibiotics and possibly antimicrobial chemicals
food production and distribution	mass production and distribution of foods globally spread food-borne pathogens
globalization	global transport of goods and services helps transmit infectious agents
human behavior	changes in sexual behavior contribute to transmission of sexually transmitted diseases (STDs)
international travel	global travel transmits disease faster than epidemiologists can detect changes in disease patterns
lack of sanitation and vaccinations	areas of high poverty and social unrest have damaged infrastructure, which inhibits the distribution of medical care
medicine	increased life expectancy contributes to existence of an older population; organ transplant advances increase number of patients who have suppressed immune systems
military conflicts	destruction of infrastructure, medical supply distribution, clean water, and safe food; mass migrations and displaced people
population growth and shifts	spread of normally localized illnesses to new areas
poverty	poor nutrition, health care, shelter, and access to clean water increase risk for disease
urban development and expansion	human encroachment into wilderness areas that are home to insect and animal vectors for disease
urban growth	increases density of human populations and increases risk of exposure to microorganisms

Important Emerging Diseases since 1980

Disease	Cause	Global Status	Probable Year of Emergence
AIDS	human immunodeficiency virus (HIV)	worldwide, epidemic in sub-Saharan Africa	1980
human monocyte ehrlichiosis	*Ehrlichia chaffeensis* (bacteria)	southeastern United States	1994
food-borne hemorrhagic fever, diarrhea	*E. coli* O157:H7	sporadic outbreaks	1983
hantavirus pulmonary syndrome	hantavirus	spread over half of the United States, originating in the Southwest	1993
Hendra virus disease	Hendra virus	Australia	1994
influenza	type A and type B influenza viruses	emerges yearly, usually originating in Asia and spreading worldwide	yearly
Nipah virus disease	Nipah virus	Southeast Asia	1999
Pfiesteria poisoning	*Pfiesteria piscicida* (algae)	U.S. East Coast	1995
Rift Valley fever	Rift Valley fever virus	Africa	2006
SARS	SARS-associated coronavirus (SARS-coV)	small outbreaks and some deaths mainly in Asia	2003
West Nile disease	West Nile virus	spread across the United States	1999

why reemerging disease seems to be on the increase. Joshua Lederberg (1925–2008), a Nobel Prize recipient in medicine, explained, in 1998, the simple reason for emergence and reemergence: "Emergence is none other than the dark side of coevolution, a typical, inexorable biological phenomenon. But we came along with marvelous vaccines and antibiotics, with sanitary water and food; for a while, some of us practiced safer lifestyles, from handwashing to discreet sex. And perhaps we thought we had licked the bugs [microorganisms] with our technology."

"But they kept and do keep evolving, and besides having let down our guard, we have contrived a world that is safer for bugs than ever before, with instantaneous travel, mass production and transport of foodstuffs, and crowded and sharply stratified urban populations." Lederberg's remarks cut to the heart of today's near-crisis in emerging and reemerging disease: Infectious microorganisms have taken advantage of changing behavior in society and their own ability to evolve quickly to outwit even the best defenses against disease that humans have invented. For example, though the development of antibiotics saved millions of lives since the 1940s, when penicillin became the first antibiotic sold, microorganisms have developed entirely new genetics to defeat the actions of most of today's antibiotics. In a way, medicine has created infectious species that are far stronger than the species that ravaged societies of centuries ago.

Current studies of emergence and reemergence combine five major elements of disease study, as follows:

- epidemiology—study of the physical and human factors that lead to disease outbreaks
- improved mortality record keeping and databases
- social and economic issues

- changes in microbial virulence and other evolutionary factors
- climate change

Each of the points listed includes numerous topics of more focused concern, so that a biology student quickly sees that the problem of emergence encompasses much more than microorganisms, transmission, and treatments. The table summarizes the key points discussed by the medical profession today in relation to emergence and reemergence.

The factors listed in the table span social, economic, biological, and environmental activities. In this way, they illustrate the complexity of studying disease emergence. As a result, emergence and reemergence have become very difficult to control, even with the best medical advances available today.

EPIDEMIOLOGY OF EMERGING DISEASES

Epidemiology is the study of the factors that cause disease outbreaks and influence the frequency and distribution of outbreaks. Specialists in emerging diseases call upon a broader type of epidemiology, which views disease in relation to global social factors and ecology: This specialization is called *systematic epidemiology*. Systematic epidemiology considers all the factors listed in the table to draw conclusions on the patterns of disease emergence worldwide.

The Centers for Disease Control and Prevention (CDC) began operations as the U.S. national public health service, in 1946. Successes followed failures, as it grew from an obscure agency staffed by microbiologists to a leader in public policy on health. The CDC led a fairly isolated life as a government agency, until the early 1980s, when an unknown illness began to devastate entire communities in the largest U.S. cities. The emergence of the acquired immunodeficiency syndrome (AIDS) virus demonstrated that disease control required participation by more individuals than doctors and medical researchers hidden away in laboratories. Public behavior and government action played a decisive role from the start of the AIDS epidemic in the United States, from the 1980s to the present. These same factors are now playing out in sub-Saharan Africa, where AIDS has virtually wiped out an entire generation of people.

The U.S. government and the CDC responded slowly to the AIDS crisis, as the 1980s unfolded. National public health agencies did not collaborate, at first, as the AIDS virus spread through the population of heterosexual men, intravenous drug users, and people who had weakened immune systems due to existing disease. In 1989, a group of American scientists who specialized mainly in viral diseases met in Washington, D.C., to emphasize the crisis developing before them as well as potential other virus diseases on the move. The meeting had been sponsored by the NIAID, the Fogarty International Center, and Rockefeller University. In that meeting, as relayed by Laurie Garrett, author of *The Coming Plague,* the medical historian William McNeill said, "It is, I think, worthwhile being conscious of the limits upon our powers. It is worth keeping in mind that the more we win, the more we drive infections to the margins of human experience, the more we clear a path for possible catastrophic infection. We'll never escape the limits of the ecosystem. We are caught in the food chain, whether we like it or not, eating and being eaten." McNeill's chilling appraisal of disease made it clear that humanity would never eradicate infectious diseases. The best hope appears to be steady and reliable surveillance of emerging diseases to prevent delay in reacting to the new disease.

Effects of Globalization on Infectious Diseases

Activity	Effect
air travel	infected travelers transmit disease to other travelers
business meetings	face-to-face meetings give opportunities for localized pathogens to spread
developing economies	commodities and raw materials from isolated regions may carry new pathogens
education	international travel for study introduces new pathogens into campus populations
new materials	materials not previously harvested become alternatives to declining materials and may contain new pathogens
oceangoing ships	ballast release in new ports introduces nonnative plants, animals, and infectious agents

Some emerging diseases, such as AIDS, change from a new medical emergency to an endemic disease, meaning the illness has become expected in a population and causes a standard number of deaths each year. The table lists some of the prevalent diseases that have emerged since 1980 and their status today. Some of these infectious agents were known in biology for many years, but they were not known to constitute a serious health threat until many years later.

Fast response by government and by the medical community to an emerging disease is supplemented with ever-improving surveillance systems. The power of today's databases and analysis of trends within them has exceeded the programs of the 1980s several hundredfold. If these systems are employed properly in spotting new diseases, medical researchers receive valuable extra time in identifying the cause of infection, learning the pathogen's characteristics, and finding a treatment.

Systematic epidemiology also relies on cooperation between national medical leaders to track diseases across the globe rather than wait for an infection to cross borders. The World Health Organization (WHO) surveys global diseases and the patterns of their emergence or reemergence. The WHO also provides guidance on sanitation, methods for preventing the development of antibiotic resistance, and special disease circumstances in world regions. The WHO and the CDC cooperate to offer guidance and policies to the public for preventing disease transmission nationally and globally.

SOCIAL AND ECONOMIC ISSUES IN DISEASE TRANSMISSION

Populations carry out many actions that give infectious agents an opportunity to spread. In many cases, people take direct control over behavior that directly or indirectly helps pathogens gain access to the population. The following voluntary activities open the door for new diseases:

- global transport and travel and immigration
- medical methods to extend lives
- desire to visit, develop, or otherwise enter previously unspoiled land
- demand for antibiotics in inappropriate uses
- increasing urbanization

Reasons for Disease Reemergence

Disease	Cause	Probable Reason for Reemergence
bubonic plague	*Yersinia pestis* (bacteria)	poverty, poor sanitation, urbanization
Chagas' disease	*Trypanosoma cruzi* (protozoa)	increased incidence of immunosuppressive conditions (transplants, human immunodefiency virus [HIV], pregnancy, increased age)
cholera	*Vibrio cholera* (bacteria)	shortage of safe water
diphtheria	*Corynebacterium diphtheriae* (bacteria)	lack of vaccination programs
dengue hemorrhagic fever	dengue virus	global warming, flooding
Ebola hemorrhagic fever	Ebola virus	encroachment into the disease reservoirs in tropical areas
malaria	*Plasmodium falciparum* (protozoa)	air travel, mass emigrations from war regions, drug resistance, changing rainfall patterns
rabies	rabies virus	encroachment into undeveloped wildlife habitats
typhoid fever	*Salmonella typhi* (bacteria)	declining food and water sanitation in poverty-stricken areas
yellow fever	yellow fever virus	global warming

Many voluntary actions that help spread disease are also attempts to improve the lives of people under stress. People who emigrate from warring countries or political persecution could hardly be blamed for seeking better circumstances.

The world has transformed from isolated national economies, in the early years of industrialization, to international trade in certain commodities, in the 1930s through the 1980s. When Web-based businesses grew in the 1980s, followed by rapid advances in telecommunications, globalized trade became the major business model, and it remains in place today. With globalized economies were intercontinental shipments of raw materials and finished products. Business travel also expanded to every part of the globe. These activities have caused obvious and discrete effects on the transmission of infectious agents. The table presents the main relationships between globalized economies and disease transmission.

Emerging diseases present the medical community with more than one moving target. Changing social behavior and economic practices contribute to the spread of many diseases that emerge in new places. At the same time, infectious microorganisms change as well, thereby further complicating the emergence of new diseases.

FACTORS IN REEMERGENCE

Microorganisms change their genetic makeup through spontaneous mutations. These mutations take place continually in microbial populations, and in the majority of these events, they lead to indistinguishable variations in the population. On occasion, a spontaneous mutation gives a microorganism a distinct advantage over all other related microorganisms. This is how resistance to antibiotics developed. A few species developed methods, some time ago, to destroy antibiotics naturally produced by other microorganisms in the environment. This small proportion of resistant cells might have grown to dominate their populations in the natural course of evolution. But the proliferation of antibiotics in medicine has encouraged the development of antibiotic-resistant microorganisms. Antibiotic resistance may be a major contributor to the reemergence of a large number of diseases that had all but disappeared several decades ago.

The combination of antibiotic resistance with factors that increase opportunities for disease transmission has caused a rebirth of many diseases. In human history, only one disease has been proven to have been completely eradicated: smallpox. All other diseases have remained in human populations to greater or lesser degrees, throughout history. Although all of the known pathogens have remained as threats to human health, the pathogens that cause a surge in new cases, at any point in time, are usually considered to be the cause of reemerging disease.

Many reemerging diseases return only to localized regions of the world, rather than recur worldwide. For example, WHO refers to tuberculosis (TB) as an epidemic, even though rates of this disease have been decreasing slowly worldwide, for several decades. But regions of Africa have seen increases in TB, especially associated with the alarming incidence rates of AIDS. A second factor in the resurgence of TB resides in new strains of the *Mycobacterium* pathogen that resist an array of antibiotics. A strain called multiple-drug-resistant TB (MDR-TB) has increased in parts of Africa that also have a high incidence of AIDS. MDR-TB has developed resistance to the antibiotics isoniazid and rifampicin and possibly other standard TB treatments. A newer MDR-TB strain called *XDR-TB* that has emerged is resistant to additional drugs: fluoroquinolone and at least one of the injectable drugs amikacin, kanamycin, and/or capreomycin. This resistance leaves people infected with XDR-TB without any sure treatment for their TB.

On April 1, 2009, China View news reported, "Although TB is preventable and treatable, when the TB bacillus becomes resistant to the two most powerful first-line anti-TB drugs, the disease develops into the multidrug-resistant TB (MDR-TB). The more serious XDR-TB, a substrain of MDR-TB that developed from the highly drug-resistant strains, has been reported in more than 50 countries—mainly Asia, Africa, and Europe, according to the WHO." TB cases have risen especially in crowded cities, such as New York City, because TB spreads easily when people are close together in places such as crowded elevators, buses and subways, and theaters. The emergence and spread of drug-resistant varieties of TB might be viewed as among medicine's most disheartening failures because TB had been under control for many decades.

Reemergence and emergence sometimes intertwine, as the new forms of TB demonstrate. The reemergence of TB has been linked to poverty in some instances and to the rise of AIDS, which weakens the body's ability to fight infection from the TB microorganism. Other diseases reemerge for the same reasons that new diseases emerge: globalization, poverty, international travel, and factors in ecology. The table presents some diseases that have reemerged because of these factors.

SOCIETY AND DISEASE

Diseases enter a society, in large part, through human actions that open the door for infection. For example, the AIDS epidemic in the United States emerged in specific subpopulations, in which the pathogen quickly spread because of sexual behavior and sharing of needles. The rabies virus has been reemerging, principally as a result of the increased expansion of housing into wildlife habitat. Some human activities alter the environment in such a way as to create a much more global shift than these two examples. Top among the global problems that have been related to new disease patterns is population growth. This event has ushered in a variety of situations that affect the emergence and spread of disease. Donald B. Louria of the New Jersey Medical School summarized the effects of world population growth on variables related to infectious disease emergence and reemergence. The following list summarizes these conditions.

- increased potential for person-to-person disease spread
- increased global warming
- larger numbers of travelers
- more-frequent wars
- increased numbers of refugees and displaced persons
- increased hunger and malnutrition
- more crowding in urban slums
- increase in the numbers of people living in poverty
- inadequate clean water supplies
- increasing dam construction and irrigation projects

Most of the important points listed here have already been discussed, except the building of new dams and irrigation projects. These types of construction cause flooding of waters upstream from dams or increase pools of standing water. Both conditions increase the breeding areas for insects that may act as vectors in carrying infectious disease. In poor regions, people using these waters for bathing, washing clothes, and cooking and drinking water will be at higher risk for infection.

Any large changes in human activities, such as housing developments built in undisturbed open space, dam construction, or poverty that puts hungry people and animals together in the search for food, can lead to increased incidence of zoonoses. Zoonoses, or zoonotic diseases, are illnesses transmissible between people and animals, for example, avian influenza. When a pathogen that has normally confined itself to an animal species begins to infect humans, the process is called a *species jump*. The health of the world's growing human population, therefore, is connected with the diseases that circulate in animal populations.

EMERGING DISEASES IN ANIMALS

Emerging diseases in animals that serve as reservoirs for human pathogens impact the overall risks to human health. A reservoir is a species that serves as a continual source of a pathogen. Human population growth affects the lives of wildlife by removing habitat, making water scarce, creating pollution, and disrupting normal breeding and social behavior in animal species. Animal populations consequently suffer increased stress, malnutrition, and poor health. These conditions give pathogens an opportunity to enter and spread throughout an animal population that does not possess a strong immune system to fight the infection. In short, human population growth has created a disease-susceptible population of wildlife. Human communities that expand into previously undisturbed areas additionally cause wildlife to find new habitat. As animals migrate into new territory, they carry infections to a new population. The veterinary pathologist Corrie Brown noted, in 1999, "Movement to a susceptible population has been a reason for emerged diseases of animals for centuries. Genghis Khan, Attila the Hun, and Napoleon all spread cattle diseases in their wake as draft animals subclinically harboring rinderpest (cattle plague) and contagious bovine pleuropneumonia were taken to new areas where cattle populations had no immunity." Domesticated animals have many of the same health risks as humans in regard to contracting a new disease.

The main factors in giving new diseases an opportunity to emerge in wild or domesticated animal populations are the following:

- migration of infected animals into a susceptible population
- disruptions to habitat, food, and water
- species crowding in smaller habitats
- changing animal husbandry practices: feedlots, antibiotics, growth stimulants, and so on

In addition to the broad categories listed here, many diseases emerge in animal populations for entirely unknown reasons or reasons that have merely been suspected. The table describes emerging diseases that are of current concern in veterinary medicine and wildlife management.

Emerging diseases create two important effects on the environment. First, increase in disease causes the loss of species biodiversity and accelerates the speed with which species become extinct. Second, dis-

Emerging Diseases in Animals

Disease	Cause	Description	Location
anaplasmosis (formerly called gall sickness)	*Anaplasma phagocytophilum* (bacteria)	anemia and death in cattle	global
bluetongue	bluetongue virus	ruminants, mainly sheep and deer	Mediterranean basin
Crimean-Congo hemorrhagic fever	Crimean-Congo virus	cattle, sheep, small mammals	Afghanistan and neighboring countries
cryptococcosis	*Cryptococcus gattii*	infections of lungs and skin and neurological disorders in marine mammals	western Canada coast
foot-and-mouth disease	coxsackievirus and other related viruses	blisters on skin and mouth, fever, neurological disorders, irregular heartbeat in cattle	sporadic, localized outbreaks
porcine respiratory and encephalitis syndrome	Nipah virus	lethargy, morbidity, and death in pigs	Southeast Asia
Pfiesteria poisoning	*P. piscicida* (algae)	skin lesions, neurological damage in fish and predators	eastern U.S. coast
Rift Valley fever	Rift Valley virus	lethargy and fever in livestock	Africa
West Nile	West Nile virus	neurological disorders and death in birds	global

eases in animals may cause increases in the incidence of zoonotic diseases in humans. In 2008, Kate Jones, research fellow at the London Institute of Zoology, said, "Emerging disease hotspots are more common in areas rich in wildlife, so protecting these regions from development may have added value in preventing future disease emergence." Jones's study team also conducted a study of data on 335 incidences of emergence, since 1940, and found that more new diseases emerged, in the 1980s, than in any other period in history, and many of them originated in animals.

The bacterial disease brucellosis in cattle has caused an ongoing controversy about the risks—real or perceived—that wildlife will transmit disease a population of domesticated animals. Brucellosis is a bacterial disease caused by species of *Brucella,* which causes spontaneous abortion, poor milk production, and infertility. Cattle farmers in the United States, in 1917, first reported brucellosis, and, over the following decades, the federal government spent substantial amounts of money to eradicate the disease from domesticated herds. In the early 1900s, the American bison had suffered through a decade of mass slaughter and numbered a few dozen animals in isolated spots, such as Yellowstone National Park. In 1966, regulations opposed to bison hunting strengthened, and the Yellowstone herd began to grow, to more than 4,000 animals. Cattle farmers on the lands surrounding the park feared that bison that wandered out of Yellowstone's boundaries to graze, especially in winter when food is scarce, would carry the disease back to domesticated cattle.

The brucellosis scare initiated a plan to shoot any bison that unwittingly wandered outside Yellowstone's borders. Animal rights proponents soon squared off against ranchers, and both groups demanded answers and action from the National Park Service. Brian McCluskey, spokesperson for the federal Animal and Plant Health Inspection Service, told the Montana *Billings Gazette,* in 2008, "Right now, the only place where there is bovine brucellosis is the greater Yellowstone area." Despite the successful eradication everywhere else in the nation, the concern that infection persists in the area's bison, elk, and other wildlife has created significant fear that brucellosis will reemerge in cattle. The ongoing dilemma of how best to protect commercial herds while preserving the lives of the American bison has created interest in the issues surrounding reemerging diseases in animals.

EMERGING DISEASES IN PLANTS

Plant susceptibility to new diseases occurs for reasons similar to those in animals: Plants' ability to fight infection is reduced when changing climates, water availability, and reduced habitat make plants more susceptible to pathogens. Nikkita Patel, officer of a project team in conservation medicine of the Wildlife Trust Consortium, said, in 2004, "The recent spate of extreme weather events in Florida and elsewhere highlights the need to protect crops and wild plants from invasive pathogens that exploit the damaged environments following hurricanes." Increased frequency and violence of large storms have been proposed as a result of shifting weather patterns due to global warming.

Plants, as can animals, can be infected by bacteria, viruses, fungi, or parasites. Emerging diseases in plants have been blamed for leading many plant species to the brink of extinction. Intensive crop cultivation methods on small pieces of land have possibly contributed to the spread of these pathogens easily from plant to plant. Individual plants already weakened from intensive cultivation practices and environmental changes become more vulnerable to any disease.

Emerging Diseases in Plants

Disease	Cause	Affected Plants
	FOOD CROPS	
cassava mosaic disease	begomovirus	cassava
citrus canker	*Xanthomonas axonopodis* (bacteria)	citrus species
High Plains disease	High Plains virus	barley, corn, grasses, wheat
karnal bunt	*Tilletia indica* (fungus)	triticale, wheat
Moko disease	*Ralstonia solanacearum* (bacteria)	banana
potato late blight	*Phytophthora infestans* (fungus)	potato
rice blast	*Magnaporthe grisea* (fungus)	barley, millet, rice, turf grasses, wheat
rice stripe necrosis	rice stripe necrosis virus	rice
sugarcane orange rust	*Pucciniania kuehnii, Saccharum officinarum* (fungus)	sugarcane
tomato yellow leaf curl	tomato yellow leaf curl virus	bean, tomato
	NON-FOOD PLANTS	
chestnut blight	*Cryphonectria parasitica* (fungus)	American chestnut
dogwood anthracnose	*Discula destructiva* (fungus)	flowering dogwood
Dutch elm disease	*Ophiostoma* species (fungus)	elms
eelgrass wasting disease	*Labyrinthula zosterae* (fungus)	marine eelgrass
Florida torreya mycosis	*Pestalotiopsis microspore* (fungus)	Florida torreya
pondberry stem dieback	various fungi	pondberry
sudden oak death syndrome	*Phytophthora ramorum*	California bay laurel, Oregon myrtle, various oaks, and other woody plant species

Global food consumption depends on four main plant staples: wheat, rice, corn, and potato. Each of these crops has, since the 1980s, been assaulted by a variety of emerging plant pathogens. The table on page 260 describes the emerging diseases of greatest concern in plant health today.

In plants, emerging diseases most often are due to infection by viruses (about 47 percent of all new diseases), followed by fungi (30 percent), bacteria (16 percent), and other rarer pathogens. Viruses, fungi, or bacteria may infect for different reasons, but in general, the main reasons behind the emergence of a new disease in a plant population are, in order, new plant introductions (including unwanted invasive species), altered weather patterns, intensive farming, changes in vector populations, recombinant varieties, and habitat disruption. Intensive farming that depletes soil nutrients, global warming, and habitat loss all stress plant species and so hurt plants' natural abilities to fight pathogens.

Emerging infectious diseases have become a serious concern in human and veterinary medicine and in food production. Pathogens that were, at one time, dormant or hidden in small locales have been released into the environment by a variety of human-caused factors. Emerging and reemerging diseases can be expected to remain or perhaps grow as a health threat in the near-future.

See also EPIDEMIOLOGY; VIRULENCE.

Further Reading

Anderson, Pamela K., Andrew A. Cunningham, Nikkita G. Patel, Francisco J. Morales, Paul R. Epstein, and Peter Daszak. "Emerging Infectious Diseases of Plants: Pathogen Pollution, Climate Change and Agrotechnology Drivers." *Trends in Ecology and Evolution* 19 (2004): 535–544. Available online. URL: http://chge.med.harvard.edu/about/faculty/journals/tree.pdf. Accessed August 21, 2009.

Brown, Corrie. "Emerging Diseases of Animals." In *Emerging Infections III.* Washington, D.C.: American Society for Microbiology Press, 1999.

Brown, Matthew. "Brucellosis Plan Suggests Special Yellowstone Area." *Billings Gazette,* 22 September 2008.

Centers for Disease Control and Prevention. Available online. URL: www.cdc.gov. Accessed August 21, 2009.

Consortium for Conservation Medicine at Wildlife Trust. "Emerging Infectious Diseases of Plants Caused by Extreme Weather Events and 'Pathogen Pollution.'" 28 September 2004. Available online. URL: www.ewire.com/display.cfm/Wire_ID/2305. Accessed August 21, 2009.

Garrett, Laurie. *The Coming Plague.* New York: Farrar, Straus & Giroux, 1994.

Keppler, Nick. "White Tail Solutions' Quiet Kill." *New Haven Advocate,* 25 December 2008. Available online. URL: www.newhavenadvocate.com/article.cfm?aid=11094. Accessed August 23, 2009.

Lederberg, Joshua. "Foreword." In *Emerging Infections I.* Washington, D.C.: American Society for Microbiology Press, 1998.

Louria, Donald B. "Emerging and Reemerging Infections: The Critical Societal Determinants, Their Mitigation, and Our Responsibilities." In *Emerging Infections I.* Washington, D.C.: American Society for Microbiology Press, 1998.

Science*Daily*. "Emerging Infectious Diseases on the Rise: Tropical Countries Predicted as Next Hot Spot." February 21, 2008. Available online. URL: www.sciencedaily.com/releases/2008/02/080220132611.htm. Accessed August 20, 2009.

U.S. Department of Agriculture. Center for Emerging Issues. Available online. URL: www.aphis.usda.gov/vs/ceah/cei. Accessed August 21, 2009.

World Health Organization. Available online. URL: www.who.int/en. Accessed August 21, 2009.

Xinhua News Agency. "WHO Calls for 'Urgent Action' against Global Epidemic of Drug-Resistant TB." April 1, 2009. Available online. URL: http://news.xinhuanet.com/english/2009-04/01/content_11112401.htm. Accessed August 21, 2009.

enrichment An enrichment is the addition of a substance to a medium for the purpose of favoring the growth of a particular microorganism. A culture of microbial cells grown on such a supplemented formula is called an *enrichment culture.*

Enrichment medium is designed to increase the numbers of a desired microorganism to detectable levels. In many cases, the substance added to a formula helps fastidious species grow. Fastidious species are varieties that are very particular about the conditions in which they grow, either in nature or in a laboratory. Enrichments are thus done to achieve any of three purposes. First, enrichment media help recover microorganisms from very unusual environments that provide very specific or unusual nutrients. Recovery is the process of removing a microorganism from its natural environment and providing the proper growth conditions for it to grow and multiply in a laboratory. Second, enrichment media aid in the separation of two microbial species that normally live in very close association. Third, some enrichment medium formulas induce microorganisms to behave in a certain manner or to produce a desired compound.

Enrichment media may be either liquid broth or solid agar. Broth is often used during the early stages of recovery of a microorganism from the environment; nutrients are more available to stressed cells in a liquid growth medium than on a solid surface. Microorganisms that are highly difficult to recover from their natural environment and grow in a laboratory may be aided by liquid preenrichment media, which give time for damaged cells to repair and allow extra time for stressed cells to begin multiplying. Preenrich-

Types of Enrichment Media

Medium Conditions	How It Works
	PHYSICAL
low-temperature incubation (32–41°F [0–5°C])	retards growth of all microorganisms except psychrophiles
high-temperature incubation (158–162°F [70–72°C])	kills all microorganisms except thermophiles
high-agar concentration (5 percent)	aids swarming growth in species such as *Clostridium tetani*
semisolid agar (0.15 percent agar)	aids in detecting motile bacteria that can swim through the agar, leaving a trail of colonies
illumination during incubation	allows growth of photosynthetic cyanobacteria
	CHEMICAL
alkali conditions (pH 8.4–8.5)	induces growth of *Vibrio* species
acidic conditions (pH 2.0)	induces growth of *Thiobacillus thiooxidans* and other extreme acidophiles
potassium tellurite (0.001 percent)	inhibits gram-positive and gram-negative bacteria except *Corynebacterium* species
mannitol salt medium containing 7.5 percent NaCl	enables only *Staphylococcus* species to grow
antibiotic media	various antibiotics added to media to allow growth of only antibiotic-resistant bacteria
Iverson medium with high sulfate and organic acids	allows growth of sulfate-reducing bacteria *Desulfovibrio* and *Desulfotomaculum*
	BIOLOGICAL
plaque formation, viruses	isolates viruses, called *bacteriophages,* that attack bacterial cells
plaque formation, bacteria	isolates *Bdellovibrio* bacteria that live as parasites in other bacteria

ment media contain all the nutrients that a microorganism would be expected to need for growth, but the nutrient concentrations are adjusted so as not to shock cells that are from nutrient-sparse environments.

Microorganisms may additionally require primary enrichment followed by a secondary enrichment. The primary enrichment medium contains ingredients to favor the initial recovery of a desired microorganism, and the secondary enrichment contains a slightly altered formula that maintains the growth of the desired isolate while suppressing the growth of other microorganisms. This primary-secondary scheme can be helpful when searching for species that make up a small proportion of microbial communities and play specialized roles within those communities.

Agar media work best for isolating particular microorganisms from others in a mixture and for studying a microorganism's distinct colony morphology (the general appearance in color, shape, and size). Microbiologists have formulated hundreds of different enrichment formulas to accomplish one of the three objectives listed earlier. Enrichment media can now be categorized by the type of microorganism they help isolate or the constituents in the enrichment formula.

TYPES OF ENRICHMENT MEDIA

Enrichment media may be categorized into two groups based on the needs of the microorganism to be recovered: elective enrichment medium and selective enrichment medium. Elective enrichment media provide unique growth factors that encourage the growth of the desired microorganism among a mixture of many other species. This type of medium takes advantage of the distinctive nutrient needs of one microorganism to differentiate it from all others. Selective enrichment medium contains substances that inhibit the growth of almost all other microorganisms but the desired one. Selective enrichment media are important in the recovery of species that naturally grow slowly or in low numbers within mixtures of more dominant species. The following scenarios provide examples of the uses of these two types of enrichment media.

- Tryptophan medium serves as an elective enrichment medium, because it encourages the growth of *Pseudomonas* species that can use the amino acid tryptophan as a sole carbon and nitrogen source.
- High-salt medium can be formulated as a selective enrichment medium by including NaCl (salt) at about 20 percent. The high salt aids the growth of halophilic (organisms that need high-salt conditions for growth) *Halobacterium* and *Halococcus*, while killing all other microbial growth.

The most common additions or deletions or other alterations to growing conditions that make up enrichment media are the following:

- use of specific carbon sources
- deletion of all carbon sources to favor the growth of autotrophs, species that get carbon from carbon dioxide in the air
- use of specific nitrogen sources
- deletion of all nitrogen to favor the growth of nitrogen-fixing bacteria that get nitrogen from the air
- deletion of all energy sources to favor photo synthetic microorganisms
- addition of specific growth factors: vitamins, minerals, amino acids, nucleic acid bases
- addition of factors derived from nature, such as marine water

In terms the alterations listed, enrichment media can be defined in four different ways: chemically defined, chemically undefined, simple, or complex. Chemically defined media contain known ingredients, all of which are at known concentrations. Chemically undefined media contain one or more ingredients of unknown composition, such as media containing a percentage of marine water. The exact composition of marine water is undefined and will, furthermore, vary from batch to batch. Simple enrichments are attained with the addition or deletion of a few constituents, such as an amino acid or a particular mineral salt. Complex enrichments contain more specialized formulas, usually made up of a variety of ingredients, many of which are uncommon for growth media.

All enrichments may be divided into types based on whether the enrichment is of a physical, chemical, or biological nature. The table provides common examples of each of these main enrichment media categories.

Microbiologists have a large selection of enrichment media that has been developed by hundreds of scientists, over several decades, to grow various microorganisms. To find a medium suitable for isolating known microorganisms, a microbiologist can refer to resources such as *Bergey's Manual of Systemic Bacteriology* or the *Difco Manual of Dehydrated Culture Media*. But before today's commonly used enrichment media grew to the current large selection, microbiologists developed formulas for specific purposes in their individual laboratories. The following passage, published, in 1977, by researchers at the Centers for Disease Control (CDC), illustrates the tedious work that went into many of microbiology's available enrichment formulas: "Forty-eight combinations of enrichment media, secondary enrichment, incubation times and temperatures, and atmospheres were examined for their efficacy in recovering different serovars of *Salmonella* that had been inoculated into ground-meat extract. . . . One-hundred and twenty-four tests were conducted for each enrichment under each condition of incubation." Development of new enrichment media remains a painstaking task for the world's most fastidious microorganisms.

USES OF ENRICHMENT MEDIA

The examples of uses for enrichment media given so far have been for the purpose of recovering one particular microorganism from a mixture of many. This procedure is valuable in several different disciplines within microbiology. The following list describes the manner in which enrichment media isolate specific microorganisms:

- soil microbiology—a microorganism from a mixture of nonliving matter, living organisms, and molds and bacteria
- clinical microbiology—a pathogen present in low levels in blood, sputum, stomach contents, and so on
- epidemiology—a potential food-borne pathogen from many other species in a fecal sample
- water quality—an indicator species that signals the presence of fecal contamination
- microbiology ecology—recovery of nitrogen-fixing bacteria (species capable of taking nitrogen from the atmosphere) from plant root nodules
- food microbiology—monitoring for potential spoilage organisms or pathogens in food products
- industrial environmental monitoring—determining the presence of mold and bacterial contaminants in a manufacturing facility

Enrichment media also serve the purpose of inducing the production of specific compounds from a microorganism. Biotechnology uses various types of enrichments to induce microorganisms to make specific end products. By altering a carbon or nitrogen source, scientists can induce microorganisms to express specific genes. *Gene expression* refers to the biological process by which a gene controls the production of a protein.

A widely used medium in recombinant deoxyribonucleic acid (DNA) studies that is based on the principles of enrichment is Luria-Bertani medium, or LB medium. Different enrichments to LB medium allow a molecular biologist to recover specific gene-containing varieties of species. For example, one derivation of LB medium provides all the nutrients required by a strain of *Escherichia coli* (*E. coli* K12) that can carry genes from other microorganisms but cannot make any of its own growth factors. By supplying the required nutrients and altering salt concentrations, molecular biologists can separate a specific *E. coli* K12 strain from all other strains.

Scientists in biotechnology have developed many medium formulas in which the addition of one or more specific compounds activates gene expression for making select enzymes, antibiotics, vitamins, and other microbial products.

The development of enrichment media has accelerated advances in microbiology, particularly in the areas of environmental microbiology and ecology, food and water quality monitoring, and clinical isolation of pathogens. This method remains an essential part of microbiology.

See also ENVIRONMENTAL MICROBIOLOGY; MEDIA; RECOMBINANT DNA TECHNOLOGY.

Further Reading

Becton Dickenson. "Difco and BBL Manual." Available online. URL: www.bd.com/ds/technicalCenter/inserts/difcoBblManual.asp. Accessed August 22, 2009.

Garrity, George M., ed. *Bergey's Manual of Systemic Bacteriology,* vol. 2, 2nd ed. New York: Springer, 2005.

Kafel, S., and F. L. Bryan. "Effects of Enrichment Media and Incubation Conditions on Isolating Salmonellae from Ground-Meat Filtrate." *Applied and Environmental Microbiology* 34 (1977): 285–291. Available online. URL: www.pubmedcentral.nih.gov/picrender.fcgi?artid=242644&blobtype=pdf. Accessed August 22, 2009.

Lindquist, John. "Principles of Enrichment and Isolation of Bacteria." Available online. URL: www.splammo.net/bact102/102enrisol.html. Accessed August 22, 2009.

enteric flora Enteric flora are microorganisms that are part of the body's normal flora and inhabit the small and large intestines. Microbiologists who study the intestinal flora of humans and other single-stomach mammals often use the term *enterics* to refer to only the bacteria in the family Enterobacteriaceae. Gram-negative facultative (live with or without oxygen) rods make up this family. In truth, bacteria in other families, as well as protozoa, also make up the broad group of enteric flora.

Enteric flora play several important roles in animal health. First, they help digest dietary nutrients. Many species of animals would not live long without their enteric flora. Second, enteric flora supply additional nutrients to the host animal; amino acids, vitamins, and small volatile fatty compounds, called volatile fatty acids, are the chief nutrients supplied. Third, enteric flora may protect the host animal from infection by other microorganisms that have been ingested. This is a known role of all the body's normal flora, not just those in the gastrointestinal (GI) tract. Fourth, the beneficial enteric flora can become dangerous health threats when they contaminate food and water. Most food-borne illnesses and many waterborne illnesses contracted by humans are caused by enteric flora.

Studies of enteric flora tend to be divided into different specialty areas based on different types of digestive systems. For instance, gastrointestinal microbiology is the scientific field concerned mainly with enteric flora of the single-stomach GI system. These species are called monogastric animals (humans, primates, pigs, dogs, cats, mice, and rats). Rumen microbiology is the study of the interrelationships between rumen flora and the host ruminant animal. Ruminant animals are species (cattle, sheep, goats, etc.) in which the stomachlike organ consists of four compartments. Some species are not true monogastrics yet are not ruminants either. These animals have a single-compartment stomach but a very large cecum containing enteric flora very similar to those found in ruminant animals. Examples are horses, rabbits, and guinea pigs. Finally, a special science focusing solely on the enteric flora of insects has emerged.

The GI system of monogastric animals consists of the stomach and the large and small intestines. This habitat provides a rich variety of nutrients from the diet, warm temperatures, and water. These areas also lack oxygen. The study of enteric flora, therefore, is a study of anaerobic microorganisms and their metabolism.

The entire digestive tract from the mouth to the anus possesses other characteristics, in addition to the amount of oxygen available. Several different environments, in fact, make up the GI tract, and in each of these places, the types of enteric flora change composition. For example, bacteria living in the mouth contend with a feast or famine supply of nutrients and surges of enzymes and saliva. Oral microbiology has become a separate specialty in microbiology. In the stomach, however, the flora there must withstand very acidic conditions. Flora in the small intestine deal with partially digested foods, intense enzyme activity, and the effects of bile. Colon flora are exposed mainly to fibrous parts of the diet. In summary, conditions change from one end of the GI tract to the other end. The main conditions that change throughout the GI tract and influence the enteric flora are nutrient supply, pH, oxygen levels, immunity, and secretions from the liver, gallbladder, and pancreas.

THE ROLES OF ENTERIC FLORA IN MICROBIOLOGY

Enteric flora have been used as study models in microbiology, for more than a century. Most of the earliest studies in microbiology used microorganisms found in water and fecal matter. Both water and feces are, after all, available in ample amounts, and many of the enteric flora proved easy to grow in laboratories. The most famous of enteric bacteria, *Escherichia coli,* remains the most studied living thing in all of biology.

In early studies of *E. coli* and other enteric bacteria, microbiologists believed the enteric flora had a commensal relationship with their host. This means the microorganisms benefited, but the host was unaffected. As the science of microbiology developed, microbiologists found the relationship between the bacteria and the animal host was more complex. Enteric flora assist the digestive enzymes secreted by the GI tract in breaking down dietary polysaccharides to sugars and proteins to amino acids. They digest a portion of the fibers in the monogastric diet that would otherwise be undigested. Enteric flora in the colon digest a portion of the fibrous compounds cellulose, hemicellulose, and pectin, and perhaps a small portion of lignin. Throughout this microbial digestion, the body receives a supply of compounds that it absorbs through the intestinal wall and into the bloodstream: B vitamins, vitamin K, amino acids, sugars, nucleic acids, and volatile fatty acids, which are used for energy. Some enteric species further contribute by capturing nitrogen within the intestines and using it to make cellular protein. This process of converting nitrogen gas to a cellular compound is called nitrogen fixation. Nitrogen-fixing bacteria build a variety of amino acids, and, by doing this, they actually improve the quality of protein available for the body's use. Meanwhile, enteric flora work in conjunction with the body's immune system to fight off pathogens. The enteric flora provide part of this protection by simply outcompeting many pathogens for living space along the GI tract's epithelia (the cells lining the GI inner surface). But the flora-host relationship is not one-sided. The host provides the comfortable environment described earlier. The relationship between enteric flora and their host is, therefore, a mutualistic one; both the host and the microorganisms benefit. Mutualism and commensalism are types of symbiosis, which is the state of two organisms living together. Enteric flora provide an example of endosymbiosis, because they live inside the host organism.

Enteric flora that have escaped the body and end up in food or water also affect human health, but not for the good. Contaminated food, drinking water, and recreational waters have all been the source of outbreaks caused by one or more enteric microorganisms. Food microbiologists and water microbiologists both study the means of disease transmission in food and water, respectively. One common illness is gastroenteritis, which is an inflammation within the GI tract. One or more of a variety of symptoms are seen in gastroenteritis: nausea, vomiting, fever, chill, malaise, and diarrhea. When enteric flora are present in food or in drinking water, they are an indication of fecal contamination that may lead to gastroenteritis or other diseases.

The enteric bacteria living in their host do not cause illness in the host. Enteric species become pathogens only when they enter the environment and infect another individual. The host is protected from its own enteric flora by a complex relationship among the mucosal lining of the GI tract, the host's immune system, and the flora themselves. Under normal, healthy conditions, enteric bacteria do not make their host sick. The GI tract's immune sys-

Important Genera of Enterobacteriaceae

Genus	Main Role in Human/Plant Health
Citrobacter	waterborne nosocomial infections, opportunistic pathogen
Edwardsiella	opportunistic pathogen causing diarrhea
Enterobacter	food-borne pathogen, nosocomial infections
Erwinia	plant pathogen, opportunistic pathogen in humans
Escherichia	food-borne pathogen
Klebsiella	respiratory illness, food-borne pathogen
Plesiomonas	gastroenteritis
Proteus	nosocomial infections, opportunistic pathogen
Salmonella	prevalent food-borne pathogen
Serratia	waterborne nosocomial infections
Shigella	food-borne dysentery
Yersinia	plague

tem relies on specialized lymphocytes to distinguish between antigens on pathogens and antigens on the normal enteric flora. This finely tuned feature of the immune system evolved as humans evolved and developed their own normal flora.

ENTERIC FLORA OF HUMANS AND OTHER MONOGASTRIC ANIMALS

Oral flora play varying roles in food digestion, depending on the type of host animal. In general, oral flora contribute less to digestion than the flora in the intestines. After the mouth, few microorganisms are found in the esophagus, and food is not digested there. Food then enters the monogastric stomach, where acidic conditions (pH 2) limit the types and amounts of flora that survive there. Some flora—both good types and pathogens—avoid being killed by stomach acids by *hiding* inside partially or undigested bits of food. The food eventually moves into the intestines, where the pH again rises. This is how bacteria are able to colonize the GI tract of infants and recolonize the GI tract of adults who have low numbers of enteric flora as a result of illness or antibiotic treatment.

All genera in Enterobacteriaceae are gram-negative facultative anaerobic straight rods. Most of the species ferment glucose and other carbohydrates. These bacteria are motile because they have peritrichous flagellae of 3–10 μm in length. Peritrichous flagellae are taillike appendages that project from all over the cell's outer surface rather than from only one end. Enteric bacteria have other projections, called fimbriae (0.3–1 μm), that help cells adhere to the inner lining of the intestines. Enteric bacteria have even smaller appendages called pili, which help the cells pass genetic information from one cell to another.

The small intestine has three parts: the duodenum, jejunum, and ileum. Numbers of enteric flora increase gradually from the duodenum to the ileum, which connects with the colon of the large intestine. The large intestine consists of the colon, cecum, rectum, and anus. The small and large intestines combined support enormous numbers of *microflora,* a term for the collection of all enteric bacteria and protozoa. In the colon alone, the concentration of bacteria reaches 10^{12} cells per gram of intestinal contents.

More than 40 Enterobacteriaceae genera have been characterized in detail. Of these, *Salmonella* and *Escherichia* are perhaps the most famous as food-borne pathogens. Other enteric bacteria are associated with many different food-borne illnesses. Their presence in food signals not only fecal contamination, but poor hygiene in general. One exception is *Yersinia pestis,* which infects the respiratory system. The genus *Erwinia* is also unique, because it is a plant pathogen and plays a lesser role in infecting mammals. Some of the important enteric bacteria are described in the table.

Most Enterbacteriaceae grow well on a variety of laboratory media. Certain types of media, called *selective media,* allow one type of microorganism to grow while inhibiting the growth of others. MacConkey's agar is a widely used selective medium for isolating Enterbacteriaceae from other kinds of bacteria because it contains bile salts. Enteric flora have adapted to live in the presence of bile excreted into the intestines from the gallbladder, but bile inhibits most other microorganisms.

Enteric bacteria can be further classified into genus and species, based on the enzymes they possess. All enteric bacteria have in common the ability to degrade sugars using the Embden-Meyerhof fermentation pathway, also known as glycolysis. Though they all share this pathway to make energy, all enteric generas make different end products during this metabolism. Microbiologists have taken advantage of the differences between genera and species of Enterobacteriaceae to devise identification tests based on sugar use and enzyme activities. In a laboratory, they first isolate all the bacteria able to grow on MacConkey's agar. In a second step, to confirm the presence of enteric bacteria, they test a sample from each colony for the presence or absence of oxidase enzyme. All enteric gram-negative rods lack the pigment cytochrome c and the enzyme that acts on it, oxidase. Enteric bacteria are, therefore, called oxidase-negative. The microbiologist, then, tests all the oxidase-negative bacteria by a scheme that distinguishes among them on the basis of sugar use, amino acid use, and a few unique enzymes. One such scheme for differentiating enteric flora is shown in the illustration and a representative test is shown in the color insert on page C-3 (top).

Microbiologists must distinguish among the enteric species for one main reason: to find the pathogen causing an outbreak of food-borne or waterborne illness. Not all enteric flora cause outbreaks, but quite a few do, and the symptoms can be very serious, even life-threatening in some circumstances. The enteric bacteria *E. coli* and *Salmonella* are perhaps the best-known causes of infection due to ingesting contaminated foods or water. But additional species cause severe illness, too. Clinical microbiologists perfect their techniques in making fast and accurate identifications of the following enteric bacteria. This information is useful for physicians treating patients and for public health officials who are tracing the cause of an outbreak.

Enterobacter

Like most of the enteric bacteria, *Enterobacter* is an opportunistic pathogen. *Enterobacter* causes some food-borne illnesses when the food is contaminated with fecal matter. This microorganism has also become a concern in nosocomial (hospital-associated) infections. Some of the more serious nosocomial infections from *Enterobacter* are meningitis, pneumonia, septicemia, and urinary tract infections. As have many other pathogens, *Enterobacter* has become resistant to several antibiotics.

Klebsiella

Klebsiella species are distinguished by a capsule of polysaccharides that surrounds the cell and aids in physical protection, nutrient uptake, and adhesion to surfaces. *Klebsiella* strains are isolated from the respiratory tract, in addition to the GI tract. The respiratory strains can cause pneumonia. Some strains of *K. pneumoniae* are rare among the enterics because they carry out nitrogen fixation in soils and water, in addition to inside the GI tract.

Shigella

Shigella dysenteriae is a serious food-borne pathogen. The illness it causes is bacillary dysentery, and the symptoms result from the shiga toxin produced by these cells. The shiga toxin is a protein that binds to the epithelial cells of the intestinal lining. After an epithelial cell absorbs a toxin molecule, its enzymes cleave the toxin into two subunits. One of the subunits blocks protein synthesis in the epithelia. The epithelial cells then die, and severe diarrhea begins. Shigalike toxins are produced by other bacteria, notably enterohemorrhagic *E. coli.* In addition, shigalike toxins damage the vascular system, which carries blood to the GI tract and carries absorbed nutrients away from it to other organs. Bacillary dysentery is feared because as few as 10 *Shigella* cells can cause the infection. The Food and Drug Administration estimates that there are 300,000 cases of *Shigella* poisoning in the United States each year.

Serratia

Serratia lives in the GI tract but also does well in water and soil, on plants, and inside insects. *S. marcescens* is a waterborne species that causes nosocomial pneumonia and urinary tract infections. Infection probably spreads on medical devices that have been exposed to contaminated water. Several species of *Serratia* are red or pink when growing as colonies on agar or on moist surfaces, such as bathroom tiles. The red color in *Serratia* and others, such as *Streptomyces,* is produced by the pigments prodigiosin and pyrimine. *Serratia* is used in industrial microbiology to produce 2,3-butanediol, which is a biomass alcohol that has been proposed as a replacement for petroleum products.

Yersinia

Yersinia is a human pathogen that has different appearances, or phenotypes, when grown in a laboratory. Most species are nonmotile below 30°C but begin regaining the ability to move as the temperature approaches 37°C. An exception is *Y. pestis,* which is always nonmotile. This species causes plague, which has assaulted humans, throughout history, in devastating epidemics. *Y. pestis* normally inhabits the digestive tract of the fleas that are parasites on rats. When a flea bites a rodent, it transmits the bacteria to the rodent, which then serves as a reservoir for the disease in humans. Although rat bites may have contributed to the plague centuries ago, today, flea bite is the main mode of transmission when humans live near large numbers of rodents.

Species other than *Y. pestis* cause food-borne illnesses. *Y. enterocolitica* and *Y. pseudotuberculosis* are two species from the GI tract of domesticated food animals. They both can cause yersiniosis when people eat contaminated meat or dairy products.

Citrobacter

Citrobacter is normal in human and animal GI tracts, but its incidence in water is so high that it is also considered a normal water microorganism. *Citrobacter* is an opportunistic pathogen that causes some nosocomial infections, and, like many other nosocomial infections, it is resistant to a growing number of antibiotics.

Strict Anaerobes

Healthy human intestines contain bacteria that are not members of Enterobacteriaceae and, in fact, can outnumber the Enterobacterieae by a wide margin. These species have specific nutritional needs, and they are strict anaerobes. Anaerobic microbiology is the field that includes the culture techniques for growing microorganisms in the absence of oxygen. Therefore, anaerobic microbiology encompasses these enteric anaerobes, as well as other strict anaerobes found in the environment. Some of the predominant genera in this diverse group of enteric flora are *Bacteroides, Clostridium, Fusobacterium, Veillonella, Streptococcus, Lactobacillus, Bifidobacterium, Eubacterium, Propionibacterium,* and *Peptococcus.* The incidence of these bacteria is higher in the large intestine compared with the small intestine.

ENTERIC PROTOZOA

The protozoa living in the monogastric GI tract are thought to play a lesser role in nutrient digestion than the protozoa in ruminant animals. Some animals share a mutualistic relationship with their intestinal protozoa. For example, the fiber-digesting protozoa in ruminant animals, horses, and insects are essential for their health. The role of protozoa in monogastrics, including humans, is less known and thought to be less vital for digesting food. In monogastric animals, protozoa are thought to be either mutualists or commensals. They may play a small role in digesting fibrous matter, but it is not known how important this role is for the animal's well-being.

The diet of ruminants, horses, and many insects is high in fibers such as cellulose and hemicellulose. Cellulose is the main plant fiber that must be digested and used for energy by herbivores and insects such as termites. Protozoa carry out a major portion of the cellulose digestion in the rumen, the cecum, or the insect digestive tract. They do this for their own energy needs, but in the course of digesting cellulose, they produce volatile fatty acids. These compounds are absorbed by the intestinal lining and carried in the blood to the host's organs. The volatile fatty acids, then, are used for energy, entering each cell's Embden-Meyerhof pathway. Protozoa in the large intestine also ingest bacteria that enter from the small intestine's ileum. This also benefits the host. By digesting bacteria, the protozoa make bacterial proteins and amino acids available for use by the host.

ENTERIC FLORA OF HORSES

Horses, guinea pigs, and rabbits are monogastric animals and strict herbivores. Their digestion depends on a large cecum that contains very high concentrations of bacteria and protozoa, comparable to the amounts in monogastric animals' large intestine. The horse has the largest, and perhaps the most active, large intestine among the domestic animals. The large intestine consists of the cecum and colon. These two organs make up 60 percent of the digestive tract and serve to digest the plant fiber in equine diets. Undigested matter from the small intestine enters the cecum, which holds a large and diverse population of microorganisms in about two gallons (7.5 liters) of water. In this organ, cellulose and other fibers begin to break down in fermentations similar to the digestion that occurs in ruminants. The partially degraded material moves out of the cecum and into the colon. Additional enteric flora further digest these materials and produce volatile fatty acids. The volatile fatty acids are absorbed by the colon and ultimately used by the horse for energy in the same way they are in ruminant animals.

The enteric bacteria of horses are predominantly anaerobic gram-negative rods and cocci. The majority of species are members of the genera *Bacteroides, Lactobacillus,* and *Streptococcus.* The horse's protozoa, however, differ from ruminant protozoa. But the protozoa in the horse cecum and those in the rumen are alike in one important way: Their types and proportions change with changes in the diet.

ENTERIC FLORA OF INSECTS

The rat flea and its role in spreading the plague are examples of an insect-enteric microorganism relationship. There are a vast number of other insects that depend on enteric flora for part or all of their digestion. For example, flies, stinging insects, and cockroaches are known to have a diverse composition of enteric bacteria. The enteric flora in these insects are studied to determine whether they serve to transmit disease in a way similar to the rat flea's transmission of plague. *E. coli* and *Salmonella* have been found in the gut of flies and cockroaches, but there is little evidence that their presence contributes to infections in humans.

Termites are known for their ability to *eat* wood. The insect's enteric flora, and not the insect itself, carry out digestion of woody fibers. Termites and other wood-eating insects contain protozoa that use celluloytic enzymes to break down wood cellulose and fibers made of lignin-cellulose mixtures. The protozoa degrade this tough polymer and break it into small strings of sugars for their own energy needs. In the process, sugars become available for absorption by the termite's gut. But there is a finer level of symbiosis working. Bacteria inside the protozoa are the actual source of cellulase enzyme. These bacteria make cellulose an available nutrient for the protozoa, and the protozoa digest cellulose to a form usable for the termite.

In termites, protozoa such as *Trichonympha* make up the greatest portion of the mass of enteric flora. The bacteria inside termite protozoa are of the Archaea domain. Anaerobic microbiologists use the termite's hindgut as a model for the production of carbon dioxide, hydrogen, methane, and nitrogen by insect microflora. The collective activities of termites and their enteric flora contribute to the carbon dioxide, hydrogen, and methane levels in the atmosphere, and so termites have joined a growing list of greenhouse gas emitters. (Rumen flora are also a significant source of greenhouse gases.) Considering that they have colonized at least two thirds of Earth's land surface, many scientists believe termites are one of many contributors to global warming.

Microbiologists have uncovered a tremendous diversity in roles that enteric microorganisms play in the environment and in health. This group of microorganisms will remain one of the most-studied in the microbiology.

See also ANAEROBE; *ESCHERICHIA COLI;* FOODBORNE ILLNESS; IDENTIFICATION; NORMAL FLORA; NOSOCOMIAL INFECTION; ORAL FLORA; PROTOZOA; RUMEN MICROBIOLOGY; *SALMONELLA;* SYMBIOSIS.

Further Reading

Drašar, Bohumil S., and Paul A. Barrow. *Intestinal Microbiology: Aspects of Microbiology 10.* Washington, D.C.: American Society for Microbiology Press, 1985.

Janda, John M., and Sharon L. Abbott. *The Enterbacteriaceae,* 2nd ed. Washington, D.C.: American Society for Microbiology Press, 2006.

Todar, Kenneth. "The Enteric Bacteria." Todar's Online Textbook of Bacteriology. Available online. URL: www.textbookofbacteriology.net. Accessed August 23, 2009.

environmental microbiology Environmental microbiology is the study of microorganisms found in various places in environments other than inside animal bodies. This discipline encompasses studies on the microorganisms of soil, water, air, and plant life. Environmental microbiologists study the types and amounts of microorganisms in different habitats in the environment and the activities each microorganism conducts in its environment.

Current environmental microbiology also involves studies of habitat and ecosystems. Microbial habitats are the places where microorganisms live. Ecosystems are complex relationships between the living things, including microorganisms, and the physical matter in a particular environment. For example, a pond ecosystem contains inanimate things such as water, gases, and sediment, and it also contains microorganisms, animals, and plant life. All of these components interconnect to make the ecosystem of a pond.

Environmental microbiology grew out of the older science of microbial ecology, which covers the interactions among microorganisms; higher multicelled organisms, such as plants and animals; and the nonliving factors in their environments. Although these two fields are now closely related, they may be distinguished as follows:

- Environmental microbiology covers the diversity and functions of microorganisms in the environment.

- Microbial ecology covers the interactions of microorganisms with their environment and the way these interactions define conditions on Earth.

Environmental microbiology has developed into specialized areas of study that have taken on increased importance with current concerns about the declining quality of Earth's environment. Environmental microbiologists who specialize in the subjects listed in the table on page 262 work with other scientists in the areas of pollution control, bioremediation, climate change, water quality, and emerging diseases.

Each focus area in environmental microbiology calls upon skills common to other types of microbiology. For instance, an environmental microbiologist must understand microscopy, identification methods, growth requirements, media, aerobic and anaerobic culture methods, sterilization, and aseptic technique. Some techniques unique to environmental microbiology are soil, water, and air sampling;

Environmental Microbiology

General Subject	Concentrated Subjects
aeromicrobiology—studies on airborne microorganisms	• aerosols • disease transmission • indoor air quality • particle movement in the atmosphere
environmental microbial metabolism	• adaptation • biogeochemical cycles • environmental pathogens • waste decomposition and treatment
plant and terrestrial life	• agriculture • disease vectors • food-borne pathogens • plant pathogens • symbiosis in plants and microorganisms
soil microbiology	• fossil fuel production • subsurface environments • surface soil microbiology
water microbiology	• deep ocean microbiology • drinking water microbiology • groundwater quality • lake and freshwater microbiology • marine and saltwater microbiology • water and wastewater treatment

sample preservation and transport; indicator microorganisms; extreme environments; pollutant degradation; and disinfection.

Environmental microbiologists also increasingly study the activities of microorganisms in the environment that may be of benefit to humans. These activities may be any of the following: enzyme production, toxic waste degradation, activities in extreme environments, gas production, nitrogen or carbon dioxide absorption, waste decomposition, and metabolism in wastewater.

THE HISTORY OF ENVIRONMENTAL MICROBIOLOGY

The earliest studies on microorganisms from the environment have been credited to the Dutch cloth merchant Antoni van Leeuwenhoek (1632–1723), who built a rudimentary microscope in the 1600s and recorded his observations on the tiny living particles he noticed. Van Leeuwenhoek examined minuscule microscopic particles in rain, well water, seawater, and snow. The prevalent theory of life, at that time, was predicated on spontaneous generation, through which living things arose from nonliving matter. In that context, van Leeuwenhoek's observations of "animacules," tiny living creatures so simple that they could easily have arisen from nonliving materials, made perfect sense. Almost two centuries would pass before Louis Pasteur and his contemporaries provided evidence that microorganisms arose from their own reproduction, and not from spontaneous generation.

Microbial ecology as a dedicated science might be traced to 1888, when the Dutch botanist Martinus Beijerinck (1851–1931) discovered nitrogen-fixing bacteria living in the bulbous nodules on the roots of certain plants. Beijerinck suspected these bacteria provided the plant with a nutrient such as nitrogen, but he was unable to uncover definitive proof. The next year, the Russian microbiologist Sergei Winogradsky (1856–1953) conducted studies on the processes by which soil bacteria derived energy from inorganic compounds. Beijerinck and Winogradsky, additionally, provided evidence for the cycling of nitrogen and sulfur through the Earth's living and nonliving matter.

Environmental microbiologists had also, perhaps unwittingly, revealed another important characteristic of microorganisms: Most microorganisms did not

Concepts in Environmental Microbiology

Concept	Description
habitat	place where a microorganism lives
ecosystem	a group of different plant, animal, and microbial species interacting with one another and with the chemical and physical factors in their surroundings
niche	role of a species, including the requirements needed to carry out that role, in an ecosystem
community	mixture of different species living together and interacting
environment	living and nonliving factors that affect an organism in its lifetime
ecology	study of the interrelationships amomg microorganisms, higher organisms, and their environment
microbial diversity	number and variety of different species in the environment
population ecology	study of the interrelationships of entire species populations with the environment
global ecology	study of the interrelationships of living things that contribute to energy and matter flow on Earth

cause disease but rather played critical and beneficial roles in the environment. Prior to these environmental studies, microorganisms were thought of mainly in terms of disease. The only good that could be extracted from the microbial world took the form of food fermentation.

Microbiologists excited by the search for environmental microorganisms began devising ways to categorize the new species. The growing collection of bacteria isolated from various points in the environment gave birth to the fields of taxonomy and systematics, which are concerned with the classification of organisms and a system for naming, respectively.

Albert J. Kluyver (1888–1956) followed in the footsteps of Beijerinck and Winogradsky to emphasize the crucial need for microorganisms in the world. He remarked, in 1924, "In a microbe-less world, the conditions for human life on earth would soon no longer be realized, so that man possesses at least as many friends as enemies in the domain of the microbes." The study of ecosystems and the interconnectedness of life on Earth cannot be done in today's science without considering microbial life.

HABITAT AND ENVIRONMENT

Environmental microbiology involves the understanding of certain concepts in biology that extend to higher organisms. As mentioned, habitat and ecosystem are important concepts in environmental microbiology, but this discipline also studies the roles of microorganisms in their habitat and in ecosystems. By learning simple relationships, microbiologists can describe larger systems such as the environment and overall ecology. The table explains important concepts needed in studying environmental microbial life.

Environmental microbiology and microbial ecology have expanded into an additional discipline within biology, called *exobiology*. Exobiology has centered on the relationships between Earth's geology and Earth's microorganisms. This field includes the evolution of microorganisms, the questions that microbial fossils can answer, and how Earth's life can be defined at the present. Exobiology has, in recent years, tried to expand the knowledge of Earth and its microorganisms to life in other places in the solar system. Just as microorganisms shaped the development of life on Earth, exobiology seeks to determine how life may be formed on other planets. At their root, future studies in exobiology can be thought of as environmental microbiology on a place other than Earth.

MACROENVIRONMENTS AND MICROENVIRONMENTS

Fundamental environmental microbiology defines the general characteristics of different places on Earth. These general places are called *macroenvironments*. Soil and ocean water are examples of macroenvironments. The definition of these environments also includes the general conditions in which microorganisms live: moisture levels, temperature, oxygen availability, pH, and more. Further study into an environment such as soil reveals additional generalized environments, which may also be described as macroenvironments: porous surface soils, deep oxygen-scarce sediments, igneous rock formations, peat bogs, and other geological layers.

Water contains its own unique set of macroenvironments. These can be saltwater or freshwater macroenvironments, or more specifically (for a freshwater lake as an example), a surface layer, a limnetic zone, a deep profundal zone, and the bottom benthic zone. The layers differ in nutrients, temperature, oxygen availability, and light, so each represents a different macroenvironment. The microorganisms that live in macroenvironments have adapted to the conditions found there.

Within many macroenvironments exist exceptionally demanding conditions for growth. Only a few species have evolved the adaptations needed to survive in these places. For example, within the ocean water macroenvironment, there exist millions of different microbial species. Descending into deeper ocean waters, where the temperature becomes chilly, light disappears, and nutrients may be limited, a more specialized macroenvironment provides a habitat for a smaller number of specialized microorganisms. At the very bottom of the sea lies a dark world of very cold temperatures, limited nutrients, and enormous pressure from the ocean waters above it. Only a few, highly specialized microorganisms have adapted to these conditions. This place may be described as a *microenvironment,* because the conditions are unique among all Earth environments. The deep ocean also represents a microhabitat, or a place found in few locations on Earth and populated by a distinct group of organisms found nowhere else on the planet.

Many microenvironments are so difficult to survive in that they are called extreme environments. These places can be populated only by microorganisms especially adapted to them, called *extremophiles.* While some microorganisms can survive outside their favored macroenvironment, species adapted to microenvironments are rarely found anywhere else.

Environmental microbiology has developed into a science that supports the larger fields of environmental science and microbial ecology. This discipline includes the specialty areas of soil, air, water, and plants, and those subjects contain further subspecialties. Environmental microbiology has become a large subject area within microbiology that touches on a diversity of topics in biology, ecology, and medicine.

See also AEROMICROBIOLOGY; BIOGEOCHEMICAL CYCLES; BIOREMEDIATION; EXTREMOPHILE; MICROBIAL ECOLOGY; SOIL MICROBIOLOGY; SYSTEMATICS; TAXONOMY; WASTEWATER TREATMENT; WATER QUALITY.

Further Reading

Jjemba, Patrick, K. *Environmental Microbiology: Principles and Applications.* Enfield, N.H.: Science Publishers, 2004.

Kerr, Richard A. "Deep Life in the Slow, Slow Lane." *Science* 296 (2002): 1,056–1,058. Available online. URL: www.geobacter.org/press/2002-05-10-science-1.pdf. Accessed August 22, 2009.

Kluyver, Albert J. "Eenheid en Verscheidenheid in de Stofwisseling fer Microben" (Unity and Diversity in the Metabolism of Microorganisms). In *Milestones in Microbiology,* translated and edited by Thomas Brock. Washington, D.C.: American Society for Microbiology Press, 1961.

Maier, Raina M., Ian L. Pepper, and Charles P. Gerba. *Environmental Microbiology.* San Diego: Academic Press, 2000.

Needham, Cynthia, Mahlon Hoagland, Kenneth McPherson, and Bert Dodson. *Intimate Strangers: Unseen Life on Earth.* Washington, D.C.: American Society for Microbiology Press, 2000.

VarnAm, Alan H. *Environmental Microbiology.* Washington, D.C.: American Society for Microbiology Press, 2000.

epidemic An epidemic disease is a disease that afflicts many people at the same time in the same geographical area. In microbiology terms, epidemic diseases are caused by infectious pathogens that spread quickly through a given population within a given area. (Epidemic diseases are often called simply *epidemics,* such as a flu epidemic.)

An epidemic is distinguished from other disease occurrences by the frequency of the disease's occurrence. If a disease occurs occasionally within various parts of a population over several years, it is called a *sporadic disease.* An outbreak is a disease occurrence in a localized place, such as a school, that affects a small number of people. An endemic disease such as the common cold does not occur sporadically within an entire population; it is always present in the population. Epidemics usually confine themselves to a geographic region, such as the U.S. East Coast or an entire country or island. When a disease spreads worldwide, it is called a *pandemic disease.* For example, acquired immunodeficiency syndrome (AIDS) began as an epidemic in the United States and a few other countries but has spread worldwide and may now be considered a pandemic disease.

Some diseases, for example, influenza, by the nature of how they develop in a reservoir and mutate inside animal cells cause frequent epidemics interspersed with pandemics. Flu epidemics occur almost every year. Over the past 300 years, global flu pandemics have occurred about every 20 years.

Three components that help define an epidemic are incidence, prevalence, and rate. Incidence is the frequency of new cases of the same disease in a population during a given period. Epidemics are characterized by having a high incidence in a specific period. Prevalence is the number of cases of the same disease in a population at any time. A high prevalence of a disease could indicate that the disease is endemic in the population. The rate of a disease equals the number of cases per a fixed number of people. For example, a rate of 10:100,000 equals 10 cases of a disease per 100,000 people.

Infectious diseases, those that are caused by a microorganism that infects the body, can be caused by any type of microorganism, but viruses and bacteria have caused the majority of history's epidemics. To begin an epidemic, a disease must be able to spread easily from person to person. A *communicable disease* is a disease that has the capacity to spread from person to person, and a *contagious disease* is one that readily moves from person to person. For example, AIDS is a communicable disease that is spread by sexual contact or in contaminated blood. Influenza (flu) is a contagious disease that can spread when people are merely in close proximity. For that reason, flu outbreaks often occur on college campuses, on sports teams, and in the workplace. Contagious diseases are the most likely candidates for starting an epidemic. Noncommunicable diseases do not spread from person to person but only cause symptoms when introduced into an individual's body. Anthrax disease that enters the body by way of inhalation or a cut is noncommunicable.

An epidemic in a population of humans behaves similarly to an illness in an individual. The entire series of events is called the *disease cycle*. The cycle begins with an incubation period, the time between infection and the first appearance of symptoms. The epidemic then spreads rapidly through the population, so that each day an increasing proportion of people become infected. As a large proportion of the population begin to seek treatment or fight the disease with their immune systems, the number of newly infected people begins to decline. The overall number of infected may still increase, but the increase occurs at a slower and slower rate. Finally, the disease begins disappearing from the population, in what is called a period of decline. If a person were to graph the number of new cases of the disease over time, an epidemic might resemble a fairly symmetrical bell curve.

Two events can change the normal characteristics of an epidemic. The first event is a latency period, which is a delay between the time of infection and the time when the first symptoms appear. A latent disease, such as AIDS, is inactive for a long time after infection; the latency period for AIDS can be 10 years or more, compared with influenza's latency period of a few days. Latency causes the bell curve to become asymmetrical because of the long, slow rise of the disease in a population. An acute disease is the second event that shifts the characteristics of an epidemic. Acute diseases develop rapidly and last only a short time. Epidemics from these diseases likewise do not last long. For example, epidemics caused by the influenza virus begin and end within a few months each year. By comparison, the AIDS epidemic in the United

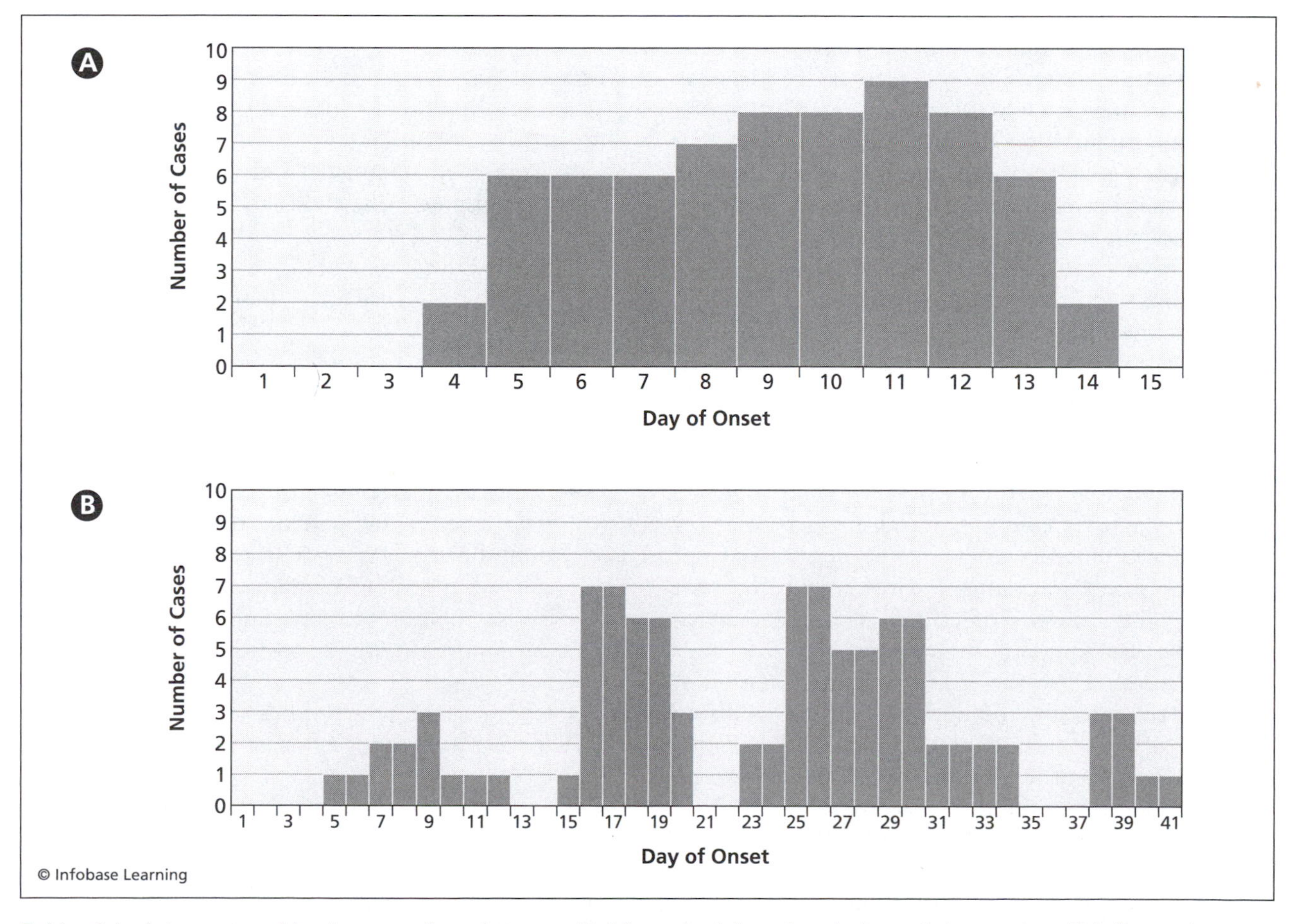

Epidemiologists create epidemic curves from data compiled from physicians, hospitals, and pharmacists. This figure shows two different patterns of epidemics: constant onset epidemic (a) and intermittent onset epidemic (b).

States began in the 1980s and did not peak until the mid-1990s, followed by a very slow decline.

RECOGNIZING EPIDEMICS

Epidemiology is the science of how disease enters and travels through a population. In order to head off an epidemic, epidemiologists monitor small, sporadic outbreaks of disease to gain clues as to whether these events will expand into the larger population. The first task of epidemiology is, therefore, to recognize when one or more isolated outbreaks turn into an epidemic.

Epidemiologists recognize two types of epidemics: a common-source epidemic and a propagated epidemic. A common-source epidemic enters a population rapidly and may reach its peak number of cases within a few weeks, followed by a rapid decline. A single infection of many people at the same time, such as food-borne illness, results in a common-source epidemic. A propagated epidemic progresses more slowly through a population, indicating that a single infected person started the epidemic. The AIDS epidemic in the United States illustrates a propagated epidemic. David Salyer of the AIDS Survival Project wrote in 2001, "In the summer of 1981, an immunologist from Los Angeles and a dermatologist from New York reported some unusual findings in the Centers for Disease Control's *Morbidity and Mortality Weekly Report*. Their articles, published a month apart, hinted at a puzzling new syndrome affecting homosexual male patients that began showing up around 1979." The AIDS epidemic inexorably spread beyond large cities to small towns and affected men and women outside the original population of victims.

Epidemiologists identify new epidemics by tracking the number of new cases occurring across a region or an entire continent. In the AIDS epidemic, the Centers for Disease Control (CDC) gathered data on an unusually high incidence of rare infections due to the breakdown of the immune system of AIDS sufferers that were emerging. In those early years of the epidemic, doctors and epidemiologists did not know the exact mechanism through which these secondary infections—infections caused by microorganisms attacking a body weakened by a primary infection—related to the AIDS virus. In addition to tracking the statistics of new cases reported from doctors' offices, outpatient clinics, and hospitals, epidemiologists gather further information on suspected epidemics by the following methods:

- examination of specimens—blood, urine, sputum, and so on—for a common microorganism
- interviews with patients to learn their habits, places of travel, occupations, and medical history
- tracking the progress of existing cases
- investigating the genetic makeup and other health characteristics of patients

The AIDS epidemic had become a particularly difficult epidemic to control, in the 1980s, because President Ronald Reagan's administration was slow to view the disease as a national crisis, in large part because of an apparent disregard for the main population of patients: homosexual men.

CONTROLLING EPIDEMICS

Effective recognition of a new epidemic and quick action to stop it cannot endure the delays that helped give AIDS its foothold in the United States. Today, epidemiologists and public health departments work quickly and in coordinated fashion to gain control of an epidemic. For example, in 2002, a health care worker in China contracted a viral disease called severe acute respiratory syndrome (SARS), which, by 2003, had spread to 27 countries and killed about 800 people. The CDC worked with the World Health Organization (WHO) to arrest further spread of SARS by calling for the following actions:

- dispatched scientists to identify the source of the pathogen (the civet cat)
- tracked the spread of new cases by country and region to identify the main mode of transmission
- set up immediate programs for developing a vaccine and testing treatments
- instituted procedures for health care workers who treat SARS patients to prevent nosocomial (hospital-associated) transmission
- issued public travel advisories for people entering countries with the epidemic
- set up extra precautions, such as quarantine for SARS patients entering the United States

The medical community controls epidemics such as SARS by establishing action plans based on the characteristics of the disease, mainly the type of pathogen that causes it and the main mode of transmission.

Modes of Disease Transmission

Mode	Description
airborne	pathogens attached to particles or moisture droplets that travel through the air
contact	direct person-to-person contact or contact from one person to another by way of an inanimate object (indirect)
food-borne	pathogens transmitted in foods
vector	transmission by way of an insect or a vertebrate animal
vehicle	a pathogen-contaminated inanimate object infects several people
waterborne	pathogens transmitted in drinking water or recreational waters

Learning the type of pathogen that has infected a population provides critical information on how to combat the microorganism and kill it. Antibiotics work against bacteria but have no effect on viruses; antibiotics that kill fungi differ from the broad-spectrum antibiotics used for killing bacterial pathogens. In addition, antibiotic or antiviral drug resistance in any of these pathogens makes the epidemic much harder to control.

Knowing the main mode of a disease's transmission is important, because health care providers can recommend to their healthy patients the best ways to avoid getting sick while researchers continue working to find a treatment. The mode of transmission also influences the manner in which a pathogen moves through a population. The table provides brief descriptions of the modes of transmission of human diseases.

Each different mode of transmission in the preceding table requires different recommendations for controlling the further spread of the pathogen. Airborne pathogens may be avoided by keeping away from people who are obviously sick, because they are coughing and sneezing. In serious epidemics, airborne transmission may be blocked by wearing a mask covering the nose and mouth. Food-borne and waterborne illnesses can be halted by avoiding the source. Food recalls initiated by the U.S. Food and Drug Administration (FDA) and boil water alerts from local water utilities help reduce infection. A boil water alert is a public warning for all customers of municipal water to heat water to a rolling boil for at least five minutes before using it. County eradication programs often serve to reduce the incidence of vector-borne diseases, and public health announcements regarding contact transmission play a role in reducing disease spread through person-to-person contact, such as sexually transmitted diseases.

Nature itself has evolved a process that helps reduce the spread of disease, and the medical community has, at times, designed vaccination programs to mimic this process called *herd immunity*. Herd immunity is the resistance to an infectious microorganism's spread that occurs when a large percentage of the population has developed immunity against the pathogen. In other words, herd immunity can halt the spread of infection even though a percentage of the population remain susceptible to the pathogen. In herd immunity, the larger the proportion of individuals who have developed immunity against the pathogen, the less chance the pathogen will have to continue spreading in the population.

Herd immunity develops as more and more people become sick and then develop antibodies against the pathogen to protect them against future infection. Vaccination programs also use the principle of herd immunity. Each flu season, health care providers vaccinate as many people as possible against the flu virus with the goal of creating herd immunity.

Highly contagious diseases may be controlled, early in their emergence, in a population, before they turn into epidemics. The process used for this purpose is called a *ring vaccination*. A ring vaccination program is set up by identifying a small number of people involved in a disease outbreak. In the first step, the immediate family members and others in close contact with the infected people receive vaccinations. In a second step, doctors vaccinate as many casual acquaintances, such as friends and coworkers, of the affected person as possible. This method develops a ring of vaccinated individuals that surrounds the infection's original source. By this approach, the pathogen becomes contained within a community protected by their immunity. Ring vaccinations can be effective, but only if a disease is diagnosed quickly and the program begins at once.

Immunity alone does not protect against all epidemics because of the role that carriers play in spreading disease. A carrier is a person who spreads a disease to others but may not always display symptoms of the disease. Four different types of carriers are as follows:

- An active carrier has obvious symptoms of disease.
- A healthy carrier harbors the pathogen but does not become sick.

- An incubatory carrier is infected and in the incubation stage and, therefore, is not yet showing symptoms.

- A convalescent carrier has recovered from the worst of the disease but continues to harbor large numbers of the pathogen.

The period in which a carrier may spread infections to others also influences how easily an epidemic will spread in a population. For instance, a casual carrier might be able to infect others for only a few to several hours, an acute carrier can infect others for days, and a transient carrier can infect other people for weeks. A chronic carrier harbors a pathogen, spreads the disease for years, and never gets sick. Perhaps the most notorious example of a chronic carrier of disease was Mary Mallon, who, between 1896 and 1906, worked as a household cook in homes in New York City and caused a rash of typhoid fever cases among the families and their guests. Nicknamed "Typhoid Mary" by the press, Mallon never exhibited outward signs that she had the disease. New York health officials caught up with Mallon, in 1907, and identified her as the cause of the typhoid fever outbreaks; city officials confined her to an island in the East River, until 1910. Mallon changed her name and slipped back into New York City, five years later, to find work as a cook again. She soon began spreading the disease, until authorities once more tracked her down and held her in custody as a means of quarantine for the next 23 years.

People who look and feel healthy might have a difficult time realizing that infectious pathogens may be hiding in their body. The *Newsday* reporter Ridgely

Epidemics in History

Epidemic	Period	Circumstances
typhoid fever	Athens in 430 B.C.E.	*Salmonella* spread in food or water to cause the Athenian plague
bubonic plague	Egypt and the Middle East in the 540s, Europe in 1400s–1700s, Far East in 1910–21	poor sanitation leads to large rat populations that serve as reservoir for *Yersinia pestis* bacteria
typhus	Ireland in 1816–19, Germany in 1942–45	rickettsia bacteria spread during the Irish potato famine and in World War II in concentration camps
various epidemics: influenza, malaria, smallpox, syphilis	1520–1600	Native Americans suffer numerous epidemics after the arrivals of European explorers
syphilis	1490s	began in Europe, and explorers spread the *Treponema* bacteria to Africa and the Far East
smallpox	1500s–1600s	numerous epidemics in England from variola virus, which then spread to the New World
yellow fever	1700s–1800s	Europe, North and South America in tropical regions, especially during the Panama Canal building in 1882–89; yellow fever virus spread by mosquito bite
cholera	1817–23, 1829–51, 1852– 59, 1863–79, 1881–96, 1899–1923, 1961–70	Europe, Middle East, and the Americas due to contaminated water carrying *Vibrio cholerae* bacteria; currently in epidemics in parts of Africa
influenza	1917–18	American soldiers returning from Europe in World War I started an epidemic, affecting 20 percent of the U.S. population
AIDS	1980–present	worldwide infection from human immunodeficiency virus (HIV) spread by sexual contact, blood transfusions, and needle sharing among drug users

Ochs revisited the tale of Typhoid Mary, in a 2009 article. "Mallon despised the moniker [Typhoid Mary] and protested all her life that she was healthy and could not be a disease carrier. She apparently could not accept that unseen and unfelt 'bugs' could infect others. As she told a newspaper: 'I have never had typhoid in my life and have always been healthy. Why should I be banished like a leper and compelled to live in solitary confinement. . . . ?'" Healthy carriers present special challenges to medical experts who are trying to find the source of an outbreak or an epidemic.

See also EPIDEMIOLOGY; INFLUENZA; LATENCY; SEXUALLY TRANSMITTED DISEASE; TRANSMISSION.

Further Reading

Biello, David. "Ancient Athenian Plague Proves to Be Typhoid." *Scientific American,* 25 January 2006. Available online. URL: www.sciam.com/article.cfm?id=ancient-athenian-plague-p. Accessed August 22, 2009.

Crosby, Alfred W. *The Columbian Exchange: Biological and Cultural Consequences of 1492.* Westport, Conn.: Greenwood Press, 1972.

Kretzschmar, Mirjam, Susan van den Hof, Jacco Wallinga, and Jan van Wijngaarden. "Ring Vaccination and Smallpox Control." *Emerging Infectious Diseases* 10 (2004): 832–841. Available online. URL: www.cdc.gov/Ncidod/EID/vol10no5/pdfs/03-0419.pdf. Accessed August 22, 2009.

Ochs, Ridgely. "Dinner with Typhoid Mary." *Newsday,* 4 April 2009.

Salyer, David. "A Look Back at the History of AIDS in the U.S." AIDS Survival Project, June 2001. Available online. URL: www.thebody.com/content/whatis/art32382.html. Accessed August 22, 2009.

Urban Rim Publications. "Infectious Diseases in History: A Guide to Their Causes and Effects." Available online. URL: http://urbanrim.org.uk/diseases.htm#Malaria. Accessed August 22, 2009.

epidemiology Epidemiology is the study of how disease distributes in and moves through a population, as well as the methods for stopping disease spread. An epidemiologist is a specialist in this field of study who monitors the patterns of disease outbreaks, new and reemerging diseases, epidemic diseases, modes of transmission, and public health programs for controlling epidemics.

Epidemiology focuses mainly on diseases at the national, regional, or global level; epidemiologists may also be called upon to help determine the cause of smaller outbreaks, such as food-borne illnesses, localized influenza outbreaks, and waterborne illnesses.

Epidemiologists rely on statistical data related to diseases in order to develop patterns of how diseases usually spread, peak, and disappear. By studying these patterns, epidemiologists improve the chance of heading off a disease that occurs in a local area before it develops into an epidemic. The table on page 278 provides the main terminology used in epidemiology for describing different types of disease occurrences in a population.

THE SCIENCE OF EPIDEMIOLOGY

Epidemiologists monitor the progress of a disease, through a community or a larger population, by collecting information from people who have become infected and sometimes from people who remain healthy. By comparing these two groups, epidemiologists may find a common thread among the infected people that is absent in healthy people. A simple example is food-borne disease in which 60 percent of the people who ate at Restaurant A last Thursday evening became ill with classic symptoms of food poisoning: nausea, vomiting, abdominal cramps, fever, and chills. An epidemiologist may learn, by interviewing many different patrons of this restaurant, that people who ate there on other days of the week or ate elsewhere did not get sick. This evidence suggests that the illness is food-borne, that it originated at Restaurant A, and that the food became contaminated only during meals served on Thursday evening.

Most diseases do not present such easy clues when they enter a population. The chief challenge for epidemiologists is the lag time that develops between entry of a new disease and the beginning of symptoms in a large proportion of the population. Often diseases are unnoticed for a time before symptoms begin, because of their natural incubation period in a person's body. Incubation period is the time between an infection by a pathogen and the first appearance of disease symptoms. As increasing numbers of people become sick, epidemiology serves as the main science for monitoring the spread of the disease, finding and handling its source, and determining why some people are at higher risk of infection than others.

Epidemiology has developed three different approaches for studying disease: descriptive, analytical, and experimental epidemiology. Descriptive epidemiology entails the collection of as much data as possible on a disease and its transmission. Usually, epidemiologists use the data they collect to perform a retrospective study, which is the process of determining a disease's source by using data to backtrack through the events leading to an outbreak or an epidemic. In analytical epidemiology, scientists study two groups of people to learn more about a disease. A healthy group is usually compared with a group of people who have the disease or have been exposed to an infectious pathogen. By collecting data on two populations that differ mainly by the presence or absence of a disease, epidemiologists gather a large amount of information quickly. Experimental epidemiology often takes place under the supervision of doctors,

Epidemiology Terminology

Term	Description
cluster	small group of cases of the same disease that can be defined by distribution, time, and place
outbreak	sudden, unexpected occurrence of a disease within a local population
sporadic disease	disease that appears occasionally in a population at irregular intervals
epidemic disease	sudden increase in a disease in a large population in a given period
pandemic disease	expansion of an epidemic over a very large region, usually worldwide
endemic disease	disease that remains in a population, causing new cases at a regular frequency
hyperendemic disease	endemic disease that gradually increases in frequency but not to epidemic levels

who prescribe a drug to one group of people with a disease and give a placebo to another group who also have the disease. As soon as the drug group shows signs of improving, doctors can prescribe the drug to the second group in place of placebo—experimental epidemiology studies are regulated by law under the authority of the U.S. Food and Drug Administration.

THE BEGINNINGS OF EPIDEMIOLOGY

The philosopher and scientist Hippocrates (about 470–400 B.C.E.) began the practice of taking notes on the signs and symptoms of people who had an obvious disease. People considered disease more mystical than clinical at the time, but Hippocrates' idea of observation and deduction would remain with medicine to the present. Epidemiology made little progress for the next 2,000 years, as sickness and health became increasingly intertwined with religion rather than science. The English physician Thomas Sydenham (1624–89) resumed the practice of recording detailed observations of sick patients, what today are called *clinical observations*.

The next 200 years introduced a variety of methods by many scientists for studying, defining, and explaining disease in relation to microorganisms. Public health officials in Great Britain began keeping records on the deaths of local citizens, in 1801. In 1838, the Englishman William Farr (1807–83) developed a national death registry. As Farr's records accumulated, he began studying the trends in different diseases and periodically published the results of his analyses to show the evolving health conditions in Great Britain.

In Vienna, from 1846 to 1848, the Hungarian physician Ignaz Semmelweis (1818–65) recorded all the births and deaths occurring at Vienna General Hospital. After studying the causes of death, Semmelweis found that one of the hospital's maternity clinics had an unusually high incidence of infection among new mothers. By backtracking, he determined that medical students assigned to the suspect clinic always spent their mornings dissecting cadavers before spending the rest of the day delivering babies: They did not bother washing their hands between the two activities! Semmelweis solved the problem by ordering all personnel to wash hands before entering the delivery rooms.

The London physician John Snow (1813–58) conducted what many historians cite as the first scientifically sound use of epidemiology. In 1848–49, a cholera outbreak ravaged London, as Snow undoubtedly noticed as increasing numbers of sick people made their way to him for relief. In 1854, another outbreak struck. Sensing a medical catastrophe, Snow gathered information from the living people who were afflicted and found information on those who had died. Snow went house to house, in the areas where his patients lived, and studied the information he collected. Many people had already fled from the sickness and death that had been overwhelming the area of Golden Square.

Studying a map of Golden Square, Snow suspected that all the illnesses began at a single source of contaminated water. After reviewing where people had eaten, ordered drinks, worked, and slept, Snow narrowed the cholera source to a water pump on Broad Street. In his 1855 book *On the Mode of Communication of Cholera,* Snow wrote, "With regard to the deaths occurring in the locality belonging to the pump, there were sixty-one instances in which I was informed that the deceased persons used to drink the pump-water from Broad Street, either constantly or occasionally. In six instances I could get no information, owing to the death or departure of everyone connected with the deceased individuals; and in six cases I was informed that the deceased persons did not drink the pump-water before their illness." After reviewing his records, Snow suggested the simplest of solutions: He had the Broad Street pump's handle removed. As a result, the number of cholera cases dropped dramatically. John Snow entered medical history as the Father of Epidemiology.

The British nurse Florence Nightingale (1820–1910) used medical statistics to demonstrate the need for better care among Britain's soldiers. During the Crimean War, Nightingale collected copious amounts of data on injuries and infections and established an organized storage and retrieval system for large amounts of medical records. She introduced her data collection and records systems to military hospitals and so demonstrated the value of a standardized method of making clinical observations. Nightingale was the first health care provider to show the importance of statistics in tracking the health of populations. Modern epidemiology could not exist without the use of advanced statistics programs.

John Snow, shown here in 1857, became known as the Father of Epidemiology for his scientific approach to finding the source of a cholera outbreak in London. ***(Wellcome Historical Medical Museum and Library)***

EPIDEMIOLOGY TODAY

Epidemiology relies on four main calculations for describing disease in a population: incidence rate, prevalence rate, morbidity rate, and mortality rate. Each calculation is a ratio of one statistic to another statistic. The calculations are called *rates* because each is determined within a set time frame.

Incidence rate is the frequency of new cases of a disease per a given population size. Public health agencies usually express incidence rate as the number of new cases per 1,000–100,000 people per year. This expression, therefore, connotes the risk of contracting a disease. For example, a person has a lower risk of contracting a disease with an incidence rate of 25:100,000, 25 new cases per 100,000 people, than a disease with an incidence rate of 650:100,000.

Incidence rate may also be expressed as a percentage: An incidence rate of 5:1,000 is equal to 0.005 percent per year. This means that of 10,000 people within a normal, healthy population, 50 can be expected to contract the disease within a year.

Incidence proportion, also called *cumulative incidence,* views incidence rate as an accumulative event over a long period. This calculation helps epidemiologists watch the trends in a disease in terms of specific factors, such as a certain age group, income level, or geographic area. For example, the incidence rate of a communicable disease in a group of 1,000 20-year-olds is 16:1,000. Sixteen individuals can expect to have the disease by the time one year has passed and 984 people will still be healthy. What is the chance that a 20-year-old will contract the communicable disease within five years? The table illustrates the calculation for the cumulative incidence of this group of people.

In the table, 78 people of the original 1,000 will contract the disease within five years. This group has a cumulative incidence rate of 0.078, or 7.8 percent.

The prevalence rate of a disease is the number of people who have a specific disease at any time. If an influenza outbreak occurs in a school of 650 students and 293 students stay home sick on Wednesday, the prevalence rate that day becomes 293 ÷ 650, or 45 percent. Morbidity rate is similar to incidence rate. Incidence rate gives an epidemiologist an idea of the ratio of new cases occurring in a population in a period. The epidemiologist can also calculate the total number of cases per population, as follows:

morbidity rate = number of cases per year ÷ number of individuals in the population

The mortality rate measures the number of people who die of a disease out of all the people who have contracted the same disease:

mortality rate = number of deaths per year ÷ number of diseased individuals

EPIDEMIOLOGY DATA

The calculations used in epidemiology require simple mathematics. A much more difficult task resides in the manner in which epidemiologists collect the data they need to study a disease. In the United States, the Centers for Disease Control and Prevention (CDC) publish a yearly list of diseases that health care

Calculation of Cumulative Incidence Rate

Year	Number Healthy	Multiply by	Number Sick (rounded to the nearest whole number)
1st	1,000	0.016	16
2nd	984	0.016	15.7 = 16
3rd	968	0.016	15.5 = 16
4th	952	0.016	15.2 = 15
5th	937	0.016	14.9 = 15
			Total sick = 78

providers report to the CDC's National Notifiable Diseases Surveillance System (NNDSS). State health regulations determine which diseases are reportable, but because the reporting is from all 50 states and the U.S. territories, the NNDSS maintains a large database on a variety of diseases. In 2009, the CDC listed about 65 infectious diseases that doctors, outpatient clinics, and other health care facilities must report or voluntarily may report to the NNDSS. By plotting of the reports of diseases for more than 10 or more years, doctors and the public learn the diseases that are on the rise in the United States, those that may be declining, and others in which prevalence is holding steady (provided the disease has been monitored for each of the years in question).

In his Nobel Prize acceptance speech, in 1905, the microbiologist Robert Koch (1843–1910) said, "The starting point in the fight against all contagious diseases is the obligation to report, because without this most cases of the disease remain unknown." The quality of the data reported to government health agencies remains key to successful epidemiology. In addition to the reporting of disease incidences to the NNDSS, epidemiologists collect data by a variety of manual methods, listed here:

- doctors' case reporting forms on all diseases
- mortality data from death certificates
- investigation of actual cases
- data collection during epidemics
- review of clinical laboratory results
- population surveys
- data on sales of antibiotics, antiviral drugs, vaccines, and other treatments

Epidemiologists may also use large databases from the U.S. Census and remote surveys with satellite-based geographic information systems (GIS) to monitor the movement of populations. Remote sensing devices on satellites collect digital images of Earth that show gains or losses of forests, water, cultivable land, wetlands, and so on. The GIS then organize the large amounts of data contained in the images to help epidemiologists make predictions on the spread of disease. For example, images that show significant flooding along the Yangtze River in China would indicate a probable rise in insect-borne diseases, especially those carried by mosquitoes.

In 2000, the public health expert Paul R. Epstein alerted the public, in a *Scientific American* article, to the effects of increasing global temperatures on infectious disease. "Diseases relayed by mosquitoes," he wrote, "such as malaria, dengue fever, yellow fever and several kinds of encephalitis, are among those eliciting the greatest concern as the world warms." By 2007, the Intergovernmental Panel on Climate Change (IPCC) had reported, "There is some evidence of climate-change-related shifts in the distribution of tick vectors of disease, of some (non-malarial) mosquito vectors in Europe and North America, and in the phenology (lifecycle and migration timing) of bird reservoirs of pathogens." Epidemiologists will contribute to the massive project of determining the effect of global change on human health.

By working with the data that scientists have accumulated on diseases throughout the world, epidemiologists now make fairly accurate predictions on the probability of contracting an infectious dis-

Does Immigration Lead to Increased Incidence of Disease?

by Anne Maczulak, Ph.D.

Immigration plays an important role in the ecology of disease. Since the earliest recorded history, explorers and conquerors have carried infectious disease from their place of origin to new lands. Tuberculosis and syphilis, two of civilization's oldest diseases, traveled with Phoenician sailors across the Mediterranean and then north along Europe's Atlantic coast, between 1500 and 300 B.C.E. The bubonic plagues that devastated Europe's population, in the 13th and 14th centuries, and that in London, in 1665–66, had followed the trade routes from Asia, across India and the Middle East, and then to the major ports of Europe. With each shipload of goods were immigrants, including soldiers and sailors, who stayed in the new land. Whatever infections they carried, they took into the taverns, shops, hostels, and gatherings they attended.

In 2005, the microbiologist Igor Mokrousov of the Pasteur Institute analyzed the genetics of more than 300 strains of *Mycobacterium tuberculosis* to trace the course of tuberculosis (TB) in populations in history. Mokrousov stated that Genghis Khan's expeditions may have been the primary cause of TB's spread, in the 13th century. The microbiology writer Jeffrey Fox noted, in *Microbe,* that *M. tuberculosis* genetics indicates that "over the past 60,000–100,000 years, its dispersal and evolution closely followed human migration patterns." Disease transmission increases when people are crowded together or have other debilitating health conditions, but no single factor has mirrored the global spread of disease as much as the migration of large groups of people from areas with infection to areas of healthy populations. This includes the international patterns of trade and travel we now call *globalization.*

Globalization has always been part of civilization. The Phoenician explorers, the Vikings, Romans dominating Europe and the Mediterranean, the eastern invasions of Europe, and Columbus's voyages are all types of globalization for their times. These movements spread culture, language, and trade goods, as well as infections. Today, the patterns of disease spread are strikingly similar but on a larger scale. Trade, business travel, international leisure travel, and military movements spread microorganisms as they have for centuries. Today, society also has large-scale migrations of people fleeing political persecution, drought and hunger, or war zones. When new groups of these travelers enter new homelands, they raise anew the fears of immigrants' causing disease in the healthy. This fear has persisted for as long as people have been mobile and contributes to the ongoing fear of large groups of new immigrants.

Because of the politics often associated with immigrant groups today, blaming immigration for emerging or reemerging disease may seem politically incorrect. Antiimmigration factions have, indeed, used the fear of new disease as a potent argument against letting immigrants into their countries. In 2003, Robert Engler, writing for the conservative social commentary publication *American Daily,* said, "The U.S. is not alone in its fight against illegal immigration and disease. Illegal Chinese immigrants to Europe are bringing to that continent malaria. Even Thailand has problems with diseases brought into that country from Burma [Myanmar]. The biological clock is ticking and a world health crisis is looming. Because the U.S. suffers more immigration than any country in the world, our health problems are growing faster. Certainly, the potential for biological disaster has come to the notice of terrorists and Al Quaeda." Although infection does move with humans, Engler fails to mention that disease also spreads with animals, insects, and changing weather patterns. Engler additionally makes the very dangerous supposition that immigration by diseased populations leads to political instability.

Making a political and social crisis out of a disease has been a common tactic of certain groups, for a long time. This tactic not only endangers the life of immigrants, but slows the progress of medical research in studying disease and so finding the best treatments. In 1989, the Dutch AIDS educator Hans Verhoef had been detained for several days, in St. Paul, Minnesota, when trying to attend a medical conference in California. Other people who are HIV-positive were denied entry at the Canadian border, even though they were scheduled for treatment in the United States. Various international medical conferences have moved out of the United States because of the ban on HIV-positive travelers. U.S. policy had evolved to favor ignorance over medical progress.

On October 30, 2009, President Barack Obama lifted this 22-year ban on travel into the country by people who have tested positive for the AIDS virus, saying the restriction was "rooted in fear rather than fact." Obama added, "If we want to be a global leader in combating HIV/AIDS, we need to act like it. Now, we talk about reducing the stigma of this disease, yet we've treated a visitor living with it as a threat." Because disease and immigration have become intertwined politically, the United States and other countries will be hard-pressed to separate the politics from the biology of disease in the future.

(continues)

Does Immigration Lead to Increased Incidence of Disease?

(continued)

AIDS was probably carried into the United States, prior to 1980, by an immigrant. But the connection between the movement of people and disease transmission is a biological fact and not a political problem or social crisis. Humans have had virtually no success at eliminating any of the diseases that have haunted them since the beginning of history, save smallpox. The rush to bar immigrants from the borders has never been an effective deterrent to disease transmission.

No easy answers exist on how to meet the medical needs of the infected, reduce the further transmission of disease, yet allow the continued globalization of world economies and cultures. As medical research has made gains in controlling some diseases, others have reemerged. This reemergence has been in large part due to globalization and immigration. For decades leading up to the AIDS epidemic's start in 1980, TB had been under control and was decreasing in most areas of the world. But, in 1996, the British epidemiologists John Porter and Keith McAdam wrote in the *Annual Review of Public Health,* "Before the HIV epidemic, the risk of TB infection in developing countries fell 1–5 percent per year. The HIV epidemic has reversed this trend. It also fell in developed countries. In the US, the risk began to increase in 1986 in New York City due to increased immigration from TB endemic countries, interaction between TB and HIV, and decay in the health infrastructure supporting TB control programs." The factors stated by Porter and McAdam are correct and highlight the complexity of dealing with diseases thought to be influenced by immigration.

No easy answers exist to the connection between immigration and infectious disease other than, yes, that in many diseases immigration certainly changes the patterns of global disease transmission. This does not suggest that governments should ban immigration from places where disease is endemic. Some disease is endemic everywhere, and such a ban essentially ends all immigration. The approach to managing diseases that move with immigration involves continued education on the hygiene methods that best prevent contamination of foods and water and block the transmission of pathogens. If vaccination programs have proved to be the best prevention for certain diseases, immigration programs must include these steps. The public, however, may be more realistic about these approaches than politicians: Education and vaccines cost money. In today's economy, the best health choices may not succeed if their costs cannot be met.

Rather than prevent the migration of sick people who by choice or by force must leave their home, true disease prevention will only succeed if targeting the source of the pathogen. Programs in every country should include, at minimum, disinfection of water supplies, proper handling and pasteurization of milk and other foods, sewage treatment, vector control, and strict monitoring of the cleanliness of ships and internationally traded goods. Vaccination, hygiene education, and the quarantine of carriers might be necessary, even though difficult, requirements for halting the global transmission of infectious disease.

See also EMERGING DISEASE; INFECTIOUS DISEASE; VACCINE.

Further Reading

Engler, Robert Klein. "Immigration and Disease: It's Enough to Make You Sick." *American Daily,* 21 November 2003. Available online. URL: www.americandaily.com/article/3364. Accessed November 9, 2009.

Medical News Today. "CDC Officials Unaware Recommendation on HPV Vaccination Would Lead to Immigration Mandate." October 2, 2008. Available online. URL: www.medicalnewstoday.com/articles/123892.php. Accessed November 9, 2009.

Porter, John D., and Keith P. McAdam. "The Re-Emergence of Tuberculosis." *Annual Review of Public Health* 15 (1994): 303–323.

Preston, Julia. "Obama Lifts a Ban on Entry into U.S. by H.I.V.-Positive People." *New York Times,* 30 October 2009. Available online. URL: http://travel.nytimes.com/2009/10/31/us/politics/31travel.html?scp=1&sq=ryan+white+act+obama&st=nyt.Accessed November 9, 2009.

ease and health factors that make some people of higher risk for disease than others. This information, in turn, helps doctors guide patients in the ways to avoid certain infections.

Epidemiology is a vital part of current health care because it gives providers historical information on disease, helps spot new diseases or epidemics on the rise, and aids in the determination of groups of people at the highest risk for given diseases. The world population will continue to rely on sound epidemiological data as a tool in protecting health.

See also CENTERS FOR DISEASE CONTROL AND PREVENTION; EMERGING DISEASE; EPIDEMIC; TRANSMISSION.

Further Reading

Epstein, Paul R. "Is Global Warming Harmful to Health?" *Scientific American,* August 2000. Available online. URL:

http://chge.med.harvard.edu/about/faculty/journals/sciam.pdf. Accessed August 22, 2009.

Intergovernmental Panel on Climate Change. "Climate Change 2007." Available online. URL: www.ipcc.ch. Accessed August 22, 2009.

Koch, Robert. "The Current State of the Struggle against Tuberculosis." Nobel Lecture presented in Stockholm, Sweden, December 12, 1905. Available online. URL: http://nobelprize.org/nobel_prizes/medicine/laureates/1905/koch-lecture.html. Accessed June 3, 2009.

Snow, John. *On the Mode of Communication of Cholera.* 1855. Reprinted by Delta Omega with introduction by Ralph R. Frerichs. Washington, D.C.: Delta Omega Honorary Public Health Society, 1988. Available online. URL: www.deltaomega.org/snowfin.pdf. Accessed August 22, 2009.

Escherichia coli *E. coli* is a gram-negative straight rod-shaped bacterium that belongs to the general group of enteric bacteria known as coliforms and is a member of family Enterobacteriaceae. The non-spore-forming, motile *E. coli* is found in the large intestine of humans and other animals, and its presence in natural waters and soils is an indication of fecal contamination. Fecal contamination of food is estimated to cause at least 73,000 food-borne illnesses each year, but because food-borne illnesses are historically underreported, cases of *E. coli* not requiring hospitalization or a visit to a doctor's office may be far greater in number.

E. coli is prevalent in the human intestines, but it is not the main contributor to microbial activities there, and it plays a small role in the earth's overall microbial metabolism. Yet *E. coli* is the most extensively studied microorganism on Earth. It has been called the "guinea pig," "lab pet," or "workhorse" of microbiology and molecular biology, and it is the microbe upon which the biotechnology industry was built. *E. coli* is a pathogen in humans and animals, and it serves also as a model for research on genes, gene sequencing, gene transfer, genetic engineering, cloning, plasmids and resistance, pathogenesis, and membrane structure and organelles.

In 1885, the German physician Theodor Escherich (1857–1911) isolated in his laboratory an unknown bacterium from fecal samples. His observations were made during a period to become known as the golden age of microbiology, which occurred from the late 1880s to the 1920s. Throughout this period, the foundations of microbial research were developed. From the dismissal of the spontaneous generation theory to the first differentiations of bacteria based on their biochemical traits, microbiologists studied a variety of microbial environments in search of new species. Digestive wastes from humans and animals were found to contain an enormous concentration of bacteria. A common fecal isolate that grew readily in the laboratory was repeatedly observed by many microbiologists, after Escherich's first observations of it. Their descriptions consistently matched his, and the microbe was soon named for him, in honor of his discovery.

The genus *Escherichia* contains species other than *E. coli,* though *E. coli* strains make up the overwhelming majority of those deposited in culture collections worldwide. A modest degree of diversity exists across the genus. *E. fergusonii* is a commensal species found in human feces, while *E. blattae* is found in the hindgut of cockroaches. *E. hermanii* has been isolated from clinical wound and skin samples, but microbiology has not yet decided whether this is a true species or a substrain of *E. coli.*

CELL STRUCTURE

An *E. coli* cell measures 2 micrometers (μm) in length and 0.5–0.8 micrometers in diameter. The cell has a typical gram-negative cell structure, consisting of an inner multilayered membrane, a thin cell wall (20–25 nanometers [nm] thickness) containing peptidoglycan, and an outer membrane. As in most living cells, the *E. coli* cell volume is predominantly water, 70 percent, with 17 percent occupied by proteins and about 7 percent nucleic acids.

A large part of our knowledge of cellular motility and adhesion to surfaces has been gained from research on the outer structures of *E. coli. E. coli*'s motility in liquids is made possible by 10 flagella, of 10–20 μm, that extend from one end of each cell. These tails, which rotate the way propellers do, demand a significant portion of cellular energy. Bacterial flagella have provided a model for studying adenosine triphosphate (ATP) hydrolysis, its regulation, and the biological mechanisms used in converting energy to motion. The cell's entire surface is also covered with shorter projections called fimbriae, and because these structures are distributed uniformly over the cell surface, they are called peritrichous fimbriae. *E. coli*'s fimbrae range from 200 to 2,000 nm in length and, as do all bacterial type I fimbriae, they help *E. coli* adhere to surfaces such as the intestinal lining. Small pili, several micrometers in length, also project from the cell surface for exchanging genetic material between two cells during conjugation. Pili are also responsible for causing the specific serological reactions in bacteria identification methods.

The *E. coli* chromosome and its genes were the first teaching tools in the early stages of molecular science, which gained momentum in the 1980s. By 1997, all of the 4,403 *E. coli* genes had been sequenced. Frederick Blattner and his team of geneticists at the University of Wisconsin were the first to report the genomic sequence, and they determined it contained 4,639,221 base pairs. The *E. coli* genome offered a foundation for the burgeoning sciences of gene transfer technology, disease-targeted gene therapy, and genome characterizations, including the Human Genome Project. By sequencing a complete genome, scientists can pinpoint the sources for all cellular activities. Sequencing is additionally valuable in determining the evolutionary history of organisms.

Prokaryotic chromosomes are double-stranded circular molecules containing segments called domains, which are created by the twisting, or supercoiling, of the DNA. Because *E. coli* contains a single strand of deoryribonucleic acid (DNA), its DNA represents the entire chromosome, excluding plasmids. Eighty-nine of *E. coli*'s total genes encode (carry the molecule-synthesizing information) for ribonucleic acids (RNAs); the remaining 4,288 genes encode for proteins.

In biotechnology, *E. coli*'s major importance relates to three components of its cellular genetics. These are restriction enzymes, operons, and gene transfer mechanisms. Microbial restriction endonucleases are enzymes capable of cutting double-stranded DNA at a specific base pair sequence called a *recognition site*. Restriction activity has been thought to have evolved as one of several protective mechanisms within microorganisms. In the case of restriction, the cell is protected against the lethal effects of foreign DNA from other species. This may be particularly critical against infiltration by phages, which are viruses that attack bacteria. Restriction endonucleases from *E. coli* became the enzymes used in biotechnology's first attempts to cut open host DNA before inserting a new, foreign gene. This process gave birth to the field of genetic engineering.

*Eco*RI and *Eco*RV from *E. coli* are two examples of the more than 300 microbial restriction enzymes used today in research and industrial microbiology. *Eco*RI acts by moving along the DNA molecule until it finds a guanine (G) base nucleic acid bound to a sequence of adenine-adenine-thymine-thymine-cytosine (AATTC) on the complementary strand. Guanine binds to cytosine in the DNA of all living things on Earth and so is bound to the C in the AATTC sequence. The recognition site for *Eco*RI is designated in genetic engineering as G/AATTC. *Eco*RV works the same way but cleaves DNA at GAT/ATC. Restriction endonucleases do not protect cells from harm as effectively as was originally thought, when they were discovered, in 1970, by Daniel Nathans, Hamilton Smith, and Werner Arber. First, some phages have mechanisms that destroy an endonuclease before it can damage their DNA. Second, certain bacterial plasmids contain genes for enzymes that destroy endonucleases. Finally, transposons often carry a resistance gene against endonuclease destruction. *E. coli* has been an integral tool in studying all of these processes. A wry observation has been made that many molecular biologists who spend their careers working with *E. coli*'s genome have never actually seen an *E. coli* cell in its entirety.

The majority of teaching activities conducted on *E. coli* use the K-12 strain, which was discovered to be a harmless representative of the species and a very cooperative grower in laboratory experiments. K-12 has been transferred repeatedly in laboratories worldwide, to the point where the lab strain probably will not grow if inoculated into the intestines. This is quite different from the dangerous *E. coli* pathogen strain O157:H7 (or, simply, O157). Today's O157 isolates and K-12 laboratory specimens differ in their DNA sequences by as much as 25 percent, making these two strains of the same species seem as if they are hardly related at all. Evolution and its related movement of genes within the genomes of these two strains, over the past few decades, account for this enormous genetic difference.

In 1961, François Jacob and Jacques Monod developed a theory to explain how gene expression is controlled to produce an orderly reproduction of proteins when they are needed and in the amounts in

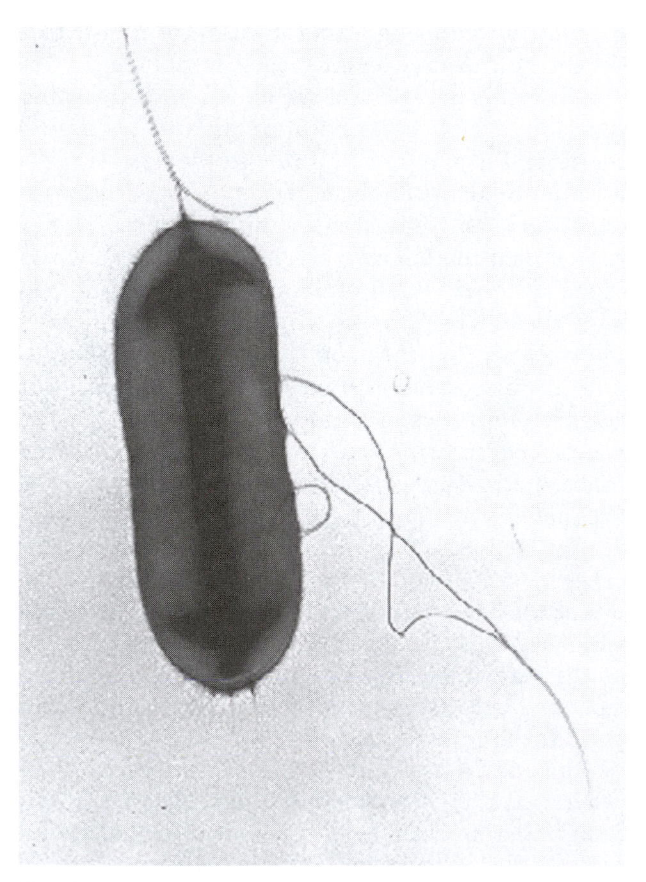

E. coli*, here showing flagella, is the most studied organism in biology.** ***(Associated Press)

which they are needed. By studying *E. coli*'s metabolism of the sugar lactose, they discovered three genes that encoded for three enzymes in a coordinated fashion. The genes were preceded by a section of DNA that initiated its replication. This section was termed the operator. The *E. coli* system of genes, the operator, and additional promoting factors have become known as the *lac* operon, and Jacob and Monod's discovery is called the operon model. Operons are studied to determine how genes are turned on or off, how genes are suppressed, and the ways in which amino acid levels affect protein synthesis. As with so many breakthroughs in microbiology, it all began with *E. coli.*

In 1983, the geneticist Barbara McClintock was awarded the Nobel Prize in physiology for her discovery of transposons in corn cells in studies that she had conducted in the 1950s. Transposons are small DNA segments that can move from one location on the DNA molecule to another location, or to another chromosome, or to a plasmid. Research on the insertion of transposons into *E. coli* DNA has improved understanding of microbial resistance to antibiotics and heavy metals, as well as toxin production. Gene transfer techniques using transposons are usually accomplished using *E. coli.*

METABOLISM

E. coli is the most famous facultative anaerobe on Earth. Facultative microorganisms use respiratory metabolism under aerobic conditions and switch to fermentative metabolism under anaerobic conditions. This versatility serves *E. coli* well, when living in its normal oxygen-free habitat in the intestines and during the times when it exits the body and lives in the environment. This versatility helps *E. coli* and other facultative microorganisms in transmission and in causing disease.

Enteric bacteria live in an environment rich in nutrients and growth factors, when compared with microbes from more austere environments. In the intestines, *E. coli* metabolizes glucose, mannitol, and lactose, among other sugars; obtains sulfur from sulfate and sulfite compounds; and can use the nitrogen from compounds containing an ammonia group. The definition of a coliform gives further detail to the growth requirements of *E. coli:* gram-negative, nonsporing, facultative bacillus that ferments lactose, with acid and gas production, within 48 hours at 98.6°F (37°C). *E. coli* carries out mixed acid fermentations, meaning that it makes more than one end product. *E. coli*'s main products are lactic, succinic, and acetic acids and ethanol from lactose, and carbon dioxide and hydrogen from glucose.

Stained enteric bacteria are similar in morphology, when they are viewed in a microscope. Therefore, biochemical tests are employed to create a simple scheme for differentiating *E. coli* from other gram-negative enterics. The first test is checking for the presence of oxidase enzyme, present in *Escherichia* and absent in other enteric genera. *Escherichia,* then, differs from the nonmotile enteric *Shigella, Yersinia,* and *Klebsiella* genera, because it is motile. Production of beta-galactosidase enzyme further separates *E. coli* from the motile *Salmonella* and *Proteus.* The genus is, then, differentiated from *Citrobacter, Enterobacter,* and *Serratia,* because it cannot use citrate in its metabolism.

Modern methods to identify bacteria to the species level use a variety of techniques that quantify the chromosomal components. There is undoubtedly more information known about *E. coli*'s chromosome than about any other species. This knowledge and simple growth reactions on selective media both serve well to confirm the identity of *E. coli.* EC medium was developed, in 1943, by Hajna and Perry for selecting coliforms from other microorganisms and, in turn, differentiating *E. coli* from the coliforms on the basis of their metabolic requirements. Bile salts in EC medium inhibit noncoliforms. If incubation is performed at 114°F (45.5°C), rather than the typical incubation temperature of 98.6°F (37°C), *E. coli* grows and other coliforms do not.

E. coli is a mesophile, meaning it grows best between 68°F and 113°F (20–45°C) but may grow with difficulty at temperatures outside that range. The cells cannot withstand extreme temperatures, acidic or basic conditions, desiccation, or high barometric pressures. The microorganism exhibits chemotaxis, whereby cells move in response to a concentration gradient. *E. coli* moves in the direction of higher concentrations of certain chemicals. When it moves from higher to lower concentrations, receptors on its surface start a series of reactions, sending the cell into erratic movements and tumbling. This irregular movement may have evolved to help the cell change course until it begins migrating toward an attractant chemical.

Several enteric bacteria produce vitamin K, which is needed by the body to aid in blood clotting. *E. coli* produces a version called vitamin K_2, which is not available in foods, and, in doing so, plays an essential role in human nutrition. As microbial cells die and break apart inside the intestines, they release vitamin K_2, which is then absorbed through the intestinal lining. *E. coli* also produces B vitamins that help meet the host's metabolic requirements.

CULTURING

The ease with which *E. coli* is grown in a laboratory explains why it has been chosen for hundreds of thousands of laboratory experiments. *E. coli* grows well on general liquid or agar media within its optimal temperature range. Specialized media are used when *E. coli* is an indicator organism for fecal contamination in water or when three is need to diagnose a presumed *E. coli* food outbreak.

In 1908, the British bacteriologist A. T. MacConkey formulated a medium for isolating coliform bacteria. MacConkey broth and agar remain in use today for isolating *E. coli* from heterogeneous food and water samples or for detecting *E. coli* in clinical specimens from medical patients. This medium became one of the first important selective media used in microbiology for separating specific microorganisms from mixtures of other species. EC medium further differentiates *E. coli* from other coliforms. In 1983, Feng and Hartman developed a revised EC medium formula to improve on *E. coli* detection. The revised formula incorporated 4-methylumbelliferyl-beta-D-glucuronide (MUG). The procedure makes use of *E. coli*'s production of the enzyme glucuronidase, which breaks apart the MUG molecule to yield a fluorescent end product that is detectable under ultraviolet light (366 nm).

Traditional aseptic techniques in a level 2 biosafety laboratory are used for broth and agar plate *E. coli* cultures. *E. coli* is grown at the conditions favored by the mesophiles: 98.6°F (37°C) in the presence of air, for at least 16–18 hours. In favorable nutrient and physical conditions, the microbe's doubling time can be as short as 20 minutes.

PATHOGENICITY

E. coli is a food-borne pathogen but can also cause infection from non-food-borne sources. In all circumstances of *E. coli*–caused illness, the microbe originated in contamination from human or animal feces.

E. coli is one of several microorganisms—bacteria, viruses, or protozoa—that are implicated in causing traveler's diarrhea. People who travel internationally are susceptible to contaminants that are present in the local water or in uncooked or improperly washed fruits and vegetables. *E. coli* is also a prevalent cause of diarrhea cases among children throughout the world, regardless of their travel history. Diarrheal diseases caused by *E. coli* are categorized according to the microbe's mode of pathogenicity, that is, the way it initiates disease and causes its symptoms.

Six mechanisms of *E. coli* food-borne illness and diarrhea have been identified (shown in the table on page 288). These groupings are collectively known as the EEC group of *E. coli*, the enterovirulent *Escherichia coli* group. EEC strains known as enterotoxigenic *E. coli* (ETEC) produce one or both of two different enterotoxins, which are toxic compounds secreted by bacteria that damage the cells lining the intestines. The heat-labile enterotoxin (HL) and the heat-stable enterotoxin (HS) each attack different receptor sites on epithelial cells. Each toxin, once it has bound to the epithelia, activates cellular membrane enzymes that promote water and electrolyte discharge into the intestines. HL binds to cellular gangliosides and activates the enzyme adenylate cyclase, which leads to increased production of cyclic adenosine monophosphate (cAMP). HS seeks glycoprotein receptors on the intestinal epithelia. This binding reaction activates cyclic quanosine monophosphate (cGMP). Both cAMP and cGMP are integral in cellular energy production by mediating phosphorylation, among other critical reactions. Their stimulation by enterotoxins initiates a wholesale release of calcium, magnesium, manganese, and other electrolytes. The result is watery diarrhea that causes significant loss of electrolytes from adults and infants, as well as some livestock. Enteroinvasive *E. coli* (EIEC), by contrast, penetrates the cells lining the intestines and multiplies there. As it produces toxins, the intestinal lining is destroyed, and bacillary dysentery ensues, characterized by diarrhea containing blood and mucus.

The remaining EEC strains cause disease by attaching to the outside of the intestinal epithelia. Enterohemorrhagic *E. coli* (EHEC) strains release toxins similar to those produced by *Shigella dysenteriae* that destroy the vascular systems of the intestines and the microvascular systems of the kidneys and the central nervous system. The resulting hemorrhagic colitis is associated with severe abdominal pain and cramps, followed by bloody diarrhea. Toxins that travel in the bloodstream, at the same time they affect the intestines, lead to hemolytic uremic syndrome, which can cause critical damage to the kidneys, especially in young children. *E. coli* O157:H7 is a particularly lethal EHEC serotype, first studied in 1982, that has caused outbreaks due to undercooked meat, contaminated dairy products and juices, and raw vegetables and fruits, recreational waters, and a variety of foods or activities in which fecal contaminants may be swallowed. Farm animals are the main reservoir of O157 infections in humans, and the contamination has been traced from food products back to farms near where the food item was produced. There are an estimated minimum of 20,000 illnesses and 250 deaths from O157 in the United States each year.

E. coli O157:H7 is the one of the more than 25 EHEC serotypes most associated with hemorrhagic colitis. Frequent food-borne outbreaks throughout the United States that have been attributed to O157 have spurred vigorous efforts to find rapid identification methods for speeding diagnosis and treatment. Clinicians have a battery of biochemical and molecular tests for focusing on this dangerous microbe. Serotype-specific probes and probes employed in the polymerase chain reaction (PCR) are used for rapid detection of specific strains of O157, such as O157:H7, in mixtures containing many diverse microorganisms and organic matter. DNA or RNA probes of 14–40 base sequences are valuable in tracing O157 outbreaks back to an individual produce farm or livestock operation. Immunochromatic assays for individual strains use antibodies, rather than nucleic acid probes, to react to specific strain antigens. The antibody-antigen reaction is linked to a visible color change, which indicates the presence of the microorganism. The latex agglutination test for *E. coli* O157 strains also bases its result on specific antibody-antigen binding. Antibodies specific for enterotoxin or other soluble fractions of O157 cells are attached to tiny latex particles. Agglutination within a mixture of antibody particles and the test sample indicates the presence of the target antibody-antigen reaction, thus the presence of the O157 strain.

Laboratory media methods are not as precise as molecular techniques for detecting *E. coli* strains such as O157 but remain useful screening tools in the initial identification of species and strains. *E. coli* O157 is unlike other *E. coli* strains, in that it cannot ferment the carbohydrate D-sorbitol. *E. coli* O157 grown on MacConkey agar supplemented with sorbitol is differentiated from other *E. coli* by a pH-linked color reaction in the colony-forming units (CFUs). Sorbitol-fermenting cells produce acid, turning their CFUs pink (most *E. coli*), while non-sorbitol-fermenting *E. coli* CFUs are colorless (O157 serotypes).

Enteropathogenic *E. coli* (EPEC) causes infantile diarrhea by attaching to and destroying the brush border of the intestines. The brush border is made of millions of microscopic projections called microvilli, which extend from the epithelial cell into the intestinal contents. The microvilli greatly increase the intestines' surface area for nutrient absorption. EPEC is a major cause of infant death in developing countries. Enteroaggregative *E. coli* (EAggEC) adheres to sections of the intestines, causing lesions. Toxin production has not as yet been found in EAggEC strains. Diffusely adhering *E. coli* (DAEC) attaches to cells of the entire intestinal lining. DAEC's mechanisms are not well known, but researchers know that malnourished children and infants are at higher risk of DAEC infection. DAEC has been implicated in some diarrhea outbreaks and may cause recurrent urinary tract infections by modes of action that have not yet been fully described.

Traveler's diarrhea

Traveler's diarrhea (TD) is an affliction in people who drink tap water or consume fresh fruit and uncooked vegetables when traveling outside their normal environment. From 30 percent to more than 60 percent of travelers from developed to developing countries experience TD, amounting to more than 15 million affected annually. The cause of TD is never identified after the disease has run its course, but assorted enteric bacteria, viruses, and protozoa have been implicated. Of the two types of *E. coli* known to cause TD, ETEC and EIEC, ETEC enterotoxins cause about 80 percent of diagnosed cases. ETEC and EIEC have different mechanisms of pathogenicity, but each results in TD's classic range of symptoms, which are characteristic of many food-borne and waterborne illnesses: mild disease with several days of loose stools to severe watery diarrhea accompanied by nausea, vomiting, abdominal cramping, bloating, fever, and/or malaise. Recent studies suggest that many patients develop irritable bowel syndrome.

TD is self-limiting in most people, within three or four days, and medical treatment is not needed to resolve mild to normal symptoms before the illness ends on its own. Severe cases may require antibiotic treatment. Dehydration is a danger in any diarrheal disease, so rehydration with fluids containing electrolytes, throughout the symptomatic period, is recommended. In addition to antibiotics for severe cases, antimotility drugs slow the activity of the intestines to reduce water loss. The risk of TD is influenced by host conditions, including age, genetic factors, and preexisting health status. Persons who experience persistent vomiting, bloody stools, high fever, or signs of severe dehydration (lack of tearing, decreased urine volume) should seek immediate medical care.

TD can be prevented, during travel, by avoiding actions that increase the risk of ingesting infectious microorganisms. The Centers for Disease Control and Prevention (CDC) and the Mayo Clinic publish precautions against contracting illness when traveling in developing countries or the Caribbean Islands, shown in the list that follows. Though certain regions of the globe are associated in Americans' minds with TD, it is also a risk, though a lower risk, in northern Europe, Canada, Japan, Australia, New Zealand, and the United States.

- Eat only meats and vegetables that are well cooked, never raw.
- Eat fruits that must be peeled rather than salads and unpeeled fruits.
- Avoid unpasteurized drinks and untreated, unsterilized water.
- Choose canned or bottled drinks in their original containers.
- Use bottled water for brushing teeth and in infant formulas.
- Avoid ice cubes or mixes made with tap water.

Types of *E. coli* Pathogenicity

Type of *E. coli*	Acute Diseases	Symptoms	Time to Symptoms	Target Population
EHEC	hemorrhagic coalitis	A, D, V, f	18–36 hours	all
ETEC	gastroenteritis (traveler's diarrhea)	Dw, A, f, N, M	24 hours–3 days	infants and adults
EPEC	infantile diarrhea	Db, Dw	18–36 hours	infants
EIEC	dysentery	D, A, V, F, C, M	12–72 hours	all
EAggEC	persistent diarrhea	Db*	8–18 hours	children
DAEC	gastroenteritis	Dw, V	unknown	malnourished children

Note: A = abdominal cramping, C = chills, D = diarrhea, Db = bloody diarrhea, Dw = watery diarrhea, Fever = fever, f = low-grade fever, M = malaise, N = nausea; V = vomiting

- Do not swim in known contaminated waters and avoid swallowing water when swimming or when showering.
- Avoid food bought from street vendors

TD has impacted the health of people on the move, since antiquity. Military campaigns, in particular, have been affected by ailments both serious and mundane. Soldiers under the Persian king Darius I suffered what historians surmise were TD-like infections while subduing rebellious Greek cities in 495–492 B.C.E. In the fourth century, the Germanic Visigoths crossed the Danube River in a series of attacks on the Roman Empire but were laid low by gastrointestinal ailments, as were the French Imperial Army, in the early 1800s, as they struggled with slave revolts in Haiti. British soldiers' battles with infection simultaneous with their war against the Turks at Gallipoli were not unlike the suffering of U.S. Marines as they fought Japan's army across the Pacific, in World War II. Though the glamour of TD is questionable, the importance of infectious disease in human history cannot be denied.

E. COLI AS AN INDICATOR ORGANISM

Coliforms have been used, for many years, as indicators of fecal contamination of rivers, lakes, and other recreational waters and in drinking water. A more accurate measure of fecal contamination, however, is the presence or absence of *E. coli,* which is normally found only in animal intestines. *E. coli* is, therefore, used as an indicator organism for fecal contaminants; its presence signals the potential contamination of water by fecal matter and other fecal pathogens.

E. coli may be detected in water samples using membrane filtration or a rapid presence-absence test. In membrane filtration, 100 ml of water is passed through a 0.45-μ pore filter with a counting grid. After filtration, the membrane is saturated with mTEC (membrane thermotolerant *E. Coli*) Agar and incubated, first, at 95°F (35°C), for two hours, to resuscitate any injured cells, then at 112°F (44.5°C), for 22 hours, to differentiate *E. coli* from other coliforms. By exposing the incubated filters to urea, *E. coli*'s urease enzyme gives the CFUs a distinctive amber color. The presence-absence test takes advantage of *E. coli*'s ability to cleave the fluorogenic compound 4-methylumbelliferone glucuronide (MUG). A water sample is added to a bottle containing broth supplemented with MUG. After a 24-hour incubation, the bottle is exposed to long-wave ultraviolet light, and the presence of fluorescence indicates that *E. coli* was in the water sample.

These indicator tests demonstrate how truly versatile *E. coli* has become as a tool to study the environment, in addition to its contributions in molecular biology, cell structure, and pathogenicity.

See also ANAEROBE; CHEMOTAXIS; CHROMOSOME; COLIFORM; ENTERIC FLORA; GENETIC ENGINEERING; GENE TRANSFER; INDICATOR ORGANISM; MESOPHILE; PATHOGENESIS; PLASMID; TOXIN.

Further Reading

Bell, Chris, and Alec Kyriakides. *E. coli: A Practical Approach to the Organism and Its Control in Foods.* London: Thomson Science, 1998.

Carey, Catherine, Judy Folkenberg, and Vern Modeland. "*E. coli* Clue to Contamination." *FDA Consumer,* 23 October 1989.

Centers for Disease Control and Prevention. "*Escherichia coli.*" Available online. URL: www.cdc.gov/nczved/dfbmd/disease_listing/stec_gi.html. Accessed August 23, 2009.

Prescott, Lansing M., John P. Harley, and Donald A. Klein. *Microbiology,* 5th ed. New York: McGraw-Hill, 2002.

U.S. Food and Drug Administration. *Bad Bug Book.* Available online. URL: www.fda.gov/Food/FoodSafety/FoodborneIllness/FoodborneIllnessFoodbornePathogensNaturalToxins/BadBugBook/default.htm. Accessed August 22, 2009.

eukaryote A eukaryote is one of two major types of cells in biology, characterized by the following features: membrane-enclosed inner structures called *organelles,* a membrane-enclosed nucleus that holds the genome (all the genetic material), deoxyribonucleic acid (DNA) associated with proteins called *histones,* and the replication of genetic material involving meiosis and mitosis. These four characteristics distinguish eukaryotic cells from the other main group, prokaryotes. But as prokaryotes do, eukaryotes adhere to the definition of a cell: A cell is the simplest collection of matter that can live.

In microbiology, bacteria and archaea make up the prokaryotes, and the eukaryotes contain every other type of microorganism: algae, protozoa, yeasts, and fungi. Higher multicellular organisms such as invertebrates, reptiles, amphibians, fish, mammals, marsupials, and birds, as well as plants, are also composed of eukaryotic cells.

Microscopic eukaryotic cells serve as excellent study tools for learning the basic components of eukaryotes in general. The most common microorganisms that serve this purpose are amoeba, *Euglena,* and paramecium.

EUKARYOTIC CELL STRUCTURE

Eukaryotic cells are typically 10–100 micrometers (μm) in diameter, contain a chemically simple cell wall composition, and may or may not have flagella. A hallmark of eukaryotic cell structure resides in the organelles, which have evolved to carry out specialized functions in the cell. The table on page 282 describes the organelles of eukaryotic cells. Only the cytoplasm and the plasma membrane are similar to the respective structures in prokaryotic cells. Eukaryotic ribosomes, the cell wall, and flagella differ in eukaryotes and prokaryotes.

The cells of animals and plants are both eukaryotic, yet they differ in the configuration of many of the organelles listed in the table. Plant cells have a much larger vacuole compared with animal cells. In plants, vacuoles serve the following functions: intracellular digestion, nutrient storage, release of wastes, secretion of enzymes that destroy free radicals, and water storage to maintain the fullness of the plant, called *turgor.* Vacuoles play a lesser role in animal cells. Plant cells also contain chloroplasts, but they have few or no lysosomes. Animal cells almost always possess lysosomes.

In addition to the difference between eukaryotes and prokaryotes in the presence of organelles, eukaryotes also organize their total DNA, or *chromosome,* in a different manner than prokaryotes, in which DNA is distributed in the cytoplasm. Eukaryotic cell DNA resides within the cell nucleus and is associated with histones—about 20 DNA base pairs per histone—that hold the large molecule in an organized, compact mass inside the nucleus. When the cell is not dividing, DNA, the histones, and other proteins called *nonhistones* are collectively called *chromatin.*

When a eukaryotic cell is ready to divide, it operates two processes in sequence: mitosis and meiosis. Mitosis is the process by which the nucleus and the chromosome distribute to opposite sides of the cell, before the cell divides in two. Meiosis is part of a sexual reproductive life cycle, in which a cell divides in two, giving each new cell a one-half copy of the chromosome.

Prior to mitosis, the cell makes a duplicate copy of its entire chromosome, a process called *DNA replication.* When mitosis starts, the chromatin changes from stringy consistency to a denser form. The nucleolus disappears, and microtubules called centrioles move to opposite ends of the cell. As the chromosome copies separate, microfilaments provide a framework for their movement in opposite directions. This framework is called the *mitotic spindle.* The cell begins pinching in the middle, until a new membrane has been formed to create a double cell, about to split into two new cells. The nucleolus begins to reappear in each new cell, a nucleur membrane reestablishes, and the chromosome starts to return to its less dense form folded around the histones.

Although the process of mitosis has been simplified for this discussion, eukaryotic chromosome and cell division is elaborate compared with the simple binary fission used when prokaryotic cells replicate. Scientists have pondered the question of how the compartmentalized eukaryotic cell developed as more complex life began to evolve on Earth.

EVOLUTION OF THE EUKARYOTE

Earth's environment, about 4.5 billion years ago, contained an atmosphere of the gases ammonia, methane, and hydrogen. The atmosphere and the

Eukaryotic Cell Organelles

Organelle	Function in the Cell
cell wall	protective covering for the plasma membrane and cell contents
centrosome	microtubule production
cilia	tiny hairlike structures for motility
chloroplasts	site of photosynthesis for energy production (absent in animal cells)
cytosol	watery environment for organelles and some metabolic systems (cytoplasm is the cytosol plus the cell organelles)
endoplasmic reticulum (ER)	protein and lipid (fats and fatlike compounds) synthesis; transport of material inside the cell; rough ER is dotted with many ribosomes; smooth ER lacks ribosomes and participates in some compound synthesis
Golgi apparatus	site for packaging and secreting materials
lysozymes	packets of enzymes that digest nutrients and break down the cell during cell death (uncommon in plant cells)
microfilaments (actin protein units) and microtubules (tubulin protein units)	maintain cell structure and participate in mitosis (cell division)
mitochondria	energy production by metabolism of sugars and amino acids
nucleolus	ribosome construction and ribonucleic acid (RNA) synthesis
nucleus	storage site for the cell's genetic material and controlling site for many cellular functions
perioxomes	packets of enzymes that protect cells against damage by free radicals, which are highly reactive by-products of oxygen metabolism
plasma membrane (also cell membrane)	control of water, nutrients, and ions into and out of the cell
ribosomes	protein synthesis in conjunction with RNA
secretory vesicles	portions of Golgi apparatus that break away and carry materials to the membrane for release
vacuole	site of nutrient digestion and storage; some function in motility

ocean contained virtually no oxygen. Violent electric storms occurred often, and lightning repeatedly pierced the atmosphere. More than a half-century ago, scientists began proposing that the energy imparted on atoms in the atmosphere might force the creation of new molecular structures, perhaps making the conditions right for producing life's building blocks: amino acids, sugars, fats, and nucleic acids.

In the 1950s, the University of Chicago professor Harold Urey and the graduate student Stanley Miller devised an experiment to test the theory that electrical energy could possibly cause simple molecules to combine and form one or more of life's building blocks. The experiment, brilliant in its simplicity, involved two flasks connected by glass tubing. One flask contained water to represent the ocean, and the second flask contained the three gases of early Earth's atmosphere. The author Joseph Panno described the Urey-Miller experiment, in his 2005 *The Cell:* "They passed an electric charge through the flask containing the atmosphere to simulate lightning and heated the water flask to produce the high temperatures of young Earth. After a week, Urey and Miller tested the contents of the flask and to their great surprise found that the water contained large amounts of amino acids. By varying the conditions of their experiment they were able to produce a wide variety of organic compounds, including nucleic acids, sugars, and fats." Miller published the findings, in 1953, in a two-page article, "A Production of Amino Acids under Possible Primitive Earth Conditions."

Once simple carbon and nitrogen compounds began forming, Earth's ocean served as a setting for larger compounds to develop. The oceans took one billion years to build up a store of proteins made from amino acids, the nucleic acids DNA and RNA, and complex carbon compounds, mainly fats and carbohydrates. The fatty compounds called *phospholipids* and various carbohydrates combined and broke apart, again and again, in the turbulent oceans of 3.5 billion years ago. The phospholipids' tendency to repel water and the carbohydrates' tendency to be miscible with water worked in conjunction: Tiny water-filled vesicles surrounded by a layer of phospholipids and carbohydrates formed. The surrounding layer was probably oriented so the most water-repellent portions of the phospholipids pointed into the middle of the layer, and more water-compatible portions plus the carbohydrates were positioned with exposure to the watery vesicle interior and the ocean on the outside. On occasion, as the vesicles formed, they enclosed bits of proteins and nucleic acids inside the vesicle's surrounding layer. The world now possessed microscopic packets of nucleic acids and proteins, protected from the environment by an organized layer that allowed some molecules inside the vesicle and repelled other molecules. As a result, a rudimentary prokaryotic cell was created.

Prokaryotes underwent evolution from simple packets of enzymes—made of proteins—to more complex structures with specialties that enabled them to survive their surroundings. A significant step in evolution took place when cyanobacteria developed. These photosynthetic bacteria possessed the ability to capture the Sun's rays for energy and to make cell constituents. Cyanobacteria began producing oxygen that slowly but steadily shifted the composition of Earth's atmosphere. By this action, the cyanobacteria bridged a gap in evolution between simple anaerobic prokaryotes and more complex aerobic cells.

In the 1960s, the University of Massachusetts biologist Lynn Margulis focused her studies of cell structure on plausible ways in which eukaryotes may have developed from prokaryotes. Margulis perhaps could not dismiss the observation that the mitochondria inside plant and animal cells looked remarkably like bacteria; they shared a similar shape and size, and mitochondria contained an efficient energy-producing mechanism. In plants, cells chloroplasts that carried out photosynthesis also seemed to represent an almost-independent energy factory tucked in the cytoplasm amid other cellular activities and organelles.

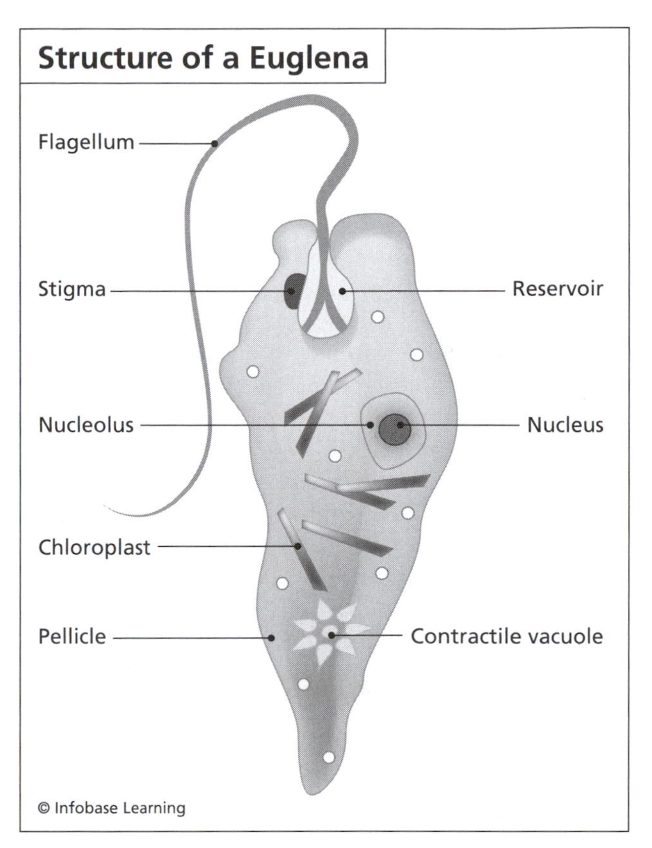

***Euglena* cells are motile, photosynthetic eukaryotes. The pellicle lies inside the outer membrane and provides flexibility. The stigma is a light-sensitive area that allows the cell to respond to light and dark. The contractile vacuole continually collects excess water inside the cell and expels it into the surroundings in order to maintain the interior pressure.**

Margulis had been a precocious scientist for her entire career—she started college at age 15 and held her own in theoretical discussions with graduate students and professors. She believed that the unique genes inside mitochondria and chloroplasts were a key piece in completing the puzzle of eukaryote evolution. In 1967, Margulis published an article explaining her theory that a fairly rapid rise in Earth's oxygen levels occurred two billion years ago, as a result of increasing photosynthetic cyanobacteria. Anaerobic cells struggled to survive, perhaps developing vulnerabilities. On occasion, a smaller aerobic bacterium may have invaded the anaerobic cell. Most of these invasions undoubtedly led to the death of many anaerobes, but, in some instances, the aerobic invader lived inside a living anaerobe as a parasite. In time, the relationship turned into symbiosis rather than parasitism: The aerobe produced energy for the anaerobe, and the host anaerobe provided a protective structure for the aerobe.

The invaders of some of the larger host cells evolved into mitochondria. The anaerobes that had been invaded by cyanobacteria probably provided the setting for the evolution of chloroplasts. Fifteen journals rejected Margulis's article, arguing that the theory had no fossil records to support it, before the *Journal of Theoretical Biology* accepted it. Scientists debated the possibility of the relationship proposed by Margulis, a relationship in which one organism lives inside another.

Advances in electron microscopy and ever-improving techniques in nucleic acid analysis amassed results to support Margulis and counter her detractors. The new studies showed that, in addition to size and shape similar to bacteria, mitochondria and chloroplasts contain circular strands of DNA, as found in prokaryotes. The organelles can, furthermore, reproduce independently of their host cell, and their ribosomes resemble prokaryotic ribosomes. By 1970, Margulis used this new evidence and results from her continued studies to publish the book *Origin of Eukaryotic Cells*. Several decades of work by many laboratories have further proven the theory of eukaryotic development now known as the *endosymbiotic theory*. The evolutionary biologist Richard Dawkins was quoted, in the 1995 book *The Third Culture: Beyond the Scientific Revolution*, as saying, "I greatly admire Lynn Margulis's sheer courage and stamina in sticking by the endosymbiosis theory, and carrying it through from being an unorthodoxy to an orthodoxy. I'm referring to the theory that the eukaryotic cell is a symbiotic union of primitive prokaryotic cells. This is one of the great achievements of twentieth-century evolutionary biology, and I greatly admire her for it."

The table below provides a capsule view of milestones in the evolution of eukaryotic cells.

Evolution of Eukaryotic Cells

Milestone	Billions of Years Ago
birth of Earth's solar system	4.6
gases escaping from molten planet form the atmosphere	4.5
oceans form	4.2
oldest known rocks appear	4.0
fossil records indicate oldest known life	3.8
bacteria begin to diversify	3.8
cyanobacteria appear	3.7
first eukaryotes appear	2.7
multi-celled organisms appear	0.9

The endosymbiotic theory clarifies many of the questions regarding how and why eukaryotic cells have developed to a greater complexity than prokaryotes. In evolution, eukaryotic cells joined into groupings in which individual cells could communicate with each other and, eventually, evolve specialized activities in a larger multicellular organism. Eukaryotic cells in higher organisms have evolved to such a degree of interdependence that a single cell would probably die if removed from the organism. Although prokaryotes, too, form communities and give evidence of different simple forms of communication, any prokaryote removed from its group could continue living as an independent entity. The table on page 285 indicates the major differences between eukaryotic and prokaryotic cells.

The eukaryotic cell represents the natural development in evolution, from simple one-celled microorganisms, to more complex cells, and finally to multicellular organisms. Eukaryotic cells make up an important part of the study of microbial cell structure and function.

See also CYANOBACTERIA; ORGANELLE; PROKARYOTE.

Further Reading

Brockman, John. *The Third Culture: Beyond the Scientific Revolution*. New York: Touchstone, 1995.

Margulis, Lynn. *Origin of Eukaryotic Cells*. New Haven, Conn.: Yale University Press, 1970.

Miller, Stanley L. "A Production of Amino Acids under Possible Primitive Earth Conditions." *Science* 117 (1953): 528–529. Available online. URL: www.issol.org/miller/miller1953.pdf. Accessed August 10, 2009.

Panno, Joseph. *The Cell*. Rev. ed. New York: Facts On File, 2010.

Sagan, Lynn. "The Origin of Mitosing Eukaryotic Cells." *Journal of Theoretical Biology* 14 (1967): 255–274.

extremophile An extremophile is any microorganism that lives in an environment characterized by extreme physical features, such as very high temperatures or very acidic conditions.

Biological cells require an optimal range of physical factors that define their environment. Cells within multicellular organisms derive these conditions from the organism's body. Cells that make up plant and animal tissue, thus, have their needs conveniently met by the organism as a whole. Parasites and pathogens that invade higher organisms also take advantage of the warm temperatures, moisture, and constant supply of nutrients to support them. But the vast majority of microorganisms on Earth live in an environment where they must cope with the physical features at hand. They must move to comfortable temperatures, regulate their pH even in very acid or alkaline habitats, and expel excess salts. Bacteria that cannot do these things cannot survive, that is, except extremo-

Differences between Eukaryotes and Prokaryotes

Characteristic	Prokaryote	Eukaryote
size	0.2–2.0-μm diameter	10–100-μm diameter
cell wall	chemically complex and with peptidoglycan	chemically simple and no peptidoglycan
flagella	composed of one fiber	composed of multiple tubules
membrane contains sterols	usually no	yes
ribosomes	70S in cytoplasm	80S, 70S in organelles
endoplasmic reticulum	absent	present
Golgi apparatus	absent	present
lysosomes and peroxisomes	absent	present
mitochondria and chloroplasts	absent	present
genetic recombination	DNA transfer between cells	meiosis and combining of gametes
cell division	binary fission	mitosis
cell differentiation	minimal	tissues and organs
	GENETIC MATERIAL:	
DNA complexed with histones	no	yes
site of DNA and its replication	cytoplasm	membrane-enclosed nucleus
number of chromosomes	one	more than one
chromosome organization	single circular chromosome	organized folding
plasmids	often present	absent

philes. Extremophiles not only seek places on Earth that present very difficult living conditions; they have evolved so that they can thrive in no other place.

Any range of conditions that a microorganism requires for life is called its *optimal range*. For example, *Escherichia coli* has an optimal temperature range of 68–113°F (20–45°C) and an optimal pH range of 6.0–8.0. If *E. coli* finds itself in conditions far outside either of those ranges, it will cease growing or even die. Other optimal conditions require a significant amount of activity by a cell. Microorganisms must work to maintain the perfect osmotic pressure inside the cell relative to the outside. In order to maintain a certain pressure within the cell membrane, the cell constantly regulates the amounts of ions, sugars, and other compounds in its cytoplasm to adjust to the pressure of the immediate surroundings. Extremophiles either make these adjustments or have evolved to contain physical features that help them survive in extreme conditions.

The main parameters that define extreme environments are the following: temperature, acidity, alkalinity, high temperature–high acidity, high temperature–high pressure, high salt concentration, toxic chemicals, osmotic pressure, hydrostatic pressure, heavy metals, low water availability, and radiation. Most extremophiles are in domain Archaea, and in fact the archaea and extremophiles are often thought of as synonymous. But the strong connections between the archaea and extremophiles may be misleading, because extremophiles may also be bacteria, yeasts, fungi, or algae.

Archaea

Archaea make up a diverse group of microorganisms that are neither bacteria nor eukaryotes but share characteristics of each of these cell types. Archaea belong to their own domain, which also contains the vast majority of extremophiles. The genetic makeup of archaea suggests that these microorganisms split from bacteria early in their evolution and subsequently developed their own distinctive characteristics, especially in their cell wall and membrane.

Archaean cell walls lack the long polymer peptidoglycan common in bacterial cell walls, but many possess a pepitodoglycanlike molecule called *pseudomurein.* Some archaea also possess glycoproteins, which are proteins with sugar side units, or a mixture of polysaccharides and proteins. These cell wall constituents might lend special durability to extremophile cells.

Archaean membranes possess unusual features that contribute to the cell's stability. First, the membrane contains a mixture of branched hydrocarbons. Some of the branches form ring structures that strengthen membranes. Side branches containing a 20-carbon compound called *isoprene* also reach completely across the membrane, from the inner membrane surface to the outer surface. No other biological membrane possesses this structure, called a *transmembrane phospholipid,* which probably helps strengthen the membrane. Second, large lipid molecules based on the 30-carbon structure of squalene also make the archaean membrane more rigid than other biological membranes. Squalene additionally withstands very high temperatures.

Because archaea live in extreme environments that would kill most other living things, they have developed unusual types of metabolism. For example, methanogenic archaea exist by using only methane gas as their carbon source, while others such as *Thermoproteus* absorb carbon dioxide. Some archaea additionally have the ability to store carbon in their cells as glycogen, the main carbon storage compound in higher multicellular organisms. Halophiles, meanwhile, operate a metabolism based on the constant balancing of sodium concentrations outside the cell with other ions inside the cell.

Archaea are more difficult to recover from the environment and study in a laboratory than other microorganisms, so many questions about their metabolism remain unanswered. But because of their special abilities, archaean enzyme systems may possess characteristics that will someday benefit industrial microbiology.

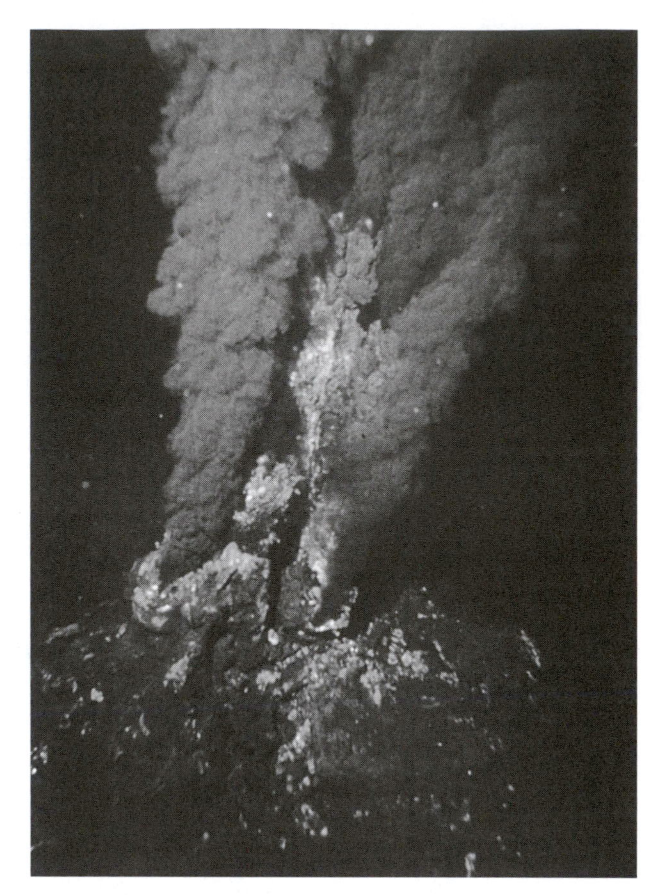

Thermal vents called black smokers on the ocean floor emit extremely hot liquids and gases. Only extremophiles can live at or near these vents. ***(NOAA/Department of Commerce)***

TEMPERATURE EXTREMOPHILES

Microorganisms that live in the same temperature range as humans are called *mesophiles.* Because mesophiles occupy a middle zone between extremes in temperature, they do not belong to the extremophiles. The extremophiles live at temperatures above or below the mesophile range, described in the table.

The extremely cold environments that favor psychrophiles may be found in glacial ice, in Arctic and Antarctic sea ice, in permanently cold climates that thaw for only a few weeks a year, or in the deepest and coldest parts of the ocean. Psychrotrophs prefer less extreme temperatures, such as the temperature inside a refrigerator. For this reason, psychrophiles represent many of the contaminants that spoil refrigerated foods or cause food-borne illness. Psychrophilic microorganisms may be further divided into the obligate psychrophiles and facultative psychrophiles. Obligate species are microorganisms that cannot survive at temperatures of greater than about 72°F (22°C). Facultative species prefer their optimal range, but, if left with no other choice, they can survive for a time at temperatures up to 100°F (38°C).

Thermophilic, or hot-temperature, microorganisms live in hot springs such as those in Yellowstone National Park (see the color insert, page C-3 [bottom]) and at the ocean bottom, where hydrothermal vents spew heated materials from deep in Earth's core.

Although many mesophiles can slow their metabolism when frozen, they do not die in the frozen state and can recover if returned to their optimal range. But elevated temperatures kill mesophiles with no opportunity for recovery, making the thermophilic survival mechanisms all the more fascinating. Scientists in research and in industry have investigated thermophiles as potential sources of new heat-resistant products. Thermophilic enzymes, for instance, offer benefits to manufacturers and biotechnology. Manufacturers often run processes that require energy input in the form of heat. By finding natural enzymes and catalysts (molecules that help enzymes run more energetically), manufacturers may cut production costs by substituting natural substances for harsh chemicals. Soap and detergent manufacturers also depend on enzymes that work at high temperatures for degrading proteins, fats, and other components of stains. Manufacturers of livestock feed additives similarly use acid-tolerant thermophilic enzymes to help nutrient digestion inside the warm and acidic confines of an animal's stomach. Finally, biotechnology has been using an enzyme from a hyperthermophile that has enabled the development of a technique called *polymerase chain reaction* (PCR).

More than 450 thermophiles have been isolated—of these microorganisms, *Thermus aquaticus* has contributed a pivotal piece to science and even crime solving. *T. aquaticus* is a hyperthermophile that lives in hot springs. In the 1980s, *T. aquaticus* was found to possess an enzyme called *Taq* polymerase, which can make millions of copies of a single gene under the correct laboratory conditions. This capability led to the development of PCR, a technique now used for multiplying the tiniest bits of deoxyribonucleic acid (DNA) into larger amounts that scientists can analyze. PCR technology contributes to disease diagnosis, scanning for certain genetic traits, and linking of suspects to a crime scene through DNA testing. The reason for *T. aquaticus*'s value in PCR resides in the ability of *Taq* polymerase to withstand temperatures up to 130°F (55°C). The PCR procedure requires a heating step in the DNA reproduction process that only a heat-resistant enzyme can perform.

Taq polymerase is one of a group of enzymes from extremophiles called extremoenzymes, because they maintain their activity even under the most challenging physical conditions. Extremoenzymes have been extracted from psychrophiles just as they have from thermophiles. Psychrophilic enzymes have allowed the development of cold-temperature laundry detergents, and cold-temperature enzymes that additionally can work in alkaline conditions give an added benefit because detergents create alkaline conditions when dissolved in water.

Some environmental microbiologists explore the remote and very unique environment of the deep ocean. One of the harshest of places on Earth for the existence of life occurs in marine trenches that lace the ocean floor. These ravines often contain deep sea hydrothermal vents called black smokers and white smokers. Cynthia Needham and her coauthors of *Intimate Strangers: Unseen Life on Earth* wrote, in 2000, "Hydrothermal vents offer a particularly interesting hunting ground [for microbiologists]. They occur at various spots along the Pacific and Atlantic sea floors, near the points where hot basalt and magma are very near the floor surface. . . . As the crust of the earth opens, underwater volcanoes form, spewing out water heated to 350°C [662°F] or more and laden with minerals." Needham pointed out that these smokers "are surrounded by a complex and bizarre community of living forms, including tube worms 2 meters [6.6 feet] in length, giant clams, bright red shrimp, mussels—and, of course, microbes." The extreme hardiness of these microorganisms and their enzyme systems may soon offer new uses in various sciences.

Types of Microorganisms That Grow in Different Temperatures

Name	Temperature Range
psychrophiles	17–65°F (-8–19°C)
psychrotrophs	32–86°F (0–30°C)
mesophiles	50–122°F (10–50°C)
thermophiles	104–162°F (40–72°C)
hyperthermophiles	150–250°F (65–121°C)

pH EXTREMOPHILES

Most microorganisms, called *neutrophiles,* prefer and can maintain an interior pH between 6.5 and 7.5 regardless of the pH surrounding the cell. Different microorganisms regulate pH by three main mechanisms: (1) ion pumps, (2) protein-based regulation systems, and (3) pH-regulating excretions.

Ion pumps consist of a system based in the cell membrane that exchanges ions from the outside to the inside or in the opposite direction for balancing intracellular pH. Neutrophiles use a potassium-proton pump, exchanging a potassium ion (K^+) from the outside with a proton (H^+) from inside the cell to raise the interior pH and pumping protons into the cell to lower the interior pH.

Protein-based regulation systems activate in very acidic conditions to shunt large amounts of protons

pH Requirements in Microbiology

Type	Description	Example Habitats	Example Microorganisms
acidophiles	require pH 0–5	acidic hot springs, mining drainage	*Sulfolobus, Ferroplasma, Lactobacillus, Aspergillus*
neutrophiles	require pH 5.5–8.0	land and water, plants, animals	*E. coli, Pseudomonas, Euglena, Paramecium*
alkaliphiles	require pH 8.5–11.5	seawater, soda lakes (high-carbonate), soaps	*Bacillus alcalophilus*

out of the cell. These systems rely on the synthesis of a protein that aids in forcing protons into an environment already high in protons. This system requires the energy in adenosine triphosphate (ATP) to move protons up a concentration, much like trying to swim upstream.

Other microorganisms excrete acidic compounds if conditions become too alkaline. The excretions may be organic acids produced naturally from the metabolism of sugars, or they may be acidic waste compounds.

Acidophiles are microorganisms that live in environments of pH 0–4. These species manage to maintain neutral conditions inside their cells by using a combination of the mechanisms described here. Pumps and protein synthesis require energy expenditure from the cell, but the cell membrane also provides a passive, non-energy-demanding system by acting as an impervious barrier to passage of protons into the interior. Alkaliphiles, conversely, need basic conditions in which the pH ranges from 8.5 to 11.5. They regulate cellular pH similarly to acidophiles, but they use a sodium-proton pump, rather than a potassium-proton pump. The table above summarizes various pH requirements among microorganisms.

Some microorganisms require more specific pH ranges than those defined by the general acidophile and alkaliphile groups. For example, the acidophile *Lactobacillus acidophilus* grows best at pH 5.5–6.0, and the acidophile *Ferroplasma* exists at pH 0. Among alkaliphiles, *Nitrosomonas* lives in seawater at pH 8, whereas *Bacillus alcalophilus* lives in places approaching pH 11.5.

SALT AND OSMOTIC PRESSURE EXTREMOPHILES

Environments having high salt concentrations also contain high osmotic pressure outside the cell. Osmotic pressure is the force needed to move a dissolved substance (solute) from a site of low solute concentration to a site with a high concentration. Microbial cell membranes are called selectively permeable membranes because they allow some substances, such as water, to pass freely across the membrane barrier but prevent other substances from leaking out of the cell or entering the cell. Osmotic pressure is the pressure needed to prevent water from flowing across the selectively permeable membrane. In environments of high salt levels, a microbial cell must spend a considerable part of its activity maintaining an osmotic pressure that enables it to survive. Microorganisms that survive in high-salt concentrations are called halophiles, and many of the same microorganisms also thrive in places of extreme osmotic pressures, called *osmophiles*.

High-salt environments contain hypertonic conditions, also called hyperosmotic conditions: situations in which the concentration of solutes is higher outside the cell than inside the cell. These conditions can be found in bodies of water such as Utah's Great Salt Lake and the Dead Sea in the Middle East, brine layers that result from water evaporation, and soils of extremely high mineral content. Halophiles such as *Halococcus* and *Halobacterium* can live in water that is almost saturated in salt (NaCl). Halophiles accomplish this specialty by constantly balancing their interior solute concentrations with the outer sodium ion (Na^+) concentrations.

Halophiles use two mechanisms for surviving hyperosmotic conditions. The first system involves holding ions or other molecules inside the cell at levels equivalent to the surrounding sodium concentration. Prokaryotes tend to use potassium ions (K^+) to maintain interior osmotic pressure; eukaryotes often use glycerol to do the same. These microorganisms retain high levels of potassium or glycerol, while pushing sodium out of the cell. In each case, the cell must expend energy to run this mechanism. Microorganisms may use other compounds to maintain osmotic pressure, compounds such as betaine, choline, glutamine, proline, or sugars. But high levels of these substances inside the cell can interfere with protein function, so halophiles maintain a precarious balance between too much and too little of inter-

Salt Tolerance in Microorganisms

Type of Halophile	Salt Range (M)	Optimal Range (M)
nonhalophile	0–1.0	less than 0.2
slight halophile	0.2–2.0	0.35
halophile	1.4–4.0	2.5
extreme halophile	2.0–5.2	more than 3.0

Note: M = molarity of a solution = number of moles of solid per number of liters of solution

nal solutes to maintain osmotic pressure. Betaine, choline, and proline are called *compatible solutes* because they cause less damage to cellular proteins than other compounds.

Some halophiles use a second mechanism to manage high-salt environments. This mechanism is based on cellular proteins that contain low proportions of nonpolar amino acids (alanine, glycine, isoleucine, leucine, methionine, phenylalanine, proline, and valine). Salt helps activate these proteins, so, in this way, the microorganism puts the high salt levels to good use in its metabolism.

Salty environments vary by the levels of salt they normally contain. The table above illustrates differences in ranges of salt tolerance among microorganisms.

Seawater contains about 40 different ions that contribute to molarity, and furthermore molarity varies with varying densities in different parts of the ocean. For this reason, biologists usually express the ocean's salt concentration as a percentage: The average salt concentration of the ocean is 3.5 percent. (Salinity equals the number of grams of a solid dissolved in 1 kg of water, so, in the case of seawater, salinity is 35 percent.) Once the salt concentration of a body of water has been calculated, the environment's waters can be grouped according to salt level, as follows: freshwater, less than 0.005 percent; brackish water, 0.005–3.0 percent; marine water or seawater, 3.0–5.0 percent; and brine, more than 5.0 percent.

OTHER EXTREME ENVIRONMENTS

Extremophiles have been found in habitats that provide living conditions just as implausible as those found at hydrothermal vents, in brine flats, or 300 feet (91 m) deep in ice. In many cases, an extremophile confronts multiple environmental challenges because extreme environments are often extreme in more than one way. For example, organisms that live near black smokers also endure very high temperatures, strong acids, as well as enormous hydrostatic pressure, which is the pressure found in deep water. Psychrophiles that live in the ocean depths but away from hydrothermal vents must live in conditions of perpetual cold and high pressure. Scientists have recently discovered an endolithic (an organism that lives in rock) halophile in a remote dry lake bed in Tibet. Many more specialized extremophiles are undoubtedly waiting to be discovered.

The table describes the diversity among other extremophiles. In these extremophiles, as in temperature or pH extremophiles, various microorganisms occupy different ranges within an extreme environment. For example, barophiles contain barotolerant species that withstand pressures slightly higher than atmospheric pressure, barophilic species survive under high hydrostatic pressure, and extreme barophiles require very high pressures.

Extremophiles use common mechanisms for regulating internal pH, tolerating extreme temperatures, and avoiding damage from chemicals, yet extremophiles also possess unique attributes adapted to the circumstances in which they live. As scientists have explored a wider selection of extreme environments, they have detected new extremophiles in these places. Studies on recognized extremophiles, as well as newly discovered species, cover a diversity of microbial activities. These studies not only help learn about how extremophiles survive, but also give researchers details on how mesophiles exist. Extremophile studies have contributed to biodiversity, biotechnology, energy systems, enzymes, evolution, membranes, metabolism, transport systems, and ultrastructure.

In 2006, scientists at Princeton University discovered a new bacterium they named *Desulforudis audaxviator,* described by the *San Francisco Chronicle* reporter David Perlman as a microorganism that "lives in complete isolation a mile and three-quarters down in a South African gold mine where darkness is total, oxygen is nonexistent and the temperature

Other Types of Extremophiles

Extremophile	Description	Example Habitats
anaerobe	• facultative: can live without oxygen • obligate: requires the absence of oxygen	sediments, bogs, deep in Earth's crust, inside animal digestive tract, oxygen-free parts of ocean
barophile (piezophile)	• barotolerant: 1–4 atmospheres (atm) • barophilic: up to 400 atm • extreme barophilic: require up to 700 atm	ocean depths or 2.5 miles (4 km) or more
endolith	• cryptendolithic: rocks at Earth's surface • subsurface: caves, caverns, underground water sources • deep-biosphere: deep mines, beneath ocean floor	inside rocks from earth surface to several miles deep
toxitolerant	• growth in presence of toxic chemicals, heavy metals, or radioactive materials	industrial waste sites, mining operations, nuclear waste sites
xerophile	• growth in extremely dry environments	deserts, Antarctica

is 140 degrees Fahrenheit [60°C]." Two years later, University of California scientists had deciphered the microorganism's genes and found *D. audaxviator* contained DNA sequences from bacteria and from archaea.

David Des Marais, a geochemist at the National Aeronautics and Space Administration (NASA), commented on *D. audaxviator:* "Where did this critter come from? Did it live on the surface millions of years ago, and if things got nasty up top, did it find its way nearly two miles [3.2 km] down and survive there on its own in total isolation? Or did it always live down there?" Scientists such as Des Marais have suggested that information learned about extremophiles may offer clues in a new science called *astrobiology,* the study of possible life on other plants. "It can teach us a lot about survival in an extreme environment that we still don't know enough about," Des Marais added, "the kind of environment that microbes on Mars might have migrated down into when the surface was no longer habitable." The relationship of extremophiles and astrobiology offers a vast territory for science to explore.

See also ARCHAEA; ENVIRONMENTAL MICROBIOLOGY; OSMOTIC PRESSURE; POLYMERASE CHAIN REACTION.

Further Reading

Brock, Thomas D. "Life at High Temperatures." *Science* 158 (1967): 1,012–1,019.

———. *Thermophilic Microorganisms and Life at High Temperatures.* New York: Springer-Verlag, 1978. Available online. URL: http://digicoll.library.wisc.edu/cgi-bin/Science/Science-idx?id=Science.BrockTher. Accessed January 29, 2010.

Kong, Fanjing, Mianping Zheng, Alian Wang, and N. N. Ma. "Endolithic Halophiles Found in Evaporite Salts on Tibet Plateau as a Potential Analog for Martian Life in Saline Environment." Presented at 40th Lunar and Planetary Science Conference, The Woodlands, Tex., March 23–27, 2009. Available online. URL: www.lpi.usra.edu/meetings/lpsc2009/pdf/1216.pdf. Accessed August 9, 2009.

Madigan, Michael T., and Barry L. Marrs. "Extremophiles." *Scientific American,* April 1997.

Microbial Life Educational Resources. "Who Are the Extremophiles?" Available online. URL: http://serc.carleton.edu/microbelife/extreme/extremophiles.html. Accessed August 9, 2009.

Needham, Cynthia, Mahlon Hoagland, Kenneth McPherson, and Bert Dodson. *Intimate Strangers: Unseen Life on Earth.* Washington, D.C.: American Society for Microbiology Press, 2000.

Perlman, David. "Microbe's DNA Code Revealed." *San Francisco Chronicle,* 11 October 2008.

fermentation The term *fermentation* has two major meanings in microbiology. First, fermentation is a type of metabolism in which enzymes degrade sugars in the absence of oxygen and with the production of organic acids and the energy storage compound adenosine triphosphate (ATP). Fermentation is, second, any industrial process in which microorganisms make a commercial product. In some cases, substances made by microorganisms are the desired end product of an industrial fermentation, and, in other cases, the microbial cells themselves serve as the product. The table describes each of these two aspects of fermentation in more detail.

FERMENTATION STUDIES IN MICROBIOLOGY

Knowledge of fermentation dates to ancient societies. People have long realized that reactions that take place in certain foods could change the consistency and taste of those foods. In some cases, the food spoiled and could not be used. In other cases, the food became altered but remained palatable and would, furthermore, last longer in the fermented form than as a fresh food. People learned that fermented dairy foods, vegetables, and fruit juices (wine) provided a good way to keep foods for long periods.

Not until the mid-1800s, did individuals begin examining fermentation reactions in a scientific way. The French microbiologist Louis Pasteur (1822–95) led this new phase in biology. Local merchants had turned to Pasteur to find the reason why their wine and beer batches sometimes spoiled and to find a method for preventing the spoilage during the shipment over long distances. By studying the actual steps and the microorganisms present in fermentations, Pasteur revealed much about fermentation that had never before been known. In the process, Pasteur initiated the science of microbiology by, first, discovering the yeasts responsible for the juice and grain fermentations and, second, studying the specific activities of these yeasts. Prior to Pasteur's work, scientists rarely linked the type of microorganism in a substance to the biochemistry taking place there.

During Pasteur's early studies on lactic acid fermentations, in 1857, he remarked, "In the same way that there exists an alcoholic ferment, the yeast of beer, which is always found wherever sugar is decomposed into alcohol and carbon dioxide, there also exists a particular ferment, the lactic yeast, which is always present when sugar becomes converted into lactic acid." Pasteur noticed also that the "lactic yeast" required certain nitrogen compounds in order to carry out its fermentation. When studying the unknown microorganism, Pasteur said, "under the microscope it is seen to form tiny globules or small objects. . . . These globules are much smaller than those of beer yeast and move actively by Brownian [random] movement." Later microbiologists deduced that Pasteur had observed *Streptococcus* bacteria.

Pasteur also became the first scientist to study alcohol fermentation by yeasts in a quantitative way. He did this by measuring the weight of the cells, in addition to their nitrogen and carbon content. Significantly, he showed that by adding a nitrogen source (compounds containing ammonium), he could increase the weight of the yeasts and enhance the fermentation process. Finally, he followed through on a request by wine and beer makers to find a way to prevent spoilage. Pasteur recommended that the merchants heat their products to kill the yeasts after

Fermentation Uses

Microbial Metabolism	Role in Biology
any biological process occurring in the absence of oxygen	anaerobic metabolism of organic matter
use of an organic compound as an electron donor and acceptor	glycolysis (conversion of glucose to pyruvic acid)
use of an organic compound to reduce (add electrons) to another compound, and the same partially degraded compound to oxidize (remove electrons) other substances	energy production in anaerobes
type of growth that depends on phosphorylation (addition of phosphate chemical groups) of sugars	ATP production
Industrial Microbiology	**Application**
production of alcohols	production of wine, beer, and liquor
production of organic acids and enzymes	fermented food production
food spoilage	food industry
large-scale production of acids, alcohols, enzymes, vitamins, antibiotics, and flavorings	biotechnology, food additives
large-scale production of drugs	biopharmaceuticals (biology-based drugs)

the fermentation process was complete. This method would later become known as *pasteurization.*

FERMENTATION IN MICROBIAL ENERGY PRODUCTION

Energy generation in prokaryotes and eukaryotes involves the transfer of electrons from one compound to the next. Electrons move from a reduced compound to an oxidized compound, which, in turn, becomes reduced when another electron is added to it. This series of reduction-oxidization steps generates energy that cells store in the compound adenosine triphosphate (ATP). The electrons, meanwhile, progress toward a final compound, which must be eliminated by the cell in order for energy metabolism to continue.

Aerobic cells uses oxygen as what biochemists call the final *electron acceptor,* or electron sink. Anaerobic microorganisms that carry out fermentation use a variety of compounds other than oxygen as a final electron acceptor. As a result, several different types of fermentation convert either organic or inorganic compounds into a wide range of end products. Industrial microbiology has taken advantage of the versatility of fermentation microorganisms for making a wide variety of end products.

The core reaction in any fermentation involves the conversion of an organic compound, called a *substrate,* mainly to acids and alcohols. Any fermentation cannot progress without an auxiliary compound called *nicotinamide adenine dinucleotide* (NAD), also referred to as coenzyme I. NAD functions as a carrier for protons (H^+), in the various reduction-oxidation reactions in fermentation. When NAD receives an electron, a proton follows, because of its attraction to the electron's negative charge. The transfer of an electron changes NAD to NADH, in which *H* represents the proton. The sum of the fermentation reactions can be expressed in the following equations:

$$\text{substrate} + \text{electron} + \text{NAD} \rightarrow \text{oxidized product} + \text{NADH}$$

$$\text{substrate} + \text{NADH} \rightarrow \text{reduced product} + \text{NAD} + \text{electron}$$

Fermentation begins with a six-carbon sugar such as glucose, which is converted through several reactions to two three-carbon pyruvate molecules. Pyruvate, then, serves as the substrate for a variety of pathways that further ferment the compound. The most-studied fermentations described in the table on page 293 are named for their main end products.

The preceding discussion suggests that much of a microorganism's metabolism is devoted to shuttling elec-

Common Microbial Fermentations from Pyruvate

Fermentation	Other Major Products	Microorganisms
acetone-butanol	butyric acid, carbon dioxide	*Clostridium*
2,3-butanediol	lactic acid, formic acid, carbon dioxide	*Enterobacter, Bacillus*
butyric acid	acetone, isopropyl alcohol, carbon dioxide	*Butyrivibrio, Clostridium, Fusobacterium*
ethanol	acetic acid, lactic acid, succinic acid, carbon dioxide, hydrogen	*Zymomonas*
ethanol (yeast)	carbon dioxide, glycerol, lactic acid, acetic acid	*Saccharomyces, Aspergillus, Mucor*
formic acid	carbon dioxide, hydrogen	enteric bacteria
homoacetate	hydrogen	*Acetobacterium, Clostridium*
isopropanol	carbon dioxide	*Clostridium*
lactic acid	trace amounts	*Lactobacillus, Streptococcus, Bacillus*
mixed-acid	acetic acid, lactic acid, formic acid, succinic acid, ethanol, carbon dioxide	enteric bacteria
propionic acid	acetic acid, carbon dioxide, hydrogen	*Propionibacterium*

trons from one compound to the next, for the purpose of producing energy the cell can use. The metabolic processes fermentation and respiration both serve this purpose, but respiration differs from fermentation by using oxygen as an external receptor of electrons, that is, its electron sink. Because fermentation does not use oxygen, the products of fermentation have the same oxidation state as the substrate. This characteristic makes fermentation a much less efficient route than respiration for supplying energy to the cell. The table below compares the main features of anaerobic fermentation and aerobic respiration. Whether using fermentation or respiration, microorganisms called *heterotrophs* can employ any of the following organic compounds as the starting substrate that will ultimately give energy to the cell: sugars, amino acids, organic acids, and the nucleic acid bases purine and pyrimidine.

Aerobic metabolism produces many times more ATP, and thus energy for the cell, compared with fermentation. In aerobic environments, aerobic microorganisms have a distinct advantage over anaerobic microorganisms.

Eukaryotic aerobic cells are slightly less efficient than prokaryotic aerobes because of the

Anaerobic Fermentation versus Aerobic Respiration

Final Hydrogen (and electron) Acceptor	Main Energy-Producing Processes	Main End Products	ATP Produced per Glucose Molecule
	FERMENTATION		
organic compounds	phosphorylation of organic molecules	organic acids, alcohols, carbon dioxide, hydrogen	2
	RESPIRATION		
oxygen	Krebs cycle and electron transport chain	carbon dioxide, water	38 (36 in eukaryotes)

Technologies Used in Industrial Fermentations

Technology	Tasks
preparation of organisms	isolation of a desired organism from others using aseptic techniques and optimal media and incubation conditions
laboratory-scale fermentations	development of continuous or noncontinuous (batch) cultures to determine optimal growth conditions, nutrients, additives, and methods for increasing cell yield
sterilization	elimination of all possible contamination of media and culture equipment
cell metabolism	understanding the metabolic pathways used for making a desired product; minimizing inhibitory conditions and maximizing conditions that increase yield
preparation and scale-up of inocula	conversion of laboratory-scale cultures to cultures of several thousand liters in volume
monitoring instrumentation	measurement of chemical and physical characteristics of bioreactor cultures using sensors, feedback and adjustment systems, mixing speed, and electronic controls
bioreactor operation	complete steps in inoculating a bioreactor, monitoring for optimal growth during the incubation, preventing contamination, and harvesting the end product

energy needed to transport molecules across organelle membranes. Prokaryotes do not possess membrane-enclosed organelles, so they are not hampered by this step. Nevertheless, it is clear that even though anaerobes have many different types of fermentations at their disposal, they live at a definite disadvantage compared with energy metabolism in aerobes.

INDUSTRIAL FERMENTATIONS

Fermentations in industry consist of any large-scale production of a desired product in a chamber called a *bioreactor* or fermenter. In industry, fermentation may not necessarily be an anaerobic process; *fermentation* simply refers to any large-scale microbial process. Industrial fermentations can involve microorganisms, animal cells, or plant cells.

The food production industry has the advantage of decades to centuries of experience in making fermented foods. Other younger industries, such as biotechnology and pharmaceuticals, have also developed products through fermentation. Lansing Prescott, John Harley, and Donald Klein wrote in *Microbiology,* in 2005, "The development of industrial fermentations requires appropriate culture media and the large-scale screening of microorganisms. Often years are needed to achieve optimum product years. Many isolates are tested for their ability to synthesize a new product in the desired quantity. Few are successful." Industrial fermentations can be difficult to perfect because they require expertise in a number of technologies within microbiology: laboratory-scale fermentations, preparation of organisms using aseptic techniques, sterilization, cell metabolism, preparation and scale-up of inocula, instrumentation of cultures, and bioreactor operation. The specialties of industrial fermentations are summarized in the table above.

Each industry that uses fermentations as part of its business includes its own specialized processes for making products. Some of the notable specialized fermentations that have been adopted by industries, primarily the food industry, are listed in the table on page 295. About 70 percent of the ingredients used in the food industry today are derived from industrial fermentations, as the table suggests.

Some of the nonfoods produced by fermentation, such as drugs, are commercial products themselves. Other nonfoods serve as components used in a particular industrial method, such as fermentation-produced protease enzymes used in turning hide into leather. Enzymes such as protease make up a large proportion of the biological substances made in industrial fermentations. These enzymes go to a variety of uses, in addition to leather processing, such as the following: laundry detergent additives, brewing, cheese production, paper bleaching, and candy making. The following list describes the main industrially made enzymes:

- Amylase made by *Aspergillus* mold or *Bacillus subtilis* bacteria digests starch.

- Cellulase made by *Trichoderma* yeast digests the fiber cellulose.

- Invertase made by *Saccharomyces cerevisiae* yeast cleaves the sugar sucrose into glucose and fructose.

- Lactase made by *Saccharomyces fragilis* breaks down lactose in dairy products.

- Lipase made by *Aspergillus* mold degrades fats.

- Oxidase made by *Aspergillus* serves as a bleaching agent.

- Pectinase made by *Aspergillus* degrades the fiber pectin found in fruits.

- Protease made by *Aspergillus* digests proteins.

- Rennin (chymosin) made by *Mucor* mold and *Escherichia coli* bacteria curdles milk proteins.

- Streptokinase made by *Streptococcus* breaks down blood clots.

Additional specialty enzymes made in industrial fermentations are phytase, used as an agricultural feed additive, and isomerase, used by chemical companies to rearrange chemical structures. These examples illustrate why enzyme production has grown into one of the most important aspects of industrial fermentation. Fermentations, furthermore, produce various enzyme inhibitors that have uses in medicine, such as the following examples:

- Acarbase inhibits alpha-amylase and in used in Type 2 diabetes treatment.

- Clavulanic acid reduces the resistance to penicillin by some bacteria.

- Fibrostatin C inhibits proline hydroxylase as a treatment for fibrosis.

Agriculture has made similar use of fermentation products as herbicides, insecticides, and fungicide.

WINE MAKING

Food production depends on many enzymes produced in commercial fermentations. Perhaps the best example of an industrial fermentation that makes a food product is the process of wine making. This fermentation's origins date to 5000 B.C.E., when farmers needed to preserve their crops. The extracts and juices could be inoculated with yeast and allowed to convert from a high-sugar liquid to one higher in alcohol that could last longer when stored. But how did ancient civilizations know to make this important leap? The wine historian Patrick McGovern said, "As intriguing and often exciting as the stories of the origins of viniculture (encompassing both viticulture—vine cultivation—and winemaking) are, this tangled 'vineyard' needs to be trod with caution. Many books on the history of wine give undue weight to one legend or another and rely on dubious translations." Although the origins of wine making remain mysterious, the basic process may have changed very little, for 7,000 years.

Early in the history of wine making, people probably realized that the entire process could be ruined if air entered the fermentation. The addition of oxygen made aerobic bacteria lingering in the fermentation mixture in low numbers (mainly *Acetobacter* and *Gluconobacter*) grow quickly and produce acetic acid from ethanol. Although the acid ruined the wine batch, it led to another usable end product: vinegar. Today's makers of vinegar products use acetic acid fermentations.

Wine making consists of two main steps: the *primary* (sense 1) fermentation and the *secondary* (sense 2) fermentation. The primary fermentation employs *Saccharomyces* yeast to convert the ample sugars present in the juice from grapes to ethanol—

Specialized Industrial Fermentations

Fermentation	Description
dihydroxyacetone	glycerol converted to dihydroxyacetone by *Gluconobacter* for the purpose of making product additives
flor fermentation	a second fermentation in the presence of air that follows primary alcohol fermentation in wine making for the purpose of making sherry
malolactic	conversion of malic acid to lactic acid by *Lactobacillus, Leuconostoc,* or *Pediococcus* in fermented foods (example: soy sauce) for the purpose of reducing acid taste
sorbose	conversion of D-sorbitol to L-sorbitol by *Acetobacter* as a step in making ascorbic acid (vitamin C)

Fermentations in Industrial Microbiology

Industry	Product	Microorganism	Type
agriculture	gibberellins—hormones that stimulate plant growth	*Gibberella fujikoroi*	fungus
biofuels	ethanol	*Zymomonas*	bacteria
	methane	*Methanothrix*	archaea
biotechnology	insulin, human growth hormone, somatostatin, interferons	bioengineered microorganisms	prokaryotic and eukaryotic cells
food production	amino acids	*Corynebacterium*	bacteria
	organic acids	*Aspergillus niger*	fungus
	vitamins	*Ashbya* (riboflavin)	fungus
		Streptomyces (B_{12})	bacteria
	dairy products	*Streptococcus, Lactobacillus*	bacteria
	sauerkraut	*Leuconostoc, Lactobacillus*	bacteria
	alcohol beverages	*Saccharomyces cerevisiae*	yeast
industrial materials	acetone	*Clostridrium acetobutylicum*	bacteria
	2,3-butanediol	*Enterobacter*	bacteria
	ethanol	*Saccharomyces cerevisiae*	yeast
	enzymes	various	bacteria, molds (fungus)
pharmaceuticals	antibiotics (penicillin)	*Penicillium*	fungus
	antibiotics (streptomycin)	*Streptomyces*	bacteria
	steroid conversions	*Arthrobacter*	bacteria
	streptokinase	*Streptococcus*	bacteria

this is an alcohol fermentation. Some varieties of grapes are known to produce high acid levels in the fermented juice. For this reason, winemakers put the liquid through a secondary malolactic fermentation. In this step, lactic acid bacteria such as *Lactobacillus* convert malic acid to lactic acid, resulting in better taste in the final product. Winemakers also add a small amount of a sulfite compound—potassium metabisulfite is common—to kill all the microorganisms that could potentially contaminate the process. Wine making thus consists of the following steps:

1. crushing grapes and recovering juice
2. primary fermentation with yeast to produce ethanol
3. secondary fermentation with bacteria to adjust acid content
4. settling and clarification
5. filtration
6. aging
7. bottling

Other industries have taken full advantage of the ability of microorganisms to produce a desired substance without the need for harsh chemicals and intense heating, both common in chemical manufacturing processes. The table describes major industrial products made by fermentations.

Fermentation is an essential aspect of microbial energy metabolism. Although fermentation is not as efficient as aerobic respiration in generating energy, it plays its own important role in the overall breakdown of organic matter on Earth.

See also ANAEROBE; BIOREACTOR; CULTURE; FOOD MICROBIOLOGY; INDUSTRIAL MICROBIOLOGY; METABOLISM.

Further Reading

Bowen, Richard. "Basic Fermentation Chemistry." Available online. URL: www.vivo.colostate.edu/hbooks/

pathphys/digestion/herbivores/ferment.html. Accessed April 10, 2009.

McGovern, Patrick. *Ancient Wine: The Search for the Origins of Viniculture.* Princeton, N.J.: Princeton University Press, 2005.

McNeil Brian, and Linda M. Harvey, eds. *Practical Fermentation Technology.* Chichester, England: John Wiley & Sons, 2008.

Pasteur, Louis. "Mémoire sur la Fermentation Appelée Lactique" (Report on the Lactic Acid Fermentation). *Comptes Rendus de l'Académie des Sciences* 45 (1857): 913–916. In *Milestones in Microbiology,* translated and edited by Thomas Brock. Washington, D.C.: American Society for Microbiology Press, 1961.

Prescott, Lansing M., John P. Harley, and Donald A. Klein. *Microbiology,* 6th ed. New York: McGraw-Hill, 2005.

filtration Filtration is a procedure in which liquid or air passes through a screenlike material perforated with tiny pores. This pore-containing material is the filter. Microbiologists use filtration for a variety of uses: separating large microorganisms from smaller ones, recovering extracellular products from cultures, removing large particulate matter from liquid medium, estimating particle size, and removing all microbial life from a liquid. When filtration is used for the latter purpose, the process is called *sterilization.*

In industry, the filtration of the air and water plays a critical part in manufacturing sterile drugs, such as injectable drugs. Specialized areas called *clean rooms* in drug manufacturing facilities rely on flawless filtration techniques for ensuring that no contaminants enter the drug or its packaging. Hospital burn units also require that the air entering the patient's room be filtered to remove all potential microorganisms that could cause infection. Microbiologists in research use filtration to separate particles of different sizes in a process called *cell fractionation,* which is the separation of cell components from each other. In cell fractionation, a microbiologist uses a combination of filtration methods and centrifugation to isolate ribosomes, mitochondria, chloroplasts, membranes, nuclei, and other intracellular structures.

A variety of filtration techniques reduce the amount of microorganisms in water, other liquids, or air for the purpose of eliminating contamination from pathogens. Filtration methods used for these purposes are as follows:

- water treatment—for removal of viruses, bacteria, spores, and cysts from drinking water
- water quality testing—collecting any organic particles in treated water for the purpose of monitoring for potential pathogens
- food production—for removal of spoilage organisms or food-borne pathogens from juices and other drinks
- laboratory studies—purification of the air entering and exiting confined workspaces called biological safety cabinets that are constructed for reducing contamination of cultures and preventing infection to the microbiologist
- wastewater treatment—for removal of organic matter that has bacteria attached to it

Manufacturers of in-home water treatment systems and portable water purification devices (devices that remove most dangerous microorganisms from water), also, use a variety of filtration methods for improving the quality of drinking water. In-home water filtration systems remove some of the bacteria normally found in water, as well as organic matter and metals that can change the taste of water. Water purification devices such as portable units used by hikers and campers remove most of the bacteria and cysts that inhabit streams and rivers. These portable devices, thus, do more than improve the water's taste; they prevent drinkers from contracting waterborne infections.

Microbiologists have also developed specific filtration methods for two different laboratory methods: enumeration and motility testing. In enumeration, a microbiologist determines the number of microbial cells present in a very large volume of water. To do this, the microbiologist filters a liter or more of water through a sterile filter. The filter captures all the cells as the water passes through. The microbiologist, then, puts the filter bacteria-side-up on agar or on a pad soaked with liquid medium. After incubating the filter and the agar, the microbiologist counts the number of colonies (called *colony-forming units,* or CFUs) on the filter's surface. This technique works because nutrients can slowly pass from the agar through the filter's pores and become available for the bacteria growing on the filter's surface.

Motility testing is the determination whether a microorganism can move under its own power. A microbiologist simply puts a drop of culture onto a filter that has been placed on top of an agar plate. Motile bacteria will swim through any pores larger than the cell's diameter to reach the nutrients in the agar.

The combination of motility and specific pore size has been employed for identifying certain bacteria. For example, treponemes are spirochete (spiral-shaped) pathogens that pass through pore sizes of 0.15 micrometer (μm) that stop almost every other type of bacteria. A microbiologist adds a drop of culture suspected of containing treponemes onto a

Types of Filtration Based on Filter Pore Size

Type	Pore Size Range	Materials Retained
conventional	5–100 μm	visible particles, dander, skin cells, hair
microfiltration	0.1–5 μm	bacteria, yeasts, smoke particles
ultrafiltration	0.01–0.1 μm	particles seen only by using electron microscopy (viruses)
reverse osmosis	0.001–0.01 μm	smallest viruses (polio), molecules, salts, metal ions
cross-flow filtration (tangential flow filtration)	micrometer (μm) to namometer (nm) levels	any of the above

0.15-μm-pore-size filter placed over agar. During incubation, the treponemes migrate through the filter into the agar, and this growth appears as a cloudy layer just below the filter. Antibiotics that kill most bacteria but not treponemes can be included in the agar medium to eliminate all species except the treponemes. A similar method has been used for isolating *Campylobacter fetus* from all other species by using 0.65-μm-pore-size filters overlaid on antibiotic-containing agar.

FILTRATION COMPONENTS

A typical filtration method for liquids contains three components: suspension, filter, and filtrate. The suspension usually begins as a cloudy or turbid liquid that transmits only a small amount of light if exposed to a light source. After the liquid passes through the filter, it becomes a clear filtrate that transmits more light. The filter retains all solid particles of a size larger than the filter's pore diameter. Depending on the objective of the procedure, a microbiologist's intent may be to collect the material in the filtrate, or the matter retained on the filter. The term *filter* is also a verb; a scientist filters a suspension by passing it through a filter.

In almost all filtrations, a flowing material passes through a stationary filter. This flow can be induced in three ways: by gravity, vacuum, or pressure. Gravity filtration allows the liquid to move through the filter on its own; concentrated suspensions usually take a long time to filter, and less concentrated suspensions usually flow quickly. An exception to this occurs when the filter contains very small pores. Filtration of even highly diluted suspensions can take many hours using gravity alone. Vacuum filtration speeds the process by adding force to the filtration system. This method involves the use of a simple apparatus holding two chambers, one on top of the other and separated by a filter. The entire apparatus can be connected by tubing to a vacuum source. The vacuum, then, pulls the liquid from the suspension in the upper chamber through the filter, and filtrate collects in the lower chamber. Pressure filtration applies the same principle of adding force to the system, but, in this case, pressure is applied to the suspension to push the liquid through the filter.

As microscopy developed during the past century, microbiologists have discovered that microorganisms inhabit a wide range of sizes. Most bacteria range in diameter from 0.5 to 2.0 μm. Viruses measure only 10–400 nanometers (nm); *Guardia* cysts and *Cryptosporidium* cysts, which are both dangerous contaminants in natural waters, measure 8–12 μm and 4–6 μm, respectively. Knowing the size of a microorganism helps a scientist select a filter of the correct pore size for a given purpose. For example, to separate viruses from a sample containing a mixture of microorganisms, a microbiologist would select a very small-pore filter, such as one with pores of 0.2-μm diameter. The tiny viruses easily pass through these pores, but bacteria and larger objects stay on the filter.

TYPES OF FILTRATION

Filtration categories relate to the pore size of the filter being used. The table above summarizes filtration categories that correlate with pore sizes, as well as the materials retained by these filters.

Water treatment and wastewater treatment plants also use a filtrationlike process called *gross filtration*. This method makes use of sand layers, called slow sand; screens; and inert materials, such as diatomaceous earth, to remove large particles and aggregates of organic matter from water. Diatomaceous earth is a material made from the cell walls of microorganisms called *diatoms*. The very slow flow of water through these materials allows the removal of many microorganisms, even though the pore sizes are very large. These gross filtration techniques remove a portion of microorganisms from water by allowing time for the cells to attach to, or adsorb to, organic matter as it slowly accumulates on the sand, diatomaceous earth, or other materials. In water and wastewater treatment, these filtration steps are referred to as *pretreatment* when they precede more specific methods for removing microorganisms.

TYPES OF FILTERS

Filter technology ranges from the very simple, as used in gross filtration, to filters that have been manufactured to hold a defined pore size. For all types of filtration other than gross filtration, the technology can be divided into two types: (1) membrane or pore filters and (2) depth filters. Membrane filters consist of a very thin (about 0.1 mm) and fragile material of a standard pore size, which can be as small as a fraction of a micrometer. Other filters can be based on pore size, but they are composed of fibrous material rather than a thin membrane. Depth filters contain thick, densely woven, or packed material that requires a suspension to travel several inches to a foot (0.3 m) from the top of the filter to the bottom. On the basis of their differences in construction, membrane filters depend on set pore sizes to perform filtration, and depth filters depend on a tortuous route for the liquid to take in order to be filtered. Some depth filters combine both technologies; they possess pores of fairly standard size and a matrix that increases the distance the suspension takes through the filter.

A wad of cheesecloth or cotton serves as microbiology's simplest filter. Wrapped fabric can be sterilized and used as an effective plug for test tubes or flasks that contain bacterial cultures. These filters remove contaminants from the air but allow the aerobic culture to receive oxygen. The French bacteriologist Louis Pasteur (1822–95) developed this elementary filter for proving the existence of microorganisms transmitted through the air. In 1861, he wrote a seminal article on the presence of airborne microorganisms that settle on surfaces and cause spoilage. "The procedure which I followed for collecting the suspended dust in the air and examining it under the microscope is very simple," he wrote. "A volume of the air to be examined is filtered through guncotton which is soluble in a mixture of alcohol and ether. The fibers of the guncotton stop the solid particles. The cotton is then treated with the solvent until it is completely dissolved. All of the particles fall to the bottom of the liquid." Pasteur rinsed the solvent off the collected particles with water, and then he studied them under his microscope. This filtration experiment led Pasteur to remark, "These very simple manipulations provide a means of demonstrating that there exists in ordinary air continually a variable number of bodies." This observation by Pasteur would eventually lead to the development of new filters specially made to remove particles from the air.

The standardized pore sizes of membrane filters work by a process called *micropore exclusion*. These filters exclude from the filtrate any particle of a certain size and larger. For example, a 0.2 µm membrane excludes all particles of about 0.25 µm and larger. But damaged cells and newly divided cells could be smaller than normal and pass through the 0.2 µm pores. To account for this possibility, a microbiologist would be more likely to define the exclusion size of a 0.2 µm filter as 0.3 µm rather than expecting the filter to catch all particles of 0.2001, or even 0.25 µm in diameter.

Membrane filters of 0.22–0.45-µm pore diameters remove most bacterial species from liquids. Smaller bacteria, viruses, and large molecules (macromolecules) must be removed using 0.01-µm pore membrane filters. Even these tiny pores cannot stop all cells because some microorganisms possess a morphology (cell structure) that enables them to pass through extremely small pores. For example, mycoplasma and spirochete bacteria can pass through most filters. Mycoplasma bacteria lack a rigid cell wall, so their pliable bodies squeeze through pores that stop most other bacteria. Spirochete bacteria, by contrast, have a long thin shape that allows them to slip through small pores.

Despite the drawbacks of using membrane filters for suspensions containing mycoplasmas or spirochetes, these filters serve a variety of uses in microbiology. Membrane filters have become common in the sterilization of aqueous solutions, growth media, drug or antibiotic formulations, and ophthalmic solutions. Membranes are also fragile and prone to tearing or puncture, which immediately ruins their capacity to exclude particles of a certain size. Laboratory workers must handle membrane filters with care and follow the manufacturer's instructions regarding whether the filter material can be exposed to heating or organic solvents, such as alcohol.

Membrane filters in use today are made of any of the following: cellulose acetate, cellulose nitrate, nitrocellulose (mixed cellulose ester filters [MCE]), polycarbonate, polyvinylidine fluoride (PVDF), nylon, or polytetrafluoroethylene (PTFE, or Teflon). Nitrocellulose filters contain a mixture of cellulose acetate and cellulose nitrate. All of the materials may be classified as either hydrophilic (water-attracting) materials or hydrophobic (water-repelling) materials. The hydrophilic or hydrophobic qualities of the membrane material make filters more suitable for some applications than others.

For example, a hydrophilic nylon material binds proteins. A microbiologist who wishes to filter a protein solution to remove all extraneous matter from the solution would not this filter, because the proteins would bind to the nylon. Filter manufacturers have developed technologies that make filter materials either hydrophilic or hydrophobic. Membrane materials that can be in either a hydrophilic or hydrophobic form are PVDF, nylon, and PFTE. Microbiologists and chemists select membrane filters according to their needs. In general, methods in microbiology require membrane filters with the following characteristics:

- compatible with a wide range of gases
- resistant to corrosion, shedding, or dissolving
- stable in a wide temperature range, -40–250°F (-40–121°C)
- withstands pressure up to 20–60 pounds per square inch (psi)
- possesses an absolute pore size (no variability of pore sizes in the same filter)

The high-efficiency particulate air (HEPA) filter also relies on pore size to remove particles, but it differs from membrane filters in two respects: HEPA filters remove particles from air and not liquids, and HEPA filters contain a thick matrix of paperlike fibers, rather than a single, thin sheet of membrane. HEPA filters remove 99.97 percent of particles that measure 0.3 μm and larger. The pore size limitations and the thick configuration of HEPA filters make them a combination of two filtration technologies: pore size and depth.

In the 1940s, the U.S. Atomic Energy Commission contracted with the Arthur D. Little company to develop a device for use in the top-secret Manhattan Project's research into making the first atomic bomb for use in World War II. The project required a device that would remove from the air radioactive particles emitted as scientists conducted their experiments. After the war, the government released the HEPA technology to the public, and, by the 1950s, companies deployed HEPA filters for removing smoke, dust, pollen, and microorganisms from the indoors of houses and businesses. Many new models of vacuum cleaners also contain HEPA filters for preventing living and nonliving microscopic particles from escaping into the air.

A HEPA filter contains a single sheet composed of tightly woven fibers. The filter manufacturer folds the sheet back and forth, until it is completely pleated. The folds in the HEPA filter produce a surface area much greater than a single flat sheet would offer. Any particle approaching a HEPA filter confronts a dense tangle of fibers, through which it must find a route in the flow of air from one surface to the filter's opposite surface. Air finds a way through the filter's maze, but the filter stops particles by three mechanisms. First, particles crash directly into a fiber, which is probable, considering the dense arrangement of the filter's

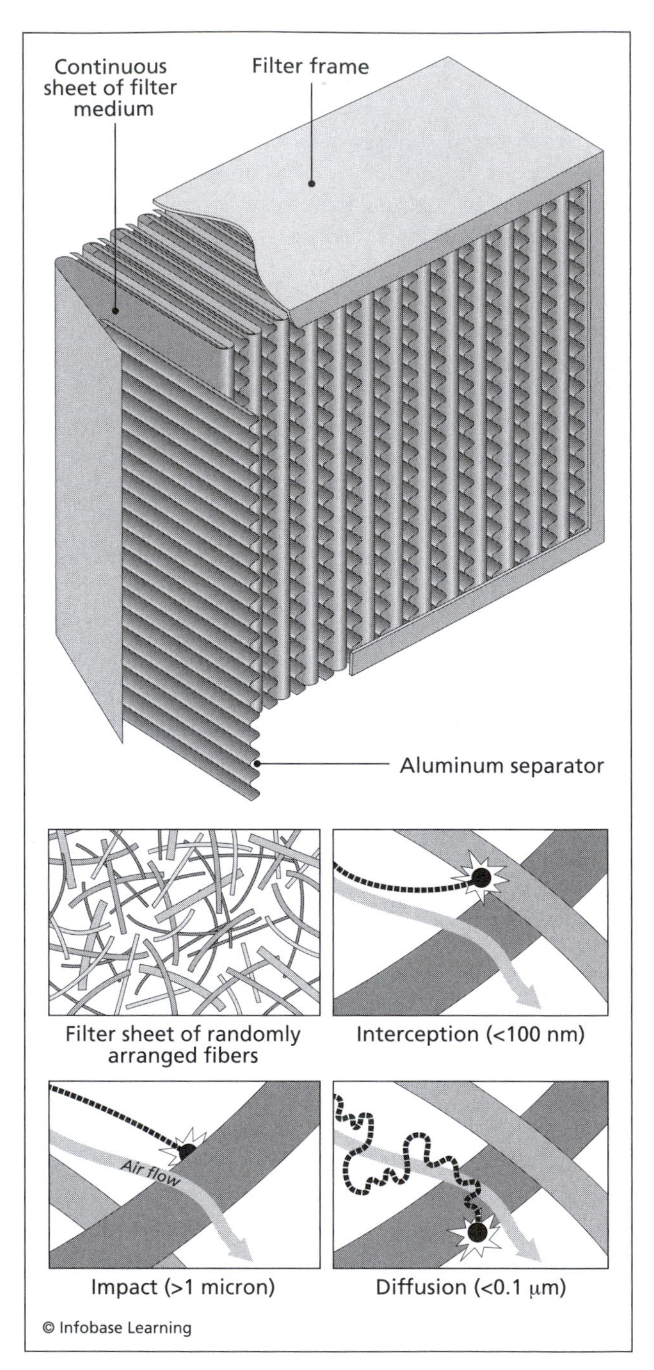

High-efficiency particulate (HEPA) filters remove almost 100 percent of infectious microorganisms from the air. Hospitals, drug manufacturing plants, and microbiology laboratories use this type of filtration. The HEPA filter configuration and materials remove particles as small as 1.0 micrometer (μm).

material. Second, a particle approaches within a very short distance of a fiber to elicit electrostatic attraction that arrests the particle on the fiber's surface. Third, the particle becomes caught in a pore too small to let it continue. The Arthur D. Little engineers named their invention an *absolute filter* because of its effectiveness in stopping particles of specific sizes.

Depth filter technology often relies on three means of stopping particles: adhesion of the particles to the filter material, small pore size, and a tortuous route of flow. Adhesion helps remove very small particles that can fit through tiny pores or make their way through a dense filter matrix. In filtration, the adhesion process is called *adsorption,* which is a process in which a particle sticks to a material's surface. Adsorptive filters remove particles in the following stepwise process:

1. macropore transport of large substances through pores of greater than 50 nm in the filter matrix

2. mesopore transport in which particles flow into smaller pores of 2–50 nm, where some adsorption occurs

3. micropore transport of small particles into tiny pores of less than 2 nm, where they adsorb to the filter

The preceding steps describe a process called *sorption.* Sorption occurs when layers of material remove particles from a flow by capturing them on the material's surface.

Most depth filters perform much less stringent filtrations than HEPA filters. The activated carbon filter has become a commonly used depth filter for removing large particles and some organic and inorganic compounds from drinking water. Activated carbon is composed of organic materials (such as coconut shells, wood, peat, or bone) that have been heated until they decompose to carbon granules. Activated carbon in granular form, as powder, or in pellets offers a tremendously large surface area for adsorbing substances from water. Drinking water treatment systems include activated carbon filters, also called *carbon cartridges,* which remove many of the following substances from drinking water: bacteria, viruses, organic matter, chemicals, and compounds that change water's odor or taste. These substances adhere to the carbon in very weak attachments, so activated carbon cannot be used as a water purification method.

Other materials have been used for filters that work in the same way as activated carbon: by binding some particles and by creating a long, tortuous route for particles to travel. Two materials have been used to substitute for carbon: silica gel, which is a matrix of silicon dioxide aggregates, and alumina, which is an aggregated form of aluminum oxide. Additional materials in depth filter technology are the following: asbestos (Seitz filter), synthetic fibers such as nylon, diatomaceous earth, cellulose, unglazed porcelain granules (Chamberland filter), or sintered glass. Sintered glass is made by heating glass pieces until they form a mass but before the glass melts completely. This process results in a porous filter rather than a solid glass block. Devices called Berkfield filters contain kieselguhr, which is a type of diatomaceous earth that is processed into coarse, medium, or fine granules.

Fiber depth filters play an important role in water microbiology, because they have the capacity to receive large volumes of water that may contain a small number of *Cryptosporidium* or *Giardia* cysts. Filtration is the main method that microbiologists use for determining the presence and concentration of these microorganisms in drinking or recreational waters.

SPECIAL APPLICATIONS IN FILTRATION

Microbiologists can recover almost any microorganism from a liquid containing an abundance of other materials by using a two-step filtration. The first step, called *cake filtration* or *prefiltration,* removes large pieces of nonliving substances and aggregates of organic matter. Cake filtration often involves use of highly porous filter paper—paper called Whatman no. 1 is most common—set inside a Büchner funnel. The microbiologist, then, filters the partially clarified filtrate through a membrane of set pore size. This method works well in recovering small amounts of materials from large volumes of dirty water. To recover viruses from water, microbiologists might use a Seitz filter in which bacteria stop in the filter of compressed asbestos but the viruses pass into the filtrate.

Water purification devices should be used by anyone planning to drink water from natural surface waters or during natural disasters that have caused possible contamination of drinking water. Water purification units state on their packaging whether the filter is "nominal 1 μm" or "absolute 1 μm." Nominal 1 μm filters contain pores that average 1.0 μm across. Absolute 1 μm filters, by contrast, have pores that are all no more than 1.0 μm across. This distinction is important for groups in the population who have high-risk health conditions, such as cancer or acquired immunodeficiency syndrome (AIDS) patients. High-risk health conditions require extra diligence in removing potentially dangerous microorganisms from the person's environment and diet. Devices containing absolute pore size filters always say the following on the package: "Absolute micron size of 1 micron or smaller" or "Tested and certified by NSF standard 53 for cyst removal." NSF refers to an organization named *NSF International,* which tests water purification devices. (The organization's name was *National Sanitation Foundation,* until 1990.)

Pressure filtration with nitrogen offers a specialized process that reduces chemical reactions in the filtrate. In this procedure, nitrogen gas pushes the suspension through a filter and thereby prevents oxygen from entering the device. By replacing oxygen with the inert gas nitrogen, a scientist can reduce the chance of oxidation reactions that might damage an important microbial end product.

Filtration also makes use of an apparatus called a *Hemming filter,* which filters a suspension as it is being centrifuged. The Hemming filter contains two chambers clamped together mouth-to-mouth with a filter in between. One chamber holds the suspension, and the entire apparatus can be fitted into a centrifuge. During centrifugation, gravitational force pushes the suspension through the filter so that the filtrate moves into the second chamber. Hemming filters speed the process of recovering small particles from very concentrated suspensions.

Any filter cannot perform as it should forever and must eventually be replaced. Sooner or later, pores clog and adsorption sites fill up, but filters have a wide range of life spans. Membrane filters about the size of a quarter last for one use only, but HEPA filters can work for a year or longer before needing replacement. In general, depth filters last longer than membrane filters.

Filtration is a simple but essential technique in microbiology. Filtration techniques have progressed so that a very specific particle size can be separated from all other particles of different sizes. This method plays an important role in the fractionation of different cell types and various cell organelles.

See also CENTRIFUGATION; CLEAN ROOM; FRACTIONATION; ORGANELLE; STERILIZATION; WASTEWATER TREATMENT; WATER QUALITY.

Further Reading

American Filtration and Separations Society. Available online. URL: www.afssociety.org/index.htm. Accessed August 10, 2009.

Millipore. "Lab Filtration." Available online. URL: www.millipore.com/lab_filtration/clf/capability. Accessed August 10, 2009.

Pall Corporation. "Sterile Filtration and Clarification." Available online. URL: www.pall.com/laboratory_1079.asp. Accessed August 10, 2009.

Pasteur, Louis. "Mémoire sur les Corpuscles Organisés qui Existent dans l'Atmosphére. Examen de la Doctrine des Générations Spontanées" (On the Organized Bodies Which Exist in the Atmosphere; Examination of the Doctrine of Spontaneous Generation). *Annales des Sciences Naturelles* 16 (1861): 5–98. In *Milestones in Microbiology,* translated and edited by Thomas Brock. Washington, D.C.: American Society for Microbiology Press, 1961.

U.S. Environmental Protection Agency. "Filtration." Tech Brief 2. September 1996. Available online. URL: www.nesc.wvu.edu/ndwc/pdf/OT/TB/TB2_filtration.pdf. Accessed April 14, 2009.

Virginia Polytechnic and State University. "Membrane Filtration Method: Fecal Coliforms." Available online. URL: http://filebox.vt.edu/users/chagedor/biol_4684/mfcoli.html. Accessed August 10, 2009.

Zajac, Ted. "HEPA Air Purifiers." ClearFlight Air Purifiers, Inc. Available online. URL: www.airpurifiers.com/technology/hepa.htm. Accessed August 10, 2009.

Fleming, Alexander (1881–1955) ***Scottish Physician, medical microbiologist*** Alexander Fleming has been credited as the discoverer of the antibiotic penicillin. Fleming's work on the effects of mold extracts on the growth of various bacteria led him to this discovery, and, in 1945, he accepted the Nobel Prize in medicine for his scientific contributions in the area of antimicrobial substances. Fleming, additionally, received at least a dozen awards for his work in bacteriology, immunology, and the treatment of diseases with chemical agents, a field now known as *chemotherapy.*

Alexander Fleming was born on August 6, 1881, in rural Ayrshire, Scotland, the seventh of eight siblings. He lived on his family's farm until his teens, when he joined three brothers and a sister in London and attended the Polytechnic School. After graduation, Fleming worked at a shipping firm, until he enlisted in a Scottish regiment at the beginning of the Boer War, in 1900, between the United Kingdom and its colonies in southern Africa. Fleming the soldier would develop lifelong interest in finding a substance that would prevent infections caused by wounds sustained on the battlefield.

After the military, Fleming followed his older brother Tom into medicine and graduated as a surgeon from St. Mary's Hospital Medical School at the University of London. The young doctor faced a career decision, in 1905, that would be determined less by science and more by his nonscientific pursuits. A 1998 PBS documentary on Fleming explained how he arrived in bacteriology rather than surgery: "If he took a position as surgeon, he would have to leave St. Mary's. The captain of St. Mary's rifle club knew that and was desperate to improve his team. Knowing that Fleming was a great shot he did all he could to keep him at St. Mary's. He worked in the Inoculation Service and he convinced Fleming to join his department in order to work with its brilliant director—and to join the rifle club. Fleming would stay at St. Mary's for the rest of his career." Sir Almroth Wright, the director, was at the time a pioneer in vaccine development.

At the start of World War I, Fleming and many of his colleagues joined the military to set up battlefield hospitals in Europe. Fleming could not help but notice how quickly a normally insignificant infection could kill, with little chance for doctors to administer treatment. Fleming became dedicated to finding new ways to fight infection effectively and rapidly. He realized that most infections received on the battlefield rarely were caused by exotic microorganisms but rather were caused by familiar species, which he had studied earlier in his laboratory. As the war progressed, Fleming experimented with new methods for treating wounds, and he made optimal use of antiseptics on all minor or major wounds.

Upon returning to St. Mary's in the 1920s, Fleming continued seeking better antiseptics. He discovered the substance lysozyme, an enzyme found in tears and other body fluids that lyses bacteria and thus defends the body against infection. Lysozyme inhibited some bacteria in laboratory experiments but did not appear to be strong enough to combat the most virulent pathogens. (A pathogen's virulence relates to the microorganism's ability to start an infection.) Fleming, perhaps consumed with the idea of discovering new antibacterial compounds, soon filled the laboratory with in-progress, aborted, and planned experiments, which each had an accompanying assortment of test tubes, flasks, and petri dishes.

The story of what happened next in Fleming's laboratory has evolved into different versions over the decades, but with but one commonality: the surprising and unexpected discovery of penicillin. In September 1928, Fleming took a several-day vacation from the laboratory. He had left behind a collection of bacterial cultures, either in need of disposal or perhaps from a half-completed experiment to which he planned to return. Cool temperatures in the unoccupied laboratory probably helped molds grow on the cultures' agar surface. Numerous microbiology historians have attributed the occurrence to either sloppy housekeeping by Fleming or careless handling of the cultures by others working in the laboratory while he was away. Whatever the circumstances, Fleming benefited from a serendipitous discovery.

Fleming described his observations of what had occurred in the *British Journal of Experimental Pathology,* the next year: "While working with staphylococcus variants a number of culture-plates were set aside on the laboratory bench and examined from time to time. In the examinations these plates were necessarily exposed to the air and they became contaminated with various micro-organisms. It was noticed that around a large colony of a contaminating mould the staphylococcus colonies became transparent and were obviously undergoing lysis." Lysis meant that a substance had caused the bacterial cells to disintegrate and die. The substance, furthermore, exerted this effect quite effectively. Fleming included in the article a picture of such a culture, in which a large mold colony dominates one side of the petri dish, normal staphylococcus colonies grow at the opposite side, and a territory between the two contains only lysed bacteria.

Fleming repeated the *Staphylococcus* experiment and also studied other bacteria such as streptococcus, gonococcus, and diphtheroids with similar results. "This simple method therefore suffices to demonstrate," he wrote, "the bacterio-inhibitory and bacteriolytic properties of the mould *[sic]* culture, and also by the extent of the area of the inhibition gives some measure of the sensitiveness of the particular microbe tested." Fleming presumed the contaminant was closely related to *Penicillium rubrum* because of its morphology. He further showed that an 800-fold dilution of a substance extracted from the *Penicillium* could remain effective in killing bacteria. He named the extracted substance penicillin.

Despite Alexander Fleming's entry into medical history, penicillin had been already been discovered by the French medical student Ernest Duchesne, in 1896, as he described in his dissertation, in 1897. Duschesne had worked out a partial purification scheme and shown that penicillin cured infections in animals, as Fleming would, more than 30 years later. But Duchesne died at only 37 years of age and never followed through on his important finding. Scholars have since searched Fleming's notes for an indication that he knew of Duchesne's study, but no evidence seems to exist to show that Fleming had any knowledge of the prior discovery of penicillin.

Fleming himself may not have grasped the entire significance of penicillin as a drug for fighting infections. He focused mainly on *Penicillium* extract's activity as a potential new antiseptic to replace weaker antiseptics. The microbiology historian Thomas Brock wrote in *Milestones in Microbiology,* in 1961, "Fleming perceived only dimly the practical applications of penicillin. He obviously was aware that such a substance would be highly useful if it would kill pathogenic bacteria in humans, but such a hope seemed much in the future in view of the low broth potencies he was obtaining. It took the development of the sulfa drugs and the impetus of World War II before the tremendous technological and medical problems could be adequately solved, which led to the development of penicillin as a practical chemotherapeutant." Fleming dabbled in penicillin experiments while remaining focused mainly on lysozyme and amassed a significant amount of information on this enzyme, although he will always be best known for his studies on penicillin.

In 1939, two pathologists at the University of Oxford teamed to reinvestigate the idea of penicillin as a drug against bacterial infection. The Australian Howard Florey and his British counterpart Ernst Chain developed a small-scale method for recovering penicillin from the liquid broth in which the mold grew. By 1940, others had realized the potential benefit of a compound that was easy to make and could kill a variety of bacterial pathogens. At that time, Britain had been immersed in the war against Nazi Germany and did not have the capacity to set up a facility for producing penicillin. The U.S. government took the lead in the large-scale manufacture of the antibiotic, which became an essential part of medics' supply kits for the war. Penicillin has since been mentioned as a factor for helping the Allied countries win World War II. At home, doctors began prescribing the drug for pneumonia, syphilis, gonorrhea, diphtheria, and scarlet fever. Diseases that had caused large numbers of deaths suddenly became curable. Wounds and infections that previously would lead to gangrene could be better controlled. Penicillin became known as a "miracle drug" or "wonder drug."

The public sought a hero for this new lifesaving drug. In a 1999 *Time* magazine article, the correspondent David Ho wrote, "Fleming alone became such an object of public adulation, probably for two reasons. First, Florey shunned the press, while Fleming seemed to revel in the publicity. Second, and perhaps more important, it was easier for the admiring public to comprehend the deductive insight of a single individual than the technical feats of a team of scientists." Florey and Chain were instrumental in developing the full potential of penicillin and of leading the efforts in penicillin's commercial production.

Fleming and Florey received knighthood, in 1944, and, the next year, they and Chain shared the Nobel Prize for the development of the miracle drug. Alexander Fleming continued to work in science until his unexpected death of a heart attack, in 1955, at which time the world mourned his passing and hailed him as a hero. The crypt of St. Paul's Cathedral in London now holds the burial place of Sir Alexander Fleming.

Fleming's story could easily have stopped with his death and his legacy kept alive in penicillin. But, for some time, medical historians have debated Alexander Fleming's place in medical history and the birth of what became known as the "Fleming myth." Many scientists who developed penicillin and made its potential a reality labored in obscurity, while Fleming's name has become forever linked to its discovery. The Fleming myth may actually teach students lessons about scientific discovery. First, discovery is based on open communication with other scientists, so that several minds can find the potential uses of the discovery. (Ernest Duchesne certainly did not follow this lesson.) Second, discovery arises from the willingness to be drawn to unexpected findings, rather than discard them, as Fleming did, when he inspected contaminated cultures. Any other scientist might well have thrown away the cultures without a thought. Finally, scientific discovery owes much to random luck, which should never be dismissed. Random luck has always been a factor in scientific discoveries and uncovering of opportunity.

See also ANTIBIOTIC; PENICILLIN.

Further Reading

Brock, Thomas. *Milestones in Microbiology.* Washington, D.C.: American Society for Microbiology Press, 1961.

Fleming, Alexander. "On the Antibacterial Action of Cultures of a Penicillium, with Special Reference to Their Use in the Isolation of *B. influenzae.*" *British Journal of Experimental Pathology* 10 (1929): 226–236.

Ho, David. "Scientists and Thinkers: Alexander Fleming." *Time,* 29 March 1999. Available online. URL: www.time.com/time/time100/scientist/profile/fleming.html. Accessed August 10, 2009.

Macfarlane, Gwyn. *Alexander Fleming: The Man and the Myth.* Oxford: Oxford University Press, 1985.

PBS. "People and Discoveries: Alexander Fleming." Available online. URL: www.pbs.org/wgbh/aso/databank/entries/bmflem.html. Accessed August 10, 2009.

fluorescence microscopy Fluorescence microscopy is the study of objects too small to see with the unaided eye by the objects' light absorption and emission. Fluorescence is the ability of a substance to absorb short-wavelength light and then emit light at a longer wavelength. Some things in nature fluoresce on their own, but other substances can be made to fluoresce by attaching a fluorescent dye or stain called a fluorochrome or a fluorophore. A microorganism stained with a fluorochrome and viewed under a microscope, using an ultraviolet (UV) light source, will appear as a bright object against a dark background. This method is referred to as *epifluorescence.* Any action in which light emits from an irradiated compound is generally known as *luminescence,* so fluorescence is a type of luminescence.

Microbiologists use techniques to make microorganisms more visible under a fluorescent microscope than they would be if left untreated. When substances in nature fluoresce without fluorochromes, the process is called *primary fluorescence* or *autofluorescence.* Usually, primary fluorescence gives too weak a light emission for studying under a microscope. Microbiologists, therefore, attach fluorescent stains to laboratory specimens to enhance the brightness when exposed to UV light. This enhancement is called *secondary fluorescence* (see the color insert on page C-4 [top, left]).

Fluorescence microscopy offers the advantages of highlighting microorganisms in greater detail than other forms of microscopy and of producing a three-dimensional image. In more conventional types of microscopy, such as bright field, dark field, or phase contrast, light illuminates the specimen from above, below, or the sides, respectively. Fluorescence microscopy uses light that is from the specimen itself, as either natural primary fluorescence or secondary fluorescence. The result is a bright and variously colored specimen showing fine details that would otherwise be invisible.

THE DEVELOPMENT OF FLUORESCENT MICROSCOPY

In microscopy, resolution is the power of an instrument to distinguish between two objects. Light microscopy and methods that depend on light's visible spectrum use a limited range of wavelengths for illuminating an object. In the 1800s, microscopists understood that resolution could be influenced by four factors: wavelength, magnification, the apertures of the microscope's objective and condenser lenses, and the geometry with which light rays met at the specimen. In the 19th century, microscope technology seemed to have reached its limits in the latter three areas, but wavelengths could be made longer or shorter than the visible light spectrum.

The German physicist Ernst Abbé (1840–1905) developed the basic principle behind all types of microscopy, in 1873, by determining that the resolution of an object is limited to about 10 times the wavelength of the light being used. This principle, to become known as Abbé's law, prompted scientists to work with wavelengths as the most powerful component in a new type of microscopy.

The British scientist George G. Stokes had first described fluorescence in physical terms, in 1852, when he noticed that the mineral fluorspar glowed red when exposed to UV light. August Koehler (1866–1948) had already been working with UV light's effect on other substances and designed an UV microscope, in 1904, to view tissue samples. But Koehler experienced immediate difficulties with biological specimens because they emitted an unexpected fluorescence. Rather than dismissing UV light as incompatible with microscopy, Koehler sensed a potential value in his finding. He began adjusting the UV light in order to improve the resolution between objects and would attain resolutions that had never before been achieved.

Koehler and his contemporaries Carl Reichert and Heinrich Lehmann stroved to enhance the power of fluorescent microscopy, but the technology did not become practical until the 1930s, when Max Haitinger perfected fluorochrome staining. Albert H. Coons and Melvin H. Kaplan developed the first important use for fluorescence microscopy, in the 1950s, by using fluorochromes attached to antibodies to highlight precise sites (acting as antigens) on cells isolated from tissue. This method, called *antibody tagging,* has become important in microbiology today for demonstrating the presence or absence of specific cellular molecules.

THE FLUORESCENT MICROSCOPY METHOD

Fluorescence and phosphorescence are both physical phenomena in which absorbed light is emitted at a different, lower-energy wavelength. Fluorescence differs from phosphorescence merely by the fact that it emits higher-energy light than phosphorescence. In fluorescent microscopy, a light source irradiates a specimen with UV light in a range of wavelengths. A component of the microscope, then, filters the light emitted by the specimen to remove any weak fluorescence that would decrease the image's sharpness. The high-energy emissions used for viewing the image are called *excitation light.* A fluorescent microscope is configured to direct only excitation light to the eyepiece for viewing the highlighted specimen against a dark background. In current microscopy, the excitation light can reach almost a million times the brightness of weak fluorescent emissions.

Excitation is created as a result of energy transitions in the fluorochrome molecule's electrons. Molecules energized by UV irradiation contain electrons that have been raised to an excited state. As the electrons return to a lower energy level, called the *ground state,* they lose energy, which takes the form of emitted light. The energy loss also results in emitted light in a different range of wavelengths than the absorbed light. In other words, a range of wavelengths shift between absorption and emission. This difference in wavelength ranges is called Stokes' shift or Stokes' law. Manufacturers of fluorescent microscopes devise a combination of filters that enhances the shift and improves the image.

The light source for a fluorescent microscope is supplied by either a mercury vapor arc lamp or a xenon arc lamp. The light passes through an exciter filter to eliminate all wavelengths but the desired range of about 400 nanometers (nm). As the light approaches the specimen—in fluorescence microscopes, the light is from behind the specimen—it passes through a dark-field condenser, a component that creates a black background. When the UV light transmits through the specimen, the fluorochrome attached to the specimen emits longer-wavelength light. Before a microbiologist sees the image produced by the transmitted light, a barrier filter removes all extraneous excitation light, which could damage the viewer's eyes. The barrier filter also removes blue and violet light, which reduce contrast in the image. A properly configured UV microscope can illuminate fine cellular structures, appendages, deoxyribonucleic acid (DNA) replication, and cell division, or it can differentiate living cells from dead.

USES OF FLUORESCENT MICROSCOPY

Fluorescence microscopy is used for diagnoses of some diseases, in public health projects that monitor for specific pathogens in the environment, and for detection of water or soil microorganisms.

Four technologies have emerged in fluorescence microscopy for medical and nonmedical studies: (1) fluorescence in situ hybridization (FISH), (2) fluorescence resonance energy transfer (FRET), (3) fluorescence lifetime imaging microscopy (FLIM), and (4) multiphoton imaging.

The FISH technique detects strands of ribonucleic acid (RNA) or DNA in biological samples. The method begins with a specimen containing nucleic acids, in which the double-stranded molecule has been separated into two strands. A microbiologist, then, exposes the sample to a probe containing a second piece of nucleic acid with a known base sequence. For example, a probe containing a section of a DNA strand from *Escherichia coli* would be able to detect *E. coli* DNA in a water sample. The probe detects its target through a process called *hybridization*, or the pairing of base units on the probe with complementary base units in the sample. By including a fluorochrome in the probe, microbiologists detect the presence of specific types of nucleic acids. The stepwise reactions that lead to a detectable fluorescent signal are the following:

1. Expose specimen to the RNA- or DNA-specific probe.
2. Allow time for the complementary strands to bind, a process called *annealing*.
3. Wash away any unbound probe.
4. Observe sample under UV light.

After the steps described here, only RNA or DNA that has hybridized with the probe becomes visible in the fluorescent microscope. *E. coli* probes detect *E. coli* cells, *Salmonella* probes detect *Salmonella* cells, and so on.

The FRET technique aids in the study of interactions between two different proteins by measuring the fluorescence they emit separately, compared with the fluorescence they emit as a protein-protein complex. In the FRET reaction, two different fluorochromes attached to proteins exchange energy when they move close to each other and are irradiated by UV light. As a result, one fluorochrome emits wavelengths of less energy than normal, and the other fluorochrome emits wavelengths of higher-than-normal energy. This unique pattern allows microscopists to detect protein interactions that have been used for detecting tumors, blood factors, and specific cell types seen in certain diseases.

In FLIM, a scientist measures the gradual decay of fluorescence from a molecule to determine its persistence in the body. FLIM has enabled biologists to track the relative concentration of one component inside a cell and watch it degrade over time. This method has become an important tool for the following applications: tracing the longevity of pollutants in the body's tissues, determining the exact intracellular location of proteins, studying in situ interactions between enzymes and substrates, identifying changes in tissue or cells due to disease, and studying the effects of pathogens on tissues.

Multiphoton imaging, also called *multiple-photon excitation fluorescence microscopy,* illuminates specimens that may not work well under normal florescence microscopy. In this technique, a specimen is exposed to light of certain wavelengths, and a second energy source, such as a laser, in which neither source alone would make a fluorochrome fluoresce. But the two energy sources together allow the fluorochrome to absorb excited light photons (light rays) and fluoresce. Multiphoton imaging has been useful for observing objects that reside in thick specimens that regular fluorescence microscopy might miss because of the scattering of emissions in thick specimens. Multiphoton imaging creates fluorochrome excitation only at the exact point of focus and eliminates images from any out-of-focus fluorochrome, thereby giving excellent focus and resolution.

FLUOROCHROMES

Regardless of the fluorescence microscopy method being used, a clear fluorescent image correlates with astute selection of a fluorochrome. Microscopists base selections of fluorochromes on the wavelength range in which the molecule absorbs light. A fluorochrome typically absorbs light in one range, emits light at a slightly longer-wavelength (lower-energy) range, and is detected by the microscopist at a narrower range of emissions, as a result of the use of filters inside the microscope.

About 210 fluorochromes have various uses in fluorescence microscopy. The table on page 307 describes 10 commonly used fluorochromes.

Most fluorochromes contain a ring structure that attaches to specific organelles (Golgi apparatus, endoplasmic reticulum, nucleus, etc.), molecules (DNA, RNA, metal ions, receptor sites, etc.), or cellular activities (protein secretion, enzyme activity, cell death [called *apoptosis*], receptor-signal interaction, etc.). Olympus Corporation, a microscope manufacturer, has explained on its Web site,

Common Fluorochromes in Fluorescence Microscopy

Fluorochrome	Main Use	Excitation Wavelength	Emission Wavelength	Emitted Color
acridine orange	detects DNA or RNA	502	526	DNA, green RNA, blue
aurimine O	*Mycobacterium* detection	460	550	yellow
calcein AM	cell viability by detecting membrane integrity	495	520	green
calcofluor white	fungus detection	440	500–520	light blue
4′,6-diamidino-2-phenylindole (DAPI)	double-stranded DNA (dsDNA)	350	470	blue
eosin	cytoplasm staining	525	545	red
ethidium bromide	dsDNA and some RNA	510	595	orange
fluorescein isothiocyanate (FITC)	antibody assays	490	525	green
Hoescht 33258	*Mycoplasma* contamination	345	485	blue
rhodamine B	antibody assays and acid-fast stain for bacteria	540	625	red

Note: wavelengths in nanometers (nm)

> The history of synthetic fluorescent probes dates back over a century to the late 1800s when many of the cornerstone dyes for modern histology [the study of tissue and cell structure] were developed. . . . Although these dyes were highly colored and capable of absorbing selected bands of visible light, most were only weakly fluorescent and would not be useful for the fluorescent microscopes that would be developed several decades later. . . . Fluorochromes were introduced to fluorescence microscopy in the early twentieth century as vital stains for bacteria, protozoa, and trypanosomes, but did not see widespread use until the 1920s when fluorescence microscopy was first used to study dye binding in fixed tissues and living cells. However, it wasn't until the early 1940s that Albert Coons developed a technique for labeling antibodies with fluorescent dyes, thus giving birth to the field of immunofluorescence.

Immunofluorescence is now a core technique in medicine for determining whether a patient has been exposed to a microbial pathogen.

IMMUNOFLUORESCENCE

Immunofluorescence uses fluorochromes that emit light when exposed to UV (200–380 nm), violet (380–435 nm), or blue light (435–500 nm). Fluorochromes couple to antibodies without interfering with the ability of the antibody to detect and attach to specific substances, called *antigens*. An antigen can be any material—microbial cells, proteins, and pollen are examples of antigens—that induces the body's immune system to make antibodies that will actively seek these foreign particles in the bloodstream or in tissue. Immunofluorescence techniques exploit the specific antibody-antigen binding to detect certain compounds. Two different techniques accomplish this: direct and indirect immunofluorescence.

In direct immunofluorescence, a scientist adds fluorochrome-linked antibody to a specimen fixed onto a glass slide and containing the target antigen. After washing off any unbound antibody, the scientist can view the specimen under a fluorescent microscope and observe the antigen's location in the specimen.

Indirect immunofluorescence detects the presence of antibodies in a person's body after possible exposure to a microorganism. In this technique, the scientist puts a known antigen onto a glass slide and then adds to it the fluorochrome-antibody complex. In this case, the antibody has been recovered from the person's blood sample. After incubating the slide to allow time for the antigen and antibody to react, then washing the slide to remove unreacted materials, the

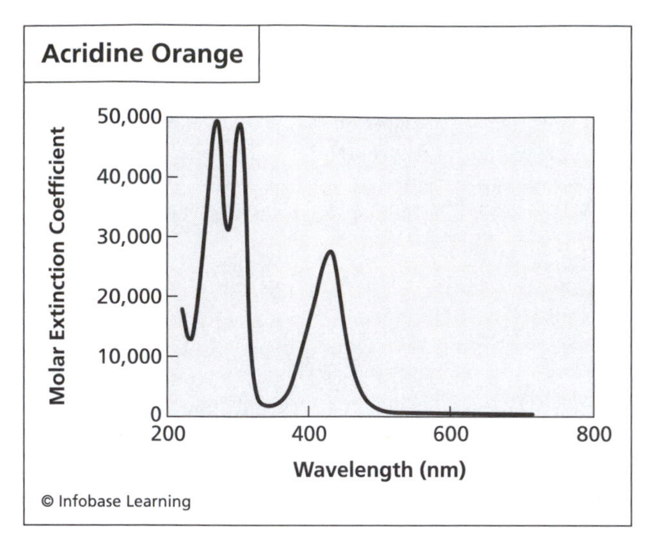

Dye molecules called fluorochromes, such as acridine orange, absorb light at specific wavelengths of fluorescent light. Fluorochromes attached to microbiological specimens make certain structures glow against a black background in a fluorescent microscope.

microbiologist views the specimen under a fluorescent microscope. Fluorescence indicates that the antibody and antigen reacted, thus proving the presence of an antibody for the microorganism in the person's blood. This is a sign that the individual had been infected by the microorganism, some time in the past, or may currently be infected.

Fluorescence microscopy has delivered excellent power of resolution and detail to microscopic studies. In addition, this technology does not require elaborate or difficult-to-use equipment. Medical microbiologists and environmental microbiologists use this technology to detect small amounts of substances, such as a few cells of a microbial species, within a heterogeneous mixture of materials. Fluorescence microscopy plays an essential role in studies of the microbial world and in more specific tasks in disease diagnosis.

See also HYBRIDIZATION; IMMUNOASSAY; MICROSCOPY; STAIN.

Further Reading

Davidson, Michael W., and Mortimer Abramowitz. "Fluorescence Microscopy: Introductory Concepts." Molecular Expressions. Available online. URL: http://micro.magnet.fsu.edu/primer/techniques/fluorescence/introduction.html. Accessed August 10, 2009.

Invitrogen. "Fluorescent Tutorials." Available online. URL: www.invitrogen.com/site/us/en/home/support/Tutorials.html. Accessed August 10, 2009.

Olympus Corporation. "Fluorescence Microscopy." Available online. URL: www.olympusmicro.com/primer/techniques/fluorescence/fluorhome.html. Accessed August 10, 2009.

Rost, Fred W. D. *Fluorescence Microscopy, vol. 2.* Cambridge: Cambridge University Press, 1995.

Spring, Kenneth R., and Michael W, Davidson. "Introduction to Fluorescence Microscopy." Nikon Microscopy U. Available online. URL: www.microscopyu.com/articles/fluorescence/fluorescenceintro.html. Accessed August 10, 2009.

food-borne illness A food-borne illness or infection is any gastrointestinal illness caused by the ingestion of food contaminated with pathogenic microorganisms—such food-borne illnesses can be caused by bacteria, viruses, protozoa, or parasites. Food-borne illness is sometimes also called *food poisoning,* but the term *food poisoning* is better reserved for illnesses caused by chemicals in food rather than live microorganisms.

Even though the United States has one of the safest food supplies in the world, food-borne illness occurs regularly and might account for up to 9,000 deaths each year. The Centers for Disease Control and Prevention (CDC) has warned, "An estimated 76 million food-borne diseases occur each year in the United States. The great majority of these cases is mild and cause symptoms for only a day or two. Some cases are more serious, and CDC estimates that there are 325,000 hospitalizations and 5,000 deaths related to food-borne diseases each year." The CDC emphasizes that these are only estimates, because many food-related illnesses are not reported.

Each year, physicians, hospitals, and outpatient clinics report to the CDC the food-borne illnesses that they diagnose in their patients. In 2008, the CDC tabulated a total of 18,499 laboratory-confirmed cases from 10 states to assess the main causes of food-borne illnesses (see the table on page 309).

In the preceding list of food-borne outbreaks, all of the microorganisms are bacteria, except the parasites *Cryptosporidium* and *Cyclospora.* Perhaps the most striking feature in these statistics relates to the vast difference between 18,499 reported cases of food-borne illness and the 76 million cases that the CDC estimates occur each year. Although the survey had been conducted in only 10 states (Connecticut, Georgia, Maryland, Minnesota, New Mexico, Oregon, Tennessee, and parts of California, Colorado, and New York), extrapolating the data to 50 states does not account for 76 million cases. This large discrepancy occurs for three main reasons. First, as mentioned, many food-borne illnesses are never reported to health care providers, so they are never reported to the CDC. Second, a large number of food-borne illness cases occur without a confirmed diagnosis as to the cause. A sick patient may be sent home with instructions to rest, drink fluids, and perhaps follow a course of a broad-spectrum antibiotic, which is an

Food-Borne Illnesses

Microorganism	Prevalence and Incidence
Salmonella	7,444 total cases; 16.20 per 100,000 people
Campylobacter	5,865; 12.68
Shigella	3,029; 6.59
Cryptosporidium	1,036; 2.25
Escherichia coli 0157:H7	513; 1.12
E. coli non-0157	205; 0.45
Yersinia	164; 0.36
Listeria	135; 0.29
Vibrio	131; 0.29
Cyclospora	17; 0.04

antibiotic that kills a wide variety of bacteria. Third, health care facilities report food-borne outbreaks on a voluntary basis and are asked by the CDC to report only the microorganisms listed here. Although this reporting helps build a picture of the nation's food-borne illness status, it ignores many infectious agents that contribute to the total cases: viruses, *Clostridium, Toxoplasma,* and other food-borne threats.

Controlling food-borne illness presents a difficult task to health care professionals and microbiologists. The CDC commented in 2009, "Despite numerous activities aimed at preventing food-borne human infections, including the initiation of new control measures after the identification of new vehicles of transmission (e.g., peanut butter–containing products), progress toward the national health objectives has plateaued, suggesting that fundamental problems with bacterial and parasitic contamination are not being resolved." Since 1998, the number of new cases of a food-borne disease per 100,000 people reported, a statistic called *incidence,* has not declined significantly for the reportable organisms in the preceding list. As the CDC has suggested, factors that have not yet been addressed appear to cause a consistent number of food-borne illnesses in the United States and worldwide.

REASONS FOR FOOD-BORNE ILLNESS

Food producers have at their disposal technologies for assuring food safety as never before in history. Food production plants use sensors to monitor constantly the food-making process to ensure temperature, moisture, and preservatives stay within specified ranges. Microbiologists use new rapid techniques to detect microorganisms rather than wait a day or more to incubate food samples. Food distributors additionally use refrigerated trucks, grocers ensure all food sold has a "sell by" date, and public health inspectors monitor the domestic food supply and imported foods. Yet with all this attention to food safety, some food microbiologists feel that the food supply has never before been as threatened as it is today. In 2004, Carol Tucker Foreman, former assistant secretary of agriculture, said, "I do think we have a serious public-health problem with regard to food-borne illness." The following factors have been cited as important contributors to food-borne illness in the 21st century:

- not enough food inspectors to monitor all the food produced domestically and internationally
- limited methods for testing food supplies before the food is shipped or sold
- mass production of foods, which increases potential spread of pathogens
- worldwide distribution of foods with varying levels of safety inspection
- lack of hygiene among untrained food handlers
- lack of hygiene, proper cooking, and storage methods in the home
- poor adherence to public health regulations by restaurants and grocers
- increase in emerging pathogens and drug-resistant microorganisms
- increase in the size of at-risk populations

Any of the factors listed here could, on its own, cause an outbreak of disease. In 2002, Patricia Griffin, head of the CDC's Food-borne Diseases Epidemiology (the study of the cause of disease outbreaks) Section, said in *National Geographic,* "Whether the overall incidence of food-borne disease has risen over the past generation is not known because we can't track all food-borne illnesses. What is clear is that the incidence is high, that some food-borne illnesses have clearly increased, and that dramatic changes in our food production system are likely to be playing a major role. Now we are more aware that the responsibility does not rest solely with the cook.

We know that contamination often occurs early in the production process—at steps on the way from farm or field or fishing ground to market." Hidden in Griffin's words is an explanation of the differences between rural, agrarian lifestyles and the lifestyles in industrialized societies, where almost every product, including food, is mass produced. Before the 1950s, in the United States, families raised much of their own meat and vegetables and prepared what they needed for feeding a single family. Prepackaged foods and items shipped from far away were rare—meat, dairy products, and fresh produce all were obtained from farms close to home. Mass production has overwhelmed much of the small, family-based food production of years past, especially in industrialized countries.

The behavior of consumers has also changed, since the 1950s, in ways that affect food handling. Some food handlers such as servers and kitchen workers do not wash their hands, sick workers who should stay home report to work in restaurants, and some obviously spoiled food items may not be discarded promptly. People in industrialized nations also eat out more than families did decades ago, creating a greater reliance on strangers to prepare a safe meal. Finally, advances in medicine have shifted the population, from 50 years ago, to a current population of more elderly people and individuals who have extended medical care and serious diseases, called a *high-risk* condition. Greater numbers of individuals who have higher-than-normal health risks mean that the chance of food-borne infection in the population will also increase. The main high-risk groups are as follows:

- elderly adults
- infants and young children
- people who have weakened immunity, such as acquired immunodeficiency syndrome (AIDS) patients
- organ transplant recipients
- people who have chronic disease
- diabetics
- pregnant women
- chemotherapy patients

National Geographic's Jennifer Ackerman pointed out, in a 2002 article on food safety, "In the name of efficiency and economy, we have also changed the way we raise our food animals. Our fish, cattle, and broiler and laying chickens are raised in giant 'factory' farms, which house large numbers of animals in tight quarters." This arrangement helps spread pathogens through animal populations and so increases the amount of pathogens in food products. In addition, massive farming operations produce large amounts of animal waste—as much as 500 million tons (454 metric tons) per year—that invariably pollutes community drinking water sources and nearby agricultural fields.

A 2006 outbreak of *E. coli* O157 infections was caused by fresh bagged spinach grown on a farm in California and shipped nationwide. After its investigation of the outbreak that had affected 27 states, the U.S. Food and Drug Administration (FDA) reported, in 2007, "The probe initially focused on the processing and packaging plant of Natural Selection Foods, LLC in San Juan Bautista, CA, where the contaminated products had been processed. The next focus of the inquiry was the source of the spinach in 13 bags containing *E. coli* O157:H7 isolates that had been collected nationwide from sick customers." The FDA investigators used molecular techniques such as deoxyribonucleic acid (DNA) fingerprinting to identify the farm where the pathogen originated and one particular field on that farm. Investigators proposed that the contamination may have been due to surface waters contaminated with manure from other farms, contaminated irrigation water, or wild animals traveling through the fields. Despite the excellent detective work in tracking down the illness, the FDA admitted, "Because the contamination started before the start of the investigation, and because of the many ways that *E. coli* O157:H7 can be transferred—including animals, humans, and water—the precise means by which the bacteria spread to the spinach remain unknown." In short, the reasons for any single food-borne illness can vary, and the chances are poor for finding a cause of an outbreak.

HYGIENE AND FOOD-BORNE ILLNESS PREVENTION

Because food-borne illness can occur almost anywhere food is served and because food-borne pathogens are invisible, people should follow good hygiene practices to protect themselves from receiving food-borne infection or spreading it. The following activities greatly reduce the chance of infection by a food-borne pathogen:

- washing hands thoroughly with soap and warm water before preparing meals or eating
- before handling food, washing hands after touching the face, touching another person, using the bathroom, or carrying, feeding, or diapering an infant or young child

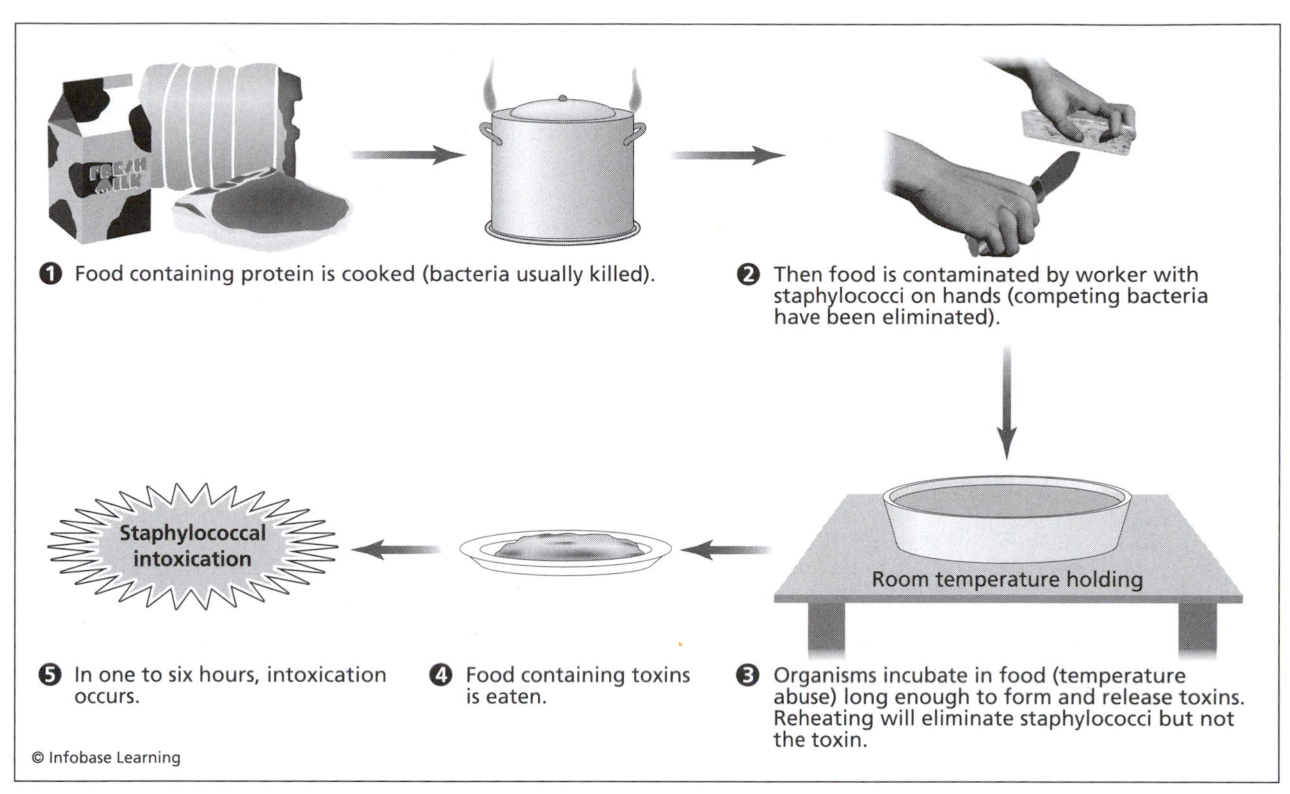

Proper handling of food prevents food-borne illness. Cooking kills many pathogens, but contamination also occurs between cooking and serving a meal. This figure shows that even if *Staphylococcus* cells are killed during cooking, the toxin may remain in the food.

- teaching children correct hand washing techniques—warm, but not hot, soapy water, for at least 20 seconds
- reducing the time that any food is outside a temperature range of 40–140°F (4.5–60°C)
- refrigerating (40°F [4.5°C] or lower) perishable foods within 90 minutes
- storing frozen foods at 0°F (-18°C) or lower
- following all cooking instructions on packaged foods
- avoiding foods past their "sell by" or "use by" dates
- avoiding meat in packages that are torn or leaking
- discarding any canned products in which the can bulges
- discarding foods with visible spoilage or bad odors
- preparing salads or other fresh foods on cutting boards or surfaces separate from those used recently for uncooked meats
- keeping raw meat, fish, and their juices away from fresh foods in the refrigerator and counters, cutting boards, and utensils used for fresh foods

Even with the many precautions that people can take to reduce the chance of getting or spreading a food-borne pathogen, food-borne illness will probably strike everyone at least once in his or her lifetime. Most food-borne illnesses are mild and do not threaten a person's long-term health. But some food-borne illnesses develop into serious conditions that call for immediate care by a doctor.

FOOD-BORNE INFECTION

A food-borne infection begins when a pathogen enters the digestive tract and begins to multiply. The pathogen then causes illness in one or more of the following ways: (1) damage of the epithelial cells that line the digestive tract, (2) penetration of the epithelial lining and entry into the bloodstream, or (3) production of toxins, which are poisons excreted by microorganisms. The resulting illness usually takes any of the following three general forms:

- intoxication—ingestion of a toxin in food or ingestion of a pathogen that produces a toxin in the body

Types and Causes of Dysentery

Type of Dysentery	Main Cause
amebic (amebiasis)	amoeba *Entamoeba histolyticus* from food or water
bacillary (shigellosis)	*Shigella* bacterial species
bacillary, general	*Salmonella, Campylobacter,* and *E. coli* 0157 in addition to *Shigella*
balantidial	*Balantidium coli* protozoa
viral	viruses, especially rotaviruses, norviruses, coronaviruses, and enteric adenoviruses

- gastroenteritis—general inflammation of the stomach and intestines
- dysentery—severe diarrhea accompanied by blood or mucus

In intoxication, toxins can damage tissue themselves, or they may increase the virulence, the ability to cause disease, of a live pathogen. Toxins may also resist heat better than the pathogen that produces them, meaning that cooking can kill the microorganism but will not destroy the toxin. (This is the reason why any food suspected of being contaminated should never be used even if it has been cooked.) Toxins from food-borne pathogens can cause any of the following: blood clotting, nerve damage, destruction of gastrointestinal (GI) epithelial lining, destruction of tissues of other organs, kidney failure, or general abdominal pain and cramps.

Gastroenteritis appears mainly as nausea, vomiting, and diarrhea and can result from infections from either bacteria or viruses. Most cases of gastroenteritis are treated with rehydration and a diarrhea treatment, but severe cases or cases in at-risk people can lead to hospitalization.

E. coli is a common cause of gastroenteritis that takes four different forms: (1) enteropathogenic gastroenteritis that damages the intestinal lining in infants with possible toxin release; (2) enteroinvasive gastroenteritis in which the bacteria destroy the intestinal lining and gain access to underlying tissue, causing dysentery; (3) enterohemorrhagic gastroenteritis associated with a toxin that causes kidney failure and lysis (breaking apart) of red blood cells, a condition called *hemolytic anemia;* and (4) enterotoxigenic gastroenteritis, in which most of the symptoms are caused by only the toxin on page 322.

Dysentery is a severe, life-threatening diarrhea associated with specific food-borne pathogens. Dysentery causes the body to lose so much water through the diarrhea that dehydration can occur rapidly and lead to prostration. The main microorganisms that cause this illness are described in the table (left).

Most food-borne illnesses never progress to life-threatening status but instead cause a variety of symptoms familiar to most people: nausea, vomiting, abdominal cramps, headache, fever, diarrhea, and lethargy. Food-borne illness, regardless of the cause, usually has more than one of these main symptoms. For this reason, food-borne illness cannot be diagnosed conclusively from the symptoms alone. Microbiologists help doctors diagnose food-borne illnesses by detecting and identifying a pathogen in a stool specimen.

FOOD-BORNE PATHOGENS

Almost any microorganism has the potential to cause an illness if a person ingests large numbers. In general, however, certain microorganisms are known to be prevalent causes of food-borne illness; the FDA lists more than 35 bacteria, viruses, and protozoa most commonly associated with food-borne illness. The table on page 313 describes the main microorganisms responsible for most food-borne infections or intoxications. Infective dose is the amount of cells known to cause symptoms, and incubation period is the time between ingestion of an infective dose of the pathogen and onset of symptoms. Infective dose is usually expressed as ID_{50}, or the amount of pathogen needed to cause symptoms in 50 percent of the population. Infective dose expressed as ID_{50} serves the following two purposes: It eliminates data from people who become infected but never get sick, and it allows epidemiologists to compare the infective dose of different diseases directly. (Toxin ID_{50} is expressed as the amount of toxin needed to make a healthy adult sick.) The wide range of infective doses and incubation periods might be due to variations in susceptibility between people in at-risk or high-risk health conditions and those at low risk.

Many food-borne pathogens originate with fecal contamination from either humans or animals. However, people who accidentally contaminate food with their own gastrointestinal (GI) bacteria or viruses do not become sick. The body's defense system eliminates most of the millions of microorganisms that a person ingests with foods each day. These defenses include the acids in the stomach, enzymes and bile acids that harm foreign microorganisms, and the GI's normal bacteria, which crowd out any incoming microorganisms. The GI tract also provides immunity against pathogens. Pathogens that attempt to attach to the GI lining are often repelled by immunity factors, and the natural motility of the intestines

Food-Borne Pathogens

Pathogen	Infective Dose	Incubation Period	Main Symptoms	Common Foods
Salmonella	20–1,000	6–48 hours	nausea, vomiting, abdominal cramps, fever, headache, mild diarrhea	meats, poultry, fish, eggs, dairy, sauces and dressings, cake mixes, cream fillings
Campylobacter jejuni	400–500	2–5 days	diarrhea, fever, abdominal pain, nausea, headache, muscle pain	raw chicken, raw milk, meats
Staphylococcus aureus	1 microgram	rapid	(T) nausea, vomiting, retching, abdominal cramping, prostration	meats, poultry, eggs, salads, dairy, cream fillings
Clostridium perfringens	several million	8–24 hours	(T) abdominal cramps, diarrhea	meats, gravy
C. botulinum	nanograms	4 hours to several days	(T) blurred eyesight or double vision, difficulty breathing and swallowing, muscle weakness	canned or bottled foods
norovirus (Norwalk virus)	100 or fewer	24–48 hours	nausea, diarrhea, abdominal pain, headache	shellfish, salads
Shigella	10	12–50 hours	(T) abdominal cramps, pain, fever, diarrhea, vomiting, lower bowel spasms	salads, raw vegetables, dairy, poultry
Cryptosporidium	1–10	2–7 days	cycles of watery diarrhea lasting 2–4 days and continuing for several weeks	vegetables exposed to contaminated water
E. coli	1 to several million	24 hours	severe watery or bloody diarrhea	any fecally contaminated foods
E. coli O157:H7	10–1,000	24 hours	(T) severe cramping, diarrhea	undercooked meats, sprouts, raw vegetables, unpasteurized fruit juices
Bacillus cereus	1 million	6–18 hours	(T) watery diarrhea, painful abdominal cramps	meat, milk, vegetables, fish
Listeria monocytogenes	1,000 or fewer	12–24 hours	fever, nausea, vomiting, diarrhea	milk, cheese, ice cream, sausages, raw meat, raw vegetables, smoked fish
Streptococcus	100–1,000	6 hours–3 days	fever, chills, dizziness, headache, nausea, vomiting, diarrhea	dairy, cheese, sausage, lobster, salads, pudding
hepatitis A	10–100	24–48 hours	fever, malaise, nausea, loss of appetite, jaundice	cold cuts, salads, vegetables, dairy, juices
rotavirus	10–100	24–48 hours	vomiting, diarrhea, low-grade fever	any fecally contaminated foods
Vibrio cholerae	1 million	6 hours–5 days	severe diarrhea, abdominal cramps, nausea, vomiting, shock	shellfish, foods exposed to contaminated water (vegetables, fruit, etc.)

Note: (T) indicates symptoms are caused by a toxin

sweeps the pathogens toward the colon, the posterior section of the GI tract. Only when very high doses of pathogens overwhelm the GI tract does immunity fall short in protecting a person from infection. The factors that determine whether a person will be sickened by food containing pathogens, therefore, depends on the relationships among three factors:

- virulence of the pathogen
- susceptibility of the person who ingests the food, called the *host*
- infective dose

These three factors determine whether a person will become sick from ingesting a specific dose of pathogens that possess a certain level of virulence.

A person normally has 400–500 different types of bacteria in the digestive tract, and these bacteria have probably evolved with humans to build a cooperative relationship called symbiosis. In one role, the GI bacteria that cling to the epithelial lining build a protective barrier against invasion by pathogens. By occupying all the receptor sites on the epithelial cells, the GI bacteria block he pathogens' access in a phenomenon called *competitive exclusion.* Normal GI microorganisms, called *gut flora,* also activate the immune system when they detect foreign microorganisms. But pathogens do not give up easily. Pathogens that enter the GI tract secrete substances that induce inflammation of the GI lining. The inflammation causes the body to secrete water into the intestines, which dislodges some of the normal gut flora—a person recognizes this response in the form of diarrhea, present in almost all food-borne illnesses. With a break in the defensive barrier, the invading pathogens, then, access the GI lining and the body's bloodstream.

Sanitation in Restaurants

by Anne Maczulak, Ph.D.

The food industry in the United States confronts greater challenges that ever before in providing safe, nutritious products to its customers. Safety in food production, processing, and preparation involves the exclusion of all unwanted substances. The main nonbiological contaminants of food are dust, dirt, metals, fibers, and chemicals, which can all endanger a consumer's health. Food microbiologists have the responsibility of preventing these items from entering food, as well as biological sources of food-borne illness, such as bacteria, fungi, protozoa, and viruses, and items that can carry these microorganisms into food: hair, dander, saliva, nasal excretions, and direct contact with the skin. The almost yearly recall by the U.S. Food and Drug Administration (FDA) of food products hints at the ongoing challenges in food safety.

Food safety begins with clean production processes on agricultural farms and slaughterhouses. The raw materials from these operations go to wholesale food processors or directly to restaurants. Both establishments must have superior programs in cleaning the facilities where food is prepared, sanitizing equipment, storing foods properly, and ensuring that food handlers follow all practices in good personal hygiene. Restaurants have been delegated an increasingly large burden in food safety for several reasons. First, an emergence of fast food restaurants, in the 1950s, increased the number of meals families ate outside the home. Second, food production in the United States evolved from small family-run businesses to large centralized production and distribution centers. These large operations made the transmission of food-borne germs to a larger population easier, as well as more difficult to trace in cases of food-borne outbreaks. Third, many foods consumed in the United States have international sources with varying commitments to safety regulations. FDA food safety inspectors cannot keep up with the large volumes of food that enter the United States, each year. Finally, consumers desire more imported foods, new ingredients, and innovative recipes, many of which affect the requirements for safe preparation.

For these reasons, sanitation makes up a critical part of the restaurant industry. The Culinary Institute of America has explained, "The importance of sanitation cannot be overemphasized. In a business based on service and hospitality, reputations and, indeed livelihoods are dependent upon the customer's good will. Few things are as detrimental to that good will as an official outbreak of a food-borne illness caused by poor sanitary practices." In this context, sanitation refers to cleaning and sanitizing of the kitchen and equipment; instituting methods that prevent cross-contamination from high-microbe foods to low-microbe foods; proper cooling, thawing, cooking, and reheating temperatures; pest control; and proper food handling by cooks and servers.

In sanitation, cleaning is the removal of all visible dirt, and sanitizing is the use of a chemical that reduces microbial numbers to safe levels. These two processes are accomplished in commercial kitchens by following a stepwise process: (1) sorting dirty dishes and utensils, (2) scraping off excess food, (3) prerinsing the items, (4) washing in detergent at 120°F (49°C), (5) rinsing off detergent in hot water, (6) sanitizing the items in water at 170°F (77°C) or in chemical sanitizer for two minutes, (7) draining, and (8) storing in clean, closed areas. Precleaning items by scraping off

In 2008, the biologist Martin Ackermann and his research team, at the Institute of Microbiology in Zurich and the University of British Columbia, reported finding a specialized subpopulation of *Salmonella typhimurium* that inhabited the digestive tract of animals and initiated the invasion of epithelial cells when a large dose of additional *S. typhimurium* entered the scene. The native *S. typhimurium* cells sacrificed themselves by setting up an invasion that would occupy the protective gut flora. The first attack, though repelled, would set up infection by the much larger army of *S. typhimurium* set to invade.

The body carries out these attacks and counterattacks between normal gut flora and pathogens hundreds, perhaps thousands, of times a day. People who are healthy and do not have underlying disease that puts them at higher risk for infection can thwart most food-borne infections before they begin. But statistics on food-borne illnesses clearly show that the pathogens often gain the upper hand. For people to prevent illness, their best defenses are conscientious handling of foods and proper hand washing. They will protect against some food-borne illnesses but probably not all. In the global business of food production, mistakes happen, and food-borne illness will be part of society forever. Food microbiologists can only try to lessen the effects and the incidence of outbreaks.

See also DNA FINGERPRINTING; *ESCHERICHIA COLI;* FOOD MICROBIOLOGY; HEPATITIS; HYGIENE; INFECTIOUS DOSE; SPECIMEN COLLECTION.

Further Reading

Ackerman, Jennifer. "Food: How Safe?" *National Geographic,* May 2002.

Ackermann, Martin, Bärbel Stecher, Nikki E. Freed, Pascal Songhet, Wolf-Dietrich Hardt, and Michael Doebeli. "Self-Destructive Cooperation Mediated by Phenotypic Noise." *Nature* 454 (2008): 987–990. Avail-

excess food is critical for the success of this process, because presence of large amounts of organic matter decreases the effectiveness of sanitization.

Cooks, food handlers, and the wait staff also have responsibility in assuring the food they serve is safe. Formal restaurants as well as fast food establishments often have a high turnover of kitchen and wait staff, so training is essential so that all employees learn good food handling techniques. These techniques, furthermore, do not have shortcuts. In other words, proper food handling must always include the following actions: (1) use of separate cutting boards for cooked and raw foods; (2) thorough hand washing with soap and warm water before starting a shift, before handling raw foods, and after using the restroom; (3) use of hair nets; (4) avoidance of touching hands to any part of the face, chewing gum, or wearing makeup; (5) use of disposable tasting spoons; (6) dating and rotating of inventory so oldest foods are used first; (7) wrapping and labeling of stored foods; and (8) preparation of foods as close to service time as possible. Well-managed restaurant kitchens also provide ample sinks, soaps, and paper towels. Toxic chemicals and cleaning products are always stored away from the food preparation area.

Restaurants take care in keeping all foods, cooked or fresh, out of a temperature range known as *the danger zone,* in which food-borne microorganisms multiply at the fastest rate. The danger zone ranges from 45°F (7°C) to 140°F (60°C). Some cooks add an extra margin of safety by using a range of 40–145°F (4–63°C). The following four practices assure that foods remain in the danger zone for a minimal period of time:

- hold in the danger zone for no more than two hours (some cooks recommend 90 minutes)
- foods prepared in advance of serving should be chilled to below 45°F (7°C) as quickly as possible
- reheated foods should be heated to at least 165°F (74°C) as quickly as possible
- thaw frozen foods in refrigerator, under running water at 70°F (21°C), or in a microwave
- cook thawed foods as quickly as possible

Even with the best practices in food preparation, germs find their way into food despite these preventions and food-borne illness will never be eliminated. People make mistakes or become distracted in busy kitchens. Not everyone's training in hygiene meets these minimal requirements. In addition, unknown microorganisms may contaminate a food through behavior that evades these established methods in food safety. Sanitation in restaurants will be expected to evolve as the demands on restaurants increase and food becomes more a global, mass-produced commodity.

See also FOOD MICROBIOLOGY; SANITIZATION; TRANSMISSION.

Further Reading

Arduser, Lora, and Douglas Robert Brown. *HACCP and Sanitation in Restaurants and Food Service Operations.* Ocala, Fla.: Atlantic, 2005.

Donovan, Mary Deirdre, ed. *The New Professional Chef.* New York: Van Nostrand Reinhold, 1995.

Jones, Timothy F., Boris I. Pavlin, Bonnie J. LaFleur, L. Amanda Ingram, and William Schaffner. "Restaurant Inspection Scores and Foodborne Disease." *Emerging Infectious Diseases* 10 (2004): 688–692. Available online. URL: www.cdc.gov/ncidod/EID/vol10no4/03-0343.htm. Accessed October 3, 2009.

able online. URL: www.vancouver.wsu.edu/fac/bishop/Teaching/A%20Biol%20403%202009/Readings/Ackerman% 20Doebli%20cooperation.pdf. Accessed August 10, 2009.

Bhunia, Arun K. *Foodborne Microbial Pathogens: Mechanisms and Pathogenesis.* New York: Springer, 2008.

Centers for Disease Control and Prevention. "Food-Related Diseases." Available online. URL: www.cdc.gov/ncidod/diseases/food/index.htm. Accessed August 10, 2009.

———. Foodnet—Foodborne Diseases Active Surveillance Network. Available online. URL: www.cdc.gov/foodnet. Accessed August 10, 2009.

U.S. Food and Drug Administration. *The Bad Bug Book.* Available online. URL:

———. "FDA Finalizes Report on 2006 Spinach Outbreak." *FDA News,* March 23, 2007. Available online. URL: www.fda.gov/NewsEvents/Newsroom/PressAnnouncements/2007/ucm108873.htm. Accessed August 11, 2009.

Wikoff, William R., Andrew T. Anfor, Jun Liu, Peter G. Schultz, Scott A. Lesley, Eric C. Peters, and Gary Siuzdak. "Metabolomics Analysis Reveals Large Effects of Gut Microflora on Mammalian Blood Metabolites." *Proceedings of the National Academy of Sciences* 106 (2009): 3,698–3,703.

food microbiology Food microbiology is a specialized discipline within industrial microbiology that exists for two purposes: food production and prevention of food spoilage and food-borne illness. This latter purpose is also called *food protection.*

Food production consists of three main areas: (1) use of microorganisms for making foods; (2) food canning, bottling, wrapping, and packaging; and (3) food preservation. Food production involves strict adherence to a food protection program that probably involves the following areas: equipment sterilization and cleaning, proper food storage, and proper handling and use of food products by consumers.

Food presents many challenges for microbiologists in assuring no unwanted microorganisms are present. This is because of the characteristics of food, as follows:

- rich supply of nutrients
- favorable moisture content
- favorable pH in most foods
- storage temperatures that are often in a microorganism's optimal temperature range
- tendency of foods left out of the refrigerator to warm quickly to room temperature
- possibility of food contamination due to handlers' poor hygiene
- international shipments of food products

International food shipment has emerged as a particularly important aspect of food microbiology. Food-borne outbreaks occur by the hundreds

Food and Microorganisms in History

Date	Event
7000 B.C.E.	Babylonians ferment grains to produce beer
4000	cheese produced, perhaps by accident in spoiled milk, in Arabia
3500	wine appears
3000	several societies learn to use salt for food preservation
2575	art in Egypt depicts use of yeast for bread making
2200	Chinese emperor Hsia Yu puts a tax on salt, an essential food preservative
1400	bread leavened with yeast is a staple of the Egyptian pharaohs
1000	Romans prepare smoked meats and use snow to preserve shrimp
1300 C.E.	Marco Polo trades spices, a food preservative, from Asia to Europe
1492	Columbus seeks a shorter route for the spice trade
1940s	homes replace iceboxes with refrigerators for food storage
1795	Frenchman Nicholas Appert invents a method for canning foods
1860s	Frenchman Louis Pasteur demonstrates microorganisms exist in spoiled milk and in wine and beer
1906	Federal Food and Drug Act passed to ensure food safety
1920s	dairy industry adopts pasteurization to prevent milk-borne diseases
1950s	food production shifts from local sources to mass production and widespread distribution

each year in the United States. Most are undetected because the source or the scope of the outbreak cannot be determined before the outbreak ends. Food microbiologists consider all of the preceding factors when they participate in national policies for keeping the food supply safe.

The preparation and preservation of foods began several thousand years ago, and, remarkably, many of the methods used in ancient history have changed very little to this day. The table provides a brief history of food microbiology.

Early civilizations sought the benefits of preservation to ensure a food supply through the autumn and winter. Food preservation also gave stability to societies that no longer needed to live a nomadic life, following crops in bloom or herds of animals. The oldest methods of food preservation are salting, smoking, fermenting, and chilling.

Preservation is crucial in maintaining a safe food supply, because food is one of the best places a microorganism can find to live. Food microbiologists emphasize the activities of microorganisms on or in food, either to make improved fermented products or to protect the food from contamination. These microbiologists must have knowledge of the types of microorganisms that prefer particular foods, factors that affect the multiplication of microbial cells in food, and the chemical and physical characteristics of foods. By knowing chemical and physical characteristics, a microbiologist can make better choices in preservatives. The chemical characteristics of food to be considered in food microbiology are oxygen content, pH, oxidation-reduction capacity, natural preservatives that may already be present, and nutrient content, which means the amount of sugars, amino acids and other nitrogen sources, fats, vitamins, and minerals. Physical characteristics of food are the following: water content, form (solid, semisolid, or liquid), and temperature the food reaches during production, packaging, shipment, and storage.

MICROORGANISMS IN FOOD PRODUCTION

Yeasts, fungi (molds), and bacteria make up the main types of microorganisms that produce healthy foods. These microorganisms participate in the preparation of almost every type of food of either an animal or a plant source. Many of these are fermented foods, meaning microbial actions take place in the food under anaerobic conditions with the main end products of organic acids or alcohols. Both of these end products serve two purposes: (1) They change the taste and sometimes the consistency of the food, and (2) they act as preservatives that inhibit the growth of spoilage microorganisms. The table on page 318 describes the main foods used today that are made entirely or in part by microbial activities.

Cheese, wine, and bread predate recorded history. Archaeological excavations and ancient art suggest that these foods were important in diets more than 4,000 years ago and may date to 7000–8000 B.C.E. Cheese serves as a good representative of foods that undergo microbial changes in order to alter the flavor, consistency, and storage. In many cases, cheese contains a better variety of nutrients than the milk from which it is made because of contribution of microbial vitamins and amino acids.

MODERN FOOD PRODUCTION

Food microbiology today must adapt to large production plants that make food products in quantities to feed many thousands of people a day. The products are packed onto trucks or ships and sent long distances, sometimes with variable temperature. Food microbiologists working in food production plants have a critical responsibility to ensure that all the food packed and shipped is of the highest quality and safety.

Food microbiologists can decrease the chance of food spoilage as well as the presence of dangerous illness-causing microorganisms by using the following techniques:

- filtration of liquids and water to remove microbial cells
- heating to kill all microorganisms and deactivate their toxins
- adjustment of water content to reduce water availability for microbial growth
- addition of chemical preservatives
- low-temperature storage to slow the growth of any microorganisms present
- contamination prevention programs

Contamination prevention consists of two components: a program called *HACCP,* for "hazard analysis and critical control point," and a cleaning and sanitation program. HACCP programs consist of all the activities of a food from the manufacturing plant to a consumer and the methods at each step along the way for ensuring the food's safety. HACCP processes help identify the places in the food production process that are most likely to receive microbial contamination; these activities or sites are called *critical points.* Microbiologists develop a HACCP program that accounts for monitoring the food production and eliminating contamination before it has a chance to enter the system. Cleaning is part of HACCP and involves rinsing

Foods Made by Microorganisms

Food	Examples	Main Microorganisms (bacteria unless noted otherwise)
cheese	• blue • Camembert • cheddar • mozzarella • Swiss	• *Brevibacterium* • *Lactobacillus* • *Leucono stoc* • *Penicillium* (mold) • *Propionibacterium*
fermented milk	• buttermilk • kefir • koumiss • yogurt	• *Bifidobacterium* • *Geotrichium* (mold) • *Lactobacill us* • *La ctoco ccus* • *Leuconostoc*
meat	• cured ham • sausages	• *Pediococcus* • *Lactobacillus*
fish	• izushi • katsuo-bushi	• *Aspergillus* • *Lactobacillus*
bread	• loaf bread • sour dough	• *Lactobacillus* • *Saccharomyces* (yeast)
vegetables and fruits	• beer • kimchi • miso • olives • pickles • soy sauce • tofu • wine	• *Aspergillus* (mold) • *Lactobacillus* • *Lactococcus* • *Pediococcus* • *Rhizopus* (mold) • *Saccharomyces* (yeast)

equipment with water and detergents. Sanitation is a more rigorous type of cleaning for removing dirt, wastes, and other possible contaminants.

Two main types of cleaning are performed in food manufacturing plants: clean-in-place (CIP) and clean-out-of-place (COP). In CIP cleaning, workers clean assembled food production equipment. The workers might alternate water rinses with soaps or stronger chemicals called *sanitizers* that kill microorganisms on inanimate surfaces. COP involves disassembling food production equipment and thoroughly cleaning and sanitizing each piece before reassembling the equipment. Common chemicals used in cleaning and sanitizing food production equipment are chlorine compounds, such as chlorine dioxide, and peracetic acid.

FOOD QUALITY AND PRODUCTION

Food packaging in clean containers is a critical component of food protection. Microbiologists must ensure that no contamination occurs when food is put into containers, a step called the *filling process*. In order to reduce contamination during the filling process, food manufacturers use aseptic packaging, which is the handling of foods and containers by methods that prevent the entry of contamination. Aseptic packaging contains the following components to achieve this:

- sterilization of packaging with heat, hydrogen peroxide, or ultraviolet irradiation
- heating of foods to a temperature that eliminates dangerous microorganisms
- use of sterilized filing equipment to put the food product into the sterile packaging
- sealed containers

Microbiologists assist the food production process with two methods: product testing and environmental monitoring. Product testing is the analysis of several representative food samples—in the final packaged form—for the presence of any contaminants known to be in a certain food. For example, meat product manufacturers test for the presence of *Campylobacter* bacteria, producers of canned vegetables test for *Clostridium* bacteria, and spice manufacturers check for the presence of molds.

Microbiologists also monitor a more general group of microorganisms that are known to contaminate all kinds of foods. But testing many different types of microorganisms is time-consuming and labor-intensive. Microbiologists depend on indicator organisms to represent many other species. The following list shows the main indicator organisms in food microbiology:

- coliforms—general group of gram-negative bacteria that indicate poor hygiene and sanitation in the manufacturing process
- fecal coliforms—bacteria such as *Enterobacter* that indicate contamination from fecal matter
- *Escherichia coli*—the most prominent fecal coliform that confirms fecal contamination
- fecal streptococci and enterococci—general group of fecal contaminants that indicate contamination of frozen foods (because fecal coliforms and *E. coli* do not usually withstand freezing)
- *Staphylococcus aureus*—indicator of contamination from people working on the production line

- *Clostridium botulinum*—indicator of contamination from anaerobic bacteria
- *Pseudomonas aeruginosa*—indicator of contamination from water sources

Microbiologists begin the test for these microorganisms by taking a sample of the food from the production line. In a laboratory, the microbiologist dilutes each sample with sterile water and, then, puts a small portion called an aliquot on growth medium. After incubation, the microbiologist inspects the medium for the presence or absence of microorganisms.

The second task of food microbiologists involves monitoring the food manufacturing facility. This process, called *environmental monitoring,* is done for the purpose of detecting microorganisms in the air, on equipment surfaces, and in the facility's water. Environmental monitoring and HACCP go hand in hand in food protection.

Anne Sherrod, an industry director of food safety, told *Food Quality* magazine, in 2007, "If our in-house testing indicates conditions are out of specification, we do a root cause analysis to find the ingredient, work in process, or finished product of concern. If the problem were in an ingredient, we would go back to the supplier to address the situation with them." As Sherrod implies, a microbiologist's job involves detective work in tracking down the source of a contamination.

Environmental monitoring and HACCP together make up a program called *quality control* (QC). QC comprises all the activities that ensure a product meets a manufacturer's specifications for a high-quality food. (Manufacturers of all products, not only foods, include QC programs in their business.) In food manufacturing, the absence of microbial contamination certainly makes up a significant part of QC, but other typical components of QC are as follows:

- taste
- color
- consistency
- odor
- separation of phases, such as oil and water
- gas production

The characteristics of a food either appeal to consumers or indicate possible spoilage by microorganisms. Some attributes of the food, such as color or the presence of gas, are easy to see. But microorganisms are invisible, and unless the food contains a huge amount of microorganisms, a consumer might not realize that a pathogen is in a food. Food microbiology is, therefore, essential in assuring food's safety.

FOOD SPOILAGE

Canning, bottling, use of soft packaging and wrappers, as well as freezing and refrigeration all take place for the purpose of reducing microbial growth in food. Despite careful packaging and monitoring, many foods contain a large number of microorganisms. A good monitoring system combined with secure packaging helps a manufacturer reduce the risk of their product's spoiling, but consumers must understand that few foods last forever without going through some chemical or microbial breakdown. The table on page 320 describes the main factors affecting microbial growth in foods.

The table on page 320 provides examples of certain conditions commonly found in food and microorganisms that grow under those conditions.

Packaging works in tandem with the characteristics of the food in food protection. For example, meat retailers package items such as steak in wrapping that prevents microorganisms from contaminating the meat's surface. In addition, they wrap meat in an atmosphere containing carbon monoxide, which maintains the meat's color and inhibits bacteria. The following list includes some additional methods used in food packaging and how they work:

- Refrigeration or freezing slows microbial metabolism.
- Drying decreases a_W.
- Adding sugar, salt, or acid kills many potential food spoilage microorganisms.
- Canning and bottling ensure sterile conditions in the container.
- Vacuum packaging reduces oxygen availability.
- Pasteurization kills the majority of microorganisms by heating.
- Gamma ray irradiation kills insects as well as bacteria, fungi, and viruses.
- Chemical preservatives can be selected to target microorganisms known to spoil specific foods.

Any of today's most common packaging methods reduces the chance of food spoiling and so increases a measurement called *shelf life,* which is the amount of time a food can last before deteriorat-

Factors Affecting Microbial Growth in Food

Factor	How It Helps Microbial Growth	How It Inhibits Microbial Growth
nutrients	provide energy, carbon, nitrogen, and other growth factors	high sugar, salt, or fats can harm or kill microorganisms
water activity (a_W)	high a_W increases water availability for maintaining cell cytoplasm and enzyme activity	low a_W stunts microbial growth
pH	pH far outside a microorganism's optimal range stops growth	pH may inhibit the growth of all microorganisms except those that thrive at high or low pH
oxidation-reduction potential (E_h)	Eh favorable to a microorganism's oxygen requirements enhances growth	higher Eh favors aerobes and inhibits anaerobes; lower Eh favors anaerobes and inhibits aerobes
natural inhibitors	eliminate competition from other microorganisms	bacteriocins, enzymes, or ability to alter the pH creates an advantage over other microorganisms
natural barriers	skins and shells provide a contained environment of optimal a_W, nutrients, pH, etc.	fruit and vegetables skins, eggshells, etc., create a protective barrier against invading microorganisms

Note: the table shows the various factors that can help or hinder microorganisms that spoil foods. The growth of any microorganism in food depends on two overall considerations: the characteristics of the food and the growth requirements of the microorganism

ing. Factors other than microorganisms can shorten shelf life, such as oxygen, chemical reactions, and moisture loss. Microorganisms, nonetheless, play a pivotal part in affecting shelf life through the following microbial activities: digestion of proteins, carbohydrates, or fats; gas production; acid production; production of pigments; and slime formation.

FOOD PRESERVATIVES

The purpose of any food preservative is to stop the growth of unwanted microorganisms before they multiply to high numbers. High concentrations of microorganisms cause infection or toxicity by secreting a toxin. That is why microbiologists speak about the microbial load of food, which is the total amount of microorganisms in a known quantity of the food.

Food preservation in the form of drying or freezing dates to the earliest civilizations. Methods also developed that combined chemical factors with physical factors to prevent microbial growth, such as smoking or salting foods or both. Fermented foods also use the end products of chemical reactions as preservatives. These end products are usually acids or

Various Food Conditions and Effects on Microbial Growth

Condition	Type of Food (example)	Microorganism It Favors (bacteria unless otherwise noted)
acid foods, pH 4.0	mayonnaise	*Lactobacillus*
nonacid foods, pH more than 5.3	seafood	*Vibrio*
high E_h, 250 millivolts (mV)	ground meat	*E. coli*
low E_h, -200 mV	canned vegetables	*Clostridium botulinum*
high a_W, 0.98	fresh poultry	*Clostridium perfringens*
low a_W, 0.80	cake	*Penicillium* (mold)

Note: E_h is a measure of anaerobic conditions; lower E_h represents conditions with less free oxygen

alcohols, and they greatly extend the shelf life of various fermented foods beyond that of the fresh food.

Today, food microbiologists contend with trying to preserve foods meant to satisfy changing consumer tastes. Consumers now seek various imported products, as well as foods with the following attributes: minimal processing; less fat, salt, and sugar; more natural and fresh ingredients; and no added chemical preservatives. "Not only do we like our foods diverse and available year-round," wrote Jennifer Ackerman for *National Geographic,* in 2002, "but we also like them convenient—prepackaged, preferably, and ready-to-eat. This means we're leaving to commercial foodmakers the peeling, chopping, and mixing of our food." The same article quoted Patricia Griffin of the Centers for Disease Control and Prevention (CDC): "The more untrained people handling food, the greater the risk of inadequate cooking or of cross-contamination of safe foods from unsafe or uncooked foods." In some cases, it seems as if food preservatives serve as the last line of defense between a consumer and food-borne illness.

Chemical preservatives added to food produced in the United States are under the jurisdiction of the U.S. Food and Drug Administration (FDA). The FDA requires that preservatives be effective against microorganisms but safe for humans. Some common chemical preservatives approved by the FDA are the following: acetic acid, ascorbic acid, benzoic acid, citric acid, sorbic acid, sodium phosphate, sodium or potassium sulfite, sodium nitrite, and sucrose. (Each of the acids, such as sorbic, may also appear on food labels with the suffix *-ate,* such as *sorbate.*)

Today's food microbiologists have adopted sophisticated techniques and instruments for monitoring food safety and protecting foods against contamination. Food microbiology continues to use techniques that have been proven, over centuries, to be effective. New technologies in packaging and chemical preservation are also at the microbiologist's disposal. The responsibilities put onto food microbiology can be expected to increase in the future.

See also ASEPTIC TECHNIQUE; FERMENTATION; FOOD-BORNE ILLNESS; HACCP; HYGIENE; PASTEURIZATION; PRESERVATION; SANITIZATION; STERILIZATION; TOXIN.

Further Reading

Ackerman, Jennifer. "Food: How Safe?" *National Geographic,* May 2002.

Banwart, George J. *Basic Food Microbiology,* 2nd ed. New York: Chapman & Hall, 1989.

Huemoeller, Jody L. "In the Lab: DIY *Staph* Testing." *Food Quality,* October/November 2007.

Microbes.info. "Food Microbiology." Available online. URL: www.microbes.info/resources/Food_Microbiology. Accessed August 12, 2009.

Roberts, Diane, and Melody Greenwood. *Practical Food Microbiology,* 3rd ed. Hoboken, N.J.: John Wiley & Sons, 2002.

U.S. Food and Drug Administration. Center for Food Safety and Applied Nutrition. Available online. URL: www.cfsan.fda.gov/list.html. Accessed August 12, 2009.

U.S. National Food Safety Programs. Available online. URL: www.foodsafety.gov/~dms/fs-toc.html. Accessed August 12, 2009.

World Health Organization. "Food Safety." Available online. URL: www.who.int/topics/food_safety/en. Accessed August 12, 2009.

fractionation Fractionation, or cell fractionation, is the physical separation of cells into their component parts. This process separates prokaryotic cells into their components, such as cell wall fragments, membranes, and deoxyribonucleic acid (DNA). Fractionation also separates eukaryotic cells into individual organelles, for example, ribosomes, mitochondria, and nuclei. Biologists use fractionation to determine cell composition by further disassembling cellular material into carbohydrates, fats, nitrogen compounds, and cell wall polymers.

The steps of cell fractionation begin with disruption, which is any method that breaks apart a cell to release its contents. After cell disruption, fractionation usually involves a series of physical methods, such as centrifugation or filtration, to remove large particles from small particles, or chemical methods, such as precipitation or chromatography to draw a molecule away from a mixture of other molecules.

The steps in fractionation vary, depending on the nature of the material that a microbiologist wishes to recover. For example, cell wall polymers behave differently than membranes in a suspension, so each requires specific steps. Fractionation steps, nonetheless, follow a general pattern described in the table.

CELL DISRUPTION

Cell disruption involves breaking apart cells to release the contents unharmed. Disruption is also called *lysis* when the cell bursts because of the action of enzymes on the membrane or changes in the environment's pressure on the cell. *Disintegration* is a general term for any physical or chemical disassembly of a cell. Cell disruption may occur with any of three different methods: (1) physical or mechanical, (2) osmotic pressure, or (3) chemical.

Physical or mechanical cell disruption methods use either of two approaches. First, special cell disruption equipment exerts pressure on microbial cells until they break open. Gram-positive bacteria, bacterial endospores, and yeasts are most resistant to mechanical disruption; gram-negative bacteria, protozoa, and animal cells are easier to disrupt by this technique. Second, abrupt changes to the cell's environment cause lysis to release the cell contents.

Mechanical cell disruption uses four main types of equipment: the French press, the Hughes press, manual tissue disintegrators, or ultrasonication. The French press, Hughes press, and some manual tissue disintegrators exert shearing forces on the cells, which cause them to break. A French press exposes cells to 10 tons of pressure per square inch (1.4 metric tons/cm^2). A related apparatus called a *Ribi cell fractionator* produces higher pressures for breaking more resistant cells and bacterial endospores. An endospore is an extremely rugged structure that certain bacteria form as a protection against the environment. The microbiologist Carl Schnaitman explained about the Ribi fractionator, "Breakage is virtually instantaneous, and the broken suspension is not subjected to additional shear forces which could damage subcellular particles." The smaller pieces collect in a cooled chamber away from the shearing action. A Hughes press acts in a similar manner on frozen suspensions, in which ice crystals help the lysis under about 30 tons of pressure per square inch (4.2 metric tons/cm^2).

Microbiologists have used simple mortar and pestles to lyse easy-to-break cell types, or they use more aggressive instruments such as a tissue homogenizer. This glass instrument holds about 10 ml of suspension, into which the microbiologist pushes a plunger that makes a tight fit inside the outer vessel. As the microbiologist pushes on the plunger, it forces the suspension to the small space between plunger and vessel, creating a shearing force.

Very hard-to-break particles, such as endospores, might require stronger forces such as the ballistic forces produced in a Braun MSK tissue disintegrator. This apparatus shakes the cells with small glass or hard plastic beads at several thousand oscillations per minute. A five-minute treatment in a Braun disintegrator can break apart bacterial endospores.

Ultrasonication produces rapidly moving bubbles within a suspension that set up shear forces that break cells. Ultrasonication has the disadvantage of often breaking apart organelles at the same time it disrupts the outer cell membrane and wall. Ultrasonication also raises the suspension's temperature and creates foam; either effect can denature proteins. *Denaturing* refers to the unraveling of proteins from their normal shape. As a result, the protein cannot function as normal.

Freeze-thawing is another easy physical method to make cells, especially gram-negative bacteria, susceptible to damage and breaking. In this technique, a microbiologist puts cells into a suspension with a pH 7.8 buffer that contains ethylenediaminetetraacetic (EDTA), the sugar sucrose, and the enzyme lysozyme. The microbiologist, then, makes the suspension's temperature drop rapidly by putting the vessel into an ice-acetone bath. The suspension is then warmed, poured into another cold buffer, and blended. The freezing-thawing cycles, followed by blending, weaken the cell walls, and the EDTA and lysozyme, then, destroy the wall and membrane. Sucrose acts as a matrix for collecting the cell organelles without causing further damage to them. A microbiologist finishes the procedure by centrifuging the mixture to remove the cell components from any unbroken cells.

Steps Used in Cell Fractionation

Step	Description	
1	disruption, lysis, or disintegration	breaking open the cell wall and cell membrane to release the cell contents
2	filtration	gross removal of large particles, especially when trying to recover a material soluble in the cyroplasm
3	centrifugation or sedimentation	separation of materials of different density by applying centrifugal force
4	washing	application of detergents to separate water-soluble components from non-water-soluble compounds
5	chromatography	separation of dissimilar substances based on their relative movement through a stationary material that is exposed to a flowing liquid
6	polyacrylamide gel electrophoresis (PAGE)	separation of various polypeptides (long chains of amino acids) from each other based on electric charge
7	precipitation	recovery of a dissolved molecule, such as DNA or ribonucleic acid (RNA), from an aqueous solution by using ethanol and salt

Osmotic lysis of cells occurs when the cells suddenly become exposed to dilute buffer or water. This method works best for lysing spheroplasts (gram-negative bacteria lacking a cell wall or with a cell wall damaged in spots) and protoplasts (gram-positive bacteria lacking a cell wall). Osmotic lysis also aids in DNA recovery from bacteria. In intact bacterial cells, DNA is folded and compacted into an area of the cytoplasm called a *nucleoid*. But RNA molecules within the nucleoid stabilize the large DNA molecule, so osmotic lysis must include a step to release the DNA from the RNA. The enzyme ribonuclease (RNase), in combination with magnesium, digests the RNA that holds the DNA molecule and allows the DNA to unfold. A microbiologist can make a simple addition of salt to the solution that causes the DNA to precipitate and settle to the bottom of the liquid. Some animal cells and bacteria are sensitive to osmotic lysis and break apart easily, but many other bacteria, such as gram-negative varieties, resist this method and require other physical methods to break the cells.

Chemical disruption methods work alone or in conjunction with physical methods, depending on the type of cell that must be broken apart. This method may also be thought of as a biological method, because the agents used for disrupting cells have biological sources. Enzymes such as lysozyme or lysostaphin both act on strong peptidoglycan molecules that make up bacterial cell walls. Lysozyme extracted from egg whites is commercially available for this purpose; *Staphylococcus* bacteria produce lysostaphin. Cell disruption procedures might employ an additional enzyme named *muramidase* made by *Staphylococcus* or the fungus *Chalaropsis*. All of these enzymatic steps usually include EDTA to help destroy the cell wall and membrane.

SEPARATION TECHNIQUES USING CENTRIFUGATION

Cell disruption may be the most difficult phase in cell fractionation. Disrupted cells need only be filtered or centrifuged to complete the separation of cell components. In some cases, centrifugation alone can separate every cell component from every other component.

The main centrifugation techniques used in cell fractionation are described in the table on page 324.

Rate-zonal and density gradient centrifugations depend on a solution in which the density increases from the top of the tube to the bottom in order to separate one type of particle from another. Sucrose often serves as the chemical used in making density gradients, because it does not destroy the fragile cell components present in a suspension. These techniques are, therefore, also called *sucrose gradients*.

Differential centrifugation alone can separate almost every cell component from every other by performing a series of centrifugations of filtrate at higher and higher speeds—at higher speed, the smaller or less dense particles will settle from the suspension. This method's drawback resides in the many different centrifugation runs that would be required to complete the cell fractionation. Differential centrifugation has played its main role in microbiology in separating whole microbial cells from cell fragments and other debris and in isolating a single organelle, such as the ribosome.

Rate-zonal and density gradient centrifugation work best in recovering many organelles of similar size, shape, and density. Rate-zonal centrifugation has been used for separating and collecting different subunits of ribosomes, nuclei, and various forms of DNA, such as nucleoid DNA and plasmid DNA. Density gradient centrifugation separates small membrane vesicles, membrane fragments, and smooth endoplasmic reticulum.

DETERGENT WASHING

Cell fractionation with chemicals depends mainly on detergents. Detergents are solutions that alter the normal surface tension that exists between aqueous liquids (water) and oily liquids. They break surface tension because they have both a hydrophilic (water-attracting) end and a hydrophobic (water-repelling) end. This structure, called an *amphipathic* molecule, enables detergents to straddle the interface between water and oily substances. All detergents belong to one of three categories, which are based on the characteristics of how they react with hydrophilic and hydrophobic substances. Ionic detergents, either anionic or cationic, contain a hydrophilic end that holds a positive or negative charge. Anionic detergents have a negative charge; cationic detergents have a positive charge. Nonionic detergents lack a charged pole.

Various detergents have different pH requirements in order to work because pH influences the molecule's overall charge. In acidic conditions (low pH), a surplus of hydrogen ions (H^+) is present. Conversely, alkaline conditions (high pH) offer a dearth of hydrogen ions but a surplus of electrons (e^-). Microbiologists conducting cell fractionation with detergents must understand the chemical requirements of these compounds in order to achieve the results they want. The detergents used in cell fractionation and other activities in biology are summarized in the table on page 325.

In cell fractionation, detergents serve three purposes. First, they lyse cells after other methods have destroyed the cell wall. Second, detergents dissolve cell cytoplasm but leave the outer membrane intact. Third, detergents can gently remove cell debris from fragile organelles.

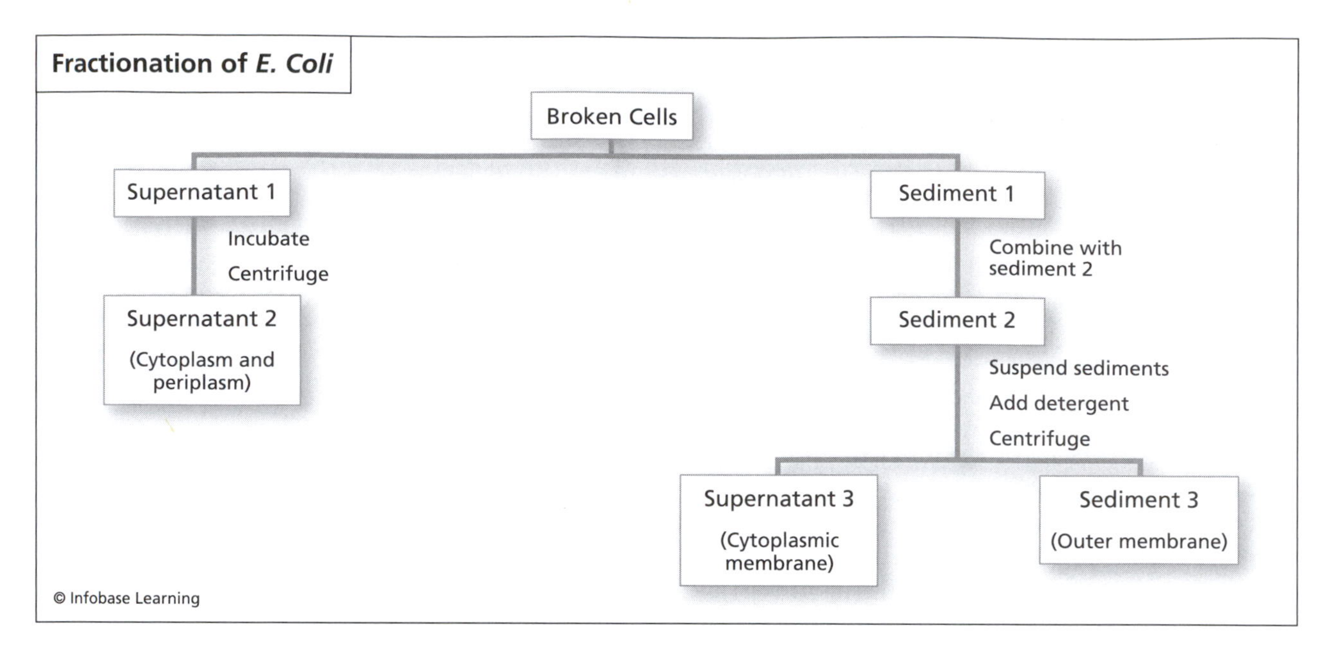

Fractionation involves a series of techniques that separate cell components. A French press can be used to break whole cells into pieces. Centrifugation separates the insoluble components, and detergents separate detergent-soluble and -insoluble substances.

CHROMATOGRAPHY

Chromatography is any method that separates molecules on the basis of how each different molecule moves between a stationary material and a flowing substance called the *mobile phase*. Chromatography mobile phases can be either gas or liquid. Various techniques in chromatography fulfill two purposes: to identify an unknown compound and to separate a compound from others. In cell biology, scientists may prefer to use a chromatography method that does not destroy the compound so that the compound may be studied in further tests.

Microbiologists use the chromatography methods described in the table on page 325 for studying cellular compounds.

Biologists also use a chromatography technique called *elution* to recover individual molecules from a mixture after the main fractionation steps have been completed. Elution involves adding the mixture to a stationary material, which is then exposed to a flowing liquid, called a *mobile phase*. The mobile liquid draws the desired molecule away from the stationary material, which continues to bind all of the other materials in the mixture. A mobile liquid of a particular pH or salt concentration usually works well in eluting specific end products of fractionation.

POLYACRYLAMIDE GEL ELECTROPHORESIS

PAGE has become the most common technique for separating different polypeptides from each other.

Centrifugation Methods in Cell Fractionation

Centrifugation Method	Description	Main Use
differential	centrifugation for a given time at a given speed	separates one particle from all others in a suspension because of the speed at which it settles out of the liquid
rate-zonal	centrifugation in a liquid that creates a density gradient; particles migrate to different depths, depending on centrifugation speed and time	separates organelles with similar density but different shape or mass
density gradient	centrifugation in a liquid that creates a density gradient; particles migrate to the region in the gradient that matches the density of the particle	separates particles on the basis of buoyant density in a liquid

Detergents Used in Cell Fractionation

Category	Activity	Examples Used in Cell Fractionation
ionic	strongly binds to proteins, resulting in protein denaturation	sodium dodecyl sulfate (SDS); cetyl trimethyl-ammonium bromide (CETAB)
nonionic	binds to hydrophobic proteins with little loss to protein function	Triton X-100; polyethylene glycol; CHAPS
bile salts	associates with fatty substances such as membrane phospholipids	commercial bile salts

Note: CHAPS = 3-[(3-cholamidopropyl) dimethylammonio]-1-propanesulfonate

The gel, called a separation or running gel, contains acrylamide, bis-acrylamide, and the detergent sodium dodecyl sulfate (SDS), also called sodium lauryl sulfate. This method is, therefore, also referred to as *SDS-PAGE*. In SDS-PAGE, the acrylamide compounds and the SDS carry out two different actions. Acrylamides coalesce in the presence of two other chemicals (persulfate and tetramethylethylenediamine) to form a gellike sheet, a step called *polymerization*. The SDS denatures the proteins into peptides so that the peptide pieces migrate through the gel matrix at different speeds.

Peptides move in SDS-PAGE gels as a result of a constant electrical current applied to the gel and the liquid buffer that carries the proteins through the gel matrix. The voltage induces the peptides, which all have an overall negative charge, to migrate toward a positive pole at the far end of the gel. Because of the winding path the peptides must traverse through the gel matrix, small peptides move faster than larger peptides. In this way, SDS-PAGE differentiates numerous peptides by two mechanisms: electrical current and pore size in the gel matrix.

The PAGE peptide bands are invisible, so an extra ingredient is added to the sample, a stain, that shows the progress of the sample as it moves through the gel. Scientists commonly use the stain Coomassie blue to locate the front edge of the migration in the

Chromatography Methods

Method	Description
adsorption	molecules separated on the basis of their relative adsorption onto a stationary material from a liquid mobile phase
affinity	specific bonding reactions, such as antigen-antibody or enzyme-substrate, to isolate a molecule
gas-liquid (GLC)	vaporized molecules partition between a stationary liquid and a gas mobile phase
high-pressure liquid (HPLC)	high-pressure forces molecules in a liquid mobile phase through a tightly packed column of small particles, which selectively adsorb the molecules
ion exchange	charged molecules in a liquid mobile phase adhere to an insoluble ion-exchange material
partition	different molecules distribute differently over a solid material such as paper when carried by a liquid mobile phase, and then exposed to another liquid mobile phase flowing at right angle to the original flow
thin-layer (TLC)	molecules rise up a thin layer of silica gel coated onto glass by capillary action, which separates the molecules by size

gel. A scientist might also use an antibody with a stain attached to it. When antibody reacts with a specific peptide, the scientist can watch the particular peptide migrate in the gel.

SPECIAL TECHNIQUES IN FRACTIONATION

Field-flow fractionation (FFF) is a technology for differentiating vaccines made of whole cells on the basis of qualities of the cell membrane. In FFF, a gravitational, centrifugal, or electromagnetic field runs perpendicular (at a right angle) to a liquid mobile phase. Separated components emerge from the flow one at a time (the elution process) and can be measured by a detector. Because FFF combines features of centrifugation, chromatography, and electrophoresis, it has become increasingly popular in separating large whole cells from small subcellular molecules, all in a single process. The cancer researcher Haleem J. Isaaq said of FFF, in 2002, "The method has now become a powerful means for the separation of polymers, colloids, and particles from a wide variety of fields, ranging from medicine to fabrication to environmental studies." FFF isolates bacteria, viruses, eukaryotic cells, macromolecules, DNA, RNA, proteins, lipoproteins, and various organelles.

Isotope fractionation is a specialized technique for isolated enzyme systems or cellular processes rather than a single cellular structure. In isotope fractionation, a scientist tags specific molecules with an isotope—isotopes have the same number of protons as the corresponding elements but a different number of neutrons in its nucleus—in a process called *labeling*. The labeled molecules participate in normal cellular activity. As a reaction progresses, a scientist measures the change in isotope composition. For instance, more isotope is associated with a starting compound (substrate) at the beginning of a reaction, but as the reaction progresses, the isotope begins appearing in reaction end products.

Fractionation serves to advance cell biology by enabling microbiologists to study specific parts of microbial cells. Cell fractionation has additionally become an integral part of DNA isolation, purification, and subsequent DNA studies. This technique is an essential part of the physical and chemical study of microorganisms.

See also CENTRIFUGATION; FILTRATION; ORGANELLE; PROTOPLAST.

Further Reading

Alberts, Bruce, Alexander Johnson, Julian Lewis, Martin Raff, Keith Roberts, and Peter Walter. "Manipulating Proteins, DNA, and RNA." In *Molecular Biology of the Cell,* 4th ed. Oxford, England: Garland Science, 2007. Available online. URL: www.ncbi.nlm.nih.gov/books/bv.fcgi?rid=mboc4.section.1522. Accessed August 9, 2009.

Isaaq, Haleem J. "Field Flow Fractionation." In *A Century of Separation Science.* Boca Raton, Fla.: CRC Press, 2002.

Schnaitman, Carl A."Cell Fractionation." In *Manual of Methods for General Bacteriology.* Washington, D.C.: American Society for Microbiology Press, 1981.

Wang, Nam Sun. "Cell Fractionation Based on Density Gradient." Available online. URL: www.eng.umd.edu/~nsw/ench485/lab10.htm#List. Accessed August 9, 2009.

fungus A fungus is an organism that belongs to domain Eukarya, lacks chlorophyll, and absorbs a variety of nutrients. Because fungi lack chlorophyll, they do not perform photosynthesis for energy production.

Fungi make up a highly diverse group of eukaryotes that range from single-celled microorganisms to large multicellular organisms. Various fungi reproduce in either of two ways: asexually by forming spores inside a structure called a *sporangium* or by the sexual formation of gametes produced in a structure called a *gametangium.* Because of the broad diversity of forms, internal structures, and reproductive idiosyncrasies of fungi, the study of these organisms makes up a dedicated science called *mycology.* Scientists who study fungi are mycologists.

About 100,000 different fungal species have been partially or completely identified, but perhaps 10 times as many species may actually exist on Earth, as yet undiscovered and unidentified. The American Society for Microbiology has explained on its Web site, "Fungi straddle the realms of microbiology and macrobiology. They range in size from the single-celled organism we know as yeast to the largest known living organism on Earth—a 3.5-mile-wide [5.6 km] mushroom." This mushroom, *Armillaria ostoyae,* began growing, about 2,400 years ago, in Oregon and now covers 2,200 acres (8.9 km^2) in Malheur National Forest in that state.

Fungi live in almost all terrestrial environments. They have three main roles on Earth in relation to humans: (1) manufacture of food, beverage, or industrial products; (2) decomposition of organic matter in the soil; and (3) as plant and animal pathogens.

Although visible fungi such as mushrooms are not microscopic, they connect to the microbial world by producing single cells, spores, and a cell structure that resembles neither plant nor animal cells. The table illustrates the main features of fungi. Of these characteristics, the most striking differences between fungi and bacteria are in cell type, membrane components, and cell wall structure. Unlike fungi, bacteria are prokaryotic cells lacking sterol compounds in their membrane, but having peptidoglycan in the cell wall. Fungi additionally possess a more complicated reproduction process than the binary fission of bacteria.

Fungi classification schemes may be the most puzzling in all of microbiology. Mycologists categorize fungi according to structure, or reproductive processes, or both. Two main groups of fungi belong to yeasts and dimorphic fungi. Yeasts are organisms that remain in a single-celled form throughout their life (mycologists currently believe) and do not resemble fungi that have more than one structure in their life cycle. Under a microscope or growing on agar, yeasts resemble bacteria more than they resemble fungi such as molds, which are the fuzzy growths such as that seen on moldy bread. Dimorphic fungi grow as either a yeast form or a filamentous form and make up the vast majority of fungal species.

Most textbooks that cover mycology describe five main divisions of fungi based on likenesses in structure, reproduction, or cell type. In mycology, a division is equivalent to a phylum in bacterial classifications. Some yeasts fit into these categories, but for simplicity yeasts usually receive separate discussion when discussing fungi. Fungal identification relies quite a bit on physical features and life cycle and has only recently included molecular characteristics. Quite a few mycologists identify fungi by examining the spores and other features under a microscope and using their experience to classify the organism. Studies of fungal ribosome structures called *18S rRNA* have been gaining importance in fungus identification and have allowed mycologists to adjust earlier classifications. The current classification scheme for terrestrial fungi is shown in the table on page 328 (top). The divisions listed hold more than 30,000 species except Zygomycota, which contains 600–700 species.

Fungi live wherever conditions are moist and offer a rich nutrient supply. For this reason, fungi usually live in moist soils and decaying matter. The mycologist Meredith Blackwell wrote, in 2009, "[Fungi] share with animals the ability to export hydrolytic enzymes that break down biopolymers, which can be absorbed for nutrition. Rather than requiring a stomach to accomplish digestion, fungi live in their own food supply and simply grow into new food as the local environment becomes nutrient depleted." Hydrolytic enzymes are enzymes that break down organic matter and add a water molecule during the reaction; biopolymers are long molecules (cellulose, for example) produced in nature. Fungi are especially valuable for degrading woody dead materials that contain the biopolymer lignin. Lignin gives plant stalks and trees their strength, but few organisms other than fungi can digest it.

Fungus identification is not always clear-cut. The Deuteromycota, known better as Fungi Imperfecti, contain organisms in which mycologists have seen no sexual phase. If a sexual phase becomes apparent, mycologists then reclassify the fungus into one of the other divisions. For this reason, Deuteromycota act, as a type of holding pen for unidentified genera. Deuteromycota is sometimes referred to as the fungal "holding division."

Additional divisions contain fungi that live all or part of their life cycle in water. These are the chytrids, slime molds, and water molds, described in the table on page 328 (bottom).

Characteristics of Fungi

Feature	Description as Found in Fungi
cell type	eukaryotic
cell membrane	sterols present
cell wall	no peptidoglycan; uses chitin for strength
spores	spores used for reproduction rather than protection
reproduction	variety of sexual and asexual processes
nutrition	chemoheterotroph (organism that uses organic compounds as a carbon and energy source)
metabolism	usually aerobic

IMPORTANCE OF FUNGI

Fungi and bacteria decompose up to 90 percent of the organic matter on Earth. In ecology, the term for this role is *decomposer.* Decomposers recycle nutrients for use by other organisms and prevent the world from filling up with natural waste. Fungi are also saprophytes (also called *saprobes*), or organisms that get nutrients from dead matter, so fungi are critical in decomposing wastes in forests, soils, and animal carcasses.

Fungi have benefited humans for millennia, since the first fermentations for making cheese, preserving vegetables, and producing wine and beer. Fungi today also produce industrial chemicals and drugs, as shown in the table on page 329.

Various fungi also degrade the lignin in wood as the first step in the papermaking process. Pigments, textile dyes, bleaching agents, and bioremediation enzymes have also been extracted from fungi by specialized industries. Because of the many benefits provided by fungi, these organisms make up an important specialty within microbiology, and mycologists will be called upon to help in the identification of new species and expanded roles for these organisms.

STRUCTURE AND IDENTIFICATION

Fungi have incredible variety of shapes and sizes, but they also possess common features. The body of

Fungi Divisions

Division	Distinctive Structure	Reproduction
Ascomycota (sac fungi)	saclike asci (singular: ascus) that release ascospores	asexual by conidiospores; sexual by haploid ascospores
Basidiomycota (club fungi)	club-shaped basidium	sexual when haploid basidiospores enter soil and develop spreading filaments
Deuteromycota (Fungi Imperfecti)	sexual cycle not observed until the fungus is allowed to begin its life cycle	asexual by conidiospores
Glomeromycota	filaments that grow inside plants and trees	asexual spores
Zygomycota	sporangium	asexual by sporangiospores; sexual by zygospores

a fungus that does not hold reproductive structures is called the vegetative body, or the *thallus*. Mushrooms make the most familiar thallus structures seen in nature, and they illustrate the broad diversity of shapes, sizes, and colors of the thallus.

Molds have been used to illustrate the next familiar feature of fungi: hyphae. Hyphae are long, thin threads, or filaments, of cells that often contain branches. A mass of hyphae often joins to form a tangled, fuzzy mass, the *mycelium*. Most people recognize mycelia when they see mold growing on fruit or bread. See the color insert on page C-4 (bottom) for a picture of a mold growing on agar.

Some fungi produce hyphae that contain one continuous thread filled with cytoplasm. These hyphae are called coenocytic hyphae. Other fungi produce hyphae that contain cross walls called septa that divide the hyphae into segments. These are septate hyphae.

Dimorphic fungi switch between the filamentous form and a single-celled yeast form, depending on the phase of their reproductive cycle. This process of turning into another form and reverting to the original form during the life cycle is called the YM shift. Factors in the environment such as nutrients, temperature, carbon dioxide levels, or moisture influence the YM shift.

Fungal identification uses features in addition to the thallus, hyphae, or mycelia (although mushrooms can be identified by the thallus alone). Without any help from 18S rRNA analysis, mycologists use two techniques to identify fungi: colony morphology on agar and microscopic examination. Mycologists record these observations and then compare them with descriptions of known fungi. Quite a few fungi, especially genera that cause disease in plants or animals, are identified this way, but this method works only if the fungus has already been identified and described in published resources.

The following features of colony morphology provide clues to fungal identification:

Divisions of Aqueous Fungi

Division	Description
Acrasiomycota (slime molds)	in high nutrient supply, they live as amoeboid cells; in low nutrients, they aggregate into a pseudoplasmodium that leaves a slime trail as it moves; the resting pseudoplasmodium produces an appendage called a *sorus* that produces new spores
Chytridiomycota (chytrids)	reproduce by asexual motile zoospore, but can use a sexual stage by forming a zygote
Myxomycota (acellular slime molds)	made of streaming masses of material, lacking a cell wall, called a *plasmodium,* which moves in the direction of food
Oomycota (water molds)	cell wall made of cellulose rather than chitin; sexual reproduction produces a large "egg" oogonium, fertilized by a sperm to form a zygote

- bacterialike colonies—irregular shapes, off-white
- yeastlike colonies—large, round, and shiny
- reverse colony morphology—the appearance of the colony from the agar plate's underside
- thermally dimorphic—displays a YM shift when incubated at one temperature, then switched to a different temperature
- thermally monomorphic—retains the same filamentous features regardless of incubation temperature

After the gross morphology examination, a mycologist takes a small amount of the mycelia and puts it into a drop of water on a glass slide. Special stains can be added that emphasize certain fungal features. The Gram stain and the acid-fast stain, normally used for bacteria, have been modified for fungi. Additional staining techniques have been developed exclusively for fungi: lactophenol cotton blue, Giemsa stain, and methenamine silver. These stains act both to highlight fungal structures and to protect the structures from distortion.

The following features of microscopic examination usually complete the identification:

- coenocytic versus septate hyphae
- shape and size of spores
- shape and size of sporangium or conidium
- presence or absence of hyphae branches
- dense or sparse hyphae branching

Methods for identifying fungi by analyzing the nucleic acid sequence of 18S rRNA have shed light on schemes that have been used for a hundred years. Sequence analysis categorizes fungi into genera and species and shows the relatedness of the divisions. Molecular biologists also use identified sequences to develop probes, which contain small pieces of deoxyribonucleic acid (DNA) or ribonucleic acid (RNA) that detect the same microorganism in nature.

Probes for common molds found in indoor air have been used for assessing air quality, and the food industry and clinical laboratories have employed similar probes to detect contaminants and pathogens, respectively. Environmental, food, and clinical microbiologists also use polymerase chain reaction (PCR) technology for finding specific fungi. PCR is a procedure that can multiply tiny bits of DNA into amounts suitable for further analysis. This sensitive technique helps microbiologists detect the presence of fungi that produce dangerous toxins that taint foods or cause severe infections.

Specific probes or the use of PCR for further nucleic acid analysis has removed some of the burden from mycologists, who had previously used only physical traits to identify fungi. In 2007, the marine biologist Mara R. Diaz wrote in *Microbe* magazine, "Diagnosing fungal infections remains a challenge because symptoms can be nonspecific, while traditional diagnostic methods tend to be slow and can prove unreliable." The traditional method of studying spores and hyphae under a microscope will be helped immensely by molecular methods. Diaz and other researchers have developed rapid methods for fungal identification based on high-throughput systems (processes that analyze 100 samples in a short period). These systems identify species by nucleotide sequences, antibodies, or antigens. But many

Important Products Made by Fungi

Fungi	Products
	FOOD
Aspergillus	cured ham, miso, soy sauce
Penicillium	blue, Roquefort, Camembert, and Brie cheeses
Rhizopus	tempeh
Saccharomyces	ale, beer, brandy, saki, whiskey, wine, breads, and baker's yeast
	INDUSTRIAL PRODUCTS
Ashbya	vitamins
Aspergillus	amylase, oxidase, protease, pectinase, lipase, tannase, xylanase, and citric acid
Bradyrhizobium	fertilizer
Metarhizium	herbicides
Mucor	rennin
Saccharomyces	invertase
various fungi	chitin and chitosan
	DRUGS
Claviceps	ergot alkaloids
Penicillium	penicillin, griseofulvin

molecular methods are expensive. Most microbiology laboratories currently do not have the resources for automated identification systems. For a while, fungal identification may depend on the experience of mycologists and the aid of simple probes.

REPRODUCTION

Reproduction in fungi can be either asexual or sexual; many species use both methods. The structures that develop in different reproductive stages of fungi also serve as an important tool for identification, especially sporangia and conidia.

Asexual reproduction occurs in three steps: (1) cell division, (2) fragmentation of hyphae, and (3) spore formation. In cell division, a fungal cell divides to form two daughter cells. This reproduction requires only the building of a new cell wall through the middle of the parent cell, a process called *transverse fission*. As cell division builds hyphae stretching longer and longer, some cells break off from the septate hyphae at the cell wall interfaces. The single hyphal cells become spores called *arthrospores* and find another place to begin growing or germinate.

Certain fungi develop derivations on simple hyphal fragmentation. In some species, hyphae stop growing as filaments at a certain point and begin producing a chain of spores at the tip of a branched bundle of hyphae. The branched bundle is called a *conidiophore*, and the spores that it grows and releases into the air are conidiophores. *Aspergillus* mold forms a round conidiophore that holds hundreds of small circular spores.

Other fungi develop sacs full of spores. For example, a sac called a *sporangium* might form at the tip of a hypha and contain hundreds of sporangiospores, which escape when the sac ruptures. The fungus *Rhizopus* uses this method to create new organisms. Other fungi, such as *Candida,* make some sacs called *chlamydospores* within the hyphae rather than at the ends and other sacs at the hyphal tips (terminal chlamydospores). The thick cell wall of the hyphae surrounds each chlamydospore, which breaks off the hypha as a single piece. Fungi that use none of these techniques for making and spreading spores may simply grow buds off their hyphae that also bud into spores. These spores are called *blastospores,* and as do chlamydospores, sporangiospores, and chlamydospores, they disperse into the air several inches to a mile (1.6 km) or more from their origin. Only arthrospores depend less on airborne travel and spread mainly through the soil or other places where the parent mold grows.

Sexual reproduction involves the union of gametes that each contains half of a complementary chromosome, a haploid structure. Some fungi produce both male and female gametes on different hyphae in the same mycelium; some fungi produce the gametes on different mycelia. Two compatible gametes that meet fuse their cytoplasm and then fuse their nuclei. The nuclear fusion creates a zygote containing the full

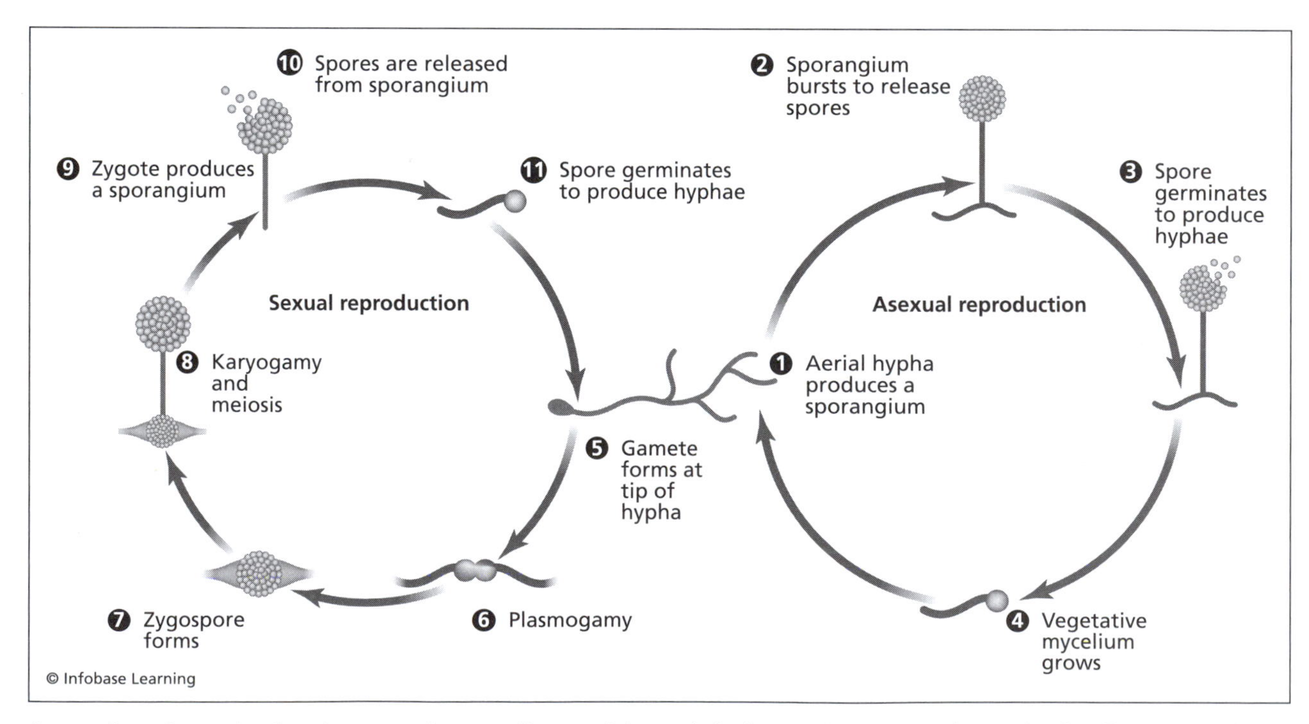

In most fungal reproduction, the vegetative mycelium participates in both sexual and asexual reproduction. Zygomycete fungi, such as *Rhizopus,* shown here, construct a sporangium that releases spores into the air. The spores germinate to form a new vegetative structure after they settle on the ground.

chromosome, a diploid structure. Zygotes develop further into the type of spores that are characteristic of the parent fungi. Thus, fungi of Ascomycota form ascospores, Zygomycota develop zygospores, and Basidiomycota produce basidiospores.

The stage of a fungus's life cycle represents more than an academic curiosity. The structure associated with certain reproduction phases often correlates with the diseases caused by fungi, called *mycoses*. For example, the yeast *Candida* usually exists in its single-celled form and causes no harm when living on the human body. When *Candida* cells develop hyphae, however, they can invade the skin to start troublesome infections.

MEDICAL MYCOLOGY

A mycosis is any fungal disease in which the organism has invaded living tissue. Medical mycology is the study of mycoses and their diagnosis. Medical mycologists and clinical microbiologists must understand the various types of fungi that infect the skin, hair, nails, and, in critical cases, the bloodstream and organs. Veterinarians also see various skin infections and more serious diseases in domesticated animals caused by fungi.

Mycoses usually have a long duration; these infections are called *chronic infections*. Chronic infections last for several weeks or longer. Mycoses are difficult to cure and last a long time in the body for two reasons. First, fungal hyphae spread and infiltrate large amounts of tissue. Second, because fungal cells are eukaryotic and related to animal cells, few antifungal drugs kill fungi without also harming the body's cells. Fungal infections in humans and domesticated animals are some of the most difficult infections to cure.

Mycoses belong to four different groups based on the pathogen's mode of entry into the body and where it causes infection, as follows:

- superficial—along hair shafts and on the outer epidermis (the skin)
- cutaneous or dermatomycoses—infections of the epidermis, hair, and nails
- subcutaneous—beneath the skin, usually by way of a puncture wound
- systemic—tissues and organs inside the body

Superficial and cutaneous infections make up a large part of dermatology, mainly because they result in rashes, pustules, redness, or itching. The fungal genera *Aspergillus, Trichophyton, Candida, Microsporum, Sporothrix,* and *Malassezia* cause most of the cutaneous infections, called *dermatomycoses,* in people living in temperate climates.

Systemic mycoses usually begin in the lungs when a person inhales spores. Oftentimes, the infection stays in the respiratory tract and causes moderately serious effects. But subcutaneous and systemic infections can have very serious consequences, if fungal cells enter the bloodstream from the lungs or the digestive tract and infest body tissues. The physician Robert Thiel explained, in his 2007 article "Systemic Mycoses: An Overview for Natural Health Professionals," that most systemic infections arise from opportunistic conditions. "Systemic mycoses can cause a tremendous variety of health problems including digestive difficulties (diarrhea, bloating, discomfort, flatulence, constipation, etc.), skin problems (rashes, eczema, psoriasis, dry skin patches, intense itching, hives, open cut-like sores, etc.), bronchopulmonary disorders, asthma, breathing difficulties, fatigue, allergies, weight loss, fever, chills, malaise, depression, and chronic sinusitis); some of them may be risk factors in developing autoimmune disorders." Systemic mycoses are difficult to diagnose, and the rather indistinct symptoms listed by Thiel have led physicians to diagnose systemic mycoses as foodborne illnesses, allergic intolerances to certain foods, or symptoms related to other diseases.

The genera *Coccidioides, Blastomyces, Cryptococcus, Candida,* and *Histoplasma* have been implicated in many systemic mycoses in humans. These organisms can infect healthy individuals, but people suffering from any condition in which their immune system has been weakened are at far more risk for contracting serious, often deadly, fungal diseases. Healthy immune systems can usually confine a fungus infection to a wound site or to the respiratory tract. Immunocompromised people have weaker defenses against the dissemination of fungal cells throughout the body, and so mycoses cause a high rate of mortalities in this subpopulation.

Fungi in the body can attack tissue in two ways: by infiltrating tissues or organs until their normal functioning cannot continue, or by excreting toxins. Toxins excreted by fungi are called *mycotoxins*. Some mycotoxins are among the deadliest biological compounds known to science.

MYCOTOXINS

Mycotoxins are compounds produced by fungi in their normal metabolism. Humans and other animals receive mycotoxin exposure when fungi contaminate foodstuffs. For example, mycotoxin poisoning that results from eating poisonous mushrooms is called *mycetism*. Systemic mycoses also introduce mycotoxins into the body by routes other than the digestive tract, that is, in contaminated wounds or by respiratory infection.

Fungi make mycotoxins for the same reason bacteria make antibiotics: These compounds help the organism compete with other microorganisms for nutrients and habitat. Fungi usually produce mycotoxins after germinating from the spore form and then growing hyphae. The toxins might help a fungus invade tissue by damaging cells. Other than mycotoxins, fungi seem to possess no means of expressing virulence.

About 20 different mycotoxins have been attributed to specific fungi, some of which are lethal at small enough doses to be considered potential bio-

Mycotoxins

Mycotoxin	Fungus That Produces It	Action in Animal Tissue
aflatoxin	*Aspergillus flavus*	liver toxicity; immune system damage; DNA replication errors leading to cancer (A, H)
asteltoxin	*Aspergillus stellatus*	respiratory impairment; paralysis of extremities (A)
aurovertins	*Calcarisporium arbuscula*	energy metabolism interference (H)
chloropeptide	*Penicillium islandicum*	disrupts cell integrity and strength (A, H)
citreoviridin	*Penicillium citreoviride*	neurotoxin causing paralysis, convulsions, respiratory arrest, death (A, H)
citrinin	*Penicillium* species	kidney damage (A)
cyclopiazonic acid	*Aspergillus* and *Penicillium* species	interferes with nerve function (A, H)
ergot alkaloids	*Claviceps, Aspergillus, Penicillium, Rhizopus* species	stimulates smooth muscles to affect blood flow (A, H)
fumonisins	*Fusarium* species	(A, H)
nivalenols	*Fusarium* species	
ochratoxins	*Penicillium* and *Aspergillus* species	damages kidneys; disrupts energy metabolism (A, H)
patulin (clavacin)	*Penicillium, Aspergillus,* and *Byssochlamys* species	disrupts protein and enzyme function, leading to death (A, H)
penitrems	*Penicillium verrucosum*	neurotoxin causing weakness, convulsions, tremors, death (A)
rubratoxins	*Penicillium rubrum*	internal hemorrhaging, liver damage, brain lesions (A)
slaframine	*Rhizoctonia leguminocola*	digestive and secretory disorders (A)
sporidesmins	*Pithomyces chartarum*	appetite loss, diarrhea, dehydration, photosensitivity, death (A)
sterigmatocystin	*Aspergillus nidulans* and *A. versicolor*	liver damage and cancer (H)
trichothecenes	*Fusarium, Trichothecium,* and *Myrothecium* species	disrupts protein synthesis and causes cell death, gastric hemorrhaging, death (A, H)
xanthomegnin	*Aspergillus, Penicillium,* and *Trichophyton*	liver damage (A, H)
zearalenone	*Gibberella zeae* and *Fusarium* species	hormonal disruption, infertility, rapid weight gain (A)

Note: A = most evidence of toxicity found in animals; H = most evidence of toxicity found in humans

Important Fungal Diseases of Trees and Plants

Disease	Cause	Plant	Effects
apple scab	*Venturia inaequalis*	apple tree	black circular spots on leaves and fruit
black rot	*Guignardia bidwellii*	grape	light brown spots that spread over the fruit
black stem rust	*Puccinia graminis*	wheat	rust-colored pustules that spread through the plant
brown rot	*Monilinia fructicola*	stone fruit	brown spots on blossoms and fruit
chestnut blight	*Endothia parasitica*	chestnut tree	brown patches on bark, spreading lesions that kill limbs
coffee rust	*Hemileia vastatrix*	coffee	orange-yellow powdery spots on underside of leaf
corn smut	*Ustilago maydis*	corn	small to large galls (outgrowths) on seedlings, leaves, and stalks
downy mildew	species of family Peronosporaceae	grapes, grasses, various vegetables	yellow spots on leaves while fungus grows on leaf underside
Dutch elm disease	*Ceratocystis ulmi*	elm tree	wilted leaves turn dull and drop off; branches die
Fusarium wilt	*Fusarium oxysporum*	tomato	stunted plant growth, poor leaf formation
late blight	*Phytophthora infestans*	potato	dark-green to black water-soaked lesions on leaves
powdery mildew	species of family Erysiphacaeae	many grasses, plants, shrubs, and trees	mildew growth covering leaves and spreading to other organs
root rot	*Heterobasidium, Thanatephorus*	various trees, plants, ornamental species	dead roots, stunted plant growth
soft rot	*Rhizopus*	flowers, fruits, and vegetables with fleshy organs	soft, water-soaked areas that spread rapidly, followed by mycelium and black spores
stinking or bunt smut	*Tilletia caries*	wheat	develops in kernels to produce a fishy-smelling puff ball in place of normal kernels
white pine rust	*Cronartium ribicola*	white pine tree	cankers on bark that exude secretions, followed by bright orange pustules
white rust	*Albugo*	chrysanthemum	white-yellow spots on upper leaf followed by pustules

logical weapons. The table lists important individual mycotoxins or mycotoxin groups.

Ingested mycotoxins originate in moldy foodstuffs, particularly grains. Mycotoxins cause either acute—rapid-onset, short-duration symptoms—or chronic illness. The major grains that present a risk of mold or mycotoxin contamination are corn, wheat, barley, and peanuts, although other grains and grasses have contributed, at times, to toxicities from mold contamination. The Technical Centre for Agricultural and Rural Cooperation in the Netherlands has identified the following mycotoxins as the most serious human health threats today:

- aflatoxin B_1 and B_2 from *Aspergillus flavus*
- aflatoxin B_1, B_2, G_1, and G_2 from *A. parasiticus*
- deoxynavalenol and nivalenol from *Fusarium graminearum, F. crookwellense,* and *F. culmorum*
- fumonisin B_1 from *F. moniliforme*
- ochratoxin A from *A. ochraceus* and *Penicillium verrucosum*
- zearalenone from *Fusarium graminearum, F. crookwellense,* and *F. culmorum*

These mycotoxins have caused health problems worldwide because they have high toxicity but also because the fungi that produce them tend to contaminate grains that have been used as staples worldwide for many years. The Technical Centre has also emphasized that aflatoxins and fumonisins will become increasing health threats in developing countries with poor food storage and distribution systems.

FUNGAL PLANT DISEASES

More than 70 percent of all plant diseases are caused by fungi; for example, there have been several instances in history in which a fungal disease of a crop caused severe hardship in society. Ireland's potato famine, in the 1840s, is the most famous example. In this event, more than one million people died of starvation or famine-related illnesses due to a *Phytophthora infestans* infection of their crops. The potato blight fungus forced 1.5 million more people to emigrate from Ireland, during the famine. Fungal diseases in crops continue to threaten populations around the world where people's lives depend on yearly harvests.

The main fungal diseases of plants are from three main groups: blights, smuts, and rusts. Blight causes a rapid degradation of plant chlorophyll that turns the leaves brown and makes them unable to support photosynthesis. Blights primarily attack potato, tomato, corn, rice, and grasses. Smut fungi infest grasses, corn, and other cereal grains by interfering with the plant's normal reproduction. Rusts live parasitically inside plants and draw their nutrients from the plant cells as they carry out their entire fungal life cycle. Rusts tend to be specific for a given fruit or plant; they attack agricultural crops, horticultural plants, and various species of natural forests.

Fungal diseases of plants encompass a large number of pathogens and symptoms—there are more than 6,000 rusts alone that infect plants. Some of the most familiar plant fungal diseases in the United States are summarized in the table on page 333. The fungi that cause these diseases are called *phytopathogenic fungi.*

In the table and in other aspects of mycology, the term *mildew* refers to any moldlike growth that tends to grow as a flat mass over a surface, which the mycelia then infiltrate. For example, the fungal growth on old leather is usually mildew because the growth does not growth outward from the leather but prefers to grow into the leather and digest the material as it grows.

Enormous diversity exists within fungi, and many species have not yet been identified. These organisms have been beneficial to humans, but under certain circumstances they can also be pathogens. For example, *Penicillium* mold is used in food production and is the source of the antibiotic penicillin, while it also produces toxins that can be deadly to humans and animals. For these reasons, mycology will remain an essential part of microbiology and medicine.

See also ASPERGILLUS; MORPHOLOGY; MYCOSES; OPPORTUNISTIC PATHOGEN; SPORE; STAIN; TOXIN; YEAST.

Further Reading

Blackwell, Meredith, Rytas Vilgalys, Timothy Y. James, and John W. Taylor. "Fungi Eumycota: Mushrooms, Sac Fungi, Yeast, Molds, Rusts, Smuts, etc." Available online. URL: http://tolweb.org/Fungi/2377#AboutThisPage. Accessed August 12, 2009.

Diaz, Mara R. "Bead Suspension Arrays for Identifying Fungal Pathogens." *Microbe,* February 2007.

DoctorFungus.org. Available online. URL: www.doctorfungus.org/index.htm. Accessed August 15, 2009.

Hamlyn, Paul F. "Fungal Biotechnology." *Northwest Fungus Group Newsletter,* April 1997. Available online. URL: http://fungus.org.uk/nwfg/fungbiot.htm. Accessed August 10, 2009.

Money, Nicholas P. *Carpet Monsters and Killer Spores.* Oxford: Oxford University Press, 2004.

Technical Centre for Agricultural and Rural Cooperation. *Mycotoxins in Grain.* Wageningen, Netherlands, 1997. Available online. URL: www.fao.org/wairdocs/X5008E/X5008e00.HTM. Accessed August 12, 2009.

Thiel, Robert J. "Systemic Mycoses: An Overview for Natural Health Professionals." *Original Internist* 14 (2007): 57–66. Available online. URL: www.healthresearch.com/yeast.htm. Accessed August 12, 2009.

Wu, Zhihong, Yoshihiko Tsumura, Göran Blomquist, and Xiao-Ru Wang. "18S rRNA Gene Variation among Common Airborne Fungi, and Development of Specific Oligonucleotide Probes for the Detection of Fungal Isolates." *Applied and Environmental Microbiology* 69 (2003): 5,389–5,397. Available online. URL: www.pubmedcentral.nih.gov/articlerender.fcgi?artid=194989. Accessed August 12, 2009.

generation time Generation time is the length of time that a population of microorganisms takes to double in number. For this reason, it is also referred to as *doubling time*. Generation time relates to microorganisms that reproduce by binary fission, which is the asexual process whereby a single cell divides to produce two identical daughter cells.

Microbiologists use generation time to evaluate cell physiology or to assess the effect of various environmental factors. For example, by adding an essential nutrient to a bacterial culture, the microbiologist might observe a shortened generation time. Adding a toxic compound could possibly lengthen generation time. These responses can be interpreted by a microbiologist as follows:

decreasing generation time =
increasing growth rate

increasing generation time =
decreasing growth rate

In the preceding relationships, growth correlates to an increase in cellular constituents and cell numbers, and growth rate is the amount of microbial material produced per hour. Bacteria do not grow as multicellular organisms grow. Growth in multicellular organisms usually equates to an increase in the organism's size. Growth in bacteria means an increase in total cell numbers, because bacteria do not grow much in size after cell division; they reach a size predetermined by their species and stay at that size. Unlike multicellular organisms, such as plants and humans, bacteria stay within a remarkably narrow size range.

In 1996, the Indiana University biologist Arthur L. Koch wrote an article titled "What Size Should a Bacterium Be? A Question of Scale," in which he considered the reasons behind standard cell sizes in bacteria. "The smallest size for a free-living organism is suggested to be largely set by the catalytic efficiency of enzymes and protein [synthesis]," he wrote. "Because of fluctuations in the environment, cells must maintain machinery to cope with various catastrophes; these mechanisms increase the minimum size of the cell. On the other hand, the largest cell is reasonably assumed to be limited by the ability of diffusion to bring nutrients to the appropriate part of the cell and to dispose of waste products." As Koch suggested, each microbial species has evolved to a size and shape that allow the cell to metabolize the most nutrients, excrete the maximal amount of wastes and products, and adapt readily to changes in its environment.

In microbiology, growth rate can more accurately be defined as specific growth rate, in which any type of microorganism can be compared to another on the basis of the mass of material it produces in a period of time, usually an hour. Measuring microbial mass also puts microorganisms on an equivalent scale with organisms that measure growth in terms of mass. Microbiologists use either cell mass or cell numbers when monitoring growth and generation times.

GROWTH RATE

Growth rate can be calculated under two different laboratory conditions: batch cultures and continuous cultures. A batch culture contains a broth (liquid) medium in which an inoculum grows with no additions or subtractions from the culture during incubation. Batch cultures contain a typical pattern of growth called a *growth curve,* made up of a slow lag phase, a rapid logarithmic (log) phase, a stationary phase, and a death phase. Continuous cultures receive a constant inflow of fresh medium containing nutrients and a constant outflow of spent medium and wastes. Bacteria in continuous cultures can be held at a constant growth rate by adjusting the rate of medium inflow (which automatically equals the outflow). In either batch or continuous culture, cells that reproduce by binary fission, such as bacteria, increase from one cell to two cells, two to four, four to eight, and so on, with each generation. The number of cells, N, derived from a single starting cell after n generations is:

$$N = 2^n$$

Most inocula contain many more cells than just a single cell. To calculate the number of cells present after n generations, from an initial number of cells in the inoculum, N_0, a microbiologist uses the following equation:

$$N = 2^n N_0$$

Bacterial cells dividing by binary fission produce a doubling in cell numbers with each generation, so microbiologists can plot the cell numbers on a $\log_2$ scale, which allows them to read the number of generations per unit time directly from the graph. Each unit on the $\log_2$ scale equals one generation.

Microbiology usually calculates growth rates from the log phase of batch cultures or the flow rate of continuous cultures. A value called the *mean growth rate constant, k,* describes this growth as follows:

$$k = n \div t$$

In this equation, t refers to a unit of time, usually hours. The $\log_2$ scale can be converted to a $\log_{10}$ scale by using the following equation:

$$\log_{10} N = 0.301 \log_2 N$$

On a $\log_{10}$ scale, the mean growth rate constant becomes

$$k = (\log N_t - \log N_0) \div 0.301\, t$$

The following example shows how this calculation works. An inoculum of 1,000 cells grows to 1×10^8 cells in seven hours. The 1,000-cell inoculum can be expressed as 1×10^3, and the mean growth rate is:

$$k = (\log 10^8 - \log 10^3) \div (0.301)(7) = (8 - 3) \div 2.107 = 2.373 \text{ generations per hour}$$

These calculations lead to a mean number of generations, which makes more sense than calculating specific growth rate, because some cells divide slightly more slowly or slightly faster than others. Specific growth rate tends to represent a theoretical calculation, while the mean growth rate depicts a real situation within a cell culture.

CALCULATING GENERATION TIME

Generation time provides information to a microbiologist for assessing the effects of nutrients, growth factors, toxins, pH, or physical conditions such as temperature on the growth of a culture.

From the previous example, generation time, g, can be determined as the following:

$$g = 1 \div (2.373 \text{ generations per hour}) = 0.42 \text{ hour per generation}$$

This calculation shows that generation time is simply the reciprocal of growth rate. The faster the growth of a microorganism, the shorter is its generation time.

Microbiologists must have an approximate idea of a microorganism's doubling time in order to set up laboratory experiments on physiology or growth requirements. Most commonly used bacteria in microbiology studies—*E. coli, Staphylococcus aureus, Salmonella, Bacillus, Pseudomonas* are examples—grow rapidly in the optimal conditions provided in a laboratory. Because laboratory conditions offer favorable incubation temperatures, pH, and oxygen availability and ample nutrients, bacteria grow at their best and generation times can be as short as 20 to 30 minutes. Generation times observed in laboratories follow general patterns. For example, aerobic bacteria tend to possess the shortest generation times, which last 20–40 minutes; anaerobic bacteria have slower doubling times of 30–60 minutes. Archaea are microorganisms that possess attributes of both bacteria and eukaryotic cells but belong to neither group; they belong to their own domain, Archaea. Some archaea have extremely long doubling times. Methanogenic archaea that produce methane gas may require several hours per generation. Some ammonia-oxidizing archaea that grow in sediments or freshwaters or associate with marine corals and sponges require from two days to a week for a new generation to appear.

The table on page 337 shows the wide variety found in bacteria generation times in laboratory studies.

Each of the generation times observed in the laboratory and shown in the table depends on the type of medium in which the bacteria grow. For example, *Streptococcus lactis* grown in milk medium takes 26

Bacteria Generation Times

Species	Generation Time (minutes)
Escherichia coli	17
Bacillus megaterium	25
Staphylococcus aureus	27
Lactobacillus acidophilus	66–87
Rhizobium japonicum	344–461
Mycobacterium tuberculosis	792–932
Treponema pallidum	1,980

Source: Kenneth Todar, *Todar's Online Textbook of Bacteriology.*

minutes per generation, whereas *S. lactis* grown in lactose broth requires 48 minutes per doubling.

THE MEANING OF GENERATION TIME

Generation time in laboratories rarely mimics the generation times found in nature. Laboratory cultures usually contain a pure collection of cells of the same species with no interference from competitors. As mentioned, laboratory conditions also provide all the best physical factors and nutrients for optimal growth. Microorganisms in nature confront much more severe conditions for growth. Some of the factors common in nature that would be likely to slow growth and thus lengthen generation times are the following:

- suboptimal temperature
- acidic or basic conditions
- limited oxygen availability for aerobes
- exposure of anaerobes to oxygen
- limited carbon and/or energy sources
- limited nitrogen, phosphorus, or sulfur compounds
- limited vitamins, minerals, and other nutrients required in trace amounts
- presence of antibiotics and other harmful products such as bacteriocins
- competition with other organisms for space and nutrients
- predation by other microorganisms

Microorganisms in their natural environment or in laboratory culture may also exhibit random delays in normal doubling times. Many of these delays can be attributed to factors that are less than optimal for the microorganism. Bacteria that normally double every 20 minutes, in broth culture incubated at 98°F (37°C), might have delayed doubling of every 30 minutes or longer, if incubated at room temperature of 72°F (22°C). Enzyme systems may slow at the cooler temperature, and the processes that use these enzymes also slow. But many other undetermined factors contribute to variable generation times. Some likely effectors of delayed cell doubling are the following:

- mutations
- faulty binary fission infrastructure
- enzyme interference
- damaged daughter cells requiring repair
- undermined cell damage requiring repair
- presence of contamination

Microbiologists assume that all cultures of the same bacteria, even if prepared by identical methods, will not grow at exactly the same rate. This variation is due to the natural variability inherent in all biological systems. Generation time, therefore, represents a generalized characteristic of bacteria. In this generalized sense, the use of generation can be helpful in following the typical characteristics of a microorganism's growth in laboratory conditions.

Generation time and growth rate serve as analytical tools for studying microorganisms' responses to their environment. Generation time helps investigations of factors that enhance or inhibit growth. Generation time measurements have been applied to several areas in microbiology, such as cell physiology, antibiotic susceptibility and resistance, nutrient requirements, and species competition.

See also CONTINUOUS CULTURE; GROWTH CURVE.

Further Reading

Koch, Arthur L. "What Size Should a Bacterium Be? A Question of Scale." *Annual Review of Microbiology* 50 (1996): 317–348.

Painter, Page R. "Generation Times of Bacteria." *Journal of General Microbiology* 89 (1975): 217–220. Available online. URL: http://mic.sgmjournals.org/cgi/reprint/89/2/217?view=long&pmid=1176952. Accessed August 15, 2009.

Todar, Kenneth. "Growth of Bacterial Populations." In *Todar's Online Textbook of Bacteriology.* Madison: University of Wisconsin, 2008. Available online. URL: http://textbookofbacteriology.net/growth.html. Accessed August 15, 2009.

gene therapy Gene therapy is a method of disease treatment in which abnormal genes in a person's deoxyribonucleic acid (DNA) are replaced with normal genes. A gene is a basic unit of the genetic material found in every living thing. This genetic material makes up a microbial, plant, or animal cell's chromosome. In viruses, the genetic material may be either ribonucleic acid (RNA) or DNA, depending on the type of virus.

Each gene contains a specific sequence of nucleotides (a purine or pyrimidine base, a five-carbon sugar, and a phosphorus-containing phosphate group). The nucleotide sequence carries information on the proteins that a cell must make to sustain its normal functioning, replication, and specialized activities, such as the type of nutrients it can use. All of the genes in an organism make that organism a unique individual in its species, but genes also define an individual in a way that leaves no doubt as to its species. Simply, human genes make humans, oyster genes make oysters, chestnut genes make chestnut trees, and so on.

Some inherited disorders in DNA lead to disease in people, animals, and plants. Other diseases are caused by infections from pathogens that interfere with normal gene functioning in the host cell. A host cell is any cell that is invaded by a foreign agent, such as a virus. As a result of either inherited or acquired disease, a body's normal gene sequences or functions become disrupted. Gene therapy acts to correct the mistakes by putting the correct gene back into the host cell's DNA. Gene therapy makes the corrections to the damaged DNA by either of two methods: ex vivo, or cell-based, gene therapy and in vivo gene therapy.

EX VIVO GENE THERAPY

In ex vivo gene therapy, a doctor removes cells from a patient, and laboratory scientists modify the cell's DNA with the correct genes. These modified cells are, then, returned to the patient's body. Ex vivo gene therapy can use almost any type of cell from a patient, but gene therapists have long preferred to use stem cells. Bone marrow and other organs produce stem cells to replenish dying or damaged cells. In the body, stem cells also have the ability to develop into many different cell types, depending on the kind the body needs. In gene therapy, technicians put genes into a patient's stem cells and, then, treat the cells in a way that will induce them to turn into a specific cell type, such as heart, kidney, or nerve cell, once they are put into the patient. This process of adapting features of a certain cell type is called *differentiation.* In the bloodstream, stem cells go to the diseased or damaged tissue type. Stem cells provide a great advantage over current methods that treat damaged tissue by transplantation. Organ or tissue transplantation is a difficult and expensive procedure, leaves the patient vulnerable to other infections, and can be hampered by a shortage of organ donors.

The Harvard University biotechnologist Arlo Miller pointed out, in 2001, the challenges in gene therapy. "While the idea of gene therapy seems simple, the actual delivery of the gene to the diseased area has proved quite troublesome. The delivery vehicle for the gene must have several characteristics: it must be safe, it must efficiently deliver the gene to a high percentage of diseased cells, it must be targeted specifically to the appropriate cells, and it should be able to regulate levels of expression of the therapeutic gene." Some gene therapies are intended to turn on certain metabolic processes in the patient's cells, and some therapies shut off errant genes that cause metabolic disorders.

Stem cell therapy may become a promising treatment for cardiovascular diseases. The National Institutes of Health has explained on its Web site, "To realize the promise of novel cell-based therapies for such pervasive and debilitating diseases, scientists must be able to manipulate stem cells so that they possess the necessary characteristics for successful differentiation, transplantation, and engraftment [attaching new tissue to existing tissue]." In order to achieve medicine's goals, stem cells must function in the body as doctors intend, they must avoid rejection by the body's immune system, and they must maintain themselves in the body for as long as a patient lives.

The major obstacle to the use of stem cells is the strong opposition that this type of medical treatment has received from the public. Stem cell therapy became a potentially new therapy for disease, in 1998, when the biologist James A. Thomson of the University of Wisconsin isolated human embryonic stem cells (eHSCs) from several-day-old embryos. As the public learned about this procedure, people soon pointed out that the use of these cells was tantamount to taking human lives, because a human embryo is equivalent to a human being. The Science and Policy division of the American Association for the Advancement of Science has said of the controversy, "Some opponents of eHSC research also argue that research on stem cells obtained from adults is just as promising and renders eHSC research unnecessary. Most scientists, however, dispute this claim, citing great potential in the field of adult stem cells but several drawbacks as compared eHSCs." On August 9, 2001, President George W. Bush confined stem cell research to only stem cells already owned by laboratories and not new stem cells from embryos. But the stem cell debate continued. In March 2009, President Barack Obama overturned a number of the previous administration's restrictions on stem cells. Medical research on stem cells is now a growing area in science, but gene therapy using human cells appears to have a difficult road ahead in order to resolve all the differences on this subject.

IN VIVO GENE THERAPY

In vivo gene therapy using viruses precludes the controversies inherent in stem cell methods. Viruses serve as a vector for introducing new genes into the body; a vector is any biological thing that carries infection from one host to another. Because viruses have evolved into expert invaders of living cells as well as masters of taking over a host cell's DNA, they act as excellent vectors in putting a foreign gene into a patient's body.

A virus, such as the cold or influenza virus, attacks a living cell in a stepwise process, simplified as follows:

1. attachment to the outer surface of a host cell
2. penetration of the virus into the cell enveloping of the virus; by the cell membrane
3. shedding of the viral coat in the cell cytoplasm
4. use of cell DNA replication system for making viral DNA

After a virus has infiltrated the host cell's DNA replication process, it injects its own genes into the host DNA. The host becomes a factory for producing more and more virus particles that can spread throughout the body. RNA viruses follow a similar attachment-penetration process, but they take over the RNA replication steps inside the cell rather than DNA replication.

Newly constructed viruses exit host cells in two ways: lysis of the cell and budding. In lysis, viruses kill the host cell by rupturing its membrane. The released viruses then seek new living cells to attack. Viruses can also exit host cells by pushing through the cell's membrane until the membrane completely encloses the virus, which then breaks free from the host, called *budding*. In budding, the host cell survives.

Because viruses inject new genetic material directly into a host cell's chromosome, scientists have used viruses to put genes into damaged DNA. In many cases, the new gene replaces a faulty gene in the patient. This type of treatment is often called *virus-mediated disease therapy*. The applications for gene therapy using viruses are fourfold: (1) a normal gene inserted into the chromosome to do the work of a nonfunctioning gene, (2) an abnormal gene removed and swapped with a normal gene, (3) an abnormal gene repaired by an enzyme in a normal process called *reverse mutation*, or (4) a new gene inserted to provide an on-off regulation for other genes.

THE STATUS OF GENE THERAPY

In September 1990, doctors at the National Institutes of Health (NIH) treated a four-year-old girl who had severe immune deficiency by intravenous infusion of white blood cells that contained the gene she lacked. The medical world considered this a qualified success: A patient could safely receive gene therapy with no obvious signs of trouble. Medical researchers began investigating other diseases that would respond to gene therapy. In 1999, University of Pennsylvania Hospital in Philadelphia prepared to treat 18-year-old Jesse Gelsinger, who had suffered all his life with a rare genetic disorder that affected his amino acid metabolism. Gelsinger entered a gene therapy trial intended to test the safety of virus-mediated treatments for future patients. Doctors injected Gelsinger with attenuated (weakened) cold virus, called *adenovirus*, which carried various corrective genes for his DNA. The researchers barely had a chance to determine whether the viruses had delivered their payload. Gelsinger soon began experiencing a dramatic inflammatory reaction to the injection. All his vital organs—liver, kidneys, lungs, brain—and his blood-clotting system malfunctioned at once, and he died, on September 17, 1999.

A devastated medical community searched for answers to the unexpected response in the young man's body. The bioethicist LeRoy Walters of Georgetown University said, "I think it's a perilous time for gene therapy. Until now, we have been able to say, 'Well, it hasn't helped many people, but at least it hasn't hurt people.' That has changed." Adenovirus treatments similar to Gelsinger's had, in fact, shown signs of danger in animal tests, but a tragic breakdown in communication between the NIH and the U.S. Food and Drug Administration (FDA), which approves drug-testing procedures, led to missed warning signs.

As of 2010, the FDA had not approved any viral-mediated therapy for patients. After a second alarming outcome in a gene therapy trial, in 2002, using a virus called a retrovirus, the FDA halted all testing on gene therapy. In Gelsinger's case and in other tests, the patient's immune response reacted in dramatic fashion to the sudden influx of viruses in the bloodstream. The stimulated immune system proceeded to destroy the foreign invader, but with tragic results. Gene therapists must learn how to suppress the immune response in order to allow the therapy to work, similarly to the way they suppress the immune system in organ transplant patients to reduce the chance of organ rejection by the body. Researchers must also determine whether gene therapy can deliver long-lasting benefits or provides only temporary effects. Finally, many of the greatest killers in the population, such as heart disease, diabetes, and Alzheimer's disease, are caused by a multitude of genes; they are called *multigene disorders.* No one has yet determined whether gene therapy can treat multigene disorders.

Stem cell research for gene therapy is regaining its momentum. In 2008, California voters approved funding for stem cell research. In March 2009, President Barack Obama made official a pledge to support the future of stem cell research. His March 9 White House press release said, "Advances over the past decade in this promising scientific field have been encouraging, leading to broad agreement in the scientific community that the research should be supported by Federal funds." Obama, on that day, signed an executive order to revoke limits that had been put on stem cell work by the previous administration.

Viral-mediated therapy has also moved forward in attempts to correct its problems of the past. The Human Genome Project has identified four types of viruses, described in the table, that are expected to yield the best results in future gene therapy trials. In all cases, the virus will be attenuated, so that it cannot cause disease on its own.

Viral-mediated gene therapy faces the risks of intentionally introducing infectious agents into a patient's body. Although researchers genetically change the virus vector to a benign form, this type of gene therapy has always raised concern that a therapy virus might revert to its virulent form once it is in the body. This concern has remained a real, if remote, drawback of viruses.

In 2009, David Schaffer of the University of California announced a novel approach to using viruses in safe gene therapy. His research team transformed a normally benign strain of AAV into a more infectious form. By this method, Schaffer thought he could develop an effective treatment for disease—his initial studies were on cystic fibrosis—but use a virus that would not present a health threat. "Both of those are situations where improvements in the properties of the vehicle can have a significant impact on the success of the therapy," Schaffer said in a campus news release. Schaffer's research team produced a new type of AAV that avoids the immune system yet is several hundred times more capable of entering lung tissue. "We devised a way to evolve viruses," he explained, "that are released from the natural constraints of evolution and have the freedom or ability to evolve toward properties that are more useful for medical application." The new virus also binds to a variety of different molecules, called *receptors,* on the surface of lung cells. This method improved the virus's ability to enter the cells.

Gene therapy researchers have tried to bypass some of the disadvantages of viruses by finding ways to inject new genes directly into a patient's cells. The following approaches to this type of gene therapy are being studied in the United States and abroad:

- liposomes—fat-based particles with an outer coat made of polymers
- nanoparticles—a specialty in nanobiology for making devices the size of molecules
- bioengineered white blood cells called lymphocytes
- direct introduction of a virus to the affected tissue

Research on gene therapy will probably develop vectors in addition to these for delivering genes safely to target organs or tissues.

Gene Therapy Viruses

Virus Type	Features	Examples
adeno-associated viruses	contains single-stranded DNA (ssDNA) and inserts genes at a specific site on chromosome 19	animal parvoviruses, human adeno-associated virus (AAV)
adenovirus	contains double-stranded DNA (dsDNA) with affinity for the mucosal membranes of the respiratory system, the digestive tract, and the eyes	common cold, conjunctivitis, pneumonia
herpes simplex	contains dsDNA and infects nerve cells	chicken pox, herpes, shingles
retroviruses	contains RNA, from which it can build dsDNA	acquired immunodeficiency syndrome (AIDS), leukemia, some sarcomas

Gene therapy is still a very new science, so its features and its flaws have not yet been completely identified. Gene therapy research is advancing, however, and may eventually emerge as a revolutionary way to repair genetic defects that previously had no cure.

See also GENETIC ENGINEERING; NANOBIOLOGY; VIRUS.

Further Reading

American Association for the Advancement of Science. "AAAS Policy Brief: Stem Cell Research." News release, March 10, 2009. Available online. URL: www.aaas.org/spp/cstc/briefs/stemcells/#ban. Accessed August 14, 2009.

Gardner, Amanda. "Gene Therapy Shows Promise for Parkinson's." ABC News/Health. October 15, 2009. Available online. URL: http://abcnews.go.com/Health/Healthday/gene-therapy-shows-promise-parkinsons/story?id=8830431. Accessed January 1, 2010.

Human Genome Project. "Gene Therapy." Available online. URL: www.ornl.gov/sci/techresources/Human_Genome/medicine/genetherapy.shtml#status. Accessed August 16, 2009.

Miller, Arlo. "Gene Therapy." *Society for Indistrial Microbiology News,* September–October 2001.

Obama, Barack. "Removing Barriers to Responsible Scientific Research Involving Human Stem Cells." The White House. Office of the Press Secretary. Executive Order, March 9, 2009. Available online. URL: www.whitehouse.gov/the_press_office/Removing-Barriers-to-Responsible-Scientific Research-Involving-Human-Stem-cells. Accessed August 16, 2009.

Sanders, Robert. "'Evolved' Virus May Improve Gene Therapy for Cystic Fibrosis." University of California news release, February 17, 2009. Available online. URL: http://berkeley.edu/news/media/releases/2009/02/17_schaffer.shtml. Accessed August 16, 2009.

Stolberg, Sheryl Gay. "The Biotech Death of Jesse Gelsinger." *New York Times Sunday Magazine,* 28 November 1999. Available online. URL: www.nytimes.com/library/magazine/home/19991128mag-stolberg.html. Accessed August 16, 2009.

genetic engineering Genetic engineering is a scientific discipline in which scientists manipulate genetic material and then manufacture a new product from it. The genetic material most often manipulated in this science is deoxyribonucleic acid (DNA). The detailed methods used within the genetic engineering specialty belong to a technology called *recombinant DNA* or genetic recombination. Recombinant DNA technology encompasses all the procedures for joining pieces of DNA from two different organisms or microorganisms to make a new DNA never before created in nature. This insertion of foreign DNA segments into a whole DNA molecule is called *gene splicing.*

Genetically engineered microorganisms have been studied and tested for the following activities:

- pollution degradation
- natural insecticides
- conferring new traits on plants and food-producing animals
- production of specific chemicals or drugs

Genetically engineered microorganisms grown in large bioreactors can produce substances that the microorganisms would not normally make. This is especially useful for products that are hard to harvest from nature: hormones, growth factors, drugs, antibiotics, and essential nutrients.

HISTORY OF GENETIC ENGINEERING

A rudimentary form of genetic engineering began, in 6000 B.C.E., when people selected certain yeasts for brewing beer. The earliest farmers seeking to perfect the quality and yield of their crops did not know about the gene or enzymes, but they nevertheless learned to enhance the traits of plants by combining two different species. The resulting plant contained the best attributes of both parent plants.

The first discovery of an enzyme has been difficult to pinpoint in history; most science historians credit the French brewers Anselme Payen (1795–1871) and Jean-François Persoz (1805–68) as the first to work with enzymes, when, in 1833, they extracted from malt a substance that could digest starch (the enzyme amylase). The Austrian monk Gregor Mendel (1822–84) laid the foundation for the field of genetics, in 1863, by showing that traits of parent pea plants carried over to progeny in discrete unidentified units, now known as genes.

Molecular genetics progressed to today's genetic engineering with the work of the British medical officer Frederick Griffith (1879–1941), who demonstrated that cells could exchange genetic material and thereby acquire new traits. James Watson (1928–) and Francis Crick (1916–2004) published a possible structure for DNA, in 1953, the first clear description of DNA's double-helical molecule, containing two strands of sugar subunits, outward-pointing phosphate groups, and complementary base pairings that connected the strands. With an accurate structure for DNA in hand, molecular biologists embarked on the new science of identifying genes and gene sets, removing some and inserting others into the DNA molecule.

Modern genetic studies and subsequent engineering can be attributed to the work of the American agricultural geneticist George Beadle (1903–89) and the biologist Edward Tatum (1909–75) of Stanford University. In 1941, they published results from experiments on the means by which *Neurospora* fungi regulate their nutrient metabolism. Beadle and Tatum wrote, in the *Proceedings of the National Academy of Sciences,* "The development and functioning of an organism consist essentially of an integrated system of chemical reactions controlled in some manner by genes. It is entirely tenable to suppose that these genes, which are themselves a part of the system, control or regulate specific reactions in the system either by acting directly as enzymes or by determining the specificities of enzymes." Biologists now understand that genes encode for enzymes that conduct the major cellular functions, the premise behind genetic engineering. *Encoding* means that the gene carries all the instructions for making a functional protein. The table on page 343 gives a brief overview of the history of genetic engineering.

Most of the discoveries that led to present-day genetic engineering have focused on the structure, functioning, activities, and repair of DNA. DNA plays the central role in genetic engineering based on four features of this genetic material. First, it must carry information, it does so in its genetic code. Second, it must replicate; the double-stranded model allows for replication. Third, genetic material must allow for genetic material to change. Mutations in the gene sequence of DNA cause this change. Fourth, DNA must be able to determine the phenotype, or the physical and chemical characteristics, of a cell. Genes do this when they are expressed as functional proteins.

Current genetic engineering builds on these principles for diagnosing disease through gene analysis and for treating some diseases by providing missing genes or replacing damaged genes. The latter two activities make up gene therapy.

Genetic engineering has created as much controversy as promise. Many people fear that genetically modified organisms (GMOs) will escape into the environment, where they can harm natural plants and animals. Ronnie Cummins of the Organic Consumers Association has warned, "When gene engineers splice a foreign gene into a plant or microbe, they often link it to another gene, called an antibiotic resistance marker gene (ARM), that helps determine if the first gene was successfully spliced into the host organism. . . . These new combinations may be contributing to the growing public health danger of antibiotic resistance." To allay these fears, microbiologists use antibiotic assays only in laboratory tests, and they now include within GMOs a set of genes that stop the microorganism's growth if it escapes into the environment.

Suicide genes have been studied, since the early 1990s, when the public took note of a rapidly expanding industry in genetically modified foods and microorganisms. Genetic engineers responded to safety concerns voiced by many critics of genetic engineering by inserting automatic safety mechanisms in GMOs. In 2004, the University of Wisconsin bacteriologist T. Kent Kirk emphasized, "Deciding whether and how to confine a genetically engineered organism cannot be an afterthought." Kirk joined a team of researchers at the National Academy of Sciences for the direct purpose of inventing multiple safety mechanisms for GMOs, in the event that one mechanism should fail. Some of the safety mechanisms that can work concurrently to stop GMOs in the environment are the following:

- suicide genes that activate when a pollutant disappears (for pollution-degrading GMOs)
- suicide genes that activate when a specific nutrient disappears
- activation genes that allow the GMO to live only under defined conditions
- nutrient requirements engineered to be higher than the nutrients normally found in nature

Since the early 2000s, safety methods for GMOs have included methods for monitoring the cells in addition to confining them to a certain area. The molecular biologists Qin Li and Wu Yi-Jun wrote, in a 2009 article in *Applied Microbiology and Biotechnology,* "One way to reduce the potential risk of genetically engineered microorganisms (GEMs) to the environment is to use a containment system that does not interfere with the performance of the GEM until activated. Such a system can be created by inserting a suicide cassette consisting of a toxin-encoding gene controlled by an inducible promoter." Qin Li and Wu enhanced this scenario by adding a way to monitor the GMO: "We constructed a GEM that can degrade organophosphorus compounds [type of pollutant], emit green fluorescence, and commit suicide when required by putting the genes that control these different functions under different promoters." These techniques have helped improve the overall value of GMOs for various tasks.

GENETIC ENGINEERING METHODS

Genetic engineering begins by identifying a gene that encodes for a certain desirable trait. For example, *Bacillus thuringiensis* contains a gene that encodes for a crystalline toxin. This toxin kills a variety of insects that devour many food crops. By putting the *Bacillus* toxin gene directly into a plant's DNA, genetic engineering creates a new plant that can ward off insect pests on its own, without the need for chemical insecticides. This elegant process begins by finding desirable genes.

Microbiologists begin by recovering a DNA molecule from a plant, animal, or microorganism of inter-

Milestones in Genetic Engineering

Date	Individual or Institution	Event
1863	Gregor Mendel	discovers traits are passed from parents to the next generation
1869	Friedrich Meischer	isolates weakly acid phosphorus-rich material from the nuclei of human white blood cells; calls it *nuclein*
1928	Frederick Griffith	demonstrates gene transfer by transformation by turning smooth, virulent *Streptococcus pneumoniae* into rough, nonvirulent *S. pneumoniae*
1941	George Beadle and Edward Tatum	propose that genes regulate cell development and function
1944	Oswald Avery, Colin MacLeod, and Maclyn McCarty	show DNA is the material that transforms cells by leaving DNA out of an experiment similar to Griffith's; transformation using any cell macromolecule other than DNA could not occur
1950	Erwin Chargaff	demonstrates adenosine and thymine are present in a 1:1 ratio; shows guanine and cytosine are also in a 1:1 ratio
1951	Rosalind Franklin and Maurice Wilkins	produce images of DNA by X-ray diffraction technology
1952	Alfred Hershey and Martha Chase	Waring blender experiment: infect bacteria with bacteriophage carrying either radioisotope-labeled DNA or protein, showing only the DNA becomes part of new generations of cells
1953	James Watson and Francis Crick	propose a double-helix structure for DNA
1958	Joshua Lederberg, George Beadle, and Edward Tatum	describe the mechanisms in genetic control of biochemical reactions
1962	James Watson, Francis Crick, and Maurice Wilkins	describe the physical structure of DNA
1968	Robert Holley, Har Gobind Khorana, and Marshall Nirenberg	identify the genetic code for amino acids
1977	Gilbert and Maxam	DNA sequencing techniques devised by using electrophoresis
1978	Genentech and City of Hope National Medical Center	first recombinant human insulin produced
1978	Daniel Nathans, Hamilton Smith, and Werner Arber	describe the action of restriction enzymes that cut DNA at specific sites
1980	Paul Berg	performs first gene splicing experiment
1981	Benjamin Hall and Gustav Ammerer	develop first genetically engineered vaccine (hepatitis B)
1985	Gary Mullis	invents polymerase chain reaction
1987	Advanced Genetic Sciences	Frostban becomes first product containing genetically engineered bacteria
1994	Calgene	'Flavr Savr' tomato becomes the first genetically engineered food
1998	James Thomson	develops first human embryonic stem cells for disease study
2000	Johann Greilhuber, Thomas Borsch, Kai Müller, Andreas Worberg, Stefan Porembski, and Wilhelm Barthlott	sequence first plant genome
2003	National Institutes of Health and U.S. Department of Energy	complete human genome sequence identification
2008	J. Craig Venter	manufactures a complete bacterial genome

est. The microbiologist then digests the DNA into small pieces with the aids of enzymes called *restriction endonucleases.* The various DNA fragments can be spliced into the DNA of various cells of the same bacterial species. By growing these cells, a laboratory can produce large amounts of cells, each of which contains a piece of the original DNA. This collection of bacterial cells that together contain the entire original DNA is called a *gene library.*

The bacteria making up gene libraries contain specific marker genes that enable microbiologists to find a few cells that contain a valuable DNA fragment among the millions of other cells in a culture that contain less valuable fragments. The gene normally found in bacteria that encodes for the enzyme β-galactosidase—this enzyme digests long galactoside molecules into small sugars—often serves as a good marker gene. By inserting the foreign DNA fragment directly into the middle of the β-galactosidase gene, a microbiologist can determine whether the bacteria received the desirable DNA. The components of this technique, called blue-white screening, are summarized in the table.

The gene called the *lacZ* gene encodes for the β-galactosidase enzyme and makes the blue-white screening test an easy method to allow microbiologists to monitor the insertion of a given DNA fragment into a microorganism. From the steps in the table, the microbiologist uses step 6 to select only white colonies, which are the only cells that received the plasmid carrying a piece of DNA from another organism. Next, the microbiologist must determine whether the foreign DNA contained by the white colonies is also the DNA that carries the particular trait that the scientist seeks.

To identify a specific gene or DNA fragment in bacteria, microbiologists use tools called *DNA probes.* A probe contains a short piece of a single DNA strand that has been made to be complementary to a specific target gene. (Some biochemical companies devote most of their business to developing probes for special industrial uses.) After mixing the probe with lysed bacteria in which the DNA has been split into single strands, a microbiologist can find the target gene. This occurs because the single-stranded probe binds only to the complementary portion of the single-stranded bacterial DNA. Probes usually contain a fluorescent molecule that allows the microbiologist easily to spot the reaction under a fluorescent microscope.

Genetic engineers have invented various methods for isolating, purifying, and producing a certain gene. The goal of any method is to produce a large amount of the pure gene, so that it can be used for making a commercial product on a large scale. Recombinant DNA technology focuses on the best methods for growing a large amount of genetically identical cells that all contain the target gene. This process is called

Creating a Marker for Genetically Engineered Bacteria

Step	Procedure	Result
1	cut original DNA and a plasmid in the same tube with restriction endonuclease enzyme that cleaves in the middle of the β-galactosidase gene	foreign DNA fragments insert into the β-galactosidase galactosidase gene of the plasmid
2	the engineered plasmid is added to a culture of bacteria resistant to antibiotic A	plasmids enter the bacteria by a process called *transformation,* and the bacteria gain resistance to antibiotic A
3	the bacteria are put on X-gal agar, a medium that contains antibiotic A and a compound called X-gal, which acts as a substrate for β-galactosidase	the antibiotic kills any bacteria that did not receive the plasmid
4	observe cells growing on the X-gal agar	only cells containing the plasmid can survive in the presence of antibiotic A
5	observe cells that produce blue colonies on X-gal agar	these cells did not receive the foreign DNA because their β-galactosidase gene remains intact (when the enzyme reacts with X-gal, it produces a blue end product)
6	observe cells that produce white colonies on X-gal agar	these cells contain foreign DNA that was inserted into β-galactosidase gene

cloning, and each identical cell (or a colony from that cell) is a clone.

Cloning relies on four main means of introducing foreign genes into a microorganism that will be used to make large amounts of the gene product. The tool used in this delivery system is called a vector, and the four types are the following:

- plasmids—small pieces of bacterial DNA separate from the main chromosome
- bacteriophages—viruses that infect bacteria
- cosmids—plasmids packaged within a bacteriophage
- artificial chromosomes—a complete chromosome identical to the recipient cell's except that it also contains the target gene

After scientists have decided on the best way of putting foreign genes into a microorganism, they then grow the genetically engineered cells in large chambers called *bioreactors*. Bioreactors can hold 1,060 gallons (4,000 l) or more of a culture that produces a commercial end product from the original target gene. The science of optimizing all of the elements that lead from an isolated DNA fragment to a new drug or other product requires a large team of scientists to contribute their specialized expertise. Andy Topping, manager of Avecia Biologics in Manchester, England, told *Genetic Engineering and Biotechnology News*, in 2009, "Our Paveway technology optimizes three elements: vectors, host strains, and fermentation protocols." Each of these manufacturing components must be perfected in order to maximize efficiency, that is, to make a large amount of product using a minimal amount of time and materials. Genetic engineering has, in this way, become a significant part of industrial microbiology.

GENETIC ENGINEERING IN INDUSTRY

The uses for genetically engineered products are in three categories: therapeutic, scientific, or agricultural. Therapeutic products are used for treating or preventing disease. Therapeutic products sold today that have been derived from genetic engineering are the following:

- hormones: epidermal growth factor, human growth hormone, insulin, relaxin, somatostatin
- blood treatments: erythropoietin, factor VIII, prourokinase, streptokinase, tissue plasminogen activator
- cancer and chemotherapy products: alpha-interferon, colony-stimulating factor, interleukins, taxol, tumor necrosis factor
- immunity affectors: interleukins, orthoclone and other monoclonal antibodies
- other disease treatments: beta-interferon for multiple sclerosis, gamma-interferon for granulomatous disease, pulmozyme for cystic fibrosis
- vaccines: hepatitis B, influenza

The yeast *Saccharomyces cerevisiae* has become the most common microorganism engineered for producing many of the listed substances. Almost 100 percent of *S. cerevisiae*'s DNA has been sequenced—that is, scientists have identified the order of its base pairs—so this yeast has been the primary choice for gene manipulations. Other cell types that are genetically engineered to produce drugs are *E. coli, Bacillus, Pichia pastoris* yeast, and various mammalian cell lines.

Many therapeutic uses for genetically engineered substances go through a difficult process of DNA isolation and sequencing and development of high-yielding clones. All new drugs must also be tested for safety and effectiveness in humans, a process called *clinical testing*. Tim Farley of the World Health Organization explained to BBC News, in 2005, the hurdles to overcome in developing a genetically engineered treatment for the acquired immunodeficiency syndrome (AIDS) virus. In this case, he discussed a genetically modified *E. coli* that would live in a person's digestive tract—*E. coli*'s natural habitat—and secrete proteins that would enter the bloodstream and block infection by human immunodeficiency virus (HIV). "Clearly there are many steps to be completed in the development and clinical testing of the product," he said, "and there may be special safety concerns over unexpected side effects due to deliberately colonizing *[sic]* the gastrointestinal tract with genetically engineered bacteria." Ideas in genetic engineering for use inside and outside medicine have obstacles to overcome before their practical application.

Scientific applications in genetic engineering involve research on better and faster methods of DNA sequencing, screening for genes in microorganisms, developing accurate probes, developing efficient microbial producers, and perfecting analytical techniques. University laboratories also play a part in genetically engineered products by developing new methods for isolating, purifying, and cloning genes.

Agriculture has, for a long time, used the principles of genetic engineering for selecting plant lines that give better results than other plants. Some of the attributes that genetic engineering can deliver in plants are the following: bruise-resistant fruits, longer-shelf-life fruits and vegetables, disease-resistant plants, natural insecticides, improved yields, drought resistance, freeze tolerance, longer growing season, and improved yields. Agricultural products made from genetically engineered microorganisms include the following:

- *Pseudomonas* bacteria that contain an insecticide gene from *B. thuringiensis*
- *P. syringae,* the ice-minus organism, which protects against ice formation on plants
- *Rhizobium* bacteria modified for enhanced nitrogen fixation
- herbicide-resistant plants containing bacterial genes
- livestock growth hormones produced by engineered *E. coli*
- cellulase enzyme made by engineered *E. coli* to make feeds more digestible
- rennin enzyme made by the engineered mold *Aspergillus niger* for forming curds in dairy products

Agriculture will probably remain a major user of new genetically engineered plants, microorganisms, and even animals.

Genetic engineering might soon benefit the development of biologically produced fuels, called *biofuels.* Modified *E. coli* and other bacteria can be made to produce a diesellike fuel. Microorganisms engineered to produce fuels would benefit the environment by forestalling the depletion of fossil fuels and slow the destruction of forests now occurring for the production of plant-based biofuels.

NATURAL GENETIC ENGINEERING

Natural genetic engineering uses techniques other than the insertion of foreign DNA into a microorganism. This type of genetic engineering, instead, involves two techniques for making microorganisms change their own genetic makeup: forced evolution and adaptation mutations.

Forced mutations occur when scientists create stresses that make microbial populations evolve faster than normal. As a hypothetical example, a bacterial species that produces an antibiotic might be induced to make extra amounts when stressed by low-nutrient conditions. This occurs because the bacteria might seek a way to outcompete other "hungry" bacteria in an environment low in nutrients.

Adaptive mutations, also called *directed mutations,* may be thought of as mutations that a microorganism purposely chooses. John Cairns of the Harvard School of Public Health reported, in 1988, an *E. coli* strain that normally could not use the sugar lactose as a carbon source but would quickly develop the ability to use it if it were the only sugar provided to it. Since Cairns's proposal became known, scientists have speculated on the reasons for this phenomenon, which seems to run counter to the idea accepted by geneticists that spontaneous mutations are due to random changes in DNA. Adaptive mutations will offer genetic engineering another approach to making microorganisms behave in a specific way.

The field of genetic engineering is a relatively young science. This technology's foundation lies in the metabolism of microorganisms and the ways they alter, share, and express genes. Genetic engineering makes use of recombinant technology, and it has been responsible for creating new industries such as biotechnology. Genetic engineering will probably produce more innovations in industrial microbiology, agriculture, and medicine.

See also BIOREACTOR; FLUORESCENCE MICROSCOPY; GENE TRANSFER; RECOMBINANT DNA TECHNOLOGY.

Further Reading

BBC News. "Bacteria Modified to Combat HIV." November 13, 2005. Available online. URL: http://news.bbc.co.uk/2/hi/health/4692905.stm. Accessed August 9, 2009.

Beadle, George W., and Edward L. Tatum. "Genetic Control of Biochemical Reactions in *Neurospora.*" *Proceedings of the National Academy of Sciences* 27 (1941): 499–506.

Cairns, John, Julie Overbaugh, and Stephan Miller. "The Origin of Mutants." *Nature* 335 (1988): 142–145.

Cummins, Ronnie. "Hazards of Genetically Engineered Foods and Crops—Why We Need a Global Moratorium." Organic Consumers Association. Available online. URL: www.purefood.org/GEFacts.htm. Accessed August 9, 2009.

Morrow, K. John. "Grappling with Biologic Manufacturing Concerns." *Genetic Engineering and Biotechnology News,* 1 March 2009.

National Academy of Sciences. "Integrated, Redundant Approach Best Way to Biologically Confine Genetically Engineered Organisms." January 20, 2004. Available online. URL: www8.nationalacademies.org/onpinews/newsitem.aspx?RecordID=10880. Accessed August 9, 2009.

Qin Li, and Wu Yi-Jun. "A Fluorescent, Genetically Engineered Microorganism That Degrades Organophosphates and Commits Suicide When Required." *Applied Microbiology and Biotechnology* 82 (2009): 749–756.

gene transfer Gene transfer is the passing of genetic material from one microbial cell to another. This transfer may take place as either horizontal gene transfer, in which genes move from one mature, independent organism to another mature organism, or as vertical gene transfer, in which a parent cell passes genes to its offspring.

Gene transfer occurs naturally in microorganisms as a way to share genes. This sharing enables microorganisms to adapt to new conditions in the environment and to confer important adaptations on the next generation. The transfer and sharing of genes make up a natural mode of deoxyribonucleic acid (DNA) recombination, which, at its simplest, is the cleaving of a strand of host cell DNA and the subsequent insertion of a gene from another cell into the gap.

To carry out natural DNA recombination, the entire process relies on a set of enzymes that perform the following steps: (1) cleavage of a gene from the donor cell, (2) cleavage of a DNA strand in the recipient host cell, (3) insertion of the new gene, and (4) reconnection of the new version of the DNA into an intact double-stranded molecule. The gene transfer step additionally requires a method of relocating a gene (or a set of genes) from one cell into another so that none of the gene's information is lost or corrupted. In biology, *recombination* usually refers to the enzymatic steps listed earlier for inserting genes into DNA, and *gene transfer* refers to the ways in which genes can be physically passed from cell to cell in the environment.

Rearrangement of genes has been critical in the evolution of species by giving rise to adaptations favorable to species survival. Gene transfer has helped build diversity in generations of microorganisms and higher organisms. Diversity, then, gives each population improved chances for surviving and for competing against other members of the population.

Recombinant technology and gene transfer have together provided the foundation for the biotechnology industry, which creates new products through the rearrangement of natural genes. These advances have helped in the development of foods with longer shelf life, plants with stronger defenses against parasites, pesticides, herbicides, drugs, and disease therapies. Biotechnologists now use many of the same techniques for transferring genes in the laboratory that microorganisms use in nature.

The four main methods that microorganisms use to transfer genes are the following:

- plasmid exchange conducted by many bacteria and some yeasts and fungi
- conjugation in bacteria and algae
- transformation in bacteria
- transduction, in which a virus transfers a gene to a host cell, or a bacteriophage transfers a gene to a bacterial cell

Once a new gene has been transferred into a recipient cell, the gene enters the DNA by one of three methods: (1) reciprocal exchange of genes between two DNA sequences on two different strands; (2) nonreciprocal transfer, whereby a new gene inserts into one strand to lengthen that strand—the result is heteroduplex DNA; and (3) site-specific recombination, in which a virus inserts a gene at a particular location in the recipient cell's DNA.

PLASMID EXCHANGE

A plasmid is a small circular doubled-stranded DNA (dsDNA) molecule present in a bacterial cell's cytoplasm and separate from the main DNA, which inhabits an area called the *nucleoid*—a few eukaryotes such as yeast can also contain plasmids. Plasmids plus nucleoid DNA make up the bacterial chromosome. The one or more plasmids in a bacterial cell operate independently of the nucleoid DNA. For instance, plasmid replication uses different replicons, which are sequences that serve as the origin point in DNA replication.

Plasmids total no more than 5 percent of the total chromosome in most bacteria, yet they play an important part in various cellular functions. Plasmids encode for a wide variety of proteins made by the cell. Encoding is the process by which a cell converts the information stored in its DNA into a protein. Plasmids encode for some of the following cellular products or activities:

- metabolic enzymes
- cell-to-cell communication
- detoxification of poisons
- resistance to chemicals or antibiotics
- bacteriocin or antibiotic production
- toxin production mechanisms
- nutrient uptake factors
- virulence factors

Bacteria use plasmid exchange for any of the listed attributes between cells of the same species or between different species. Plasmids, thus, play a role in transferring traits from one type of microorganism to another, unrelated type of microorganism. As a result, many different species in the microbial world can evolve with new traits that enhance their survival. By contrast, multicellular organisms pass their traits from one generation to the next only by transferring genes from one generation to the next.

CONJUGATION

Conjugation occurs through physical contact between cells. Sometimes called *bacterial mating*, conjugation was discovered, in 1946, by the American molecular biologists Joshua Lederberg (1925–2008) and Edward Tatum (1909–75). In their experiments, Lederberg and Tatum mixed together two different auxotrophs—organisms lacking the ability to synthesize a certain essential nutrient—each with distinct nutrient requirements. The biologists selected strains that needed an external supply of more than one nutrient, such as biotin or cysteine. Because of the very specific nutrient needs of these microorganisms, they were called *strict auxotrophs*. After the two microorganisms incubated together in the same culture, they produced a new generation of cells that could grow on medium lacking nutrients they normally required for growth. Lederberg and Tatum showed that the two auxotrophs had somehow exchanged genetic information that would benefit both species.

In 1950, the Harvard Medical School microbiologist Bernard Davis (1916–94) demonstrated that the gene transfer of the Lederberg-Tatum experiment took place through physical contact between cells. Davis devised an apparatus in which two auxotrophs would be in individual chambers separated only by a filter that allowed media to pass through but halted the movement of the cells. Separated cultures never exhibited gene transfer in a Lederberg-Tatum test, but when Davis allowed the two cultures to mix, the cells readily exchanged genes.

Lederberg next proposed the concept of specialized donor cells and recipient cells. In laboratory notes compiled in 1955, Lederberg wrote, "Until fairly recently, all strains of *E. coli* K-12 were believed to be mutually compatible. It has now been established that two 'mating types' exist, designated F^+ and F^-, so that $F^- \times F^-$ is sterile, while $F^+ \times F^-$ and $F^+ \times F^+$ are fertile." The table provides details of the mechanisms of bacterial conjugation described by Lederberg.

In all three types of bacterial conjugation, a fertility factor, or F factor, serves as the main mechanism for transferring genetic material from one cell to another. Depending on the type of conjugation, the F factor can be a plasmid, or it can be a portion of a donor cell's chromosome.

Conjugation in protozoa and algae proceeds in a different manner than bacterial conjugation. In these eukaryotes, two cells fuse, and each exchanges a haploid nucleus containing half of the chromosome. The haploid nucleus, called a micronucleus, of a donor, then, fuses with the haploid nucleus of the recipient to form a macronucleus. When each half of the fused cells has a macronucleus with a full complement of DNA, they cells are said to be fertilized, and they separate.

TRANSFORMATION

Genes sometimes transfer from one cell to another as "naked DNA" through laboratory medium or in the environment. The English military medical officer Frederick Griffith demonstrated this process, called transformation, in 1928, with two different strains of *Streptococcus pneumoniae*. In what is now known as Griffith's transformation experiment, Griffith used one encapsulated virulent (disease-causing) *S. pneumoniae* strain and one nonencapsulated avirulent (not disease-causing) strain. As expected, mice injected with the virulent strain died, and mice injected with avirulent cells lived. But, when Griffith killed virulent cells and mixed them with living avirulent cells, then injected mice with the mixture, the mice died. He concluded that the live avirulent bacteria had taken in an unknown factor from the killed cells that made the normally harmless bacteria virulent as well. A

Bacterial Conjugation

Type	Symbol	Description
polar	$F^+ \times F^-$ mating	an F^+ strain (donor) forms a pilus, a thin appendage that extends to an F^- strain (recipient) cell, and genetic material (a plasmid) passes between the cells
high-frequency recombination	Hfr	chromosome replication takes places in an Hfr cell, and a large piece of the replicated material transfers to an F^- cell as the replication progresses
F′ plasmid	$F' \times F^-$ mating	an Hfr cell produces a plasmid from a part of its chromosome, and as it replicates, the plasmid, it also sends a copy to an F^- cell; the F^- cell becomes an F′ cell with its own F′ plasmid

mechanism other than conjugation or other cell-to-cell association seemed to have occurred. DNA served as the so-called transforming factor that turned ordinary cells into deadly pathogens. Griffith further showed that the live cells had taken in the virulence factor directly from the medium as virulent cells lysed and released their contents into the liquid medium.

Microbiologists have built upon Griffith's transformation experiment to show that certain cells, called *competent cells,* absorb DNA or DNA fragments directly from the environment. Only about one cell in 1,000 is competent and able to incorporate new DNA into its native chromosome, but this is sufficient to enable a population to adapt new traits over several generations.

In transformation, a competent cell attaches to dsDNA that has been released by a lysed cell in the environment. The competent cell, then, secretes the enzyme endonuclease to break the DNA into smaller fragments. DNA uptake occurs in a different manner for each of the two DNA strands. Each uptake mechanism operates independently of the other. In the first mechanism, a competent cell envelops a free DNA strand with its membrane. The second mechanism involves a membrane protein that connects to a DNA strand and carries it into the cytoplasm in an energy-demanding process. Regardless of the method used for entering the cell, each single strand of the foreign DNA aligns with a homologous section (a section with complementary base pairs) of the recipient DNA.

Certain bacterial species have adapted specializations in their DNA uptake mechanisms. For example, the transformation system of *Bacillus subtilis* contains a structure called a *pilin complex.* The pilin complex creates an infrastructure in the membrane made of DNA-binding proteins, channel-forming proteins, and enzymes. These components work together to place DNA in the *Bacillus* cell.

In laboratories, competency occurs only if specific conditions exist, depending on the bacterial species. Microbiologists must optimize these conditions in order to elicit gene transfer. The main competency factors are phase of bacterial growth, presence or absence of certain nutrients, and adverse environmental conditions, such as high temperature or strong salt concentration. Microbiologists have discovered that the following genera are most likely to use transformation in laboratory gene transfer studies: *Acinetobacter, Azotobacter, Bacillus, Helicobacter, Moraxella, Neisseria, Pseudomonas,* and *Streptococcus.*

Transformations in laboratory experiments can be difficult and inefficient. Microbiologists overcome some of these drawbacks by exposing bacteria to very high concentrations of DNA to favor the transformation process.

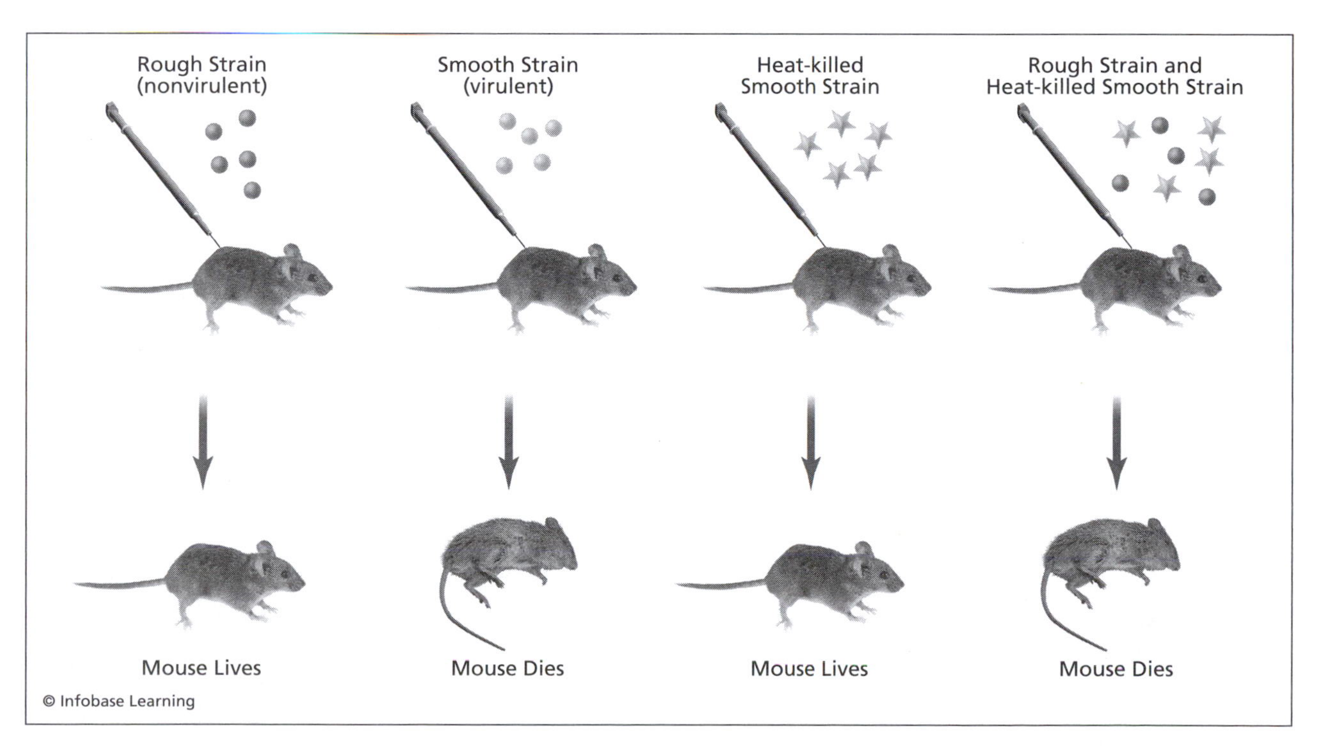

In 1928, Frederick Griffith showed that bacteria transfer genetic information between cells. By mixing killed virulent bacteria with live nonvirulent bacteria, Griffith showed that the supposedly nonvirulent cells could kill the host. The nonvirulent cells had absorbed virulence "transforming factor" from the suspension. In 1944, Oswald Avery, Colin MacLeod, and Maclyn McCarty showed that the transforming factor was DNA.

TRANSDUCTION

Transduction takes place when a virus inserts its genes into a host cell. This type of gene transfer occurs when a plant virus infects a plant cell or an animal virus infects an animal or human cell. Bacteria are also infected by a special class of viruses called *bacteriophages,* which attack only prokaryotic cells.

In 1952, Lederberg teamed with a fellow University of Wisconsin geneticist, Norton Zinder, to describe a new type of gene transfer they had detected in *Salmonella.* After a series of experiments using bacteriophages and *Salmonella* cultures of antibiotic-resistant and nonresistant strains, Zinder concluded, "The mechanism of genetic exchange found in these experiments differs from sexual recombination in *E. coli* in many respects so as to warrant a new descriptive term, *transduction.*" These experiments not only identified another method by which bacteria exchanged genetic information, but also helped elucidate the manner in which viruses infect healthy cells.

Transduction occurs in the following steps:

1. A bacteriophage attaches to a bacterial cell.
2. The bacteriophage injects its DNA into the cell.
3. Enzymes break apart the host DNA and insert pieces of the bacteriophage's DNA.
4. When new bacteriophages develop inside the cell, they contain some bacterial DNA.
5. The cell lyses and releases bacteriophages that can attack other bacteria.
6. When the bacteriophage infects a new bacterial cell, it introduces DNA from the originally infected bacteria.

By the steps listed here, genetic traits of one type of bacteria can move into another population of bacteria, provided they are both susceptible to the same bacteriophage. This process is called *generalized transduction;* in it, parts of bacterial DNA are packaged into new bacteriophage particles in a somewhat random manner. A second type of transduction, called *specialized transduction,* puts only specific bacterial genes into the new virus particles. When the host cell lyses, the bacteriophages carry the specific traits to other bacteria. Specialized transduction carries genes for toxins and virulence factors, so it plays a critical role in pathogenesis, the origin and development of disease.

These findings on various modes of gene transfer in bacteria opened up new studies in the genetics of pathogens, host cell susceptibility, and disease at the molecular level of DNA. Gene transfer studies, therefore, laid the foundation for current genetic testing for disease and for gene therapy in fighting certain diseases. In a 1996 interview, Lederberg recalled his accomplishments, saying, "Well, I think objectively there's no doubt that opening up genetics of bacteria had an impact. It was part of a movement. This is not the chemical side of it, but it's the cell biological aspect of it. It was not only the experimental results, it was a style of experimentation. . . . If you look at how we did go about discovering conjugation, discovering transduction, the experimental designs; those are very familiar today, but I think we more or less invented them." Studies by Lederberg, Tatum, Davis, and other geneticists truly started a new era in biology at the molecular level. As geneticists further investigated the workings of the chromosome and its genes, they uncovered additional surprises. One of the most interesting discoveries was the transposon, an entity in biology so astounding that the scientific community refused to believe in its existence until 30 years after its discovery.

TRANSPOSITION

Transposons are segments of DNA that move—they are transposed—from one part of a DNA molecule to another part. Transposons may move from one site to another on the same DNA, from one DNA molecule to another, or from DNA to a plasmid. These segments can range from 700 to 40,000 base pairs in length, and the process of moving from one site to another is called *transposition.*

The American molecular biologist Joshua Lederberg shared the 1958 Nobel Prize in medicine with Edward Tatum and George Beadle for discovering how bacteria exchange genetic material. ***(U.S. National Library of Medicine, National Institutes of Health)***

The American geneticist Barbara McClintock (1902–92) studied the details of transposition, in the 1940s and 1950s. In 1950, she published results showing that transposons—they were first called "jumping genes"—existed in corn cells. Almost immediately, McClintock received harsh criticism for her proposal; scientists could not believe that cells would permit survival of a system that scrambled the genetic code and would wreak havoc with cell metabolism. With declining support from the scientific community and mounting criticism, McClintock stopped working on transposons, in 1953.

By the 1970s, scientists had clarified many activities in gene control within cells and had begun to understand how transposons could exist, although they may not have grasped why such a phenomenon occurred. Experiments on gene regulation in eukaryotes, prokaryotes, and viruses built a body of evidence that supported the existence of transposons. In 1983, McClintock received the Nobel Prize in physiology for her discovery of transposons, 30 years earlier. In her Nobel acceptance lecture, she expanded on the "why" of transposons: "It is the purpose of this discussion to consider some observations from my early studies that revealed programmed responses to threats that are initiated within the genome itself, as well as others similarly initiated, that lead to new and irreversible genomic modifications. These latter responses, now known to occur in many organisms, are significant for appreciating how a genome may reorganize itself when faced with a difficulty for which it is unprepared." McClintock explained, "Our present knowledge would suggest that these reorganizations originated from some 'shock' that forced the genome to restructure itself in order to overcome a threat to its survival." McClintock further proposed that some transposon movements in microorganisms may have led to the development of new species.

Transposons seem likely to contribute to the development of adaptations within species in ways similar to spontaneous mutations in DNA. The frequency of transposition in bacteria is equivalent to the rate of spontaneous mutations: from about one in 10,000 to one in one million DNA replications.

Microbiologists may not yet have uncovered all of the information that tranposons carry, but they do know that some of the known transposon genes encode for the following:

- mechanism for transposition
- DNA repair after transposition
- toxin production
- antibiotic resistance

Cells, plasmids, or viruses can each carry transposons between microorganisms of the same species or different species. As do other methods of gene transfer, transposition probably helps microorganisms gain adaptations that aid in survival and evolution.

Gene transfer has grown into one of the most vital areas of study in microbiology, genetics, and disease therapy. In his 1958 Nobel lecture for the prize he shared with Lederberg and George Wells Beadle, Edward Tatum said, "As a biologist, and more particularly as a geneticist, I have great faith in the versatility of the gene and of the living organisms in providing the material with which to meet the challenges of life at any level." The science of gene transfer now leads a new generation of gene therapies for disease and stem cell research.

See also GENE THERAPY; MUTATION RATE; PLASMID; RECOMBINANT DNA TECHNOLOGY; RESISTANCE; VIRUS.

Further Reading

Lederberg, Joshua. "Instructions for Crossing Biochemical Mutants of *Escherichia coli* K-12." Laboratory notes, January, 1955. Available online. URL: http://profiles.nlm.nih.gov/BB/B/D/B/B/_/bbbdbb.pdf. Accessed August 10, 2009.

———. "Interview with Prof. Lederberg, Winner of the 1958 Nobel Prize in Physiology and Medicine." By Lev Pevzner. March 20, 1996. Available online. URL: http://almaz.com/nobel/medicine/lederberg-interview.html. Accessed August 10, 2009.

Lederberg, Joshua, and Edward L. Tatum. "Gene Recombination in *Escherichia coli.*" *Nature* 158 (1946): 558.

McClintock, Barbara. "The Significance of Responses of the Genome to Challenge." Nobel lecture, December 8, 1983, Stockholm, Sweden. Available online. URL: http://nobelprize.org/nobel_prizes/medicine/laureates/1983/mcclintock-lecture.pdf. Accessed August 10, 2009.

Sardesai, Abhijit A., and J. Gowrishankar. "Joshua Lederberg—a Remembrance." *Journal of Genetics* 87 (2008): 311–313. Available online. URL: www.ias.ac.in/jgenet/Vol87No3/311.pdf. Accessed April 17, 2009.

Tatum, Edward. "A Case History in Biological Research." Nobel lecture, December 11, 1958, Stockholm, Sweden. Available online. URL: http://nobelprize.org/nobel_prizes/medicine/laureates/1958/tatum-lecture.html. Accessed August 10, 2009.

Zinder, Norton D., and Joshua Lederberg. "Genetic Exchange in *Salmonella.*" *Journal of Bacteriology* 64 (1952): 679–699. Available online. URL: http://profiles.nlm.nih.gov/BB/A/B/F/L/_/bbabfl.pdf. Accessed August 10, 2009.

genomics A genome is the entire set of genetic information in any cell, prokaryotic or eukaryotic, or in a virus. The discipline of genomics is the study of genome organization, the information genes contain, and the products formed when genes are expressed.

Gene expression is the conversion of the information contained in a gene into a functioning protein.

Genomics represents the newest era in microbiology's development. Prior to the mid-1800s, biologists had little knowledge of the microbial world. Some scientists, such as Antoni van Leeuwenhoek (1632–1723) and Robert Hooke (1635–1703), had glimpsed microscopic cells, but they used weak and rudimentary microscopes to make their observations. In the mid-1800s, microscopy and the germ theory advanced. Many biologists, principally the Frenchman Louis Pasteur (1822–95), led a period of rapid discovery in microbiology. This period, from about 1850 to 1920, became known as the Golden Age of Microbiology. Not until the 1940s, did biologists begin characterizing the genetic material inside cells. In 1944, Oswald Avery (1877–1955), Colin MacLeod (1909–72), and Maclyn McCarty (1911–2005) identified the genetic material that confers the characteristics of parents to the next generation as deoxyribonucleic acid (DNA). In 1946, James Watson (1928–) and Francis Crick (1913–2004) proposed the double-helix structure of DNA. About six years later, François Jacob (1920–) and Jacques Monod (1910–76) explained the mechanisms by which genes control protein synthesis. Molecular studies advanced rapidly into a new molecular age of microbiology.

Genomics scientists now endeavor to identify the DNA base sequence of all the genes of many microorganisms. Mark Hahn of Woods Hole Oceanographic Institute (WHOI) in Massachusetts wrote in *Oceanus,* in 2005, "Genomics is more than simply determining the sequence of nucleotides in an organism's genome. . . . It is a new approach to questions in biology, distinguished from traditional approaches by its scale. Rather than studying genes one by one, genomic approaches involve the systematic gathering and analysis of information about multiple genes and their evolution, functions, and complex interactions within networks of genes and proteins." The first step in analysis of the genome is a process called *DNA sequencing.*

The first DNA sequencing of an entire genome took place, in 1976, in bacteriophage MS2, a specific type of virus that infects bacteria. In 1995, a team of molecular biologists at Johns Hopkins University, led by Robert D. Fleischmann, completed sequencing the entire genome of the bacterium *Haemophilus influenzae.* Since then, the genome of *Escherichia coli* has become the most studied microbial DNA in science. Specialized depositories also store thousands of individual gene sequences for hundreds of microorganisms. The National Human Genome Research Institute (NHGRI), part of the National Institutes of Health, sequenced the entire human genome, by the early 2000s. Scientists are presently taking steps into a new area, in which they intend to create a new genome from a collection of preexisting sequences.

THE SCIENCE OF GENOMICS

Genomics includes three main areas of study: (1) structural genomics, (2) functional genomics, and (3) comparative genomics. Experts in structural genomics study the physical structure of the genome plus the actual subunit—called *bases*—sequence of DNA. Scientists in functional genomics study the manner in which the genome works. This includes the transfer of information from genes to the proteins they encode and all the steps within this process. Comparative genomics involves the comparison of genomes from different species. The differences and similarities provide information on evolution, relatedness, and mutation.

Genomics begins with the extraction of pure DNA from an organism, so that its base composition can be analyzed. DNA extraction techniques have been refined over the past few decades but adhere to the following general steps, in this case, for *E. coli* bacteria:

1. Incubate *E. coli* in Luria broth, which supplies nutrients needed for gene replication.

2. Recover the bacteria and lyse the cells in detergent.

3. Add an enzyme (a protease) that releases the DNA from any attached proteins.

4. Add chilled alcohol, which dissolves all of the constituents except DNA.

5. Allow DNA to settle gradually out of the liquid by gravity.

6. Collect the whitish DNA material by spooling it onto a glass rod.

Scientists then prepare the DNA for sequencing. The English biologist Frederick Sanger (1918–) developed a technique for DNA sequencing, in 1975, that provided the template for today's methods. Scientists initially carried out the steps manually, but automated DNA analyzers now do the work, hundreds of times faster, on many more samples. In manual sequencing, a scientist can determine the nucleotide base—adenine, thymine, cytosine, and guanine—sequence directly from a material called polyacrylamide gel electrophoresis (PAGE), which contains a variety of DNA strands of differing length. This mixture of DNA strands from a starting quantity of DNA is called a *DNA library.* By recording each unique-length strand, from the shortest to the longest, in the DNA library of a microorganism, the scientist deter-

Microbial Genomes

Microorganism	Genome Size (Mb)
BACTERIA	
Bacillis subtilis	4.20
E. coli	4.60
Mycobacterium tuberculosis	4.40
Staphylococcus aureus	2.80
Streptococcus pyogenes	1.90
ARCHAEA	
Halobacterium species	2.57
Methanococcus jannaschii	1.66
Thermoplasma acidophilum	1.56
EUKARYOTES	
Saccharomyces cerevisiae	13.0

mines the exact base sequence of DNA. An example sequence would be written as follows:

AATCAAAGCTTAGACAAT

In this DNA sequence, A represents adenine, T is thymine, C is cytosine, and G equals guanine. The Dolan DNA Learning Center Web site provides excellent animation that explains the step-by-step process of DNA sequencing method.

Automated DNA sequencers use a method called *capillary gel electrophoresis*. This method adapts traditional PAGE to a system that analyzes very small-volume samples of DNA. By reducing the sample volume, automated methods can determine the sequence of almost 100 samples in a single instrument run.

A single chromosome contains thousands of base pairs—the bases bind to each other in pairings to create a DNA double-strand molecule. As geneticists tried to sequence ever larger genomes, even the automated modification of Sanger's method reached its limit, of no more than 1,000 base pairs. The table shows the genome size of cell types that have been fully sequenced and are common in genomics studies. The genome sizes are quite a bit larger than 1,000 base pairs; one megabase (Mb) equals 1 million base pairs. (The human genome contains about 300 Mb.)

In 1986, the U.S. Department of Energy began funding the Human Genome Project, to initiate a 15-year multilaboratory program to sequence the entire human genome. The yeast *Saccharomyces cerevisiae* and the archaean *Methanococcus jannaschii* served as a model for sequencing an entire genome. Project scientists completed the entire sequencing of both microorganisms in 1996; the next year, scientists completed the genome of *E. coli*. The *E. coli* team leader Frederick Blattner of the University of Wisconsin said, in 1997, "Determination of the complete inventory of the genes of organisms is one of the major goals of biology—analogous to development of the periodic table of elements in chemistry. Once they are all known and relationships between them become evident, a classification system for understanding the basic functions of life can be erected." The project was highly successful: The first human genome sequence was completed in 1999, and, by 2006, the project scientists completed the entire human genome ahead of schedule.

Rapid completion of DNA sequencing, such as that demonstrated by the Human Genome Project, owes its success to a method called *whole-genome shotgun sequencing*. The geneticists J. Craig Venter and Hamilton Smith developed this method, in 1996, to accommodate large genomes found in bacteria. The shotgun method consists of the following five main steps:

1. library construction—Enzymes break bacterial DNA into random-sized DNA fragments that are inserted into plasmids (small circular segments of DNA), and the plasmids are put into *E. coli* cells by a process called *transformation*.
2. sequencing—Automated sequencers determine the base sequence of all the fragments.
3. fragment alignment—A computer assembles a hypothetical genome sequence based on the fragment sequences and where they overlap.
4. gap filling—Analysis of previously sequenced bacterial DNA sections supplies information for the computer program to fill in any gaps in the newly constructed sequence.
5. editing—A computer program proofreads the entire sequence to find any inconsistencies by comparing with databases of known sequences.

Geneticists often perform sequencing studies with the goal of learning more about a microorganism's specific genes. In this case, they add a sixth step to the procedure, called *annotation*. Annotation is the process of determining the location of a specific gene in a genome. Some of the gene-associated processes that microbiologists have located on bacterial genomes are the following:

- cell membrane components
- nutrient transport proteins
- DNA replication enzymes
- protein synthesis enzymes
- amino acid synthesis
- fat metabolism

The main steps of shotgun sequencing include many more details than described here. Because of the complexity of sequencing large genomes, scientists have depended on computers to create the sequence from the analysis data, fill any gaps, proofread, and fix errors. A new field, called *bioinformatics,* uses computers to organize and analyze all the data that make up individual DNA sequencing tasks. To help with the amount of data accumulating on DNA sequencing taking place at many different laboratories, the National Institutes of Health has formed GenBank. GenBank is a depository of all of the annotated DNA sequences currently in existence. One of the newest sequences, added in 2009, belonged to the H1N1 influenza virus, which threatened to spread into an epidemic in that year. By quickly determining the DNA sequence for pathogens such as H1N1, scientists can develop effective vaccines.

DNA sequencing and annotation, together, provide scientists with information on the number of genes required to carry out certain cellular activities. For example, the pathogen *Mycoplasma genitalium* has been fully sequenced—it contains a small genome of only 0.58 Mb—with about 50 percent of the annotated genes correlated to various cell functions. An organism with such a small genome may give evidence as to the minimal amount of genes an independently living cell needs to survive. Scientists have developed an overview of *M. genitalium*'s genome, summarized in the table.

The genome of *M. genitalium* provides interesting clues about its survival. Protein synthesis takes up most of the genetic information carried by this bacterium; regulation of conditions inside the cell includes few genes. Membrane proteins are controlled by a large number of genes, which are important because *M. genitalium* is a pathogen that infects by, first, attaching to a host cell. By studying the mechanisms for making membrane proteins, microbiologists might be able to determine the cell's virulence factors and develop a vaccine or treatment for it. Virulence factors are features on a bacterial cell that determine whether the cell is a pathogen, as well

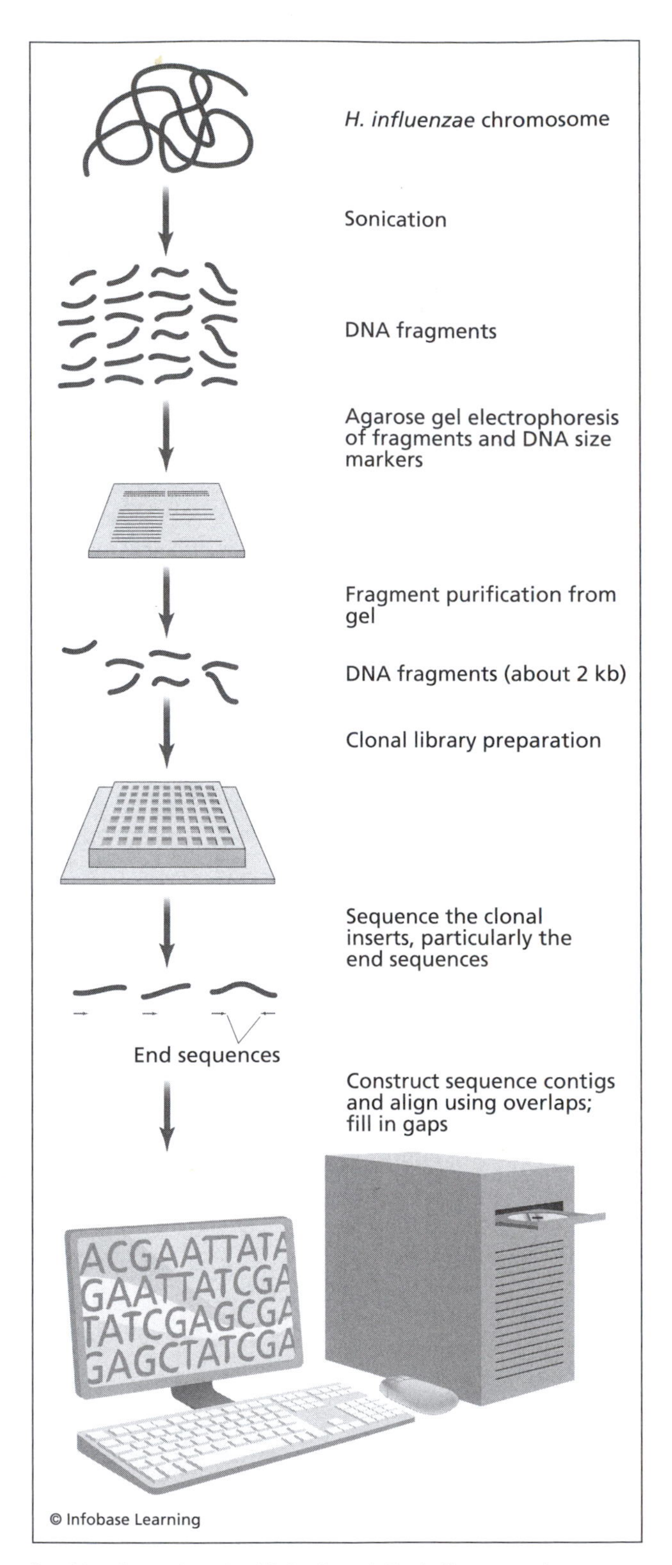

In 1995, the molecular biologists J. Craig Venter and Hamilton Smith developed whole-genome shotgun sequencing to shorten the time required for genome sequencing. The main steps are (a) breaking the DNA molecule, (b) attaching the fragments to *E. coli* plasmids, (c) putting fragments in order using automated sequencers and thousands of known templates, and (d) using computer analysis to close the gaps.

Genome Characteristics of *M. genitalium*

Characteristic	Value
total number of DNA molecules	1
total DNA size	0.58 Mb
total number of genes	521
number of protein coding genes	482
number of genes coding for membrane proteins	140
number of genes for mechanisms to evade the immune system	22
number of genes for cell regulatory functions (maintaining pH, water content, salt content, etc.)	5
number of genes with unknown roles	70

Source: J. Craig Venter Institute

as the degree to which the pathogen causes harm to the host.

Geneticists usually find that creating a diagram of an entire genome helps visualize where certain genes are located on the DNA. This process is called *gene mapping*, and the diagram of a DNA sequence is the *gene map*.

Gene mapping provides information for geneticists to use in cloning specific genes associated with disease. Cloning is the creation of many identical copies of the original gene. Geneticists, then, study the sequence of the cloned gene to see whether they can find any flaws or idiosyncrasies in the way it works that might be the cause of the disease. In some fortunate instances, scientists can use gene therapy to replace the defective gene with a fully functioning gene.

APPLICATIONS OF GENOMICS

Genomics has offered the best promise as a tool in medicine for two main uses: (1) detecting genetic diseases and curing them by gene therapy and (2) creating vaccines more effective than current vaccines. Microorganisms often serve as a model for new techniques in gene analysis before using the method for human genetic testing. The result is called personalized medicine, in which a treatment developed for a patient is based on that patient's unique genetic makeup. The physician Leslie Biesecker, chief genetic disease investigator at NHGRI, explained, in 2009, "[The project's goal is to] understand the genetic architecture of disease." Scientists are now using genomics to find and treat genetic origins of cardiovascular diseases, cystic fibrosis, muscular dystrophy, and schizophrenia. The powerful computer programs used in bioinformatics also search genes for any variations that correlate with a disease.

Outside medicine, researchers are analyzing microorganisms from the environment for new uses. In the mid-2000s, scientists determined the DNA sequence of a variety of marine microorganisms. Unusual pigments were detected in some microorganisms that carry out a different type of photosynthesis than found on land. WHOI's Hahn wrote, "Genomic studies of uncultured bacterial samples from oceanic waters showed that these pigments are much more diverse and widespread than expected. These findings are changing the way we think about the importance of marine bacteria in the flow of carbon and energy in marine ecosystems, including their possible role in the uptake of atmospheric carbon dioxide." Genomics is the most powerful way to study processes in the environment.

The genomics pioneer J. Craig Venter has already embarked on the next generation of genomics. In 2006, Venter told the *Washington Post* writer Michael Rosenwald, "Genomics is going to do for energy and [the] chemical field what it did in the early 1990s for medical biotechnology." Venter is now constructing a new bacterial genome to create a synthetic cell for the purpose of absorbing carbon dioxide and producing methane, a gas that can be burned for energy. His team started the experiment with known DNA sections from *M. genitalium* and added bases until they built the species's entire 582,970-base-pair genome. Venter now plans to move forward in the construction of new genomes. But not everyone thinks the sky is the limit for genomics. In 2008, Jim Thomas of the ETC Group, a Canadian organization that monitors biotechnology, warned, "The fact that he's [Venter] pushing ahead with this without any societal oversight is very worrying." Genomics is, indeed, advancing faster than most other technologies developed in the history of science. This new science still needs input from the public, scientists, and the government regarding safety. The medical community will decide the best uses for new products made through genomics.

See also CHROMOSOME; GENE THERAPY; GENE TRANSFER; PLASMID.

Further Reading

Davidson College. "Sequencing Whole Genomes: Hierarchal Shotgun Sequencing v. Shotgun Sequencing." Available online. URL: www.bio.davidson.edu/courses/GENOMICS/method/shotgun.html. Accessed August 22, 2009.

Dolan DNA Learning Center. "Early DNA Sequencing." Available online. URL: www.dnalc.org/ddnalc/resources/sangerseq.html. Accessed August 22, 2009.

GenBank. National Institutes of Health. Available online. URL: www.ncbi.nlm.nih.gov/Genbank. Accessed August 22, 2009.

Glazer, Vicki. "Linking Disease to Gene Variations." *Genetic Engineering and Biotechnology News,* 1 January 2009.

Hahn, Mark E. "Down to the Sea in (Gene) Chips." *Oceanus,* 25 May 2005. Available online. URL: www.whoi.edu/oceanus/viewArticle.do?id=4944. Accessed August 22, 2009.

Human Genome Program. "Complete *E. coli* Genome Sequence in Public Databases." *Human Genome News,* January 1998. Available online. URL: www.ornl.gov/sci/techresources/Human_Genome/publicat/hgn/v9n1/07ecoli.shtml. Accessed August 22, 2009.

J. Craig Venter Institute. Available online. URL: www.jcvi.org. Accessed August 22, 2009.

National Human Genome Research Institute. National Institutes of Health. Available online. URL: www.genome.gov. Accessed August 22, 2009.

Pollack, Andrew. "Genome Creation Lurches Science toward Man-Made Life." *San Diego Union-Tribune,* 25 January 2008.

Rosenwald, Michael S. "J. Craig Venter's Next Little Thing." *Washington Post,* 27 February 2006. Available online. URL: www.washingtonpost.com/wp-dyn/content/article/2006/02/26/AR2006022600932.html. Accessed August 22, 2009.

University of Michigan. "How Do We Sequence DNA?" Available online. URL: http://seqcore.brcf.med.umich.edu/doc/educ/dnapr/sequencing.html. Accessed August 22, 2009.

germ theory The germ theory is the principle that states that microorganisms cause disease. Although the concept that infectious diseases are caused by microorganisms is accepted today, the germ theory created a radically new way of thinking about disease when it emerged in science, in the 1800s. Previous to development of the germ theory, scientists and nonscientists believed that disease descended on individuals as punishment from the heavens for crimes or other bad behavior. The germ theory would overturn this philosophy by explaining facts in biology that are difficult concepts even today.

The germ theory offers the simplest explanation of infectious disease. Infectious diseases are diseases in which a microorganism, called a *pathogen,* infects a living body and lives all or part of its life cycle in the body. A pathogen is, thus, any microorganism that causes disease. The germ theory applies to infectious diseases; many other diseases that exist do not originate with an infection (neurological, genetic, some cancers).

The idea that some diseases are caused by a microscopic organism would be a difficult argument to win prior to the Golden Age of Microbiology, which began with the French microbiologist Louis Pasteur's studies, in the 1860s, and continued into the early 1900s. This period would become known as the golden age because it was a time of rapid advances in the study of microorganisms, hygiene, infection, and microscopy. To the untrained, however, the germ theory would ask people to accept some difficult concepts, such as the following:

- Diseases have a biological explanation.
- Infectious disease is a process in the body that occurs when the body interacts with a pathogen.
- The symptoms of disease are related to the body-pathogen interactions.
- Pathogens are microorganisms, which are invisible yet real entities that have structure and functions.

Perhaps the proponents of the idea of *germs*—a generic term for microorganisms, usually harmful—had their most difficult challenge in convincing others that disease could be started by a living thing that was invisible. How could so much harm be perpetrated by something no one could see? The germ theory and the rest of microbiology could not have advanced without the simultaneous advancements in microscopy that allowed scientists to study things that could not be seen by the unaided eye.

HISTORY OF THE GERM THEORY

The germ theory took root in Europe at a time in history in which medical practitioners adhered to the ancient medical doctrine espousing the three humors: blood, phlegm, and yellow or black bile. Greek, Egyptian, and Indian cultures developed similar theories on how an imbalance of these bodily substances led to disease. The Greek physician Hippocrates (460–377 B.C.E.) included the issue of balancing humors when he discussed the processes of disease. The humors not only determined a person's health, but also his or her personality: Blood accounted for feelings of joy, phlegm exerted its effect by causing melancholy, and bile controlled temperament. The medical practice of bloodletting developed for the purpose of releasing humors that caused evil in a human's soul. The science of bloodletting persisted well into the 17th century and aroused debates not on its merit but on the best ways to free bad humors from the body: by bleeding close to a lesion on a person's body or far from the lesion and the type of bloodletting instrument to use.

The earliest societies also had an inkling of the contagiousness of disease. They separated the sick from their communities, often in a cruel manner, and seemed to understand that crowding in dense populations let disease roam through a population. Philosophers, the church, and monarchs added their own stamp to the meaning of disease. The medical historian Fielding Garrison wrote, in 1921, in his book *Introduction to the History of Medicine,* of Roman citizens "who had a household god for nearly every disease or physiological function." Disease theory progressed very little, in the 1,600 years from ancient Rome to the Renaissance. The French medical philosopher Symphorien Champier (1472–1539) advised that the sick need only rely on the soil of France to cure them and implied that the soil from any other country would be useless! In this medical climate, the Italian poet Girolamo Fracastoro (1484–1553), who also had training in medicine, proposed a startlingly enlightened new view of disease and the idea of *contagion*.

In the Middle Ages, contagions were thought of as amorphous entities that carried sickness from one person to another. Fracastoro did not dispute this idea, but he proposed an idea about contagious disease that had probably never before been expressed. In 1546, he wrote, "There are, it seems, three fundamentally different types of contagion: the first infects by direct contact only; the second does the same, but in addition leaves fomes, and this contagion can spread by means of that fomes. . . . By fomes I mean clothes, wooden objects, and things of that sort, which though not themselves corrupted can, nevertheless, preserve the original germs of the contagion and infect by means of these: thirdly, there is a kind of contagion that is transmitted not only by direct contact or by fomes as intermediary, but also infects at a distance." Fracastoro wrote in Latin, and modern translations have used the word *germ* to convey Fracastoro's meaning. It is unlikely the poet used the term with the understanding people today associate with germs. He, nonetheless, published a theory for disease transmission and germs that was three centuries ahead of its time.

Physicians struggled to define the true nature of contagion, but until the evolution of the microscope, they lacked a way to connect the idea of invisible germs with disease. Many people took Fracastoro's theory that infection can occur "at a distance" to mean that contagion was a cloudlike, invisible force that moved at will. Antoni van Leeuwenhoek (1632–1723) had first observed bacteria and protozoa under a microscope and described them as living entities, but microbiology and medicine followed separate, parallel paths until the golden age of microbiology, when the concepts of contagion and germs would finally intersect.

At the same time that the most insightful scientists were puzzling over the existence and the role of cells in biology, they confronted a second dogma that had survived for centuries: spontaneous generation. Even respected scientists believed that life arose spontaneously from nonliving matter; some developed fairly sophisticated experiments that seemed to prove this theory. The concept of biogenesis—cells arise only from preexisting living cells—would repudiate spontaneous generation in time. The notion that germs cause disease would be a blow to proponents of spontaneous generation.

The German histologist Jacob Henle (1809–85) applied the cell theory to his work in studying and describing tissues of the human body. As a histologist, he sought ways to examine cell organelles by staining them and using magnification. The rapidly advancing field of microscopy aided Henle in becoming one of the greatest histologists in history. In addition, he was a brilliant scientist, who perhaps was the first to connect microorganisms to disease in scientific terms. Henle proposed, in 1840, "The first hypothesis [of disease] was that the contagium contained lower plants or animals. The second possibility is that the contagium consists of animal elements which can be cultured and isolated from the sickness and which can become free-living entities, from which the infection can spread." He noted that the second possibility seemed improbable because no one had seen these "elements which can maintain their energy of life for so long a time after being separated from the body." But many microbiologists were already gathering evidence on the germs that could live on their own but also cause disease.

In 1836, the Italian entomologist Agostino Bassi (1773–1856) showed that a fungus caused silkworm disease. This was not a trivial observation, because the disease had been threatening Europe's lucrative silk industry at the time. Silk producers and traders called upon a microbiologist in France who had been gaining fame for unraveling many secrets of the microscopic world: Louis Pasteur.

Pasteur has been linked to the development of the germ theory as no other scientist was before or since. Pasteur developed a reputation as a talented microbiologist, who had demonstrated the processes behind alcohol fermentations, food spoilage, pathogen transmission, and vaccines. He had built a store of knowledge greater, perhaps, than anyone else at the time on the life cycles and end products of bacteria and yeasts.

To disprove spontaneous generation, which many microbiologists of the period cast in serious doubt, in 1861, Pasteur devised a simple experiment to show whether microbial growth can arise spontaneously from the air. Pasteur filled a round long-neck flask with a nutrient-containing broth (hay infusion) that could support the growth of microorganisms. He, then, boiled the liquid for a period that would kill any microorganisms that might have been in the liquid. Finally, Pasteur heated the elongated neck of the flask and formed the neck into an S shape. The neck allowed air into the flask, but the S shape prevented microorganisms from falling out of the air into the broth. The broth remained clear and sterile for the entire time Pasteur watched it. One of Pasteur's flasks is in the Pasteur Institute in Paris and has not grown any microorganisms since Pasteur set up his experiment, 150 years ago. But when he broke the neck off similar flasks and exposed the broth to the air directly above the opening, airborne microorganisms entered the broth and began to turn the clear liquid cloudy. Pasteur opined, "I do not know any more convincing experiments than these, which can be easily repeated and varied in a thousand ways." Other scientists did repeat Pasteur's experiment and agreed that it provided a reasonable explanation for the presence of microorganisms in the environment.

In England, the physician Joseph Lister (1827–1912) took note of Pasteur's work. Lister combined the information from Pasteur's studies with his own understanding that disinfection played a role in eliminating germs from medical settings. The Hungarian physician Ignaz Semmelweis (1818–65) had, in 1850, proven his theory that hand washing, or the lack of it, related directly to the chance of surgery patient's developing infections. (Semmelweis became increasingly strident in defense of hygiene in obstetrics and wrote angry letters to the European medical community—sometimes accusing obstetricians of murder—to convince them that their casual methods were harmful or deadly to their patients. The Semmelweis Society has alleged on its Web site, "His contemporaries, including his wife, believed he was losing his mind and he was in 1865 committed to an asylum [mental institution]." Semmelweis died there only 14 days later, possibly after being severely beaten by guards.)

The surgeon Joseph Lister carried Semmelweis's ideas forward, however, by applying an antiseptic (phenol) to all his patients' surgical wounds. Lister, further, insisted on good hygiene, including hand washing and cleaning of the surgery room. Unlike Semmelweis, Lister drew a connection between infection and the minute particles that Pasteur had shown travel in the air. At last, the concept of germs and the concept of infection merged.

The germ theory needed only a final definitive proof by scientific study to show that germs cause disease. The physician Robert Koch (1843–1910) had traveled from Germany's mining highlands to the University of Göttingen, in 1862, to study medicine. A precocious learner since he was young, Koch soon devised elegant experiments to study the mechanisms of anthrax and tuberculosis. In practice in Berlin, Koch collaborated with other scientists, including R. J. Petri (usually known as R. J. Petri) (1852–1921), in improving culture and staining techniques for bacteria. Koch introduced the method of obtaining a pure culture in which one microorganism is isolated from all others so that it can be studied by itself. With the ability to isolate pure cultures of certain bacteria, Koch applied the new technique to the study of infection and disease. He devised a set of principles to be known as Koch's postulates, which provided a way for medical researchers to prove that a specific microorganism has caused a specific disease in a patient. Derivations of Koch's postulates remain in use in medical research today.

THE GERM THEORY TODAY

The germ theory influences medical care of patients in direct and indirect ways. Knowledge of disease-causing germs, or pathogens, is the reason for hand washing, surgical scrubbing, sterilization of medical instruments, and sterilization of items in surgery rooms. The germ theory also led to an expanded list of chemicals, since Lister's time, that can be used as antiseptics on the skin and as disinfectants of inanimate items. The germ theory additionally led to the development of aseptic techniques in medicine and in microbiology, so that a scientist can prevent contamination from unwanted microorganisms. Pasteur's proof of the germ theory and the refutation of spontaneous generation have been identified by most historians as the beginning of microbiology.

The germ theory required a significant leap of faith for many people, and it would not have developed as quickly as it did without advances in microscopy. Even today, many people have a difficult time understanding germs because they cannot be seen. Hand washing, bathing, covering the mouth when sneezing are all actions that relate to the acceptance of the germ theory. In 1878, Pasteur wrote in his seminal publication "Germ Theory and Its Applications to Medicine and Surgery," "All things are hidden, obscure and debatable if the cause of the phenomena be unknown, but everything is clear if this cause be known." These words can be used to describe all of science.

See also ANTISEPTIC; KOCH'S POSTULATES; LISTER, JOSEPH; MICROBIOLOGY; MICROSCOPY; PASTEUR, LOUIS.

Further Reading

Brock, Thomas, ed. *Milestones in Microbiology.* Washington, D.C.: American Society for Microbiology Press, 1961.

Garrison, Fielding H. *An Introduction to the History of Medicine.* Philadelphia: W. B. Saunders, 1921. Available online. URL: http://books.google.com/books?id=JvoIAAAAIAAJ&pg=PA82&lpg=PA82&dq=three+humors&source=bl&ots=LqpRb55HSb&sig=jfIKc-nEiT9C6vFCLKgAD4J2Mrs&hl=en&ei=jF0kSpPaJozstQOu0tH9Aw&sa=X&oi=book_result&ct=result&resnum=4#PPA3,M1. Accessed August 22, 2009.

Harvard University. "Concepts of Contagion and Epidemics." Available online. URL: http://ocp.hul.harvard.edu/contagion/concepts.html. Accessed August 22, 2009.

———. "Germ Theory." Available online. URL: http://ocp.hul.harvard.edu/contagion/germtheory.html. Accessed August 22, 2009.

Pasteur, Louis. 1848. "Germ Theory and Its Applications to Medicine and Surgery." Translated by H. C. Ernst. Available online. URL: www.fordham.edu/halsall/mod/1878pasteur-germ.html. Accessed August 22, 2009.

Semmelweis Society International. "Dr. Semmelweis." Available online. URL: www.semmelweis.org/about/dr-semmelweis-biography. Accessed August 22, 2009.

glycolysis Glycolysis, also called the *Embden-Meyerhof pathway,* is a series of enzyme-driven reactions that splits the sugar glucose into two pyruvate molecules. Glycolysis is the only metabolic pathway that occurs in all living things.

In microorganisms, the pyruvate from glycolysis becomes the starting point for fermentation or for respiration. Fermentation is a hallmark of anaerobic species, and respiration refers to the metabolism of aerobic species. In aerobic cells, the pyruvate enters the Krebs cycle, which produces compounds that, then, donate electrons to the electron transport chain. While fermentation is the main energy-producing pathway in anaerobes, aerobes use a more efficient system with the combined activities of the Krebs cycle and the electron transport chain.

The German chemists Gustav Embden (1874–1933) and Otto Meyerhof (1884–1951) both pursued their interest in the energy metabolism of muscle. Embden focused on carbohydrate metabolism and, in the late 1800s, became the first to describe all the steps leading from glucose to lactic acid. The author William Bechtel wrote, in his 2006 book *Discovering Cell Mechanisms,* "This work by Embden, together with investigations by Otto Meyerhof demonstrating that very similar coferments were required in alcohol fermentation and lactic acid fermentation, pointed strongly to a close connection between the two processes." Nicole Kresge, Robert Simoni, and Robert Hill added in the *Journal of Biological Chemistry,* "Meyerhof was also interested in analogies between oxygen respiration in muscle and alcoholic fermentation in yeast and proved, in 1918, that the coenzymes involved in lactic acid production were the same as the yeast coenzymes discovered by [Arthur] Harden and [William] Young, revealing an underlying unit in biochemistry." Coenzymes are small molecules that help enzymes work efficiently. Meyerhof coined the term *glycolysis* to identify the pathway used by both fermentation and respiration. Meyerhof was awarded the Nobel Prize in physiology, in 1922, for his discoveries in biochemistry, including glycolysis.

GLUCOSE

Glycolysis is the oxidation of six-carbon glucose ($C_6H_{12}O_6$) to two three-carbon pyruvic acids. The word *glycoloysis* means the "lysing of a sugar." Glycolysis can be divided into two stages: the preparatory stage and the energy-conserving stage. The preparatory stage involves the modification of glucose to related six-carbon sugars before being split in two. The energy-conserving stage comprises all the steps from the formation of the first three-carbon compound, glyceraldehydes-3-phosphate, to pyruvic acid. (The term *pyruvate* refers to the ionized form [CH_3COCOO^-] of the compound pyruvic acid [$CH_3COCOOH$]).

Glucose serves as the main energy source for most living things. Nature stores glucose mainly in the long glucose chains of starch or in cellulose. (The storage form for glucose in the human body is glycogen.) Microorganisms get glucose by digesting starch, cellulose, and other polymers to release the glucose units. The principal difference between these two starch and cellulose lies in the structure of bonds that link each glucose molecule to the next. Most microorganisms using the enzyme amylase break apart the alpha-1–4 linkages (α-1–4) in starch move easily than they break apart the beta-1–4 (β-1–4) linkages that hold cellulose together. Starch is, therefore, termed a *readily available glucose source.*

Cellulose provides much more strength to plants than starch. For example, a celery stalk contains a high proportion of cellulose, and a potato predominantly contains starch. Amylase is common among microorganisms, but the enzyme cellulase occurs rarely in nature; humans do not produce any cellulase to digest the fibrous portions of the diet. In the microbial world, certain bacteria and many fungi contain cellulase, so these microorganisms, play a pivotal role in the decomposition of organic matter. Two types of animals have little trouble digesting cellulose because of the diverse population of bacteria in their digestive tracts. The bacteria and protozoa in the rumen of ruminant animals and in the cecum of horses and rabbits digest dietary cellulose. Examples of ruminant animals are cattle, sheep, goats, deer, and giraffes. Humans possess some cellulolytic bacteria in the colon, which contribute a little to cellulose digestion.

Glucose also combines into chains shorter than starch or cellulose. Fairly long strings of glucose units belong to a group of compounds called dextrins. Dextrins contain many glucoses but fewer than starch does. Both starch and dextrins are called *polysaccharides.* Glucose also combines into two- or three-unit short chains, called disaccharides and trisaccharides, respectively. Glucose by itself is a monosaccharide. The following list shows how common monosaccharides combine to form di- and trisaccharides.

- sucrose = glucose + fructose
- maltose = glucose + glucose
- lactose = glucose + galactose
- maltotriose = glucose + glucose + glucose
- raffinose = glucose + fructose + galactose

Sucrose is common table sugar, which is also found in nature in fruits and vegetables. Lactose is the sugar in breast milk and cow's milk.

Microorganisms belong to two groups in relation to glucose metabolism: glucose degraders or glucose synthesizers. Species that break down glucose by glycolysis belong to the first group and contain many aerobic bacteria and organisms that perform fermentations. Other species can make glucose from small starting compounds. For example, the process of making glucose from two pyruvic acids is called *gluconeogenesis*. Glycolysis and gluconeogenesis, thus, run in opposite directions. In fact, seven of the enzymes in gluconeogenesis are exactly the same in glycolysis. This can occur because several reactions in glycolysis are reversible reactions: They proceed in either direction, depending on the needs of the cell.

GLYCOLYSIS AND ENERGY PRODUCTION

Reversibility of some of the reactions in glycolysis may have evolved as a safety mechanism for times when food became scarce. For example, the human liver runs either glycolysis or gluconeogenesis, depending on whether the body needs fuel to perform work (glycolysis) or needs to hold more glucose in reserve (gluconeogenesis).

The irreversible reactions in glycolysis play a key role in regulating a cell's metabolism. These reactions, called *committed steps*, take place in the preparatory stage, when the enzyme hexokinase puts a phosphate chemical group (PO_4^-) on the glucose. In addition to hexokinase, glycolysis contains two more committed steps by the enzymes phosphofructokinase and pyruvate kinase. As regulators, these steps assure that pathways do not run when not needed. The table on page 361 describes the step-by-step procession of glycolysis reactions.

Steps 1–3 in the table contain six-carbon compounds. In step 4, fructose splits into two three-carbon compounds, and the remaining steps also contain three-carbon compounds.

Glycolysis generates energy in addition to preparing substrates to the Krebs cycle or fermentation. Energy in biological systems often takes the form of the phosphate bond in adenosine triphosphate (ATP). Glycolysis generates two ATPs in total. One ATP must be used to put a phosphate on glucose in step 1, and a second ATP enters at step 3. Steps 7 and 10 generate one ATP each, and since the glycolysis pathway divides

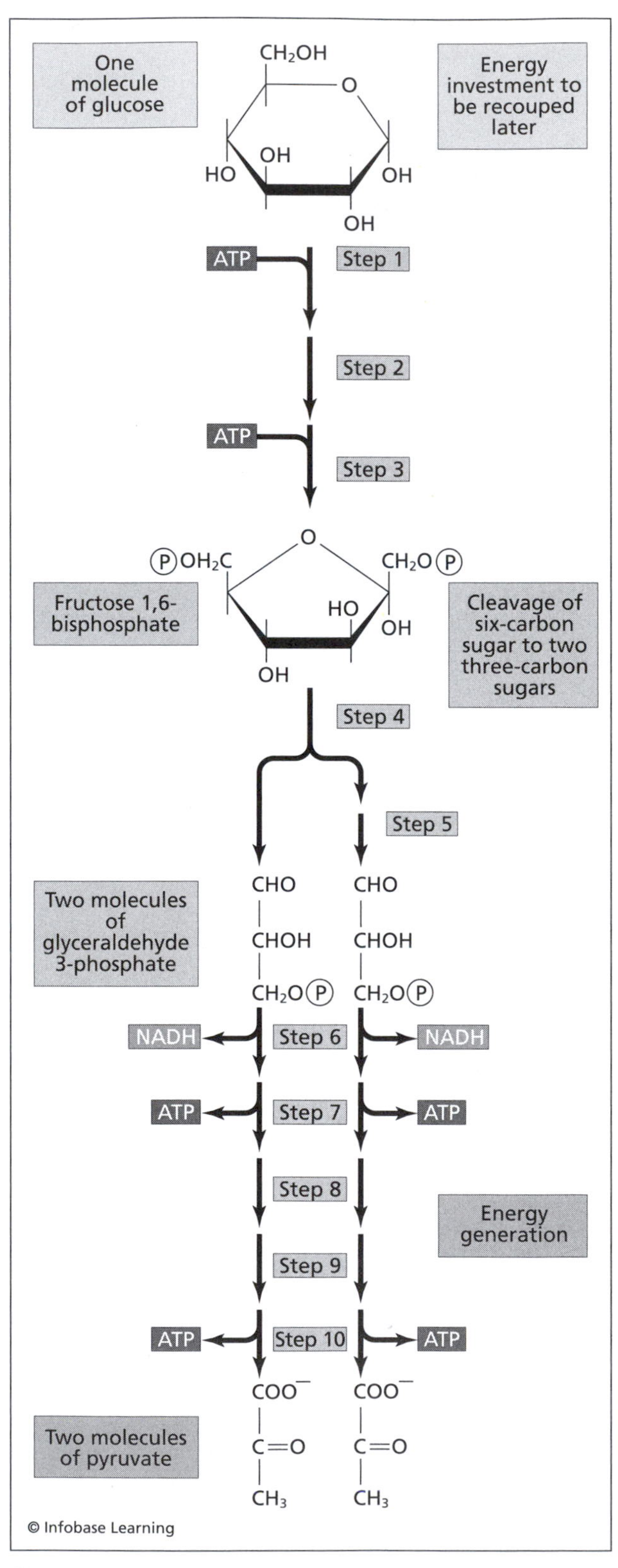

Glycolysis is a universal metabolic pathway in biology. This anaerobic process that converts glucose to pyruvate precedes the Krebs cycle, fermentation, and other major energy metabolism pathways.

Glycolytic Reactions

Step	Substrate	Product	Enzyme
1	glucose	glucose-6-P	hexokinase
2	glucose-6-P	fructose-6-P	phosphoglucose isomerase
3	fructose-6-P	fructose-1,6-bis-P	phosphofructokinase
4	fructose-1,6-bis-P	dihydroxyacetone-P + glyceraldehyde-3-P	aldolase
5	dihydroxyacetone-P	glyceraldehyde-3-P	triose phosphate isomerase
6	glyceraldehyde-3-P	glycerate-1,3-bis-P	glyceraldehyde 3-phosphate dehydrogenase
7	glycerate-1,3-bis-P	3-phosphoglycerate	phosphoglycerate kinase
8	3-phosphoglycerate	glycerate-2-P	phosphoglycerate mutase
9	glycerate-2-P	phosphoenolpyruvate	enolase
10	phosphoenolpyruvate	pyruvate	pyruvate kinase

Note: P = phosphate group (PO_4^{-3})

into two pathways (step 4), the total ATP output is four. By accounting for the two ATPs required in glycolysis's preparatory stage, the net ATP production is:

- step 1 (-1 ATP)
- step 3 (-1 ATP)
- step 7 (+1 ATP) × 2 = 2 ATP
- step 10 (+1 ATP) × 2 = 2 ATP
- net ATP produced = 2

Put another way, for each glucose that enters glycolysis, the cell produces two ATPs. In all of biology, glycolysis, which does not use oxygen, generates much less energy than aerobic pathways such as the Krebs cycle.

Aerobic microorganisms shunt the pyruvic acid made from glycolysis into respiration, which generates a net of 36 ATPs, as glucose oxidizes to carbon dioxide and water. Anaerobic microorganisms do not oxidize carbon compounds fully and instead make a mixture of fermentation products, namely, alcohols and organic acids. Despite the poor energy production compared with that of other metabolic pathways, glycolysis plays a central role in biology. Not surprisingly, glycolysis has been called "the universal pathway."

See also FERMENTATION; METABOLIC PATHWAYS; METABOLISM.

Further Reading

Bechtel, William. *Discovering Cell Mechanisms: The Creation of Modern Cell Biology.* Cambridge: Cambridge University Press, 2006. Available online. URL: http://books.google.com/books?id=WrEquK3hoDwC&printsec=frontcover&source=gbs_summary_r&cad=0#PPP1,M1. Accessed August 22, 2009.

Kresge, Nicole, Robert D. Simoni, and Robert L. Hill. "Otto Fritz Meyerhof and the Elucidation of the Glycolytic Pathway." *Journal of Biological Chemistry* 280 (2005): 124–126. Available online. URL: www.jbc.org/cgi/content/full/jbc;280/4/e3. Accessed August 22, 2009.

Prescott, Lansing M., John P. Harley, and Donald A. Klein. *Microbiology,* 6th ed. New York: McGraw-Hill, 2005.

Tortora, Gerard J., Berdell R. Funke and Christine Case. *Microbiology: An Introduction,* 8th ed. San Francisco: Benjamin Cummings, 2004.

Gram stain The Gram stain is the most widely used method in microbiology as an early step in identifying an unknown bacterium. The Gram stain differentiates bacteria into groups, gram-positive and gram-negative, so it is called a *differential stain.* Gram-positive species produce a dark blue to purple color when stained, and gram-negative species turn pink to red when stained.

The Danish doctor Hans Christian Gram (1853–1938) invented the Gram stain, in 1884, while working at Berlin's General Hospital. Gram had been attempting to isolate cocci from the lung tissue of patients

Gram Stain Components

Reagent	Formula	Action
crystal violet (solution A)	1.0 g per 100 ml distilled water	primary stain
solution B	1.0 g sodium bicarbonate per 100 ml distilled water	a few drops adjusts pH of the specimen
iodine mordant	1.0 g iodine + 2.0 g potassium iodide per 100 ml distilled water	intensifies primary stain by trapping dye in the cell
alcohol	ethanol, 95 percent in distilled water	removes all unreacted primary stain
safranin	2.0 g safranin O per 100 ml distilled water	counterstain

who had died of pneumonia. For this, he needed a stain that could make the microorganisms easy to find among the tissue constituents. He had already taken note of useful techniques developed by the German physician Robert Koch (1843–1910) and the hematologist Paul Ehrlich (1854–1915) for the tuberculosis pathogen, but these had never been tried on the group of microorganisms called *pneumococci* that invaded lung tissue. Gram specifically needed a way to find *Schizomycetes* bacteria in the tissue samples.

The two keys to Gram's method resided in the development of a dye that would be trapped as a precipitate inside bacterial cell walls. The second factor related to a second stain, called a *counterstain,* to highlight cells that did not react with the initial stain. This counterstain has formed the basis for today's modifications of Gram's original method. Almost all bacteria are either gram-positive or gram-negative. Gram, furthermore, pointed out in his summary of the procedure, in 1884, "This method is very quick and easy. The whole procedure takes only a quarter-hour, and the preparations can remain many days in clove oil without the *Schizomycete* cells becoming decolorized." Those words remain true today, although microbiologists no longer use clove oil to preserve samples.

GRAM STAIN METHOD

Microbiologists in different laboratories make slight modifications to the Gram stain method, but the basic following steps remain:

1. Put a drop of bacterial culture on a glass microscope slide.
2. Heat the slide gently over the flame of a Bunsen burner, until the specimen adheres to the glass; this step is called heat fixing.
3. Apply a few drops of crystal violet (a purple dye) to the specimen.
4. After about one minute, rinse off the crystal violet with water.
5. Cover the specimen with an iodine solution and wait about another minute.
6. Rinse off the iodine solution with alcohol or alcohol-acetone solution; this step is called *decolorizing.*
7. Rinse off any remaining alcohol and cover the specimen with safranin (a red dye).
8. Rinse off the safranin with water after one minute, pat the slide dry, and observe under a microscope.

This procedure works because of four main components. Crystal violet acts as the first component, called the *primary stain.* In this step, all the cells are stained purple-blue. The iodine solution is the second component, called a *mordant.* A mordant is any substance that, when added to a stain, makes the final color more intense. Third, alcohol acts as a decolorizing agent, which removes all the crystal violet–iodine from cells that cannot retain it. In this step, a microbiologist rinses away any primary stain–mordant that has not reacted with the cells. Fourth, safranin acts as the counterstain to color all the cells that did not retain the primary stain. As a result, a microbiologist will observe either purple-blue gram-positive cells or reddish pink gram-negative cells (see the color insert on page C-5 [top, left]).

The table above provides an example of reagents typically used in the Gram stain method.

The Gram stain method may yield variable results if not run consistently every time it is used. Once a

Gram-Positive versus Gram-Negative Characteristics

Feature	Gram-Positive Bacteria	Gram-Negative Bacteria
peptidoglycan layer	20–80 nanometers (nm) thick	2–7 nm thick
cell shape	usually spheres or rods	spheres, ovals, straight or curved rods, spirals
motility	rare	common
appendages	usually lacking	pili, fimbrae, stalks
endospore formation	present in some species	lacking
capsule formation	lacking	present in some species
effect of penicillin and cephalosporin antibiotics	susceptible	generally resistant

laboratory finds the best modification for its purposes, this method should not change. Some factors that affect the Gram stain outcome are:

- specimen thickness
- moisture in the specimen
- incomplete or excessive decolorization
- variable rinsing procedures
- variable dye concentrations in solution

As Christian Gram said, the Gram method is simple, but the technique, nevertheless, requires practice to get the best results.

THE GRAM REACTION

The reason for the Gram stain's differential result has been a subject of study for a long time. Microbiologists generally agree, at this time, that a major difference in cell wall structure distinguishes gram-positive cells from gram-negative. When cell walls are removed from gram-positive cells and the cells restained, they become gram-negative.

Almost all bacteria have either of two major structures surrounding the cytoplasm, the watery insides of the cell. Gram-positive cells contain the following layers from the outside inward: peptidoglycan, a periplasmic space, and the cell membrane. Gram-negative cells differ by having the following layers: lipopolysaccharides (long sugar chains with intermittently attached fats), outer membrane, outer periplasmic space, peptidoglycan, inner periplasmic space, and the inner cell membrane. Also noteworthy, the peptidoglycan layer of gram-positive bacteria is much thicker than the same layer in gram-negatives. These different outer structures determine the two main gram reactions.

During exposure to crystal violet, the strong peptidoglycan of bacteria does not become stained but rather acts as a permeable layer that lets the dye through. The iodine, then, passes through the peptidoglycan to react with crystal violet to form what is called a CV-I complex. Two theories explain what happens next. One theory proposes that the alcohol wash shrinks the pores of peptidoglycan. Because gram-positive cells have a thick peptidodoglycan layer, the CV-I complex cannot escape and remains in the cell wall. By contrast, the CV-I complex easily washes away through the thin gram-negative peptidodoglycan layer with alcohol. A second theory proposes that the CV-I complex does wash away in gram-positives because it is a much larger substance than either the crystal violet or iodine molecules are separately. Again, the thickness of the gram-positive cell wall affects the Gram reaction. Gram-negative cells end up colorless after the exposure to CV-I. For this reason, microbiologists put safranin on the specimen as the last step so that gram-negatives will be as visible as gram-positives when viewed under a microscope.

GRAM STAIN RESULTS

The Gram stain method works best when used on cells from a young, growing culture; older cultures stain poorly and give variable results. One reason for

Green Bacteria

Group	Phylum	Metabolism	Main Electron Donor	Main Feature
nonsulfur	Chloroflexi	photoheterotrophy	organic compounds	thermophilic
sulfur	Chlorobi	photoautotrophy	hydrogen sulfide H_2S)	growth in marine water, freshwater, or sulfur springs

this result may be that the cellular structure in older cultures is beginning to break down or have flaws due to mutations.

Gram stain results give an indication of common characteristics of gram-positives compared with gram-negatives. The table on page 363 lists the major differences in morphology (the physical form of organisms) and metabolism, but these are generalizations, and biology contains many exceptions.

The Gram stain is the first step in identifying any unknown bacteria. But microbiologists cannot use this method alone for identification. After determining whether a bacterium is gram-positive or gram-negative, a microbiologist conducts various physical and biochemical tests to obtain a conclusive identification of an unknown species.

Some bacteria normally give intermediate results when Gram stained. These microorganisms are called *gram-variable.* In this case, some of the cells look gram-positive and others look gram-negative in the same specimen. Assuming that the microbiologist has not committed a staining error, the gram-variable reaction may be because bacteria have a unique chemical composition that resists the Gram stain. Other species lose their capacity to hold on to the dye in an old culture, meaning a culture that is in stationary or late stationary stage. These bacteria contain a mixture of gram-positive and gram-negative cells. For example the gram-positive genera *Bacillus* and *Clostridium* become gram-variable in older cultures.

The Gram stain is one of the fundamental methods used in microbiology. New stains have been invented for special types of cells or for highlighting of individual cell features. The Gram stain method will, however, continue as an important method in microbiology.

See also CELL WALL; IDENTIFICATION; MORPHOLOGY; STAIN.

Further Reading

Gram, Christian. "Ueber die isolirte Färbung der Schizomyceten in Schnitt-und Trockenpräparaten" (The Differential Staining of *Schizomycetes* in Tissue Sections and in Dried Preparations). *Fortschritte der Medicin* 2 (1884): 185–189. In *Milestones in Microbiology.* Washington, D.C.: American Society for Microbiology Press, 1961.

Smith, Ann C., and Marisse A. Hussey. "Gram Stain: Gram-Variable Rods and Cocci." Available online. URL: www.microbelibrary.org/asmonly/details.asp?id=2028. Accessed June 4, 2009.

Sutton, Scott. "The Gram Stain." Microbiology Network. Available online. URL: www.microbiol.org/white.papers/WP.Gram.htm. Accessed June 4, 2009.

green bacteria *Green bacteria* is a general term for two types of gram-negative anaerobic species. The first group, called the green nonsulfur bacteria, use light as their primary energy source in photosynthesis. The second group, green sulfur bacteria, capture atmospheric carbon dioxide (CO_2) as their carbon source. Both groups are called green bacteria because of the green pigment chlorophyll that functions in the cell's photosynthesis. Analysis of the ribosomal ribonucleic acid (rRNA) in these bacteria has shown that they are not closely related and, thus, evolved along different routes.

The two main groups of green bacteria differ by molecules they use in the process of taking CO_2 from the air, called *fixation.* Green nonsulfur bacteria need organic compounds, as in CO_2 fixation, and green sulfur bacteria rely on reduced (having an excess of electrons) sulfur-containing compounds. The table above illustrates the main features of and differences between the two groups.

The table also indicates that green bacteria have some unique attributes. Green nonsulfur bacteria use a type of metabolism called photoheterotrophy, which means these bacteria use light as an energy source *(photo)* and various organic compounds other than CO_2 as the carbon source *(hetero).* Green sulfur bacteria live by photoautotrophy, in which light is the energy source, but the microorganism can make its own carbon compounds *(auto)* by fixing CO_2 from the air with the help of an inorganic type of compound called an *electron donor.* The gas H_2S serves as the overwhelming favorite electron donor for green sulfur bacteria.

Microbial ecologists classify green bacteria in terms of their preference for energy-carbon sources. In this case, green bacteria relate to purple bacteria, which are so called because their combination of carotenoid and chlorophyll pigments makes them brownish, reddish, or orange-purple. Green nonsulfur and purple nonsulfur bacteria use light for energy, but they cannot use CO_2 to build cell constituents. Instead, they must use small organic compounds, such as alcohols, fats, carbohydrates, or organic acids. By contrast, green sulfur and purple sulfur bacteria use light and sulfur-containing compounds. Purple sulfur bacteria belong to an entirely different group called proteobacteria, based on earlier classification systems.

Green bacteria (and purple bacteria) possess one major difference in photosynthesis from other photosynthetic organisms such as cyanobacteria and higher plants: Green bacteria photosynthesis is anoxygenic. Anoxygenic photosynthesis does not consume water or produce oxygen (O_2). Other photosynthetic organisms are almost always oxygenic, meaning they produce oxygen.

GREEN BACTERIA AND EVOLUTION

Earth four billion years ago held a habitat characterized by thermal springs, warm oceans, and erupting volcanoes. The atmosphere was heavy with volcanic ash and gases such as hydrogen, methane, and carbon dioxide. A limited amount of sunlight filtered through the haze to reach Earth's surface. The earliest bacteria developed under these conditions, and by necessity they were thermophiles, organisms that can live in extremely hot environments. The ancient cells existed on simple carbon compounds from the atmosphere. A great step forward in evolution occurred about 3.5 billion years ago, when some cells developed the ability to convert the energy from sunlight into energy that could be held in the cell's chemical bonds. Scientists have not identified all the processes through which this step took place, but green bacteria could well have been one of the first organisms to inhabit Earth.

The Cornell University bacteriologist Betsey Dexter Dyer wrote, in 2003, "Green nonsulfur bacteria are descendants of the first photosynthetic bacteria. Their branch on the family tree appeared soon after that of the ancient hyperthermophiles. Green nonsulfurs love hot water too, although they are not true extremophiles (lovers of extreme conditions)." These bacteria now inhabit hot springs of 140–176°F (60–80°C).

Scientists cannot yet agree on what type of metabolism preceded all other types in evolution, but the green nonsulfur genus *Chloroflexus* provides fossil evidence that it emerged on Earth before most other bacteria. The bacteriologist Dyer has suggested that the immediate ancestor of *Chloroflexus* ran only fermentation, an anaerobic energy-generating system that converts sugars to other organic compounds such as organic acids and alcohols. "According to this scenario," Dyer wrote in 2003, "it is as if during evolution, *Chloroflexus* kept sets of old genes (like old furniture) in its attic or basement, rather than getting rid of them. . . . Sometimes items were restored to use, combined in some new way, or even modified through mutation. This is how two new metabolic processes—photosynthesis and aerobic respiration—might have evolved in *Chloroflexus*. Indeed, some aspects of photosynthesis in *Chloroflexus* look like parts of fermentation run backward." Scientists still have the task of comparing the ages of green bacteria with those of other known ancient types such as cyanobacteria to determine how today's metabolic pathways evolved.

GREEN NONSULFUR BACTERIA

The green nonsulfur bacteria of phylum Chloroflexi use bacteriochlorophyll as the main photosynthetic pigment. This molecule differs in structure from the plant chlorophylls *a, b,* and *c* in containing an additional four-carbon side chain. Different species use variations of bacteriochlorophyll, which differ by absorbing light at slightly different wavelength ranges.

Chloroflexus is a major representative of green nonsulfur bacteria. Species change their metabolism according to environment, from anoxygenic photosynthesis to aerobic heterotrophy. In anoxygenic photosynthesis, *Chloroflexus* relies on small carbon compounds such as two-carbon acetate or three-carbon pyruvate. In the microorganism's normal environments, these compounds are probably end products from other microorganisms.

Chloroflexus produces filamentous (long and stringy) cells that glide over watery surfaces. Despite the name of this microorganism's group, *Chloroflexus* forms mats that range from yellow to orange-red on the surface of alkaline hot springs such as found in Yellowstone National Park. The bacteria also form green mats that tend to be difficult to see in habitats that also encourage the growth of cyanobacteria, which are also greenish. *Chloroflexus* often grows as a mat just beneath another mat of cyanobacteria. (The green nonsulfur *Oscillochloris* has, in fact, been mistaken for cyanobacteria in mixed mats.) In this situation, the *Chloroflexus* lives as a chemoheterotroph and depends on the photosynthetic cyanobacteria to absorb CO_2 from the atmosphere. The green nonsulfur microorganisms in such a layered mat may resort to being facultative anaerobes, meaning they use oxygen if it is available but can also shift their metabolism to live without oxygen. An actively photosynthesizing mat of cyanobacteria probably provides enough O_2 for the green bacteria, making facultative systems less common.

Green nonsulfur bacteria are difficult to grow in a laboratory as a pure culture—pure cultures contain a single microorganism. Rather, these bacteria thrive in complex mixtures with other organisms and even some nonliving constituents from the environment. Microbial ecologists create these very conditions by constructing a culture mixture called a Winogradsky column. In a Winogradsky column, several different environmental microorganisms taken in a single sampling grow together in a laboratory by using their natural interactions.

GREEN SULFUR BACTERIA

Green sulfur bacteria make up a small group of anaerobes that can live in conditions of low light as long as some light is provided to run photosynthesis. The anoxygenic photosynthesis of these bacteria is represented by the following equation:

$$CO_2 + H_2S + \text{light} \rightarrow C_6H_{12}O_6 + S^0$$

In this equation, the six-carbon compound is a sugar. Green sulfur bacteria also rid the cell of the buildup of elemental sulfur (S^0), by excreting sulfur granules into the surroundings. But excess sulfur produced by green bacteria is not wasted in the earth. Within a biogeochemical cycle called the *sulfur cycle,* other bacteria readily use the sulfur given off by green sulfur bacteria in their own metabolism. Species such as *Desulfuromonas acetooxidans* participate in the sulfur cycle by using S^0 in reactions with small carbon compounds and water. This step is called *sulfur oxidation.*

The sulfur cycle is a critical process for recycling the 10th most abundant element in the earth's crust. Living organisms require sulfur in protein synthesis, enzyme activity, hormones, and various cellular processes. Microorganisms that drive the sulfur cycle play a crucial role in making this nutrient available to other bacteria and higher organisms. Interestingly, the sulfur cycle is a rare biogeochemical cycle that can function entirely in moist soils or water sediments and without O_2.

Green sulfur bacteria live wherever there are light and a supply of reduced sulfur compounds. These microorganisms have the advantage of surviving on very dim light, so they can live in zones where other photosynthetic organisms cannot. Stagnant ponds that receive light, hold low oxygen levels, and emit a telltale rotten-egg smell of H_2S present microbiologists with very likely habitats for finding these bacteria. Green sulfur bacteria live in marine waters, especially in shallow estuaries, where light penetrates the upper layers. They also inhabit stagnant freshwater bodies and waste ponds, which are sites that allow for slow digestion of wastewater.

Green sulfur bacteria cannot withstand hot temperatures—they are mild thermophiles—so in hot springs, they must live along the shallow edges, where the temperatures are cooler. They also require a high concentration of reduced sulfur, so in mixed communities, green sulfur bacteria often occupy a spot near sulfate-reducing bacteria. In layered mats covering ponds, the green bacteria lie directly atop the sulfate reducers, where they capture sulfur from below and light from above.

The genus *Chlorobium* usually represents all other green sulfur bacteria in laboratory studies. *C. tepidum* can be recovered from the upper layer of microbial mats that cover hot springs, and much of the information amassed on green sulfur bacteria has been gained from this species. This microorganism excretes globules of yellow-white sulfur as it grows and builds long coiled chains of C-shaped cells during incubation.

Scientists have proposed different theories as to the possible role of green sulfur bacteria in bacterial evolution. These bacteria can thrive in very dim light and low oxygen, but high levels of reduced sulfur, similar to conditions on Earth four billion years ago. The *C. tepidum* genome was the first of any green sulfur bacteria to be sequenced. By analyzing the genetic makeup, microbiologists found clues to the way photosynthesis evolved. The biologist Jonathan Eisen of the Institute for Genomic Research in Maryland said, in 2002, shortly after completing the sequencing, "Because of their unusual mechanisms of harvesting and using the energy of light, the green sulfur bacteria are important to understanding the evolution and the mechanisms of both photosynthesis and cellular energy metabolism. The ability to carry out photosynthesis in the absence of oxygen is particularly important to evolutionary studies since it is believed that the early atmosphere of Earth had little oxygen. That is why some scientists have suggested the green sulfur bacteria were the first photosynthetic organisms." The next year, a surprising finding revealed more clues about the lifestyle of green sulfur bacteria.

The photosynthesis expert Robert Blankenship of Arizona State University announced, in 2003, that he had found green sulfur bacteria from very hot vents that occur at the deepest, blackest parts of the ocean floor. These vents, called black smokers, spew a mixture of compounds, including reduced sulfur compounds, so it might not be surprising to find green sulfur bacteria. But how would a photosynthetic microorganism exist in such lightless conditions? Blankenship said of the deepwater bacteria, "These guys are really scrounging for every photon that is there. So in a way it makes sense that that might be the sort of organisms we would find down there at the vents." After finding green sulfur bacteria in samples taken by robot from very deep sites, scientists determined that chemical reactions occurring at the vent opening may periodically emit tiny amounts of light. Blankenship explained how photosynthetic bacteria make use of these extremely low light intensities: "They have this wonderful device called a chlorosome." The chlorosome accumulates bits of energy from the low light in chemicals bonds until enough energy is available to run the bacteria's own chemi-

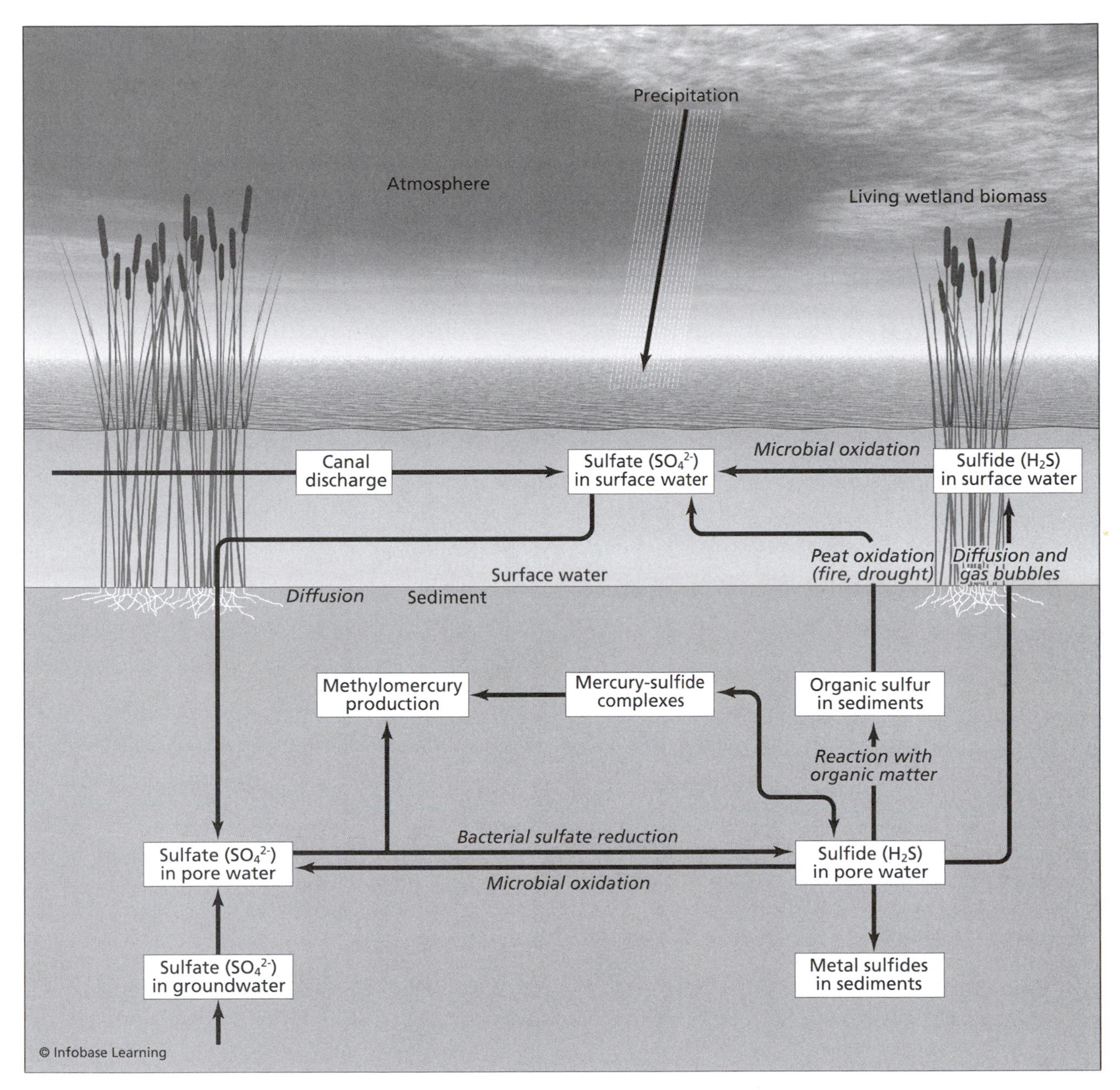

The sulfur cycle is a complex biogeochemical cycle in which sulfur compounds are either reduced or oxidized, or sulfur may be mineralized (released from organic form). Anaerobic green sulfur bacteria use hydrogen, hydrogen sulfide, and elemental sulfur in this process to make sugars.

cal reactions. Scientists now study how humans can develop energy systems that can approach the efficiency displayed by the green sulfur bacteria.

GREEN BACTERIA IN INDUSTRY

Green bacteria provide a glimpse into how life evolved on Earth, and in recent years, scientists have investigated the use of green bacteria for new energy systems. The Pennsylvania State University scientist Donald Bryant has led a research team in defining the structure of green bacteria chlorophyll. He explained the importance of their work as follows: "We found that the orientation of the chlorophyll molecules makes green bacteria extremely efficient at harvesting light." For this reason, green bacteria may provide a contribution to new energy sources that replace fossil fuels. Bryant's international team of scientists suggested that the chlorophyll of green bacteria might become a platform for artificial photosynthetic systems similar to the units that presently convert sunlight to electricity. Bryant said, "The ability to capture light energy and rapidly deliver it to where it needs to go is essential to these bacteria, some of which see only a few photons of light per chlorophyll per day." This efficiency may be the best reason for pursuing solar energy to electrical energy conversions provided by green bacteria.

See also BIOGEOCHEMICAL CYCLES; EXTREMOPHILE; HETEROTROPHIC ACTIVITY; MICROBIAL ECOLOGY; PURPLE BACTERIA.

Further Reading

Dyer, Betsey Dexter. *A Field Guide to Bacteria.* Ithaca, N.Y.: Cornell University, 2003.

Hart, Stephen. "Photosynthesis in the Abyss." *Astrobiology Magazine,* 5 May 2003. Available online. URL: www.astrobio.net/news/article451.html. Accessed August 12, 2009.

Institute of Genomic Research. "Clues to the Evolution of Photosynthesis." July 1, 2002. Available online. URL: http://news.bio-medicine.org/biology-news-2/Clues-to-the-evolution-of-photosynthesis-7272-1. August 12, 2009.

United Press International. "Green Bacteria Can Harvest Light Energy." May 4, 2009. Available online. URL: www.upi.com/Science_News/2009/05/04/Green-bacteria-can-harvest-light-energy/UPI-275 4124 1470800. Accessed August 15, 2009.

growth curve Growth in cells is any increase in cellular constituents and is measured by an increase in cell numbers. A growth curve is a graph that depicts this change in microbial cell number over time. Microbiologists study growth curves by inoculating a small number of cells to liquid medium, called *broth,* and measuring cell density in the liquid from that point forward.

Bacterial growth curves are made possible by the cells' reproduction by binary fission. Binary fission is an asexual method of cell division used by prokaryotic cells in which a parent cell splits in half to make two identical daughter cells. The resulting growth dynamics that define a growth curve take place under laboratory conditions in liquid batch cultures. No fresh medium flows into a batch culture, and no waste products leave it. This static condition gives rise to different phases of growth within the culture, which together make up the growth curve.

The four phases of growth are described in the table in the order in which they occur. The growth curves of almost all bacterial species progress over several hours to several days.

The logarithmic phase of growth, also called the *log phase* or exponential phase, is a period of peak metabolism in batch cultures. In the 1950s, the microbiologist Jacques Monod (1910–76) developed a vessel called a *chemostat* that could continually take in fresh sterile medium and expel spent medium of nutrient-depleted broth and excess cells. By adjusting the rate of flow, a researcher could keep microorganisms growing inside the chemostat in a never-ending log phase. This apparatus became the basis for the continuous culture method used today.

Microbiologists study the log phase principally to learn about nutrient requirements, nutrient uptake, optimal growth conditions, the effects of antimicrobial substances, and production of secondary metabolites, which are compounds produced by bacteria outside their normal energy metabolism.

In the log phase, microorganisms can grow to very high numbers in a short period, depending on the species's characteristic generation time or doubling time. *Generation time* is the time between each cell divisions. A short generation time of 20–30 minutes results in a rapid increase in cell numbers during the log phase. Longer generation times, of about an hour, cause a less rapid increase in cell numbers over the same period.

A single cell of *Escherichia coli* will expand to more than two million in seven hours. Fresh cultures that have been incubated for several hours or overnight can, thus, contain many millions of cells in a

Phases of the Growth Curve

Phase	Description	Culture Conditions
lag	cells synthesize new components to get ready for cell division	ample nutrients in relation to a low concentration of cells
logarithmic (log)	cells grow and divide at a constant maximal rate, depending on the microorganism's genetics	each cell divides to make two new cells so that the entire population increases exponentially (logarithmically), producing new cells and consuming nutrients at a rapid rate
stationary	cell deaths begin to equal the rate at which new cells form	nutrients begin depleting and wastes begin accumulating
death	number of deaths exceeds the number of new cells formed	buildup of wastes becomes toxic and accelerates the population's death

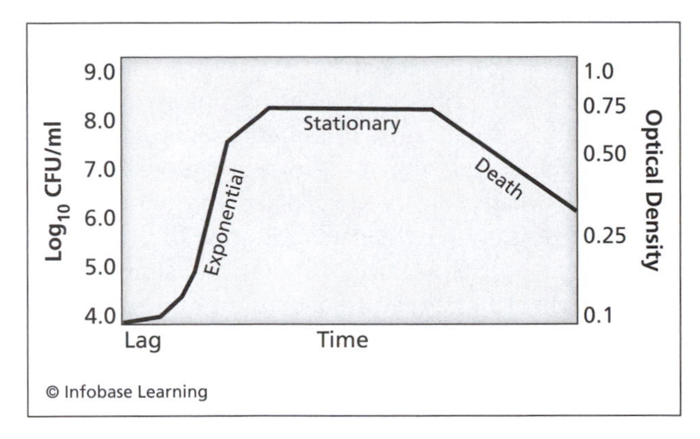

Bacteria grown in batch culture conditions follow a typical growth curve characterized by explosive growth in the exponential (logarithmic) phase, gradual slowing, and death. Microbiologists track these phases by measuring the culture's optical density.

single milliliter (ml). Microbiologists manage these very large numbers of cells per volume of medium by converting the cell number to its logarithmic_{10} value; for example, 4.2×10^{10} becomes 10.623.

THE PHASES OF GROWTH

Cells in lag phase do not divide right away. These cells need to adjust to the conditions in the medium, synthesize cell components, build up their normal array of enzymes and proteins, and repair any cell damage that may have occurred. After all of these activities have been completed, each cell starts the process of dividing into two new cells. The length of a lag phase changes within a species, depending on factors associated with the inoculum or the medium at the time of inoculation. The following factors can lengthen the lag phase in bacteria:

- inoculum from an old culture
- inoculum from a refrigerated culture
- inoculation of a medium chemically different from the inoculum's medium

The following factors can shorten the lag phase:

- inoculum from a culture in log phase
- inoculum to prewarmed medium
- inoculation of the same medium

The transition from lag phase to log phase begins the dramatic increase in the number of cells and the rate of cellular replication until the rate stabilizes and becomes constant. In logarithmic growth, cells divide and metabolize nutrients at the fastest rate that their genetics allow. When the growth rate becomes constant, the situation is called *balanced growth,* because all cell components are being manufactured at constant rates relative to each other.

A batch culture cannot sustain the log phase's enormous increase in population size forever. Eventually, conditions in the culture change and start affecting the microorganisms' growth. The growth rate begins to slow because some nutrients become depleted even though other nutrients might still be in good supply. This situation puts the culture in a state of unbalanced growth, wherein cell component synthesis rates vary in relation to each other.

When the log phase ceases and metabolism slows, the culture enters stationary phase. The stationary phase continues for a period in which there is no net increase in the population, because cells die at about the same rate as new cells form. As the stationary phase progresses, nutrients disappear, wastes and cell debris accumulate, and the pH changes. For example, bacteria that produce lactic acid cause the buildup of acid in a batch culture until the acid inhibits further growth.

After the stationary phase, cells enter the death phase, in which the number of cell deaths exceeds the number of new cells formed. The number of living cells drops quickly, so quickly, in fact, that this phase has also been called the *logarithmic decline phase.* Within a few hours, many millions of cells can die and lyse. The cell debris, then, settles to the bottom of the culture tube.

Some survivors of the death phase may go into a state of dormancy, in which they remain alive but do not reproduce. A dormant cell left in a culture that has completed its death phase would not grow because all the required nutrients have been used up, and the broth's conditions do not favor growth. But if a microbiologist takes a small volume from this culture and transfers it to fresh medium, the dormant survivors have a chance to become active and replicate. Microbiologists do not usually let a culture continue to the death phase. They either prepare a new culture from cells in log phase or simply discard any culture that has reached death phase.

MEASURING GROWTH CURVES

Growth curves can be monitored in two ways: manual culture techniques or an instrument that measures the culture's cell density. The manual method is called a *plate count;* in it, a microbiologist takes a small sample (an aliquot) from the growing culture, at periodic intervals. The microbiologist immediately dilutes each aliquot and inoculates it to agar plates. After incubating the plates, the microbiologist counts the visible colonies, called colony-forming units (CFUs),

Growth Curve Plate Counts

Time of Sampling (hours)	Expected Growth Phase	Plate Counts (CFU/ml)
0, inoculation	start of lag	50
2	lag	200
6	early log	7.0×10^3
10	midlog	5.0×10^6
14	midlog	4.0×10^7
18	late log or early stationary	9.4×10^7

on the agar. The number of CFUs can be extrapolated back to the concentration of cells at the time when the microbiologist took an aliquot from the culture.

A typical growth curve sampling may include the following aliquot intervals: inoculation, lag phase, two or more in log phase, and a sample to confirm the culture has entered stationary phase. Most microbiology laboratories report plate counts as the number of CFUs per milliliter (ml) of culture. A typical set of plate counts from a growth curve might look like the example in the table above.

Plate counts offer an advantage to the microbiologist because they indicate the number of cells present in the culture at any time in the growth curve. The disadvantages of plate counts relate to their labor-intensive procedures and the fact that the plate counts require about 24 hours of incubation before giving results. By that time, the culture has probably entered death phase.

Various instruments make the monitoring of growth curves much easier and faster than plate counts, although these methods do not give an exact value in CFU/ml. The two main instruments used for growth curve studies are the spectrophotometer and the turbidimeter. Both instruments measure the amount of cloudiness, called *turbidity,* that forms in a culture as cell density increases. (A newly inoculated culture appears as clear as uninoculated liquid medium. As the growth curve enters log phase, the liquid becomes visually cloudy, and cloudiness increases as the log phase continues.) Both instruments work by directing a light beam at the culture to measure the culture's effect on the light's path. A density of cells can redirect the light path, and this action equates to a value called *optical density.*

Optical density is a general term for the manner in which a turbid liquid affects the path of light. Spectrophotometers and turbidimeters are almost equivalent in the way they measure optical density. Spectrophotometers use light absorbance (A) to define the degree of light scattering caused by a culture; spectrophotometers also have capabilities to measure light of different wavelengths correlating to colors in the visible light spectrum. Most new models of spectrophotometers also give results as light transmittance (T), which is the inverse of absorbance. Percent transmittance is the portion of light that a suspension allows to pass through. These relationships are shown as follows:

high cell density = increased A or decreased T

low cell density = decreased A or increased T

Optical density (OD) as measured on a spectrophotometer relates to the percent of light transmittance using the following equations:

$$OD = - \log_{10} \times T$$

$$T = 10^{-OD}$$

The table below provides some conversions of optical density to percent transmittance:

These relationships demonstrate that the larger the number of bacteria in a spectrophotometer's light path, the lower the intensity of light will get through the culture. Microbiologists may replace optical density with values for light absorbance that relate directly to concentration. In spectrophotometry, this relationship derives from an equation called Beer's law:

$$A = \varepsilon\, l\, c,$$

where ε equals the absorptive capacity of a molecule—the value used to be called the *extinction coefficient*—*l* is the length of the sample the light passes through, and *c* is the concentration of the substance that absorbs the light. (A similar law, called the *Beer-Lambert law,* can be used for the absorption of light

Optical Density Relationship to Percent Transmittance

OD	Percent T
0.0	100.0
0.25	56.2
0.75	17.8
1.00	10.0
1.50	3.2

by colored samples.) A typical growth curve measured on a spectrophotometer gives a range of A values between 0 and 1.0, where 0 is a clear sample that does not absorb light, and 1.0 is a very turbid sample that absorbs a large amount of light.

Turbidimeters measure the angle of light scattering rather than transmittance-absorbance. The greater the angle of light scatter, the higher the culture's cell density. A turbidimeter works on a scale from 0 to 1,000 NTUs (nephelometric turbidity units). An NTU correlates with the amount of light that a sample can scatter to a given angle. For example, a turbidimeter can be set to measure only the light that a culture redirects at a 90° angle. A value of 1 NTU is equivalent to a low-turbidity liquid and values approaching 1,000 NTU indicate a highly turbid liquid.

Microbiologists have designed procedures to equate turbidity value with cell concentration of a given bacterial species. The Swedish microbiologist Roland Lindqvist wrote in the journal *Applied and Environmental Microbiology,* in 2006, "Estimation of microbial growth parameters from measurements of turbidity has the advantages of being rapid, nondestructive, and relatively inexpensive compared to many other techniques, e.g., classical viable count methods. However, turbidity methods also have limitations, such as being applicable only to liquid cultures and having high detection limits in the range of 10^6 to 10^7 bacteria ml^{-1} [per ml] and consequently yielding little direct information on the lag phase." Although standards have become commercially available for turbidimeters, as Lindqvist suggests, the relationship between NTU and plate counts is difficult to assess.

The McFarland procedure requires a set of barium chloride–sulfuric acid solutions called McFarland standards. In 1907, the medical microbiologist and physician Joseph McFarland developed these standards and the scale shown in the table below for equating turbidity with the concentration of gram-negative bacteria. *E. coli* served as the study model for the McFarland scale. Because *E. coli* remains the primary bacteria in research today, many labs continue to use the McFarland scale for adjusting cultures to an approximate concentration. Plastic beads have now replaced the strong chemicals that McFarland used. The density of the beads correlates to the chemical McFarland standards shown here.

Each species has its own McFarland scale, and the numbers are not interchangeable between species. The McFarland scale can only be used for these species. Even with that limitation, McFarland standards remain a useful tool in clinical, food, and research microbiology.

GROWTH CURVE APPLICATIONS

Growth curves offer insight into the way a particular microorganism responds to conditions in its environment. Microbiologists who study new microorganisms

McFarland Standard Preparation and the McFarland Scale for *E. coli*

McFarland Standard	Sulfuric Acid (ml)	Barium Chloride (ml) (10^8 per ml)	Density of Bacteria
1	9.9	0.1	3.0
2	9.8	0.2	6.0
3	9.7	0.3	9.0
4	9.6	0.4	12.0
5	9.5	0.5	15.0
6	9.4	0.6	18.0
7	9.3	0.7	21.0
8	9.2	0.8	24.0
9	9.1	0.9	27.0
10	9.0	1.0	30.0

Note: Sulfuric acid (H_2SO_4) and barium chloride ($BaCl_2$) are each a 1 percent solution in water

Typical Growth Curve Setups

Parameter Tested	Tube 1	Tube 2	Tube 3	Tube 4	Tube 5
temperature (°C)	20	25	32	37	45
pH	4.0	5.0	6.0	7.0	8.0
carbon source	glucose	glycerol	maltose	amylose	cellulose
nitrogen source	ammonia	nitrates	amino acids	purines, pyrimidines	yeast extract
vitamins (thiamine; mg/l)	0.001	0.01	0.1	1.0	2.0
manganese ($MnCl_2 \cdot 4H_2O$; mg/l)	0.0005	0.001	0.05	0.10	0.5

Note: Yeast extract is a heterogeneous material from dried, lysed yeast cells containing amino acids, vitamins, minerals, and organic carbon

prepare growth curves to study the microorganism's optimal growth condition in a laboratory. A set of broth medium tubes can be prepared, each containing a different condition for growth. By varying a single growth factor at a time in the medium, pH, for example, a researcher can develop a profile on the microorganism's preferred growth conditions. The table gives five examples of typical conditions for studying a pure culture of a microorganism.

In addition to helping evaluate the growth and nutrient requirements of a microorganism, growth curves help define the response of any microorganism to antibiotics. The main antibiotic test using growth curves is the *minimum inhibitory concentration* (MIC) test; in it, a microorganism is exposed to increasing levels of a single antibiotic. Growth curves identify two important pieces of information in this type of testing: the minimal concentration of an antibiotic needed to kill the microorganism and the minimal concentration needed to inhibit the microorganism's growth. MIC testing has been useful for evaluating the effectiveness of new antibiotics against a variety of bacteria. Automated methods now enable microbiologists to run many MIC tests at the same time using microwell plates, plastic trays, each of which holds 96 wells of about 0.3 ml each. This small-volume testing is called *microdilution*, and because this testing can analyze many samples in a short period, it is a type of high-throughput testing.

The disadvantages of growth curves relate to their use in only defined laboratory conditions and batch cultures, neither of which represents the normal habitat of bacteria. But growth curves, nonetheless, provide useful information on optimal conditions for growing bacteria in a laboratory. This method has been a standard way of determining the nutrients and physical factors that enhance bacterial growth and chemicals that can inhibit growth. Because growth curves are easy and inexpensive to run, they will continue to play a role in microbiology.

See also BINARY FISSION; CONTINUOUS CULTURE; GENERATION TIME; LOGARITHMIC GROWTH; MINIMUM INHIBITORY CONCENTRATION.

Further Reading

Koch, Arthur L. "Growth Measurement." In *Methods for General and Molecular Bacteriology.* Washington, D.C.: American Society for Microbiology Press, 1994.

Lindqvist, Roland. "Estimation of *Staphylococcus aureus* Growth Parameters from Turbidity Data: Characterization of Strain Variation and Comparison of Methods." *Applied and Environmental Microbiology* 72 (2006): 4,862–4,870. Available online. URL: http://aem.asm.org/cgi/reprint/72/7/4862.pdf. Accessed August 20, 2009.

Sutton, Scott. "Measurement of Cell Concentration in Suspension by Optical Density." *PMF Newsletter,* August 2006. Available online, URL: www.microbiol.org/white.papers/WP.OD.htm. Accessed August 20, 2009.

Todar, Kenneth. "The Growth of Bacterial Populations." Todar's Online Textbook of Bacteriology. Available online. URL: www.textbookofbacteriology.net/growth.html. Accessed August 20, 2009.

H

Hazard Analysis Critical Control Points (HACCP) HACCP is a process used in manufacturing to identify procedures or places with the highest risk of contamination. HACCP guards against either contamination from microorganisms or nonliving particles and consists of two components. The first component involves hazard analysis, which is the identification of all the factors that could lead to contamination. Hazard analysis may be used for raw materials, product containers, air circulation systems, and even the manufacturing plant's employees. The second component relates to the critical control points, which are places in the manufacturing plant easy to monitor and where contamination can be eliminated or controlled.

The Pillsbury food company developed the HACCP concept, in the 1960s, to help the National Aeronautics and Space Administration (NASA) and the U.S. Army develop safe foods for astronauts to take into space. Spoiled food or food-borne illness would be equally disastrous in space travel, so the food production system needed extra safety measures to prevent food contamination. The only sure way to confirm that a food contains no contaminants is to sample each food item and check for contamination, a means that is not practical. Scientists, therefore, developed HACCP to serve as a process that would provide maximal safety in food manufacturing by focusing only on the most likely sources of contamination.

By the 1970s, food microbiologists, statisticians, and other specialists in industry had devised a plan for safeguarding products by applying the principles of HACCP. The food production industry adopted HACCP, because it delivered a twofold benefit: prevention of the spread of food-borne contaminants to consumers and reduction of money lost in spoiled food. Today, the following agencies endorse HACCP as an important part of safe food production: the Food and Agricultural Organization (FAO), U.S. Department of Agriculture (USDA), U.S. Food and Drug Administration (FDA), National Advisory Committee on Microbiological Criteria for Foods, and World Health Organization (WHO).

Before HACCP was invented, the U.S. government sent inspectors to manufacturing facilities, where they selected random food products and tested them for contamination. The process was time-consuming, costly, and ineffective. But as the food industry increasingly made HACCP part of their standard operations, the FDA noted the improved conditions in food production and quality. In 1995, the FDA began requiring HACCP for fish and seafood products; in 2001, the agency added HACCP to juice production facilities. The FDA has since mandated HACCP programs for other foods, and makers of personal care products such as lotions and shampoos and some drugs have also adopted HACCP principles.

The Association of Food and Drug Officials endorses the following seven principles that make up the basis of HACCP programs.

1. Analyze hazards. Conduct analysis to identify potential contamination hazards and propose measures to control them.

2. Identify CCP. Evaluate all the processing steps and then identify the critical control points (CCPs).

3. Establish preventive measures. Establish an acceptable limit of contamination for each CCP.

4. Establish monitoring procedures. Establish systems for monitoring each CCP to assure it remains within acceptable limits.

5. Establish corrective actions. Develop a corrective action plan for instances of unacceptable levels of contamination.

6. Verify. Establish a verification procedure to assure the corrective actions were performed and gave the desired result.

7. Document. Keep records to document all activities in the program, monitoring results, corrective actions, and corrective action outcomes.

Every industry that was required to use HACCP or that volunteered to use it follows these principles. The FDA monitors seafood and juice manufacturers, and the USDA oversees HACCP in the meat and poultry industries. The U.S. dairy industry and other sectors in food production follow the program on a voluntary basis.

A top-notch HACCP system alone does not guarantee 100 percent of food products will be safe. A 2008 article on HACCP in *Food Manufacturing* pointed out the following:

- No system is 100 percent fool proof. Regardless of what measures the industry has in place, there is always some threat that could slip through the cracks.

- For HACCP principles to work, they would have to be adopted by the entire supply chain—starting as early as with the growers.

- Having systems in place does not always mean they will be followed. Plants must be 100 percent dedicated to HACCP for it to be successful.

HACCP is one weapon against contamination, but even the best programs have a difficult time achieving the very difficult goal of keeping dangerous substances out of food.

In early January 2009, the Centers for Disease Control and Prevention (CDC) notified the public that 388 people in 42 states had been made ill from peanut butter contaminated with *Salmonella typhimurium* bacteria. Epidemiologists tracked the cause of the outbreak to King Nut brand peanut butter, which used the Peanut Corporation of America (PCA) as their manufacturer. The disease outbreak experts further traced the problem to a PCA manufacturing plant in Plainview, Texas, and to one five-pound (2.3 kg) container of King Nut brand in particular. By February, PCA announced a nationwide recall of its products shipped from its Blakely, Georgia, plant. In retrospect, the public learned of filthy conditions at PCA plants, no sanitation procedures, and no HACCP program. By the time of the recall, tainted products had sickened at least 700 people in 43 states and killed nine. Even more alarming, CDC inspectors found evidence that contaminated products had been shipped from the Georgia plant since 2006. A HACCP system might have prevented this outbreak.

Microbiologists and plant managers who develop HACCP systems must be able to distinguish between two necessary processes: sanitation and sanitization. Sanitation is the promotion of good hygiene in a facility by conducting regular cleanings and removing trash. Sanitization is a microbiological method in which a chemical lowers the number of microorganisms to safe levels. In both cases, HACCP systems are most effective when the sanitation and sanitization programs are designed specifically for the manufacturing plant.

HACCP PROCEDURES

Some manufacturers may believe that a HACCP program would be too costly to implement or take too much time setting up. A recall such as the one that forced PCA into bankruptcy and required the company's president to testify to Congress—the company president, Stewart Parnell, refused to answer questions, citing his Fifth Amendment right—is worse than the extra work HACCP demands. Once implemented, a HACCP program can streamline the quality control steps a manufacturer takes to ensure a high-quality product.

Each company sets up a HACCP plan that is appropriate for the product it makes and the conditions inside the manufacturing plant. One generic HACCP plan does not make sense for the hundreds of food companies in the nation. Most HACCP plans do, however, have aspects in common. The table describes the main components of food industry HACCP programs.

HACCP specialists have adopted new technologies to help in monitoring large volumes of foods. Many of these methods are called *real-time methods,* because they produce information about a product as it is manufactured. This offers a marked advantage over methods in which technicians removed samples from the production process and then analyzed them in a laboratory. The results from these analyses often arrived after the product had already been packaged and shipped to a ware-

Components of HACCP Programs

Component	Topics within HACCP
plant operations	• sanitation program • water supply • ventilation • general cleaning • pest control • sanitizers and disinfectants
equipment	• routine maintenance • repairs • testing and calibration • sterilization • cleaning and sanitizing
food handling	• refrigeration and freezing • raw material receipt records • raw material storage methods • cooking temperatures • handling, noncooked, fresh foods • batch records • aseptic packaging • storage and shipment requirements
microbial monitoring	• water quality • equipment • raw materials • containers • finished product • plant environment: air, floors, walls, ceiling, etc.
nonmicrobial monitoring	• dust, dirt, fibers, and particles • finished product: color, consistency, odor, taste, etc.
employees	• good hygiene • proper hand washing techniques • proper attire: hairnets, gloves, protective overalls, etc. • sick employees excused from plant operations
management•	• assure all employees properly trained • assure HACCP programs followed • assign a quality control group • provide ongoing training in new technologies

house. The main technologies used in nonmicrobial food analysis are the following:

- gas chromatography/mass spectrometry—detects minute amounts of pesticides in the food
- oxygen monitors—measure food exposure to oxygen, a spoilage accelerator
- temperature monitoring and feedback systems—continuous temperature measurements in refrigerators and freezers and automatic adjustments for remaining within specified ranges
- inductively coupled plasma emission—metal analysis based on emitted spectrum when heated
- spectrophotometry—analysis of molecules based on light absorption
- turbidimetry—analysis of clearness or cloudiness of the product

Companies may insert additional tests for a particular product. For example, many products change consistency over time and separate into water and oil phases. Companies design special tests to monitor these physical processes in their products, whether the product is a salad dressing, paint, shampoo, or various other common items.

In addition to testing with electronic instruments, the food chemistry includes special tests for specific compounds. Some of these tests belong to an area known as *wet chemistry,* in which laboratory procedures using solutions and reagents replace sophisticated instruments. Wet chemistry can be used to detect minute amounts of contaminants or to test for the correct level of nutrients.

HACCP microbiology also involves a combination of instrumentation and manual methods. These methods detect the presence of all the major food-borne pathogens that cause the greatest health threats in food products. Routine analyses performed by food microbiologists for detecting these microorganisms are summarized in the table.

HACCP programs also include strict monitoring of the water used in rinsing equipment and in making food products. Food microbiologists who conduct HACCP programs use the same testing done by water treatment plants to confirm that no pathogens exist in drinking water. These water quality tests focus mainly on fecal coliforms, fecal streptococci, and *E. coli.*

An emerging technique in HACCP involves predictive microbiology. Predictive microbiology uses mathematical calculations to determine how a certain microorganism will behave in a food under specific conditions. For example, a microbiologist would use predictive microbiology to assess the growth of a small amount of *Salmonella* cells, about 100 cells per gram of food, in gravy when the gravy is cooked at temperatures 5°F (2.8°C) lower than required and stored in a refrigerator set 4°F (2.2°C) higher than required.

Although predictive microbiology seems to be a refined way of projecting microbial growth under a

HACCP Microbiology

Test or Target Microorganism	Description
aerobic mesophilic spore count	enumerates the endospore-forming bacteria that grow at 50–122°F (10–50°C)
anaerobic spores and total anaerobic count	detects presence of endospore-forming *Clostridium botulinum* and the total number of clostridia
Bacillus cereus	detects aerobic endospore-forming *B. cereus*
Clostridium perfringens	detects anaerobic endospore-forming *C. perfringens*
Escherichia coli and *E. coli* 0157:H7	detects total *E. coli* and the pathogenic strain 0157:H7
Enterobacteriaceae	detects bacteria in this family, which includes fecal bacteria
fecal coliforms and streptococci	detects specific groups of bacteria within Enterobacteriaceae
lactic acid bacteria	detects spoilage microorganisms that convert food components into acid
Listeria monocytogenes	detects *L. monocytogenes* in refrigerated dairy products
mold	detects a variety of molds that contaminate food
psychrophile count	enumerates microorganisms that grow at cold temperatures
Salmonella	detects various species
Shigella	detects toxin-producing species
Staphylococcus aureus	detects *S. aureus,* which usually indicates contamination from people's hands
total plate count	enumerates total aerobic bacteria of all types
yeast	detects various species

wide variety of conditions, it began in microbiology as a trial-and-error technique. The food safety experts Greg Burnham, Donald Schaffner, and Steven Ingham wrote in a 2008 issue of *Food Quality,* "Early in the 1800s, [the French inventor] Nicholas Appert [1749–1841] had discovered that food heated in sealed containers would not spoil during extended storage. His discovery earned Appert a large cash award and allowed for the feeding of Napoleon's vast armies. Appert didn't understand why food spoiled, and the causes of spoilage remained unknown until the discoveries of Louis Pasteur some 50 years later." Predictive microbiology now defines the conditions that Appert discovered, during 15 years of successes and failures. Microbiologists consider factors such as the heat resistance of specific bacteria, the heat-transfer properties of food, and the time a food will spend at temperatures outside its recommended range. Predictive microbiology has just begun to transform food microbiology so that consumers receive greater assurance of safe products.

THE EFFECTIVENESS OF HACCP

Incidences such as the one at PCA that led to a nationwide food-borne illness outbreak demonstrate the risks of ignoring HACCP principles. But even with HACCP, some manufacturers conduct their food safety activities better than others. Many companies keep their HACCP systems running optimally by using an additional procedure called an *audit.* An audit is an inspection conducted by a company employee or a contractor from outside the company to observe each part of the HACCP system. This is done for the purpose of identifying flaws, missing steps, overlooked sources of error, and hidden places where contamination may occur. Audits also help employees develop corrective actions when conditions go wrong. The person conducting the audit must be completely honest and follow a detailed inspection checklist. Company managers must receive these audit results and ensure that any problems the inspector identifies are quickly fixed.

HACCP programs and their corresponding audits have been under increasing pressure as the food industry evolves. The following trends in food production have made HACCP more difficult, but they also make HACCP vital for protecting food's safety:

- increasing supplies of raw ingredients from various domestic and international sources
- faster production speeds
- inventory movement, meaning products do not remain at the plant long before reaching consumers
- demand by consumers for new and exotic foods that have little history in HACCP programs

Additional trends toward local farm-grown foods, preservative-free products, and organic products also add stress to a program intended for assuring food safety.

In 2001, the food and health authority Eunice Taylor of the University of Central Lancashire, United Kingdom, noted that HACCP programs have been slow to enter the world of small companies and specialty food producers. Taylor wrote, "There is increasing evidence that whilst HACCP is widespread in large food operations its use is limited within small companies. This is reflected in recent studies in the U.K. and Europe which have found that small companies are less likely to invest in hygiene and food safety than larger companies and are less likely to have HACCP in place." In addition, the smaller the company, Taylor found, the less likely it used HACCP.

Food-borne pathogens seem able to find the smallest weaknesses in any HACCP system, and food-borne illnesses will probably never vanish. Diligent adherence to HACCP principles will, however, greatly reduce the risk of selling a food that contains a pathogen. Equally important, HACCP provides a mechanism for learning from past mistakes and correcting those mistakes. As a result, a food manufacturer using HACCP should improve over time and reach almost 100 percent assurance that the food it produces is safe for consumers.

See also FOOD-BORNE ILLNESS; FOOD MICROBIOLOGY; HYGIENE; SANITIZATION; WATER QUALITY.

Further Reading

Associated Press. "Peanut Plant Inspector Missed Conditions." March 20, 2009. Available online. URL: http://archives.chicagotribune.com/2009/mar/20/health/chi-food-safety_frimar20. Accessed August 20, 2009.

Burnham, Greg M., Donald W. Schaffner, and Steven C. Ingham. "Predict Safety." *Food Quality,* April/May 2008.

FoodHACCP.com. Available online. URL: www.foodhaccp.com. Accessed August 20, 2009.

Food Manufacturing. "Market Update: HACCP in the Food Industry." October 2008. Available online. URL: www.foodmanufacturing.com/pdfs/fmg_10%20october.pdf. Accessed August 20, 2009.

Pavlov, Alexander, and P. Chukanski. "Critical Control Points in Assessment of Microbiological Risk in a Small Meat-Processing Enterprise." *Bulgarian Journal of Veterinary Medicine* 7 (2004): 167–172. Available online. URL: www.uni-sz.bg/bjvm/vol7-no3-05.pdf. Accessed August 20, 2009.

Taylor, Eunice. "HACCP in Small Companies: Benefit or Burden?" *Food Control* 12 (2001): 217–222. Available online. URL: www.nutricion.org/publicaciones/pdf/taylor.pdf. Accessed August 20, 2009.

U.S. Food and Drug Administration. "Hazard Analysis and Critical Control Point." Available online. URL: www.foodsafety.gov/~fsg/fsghaccp.html. Accessed August 20, 2009.

hemolysis Hemolysis is the action of breaking apart red blood cells so that they release hemoglobin and no longer function in the body. Several species of bacteria possess hemolytic activity during infection or in a laboratory, when grown on medium containing blood. Microbiologists observe the presence of hemolysis and certain characteristics of the reaction on agar to identify various gram-positive cocci. Evidence of hemolysis in laboratory tests may also indicate how a pathogen works in the body to further an infection.

Hemolysis caused by pathogens can create a condition called hemolytic anemia, which is a lowered concentration of red blood cells. Specific pathogens, such as *Escherichia coli* O157:H7, also cause hemolytic uremic syndrome, in which a toxin produced by the bacteria leads to anemia and permanent damage to kidney cells.

The toxins excreted by bacteria that lyse blood cells form one important means of virulence, which is the degree of disease-causing activity possessed by a pathogen. The hemolysins join other unrelated substances made by pathogens called virulence factors, that is, factors that make a pathogen more virulent. Any toxin that destroys membranes in a host cell aids the spread of an infection in the body by one or more of the following processes: destruction of tissue for the release of nutrients, penetration of organs, combat of immune system cells meant to kill infections, and escape of pathogens from structures in the body designed to arrest infection.

Membrane-Disrupting Type II Exotoxins

Toxin	Target Membrane	Producer
hemolysin	red blood cells	*Streptococcus*
leukocidin	white blood cells	*Staphylococcus*
alpha toxin (lecithinase)	capillaries	*Clostridium*
kappa toxin (collagenase)	connective tissue	*Clostridium*
mu toxin (hyaluronidase)	tissue matrix	*Clostridium*
elastase	lung tissue	*Pseudomonas*

The physician James H. Brown (1884–1956) of the Rockefeller Institute for Medical Research first experimented with agar formulations containing blood, in 1919, for the purpose of differentiating among streptococci. Although Brown sought a technique for making bacterial colonies of different species display differing colony appearance, he also discovered the hemolysis reaction.

HEMOLYSINS

A hemolysin, in a general sense, is any substance that destroys red blood cells. The process is called *hemolysis*. Many pathogenic bacteria produce toxic proteins that destroy the membranes of either white blood cells or red blood cells in humans and animals. Proteins that act only on white blood cells, or leukocytes, are often referred to as *leukocidins*.

Both leukocidins and hemolysins are proteins excreted by bacteria and work in the environment even if the bacteria are no longer present. This type of toxin is called an *exotoxin*, because it works outside the cell—endotoxins stay inside bacteria and enter the environment only when the bacteria dies and lyses. Hemolysins are called pore-forming toxins, membrane-disrupting toxins, or type II toxins. As described by the microbiologist and author Alistair Lax, "Many toxins damage the cell membrane . . . by organizing their molecules into a ring shape and inserting this into the membrane to form a hole, or pore, in it." This process as described here seems simple, but hemolysins excreted by pathogens must possess the following characteristics in order to attack host cells:

- excretion from the cell in active form
- maintenance of activity outside the bacterial cell
- ability to congregate with other toxin molecules to create a pore-forming structure

Lax also pointed out that the pore structure "has to display hydrophobic patches on its outside so that it can breach the hydrophilic and hydrophobic of the membrane to insert into it. However, in the time between leaving the bacterium and inserting into a membrane, the toxin monomers [subunits] . . . have to stay dissolved in the fluid bathing the cell." Lax added, "Not only that, these toxins have to move into the membrane and stop after insertion without going right through into the [red blood] cell. All this information is coded into the toxin molecule—nothing else helps it insert into the membrane."

Other type II exotoxins produced by bacteria destroy the membranes of a host cell. The table summarizes the main types of type II toxins, including hemolysins.

Two additional types of toxins belong to the bacterial exotoxins. The first type contains the AB toxins, which bind and enter host cells in a complex two-step procedure. The second type, called *specific host site exotoxins*, contains specialized toxins that direct their activities against certain tissue types. For example, neurotoxins attack nerve cells and enterotoxins attack intestinal cells.

Hemolysins probably do most of their damage by forming pores through the cell membrane. Because red blood cells lack a nucleus, they do not have the cell repair mechanisms found in other types of cells in the body. A hemolysin causes damage by inserting into the cell membrane and reaching from the red blood cell membrane's outer surface across to the interior surface. The hemolysin protein develops a pore in the membrane through which water from the environment rushes into the cell. As a consequence, red blood cells swell and then burst, spilling hemoglobin and ions. Each hemolysin molecule creates one pore, so a high toxin concentration can decimate a single red blood cell.

The three most-studied groups of hemolysins are (1) streptolysin-O and streptolysin-S, produced by *Streptococcus pyogenes;* (2) various hemolysins produced by *Staphylococcus aureus;* and (3) hemolysin II, made by *Bacillus cereus*. Streptolysin-O is inactivated by oxygen and acts only in anaerobic conditions. Streptolysin-S is stable in the presence of oxygen, so it works in aerobic conditions.

S. aureus produces four different types of hemolysin: alpha, beta, gamma, and delta. These substances are toxic to red blood cells and various immune system cells, and all are credited with giving *S. aureus* its virulence. These hemolysins cause destruction by a variety of actions on the membrane. In addition to the common hemolysin action of pore formation in membranes, *S. aureus* hemolysins cause other changes in membrane structure such as ripples. Almost all strains of *S. aureus* produce gamma and delta hemolysins. Alpha hemolysin exerts more activity on white blood cells than red blood cells, and beta hemolysin tends to associate with *S. aureus* strains that are more pathogenic in animals than in humans.

TYPES OF HEMOLYSIS

Microbiologists use evidence of hemolysis on agar plates for identifying different cocci. This testing begins with differential agar medium, which is medium that distinguishes among different groups of bacteria. For hemolysis testing, blood agar containing 5–10 percent blood from sheep or horses serves as the differential medium. By observing the visible hemolysis reaction caused by bacteria on the blood cells in the agar, a microbiologist can identify an unknown coccus.

Hemolysis testing is used in two main endeavors: classification of new cocci that have not previously been studied and clinical microbiology. Clinical microbiologists routinely use blood agar incubations to examine samples from throat swabs—these samples often contain high numbers of streptococci or staphylococci in respiratory infections. Incubated blood agar plates that contain microorganisms possessing hemolysin display one of three main types of hemolysis:

- alpha-hemolysis—greenish discoloration of the agar surrounding the bacterial colony
- alpha-prime hemolysis—small zone of complete hemolysis surrounded by a zone of partial hemolysis
- beta-hemolysis—complete clearing of the agar surrounding the bacterial colony
- gamma-hemolysis—no discoloration or clearing and possible brown discoloration

Gamma-hemolysis is a lack of any red blood cell lysis. The unability to lyse red blood cells serves as a means to identify some nonhemolytic species.

Alpha-hemolysis occurs when bacteria possess an enzyme that reduces hemoglobin to methemoglobin. Methemoglobin leaches from the cells and surrounds the colony on the agar surface. This causes a greenish discoloration of the agar close to the colony. Alpha-prime hemolysis indicates the production of peroxidase enzyme, in addition to hemolysin, resulting in a double-zone appearance. Beta-hemolysis results from the complete destruction of red blood cells to make the agar appear clear rather than discolored.

Hemolysis Reactions of Common Cocci

Bacteria	Type of Hemolysis	Test
Streptococcus pyogenes	beta	food contamination or other infections
Streptococcus pneumoniae	alpha	respiratory infection
Streptococcus oralis	alpha	sore throat, throat infection
Staphylococcus aureus	beta	skin infections or food contamination
Enterococcus faecium	alpha	food contamination
Enterococcus faecalis	gamma	food contamination
Lactococcus	gamma	food spoilage

Hemolysis reactions have been most useful in differentiating various cocci. For example, a microbiologist can partially identify different genera of cocci with the following known reactions:

- *Streptococcus*—alpha or beta
- *Enterococcus*—alpha, beta, or gamma
- *Lactococcus*—gamma
- *Staphylococcus*—beta

In the case of *Streptococcus,* oral infections (strep throat) cause alpha-hemolysis, and other streptococcus infections carry out beta-hemolysis. The table on page 379 describes hemolysis reactions that help identify cocci in clinical microbiology.

Hemolysis reactions may be fine-tuned to gain further information on a species. For example, strains of *Streptococcus* called group D streptococci are beta-hemolytic on horse or rabbit blood agar but alpha-hemolytic on sheep blood agar. Growing an unknown coccus on both types of blood agar can provide a tentative identification of group D streptococcus. Furthermore, microbiologists have found that the best incubation conditions for examining hemolysis on blood agar use a temperature of 98.6°F (37°C), for 48 hours, for both aerobic and anaerobic bacteria.

The hemolysis reaction is a common and valuable aid in the quick identification of bacteria and has long been part of clinical microbiology.

See also ESCHERICHIA COLI; STREPTOCOCCUS; TOXIN; VIRULENCE.

Further Reading

Bownik, Adam, and Andrzej K. Siwicki. "Effects of Staphylococcal Hemolysins on the Immune System of Vertebrates." *Central European Journal of Immunology* 33 (2008): 87–90.

Lax, Alistair J. *Toxin: The Cunning of Bacterial Poisons.* Oxford: Oxford University Press, 2005.

Noble, Robert C., and Kenneth L. Vosti. "Production of Double Zones of Hemolysis by Certain Strains of Hemolytic Streptococci of Groups A, B, C, and G on Heart Infusion Agar." *Applied Microbiology* 22 (1971): 171–176. Available online. URL: http://aem.asm.org/cgi/reprint/22/2/171.pdf. Accessed August 20, 2009.

University of South Carolina. "Streptococci." In Microbiology and Immunology Online. Available online. URL: http://pathmicro.med.sc.edu/fox/streptococci.htm. Accessed August 20, 2009.

hepatitis Hepatitis is any inflammation of the liver. This condition may be caused by poisons (toxins), drugs, or viruses; when caused by a virus, the disease is called *viral hepatitis.* The viruses that cause inflammation of the liver are given the generic name of *hepatitis viruses.* The specific viruses that cause hepatitis are herpes virus (two different types), cytomegalovirus, Epstein-Barr virus, and nine viruses in a general group called *hepatotropic viruses.* The hepatotropic viruses are hepatitis A, B, C, D, E, F, and G. Hepatitis A and B were characterized in the 1960s, hepatitis D in 1977, and hepatitis C and E by 1990. A more recently discovered virus, called *transfusion-transmitted virus* (TTV), has also been associated with cases of hepatitis.

VIRAL HEPATITIS

Acute cases of hepatitis have a fast onset of symptoms, and the disease may have a short duration in the body. Acute hepatitis has been most associated with the hepatitis A (HAV), B (HBV), and C (HCV) viruses. These three unrelated viruses can cause nausea, lethargy, abdominal pain, and a condition called *jaundice,* which is a yellow coloration in body tissues. Jaundice results from excess levels of the compound bilirubin, released from the damaged liver into the bloodstream.

During acute hepatitis, liver cells cannot function as they normally do in detoxifying dangerous compounds and wastes that the liver removes from the bloodstream. The injured hepatic cells also falter in other important liver functions, such as the production of clotting factors, cholesterol metabolism, glycogen (the body's storage compound for the sugar glucose) storage, and fat-soluble vitamin (A, D, E, and K) storage. In instances where no vaccine is available to prevent hepatitis virus infection, doctors recommend good sanitation and good personal hygiene as the best prevention against infection. Sanitation involves cleaning surfaces that may harbor virus particles put there by others, and hygiene refers to proper hand washing techniques, avoidance of touching hands to the face, and avoidance of direct contact with known carriers of the virus.

In the United States, HAV and HBV do not exceed an incidence rate of two cases per 100,000 people, the lowest rates the Centers for Disease Control and Prevention (CDC) has ever recorded for this disease. HCV reaches only 0.3 case per 100,000, but infections from this virus are not declining and may be increasing in certain age groups and socioeconomic groups. The CDC has attributed the success in arresting HAV and HBV to effective vaccines against the viruses, introduced since the late 1990s.

The table on page 381 describes important features of the hepatitis viruses. Some of these viruses have been studied only recently, and less information has been accumulated on them (hepatitis F and G and TTV) than on A, B, and C. As do all viruses, the hepatitis viruses carry either deoxyribonucleic acid (DNA) or ribonucleic acid (RNA), which they inject into a host cell to make new virus particles. Fecal-oral transmission is fecal contamination in food, water, or objects contaminated with fecal matter. The World Health Organization (WHO) distinguishes hepatitis A and E as viruses transmitted mainly in contaminated food or water and hepatitis B, C, and D as forms usually transmitted parenterally (routes other than by ingestion, such as blood transfusions). Hepatitis B is also a sexually transmitted disease.

HAV, HBV, and HCV are the most common causes of hepatitis worldwide. Hepatitis F (HFV) has been implicated as an additional, distinct hepatitis virus that causes fulminant (extremely rapid onset) inflammation, often after a blood transfusion. In 1993, however, the Japanese virologist Toshikazu Uchida completed genetic analysis on HFV and produced evidence that it may be a mutant of HBV. Medical researchers have not, as yet, completely characterized HFV and HGV or confirmed that these viruses are distinct from the other hepatitis viruses. Any new hepatitis viruses discovered in the future will receive the name *hepatitis H, hepatitis I,* and so on.

HEPATITIS A

HAV is most commonly contracted through exposure to contaminated food, especially shellfish, or water. HAV outbreaks tend to be epidemics among children and young adults in confined institutional settings, such as mental hospitals, boarding schools, or juvenile correction facilities. This form of hepatitis also occurs among travelers visiting places with minimal sanitation. A carrier, human or animal, has never been identified for HAV.

Hepatitis A is usually a mild infection with an incubation period of two to six weeks with symptoms lasting two to 12 weeks. Full recovery requires several weeks to months. More severe cases may cause fever and gastrointestinal upset. HAV reproduces in liver cells and reaches the intestines through the bile duct. A person infected with HAV sheds active virus particles in the feces, and the peak time of virus shedding takes place during the latter part of the incubation period, when symptoms have not yet developed. In this scenario, infected people do not know they are sick but can excrete the virus. A hepatitis A outbreak can spread through a population before any signs of disease have emerged. This explains why good sanitation and personal hygiene are critical for reducing infection.

No drugs specifically treat HAV infection. A vaccine (trade name *Havrix*) contains killed HAV particles and can prevent the disease when given before or immediately after a supposed exposure.

HEPATITIS B

Hepatitis B infection is sometimes referred to as serum hepatitis, as it occurs only by direct inocu-

Hepatitis Viruses

Virus	Form of Genome	Transmission	Prevention
hepatitis A	RNA	fecal-oral	HAV vaccine
hepatitis B	DNA	blood, body secretions, sexual, contaminated needles	HBV vaccine
hepatitis C	RNA	blood, sexual	blood screening
hepatitis D (delta virus)	RNA	blood, sexual	same vaccine as for HBV
hepatitis E	RNA	fecal-oral	good hygiene
hepatitis F	not yet proved to exist as a distinct virus	blood transfusion	unknown
hepatitis G	RNA	sexual, parenteral	unknown
transfusion transmitted virus	DNA	blood, blood transfusion, possible fecal-oral	unknown
cytomegalovirus	DNA	body secretions, blood	good hygiene
Epstein-Barr virus	DNA	saliva	good hygiene

lation or inoculation through the body's mucous membranes. Mucous membranes occur at the body's passages or cavities that are exposed to the air: nasal passages, mouth, and urogenital tract. Humans serve as the main carrier in infecting other people. Intravenous drug users also have a higher-than-normal incidence of HBV infection. The virus is not spread in food or water or by casual contact with others. The virus can, however, remain active outside the body on inanimate surfaces for up to a week. Hand-to-mouth inoculation, after touching a contaminated surface, may be possible, but it is not the main route of transmission for disease.

The WHO estimates at least two billion people worldwide carry HBV, and many people do not know they harbor the virus. The illness's acute symptoms are jaundice, dark urine, fatigue, nausea, vomiting, and abdominal pain. HBV also causes a chronic form, which infects about 400 million people and is more difficult to detect and may be undiagnosed for a person's lifetime. Chronic HBV infection is often discovered many years after exposure, through the presence of a jaundiced liver. Acute and chronic forms of the disease kill about 5,000 people in the United States every year.

Developing countries currently have higher risk for HBV transmission. The infection is endemic (prevalent in the population) in China and other parts of Asia—one in 10 Asians may be infected—and high rates also occur in the Amazon region, parts of Europe, the Middle East, and India. Western Europe and North America have chronic infections in less than 1 percent of their populations.

Medical agencies such as the Centers for Disease Control and Prevention (CDC) and WHO have identified the following main modes of transmission for HBV infection:

- perinatal—mother to baby at birth
- unsafe injections
- blood transfusions
- sexual contact

In developed countries, HBV infection has become a growing concern among illegal drug users and young adults who have high sexual activity. The U.S. Department of Health has estimated that by age 17, about half of all adolescents (ages 10 to 18) have had sexual intercourse, a rate that makes adolescents and young adults a subpopulation of higher risk for infection compared with the general population. Unsafe injections using dirty needles have created another growing problem in young adults. Rather than injections from illegal drug use, tattoos and piercings have been implicated in a rise in HBV infection in some parts of the world. The hepatologist (liver specialist) Graeme Macdonald of the Princess Alexandra Hospital in Brisbane, Australia, commented, in 2008, "There may be tattoo artists who are using the same needles and also not changing the ink pots, which is a risk. I see a large number of people waiting for liver transplants as a result of hepatitis, and that number is growing." Without strong preventive measures, hepatitis B infection may become a reemerging health problem in the coming years.

People at high risk of contracting hepatitis B should receive the HBV vaccine. These people include health care workers, young adults who are sexually active, hemodialysis patients, and intravenous drug abusers. The HBV vaccine contains proteins from the surface of the virus (antigens) that induce the body's immune system to build antibodies to the live virus. Vaccine manufacturers prepare the HBV vaccine (trade names *Recombivax HB* and *Engerix-B*) by using recombinant DNA techniques to put the antigens on cells of baker's yeast. Vaccination against HBV infection provides protection for life in most individuals. The presence of HBV antibodies in a person's blood also serves as a good test for doctors to determine whether a person has been exposed to HBV or is a carrier.

HEPATITIS C

HCV is carried mainly in blood (called a *blood-borne infection*) and until recently was known as non-A, non-B hepatitis (NANB), because little information was available on the virus's makeup and transmission. HCV transmits mainly through blood transfusions with lesser numbers of cases resulting from sexual or perinatal transmission. Similarly to HBV, HCV can survive outside the body for several hours to a day.

Hepatitis C has an incubation period of four to 12 weeks, but few symptoms develop. Doctors usually diagnose HCV infection by detecting higher-than-normal liver enzyme levels in repeated blood tests. The body makes antibodies to HCV, but these molecules do not destroy the virus and do not provide strong immunity. One problem in gaining immunity against HCV may be a result of the many different subtypes of HCV that infect humans. Six genotypes (specific gene compositions) and more than 60 subtypes of HCV have been discovered. Antiviral agents such as interferon have been tried for the treatment of hepatitis C, but the success of these drugs is limited, and they often produce worse symptoms than the disease.

Hepatitis C has an acute form and a chronic form. Because treatments for the infection are limited, acute hepatitis C often turns into a chronic infection. No vaccine exists, at present, for HCV, and no treatments are available. The CDC recommends that a person who has HCV infection follow doctors' orders to rest, drink plenty of fluids, and get adequate nutrition. HCV leaves the body in about 25 percent of the people who contract it without causing a chronic infection.

HEPATITIS D

HDV, known as the delta virus, is unusual because it causes infection only if a person already has a HBV infection or as a simultaneous infection with HBV, called *coinfection*. For this reason, the best preventive measure against the delta virus is vaccination with the HBV vaccine. These infections occur rarely in the United States.

Delta virus infections may be acute or chronic, and the presence of the virus probably increases the effects of an HBV infection. Delta virus uses a portion of the HBV replication mechanism in liver cells to produce more delta particles, but this mechanism is not completely understood. The virus is referred to as a *defective virus,* because it cannot multiply in the body on its own. As delta virus commandeers part of HBV's replication process, it decreases the amount of new HBV particles that the liver sheds. Acute delta-HBV infections are referred to as biphasic infections because they consist of two distinct phases. In the first phase, a regular HBV infection ensues, but about a week after hepatitis B symptoms develop, the second phase begins, marked by a second surplus of liver enzymes in the blood. This second occurrence is probably due to the effects of the delta virus.

Researchers have not clarified all the details of how delta virus enters liver cells, but once inside a hepatocyte, delta virus replicates similarly to other viruses that invade cells. Only at the final steps, in which the virus must be packaged before exiting the host cell, does delta virus require help from the HBV packaging enzymes.

The most severe acute cases of coinfection are called superinfections. In 90 percent of patients in whom a delta-HBV superinfection develops chronic hepatitis also develops. These infections cause more severe cirrhoses than other hepatitis infections, as well as more fatalities. HDV cases are rare in the United States but have developed into a greater concern in Russia, Romania, Mediterranean regions, and parts of Africa and South America among all age groups.

Diagnosed hepatitis B cases that suddenly progress to more serious symptoms may be evidence of a delta virus infection. The symptoms are more severe than found in HBV infection and lead to cirrhosis of the liver (a diseases state that causes loss of liver function). Mortality rate may reach 10 percent in these situations, because delta virus infections have no drug treatments. The WHO has stated that only liver transplantation assures treatment for the infection that has progressed to severe cirrhosis of the liver. In a few instances, the drug alpha-interferon has helped lessen the severity of the symptoms.

HEPATITIS E

HEV transmits similarly to HAV in food and water, but HEV is much more prone to cause an acute disease. The incubation period ranges from three to eight weeks. Within eight weeks of onset of symptoms, a person can suffer severe liver damage, jaundice, and encephalopathy (brain dysfunction).

Hepatitis E has caused several epidemics in developing parts of Asia, Africa, and Central and South America. (Hepatitis E incidence in the United States is low.) These epidemics have followed seven- to 10-year cycles that correlate with seasonal conditions that may affect water contamination with fecal matter. Pregnant women have an especially high fatality rate (15 to more than 30 percent) compared with nonpregnant women. Acute infection leads to liver failure, bile duct injury, inflammations, and kidney failure. Nonpregnant women and the rest of the general population experience less severe general symptoms of hepatitis and recover.

Researchers have not found a definitive reason for the cyclical nature of HEV epidemics. Some scientists have proposed that the epidemics follow a general trend, in which water purification and distribution infrastructure ages. Others have suggested that an unknown animal reservoir may contribute to higher and lower incidences of outbreaks, over several years. The CDC scientists Michael Favorov and Harold Margolis wrote, in 1999, in *Emerging Infections,* "Infection with HEV, a recently characterized hepatitis virus, produces epidemics of disease associated with fecal contamination of drinking water. Although clearly an enterically transmitted virus, there appears to be a low rate of person-to-person transmission and low rates of infection among children. Serologic evidence of HEV infection in domestic animals, the recent discovery of an HEV strain in swine, and the isolation of a swine-like HEV strain from a person with hepatitis E indicate the need to determine the role that animals play in the epidemiology of HEV infection." Since Favorov and Margolis proposed this theory, scientists have detected antibodies to HEV in primates, pigs, and chickens, and hepatitis E is now defined as a zoonotic disease, a disease that humans contract from animals.

OTHER HEPATITIS VIRUSES

TTV is one of several viruses that can be transmitted from an infected to a healthy person in a blood transfusion. In the late 1990s, virologists began implicating this poorly characterized virus as the cause in acute liver disease and liver failure in patients who did not show evidence of infection by the hepatotropic viruses A through E. TTV is found in much higher concentration in the liver than in other tissues in people who have hepatitis. In addition, TTV may be present in more than 50 percent of patients who have an existing case of chronic hepatitis caused by any of the other hepatotropic viruses.

The Polish pediatrician Dariusz Lebenzstein reviewed the state of the research on TTV in the *Medical Science Monitor,* in 2000, and reported, "The review of literature suggests that TT virus is a wide-spread pathogen. The variety of transmission routes of the virus (parenteral route, particularly fecal-oral route) is probably the reason." Lebenzstein also pointed out that the mode of action of TTV on liver cells remained unknown at that time. Virology studies, since 2000, have uncovered several different genotypes of TTV, some of which might be implicated in causing liver cirrhosis and others that do not appear to be connected to liver disease. TTV's role in hepatitis has not yet been completely characterized.

Cytomegalovirus (CMV) is a common, highly contagious virus that infects the mucous membranes of the eye and other mucous membranes. CMV infections, on occasion, spread to the liver, where they cause mild symptoms that rarely lead to jaundice. Individuals who have acquired immunodeficiency syndrome (AIDS) and other immunocompromised people such as organ transplant recipients and cancer chemotherapy patients have a higher risk of infection, and, in some cases, the CMV infection can be fatal.

AIDS patients or individuals who are positive for the human immunodeficiency virus (HIV) without AIDS may contract pneumonia, gastrointestinal disease, nervous system injury, and/or retinitis (inflammation of the retina of the eye) as a result of CMV infection. CMV hepatitis has also become a prevalent postoperative condition arising from liver transplantation. Patients receive treatment with ganciclovir to stop the CMV infection.

Epstein-Barr virus (EBV) is a herpes virus that causes mononucleosis. Prolonged EBV infections cause the virus to spread to various organs, including the liver, for replication. This infection leads to a mild increase in blood levels of liver enzymes but few or no symptoms of hepatitis. Mononucleosis is also associated with hepatomegaly, or enlargement of the liver, but not jaundice. This reaction is due in part to inflammation in the liver and a possible autoimmune response of the body to the presence of EBV. Autoimmune responses are actions in which the body sets up an immune reaction against its own cells or tissue.

The various viral hepatitis diseases collectively have a high incidence worldwide. Although these diseases cause significant debilitation, relatively few deaths result when compared with the large number of people who have been infected with a hepatotropic virus. Despite the prevalence of hepatitis in society, many of the mechanisms of the viruses remain unknown. The exact causes, modes of action on liver cells, preventions, and treatments all need further research and testing.

See also SEXUALLY TRANSMITTED DISEASE; TRANSMISSION; VIRUS.

Further Reading

Centers for Disease Control and Prevention. "Viral Hepatitis." Available online. URL: www.cdc.gov/hepatitis. Accessed August 20, 2009.

Daniels, Danni, Scott Grytdal, and Annemarie Wasley. "Surveillance for Acute Viral Hepatitis—United States, 2007." *Morbidity and Mortality Weekly Report* 58 (2009): 1–29. Available online. URL: www.cdc.gov/mmwr/PDF/ss/ss5803.pdf. Accessed August 20, 2009.

Davies, Hannah. "Tattoo Craze Causes Surge in Hepatitis Cases." Queensland, Australia, *Courier Mail,* 18 May 2008. Available online. URL: www.news.com.au/couriermail/story/0,23739,23713348-23272,00.html. Accessed August 20, 2009.

Favorov, Michael O., and Harold S. Margolis. "Hepatitis E Virus Infection: An Enterically Transmitted Cause of Hepatitis." In *Emerging Infections 3*. Washington, D.C.: American Society for Microbiology Press, 1999.

Hafez, Mohamed M., Sabry M. Shaarawy, Amr A. Hassan, Rabab F. Salim, Fatma M. Abd El Salam, and Amal E. Ali. "Prevalence of Transfusion Transmitted Virus (TTV) Genotypes among HCC Patients in Qaluobia Governorate." *Virology Journal* 4 (2007): 135–140. Available online. URL: www.virologyj.com/content/pdf/1743-422X-4-135.pdf. Accessed August 20, 2009.

Hunt, Richard. "Hepatitis Viruses." In *Virology*. Columbia: University of South Carolina School of Medicine, 2004. Available online. URL: http://pathmicro.med.sc.edu/virol/hepatitis-virus.htm. Accessed August 20, 2009.

Immunization Action Coalition. "Unusual Cases of Hepatitis B Virus Transmission." Available online. URL: www.immunize.org/catg.d/p2100nrs.pdf. Accessed August 20, 2009.

Lebenzstein, Dariusz M. "Is TT Virus (Transfusion Transmitted Virus) a Novel Hepatotropic Agent Causing Hepatitis?" *Medical Science Monitor* 6 (2000): 823–826. Available online. URL: www.medscimonit.com/fulltxt.php?ICID=508346. Accessed August 20, 2009.

World Health Organization. "Hepatitis." Available online. URL: www.who.int/topics/hepatitis/en. Accessed August 20, 2009.

heterotrophic activity Heterotrophy is a mode of metabolism in which a microorganism uses preformed organic compounds as the main carbon source and usually as an energy source. A *heterotroph* is the general name given to any microorganism that lives using heterotrophy.

The diverse group of heterotrophic microorganisms contains pathogens, food production contaminants, spoilage microorganisms, and a large portion of environmental species.

Heterotrophy is one of the two major mechanisms of nutrient use among microorganisms; the other is autotrophy. These two types of nutrient use are distinguished as follows:

- heterotrophy—use of a variety of preformed, highly reduced organic compounds made by other organisms

- autotrophy—use of carbon dioxide (CO_2) as the main or the sole carbon source for building all other organic compounds within the cell

Heterotrophy evolved as an advantage to the microorganisms that use it because of its energy efficiency. Most microorganisms would probably use CO_2 as a main carbon source simply because it is abundant in all aerobic and anaerobic environments. But in order to incorporate this gas into cellular metabolism, an organism must, first, reduce the compound, by adding negatively charged electrons in a process that also hydrogenates the carbon atom. An example of one such reduction performed by methanogenic bacteria is shown in the equation. Methanogenic bacteria are species that produce methane gas.

$$CO_2 + electrons^- + H^+ \rightarrow CH_4 + H_2O$$

A problem arises in this type of metabolism because CO_2 does not supply energy or hydrogen to the cell. Reduction requires energy from the cell to complete the reaction, putting demands on autotrophs to find an energy source. Some autotrophs solve this dilemma by using photosynthesis to capture energy from the Sun. Heterotrophic microorganisms do not confront this challenge.

Heterotrophs also have an advantage over autotrophs in being very flexible in their choice of nutrients. A wide range of carbon compounds can be used in heterotrophy for energy production: sugars, organic acids, polysaccharides, fats, and amino acids. A heterotroph such as *Pseudomonas* bacteria can use well over 100 different compounds as carbon and energy sources. The heterotrophs symbolize the great diversity that exists in the microbial world.

TYPES OF METABOLISM

Metabolism, the collective reactions used for converting carbon compounds to energy and cellular building blocks, can be accomplished in four main ways, as summarized in the table on page 386.

Chemoheterotrophs live in a wide range of habitats: soil, food, freshwater, marine water, and decaying matter and as pathogens in plants and animals. They can be bacteria or fungi and thrive in either extreme environments or more moderate environments. Many heterotrophs, furthermore, adjust their metabolism to the compounds available to them.

Because heterotrophic microorganisms depend on organic compounds, chemoheterotrophs have also been called *chemoorganoheterotrophs,* and photoheterotrophs have been called *photoorganoheterotrophs.* In more general terminology, organotrophs are any microorganisms that use organic compounds as an electron source for energy metabolism. Chemotrophs obtain energy by metabolizing either organic or inorganic compounds. Finally, mixotrophs combine different types of metabolism: They use inorganic energy sources and organic compounds for all other metabolism.

Heterotrophs also contain two types of organisms classified by whether they subsist on living matter or dead nonliving organic matter. Biotrophs live on living organisms. Parasites and pathogens are biotrophs. Saprophytes derive their nutrition from nonliving organic matter. For example, fungi that degrade decaying woods and plants are saprophytic organisms.

TYPES OF HETEROTROPHY

Chemoheterotrophs include all fungi, protozoa, invertebrates, and higher organisms, in addition to bacteria. Among these life-forms, chemoheterotrophy contains two of biology's basic types of energy and carbon metabolism: fermentation and respiration. Both types of metabolism shuttle electrons between compounds in a sequence run by enzymes for the purpose of releasing energy from chemical bonds. Fermentation is the anaerobic breakdown of carbohydrates in which organic compounds serve as a final electron acceptor. Respiration is also a sequence of enzymatic reactions, but it can occur either aerobically or anaerobically, it generates more chemical energy than fermentation, and an inorganic compound serves as the final electron acceptor.

Fermentation is less efficient than respiration because no net oxidation and reduction take place. In fermentation, enzymes shuttle electrons among different organic compounds. The end products of fermentation vary by the type of microorganism and the type of substrate (the starting compound). A common fermentation pathway is called *glycolysis,* which is the degradation of the sugar glucose to pyruvic acid. Fermentation end products from pyruvic acid are typically organic acids, alcohols, and CO_2. Respiration, by contrast, involves enzyme systems that, in

Metabolic Categories of Microorganisms

Metabolism	Main Carbon Sources	Main Energy Sources
chemoheterotrophy	various, usually sugar such as glucose	often the same as the carbon source
photoheterotrophy	various organic compounds	light
chemoautotrophy	CO_2	inorganic compounds such as hydrogen sulfide (H_2S), sulfur (S), ammonia (NH_3), nitrite (NO_2^-), or metals
photoautotrophy	CO_2	light

sequence, degrade a substrate completely to CO_2 and H_2O with the generation of about 18 times the energy that fermentation generates. Both types of metabolism store the energy they generate mainly in the chemical bonds of adenosine triphosphate (ATP). Respiration also begins by breaking down glucose by the glycolysis pathway, then uses the Krebs cycle to degrade pyruvic acid. Finally, respiration employs a series of reactions called the *electron transport chain* to generate energy held in ATP and produce CO_2 and H_2O.

TYPES OF HETEROTROPHS

Heterotrophs include all fungi and other eukaryotic microorganisms such as protozoa. Among bacteria, several groups use heterotrophy. A method in microbiology called a *heterotrophic plate count* is the sampling and incubation of diverse microorganisms from the environment that can use a variety of substrates for growth on laboratory medium. The table below highlights the main heterotrophic groups of bacteria.

The bacteria described in the table illustrate important steps in the evolution of simple single-celled organisms to multicellular organisms, especially in regard to a shifting away from dependence on simple molecules to more complex energy and carbon sources. Evolution on Earth began with fermentative bacteria that used only simple substrates in anaerobic conditions. The development of cyanobacteria marked a point at which an alternative type of energy metabolism, photosynthesis, emerged to replace fermentation. Cyanobacteria and other photosynthetic microorganisms put oxygen into the atmosphere over time, and aerobic respiration began to develop.

The microbiologist Andrew White of Staffordshire University in the United Kingdom explained on his Web site Earth's biology of 3.5 billion years ago: "The first cells were probably chemoautotrophs synthesizing ATP by oxidation of hydrogen

Heterotrophic Bacteria

Group of Bacteria	Description	Energy Source	Examples
ammonia-oxidizing	soil species that convert ammonium (NH_4^+) to nitrite (NO_2^-) in nitrification	organic compounds	*Nitrosomonas*
cyanobacteria	plantlike organisms; some anaerobic	light	*Anabaena*
enteric	facultative anaerobic species that ferment sugars in the intestines	organic compounds	*Escherichia*
green nonsulfur	extremophile users of organic compounds but can be autotrophic	light	*Chloroflexus*
nitrite-oxidizing	soil species that convert nitrite (NO_2^-) to nitrate (NO_3^-)	organic compounds	*Nitrobacter*
purple nonsulfur	aquatic species, usually anaerobic	light	*Rhodospirillum*

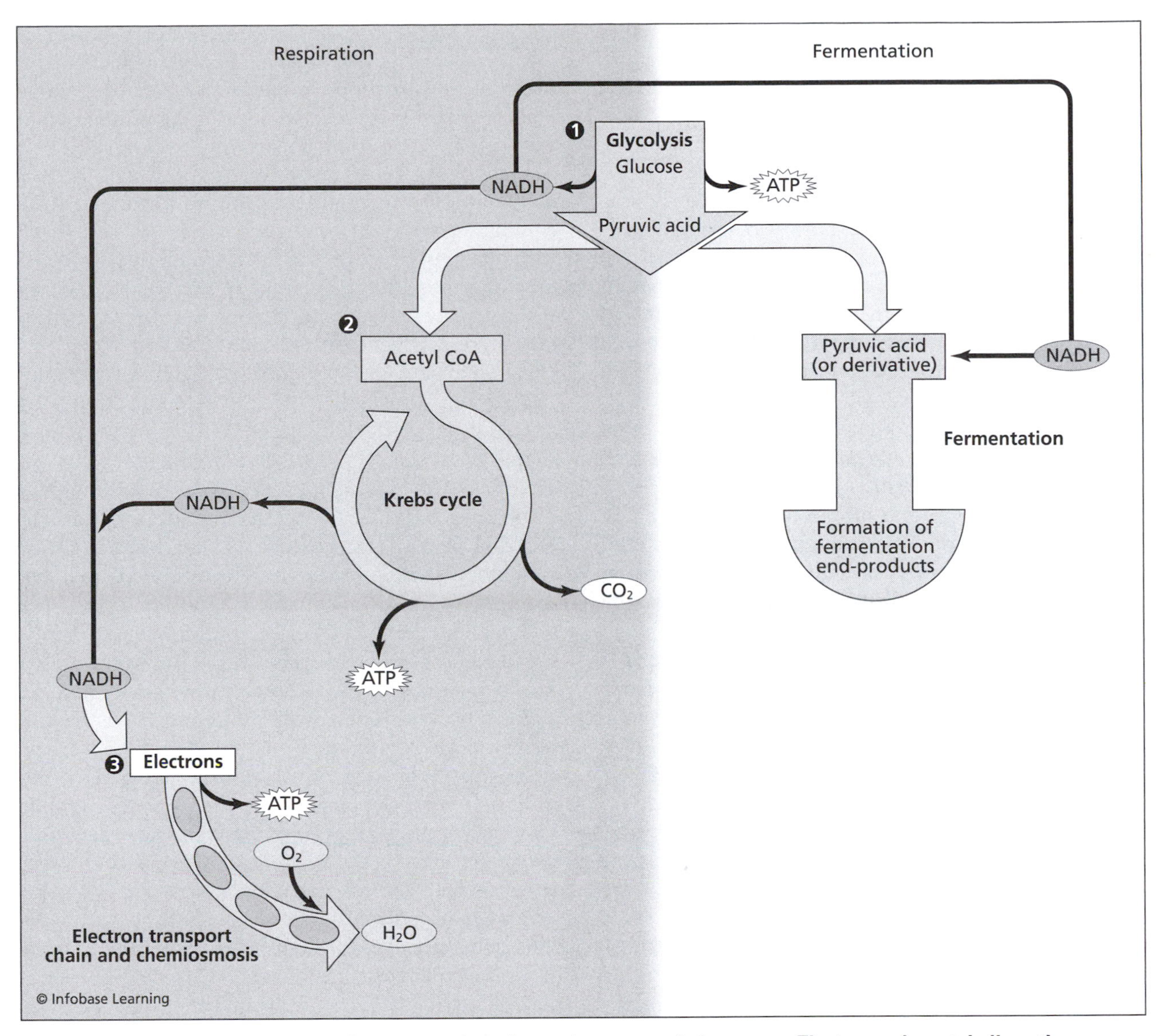

Heterotrophs, also called organotrophs, use a variety of organic compounds for energy. The two main metabolic pathways in heterotrophy are aerobic respiration, which uses oxygen as an electron acceptor, and fermentation, which uses organic compounds as electron acceptors.

sulphide and iron (II) compounds then abundant in the environment. The released energy could have been harnessed via production of a proton gradient, stimulating evolution of electron transport chains, and the reducing equivalents (electrons) generated used in carbon dioxide fixation and thence biosynthesis. There may also have been opportunistic absorption of abiotically produced organic compounds from the environment (i.e., chemoheterotrophy), although these were not present in sufficient quantity to make anything more than a small contribution." White's statement summarizes the significant contributions that heterotrophs have made to energy metabolism on Earth and the evolution of today's biodiversity.

See also CYANOBACTERIA; FERMENTATION; GREEN BACTERIA; METABOLISM; PURPLE BACTERIA.

Further Reading

Todar, Kenneth. "The Origin, Evolution and Classification of Microbial Life." In *The Microbial World*. Madison: University of Wisconsin, 2008. Available online. URL: http://bioinfo.bact.wisc.edu/themicrobialworld/origins.html. Accessed August 20, 2009.

White, Andrew J. "Evolution." Available online. URL: www.staffs.ac.uk/schools/sciences/biology/Handbooks/evolearly.htm. Accessed August 20, 2009.

Hooke, Robert (1635–1703) ***British Inventor, naturalist, physicist*** Robert Hooke has been credited with originating the cell theory, the doctrine that states that all living things are made of one or more organized units called *cells*. The concept of the cell now makes up one of several themes that unify modern biology. The cell theory states that cells are every organism's basic unit, which forms the organism's structure and makes functions possible. Biology is divided into two main cell types: prokaryotic cells containing bacteria and archaea and eukaryotic cells containing protists (single-celled eukaryotic species), fungi, plants, and animals.

Born on July 18, 1635, Robert Hooke was the son of a church curator in Freshwater on the Isle of Wight, England. As did many youngsters of the day, Robert began his early years in poor health and suffered debilitating headaches. His parents had expected to steer Robert toward a life in the church to further the family tradition, but because of their son's poor health, they may have assumed the youngster would not live to adulthood. Hooke's parents gave up on his further education by the time Robert was 10, and the boy began a life of self-education.

Hooke found himself drawn to both science and art. He observed the nature and geography around him and made detailed notes on his observations, but few of Hooke's writings have survived. Simply observing nature would not provide a livelihood, so Hooke turned to another of his many talents, drawing and painting, to earn a living. After his father's death, in 1648, Hooke went to London to apprentice with a portrait painter. Despite Hooke's proficiency in art, he grew restless with the training and left his apprenticeship to enroll at Westminster School, in 1648. Hooke proved to be a good student and before long devoured lessons in chemistry, geometry, engineering, and Latin. By 1653, Hooke's teachers thought their student had reached the boundaries of their resources, and they sent Hooke with their blessings to the University of Oxford.

Hooke pursued interests in mechanical instruments, mathematics, and engineering. During his time in Oxford, he tried various inventions: a flying machine, spring-controlled clocks, and construction of an air pump that was the predecessor for today's air pumps. Hooke had begun to build another reputation for himself as a brilliant scientist with so many interests he could scarcely stay focused on any one at a time.

In the mid-1600s, Great Britain experienced a period of shifting political philosophies that influenced religion, the arts, and science. In 1660, a consortium of college professors petitioned King Charles II to approve the formation of the Society for the Promoting of Physico-Mathematical Experimental Learning. The society intended to promote the pursuit of scientific ideas through experimentation rather than building concepts solely on theory, many of which were tightly bound to religious dogma. Charles II granted the royal charter for the society, in 1662, to be called the Royal Society of London. The society members appointed Robert Hooke as the organization's curator of experiments. One of Hooke's new responsibilities as curator was to follow a rather demanding schedule of introducing new experimental methods to the society. Hooke's rampant energies in science were perfectly suited to the responsibility. In the next several years, Hooke's rich supply of ideas helped raise the Royal Society of London to a preeminent position in the scientific community. The historian Howard Gest wrote in *Microbe* magazine, in 2007, "Hooke became a commanding intellectual presence in the Society, and as curator provided the main substance of meetings. His interests ranged over physics, mechanics, astronomy, chemistry, geology, and biology. Moreover, he was a prolific inventor, especially in connection with microscopes and telescopes." When Hooke built a simple microscope from a tube holding two convex lenses, he helped launch the science of microbiology.

Hooke began studying the microscopic features of wood, leaves, hair, feathers, vegetables, animals, and other samples from nature. He also used his drawing skills to document his observations in a series of detailed sketches. The Royal Society published the notes and sketches in a large tome called *Micrographia,* in 1665. After examinations of cork, feather, or vegetable under his microscope, Hooke wrote in *Micrographia* [spelling corrected to modern form], "The pith also that fills that part of the stalk of a Feather that is above the Quill . . . so that I guess this pith which fills the feather, not to consist of an abundance of long pores separated with Diaphragms, as Cork does, but to be a kind of solid or hardened [collection] of very small bubbles consolidated into that form, into a pretty stiff as well as strong concrete, and that each Cavern, Bubble, or Cell, is distinctly separate from any of the rest. . . . In several of those Vegetables, whilst green, I have with my Microscope, plainly enough discovered these Cells or Holes filled with juices." This may have been the first use of the word *cell* as it is used today. Hooke drew striking, detailed sketches of the subjects he studied, including what may be the first published image of a microorganism: the reproductive structures called *sporangia* on a "blue mould" that scientists, generations later, surmised was the common mold *Mucor.*

THE CELL THEORY

The development of the cell theory could not have occurred without concurrent advances in microscopy. Knowing Hooke's personality, it would not be hard to imagine that he constantly tinkered with microscopy to inspect the natural world. In truth, the theory of cells would not have grown into a science of its own without the contributions of individuals in addition to Hooke in later years. One contemporary of Hooke's, the Dutch cloth merchant

Development of the Cell Theory

Person	Date	Contribution
Robert Hooke	1665	publishes his observations of microscopic things and first used the term *cell*
Antoni van Leeuwenhoek	1673	described small objects under his microscope; first person to describe bacteria and protozoa
Lorenz Oken	1805	German naturalist and philosopher, proposed that all living things consist of cells
Robert Brown	1833	Scottish botanist, first to use the microscope for the study of plants; first to recognize the nucleus and to use the term *nucleus*
Theodor Schwann and Matthias Schleiden	1838	German biologists who proposed that the cell is the one unit upon which all plants and animals begin their development; the introduction of modern cell theory
Rudolf Virchow	1858	proposed that every cell must result from a preexisting cell

Antoni van Leeuwenhoek, used his own rudimentary microscopes to observe single-celled living things in samples he collected from the environment and from the human body. Hooke's use of a microscope to observe nature at the small scale and van Leeuwenhoek's actual description of tiny "animalcules" in tandem created a new science: microbiology, the study of independent single-celled organisms.

The observations of the microbial world did not lead to an automatic leap to the cell theory. In fact, the Italian biochemist Paolo Mazzarello explained, in 1999, that Hooke's and van Leeuwenhoek's discoveries may well have fostered an entirely different doctrine than the modern cell theory. "The existence of an entire world of microscopic living beings was seen as a bridge between inanimate matter and living organisms that are visible to the naked eye," Mazzarello wrote in *Nature Cell Biology*. "This seemed to support the old Aristotelian doctrine of 'spontaneous generation', according to which water or land bears the potential to generate, 'spontaneously', different kinds of organism." Generations of scientists would put the theory of spontaneous generation to rest, as the cell theory emerged.

History has recognized the contribution of Hooke's first observations of cells as the beginning of the cell theory. The table above describes how the cell theory developed, over two centuries, through the contributions of a parade of scientists who built on the concept of the cell, improvements to the microscope, or both.

Over the next two centuries, scientists unraveled mysteries of how cells live on their own, reproduce, manage their genetic material, and communicate with each other. The modern cell theory that is the foundation for biology adheres to the following six principles:

- All known living things are composed of cells.
- Cells are the most basic structural and functional units of living things.
- Cells result only from preexisting cells and not from spontaneous generation.
- Cells contain hereditary information that is passed to new cells during cell division.
- All cells contain similar chemical composition.
- All energy metabolism of living things takes place inside cells.

The cell theory is one of the unifying concepts of biology. Cells represent all the cells that preceded them; that is, cells are related to their ancestors. Yet present-generation cells also differ in either major or subtle ways from their ancestors. These differences have been mainly due to adaptations to changing conditions in an environment, which are the basis of evolution.

The cell theory has since merged with other tenets of biology to explain how the living world works. Biologists have developed 12 themes that run through all of biology, which are described in the table on page 390.

Biologists will certainly add new concepts to the list in the table as technologies advance. Mazzarello wrote, in 1999, "With the theory of evolution, the cell theory is the most important generalization in biology. There is, however, a missing link between these theories that prevents an even more general and unifying concept of life. This link is the initial

Twelve Unifying Themes in Biology

Theme	Main Principle
1. cell theory	cells are the basic unit of every living thing
2. heredity	living things maintain continuity through generations by passing on genetic information held in deoxyribonucleic acid (DNA)
3. organization	the living world has a hierarchal organization that contains all biota
4. regulation	feedback mechanisms in biology help living things maintain a steady state
5. interactions with environment	organisms respond to and interact with stimuli in their environment
6. energy maintains life	the Sun's energy is captured by living things on Earth and transferred to all other organisms
7. diversity	the biological world can be categorized into three domains: Bacteria, Archaea, and Eukaryota
8. unity	even with diversity, all biota also possess some genetic characteristics in common
9. evolution	the adaptation of populations to their environment is due to natural selection, which creates both diversity and unity
10. structure and function	form and function are connected in some way in all of biology
11. scientific inquiry	progress derives from discoveries based on observation and on the testing of hypotheses in a way that others can repeat
12. science, technology, and society	science strives to develop new technologies that have direct bearing on the well-being of society

passage from inorganic matter to the primordial cell and its evolution—the origin of life." Biologists are nearing a milestone when this link progresses from conjecture to proof, but no one knows how far in the future that milestone exists.

Robert Hooke's powers of inspection and his diligence in recording fine details created the modern biology studied in schools today. Until 2006, most historians relied mainly on writings by Hooke to the Royal Society. But, in that year, hundreds of pages of Hooke's laboratory notes were discovered in a privately owned house in England. The Royal Society now houses these papers, called the Hooke Folio. In it, letters between Hooke and van Leeuwenhoek have revealed the thoughts of both men as they mused over the nature of microscopic creatures. They also maintained a detailed correspondence on microscopy and ideas for improving their instruments.

Hooke spent the rest of his career following a characteristically disjointed path. The older Hooke's enthusiasm for diverse subjects sometimes appeared as a lack of focus. He grew irritable with many of his contemporaries for being slow to see the world as he saw it. Much of Hooke's energy also went into a decades-long rivalry with Isaac Newton over who had first proposed the theory of gravity. His biographer Allan Chapman wrote, in 2005, "It is unfortunate that Robert Hooke's controversy with Newton, which undoubtedly embittered the last 16 years of his life, has somehow been allowed to become so formative in history's wider judgement *[sic]* of the man." Chapman described Robert Hooke as one of history's greatest scientists, particularly in the manner in which Hooke approached his work: by meticulous examination, detailed record keeping, demonstration, and communication to his peers.

In 1678, Robert Hooke's brother committed suicide, possibly over poor finances. Because the act of suicide was by law a felony in England, immediate family members were forced to turn over a large portion of their estate to the Crown. Robert's family may have faced this obstacle, as historians such as Chapman have suggested that, in Hooke's final years, he became miserly with his money. Hooke died in London of undetermined causes, in 1703.

Even trained microbiologists may not realize all of the areas of science to which Robert Hooke contributed his talents. He will certainly be remembered for two critical precepts taught to all science students. First is the concept of the cell as the basic unit of all living things. Hooke's second contribution related to his superb demonstration of scientific inquiry, in which facts are based on experimentation and confirmation rather than faith.

In 1664, Hooke addressed the Royal Society of London with the following: "The rules you have prescrib'd yourselves in your philosophical progress do seem the best that have ever yet been practis'd. And particularly that of avoiding dogmatizing and the espousing of any hypothesis not sufficiently grounded and confirm'd by experiments. This way seems the most excellent, and preserve both philosophy and natural history from its former corruptions." Hooke's simple proposal of scientific inquiry would take years for acceptance, but history will repeatedly show the soundness of Hooke's approach to science.

Robert Hooke died, on March 3, 1703, of natural causes in London. Although his legacy had been overlooked for many years after his death, Hooke has now been accepted as one of the leading minds in the history of scientific discovery.

See also LEEUWENHOEK, ANTONI VAN, MICROSCOPY.

Further Reading

Chapman, Allan. *England's Leonardo: Robert Hooke and the Seventeenth-Century Scientific Revolution.* Bristol, England: Institute of Physics Publishing, 2005.

Gest, Howard. "Fresh Views of the 17th-Century Discoveries by Hooke and van Leeuwenhoek." *Microbe,* October 2007.

———. "The Remarkable Vision of Robert Hooke (1635–1703): First Observer of the Microbial World." *Perspectives in Biology and Medicine* 48 (2005): 266–272.

Hooke, Robert. *Micrographia: Or Some Physiological Descriptions of Minute Bodies Made by Magnifying Glasses with Observations and Inquiries Thereupon.* London: Council of the Royal Society of London, 1664. Available online. URL: www.gutenberg.org/files/15491/15491-h/15491-h.htm. Accessed August 9, 2009.

Mallery, Charles. "Cell Theory." University of Miami, 2003. Available online. URL: www.bio.miami.edu/~cmallery/150/unity/cell.text.htm. Accessed August 9, 2009.

Mazzarello, Paolo. "A Unifying Concept: The History of Cell Theory." *Nature Cell Biology* 1 (1999): E13–E15. Available online. URL: www.nature.com/ncb/journal/v1/n1/full/ncb0599_E13.html. Accessed August 9, 2009.

Panno, Joseph. *The Cell.* New York: Facts On File, 2005.

Waggoner, Ben. "Robert Hooke (1635–1703)." 2001. Available online. URL: www.ucmp.berkeley.edu/history/hooke.html. Accessed August 9, 2009.

human immunodeficiency virus The human immunodeficiency virus (HIV) is the pathogen that causes acquired immunodeficiency syndrome (AIDS). HIV is a retrovirus, meaning it carries its genetic information in ribonucleic acid (RNA) and must infect living cells in order to make deoxyribonucleic acid (DNA) before replicating.

Since the birth of the AIDS epidemic in the United States and abroad in the early 1980s, this virus has become one of the most studied pathogens in medical history. Unlike most other pathogens, the virus and the disease it causes have raised societal, industry, and government issues in the area of public health.

HISTORY OF HIV RESEARCH

The history behind the AIDS epidemic and the discovery of HIV demonstrates the contentiousness that can result from new and frightening outbreaks. The AIDS disease and even HIV infection that has not developed into disease have engendered fear, political maneuvering, religious discourse, and misinformation. The AIDS epidemic arose in such a rapid and unexpected manner, in the United States, that public health officials may have been confused and disorganized as the numbers of deaths mounted.

The American author Laurie Garrett wrote in 1994's *The Coming Plague,* "The shock of the AIDS epidemic prompted many more virus experts in the 1980s to ponder the possibility that something new was, indeed, happening. As the epidemic spread from one part of the world to another, scientists asked, 'Where did this come from? Are there other agents out there? Will something worse emerge—something that can be spread from person to person in the air?'" The questions far outnumbered the answers, and although the AIDS pathogen emerged, in 1980 or 1981, in the United States, the medical community did not organize concerted efforts in research, diagnosis, treatment, or prevention until the mid-1980s. The first antibody test to detect HIV in the blood became available in 1985. By then, more than 20,000 cases of AIDS had been reported to the World Health Organization (WHO), with perhaps many times that number infected. Some of the most basic biology of AIDS has been revealed only recently; desperate attempts at diagnosis and therapy emerged first.

The Centers for Disease Control and Prevention (CDC) has proposed a subspecies of a chimpanzee native to western equatorial Africa as the probable source of HIV. The CDC Web site has explained, "The virus most likely jumped to humans when humans hunted these chimpanzees for meat and came into contact with their infected blood. Over several years, the virus slowly spread across Africa and later into other parts of the world." Modern DNA analysis methods determined that the earliest case of HIV occurred in a man in Republic of

Congo, in 1959. From that point, the virus spread to North and South America, Europe, and Australia.

In 1981, doctors in New York City, Los Angeles, and San Francisco noticed an increase in a rare skin cancer called Kaposi's sarcoma, as well as higher incidences of a lung infection caused by the obscure protozoan *Pneumocystis carinii.* Neither of these infections would have been commonplace in the young men who were visiting the doctors' offices. Medical records indicate that a few cases of these illnesses had occurred in New York and Los Angeles, in the 1960s, without raising suspicion.

The year 1981 marked the beginning of a steady increase in the number of men consulting doctors for Kaposi's sarcoma, *P. carinii,* and other rare maladies. Information from medical advisers would be replaced, six months later, with new, completely opposite medical advice. The nation understood only that the infection seemed to spread among homosexual men. Newspapers referred to the disease as gay-related immune deficiency (GRID).

By 1983, researchers identified blood as the source of an infective virus in humans that killed the body's immune system. The news filled with stories of "killer blood." The new infection caused a curious pair of responses. In some quarters, GRID caused concern of almost panic proportions, yet the U.S. government all but ignored the growing problem. President Ronald Reagan would delay referring to the disease by name—called AIDS since 1982—for a full five years after the CDC became aware of it and more than 2,000 Americans had died.

In 1983, virologists at the Pasteur Institute in France isolated a virus from lymph nodes and proposed in *Science* magazine that this new virus caused AIDS. The researchers, led by Françoise Barré-Sinoussi, named it *lymphadenopathy-associated virus* (LAV), because of its harm to the lymphatic system. In May of that year, the National Cancer Institute (NCI) in the United States announced that it had found the true pathogen in a blood sample, a virus they called HTLV—for human T-cell lymphotropic virus—the third HTLV that had been discovered. The *New York Times* reported, "Evidence of an unusual virus has been found in up to one-third of blood samples from some victims of acquired immune deficiency syndrome, or AIDS, an immunological disorder that has killed about half of the 1,352 people known to have acquired it, according to the Federal scientist in charge of investigating the epidemic." The epidemiologist James Curran was the government scientist, and an NCI team led by Robert Gallo possessed the virus in their laboratory; they had isolated it from the blood of a diseased patient.

At the same time the United States announced the potential discovery of the AIDS virus, Barré-Sinoussi and Luc Montagnier of the Pasteur Institute thought they had the correct, different virus. A bitter dispute would begin over the AIDS virus discovery. Each side accused the other of faulty science and even of data based on a contaminant. Years later, Gallo would admit that the virus in his studies was probably contaminated, but he remained dedicated to finding a cure.

The competition between the two groups calmed slightly, as AIDS research progressed in hundreds of laboratories worldwide. In truth, once the virus became known, these diverse laboratories began sharing information on developing blood detection tests, defining the immune reactions against HIV, and examining the structure of the virus. With each improvement in technology, scientists accumulated more information on HIV.

In 1987, the United States and France signed an agreement that gave each country an equal share in royalties from an AIDS detection test for blood. One year later, Gallo and Montagnier coauthored an article in *Scientific American* on the knowledge about HIV, to that point. They recounted the early years of HIV research, writing, "By the mid-1970s no infectious retroviruses had been found in human beings, and many investigators firmly believed no retrovirus would ever be found. . . . Many excellent scientists had tried and failed to find such a virus." Work on HIV slowly accelerated, from that point onward, through government and private funding to research laboratories.

Also in 1987, a group of researchers detected a similar retrovirus in AIDS patients in West Africa. The two viruses contained similar genomes and structures. The HIV isolated from lymph nodes at the Pasteur Institute became known as HIV-1 and the West African virus was called HIV-2.

In 2008, Luc Montagnier and Françoise Barré-Sinoussi were recipients of the Nobel Prize in medicine for their work in discovering the AIDS virus. Somewhat surprisingly, the Nobel committee excluded Robert Gallo. The *New York Times* reported, "Dr. Gallo told The Associated Press on Monday that it was 'a disappointment' not to have been honored with the French team. Later, Dr. Gallo issued a statement congratulating this year's Nobel Prize winners and said he 'was gratified to read Dr. Montagnier's kind statement this morning expressing I was equally deserving.'" A generation of scientists had devoted extraordinary energy to the HIV research that culminated in the Nobel Prize.

THE VIRUS

HIV is a retrovirus, a type of virus that carries ribonucleic acid (RNA) as its genome and infects human cells in order to make DNA. The genome is the entire collection of genetic material of a cell or virus.

HIV Surface Structures

Structure	Composition	Activity
gp120	glycoprotein (protein with sugars attached; molecular weight of 120,000)	grasps a receptor site on the surface of a specific immune cell, facilitates virus entry
gp41	glycoprotein	supports the gp120 binding site
major histocompatibility complex, class I (MHC-1)	protein	mimics structure of human cell surface to prevent immune response to the virus
MHC-2	protein	same as MHC-1

Retroviruses also carry an enzyme called reverse transcriptase that controls the process of making viral DNA out of RNA. This enzyme is referred to as reverse because the normal direction of genome replication is from DNA to RNA in a process called *transcription*.

The main body of the virus, called the *virion*, measures about 100 micrometers (μm) in diameter. The virions are spherical and possess an outer surface made of a lipid bilayer, a layered surface in which the hydrophilic (attracted to water) ends of fatty compounds (lipids) point toward the outside and the interior of the virion and hydrophobic (repelled by water) point into the bilayer's interior. The virus also contains a variety of compounds dispersed across its outer surface, and these compounds are instrumental in helping the virus infect human cells. The normal HIV structure contains 72 of these surface compounds, which spike out from the virion. The table describes the main surface compounds that play one or more roles in HIV infection.

The surface glycoproteins gp120 and gp41 bind together to form a spike. Three gp41 strands attach to each other, forming an appendage that juts out from the virion surface, and three gp120 molecules attach to the most distal end of these strands. The gp120 structure has received considerable attention in AIDS research because of its crucial role in HIV infection. Researchers have proposed that gp120 carries out at least the three following steps in infection.

- searches for specific immune system cells called T-lymphocytes (T cells) that contain a surface protein called CD4
- attaches to the CD4 receptor site
- helps inject viral contents into the T cell

The relationship between HIV's gp120 and the T-cell's CD4 is one of the most critical stages in developing AIDS other than the initial infection of a healthy person by the virus. This association, discussed in further detail later, has been the target of research into the following subjects in HIV treatment:

- immune responses to HIV
- natural HIV immunity
- HIV resistance to the immune system or to antiviral drugs
- receptor site blocking
- vaccines

Much of the resistance that HIV has toward the immune system and against AIDS drugs results from the ability of gp120 to add or remove sugar subunits. This ever-changing appearance prevents immune cells or drugs from recognizing the HIV.

The interior of the virus contains a simple triangular or cone-shaped core. The core contains two main components: (1) two strands of RNA and (2) at least two different proteins that protect the inner core and the RNA. Protease enzymes reside just outside the core and break down the outer viral coat and the inner core's coat to release HIV contents into a human cell.

In 2006, a news release written by Wai Lang Chu for *Drug Researcher* said that a HIV research team "consisting of English and German scientists, took a vast number of images of the virus, which is 60 times smaller than red blood cells, and used a complex computer program to combine them." Studies such as this one are providing additional insight into HIV structure and function, the relationship between the way a virus is built and the way it carries out its activities.

HIV researchers have tried to find the history of the virus and the way it first entered the human popu-

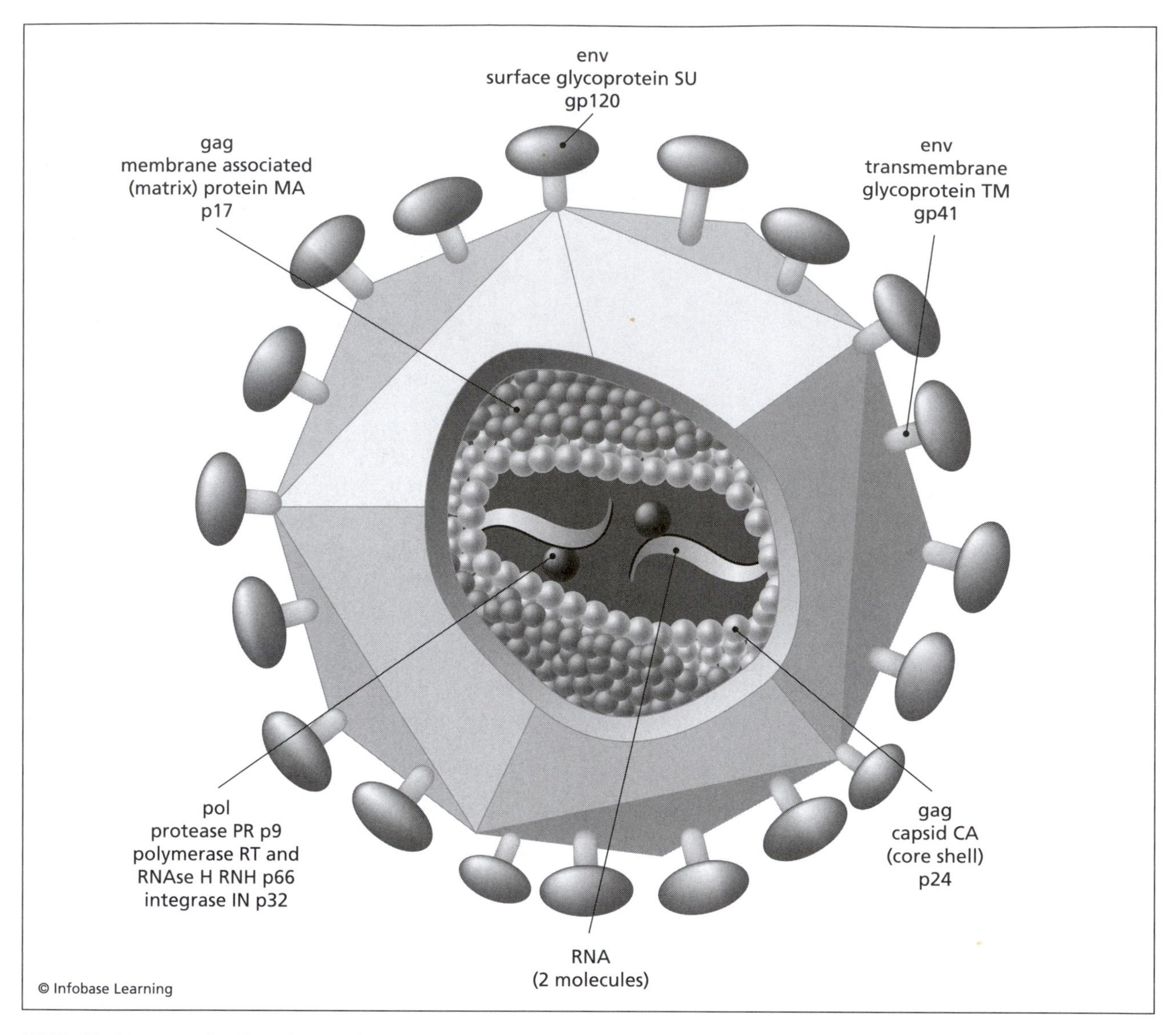

HIV holds 72 external spikes that are formed from two main proteins: gp120 and gp41 (Gp = glycoprotein). The spikes bind with a glycoprotein called CD4 on the surface of the immune system's T cells before invading these cells and possibly other cells containing CD4.

lation. Most primates (apes, chimpanzees, and monkeys) carry a simian immunodeficiency virus (SIV) that relates closely to HIV but causes no symptoms in the animals. In the mid-2000s, molecular biologists who compared human and chimpanzee chromosomes found evidence of different defense mechanisms against the same retrovirus. Millions of years ago, humans probably had superior resistance to an ancient form of retrovirus to which primates were susceptible. As modern primates evolved, their defenses to certain retroviruses improved. During the same time period, the genetic studies showed that humans lost an antiviral protein that had protected them, eventually leaving humans more susceptible to modern versions of HIV.

Despite discoveries related to HIV's history, the genetic development of HIV into today's pathogen remains speculative. In 2009, the University of Alabama-Birmingham medical researcher Beatrice Hahn said, "If we could figure out why the monkeys don't get sick, perhaps we could apply that to people." Hahn led a nine-year study on SIV carriers that found that the infected animals' death rates were 10 to 16 times higher than those of uninfected animals. Necropsies on the dead animals showed unusually low T-cell white blood cells, the same as found in human AIDS (acquired immunodeficiency syndrome) patients. The particular strain of SIV that had been the focus of this study was almost identical to HIV. Daniel Douek of the National Institute of Allergy and Infectious Diseases said of ther finding, "From an evolutionary and an epidemiological point of view, these data can be regarded as a 'missing link' in the history of the HIV pandemic." Learning more about the way HIV has adapted to survive in different hosts might reveal opportunities for blocking the virus's infective mechanisms.

CD4 Cells

CD4 cells are any cells in the human body that contain CD4 proteins on their surface. HIV can infect any cell that contains CD4 proteins, but most of its damage results from attacking the white blood cells called *T-helper lymphocytes* (or T-4 cells, CD-positive cells, or helper cells). T-4 cells contain CD4, which offers the virus a type of "docking station" to start an infection of the cell. As HIV infects more and more of the body's T-4 cells that are specifically made for fighting infection, the numbers of these cells start to decrease in the blood. Every type of CD4 is made by the body to fight different types of infectious agents, whether they are bacteria, viruses, or other microorganisms. When HIV reduces the number of CD4-containing cells, especially T-4 cells, the body progressively loses its ability to repel infections by other pathogens. Patients who have developed AIDS are at risk for contracting any number of secondary infections, meaning infections that follow the initial HIV infection. Doctors monitor the white blood cell count and the T-4 cell count when studying the progress of an HIV infection into AIDS.

Normal CD4 cell counts are between 500 and 1,600 cells per cubic millimeter (mm^3) of blood. In people infected with HIV, the count can drop to 0 per mm^3. In this precarious health situation, the secondary infections that attack the body may also be caused by opportunistic pathogens. These microorganisms are normally benign on the human body, but given the right set of circumstances, such as lowered T-4 cell counts, they, too, can cause a dangerous infection—they take the opportunity given to them.

CD4 counts can vary from person to person and in the same person if fighting a minor cold or allergy. Clinical microbiologists, therefore, prefer to express CD4 cell counts as a percentage of a person's number of total lymphocytes. The CDC defines AIDS as the condition in which a person has a CD4 cell count of less than $200/mm^3$ or a CD4 percentage of less than 14 percent. Doctors usually begin anti-HIV procedures (called antiretroviral therapy [ART]) if a patient's blood begins approaching a low CD4 percentage rather than wait for the HIV-infected person to reach the CDC's parameters.

TRANSMISSION AND INFECTION

In medicine, transmission is the passing of a disease from one person or animal to another. Transmission of HIV-1, the variant that causes disease in humans, occurs by three main mechanisms: (1) sexual contact, (2) blood transfusions, or (3) perinatal (mother-to-fetus) transfer. Because of the sexual contact mode of transmission, AIDS is one of many sexually transmitted diseases (STDs).

HIV-1 also belongs to a group of viruses called *lentiviruses.* These viruses take a long time to develop a disease; the infections are often called *slow virus diseases.* HIV usually remains latent or dormant (not active) for one to 10 years. A person known to harbor HIV but not yet showing symptoms of AIDS is said to be in a *latency period.* During the latency period, the virus is scarce in blood but persists in the lymph nodes.

In the early history of the world AIDS epidemic, no one understood the characteristics of the virus and how it was transmitted by infected to healthy individuals. Many questions remain open about the full nature of HIV transmission, but more than two decades of the epidemic plus extensive research have accumulated the following points about transmission:

- occurs through sexual contact, blood transfusions, and use of infected needles
- can be transmitted from mother to fetus (perinatal transmission) or in breast milk
- highest HIV concentrations are in (in order from most to least) blood, semen, vaginal fluids, breast milk
- saliva, sweat, tears, urine, or feces does not carry HIV

Inside the body, HIV follows the conventional processes of infection carried out by other viruses. After the gp120 spikes of the virion connect with CD4 protein on a T-4 cell, the virus's outer jacket, called the *envelope,* fuses with the cell's membrane. The HIV core enters the cell cytoplasm. Viral proteases dissolve the core to release the viral RNA, and then the HIV's reverse transcriptase assembles double-stranded DNA (dsDNA) from the information contained in the HIV RNA. The RNA-to-DNA is a two-step process, in which, first, the enzyme RNA/DNA-dependent DNA polymerase copies the single strand of RNA into a single strand of DNA, and, second, the reverse transcriptase builds a dsDNA from the single strand. At this point, the HIV genome is ready to take the critical step of infiltrating the human cell's genome.

The new dsDNA moves across the nucleus membrane by a mechanism that has not been fully explained yet in HIV research. Once inside the nucleus, an enzyme carried by the virus called *integrase* inserts the HIV DNA into the host cell's DNA. This complex of viral dsDNA and the viral integrase that infiltrates the host DNA is called the HIV *provirus.* Although no new viruses have yet been made, this step leads to the inevitable production by the T-4 cells of new viruses. As a result, the amount of HIV increases in the body, and the amount of T-4 and other CD4-containing cells decreases. An infected person who does not receive ART will be likely to progress from an HIV infection to AIDS.

Part of the confounding features of HIV is contributed by the great variability of the progress of HIV infection. Latency periods can vary from several months to more than a decade. The HIV-1 type that infects humans also follows an inconsistent course, once it infects the body. Humans experience four main outcomes from HIV-1 infection.

AIDS

The first form of AIDS was originally called *AIDS-related complex* (ARC), in the 1980s. ARC causes mild symptoms such as fever, fatigue, headaches, rashes, diarrhea, weight loss, and lymph node enlargement. Infected persons might contract opportunistic infections such as oral candidiasis caused by the yeast *Candida*. ARC persists for a few weeks and recurs periodically. In some people, ARC develops into a full case of AIDS. Other individuals continue suffering milder or more serious symptoms that persist for years without turning into AIDS. The term *ARC* has been replaced, since 2000, with the name *chronic symptomatic HIV infection* to include HIV-positive individuals who have the virus for many years without developing AIDS.

The second and most prevalent form of AIDS is a full case with classic symptoms and secondary infections. AIDS patients often experience a period after infection but before symptoms appear when HIV infection is latent. This period is called asymptomatic HIV infection and lasts as long as 10 years after the initial infection if the person was infected only once. Anti-HIV-1 antibodies appear in an infected person's blood during this period, but the virus persists in the body by hiding inside T-4 and other CD4 cells and accumulating in the lymph nodes. CD4 cells may actually rise in the blood, initially, in response to the infection, then, when the virus starts replicating, the CD4 cell numbers drop dramatically. Other components of the human immune system depend on CD4 cells for activity, and so a decline of the entire immune system ensues.

With immunity lost or severely damaged, a person has poor defense against other pathogens. Most disturbingly, AIDS patients cannot repel microorganisms that the healthy body defends against every day. A litany of opportunistic infections can occur in AIDS patients, and these infections, rather than HIV itself, debilitate the body and very often lead to death. The table lists the most prominent opportunistic infections that occur in HIV infections.

A third manifestation of AIDS attacks the central nervous system (CNS). When HIV-1 infects immune system cells called macrophages, it can enter the CNS, because macrophages cross the blood-brain barrier, which blocks most other cell types. Patients suffer from headaches, fever, and symptoms collectively known as *AIDS-dementia complex:* abnormal reflexes, irregular muscle action, and cognitive difficulties such as memory loss, mood disorders, and inability to concentrate. All of these symptoms worsen, over time.

The fourth consequence of AIDS can be cancer. Various cancers become common in AIDS progression. The main AIDS-related cancers are:

- Kaposi's sarcoma
- rectal or cervical cancer
- oral cancers
- lymphomas

Some of these cancers are associated with secondary infections. For example, a herpes virus known as human herpesvirus 8 (HHV-8) is almost always present with Kaposi's sarcoma. AIDS that has received no treatment can lead to any of the four types described here with a larger list of more subtle symptoms.

In 2008, the Joint United Nations Programme on HIV/AIDS (UNAIDS) published the statistics, to date, on the infection and disease. Various therapies have been developed for AIDS, and the disease has changed in its demographics, that is, the places and the populations where it is prevalent. People in the United States often misinterpret these changes as indications that AIDS has been defeated. In fact, there still is no complete cure for AIDS. UNAIDS published the following statistics on the global status of this disease:

- Since 2007, there have been about 2.7 million new HIV infections and 2 million HIV-related deaths per year.
- The global percentage of adults living with HIV has leveled off, since 2000.
- New HIV infections are increasing in the Russian Federation, parts of Eastern Europe, and Indonesia.
- Sub-Saharan Africa accounts for 67 percent of all HIV infections and 75 percent of AIDS deaths.
- About 50 percent of women are living with an HIV infection.

Opportunistic Infections Associated with AIDS

Microorganism Type	Opportunistic Pathogens
bacteria	• *Mycobacterium avium* • *M. tuberculosis* and other species • *Salmonella*
protozoa and cysts	• *Coccidioides immitis* • *Cryptococcus neoformans* • *Cryptosporidium* • *Cyclospora* • *Histoplasma capsulatum* • *Isospora* • *Toxoplasma*
viruses	• cytomegalovirus • herpes simplex
yeasts and other fungi	• *Candida* • *Pneumocystis carinii*

- Since the beginning of the global epidemic in 1980, 25 million people have died of AIDS.
- New HIV infections now outnumber AIDS deaths.

Some of these statistics suggest that the world has turned a promising corner toward getting control of AIDS. While the percentage of the world's population living with HIV has leveled off, the overall increase in population means that the number of people, including children, living with HIV is increasing. Some portions of the population have entered a very high-risk health situation because of HIV. "In countries most heavily affected," the UNAIDS report *08 Report on the Global AIDS Epidemic* said, "HIV has reduced life expectancy by more than 20 years, slowed economic growth, and deepened household poverty. In sub-Saharan Africa alone, the epidemic has orphaned [lost one or both parents] nearly 12 million children aged under 18 years." HIV infection this profound has shifted the age, income, and other demographics in entire populations. Many scientists have called AIDS the "modern Black Death" to compare it to the massive plagues that afflicted Europe in the Middle Ages.

ANTI-HIV TECHNOLOGIES

Three types of antiviral drugs are available to fight HIV infection: (1) reverse transcriptase inhibitors that work at the DNA synthesis step (called *nucleoside reverse transcriptase inhibitors*), (2) reverse transcriptase inhibitors that interfere with the building of DNA from the virus's RNA molecule (called *nonnucleoside reverse transcriptase inhibitors*), and (3) protease inhibitors that inhibit the enzyme the virus needs to build new virions. Azidothymidine (AZT) entered the market as the first AIDS drug, in 1987, as a reverse transcriptase inhibitor. But AZT caused dangerous side effects. Other drugs have since been developed to destroy HIV in the body or at least prevent the HIV in the body from replicating.

People who are infected with HIV, called *HIV-positive,* have a variety of drugs at their disposal. No single treatment regimen rids the body of HIV; doctors develop unique drug combinations for a patient's individual needs. From the dozens of medications that AIDS patients took earlier in the epidemic, drug companies are now developing a single pill that will work for a wide variety of AIDS sufferers and HIV-positive people. The table below provides an overview of antiviral drugs that have been tried in HIV treatment.

Some of the drugs in the table work best in combination. In addition, drug companies have offered new classes of drugs in the following categories: integrase inhibitors, virus entry inhibitors, and multiclass combination drugs.

Twenty years of research have provided a large volume of information on HIV, yet this virus remains enough of a mystery to confound attempts at eliminating it from the body. HIV infections and the disease that HIV causes have developed into one of the world's most critical public health concerns. AIDS

Antiviral Drugs for HIV Infection or AIDS

Type	Generic Name
nucleoside reverse transcriptase inhibitors	abacavir azidothymidine didanosine emtricitabine lamivudine stavudine zalcitabine zidovudine
nonnucleoside reverse transcriptase inhibitors	delaviridine efavirenz etravirine nevirapine rilpivirine (experimental)
protease inhibitors	amprenavir indinavir nelfinavir ritonavir saquinavir

continues to kill millions of people worldwide, each year, so HIV research will be a topic of study in virology for years to come.

See also EPIDEMIC; INFECTION; LATENCY; OPPORTUNISTIC PATHOGEN; VIRUS.

Further Reading

AIDS.org. Available online. URL: www.aids.org. Accessed August 22, 2009.

Altman, Lawrence K. "Rare Virus May Have Link with Immunological Illness." *New York Times*, 1 May 1983. Available online. URL: www.nytimes.com/1983/05/01/us/rare-virus-may-have-link-with-immunological-illness.html?&pagewanted=1. Accessed August 22, 2009.

Barré-Sinoussi, Françoise, et al. "Isolation of a T-Lymphotropic Retrovirus from a Patient at Risk for Acquired Immune Deficiency Syndrome (AIDS)." *Science* 220 (1983): 868–871. Available online. URL: www.sciencemag.org/cgi/content/abstract/220/4599/868. Accessed August 22, 2009.

Bazell, Robert. "Dispute behind Nobel Prize for HIV Research." NBC News commentary, October 6, 2008. Available online. URL: www.msnbc.msn.com/id/27049812. Accessed August 22, 2009.

Why AIDS Is Not Going Away

by Carlos Enriquez, Ph.D., Chabot College, Hayward, California

The first documented cases of acquired immunodeficiency syndrome (AIDS) occurred in Africa, in the late 1950s, in Kinshasa, Republic of Congo, in 1959, then, in 1976, in the Western Hemisphere. However, the AIDS epidemic did not became apparent until 1981, in the Los Angeles area, where unusual severe infections caused mostly by fungi were observed in young male homosexuals. It took researchers two more years to identify the causative agent, a retrovirus, which was named the *human immunodeficiency virus* (HIV). This new virus is believed to have originated in monkeys in Africa, around 1930.

AIDS became a pandemic, soon affecting most regions of the world and, more severely, sub-Saharan Africa. The modes of transmission, in order of importance, from highest to lowest are male-to-male sexual contact, intravenous drug use, and heterosexual contact. HIV destroys a very important type of cell of the immune system, the T-helper lymphocyte, thus leading to immunosuppression. Developing a vaccine against HIV has been very challenging because this virus has the ability to change rapidly. In addition to the practice of safe sex and the use of clean syringes by intravenous drug users, combatting HIV involves the use of many drugs simultaneously, in an effort to prevent the virus from becoming resistant to these antiviral drugs. HIV/AIDS is a sober example of how vulnerable we are when encountering a new pathogenic microorganism. Only a determined and educated society would have a chance of fighting successfully such a formidable enemy.

Why is AIDS not going away? In 1981, the *New England Journal of Medicine* published an article authored by Michael S. Gottlieb, a 33-year-old assistant professor at the University of California–Los Angeles (UCLA) Medical Center, describing a hitherto unknown syndrome in otherwise healthy male homosexuals. The syndrome was characterized by the presentation of unusual fungal and viral infections consistent with an impaired immune system. The condition was named *acquired immunodeficiency syndrome* (AIDS). Soon the condition developed into a pandemic, spreading all over the world. Even before the identification of the cause of AIDS, epidemiologists determined that the disease was mainly transmitted by male-to-male sexual contact and by exposure to blood from those who have AIDS. The early insight from epidemiologists resulted in the first AIDS control guidelines, such as the practice of safe sex and avoidance of exposure to blood. However, fear and ignorance, in the early years of AIDS, led to extreme actions that often discriminated against those afflicted by the disease, including children. Frantic work by scientists in many countries began, trying to identify the cause of AIDS.

In 1983, a group of researchers in France led by Luc Montagnier identified the virus that causes AIDS. This agent, a retrovirus, was later dubbed *human immunodeficiency virus* (HIV). When, in 1983, HIV was identified as the virus that causes AIDS, many scientists thought that the development of a vaccine to prevent the disease would soon follow. Twenty-five years later, the availability of a vaccine against HIV is still elusive. Why? To answer this question, we need, first, to understand the nature of HIV. This virus is believed to have originated in Africa. The first documented human cases of HIV/AIDS occurred in Kinshasa, Republic of Congo, in Africa, in 1959. But the virus is likely to have caused the first cases of AIDS in isolated small villages in Africa, in the 1930s, when it began to spread slowly to other places. Most scientists agree that HIV originated from the *simian immunodeficiency virus* (SIV) in Mangabey monkeys that passed it to chimpanzees, and these in turn to humans, through a series of mutations. Humans probably were first exposed to the virus through handling, preparing, and consuming monkey meat. The first case of AIDS

Borenstein, Seth. "HIV's 'Missing Link' Detected in Ailing Chimps." *San Francisco Chronicle*, 23 July 2009. Available online. URL: http://articles.sfgate.com/2009-07-23/news/17219585_1_chimps-virus-infected. Accessed January 29, 2010.

Centers for Disease Control and Prevention. "HIV/AIDS." Available online. URL: www.cdc.gov/hiv. Accessed August 22, 2009.

Gallo, Robert C., and Luc Montagnier. "AIDS in 1988." *Scientific American*, October 1988. Available online. URL: www.scientificamerican.com/article.cfm?id=aids-in-1988-gallo-montagnier. Accessed August 22, 2009.

Guyader, Mireille, Michael Emerman, Pierre Sonigo, François Clavel, Luc Montagnier, and Marc Alizon. "Genome Organization and Transactivation of the Human Immunodeficiency Virus Type 2." *Nature* 326 (1987): 662–669. Available online. URL: www.nature.com/nature/journal/v326/n6114/pdf/326662a0.pdf. Accessed August 22, 2009.

Internet Pathology Library. "AIDS Pathology." Available online. URL: http://library.med.utah.edu/WebPath/TUTORIAL/AIDS/AIDS.html. Accessed August 22, 2009.

Joint United Nations Programme on HIV/AIDS. "08 Report on the Global AIDS Epidemic." Geneva, Swit-

in the Western Hemisphere was reported, in 1979, in a sailor who frequented the coasts of Africa. By the mid-1980s, AIDS had become a major public threat worldwide. In 1988, most recorded cases of HIV affected men in the United States. Today, however, HIV is found all over the world, and half those living with HIV are women. According to the 2007 AIDS epidemic update, published by the World Health Organization (WHO), every day more than 6,800 persons are infected with HIV and more than 5,700 persons die of AIDS. The most affected area in the world is sub-Saharan Africa, where, in 2007, approximately 22.5 million adults and children were living with HIV, 1.7 million became infected, and 1.6 million died of AIDS.

Retroviruses, including HIV, have a peculiar replication strategy; the genetic material of these viruses consists of two identical copies of singled-stranded ribonucleic acid (ssRNA), which are transcribed into complementary deoxyribonucleic acid (cDNA) by the enzyme reverse transcriptase (RT), which is carried by the virus itself. The cDNA, then, integrates into a chromosome of the infected cell, where it may remain dormant for long periods, or can take over the metabolic machinery of the cell to manufacture more viruses, resulting in the destruction of the infected cells. HIV has a marked predilection to infect a very important subset of cells of the immune system, the helper T cells. As the infection progresses and more helper T cells are destroyed, the immune system starts to collapse, resulting in AIDS. HIV is a virus that mutates readily. This is mainly due to the nature of the HIV enzyme RT, which is relatively inefficient in the process of copying the viral RNA into DNA. RT sometimes causes mutations, when it incorporates the wrong nucleotides to the new DNA molecule during the transcription process. As a result, the HIV offspring may be composed of many mutant viruses with as many new traits. This characteristic of HIV has represented a formidable challenge to scientists who have tried to develop vaccines, or drugs against HIV. The U.S. Food and Drug Administration (FDA) lists more than 200 HIV-vaccine related clinical trials; unfortunately, a safe and effective vaccine against HIV has not been found yet. A recent and promising trial had to be terminated prematurely because it appeared that the vaccine itself made those who received it more susceptible to HIV infection. Although the process of developing a vaccine against HIV has been mostly unproductive, the development of drugs against the virus has been more rewarding. Currently, a person infected with HIV may be able to live a relatively normal life if given the proper antiretroviral therapy, but this therapy is costly and complicated. The ability of HIV to evade the immune system, by constantly changing by mutation, also poses a problem with the use of antiretroviral drugs. HIV often becomes resistant to the drugs after a period of use. To reduce the chance of HIV's becoming resistant to a single drug, several antiretroviral drugs are given simultaneously. This approach, called highly active antiretroviral therapy (HAART), may include the combination of several drugs. Patients under HAART treatment often have to take up to 40 pills every day. Although prevention should be one of the best strategies to prevent HIV infections, it is a formidable challenge. Controlling AIDS would involve engaging the society in a number of social and developmental issues, including education, drug use, and homophobia.

See also EPIDEMIOLOGY; VIRUS.

Further Reading

Centers for Disease Control and Prevention. "HIV/AIDS." Available online. URL: www.cdc.gov/hiv. Accessed November 8, 2009.

Levy, Jay A. *HIV and the Pathogenesis of AIDS*, 3rd ed. Washington, D.C.: American Society for Microbiology Press, 2007.

U.S. Department of Health and Human Services. "HIV/AIDS 101." Available online. URL: http://aids.gov. Accessed November 8, 2009.

zerland: UNAIDS, 2008. Available online. URL: www.unaids.org/en/KnowledgeCentre/HIVData/GlobalReport/2008/2008_Global_report.asp. Accessed August 22, 2009.
Lang Chu, Wai. "3D HIV Structure Reveals New Insights." January 26, 2006. Available online. URL: www.drugresearcher.com/Emerging-targets/3D-HIV-structure-reveals-new-insghts. Accessed August 22, 2009.
Russell, Sabin. "Ancient Viral Battle Opened Humans to HIV." *San Francisco Chronicle,* 22 June 2007.
Santora, Marc, and Lawrence K. Altman. "Rare and Aggressive H.I.V. Reported in New York." *New York Times,* 12 February 2005. Available online. URL: www.nytimes.com/2005/02/12/health/12aids.html?_r=2. Accessed August 21, 2009.
Shilts, Randy. *And the Band Played On: Politics, People, and the AIDS Epidemic.* New York: St. Martin's Press, 1987.

hybridization Hybridization is the action of combining genetic material from one organism with the genetic material of a different organism. For this reason, it can also be called *nucleic acid hybridization,* referring to the molecules that hold biology's genetic information: deoxyribonucleic acid (DNA) and ribonucleic acid (RNA). In microbiology, hybridization takes place between two different microorganisms, and microbiologists use three different types of hybridization: (1) DNA-DNA hybridization, (2) DNA-RNA hybridization, and (3) plasmid hybridization.

The geneticist Alexander Rich of the Massachusetts Institute of Technology developed the method for hybridizing DNA molecules, in 1960. During that time, the double-helix structure of DNA was still a new concept among even the world's preeminent scientists. In 2006, Rich recalled, "In the 1950s it was widely assumed that 'DNA makes RNA, RNA makes protein.' This was not based on experimental evidence that DNA and RNA could combine but was more in the nature of an intuitive belief. However, by early 1960 I was finally able to carry out a direct experiment, the first DNA-RNA hybridization." Rich and others used hybridization to clarify the structure of nucleic acids and their relationships to each other in genetic transfer. Hybridization studies now serve in the following additional ways:

- determination of relatedness of the genetic material from different microorganisms
- identification of unknown microorganisms by hybridizing with nucleic acids of known species
- determination whether two proposed species in the same genus are actually the same microorganism
- composition of new DNA to be transferred into a microorganism
- detection of a specific cloned gene in a method called *colony hybridization*

The relatedness between bases in the DNA of two different microorganisms or two different sections of the same DNA molecule is called *DNA homology.* In order to determine DNA homology, a microbiologist uses the hybridization method.

HYBRIDIZATION TECHNIQUES

Hybridization takes advantage of two characteristics of the double-stranded DNA molecule (dsDNA). First, when a dsDNA solution is heated, the two strands detach from each other, creating a solution of single-stranded DNA (ssDNA). This step is called *denaturation,* and the temperature at which DNA separates into two strands is called the *melting temperature.* Melting temperature varies slightly, depending on the origin species of the DNA, but, in general, it occurs at between 176 and 194°F (80–90°C). The second feature of DNA becomes apparent when the solution is cooled and held at a temperature of 77°F (25°C): The single strands reattach, in a process called *reannealing.* In reannealing or renaturation, the various single strands of DNA recombine along stretches of complementary bases: That is, adenine binds with thymine and guanine binds with cytosine. This recombination of complementary bases creates the foundation of homology studies. DNA-RNA combinations can be accomplished the same way, with two exceptions: (1) RNA is already a single-stranded molecule, and (2) RNA molecules pair adenine with uracil instead of thymine.

If the goal of DNA-DNA hybridization is to create a hybrid DNA molecule, a microbiologist adjusts the heating and cooling process to favor the binding of two unrelated DNA molecules. Strands with similar but not identical DNA base sequences bind, but they do not form as strong a connection as do two completely related strands. In this circumstance, the microbiologist might use cooler temperatures for denaturation and renaturation to help the hybrid's two strands stay together.

Homology studies on two presumably different microorganisms indicate whether the microorganisms are closely related or distantly related. This helps determine whether the two species evolved at approximately the same time or evolved at different times or along different paths. Two microorganisms are considered members of the same species if they meet two criteria in DNA-DNA hybridization:

They share at least 70 percent of the same DNA base sequences, and their melting temperatures differ by less than 5 percent.

Hybridization methods require standardized conditions that help the DNA strands denature and renature. Three main factors contribute to the success of a hybridization method:

- base composition—The dsDNA with a high percentage of guanine and cytosine pairs is more difficult to separate than dsDNA with a low percentage of these bases because guanine-cytosine contains a stronger chemical bond than adenine-thymine.
- chemical conditions—High levels of cations (positively charged elements) such, as sodium (Na^+), stabilize dsDNA, and other chemicals, such as urea, make dsDNA more prone to denature.
- strand length—Long dsDNA strands may require extra energy as heat to denature into two single strands.

In studies of guanine-cytosine (G + C) amounts, a high level would be expressed as 50–75 percent G + C, and a low level would be expressed as 20–40 percent G + C. The ratios of G + C to adenine-thymine (A + T) are characteristic of bacterial groups. For example, *Streptomyces* is a filamentous (string-like) bacterium with a high G + C content of about 70 percent, whereas *Staphylococcus* has a low G + C content of about 36 percent. When hybridization is being used to identify unknown bacteria, microbiologists often analyze the G + C content, as well. The G + C content alone cannot identify a species; hybridization remains the best method for determining the genetic makeup of a microorganism.

Microbiologists monitor the hybridization process by attaching a radioactive element such as phosphorus (^{32}P), carbon (^{14}C), or hydrogen as tritium (^{3}H) to the nucleic acid pieces. Thin filters made of the material nitrocellulose serve as a good platform for holding nonradioactive DNA strands. A microbiologist, then, incubates the filters immersed in a solution of strands tagged with the radioactive element. During incubation, the radioactive ssDNA has the chance to bind with complementary strands of ssDNA stuck to the filter. After incubation, the microbiologist rinses off the filter to remove any DNA that did not hybridize and, then, puts the filter, now holding radioactive dsDNA, into an instrument that counts radioactivity. (The filters must be immersed in a special solution that aids in the detection of radioactive emissions; an instrument called a *scintillation counter* detects each emission and records it.) The higher the amount of radioactivity collected on the filter, the greater the homology between two microorganisms.

Sometimes, two microorganisms share so few common DNA sequences that DNA-DNA hybridization does not work. In this case, scientists turn to RNA-DNA hybridization to study the relationship between distantly related species. RNA-DNA hybridization compares the base sequences in either ribosomal RNA (rRNA) or transfer RNA (tRNA) with the sequences in a molecule of DNA. Ribosomal RNA is a single strand in the ribosome organelles of cells that participates in protein synthesis. Transfer RNA is a folded RNA molecule that has some double-stranded sections and acts in protein synthesis by delivering amino acids to the protein-building mechanism in the cell. The genes in rRNA and tRNA tend to evolve more slowly than a cell's remaining genes. For this reason, RNA-DNA may be able to detect small degrees of relatedness between microorganisms that did not evolve at the same time or on similar paths.

The purpose of plasmid hybridization usually differs from that of DNA-DNA or RNA-DNA methods that mainly study the relationship between two different microorganisms. A *plasmid* is a small circular piece of DNA that is separate from a microorganism's main DNA. Plasmid hybridization involves the insertion of new genes into a plasmid to create a new trait in a cell. In other words, plasmid hybridization is part of genetic engineering, the process of combining genetic material from two different organisms to create a new organism. As a result of plasmid hybridization, a scientist can produce a new organism containing recombinant DNA that contains the organism's normal genes plus new genes that provide a certain benefit. The original DNA belonging to the recipient microorganism is called *native DNA* or native plasmid DNA, and the piece to be inserted into it is called *foreign DNA*.

Plasmids have become useful tools in gene transfer, the process of putting a new gene into an existing microorganism. Many bacteria in the environment or in a laboratory culture absorb plasmids from other microorganisms into their own cells. This process often serves the recipient cell well by giving it a new trait that improves its survival.

PROBES

A valuable invention in microbiology that has developed from hybridization technology is the nucleic acid probe. The nucleic acid probe is a type of biosensor that detects a specific DNA or RNA in mixtures on the basis of hybridization technology. Sometimes, the mixture can be very complex, such as soil or food, either of which contains many other microorgan-

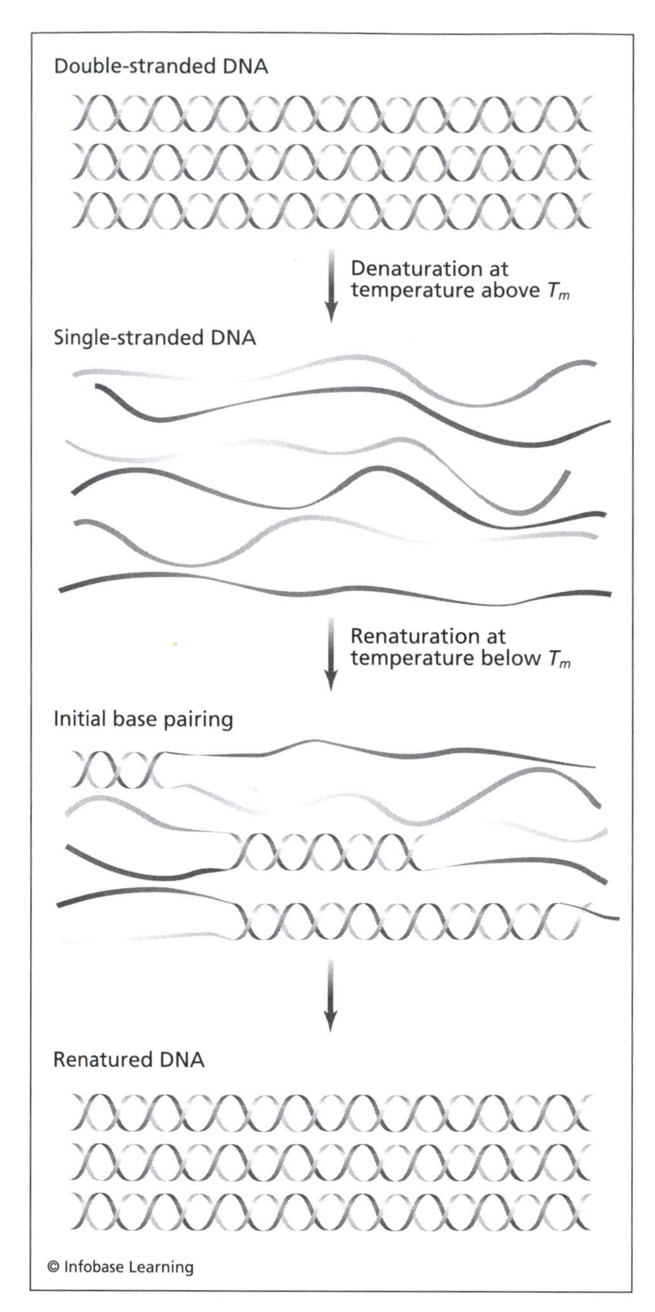

In DNA hybridization, double-stranded DNA is heated to its melting temperature, T_m. The strands separate, and then, when the suspension cools, complementary strands bind together, in a process called renaturation. By using the DNA from two different microorganisms in the mixture, a scientist can determine how closely the species are related by calculating percent renaturation.

isms. Because hybridization is very specific in finding nucleic acid base sequences, it is the best technique for finding a low concentration of a given organism within a mixture or other living and nonliving matter.

Nucleic acid probes may be referred to also as DNA probes or RNA probes, depending on the type of nucleic acid they contain, or simply as gene probes. Probes that detect gene sets with control over an entire process, such as nitrogen fixation (capture of nitrogen gas from the air), are called functional gene probes.

Companies now specialize in making gene probes containing specific genes from microorganisms such as *Escherichia coli, Staphylococcus, Salmonella,* and many other species in medical, food, and environmental microbiology. These companies also prepare made-to-order probes that contain a gene from a rare microorganism. DNA synthesis methods in laboratories have made possible this ability to craft a new piece of DNA with exactly the correct base sequence as found in the original microorganism. Elizabeth Marlowe, Karen Josephson, and Ian Pepper described the process in *Environmental Microbiology,* in 2000: "The basic strategy in the construction of a gene probe is to obtain the sequence of the target gene and then to select a portion of this sequence for use as a probe. The size of the probe can range from 18 base pairs to as many as several hundred base pairs. The probe is then synthesized and labeled in such a way that it can be detected after it hybridizes to the target sequence." Labels used for monitoring the probe in the mixture being analyzed may be radioactive molecules, antibodies, or other chemicals that give off a measurable signal.

As mentioned, radioactive labeling helps monitor a nucleic acid simply by measuring the radioactivity in a sample using a scintillation counter. An alternate method for detecting the radioactivity is autoradiography. In this technique, a microbiologist places a sheet of photographic film on top of the nitrocellulose filter and incubates them, for up to a week, in the dark. The radioactive emissions given off by labeled nucleic acid on the filter turn the exposed film dark. All places on the film not adjacent to a radioactive gene turn white.

Other labels avoid the use of radioactive materials altogether. Probe makers can incorporate a compound into the nucleic acid. This compound, then, will react with a second material, which emits a visual signal. For example, a fluorescent dye called fluorescein attached to a gene probe can be bound to a nitrocellulose filter. A scientist, then, exposes the study sample to the probe to allow complementary bases to bind. After the incubation period, the scientist further exposes the filter to an antibody specific for the fluorescein-containing molecule. When the antigen and antibody react, they cause the fluorescein to emit fluorescence, which can be detected by an instrument called a fluorometer. Other, similar labels include biotin reacting with streptovidin to produce luminescence and digoxigenin that reacts with an antibody to produce a purple color.

Gene probes have been valuable in detecting food-borne pathogens. In some cases, the probe has been used to identify the source of a pathogen, and, in other cases, probes help detect pathogens in food before they can cause infection. The U.S. Food and Drug Administration has explained, "The relatively short length of synthetic oligonucleotide [a small piece of DNA] probes means that they are specific for particular regions of DNA. There is only about 1 chance in 15,000 that a sequence length of 18 bases would appear more than once in the *E. coli* genome. With a 22-base probe, the chance drops to about 1 in 4 million." With this power, gene probes can find entities much more refined than a bacterial cell: They can search for particular strains of species or even the virulence genes in pathogens.

HYBRIDIZATION-RELATED METHODS

Scientists have developed techniques that gather more information from a single hybridization than simply whether a gene is present or absent. These techniques are colony hybridization, Southern hybridization (usually called a Southern blot), and Northern hybridization (or Northern blot).

Colony hybridization enables a microbiologist to find a specific gene sequence in a bacterial colony growing on an agar plate with other bacterial colonies. This process has been useful for finding clones of cells that contain a certain gene. In this method, a microbiologist presses a round filter paper on top of the plate's colonies to collect cells from each colony on the paper. The paper is, then, lifted off the plate; the papers with cells are sometimes called lifts. The following steps lyse (break apart) the cells and bind the cells' DNA directly to the paper. By using a gene probe, the microbiologist finds all the cells that contain the target gene. The original agar plate with the remaining colonies can be used as a source of pure colonies that contain the target gene.

Southern blotting provides a technique for finding a gene in a mixture of DNA pieces. The DNA pieces can be separated from each other in a jelly-like plane of material by using a method called gel electrophoresis. After the different-size pieces separate from each other in the gel, a scientist presses a special filter on top of the gel to collect all the pieces. This filter sheet can be analyzed with a gene probe to find the target gene.

The Northern blotting technique resembles the Southern blot, but it is used for RNA rather than DNA. These methods are called *blotting,* and the final results are a blot, because a scientist picks up nucleic acids by blotting them up as a sponge would absorb water from a floor. The Southern blot is named after its inventor, the British biologist Edwin M. Southern (1938–), who developed the method, in 1975. Two years later, the Stanford University biologists James Alwine, David Kemp, and George Stark named their method for studying RNA the *Northern blot* to distinguish it from the Southern.

Hybridization techniques have been among the most useful products of genetic engineering. In turn, they have been helpful in the development of very specific probes that can find a small amount of substance in a large volume of material. Hybridization combines the skills of whole-cell microbiology with new methods in molecular biology.

See also BIOSENSOR; GENETIC ENGINEERING; GENE TRANSFER; PLASMID; RECOMBINANT DNA TECHNOLOGY.

Further Reading

GeneProbes.org. Available online. URL: www.geneprobes.org. Accessed June 8, 2009.

Hill, Walter E., Atin R. Datta, Peter Feng, Keith A Lampel, and William, L. Payne. "Identification of Bacterial Food-borne Pathogens by Gene Probes." In *Bacteriological Analytical Manual.* Washington, D.C.: Food and Drug Administration, 2001. Available online. URL: www.fda.gov/Food/ScienceResearch/LaboratoryMethods/Bacteriological AnalyticalManualBAM/ucm072659.htm#authors. Accessed August 22, 2009.

Marlowe, Elizabeth M., Karen L. Josephson, and Ian L. Pepper. "Nucleic Acid–Based Methods of Analysis." In *Environmental Microbiology.* San Diego: Academic Press, 2000.

Rich, Alexander. "Discovery of the Hybrid Helix and the First DNA-RNA Hybridization." *Journal of Biological Chemistry* 281 (2006): 7,693–7,696. Available online. URL: www.jbc.org/cgi/reprint/281/12/7693. Accessed August 22, 2009.

Strachan, Tom, and Andrew P. Read. "Nucleic Acid Hybridization Assays." In Human *Molecular Genetics* 2. New York: Wiley & Sons, 1999. Available online. URL: www.ncbi.nlm.nih.gov/books/bv.fcgi?rid=hmg.chapter.457. Accessed August 22, 2009.

hygiene Hygiene is the observance of rules regarding cleanliness for the purpose of preventing infection and maintaining health. Hygiene can be grouped into industrial or occupational hygiene, community hygiene, and health hygiene. Industrial hygiene protects workers from infections they could receive on the job. In many cases, proper industrial hygiene helps manufacturers produce safe products that are free of contaminating microorganisms, so both workers and consumers receive protection. For example, industrial hygiene is very important in food manufacturing plants in order to assure customers that the food they buy will not make them sick. Community hygiene consists of the actions that protect the health of a large group of people in a town, state, or nation, with particular attention to communicable diseases. Communicable diseases are diseases that pass from one host to another. Health hygiene consists of two main areas: personal hygiene and oral hygiene, also called *dental hygiene.* Health hygiene includes all the activities that people do to maintain their body's health by preventing infection.

The activities that collectively make up a hygiene program are referred to as *hygiene practices.* In general, these practices involve the following topics: sanitation, cleaning, disinfection or sanitization, and general health-promoting activities such as regular medical care, good nutrition, adequate rest, and exercise. All of these health-promoting activities indirectly affect the risk a person has of contracting an infection. Regular monitoring of a person's health by a trained medical provider helps identify new infections before they develop into more serious conditions. Underlying diseases or preexisting health conditions increase the chance of infection. People who have these conditions are called *at-risk* or high-risk health individuals. At-risk groups have higher-than-normal risk of infection because their immune systems are already being challenged by the preexisting health condition. This immunocompromised condition gives infecting microorganisms better opportunities to invade the body. Nutrition, rest, and exercise support a healthy immune system, and this gives a person the best chance of fighting an infection.

ANTIMICROBIAL PRACTICES IN HYGIENE

The practices used in all categories of hygiene rely on sanitation, cleaning, and the use of microorganism-killing products called biocides. Biocides are chemicals that kill specific microorganisms known to be in a person's environment. The two main types of biocides are disinfectants and sanitizers, which are intended only for use on inanimate surfaces. Disinfectants kill all microorganisms except bacterial spores, such as the spores produced by *Bacillus* or *Clostridium.* Sanitizers kill a portion of all microorganisms to lower their numbers to safe levels. For example, assuming no bacterial spores are present, a disinfectant kills 100 percent of the microorganisms on a surface such as a bathroom sink, but a sanitizer may kill 999 of 1,000 microorganisms present on the sink.

Sanitation applies various activities to keeping an area safe from most dangerous microorganisms. In industrial and community hygiene, sanitation programs usually contain written steps that describe the actions people are to take for keeping an area clean and healthy. A typical sanitation program contains the following elements:

- regular trash removal
- secure trash storage until pickup
- safe drinking water supply
- adequate disposal for human wastes
- vermin and insect control
- regular cleaning to remove dirt, dust, allergens, and pollen
- regular monitoring for cleanliness
- periodic facility shutdowns for full cleaning of equipment
- medical care on call for emergencies
- adequate hand washing and showering facilities, changing rooms, and break rooms
- periodic monitoring for adequate ventilation

The World Health Organization (WHO) includes activities as part of sanitation that aid the overall well-being of communities. In other words, WHO combines antimicrobial activities with cleaning, monitoring, and maintenance of good health as all part of sanitation. The WHO has explained on its Web site: "Sanitation generally refers to the provision of facilities and services for the safe disposal of human urine and faeces *[sic].* Inadequate sanitation is a major cause of disease world-wide and improving sanitation is known to have a significant beneficial impact on health both in households and across communities. The word 'sanitation' also refers to the maintenance of hygienic conditions, through services such as garbage collection and wastewater disposal." This statement emphasizes the relationship between hygiene and sanitation.

Cleaning is a specific activity within any sanitation program. In hygiene, it can mean the cleaning of households or manufacturing equipment or personal cleaning by hand washing, showering, and dental care. Cleaning usually refers to the removal of gross dirt, that is, dirt that can be seen or felt.

Cleaning dates to the beginning of civilization, when a nearby body of water served as a washing place for a community. Soaplike materials have been found in the ruins of ancient Babylon of 2800 B.C.E. The Egyptians, in about 1500 B.C.E., invented soap that resembled modern soaps by combining animal tallow or vegetable oils with alkaline salts. The ancient Greeks and Romans used cleansing materials made of ash, sand, pumice, or clay physically to scrape dirt off the skin. Later in the Roman Empire, starting about 300 B.C.E., the Romans built public baths and introduced the use of herbal and flower oils as deodorants. The fall of Rome, about 470 C.E., led to a concurrent decline in hygiene.

From the fall of Rome through the Middle Ages, people did not make community or personal hygiene a priority for their well-being. As a consequence, the Middle Ages endured several devastating epidemics of bubonic plague that reduced populations by up to two thirds of the population before the plague. In the Middle Ages, several different hygiene mistakes combined to create perfect conditions for the *Yersinia pestis* pathogen to spread. Fleas carry this bacterium, and when fleas infest rats, the rats provide a constant source of the pathogen for infiltrating human populations. Because of the terrible hygiene through the 16th century, rat populations grew out of control in almost all large cities and smaller towns and carried the plague with them. People caught the disease when a flea bit them or, on rare occasions, from a rat bite. The hygiene mistakes made during this period of history were the following:

- garbage left in the open and not burned or buried
- sewage disposal in the open street
- unsafe water for drinking and cooking
- inadequate burial of the diseased
- dense living conditions in cities
- close association with domesticated and wild animals
- large populations of rats and other vermin

Soap makers made their products available in Europe as early as the seventh century, but soap alone could not counteract all the other flaws in hygiene. The modern formula for soap developed, in about 1810, when the French chemist Nicolas Leblanc (1742–1806) created formulas that balanced a mixture of fatty acids and glycerin. In the mid-1800s, the Belgian chemist Ernest Solvay (1838–1922) invented a soap making process using sodium chloride (table salt) to improve the solubility and cleaning power of soap. Soap has become an under-appreciated part of hygiene. Because it mixes with water and breaks up greases and oils, soap removes dirt from hard surfaces or from skin much better than water alone. Dirt on inanimate objects or on the body carries various microorganisms that can cause disease.

INDUSTRIAL HYGIENE

Industrial hygiene programs protect the safety of employees from hazards that occur in the workplace. These hazards may be chemicals, physical hazards, or biological hazards, such as pathogens. The following occupations expose workers to pathogens: doctors and nurses, hospital orderlies, phlebotomists, clinical microbiologists, and biotechnologists. These occupations require extra safety measures to prevent workers from being infected. Poor hygiene in health care settings contributes to nosocomial infections, which are infections due to hospital stays. In the case of health care professions, workers must avoid exposure to pathogens so that they do not contract an infection from patients. The main hazards among health care workers are the following:

- needle-sticks with needles containing pathogen-contaminated blood
- blood and body fluids sprayed from patients in trauma units or in surgery
- airborne transmission of pathogens from patients by coughing, sneezing, or talking
- direct contact transmission
- infection transmitted by bedding, eating utensils, or bedpans
- exposure to pathogens during identification procedures in clinical microbiology laboratories

In biotechnology, workers are more likely to be exposed to bioengineered microorganisms. These are microorganisms containing genes from other organisms.

Food microbiologists adapt industrial hygiene to their circumstances, which differ from those in the jobs listed earlier. In food microbiology, workers must ensure that food products do not contain microorganisms that spoil the food or cause illness. Manufacturers of sterile drugs, such as an injectable vaccine, also have strict industrial hygiene programs. The goal of hygiene in these industries is protection of the product.

Industrial hygienists help facilities develop programs for protecting workers and products. The main responsibilities of this profession are the following:

- investigating the workplace for potential hazards

- developing techniques to prevent exposure to pathogens
- developing environmental monitoring
- evaluating indoor air quality
- implementing government requirements on worker safety
- training employees in safety procedures
- supplying the appropriate safety equipment for workers
- developing emergency procedures in event of an infection
- developing emergency cleanup procedures for spills of pathogens
- overseeing proper disposal of biohazardous wastes (wastes containing pathogens or their toxins)

Environmental monitoring in the preceding list refers to the determination of harmful microorganisms in a person's surroundings. In order to perform the responsibilities listed, industrial hygienists work together with microbiologists to detect and eliminate pathogens.

COMMUNITY HYGIENE

Community hygiene covers all the activities that affect the overall health of a population. This discipline includes waste disposal, wastewater collection and treatment, drinking water distribution, insect and rodent control, and emergency health care for people who have been exposed to pathogens.

The plagues of the Middle Ages and other periods in history (ancient Rome, Greece, and China, and the plague of London) were the result of poor community hygiene. City or county public health departments have long since corrected the problems that gave rise to the plagues, yet epidemics still occur, and even the bubonic plague has not disappeared. Community hygiene requires constant diligence to prevent the spread of contagious diseases.

The WHO has identified the following main focus areas in global community hygiene:

- drinking water source protection
- proper solid waste disposal
- wastewater drainage
- control of animal wastes from farms and feedlots
- open market hygiene

Open air markets create special circumstances where pathogens can be transferred among many people and onto a variety of foods in a short period. The WHO has explained, "Markets often represent a health hazard because foodstuffs may not be stored properly and because markets may lack basic services, such as water supply, sanitation, solid waste disposal and drainage." Food vendors must understand hygiene basics to make open air markets a safe place to purchase food. For example, a vendor might rinse fruits with water to keep away flies. Water contaminated with pathogens, however, creates a health problem by putting dangerous microorganisms onto the food.

Farmers' markets and restaurants abide by the same principles in order to protect community health. Poor food handling spreads food-borne illness, so workers in the food industry must follow a number of steps in protecting food: proper refrigeration, proper cooking temperature, inspection of fresh produce for dirt, rinsing of all fresh vegetables and fruits, disposal of spoiled food, and insect and rodent control. Food handlers additionally must have excellent health hygiene habits in order to prevent spreading microorganisms from their bodies to food.

HEALTH HYGIENE

Health hygiene involves the personal hygiene activities of hand washing, bathing, and laundering clothes plus good dental hygiene. Health hygiene conducted properly protects a person from infection and protects others from the spread of pathogens between people.

Hand washing removes dirt and transient microorganisms from the skin. Transient microorganisms stay on the skin for short periods and do not make up the normal native microorganisms of skin. Pathogens can make up a portion of the transient microorganisms at any time, so people should always wash their hands in the following circumstances:

- after using a bathroom
- after diapering an infant
- after touching pets or livestock
- before eating or preparing meals

In order to remove transient microorganisms from hands, hand washing should involve the following:

- use of soap and warm, running water
- avoidance of hot water, which can burn the skin
- scrubbing all parts of the hand, including between the fingers and the fingertips
- washing for 20–30 seconds
- thorough drying to prevent chapping

The preceding steps show that hand washing involves more than a brief rinse—a University of Arizona microbiologist has determined most people rinse their hands for only four seconds. People should avoid scalding and chapping the skin because injuries to the body's protective skin barrier create an opportunity for infection. Regular bathing and clothes laundering both serve to reduce exposure to harmful fungi, bacteria, and viruses.

Washing also breaks a mode of passing pathogens from person to person called indirect transmission. In this type of transmission, a pathogen moves from one person to the next by way of an object such as a doorknob, gym equipment, or a towel. To break this transmission, a person should wash the hands as soon as possible after touching surfaces that others touch repeatedly.

Dental hygiene involves all of the preventive measures to combat disease or infection of the teeth and the mouth. The following actions make up the basics of good dental hygiene:

- regular brushing between the teeth
- regular use of floss
- removal of impacted food debris
- plaque removal by a dental hygienist twice a year

These dental hygiene activities reduce the chances of dental caries and periodontal disease, which are both caused by oral bacteria.

THE HYGIENE HYPOTHESIS

The hygiene hypothesis is a theory developed, in 1989, by the German pediatrician Erika von Mutius stating that regular exposure to microorganisms during youth helps build stronger immunity than little exposure to germs. As a specialist in childhood allergies, von Mutius sought a reason for increasing rates of asthma and allergies in industrialized nations. She compared the health status of youngsters living in clean houses with that of children who lived in tenements among vermin and dirty conditions. When von Mutius recorded fewer cases of asthma and allergies among the poor children, she concluded that a steady exposure to various microorganisms helps immune system development.

Von Mutius received little initial support of the hygiene hypothesis. The University of Arizona's Charles Gerba has opined that the hygiene hypothesis may lead to complacency about germs. "In some ways," he told the *Chicago Tribune* reporter Bob Condor, in 2001, "you can never be too clean." The microbiologist Stuart Levy of Tufts University has countered that the hygiene hypothesis makes sense. Levy said in the same article, "It's just like a child needs exercise to build strong bones and muscles. A child's immune system needs its own workout to develop a normal resistance to infections." Scientists have increasingly supported the hygiene hypothesis, but this area needs many more studies to determine the relationship between hygiene and immunity.

Hygiene plays a pivotal role in prevention of infection and contamination of food and drug products. Hygiene also is a critical part of health care so that hospitals and health care providers do not transmit pathogens from sick patients to healthy people.

See also DISINFECTION; FOOD MICROBIOLOGY; NOSOCOMIAL INFECTION; SANITIZATION.

Further Reading

American Industrial Hygiene Association. Available online. URL: www.aiha.org/Content. Accessed June 18, 2009.

Condor, Bob. "The Dirt on Being Skeaky (sic) Clean." *Los Angeles Times,* 30 April 2001. Available online. URL: http://articles.latimes.com/2001/apr/30/health/he-57460. Accessed June 19, 2009.

Gelfand, Erwin W. "The Hygiene Hypothesis Revisited: Pros and Cons." *MedScapeCME,* 10 April 2003. Available online. URL: http://cme.medscape.com/viewarticle/452170. Accessed June 19, 2009.

Soap and Detergent Association. Available online. URL: www.sdahq.org/index.cfm. Accessed June 18, 2009.

U.S. National Library of Medicine. "Germs and Hygiene." Available online. URL: www.nlm.nih.gov/medlineplus/germsandhygiene.html. Accessed June 18, 2009.

World Health Organization. "Sanitation." Available online. URL: www.who.int/topics/sanitation/en. Accessed June 18, 2009.

I

identification Identification includes all the activities used in microbiology for determining the species of an unknown microorganism. The importance of identifying microorganisms varies with different disciplines in microbiology. For example, environmental microbiologists may be interested in the actions of entire microbial communities, regardless of the species in the communities. By contrast, clinical microbiology requires that a microbiologist identify any potential pathogen in a patient's specimen quickly and accurately, so that a physician can prescribe an effective treatment.

Identification methods belong to three categories: (1) physical, (2) biochemical, and (3) genetic. Microbiologists call upon all three of these categories to identify unknown bacteria and fungi. Physical methods use the morphology (the appearance) of the microorganism by observing the cells under a microscope, using staining techniques, looking for distinctive external features such as flagella, and noting the morphology of colonies growing on agar. Biochemical tests look for the presence or absence of enzyme systems indicated by the ability to convert one substance to another substance. For example, a starch hydrolysis test determines the presence or absence of the enzyme amylase, which degrades starch into sugars. To test for this, a microbiologist grows the unknown microorganism on starch-containing agar. Iodine reacts with the intact starch molecule, so adding an iodine solution to the agar plate after incubation, causes amylase-positive colonies to become surrounded by clear zones indicating starch breakdown. The agar surrounding amylase-negative colonies turns blue. A battery of simple visual tests identifies bacteria to the genus or species level by process of elimination. The microbiologist accumulates test results and the physical features of the unknown cells and colonies and compares them with known microorganisms until finding a match.

Physical and biochemical methods inspect the phenotype of a microorganism, that is, the way the genes are expressed. Genetic analysis inspects a microorganism's genotype, its genetic makeup or the genes themselves.

Genetic identification developed within the last 20 years; physical and biochemical methods date back 100 years. Genetic analysis of microorganisms involves characterization of the genome, which is the entire collection of genetic material inside a cell or virus particle. This technology is advancing rapidly, and the analysis of deoxyribonucleic acid (DNA) and ribonucleic acid (RNA) has already begun to overtake older methods. Identification by genetic testing is changing the way microorganisms are classified.

Systematics is the science of classifying all living things into organized groups by the ways they are related to one another. The earliest systematics, begun in the late 1600s, relied on the observation of physical features of plants and animals with the help of microscopes to study cells. Biochemical tests evolved during the Golden Age of Microbiology, from the mid-1800s to early 1900s, when Louis Pasteur (1822–95) and his contemporaries began learning about metabolism and microbial end products such as alcohols, acids, and gases. The emergence of genetic engineering, in the 1970s, enabled biologists to study cells at the molecular level. The structure and function of DNA and RNA led to methods for analyzing the relatedness of all microorganisms. A new systematics scheme for the 21st century might soon revise today's classifications.

Bacterial Identification Methods

Method	Description
	PHYSICAL
microscopy	inspection of microscopic internal and external features of cells
colony morphology	color, shape, and size of colonies on specific agar types
cell morphology	shape (cocci, rods, spiral, endospores) and size
staining	gram-positive versus gram-negative
motility	ability of a cell to move under its own power
serotype	a unique type of cell within a species, differentiated by surface substances
	BIOCHEMICAL
catalase	an enzyme that converts hydrogen peroxide (H_2O_2) to water
oxidase	an enzyme that is part of aerobic energy metabolism
carbohydrate fermentation	presence of acid and/or gas when grown on sugars or alcohols
nitrate reduction	ability to use nitrate in energy metabolism
coagulase	an enzyme that clots blood
IMViC	indole, methyl red, Voges-Proskauer, citrate; combined test for the presence of indole, acid, and acetoin production and ability to use sodium citrate as the sole carbon source (see the color insert, page C-4, center)
gelatin liquefaction	ability to produce protease that degrades gelatin
urease	an enzyme that splits urea to ammonia and carbon dioxide
esculin hydrolysis	ability to cleave sugar-containing compounds
	GENETIC
homology	relatedness of unknown deoxyribonucleic acid (DNA) to known DNA
hybridization	use of probes containing specific pieces of DNA that will only bind to complementary DNA
rRNA	analysis of ribosomal ribonucleic acid (rRNA) unique base sequences

IDENTIFICATION OF BACTERIA

Among all different types of microorganisms, bacterial identification makes the broadest use of physical, biochemical, and genetic methods. Physical differentiation of bacteria begins with the morphology of bacterial colonies on general growth agar and differential agar. See the color insertions on pages C-2 (lower right) and C-5 (bottom) for colony and cell morphology.

Colonies of a given genus look the same if grown on the same type of agar. This is true whether two different cultures of a species incubate a day apart or a year apart. Microbiologists observe the characteristics of colonies on generic agar that provides a wide variety of nutrients. On this type of agar, species produce a variety of colony types: smooth versus rough, small versus large, dome-shaped versus flat, white versus gray, and so on. Microbiologists, then, supplement this finding by growing the bacteria on differential agar media, which do a better job at distinguishing between major groups of bacteria. For example, eosin methylene blue (EMB) agar is a differential medium containing the dyes eosin and methylene blue. The use of these dyes distinguishes between bacteria that ferment the sugar lactose and bacteria that do not. EMB agar contains lactose as part of its formula, along with a protein, vitamins, the dyes, and the sugar sucrose. A typical result on EMB agar would be the following:

- *Escherichia,* a lactose fermenter, produces black colonies on EMB agar.
- *Salmonella,* not a lactose fermenter, produces colorless colonies.

This example illustrates the main difference between generic agar formulas and differential formulas: Generic agar produces fairly generic-looking colonies, and differential agar uses biochemical differences between species to give much more information. A microbiologist can further identify each microorganism by an additional test, as follows:

- *Salmonella* can be further differentiated to species serotyping, that is, identifying bacteria on the basis of the outer structures called antigens.
- *E. coli* strain O157:H7 can be distinguished from most other *E. coli* strains by growing it on MacConkey sorbitol agar, containing, the sugar sorbitol. Sorbitol-metabolizing *E. coli* produces dark pink colonies, and non-sorbitol-metabolizing *E. coli* O157:H7 produces colorless colonies.

Microbiologists have developed several hundred different types of media to separate species by morphology and biochemistry at the same time. The table describes some of the common identification techniques used by most laboratories for bacteria.

In hospitals, the correct and rapid identification of a bacterial pathogen may save someone's life. Many of the methods listed in the table are accurate but require one or more days to complete. Since the early 1990s, biological engineers have invented instruments that provide "rapid identification." Three main types of rapid systems are (1) manual biochemical kits, (2) fatty acid analysis, and (3) serology. These systems are called rapid because an unknown bacterium can be identified in less than an hour. However, the methods still require time to grow a culture before the test begins.

Manual biochemical tests include the examples in the table, plus additional few that can be run in very small wells rather than test tubes. Cardboard strips that hold 20 wells, each containing a few drops of a biochemical test broth, can be inoculated with bacteria, be incubated, and provide an identification within 24 hours. These strips are called *identification kits.*

Fatty acid analysis involves determining the unique collection of fatty acids in bacterial cells using an instrument called a *gas chromatograph.* Fatty acids are carbon chains ranging from two to 20 carbons and containing a carboxyl group (COOH) at one end. In this method, technicians break apart the bacteria and recover the fatty portion of the cell contents. They, then, convert the fatty acids into a volatile chemical form called a *fatty acid methyl ester.* The gas chromatograph contains a tubelike component filled with an inert gas that the instrument slowly pumps through the tube. Fatty acid methyl esters injected into the gas flow gradually separate from each other, the smallest moving more quickly through the tube than large compounds. A detector at the end of the line determines the type of fatty acid exiting the flow and its amount relative to the other fatty acids. This analysis results in a unique pattern of fatty acids called a *profile,* printed on a chart called a *chromatogram.* By comparing the profile to a database of known fatty acid profiles, a microbiologist can identify the species.

Serology identifies bacteria by examining the reaction of bacterial antigens, which are compounds on the cell surface, with specific antibodies. The antigen-antibody reaction is very specific and identifies bacteria to species as well as subspecies, called *strains* or *serovars.* This method of identification is serotyping.

16S rRNA

Analysis of 16S rRNA by ribotyping has become the most powerful way to identify microorganisms. This method makes use of the RNA associated with ribosomes, which are small components of cells that carry out protein synthesis. Ribosomes are built in subunits; bacteria possess one subunit identified as 16S based on its mass. (*S stands for the Svedberg unit used in centrifugation.) Microbiologists determine unique base sequences present in 16S rRNA to identify bacteria.*

The University of Virginia microbiologist Jill Clarridge explained in *Clinical Microbiology Reviews,* in 2004, "The traditional identification of bacteria on the basis of phenotypic characteristics is generally not as accurate as identification based on genotypic methods. Comparison of the 16S rRNA gene sequence has emerged as a preferred genetic technique. 16S rRNA gene sequence analysis can better identify poorly described, rarely isolated [bacteria . . . and] can lead to the recognition of novel pathogens and noncultured bacteria." The 16S rRNA analysis requires a small sampling of bacteria, so it is

Fungal Identification by Microscopy

Structure	Description	Use in Identification
apophysis	a swelling of the sporangiophore stalk where the sporangium attaches	shape, size
cells	single-celled phase of a multiphase life cycle; any yeast	shape, size, color, color of colonies on agar
conidiophores	hyphae that extend upward into the air and hold a conidia at their tips	shape, visible walls, clustered or not clustered
filaments	long threadlike tips of elongating hyphae	easier to see in some species than others
hyphae	main tubular structure that elongates during growth and supports reproductive structures	septate (having cross walls) or nonseptate, size of cells, thickness
rhizoids	rootlike, branched hyphae that grow downward into the growth medium	presence or absence
spherules	large, round structures containing spores	presence or absence
sporangia	saclike structures containing asexual spores (sporangiospores)	presence or absence, shape, size
sporangiophores	asexual spores	density in the sporangium, shape of the clusters

useful for species that are difficult to grow in laboratory conditions. For very small samples of bacteria, microbiologists can increase the amount of rRNA they need for analysis by using the polymerase chain reaction (PCR). PCR rapidly creates millions of exact copies of gene sequences in a few hours.

The general steps in the 16S rRNA identification method are the following:

1. lysis of bacterial cells to release DNA
2. use of nucleic acid probe to locate the 16S rRNA gene on the DNA
3. production of multiple copies of the gene using PCR
4. sequencing of the 16S rRNA gene

The preceding steps show that 16S rRNA identification methods include two additional microbiology techniques: hybridization and PCR. Hybridization is the bonding of a DNA strand from an unknown microorganism with a complementary strand from a known microorganism.

IDENTIFICATION OF FUNGI

Until molecular methods became available, mycologists (fungus specialists) identified fungi by staining them and observing the cell's structures under a microscope. Staining remains a handy way to highlight features when nucleic acid analysis is not available. The following stains are used for fungal identification:

- acid-fast—positive reaction is pink to red
- ascospore—reproductive ascospore structures stain green and vegetative (nonreproductive) structures stain red
- Gram—gram-positive cells are dark blue
- lactophenol cotton blue—acts as a preservative and highlights some structures in blue
- methenamine silver—structures of the fungus delineated in black against a pale green background

Mycologists can also identify some fungi without using stains by preparing a wet mount of the specimen. A wet mount contains the fungus in water placed onto a glass microscope slide. Since fungal cells are bigger than bacteria, mycologists easily observe features of reproductive and vegetative structures. In many cases, a wet mount alone leads to identification. The table above presents the main fungal structures used in identification.

Virus Identification Methods

Method	Description
biochemical analysis	determination of the structure of proteins or polypeptides (long amino acid chains) from the outer coat
electron microscopy	use of an extremely powerful microscope that can resolve structures at 1 nanometer (nm)
genetics	DNA or RNA sequencing
immunology	use of antibodies specific for known viruses against the unknown virus as the antigen
X-ray diffraction	determination of structure by photographing a crystallized compound

Molds are fungi that grow in furry masses on agar. Some molds produce distinctive colors, shapes, or vertical formations. Characteristic growth on agar can help with identification with two cautions: Many molds look alike on agar, and mold cultures change appearance with prolonged incubations times.

Identification of fungi by their morphology even by an experienced mycologist can be unreliable. Fungi also need two days to two weeks to mature on agar. This timing would be catastrophic in clinical microbiology, in which speed is a priority. The biologist Mara Diaz explained in *Microbe,* in 2007, "Clinical microbiologists who are called on to identify fungal pathogens need to have multiplex, high-throughput diagnostic tools at their disposal." Ribosomal RNA gene sequences identify fungi today in a manner similar to 16S rRNA methods used for bacteria.

IDENTIFICATION OF VIRUSES

Viruses reproduce only inside living cells. For this reason, viral identification has been difficult because a virus cannot be grown on agar as bacteria or fungi can. Virologists have developed in vitro techniques for keeping animal or plant cells alive in the laboratory. These cell cultures have aided virus detection and identification.

The main techniques for identifying viruses are described in the table above. These methods encompass physical, biochemical, and genetic tests.

High mutation rates in some viruses hamper genetic analysis. Despite sophisticated DNA or RNA analysis, mutations have made virus identification a field in need of new and rapid technologies.

Identification is the cornerstone of microbiology. Few activities in microbiology proceed without an accurate assessment of a pathogen, a food spoilage species, or a microorganism that performs a crucial role in the environment. For this reason, students of microbiology learn identification techniques early in their education.

See also CLINICAL ISOLATE; HYBRIDIZATION; MEDIA; MORPHOLOGY; POLYMERASE CHAIN REACTION; SEROLOGY; STAIN; SYSTEMATICS.

Further Reading

Clarridge, Jill E. "Impact of 16S rRNA Gene Sequence Analysis for Identification of Bacteria on Clinical Microbiology and Infectious Diseases." *Clinical Microbiology Reviews* 17 (2004): 840–862.

Clem, Amy L., Jonathan Sims, Sucheta Telang, John W. Eaton, and Jason Chesney. "Virus Detection and Identification Using Random Multiplex (RT)-PCR with 3′-Locked Random Primers." *Virology Journal* 4 (2007): 65–76. Available online. URL: www.virologyj.com/content/4/1/65. Accessed August 22, 2009.

Diaz, Mara R. "Bead Suspension Arrays for Identifying Fungal Pathogens." *Microbe,* February 2007.

Kaiser, Gary E. "Bio 230 Microbiology Lab Manual." 1999. Available online. URL: http://student.ccbcmd.edu/courses/bio141/labmanua/toc.html. Accessed August 22, 2009.

Reier-Nilsen, Tonje, Teresa Farstad, Britt Nakstad, Vigdis Lauvrak, and Martin Steinbakk. "Comparison of Broad Range 16S rRNA PCR and Conventional Blood Culture for Diagnosis of Sepsis in the Newborn: A Case Control Study." *Biomed Central Pediatrics* 9 (2009): 5–13. Available online. URL: www.biomedcentral.com/content/pdf/1471-2431-9-5.pdf. Accessed August 22, 2009.

Todar, Kenneth. "Online Textbook of Bacteriology." 2008. Available online. URL: www.textbookofbacteriology.net. Accessed August 22, 2009.

immunity Immunity is the body's process for resisting disease. The immune system consists of white blood cells, lymphocytes (cells produced by the lymph system), lymph nodes, the thymus gland, the spleen, and bone marrow. These components complement each other in creating an immune response, which is the detection and destruction of foreign matter in the body. Foreign matter that may activate an immune response are the following: microorganisms, fungal spores, pollen, proteins, some chemicals, or animal or plant tissue. Immunity against microorganisms serves as the main defense that people have against infection.

The immune system recognizes the first occurrence of foreign matter, such as a pathogen, in the blood or in the body's tissue, and then destroys or neutralizes it. Anytime the same pathogen returns in subsequent infections, the body should be better prepared to fight the invader, whether it is a virus, bacterium, fungus, protozoan, or parasite. Overall, a person's properly functioning immune system repels infections from microorganisms that have contact with the body every day.

DISCOVERY OF THE IMMUNE RESPONSE

In 1901, the German microbiologist Emil von Behring (1854–1917) studied cultures of the bacteria that cause diphtheria (*Corynebacterium diphtheriae*) and tetanus (*Clostridium tetani*), for the purpose of finding a substance that could inactivate these pathogens' toxins. He and the Japanese bacteriologist Shibasaburo Kitasato (1852–1931) showed that sterilized cultures of the bacteria—sterilization killed the bacterial cells—injected into healthy animals caused the formation of substances in the blood that neutralized the toxins in test tubes. These so-called antitoxins, furthermore, cured animals that had the symptoms of disease. Von Behring referred to the phenomenon as *humoral immunity* because medicine at the time identified the body's fluids as humors. In his speech when he was awarded the Nobel Prize in medicine, von Behring said, "It is a humoral therapy, because its activity develops only within the fluid and [dissolved] components of the individual who is ill or threatened with illness." Von Behring continued applying the theory of toxin neutralization to the treatment of tuberculosis in cattle. "I need hardly add," he said upon winning the Nobel, "that the fight against cattle tuberculosis only marks a stage on the road which leads finally to the effective protection of human beings against the disease." While von Behring introduced the concept of immunity and antitoxins, other scientists were unraveling the complex nature of immunity.

The Russian biologist Elie Metchnikoff (1845–1916) demonstrated the activity of cells called *phagocytes,* which are immune system cells that engulf and digest microorganisms, in tissue and blood. Metchnikoff, who won the 1908 Nobel Prize in medicine for his research in immunology, showed that phagocytes reacted more efficiently to foreign matter in animals that already had some immunity to a microorganism. This finding showed that phagocytes provided two kinds of immunity: specific and nonspecific.

Specific immunity arises from antibodies that the body makes in response to a specific microorganism. For example, antibodies made by a child infected with the chicken pox virus (varicella) provide immunity against varicella in subsequent years. Nonspecific immunity arises from other factors in the body that enables a person to fight infection. This nonspecific response, called *innate immunity,* can be inherited, arise from fever, or result from certain of the body's cells that resist being infected. For example, dogs have innate immunity to humans' common cold.

Immunologists have since discovered many of the mechanisms of specific and nonspecific immunity, and yet some questions remain. Sometimes, a disease such as acquired immunodeficiency syndrome (AIDS), which attacks the immune system, creates a public health crisis that it accelerates the research on immunity. Several heretofore unknown structures on immune system cells were discovered because of AIDS.

ORGANIZATION OF THE IMMUNE SYSTEM

The immune system is made up of two main branches: innate immunity and acquired immunity. Innate, or nonspecific, immunity consists of protections against infection that a person receives upon birth. These innate defenses include physical barriers such as the skin and mucous membranes, phagocytosis (the ingestion and destruction of foreign matter by a cell), and inflammation. Innate immunity may also involve factors that have not been discovered, such as specialized structures on the outside of cells that repel pathogens. (Immunity does not develop against only pathogens; any foreign material elicits an immune response, even if the foreign matter is harmless.)

Acquired or *specific immunity* refers to defenses that the body develops on its own. Two types of acquired immunity exist: natural and artificial. Both of these types may be in active form or passive form. These alternate forms of immunity make up what is referred to as "the duality of the immune system." This means that if an infectious agent finds its way past any component of a healthy immune system, the body has one or more alternate responses against the invader.

Acquired immunity develops throughout a lifetime, and it may last an entire lifetime or for shorter periods. For example, acquired immunity against measles lasts a lifetime, but a person can become sick by ingesting *Escherichia coli* many times because the acquired immunity against *E. coli* is short-lived.

Naturally acquired immunity is the production of antibodies against a foreign particle, called an *antigen.* The active form of naturally acquired immunity develops when the body produces antibodies against the specific antigen. For instance, a measles infection will induce the body to produce antibodies against the measles virus in case of subsequent exposure, but this antibody is powerless against a *Staphylococcus* infection. The passive form of naturally acquired immunity is gained from the mother, when antibodies are passed to the fetus, before birth and shortly after birth in mother's milk. Nursing cows give newborn calves passive immunity in colostrum, a thick milk filled with antibodies that the calf drinks during its first three days of life.

Artificially acquired immunity is produced by vaccination. The active form develops when a person receives a vaccine containing an antigen and, then, produces antibodies to that antigen. The passive form occurs when a vaccination involves injecting preformed antibodies into the body.

Antibodies are proteins called immunoglobulins. Different types of antibodies function in the human

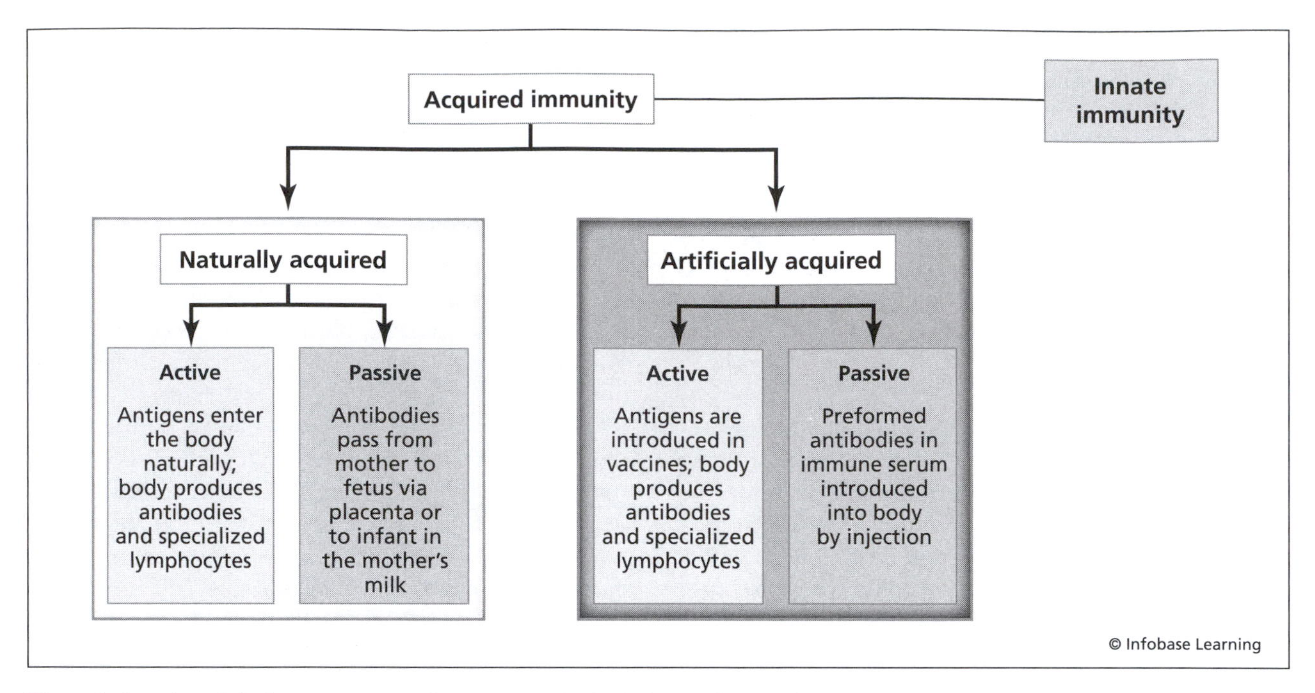

The main function of the immune system is to protect the body from infection from microorganisms, the growth of tumor cells, and the infiltration of foreign matter, such as pollen. The duality of the immune system assures that backup systems support the first lines of defense provided by innate immunity.

body, distinguished from each other by molecular weight and, in some cases, by the molecule's structure.

The duality of the immunity system allows for a second classification scheme based on antibody defenses versus cellular defenses. Antibody-mediated immunity (humoral immunity) is the production of antibodies against infection. Cell-mediated immunity involves specialized cells from the lymphatic system, or lymphocytes, called *T lymphocytes.* T cells regulate an array of further reactions that build the immune response to foreign matter.

ANTIBODY-MEDIATED IMMUNITY

All types of acquired immunity—natural or artificial, active or passive—are based on the crucial reaction of antibody to antigen. Most antigens are proteins or long sugar chains called *polysaccharides.* These compounds can live free in the bloodstream, but they are usually attached to something else, such as a microbial cell's surface or the outer coat of a virus.

Five different types of immunoglobin (Ig) make up the body's system of antibodies. A globulin is a large protein (more than 150,000 molecular weight) that dissolves in watery fluids such as serum. The proteins in serum contain three fractions, named the alpha, beta, and gamma proteins. The gamma fraction, or gamma globulin (IgG), contains most of the antibodies that provide protection against infection.

The simplest structure in a globulin molecule holds four proteins, which together form a unit called a monomer. The monomer proteins contain two identical short chains, called *light chains,* and two identical long chains, called *heavy chains.* The four chains bind to create the Y-shaped monomer. The tips of the two arms of the Y act as binding sites for an antigen, so that each antibody can bind to two antigens. Immunologists designate the antigen-binding sites as an antibody's variable region and the longer stem of the Y as the antibody's constant region.

B lymphocytes, or B cells, made in stem cells inside bone marrow or the liver of fetuses, enter the lymphatic system and produce the five classes of antibodies, described in the table on page 415. B cells detect specific antigens through receptors on their cell surface that complement structures on the antigen. The body must contain a variety of B cells to recognize the many different antigens a person is exposed to over time. When a B cell detects the antigen, it becomes an activated B cell and then divides into many clones, called *plasma cells,* that recognize the same antigen.

The microbiologists Lansing Prescott, John Harley, and Donald Klein have described plasma cells as "literally protein factories that produce about 2,000 antibodies per second in their brief five- to seven-day life span." A second type, called a *memory B cell,* operates much longer (years to decades) and reacts to specific antigens that have invaded the body previously. Memory B cells give a person a rapid immune response to the same antigen in subsequent exposures. The complementary actions of plasma cells and memory B cells, therefore, give the body a quick

Immunoglobulin Classes in Humans

Class	IgG	IgA	IgM	IgD	IgE
Percent of Total	80	10–15	5–10	0.2	0.002
Structure	monomer	dimer	pentamer	monomerr	monomer
Location of Activity	blood, lymph, intestines	blood, lymph, tears, saliva, mucus, milk, intestines	blood, lymph, B cell surface	blood, lymph, B cell surface	cell-bound
Half-life (days)	23	6	5	3	2
Main Functions	helps phagocytosis, targets toxins and viruses	protects mucosal surfaces	targets microorganisms in early infection	B cell recognition of antigens	allergic reactions, targets parasitic worms

Note: Half-life equals the time for half of the compound to lose activity

antibody response to infection and a long-term protection against infection, respectively.

In the table describing the immunoglobulins, two antibodies contain more than one monomer. IgA contains two monomers, so it is a dimer structure, and IgM contains five monomers, making it a pentamer structure.

The key reaction in the antibody-mediated immune response is the formation of an antigen-antibody complex. When the variable regions of an antibody bind to the specific antigen recognized by the antibody, the resulting complex neutralizes the antigen, so it cannot cause further infection. The complex holds the antigen until other immune cells arrive to destroy it.

The antibody response to antigens takes place in two types of responses, in the event of two consecutive exposures to an antigen: the primary antibody response and the secondary antibody response. In the primary response, after detection of the presence of an antigen, IgM appears in the bloodstream, within a few days, followed quickly by IgG. Both IgM and IgG circulate for about one and two weeks, respectively. During this time, the primary antibody response also helps the immune system set up memory against future invasions by the same antigen. This memory will be held in clones of the earliest memory B cells. A second exposure to the same antigen, from one to two months after the first exposure, gives rise to the secondary antibody response. In this response, IgM levels again increase above their normally low circulating levels in blood. IgG, however, makes a rapid rise and reaches levels in the blood much higher than at any previous time. Memory B cells also modulate this secondary antibody response, called the *heightened secondary*, or anamnestic, response. The primary and secondary events, therefore, assure that although a person may be harmed once by a pathogen, the body will be prepared to fight subsequent exposures to the same pathogen.

CELL-MEDIATED IMMUNITY

Part of the duality of the immune system is rooted in the coordination of antibody-mediated response and cell-mediated response. Cell-mediated immunity involves specialized T lymphocytes, also called *T cells,* plus a variety of white blood cells that play defined roles during an immune response. The cells of the immune system are described in the table on page 429.

T-cells are the major component of cell-mediated immunity. These lymphocytes provide the following general benefits to the immune systems:

- participate in B cell activation
- carry memory to specific antigens
- bind with antigens
- interact with macrophages
- release specialized immune chemicals that enhance the immune response

As a whole, the immune response is one of biology's most elegantly coordinated processes.

THE IMMUNE RESPONSE

The immune response has evolved into a complex, yet elegant series of complementary actions meant

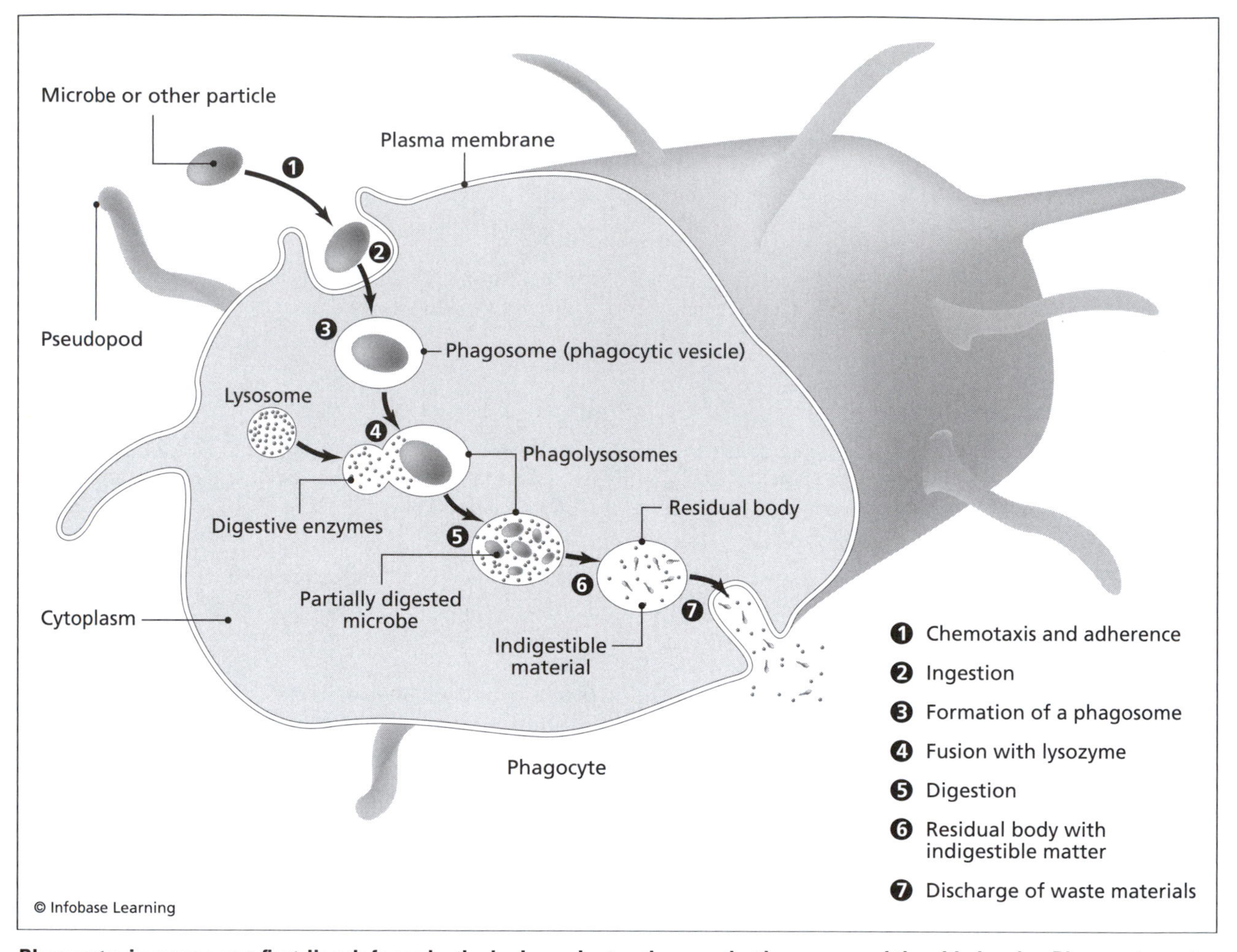

Phagocytosis serves as a first-line defense by the body against pathogens that have crossed the skin barrier. Phagocytes act in a nonspecific way to envelop and destroy any particles they identify as foreign to the body.

to defeat almost any invader. The cascade of events that make up a response begin when foreign matter enters the body. The antibody-mediated steps and cell-mediated activities, then, begin coordinated series of actions to eliminate the invasion. The steps are listed in the general order in which they occur:

- Macrophage and other cells find foreign matter, ingest it, and break it apart inside the cell. The macrophage, then, puts fragments of the antigen on its own cell surface. This is now called an *antigen-presenting cell.*
- A T helper (also called *helper T*) cell attaches to the antigen-presenting cell by binding to the antigen fragment.
- The antigen-presenting cell releases interleukin-1 and the T helper cell releases interleukin-2 into the bloodstream.
- The interleukins activate growth of T-cells and B-cells.
- Protein called *complement* helps carry out phagocytosis of the antigens by stationary and blood-migrating macrophages.

After the steps outlined, the immune response follows two routes, initiated by T cells and B cells. Cytotoxic T-cells find any cell that has ingested antigen, binds to the cell, and then kills it along with the antigen. B-cells, meanwhile, differentiate into the two types that will multiply and travel the bloodstream. B cells bound to antigen sometimes cannot become activated without aid from T helper cells. Antigens called T-dependent antigens require this extra help. Common T-dependent antigens are bacteria, foreign red blood cells, and certain proteins. The entire T-dependent antigen reaction requires the helper T cell, the B cell that has bound to an antigen, and macrophages. The interaction of these three components stimulates T cells to release compounds that support and enhance immunity.

Proteins called *cytokines* act as the main chemical messenger within the immune system. Immunologists have gathered new information about these

Cells of the Immune System

Cell	Source of Mature Cell	Functions	Special Features
B cell	bone marrow	antibody production	works in the spleen and lymph nodes
T cell	thymus gland	bind antigens until other immune cells arrive	contains four different types
monocytes	bone marrow stem cells	phagocytosis	after eight hours in blood, mature into macrophages
macrophages	monocytes	phagocytosis	migrate to specific tissues
basophils	bone marrow stem cells	release substances that regular blood flow to an infection	become coated with IgE to trigger inflammatory response
eosinophils	bone marrow stem cells	destroy protozoa and other parasites	mobile cells that pursue their targets
neutrophils	bone marrow stem cells	phagocytosis and antimicrobial action at infection sites	carry a variety of enzymes that digest microorganisms
mast cells	bone marrow	build the inflammatory response	create allergies and hypersensitivities
dendritic cells	monocytes	distinguish between pathogens and native microorganisms	after destroying infectious agent, joins other immune cells against infection
natural killer cells	lymph precursor cells from bone marrow	destroy tissue infected with pathogens and tumor cells	surface receptors that recognize "self" from "non-self"

compounds, which were largely unknown a few decades ago. The main cytokines known to affect immunity are the following:

- gamma-interferon—inhibits virus replication and increases macrophage activity
- interleukin-1—stimulates T helper cells and attracts phagocytes during inflammation
- interleukin-2—stimulates the proliferation of B cells and various T cells
- interleukin-8—attracts phagocytes to an inflammation site
- interleukin-12—helps in differentiation of T cells that contain the protein CD4 (CD4 cells)
- tumor necrosis factor—kills tumor cells and enhances activity of phagocytes
- granulocyte-macrophage colony-stimulating factor—stimulates formation of white and red blood cells from stem cells

Entire specialties within immunology focus on the cytokines and their interaction with cells, especially T cells. The acquired immunodeficiency syndrome (AIDS) epidemic that emerged in the United States, in the early 1980s, stimulated research on immunity factors as well as T cells because of the manner in which the AIDS virus attacks these components. The immune system is known to contain the following types of T cells:

- cytotoxic T cells—destroy foreign cells on contact
- suppressor T cells—possibly suppress B cell and T cell activity after an infection ends
- delayed hypersensitivity T cells—involved in specific immune responses, such as rejection of foreign (transplanted) tissue
- CD4 cells—primarily T helper cells with the CD4 protein on their surface
- CD8 cells—include mainly cytotoxic and suppressor T cells

Each of the T-cells has additional jobs that make these critical to maintaining a healthy immune response. Any disease that primarily attacks the T cells poses a greater risk to the body than almost any other illness.

VACCINES AND HERD IMMUNITY

Vaccines confer artificially acquired immunity on the body. Vaccination, or immunization with a deactivated antigen, causes the body to make antibodies against the antigen, so this is the active form of artificially acquired immunity. Some vaccinations contain antibodies rather than antigen. When these antibodies are injected into the blood, the result is passive artificially acquired immunity.

Vaccination has been crucial in limiting the spread of several highly communicable diseases. In a single population in a defined area, vaccination can lead to a phenomenon called *herd immunity*. Herd immunity is the resistance of a population to a disease that occurs because a large percentage, but not everyone, in the population has immunity. The larger the proportion of immune people, the smaller the chance the uninfected will contract the disease. This process works because in a predominantly immune group of people, the number of susceptible people decreases. Without susceptible individuals, a pathogen cannot spread through a population.

The dynamics of how herd immunity works may change from population to population and with different pathogens. However, herd immunity helps protect the health of entire communities.

Immunity is an important part of how successful a pathogen will be in setting up an infection in an individual and spreading infection through a population. An understanding of infectious disease requires a basic understanding of the human immune system.

See also IMMUNOASSAY; INFECTION; SEROLOGY; VACCINE.

Further Reading

DeFranco, Anthony L., Richard M. Locksley, and Miranda Robertson. *Immunity: The Immune Response in Infectious and Inflammatory Disease*. Corby, England: New Science Press, 2007. Available online. URL: www.new-science-press.com/browse/immunity/info. Accessed August 10, 2009.

Kaufmann, Stefan H. E., Alan Sher, and Rafi Ahmed, eds. *Immunology of Infectious Diseases*. Washington, D.C.: American Society for Microbiology Press, 2002.

Prescott, Lansing M., John P. Harley, and Donald A. Klein. *Microbiology*, 6th ed. New York: McGraw-Hill, 2005.

Russell, David G., and Siamon Gordon. *Phagocyte-Pathogen Interactions: Macrophages and the Host Response to Infection*. Washington, D.C.: American Society for Microbiology Press, 2009.

Seeley, Rod R., Trent D. Stephens, and Philip Tate. *Anatomy and Physiology*, 7th ed. Boston: McGraw-Hill, 2006. Available online. URL: http://highered.mcgraw-hill.com/sites/0072507470/student_view0. Accessed August 10, 2009.

Von Behring, Emil. "Serum Therapy in Therapeutics and Medical Science." Nobel lecture presented in Stockholm, Sweden, December 12, 1901. Available online. URL: http://nobelprize.org/nobel_prizes/medicine/laureates/1901/behring-lecture.html. Accessed June 20, 2009.

immunoassay An immunoassay is any procedure that uses an antibody to detect the presence of an antigen in the body. An antigen is any particle not belonging to the body, such as a pathogen, pollen, or transplanted tissue. The antibody is a protein produced by the immune system with the goal of finding and attaching to an antigen. Because antibodies are very specific for molecules on the surface of their target antigens, immunoassays provide a valuable way to detect one microorganism in a heterogeneous mixture of living and nonliving substances.

Clinical microbiologists use immunoassays as one of several methods for identifying unknown pathogens recovered from hospital patients. In this use, immunoassays belong to three groups: end-point, enzyme, and sandwich. End-point immunoassays produce a signal when all of the antigen and antibody molecules have bonded into an antigen-antibody complex. Enzyme immunoassays use the properties of enzymes to carry out specific reactions as a means of detecting compounds. A sandwich immunoassay uses a compound bound to a solid material and attached to another compound, which produces a signal when it connects to an antibody.

Scientists have developed many derivations in each main category. The most common immunoassays used in microbiology are the following:

- end-point immunoassays—agglutination, precipitin ring test, immunodiffusion
- enzyme immunoassay—enzyme-linked immunosorbent assay (ELISA), enzyme-multiplied immunoassay technique (EMIT)
- sandwich immune assays—fluorescent antibody test

Disease diagnosis includes immunoassays as an exact way to identify a pathogen or a toxin in a patient. These assays have been used for detecting the following health conditions: adrenal and pituitary function, allergies, anemia, bone disease, cardiovascular disease, diabetes, infectious disease, inflammation, reproductive hormones, thyroid disorders, and tumors. In infectious disease diagnosis, some of the viruses and bacterial infections that can be detected in the blood by immunoassays are human immunodeficiency virus (HIV), cytomegalovirus (CMV), papilloma virus, food-borne pathogens, sexually transmitted diseases (gonorrhea, chlamydia, syphilis), brucellosis, and hepatitis.

The German microbiologist Robert Koch (1843–1910) discovered the first immunoassay against tuberculosis-causing *Mycobacterium tuberculosis,* during studies in which he sought a vaccine against tuberculosis. Koch noticed a reaction at the injection site of guinea pigs that received a vaccine made of a *M. tuberculosis* cell suspension. A red and swollen area formed on the skin, one to two days after the injection. Medical science had not yet delineated all of the stages in a typical immune response; Koch had unwittingly developed and observed a classic inflammation reaction. Koch's procedure evolved into the tuberculosis test used today, one of many such tests that take advantage of the specific binding of antibody to antigen.

Antibodies are invisible to the unaided eye. In order to make any immunoassay practical, microbiologists must devise a way of connecting the antigen-antibody reaction to a visible signal. The most commonly used visible signals are the following: formation of an aggregate or precipitate, fluorescence, and color change.

AGGLUTINATION AND PRECIPITATION

Agglutination and precipitation reactions both create a solid aggregate when an antigen-antibody complex forms in a test tube. Agglutination reactions occur when the antigen is an insoluble particle, such as a bacterial cell, or a soluble antigen attached to an insoluble particle. When the antigen and antibody levels become almost equal, these molecules link together to form a mat that sinks to the bottom of the reaction mixture. Precipitation reactions, by contrast, involve soluble antigens that become insoluble when bound with an antibody.

Microbiology uses three types of agglutination tests that give results in less than an hour: (1) direct agglutination, (2) indirect agglutination, and (3) hemagglutination. In direct agglutination, a scientist adds increasing amounts of antibody to replicate tubes containing antigen, until agglutination occurs. This test not only demonstrates the presence of antigen, but also estimates the amount of antigen present. Indirect agglutination measures antigens that have been attached to microscopic beads. Most laboratories use beads made of latex; the test is often called a *latex agglutination test.* Adding

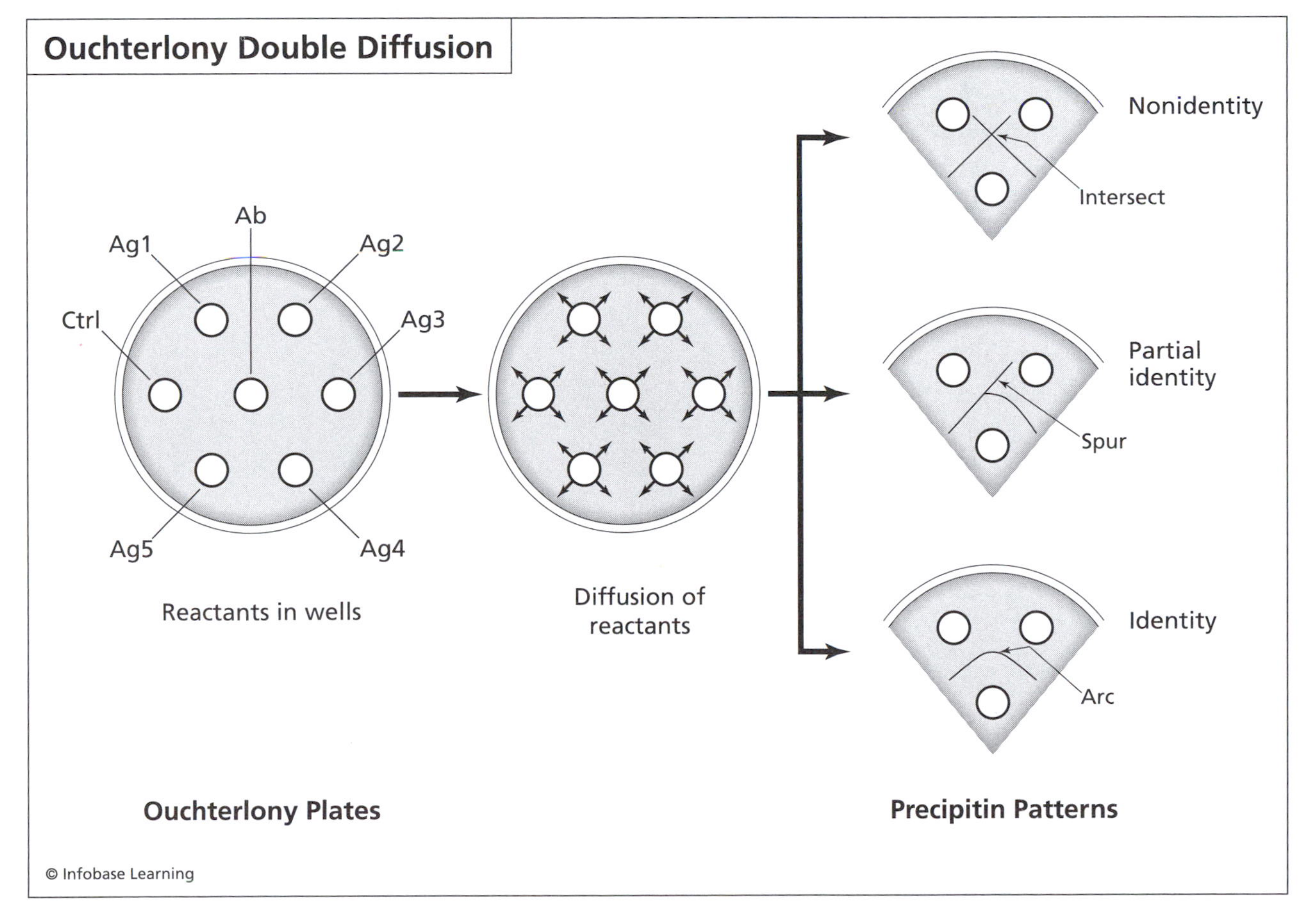

In double diffusion immunoassay, microbiologists test the activity of an antibody (Ab) to several different antigens (Ag), simultaneously (Ctrl = water control). The Ab and Ag fill wells in the agar, then diffuse outward from their wells. The place where an Ab-Ag complex precipitates, visible as a line in the agar, identifies the Ag with which the Ab reacts.

antibody to a suspension of antigen-coated beads causes visible aggregates to form, similar to the mats made by direct agglutination. Laboratories can also set up the test in reverse by attaching antibodies to the beads, then adding a suspension of antigen.

Hemagglutination clumps red blood cells together by reacting them with antibodies specific to the surface molecules on the red blood cells. This test is used in blood typing. Microbiologists have adapted the hemagglutination test for the detection

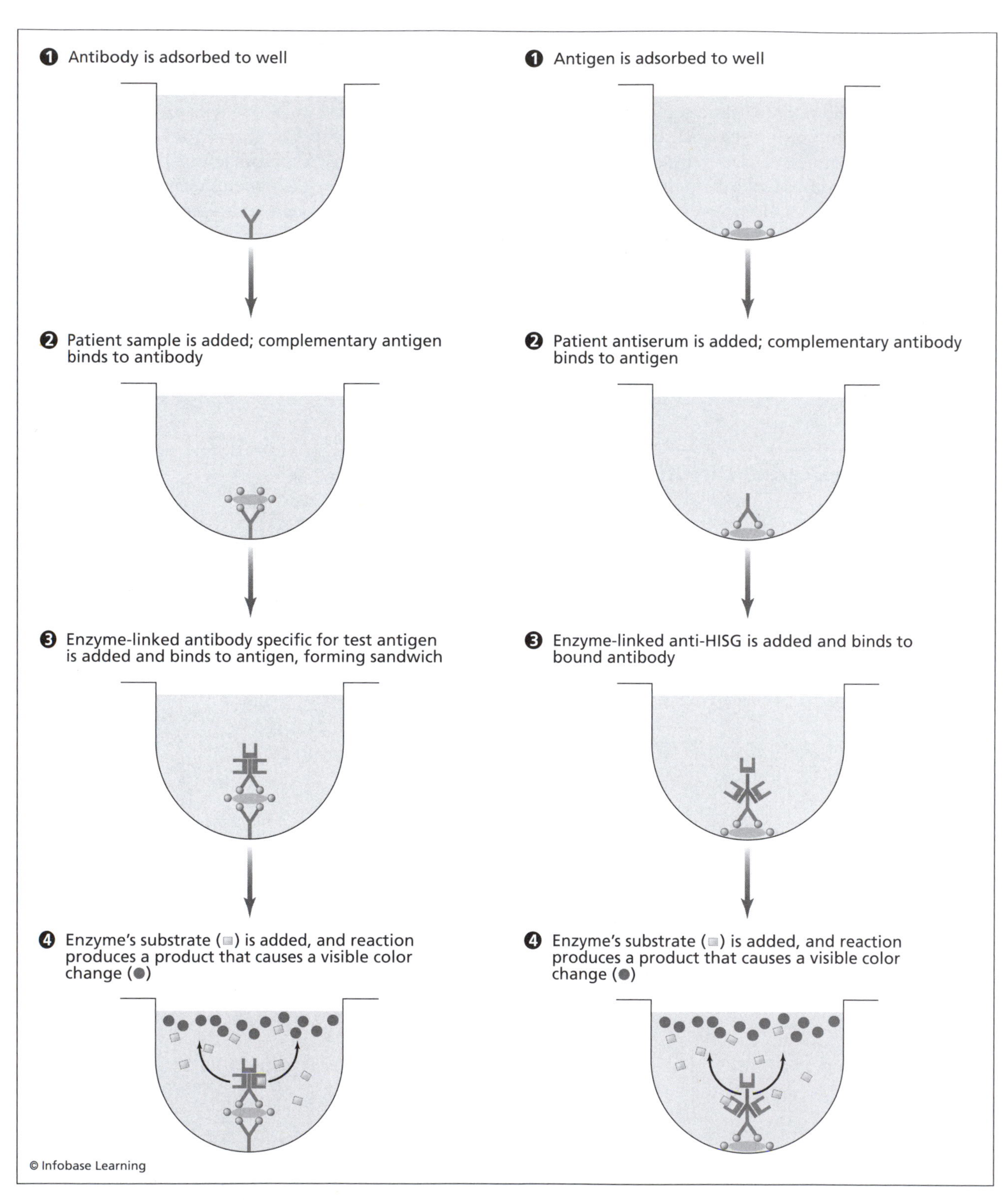

The ELISA test is one of the most widely used immunoassays. The assay is a sandwich method for detecting either antigen or antibody. The direct ELISA determines the identity of an unknown antigen by reacting it with a known antibody. The indirect ELISA identifies an unknown antibody.

of viruses. Several viruses can form aggregates of red blood cells in the same manner as antibodies. Hepatitis, influenza, measles, mumps, herpes simplex, and CMV have been detected and quantified by using viral hemagglutination tests.

Precipitation tests—also called immunoprecipitation—rely on the ability of the antibody to bind enough antigen to form an insoluble aggregate that can be seen in a test tube. The precipitin ring test involves the mixing of soluble antigens and antibodies in the same tube. Antigen-antibody complexes form within minutes and, then, arrange into a lattice structure, over the next few hours. When the lattice structure grows large enough, it is visible as a cloudy band (or ring) in the middle of the tube.

Two types of immunodiffusion tests involve precipitation in agar at a location where antigen and antibody bind. In a single radial immunodiffusion (RID) assay, antibody is added to molten agar, during agar plate preparation. After the agar solidifies, a microbiologist cuts small wells, about one quarter inch (0.6 cm) in diameter, in the agar, then adds an antigen solution. Over 24 hours of incubation, the antigen diffuses out of the well into the agar. Similar to the precipitin test, when antigen and antibody reach equilibrium, a ring of precipitant forms around the well. The second method, called the double diffusion agar assay, or the Ouchterlony technique, uses antigen in more than one well and antibody in another well on the same agar plate. The substances diffuse into the agar during incubation, and the pattern of the precipitate in the area between wells indicates whether the antigen and antibody are specific for each other.

ENZYME-LINKED ASSAYS

Enzyme immunoassays (EIAs) are widely used in microbiology in two ways: direct assays that detect antigens and indirect assays that detect antibodies. The ELISA test is one of the most widely used assays in biology, and it can be used either as a direct or an indirect assay.

The following steps explain the direct ELISA test:

1. add antibody to replicate wells in a plastic microtiter plate (a flat plate containing small wells of about 0.3 ml capacity)
2. the antibody attaches to the walls of the wells
3. add the unidentified antigen
4. the antibody binds the antigen to it
5. rinse off excess, unbound antigen
6. add a second enzyme-linked antibody specific for the presumed antigen
7. the second antibody attaches to the antigen that is already attached to the first antibody
8. rinse off excess, unbound second antibody
9. add the enzyme's substrate (the compound with which an enzyme reacts) to the wells
10. the enzyme-substrate reaction develops a visible color

Two critical steps that give this procedure its precision are, first, the very specific binding between an antibody and the antigen and, second, the enzyme-substrate reaction.

An indirect ELISA takes advantage of the same antibody-antigen and enzyme-substrate reactions as the direct assay, but for the purpose of detecting unknown antibody. In indirect ELISA, an unknown antigen attaches to the plastic well, and a technician adds to it a patient's antiserum, which might contain certain antibodies. Next, the technician adds a specially developed antibody called antihuman immune serum globulin (anti-HISG), which reacts with any human antibody. The anti-HISG also carries an enzyme. The entire indirect ELISA sandwich, thus, consists of the following substances attached in order to the plastic well:

1. a known antigen for the antibody to be detected
2. antiserum from the patient
3. enzyme-linked anti-HISG

If a person's blood contains the target antibody, the three components will bind. By adding in the enzyme's substrate, the enzyme-substrate reaction emits a color that can be measured in a spectrophotometer (any instrument that detects color of specific wavelength and measures its intensity). The physician Niel Constantine of the University of Maryland described the value of this assay for screening patients for the human immunodeficiency virus (HIV): "For the laboratory diagnosis of HIV, the mere presence of specific antibodies signals that infection has occurred. . . . ELISA is the most commonly used type of test to screen for HIV infection because of its relatively simple methodology, inherent high sensitivity, and suitability for testing large numbers of samples, particularly in blood testing centers." The U.S. Food and Drug Administration has allowed the use of about 10 different variations of HIV-testing ELISA kits that contains all the needed ingredients except the antiserum.

Both types of ELISA tests offer examples of a sandwich immunoassay, in addition to an enzyme-linked assay. All of the reactions take place attached to a solid surface, the wall of the plastic well.

The method called EMIT is a nonsandwich method done in a test tube. Most EMIT procedures today are for detecting the presence or absence of drugs in blood or urine. EMIT has been used for detecting the presence of antibiotics in these body fluids. In brief, the EMIT procedure involves the following steps:

1. mix urine or blood sample with a drug-specific enzyme and antibody
2. antibody attaches to the drug, and the drug reacts with the enzyme
3. add the enzyme's substrate
4. use a spectrophotometer to measure color change in the reaction mixture

The EMIT color change occurs, over time, as the drug and enzyme reaction progresses. EMIT offers the advantage of testing for more than one drug in the same assay.

FLUORESCENCE ASSAYS

Fluorescent-antibody (FA) methods are used for detecting unknown pathogens in patient specimens and for detecting specific antibodies in serum. Because these immunoassays use fluorescence to signal a specific reaction, the technology is called *immunofluorescence.* These tests take less than an hour to run and use only a glass microscope slide to hold the reaction.

Direct FA methods test for the presence of an unknown microorganism, which acts as the antigen. For example, a patient suspected of having tuberculosis would have *M. tuberculosis* in sputum. A clinical microbiologist tests the sputum sample for the presence of *M. tuberculosis,* by using the antibody specific for this species. The antibody, also, has a substance called fluorescein attached to it. After a brief incubation to let the antibody react with the bacteria, the microbiologist rinses off excess antibody and views the sample under a fluorescent microscope. Yellow-green fluorescence indicates the presence of *M. tuberculosis.*

Indirect FA detects the presence of an antibody in the serum after a person has been exposed to a microorganism. This assay uses fluorescein-labeled anti-HISG to make the antigen-antibody reaction easy to see under fluorescent light. In indirect FA, a known microorganism can be mixed with the patient's antiserum. After several minutes for the antibodies to bind with the microorganism, a technician adds the labeled anti-HISG. Presence of fluorescence indicates that the patient has the target antibodies. For example, when a person tests positive for *Streptococcus* antibodies, that person currently has or at one time had a *Streptococcus* infection.

Immunoassays give microbiologists the ability to use highly specific unions between antibodies and antigens for a variety of tasks. Today, immunoassays help diagnose disease, detect antibiotics, find food-borne pathogens, and identify unknown microorganisms.

See also DIFFUSION; IMMUNITY.

Further Reading

Constantine, Niel. "HIV Antibody Assays." HIV Insite. May 2006. Available online. URL: http://hivinsite.ucsf.edu/InSite?page=kb-02-02-01. Accessed June 23, 2009.

Shimeld, Lisa Anne, ed. *Essentials of Diagnostic Microbiology.* Florence, Ky.: Delmar Cengage Learning, 1999.

University of Arizona. "ELISA Activity." Available online. URL: www.biology.arizona.edu/immunology/activities/elisa/elisa_intro.html. Accessed June 20, 2009.

Wild, David, ed. *The Immunoassay Handbook,* 3rd ed. Philadelphia: Elsevier, 2005.

indicator organism An indicator organism is a microorganism whose presence suggests the probable presence of other microorganisms, usually pathogens. In its role, the indicator organism acts as a substitute or a surrogate for a different microorganism. Bacteria serve as the major indicator organisms.

Microbiologists use indicator organisms in instances when it would be too time-consuming, tedious, or impractical to examine all of the microorganisms in a sample. The greatest need for indicators arises in the following areas: water microbiology, food microbiology, preservative testing, environmental microbiology, and biotechnology. The table explains how indicator organisms are used in these areas.

Microbiologists often select more than one indicator microorganism because no single species fulfills all the requirements of a perfect indicator organism, described in the following list. This list summarizes the criteria for an ideal indicator organism for environmental pathogens:

- The indicator should be able to live in the environment.
- The indicator should be present whenever the pathogen is present.
- The indicator should survive longer in the environment than the pathogen.
- The indicator should not grow to large numbers in the environment.
- The concentration of the indicator should relate to the pathogen's concentration.
- The method for detecting the indicator should be easy to perform.

Each discipline in microbiology—water (see table below), food, biotechnology, and others—might include a few additional criteria for selecting an indicator organism. For example, indicators used in water microbiology to test for the presence of fecal contamination should be microorganisms native to the digestive tracts of warm-blooded animals. Indicators should have some relationship to the microorganisms being targeted. For example *Escherichia coli* serves as an indicator of fecal contamination in water microbiology, but *Salmonella* might be a better choice as an indicator in food microbiology. This is because *E. coli* indicates fecal contamination of water sources, and *Salmonella* is a more accurate indicator of poor food storage, handling, or cooking.

For many years, another criterion of an ideal indicator specified that the indicator not be pathogenic, mainly because nonpathogenic microorganisms are safer to study in laboratories than pathogens. But microbiologists sometimes prefer using indicators that give the most accurate assessment possible of the true conditions in the environment or food. That is why bacteria such as *E. coli* and *Salmonella,* themselves pathogens, serve as good indicators.

DISADVANTAGES OF INDICATOR ORGANISMS

Molecular probes highlight a weakness of indicator organisms: Indicators often do not give information on an exact species. Before molecular techniques became available, scientists used two assumptions about indicators: The presence of an indicator correlated with the presence of pathogens, and the pathogens correlated with the probability of illness. Several studies on indicators, in the past few decades, have detected the following two common errors in results from water indicators, particularly coliforms:

- false positives—presence of an indicator identifies water as a health risk when the water actually is safe
- false negatives—presence of an indicator is undetected even though the water contains pathogens

The risk of receiving false positive or false negative results in any test method has prompted microbiologists to question the usefulness of indicator organisms. In 1998, the EPA called together experts in the use of indicators to determine the disadvantages and methods to correct these flaws. The meeting participants agreed that new *E. coli* tests were a better indicator of pathogens than coliforms, and even those *E. coli* tests needed a confirmation assay.

In 2006, the microbiologists Michael Doyle and Marilyn Erickson discussed the problem of false positive and false negative results from coliform, fecal coliform, and *E. coli* tests on foods. The authors cited an "instance where fecal coliform data have been inappropriately interpreted [involving] two Canadian recalls of sprouts where high levels of fecal coliforms were later identified to be *K. [Kleb-*

Uses for Indicator Organisms

Area	Use	Description
water microbiology	water quality testing	detection of pathogens in drinking water, treated wastewater, recreational waters, or agricultural waters
food microbiology	contaminant detection, testing for proper storage and packaging	detection of spoilage microorganisms or pathogens
preservative testing	food preservatives, consumer product preservatives	challenging the effectiveness of a preservative in a food or formula
consumer products	disinfectant testing	testing the effectiveness of a disinfectant against *Staphylococcus, Salmonella,* and *Pseudomonas* to represent other bacteria
environmental monitoring	air, water, or surface contamination	detection of microorganisms that indicate poor sanitation
environmental microbiology	soil and water testing	detection of pathogens in natural environments
biotechnology	bioengineering	tracking the movement of microorganisms released into the environment

Indicators of Microbial Contamination of Drinking Water

Microorganism or Group	Microorganisms It Represents	Type of Contamination
total coliforms	general gram-negative bacteria	water treatment process
fecal coliforms	fecal microorganisms	fecal matter
E. coli	fecal pathogens	fecal matter
Cryptosporidium	small cysts	human and animal fecal wastes
Giardia lamblia	large cysts and protozoa	human and animal fecal wastes
Legionella	bacteria that multiply in water	water distribution systems
enteric viruses	pathogenic viruses	human and animal fecal wastes
heterotrophic plate count	general bacteria	environmental microorganisms

siella] pneumonia [common in natural waters]. In the health hazard alert accompanying these results, a warning was issued that this organism could cause gastrointestinal illness in humans. While this strain is an opportunistic pathogen outside the intestinal tract causing respiratory and urinary tract infections, gastrointestinal illness rarely occurs. Hence, the overly cautious warning was likely due to the association of this bacterium with the fecal organism group." Doyle and Erickson have joined others in pointing out the hazards of erroneous indicator results from food and water. The *E. coli* test has been proposed as more accurate in predicting pathogens than the coliform or fecal coliform tests.

Current EPA-approved *E. coli* tests call for confirmation tests to eliminate false positives and negatives. The additional testing also rules out fecal contamination when positive fecal coliform test results are obtained from nonfecal bacteria. New tests also distinguish *E. coli* from nonfecal coliforms.

Laboratories that monitor fecal streptococci with other indicator organisms often use the ratio of fecal coliforms to fecal streptococci as an indication of fecal contamination's origin. In general, water coliform-streptococcus ratios indicate the following:

- more than 4.0—presence of fecal contamination of human origin
- 4.0–0.7—a mixture of human and animal fecal matter
- less than 0.7—presence of fecal contamination of animal origin

Water microbiologists have devised ratios that give more detailed information on the fecal contamination of water, as follows: humans, 4.4; ducks, 0.6; sheep or pigs, 0.4; and cattle, 0.2. The use of any coliform-streptococcus ratio requires the following precautions:

- The pH of the sample influences the relative populations of coliforms and streptococci.
- Samples must be tested quickly because fecal streptococci die faster than fecal coliforms.
- Mixed pollution origins can complicate the results.
- The ratio does not apply to bay, estuary, or ocean water.
- Samples containing low amounts of fecal streptococci give inaccurate results.

The EPA does not, at present, require fecal coliform–fecal streptococcus ratios for water quality assessment.

Current indicator organisms, including *E. coli*, do not assess the presence of protozoa or viruses. These microorganisms require specialized laboratory techniques and special microbiology training. Viruses involve time-consuming tests, and, unlike with bacteria, one virus does not necessarily represent a group of viruses. Microbiologists have confronted the issue of viruses in two ways. First, they conduct laboratory studies to find bacteria that also serve as good indicators for the presence of viruses. For example, fecal streptococci have been identified as better virus indicators than fecal coliforms. Second, microbiologists use bacteriophages, viruses that infect only bacteria, as indicators of virus pollution. Bacteriophage behavior in the environment

Disinfectant Product Indicator Organisms

Microorganism	Microorganisms It Represents
Salmonella typhimurium	gram-negative bacteria
Staphylococcus aureus	gram-positive bacteria
Pseudomonas aeruginosa	hospital-associated bacteria
Aspergillus niger	molds
Trichophyton mentagrophytes	fungi

mimics enteric (from the digestive tract) viruses; therefore, the bacteriophage is a good indicator. Bacteriophages are also easier to study in laboratories and are not a health threat to microbiologists.

Microbiologists will continue to seek new indicator organisms that meet all necessary criteria. The advent of molecular techniques makes microbial testing more accurate than culture methods, but many laboratories are not yet equipped to use molecular probes. The culturing of indicator organisms, therefore, remains the main method used as long as microbiologists understand the potential drawbacks.

APPLICATIONS FOR INDICATOR ORGANISMS

Indicator organisms play an important part in water testing, partly because of their ease of use and partly because of EPA requirements. (The EPA evaluates new technologies that help microbiologists make better water quality assessments. The *E. coli* test kits used today resulted from cooperation between the EPA and water testing laboratories.) In addition to the water quality indicators already required, water microbiologists study other candidates as good indicators. Representatives of the bacterial genera *Aeromonas, Mycobacterium, Pseudomonas,* and *Staphylococcus* receive some attention as alternate indicators.

Makers of biocides such as disinfectants also rely on indicator organisms to streamline the testing of new product formulations. The EPA has responsibility for evaluating test results and approving the sale of new disinfectants. To do this, EPA scientists review the effectiveness of a disinfectant against the indicator organisms described in the table.

In disinfectant testing, no single virus represents all other viruses. That is, each individual virus—influenza virus, respiratory syncytial virus, rhinovirus, and so on—must be tested separately.

The food industry and consumer product manufacturers use indicator organisms primarily for testing preservative systems. Packaged, processed foods receive testing, as well as nonfood products, such as shampoos, cosmetics, lotions, and skin creams. Microbiologists test preservative effectiveness by a method called challenge testing. In this procedure, the microbiologist adds a mixture of several different bacteria and molds directly to a finished product. The microbiologist then withdraws samples from the product at intervals, typically at the time of inoculation and at one, three, seven, 21, 90, 180, and 365 days. An effective preservative will kill all of the indicator organisms, also called *challenge organisms,* over a period. Weak preservatives require longer periods to kill the challenge organisms, and strong preservatives take a shorter time.

The indicator organisms selected for challenge testing include microorganisms that historically contaminate the specific product. For example, mayonnaise would be challenged with *Lactobacillus,* which lives in low-pH conditions provided by this food, and canned vegetables, at neutral pH, would be challenged with *Clostridium,* which can contaminate this type of product. The rest of the challenge organisms include common contaminants such as *Staphylococcus, Bacillus, Pseudomonas,* and *Aspergillus.*

In biotechnology, microbiologists study the behavior of indicator organisms released into the environment before releasing a bioengineered microorganism that contains foreign genes. Indicators selected for this task can be generic microorganisms such as *Bacillus.* A microbiologist can also select the same species but without any new genes inserted into its DNA. These natural microorganisms are called wild type. For example, the indicator for a bioengineered *Bacillus* species would be wild type *Bacillus,* or wild type *Pseudomonas* for bioengineered *Pseudomonas.* Regardless of the species, a microbiologist must be able to track the microorganism in the environment. This can be done with probes specific for a certain gene carried by the wild type bacteria.

Indicators for bioengineered bacteria help biotechnologists evaluate the following factors:

- time the bacteria stay alive in the environment
- distance the bacteria travel from the release location
- probability of mutation to a virulent form

The behavior of indicators in these tests helps answer questions the public may have about bioengineering. The test results also help biotechnologists improve on bioengineering techniques.

Indicator organisms have been an important aid in microbiology for more than 100 years, and they continue to play a part in various industries. They have disadvantages, but with an understanding of those disadvantages, indicator organisms are essential in learning about many microorganisms by studying a few.

See also BACTERIOPHAGE; BIOSENSOR; COLIFORM; PRESERVATION; WASTEWATER TREATMENT; WATER QUALITY.

Further Reading

BusinessWire. "EPA Approves Colitag *E. coli* Detection Technology for Public Drinking Water Safety." February 16, 2004. Available online. URL: www.thefreelibrary.com/EPA+Approves+Colitag+E.+coli+Detection+Technology+For+Public+Drinking...-a0113293789. Accessed August 9, 2009.

Doyle, Michael P., and Marilyn C. Erickson. "The Fecal Coliform Assay, the Results of Which Have Led to Numerous Misinterpretations over the Years, May Have Outlived Its Usefulness." *Microbe,* April 2006. Available online. URL: www.woodsend.org/pdf-files/MicrobeNews.pdf. Accessed August 9, 2009.

Hach Company. "The Use of Indicator Organisms to Assess Public Water Safety." *Technical Information Series* 13 (2000): 1–47. Available online. URL: www.hach.com/fmmimghach?/CODE%3AL7015547%7C1. Accessed August 9, 2009.

National Academy of Sciences. *Indicators for Waterborne Pathogens.* Washington, D.C.: National Academies Press, 2009.

U.S. Environmental Protection Agency. "Drinking Water Contaminants." Available online. URL: www.epa.gov/safewater/contaminants/index.html#micro. Accessed August 9, 2009.

industrial microbiology Industrial microbiology involves the processes leading to a product made by a microorganism or a microorganism that itself is a product. This vast field contains the following major areas: bioremediation, biotechnology, brewing, consumer goods, diagnostics, food and food additives, industrial enzymes, metals and mining, microbiology products, and pharmaceuticals. An emerging interest in microorganisms as alternative energy sources represents a new discipline in industrial microbiology, called *bioenergy.*

Industrial microbiology can also be called *applied microbiology* because it relates to the development of commercial applications for microorganisms. Basic microbiology, by contrast, consists of studies on the mechanisms of microorganisms for gaining knowledge of the microbial world, not for commercial gain. Applied microbiology takes place primarily in industries, and basic microbiology takes place mainly at universities.

In industries that make a microbial product, the term *fermentation* refers to any growth of a microorganism in large volumes (several thousand gallons). This terminology differs from that related to fermentation as a biochemical process, in which an anaerobic microorganism makes alcohol or acid from sugar. Industrial fermentations are, therefore, any large-

Industrial Microbiology

Area	Description
bioremediation	microorganisms for cleaning up pollution
biotechnology	drugs from bioengineered microorganisms; microorganisms for agriculture (growth factors, natural pesticides, freeze protection)
brewing	yeasts and bacteria for beer and wine production
consumer products	preservative development and testing for personal care products, paints, wood, leather products, plastics, biocides
food and food additives	cheeses, fermented foods, enzymes, vitamins, amino acids; preservative development and testing
industrial enzymes	amylases, proteases, lipases, cellulases
metals and mining	recovery of metals from ore and cleanup of mine tailings
microbiology products	agar and broth media; clean rooms, equipment (petri dishes, test tubes, inoculating loops, Bunsen burners, biohazard containers, incubators); identification kits; microscopes; stains; sterilizers
pharmaceuticals	antibiotics, gene therapy, hormones, therapeutic enzymes

Milestones in Industrial Microbiology

Date	Event	Person
1916	development of fermentation process for making acetone, the largest industrial fermentation product after ethanol	Chaim Weizmann (1874–1952), England
1926	demonstration that enzymes are proteins	James Sumner (1887–1955), United States*
1933	microbial conversion of D-sorbitol to L-sorbose for vitamin C synthesis, the Reichstein-Grüssner process, still in use	Tadeus Reichstein (1897–1996), Poland*
1940s	development of large-scale manufacture of penicillin	Howard Florey (1898–1668), Australia, and Ernst Chain (1906–79), Germany*
1941	determination of relationship between genes and enzymes	George Beadle (1903–89) and Edward Tatum (1909–75), United States*
1953	determination of the structure of the DNA molecule	James Watson (1928–), United States; Francis Crick (1916–2004), England; and Maurice Wilkins (1916–2004), New Zealand
1956	first commercialization of amino acid (glutamic acid) production by microorganisms	Japan
1973	invention of recombinant DNA method	Stanley Cohen (1922–)* and Herbert Boyer (1936–), United States
1980s	application of fungal enzymes to large-scale production of high-fructose corn syrup from starch	United States
1985	invention of polymerase chain reaction	Kary Mullis (1944–), United States*
1995	beginning of genomics, first full microbial genome published for *Haemophilus influenzae*	Hamilton Smith (1931–), United States
2002	beginning of large-scale production of biofuels	several countries
2009	expansion of nanotechnology	United States

Note: *denotes a Nobel Prize winner

scale growths of microorganisms. Industrial vessels called *bioreactors* allow microbiologists to adjust growth conditions in the culture to increase the yield (amount produced) or purity of the final product.

Industrial microbiology emphasizes the following specialties: monitoring for contamination, preservative testing, sanitation, maintenance of stock cultures, aseptic technique, and new product research and development. New product research often involves environmental microbiology, in which scientists seek new species or familiar species with new traits. The table on page 426 describes the main areas within today's industrial microbiology.

Industrial microbiology has increasingly made use of bioengineering to create new microbial products, or microorganisms with useful traits. For example, the pharmaceutical industry works closely with biotechnology, an industry that grew out of bioengineering. Biotechnology also contributes to bioremediation, enzyme production, and the microbiology products industry. Many microbiology product companies specialize in making biosensors, probes, immunoassay kits, and deoxyribonucleic acid (DNA) sequencing instruments.

Recovering a microorganism from the environment, whether from surface soils, subsurface sediments, the ocean, or freshwater, requires added skills in pure culture and identification. It is also important that good personal hygiene be maintained in many industries for the following reasons: (1) prevention of adding pathogens to a product, (2) prevention of putting spoilage organisms into a product, and

Industrial microbiology and biotechnology depend on large fermentation facilities to make microbial products with commercial value. ***(Bioprogress Engineering Department, Faculty of Chemical and Natural Resources Engineering, University of Malaysia)***

(3) prevention of contaminating an industrial microorganism that is used for making a product. Poor personal hygiene of workers puts both consumers and the company at risk.

HISTORY OF INDUSTRIAL MICROBIOLOGY

The commercial use of microorganisms began, before 1000 B.C.E., in China, Egypt, and Greece, when street market vendors sold brewed beverages, vinegar, and leavened bread. Beer and wine production, cheese making, and baking developed into industries that continue today. Leather makers also applied microorganisms from dung to remove flesh from fresh hides, a precursor to the industrial uses of enzymes.

Major breakthroughs in the manipulation of microorganisms for commercial uses did not take shape until the Golden Age of Microbiology, from the mid-1800s to the early 1900s, when the germ theory and advances in microscopy helped scientists study single-celled organisms in detail. Scientists such as the Frenchman Louis Pasteur (1822–95) studied basic metabolic pathways such as fermentation by yeasts, and Pasteur also investigated the nutrients and substrates used by the yeasts and the end products they made. Pasteur's work provided early scientific evidence of microbial substances that convert specific compounds into other compounds. Ethanol from yeast fermentations would become the first major industrial product made by a microorganism.

By 1897, three scientists in Germany collaborated to demonstrate the action of enzymes. Eduard Büchner (1860–1917), awarded the 1907 Nobel Prize in chemistry; his brother, Hans Ernst Büchner (1850–1902); and Martin Hahn (1865–1934) showed that extracts from ground yeast cells could catalyze biological reactions, such as the fermentation of the sugar glucose to the alcohol ethanol. Catalysis is the action of making a reaction energetically favorable for a living cell. But the Büchners and Hahn had also shown that these substances, enzymes, worked outside the cell. This finding launched the new science of enzymology, which remains important in industries today. The table on page 427 reviews the major advances in industrial microbiology since the discovery of enzymes.

The table shows how industrial microbiology contains the following stages, which have highlighted its growth:

- discovery of enzymes
- molecular studies of DNA
- genomics

Nanobiology, the use of biological systems at the atomic level, might represent the next major advance in industrial microbiology. The emphasis in industrial microbiology will still include medicine and food production, but the future will also expand in alternative energies and environmental sciences.

The industrial biochemical engineer Peter Michels and the chemist John Rosazza wrote for the *Society of Microbiology News,* in 2009, "The astonishing diversity of microbial natural products reflects the enormous range of metabolic chemistry that microorganisms have at their disposal." In that light, the number of industries that use microbial activities may be far below the full potential that microorganisms offer. Industrial microbiology can be expected to grow several-fold in the near future.

INDUSTRIAL MICROORGANISMS

Any microorganism that can be harnessed for making a product, safely and efficiently, has potential in industry. Many microorganisms have already been put to use in the industries listed in the earlier table. Bacteria, molds, and yeasts currently carry out most industrial processes—molds and yeasts are both classified as fungi.

Each industry tends to rely on specific microorganisms to satisfy its particular needs, but a few microorganisms work across industries. For example, the yeast *Saccharomyces* contributes to beer brewing, wine making, and baking. The mold *Aspergillus* performs even more diverse jobs: ham curing, soy fermentations, cheese production, digestive aids, laundry detergent enzymes, paper bleaching, and leather tanning. The table on page 429 describes the prevalent microorganisms in industrial microbiology.

Microbiology continually uncovers more uses for the microorganisms listed here and for other genera. Studies on microbial genomes can be expected to help microbiologists work with known species and to design new recombinant species for making new microbial products. For example, most microorganisms lend themselves to the development of specific probes. These probes now help in the detection of pathogens in food, patient specimens, and the environment. Biosensors, deoxyribonucleic acid (DNA) fingerprinting techniques, and genome studies represent fast-growing areas for advancing industrial microbiology, and alternative fuel produced by microorganisms is a rapidly advancing area within industry.

BIOENERGY

Energy produced from biological activities is called *bioenergy,* and microorganisms offer a variety of ways in which alternative biofuels may replace conventional fossil fuels. Microbiologists in energy industries now search for microorganisms that can efficiently convert plentiful substances into clean fuels that do not harm the environment. These bioenergy sources belong to three main categories: (1) biofuels from biomass, (2) hydrogen fuel production, and (3) microbial fuel cells that produce electricity.

Industrial Microorganisms

Microorganism	Uses
	BACTERIA
Acetobacter	acetic acid production
Bacillus	amylases for brewing, lipases for drain cleaners, vitamin production
Clostridium acetobutylicum	production of solvents acetone and butanol
cyanobacteria	energy production from photosynthesis
Escherichia coli	genetic engineering
Lactobacillus	acid production in food preservation, lactic acid production in dairy products and pickling
Leuconostoc	lactic acid fermentations for pickling and soy sauce
Methanobacterium	methane production
Pediococcus	sausage production
Streptococcus	streptokinase for blood clot lysis, milk curdling
Streptomyces	antibiotics, growth promoters for meat animals, penicillinase to combat penicillin resistance
Xanthomonas	production of xanthan gums as thickeners
	MOLDS
Ashbya	vitamin production
Aspergillus	citric acid, meat tenderizer proteases, digestive aid production, meat curing, plastic raw materials, fungicides and insecticides, paper bleaching, proteases for leather tanning, lipases for detergents, cholesterol-reducing drugs (Lovastatin)
Gibberella	plant growth promoters (gibberellins)
Mucor	enzymes for cheese production
Penicillium	antibiotics, cholesterol-reducing drugs (Pravastatin), meat curing
	YEASTS
Saccharomyces	amulase for brewing and bread leavening; invertase for candy manufacture; lactase for digestive aids, wine fermentation, brandy and whiskey production
Trichoderma	cellulases for fruit juice, coffee, and paper production

Components of Environmental Monitoring

Component	Description
requirements	minimal requirements for cleanliness or sterility published by the FDA or equivalent government agency
baseline monitoring	amount of microorganisms and types normally found in the manufacturing environment
cleaning and disinfecting program	processes that will be used to prevent contamination
data collection and analysis	methods for determining the presence/absence of microorganisms, types, and numbers and places where they occur most frequently
trending	establishment of any correlations between contamination and certain plant activities: season, vacations, change in raw materials, change in cleaning/disinfecting schedule, night shifts, and so forth
alert and action levels	setting levels for an alert; microbial numbers that indicate an imminent problem, or action; microbial numbers that require immediate corrective action
corrective action plan	procedures for handling excursions outside the acceptable limits for microbial numbers
inspections	periodic checks to assure employees are following required steps for preventing contamination
reporting	maintenance of detailed records of monitoring results, trends, corrective actions, and suggestions for improvement

Biomass is plant waste that gives off energy when burned. Burning wastes some of the energy within biomass and contributes to air pollution. Microorganisms offer an alternative way to convert biomass to a cleaner energy source by fermenting plant fibers to produce ethanol. The fibers have been called lignocellulosic materials because they contain a combination of lignin and cellulose fibers that humans cannot use for food. Some microorganisms break down these tough, woody fibers into sugars. Algae, then, convert the sugars to an oily by-product that can be refined into biofuel for cars or for heating.

Photosynthetic microorganisms contribute to fuel production by using the sun's energy to convert water to hydrogen. Cyanobacteria and algae have been investigated as the most promising microorganisms to perform this conversion. In *Microbe* magazine, in 2009, the energy researcher Pin-Ching Maness wrote, "Photosynthetic green algae and cyanobacteria provide a more promising pathway [than nonphotosynthetic hydrogen-producing microorganisms] for generating hydrogen on a large scale. Hydrogen production by these microorganisms depends on the availability of plentiful resources, namely water as a substrate and solar energy as the energy source. Moreover, the oxygen and hydrogen that such cells produce could be used in a fuel cell to generate electricity." Hydrogen gas produced this way can serve as a fuel itself, or, as Maness suggested, be incorporated in biological fuel cells.

Some industries have already designed fuel cells for focused uses. The University of California microbiologists explained, in 2009, in *Microbe,* "Breweries are taking their wastewater, which is rich in organic material, and turning it into electricity with bacteria in microbial fuel cells (MFC). MFCs can generate valuable commodities from a variety of organic wastes that are abundant and essentially free—bacteria have generated electricity from industrial wastewater, sewage, and even sediment." In selected circumstances, MFCs have helped companies save on energy use, but they have not yet been developed to a large scale in industry, mainly because of relatively low energy outputs compared with a manufacturing plant's energy needs. When MFCs become practical in the future, they will benefit industry in two ways: by, first, helping established industries use renewable energy sources and, second, introducing a new industry that produces microbial energy.

ENVIRONMENTAL MONITORING

Industrial microbiologists help their companies develop new products from microorganisms. A second important responsibility involves the monitoring of a manufacturing facility so that contamination does not ruin a product. This specialty, called *environmental monitoring,* is critical in the following areas: food production, drug manufacturing, and consumer goods manufacturing.

Environmental monitoring has high importance in the manufacture of sterile drugs, such as vaccines that are injected into the body. These types of drugs have zero tolerance for the presence of a contaminant because of the severe health effects that a microorganism can cause in the bloodstream. The U.S. Food and Drug Administration enforces strict regulations on how sterile drugs must be made and monitored. Sterile drug manufacturing requires the use of clean rooms, which are designed to keep out all extraneous microorganisms.

Of the components listed in the table, two require microbiological methods in a laboratory: (1) establishment of baseline data and (2) data collection and analysis. Both components require the same activities. The main difference is that baseline monitoring takes place before manufacturing begins, so that a microbiologist understands the normal background amount of microorganisms in a facility. In clean rooms, this background level must be reduced to zero before any drug manufacturing can begin. In other industries, a low level of microorganisms can be tolerated as a background level. When the manufacturing process begins, the microbiologist starts collecting data on microorganisms present; this step represents the primary component in environmental monitoring.

Identification of the microorganisms provides three important pieces of information: (1) identification of microorganisms that are always present in the manufacturing environment, (2) identification of microorganisms that sometimes occur in very large numbers, and (3) presence of pathogens. Microbiologists must make decisions on each of these occurrences about how they affect a consumer's health risk from the product. In 2003, the microbiologist Johanna Maukonen described situations common in manufacturing: "Microorganisms in food and industrial environments are distributed unevenly; and there is a great variation in the cell density and composition of microbial population over space and time. Typically, the microbial cells are located in the surfaces of the food matrix and process equipment; and the cell density and species distribution may vary in different parts of a food product." In short, industrial microbiology is seldom a static situation.

Industrial microbiology is the most diverse area within the science of microbiology. Industrial microbiology encompasses every specialty within the science, and it has contributed new technologies throughout its history. Industrial and basic microbiology complement each other in examining the role of microorganisms and the services microorganisms can provide to humans.

See also *ASPERGILLUS;* BIOREACTOR; CLEAN ROOM; CULTURE; FERMENTATION; FOOD MICROBIOLOGY; GENETIC ENGINEERING; GENOMICS; GERM THEORY; HAZARD ANALYSIS CRITICAL CENTRAL POINTS (HACCP); IDENTIFICATION; MICROBIOLOGY; NANOBIOLOGY; *SACCHAROMYCES;* SAMPLE.

Further Reading

Govind, Nadathur S., and Arup Sen. "Combining Agriculture with Microbial Genomics to Make Fuels." *Microbe,* June 2009. Available online. URL: www.microbemagazine.org/index.php?option=com_content&view=article&id=308:combining-agriculture-with-microbial-genomics-to-make-fuels&catid=132:featured&Itemid=196. Accessed August 9, 2009.

Maness, Pin-Ching, Jianping Yu, Carrie Eckert, and Maria L. Ghirardi. "Photobiological Hydrogen Production—Prospects and Challenges." *Microbe,* June 2009. Available online. URL: www.microbemagazine.org/index.php?option=com_content&view=arti cle&id=309:photobiological-hydrogen-productionprospects-and-challenges&catid=132:featured&Itemid=196. Accessed August 9, 2009.

Maukonen, Johanna, Jaana Mättv, Gun Wirtanen, Laura Raaska, Tiina Mattila-Sandholm, and Maria Saarela. "Methodologies for the Characterization of Microbes in Industrial Environments: A Review." *Journal of Industrial Microbiology and Biotechnology* 30 (2003): 327–356.

Michels, Peter C., and John P. N. Rosazza. "The Evolution of Microbial Transformations for Industrial Applications." *Society for Industrial Microbiology News,* March/April 2009.

Waites, Michael J., Neil L. Morgan, John S. Rockey, and Gary Higton. *Industrial Microbiology: An Introduction.* Malden, Mass.: Wiley-Blackwell, 2001.

Wrighton, Kelly C., and John D. Coates. "Microbial Fuel Cells: Plug-in and Power-on Microbiology." *Microbe,* June 2009. Available online. URL: www.microbemagazine.org/index.php?option=com_content&view=article&id=307:microbial-fuel-cells-plug-in-and-power-on-microbiology&catid=132:featured&Itemid=196. Accessed August 9, 2009.

Yan, Yajun, and James C. Liao. "Engineering Metabolic Systems for Production of Advanced Fuels." *Journal of Industrial Microbiology and Biotechnology* 36 (2009): 471–479.

infection Infection is the initial invasion of the body by microorganisms. Infections may develop into disease or remain localized in the area where the microorganism entered. For example, *Mycobacterium tuberculosis* invades the lungs and leads to a disease, tuberculosis. A cut on the skin may lead to an infection of *Staphylococcus aureus* confined to the injury, but the infection never spreads beyond the cut.

An infectious disease is an illness caused by a microorganism, spread between people or from animals to people, and caused by bacteria, fungi, protozoa, viruses, or cysts. Helminths (worms) also cause infection, often when they are in a microscopic immature stage. Bacteria, fungi, and viruses cause most infections in plants and trees.

Infections are classified by the infectious microorganism, by the tissue affected, or by the source of the infectious microorganism. The table on page 433 shows the common classifications describing infections; however, a particular infection may belong to more than one category at the same time. For example, if a person becomes sick from ingesting food contaminated with *Salmonella,* the person has a bacterial, a food-borne, and a gastrointestinal infection. Any of these three descriptions is accurate in describing the illness.

The classifications listed in the table also contain subcategories. For example, a respiratory infection may be specifically described as a bronchial infection. A yeast infection might also be a *Candida* infection, a viral infection might be a human immunodeficiency virus (HIV) infection, and so on.

The sources listed in the table also describe modes of transmission, the way in which a person contracts an infectious microorganism. Infection, infectious disease, and disease transmission are interrelated and depend on the type of microorganisms causing illness. Discussions about infection also must include the role of the immune system in fighting infection in the body. A strong immune response to a pathogen can stop an infection before a person realizes he or she has it.

INFECTIONS AND DISEASE

Infections present a wide variety of outcomes from severe disease to a mild event. Two major factors determine the seriousness of an infection and its outcome: the health of the host and the pathogenicity of the microorganism. A pathogen is a disease-causing microorganism, and pathogenicity is the ability of a pathogen to cause disease after it infects the body.

Pathogens may be thought of as parasites, once they invade the body, because they benefit from the relationship, but the host (the infected person) is harmed. The host-parasite relationship describes a complex group of factors that either give an advantage to the host in warding off infection or give an advantage to the pathogen in sustaining an infection. The strength of the host's immune system is a major factor in fighting infection. Pathogens gain an advantage through their pathogenicity, their ability to cause disease, and their virulence, the intensity of the pathogenicity.

The more scientists learn about the mechanisms that various pathogens use to start infections, the more complex the host-pathogen relationship is discovered to be. Medical microbiology devotes an entire area of research to pathogenesis, the manner in which diseases develop from an initial infection.

HOST MECHANISMS AGAINST INFECTION

The body possesses multiple barriers that have evolved for the purpose of preventing infection. The immune system may be thought of as the primary line of defense, but, in fact, the immune system should be thought of as the last line of defense against a microorganism that has entered the body. The human body has barriers that prevent many pathogens from ever entering the body or staying for long on the skin or in the digestive tract.

The outside of the body challenges invading microorganisms with an array of physical, chemical, and biological barriers. The skin is a contiguous covering over almost the entire outer surface of the body. The skin and mucous membranes—membranes that line body openings—have together been called the body's first line of defense. Only when these barriers are damaged by a cut, scrape, rash, or burn, does the barrier allow access for pathogens to enter the body.

Chemical defenses work on the surface of the body and inside the body. On the skin, oils in sebum prevent the skin from drying out and prevent chapping that could give pathogens entry to the blood. Perspiration flushes pathogens from the skin and contains the enzyme lysozyme, which damages bacterial cells by degrading the cell wall. The vaginal area also maintains an acidic pH, which inhibits the growth of bacteria. Ingested pathogens that enter the stomach must survive the very acidic conditions caused by hydrochloric acid secretion from the stomach lining, as well as digestive enzymes.

The native bacteria on the skin and in the digestive tract, called *normal flora,* meanwhile, provide biological defense against infection. Normal flora combat pathogens in four main ways:

1. Normal flora that have adhered to the skin take up space and do not give transient microorganisms much room to settle.

2. Normal flora similarly outcompete new microorganisms for nutrients on the skin or on the intestinal lining by being adapted to those places.

3. Some native microorganisms change the conditions on the body to make a region inhospitable for other species. Some bacteria, for instance, lower the pH on

Types of Infections

Method of Classification	Description
	CAUSATIVE MICROORGANISM
bacterial	caused by bacteria
fungal	caused by fungi
opportunistic	caused by a normally harmless microorganism
protozoan	caused by protozoa
viral	caused by viruses
yeast	caused by yeasts
	TISSUE OR SYSTEM AFFECTED
bladder	in the bladder
ear	in the ear
eye	in the eye or associated mucous membranes
gastrointestinal	in the stomach, small intestine, or large intestine
respiratory	in the trachea or lungs
skin	on or in the epidermis
urinary	in the urethra
	SOURCE
airborne	transmitted by air and usually inhaled
blood-borne	transmitted through contact with infected blood
food-borne	ingested with food
nosocomial	transmitted during a stay in a hospital
waterborne	ingested with drinking or recreational waters
	ONSET OR TRANSMISSION
acute	appears suddenly and may be brief in duration
chronic	having a long duration
fulminating	microorganisms multiply rapidly to high numbers
local	has not spread and remains near the entry site
subacute	intermediate between acute and chronic
subclinical	confirmed by immunological testing but does not produce signs or symptoms
systemic	microorganism has spread throughout the body
primary	caused by the first microorganism that invades the body
secondary	caused by one or more microorganisms that follow the primary infection

the skin, as they degrade sebum into fatty acids. This acidic pH holds other microorganisms in check.

4. Many native microorganisms produce either antibiotics or bacteriocins that inhibit nonnative microorganisms. Antibiotics inhibit species that are not closely related to the producer; bacteriocins inhibit closely related microorganisms, as when a *Streptococcus* species produces a bacteriocin that inhibits other *Streptococcus* species.

When these normal body defenses break down, even normally harmless native flora can cause infection. This event, called an *opportunistic infection,* occurs when conditions of the body change in a way that allows normal flora to infect. For example, antibiotic treatments can reduce the number of native bacteria on the skin. When this happens, fewer bacteria are available to exert their inhibitory effects. *Candida* yeast infections in women are common when antibiotic treatment eliminates the bacteria that control vaginal pH. A shift in pH allows the *Candida* to multiply and cause a vaginal yeast infection.

If a microorganism evades the first line of defenses and enters the body, the immune system activates its own first, second, and third responses to keep the infection confined and then destroy the pathogen. The subject of immunology covers the array of defenses that the human immune system releases to eliminate any invader in the body. In general, the immune system combats infection by three defenses: inflammation, phagocytosis (the engulfing of pathogens by a body's cell and destruction of the pathogen), and the production of antibodies.

Inflammation is a generalized response to infection characterized by redness, swelling, heat, and pain at the infection site. These actions signal the immune system to rally certain types of cells to the infection site. Heat alone helps kill some of the weakest pathogens. Redness and swelling indicate a three-step process taking place: (1) vasodilation and increased vessel permeability, (2) migration of phagocytes to the site, and (3) tissue repair.

Vasodilation increases blood flow to the infection site and causes the characteristic redness and warmth. Blood vessels also become more permeable at this time, meaning they allow more substances in the blood to pass through the vessel wall and move to the infected area. This movement of substances can lead to inflammation's swelling, called *edema.* When the body's cells are injured, perhaps by a cut that becomes infected, they release compounds called *histamines* that cause the increased permeability. Other compounds, called *kinins,* help with permeability and with vasodilation. But the immune system is complex, and this complexity helps it avoid the counterattacks by a pathogen that is trying to spread into the body. Additional compounds alert phagocytes to the site, and clotting factors in the blood move in the direction of the infection to form clots that will wall off the infection from the rest of the bloodstream. An infected cut exudes pus, which is a mixture of body fluids, immune system compounds, dead cells, and the pathogen. This conglomeration signals that the immune response is working hard to confine the infection.

Inflammation must eventually subside, or it will damage the body's cells. As inflammation decreases, phagocytes populate the area around the site and destroy foreign matter inside or outside the blood. Phagocytes travel through tissue by a process called *amoeboid movement,* because it resembles the fluid motions of amoeba cells. Several additional immune cell types play specific roles at the infection site as the phagocytes work. Large cells called macrophages arrive for the task of engulfing and degrading damaged tissue as a precursor to repairing the damage.

If the inflammation described here has defeated the infection, the third stage, tissue repair, begins. The tissue itself builds new tissue in two parts: stroma, or supporting connective tissue, and parenchyma, or the functioning part of the tissue.

In events when the inflammation cannot contain the infection, the body must develop backup responses. These responses support the ongoing activities of the immune system that continue to battle the infection. Three support systems work in conjunction with the body's immune system: (1) a fever response, (2) the complement system, and (3) interferons.

The hypothalamus, in the brain, controls the development of fever as a reaction to infection. The hypothalamus normally keeps the body's temperature at 98.6°F (37°C). When phagocytes begin attacking pathogens, they also release a compound called cytokine interleukin-1 (IL-1), which activates the fever mechanism in the hypothalamus. The brain, then, raises the temperature to about 102.2°F (39°C). The body adjusts to the rise in temperature by constricting blood vessels and increasing the rate of metabolism, both leading to the shivering reflex. Shivering continues until the body reaches the higher temperature, and that high temperature persists until the levels of IL-1 decline. High temperature may inhibit pathogens, to some extent, but its main objective is to intensify the body's infection-fighting mechanisms, such as release of lymphocytes from the lymph nodes and antiviral compounds called *interferons*. The peak of the fever is called the *crisis* stage, and, from there, the temperature begins dropping to normal, and the person begins to sweat, a process that helps cool the body.

The complement system consists of more than 30 proteins made by the liver. Discovered, in 1895, by the Belgian Jules Bordet (1870–1961), this system helps fight infection by contributing to the destruction of pathogens, inflammation, phagocytosis, and tissue repair. The complement system proceeds in a cascade of events; that is, one protein's reaction against infection triggers the next protein in the complement sequence. As the cascade continues, the overall effect becomes an increasingly stronger response to infection.

Complement proteins named C1 through C9 represent the backbone of the complement system. Products from these proteins, after being activated, constitute the remainder of the system. Three different processes activate the complement system, so that if one system cannot function, an alternate can take over the fight against the infection. The table describes the three activation processes.

Medical research continues to investigate the details of the complement system in the hope of using complement proteins as disease therapy.

Various research groups discovered interferons, in the 1950s, during a time when the molecular basis of disease dominated medical research. In 1957, the virologists Alick Isaacs (1921–67) of Great Britain and the Swiss Jean Lindenmann (1924–) coined the term *interferon* for a protein released from cells infected in vitro with the influenza virus. Subsequent research has added details to knowledge of the nature of interferon and its activity.

Interferon belongs to the nonspecific portion of the immune system that defends against any foreign particle it detects in the body. Various cells secrete interferon when a virus infects the cell. Three types of interferons exist, belonging to two groups, interferons I and II. Interferon I contains alpha (IFN-α) and beta interferon (IFN-β), released by virus-infected leukocytes and virus-infected fibroblasts (connective tissue cells), respectively. Interferon II contains gamma interferon, which is made by the immune system's T lymphocyte (T cell) and natural killer (NK) cells. The cells that make and release interferon do so only when infected by a virus. Once in the bloodstream, interferons enhance the activity of the immune system, particularly T cells and NK cells.

Interferon has been studied as a possible drug against infection, but some characteristics of the protein must, first, be overcome. First, injected interferon has caused nausea, vomiting, weight loss, and fever in some patients and can be toxic at high levels. Second, interferon does not usually stay active in the body long enough to fight an established infection. Third, viruses might develop resistance to interferon if it becomes a commonly used treatment for infection. In 2006, the molecular geneticist and interferon expert Sidney Pestka said, "The great promise of interferon as an antiviral agent was evident from the moment of its discovery." But production of a stable, effective interferon has been elusive.

MICROBIAL MECHANISMS OF INFECTION

Pathogens would not have evolved with humans if they did not have some tricks for avoiding the body's defenses. All of the evasive actions used by pathogens contribute to their pathogenicity and virulence. Various pathogens possess one or more of the following countermeasures to the body's defenses:

- Mutation makes the pathogen difficult to be recognized as a foreign particle.
- Capsules inhibit the ability of phagocytes to engulf the cell.
- An M protein on a pathogen's surface makes it difficult for a phagocyte to attach to the cell.
- Leukocidins released by an engulfed pathogen kill the phagocyte.
- Some pathogens survive for long periods inside immune system cells.
- Coagulase enzyme made by a pathogen breaks down the clot surrounding the infected site, allowing cells to access the bloodstream.

Pathogens that have plagued humans through history have developed additional mechanisms for causing and prolonging infections.

INFECTION CONTROL

The British surgeon Joseph Lister (1827–1912) began the discipline of infection control, by being the first physician to stress cleanliness and the use of antiseptics when treating patients. Infection control has since grown into a specialty of both microbiology and public health.

Infection control begins with an awareness of microorganisms in the environment, some of which may be pathogenic, as well as the best ways to prevent infection. Infection control in health care settings and at home can be divided into the following three main categories: (1) personal hygiene, (2) cleaning, and (3) medical attention to new infections. Personal hygiene such as hand washing and covering the mouth when sneezing or coughing greatly reduces the spread of pathogens to others. *Cleaning* is a general term for the appropriate use of cleaning products, disinfectants, and antiseptics to remove potential pathogens from a person's environment. For example, a disinfectant removes microorganisms from inanimate surfaces, an ability that can be valuable in a kitchen to prevent food-borne infections. Antiseptics work directly on the skin and should be used if a person receives a break in the skin.

Medical attention ranges from a bandage applied over a cut to a physician's treatment of a bronchial

Complement Activation Systems

Name	How It Works
classical pathway	initiated by antibody reaction with a pathogen; this antigen-antibody complex activates C1
alternative pathway	activated by direct contact between a pathogen and complement protein C3
lectin pathway	when macrophages encounter pathogens, they induce the liver to produce lectin proteins; the mannose-binding lectin binds to pathogens, activating C2 and C4

infection. Because infections can be minor or severe, medical attention should be appropriate to the circumstances. Certain groups within the population require diligent attention to even minor infections because these people are at higher-than-normal health risk. This high-risk population consists of people who have weakened immune systems (acquired immunodeficiency syndrome [AIDS] patients, organ transplant recipients, chemotherapy patients), the elderly, the very young, pregnant women, and people who have a preexisting disease. In all of the high-health-risk situations, the immune system cannot cope with the invasion of large amounts of pathogens. Even small infections such as a cut on the hand can turn serious in high-risk people. Infection control has become an important part of health care in both high-risk groups and the general population.

Infections represent the dangers of the microorganisms in a person's environment. Most exposures to microorganisms do not result in infection, because pathogens are rare relative to the other microorganisms in the environment, and the body has defenses against infection. The body's barriers against infection usually eliminate potential trouble without a person's knowledge. Only when a pathogen evades the barriers does infection occur. This event is the central theme of infectious disease medicine.

See also HYGIENE; IMMUNITY; INFECTIOUS DISEASE; LISTER, JOSEPH; NORMAL FLORA; NOSOCOMIAL INFECTION; PATHOGEN; PATHOGENESIS; TRANSMISSION; VIRULENCE; VIRUS.

Further Reading

Centers for Disease Control and Prevention. "Infection Control in Healthcare Settings." Available online. URL: www.cdc.gov/ncidod/dhqp. Accessed June 27, 2009.

Hunt, Margaret. "Interferon." Microbiology and Immunology Online. Columbia,: University of South Carolina School of Medicine, 2006. Available online. URL: http://pathmicro.med.sc.edu/mhunt/interferon.htm. Accessed June 27, 2009.

Infection Control Today. Available online. URL: www.infectioncontroltoday.com. Accessed June 27, 2009.

Isaacs, Alick, and Jean Lindenmann. "Virus Interference. 1, The Interferon." *Proceedings of the Royal Society of London, Biological Sciences* 147 (1957): 258–267. Available online. URL: http://caonline.amcancersoc.org/cgi/reprint/38/5/280. Accessed June 27, 2009.

Massachusetts Institute of Technology. "Eminent Scientist Receives $100,000 Lemelson-MIT Lifetime Achievement Award: Dr. Sidney Pestka Recognized for Groundbreaking Research on Anti-Viral Treatments." May 3, 2006. Available online. URL: http://web.mit.edu/invent/n-pressreleases/n-press-06LAA.html. Accessed June 27, 2009.

infectious disease An infectious disease is an illness caused by viruses, bacteria, fungi, or protozoa and can be transmitted between people or from animals to people. Infectious diseases strike every organ; some affect more than one organ at the same time. Depending on the pathogen, a microorganism may cause a few defined symptoms or produce a group of general symptoms. Infectious diseases also range from mild to deadly, and, in some parts of the world they are the major causes of death, especially among children.

Some diseases can be caused by a microorganism but do not qualify as infectious diseases because they do not spread from host to host. For example, anthrax disease develops when soil contaminated with *Bacillus anthracis* enters the body in a wound, by inhalation, or by ingestion but is not transmitted from person to person.

The main threat from infectious disease results from an *epidemic*. An epidemic is a disease that affects a large number of people at almost the same time. Epidemics sweep across populations or geographic regions. Other infectious diseases remain localized in an area and emerge periodically in smaller groups of people, a situation called *sporadic disease*.

TYPES OF INFECTIOUS DISEASES

Physicians diagnose any disease by referring to the illness's signs and symptoms. Signs are tangible clues that a disease is present, such as fever, rash, muscle spasms, or paralysis. Symptoms are intangible events experienced by a patient: pain, lethargy, nausea, and weakness. A group of signs and symptoms that together characterize a certain illness is called a *syndrome*.

Infectious diseases can, additionally, be defined as communicable or contagious. Communicable diseases transmit either directly from person to person or indirectly, that is, spread on common objects, such as an improperly sterilized dental instrument. A contagious disease is a communicable disease that spreads very easily from person to person, such as influenza.

Infectious diseases can be acute, having a fast onset and short duration, or chronic, developing slowly and persisting for a long time. Influenza provides an example of an acute disease, whereas hepatitis often develops as a chronic disease. Infectious diseases intermediate between these two types are called *subacute*. Infectious diseases may be latent, meaning the onset of symptoms does not occur until very long after the initial infection, which is the first entry of a pathogen into the body. Herpes and acquired immunodeficiency syndrome (AIDS) provide examples of latent infectious diseases.

DEVELOPMENT OF INFECTIOUS DISEASE

A disease from a microorganism begins with an infection. An infection is any establishment of illness-causing microorganisms in or on the body. The routes that microorganisms take to cause disease vary with the specific pathogen. In infectious disease, these routes, called *portals of entry*, are usually the skin, mucous membranes, lungs, or digestive tract.

After an infection, a pathogen must overcome the body's defenses in order to cause disease. The body produces physical, chemical, and biological factors that fight infection, led by the immune system, a multitiered set of defenses that halts most infections. If a pathogen evades these defenses, a disease develops, following a standard set of stages whether it is acute or chronic. The development of infectious disease is described in the table on page 438.

Any person who has signs or symptoms of a disease is capable of transmitting the disease to others. In some instances, a person carries a pathogen and transmits it to others but never shows any outward indication of the disease. The Irish immigrant Mary Mallon (1869–1938), better known as Typhoid Mary, worked as a cook in New York City, for several years, during which she transmitted *Salmonella* that caused typhoid fever. Throughout her life, Mallon never showed any signs or symptoms of typhoid fever and entered medical history as an example of an asymptomatic carrier. Carriers are people who act as a reservoir of disease and spread it to others. Some carriers have signs or symptoms, but others like Mary Mallon can be thought of as silent carriers.

Infectious disease can develop in any individual in good health, but infection becomes a higher risk in people with predisposing factors, which are any conditions that make the body more susceptible to invasion by a pathogen. For example, smokers have predisposing factors that increase their chance of respiratory diseases. This predisposing factor is fairly easy to define, but many such factors remain only partly understood by medical researchers. The following predisposing factors are known to increase the risk of contracting an infectious disease: smoking, poor nutrition, lack of access to clean water, poor hygiene, age, preexisting illness, chemotherapy, recent organ transplantation, chronic diseases, and certain occupations, such as hospital jobs. The following other factors may also contribute to increased susceptibility to infectious disease: stress, lack of sleep, fatigue, and emotional disturbances. Certain deleterious factors in the environment have also been implicated in increasing risk, such as contaminated tap water or residence near a dense industrial center.

GLOBAL INFECTIOUS DISEASES

Infectious disease accounts for 16.2 percent of mortalities worldwide (as of 2008), second to heart disease, which kills almost 30 percent of people. (Cancer is the third-leading killer, claiming 12.6 percent.) The portion of deaths due to infectious disease translates to 1.08 billion people annually.

Development of any infectious disease involves all of the factors discussed here in a complex relationship. For example, a healthy person can be susceptible to disease if exposed to a very high dose of pathogens, but a chemotherapy patient might contract an illness by being exposed to a few cells of the same pathogen. As a population changes—the average age increases or decreases, income increases or declines, more people move to cities—changes in infectious disease patterns also change. As of 2009, the World Health Organization (WHO) has reported in "World Health Statistics 2009" that three of every 10 deaths are due to communicable diseases. Although medical care is improving worldwide, other factors, such as growing poverty, overpopulation, and wars, can override medicine's progress.

The WHO and other health agencies such as the Centers for Disease Control and Prevention (CDC) admit that determining an accurate number of people with a given disease is difficult. In its 2009 report, the WHO stated, "Some diseases—malaria and yellow fever, for example—are endemic [established in the population] to certain geographical regions, but are extremely rare elsewhere. Diseases such as plague are prone to outbreaks which cause case numbers to fluctuate wildly over time. Some diseases are best tackled with preventive measures such as mass drug treatment, so reporting the number of cases is a lower priority than estimating the population at risk. For vaccine-preventable diseases, case numbers are affected by immunization rates." Global disease diagnosis can be further complicated by a lack of diagnostic equipment in poor regions or migrations of people due to famine or war. Finally, high mortality rates from a disease may not be a good indicator of the disease's overall threat to a population. Ebola virus has a high mortality rate, but it remains very rare worldwide.

The table on page 439 summarizes the major infectious diseases currently tracked by the WHO. In most updates from the WHO on infectious diseases, the agency must wait for individual countries to confirm disease cases and, then, compile the data. Ties Boerma, WHO statistician, explained in 2008, "Countries have a backlog of two, three years in publicizing their own information." In nations that do not compile death registration data, epidemiology studies provide an idea of an individual disease's global status.

ZOONOTIC INFECTIOUS DISEASES

Zoonotic diseases, also called *zoonoses,* are infectious diseases transmitted between humans and animals. The medical community and the agriculture industry focus mainly on diseases that people can catch from animals, but some diseases are transmitted from humans to animals.

Veterinary medicine in the United States identifies about 50 zoonotic diseases that present the greatest risk to human health. The WHO has cited about 200 zoonoses worldwide. These can range from mild foodborne illnesses to pathogens so deadly they are considered bioterrorism threats, for example, the Nipah virus. Most zoonotic infections result from either an animal bite or close association with and handling of live animals or animal products. From two million to almost five million animal bites occur annually in the United States; the wide range in this estimate is due to the many bites that probably are unreported. The main animal bites that cause medical concern are by dogs, cats, reptiles, and rodents, and these may be either domesticated animals or wild animals. Wildlife such as coyotes, squirrels, and fish also account for a portion of serious bites in the United States.

Dog and cat bites account for the majority of infections, and these are usually caused by the following bacteria: *Staphylococcus, Streptococcus, Pasteurella, Bacteroides, Clostridium,* and *Klebsiella.* These infections may cause local inflammations at the bite site, with the exception of *Pasteurella,* which leads to a more serious systemic infection. Handling live animals or animal products from killed animals, such as hides, also increases the risk of certain zoonotic diseases. Anthrax is an example of a disease that historically has resulted more often from the handling of animal products than from handling of live animals. On rare occasions, a person can contract a zoonotic disease by inhaling an infectious agent. Inhalation anthrax is one such example. The rabies virus has also caused infections when a person is in a confined area with a dense population of disease-carrying animals. An example of this circumstance would be a cave that serves as a roosting place for bats.

The table on page 440 describes the prevalent zoonotic diseases that occur worldwide.

Zoonoses remain a significant health problem in many parts of the world where people work closely with food animals and nonfood animal products. New zoonotic diseases may be emerging as the human population encroaches into areas where animals have lived undisturbed. With poverty, people also farm areas that were formerly jungle or savannah, and these activities can also be expected to uncover new zoonoses.

CURRENT ISSUES IN INFECTIOUS DISEASE

International programs in which countries work together to decrease the incidence of infectious disease have achieved notable successes, such as progress made against malaria and childhood deaths due to measles. In 2008, the WHO reported that worldwide deaths from measles declined by 74 percent, between 2000 and 2007. Peter Strebel of the WHO's Department of Immunizations, Vaccines, and Biologicals said in *HealthDaily News,* "The progress made today [against measles] is a major contribution to achieving the United Nations Millennium Development target of reducing child mortality by two-thirds by 2015." Other diseases have been more difficult to control. For example, HIV/AIDS in sub-Saharan Africa is an epidemic; HIV infected 1.9 million people in 2007, and this region contains 22 million people who are HIV-positive (HIV detected in the body). Meanwhile, new health

Stages of an Infectious Disease

Stage	Description
infection	initial entry of the pathogen; no signs or symptoms
incubation	pathogen establishes itself in the body and begins multiplying; no or mild signs or symptoms
prodromal period	short period after incubation with characteristic, but mild signs or symptoms
illness	most serious phase of the disease, with overt signs and symptoms; immune system in full activity
decline	period of one day to several days in which signs and symptoms subside; overworked immune system leaves patient at risk for secondary infections by pathogens
convalescence	recuperation of normal body strength

problems emerge before the old ones have been solved. In 2008–09, a new influenza strain, named H1N1, spread throughout the world. International health agencies reacted quickly to the imminent dangers from this atypical flu strain. The emergence of H1N1 showed how quickly a new pathogen can transmit through populations.

The WHO has described in its "World Health Report 2007" the changes in global behavior that have affected the threats from infectious diseases. An increased population with general migration to crowded urban centers increases the risk of disease transmission. Global travel on long flights has always increased the chance of transmission, but travel includes additional stressors: overcrowded flights, delays, limited meals, and sleep-deprived travelers. "In today's world," the report said, "public health security needs to be provided through coordinated action and cooperation between and within governments, the corporate sector, civil society, the media and individuals. No single institution or country has all the capabilities needed to respond to international public health emergencies caused by epidemics, natural disasters, environmental emergencies, chemical or biological attacks, or new and emerging infectious diseases. Only by detecting and

Infectious Diseases Worldwide

Disease	Cause	Approximate Incidence, New Cases Annually
hepatitis B	hepatitis B virus	2 billion (new plus preexisting cases)
sexually transmitted diseases	various bacteria, viruses, protozoa, and yeasts	340 million
malaria	*Plasmodium* species (P)	247 million
HIV/AIDS	human immunodeficiency virus (HIV)	33 million
influenza	influenza virus, type A and type B	4 million
tuberculosis	*Mycobacterium tuberculosis* (B)	2.6 million
hepatitis A	hepatitis A virus	1.4 million
mumps	paramyxovirus	409,000
measles	morbillivirus	281,000
leprosy	*Mycobacterium leprae* (B)	213,000
rubella	rubella virus	196,000
cholera	*Vibrio cholerae* (B)	178,000
pertussis (whooping cough)	*Bordetella pertussis* (B)	162,000
meningitis	various bacteria, viruses, and fungi	31,000
tetanus	*Clostridium tetani* (B)	20,000
Japanese encephalitis	Japanese encephalitis virus	10,000
diphtheria	*Corynebacterium diphtheriae* (B)	4,300
bubonic plague	*Yersinia pestis* (B)	2,300
poliomyelitis	polio virus	1,400

Note: B = bacteria; P = protozoa

reporting problems in their earliest hours can the most appropriate experts and resources be deployed to prevent or halt the international spread of disease." In addition to these factors, climate change must be considered in any discussion of emerging infectious diseases.

Major Zoonotic Infectious Diseases

Disease Agent	Type of Pathogen	Main Animal Carrier	Disease
anthrax *(Bacillus anthracis)*	bacteria	cattle, sheep	anthrax
Brucella *(Brucella melitensis)*	bacteria	pigs	brucellosis (undulant fever or Malta fever): intermittent, undulating fever with weakness, weight loss, and headaches
Chagas' disease *(Trypanosoma cruzi)*	protozoa	reduviid bugs that live in burrows in the soil	trypanosomiasis with fever, lethargy, liver damage, and lymph system disruption
fungi, various	fungi	domesticated animals	dermatomycoses with skin rashes, inflammations
equine encephalitis virus	virus	horses	encephalitis with fever, headache, nausea, paralysis, death
hantavirus	virus	wild rodents	pulmonary syndrome
influenza	virus	pigs, birds	influenza
Japanese encephalitis virus	virus	pigs	acute encephalitis
Leptospira interrogans	bacteria	pigs	leptospirosis with damage to kidneys or liver, meningitis, pulmonary hemorrhage, skin lesions
Lyme disease *(Borrelia burgdorferi)*	bacteria	deer (transmitted by ticks)	Lyme disease with fever, headache, joint pain, malaise, and muscle pain
Nipah virus infection	virus	pigs	encephalitis
plague *(Yersinia pestis)*	bacteria	rats, prairie dogs, squirrels (transmitted by fleas)	bubonic plague with high fever, swollen and painful lymph nodes, malaise, hemorrhages in skin
Q fever *(Coxiella burnetii)*	bacteria	birds, rodents (transmitted by ticks)	acute fever, headache, chills, sweats, and heart, liver, or nerve damage
Rocky Mountain spotted fever *(Rickettsia rickettsii)*	bacteria (transmitted by ticks)	rodents and other small mammals	inflammation and leaking of blood vessels, shock
Toxoplasma gondii	protozoa	mainly cats	toxoplasmosis with fever, destruction of lymph nodes, rash, muscle pain, and enlarged liver and spleen
tuberculosis *(Mycobacterium)*	bacteria	cattle, birds	tuberculosis
tularemia *(Francisella tularensis)*	bacteria	small mammals, especially rabbits and hares	tularemia with fever, headache, chills, and symptoms related to site of infection (eyes, throat, lungs)
West Nile virus	virus	rodents (transmitted by mosquitoes)	West Nile disease with fever, rash, muscle weakness, nausea, vomiting, seizures

Global warming experts have cited a number of actions that will begin or have already begun as a result of temperature rise. The following global warming–associated events affect infectious disease transmission:

- ocean level rise will increase flooding in lowlands, leading to increased populations of mosquitoes, carriers of disease
- flooding will increase risk of water contamination
- droughts will drain groundwater sources, causing possible contamination by drawing in water from surrounding soils
- increased numbers of rats and other disease-carrying pests
- disruption of food and water supplies will increase human susceptibility to infection
- tropical diseases will spread to temperate areas with increased temperatures
- increased pollen-producing vegetation will irritate respiratory conditions

International Health Regulations (IHR), drawn up in 2005, is a set of laws for the prevention of cross-border spread of infectious diseases—the IHR became fully active in 2007. The IHR contains the following main emphasis areas related to the spread of disease: international trade and shipping; monitoring of international ports, airports, and border crossings; travelers' health; global outbreak alerts; national health surveillance reports; and coordinated response activities to pandemics. The IHR may become increasingly important as the global warming–related health threats increase, as many experts expect.

In conjunction with the IHR, world health agencies such as the WHO have made the following communicable infectious diseases priorities for control: HIV/AIDS, tuberculosis, malaria, and blood transfusion–transmitted infections, such as those caused by viruses: hepatitis, cytomegalovirus, Epstein-Barr virus, herpes, and West Nile. Diseases targeted for eradication are leprosy, dengue fever, and parasitic worm infections. Leprosy is a bacterial disease caused by *Mycobacterium leprae,* and dengue fever results from infection by the dengue virus.

Infectious disease prevention and control are major parts of medical microbiology because these diseases collectively are the second leading cause of death worldwide. Infectious diseases vary by the agent that causes them, the reservoir of that agent, and the factors occurring between the pathogen and the host. Because of the vast amount of information on this subject, each infectious disease, or group of related diseases such as sexually transmitted diseases, constitutes a specialty within medical microbiology.

See also ANTHRAX; INFECTION; INFLUENZA; SEXUALLY TRANSMITTED DISEASE; TRANSMISSION; VIRUS.

Further Reading

Associated Press. "Heart, Infectious Diseases Top Killers Worldwide." October 27, 2008. Available online. URL: www.msnbc.msn.com/id/27402241. Accessed July 2, 2009.

AVERT. Available online. URL: www.avert.org. Accessed July 2, 2009.

Centers for Disease Control and Prevention. "Infection Control in Healthcare Settings." Available online. URL: www.cdc.gov/ncidod/dhqp. Accessed June 30, 2009.

Global Health Council. "Mortality and Morbidity of Infectious Diseases." Available online. URL: www.globalhealth.org/infectious_diseases/mortality_morbidity. Accessed July 2, 2009.

Infectious Diseases Society of America. Available online. URL: www.idsociety.org. Accessed June 28, 2009.

Kaur, Paramjit, and Sabita Basu. "Transfusion-Transmitted Infections: Existing and Emerging Pathogens." *Journal of Postgraduate Medicine* 51 (2005): 146–151. Available online. URL: www.jpgmonline.com/article.asp?issn=0022-3859;year=2005;volume=51;issue=2;spage=146;e page=151;aulast=Kaur. Accessed July 1, 2009.

National Institute of Allergy and Infectious Diseases. Available online. URL: www3.niaid.nih.gov. Accessed June 30, 2009.

Paediatric Infectious Diseases Web. Available online. URL: www.paediatric-infectious-diseases.com/index.htm. Accessed July 2, 2009.

Reinberg, Steven. "Worldwide Measles Deaths Drop Dramatically." MedicineNet, December 4, 2008. Available online. URL: www.medicinenet.com/script/main/art.asp?articlekey=94705. Accessed July 2, 2009.

University of Wisconsin School of Veterinary Medicine. "Zoonotic Diseases Tutorial." Available online. URL: www.vetmed.wisc.edu/pbs/zoonoses. Accessed August 4, 2009.

U.S. Department of State. "Infectious and Chronic Disease." Available online. URL: www.state.gov/g/oes/id. Accessed July 2, 2009.

World Health Organization. "Communicable Disease Surveillance and Response: International Health Regulations (IHR)." Available online. URL: www.searo.who.int/en/Section10/Section2362.htm. Accessed July 1, 2009.

———. "Infectious Diseases." Available online. URL: www.who.int/topics/infectious_diseases/en. Accessed June 28, 2009.

———. "World Health Report 2007: A Safer Future." Available online. URL: www.who.int/whr/2007/en. Accessed July 1, 2009.

infectious dose An infectious dose is the number of microorganisms that will infect 50 percent of a population within a specified period. Infectious dose, written in medical literature as ID_{50}, is determined by injecting laboratory animals with a pathogen and measuring the response of the animals by observing signs of disease. The ID_{50}, therefore, also names the test method used for comparing the strength of pathogens.

Infectious dose determinations provide medical researchers with information on the pathogenicity of a particular microorganism, that is, the ability of the pathogen to cause disease. Comparing two hypothetical pathogens, pathogen A with an ID_{50} of 15 has more capacity to cause disease than pathogen B, with an ID_{50} of 30. In this experiment, half of the animals receiving a dose of 15 of pathogen A contract disease. In order to make half of the animals sick with pathogen B, the dose must be raised to 30.

THE INFECTIOUS DOSE TEST METHOD

Scientists usually use rats or mice to determine infectious dose. This dose is either an actual number of microorganisms or the number per 100 grams of the animal's body weight.

A generalized ID_{50} test method consists of the following steps, designed to minimize the number of animals used in each test:

1. Prepare a suspension of pure microorganisms at a known concentration.
2. Administer a dose volume to give a known dose of microbial cells to the animals by either ingestion or injection.
3. Observe animals at periodic intervals for signs of infection.

Signs and symptoms of infection or disease vary by the type of pathogen. Some examples of the signs/symptoms used in infectious dose testing are in the following list. *Morbidity* is a term used for any state of being diseased, so ID_{50} test results are often reported as "signs of morbidity."

- fever
- altered heart rate
- neurological signs: tremors, paralysis, and others
- lethargy
- prostration
- death
- evidence of infection in biopsy specimens
- evidence of infection during necropsy

At the end of a specified period—one day to several weeks—the scientist records the number of animals showing morbidity. Doses may be increased, until 50 percent of the animals have signs of morbidity.

Laboratories have invented more rapid tests that act as confirmation of infectious dose without using additional animals, which also takes more time. A rechallenge ID_{50} test begins the same way as a conventional ID_{50} test, but at its completion, the scientist redoses all animals that did not show morbidity during the first test. All or a portion of the redosed or rechallenged animals may then develop morbidity. Some tests gather morbidity and death data in this manner, by determining the percentage of animals in the challenged group that died upon receiving the second dose.

INTERPRETATION OF ID_{50} TEST RESULTS

Infectious dose calculations provide helpful information by comparing the pathogenicity or virulence, the intensity of pathogenicity, of two different pathogens or two different strains of the same pathogen. A strain is a subpopulation of cells with unique traits descended from a single microorganism. For example, *Escherichia coli* is a bacterial species that lives in the digestive tract of warm-blooded animals. *E. coli* can cause illness when it is ingested. *E. coli* O157:H7 is a more virulent strain of *E. coli* from the digestive tract of cattle.

Infectious dose determinations must be used with caution because of the features of this test method. The main precautions of the ID_{50} test are as follows:

- in vitro test that may not represent in vivo conditions
- uses laboratory animals that may not be representative of human health conditions
- only useful in susceptible animals
- results do not account for variability in the pathogen

Microbiologists have learned to interpret ID_{50} test results with the preceding precautions in mind. Additional factors, detailed in the table on page 460, affect the way a pathogen behaves in vivo. These factors would not always be detectable by the in vitro ID_{50} test.

The factors described in the table on page 446 illustrate that the ID_{50} test may be best used in combination with other information about a pathogen, such as any epidemiological information known about a disease caused by the pathogen. Epidemiology is the study of how diseases originate and spread in a population. For instance, doctors know that the infectious dose of a pathogen in a healthy population

Will Global Warming Influence Emerging Infectious Diseases?

by Kelly A. Reynolds, Ph.D., Zuckerman College of Public Health, University of Arizona, Tucson, Arizona

According to the World Health Organization, infectious diseases cause approximately 26 percent of deaths worldwide. Ninety percent of those deaths are due to a mere six agents, at least half of which are known to be influenced by climate changes. Unquestionably, global warming will influence emerging diseases as the survival, replication and persistence of infectious microbes and their vectors are highly dependent on related climatic factors. It is not known, however, to what extent diseases will increase, or decrease, around the globe or in specific regions. Predicting human health effects related to climate change requires complex and holistic consideration of the many socioeconomic, ecological, environmental, and biological variables. The extent of climate change impacts on humans depends largely on the condition and preparedness of the populations exposed and their ability to adapt to the introduction of new insults.

The term *emerging diseases* refers to illnesses that suddenly appear in the population (emergence) or have been previously documented and are rapidly increasing in the number of new cases (resurgence) or geographical range (redistribution). From 1940 to 2004, 335 human infectious diseases reportedly emerged on Earth. The rate of these events has risen significantly over time, with many corresponding to documented climate changes, especially regarding vector-borne diseases.

Vector-borne diseases spread primarily via arthropods (i.e., mosquitoes, ticks, snails, fleas, tsetse flies) carrying infectious pathogens (i.e., viruses, bacteria, protozoa) from one individual to another, sometimes with an intermediary animal host. Nearly half of the world's population is estimated to be infected by a vector-borne disease, such as malaria, babesiosis, encephalitis, leishmaniasis, schistosomiasis, rickettsioses, Chagas' disease, onchocerciasis, yellow fever, chikungunya fever, dengue, plague, Lyme disease, sleeping sickness, leptospirosis, and West Nile virus illness. All of these illnesses are important public health threats and known to be influenced by climatic factors such as temperature, humidity, and precipitation. Malaria, caused by a protozoan parasite and transmitted by mosquitoes, infects an estimated 300–500 million people, resulting in one to three million deaths per year.

Zoonotic infections, due to pathogens transmitted between vertebrate animals and humans, are also linked to climatic events. Climate irregularities contribute to a series of ecosystem transformations, including increases or losses of vegetation affecting rodent populations and their predators. Severe and fatal hantavirus infections, for example, emerged in the Four Corners region of the United States (New Mexico, Arizona, Utah, Colorado), in 1993, after unseasonable El Niño–induced summer rains that led to a boom in deer mice populations—carriers of hantavirus. Plague, tularemia, leptospirosis, Lassa fever, and many other diseases, are transmitted by rodents and other wild animals, whose populations follow climatic trends. Domestic animal zoonoses, sensitive to climate, include tuberculosis, Rift Valley fever, avian influenza, and a variety of helminth (worm) infections. Indirect contact with animal wastes occurs via runoff from agricultural farms into marine and freshwater environments, impacting recreational and drinking water quality, as well as food safety.

Naegleria fowleri, Vibrio cholerae non-01/non-0139, *Pseudomonas aeruginosa, Legionella* spp., and *Aeromonas* are examples of generally harmless, naturally occurring surface water organisms that increase in numbers as temperatures and nutrients increase in their habitat. *Naegleria fowleri* (fatal brain-eating amoeba) feeds on algae in warm surface waters, increasing in number along with available food sources. In addition, algal blooms often produce toxins that cause skin rashes, respiratory ailments, and paralytic shellfish poisoning via consumption of shellfish harvested from contaminated areas.

Another shellfish contaminant, *Vibrio parahaemolyticus,* is an important food-borne pathogen. An outbreak from oysters farmed in Alaska followed record summer heat resulting in water temperatures above 59°F (15°C), initiating growth of the organism. This marks the most northerly outbreak of *V. parahaemolyticus* reported to date.

Enteric pathogens, such as *Cryptosporidium, Giardia,* norovirus, and *Vibrio cholerae,* are spread by food and water and typically cause diarrhea in humans. Worldwide, diarrhea is a major cause of mortality, resulting in 3.5 million deaths per year, primarily in developing regions. Cholera epidemics are commonly associated with climatic events such as El Niño, with peak incidence lagging a predictable amount of time (as little as three weeks) behind ambient temperature increases. *Vibrio cholerae* is a bacterium sensitive to long- and short-term climate changes, largely because of their association with marine plankton and copepod populations.

Many enteric illnesses follow a seasonal pattern and correlate significantly with climatic events. Studies in the Pacific Islands showed a 3 percent increase in diarrhea incidence for every 1.8°F (1°C) increase in temperature. Other studies found that a temperature increase of 1 percent results in a proportional 8 percent increase in hospital admissions. In the United States, more than 50 percent of waterborne disease outbreaks are preceded by heavy precipitation.

(continues)

Will Global Warming Influence Emerging Infectious Diseases?

(continued)

The term *climate change* is broadly used and not quantitatively defined, partly because areas around the world will be affected differently by global warming. Some regions will experience more droughts and others more floods. Droughts lead to poor hygiene practices, decreased food supplies, and famine, whereas floods result in an increase in disease-carrying insects and rodents, water contamination, and damage to sanitary sewers and drinking water distribution infrastructure.

Even slight increases in average daily temperatures can alter the earth's hydrologic cycle, leading to extreme weather events. Some of these effects can be predicted, while others remain uncertain. The global mean surface temperature is expected to increase at a rate of 0.36–0.9°F (0.2–0.5°C) per decade. Warming and precipitation claim 150,000 lives per year, and this number is expected to double by 2030. While some level of global warming is certain, identifying the impact of climate change on the emergence and reemergence of human disease is a challenge because of the many complex and interrelated processes.

Numerous articles and reports have been published related to the assessment of climate impacts on infectious diseases, including those by the Intergovernmental Panel on Climate Change (IPCC), the National Research Council (NRC), and the United States Global Change Research Program (USGCRP). Although specific reports vary in their impact assessments and priority issues, all generally agree that current predictive models lack the certainty and detail needed to evaluate the complexity of the climate-disease relationship adequately and that more empirical data are needed to improve the models. In addition, how well populations are able to predict, recognize, and adapt to changes in disease patterns should not be discounted.

Global efforts in education, research, water treatment, and dissemination of medical supplies/vaccinations will be necessary to reduce the impact of global warming. Individual controls such as the use of insect repellents, avoidance of contaminated recreational waters and shellfish, boiling water, and limiting travel can be highly effective. Community level controls such as the widespread use of pesticides, flood control infrastructure, animal waste management procedures, and policies related to recreational water and shellfish harvesting quality standards are also essential.

Current and future research should focus on improving the communication and response of public health systems, monitoring disease and climate patterns, and developing rapid detection systems for increasing response time to problems of prevention, control, and treatment. Predictive modeling will determine the range of possibilities for future outcomes. Understanding the likelihood of events and where they are expected to occur will help communities prepare for managing such events. If diseases are shifted to more

of people differs markedly from the infectious dose of the same pathogen in at-risk populations, groups who have a higher than normal risk for infection. The main at-risk or high-risk populations are the following:

- elderly adults
- infants and young children
- pregnant women
- immunocompromised individuals: acquired immunodeficiency syndrome (AIDS) patients, chemotherapy patients, organ transplant recipients
- people with preexisting or chronic health conditions such as diabetes
- smokers

Certain occupations affect infectious dose, also, such as jobs that expose workers to stressors: chemicals, long hours, lack of sleep, or loud noise.

MINIMAL DOSE AND LETHAL DOSE

Medical research on pathogens uses variations on infectious dose to add to the knowledge about the pathogen's capacity to infect and cause disease. Minimum infective dose (MID) or minimal infectious dose equals the minimal number of pathogenic cells needed to elicit an immune response in an animal. Put another way, an immune response indicates that the body detects the presence of infection. A MID may not lead to disease, even though it does cause an infection.

MID serves as an alternate way to assess pathogenicity. Some pathogens require only one cell to start an infection, multiply in the body, and initiate disease. Studies on MID help in determining the actual number of pathogens that could threaten the health of humans or animals. MID studies are also used for two additional determinations: (1) differences between the route of infection—blood, ingestion, inhalation, or other—and development of disease and (2) relationship between dose and the severity of the disease. The veterinary pathologist Gillian Dean explained in a 2005 article, "Minimum Infective Dose of *Mycobacterium bovis* in Cattle," of bovine tuberculosis,

affluent regions, systems may be in place to deal better with controls, compared to those of the previously stressed region, resulting in, perhaps, a net decline of illness. Unfortunately, many of the predicted hotspots for climate-related illnesses are in some of the poorest countries in the world.

Global warming will influence emerging diseases, but the degree of impact those diseases have on public health will depend on a variety of factors, including variations in the health status of the population exposed and how well prepared the population is to respond, control, and adapt to changing conditions. Adaptation is key to survival. Microorganisms provide clear examples of how populations can adjust to a range of environmental conditions. Few places on Earth are void of microbial life, including extreme environments such as volcanic vents and the arctic cold. Humans, too, have the ability to prepare and adapt to climate change effects.

See also EMERGING DISEASE; PATHOGEN.

Further Reading

Checkley W., L. D. Epstein, R. H. Gilman, D. Figueroa, R. I. Cama, J. A. Patz, and R. E. Black. "Effect of El Niño and Ambient Temperature on Hospital Admissions for Diarrhoeal Diseases in Peruvian Children." *Lancet* 355 (2000): 442–450.

Curriero, F. C., J. A. Patz, J. B. Rose, and S. Lele. "The Association between Extreme Precipitation and Waterborne Disease Outbreaks in the United States, 1948–1994." *American Journal of Public Health* 91 (2001): 1,194–1,199.

Jaenisch, T., and J. Patz. "Assessment of Associations between Climate and Infectious Diseases: A Comparison of the Reports of the Intergovernmental Panel on Climate Change (IPCC), the National Research Council (NRC), and United States Global Change Research Program" *Global Change and Human Health* 3 (2002): 67–72.

Jones, K. E., N. G. Patel, M. A. Levy, A. Storeygard, D. Balk, J. L. Gittleman, and P. Daszak. "Global Trends in Emerging Infectious Diseases." *Nature* 451 (2008): 990–993.

Lafferty, K. D. "The Ecology of Climate Change and Infectious Diseases." *Ecology* 90 (2009): 888–900.

Martinez-Urtaza, J., B. Huapaya, R. G. V. Gavilan, J. Blanco-Abad, C. Ansede-Bermejo, Cadalso-Suarez, A. Figueiras, and J. Trinanes. "Emergence of Asiatic *Vibrio* Diseases in South America in Phase with El Niño." *Epidemiology* 19 (2008): 829–837.

McMichael, A. J., R. E. Woodruff, and S. Hales. "Climate Change and Human Health: Present and Future Risks." *Lancet* 367 (2006): 859–869.

Parkinson, A. J. and J. C. Butler. "Potential Impacts of Climate Change on Infectious Diseases in the Arctic." *International Journal of Circumpolar Health* 64 (2005): 478–486.

Patz, J. A., D. Engelberg, and J. Last. "The Effects of Changing Weather on Public Health." *Annual Reviews in Public Health* 21 (2000): 271–307.

Singh, R. B. K., S. Hales, N. de Wet, R. Raj, M. Heamden, P. Weinstein. "The Influence of Climate Variation and Change on Diarrheal Disease in the Pacific Islands." *Environmental Health Perspectives* 109 (2001): 155–159.

World Health Organization. "Climate Change and Human Health 2003." 133–158. Available online. URL: www.who.int/globalchange/environment/en/ccSCREEN.pdf. Accessed June 5, 2010.

"In the intranasal model, 5×10^5 to 10^6 CFU [colony-forming units, which equals number of cells] resulted in multiple respiratory lesions, while 5×10^2 to 10^4 CFU resulted in a more variable pathology (some animals had multiple lesions, and some animals had no lesions) and 10^2 CFU resulted in no visible lesions at all." MIDs have been determined for bacterial, viral, fungal, and protozoan pathogens.

Lethal dose (LD_{50}) is an amount of pathogens that destroys 50 percent of the host cells or host animals. Virus lethal dose determinations often use mammalian cells in in vitro tissue culture methods to observe actual cell death, when the tissue has been infected by viruses. Virologists have developed two different in vitro tests for measuring specific virus activities. The first test, the tissue culture infectious dose ($TCID_{50}$), measures the activity of viruses inside mammalian cells grown in vitro. The second test, called a plaque assay, measures the amount of virus particles needed to destroy bacteria grown as a film on agar. Holes in the bacterial film called *plaques* indicate places where a virus particle has destroyed the surrounding bacterial cells.

In animal tests, the LD_{50} test works identically to the ID_{50} test, but death, rather than signs of disease, is measured as the end point of the test (An end point is any reaction that indicates a test method has achieved its desired result and the test is complete.) Scientists sometimes set up LD_{50} tests to determine a lethal dose of pathogen and, then, prepare various dilutions of the dose to determine the infectious dose. Minimum lethal dose (MLD) equals the minimal dose needed to kill 100 percent of test animals. As mentioned, scientists have also designed experiments that gather both infectious dose and lethal dose data using a rechallenge of pathogens in LD_{50}: ID_{50} tests.

Lethal dose tests have become common indicators of pathogenicity because the end point, death, is much easier to assess than the signs and symptoms of disease that require careful observations or measurements. Each method, however, belongs to a group of test methods called end point tests because the test continues until a predefined event, the end point, has occurred. End points may be death, initiation of fever, increased heart rate, or other physiological responses.

Enzyme-linked immunoassays have been combined with ID_{50} tests to receive faster and potentially more accurate information on the infection. One such test is the enzyme-linked immunosorbent assay infectious dose test, or the ELISA-ID_{50} test. This method checks the blood for the appearance of antibodies to the pathogen after it has been dosed into the animal. The presence of antibody specific to the pathogen serves as the end point in this test.

APPLICATIONS OF INFECTIOUS DOSE

Many factors cause large variability in infective dose so that its main benefits are twofold: (1) for comparison of relative pathogenicity of two microorganisms and (2) determination of the dose of pathogens that can make a person sick, but only when a large amount of information is already known about the pathogen. For many years, medical researchers used the following assumption before studying infectious dose in animals: Minimal animal infectious doses equal minimal human infectious doses. Researchers now know this is an incorrect assumption. Some pathogens that infect animals are, furthermore, harmless in humans, or vice versa. An exact number to describe any infectious dose must be used with caution.

Infectious dose has been most useful in studying food-borne illnesses. These numbers vary, but they can still help assess the risks associated with eating contaminated food. Food-borne pathogens with a low infectious dose (less than 100) present much higher risks in food than pathogens that have a very high infectious dose (several thousand to more than one million). But even pathogens with a high infectious dose can be dangerous if they produce toxins. Cooking may kill all of the pathogens, but cooking does not destroy toxins in the food that have been produced by many thousands of cells. For this reason, food microbiologists also use minimal toxic dose to determine the number of pathogens that make most people sick from either the pathogen or the pathogen's toxin.

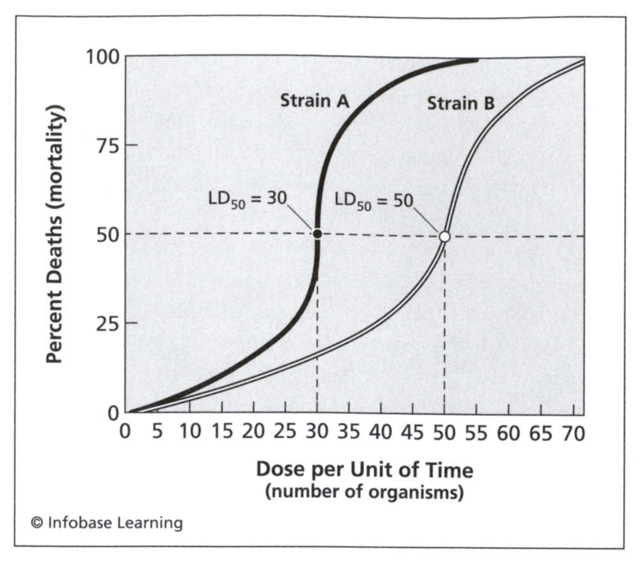

The lethal dose assay indicates the dose of a toxin or pathogen that kills 50 percent of a population. In this example, strain A is more lethal than strain B because strain A requires a dose of 30 cells to kill 50 percent of the test animals, whereas strain B needs 50 to do the same.

Factors in Interpreting ID_{50} Test Results

Factors	Description
pathogen variables in vitro	• genetic variability • potential stresses to the pathogen before dosing
pathogen variables in vivo	• interactions with food, tissue, or blood • susceptibility to pH • unique immunological factors • interaction with other microorganisms
host variables	• age, pregnancy, general health • unidentified stressors • preexisting health conditions • genetic metabolic variability • nutrition • immune system competence

In food microbiology, the term *infective dose* is more common than *infectious dose*. Food-borne infective dose refers to the number of microorganisms needed to make a normal person sick; food microbiology does not use the 50 percent standard for most descriptions of food-borne illness. The table on page 477 shows infective doses of common food-borne pathogens.

Infectious dose can be thought of in variable terms only because of the many factors that create infection. In addition to the virulence of a pathogen, each person has preexisting health conditions that greatly affect infectious dose. No one should make the mistake of dismissing a small number of pathogens as a safe number.

See also FOOD-BORNE ILLNESS; IMMUNOASSAY; PLAQUE.

Further Reading

Dean, Gillian S., Shelley G. Rhodes, Michael Coad, Adam O. Whelan, Paul J. Cockle, Derek J. Clifford, R. Glyn Hewinson, and H. Martin Vordermeier. "Minimum Infective Dose of *Mycobacterium bovis* in Cattle." *Infection and Immunity* 73 (2005): 6,467–6,471. Available online. URL: http://iai.asm.org/cgi/reprint/73/10/6467. Accessed July 1, 2009.

Infective Doses of Food-Borne Pathogens

Pathogen	Dose (number of cells per 100 grams food)
Bacillus cereus	10^5–10^{11}
Campylobacter jejuni	less than 500
Clostridium botulinum	nanogram amounts of toxin
Clostridium perfringens	10^6–10^{10}
Cryptosporidium parvum	less than 30 (cysts)
E. coli	varies with person from 10–10^8
E. coli 0157:H7	as low as 10
Listeria monocytogenes	varies with person from about 100 to 10^6
Salmonella species	15–1,000
Staphylococcus aureus	10^6

Mara, Duncan, and Nigel Horan, eds. *Handbook of Water and Wastewater Microbiology.* San Diego: Academic Press, 2003.

Schiff, Gilbert M., Gerda M. Stefanovic, Betsy Young, and Julia K. Pennekamp. "Determination of Minimal Infectious Dose of an Enterovirus in Drinking Water." U.S. EPA Health Effects Laboratory, Research and Development, June 1983. Available online. URL: http://nepis.epa.gov/Exe/ZyNET.exe/20016T19.txt?ZyActionD=ZyDocument&Client=EPA&Index=1981%20Thru%201985&Docs=&Query=&Time=&EndTime=&SearchMethod=1&TocRestrict=n&Toc=&TocEntry=&QField=pubnumber%5E%22600S183004%22&QFieldYear=&QFieldMonth=&QFieldDay=&UseQField=pubnumber&IntQFieldOp=1&ExtQFieldOp=1&XmlQuery=&File=D%3A%5CZYFILES%5CINDEX%20DATA%5C81THRU85%5CTXT%5C00000013%5C20016T19.txt&User=ANONYMOUS&Password=anonymous&SortMethod=h%7C-&MaximumDocuments=10&FuzzyDegree=0&ImageQuality=r75g8/r75g8/x150y150g16/i425&Display=p%7Cf&DefSeekPage=x&SearchBack=ZyActionL&Back=ZyActionS&BackDesc=Results%20page. Accessed August 22, 2009.

Schmitt, Ronald H. "A Reference Guide for Food-Borne Pathogens." 2005. Available online. URL: http://edis.ifas.ufl.edu/pdffiles/FS/FS12700.pdf. Accessed August 20, 2009.

U.S. Food and Drug Administration. "Bad Bug Book." Available online. URL: www.fda.gov/Food/FoodSafety/Food-borneIllness/Food-borneIllnessFood-bornePathogensNaturalToxins/BadBugBook/default.htm. Accessed August 22, 2009.

influenza The term *influenza* describes a group of viruses that cause sporadic and epidemic outbreaks of influenza (the flu) in mammals. The human disease influenza is a respiratory illness characterized by fever and chills, headache, and muscle pain. The disease is not normally fatal in healthy individuals in the general population, but it, nevertheless, causes a significant number of deaths worldwide each year, especially among the elderly, the young, and the infirm. In years with severe epidemics, fatalities from the virus infection increase dramatically. The influenza virus has been, throughout history, one of the most difficult pathogens to keep out of human populations.

Viruses in the genus *Influenzavirus* contain a single strand of ribonucleic acid (RNA) that represents the virus's entire genetic material, called its genome. As with all viruses, influenza must infect a healthy cell in order to replicate and build new virus particles. Three different types of influenza infect humans: types A, B, and C. These three forms have the following characteristics:

- type A—most common form, which causes most of the severe flu epidemics
- type B—can cause epidemics, but the disease is milder than type A flu
- type C—not associated with serious outbreaks

Type A causes the majority of severe flu cases in humans and is complicated by seasonal changes in the components of the virus's outer coat. These small changes in the composition of the virus surface are collectively called *antigenic shifts,* or antigenic drift. In other words, the virus, which acts as an antigen in the body, presents constantly changing structures for the body's immune system to recognize. Antigenic shift presents two challenges to preventing the flu: First, vaccines prepared for a flu outbreak may not be effective against other versions of the virus in the next flu season or in a current flu outbreak, and, second, a person's immunity against one version of the virus may not be effective against another version. The variations within influenza have made this pathogen one of the most difficult problems in health care today.

INFLUENZA VIRUS

The influenza virus genome contains eight RNA segments that encode for (carry information for synthesis of) 10 viral proteins. The genome resides inside a protein coat called the *capsid,* which is covered by an outer fatlike lipid layer. The lipid layer, called the *envelope,* of type A influenza holds two types of glycoprotein (protein with sugar attached to it) projections on the outer particle that play important roles in infectivity. The table below describes the differences between these two projections, called the hemagglutinin spikes (H spikes) and the neuraminidase spikes (N spikes). Influenza types B and C do not contain these structures. The H and N spikes play an important role in causing disease because they give the virus a new structure that the body's immune system may not recognize, and, therefore, it does not readily set up a defense. The variety in H and N spikes and subsequent antigenic drift has made influenza one of the most difficult virus infections to control.

The H spikes and N spikes on the influenza type A virus differ slightly in structure, so that the different forms have been named *H1* (or H_1), *H2, H3;* and *N1* (or N_1), *N2, N3;* and so on. Certain flu epidemics, in history, have been caused by particular combinations of H and N spikes; these combinations are called *antigenic subtypes.* Certain antigenic subtypes have made the virus unusually virulent, in some years, and caused dangerous flu epidemics. Virulence is the capacity to cause severe disease. The table on page 449 describes major flu outbreaks in history and the antigenic subtypes associated with these epidemics. Epidemics are outbreaks that rapidly spread through a large, defined population, such as the mid-Atlantic United States. Epidemics that spread into a much larger global population are called *pandemics.* In modern medical history, influenza has been the cause of several pandemic outbreaks, with no end in sight.

Humans become infected mainly by viruses containing H1, H2, or H3 spikes, although sporadic flu cases have been caused by strains with H5 and H9 spikes. Animals tend to become infected with viruses containing H4 and higher spikes. Currently, N1 and N2 combine with H spikes in most human and animal flu outbreaks.

Flu outbreaks normally occur annually, in the winter months. No one has definitively proven why winter produces more severe outbreaks, but two theories have been proposed as the most likely reasons. First, people crowd together indoors during the winter months (in temperate climate regions) and thus aid the spread of the virus. Second, factors in the upbringing of young animals may affect the animal-to-human transmission of the virus on a cyclical basis that coincides with the months November through February.

A third factor clearly also contributes to making some years' viruses more virulent than those of others, leading to the outbreaks described in the table. Scientists have not completely discovered all the reasons why the influenza virus contains antigenic subtypes of greater virulence in some years than other years. Although vaccine manufacturers have developed methods to prepare for each season's most probable new influenza virus, the system is not 100 percent foolproof, and unique viruses will always get through and cause infection.

The biology behind flu outbreaks relates to the virulence of the virus and the body's ability to combat the virus; both factors are determined, in large part, by the virus's antigenic subtype. A third biological factor arises from the ease with which influenza viruses transmit from animals to humans. The flu is a zoonotic disease, meaning it originates in animals and, then, moves to the human population, where it causes illness. Influenza viruses possess weak attraction for specific animal tissues, explaining why influenza transmits between species more easily than many other viruses. Flu outbreaks that have entered human populations have originated principally from swine and birds, but horses, cats, and marine mammals also harbor the influenza virus from time to time.

Pigs serve as a reservoir for new antigenic subtypes of influenza viruses by acting as a place where more than one virus subtype exists and exchanges genetic information. Two different influenza strains that have infected a pig may share their genes in a process called *reassortment,* so that a new virus that is more virulent than either of the originals emerges. Humans and even pigs lack immunity to this new strain, so they are very susceptible to infection. The University of Wisconsin School of Veterinary Medicine has explained on its Web site, "Genetic reassortment is the basis for one form of influenza variation called 'antigenic shift.' An antigenic shift occurs when an influenza virus of a new subtype enters a population and, because the animals/humans lack any preexisting immunity to the new virus, it spreads rapidly throughout the population. 'Pandemics' of human influenza tend to occur ~every 10–40 years." People also can be infected by the influenza virus that sickens pigs, even without any reassortment prior to the transmission from pigs to humans.

Influenza experts had long believed that the transmission of influenza from birds to humans

Influenza Antigens

Type of Spike	Approximate Number per Virus Particle	Activity
H	500	enables virus to recognize target cells before attaching to them
N	100	helps new virus particles exit the host cell after replication

occurred only rarely. Evidence has been mounting, however, to show that avian influenza viruses cause considerable harm to human populations. The 1957 and 1968–69 pandemics resulted from reassortment between an avian virus and a human influenza virus that had already caused sporadic cases of flu. The new strain, called a *variant,* was much more virulent than either strain alone. The most severe flus have, in fact, been those in which two or three different influenza viruses exchange genes inside an animal cell and create a brand-new strain. People living or working with infected animals will become infected and begin the flu's spread in the human population. The University of Wisconsin School of Veterinary Medicine stated, "A very unique outbreak of influenza occurred in Hong Kong in 1997. It was unique because it involved an H5N1 avian influenza virus, because infection in people was epidemiologically linked to direct contact with infected chickens in the live-bird markets, and because of the severity of the disease induced. In fact, 6 of 18 infected people died." Similar sporadic and deadly outbreaks have been found, since then, caused by bird-to-human transmission.

The world health care community contributes, in its own way to additional cases of flu, when it develops a vaccine that does not work against the season's main influenza strains. Even with the best of decisions that lead to an effective vaccine, societal factors have had an impact on the way flu travels through world populations. The main factors are the following:

- inadequate funding for vaccination programs
- inadequate distribution of vaccines
- incomplete animal vaccination efforts
- lack of quarantine of sick people or animals
- globalization that increases the worldwide spread of the virus
- alteration of bird habitats that affect migration and breeding grounds
- global trade in live pet birds

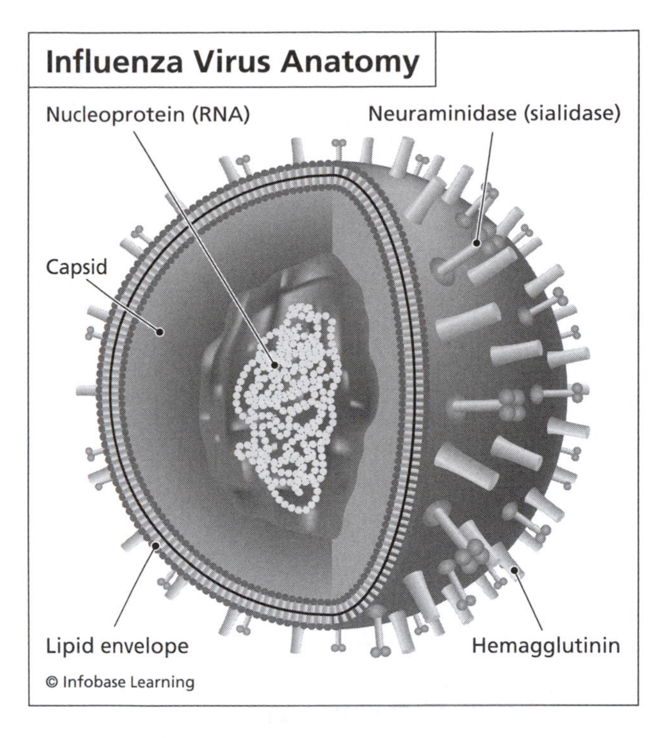

Influenza virus is not usually lethal to healthy individuals with strong immune systems, but it, nonetheless, causes thousands of deaths worldwide each year, mainly of young children.

In 2008, a new H1N1 influenza virus emerged and took advantage of several of the factors listed, especially globalization. The *Time* reporter Bryan

Major Type A Influenza Outbreaks

Subtype	Year of Outbreak	Probable Origin	Consequences
H3N2	1889	southern China	first confirmed modern pandemic
H1N1 (Spanish flu)	1918–19	China	deadliest flu epidemic in history, killing 20 million people
H2N2 (Asian flu)	1957	East Asia	pandemic that killed at least two million people
H3N2 (Hong Kong flu)	1968–69	Hong Kong	pandemic that killed almost one million people
H1N1 (Russian flu)	1977	Russia	possible escape of the pathogen from a laboratory
H5N1 (avian flu)	2005	Asia, Europe, Near East, Africa	unusual, by spreading more readily from birds to humans than from humans to humans
H1N1 (swine flu)	2008–09	Mexico	pandemic and unusual, by spreading during summer months

Walsh wrote, in May 2009, "H1N1 makes clear how vulnerable our interconnected globe is to emerging diseases. As a result of jet travel and international trade, a new pathogen managed to seed itself in more than 20 countries in less than two weeks." Good medical care in most places in the world may make the flu incapable of killing large numbers of people in a single outbreak, but the speed with which pathogens move, as Walsh noted, has become more troublesome than it was 30 years ago.

INFLUENZA EPIDEMICS

The influenza virus has caused some of history's most deadly epidemics. Hippocrates recorded the earliest known flu outbreak, in 412 B.C.E., but flu outbreaks have plagued every generation.

The worst flu epidemic in history began, in 1918, during World War I. Soldiers at Fort Riley, Kansas, started to visit the infirmary in increasing numbers, complaining of headache, chills, and fever. Some had nausea and a persistent dry cough. All of these were typical flu symptoms, but this outbreak would prove to be atypical. The healthiest and strongest young adults were as vulnerable as the elderly, and the disease often progressed into a devastating lung infection and death. The epidemic to become known as "the Great Influenza" spread throughout the United States and the world. In the next two years, the pandemic had claimed 22 million lives. The occurrence of a world war, from 1914 to 1918, undoubtedly helped in transmitting the virus worldwide.

Medical historians have studied medical records to find clues as to why the great influenza was the worst epidemic of all time. Researchers have even reconstructed the 1918 influenza virus's genes. The unusually virulent 1918 influenza virus elicited a dramatic immune response in the lungs of the infected. The virologist Yoshihiro Kawaoaka led the research team that studied the 1918 virus's unique virulence. In 2007, Kawaoaka told *ScienceDaily*, "[The study] proves the 1918 virus was indeed different from all other flu viruses we know of." The virus initiated an exaggerated immune response that could spiral out of control and attack the victim's organs and kill the host. In addition, medicine knew almost nothing of viruses, in the early 20th century—the influenza virus would not be discovered until 1933. Researchers today have likened the 1918 flu response to toxic shock syndrome, in which a violent reaction by the body to an infection interferes with organ function and causes death.

The 1918 influenza virus has since been identified as a H1N1 subtype. In 2005, another H1N1 influenza appeared in Mexico. This strain of influenza contained a blend of bird flu, swine flu, and a human-type virus. By 2008, the virus had spread globally, and the World Health Organization (WHO) labeled it a new flu pandemic. The continued H1N1 pandemic of 2009 differs from other yearly flu outbreaks in the following three ways:

- H1N1 occurs year-round.
- The virus moves as easily from humans to pigs as from pigs to humans.
- The virus has been aided as no previous influenza has by worldwide travel and globalization of economies.

These factors indicate that H1N1 has increased opportunity to transmit and cause disease.

Modern medicine uses a variety of techniques to monitor the spread of flu through a population or around the world. Epidemiologists, people who study the spread of disease in a population, gather and analyze large amounts of data on the incidence of flu cases and the type of influenza virus causing most illnesses each year. Similar epidemiology also helps vaccine manufacturers predict which influenza will be the most likely pathogen in next year's flu season. Even with epidemiological assistance, doctors can help determine whether a new flu epidemic is growing simply by monitoring the illnesses they see among patients. Doctors and epidemiologists monitor flu outbreaks by the following methods:

- monitoring online queries from the public on influenza or influenza symptoms
- recording increases in physician visits and the symptoms being presented to doctors
- monitoring the drugstore sales of medicines that treat flu symptoms

These monitoring techniques help gain perspective on in-progress flu epidemics, but, as in years past, medical care providers often learn about a new outbreak after the outbreak has affected many people in a population.

THE FLU

Influenza virus can infect in epidemic proportions because it transmits easily from person to person. The flu is called a contagious disease because people can catch it from others. The virus travels between people in tiny moisture droplets called *bioaerosols* that are expelled with sneezing, coughing, and perhaps talking. As a result, a person becomes infected by inhaling the infectious virus, and, so, the flu may be described as a contagious respiratory disease.

Flu carriers can transmit the virus beginning one day before their own symptoms start and up to five days after the start of the flu symptoms. The symptoms begin from one to four days after the virus has entered the body.

The flu symptoms begin suddenly compared with those of the common cold, which develops over a few days. The typical flu symptoms are the following:

- high fever
- headache
- weakness and tiredness
- dry cough
- sore throat
- runny or stuffy nose
- muscle aches

More serious cases of the flu may develop in the elderly or the young or in people with weakened immune systems. In addition to the symptoms listed here, adults in a higher-risk health condition may experience severe respiratory distress and death. In children, the flu can cause severe diarrhea and vomiting, particularly among children who receive inadequate nutrition.

Although most normal adults can recover from flu in about two weeks, severe cases can lead to hospitalization. The CDC has stated that 5–20 percent of people in the United States contract flu each year. Of these individuals, more than 200,000 people must be admitted to a hospital because of life-threatening conditions; about 36,000 people in the United States per year die of the flu.

People can reduce their chance of catching the flu by following the basic practices of good personal hygiene, highlighted by the following actions:

- washing hands often with soap and water, especially after sneezing or coughing
- covering the mouth while sneezing or coughing to prevent spread of the flu to others
- avoiding touching the eyes, nose, or mouth with fingers or hands
- avoiding people who have flu symptoms
- staying home from work or school while sick with the flu
- receiving a flu vaccination

The CDC recommends the following three actions for flu prevention: (1) take time to be vaccinated, (2) take everyday preventive actions, and (3) take flu antiviral drugs if recommended. Everyday preventive actions include the hygiene practices listed here and known to prevent the spread of the flu, but vaccines and other drugs have had a more difficult time achieving 100 percent effectiveness against flu.

INFLUENZA VACCINES

Influenza vaccines are of two types: injection and nasal spray. The injectable vaccine, known as the flu shot, and nasal sprays contain influenza virus particles that have been treated so they cannot cause disease. Adults who have no preexisting or chronic illnesses usually receive a flu shot, which must be given yearly.

Makers of injectable flu vaccines mix together the components of more than one influenza virus that researchers expect to cause the main health threat in the upcoming flu season. The predictions of the next major influenza threats must be based on global surveillance studies on the current emerging antigenic subtypes. The CDC and the WHO lead the research efforts to make the most accurate predictions of the causative agents of the next flu outbreak. The researchers usually select three subtypes most likely to infect large numbers of animals in regions of the world where the flu usually originates. Because the vaccine contains three different influenza strains, it is called a *trivalent vaccine*. Today's flu vaccines no longer use whole virus particles, as they did in previous years. The vaccines, instead, are called *split-virus vaccines* because they contain only pieces of the viruses and not the whole virus particle. Split-virus vaccines produce fewer side effects than previous whole-virus versions.

Vaccine producers, then, grow the selected viruses in chicken eggs for several days to two weeks; purify the virus; then expose it to chemicals to inactivate it. The entire process, from virus selection to the manufacture of a vaccine, takes about 11 months, so if the scientists err in predicting the main influenza strains, that year's vaccine may not be effective against most influenza infections.

For many years, beginning in 1945, scientists used mercury-containing compounds to inactivate the influenza virus for vaccines. Small amounts of mercury injected into people with each flu shot created a health threat much worse than the effects of the flu itself, so medical researchers began testing a variety of alternative inactivation methods. Some of the following methods have been used instead of mercury compounds in making flu vaccines: heating (pasteurization), low pH, chemicals such as chlorine, and organic solvents.

The National Network for Immunization Information (NNII) in Galveston, Texas, has stated that less than 1 percent of adults experience side effects from flu shots. The main known side effects are:

- soreness or tenderness at the injection site
- fever
- chills
- general weakness

In 1976, a batch of swine flu vaccine caused an unexplained increase in the incidence of a nerve disorder called Guillain-Barré syndrome (GBS). The incident threw the safety of the flu shot into doubt, and medical researchers soon set up studies to find any connection between flu shots and GBS. Medical researchers have since compiled data on the vaccine's side effects, including the possibility of a connection between flu shots and GBS. In 2004, the public health specialist Penina Haber was the lead author of a study to assess the flu shot–GBS connection. She wrote in the *Journal of the American Medical Association,* "Guillain-Barré syndrome remains the most frequent neurological condition reported after influenza vaccination to the Adverse Events Reporting System (AERS) since its inception in 1990." The antigenic subtypes of viruses in each year's flu vaccine vary, yet a connection seemed to exist between the virus or the vaccine and GBS. Additional research that has been completed implicates the influenza virus as the link to GBS. The French virologist Elyanne Gault told Reuters in, 2009, "that there was virological evidence that influenza infection is a trigger for Guillain-Barré syndrome, with a frequency related to the level of influenza epidemics." Even with additional confirmations of a link between the virus and GBS, GBS remains a very rare occurrence. The NNII has estimated that about one person in one million immunized with swine flu vaccine, for example, will develop the illness.

A second type of influenza vaccine, the nasal spray vaccine, uses live attenuated influenza viruses, meaning the virus has not been killed but has been passed through several generations to remove its virulence. Nasal spray flu vaccine has been developed mainly for healthy people older than two years who are not elderly and who have no other chronic health conditions.

In 2008, a new H1N1 influenza strain spread worldwide, and, by mid-2009, it had caused 1,200 deaths. Health agencies and drug manufacturers reacted quickly in order to arrest the pandemic. Part of the prevention program included a rush to make an effective vaccine available to combat what many experts expected to be a deadly pathogen. The reporter David Batty wrote for the United Kingdom's *Guardian,* "To ensure the vaccine is available as soon as possible, the EMA [European Medicines Agency] is allowing companies to bypass large-scale human trials." Drug companies tried to stay ahead of a poten-

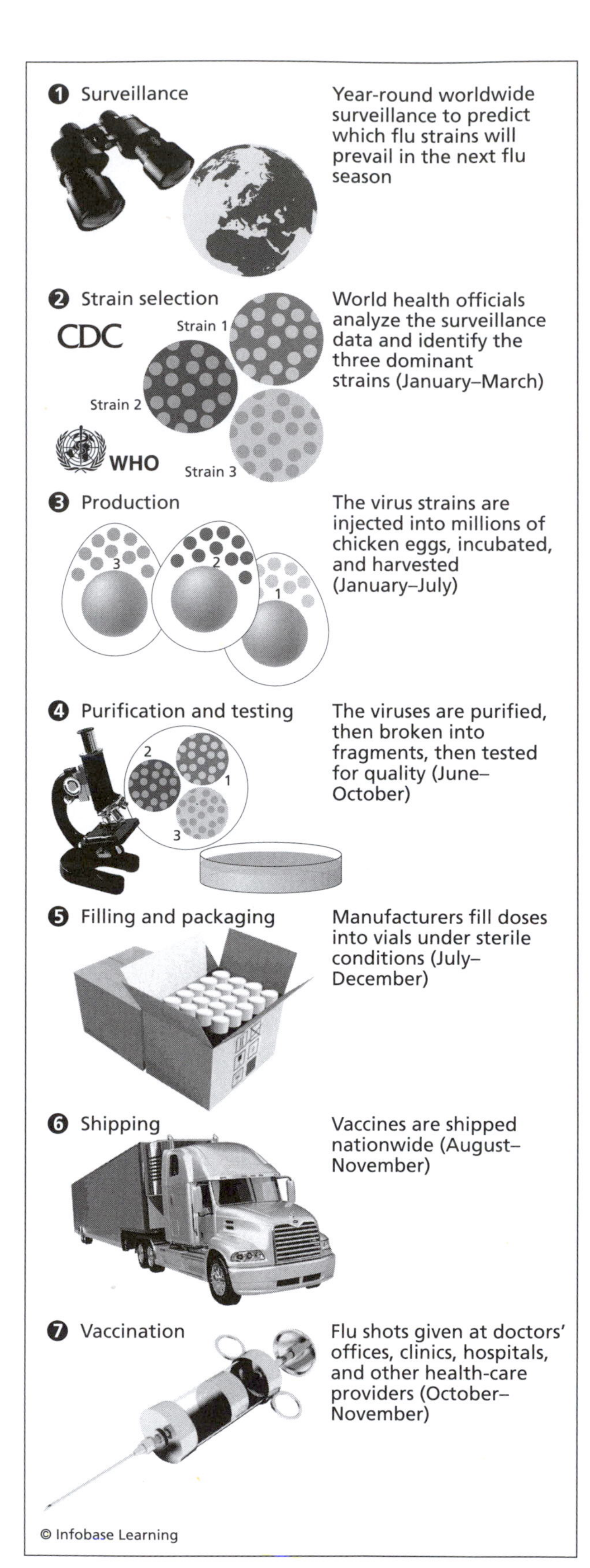

Vaccine preparation for each flu season takes an entire year. If incorrect strains are used for making the vaccine, then that year's batch of flu shots will be ineffective.

tial flu pandemic by accelerating their testing and decision-making process for the H1N1 vaccine. Batty wrote, "The idea was to do most of the testing [on a similar avian influenza virus] before a pandemic, so that when it hit, the drug companies could insert the pandemic virus into the vaccine. When the first doses are ready, the EMA will approve them largely based on data from the bird flu vaccine, since both will have the same basic ingredients." Scientists in the United States, meanwhile, decided to test the new H1N1 vaccine in humans before selling it. Describing the fast timing, Anthony Fauci, director of the National Institute of Allergy and Infectious Diseases, said in 2009, "It's [completion of the testing] going to be close. I believe it can be (ready by October) if things run smoothly. We hope they will, but you never can tell when you're dealing with biological phenomena like making vaccines and administering them." Even if the H1N1 threat is halted with minimal effects on human health, influenza viruses will appear each year with unique traits that will make future influenza strains just as difficult to control.

TREATMENTS AGAINST INFLUENZA

The use of vaccines and other preventative measures provide the safest approach to avoiding the flu. Some treatments are available, however, for instances in which influenza becomes a serious threat to a person's life. The CDC lists four different drugs that are effective antiviral treatments against influenza infection: the products amantadine, ostelamivir (product name *Tamiflu*), rimantadine, and zanamivir (product name *Relenza*). The CDC currently recommends only ostelamivir or zanamivir, alone or in combination, as the best treatments for the flu if taken within 48 hours of the appearance of flu symptoms. Each antiviral drug is prescribed by doctors, who take into account the following information in order to select the most effective treatment:

- type of influenza strain suspected of causing the flu
- known information on strains resistant to antiviral drugs
- age
- preexisting health conditions
- allergies to the drugs

Of the factors listed here, resistance of the influenza virus may become the most important consideration when prescribing the antiviral treatment. As of 2009, the CDC reported that more than 92 percent of H1N1 strains isolated from patients had developed resistance to ostelamivir.

Ostelamivir and zanamivir belong to a group of drugs called neuraminidase inhibitors. Viruses use the enzyme neuraminidase to damage host cells, allowing the new virus particles to escape and infect other cells in the body. By inhibiting this action, the drug prevents the virus particles from spreading infection. For this reason, these two drugs have become the most accepted treatments for fighting influenza infections.

Antiviral treatments of influenza infections are usually given to people older than two years who are otherwise healthy. People who have any of the following conditions should not use antiviral influenza treatments: allergies to the drug, chronic respiratory diseases, any kidney disease, or chronic debilitating diseases such as diabetes. Workers in health care professions usually take influenza treatments so that they recover completely from infection and cannot transmit the virus to patients. The antiinfluenza treatments currently in use reduce the duration of flu symptoms by one day if taken within 48 hours of the symptoms' onset.

Some physicians have recommended taking antiinfluenza drugs as a preventative against catching the flu, during times of a local outbreak. The following guidelines should be considered when using antiviral treatments as a flu preventative:

- taking the drugs for at least two weeks during a local flu outbreak
- if living in close contact with someone who is sick with the flu
- if already vaccinated but also at high risk for exposure to the virus (health care workers)
- if vaccinated within two to four weeks ago, in which time antibodies to influenza have not yet developed

As in the case of influenza vaccines, medical research has a long road ahead in finding the best and safest means of treating an ongoing influenza infection.

INFLUENZA IN ANIMALS

Influenza is a zoonotic disease, meaning it is transmitted between animals and humans. Human medicine concerns itself with the modes of transmission from animals to human populations, but the flu can be as devastating a health problem in veterinary medicine. The main animals that contract the influenza virus are swine, birds, horses, and marine mammals. Other wildlife and migrating birds may also carry and spread influenza viruses that can infect humans, but all of these flu carriers may not have yet been identified.

Swine influenza virus (SIV) causes hog flu or pig flu. This disease is an acute infection and highly contagious among swine. SIV is a type A influenza virus with mainly a H1N1 antigenic subtype. Rarely, H1N2 and H3N2 have been discovered in veterinary studies. Pigs also contract the type B and type C influenza viruses, but these are not thought to cause illness. Swine that are infected with SIV often contract additional bacterial infections, called *secondary infections,* which do most of the harm to the animal. As in humans, the virus infects the lungs and can cause lesions in bronchial tissue in the most serious infections.

SIV stays active outside the animal for up to two weeks, except in cold weather. In North America, outbreaks usually occur in the fall or winter, but in warmer climates, the infection can occur at any time of the year. No effective treatment exists for SIV infections, but veterinarians can lessen the severity of the disease by administering antibiotics to stop any secondary infections. The best control for swine flu is vaccination of domesticated pig herds, which reduces the income loss of meat-producing animals and lessens the chance of spread to humans.

Until the 1990s, bird flu (avian flu or fowl plague) was not thought to transmit readily from birds to humans. A spread of H5N1 influenza from birds to humans, from the mid-1990s to 2004, showed that bird flu may be almost as great a threat to people as swine flu. The H5N1 virus spread from Asia to India, Europe, and Africa, carried mainly by migrating waterfowl (geese and ducks) and spread by infected poultry farms. Various bird influenza viruses that have low pathogenicity (disease-causing ability) have also been recovered from wild shorebirds, pet birds, and flightless birds such as ostrich and emu.

The type A bird influenza may have subtypes of H1–15 and N1–9. Birds spread the virus by either ingestion or inhalation, and the symptoms appear within one week, sometimes as quickly as one day after infection. Birds infected with influenza develop both respiratory and digestive illnesses, which can be prevented with vaccination programs.

In horses, equine influenza is a highly contagious disease started by either of two type A virus subtypes: equi-1 and equi-2. In addition to domesticated horses, donkeys and zebras can be infected by either virus. The symptoms include mild to high fever, nasal discharge, dry cough, depression, loss of appetite, weakness, and swollen lymph nodes.

Equine flu can be prevented by keeping sick horses away from healthy ones and by practicing good barn housekeeping: minimization of dust and dirt, good ventilation, removal of dirty bedding, and clean hays and other feeds. Veterinarians also give domesticated horses vaccinations for each coming flu season. Equine influenza is not known to infect humans, although, in 2004, the virus surprisingly appeared in dogs at Florida greyhound tracks and caused canine flu. By the next year, the flu had spread to nine more states. Ed Dubovi of Cornell University's animal health laboratory said at the time, "Of all animals, dogs have the most intimate contact with humans on a daily basis, so the potential for human infection has to be in the back of our minds." People have not been known to contract equine flu, but they can spread it to healthy horses on their hands and clothing. The virus stays active on hard, dry surfaces for two to three days and for up to three weeks in water.

Type A influenza virus also infects marine mammals such as harbor seals and pilot whales. These findings have been discovered by diagnosing samples taken from animals stranded on beaches. Influenza virus can spread rapidly in dense populations of seals and possibly of other mammals that congregate in large numbers. Unseasonably warm temperatures and stresses from poor nutrition or other diseases or injuries have exacerbated flu epidemics in marine mammals, in recent years. The National Oceanic and Atmospheric Administration has established that many current strandings of marine mammals are due to viral diseases such as herpes, papilloma virus, hepatitis, and influenza. The effect of marine mammal illness on human health has not been determined.

Because influenza is an unpredictable pathogen, it may never be completely eradicated from humans or animals. The *Newsweek* reporter Laurie Garrett wrote, in 2009. At the viral level, influenza is an awfully sloppy microbe that is in a constant state of mutation and evolution. . . .

> When a virus infects a cell, its chromosomes essentially fall apart into a mess, which is copied to make more viruses that then enter the bloodstream to spread throughout the body. Along the way in this copying process any other genetic material that may be lying about the cell is also stuffed into the thousands of viral copies that are made. If the virus happens to be reproducing this way inside a human cell, it picks up *Homo sapiens* genetic material; from a chicken cell it absorbs avian genes; from a pig cell it garners swine RNA. The jackpot events in influenza evolution occur when two different types of flu viruses happen to get into an animal cell at the same time, swapping entire chromo-

somes to create 'reassorted' viruses. This occurrence is so common in influenza that it is a yearly event. Trying to prevent influenza is, essentially, trying to prevent an emerging disease every year.

A significant amount of information has been compiled, that indicates that influenza is a persistent threat to human and animal health. Devastating flu epidemics may, perhaps, not return to cause the deaths that have occurred in the past because of advances in flu prevention and treatment. But the influenza virus remains an ever-present danger in human and veterinary medicine, demanding constant vigilance and good hygiene.

See also COMMON COLD; EPIDEMIC; IMMUNITY; INFECTIOUS DISEASE; TRANSMISSION; VIRUS.

Further Reading

Batty, David. "Fast-Tracked Swine Flu Vaccine Will Be Safe, Officials Insist." *Guardian,* 26 July 2009. Available online. URL: www.guardian.co.uk/world/2009/jul/26/fast-tracked-swine-flu-vaccine. Accessed June 4, 2010.

Billings, Molly. "The Influenza Pandemic of 1918." 2005. Available online. URL: http://virus.stanford.edu/uda. Accessed July 20, 2009.

Cable News Network. "U.S. Trials for H1N1 Vaccine Announced." June 6, 2010. Available online. URL: www.cnn.com/2009/HEALTH/07/22/swine.flu.vaccine.trials. Accessed August 4, 2009.

Centers for Disease Control and Prevention. "Influenza." Available online. URL: www.cdc.gov/flu. Accessed August 3, 2009.

Garrett, Laurie. "The Path of a Pandemic." *Newsweek,* 18 May 2009.

Ginsberg, Jeremy, Matthew H. Mohebbi, Rajan S. Patel, Lynnette Brammer, Mark S. Smolinski, and Larry Brilliant. "Detecting Influenza Epidemics Using Search Engine Query Data." *Nature* 457 (2009): 1,012–1,014. Available online. URL: www.nature.com/nature/journal/v457/n7232/full/nature07634.html. August 10, 2009.

Haber, Penina, Frank DeStefano, Frederick J. Angulo, John Iskander, Sean V. Shadomy, Eric Weintraub, and Robert T. Chen. "Guillain-Barré Syndrome Following Influenza Vaccination." *Journal of the American Medical Association* 292 (2004): 2,478–2,481. Available online. URL: http://jama.ama-assn.org/cgi/content/full/292/20/2478. Accessed August 4, 2009.

The Medical News. "Equine Flu Virus Jumps to Dogs." September 27, 2005. Available online. URL: www.news-medical.net/news/2005/09/27/13406.aspx. Accessed August 5, 2009.

National Centers for Coastal Ocean Science. National Oceanic and Atmospheric Administration. "Coastal Marine Mammal Stranding Assessments (2009)." Available online. URL: www8.nos.noaa.gov/nccos/npe/projectdetail.aspx?id=57&fy=2009. Accessed August 5, 2009.

National Institute of Allergy and Infectious Diseases. *NIAID Influenza Research: 2009 Progress Report.* Washington, D.C.: Department of Health and Human Services, 2009. Available online. URL: www3.niaid.nih.gov/topics/Flu/PDF/fluResearch09.pdf. Accessed August 5, 2009.

———. "Understanding Flu." Available online. URL: www3.niaid.nih.gov/topics/Flu/understandingFlu. Accessed August 5, 2009.

National Network for Immunization Information. Available online. URL: www.immunizationinfo.org. Accessed August 3, 2009.

Reuters. "Influenza May Trigger Guillain-Barré Syndrome." January 28, 2009. Available online. URL: www.reuters.com/article/healthNews/idUSTRE50R6IK20090128. Accessed August 4, 2009.

ScienceDaily. "Lethal Secret of 1918 Influenza Virus Uncovered." January 17, 2007. Available online. URL: www.sciencedaily.com/releases/2007/01/070117134419.htm. Accessed August 10, 2009.

Stanford University. "The Influenza Pandemic of 1918." Available online. URL: www.stanford.edu/group/virus/uda. Accessed August 9, 2009.

Taubenberger, Jeffery K., Ann H. Reid, and Thomas G. Fanning. "Capturing a Killer Flu Virus." *Scientific American,* 27 April 2009. Available online. URL: www.scientificamerican.com/article.cfm?id=capturing-a-killer-flu-virus. Accessed August 5, 2009.

University of Wisconsin School of Veterinary Medicine. "Influenza as a Zoonotic Disease." Available online. URL: www.vetmed.wisc.edu/pbs/zoonoses/influenza/influenzaindex.html. Accessed August 3, 2009.

Jenner, Edward (1749–1823) ***British Physician, surgeon, advocate for aseptic techniques in medicine*** Edward Jenner takes a place in the history of microbiology as the discoverer of the process of vaccination to prevent infection and as a scientist who helped launch the Golden Age of Microbiology, which began in the mid-1800s and lasted to the early 1900s. Rapid advances occurred in microbiology, during this period, in which germs such as bacteria were definitively linked to disease and other processes in nature. The Golden Age of Microbiology also contributed advances in medical microbiology in the areas of disease diagnosis, sanitation, disinfection, and immunity. Edward Jenner made the chief contributions in the area of immunity when he developed a vaccine to protect people from smallpox.

Edward Jenner was born the eighth of nine children in county Gloucestershire, England, on May 17, 1749. Jenner's parents had died by the time he turned five, leaving him in the care of his older sister. Young Jenner developed an interest in the natural world and would begin a career in medicine when he became apprentice to a surgeon working not far from his birthplace. Jenner completed medical training at St. George's Hospital in London, in 1770, and established a medical practice of his own in his hometown of Berkeley in Gloucestershire, in 1772.

As a family doctor and surgeon, Edward Jenner practiced the medical treatments of the day. Severe infections would often require treatment by amputation—without anesthesia! Diagnostic advances such as the use of a stethoscope and even taking body temperature did not arrive until very late in Jenner's medical career. Much of the best medical advice dispensed by Jenner and his contemporaries often resulted from astute intuition. That intuition would serve Jenner well in his work in finding a defense against smallpox.

Smallpox had been one of the most feared diseases, since antiquity. The virus had been transmitted throughout Europe, in the 18th century, and would also travel with European settlers to North and South America, killing entire populations of Native people. Healers in Asia had discovered the usefulness of inoculating healthy people with scabs from a sick person for the purpose of warding off infections. Lady Mary Wortley Montagu (1689–1762), the wife of Britain's ambassador to Turkey, probably learned about the idea of inoculation during her time in the Ottoman Empire and, in 1717, applied the theory by inoculating her children with the smallpox virus (known at the time as cowpox) to save them from the fate that she and her brother had suffered. Montagu's brother had died of the pox, and Mary had been permanently scarred from a smallpox infection, when she was young. Montagu used a technique called *variolation,* in which she took a small specimen from the pox lesions of a smallpox patient and inserted it into her children, probably by telling them to inhale it. By this method, the virus would circulate through the body and elicit the immune system to build defenses, and immunity, against it.

Jenner developed an alternative to variolation's introduction of a live, virulent pathogen into a healthy body, which may have caused as many deaths as the lives it saved. He began his work against cowpox, in the late 1700s, in his own rural community, where the pox virus had devastated local farms as well as families. In 1798, Jenner wrote, "There is a disease to which the Horse, from his state of domestication, is frequently subject. The Farriers have termed it *the Grease.* It is an inflammation and swelling in the heel, from which issues matter possessing properties of a very peculiar kind, which seems capable of generating a disease in the Human Body . . . which bears so strong a resemblance to the Small Pox, that I think it highly probable it may be the source of that disease." The farms surrounding Berkeley had seen instances in which horse disease would transmit to the community's dairy cows. Jenner had, at the time, made the insightful deduction that a pathogen was probably being spread on the hands of farmers, from animal to animal, farm to farm. The same pathogen would be transmitted from sick cows to people during milking.

A young woman who visited Jenner's practice mentioned that she milked cows every day but had no fear of contracting cowpox, because she had had the disease years earlier and knew that anyone who had the disease would never develop it again. The name of this "dairymaid" has, unfortunately, been lost to history; she probably was the first person to define the process of vaccination accurately. Edward Jenner did not fail to recognize the importance of her theory, however, and began collecting specimens from cowpox blisters and scabs.

Jenner described an experiment, in 1796, using the cowpox scrapings—he called the substance "variolous matter"—as a mechanism for preventing a smallpox infection: "The more accurately to observe the progress of the infection, I selected a healthy boy, about eight years old, for the purpose of the inoculation for the Cow Pox." Jenner inoculated the boy by making small incisions on the arms and injecting the pox material. "The material from this boy was used to inoculate another person, with the same results [mild side effects but no disease]. The material was passed from one person to another through five passages from the cow." Jenner's material fulfilled two major requirements of modern vaccines: (1) It did not cause disease in the recipients, and (2) it prevented an infection by the specific disease-causing agent.

Edward Jenner's legacy did not develop overnight. Jenner had long been convinced of a relationship among smallpox, cowpox, and swinepox, although most other practitioners, at the time, found this idea to be unsound. After Jenner published his case studies of successful vaccinations, the medical community criticized his theories, science, and conclusions. The critics probably had difficulty understanding the logic of introducing material from a sick person into a healthy person, even though variolation had been practiced for centuries. Even political satirists derided Jenner and warned the public that if a person did not have smallpox, he surely would under Jenner's care. The physician Christian Charles Schieferdecker published, in his 1856 book *Evils of Vaccination,* an imaginary conversation between a mother and the family doctor:

> *Doctor.*—Bring your babe now for vaccination!
> *Mother.*—How you have frightened me, I tremble all over!
> *Doctor.*—Why? What is the matter?
> *Mother.*—Pardon me; but I feel a perfect horror creeping over me at the mere thought on vaccination, since my poor Charles has died in consequence of it.
> *Doctor.*—O nonsense! Charles did not die on vaccination; do not believe such a thing. He got dysentery while teething.
> *Mother.*—I beg you to wait another year; Tom is now very delicate indeed.
> *Doctor.*—Only the better! and I have this very moment excellent, fresh vaccine-lymph.
> *Mother.*—In God's name be it done! But, doctor, the responsibility rests on your shoulders!
>
> Nine days after this conversation and following vaccination, Tom was a corpse, with two vaccine-blisters on each arm.

Jenner continued in his path toward a smallpox vaccine and, for the next two years, after the publication of his first vaccination case studies, he conducted additional successful vaccinations. By three years after the publication of Jenner's vaccination case studies, doctors in Europe had begun using the cowpox vaccination technique to prevent their patients' contracting smallpox. More than 100,000 people would be vaccinated, by the end of 1801.

Jenner's reputation in the medical establishment took another blow when complications arose in a small percentage of vaccine recipients. Some developed smallpox or other infections. These infections had probably been the result of cowpox vaccines contaminated either with virulent smallpox virus or with other pathogens. Occasionally, a vaccinated person would have smallpox several years after being vaccinated against it. Jenner and his peers did not realize at the time that smallpox vaccination would not last forever and needed booster injections every five to 10 years. In 1801, Jenner published an update on his state of knowledge on the smallpox vaccine, in the *Origin of the Vaccine Inoculation.* The publication, included additional proof of how and why the vaccine he developed worked against the smallpox virus. "The distrust and skepticism that naturally arose in the minds of medical men, on my first announcing so

The British physician Edward Jenner (1749–1823) helped reform the care of surgery patients by calling for the use of antiseptics and sterile instruments. ***(Dibner Library of the History of Science and Technology)***

unexpected a discovery, has now nearly disappeared." A new generation of doctors would, soon, follow to perfect the smallpox vaccine.

Starting in the mid-20th century, the Centers for Disease Control and Prevention (CDC) teamed with international health agencies to launch a worldwide vaccination program against smallpox. By 1949, the CDC had eradicated smallpox in the United States, and the last known case anywhere on the globe was recorded in Somalia in 1977.

VACCINATION

About a century after Edward Jenner tried his revolutionary technique for preventing infections, the microbiologist Louis Pasteur (1822–95) delved into the mechanism of how the vaccine worked. Pasteur actually coined the term *vaccination,* from the Latin *vacca,* for "cow," and as a nod to Jenner's early studies on cowpox. Subsequent scientists would learn more about the immune system and find the reason why vaccination works: Vaccines give the body acquired immunity. Acquired immunity is the body's defense mechanism against infection through the production of antibodies.

The cowpox virus used by Jenner causes a disease different from the smallpox that had ravaged populations for centuries. But because the two viruses possess very similar features, each can induce the formation of immunity against the other. Although Jenner used a live virus, as did Mary Montagu, Jenner concentrated on the actual pathogen, rather than the heterogeneous exudates emerging from pustules in diseased people. Jenner's experiments marked the first time a live virus had been prepared specifically to kill invading pathogens of future infections. When Jenner observed that the cowpox virus continued to confer immunity, even after passing through several different people, he had proposed the concept of making effective, but weakened strains. These would develop into today's process of making vaccines from weakened viruses. These vaccines are now called *attenuated live vaccines.*

EDWARD JENNER IN SCIENCE

Edward Jenner received harsh criticism for introducing the revolutionary idea of introducing a known pathogen into a healthy person. Aside from purely medical concerns regarding vaccination with live viruses, Jenner confronted criticism that had nothing to do with his laboratory work. Jenner faced the prejudices that the London medical community held about country doctors. Perhaps, the conservative members of the medical world believed that the greatest medical discoveries could only be born in university research laboratories. Edward Jenner's curiosity about all aspects of nature, fortunately, overcame the humiliation that many tried to heap on him regarding vaccination.

Edward Jenner made contributions to science that reached far beyond viruses. The following list describes some of the main areas of interest and discovery by Edward Jenner.

- developed one of the first free clinics for medical treatment of the poor
- developed a method for reducing the toxicity of a common treatment for parasites, called *tartar emetic*
- contributed to research on heart disease and noted the link between damaged heart valves and rheumatic heart disease
- recovery and study of fossils from ancient plesiosaurs
- studies on geological changes in England
- studies on the physiology of animal hibernation
- developed theories on undiscovered bird migration routes

- gathered information on bird (the cuckoo) behavior and survival mechanisms

Edward Jenner's contribution to science will always be most closely linked to the development of vaccines. Jenner's work can, indeed, be thought of as the very first step toward the elimination of smallpox from the earth, a feat officially recognized by the World Health Organization, in 1980. In addition to opening a new science in vaccine development, Jenner initiated the study of the body's immune system and the important relationship between immunity and infectious disease.

In his later years, Jenner provided advice to a new generation of doctors seeking guidance on safe handling and production of the cowpox vaccine. In retirement, he returned to his hobbies: fossil collecting and gardening. Even then, Jenner continued innovating by propagating new breeds of berries, figs, and grapes.

Edward Jenner died of stroke, on January 26, 1823, one day after making a house call on a medical colleague who had himself suffered a stroke. He is buried in his family tomb in Berkeley, England.

See also IMMUNITY; MICROBIOLOGY; VACCINE.

Further Reading

Dugdale, Sarah, Susan Lewis, Mike Gold, and Helen Marshall. "Vaccine Safety and Community Attitudes in SA." *Public Health Bulletin* 4 (2006): 25–29. Available online. URL: www.health.sa.gov.au/PEHS/publications/PHB-comm-disease-ed4-06v2.pdf. Accessed August 9, 2009.

Jenner, Edward. *An Inquiry into the Causes and Effects of the Variolae Vaccinae, a Disease Discovered in Some of the Western Counties of England, Particularly Gloucestershire, and Known by the Name of The Cow Pox.* 1798. Available online. URL: www.foundersofscience.net. Accessed August 9, 2009.

———. *The Origin of the Vaccine Inoculation.* London: D. M. Shury, 1801. Available online. URL: http://pyramid.spd.louisville.edu/~eri/fos/jenner_Origin.pdf. Accessed August 10, 2009.

The Jenner Museum. Available online. URL: www.jennermuseum.com. Accessed August 10, 2009.

Riedel, Stefan. "Edward Jenner and the History of Smallpox and Vaccination." *Proceedings (Baylor University Medical Center)* 18 (2005): 21–25. Available online. URL: www.pubmedcentral.nih.gov/articlerender.fcgi?artid=1200696. Accessed August 10, 2009.

Scheiferdecker, Christian C. *Evils of Vaccination.* Philadelphia: Henry B. Ashmead, 1856. Available online. URL: http://books.google.com/books?id=6CUaAAAAYAAJ&pg=PA23&lpg=PA23&dq=Christian+Charles+Schieferdecker&source=bl&ots=Z5srgZkGQa&sig=6LVm_5rpYtShVNNyMpnRbr_uMEE&hl=en&ei=0JOAStiEGZHuswOn3t2JCQ&sa=X&oi=book_result&ct=result&resnum=5#v=onepage&q=&f=false. Accessed August 9, 2009.

Koch's postulates Koch's postulates are a list of criteria developed by the German physician Robert Koch (1843–1910) (pronounced kōk) for proving that a specific microorganism is the cause of a specific disease. According to Koch, four conditions must be met in order to prove that a microorganism causes a given disease.

The Koch's postulates are the following:

1. The same microorganism must be present in every case of the disease.

2. The microorganism must be isolated from a patient with the disease and then grown in pure culture (no other microorganisms present).

3. If inoculated into a healthy person, the same microorganism must cause the disease.

4. The microorganism must again be isolated from the new diseased person and shown to be the same microorganism as the original.

Robert Koch could not use humans to prove these criteria for identifying a pathogen (a disease-causing microorganism), so he substituted mice in his experiments.

Koch's postulates contributed to the advancement of microbiology in two main ways. First, the criteria gave the definitive proof needed to confirm the germ theory, a doctrine that proposed that all infectious diseases are started by specific microorganisms associated only with that disease. Second, the postulates provided a sound scientific method for diagnosing an infectious disease.

Koch's postulates also contributed to the overall progress of science by offering an example of the correct way to approach scientific inquiry. The table on page 462 outlines the basic steps in scientific inquiry and how Koch fulfilled each of them.

Robert Koch made other crucial findings in his laboratory that built a foundation for medical microbiology. First, he demonstrated that individual microorganisms can cause individual diseases. Second, Koch worked out a method of preparing pure cultures of bacteria. Although Koch's cumbersome method would soon be improved with the discovery of agar media for growing microorganisms, the pure culture method remains a cornerstone of microbiology. Third, Koch identified the causative agent of tuberculosis, during a time in history that was losing many lives to the disease. Fourth, Koch accumulated a vast amount of information on the anthrax microorganism that had never before been known.

Koch conducted the majority of his work on the microorganisms that cause tuberculosis and anthrax. To prove his hypotheses further, he varied his methods, by using different microorganisms, trying blood as an inoculum, and inoculating pathogens from one species into another. Koch concluded his 1884 presentation, "The process outlined above [anthrax microorganisms in mice], which has been successful in proving the parasitic nature of anthrax, and which has led to inescapable conclusions, has been used for the basis for my studies on the etiology of tuberculosis. These studies first concerned themselves with the demonstration of the pathogenic organism, then with its isolation, and finally with its reintroduction." Koch's scientific principles provided a critical step in the etiology of disease, that is, the study of the cause of disease.

As microbiologists have learned more about the life cycles of pathogens, they have challenged, confirmed, or expanded on Koch's postulates. Some microorganisms have a life cycle that does not conform with all of Koch's criteria. These exceptions to Koch's postulates do not refute the principles; they illustrate the variability that exists in the biological world and the difficulty of applying one theorem to all microorganisms. Because of this variability, microbiologists must consider several exceptions to Koch's postulates.

The first complication to Koch's postulates occurs when a pathogen does not grow in laboratory conditions or microbiologists have not yet discovered a way to grow them. In some cases, a pure culture cannot be obtained. For example, many viruses grow only in the body, and culture techniques for them do not exist. Microbiologists have circumvented this problem by keeping the pathogen alive in blood or living tissue. Sometimes, a microorganism that cannot be detected on culture medium must be detected in the test animal's blood, by reacting it with antibodies. This process is called an *immunoassay.*

Medical researchers must also understand that animal models used for identifying a pathogen through Koch's postulates do not always behave the same way as humans. For example, the human immunodeficiency virus (HIV) that causes acquired immunodeficiency syndrome (AIDS) in humans is not known to grow in other animals. Other viruses, the rickettsia bacteria, and sexually transmitted diseases have no animal model to represent the human body. There have been occasions in history in which humans were used as experimental models—sometimes without their knowledge—for studying disease. People, today, believe this type of testing on humans is unethical and prefer that researchers use animal models. Other people oppose testing on animals and prefer that all disease studies take place in vitro (laboratory conditions), rather than in vivo (inside a living body).

A third exception to Koch's postulates occurs when the symptoms of a disease are not easy to pinpoint. For example *Streptococcus pyogenes* can cause sore throat, inflamed joints, or fevers such as scarlet fever or rheumatic fever. This type of variability would be difficult to extrapolate to Koch's postulates.

A fourth exception, similar to the third example mentioned, relates to diseases that have no set symptoms at all. Some inflammations exhibit a variety of symptoms. Food-borne illnesses can all have similar signs and symptoms, even when caused by unrelated bacteria: nausea, vomiting, abdominal pain, possible fever, and diarrhea.

Fifth, diseases called *polymicrobial diseases* begin as a result of the action of more than one microorganism. Septicemia (infection of the bloodstream) caused by an injury to the intestinal lining is an example of a polymicrobial disease state. In this case, Koch's requirement for isolating a pure culture of a single pathogen does not apply.

Sixth, microorganisms with complex life cycles inside and outside the body present a problem. The waterborne protozoan *Cryptosporidium,* for example, forms different stages after infecting the human intestinal tract. These stages appear and disappear as the organism goes through its life cycle. The infected person also suffers recurring bouts of symptoms (severe diarrhea, painful abdominal cramping) separated by weeks of fairly good health.

Microbiologists have revised Koch's postulates to make them more generic to fit a broad range of circumstances. These so-called simplified Koch's postulates are the following:

1. Demonstrate the universal presence of the microorganism in diseased individuals.

2. Isolate the microorganism in a pure culture or in a mixed culture (more than one microorganism present).

3. Use the isolated microorganism to recreate the disease process, either in vivo or in an in vitro model.

4. Observe the disease conditions and reisolate the microorganism.

The revised version of Koch's postulates helps microbiologists meet the criteria when working with out-of-the-ordinary conditions. The revisions allow them to devise new models if an animal model does not work. The revisions also make allowances for diseases that are the result of more than one microorganism.

ROBERT KOCH

Robert Koch was born, on December 11, 1843, in Clausthal, Germany, the son of a miner. He taught himself to read, by age five, and showed promise for a career outside the mines. With an interest in biology and medicine, Koch entered the University of Göttingen, at 19. He received his medical degree in 1866. After working for a short period in Berlin with Rudolf Virchow (1821–1902), an early developer of the cell theory, Koch began his own medical practice. Only three years later, in 1870, he volunteered for service in the Franco-Prussian War. That post would lead to his assignment as a district medical officer, in 1872.

The Steps of Scientific Inquiry

Step	Action	How It Relates to Koch's Postulates
1	develop a hypothesis causing a disease	Koch presumed an unknown microorganism is causing a disease
2	select instruments or other items for testing the hypothesis	Koch selected a mouse experimental model and used a diseased animal
3	collect information by experimentation and observation	He isolated a microorganism from a dead animal, grew it in a laboratory, and injected it into a healthy animal and observed the results
4	accept the hypothesis	He concluded that the original microorganism caused the disease in healthy animals
5	confirm with more testing	Koch reisolated the same microorganism from the second diseased animal
6	communicate the observations and conclusions	He published his findings, in 1884, in the article "The Etiology of Tuberculosis"

At the height of Koch's career, the diseases anthrax and tuberculosis affected a large portion of the population. Anthrax, caused by a *Bacillus* spore, created a particular threat for workers who handled hides, sheared sheep, or farmed; anthrax spores are found in soil. Tuberculosis, additionally, endangered the general population, because it was contagious, and doctors had almost no defenses against it or means of prevention. Koch's natural inquisitiveness led him to set up studies on the manner in which anthrax infected mice. He soon discovered that *Bacillus* bacteria could withdraw into a small, indestructible spore form when stressed. No one had observed this feature before, and Koch surmised that the spore leant *Bacillus* an extra weapon as a pathogen and protected it in the environment. Koch, furthermore, showed that anthrax cells grown for generations in his laboratory retained the ability to cause anthrax disease after he put them back into mice. "It may be assumed," he wrote in 1877, "that when these spores in some way reach the blood stream of a sensitive animal, a new generation of bacilli will be produced." Proof of this hypothesis by experimentation made Koch widely respected among his peers.

In the 1880s, Koch devised new methods for obtaining pure cultures of bacteria, first, by growing them on sliced potato, and, then, by trying a newly discovered substance called *agar* in a small, shallow dish invented by Julius R. Petri (1852–1921). All microbiologists now use this familiar petri dish in their laboratories. Koch also developed biological stains to make the bacilli easier to observe under a microscope. With pure cultures in hand and an easy way to monitor the microorganisms, Koch, next, set up an experiment to prove whether a single type of organism could be correlated with a specific disease. By 1884, Koch published the results of his anthrax and tuberculosis experiments and proposed the criteria for linking a microorganism to a disease.

Koch was soon appointed surgeon general stationed in Berlin. In this position, Koch led research teams that helped control cholera epidemics in Europe and a parasite outbreak in livestock in South Africa. Koch continued to study the characteristics of the tubercle bacilli that cause tuberculosis, although his assistants now spent more time in the laboratory than Koch. The tuberculosis organism had been maddeningly difficult to study, throughout his research career. Perhaps that explains why Koch made a career shift, as noted by the historian Thomas Brock in his biography of Robert Koch, "But then, sometime in late 1889, Koch's behavior changed. Suddenly he was back in the laboratory again doing work with his own hands. But he was working completely alone, behind closed doors, and for days at a time he would talk to no one. . . . And then, in August 1880, Koch implied that he had found a cure for tuberculosis! The medical world was thunderstruck, and the news spread rapidly around the world. The great Koch had triumphed again, but this time in an even more miraculous way." The discovery, if true, would remove from the world one of its most feared diseases.

Koch called his tuberculosis treatment *tuberculin,* which was a mixture of the bacillus, to be injected directly under the skin. In guinea pigs, this treatment had shown remarkable results in resolving symptoms of the disease. Although Koch's reasoning was correct—tuberculin could act as a substance that would activate an immune response against the pathogen—the treatment caused problems in human subjects. Swelling, rashes, and other sometimes-painful reactions to the tuberculin injection occurred. (Koch had first tested tuberculin in his own body and had experienced a series of effects, from nausea and vomiting to fever and joint pain.) Through it all, Koch tried to keep his remedy a secret from colleagues and the public, because the country had few restrictions on who could distribute drugs. He probably realized that anyone could appropriate tuberculin as his own and sell it for a variety of ills.

The tuberculosis cure created a frenzy of excitement worldwide. Louis Pasteur sent congratulations from France; Joseph Lister visited Koch's laboratory to view treatments close up. Sherlock Holmes's creator, Arthur Conan Doyle, also a physician, visited Koch's laboratory to write a human interest piece for an English newspaper.

Less than a year after Koch's dramatic announcement, more than 2,000 tuberculosis patients had received a total of 17,500 tuberculin injections. Tuberculin had less than spectacular results in humans and produced uncomfortable side effects. In most quarters of the medical community, Koch remained well regarded, but he may have sensed that he had lost some luster. He returned to studies in a newly built research facility. A seasoned researcher, Koch drew the conclusion that the tuberculosis microorganisms in different animals were not exactly alike, and the disease did not transmit from animals to people. He announced this opinion, in 1901, and again a few years later, to a medical group and received vehement criticism from many in the audience. Years later, microbiologists would realize that Koch had been correct.

In 1905, the Nobel Prize Committee awarded Robert Koch its prize for his work in medicine. In his remarks to the Nobel audience, he retraced his career studying the tuberculosis pathogen and his years as a medical officer fighting epidemics. "The struggle against tuberculosis," he said that day in December, "is not dictated from above, and has not always developed in harmony with the rules of science, but it has originated in the people itself, which has finally correctly recognized its mortal enemy. It surges forward with elemental power, sometimes in a rather wild and disorganized fashion, but gradually more and more finding the right paths." Robert Koch died of a heart attack, in 1910, in Baden-Baden, Germany. Robert Koch entered history as a brilliant scientist, microbiologist, and epidemiologist and can be considered one of the most important contributors to medical microbiology.

ADVANCES IN KOCH'S POSTULATES

Koch's postulates may now be extended into other areas of disease study. Molecular biologists can identify specific genes from pathogens that cannot be grown reliably in laboratory culture. These genes are called *virulence genes,* because they are responsible for giving a microorganism tools for infecting a body and propagating within the body. Molecular biologists have applied Koch's postulates to virulence genes, similarly to the way Koch originally proposed them for whole bacteria. The following new postulates apply to diseases that molecular biologists study by analyzing a microorganism's virulence genes responsible for creating a disease:

1. The virulence genes should be more prevalent in strains of the microorganism known to cause the disease than in nonpathogenic strains.

2. Inactivation of the virulence genes should lessen or eliminate the pathogen's ability to infect a host.

3. Replacement of an inactivated or mutated virulence gene with a fully functional virulence gene should restore the pathogen's ability to infect.

4. The gene should be expressed (become activated and control protein synthesis) at some point during the disease.

5. An immune response developed by the infected host should develop specifically against the virulence gene products (proteins, enzymes, toxins, etc.).

Other areas of science in addition to molecular biology have adapted scientific processes that follow Koch's original postulates. The provost of the University of Southern Mississippi wrote in *Microbe* magazine, in 2006, "When Koch framed his postulates, they were bold and visionary. Now, these same principles are being applied to many other types of problems as microbiologists continue to use them, either knowingly or not, to address problems far outside the realm in which Koch once worked." Koch's postulates remain a model for scientific research.

See also GERM THEORY; PATHOGEN.

Further Reading

Brock, Thomas D. *Robert Koch: A Life in Medicine and Bacteriology.* Washington, D.C.: American Society for Microbiology Press, 1999. Available online. URL: http://books.google.com/books?id=HIMB2Ahkmo0C&printsec=copyright. Accessed August 22, 2009.

Garrison, Fielding H. *An Introduction to the History of Medicine.* Philadelphia: W. B. Saunders, 1921. Available online. URL: http://books.google.com/books?id=JvoIAAAAIAAJ&pg=PA82&lpg=PA82&dq=three+humors&source=bl&ots=LqpRb55HSb&sig=jfIKc- nEiT9C6vFCLKgAD4J2Mrs&hl=en&ei=jF0kSpPaJozstQOu0tH9Aw&sa=X&oi=book_result&ct=result&resnum=4#PPA3,M1. Accessed August 22, 2009.

Grimes, D. Jay. "Koch's Postulates—Then and Now." *Microbe,* May 2006.

Koch, Robert. "Die Aetiologie der Tuberkulose" (The Etiology of Tuberculosis). *Mittheilungen aus dem Kaiserlichen Gesundheitsamte* 2 (1884): 1–88. In *Milestones in Microbiology,* Washington, D.C.: American Society for Microbiology Press, 1961.

———. "The Current State of the Struggle against Tuberculosis." Nobel Lecture presented in Stockholm, Sweden, December 12, 1905. Available online. URL: http://nobelprize.org/nobel_prizes/medicine/laureates/1905/koch-lecture.html. Accessed August 22, 2009.

———. "Untersuchungen über Bakterien V. Die Aetiologie Milzbrand-Krankheit, Begründet auf die Entwwicklungsgesschichte des Bacillus anthracis" (The Etiology of Anthrax, Based on the Life History of *Bacillus anthracis*). *Beiträge zur Biologie der Pflanzen* 2 (1877): 277–310.

lactic acid bacteria Lactic acid bacteria (LAB), or lactobacteria, is a general group of various species common in nature and having three characteristics: (1) gram-positive, (2) non-endospore-forming and usually nonmotile, and (3) metabolism of the sugar glucose by lactic acid fermentation. Lactic acid fermentation is an anaerobic pathway in which the sugar glucose is converted to lactic acid.

LAB are important for making fermented dairy products, meat products, and breads. Agriculture, additionally, uses lactic acid fermentation to turn plants into a more digestible foodstuff for livestock. Fermentations by LAB have been used, for centuries, in food production and preservation, but these same fermentations in certain foods cause spoilage or bad odors and off-flavors. For these reasons, LAB have become indispensable for the food industry and important in food microbiology.

LACTIC ACID FERMENTATION

Lactic acid fermentation, as do all types of fermentation, occurs in low-oxygen conditions. Fermentation is any metabolic pathway that uses organic compounds rather than oxygen to allow energy-generating reactions to operate. Lactic acid fermentations carried out by LAB can be either homolactic or heterolactic. Homolactic fermentation produces lactic acid as its sole end product. Heterolactic fermentations produce a compound, usually the alcohol ethanol, in addition to lactic acid.

Homolactic fermentation begins with glycolysis, a metabolic pathway found throughout nature that turns a molecule of glucose (six carbons) into two molecules of pyruvate (three carbons). LAB convert all of the pyruvate to the three-carbon compound lactic acid.

Heterolactic fermentation also follows glycolysis to make lactic acid from pyruvate, but the LAB additionally take one of the intermediate compounds

Cheese Production by Lactic Acid Fermentation

Cheese	Flavor-Producing Fermentation Products	Other Microbial Factors
Camembert	lactic acid and carbonyl compounds	produced by LAB in combination with yeast and *Penicillium* mold
cheddar	primarily lactic acid	additional LAB added during ripening
Roquefort	lactic acid, fatty acids, and methyl ketones	produced in combination with *P. roqueforti* mold
Swiss	lactic acid, propionic acid, and carbon dioxide	produced by LAB in combination with propionic-producing bacteria

from glycolysis and use it in another pathway (phosphoketolase pathway) to make ethanol. Some LAB produce acetic acid instead of ethanol.

Lactic acid fermentations of foods have been used since antiquity and were probably discovered by accident in spoiled food. People soon learned that a properly controlled fermentation could change the texture and flavor of foods, as well as help preserve them. The lactic acid produced by LAB in the food causes the food's pH to drop. These acidic conditions inhibit the growth of most microorganisms other than the LAB. The following nondairy food items are preserved this way:

- fresh meats to make sausages and cold cuts
- cured olives
- pickles from fresh cucumbers
- sauerkraut from fresh cabbage

Pickles and sauerkraut are referred to as high-acid foods, with a pH of 4.0 or lower. Olives have a pH of 4.5–4.8, and the pH of cured sausages ranges from 4.4 to 5.6, and they are called *acid foods.*

Lactic acid fermentations of dairy products predate microbiology. The food science experts Jeffery R. Broadbent and James L. Steele wrote in *Microbe,* in 2005, "Because fermented dairy foods developed before the emergence of microbiology as a science, manufacturing processes for all varieties long relied upon naturally occurring LAB to acidify milk. It was not until the discovery of lactic acid fermentation by [Louis] Pasteur in 1857, and development of pure LAB starter cultures later that century, that the door to industrialized cheese and milk fermentations opened." Starter cultures are the freshly prepared growth of microorganisms used for growing larger quantities of the same species. Some microbiologists specialize only in LAB research in order to define the genetics of these important bacteria.

Microbiologists have determined the gene sequence in the deoxyribonucleic acid (DNA) of LAB. Food manufacturers want to find the genes that correlate with certain flavors in cheese, good growth in starter cultures, and tolerance to refrigeration or freezing. The gene sequences of the majority of LAB have now been determined.

Cheese production provides an example of the contributions LAB make in preparing a high-quality food. The final taste, odor, and texture of cheese result from many factors: the type and composition of the milk, the manufacturing process, and the ripening process. The bacteria affect these factors by secreting enzymes that break down the milk proteins and lipids (fatlike compounds). As cheese ripens, LAB might also act on carbohydrates and the cheese curd. Other bacteria may be added to give certain cheeses a distinctive flavor or to intensify existing flavors. The table on page 465 shows the different fermentation end products of various bacteria in cheese production.

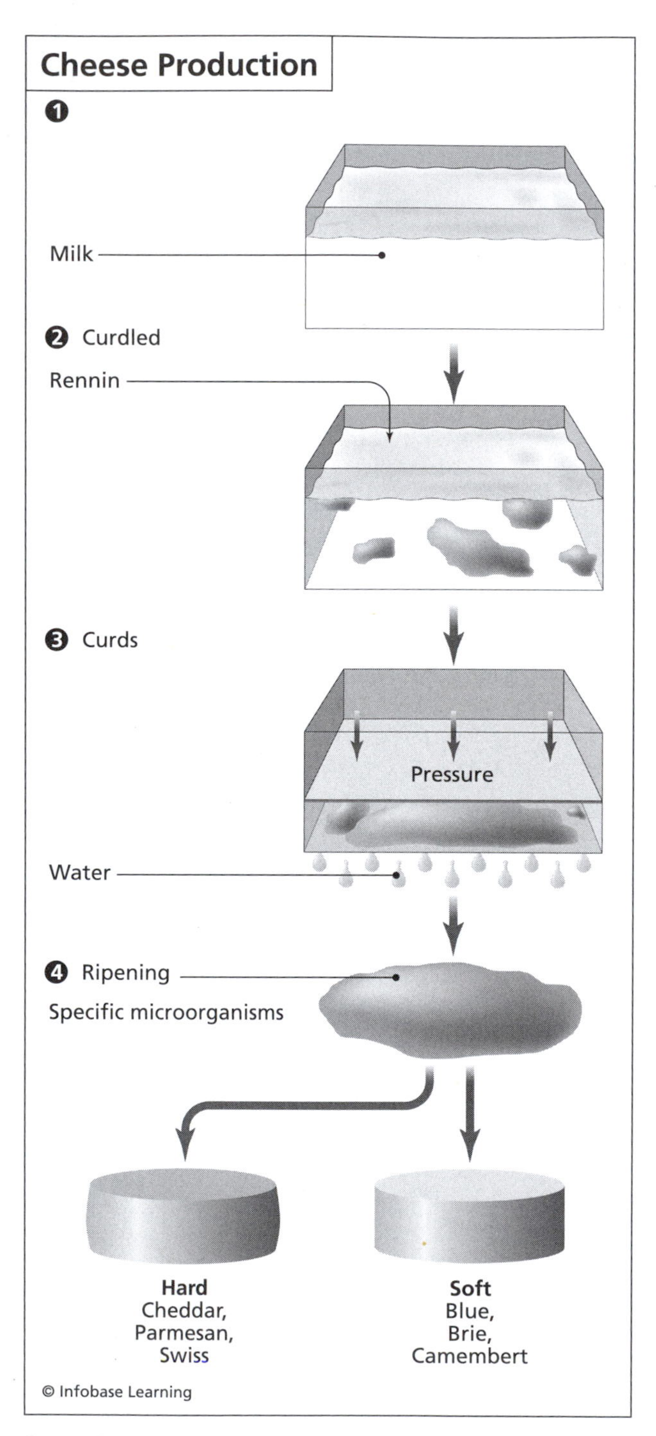

Several hundred different types of cheeses follow this general production path. *Penicillium* mold, the bacteria *Propionibacterium and Brevibacterium,* and the lactic acid bacteria predominate in cheese making.

Brown algae of the genus *Sargassum* form huge floating masses in the Atlantic Ocean's Sargasso Sea. This seaweed mixes various multicellular organisms and microorganisms to create a unique ecology. *(NOAA/Department of Commerce)*

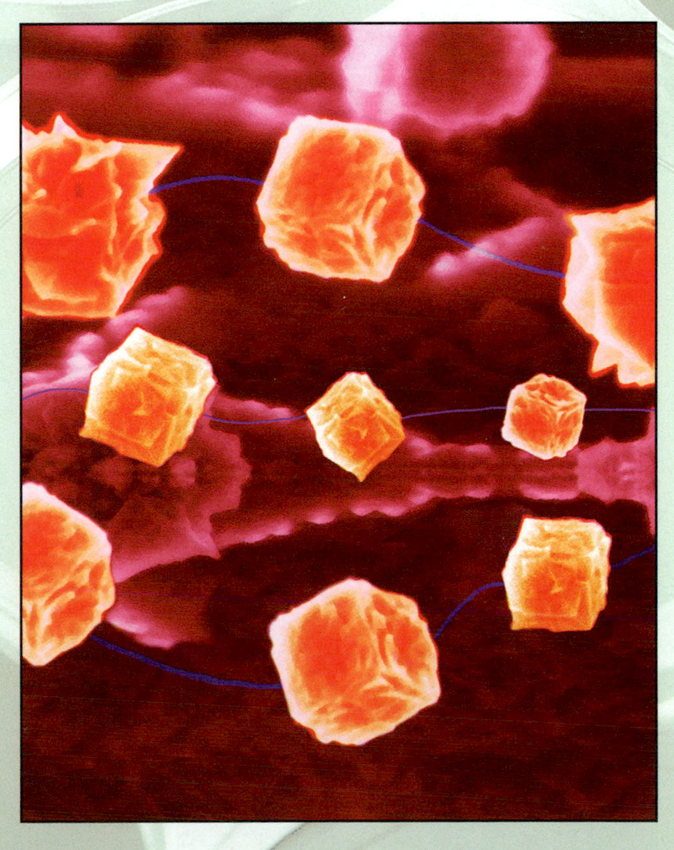

Parts of bacteria or viruses might be used in future nanoscale biosensors. This scanning electron microscope (SEM) colorized image shows cube-shaped blood glucose biosensors connected by carbon nanotubes that transmit information. *(Jeff Goecker, Discovery Park, Purdue University)*

Euglena is a motile alga that can create blooms on freshwater surfaces. *(Bill Henley)*

Lichens are composed of a fungus and either algae or cyanobacteria. Pigments from the non-fungus portion give a variety of colors, shown growing on this boulder. *(Dave Powell, USDA Forest Service)*

Mixed cultures contain many different species of bacteria, obvious by a variety of sizes, shapes, and colors of the colonies on agar. *(David B. Fankhauser)*

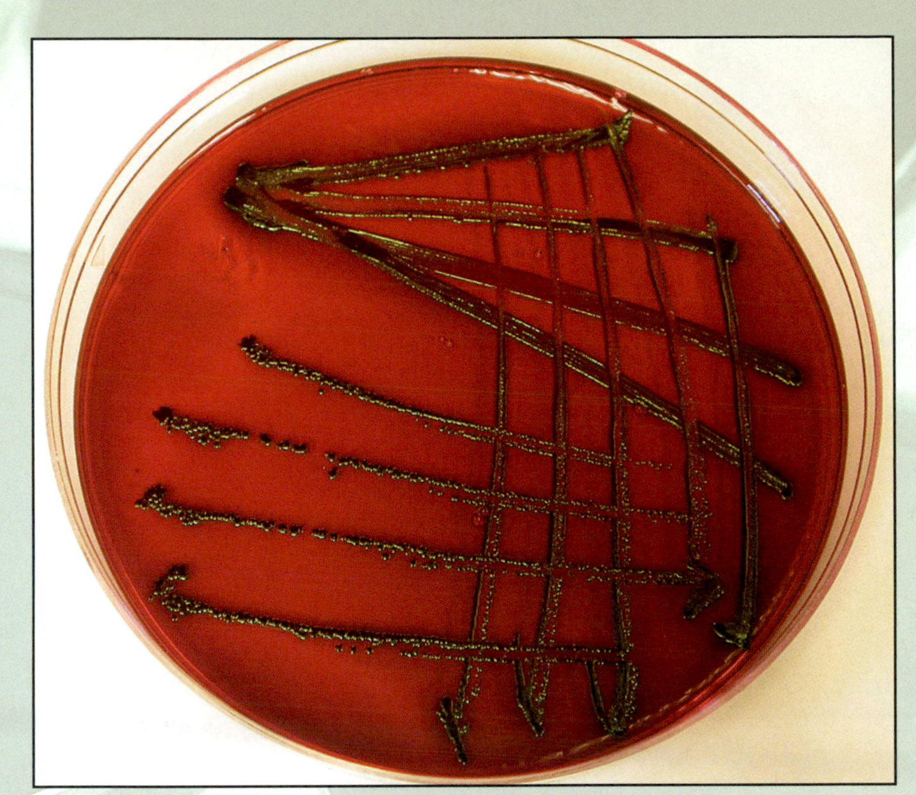

Eosin Methylene Blue (EMB) Agar differentiates *E. coli*'s dark colonies with a metallic green sheen from other enteric bacteria. *(Stephen Spilatro)*

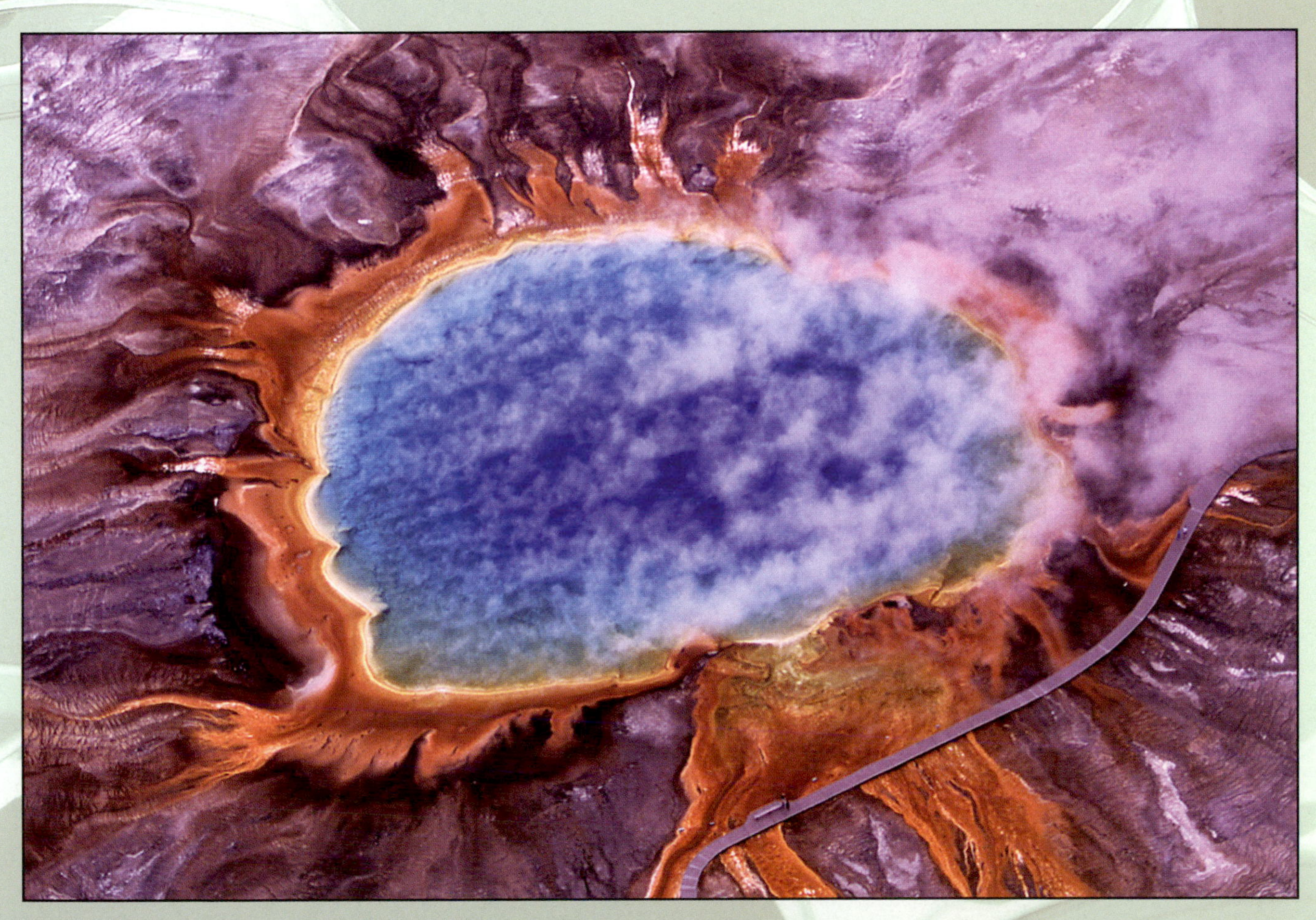

The extremophiles recovered from Yellowstone National Park's Grand Prismatic Spring are usually archaea, which make up a large proportion of thermophiles. For an idea of the spring's scale, note the pedestrian boardwalk at the bottom. *(Jim Peaco, National Park Service)*

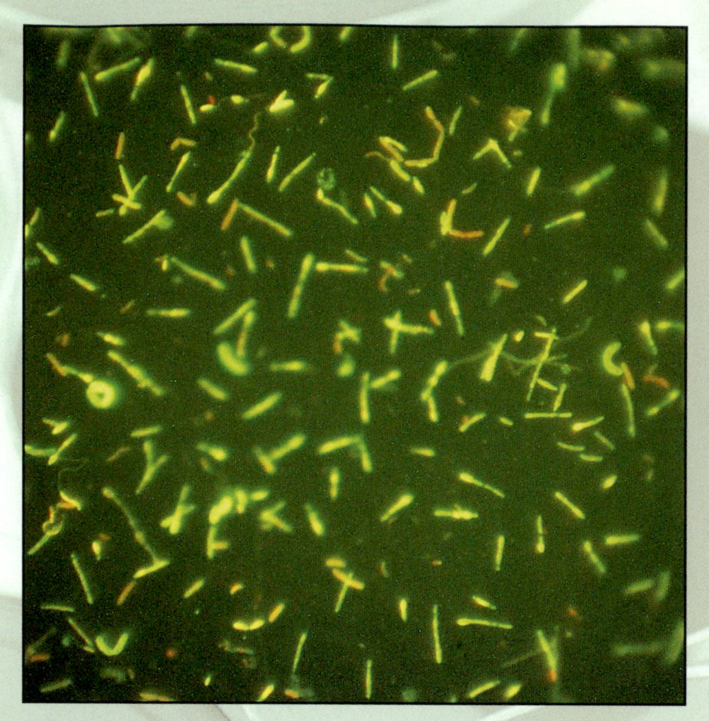

Fluorescence microscopy aids in microbial identification. Antibodies linked with a fluorochrome dye attach to a specific microorganism and the dye fluoresces in ultraviolet light. *(Stephen Spilatro)*

Some bacteria degrade tryptophan to indole. Kovac's reagent forms a red color on the surface, a positive result (tube on the left). Orange-yellow color (tube on the right) is a negative result. The indole test differentiates *E. coli* (positive) from Enterobacter (negative). *(Chiang Mai University, Division of Clinical Microbiology)*

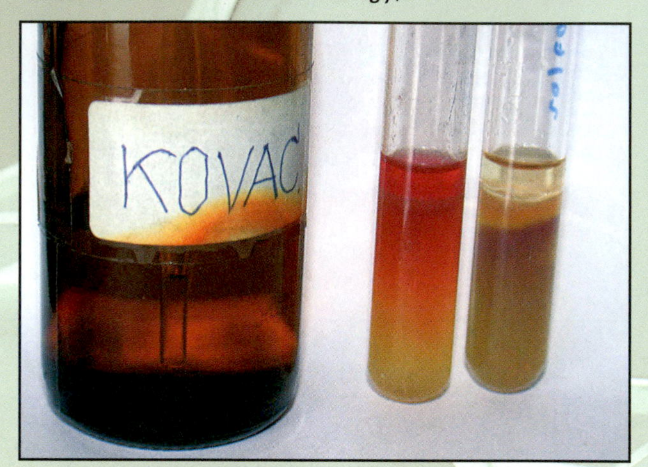

Molds such as this *Penicillium* are fungi composed of filaments and grow on surfaces. Molds are usually fuzzy, powdery, woolly, or velvety. *(Michael Gregory)*

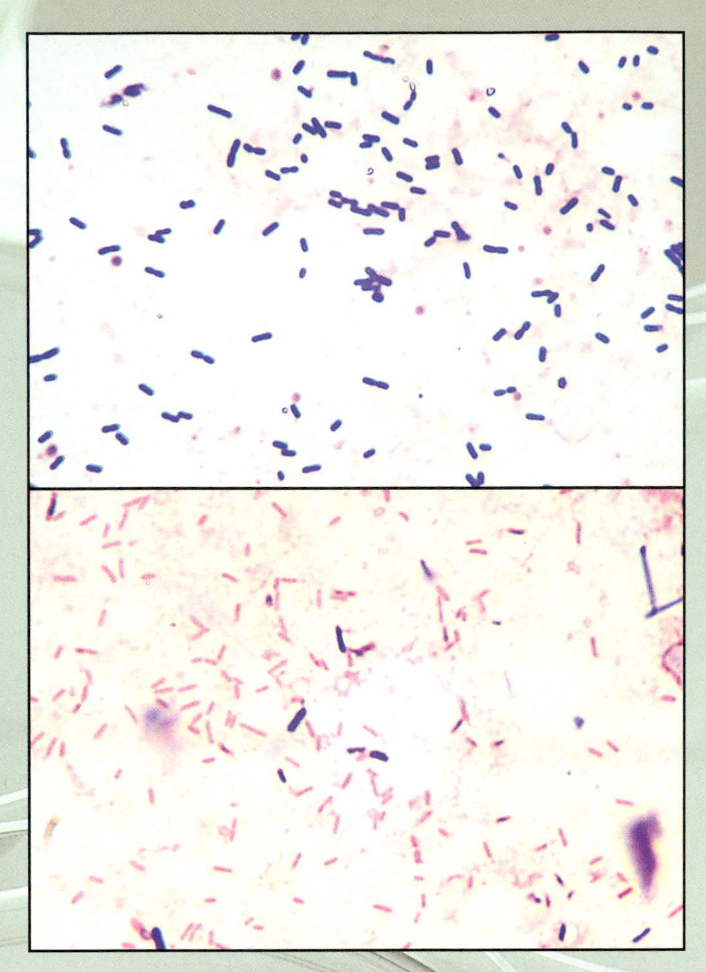

(Top) Gram-positive rods appear dark blue in a light microscope. *(Heidi Hoefer)*
(Bottom) Gram-negative rods appear pink. This specimen includes a few gram-positive cells. *(Heidi Hoefer)*

The Winogradsky Column supports microorganisms as they interrelate in nature. These columns show algae and cyanobacteria in the green, oxygenated layer, above a sulfur-rich black layer, a large red-orange zone containing photoheterotrophs, and deep anaerobic sediments. *(Mona Eilertsen)*

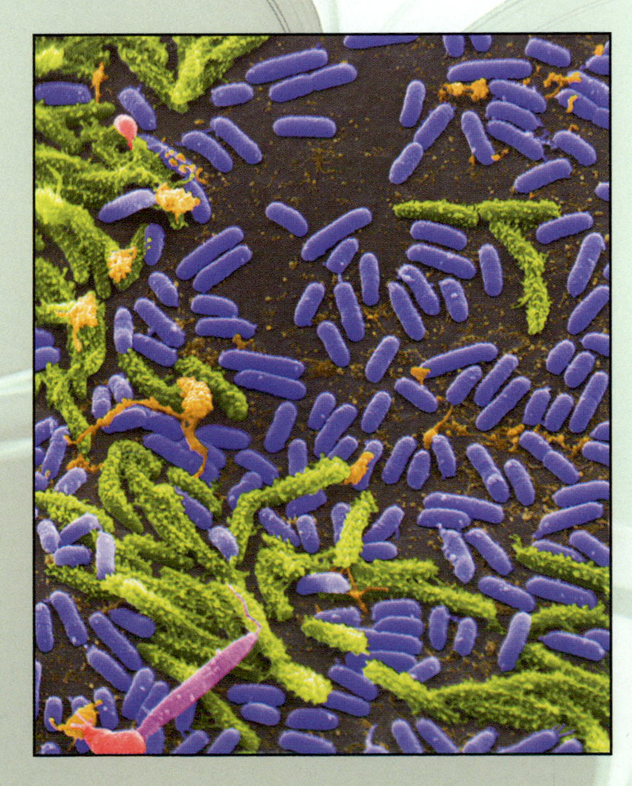

Cell shape is the first characteristic of bacteria identification. The blue colorized cells are slightly curved rods, typical of the genus *Vibrio*. *(Tina Carvalho, University of Hawaii at Manoa)*

Dark-field images come from light that has been reflected or refracted by the specimen and have a dark background. This dark-field image shows a colony of algae in a wet mount. *(University of Cambridge, Goldstein Laboratory)*

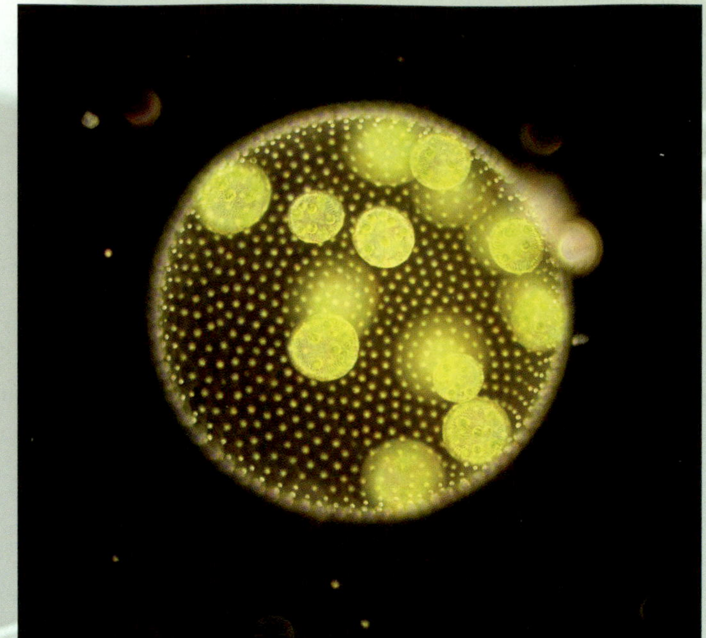

Differential interference contrast microscopy (DIC) splits light beams to give slightly different paths through a transparent specimen. DIC enables microbiologists to observe detailed cellular structures, such as these *Euglena* cells, without staining the specimen. *(Seth Price)*

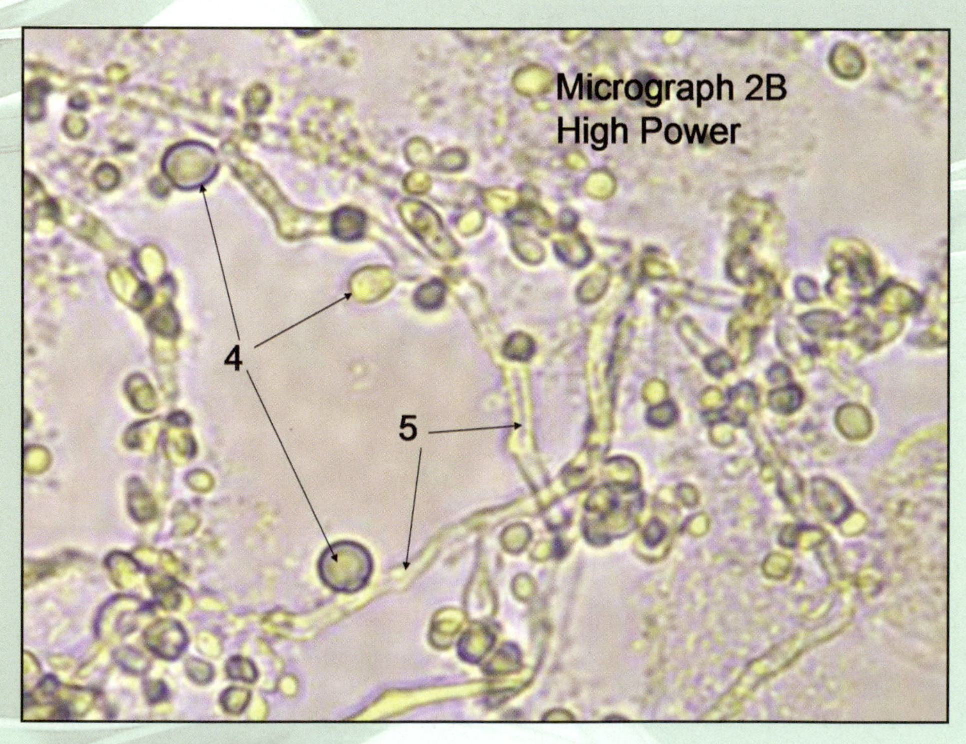

A wet mount of *Saccharomyces* yeast helps in viewing structures such as yeast cells (no. 4 arrows) and hyphae (no. 5 arrows). *(State of Michigan Department of Community Health)*

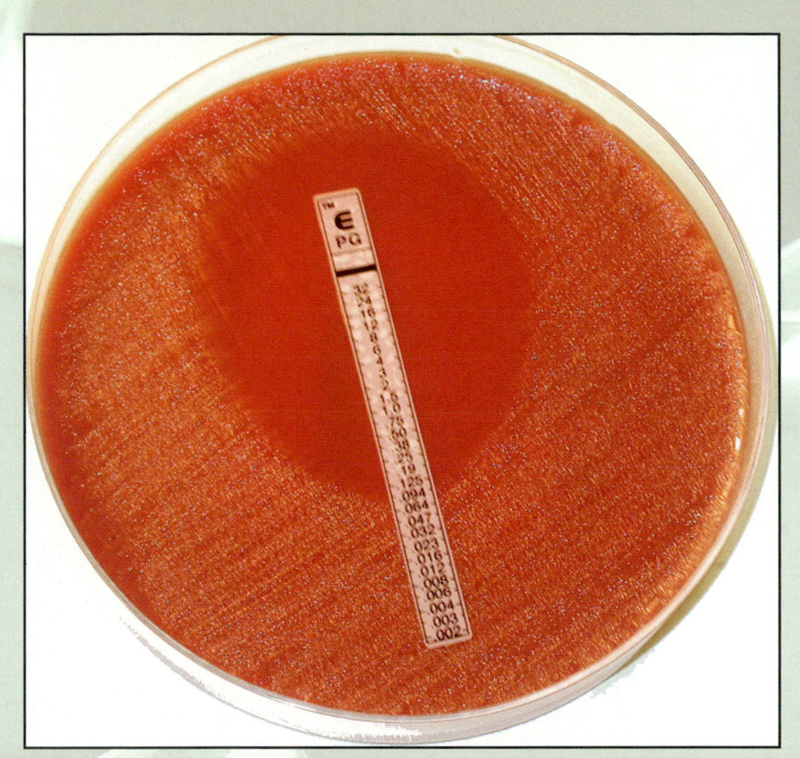

The E test strip shows the minimum inhibitory concentration (MIC) of an antibiotic in a lawn of bacteria. A microbiologist reads the MIC at the intersection of the clear zone of inhibition and the strip. *(Jean Phillips)*

Protoplasts are bacterial or plant cells that lack a cell wall. These protoplasts are from *Arabidopsis*, a plant related to mustard. *(Tsuneaki Asai)*

Halobacterium growing in these salt evaporation ponds produce the membrane protein bacteriorhodopsin, which participates in energy metabolism. The protein also produces a characteristic purple color. *(State of Michigan Department of Community Health)*

Yeast cells are larger than bacteria and usually produce larger colonies on agar. This yeast has contaminated a liquid in the same way molds contaminate foods and beverages. *(Wikimedia)*

Other dairy products made by the fermentative action of LAB on milk are butter, buttermilk, cottage cheese, sour cream, and yogurt. Nondairy products include beer, wine, and ciders.

Sourdough bread, invented by California gold miners in the mid-1800s, owes its unique sour flavor to two specific microorganisms. The yeast *Saccharomyces inusitatus* leavens the dough, and the LAB *Lactobacillus sanfrancisco* gives the bread its sour taste from lactic acid.

LACTIC ACID BACTERIA

The LAB group contains five main genera and several hundred species. The largest genus, *Lactobacillus,* alone contains more than 100 species. The LAB, in order of importance in food production, are:

- *Lactobacillus*—both hetero- and homofermentative
- *Lactococcus*—mainly homofermentative
- *Streptococcus*—homofermentative
- *Leuconostoc*—heterofermentative
- *Pediococcus*—homofermentative

Most acid foods are created by using more than one LAB. The blend of lactic acid from either type of fermentation or various end products of heterofermentation contributes to a food's distinctiveness.

Lactobacillus contains several species that produce various foods. The prominent species that make fermented dairy products are *L. acidophilus, L. brevis, L. casei, L. dulbrueckii, L. helviticus, L. johnsonii, L. plantarum,* and *L. rhamnosus.* These species grow well in acidic conditions of pH 3.5–5.0. These bacteria can also live in three different levels of oxygen availability in their environment: (1) anaerobic, lacking oxygen; (2) microaerophilic, requiring low levels of oxygen; or (3) facultative anaerobic, able to convert between anaerobic and aerobic metabolism.

Lactobacillus lives in the digestive tract of humans and other animals, but it is not a pathogen and rarely causes food-borne illness. The food industry's main concern regarding *Lactobacillus* is the LAB's ability to cause spoilage in milk, meat, and wine products. *Lactobacillus* and some species of *Leuconostoc* and *Pediococcus* carry out a metabolism called *malolactic fermentation.* In this metabolic pathway, the LABs convert a compound from the Krebs cycle, malic acid, to lactic acid. The Krebs cycle is an important energy-generating pathway from pyruvate and found in aerobes. Malolactic fermentations can alter the desired acidity of acidic foods, with the potential to produce a poor flavor or completely spoil the food.

FOOD SPOILAGE BY LACTIC ACID BACTERIA

LABs cause meat, milk, or wine spoilage if the conditions in the food have been accidentally altered. Almost all spoilage conditions result from a variety of nonmicrobial factors, as well as microorganisms that are not LABs. However, food microbiologists must include *Lactobacillus* as one of the main spoilage factors in food production and storage.

Meat spoilage depends on the number of microorganisms in the meat, plus physical factors, such as temperature, exposure to the air, and the nature of the meat itself. Ground meats with a large surface area are vulnerable to a high amount of aerobic and anaerobic bacteria. Minimally trimmed steaks and chops have a lower risk of contamination. *Lactobacillus* usually spoils packaged and chilled red meats of either variety. Packaging of meats in oxygen-impermeable wrapping or preparation under vacuum packing inhibits aerobic bacteria, which give *Lactobacillus* a chance to grow. *Lactobacillus* also resists the inhibitory effects of smoking or curing with nitrate compounds. For this reason, *Lactobacillus* spoils cured meats such as bacon and ham. In ham, the microorganism causes a greening effect by producing hydrogen peroxide, which reacts with blood's heme pigments to create a green color.

Milk spoilage or souring is caused by the utilization of milk proteins by *Lactococcus. Lactococcus lactis* converts amino acids from degraded proteins into the compound aldehyde, which contributes to off-flavors and odors. *Lactococcus*'s action of curdling proteins creates nonacidic environments that promote the growth of other non-LABs. These bacteria continue digesting the milk components, and the familiar odor and color of sour milk result.

In wines of low acid content, LABs carry out malolactic fermentation, which gives certain wines a biting, poor flavor. If the fermentation is allowed to continue, LABs can produce more lactic acid, which alters the wine's taste, usually referred to as *smooth.* If a winemaker does not want malolactic fermentation to occur, increasing the alcohol content inhibits the LABs. LABs also produce a condition called *ropiness.* Ropiness in wine occurs when *Leuconostoc* or *Lactococcus* builds long sugar chains, called *polysaccharides,* that become a slimy material in the liquid.

Food microbiologists now work with LABs for the purpose of giving foods distinct flavors. Research laboratories may develop strains of LABs that produce special enzymes or convert sugars and amino acids to desirable flavor compounds. Molecular biology may contribute increasingly to the benefits of LABs in food production.

See also FERMENTATION; FOOD MICROBIOLOGY; GLYCOLYSIS; METABOLIC PATHWAYS; METABOLISM.

Further Reading

Broadbent, Jeffery R., and James L. Steele. "Cheese Flavor and the Genomics of Lactic Acid Bacteria." *Microbe*, March 2005.

Medical News Today. "Lactic Acid Bacteria—Their Uses in Food." September 26, 2004. Available online. URL: www.medicalnewstoday.com/articles/14023.php. Accessed August 16, 2009.

Romero-Garcia, Susana, Claudia Hernandez-Bustos, Enrique Merino, Guillermo Gosset, and Alfredo Martinez. "Homolactic Fermentation from Glucose and Cellobiose Using *Bacillus subtilis*." *Microbial Cell Factories* 8 (2009): 23–31. Available online. URL: www.microbialcellfactories.com/content/8/1/23. Accessed August 16, 2009.

latency Latency is the condition of some diseases in which the disease is undetected in the body for a period (latent period), before developing into symptoms. In virus infections, latency is the period after infection when the virus stops reproducing and before it becomes active again. When an infectious agent leaves the latency period and begins multiplying, symptoms develop, and the condition is said to be an *active disease*. This transition from latent period to active stage is called *reactivation*. Once reactivation takes place, the pathogen can spread to others. For that reason, latency can be defined as the period between infection and disease transmission.

Latent diseases take two forms: (1) diseases that have a latency period before the first symptoms appear and (2) diseases that produce symptoms intermittently, with a latent period between episodes. Some pathogens, furthermore, are active but not producing symptoms (asymptomatic), and these microorganisms have the potential to spread from person to person, before any symptoms appear.

Latent infections and chronic infections together are known as *persistent infections*, which are any infections that last for years. Chronic infections differ from latent infections because chronic disease produces constant, detectable signs and symptoms. A subclinical infection, by contrast, is one in which symptoms have developed but have not been detected.

Latency complicates the work of epidemiologists when they are trying to determine the extent of a disease in a population, especially when estimating the disease's incidence and prevalence. Incidence is the frequency of new cases of a disease in a population; prevalence is the number of cases in a population at any given time. When signs or symptoms of disease do not appear soon after infection, epidemiologists risk underestimating the true number of infected people. Underestimations make it difficult for health care providers to anticipate new outbreaks and for drug companies to produce adequate amounts of vaccines, antibiotics, or other therapies.

LATENCY, INCUBATION PERIODS, AND SLOW DISEASES

The latency period differs from a disease's incubation period. An incubation period is the time in which the infectious microorganism reproduces, leading up to the first appearance of symptoms. The number of pathogens that start the initial infection, called an *infectious dose*, often affects the length of the incubation period. In general, however, incubation periods are short and fairly constant, compared with latent periods, which tend to be longer and variable from person to person and are not necessarily affected by infectious dose. The following factors are believed to affect the latency period, but the exact mechanisms are only partially known:

- immunity
- genetic factors
- diet and nutrition
- age
- presence of preexisting disease

Epidemiologists have studied certain latent diseases in detail to determine how the factors listed here affect the latency period. Researchers have compiled large amounts of data on the human immunodeficiency virus (HIV) that causes acquired immunodeficiency syndrome (AIDS), in an effort to predict the period between infection and the first signs of AIDS. Doctors, then, make predictions of when an infected person might start to show AIDS symptoms. The predictions are based on probability: That is, they are estimates drawn from large amounts of data on the disease.

Some latent viral diseases can be extremely difficult to track because viruses are undetected in the blood, antibodies against the virus do not circulate, or the infected person is asymptomatic for months to years. A group of viruses that cause these conditions are called *slow viruses*. Slow virus diseases have a very long incubation period, but they do not undergo a latency period. Various conventional viruses that attack the nervous system usually behave as slow viruses, such as progressive multifocal leukoencephalopathy (PML) and subacute sclerosing panencephalitis (SSPE). In addition, the protein responsible for the neurological disorder Creutzfeldt-Jacob disease also behaves as a slow virus does, though it is not a virus.

LATENT DISEASES

Tuberculosis (TB) is an infectious communicable disease caused by *Mycobacterium* bacteria. TB has also grown into a worldwide health threat, killing two million people annually. Because TB is a latent disease, it has been difficult to monitor and control in many parts of the world. People infected with *M. tuberculosis* in latency are not contagious, and only 5–10 percent of people who have TB in latency will develop the active disease. But as the molecular biologists JoAnne Flynn and John Chan wrote in *Infection and Immunity,* in 2001, "Up to one-third of the world's population is infected with *M. tuberculosis,* and this population is an important reservoir for disease reactivation. Understanding latent and reactivation tuberculosis, at the level of both the host and the bacillus, is crucial to worldwide control of this disease." TB presents an especially alarming challenge to doctors, because the body's own immune system protects the pathogen and helps latency.

Latent tuberculosis is caused by the immune response of the pathogen in the small air-exchanging pockets in the lungs called *alveoli.* The bacteria enter immune cells called *macrophages* in the alveoli, and the macrophages, then, signal the immune system for further action. The immune system encloses the entire infected area of the lung in a tubercle (a cyst-like vesicle) to contain the infection. This protection allows *M. tuberculosis* to remain alive and infectious for years, before beginning the disease process. *M. tuberculosis* cells inside the lung tubercles, meanwhile, evade any further attacks by the immune system and can live there for decades.

Gary Schoolnik, professor of medicine at Stanford University, concluded, in 2003, "This is an extraordinarily successful survival strategy." Schoolnik's laboratory identified genes that control the way the bacteria change their metabolism once inside a tubercle so that they can exist in low oxygen levels and in the presence of nitric oxide that the immune cells excrete. "Latency is like a contract between the host and the bacteria. As long as the levels of nitric oxide remain constant, the bacteria will not divide. It's a kind of mutualism." TB is one of the oldest known infectious diseases; its method of infection may have evolved as humans evolved.

Many different viruses also set up latent infections, the most notorious of which is HIV, which can enter a latent period of up to 10 years before developing the disease. As with TB, viruses have evolved with humans so that their mechanisms of evading the human immune response are now finely tuned. Viruses use one of two general mechanisms for establishing latency:

- using systems that evade or escape the immune system
- infecting cells that are inaccessible to the immune system

Viruses accomplish the first method by putting compounds on their outer coat that make them unrecognizable by the immune system. Viruses can evade immune cells called *T-cells,* for example, by

Latent Viral Infections

Virus	Disease	Latency Mechanism	Latency Period
cytomegalovirus	congenital CMV infection	stays dormant in cells of lung, heart, and bone marrow	30 years or longer
Epstein-Barr	mononucleosis	stays dormant in cells of the immune system	20 years or longer
hepatitis B	hepatitis	invades liver cells	variable
herpes simplex 1	genital herpes	in nervous system, then along nerves to the skin, when reactivated	lifetime
HIV	acquired immunodeficiency syndrome (AIDS)	stays dormant in lymph nodes	10 years or longer
varicella-zoster	chicken pox, herpes zoster (shingles)	in nervous system, then along nerves to the skin, when reactivated	lifetime

synthesizing proteins that bind and neutralize a T cell meant to destroy them. Viruses that use the second method of evasion do so, usually, by infecting nerves or the brain. The table on page 469 lists prevalent viruses that cause latent infections.

Medical researchers investigate the factors that make a dormant virus become reactivated to begin disease. For many latent diseases, the details of this critical step are not fully known. Some of the proposed reasons for reactivation are:

- reinfection by the same pathogen
- infection by a different pathogen
- weakened immunity
- chronic disease

Several other factors have been suggested as possible triggers for reactivation, including stressors of a physical, physiological, or psychological nature. In 2008, the National Aeronautics and Space Administration (NASA) reported that it had been studying the stress of spaceflight on reactivation of latent viruses in infected astronauts. NASA reported, "Saliva samples collected from crew members travelling on the shuttle to and from ISS [International Space Station], since 2000, have provided preliminary results for the Latent Virus Investigation." Epstein-Barr virus and varicella-zoster virus had become reactivated in some of the shuttle crew members during their missions. Spaceflight studies on virus reactivation have continued for the purpose of answering questions about reactivation in normal lifestyles.

Many factors that contribute to latency remain unknown. Medical researchers have continued to explore the means by which pathogens hide in the body and, most important, why and how pathogens reactivate and start disease. Latency is a crucial area of study in medicine.

See also HUMAN IMMUNODEFICIENCY VIRUS; IMMUNITY; VIRUS.

Further Reading

Flynn, JoAnne L., and John Chan. "Tuberculosis: Latency and Reactivation." *Immunity and Infection* 69 (2001): 4,195–4,201. Available online. URL: www.pubmedcentral.nih.gov/articlerender.fcgi?artid=98451. Accessed August 15, 2009.

Hunt, Richard. "Human Immunodeficiency Virus and AIDS—the Course of the Disease." In Microbiology and Immunology Online. Available online. URL: http://pathmicro.med.sc.edu/lecture/HIV3.htm. Accessed August 15, 2009.

Pierson, Duane L., and Satish K. Mehta. "Incidence of Latent Virus Shedding during Space Flight." International Space Station Fact Sheet. June 27, 2008. Available online. URL: www.nasa.gov/mission_pages/station/science/experiments/Latent-Virus.html. Accessed August 15, 2009.

Science*Daily*. "How a Latent Virus Eludes Immune Defenses." April 8, 2005. Available online. URL: www.sciencedaily.com/releases/2005/03/050328173659.htm. Accessed August 15, 2009.

"Stanford Researchers Identify Genes Involved in Tuberculosis Latency." February 16, 2003. Available online. URL: www.eurekalert.org/pub_releases/2003-02/sumc-sri021203.php. Accessed August 15, 2009.

Leeuwenhoek, Antoni van (1632–1723) ***Dutch Scientist and tradesman*** Antoni van Leeuwenhoek has been called the Father of Microbiology because he was the first to observe microbial cells through a microscope and draw conclusions about the microscopic living things he saw. Van Leeuwenhoek's three main contributions to the development of microbiology were (1) assembly of a practical microscope for observing things too small to be seen with the unaided eye; (2) experimentation with various magnifications, eventually reaching a magnification of about 300 times for his studies of microorganisms; and (3) drawings of the microorganisms he observed.

Van Leeuwenhoek was born in Delft, the Netherlands, into a large family with backgrounds in brewing and basket making. In his early teens, he sought an apprenticeship as an accountant with a cloth merchant in Amsterdam. During his training, the apprentice became interested in the simple microscopes (single-lens assemblies similar to a magnifying glass) that cloth merchants used to inspect the quality of fabric threads. At age 22, after six years apprenticing, van Leeuwenhoek returned to Delft to open his own textiles business and took with him a microscope he had purchased in Amsterdam.

As van Leeuwenhoek's business developed, he spent his spare time in perfecting his skills in shaping and grinding glass to make lenses. By experimenting with the convex shape of blown glass, he could affect the level of magnification of each different lens. The lenses measured only about a few millimeters across. Van Leeuwenhoek fixed a single lens between brass plates and held the instrument up to his eye to view specimens that he steadied by mounting onto a pin. Van Leeuwenhoek put each specimen in focus by turning screws on the plates, which adjusted the exact position of the lens relative to the specimen.

Others had experimented with rudimentary microscopes before van Leeuwenhoek. The Dutch eyeglass maker Zacharias Janssen (1580–1638) had already discovered the results of placing more than one lens in sequence inside a tube, thus building the first compound microscope. Janssen had not, however, applied his invention to biology but instead concentrated efforts on improving eyeglass technology. Van Leeuwenhoek's contemporary the British scientist Robert Hooke (1635–1703) may have also developed a compound microscope, but historians have been perplexed as to whether Hooke used van Leeuwenhoek's instruments as a model for further improvement, or vice versa. Both men are known to have corresponded with each other, for a time, on the subjects of microscopes and magnification.

The microbiology historian Howard Gest studied recently discovered writings by Hooke and compared them to the scant notes available from van Leeuwenhoek. In 2007, Gest referred to Hooke's "description of how to make single-lens, hand-held miniature microscope of the kind Leeuwenhoek later improved and used extensively." Gest added, "Many microbiologists mistakenly believe that Leeuwenhoek invented this kind of hand-held microscope. However, in contrast to Hooke's lengthy published descriptions of such devices, Leeuwenhoek was notoriously secretive about his own methods and microscopes. He never disclosed the techniques that he used for grinding lenses or his conditions for illuminating samples." Antoni van Leeuwenhoek may have used Hooke's exquisitely illustrated book *Micrographia* as resource microscopic studies of nature. Historians have yet to determine the roles played by Hooke and van Leeuwenhoek in inventing the microscope. Few could doubt, however, that van Leeuwenhoek used the invention mainly for viewing microorganisms.

Van Leeuwenhoek possessed a keen intuition about places where microorganisms live. He took samples of soil, water, souring milk, saliva, and the plaque he scraped from the surface of his teeth and from others'. He is known to have inspected lake water, rainwater, and melted snow. No evidence exists to suggest van Leeuwenhoek speculated on the microscopic creatures he observed or even proposed how or why they existed in his specimens, but, as did Hooke, van Leeuwenhoek possessed superior powers of observation. His findings were all the more impressive because of the single lens he used, rather than compound assemblies. Lens making started as a hobby for van Leeuwenhoek, but his skill at making lenses exceeded anything that previously existed, and he gained a reputation throughout Europe. Scientists, merchants, and the curious traveled to Delft to see the instruments. Little evidence exists that van Leeuwenhoek built microscopes or ground lenses for profit.

Fortunately for historians of microbiology, van Leeuwenhoek documented his results in a series of letters he sent to the Royal Society of London, a newly formed association of scientists that would rise to lofty stature in Europe, in later years. The society did not immediately accept van Leeuwenhoek's information. Society members dismissed one of van Leeuwenhoek's earliest descriptions of his microscope, circa 1673, as a hoax because few of the scientists believed a person could magnify objects 200 times with a single lens.

Van Leeuwenhoek is known to have addressed about 100 letters to the Royal Society, which translated them from his native Dutch to English and Latin, but he may have written a total of more than 200. The collection of subjects in the surviving letters gives evidence of the cloth merchant's insatiable curiosity: red blood cells, protozoa, insects, hair, mineral crystals, and various powders. The writings provide evidence that van Leeuwenhoek discovered lymphatic capillaries and may have been the first person to study the cell nucleus and observe the lysis of red blood cells in water.

In 1677, van Leeuwenhoek wrote, "May the 26th, I took about 1/3 of an ounce of whole pepper and having pounded it small, I put it into a Thea-cup with 2 ½ ounces of Rainwater upon it." He let the pepper settle and, then, examined the rainwater in his microscope. Van Leeuwenhoek saw microorganisms, but he assumed they had died within a week when he could no longer detect them swimming in the water. Late on the evening of June 2, van Leeuwenhoek noted that a small number of cells appeared to be living and moving. "But the 3rd of June," he wrote, "I observed many more which were very small, but 2 or 3 times as broad as long. . . . The 4th of June in the morning I saw great abundance of living creatures; and looking again in the afternoon of the same day, I found great plenty of them in one drop of water, which were no less than 8 or 10000." Van Leeuwenhoek described microorganisms, perhaps bacteria as well as protozoa, and unknowingly had monitored the progress of cell multiplication.

The various letters describing microorganisms often included illustrations with sufficient detail for modern microbiologists to recognize the types that van Leeuwenhoek probably had seen. Although van Leeuwenhoek would famously coin the phrase "little animalcules" to describe microscopic living things, he had probably, first, applied this description to spermatozoa recovered from semen rather than microorganisms. When he reported his observation to the Royal Society, its members assumed that the spermatozoa were parasites rather than components of the reproductive system.

Compelling evidence for van Leeuwenhoek's place as history's first microbiologist can be gleaned from a letter he sent to the Royal Society, in 1684, which included sketches of rod-shaped cells and cocci, as well as a description of cell motility. (The historian Brian Ford has suggested that van Leeuwenhoek may have commissioned an artist to redraw his crude sketches.) Studying saliva, van Leeuwenhoek noticed "very many living Animals, which moved themselves very extravagantly. The biggest sort had the shape of [a thin rod]. Their motion was strong and nimble, and they darted themselves thro[ugh] the water or

Antoni van Leeuwenhoek gave birth to the science of microbiology when he observed live microorganisms in a microscope. ***(Dibner Library of the History of Science and Technology)***

spittle, as a Jack or a Pike does thro[ugh] the water." Van Leeuwenhoek continued by describing a second, shorter rod that spun through the liquid "like a Top." The descriptions also included ovals and cocci. Most important, van Leeuwenhoek conducted a simple experiment on the microorganisms.

After gargling with wine vinegar, van Leeuwenhoek sampled the contents of his mouth. Among his observations, he saw a noticeable decline in rod-shaped cells but no reduction in the cells that lodged in the scurf of his teeth (plaque). "I took a very little wine-Vinegar and mixt it with the water in which the scurf was dissolved, whereupon the Animals dyed presently. From hence I conclude, that the Vinegar with which I washt my Teeth, killed only those Animals which were on the outside of the scurf, but did not pass thro[ugh] the whole substance of it." Few students may presently appreciate the importance of van Leeuwenhoek's discoveries: He had made a crucial observation on the protective nature of biofilm against chemicals and had performed one of the earliest experiments on the effect of disinfectants on microorganisms.

Antoni van Leeuwenhoek has received only modest appreciation for his ability to make accurate measurements of microscopic objects. For instance, he estimated the approximate size of red blood cells and described microorganisms as being "no bigger than 1/100th part of a sand." Even van Leeuwenhoek's guess at the number of microorganisms in existence was closer to the truth than anyone at the time imagined. He said that microorganisms "are so many that I believe they exceed the number of Men in a kingdom." Scientists of the 18th and 19th centuries continued to use van Leeuwenhoek's writings and painstaking depictions of cell shape, size, and movement as references for their own bacteriological studies. Van Leeuwenhoek's microorganisms would lay the foundation for later scientists, such as Louis Pasteur, to disprove the theory of spontaneous generation, and van Leeuwenhoek himself dabbled in experiments related to this hypothesis, that life arose automatically from nonliving matter.

In the decade after van Leeuwenhoek published his first description of a microscope, Royal Society members followed his notes to build their own models, and the power of magnification that the cloth merchant had achieved astounded them. In 1680, they elected van Leeuwenhoek to join the society, where he remained for the rest of his career. He devoted increasing time to science and built a reputation as, perhaps, the world's expert on microscopy. Royalty and other world leaders visited him in his London laboratory, and the admiration of the public caused van Leeuwenhoek to request that visitors schedule appointments with him in advance.

As did Robert Hooke, van Leeuwenhoek eventually developed a method of illuminating the specimen to enhance the magnified image, but the method by which he did this has been lost to history. Upon his death in 1723, van Leeuwenhoek bequeathed 26 of his microscopes to the Royal Society.

Antoni van Leeuwenhoek has been given credit as the originator of microbiology and microscopy. He also epitomized science through his inquisitiveness, powers of observation, record keeping, and experimentation.

See also BIOFILM; HOOKE, ROBERT; MICROBIOLOGY; MICROSCOPY.

Further Reading

Committee of Dutch Scientists. *The Collected Letters of Antoni van Leeuwenhoek.* Berwyn, Pa.: Swets & Zeitlinger, 1996.

Ford, Brian J. "From Dilettante to Diligent Experimenter, a Reappraisal of Leeuwenhoek as Microscopist and Investigator." *Biology History* 5 (1992): December. Available online. URL: www.brianjford.com/a-avl01.htm. Accessed January 3, 2010.

Gest, Robert. "Fresh Views of 17th-Century Discoveries by Hooke and van Leeuwenhoek." *Microbe,* 10 November 2007.

van Leeuwenhoek, Antoni. "An Abstract of a Letter from Mr. Anthony Leeuwenhoeck at Delft, Dated Sept. 17, 1683. Containing Some Microscopical Observations, about Animals in the Scurf of the Teeth." *Philosophical Transactions of the Royal Society of London* 14 (1684): 568–574. In *Milestones in Microbiology.* Washington, D.C.: American Society for Microbiology Press, 1961.

———. "Observations Communicated to the Publisher by Mr. Antony van Leeuwenhoek, in a Dutch Letter on the 9th of Octob. 1676 Here English'd: Concerning Little Animals by Him Observed in Rain- Well- Sea- and Snow-water; As also in Water Wherein Pepper Had Lain Infused." *Philosophical Transactions of the Royal Society of London* 11 (1677): 821–831. In *Milestones in Microbiology.* Washington, D.C.: American Society for Microbiology Press, 1961.

New York Times News Service. "'Little Animalcules' Discovered 300 Years Ago." Eugene, Oreg. *Register-Guard,* 25 December 1975. Available online. URL: http://news.google.com/newspapers?nid=1310&dat=19751224&id=j7oUAAAAIBAJ&sjid=QOADAAAAIBAJ&pg=6678,477755. Accessed January 3, 2010.

lichen A lichen is a composite organism containing a fungus living in symbiosis with either an alga or a cyanobacterium. Symbiosis is any relationship in which two organisms live in a specific association. Lichens had, at one time, been thought of as a mutualistic pairing in which both organisms benefited. Additional studies of lichens have provided evidence that lichens follow a type of symbiosis called *parasitism,* in which one organism gains an advantage from the relationship and the other may be harmed.

Lichenization is the process by which the long, stringy appendages made by fungi grow together with a microorganism. The fungus portion of a lichen is called the *mycobiont,* which is usually a member of class Ascomycetes in kingdom Fungi. The microbial portion is called the *phycobiont* (also called a *photobiont* because of its photosynthesis), whether it contains algae or cyanobacteria. In most lichens, the mycobiont dominates the relationship so that lichens are often categorized simply as fungi or lichenized fungi. A minority of lichens contain an additional member, either a phycobiont that works in association with the other two organisms or another fungus that acts as a parasite.

The Finnish botanist William Nylander (1822–99) contributed most of mycology's early knowledge about lichens, and the German biologist Wilhelm Friederich Zopf (1846–1909) described many of the chemical reactions within lichens. Lichens received little attention, however, for decades, compared with other fungi. In botany a century ago, and even today, lichens have been mistaken as moss, which is a small nonvascularized plant and not a fungus. The worth of various lichens is now becoming appreciated as sources of industrial dyes, medicines, indicators of air or soil pollution, and food.

Lichens are known to be among the oldest plantlike organisms on Earth. The University of Kansas botanist Thomas Taylor reported, in 1995, finding the oldest fossil record of a lichen, near Aberdeen, Scotland, and estimated its age at 400 million years. In 2005, Xunlai Yuan of the Nanjing Institute of Technology found even older lichens in South China, possibly 600 million years old. The earliest lichens probably predated these discoveries, when aquatic fungi first developed a symbiotic relationship with aquatic algae. Shuhai Xiao of the Nanjing Institute said, in 2005, "The ability to form a symbiotic relationship between fungi and algae may have evolved long before the colonization of land by land-based lichens and green plants, which also form symbiotic relationship with various fungi." When plant life, then, started to dominate Earth's land surfaces, lichens made up the predominant life-form. Earth probably resembled a colorful palette produced by the bright pigments that protected ancient lichens from the early atmosphere's high doses of ultraviolet light.

Lichens, now, inhabit almost every place on the earth's surface, and almost 19,000 species have been documented thus far. (See color insert, page C-2.) They continue to be valuable, both as a study tool for early life on Earth and for the useful roles they play, such as the following:

- indicators of air pollution
- production of dyes for clothing
- production of the antimicrobial compound usnic acid
- secretion of scents used in perfumes
- secretion of antitumor compounds
- food for animals living on the tundra
- human foods

In addition to the value of lichens as studies on early plantlike life on Earth, these organisms also provide a model for symbiosis.

LICHEN STRUCTURE AND SYMBIOSIS

The mycobiont and phytobiont of every lichen grow together to form the main body, called the *thallus,* which, in looks and behavior, is completely differently in either of the species grown alone. Fungal hyphae form a strong protective outer covering called the *cortex,* which is exposed to the air. The phycobiont usually forms the layer directly under the cortex. Cyanobacteria or algae cells intersperse within a tangle of hyphae. Below the tangle, a second cortex rests against the surface of a tree or rock. The region between the two cortexes is called the *medulla.* Some hyphae burrow through the lower cortex and extend to the tree or rock. These hyphae, called *holdfasts,* help the lichen stay attached to the surface.

Lichens belong to three different categories, which are based on the morphology of the thallus. Crustose lichens form a crust that covers a surface in nature. Foliose lichens, by contrast, have a more leafy structure, and fructicose lichens grow finger-like appendages.

When each biont of a given lichen grows independently but in the same culture vessel, algae excrete only about 1 percent of the carbohydrates they produce by photosynthesis and retain the rest. When algae grow in association with fungi, however, the algal cell wall allows more carbohydrates to pass out of the cell, and the fungus receives up to 60 percent of the algae's carbohydrates. The microorganisms also receive two benefits from the fungus: protection from drying out (desiccation) and attachment to a stable environment.

Lichens fulfill their nitrogen requirements by either absorbing nitrate compounds from the environment or depending on certain types of cyanobacteria that pull nitrogen from the atmosphere, a process called *nitrogen fixation.* The bacteria probably also supply the mycobiont with vitamins and trace nutrients (nutrients needed in very small amounts), and the mycobiont participates in the symbiosis by absorbing water from the surroundings.

LICHEN HABITATS

Lichen habitats range from the poles to tropical rain forests. They usually grow on the firm surfaces of rocks, tree bark, soil, leaves, and mosses. Lichens often participate in the gradual breakdown of these surfaces. For example, the disintegration of rock, called *weathering,* is partly due to lichens growing on the rock. Depending on the species, the lichen can perform endolithic growth, meaning the organism grows inside rock, or epilithic growth, which is growth on the rock's surface.

Some lichens live in extreme environments, in which temperature or dryness inhibits most other life. Scientists have found lichens growing at the earth's poles, in deserts, on the tundra, and at the top of high mountain ranges. In northern climates, reindeer and caribou dig through the snow to find lichen as a food source. Lichens' ability to scratch out a life in harsh environments may also explain why they serve as primary life in biological succession. Succession is a phenomenon in nature in which a barren landscape becomes populated, over time, with plant life, the smallest and simplest plants growing first, then successively replaced by larger, more complex plants and trees. On lands destroyed by fire, mining, or grazing, lichens may be the first organisms to return life to the area. The ability of lichens to survive dry and nutrient-poor conditions allows them to precede all other plant life in regenerating a damaged environment.

LICHEN PRODUCTS

Lichens produce a wide variety of useful substances, many of which are being investigated for further uses in medicine or industry. Most substances produced by lichens are organic chemicals that contain a ring structure, called *phenolic compounds.* From this basic structure, lichens produce at least 100 derivations that serve multiple purposes in nature. The biologists Lucia Muggia, Imke Schmitt, and Martin Grube explained, in a 2009 article, "Lichens as Treasure Chests of Natural Products," "Besides compounds common to all or most lichen groups, some species contain unique products. Arthogalin, for example, is a cyclic [peptide] known only from a species endemic to the Galapagos Islands." Mycologists have probably discovered only a fraction of the unique substances made by lichens that live only in isolated places.

A number of lichen extracts have been shown to have value for medical purposes. The main benefits are the following:

- antibiotics against bacteria
- antibiotics against fungi
- antiviral treatments
- antitumor treatments
- antimutagenic agents (inhibitors of mutations in cells)
- enzyme inhibitors
- antioxidants

A benefit of lichens that has been appreciated, for 4,000 years, lives in the variously colored pigments these organisms produce. Lichen extracts for making fabric dyes have been used since antiquity. Because many lichens grow only in single geographic regions, dyes recovered from lichens have been categorized by source instead of color, for example, Norwegian Korkje, Irish lichen dyes, and Scottish lichen dyes. Some of the most common and oldest lichen dyes are the following:

- purple orchil
- cudbear (red)
- orcein (purple-red)
- Parmelia (reddish brown)
- vulpinic acid (yellow)
- Peltigera (auburn)

Environmental scientists have devised methods for using lichens as indicators of pollution. Because lichens readily absorb metal cations (positively charged ions), scientists can harvest the lichens and analyze their constituents to determine the presence of metal pollution. Lichens also absorb compounds from the atmosphere, so similar harvesting and analysis have been adapted as a method for assessing air pollution.

In the mid-1800s, the Finnish botanist William Nylander (1822–99) noticed differences between the lichens growing on trees in the countryside outside Paris and the lichens in the city. For the next century, research progressed steadily in drawing a relationship between lichens and the chemicals in air. In the 1980s, the U.S. Forest Service (FS) began using lichens as a standard method for measuring air pollution. The FS Lichens and Air Quality Workgroup has identified the following four values of lichens in environmental studies: They (1) act as valuable components of ecosystems, (2) can be damaged or killed by air pollution, (3) provide useful indicators of air quality, and (4) can possibly be restored by reducing air pollution. The University of Minnesota botanist Clifford Wetmore explained, in 1987, the usefulness of lichens detecting advancing pollution in Lake Superior: "Isle Royale is a pristine environment . . . but from the lichen damage in certain areas, I can tell that some pollution is coming in, probably from Thunder Bay in Canada, where there is a paper plant and other industries and no scrubbers on the smokestacks." (A scrubber is a device that cleans smokestack emissions.) Lichens have been used in pollution assessment ever since.

The link between lichen health and air pollution has grown in importance because of the many pollutants that affect lichen growth, such as the following: nitrogen emissions, sulfur dioxide emissions, acid rain, and fluorine gas. Lichens also have variable sensitivity to metals from polluted air or soil. Depending on the specific metal, the effects on lichen health can be moderate to severe, when they are exposed to cadmium, copper, lead, mercury, or zinc. Lichens do not give an exact measurement of pollutants, but their high sensitivity as pollution indicators makes them valuable in environmental studies.

Lichen biology and genetics have lagged behind similar research on bacteria and fungi, but the enormous variety of lichens undoubtedly offers benefits for the future.

See also ALGAE; CYANOBACTERIA; FUNGUS; INDICATOR ORGANISM; SYMBIOSIS.

Further Reading

Brody, Jane E. "Is the Air Pure or Foul? Lichens Can Tell the Tale." *New York Times*, 18 August 1987. Available online. URL: www.nytimes.com/1987/08/18/science/is-the-air-pure-or-foul-lichens-can-tell-the-tale.html?scp=4&sq=lichens&st=nyt. Accessed August 18, 2009.

Casselman, Karen Diadick. *Lichen Dyes: The New Source Book*. Toronto: General, 2001.

Muggia, Lucia, Imke Schmitt, and Martin Grube. "Lichens as Treasure Chests of Natural Products." *Society for Industrial Microbiology News*, May–June 2009.

Nash, Thomas H., ed. *Lichen Biology*, 2nd ed. Cambridge: Cambridge University Press, 2008.

National Lichens and Air Quality Database and Clearinghouse. U.S. Forest Service. Available online. URL: http://gis.nacse.org/lichenair/index.php. Accessed August 18, 2009.

Sci-Tech, China.org.cn. "Fossils Shed Light on Lichen Evolution from Sea to Land." May 13, 2005. Available online. URL: http://en.invest.china.cn/english/scitech/128654.htm. Accessed August 18, 2009.

Taylor, Tom N., H. Haas, W. Remy, and H. Kerp. "The Oldest Fossil Lichen." *Nature* 378 (1995): 244. Available online. URL: www.uni-muenster.de/GeoPalaeontologie/Palaeo/Palbot/nature.html. Accessed August 17, 2009.

University of California Museum of Paleontology. "Introduction to Lichens: An Alliance between Kingdoms." Available online. URL: www.ucmp.berkeley.edu/fungi/lichens/lichens.html. Accessed August 18, 2009.

Lister, Joseph (1827–1912) ***British Surgeon*** Joseph Lister has been called the father of modern surgery for his discovery of the need for sterile conditions during medical treatment. Prior to Lister's work, hospitals followed none of the safety precautions known today. In many cases, staying in a hospital was more of a health threat than seeking treatment at home. Today's use of sanitation, antiseptics, and disinfectants in clinical settings results directly from the contributions of Joseph Lister to medical science and microbiology.

Joseph Lister was born on April 5, 1827, in Upton, England, the son of a physicist. A good student, Lister consistently advanced ahead of others in school, and he enrolled at University College, London, to study medicine before turning 20. Lister was fortunate to complete his studies at a time when the surgical profession was undergoing major changes. Medical schools allowed students to learn by using cadavers, a practice that had been thought of as reprehensible just a few years earlier, and surgeons also began using ether to anesthetize surgery patients. The public had never held surgeons in the high esteem they had long held for physicians, but new surgical techniques developed midcentury convinced some that surgery had grown into a refined skill. Unfortunately, hospitals did not offer a safe, clean haven for patient recovery from illness or surgery. "Hospitalism" was the main cause of infections, fever, and septicemia, an infection spread systemically into the bloodstream. Medical historians have described hospitals of the period as places to go to die, not to be cured. Joseph Lister would take note of the deplorable conditions in hospitals to launch his greatest gift to science.

After graduating, Lister joined the university faculty, in 1848, and enhanced his skills as a surgeon under the guidance of the renowned physiologist William Sharpey (1802–80). Lister moved on to become assistant surgeon, in 1856, at the Edinburgh Royal Infirmary in Scotland, and, then, was appointed main surgeon in Glasgow, in 1861. The Glasgow Royal Infirmary had just completed building a new surgical building, which gave Lister hope that the clean surroundings would decrease the mortality rate of surgical patients. Hospitals promised only a 50 percent survival rate among surgical patients when Lister joined the staff.

Lister had already developed a consuming interest in the problem of postsurgery mortality. He studied articles by Louis Pasteur (1822–95) describing how tiny microorganisms could be carried through the air and contaminate anything they landed on. If microorganisms set down on an open wound, the infection could quickly lead to sepsis, the presence of microorganisms inside the body. Antibiotics were almost a century in the future, but Lister vowed to prevent infection of wound injuries, particularly compound fractures, and surgical incision sites. This precaution, he believed, was paramount to a patient's recovery. "I arrived several years ago at the conclusion," he wrote, in 1867, "that the essential cause of suppuration [formation of pus] in wounds is decomposition brought about by the influence of the atmosphere upon blood or serum retained within them." From this observation, Lister would recommend the use of a new chemical, called an *antiseptic*, for the protection of open wounds from infection. He began a search for "some material capable of destroying the life of the floating particles."

Lister taught medical students under his tutelage that wounds required a sterile dressing after being treated and that the area around the wound should be also be made sterile. He wrote, in 1867, in the medical journal *Lancet*, "Bearing in mind that it is from the vitality of the atmospheric particles that all mischief arises, it appears that all that is requisite is to dress the wound with some material capable of killing these septic germs. . . . Carbolic acid proved in various ways well adapted for the purpose." But Lister's colleagues had not been convinced, especially since the acid vapors burned their eyes and burned patients' skin. They, furthermore, did not see any value in clean dressings for wounds or sterile surgical instruments, or for spraying a portion of the acid into the air as Lister had to prevent airborne contaminants from reaching the wound.

Clean and sterile conditions for medical care had been advancing on other fronts, so that, unbeknownst to Lister, others in the medical community were beginning to connect cleanliness with health. The Hungarian physician Ignaz Semmelweis (1818–65) had begun insisting that all medical students wash their hands between patient visits to reduce the spread of streptococcus infections. Although Semmelweis presented a lecture on this concept in Germany, in 1850, no evidence exists that Lister knew of him. Florence Nightingale (1820–1910) led her nurses in the proper techniques for maintaining sanitary conditions in the vicinity of soldiers in field hospitals during the Crimean War. She had been every bit as appalled by the terrible conditions in military hospitals as Lister was by city hospitals.

Newly graduated doctors entering the profession would be more open to Lister's advice than older doctors, who may have taken Lister's criticism of their methods as a personal insult. The idea of using an antiseptic on wounds traveled to Germany, and German doctors realized they could save soldiers' lives on the battlefield during the Franco-Prussian War by applying carbolic acid. By 1878, the German microbiologist Robert Koch (1843–1910) would advocate the use of sterile dressings and instruments for all surgeries. Koch would devote much of his career to developing the principles of infection and connecting an infection with a specific pathogen. But Lister still needed to convince Britain's medical community that any microorganism presented potential danger and should never be allowed to contaminate patients.

In 1877, Lister took a position as chair of surgery at King's College, London. At the time, doctors treated simple bone fractures by wiring together the ends, but this called for deliberate breaking to create a compound fracture, which protruded through the skin. Patching up a patient with no thought of sterile conditions invariably led to infection and sepsis. Death soon followed. In short, a simple fracture could easily kill a person. Lister's reputation preceded him at King's College, including his odd ideas on putting harsh chemicals directly on a person's skin. Doctors on staff were as skeptical as Lister's previous colleagues, assuming that the chemicals did far more harm than good to the patient's skin. But, shortly after his hiring, Lister found an opportunity to convince others of the principle of asepsis, the lack of contamination.

In 1877, a man entered the hospital with a broken kneecap. Lister shocked the other doctors by intentionally opening the man's knee in order to repair the patella. A kneecap was not the same as a simple break of a long bone, and Lister's peers saw no future for the patient. Throughout the surgery, Lister used sterilized scalpels and other instruments and sterile dressings, and he wiped the area around the surgical site with carbolic acid antiseptic. Lister also had altered the carbolic acid formula to lessen its irritating effects on the skin. He explained in the article "On the Antiseptic Principle in the Practice of Surgery," "This difficulty [damage to the skin] has, however, been overcome by employing a paste composed of common whitening (carbonate of lime) mixed with a solution of one part of carbolic acid in four parts of boiled linseed oil, so as to form a firm putty." After the patient recovered, Lister had performed six additional successful surgeries. Surgeons who had been giving their patients a mere 50-50 chance of recovery took note of Lister's record of 100 percent recovery. The use of antiseptics would soon become standard procedure for treatment of hospital patients.

Joseph Lister continued innovating throughout his career. He constantly sought better ligatures, bandages, and surgical methods. Carbolic acid would be replaced by other antiseptics that were much safer on the skin but just as effective. As early as 1867, Lister proudly wrote of the improved conditions in hospitals that used aseptic techniques, "Since the antiseptic treatment has been brought into full operation, and wounds and abscesses no longer poison the atmosphere with putrid exhalation, my wards, though in other respects under precisely the same circumstances as before, have completely changed their character; so during the last nine months not a single instance of pyaemia *[sic]*, hospital gangrene, or erysipelas has occurred in them." *Pyemia* is another term for septicemia, which can create abscesses throughout the body. Erysipelas is infection of the skin. Lister had convinced his professional peers: A hospital need only adopt the use of antiseptics and change no other routine in order to improve patient health.

Joseph Lister might never have found an audience willing to listen to him if it were not for the germ theory proven by Louis Pasteur, just a few years earlier in France. Prior to Pasteur's work, doctors did not believe microorganisms were the cause of postsurgical infections. They often did not recognize the signs of infection at all and attributed poor recoveries to adverse reactions to anesthesia. Chemicals were something to be avoided when working with patients, so Lister's theory of chemical antiseptics would be forced to overcome considerable resistance as a radical and foolish idea.

After retiring in 1896, Lister accepted an appointment at the respected Royal Society of London organization of scientists. He died of pneumonia in London, on February 10, 1912. Lister's legacy became today's strict rules for cleanliness and sterile conditions in hospitals, proper antiseptic care of open wounds, and sterilization of instruments that touch the body.

See also ANTISEPTIC; GERM THEORY; KOCH'S POSTULATES; PASTEUR, LOUIS.

Further Reading

Gariepy, Thomas P. "The Introduction and Acceptance of Listerian Antisepsis in the United States." *Journal of the History of Medicine and Allied Sciences* 49 (1994): 196–206.

Lister, Joseph. "On a New Method of Treating Compound Fracture, Abscess, and So Forth; with Observations on the Conditions of Suppuration." *Lancet* 1 (1867): 326, 357, 387, 507. Available online. URL: www.kendallasmith.com/launch/suppuration.pdf. Accessed August 19, 2009.

———. "On the Antiseptic Principle in the Practice of Surgery." *British Medical Journal* 2 (1867): 246–248. Available online. URL: www.pubmedcentral.nih.gov/picrender.fcgi?artid=2310614&blobtype=pdf. Accessed August 19, 2009.

University of Dayton. "Sir Joseph Lister. 'The Father of Modern Surgery.' " Available online. URL: http://campus.udayton.edu/~hume/Lister/lister.htm. Accessed August 19, 2009.

logarithmic growth Logarithmic growth, also called *exponential growth,* is the phase of bacterial growth in which cells reproduce at a constant, maximal rate. Generation time (also called *doubling time*) is the time required for a cell to divide into two new cells, and this determines the rate in which cell numbers increase during logarithmic growth. A doubling time of only 20–30 minutes causes a dramatic increase in cell numbers over a short period. Longer doubling times slow the logarithmic growth rate.

LOGARITHMS

A logarithm is a mathematical value that corresponds to a number's exponent. For example, the number 1×10^4 (sometimes abbreviated 10^4), where 4 is the exponent, has a logarithm equal to 4.00. Numbers other than 1 correspond to values on a conversion scale called the base 10 logarithm, or $\log_{10}$. On this scale, each whole number change represents a change of 10 times. An increase from 1×10^4 to 1×10^5 equals an increase by a factor of 10, or one logarithm.

$$\log_{10} 10 = 1, \text{ because } 10^1 = 10$$

$$\log_{10} 100 = 2, \text{ because } 10^2 = 100, \text{ or } 10 \times 10$$

$$\log_{10} 1{,}000 = 3, \text{ because } 10^3 = 1{,}000, \text{ or } 10 \times 10 \times 10$$

Calculators or published $\log_{10}$ tables provide the logarithm for numbers other than 1, for example:

$$\log_{10} 150 = 2.1761$$

$$\log_{10} 63{,}500 = 4.8028$$

The usefulness of logarithms becomes clear when trying to express the numbers of bacteria in arithmetic numbers rather than logarithmic numbers. Doing calculations such as 1.48×10^6 minus 8.61×10^5 becomes very cumbersome compared with the same calculation using the logarithms, 6.17 - 5.94.

The second advantage of logarithms is the ease with which $\log_{10}$ numbers can be shown on a graph. Linear graphs follow a number sequence such as 2, 4, 6, 8, 10, and so on. Logarithmic sequences follow a sequence like that shown in the table on page 479: 2, 4, 8, 16, 32, . . . , where the difference between numbers in sequence increases at an increasing rate. The table illustrates the rapid increase in numbers that a bacterial culture reaches, when each cell divides to make two new cells. Without a conversion to $\log_{10}$ numbers, a graph of bacterial growth would be impractical.

BACTERIAL GROWTH IN LOG PHASE

Bacterial growth in static conditions, called a *batch culture,* progresses in four phases: (1) lag phase, (2) log or logarithmic phase, (3) stationary phase, and (4) death phase. Lag phase is a period of slow growth in which cells build cellular components. Stationary phase and death phase are the slowing of logarithmic growth and the decline of the cell population, respectively.

Microbiologists usually study growth rates in liquid broth cultures rather than cultures on solid agar. Bacteria grown in a single flask or test tube of broth represent a batch culture. Eventually, batch cultures run out of nutrients and are inhibited by buildup of end products. The culture, then, ceases logarithmic growth and enters a stationary phase, in which new cells develop at the same rate older cells die. The second growth method is called *continuous culture.* In this method, the culture receives a constant inflow of fresh medium and a constant outflow of spent medium containing cells. By adjusting the rate of this flow, a microbiologist can keep a culture in logarithmic growth indefinitely.

A tube of broth medium containing one cell with a doubling time of 20 minutes illustrates the dynamics of logarithmic growth. After 20 minutes in the medium, the cell divides by binary fission into two cells. Twenty minutes later, the two divide into four; 20 minutes after that, the four cells divide into eight cells, and so on. Although the numbers of microbial cells are small, at first, they increase at increasingly higher rates, every 20 minutes. The table shows how the slow initial growth of a culture explodes into logarithmic growth, within a few hours. Because the population doubles every 20 minutes, in this example, the increase in the population is always 2^n, where n is the number of generations. The number of cells, at any point in the growth curve, is also called the population size.

The table shows how a single cell grows into a population containing 3.28×10^4 ($\log_{10} = 4.516$) cells, in five hours. Most inocula do not contain a mere single cell, but at least 100–1,000 cells. A fairly small inoculum of 100 cells leads to a population of more than 3 million (3.0×10^6 or $\log_{10}$ 6.48), in five hours.

Biology cannot sustain the frenetic reproduction rate that occurs in a bacteria culture in logarithmic growth. Eventually, the system must slow down. During the constant part of the log phase, the cells experience balanced growth, wherein all cellular constituents are made at constant rates relative to each other. When nutrients begin to be used up and wastes accumulate, the culture enters a period of unbalanced growth, which leads to a slowdown of reproduction and the end of the log phase. A microbiologist can add nutrients to the culture, before this happens, and so initiate a new log phase, but this too will end, when nutrients again become depleted.

CELLULAR ACTIVITIES IN LOGARITHMIC GROWTH

In nature, logarithmic growth becomes more rare than in laboratory conditions because of factors in the environment that hamper microbial growth: limited nutrients, temperature variations, competition from other organisms, and drought, flooding, and freezing. Logarithmic growth takes place in nature only in short bursts, such as in a still pond that suddenly receives an influx of runoff carrying readily degradable substances.

In laboratory studies, microbiologists use logarithmic growth in four main study areas: (1) protein and deoxyribonucleic acid (DNA) synthesis, (2) nucleic acid replication and gene regulation, (3) antimicrobial testing, and (4) optimal growth conditions.

Microbial Growth

Time (minutes)	Division	2^n	Number of Cells
0	0	2^0	1
20	1	2^1	2
40	2	2^2	4
60	3	2^3	8
80	4	2^4	16
100	5	2^5	32
120	6	2^6	64
140	7	2^7	128
160	8	2^8	256
180	9	2^9	512
200	10	2^{10}	1,024
220	11	2^{11}	2,048
240	12	2^{12}	4,096
260	13	2^{13}	8,192
280	14	2^{14}	16,384
300	15	2^{15}	32,768

During the slow lag phase of bacterial growth, cells prepare for reproduction by building cellular constituents. This phase involves DNA replication, ribonucleic acid (RNA) activity and synthesis, and protein synthesis. Entering logarithmic growth, RNA content and the RNA-DNA ratio begin to correlate with growth rate. In general, an increased growth rate corresponds with increased RNA content and increased RNA-DNA ratio. Protein synthesis, similarly, increases in logarithmic growth, as new cells expand in size, make enzymes, and build the infrastructure for the next cell division.

Gene regulation plays a central role in controlling events, during the log phase. Many genes turn on during logarithmic growth to encode for proteins needed for active metabolism. A variety of genes also work in a complex manner to turn on some systems and turn off others inside the cell, all for the purpose of enabling a cell to take advantage of favorable growth conditions. Logarithmic cells regulate their growth by three major mechanisms: (1) regulation of messenger RNA (mRNA) synthesis, (2) control of enzyme activity, and (3) various multigene systems called *global regulation.* Messenger RNA is one of three RNA forms involved in gene expression, that is, the conversion of a gene contained in DNA into a functioning protein. Messenger RNA is a single strand of RNA that has been synthesized on the basis of corresponding base units in DNA. After mRNA has assembled the base sequence that defines a gene, it moves to the cell's ribosomes to hand the instructions off to another type of RNA (transfer RNA), which then participates in protein synthesis from single amino acids. By either inducing the activity of mRNA to increase or repressing mRNA activity, a cell can regulate its protein synthesis.

Many of the proteins produced by a microorganism are enzymes, which are essential compounds that run the machinery of the cell. Controlling enzyme activity, thus, controls growth rate. Microorganisms control enzyme activity in the following three main ways:

- allosteric regulation—regulation of cellular enzymes by a specialized subgroup of enzymes called allosteric enzymes that can activate or inactivate another enzyme by binding to it
- enzyme modification—the attachment or removal of a certain chemical group, phosphate (PO_4^-), for example, that modulates an enzyme's activity
- feedback inhibition—control of an enzyme based on the amount of that enzyme's end product

Bacteria also employ regulatory systems that are more complex than single-enzyme and mRNA processes. These systems are multigene systems, meaning they depend on many genes working at the same time. The genes that make up this global regulation are together called a regulon. Two basic components of the regulon are an operon and a regulatory protein. An operon is a set of genes that work together, with their expression controlled by a segment of DNA called an operator. The regulatory protein binds to the operator either to turn on or to turn off the operon. By controlling operons, an entire regulon might be turned on or off, that is, *up-regulated* or *down-regulated,* respectively.

The regulation of genes and functions pertaining to cell growth still contains unknowns. Overall, however, global regulation enables microorganisms to respond to the environment by controlling many genes at once. Some of the environmental factors known to elicit global regulation are the following:

- heat-shock and cold-shock response
- response to acid

- osmotic stress (pressure inside cells relative to the outside pressure)
- virulence and pathogenicity
- catabolite repression (inhibition of the breakdown [catalysis] of a substrate when end product builds up
- sporulation (the formation of a protective endospore by some bacteria)

The characteristics of a microorganism's logarithmic growth also aid in determining two characteristics: the effect of antimicrobial compounds on growth and optimal growth conditions. Antimicrobial substances such as antibiotics can greatly reduce growth rate. Exposing a susceptible species to an antibiotic will decrease the slope of a plotted log growth line. (Antibiotics usually also lengthen the lag phase.) A method called minimum inhibitory concentration uses this altered logarithmic growth to assess the effectiveness of antimicrobial agents.

Many bacteria are most susceptible to antibiotics during logarithmic growth. The group of antibiotics called *beta-lactam antibiotics* act by disrupting the construction of new cell walls in growing bacteria. These antibiotics, thus, exert their greatest effect on rapidly growing and dividing cells.

Microbiologists must study the best conditions for growing microorganisms in a laboratory. Each microorganism has certain optimal conditions that enhance its growth in a test tube or on an agar plate. The growth rate during the log phase gives an indication of optimal and suboptimal growth conditions. Logarithmic growth has been used to determine a species's optimal conditions for temperature, oxygen level, pH, and individual nutrients. In biotechnology, microbiologists determine these optimal conditions for the purpose of making microbial end products that have commercial value. Biotechnologists must, furthermore, pay attention to the end products microorganisms make during optimal and suboptimal conditions. Understanding the products associated with logarithmic growth helps biotechnology companies produce the greatest yield of a desired cellular product.

Microorganisms make some substances during the log phase but may produce different substances when logarithmic growth begins to slow. Primary metabolites are substances made during a microorganism's logarithmic growth phase. These substances usually relate directly to the growth and survival of the microorganism. Many microbial products made in biotechnology, however, are

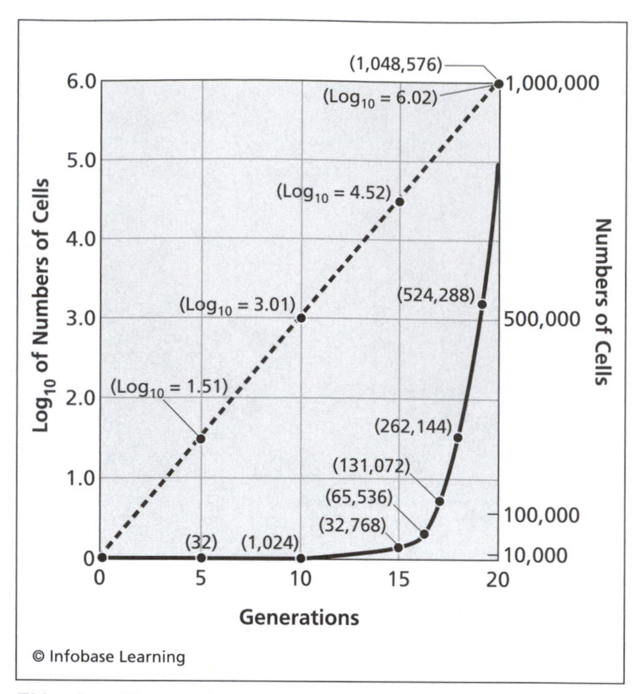

This plot of $\log_{10}$ values of corresponding arithmetic numbers shows the usefulness of logarithms when handling bacterial numbers.

secondary metabolites, which are substances produced after the period of most active growth. The following substances are secondary metabolites: antibiotics, enzyme inhibitors, pigments, and toxins. Companies that make any of these substances must maintain control of the microbial log phase.

Logarithmic growth is a hallmark of microbial metabolism, particularly in bacteria. The log phase is the period of greatest cellular activity within the growth curve. For this reason, many of the conclusions drawn about microbial metabolism have been derived from cultures in logarithmic growth.

See also GENERATION TIME; GROWTH CURVE; MINIMUM INHIBITORY CONCENTRATION.

Further Reading

Dale, Jeremy, and Simon Park. *Molecular Genetics of Bacteria,* 4th ed. Chichester, England: John Wiley & Sons, 2004.

Painter, Page R. "Generation Times of Bacteria." *Journal of General Microbiology* 89 (1975): 217–220. Available online. URL: http://mic.sgmjournals.org/cgi/reprint/89/2/217?view=long&pmid=1176952. Accessed August 15, 2009.

Todar, Kenneth. "Growth of Bacterial Populations." In *Todar's Online Textbook of Bacteriology.* Madison: University of Wisconsin Press, 2008. Available online. URL: http://textbookofbacteriology.net/growth.html. Accessed August 15, 2009.

M

malaria Malaria is a worldwide infectious parasitic disease of the liver and red blood cells caused by protozoa of genus *Plasmodium*. The pathogen contains various stages that take place inside and outside the human host, with a portion of the cycle occurring in an insect. Female mosquitoes of the genus *Anopheles* act as the vector in malaria's transmission from infected to uninfected people. When living in a human host, *Plasmodium* behaves as a parasite, meaning it receives a benefit from its host while also harming the host.

Several different *Plasmodium* species have been implicated as causing malaria, mainly *P. falciparum*, *P. malariae*, *P. ovale*, and *P. vivax*. These pathogens belong to a group of eukaryotes called the *apicomplexa*, which are microorganisms that possess a specialized organelle at the apex or tip of their cells, containing enzymes. The enzymes eat through host tissue during infection, and, thus, allow the infective form of the protozoa to enter liver and red blood cells. Apicomplexa organisms possess two additional characteristics that make them difficult pathogens to kill. First, apicomplexa such as *Plasmodium* have a complicated, multistage life cycle that involves transmission between several hosts. Second, apicomplexa spend part of their life cycle inside the host's cells.

The *Anopheles* mosquito plays a crucial role in the malaria pathogen's development. *Plasmodium* undergoes sexual reproduction inside the insect's gut and develops into the infective form, which can be transmitted to humans in a mosquito bite. The World Health Organization (WHO) advocates programs to eradicate this mosquito as one of the principal malaria preventions. Methods of controlling diseases by reducing insect populations that transmit the disease are collectively called *vector control*.

The WHO estimates that malaria today infects 250 million people each year and causes about one million deaths, mostly of African children. This organization has also pointed out, "Malaria takes an economic toll—cutting economic growth rates by as much as 1.3 percent in countries with high disease rates." Countries in sub-Saharan Africa contain malaria hot spots, places where the incidence of malaria is higher than in any other place in the world. The following countries have been identified by WHO as malaria hot spots: Angola, Bangladesh, Burkina Faso, Brazil, Cambodia, Cameroon, Chad, Côte d'Ivoire, Colombia, Democratic Republic of the Congo, Ethiopia, Ghana, India, Kenya, Madagascar, Malawi, Mali, Mozambique, Myanmar, Niger, Nigeria, Pakistan, Papua New Guinea, Senegal, Sudan, Tajikistan, Turkey, Uganda, United Republic of Tanzania, and Uganda. These countries have in common either tropical areas or regions with flooding, which increases breeding opportunities for mosquitoes.

The United States eradicated malaria in most regions, in the 1950s, although local outbreaks continue and cause about 1,000 cases annually. The Centers for Disease Control and Prevention (CDC) and the WHO have estimated that malaria is increasing worldwide and may reemerge in areas formerly free of the disease. CDC and WHO scientists agree that drug-resistant *Plasmodium* species probably contribute to malaria's return.

A second, more controversial reason for malaria's rise has been attributed to global warming. As global temperatures increase, ocean levels rise, and flooding increases; both events produce more bodies of stagnant water that provide breeding grounds for mosquitoes. The University of Michigan ecologist Mercedes Pascual supported the theory of malaria

The *Anopheles* mosquito is a vector for malaria. It feeds on human hosts primarily in the evening and early morning before sunrise. *(CDC)*

related to global warming but conceded that other factors such as drug and pesticide resistance, changing patterns of land use, and human migrations probably affect the incidence of malaria. Pascual said, in 2006, in Science*Daily*, "Our results [on malaria and temperature rise] do not mean that temperature is the only or the main factor driving the increase in malaria, but that it is one of many factors that should be considered." Others have argued that global warming is too big and ill-defined an event ever to be definitively linked with the incidence of a single disease. The biologist Paul Reiter of the Pasteur Institute in Paris has led the criticism of the global warming theory. In 2008, Reiter wrote in the *Malaria Journal,* "Simplistic reasoning on the future prevalence of malaria is ill-founded; malaria is not limited by climate in most temperature regions, nor in the tropics. . . . Future changes in climate may alter the prevalence and the incidence of the disease, but obsessive emphasis on 'global warming' as a dominant parameter is indefensible." Reiter proposed that malaria will be influenced by factors "linked to ecological and societal change, politics and economics." These factors present as much if not more difficulty for study as global warming.

At present, infectious disease experts agree that malaria is the world's most important tropical disease caused by a parasite, and that the disease can and does exist in some temperate regions. Because more than 40 percent of the world's population live in places where malaria is prevalent, this disease continues to cause a worldwide health crisis.

LIFE CYCLE OF *PLASMODIUM VIVAX*

Of the four protozoa known to cause malaria, most of the information on the pathogen's life cycle has been gathered from *P. vivax*. *P. vivax* has eight different stages, each with a unique morphology, as the microorganism moves from the mosquito to humans and back to the insect.

The table on page 483 describes the stages of the *P. vivax* life cycle in malaria.

In the malaria life cycle, young trophozoites develop a ring shape resembling a small diamond ring inside the red blood cell, which can be seen in a microscope. Alphonse Laveran saw this form when he discovered the malaria pathogen. The table also indicates that *Plasmodium* undergoes sexual reproduction inside mosquitoes and asexual reproduction when inside humans. Because mosquitoes serve as the site for the sexually reproducing form of *Plasmodium,* the insect is called a *definitive host*. Humans play the role of intermediate host, in the malaria life cycle.

As in Laveran's day, doctors currently diagnose malaria by inspecting blood specimens for the presence of *Plasmodium*. *Plasmodium*'s life cycle can complicate diagnosis because patients develop intermittent fevers that reappear about every 24 hours, when a new generation of merozoites enter the bloodstream from the liver. This cyclical rhythm puzzled doctors until medical researchers worked out the entire life cycle and realized that the parasite's timing ensures that gametocytes are at their highest levels in the blood at night, when mosquitoes bite, thus helping *Plasmodium*'s survival.

The species *P. ovale* and *P. vivax* include an additional stage, in which sporozoites enter a dormant, or nonmetabolizing, stage called hypnozoites. Hypnozoites may remain dormant in the liver, for months or years, before again becoming active.

TYPES OF MALARIA

Malaria's general symptoms of fever, chills, headache, and vomiting begin 10–30 days after a bite by a mosquito carrying the pathogen. As with most infectious diseases, people who have weak immune systems have a higher risk of infection and serious harm from malaria than the general population. These high-risk groups are acquired immunodeficiency syndrome (AIDS) patients, organ transplant patients, people who have long-term debilitating disease, elderly adults, young children, and pregnant women.

The species of *Plasmodium* causing an infection relates to the severity of the disease. Severity can be divided into two general groupings: uncomplicated malaria and severe malaria. Uncomplicated malaria, also called *benign malaria,* presents the general symptoms noted in the previous paragraph and can be cured with drug therapy. Benign malarias have been associated mainly with infections from *P. ovale, P. malariae,* or *P. vivax* and seldom develop life-threatening conditions. Severe, or malignant,

Life Cycle of *P. vivax* in Malaria

Stage Morphology	Location	Description
sporozoites	mosquito salivary gland, then bloodstream to liver cells	injected into human in a mosquito bite, then penetrate liver cells and multiply in a process called *schizogony*
merozoites	red blood cells (RBCs)	in RBCs early merozoites develop into either mature trophozoites, called a *schizont,* to make more merozoites or into gametocytes
gametocyte	RBCs	male or female gametes (contain one-half genome) taken into a mosquito with a bite
ookinete	mosquito gut	a zygote (contains the full genome) forms by the union of male and female gametocytes, called *gametogenesis*
oocyst	mosquito gut wall	oocysts form after ookinetes penetrate the gut wall
sporozoites	mosquito salivary gland	in sporogony, oocysts develop many sporozoites inside, before bursting and releasing the new generation

malaria from *P. falciparum,* by contrast, causes additional symptoms such as abdominal and muscle pains, anemia, swelling of the spleen, and convulsions. A severe form of *P. falciparum* malaria, called *cerebral malaria,* leads to coma and death. This form of the disease, called *falciparum malaria,* has a mortality rate of 20 percent in adults and 15 percent in children.

The cyclical characteristic of malaria has led medical researchers to define the three following distinct stages as it progresses in the human body:

- stage 1—shaking chills lasting from minutes to hours
- stage 2—sweating and high fever, cough, headache, backache, abdominal pain, vomiting, diarrhea, and altered consciousness, lasting several hours
- stage 3—profuse sweating, lowered fever, fatigue, starting two to six hours after stage 2

Survivors of cerebral malaria can experience permanent brain damage. In rare cases of falciparum malaria, hemolytic anemia (anemia caused by destruction of red blood cells) develops into a condition called *blackwater fever,* in which protein levels from hemoglobin increase in the urine.

Mixed malaria infections, in which more than one *Plasmodium* species infects a host, are common. The WHO reported, in 2010, "In Thailand, despite low levels of malaria transmission, one third of patients with acute *P. falciparum* infection are co-infected with *P. vivax.*" Routine diagnosis by studying blood specimens in a microscope often misses a coinfection by a second *Plasmodium* species.

Treatment for malaria works best if administered within 24 hours of its diagnosis. Chloroquine (as chloroquine phosphate) served as the main treatment for malaria, for many years, until parasites resistant to this compound began to emerge, in the recent past. The WHO now recommends alternative treatments based on the disease's severity and the causative species. Some of the main malaria treatments in use worldwide are the following:

- artemisinin-based combination therapy (ACT)
- quinine sulfate
- mefloquine
- sulfadoxine
- pyrimethamine
- any combination of these drugs

Artemisinins are various antibiotics based on the four-ring structure of the compound artemisinic acid extracted from the plant sweet sagewort.

Nonantibiotic treatments have also been used in malaria patients. Blood exchange transfusions provide an example of this method of treatment to rid the body of the parasite.

THE WORLDWIDE MALARIA CRISIS

Although preventable and treatable, malaria is a public health problem in more than 100 countries and affects 40 percent of the world's population. Malaria is endemic in large parts of Africa, Central and South America, the South Pacific, the Indian subcontinent, and Southeast Asia. Social factors have also caused increases in malaria's transmission in eastern Europe, Hispaniola (Haiti and the Dominican Republic), and Jamaica. The Malaria Foundation International has cited the following factors as causes of recent increases in malaria in new areas:

- preexisting high-risk health conditions, especially AIDS in sub-Saharan Africa
- global travel
- refugees, displaced persons, or labor forces moving into endemic (prevalent throughout the population) areas

Population growth into low-lying areas, where water accumulates could add to the threat of increased malaria cases.

MALARIA CONTROL PROGRAMS

New vaccine research and the search for newer, more effective treatments for malaria make up two parts of a worldwide three-pronged effort to eradicate this disease. Vector control constitutes the third approach.

The WHO stated, in 2006, "Vector control remains the most generally effective measure to prevent malaria transmission and is therefore one of the four basic technical elements of the GMCS [Global Malaria Control Strategy]." The other three elements proposed by WHO in malaria control are (1) early diagnosis and prompt treatment, (2) early detection and containment of epidemics, and (3) increased research into the economic, social, and ecological factors in malaria's transmission.

Several complementary actions contribute to current vector control initiatives. The following have been advocated by the WHO as being most effective, especially when these control measures are used in combination:

- indoor residential spraying (IRS) with pesticides, called *adulticides,* which kill the mature insect
- alternate sprayings with DDT and pyrethroid pesticides to slow the development of pesticide resistance
- drainage of small-scale bodies of standing water
- use of pesticides called *larvicides* on standing water to kill mosquito larvae

The steps listed here complement additional measures that people can take to prevent mosquito bites. Preventive measures involve reducing the proximity of people to mosquitoes and, thus, decrease the probability of bites. Measures such as insect-repelling netting over beds, window and door screens, repellent clothing, and small-volume use of insect repellents indoors help break the mosquito-human connection.

Malaria has been a persistent health threat, throughout civilization. A sound prevention method has yet to be developed, and this problem has been made worse by the emergence of both antibiotic-resistant *Plasmodium* and pesticide-resistant mosquitoes. Malaria, an ancient disease, appears to present an ongoing health crisis for the near-future.

See also PROTOZOA; TRANSMISSION.

Further Reading

Azzu, Vian. "Malaria Vaccine Holds Out Eradication Hope." *NewScientist,* 14 August 2009. Available online. URL: www.newscientist.com/article/mg20327213.800-malaria-vaccine-holds-out-eradication-hope.html. Accessed August 30, 2009.

Centers for Disease Control and Prevention. "Malaria: Topic Home." Available online. URL: www.cdc.gov/malaria/index.htm. Accessed August 29, 2009.

Hippocrates. *The Genuine Works of Hippocrates.* Translated by Charles Darwin Adams. New York: Dover, 1868. Available online. URL: www.chlt.org/sandbox/dh/Adams/index.html. Accessed August 29, 2009.

Malaria Foundation International. Available online. URL: www.malaria.org/index.php. Accessed August 30, 2009.

National Institutes of Health. "Malaria." Available online. URL: www.nlm.nih.gov/medlineplus/malaria.html. Accessed August 30, 2009.

Nchinda, Thomas C. "Malaria: A Reemerging Disease in Africa." *Emerging Infectious Diseases* 4 (1998): 398–403. Available online. URL: www.cdc.gov/ncidod/eid/vol4no3/nchinda.htm. Accessed August 30, 2009.

Reiter, Paul. "Global Warming and Malaria: Knowing the Horse before Hitching the Cart." *Malaria Journal* 7 (2008): S3–S11. Available online. URL: http://malariajournal.com/content/7/S1/S3. Accessed August 29, 2009.

Science*Daily*. "Warming Trend May Contribute to Malaria's Rise." March 22, 2006. Available online. URL: www.sciencedaily.com/releases/2006/03/060322142101.htm. Accessed August 29, 2009.

World Health Organization. "Guidelines for the Treatment of Malaria, 2nd ed." 2010. Available online. URL: http://apps.who.int/malaria/docs/TreatmentGuidelines2006.pdf. Accessed June 6, 2010.

———. "Malaria." Available online. URL: www.who.int/topics/malaria/en/index.html. Accessed August 29, 2009.

———. "Malaria Vector Control and Personal Protection." Available online. URL: http://apps.who.int/malaria/docs/WHO-TRS-936s.pdf. Accessed August 30, 2009.

———. "World Malaria Report 2008." Available online. URL: http://apps.who.int/malaria/wmr2008. Accessed August 29, 2009.

marine microbiology Marine microbiology is the study of microorganisms that use the marine environment as their habitat. The main types of marine habitats are the ocean, coral reefs, intertidal flats, estuaries, and rocky and sandy shores. Each of these places contains several more defined habitats, such as layers of the ocean, from the surface to the sea bottom, as well as microbial communities attached to marine surfaces.

Marine microorganisms belong to the archaea, bacteria, fungi, protozoa, and algae, a group of bacteria called *cyanobacteria* belong to a larger and diverse group of the microscopic organisms, called *plankton.* Plankton serves as a foundation in almost all marine food chains, processes by which small plant life or animals are eaten by larger and more complex animals, at each step up the chain, until reaching a top predator. A typical marine food chain begins with plankton and might progress upward as follows: invertebrates, small fish, larger fish, marine mammals, and shark.

Marine microbiology contains the following areas of specialization: marine microorganisms, extremophiles (microorganisms that live in extreme conditions that most other life cannot survive), plankton, biogeochemical cycles in marine environments, ocean ecology, nutrient use and end products by marine microorganisms, microbial populations of marine animals, and marine fish or mammal infections. Marine microbiology also focuses on the corrosive effects of aquatic microorganisms on ships and underwater structures.

Marine biologists have adapted molecular techniques for studying the relationship of aquatic species to each other and to terrestrial microorganisms, the species that play roles in coral reef health, and marine pathogens. These molecular techniques usually include tools called *biosensors* that contain a section of microbial deoxyribonucleic acid (DNA). By examining marine samples with biosensors, scientists can locate specific genes within a heterogeneous mixture of substances.

Marine microbiology has expanded into technologies in addition to biosensors for studying aquatic genes, especially the search for new substances made by marine organisms that will serve as antibiotics or disease therapies. Because microbiologists have estimated that the number of marine microbial species far exceeds the number on land, the marine environment represents an untapped source for new products. A professor of infectious diseases at the University of California–San Diego told *Drug Discovery News,* in 2006, "Most of these marine species are still waiting to be discovered, providing a nearly unfathomable resource for potential antibiotic discoveries." The increasing rate of antibiotic resistance in known pathogens has made marine microbiology a higher priority in new drug development.

Marine Microbiology Research Centers

Center	Location
Hatfield Marine Science Center, Oregon State University	Newport, Oregon
Max Planck Institute for Marine Microbiology	Bremen, Germany
Monterey Bay Aquarium Research Institute	Moss Landing, California
Scripps Institution of Oceanography	La Jolla, California
University of California–Santa Cruz Institute of Marine Sciences	Santa Cruz, California
Woods Hole Oceanographic Institution	Woods Hole, Massachusetts

THE MARINE ENVIRONMENT

The marine environment takes up most of Earth's biosphere, which is the portion of land and atmosphere that supports life. Marine waters make up 97 percent of all of Earth's water. Despite the clear importance that this massive environment has in all of biology, much in marine microbiology remains undiscovered for two main reasons. First, the sheer size of the marine environment and the diversity of its life have created challenges in studying the scope of marine ecology, and, second, many sites in the marine environment have been almost inaccessible.

The marine environment contains three major characteristics: (1) depth, (2) extremes in temperatures, and (3) darkness. Depth has made some areas nearly impossible for detailed study. For example, the undersea canyon called the Mindinao Deep reaches 35,400 feet (10,790 m) below sea level, or about 6,400 feet (1,951 m) more than Mt. Everest's height.

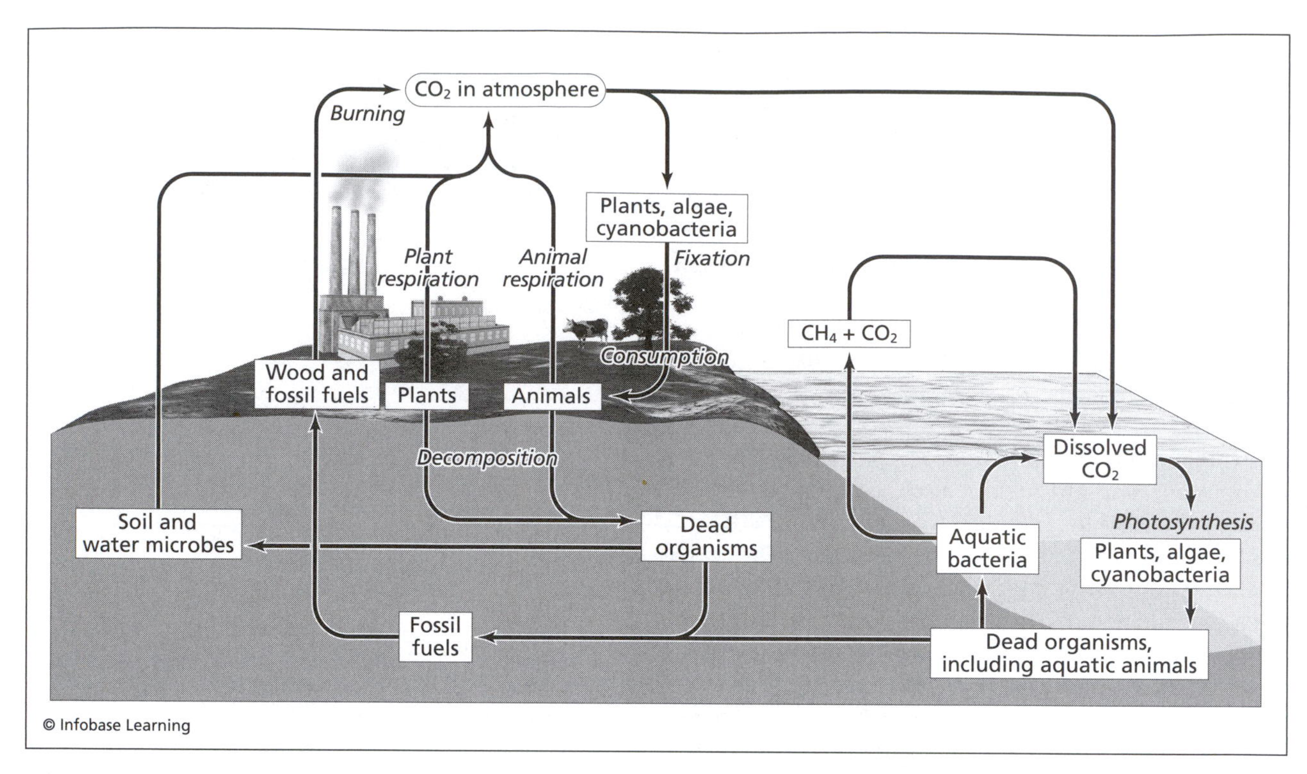

Marine microorganisms play a significant role in cycling Earth's carbon. Photosynthetic algae and cyanobacteria take in more atmospheric carbon than all terrestrial organisms combined. The assimilated carbon, then, feeds marine and terrestrial food chains.

At this depth, water produces intense pressures that many organisms cannot withstand. Barometric pressure increases gradually with depth, which in turn affects the types of microorganisms living at different depths. The ocean surface where sunlight penetrates is called the *photic zone* and contains microorganisms that do not tolerate high pressures. Greater depths increase the barometric pressure to a maximum of about 1,100 atmospheres, at the deepest places. The following types of microorganisms live at different depths based on the barometric pressures they tolerate:

- barotolerant—33–13,123 feet (10–4,000 m)
- moderate barophilc—4,921–18,450 feet (1,500–5,500 m)
- extreme barophilic—16,404–36,089 feet (5,000–11,000 m)

Several processes within barophiles (also termed *piezophiles*) differ from metabolism on the surface. Reproduction, DNA replication, protein synthesis and functioning, and membrane transport of nutrients in and wastes out of cells must withstand intense pressures by mechanisms not completely understood. Barophiles may contain a different proportion of fatty compounds called lipids and proteins compared with other microorganisms. These microorganisms may additionally incorporate extra processes whereby they can quickly repair damage to the cell due to the high pressures. Researchers have found specific genes in barophiles that turn on or off, depending on the barometric pressure of the surroundings. The microorganisms that live in marine waters as well as microorganisms living inside marine animal digestive tracts both must tolerate elevated pressures of the deep sea.

Temperature in most of the ocean surface ranges from 59°F to 86°F (15–30°C). Less than 30 percent of Earth's ocean surface waters have temperature below 50°F (10°C). At depths greater than 3,280 feet (1,000 m), temperatures reach a range of 41–29°F (5 to -1.5°C), where psychrophiles grow best. Many hydrothermal vents also dot regions of the Pacific Ocean floor, releasing pressurized water from the earth's core. The microorganisms living in areas near vents are hyperthermophiles, which withstand temperatures up to 750°F (400°C).

Temperature varies in gradients in the ocean, with the water becoming colder at increasing depths. Unlike pressure, which increases steadily, ocean temperatures undergo stratification, in which layers in a temperature range lie above or below each other. The area that separates each warmer layer above from a cooler layer below is called a *thermocline*.

The marine environment also contains high salt concentrations not found in the earth's freshwaters. The salt levels of water, called salinity, relate to the amount of salt as sodium chloride (NaCl) dissolved in 1,000 grams of water. A salinity of 1 equals 1 part salt in 1,000 parts water, or 1 part per trillion (ppt). The average ocean salinity is 35 ppt. Estuaries range from about 0.5 ppt to 17 ppt. Microorganisms living in these high-salt conditions constantly expend energy to pump salt out of their cells in order to survive.

The marine environment consists of various habitats with narrower characteristics of temperature, pressure, and sunlight penetration. The table below describes the marine habitats of the ocean.

Oceanographers sometimes use additional classifications of ocean zones to define the conditions found there further. For instance, the dysphotic zone, or twilight zone, starts beneath the photic zone and reaches to about a depth of 3,300 feet (1,006 m). A small amount of sunlight penetrates the dysphotic zone, so scientists long believed no photosynthesis occurred there. In 2005, the biochemist Robert Blankenship of Arizona State University discovered photosynthetic bacteria living near hydrothermal vents 7,875 feet (2,400 m) deep. The bacteria belong to a group called *green sulfur bacteria*. These bacteria use dim light that emerges from the hydrothermal vents and confine themselves to the narrow region between the extremely hot waters from the vent and the very cold water at the ocean floor. Blankenship said, "This is startling in the sense that you do not expect to find photosynthesis in a region of the world that is so completely dark." The discovery provided insight into a process called *low-light photosynthesis,* which enables certain microorganisms to adapt photosynthesis to an extreme environment.

In addition to the zones described in the table, microorganisms might live in two additional habitats: the epibiotic zone or the endobiotic zone. The epibiotic zone comprises underwater surfaces such as rocks, plant life, seaweed, coral, or human-made structures. This zone provides habitat to microorganisms that attach to surfaces and form communities. The endobiotic zone refers to the habitat inside fish or other marine life, usually in the digestive tract of these animals.

Each of the ocean habitats contains specific ranges of organic matter and other nutrients and sunlight, in addition to temperature and pressure. These four main factors determine the types of microorganisms that might be expected in these zones. Other factors that affect the types and amounts of microorganisms in marine waters are the following:

- tides
- distance from land
- diurnal (day-night) fluctuations in microbial populations
- season
- beneficial or antagonistic effects between microorganisms
- bacteriophages (viruses that infect bacteria)
- composition of sediments
- dissolved oxygen levels

In addition to the natural characteristics listed here, marine waters contain factors related to human activities that influence microbial populations. Pollution in runoff, oil spills, industrial chemicals, and

Marine Habitats in the Ocean

Zone	Depth	Description
photic	to 656 feet (200 m)	layer where sunlight penetrates
aphotic (midnight zone)	below 656 feet (200 m)	no sunlight penetration
neuston pelagic zones	surface	air-water interface zone
epipelagic	upper 328–656 feet (100–200 m)	upper region of the water column (open ocean, or blue water)
mesopelagic	656–3,281 feet (200–1,000 m)	bioluminescence common, nutrients scarce
bathypelagic	3,281–13,123 feet (1,000–4,000 m)	limited nutrients
abyssopelagic	below 13,123 feet (4,000 m)	extreme pressure and near-freezing temperatures
benthic	sediments on ocean floor	nutrients from particles that settled from above

acid rain affects the type and amount of marine microorganisms in any particular area of the ocean.

CHARACTERISTICS OF MARINE MICROORGANISMS

Marine microorganisms living in the aphotic zones either must be extremophiles or at least be able to tolerate harsh conditions. The photic zone, by contrast, contains a massive amount of microbial and nonmicrobial life that live in more temperate conditions. Cyanobacteria and algae dominate the photosynthetic microbial life in the photic zone. Nonphotosynthetic species and extremophiles begin to dominate below 656 feet (200 m).

The topic of how many microorganisms exist in the marine environment has interested scientists, since the first studies in marine microbiology. In 1998, the biologists William Whitman, David Coleman, and William Wiebe of the University of Georgia calculated the following numbers of marine prokaryotes:

- continental shelf, 1.0×10^{26}
- open ocean above 656 feet (200 m), 3.6×10^{28}
- open ocean below 656 feet (200 m), 6.5×10^{28}
- sediments in top 4 inches (10 cm), 1.7×10^{28}

These estimates total 1.2×10^{29} prokaryotes in the earth's marine environment. In 2001, Markus Karner, Ed DeLong, and David Karl used a fluorescent hybridization technique to estimate DNA and ribonucleic acid from cells' ribosomes (ribosomal ribonucleic acid [rRNA]). From their results, they extrapolated the number of microorganisms found at different depths to the total ocean and concluded the world's oceans contain 1.3×10^{28} archaea and 3.1×10^{28} bacteria. By using nucleic acid studies, researchers can better assess all the microorganisms present, both culturable and nonculturable, in laboratories.

Marine microbiologists continue to discover new microbial life in marine waters. Limitations in culture methods have hampered some of the progress in marine microbiology. Despite advances in studying the molecular components of microbial cells, down to the smallest gene, microbiologists still obtain most of the information about a microorganism's metabolism by culturing it in a laboratory. This requires growth on solid agar media. Development of new media for growing laboratory cultures of as-yet-unknown and noncultured microorganisms has been a challenge. The microbiologist Stephen Giovannoni of Oregon State University suggested, in 2007, in *Oceanography* that techniques for growing marine species should begin with current culture methods to include the following adjustments:

- longer incubation times
- lowered nutrient concentration
- replication of marine temperature, light, and nutrient conditions
- development of specialized growth chambers

Many marine microorganisms might belong to communities, such as biofilms, which colonize inanimate or living surfaces in the ocean. For this reason, Giovannoni and his colleagues have proposed that the best way to recover difficult-to-grow ocean species would be to grow more than one species together.

As Karner's study showed, information on marine microorganisms might increasingly depend on analysis of microbial nucleic acids, namely, 16S rRNA. This RNA derives from a component of the cell's ribosomes, the site of protein synthesis based on instructions held in the cell's DNA.

MARINE BIOGEOCHEMICAL CYCLES

The oceans contribute significantly to the cycling of nutrients on Earth. The major nutrient or biogeochemical cycles in which a portion of the cycle occurs in marine waters are the carbon, hydrogen, nitrogen, oxygen, phosphorus, and sulfur cycles, in addition to the cycling of various minerals. The ocean participates in absorbing compounds containing these elements from the atmosphere, converting the elements into a different form, and then releasing them back to the atmosphere. Plant and animal life in the ocean and on land plays intermediary roles in these nutrient conversions.

The ocean plays a crucial role in carbon cycling and in modulating global warming by absorbing at least one quarter of all the carbon dioxide emitted into the atmosphere and storing 50 percent more carbon dioxide than does the atmosphere. Oceans absorb this gas from the air in order to establish equilibrium between the water and the atmosphere, but because the exchange takes several hundred years, the oceans serve as a continual absorption site for this greenhouse gas. Once the ocean has absorbed a carbon atom, that atom will remain in the ocean for at least 500 years.

The ocean's carbon cycle consists of a physical feature and a biological feature. In the physical portion, the water's temperature and its acidity, which is regulated by the dissolution of shells and release of carbonate into the water, affect the absorption of carbon dioxide. Increasing acidity in marine waters, due in part to acid rain, decreases the amount of carbon dioxide the water can absorb. Regarding temperature, carbon dioxide levels increase in colder, deeper waters that can dissolve more of the gas. This is called the *soda bottle effect,* in which cold soda holds more dissolved carbon dioxide than warm soda. When marine waters warm, they release carbon dioxide back into the atmosphere.

The biological portion, called *the marine biological pump,* operates by the action of plankton of plant origin, called *phytoplankton.* Algae and cyanobacteria take part in this biological pump because of their ability to absorb carbon dioxide from the atmosphere during photosynthesis, as do terrestrial plants. Microorganisms move marine carbon dioxide in two directions. First, the photosynthetic microorganisms absorb carbon dioxide in the photic zone and convert it into organic matter in their cells. But when microorganisms and higher marine organisms die, the dead matter drifts slowly to the sea bottom. There, other microorganisms decompose the organic matter and release carbon dioxide back into the water.

Damage to the environment impacts the ability of the oceans to regulate carbon cycling. The National Aeronautics and Space Administration's (NASA's) Web site has explained, "Prior to the Industrial Revolution, the annual uptake and release of carbon dioxide by the land and the ocean had been on average just about balanced. In more recent history, atmospheric concentrations have increased by 80 ppm (parts per million) over the past 150 years." The ocean and its microbial life play a role as a huge carbon repository called a carbon sink. This role has become increasingly important, as NASA suggests. Marine microbiology has an exciting future in learning more about marine microorganisms' role in the environment and in all other aspects of biology.

See also BIOFILM; BIOGEOCHEMICAL CYCLES; BIOSENSOR; EXTREMOPHILE; GREEN BACTERIA; PLANKTON.

Further Reading

Center for Microbial Oceanography: Research and Education. Available online. URL: http://cmore.soest.hawaii.edu/index.htm. Accessed August 30, 2009.

Derra, Skip. "Researchers Find Photosynthesis Deep within Ocean." *ASU [Arizona State University] News,* 21 June 2005. Available online. URL: www.asu.edu/feature/includes/summer05/readmore/photosyn.html. Accessed August 30, 2009.

Karner, Markus B., Ed F. DeLong, and David M. Karl. "Archaeal Dominance in the Mesopelagic Zone of the Pacific Ocean." *Nature* 409 (2001): 507–509.

Morita, Richard Y. "In Memoriam Dr. Claude E. ZoBell." *Marine Ecology Progress Series* 58 (1989): 1–2. Available online. URL: www.int-res.com/articles/meps/58/m058p001.pdf. Accessed August 30, 2009.

Sarmiento, Jorge L., and Nicolas Gruber. *Ocean Biogeochemical Dynamics.* Princeton, N.J.: Princeton University Press, 2006.

Science*Daily*. "Marine Bacteria Can Create Environmentally Friendly Energy Source." January 15, 2007. Available online. URL: www.sciencedaily.com/releases/2007/01/070115102154.htm. Accessed August 31, 2009.

Vosseler, Julius. *Amphipodan der Plankton-Expedition* (Amphipods of the Plankton Expedition). Kiel, Germany: Verlag von Lipsius and Tischer, 1901. Available online. URL: http://books.google.com/books?id=h9wNAQAAIAAJ&pg=RA2-PA5&lpg=RA2-PA5&dq=%22Julius+Vosseler%22&source=bl&ots=7sxrVc0GsT&sig=Upa8FZ7oYyDN5uR9BVJe9P5niSI&hl=en&ei=MQybSpuZFYjSNb2PrLYB&sa=X&oi=book_result&ct=result&resnum=2#v=onepage&q=&f=false. Accessed August 30, 2009.

Willis, Randall C. "Scraping Bottom: UCSD Looks to Ocean for New Antibiotics." *Drug Discovery News,* August 2006.

Whitman, William B., David C. Coleman, and William J. Wiebe. "Prokaryotes: The Unseen Majority." *Proceedings of the National Academy of Sciences* 95 (1998): 6,578–6,583. Available online. URL: www.bren.ucsb.edu/academics/courses/219/Readings/Whitman%20et%20al%201998%20Unseen%20Majority.pdf. Accessed August 31, 2009.

ZoBell, Claude E. *Marine Microbiology: A Monograph on Hydrobacteriology.* Waltham, Mass.: Chronica Botanica, 1946. Available online. URL: www.archive.org/stream/marinemicrobiolo00zobe/marinemicrobiolo00zobe_djvu.txt. Accessed August 31, 2009.

media Medium is any formulation of nutrients in water that promotes the growth of microorganisms. Media may be solid, semisolid, or liquid. Solid media contain a gellike material called *agar* that is liquid when heated but solidifies after cooling. Microbiologists pour molten agar media into either petri plates, to make a horizontal surface, or into test tubes, tilted to create to a slanted agar surface (called a *slant*) or not tilted for a horizontal surface (called a *stab, deep,* or *butt*). Liquid media, called *broth,* contain similar ingredients to solid media but lack agar or any other solidifying agent. When microorganisms have been inoculated to media and incubated so that the cells reproduce, the resulting population of cells is called a *culture.*

Hundreds of different media formulations now help microbiologists grow thousands of species of microorganisms, from easy growers such as *Escherichia coli* to species such as *Legionella pneumophila,* which has very strict nutrient requirements. All media must meet a microorganism's nutrient requirements, or else the subsequent growth will be poor or nonexistent. Media provide specific ranges of the following parameters to meet the optimal growing conditions of given bacteria, archaea, fungi, yeasts, or algae:

- nutrients—supplies all the nutrients needed by the cell that the cell cannot make itself or absorb from the air

- pH—must be adjusted to a range that suits the acidic, neutral, or basic requirements of the microorganism

- oxygen—anaerobic bacteria require that oxygen be purged from media before inoculation

- water—some species grow only in broth or semisolid media

The nutrients supplied by most media meet a species's growth requirements for carbon, nitrogen, sulfur, inorganic ions (salts and metals), vitamins, fats, purines and pyrimidines (the building blocks of deoxyribonucleic acid [DNA], and ribonucleic acid [RNA]), and certain ingredients that provide miscellaneous growth factors. More than a century of studies on microorganisms has clarified the needs of groups of bacteria and other microorganisms. In other words, most enteric bacteria can be grown on the same type of medium, while a variety of common fungi all grow well on another formula, developed for these eukaryotes.

MEDIA COMPOSITION

The media that support growth of the most-studied microorganisms contain certain ingredients in common. Microbiologists can vary the formulas by adding or subtracting ingredients. The table below offers an example of a generic formula for aerobic bacteria. Most media do not contain all of these ingredients but contain various combinations of ingredients other than glucose.

Broth media such as the example in the table can be made into a solid medium by including 15.0 grams of agar powder per liter. Alternative substances that have been used instead of agar are alginate, gelatin, gellan gum, Pluronic polyol F127, and silica gel.

Some microorganisms are from environments that supply only low amounts of nutrients. For example, bacteria that grow in purified water have adapted to very low levels of carbon, nitrogen, and all other factors for growth. Inoculating these bacteria to a rich medium, that is, a medium that supplies ample amounts of many different nutrients, would cause the bacteria to enter nutrient shock and die. Some media, therefore, have defined, low levels of ingredients that would not support the growth of most common laboratory bacteria.

A Generic Medium Formula

Ingredient	Amount (grams per liter distilled water)	Purpose
glucose	1.0	carbon and energy source
glycerol	5.0	fat synthesis
salts: ammonium sulfate, potassium nitrate, potassium phosphate, sodium chloride	0.5 each	enzyme and membrane function
trace salts: calcium chloride, iron sulfate	0.1 each	enzyme and membrane function
trace elements: cobalt, copper, manganese, molybdenum, zinc	0.001 each	enzyme function and DNA replication
vitamins: thiamine, vitamin B_{12}	micrograms	enzyme function in energy generation
beef extract	1.0	source of miscellaneous nitrogen compounds
yeast extract	3.0	source of B-complex vitamins, inorganic salts, and amino acids
hydrolyzed casein (milk protein) or gelatin	1.0	source of amino acids
peptone	3.0	source of peptides
tryptone	5.0	source of peptides
hydrochloric acid or ammonium hydroxide	drops	pH adjustment

Types of Media for Growing Bacteria

Medium	Purpose
maintenance	routine growth and storage of common laboratory bacteria
recovery (also resuscitation or preenrichment)	stimulates the growth of bacteria that have been stored frozen by supplying extra nutrients
defined	nutrient requirement studies; all ingredients known
semidefined	determining growth requirements; all ingredients known except for the addition of yeast extract, peptone, and so on
undefined	growth of unknown bacteria or new bacteria with unknown growth requirements; most ingredients unknown in levels and composition
minimal	growth of bacteria from low-nutrient environments
selective	isolating particular bacteria by suppressing the growth of all others
differential	distinguishing among different groups of bacteria
enrichment	promoting growth of a physiological type of bacteria in a mixture of diverse species

TYPES OF MEDIA

Microbiology contains a rich variety of medium formulations for specific purposes. The table above describes the main types of microbiological media for bacteria.

Any medium that contains substances such as yeast extract, beef extract, casein hydrolysates, and so on, represents a complex medium, which is any formula containing some ingredients of unknown chemical composition. Extracts and similar additions have inexact compositions that vary from batch to batch, but these ingredients can be very helpful in microbiology because they supply a wide range of nutrients. Without complex media, microbiologists would spend a large amount of time seeking the best formula for a new microorganism.

SPECIALIZED MEDIA

Selective, differential, and enrichment media have become useful specialized media for two purposes. First, these media enable microbiologists to separate one type of microorganism out of a heterogeneous mixture of others. Second, these media help in the identification of an unknown microorganism. Microbiologists rely on specialized media for early recoveries or identifications of a microorganism but turn to maintenance or defined medium for pure cultures.

Selective media contain one or more ingredients that encourage the growth of one microorganism in preference to all others. Microbiologists use any of the following additions to make medum selective for a given microorganism: antibiotics, bile salts, dyes, potassium tellurite, unique carbon-energy sources such as cellulose, and pH. Many more types of selective media exist in microbiology for favoring the growth of one microorganism over all others. The table at the top of page 492 describes important selective media used in general, industrial, or clinical microbiology laboratories.

Selective media have a variety of uses in different disciplines within microbiology. Clinical microbiologists, for example, may use selective media to differentiate a pathogen, such as *C. diphtheriae,* from other bacteria on the body. Industrial microbiology uses selective media for finding contaminants, and water microbiologists might use selective media to detect the presence of certain microorganisms, such as *E. coli,* in the environment.

Some selective media serve also as differential media. MacConkey agar can select or differentiate bacteria. The inclusion of lactose sugar and neutral red dye in MacConkey agar makes lactose-fermenting bacterial colonies appear pink to red. Non-lactose-fermenting bacteria either do not grow or produce colorless colonies.

Differential media, shown in the table at the bottom of page 492, play an important role in clinical microbiology because they separate closely related gram-negative-staining enteric bacteria from each other. This characteristic may be of importance in diagnosis of infection or food-borne illness.

Enrichment media emphasize general capabilities of one group of bacteria compared with another group. For example, media made by purging all oxy-

Commonly Used Selective Media

Medium	Selective Agent	Microorganism It Favors
brilliant green bile agar	brilliant green dye	*Salmonella* species (bacteria)
eosin methylene blue (EMB) agar	eosin and methylene blue dyes	*Escherichia coli* bacteria
MacConkey agar	lactose	lactose-fermenting enteric bacteria
mannitol–egg yolk–polymixin	mannitol carbon source,	*Bacillus cereus* bacteria
(MYP) agar	lecithinase enzyme, and polymixin antibiotic	
mannitol salt agar	mannitol carbon source and high salt concentration	*Staphylococcus* species (bacteria)
potato dextrose agar	potato infusion and pH 5.6	yeasts and molds
Pseudomonas isolation agar	Irgasan antimicrobial	*Pseudomonas* colonies appear green to blue-green; all other bacteria are inhibited
tellurite blood agar	potassium tellurite	*Corynebacterium diphtheriae* bacteria

gen enrich conditions for anaerobic bacteria and aerobic microorganism do not grow. Enrichment can be either physical or biochemical. Physical enrichments include altered incubation temperatures and adjusted oxygen levels. The common biochemical enrichments, which are part of the medium's formula, are obtained from pH adjustments, specific inhibitors, salts, nutrient dilution, unique carbon-energy sources, and unique nitrogen sources. Environmental microbiologists also develop enrichment media containing pollutants such as toluene in order to provide enrichment for bacteria that can grow on this organic solvent. Other

Commonly Used Differential Media

Medium	Differentiation	Microorganism Reactions
blood agar	hemolysis of blood	hemolytic *Staphylococcus* produce alpha (complete) or beta (incremental) zones of hemolysis around each colony; all other bacteria produce no hemolytic zones
hektoen enteric agar	ferric ammonium citrate, brom thymol blue and acid fuchsin dyes	*Salmonella typhimurium* produces black colonies due to hydrogen sulfide generation from the ferric compound; *Shigella* produces greenish blue colonies; *E. coli* produces salmon-orange colonies
MacConkey agar	neutral red dye	*E. coli* produces pink colonies; *S.* typhimurium produces translucent; *Proteus mirabilis* produces colorless
triple sugar iron (TSI)	iron salts and phenol	*E. coli* produces gas (bubbles); *Salmonella*
agar	red dye	produces gas and hydrogen sulfide (turns agar black); *Shigella* produces red colonies but not gas or hydrogen sulfide
violet red agar (with 4-methyltumbelliferyl-β-D-glucuronide [MUG])	bile salts and crystal violet dye	under fluorescent light, *E. coli* fluoresces, *Enterobacter* does not fluoresce, and other bacteria are inhibited

enrichment media have been developed using blood, synthetic seawater, soil extract, and rumen liquid for microorganisms with distinctive growth requirements.

Enrichment media have been divided into two main groups: elective and selective. Elective enrichment media grow limited types of microorganisms in terms of their unique nutritional or physiological traits. For example, toluene-enriched media grow only bacteria that can metabolize toluene. Selective enrichment media contain an ingredient that inhibits most types of microorganisms, except the desired microorganism. Selective enrichment media and selective media are very nearly the same and are usually used interchangeably. While selective enrichments inhibit all growth except growth of the desired microorganism, selective media create favorable conditions for one microorganism over all others.

SPECIAL-USE MEDIA

Some media have been developed for specific tasks in microbiology. These special formulas are described in the table below.

Special-use media have become valuable in several different disciplines within microbiology. For example, biphasic media have allowed the growth of fastidious or slow-growing microorganisms or growth of microorganisms sensitive to end product buildup. Biphasic media confine all microbial growth to liquid overlay. This layer may be dilute broth medium or water. The microorganisms survive by drawing nutrients from the agar layer as compounds slowly diffuse from the solid to the liquid. Biphasic media can also be useful for recovering microorganisms from environments of low nutrient content.

Clinical microbiologists and laboratories with the main task of identifying unknown microorganisms employ special-use media to observe characteristics of an isolate, for instance, motility, enzyme activity, and gas production, of specific compounds. These same microbiologists also rely on transport media to keep microorganisms alive during transport from a patient or environmental site to the laboratory.

MEDIA FOR MICROORGANISMS OTHER THAN BACTERIA

Filamentous fungi (produce fuzzy or hairy colonies) and yeasts use similar growth media containing agar. Many common molds, such as *Aspergillus* and *Penicillium*, can grow on maintenance media that microbiologists use for bacteria. Media specially formulated for fungi typically have a lower pH (about 5.6) than bacterial media. The main agar media used for fungi and yeasts are potato dextrose agar and malt extract agar.

Algae can be grown in broth media that have been supplemented with a variety of salts and vitamins. Media differ for freshwater algae and marine algae, mainly in salt content. Marine algae media may contain more than 30 grams of mixed salts per liter of water.

All types of media for any use in microbiology can be altered to suit a particular microorganism best. In this way, media provide the foundation for all activities using identified, unidentified, and new microorganisms.

See also AGAR; COLONY; CULTURE; GRAM STAIN; IDENTIFICATION; MOTILITY.

Special-Use Media

Medium	Description	Use
biphasic (or diphasic)	a solid agar layer under a liquid or semisolid layer called an *overlay*	agar layer provides a reservoir of nutrients for microorganisms requiring long incubation times (10–30 days)
prereduced (or reducing) semisolid	prepared to contain almost no free oxygen contains low concentration of agar to give the cooled medium a soft, jellylike consistency	growth of anaerobes for observing motility or chemotaxis (motility in the direction of a stimulus)
test	incorporates special substances in addition to normal ingredients for demonstrating a microorganism's unique trait	for identifying genus or species
transport	low nutrient levels	maintaining of viability of microorganisms for short-term storage or transport of a sample to a laboratory

Further Reading

Difco and BBL Microbiology. Available online. URL: www.vgdllc.com/microbiology.htm. Accessed September 1, 2009.

Eddleman, Harold. "Ingredients Used in Microbiological Media." Available online. URL: www.disknet.com/indiana_biolab/b041.htm. Accessed September 1, 2009.

Tortora, Gerard J., Berdell R. Funke, and Christine Case. *Microbiology: An Introduction,* 8th ed. San Francisco: Benjamin Cummings, 2004.

mesophile A mesophile is a microorganism having an optimal temperature for growth in the range of 68–113°F (20–45°C). Mesophiles generally have the ability to grow, but more slowly than normal, at temperatures down to 59°F (15°C) but rarely survive above 113°F (45°C). The lowest end of the optimal temperature range is called the *minimal growth temperature,* and the highest end of the range is the *maximal growth temperature.*

Mesophilic growth makes up part of a spectrum of microbial growth based on the optimal temperature range of microorganisms. The optimal range is the span of degrees in which a microorganism grows fastest, equivalent to its growth rate. Growth rate is the number of cells produced in a given period. Microbiologists often use a more exact definition for this rate of growth, called the *specific growth rate,* μ, which is:

number of grams of microbial mass formed ÷
grams of microbial mass formed per hour

The unit for μ, thus, becomes $hour^{-1}$.

Since ranges of optimal temperatures of different microorganisms overlap, there exist no clear-cur dividing lines between mesophiles and psychrophiles (cold-loving organisms) or thermophiles (heat-loving organisms). But because reproductive rate drops quickly when temperature leaves a microorganism's optimal range, microbiologists must be aware of the range, in order to keep cells alive.

Microorganisms with optimal ranges outside the mesophilic range are called temperature extremophiles to distinguish them from extremophiles that survive harsh conditions caused by factors other than temperature, such as acids or high salt concentration.

MESOPHILIC METABOLISM

Mesophiles differ from temperature extremophiles in their enzyme systems and composition. Mesophiles' enzymes, structural proteins, and protein synthesis apparatus cannot function at temperatures outside the optimal range. The table on page 495 describes the features of mesophile cell composition by comparing them with those of psychrophiles and thermophiles.

Mesophiles live on terrestrial mammals, in the soil and waters of temperate climates, and on plants. Human pathogens are also mesophiles; their optimal temperature correlates with the temperature of the human body. The following additional mesophiles have significance in industry:

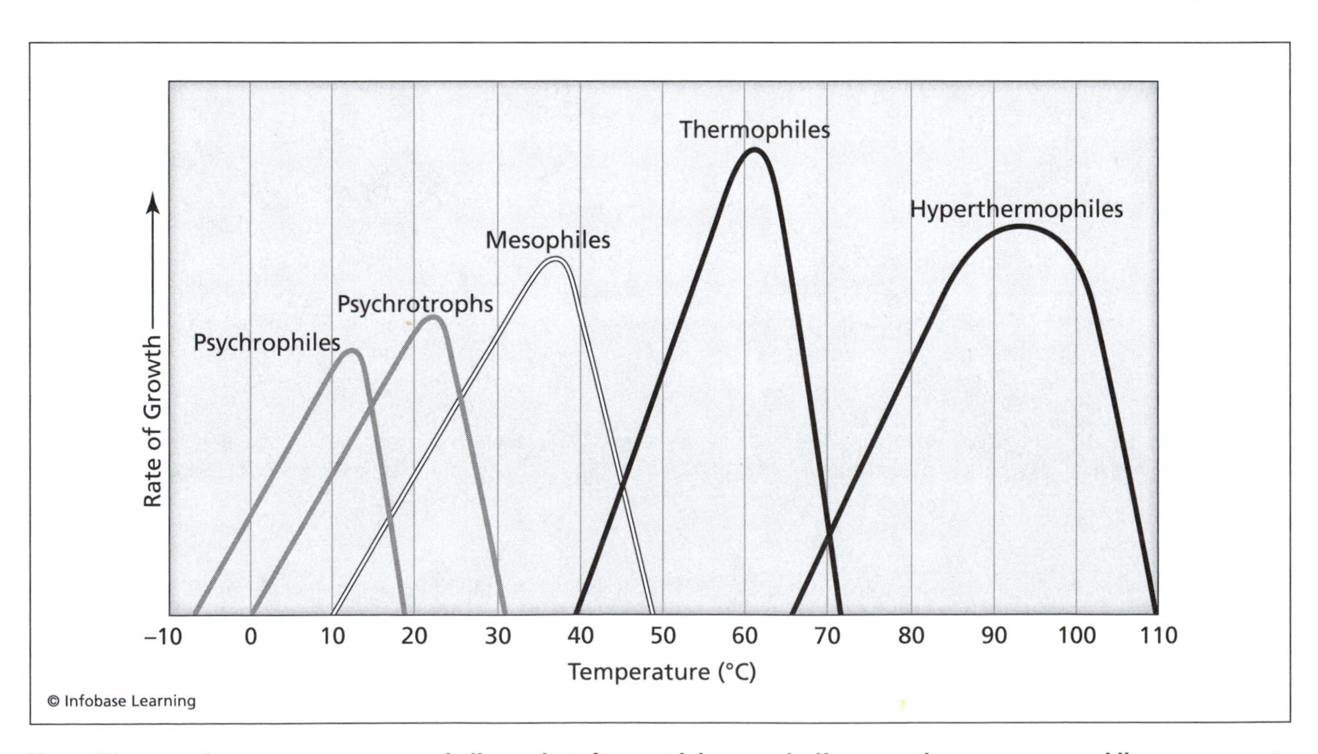

Mesophiles grow in a temperature range similar to that of terrestrial mammals. Human pathogens are mesophiles, as are most food-borne microorganisms.

Mesophiles, Psychrophiles, and Thermophiles

Features Compared with Mesophiles	Psychrophiles	Thermophiles
cell wall	similar	endospore formation in some species
enzymes	function at low temperatures	heat-stable
membranes	contain high levels of unsaturated fats that remain semifluid when cold	fatty acids more saturated, more branched, and higher in molecular weight
proteins	more heat-sensitive	extra amino acid proline for rigidity; more bonds for stability
transport systems	function at low temperatures	heat-stable

- food industry—spoilage microorganisms and food-borne pathogens
- food production—microorganisms used in fermentations and enzyme and vitamin production
- pharmaceuticals—microorganisms that produce drugs, antibiotics, and antibodies
- chemistry—microorganisms that replace strong chemical reactions with biological reactions
- biotechnology—gene transfer microorganisms and recombinant deoxynucleic acid (DNA) methods

Most academic microbiology laboratories also use mesophiles for teaching or for research. These microorganisms grow well in laboratory conditions, if provided with their required nutrients. A key aspect in growing mesophiles is the use of an incubator, which provides the cultures with the optimal temperature for growth.

INCUBATION

Many, but not all, mesophiles grow well at room temperature. Microbiologists prefer to grow mesophilic cultures in an incubator for two reasons: to provide a secure place that reduces the chance of contamination and to provide a constant, optimal temperature for growth.

Incubation temperature can dramatically affect the rate of growth, because many mesophiles have a narrow temperature range for optimal growth. Incubators must operate so that they maintain a preset temperature without varying more than 2°C. Microbiology universally uses the Celsius scale for incubation temperatures. In general, microbiologists incubate mesophilc bacteria and yeasts at 35–37°C and mesophilc fungi at 25–32°C. Bacteria isolated from soil or water may grow better in a range from room temperature (22–23°C) to 25°C.

Environmental microbiologists often take samples from nature to a laboratory to isolate bacteria or fungi. Bacteria and fungi can be separated from each other by using two different types of growth media and different incubators set at the optimal temperature for each microorganism. The samples containing bacteria are likely, however, to contain bacteria that have different optimal ranges. Microbiologists recover the maximal number of unknown bacteria from these samples by incubating, first, at a low temperature, such as 23°C, then, moving the cultures to an incubator set at a higher temperature, such as 37°C. As a result, more different types of bacteria have a chance to grow in their optimal conditions.

Dual-temperature incubations described here always start at the lower temperature and proceed to higher temperatures. Most bacteria can withstand temperatures cooler than their optimal range but not temperatures much above that range. Starting at cooler

Optimal Cooking Temperatures for Killing Mesophiles

Food	Cooking Temperature
precooked meats	140°F (60°C)
steaks, roasts, chops	145°F (63°C)
fresh pork, fresh ham, ground meats, eggs	160°F (71°C)
poultry, casseroles, leftovers	165°F (74°C)

Refrigeration Storage Times

Food	Storage Time
raw sausage, ground meat, fresh poultry	1–2 days
soups and stews, cooked meats, pizza	3–4 days
salads, cold cuts, uncooked steaks, chops, and roasts	3–5 days
processed meats (opened package)	1 week
processed meats (unopened package)	2 weeks

temperatures and then raising the incubation temperature gives cool-temperature mesophiles a chance to grow before the incubation temperature rises.

PREVENTING MESOPHILIC GROWTH

Clinical and food microbiologists have a special interest in mesophiles because these microorganisms present the greatest risks to patients or products, respectively. Pathogens use the human body as an incubator, which facilitates their growth after they infect by way of the skin, digestive tract, or lungs. Food pathogens and spoilage microorganisms also take advantage of foods that are not refrigerated, especially in hot weather. These microorganisms grow to large numbers very quickly in food production facilities, restaurants, or homes, if precautions to prevent growth are not taken.

People can control mesophilic growth in food by reducing the time that food spends in what food microbiologists call the danger zone. In this zone of about 40–140°F (4.4–60°C), mesophilic bacteria grow fastest and can multiply to hundreds of millions of cells in a few to several hours. To prevent foods from producing high numbers of bacteria, the food industry and professional food preparers follow three practices to avoid the danger zone: (1) refrigeration below 40°F (4.4°C), (2) cooking until food reaches an internal temperature of 145°F (63°C), and (3) refrigerating or discarding all leftovers that have been left at ambient temperature (the temperature of the surroundings) longer than 90 minutes. The Food Safety program of the U.S. Department of Health and Human Services publishes more specific guidelines for cooking temperatures to kill mesophiles, shown in the table at the bottom of page 495.

Because mesophiles live on the skin and in the digestive tract, people should wash their hands before preparing meals or having contact with anyone at higher than normal health risk, such as the elderly or people who have a weakened immune system. Hand washing also rids the skin of fecal bacteria that can contaminate common surfaces, particularly in places that have infants, young children, the incapacitated, and hospital patients.

Mesophiles can grow slowly at temperatures below their optimal range, explaining why refrigerated foods do not last indefinitely. The Food Safety program also proposes guidelines for the maximal time foods should be refrigerated before discarding. The table (left) summarizes some of these guidelines.

Food preservatives have added a measure of safety for consumers by inhibiting the growth of mesophiles. But preservatives do not provide foolproof protection, for the following reasons. First, preservatives can lose effectiveness over time. Second, large numbers of bacteria in food during its preparation use up the preservative's capacity to work for long periods. Third, preservatives do not usually kill bacteria and fungi but only inhibit their growth. For this reason, preserved foods stored for a long time may have large numbers of mesophiles.

Mesophiles make up the vast majority of microorganisms that people have contact with every day. Good hygiene and proper food handling, therefore, play an important role in preventing illness from these microorganisms.

See also EXTREMOPHILE; FOOD MICROBIOLOGY.

Further Reading

Chan, May, Richard H. Himes, and J. M. Akagi. "Fatty Acid Composition of Thermophilic, Mesophilic, and Psychrophilic Clostridia." *Journal of Bacteriology* 106 (1971): 876–881. Available online. URL: www.pubmedcentral.nih.gov/picrender.fcgi?artid=248714&blobtype=pdf. Accessed September 1, 2009.

FoodSafety.gov. Available online. URL: www.foodsafety.gov. Accessed September 21, 2009.

Morita, Richard Y. "Psychrophilic Bacteria." *Bacteriological Reviews* 39 (1975): 144–167. Available online. URL: www.pubmedcentral.nih.gov/picrender.fcgi?artid=413900&blobtype=pdf. Accessed September 1, 2009.

metabolic pathways Metabolic pathways comprise the succession of enzymatic steps that lead from a starting compound to end products and, in this process, produce energy for a cell and cellular constituents. Metabolic pathways make up the larger process called *metabolism*, which is the sum of all chemical reactions that take place in a living cell. The pathways of metabolism can be divided into two major types: catabolism and anabolism. Catabolism involves all of the pathways that break down large compounds into simpler compounds and molecules. Anabolism consists of all the pathways that cells use for synthesizing complex compounds from simpler compounds and molecules.

Metabolic pathways begin with the smallest unit of matter in chemistry, the atom, which contains a nucleus made of positively charged protons plus uncharged neutrons and negatively charged electrons that orbit the atom's nucleus. All atoms with the same number of protons behave the same way chemically, so they belong to the same chemical element. Chemical elements, such as hydrogen, carbon, nitrogen, and sulfur, are the building blocks of all matter on Earth.

Atoms combine by using three basic types of bonds to form molecules. Molecules can be made of one element or a combination of elements, such as hydrogen (H_2) or hydrogen sulfide (H_2S), respectively. The three types of bonds that hold molecules together are ionic, covalent, and hydrogen:

- ionic bond—created by an atom that has lost an electron to form a positive charge and an atom that has gained an extra electron to form a negative charge
- covalent bond—formed by two atoms that share one or more pairs of electrons
- hydrogen bond—a hydrogen atom forms a covalent bond with oxygen or nitrogen and is also attracted to, but does not bond with, another oxygen or nitrogen atom

Chemical reactions inside cells occur by the action of enzymes, which are complex compounds (proteins) that break bonds, rearrange chemical structures, and build new bonds. Enzymatic reactions either use up energy as they progress, called *endergonic reactions,* or release excess energy, called *exergonic reactions.* Compound synthesizing reactions of anabolism usually consume energy, and the decomposition reactions of catabolism are usually exergonic. Living cells must, therefore, use some energy to build cellular constituents, including energy storage compounds, then liberate energy for their use by breaking down (catabolizing) cellular constituents.

The main inorganic compounds involved in cellular catabolism and anabolism are water, acids, bases, and salts. The main organic compounds involved in metabolism are explained in the table below

Amino acids are the building blocks of proteins that act either as enzymes or as structural components of the cell. A typical amino acid structure contains a central carbon atom bound to an amino group ($-NH_2$), a hydrogen atom, a carboxyl group, and a side group that is unique to each amino acid. Only 20 amino acids make up the proteins found in nature, most containing two mirror image structures, called an L-form (left-handed) and a D-form (right-handed). Proteins of living cells are usually in the L-form. An exception occurs in bacterial cell walls, which contain some D-forms. Antibiotics also contain some D-amino acids.

The four most abundant elements in microorganisms are hydrogen, carbon, nitrogen, and oxygen, which are supplied by complex molecules, such as proteins, or simple molecules, such as amino acids or gases. Microorganisms need an array of additional nutrients to help enzymes run the metabolic pathways that recycle the four main elements for microbial use. These nine nutrients are calcium, chlorine, iodine, iron, magnesium, phosphorus, potassium, sulfur, and sodium. Some microorganisms require additional nutrients in very small amounts, called *trace elements,* for example, cobalt and manganese.

Much of the energy stored and released by cells resides in the compounds adenosine triphosphate (ATP) and adenosine diphosphate (ADP). Both compounds are composed of carbon, nitrogen, hydrogen, oxygen, and phosphorus and store the excess energy released by exergonic reactions and save it for the energy demands of endergonic reactions. The energy resides in the bonds between the three phosphate groups ($-PO_4^-$) in ATP or the two phosphate groups of ADP. The storage-release role of energy by ATP-ADP is the reason these compounds are sometimes called the energy currency of living cells.

Every metabolic pathway in microorganisms functions either to use or to produce energy held in

The Main Organic Compounds of Metabolic Pathways

Compound	General Structure	Major Anabolic Pathways	Major Catabolic Pathways
carbohydrates	carbon, hydrogen, and oxygen as $C_xH_{2x}O_x$	glucogenesis, photosynthesis	glycolysis
lipids	long carbon chains with hydrogen attached, ending in carboxyl (-COOH)	lipogenesis	lipolysis
proteins	chains of amino acids	protein synthesis	proteolysis
nucleic acids	deoxyribonucleic acid (DNA) and ribonucleic acid (RNA)	DNA or RNA synthesis	purine and pyrimidine catabolism

ATP. To do this, each pathway must also adhere to the two following laws of thermodynamics:

- first law of thermodynamics—energy can neither be created nor be destroyed

- second law of thermodynamics—physical and chemical processes proceed in a direction of increasing randomness or disorder, a state called *entropy*

Metabolic pathways that synthesize large complex compounds from simpler compounds decrease the entropy in a system. In order to accomplish this, these reactions consume energy in order to reverse the natural flow of nature toward disorder. For this reason, bacteria, protozoa, trees, humans, and every other living thing require energy to grow, repair injuries, reproduce, and move. Although living things can reverse the natural progression toward entropy, they can do this only temporarily. Ultimately, the universe follows the second law of thermodynamics.

The major metabolic pathways in microorganisms deal with the synthesis or breakdown of carbohydrates, usually glucose, and proteins and lipids (fats). In 1959, the Dutch microbial ecologist Albert Kluyver (1888–1956) of the University of Delft in Holland summarized metabolism in all types of cells: "The most specific characteristic of living matter resides in its metabolic properties. It is the empirical fact that the maintenance of life requires a continuous supply of special chemical substances which, in the living cell, undergo transformations that lead to their partial excretion in altered form." The pathways that control these mechanisms have been adapted by microbiologists to degrade certain substances or to produce other substances that have commercial value.

The two major factors that influence the types of pathways a microorganism can run relate to its requirements for oxygen and its ability to carry out photosynthesis. Oxygen-requiring microorganisms, called aerobes, use oxygen in their energy-generation pathways, while anaerobes, which live without oxygen, use simple organic compounds in their energy pathways. Both aerobes and anaerobes depend on an organic or inorganic compound as an energy source. Photosynthetic microorganisms differ, in depending on sunlight as their energy source. The photosynthetic microorganisms then use the Sun's energy to power the endergonic reactions needed to build cellular constituents.

The explanation offered in the preceding paragraph does not illustrate the breadth of diversity in microbial pathways. Metabolism in the microbial world is much more diverse than the metabolism of higher organisms. In microorganisms, furthermore, the types of metabolic pathways a species uses can characterize that species. Kluyver explained, in 1924, "The most specific characteristic of living matter resides in its metabolic properties." Microbiologists, today, look for the absence or presence of certain metabolic pathways to help identify unknown microorganisms.

CARBOHYDRATE METABOLISM

Many microorganisms depend on glucose as a source of carbon and energy. Three metabolic pathways represent the vast majority of routes by which microorganisms break down glucose and capture its energy and retain its carbon. These pathways are (1) glycolysis, (2) the pentose phosphate pathway, and (3) the Entner-Doudoroff pathway. Of these three, glycolysis is, by far, the most common and operates in all aerobic and anaerobic cells.

Glycolysis—also called the Embden-Meyerhof-Parnas or glycolytic pathway—starts with a six-carbon glucose molecule and converts it to two three-carbon molecules of pyruvate. The initial steps of glycolysis use two ATP molecules per glucose, but the entire pathway replaces these ATPs and produces two more.

Some microorganisms run the pentose phosphate pathway at the same time they carry out glycolysis. As does glycolysis, the pentose phosphate pathway leads to the production of pyruvate, but the intermediary compounds it produces differ from those in glycolysis. These intermediaries range from three- to seven-carbon compounds. The pentose phosphate pathway produces some energy for the cell, but this series of reactions tends to become more important in certain synthetic pathways rather than energy-yielding pathways.

The Entner-Doudoroff pathway combines features of glycolysis with the pentose phosphate pathway to produce one ATP. Most bacteria rely on glycolysis and can simultaneously operate the pentose phosphate pathway. The Entner-Doudoroff pathway is confined mainly to gram-negative bacteria, especially the genera *Agrobacterium, Azotobacter, Pseudomonas,* and *Rhizobium.* The enteric species *Enterococcus faecalis* is a rare gram-positive species that uses this pathway.

After completing glycolysis, aerobes and anaerobes diverge in the manner in which they produce energy from the starting glucose molecule. Aerobes use aerobic respiration, which characterizes the pathways that follow glycolysis, consume oxygen, and produce large amounts of energy relative to anaerobic species. Anaerobes, by contrast, depend on either fermentation or anaerobic respiration. Each of these anaerobic pathways produces less ATP, and thus energy, for the cell's use than aerobic pathways.

The main reason for the energy production disparities between aerobes and anaerobes is complex but can be simplified by considering the role of electrons in energy production. In addition to the ATP

Electron Acceptors in Major Anaerobic Metabolic Pathways

Pathway	Electron Acceptor	End Product
dissimilatory nitrate reduction	nitrate (NO_3^-)	nitrite (NO_2^-)
denitrification	nitrate (NO_3^-)	nitrite (NO_2^-), nitrous oxide (N_2O), or nitrogen (N_2)
methanogenesis	carbon dioxide (CO_2)	methane (CH_4)
assimilatory sulfate reduction	sulfate (SO_4^{2-})	hydrogen sulfide (H_2S)
sulfur respiration	elemental sulfur (S^0)	hydrogen sulfide (H_2S)
iron reduction	ferric iron (Fe^{3+})	ferrous iron (Fe^{2+})

produced in glycolysis, ATP and alternate energy storage compounds are produced by cells when enzymes transfer electrons or protons from one compound to the next in a metabolic pathway's series of steps. Aerobes use two processes for this, the *electron transport chain* and the *oxidative phosphorylation pathway,* respectively. Electron transport operates in membranes, where a series of compounds called *cytochromes* transfer electrons from one pigment to the next until an oxygen molecule at the end of the line accepts the electrons in the following reaction:

$$2 \text{ electrons} + \tfrac{1}{2} O_2 + 2 H^+ \rightarrow H_2O$$

Overall, an aerobe using electron transport can convert one glucose molecule into cellular carbon, energy, and water.

Oxidative phosphorylation is similarly efficient. This process uses a part of the energy produced by electron transport to create a charge difference, called a *gradient,* across a cell's membrane. A mechanism called the proton motive force creates this unequal distribution of charges, from one side of a membrane to the other side. The proton motive force induces protons (H^+) to migrate across the membrane to reestablish equilibrium; in so doing, the process produces ATP. Every three protons that move across the membrane produce one ATP molecule.

Common Fermentation Pathways

Type of Fermentation	End Products	Example Microorganisms
acetone-butanol	acetone, butanol, butyric acid, carbon dioxide	*Clostridium*
alcohol	ethanol, carbon dioxide	*Saccharomyces*
butandiol	2,3-butanediol, ethanol, formic acid, carbon dioxide, hydrogen	enteric bacteria
butyric acid	acetic acid, butyric acid, carbon dioxide, hydrogen	*Clostridium, Butyrivibrio*
homoacetate (acetogenesis)	acetic acid	*Acetobacterium, Clostridium*
lactic acid	lactic acid	*Lactobacillus, Streptococcus*
malolactic	lactic acid, carbon dioxide	*Lactobacillus, Leuconostoc, Pediococcus*
mixed acid	ethanol, lactic acid, succinic acid, acetic acid, carbon dioxide, hydrogen	*Escherichia coli, Salmonella, Shigella*
propionic acid	propionic acid, acetic acid, carbon dioxide, hydrogen	*Propionibacterium*

Glucose Sources for Metabolic Pathways

Substrate	Enzyme	End Products
cellobiose	cellobiose phosphorylase	alpha-D-glucose-1-phosphate + glucose
cellulose	cellulase	cellobiose
galactose	multiple enzymes	glucose-6-phosphate
lactose	beta-galactosidase	galactose + glucose
maltose	maltase	2 glucoses
sucrose	sucrase	glucose + fructose

The electrons that feed an aerobe's electron transport chain have three different sources: (1) glycolysis, (2) fat metabolism, and the tricarboxylic acid (TCA) cycle. The TCA cycle—also called the Krebs cycle or the citric acid cycle—degrades pyruvate from glycolysis, fatty acids from lipid breakdown, and amino acids from protein breakdown to carbon dioxide (CO_2), water, electrons, and energy storage compounds. The TCA cycle's main energy storage compounds are the following electron-carrying compounds:

- nicotinamide adenine dinucleotide (NAD)
- flavin adenine dinucleotide (FAD)

These two compounds participate in ATP production during specific steps in the TCA cycle. NAD, for example, produces the energy to form an ATP molecule and does this by taking advantage of charge gradients in some of the TCA cycle's steps. The simplified process proceeds as:

$$NADH + 2 \text{ electrons} \rightarrow \tfrac{1}{2} O_2 + 1 \text{ ATP}$$

By taking advantage of NAD's or FAD's participation in the TCA cycle, aerobic bacteria can produce a sum total of 38 ATP molecules from one glucose that enters glycolysis. It manufactures these ATPs at various steps in glycolysis, the TCA cycle, the electron transport chain, and oxidative phosphorylation. Some bacteria, algae, fungi, and protozoa use a modified TCA cycle, called the *glyoxalate cycle,* that begins with acetate as the carbon source.

Anaerobic microorganisms use either anaerobic respiration or fermentation to produce energy to power all the cell's functions. Neither metabolic pathway is as efficient in energy production as aerobic pathways. While aerobes degrade glucose completely to carbon dioxide, water, and energy, anaerobes cannot use oxygen as an electron acceptor but rely on organic compounds to play the role of electron acceptor. Because of the energetic of the anaerobic steps, anaerobes never have the same maximal energy from glucose as aerobes.

Anaerobic respiration uses molecules other than oxygen as the metabolic pathway's final electron acceptor. Fermentation, similarly, does not make use of oxygen but, instead, uses simple organic compounds only as the electron acceptor. The table at the top of page 499 provides examples of electron transport and, thus, energy generation in anaerobic pathways.

Some anaerobic bacteria use trace elements as part of the energy production pathway, similar to those listed in the table. The following bacteria provide examples of the use of more electron acceptors:

- arsenic—*Desulfotomaculum*
- copper—*Thiobacillus*
- mercury—*Desulfovibrio*
- selenium—*Aeromonas*

Fermentation relies on similar transfer of electrons from compounds in sequence, but the end product, the final electron acceptor, is always another organic compound. Many different types of fermentations starting with pyruvate or other organic compounds exist in nature, each leading to a different mixture of end products. The table at the bottom of page 499 summarizes fermentations that have been studied most in microbiology.

Microorganisms have evolved to be flexible in the types of carbohydrates they can use for producing energy and building new cells. Although glucose is the most studied substrate (the compound upon which an enzyme acts), microorganisms can use inorganic compounds for energy, as shown earlier, or glucose in forms other than the free sugar. The table on this page shows the variety of substrates from which microorganisms liberate glucose by using specialized enzymes.

Each of the end products of the nonglucose substrates shown in the table feeds into glycolysis. Microorganisms also degrade polysaccharides of varying lengths into shorter sugar chains, such as trisaccharides (three sugars) or disaccharides (two sugars). Specialized enzymes degrade these substrates to supply sugars for glycolysis. For example, amylase enzymes break down the polysaccharides starch and glycogen, each composed of glucose, maltose, and shorter polysaccharides.

The ultimate goal of the carbohydrate catabolism pathway is energy production. Regardless of how a microorganism degrades sugars or other organic compounds inside the cell, the resulting ATPs help keep the cell's energy cycle running. ATP stores the energy that will be needed for all cellular activities: reproduction, cell growth, injury repair, compound synthesis, nutrient transport, and movement.

Microorganisms receive glucose in one of two ways: absorption through the cell membrane and photosynthesis. Photosynthesis converts light energy into chemical energy by absorbing sunlight and using this energy to drive the glucose synthesis reactions. The photosynthetic pathway also generates ATP, which the cell depends on as an energy source for synthesizing other cellular constituents, such as proteins.

In addition to supplying their own energy needs, photosynthetic microorganisms serve as a foundation for a multitude of food chains. By converting the Sun's energy to a form that can be used by other living organisms, photosynthetic microorganisms play a vital role in the earth's energy balance. In addition to higher plants, the following microorganisms carry out photosynthesis: green, brown, and red algae; unicellular algae such as *Euglena,* dinoflagellates, and diatoms; cyanobacteria; green sulfur and nonsulfur bacteria; and purple sulfur and nonsulfur bacteria. Microorganisms that cannot carry out photosynthesis make glucose by the gluconeogenesis pathway, which resembles glycolysis in reverse.

PROTEINS AND NITROGEN METABOLISM

Nitrogen is an essential part of amino acids, nucleic acids, and coenzymes (compounds that play an auxiliary role in enzymatic reactions). Some microorganisms participate in a biogeochemical cycle called the nitrogen cycle, by capturing nitrogen gas from the atmosphere and converting it to a form that another microorganism (or a plant) can use for making amino acids and, thus, cellular proteins. The three main pathways that accomplish the conversion of inorganic nitrogen gas to proteins are:

- nitrogen fixation—reduction of atmospheric nitrogen to ammonia (NH_3)
- nitrogen assimilation—incorporation of ammonia nitrogen into organic compounds
- amino acid synthesis—conversion of various organic compounds, called *amino acid skeletons,* into amino acids
- protein synthesis—assembly of amino acids in specific sequence, guided by ribonucleic acid (RNA) and the genes held in deoxyribonucleic acid (DNA)

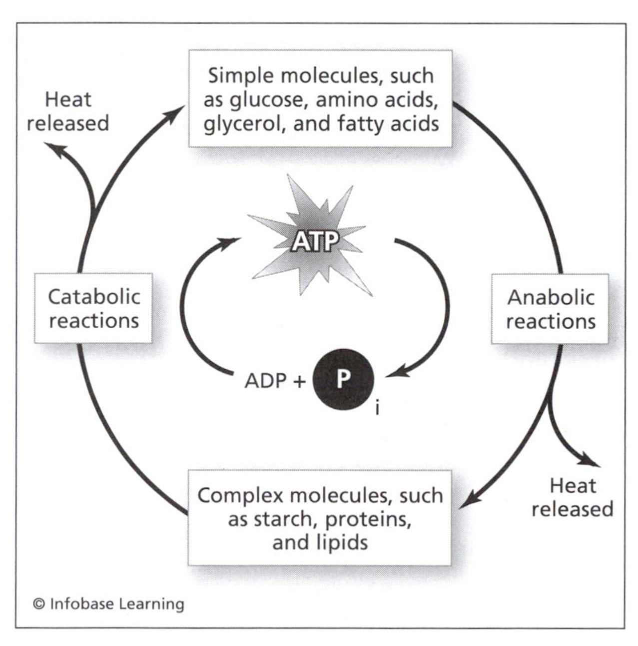

Cells contain their own energy cycles, consisting of energy flow into and out of the phosphate bonds of adenosine triphosphate (ATP). Nutrients and organic or inorganic energy sources feed the continuation of this cycle.

The purine and pyrimidine compounds that make up RNA and DNA follow complex synthesis pathways. Seven different molecules contribute part of the purine skeleton for adenine or guanine, including amino acids, the five-carbon sugar ribose, and the B-complex vitamin folic acid. The synthesis of the pyrimidines uridine, cytosine, and thymine depends, in large part, on ribose to build the molecules' skeletons.

Microorganisms also decompose nitrogen compounds for two purposes: for retrieval of nitrogen for use in building new nitrogen-containing compounds or for energy. Microbial enzymes called *proteases* cleave proteins into amino acid chains called *peptides.* Additional proteases, then, break individual amino acids from peptides, which the microorganism can use as energy by feeding amino acids in the cell's TCA cycle. Various amino acids can enter the TCA cycle at different places. The intermediate compounds in glycolysis and the TCA cycle, likewise, serve as starting structures for making new amino acids. For example, the pyruvate formed in glycolysis can either enter the Krebs cycle for energy production or serve as the starting point for making certain amino acids. The table on page 502 shows how amino acids, and thus proteins, relate in this way to energy metabolism systems from carbohydrates.

By connecting amino acid and protein metabolism with carbohydrate metabolism, microorganisms can take advantage of alternate routes for obtaining energy during times of scarce nutrients.

LIPID METABOLISM

Like carbohydrates and proteins, lipid synthesis and breakdown, called *beta-oxidation,* proceeds in either direction in microorganisms, depending on whether a cell needs to store or use up an energy source. Microorganisms contain a variety of lipids, most of which use fatty acids as the starting compound or the backbone. A fatty acid is a carbon chain of one to more than 30 carbon molecules. The carbons of the chain connect to at least one hydrogen molecule, except for one of the terminal carbons, which forms a carboxyl group (-COOH). A typical fatty acid would have the following structure:

$$H_3C\text{-}CH_2\text{-}CH_2\text{-}CH_2\text{-}CH_2\text{-}COOH$$

The six-carbon fatty acid shown in the formula is called a saturated fatty acid, because all of the available carbon binding sites are filled with hydrogen molecules, except the terminal carbon. The same fatty acid converted to an unsaturated form would contain a double bond between two of the carbons, such as:

$$H_3C\text{-}CH_2\text{-}CH{=}CH\text{-}CH_2\text{-}COOH$$

The thousands of species of microorganisms produce a diversity of fatty acids, many with side branches of additional carbon chains or odd-numbered chains. The table on page 503 describes the naming conventions for some fatty acids that are widespread in nature and are composed only of straight, nonbranched chains.

Simple lipids, or fats, consist of saturated and unsaturated fatty acids and triglycerides, which are three-carbon glycerol molecules with a fatty acid attached to each of the glycerol's carbons. Glycerol containing three fatty acids is called a *triglyceride,* glycerol with two fatty acids is a *diglyceride,* and one fatty acid makes a *monoglyceride.* The rest of the lipids are called *complex lipids;* they consist mostly of phospholipids and steroids. A phospholipid contains glycerol, two fatty acids, and a phosphate group ($-PO_4$) in place of the third fatty acid of a typical triglyceride. Steroids possess a dramatically different structure from other lipids; they contain four carbon rings plus a long fatty acid chain.

Microorganisms build their own lipids from a compound that also acts as the bridge between glycolysis and the TCA cycle, acetyl coenzyme A (acetyl-CoA). A microorganism can add many or few links to acetyl-CoA to make fatty acids of various lengths, or it can use acetyl-CoA as a starting point in steroid synthesis. Phospholipids, fatty acids, and steroids occur in microbial membranes and play critical roles in maintaining the functions and structures of membranes.

For energy from lipids, microorganisms begin degrading fatty acids carbon by carbon, splitting off acetyl-CoA units. An enzyme cleaves off the coenzyme portion of acetyl-CoA to make a two-carbon compound, which can, then, enter the TCA cycle for energy production.

PEPTIDOGLYCAN SYNTHESIS

The long polysaccharide (also called a *polymer*) peptidoglycan acts as a protective outer covering in most bacteria. In gram-positive bacteria, a thick peptidoglycan layer lies on the outside of the cell. Gram-negative bacteria differ in having a thinner peptidoglycan layer between two membranes. In both conditions, the complex structure of peptidoglycan is essential for cell survival.

Peptidoglycan synthesis begins with a backbone of two alternating compounds: N-acetylmuramic

Protein Anabolism from Carbohydrate Catabolism

Intermediate Compound	Carbohydrate Pathway	Amino Acids
glucose-6-phosphate +	glycolysis	phenylalanine, tyrosine, tryptophan
3-phosphoglycerate	glycolysis	serine serine to cysteine and glycine glycine to histidine
pyruvate	glycolysis	alanine, isoleucine, leucine, lysine, valine
alpha-ketoglutarate	TCA cycle	glutamic acid glutamic acid to arginine, glutamine, proline
oxaloacetate +	TCA cycle	aspartic acid aspartic acid to asparagines, isoleucine, lysine, methionine, threonine

Note: A "+" symbol indicates that additional intermediate compounds exist in this pathway

Straight-Chain Fatty Acids

Common Name	Scientific Name	Number of Carbons
	SATURATED	
propionic	propanoic	3
butyric	butanoic	4
valeric	pentanoic	5
caproic	hexanoic	6
caprylic	octanoic	8
capric	decanoic	10
lauric	dodecanoic	12
myristic	tetradecanoic	14
palmitic	hexadecanoic	16
stearic	octadecanoic	18
arachidic	eicosanoic	20
	UNSATURATED	
palmitoleic	cis-9-hexadecanoic acid	16:1
oleic	cis-9-octadecanoic acid	18:1
linoleic	9,12-octadecadienoic acid	18:2
arachidonic	5,8,11,14-eicostetraenoic acid	20:4

Note: unsaturated fatty acid shorthand = number of carbons:number of double bonds

acid and N-acetylglucosamine. As the microorganism builds the peptidoglycan chain, it also extends branches between chains, until the layer takes on a lattice structure. Many of the details of this complex, multistep process have not been fully detailed. At least eight major steps occur in peptidoglycan synthesis. Newly growing cells become vulnerable during this process for the following three reasons: (1) Peptidoglycan synthesis demands energy from the cell, (2) the process must proceed in an exact stepwise fashion, and (3) factors that interfere with the synthesis before it is complete will result in the cell's death.

Because of the energy demands of peptidoglycan synthesis and its need for carbon, nitrogen, oxygen, and phosphorus, plus the factors needed by the enzymes, this cellular process is a culmination of almost all other metabolic pathways in bacteria.

See also FERMENTATION; GLYCOLYSIS; METABOLISM; NITROGEN FIXATION.

Further Reading

Gottschalk, Gerhard. *Bacterial Metabolism,* 2nd ed. New York: Springer-Verlag, 1986.

Kluyver, Albert J. "Unity and Diversity in the Metabolism of Micro-organisms." In *Albert Jan Kluyver, His Life and Work.* Translated by Thomas Brock. Amsterdam, Netherlands: North-Holland, 1959.

Prescott, Lansing M., John P. Harley, and Donald A. Klein. *Microbiology,* 6th ed. New York: McGraw-Hill, 2005.

Todar, Kenneth. "Microbial Metabolism." The Microbial World. Madison, Wisc.: University of Wisconsin, 2008. Available online. URL: http://textbookofbacteriology.net/themicrobialworld/metabolism.html. Accessed June 6, 2010.

metabolism Metabolism is the sum of all chemical reactions that occur in a living cell. In metabolism, catabolism involves all pathways in which large compounds degrade into simpler compounds and molecules, and anabolism consists of all pathways for synthesizing complex compounds from simpler compounds and molecules. In all instances of metabolism, microorganisms use the energy produced in catabolism to run the reactions of anabolism and thus build new cell constituents. These two processes operate similarly in prokaryotes as well as complex organisms.

Metabolism in any organism must satisfy two main requirements: energy production to run the cell's processes and carbon compounds for building new cell constituents. Microbiologists, often, classify microorganisms in terms of the modes of energy production and carbon used by the cell. Other nutrients, such as nitrogen, phosphorus, sulfur, oxygen, water, and minerals, support the main metabolic routes of energy production and carbon use.

Microorganisms are the most diverse group of living things on Earth. Consequently, microorganisms have a wide variety of mechanisms for obtaining energy from organic or inorganic compounds and many different end products from those metabolic pathways. To make sense of the vast diversity of microbial metabolism, biologists group microorganisms into four main metabolic groups, shown in the table on page 504.

The table illustrates the range of cellular metabolism among the earth's microorganisms. Chemoheterotrophic metabolism resembles that in mammals, in which organic compounds play the central role in energy and carbon metabolism. Yet other microorganisms can live in austere conditions, existing only on sunlight and carbon dioxide gas to meet their energy-carbon needs. In much of the microbial world, microorganisms work together as communities, in which the end products of one species serve as an energy or carbon source for other microorganisms. In fact, many chemoheterotrophs and photoheterotrophs use the end products of other microorganisms. In general, the four types of metabolism can be found in the following broad classes of microorganisms:

- chemoheterotrophy—bacteria, fungi, and all fungi and animals
- chemoautotrophy—nitrogen-, hydrogen-, sulfur-, and iron-utilizing bacteria, and some archaea
- photoheterotrophy—green and purple nonsulfur bacteria
- photoautotrophy—cyanobacteria, green and purple sulfur bacteria, algae, and plants

Within each of the metabolic groups, various metabolic pathways exist for keeping a cell alive and allowing it to function and reproduce. No microorganism possesses every metabolic pathway in nature, but one pathway, called *glycolysis,* is universal to all living things. Glycolysis is the breakdown of a six-carbon glucose sugar to two three-carbon molecules of pyruvate, which then feed various other forms of energy-producing catabolism.

In addition to glycolysis, microbiologists study metabolism as it relates to the use of oxygen. After glycolysis—glycolysis does not require oxygen and is therefore an anaerobic pathway—aerobic microorganisms use different pathways for energy production and carbon use than anaerobic microorganisms. The dividing line between these two groups is not clear-cut, as many microorganisms contain pathways that use oxygen when oxygen is present and turn to alternate pathways in the absence of oxygen. In general, however, the main metabolic pathways can be described in terms summarized in the table on page 506.

Each of the metabolic pathways listed in the table contains specific enzymes that modulate the pathway's progress. The regulation of any type of metabolism is critical to cell survival for two main reasons. First, microorganisms must not run their metabolism so quickly that they use up all available nutrients and, thus, cannot sustain themselves. Second, microorganisms must regulate metabolism to prevent the buildup of end products, some of which are toxic to the cells.

Studies on microbial metabolism began in earnest, from the mid-1800s to the early 20th century, during what is now called the Golden Age of Microbiology. By the beginning of this period in history, biologists had made inroads into the behavior of bacteria and yeasts, their ability to change the properties of foods, their ubiquity in nature, and the structure of microbial cells, the latter aided by advances in microscopes. By the turn of the 20th century, microbiologists had delved into the ways by which microorganisms convert sugars and other substances, primarily into alcohols and acids.

Early studies on metabolism took place in the laboratory of the French microbiologist Louis Pasteur (1822–95), who began with investigations of fermentation. In 1857, Pasteur wrote, "It is known [by chemists] that it is necessary only to take a solution of sugar and add chalk, which keeps the medium neutral, a nitrogenous material, such as casein . . . in order to have the sugar transformed into lactic acid. But the explanation of this phenomenon is quite obscure, since the way in which the decompostable nitrogenous material acts is completely ignored. . . . Careful studies up to the present have not revealed the development of organized beings during the fermentation process." Pasteur's subsequent studies showed that microorganisms ferment sugars to alcohols and acids, such as lactic acid.

In the late 1800s, the Dutch biologist Martinus Beijerinck (1851–1931) and the Russian soil scientist Sergei Winogradsky (1856–1953) studied the nitrogen and sulfur utilization of specific types of bacteria. Beijerinck had taken on the difficult task of culturing the bacteria he discovered living in the root nodules of plants called *legumes.* Beijerinck's procedures, using bacteria that would later be classified in the genus *Rhizobium,* led to the complex study

The Main Types of Microbial Metabolism

Type of Metabolism	Energy Source	Carbon Source
chemoheterotrophy	organic compounds	organic compounds
chemoautotrophy	inorganic compounds	carbon dioxide (CO_2)
photoheterotrophy	light	organic compounds
photoautotrophy	light	carbon dioxide

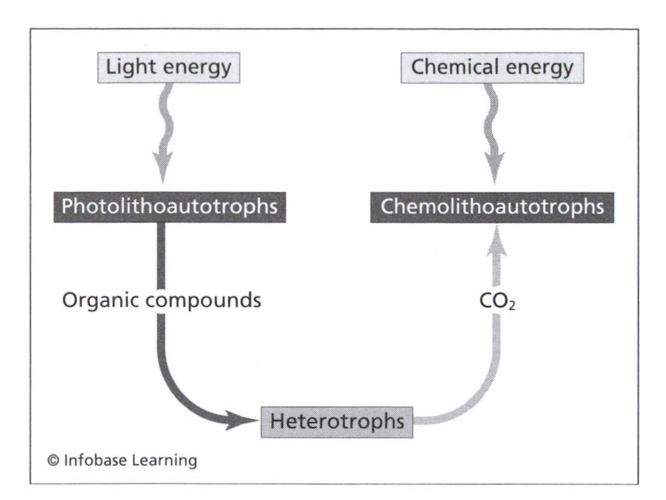

Energy flow on Earth moves through every type of microorganism. Bacteria are the principal drivers of energy capture from the sun, the decomposition of organic matter, the return of carbon to the atmosphere, and the transfer of energy to food chains.

of microbial nitrogen metabolism. Microbiologists would, eventually, recognize the bacteria Beijerinck studied as nitrogen-fixing bacteria—nitrogen fixation is the capture of atmospheric nitrogen.

At the same time as Beijerinck's work, Winogradsky developed methods for learning about microorganisms that used sulfur in their respiration. "In 1887," he said, "I published the results of my research on the physiology of the organisms of sulfur-containing waters. In another extensive work I have shown further that these unusual organisms, so numerous that I can distinguish 15 genera and more than 25 species, form a distinct physiological group that is characterized by the role that sulfur plays in their economy." Winogradsky would call the microorganisms *sulfur bacteria,* a name that persists today for the microorganisms that need sulfur-containing compounds for energy metabolism. Winogradsky also elucidated the process by which bacteria use and reuse nitrogen, in what are now known as *biogeochemical cycles.*

Environmental microbiologists and ecologists have gained a much greater understanding, since the turn of the century, of the vital role played by microbial metabolism in nutrient cycling. The cycling of nutrients through living and nonliving matter is essential to maintaining life on Earth and even contributes to climate and natural resources. For example, carbon cycling controls carbon dioxide levels in the atmosphere, which affect climate change. Natural resources, such as forests and wildlife, also depend on nutrient cycling, which begins when microorganisms make carbon, nitrogen, and other elements available for higher life-forms. In general, the metabolism of Earth's biota would not function without the microbial metabolism that takes place in soil and water.

CHEMOHETEROTROPHY

Chemoheterotrophy lacks a clear distinction between energy sources and carbon sources. Many chemoheterotrophs use an organic compound, such as glucose, to serve both needs. This metabolism has also been called *chemoorganotrophy* or *heterotrophy.* Chemoheterotrophs (or organotrophs) make up some of the most recognized names in microbiology, including pathogens.

When metabolizing glucose, chemoheterotrophs take glucose's carbon molecules and use them for building other carbohydrates or proteins, lipids, nucleic acids (deoxyribonucleic acid [DNA] and ribonucleic acid [RNA]), or vitamins. These microorganisms get energy from glucose by using the electrons from glucose's hydrogen atoms and passing the electrons into energy-generating processes.

Chemoheterotrophs play an important part in ecology, because of their degradation of organic matter. By degrading organic compounds, chemoheterotrophs make carbon available for other uses. The overall process is called the *carbon cycle.* Two types of chemoheterotrophs conduct this carbon recycling: saprophytes and parasites. Saprophytes live on dead organic matter and, so, serve a crucial role in decomposing substances; such as leaves, trees, animal carcasses, and waste matter. Parasites derive nutrients from living tissue, so these microorganisms obtain their carbon and energy from higher animal or plant life.

Chemoheterotrophs use glycolysis in conjunction with the carbon metabolism pathways of the tricarboxylic acid cycle or fermentation. These microorganisms can, therefore, be either aerobes or anaerobes, and they inhabit almost every place on Earth.

CHEMOAUTOTROPHY

Chemoautotrophs use a variety of compounds for energy generation but can use only carbon dioxide for carbon. The energy sources must be reduced inorganic compounds, meaning the molecules have an excess of electrons. The following bacteria provide examples of chemoautotrophs and their main energy sources:

- *Nitrosomonas*—ammonia (NH_3)
- *Nitrobacter*—nitrite (NO_2^-)
- *Beggiatoa*—hydrogen sulfide (H_2S)
- *Thiobacillus thiooxidans*—elemental sulfur (S^0)
- *T. ferrooxidans*—ferrous iron (Fe^{2+})
- *Pseudomonas carboxydohydrogena*—carbon monoxide (CO)

Microbiology's Main Metabolic Pathways

Pathway	Purpose	Oxygen Requirement
glycolysis	degrades glucose into a usable form (pyruvate)	anaerobic
Entner-Doudoroff	degrades glucose into a usable form (pyruvate)	anaerobic
pentose phosphate	degrades glucose into a usable form (various compounds)	anaerobic
tricarboxylic acid cycle	energy from pyruvate, amino acids, and fat breakdown products	aerobic
fermentation	degrades organic compounds into other organic compounds	anaerobic
anaerobic respiration	uses nitrogen, sulfur, iron, or carbonate (CO_3^{2-}) in place of oxygen	anaerobic
photosynthesis	converts light energy into chemical energy	anaerobic with oxygen as an end product

As all metabolic types do, chemoautotrophs contribute to Earth's carbon cycle, but their need for specific inorganic energy sources confines them to certain habitats. Nitrogen-utilizing chemoautotrophs live in soils, water, and legume root nodules. Depending on the type of nitrogen metabolism a chemoautotroph possesses, these microorganisms must reside where other microorganisms put nitrogen into a usable form. In the process, the entire transfer of nitrogen through various organic and inorganic forms makes up the nitrogen cycle.

Sulfur-utilizing chemoautotrophs play a similar role to nitrogen utilizers, but in the sulfur cycle. Many of the steps in sulfur and nitrogen cycling performed by chemoautotrophs are part of anaerobic respiration.

PHOTOHETEROTROPHY

Photoheterotrophs convert light energy to chemical energy, but they cannot convert carbon dioxide to sugar. These microorganisms depend on organic compounds, such as carbohydrates, fats, or alcohols, to supply carbon. Photoheterotrophs differ from photosynthetic microorganisms in three ways, beginning with their inability to use carbon dioxide. Photoheterotrophs also do not produce oxygen as more familiar photosynthetic microorganisms, such as algae, do, and they are thus called *anoxygenic* microorganisms. Third, photoheterotrophs do not use hydrogen from water in their photosynthesis.

Two major groups of bacteria use photoheterotrophy: green nonsulfur bacteria and purple nonsulfur bacteria. Each group of genera gets its name from the type of pigments found in its cells, though they are not necessarily colored green or purple. Neither group requires sulfur compounds for energy.

Many of the green nonsulfur bacteria are also thermophiles, meaning they thrive at very high temperatures. Green nonsulfur bacteria often form mats on top of the alkaline waters of hot springs that reach 176°F (80°C). Some green nonsulfur bacteria also live in more temperate conditions in marine waters or high-salt waters such as salt lakes. In these habitats, green nonsulfur bacteria are facultative anaerobes, meaning they can live even when oxygen becomes depleted. In aerobic conditions, green nonsulfur bacteria shut down their photosynthesis and live as heterotrophs. When oxygen becomes scarce, the bacteria use photosynthesis for generating energy.

Purple nonsulfur bacteria also switch among different types of metabolism, depending on oxygen availability. They live primarily in the mud beneath shallow bogs and ponds, even temporary puddles, where oxygen is scarce but sunlight filters through. Although these bacteria can live aerobically, they enjoy an advantage in these habitats, where other microorganisms cannot tolerate the limited oxygen and nutrients.

PHOTOAUTOTROPHY

Photoautotrophs include the organisms that use carbon dioxide for carbon and light for energy by running photosynthesis. Among microorganisms, these are mainly cyanobacteria, green sulfur bacteria, purple sulfur bacteria, and algae. Because these microorganisms give off oxygen, they are called *oxygenic.*

Cyanobacteria and algae perform the traditional photosynthesis that most biology students recognize.

The general sugar CH_2O formula most often represents glucose, $C_6H_{12}O_6$.

$$CO_2 + H_2O + \text{light} \rightarrow CH_2O + O_2$$

Green sulfur and purple sulfur bacteria have evolved a similar but alternative pathway that dispenses with water:

$$CO_2 + H_2S + \text{light} \rightarrow CH_2O + S$$

Cyanobacteria, algae, and green plants replenish almost all of the earth's atmospheric oxygen. Green sulfur and purple sulfur bacteria, by contrast, produce yellow elemental sulfur, which does not easily degrade. The biologist Betsey Dexter Dyer wrote of the sulfur bacteria, in her 2003 book *A Field Guide to the Bacteria,* "Globules of sulfur often remain attached to the outside of the cells, although in some environments the sulfur is converted to the more soluble sulfate (SO_4), and thus the bacteria are freed from carrying around globules of their own wastes." Sulfate-reducing bacteria use the sulfur released by the green sulfur and purple sulfur bacteria in their own energy metabolism, and this process helps recycle the earth's sulfur.

LITHOTROPHY

Specialized bacteria that live in nutrient-sparse habitats have been nicknamed "rock eaters" because they seem to exist on nothing at all. A lithotroph is any microorganism that derives energy from simple, reduced compounds: CH_4 (methane), CO, Fe^{2+}, H_2, H_2S, Mn^{2+}, NH_4, NO_2, and N_2O. The microorganism may or may not use oxygen in energy metabolism, and it usually relies on carbon dioxide for carbon. Lithotrophs may be either chemoautotrophs or photoautotrophs. Some of these lithotrophs belong to additional specializations. For example, the bacteria that use methane, called methanotrophs, are part of methane metabolism that occurs only in specific anaerobic environments, such as the digestive tract of ruminant animals.

Lithotrophs have been found in places that most other organisms avoid. In 2008, *National Geographic* quoted the Japanese biotechnologist Masaru Tomita's explanation of how industry has put certain lithotrophs to work: "Certain microbes react to metal ions and help copper be leached out of low-quality ore." The ultimate goal is for biotechnology to develop bacteria with enhanced ability to carry out this leaching.

Bacteria that live on or in rock contribute a significant amount to the natural breakdown of rock called *weathering.* The numbers of rock-dwelling bacteria may be small compared with those of bacteria in nutrient-rich habitats, but because these bacteria have little competition for nutrients from other organisms, they persist. By slowly oxidizing the minerals in rock, these microorganisms participate in a process called the *rock cycle,* in which the earth's sediments rise to the surface, weather, sink into the earth's mantle, and eventually reemerge.

Microbial metabolism influences the studies of both the living and the nonliving earth. Scientists have, undoubtedly, not yet discovered all of the unique types of metabolism that may exist in extreme habitats, such as the sediments below the ocean. Metabolism will continue to be pivotal in all aspects of microbiology.

See also BIOGEOCHEMICAL CYCLES; GLYCOLYSIS; HETEROTROPHIC ACTIVITY; METABOLIC PATHWAYS; MICROBIAL ECOLOGY; NITROGEN FIXATION; RUMEN MICROBIOLOGY.

Further Reading

Beijerinck, Martinus W. "Die Bacterien der Papilionaceenknöllchen" (The Root-Nodule Bacteria). *Botanische Zeitung* 46 (1888): 725–804. In *Milestones in Microbiology,* translated by Thomas Brock. Washington, D.C.: American Society for Microbiology Press, 1961.

Dyer, Betsey Dexter. *A Field Guide to the Bacteria.* Ithaca, N.Y.: Cornell University Press, 2003.

Hornyak, Tim. "Rock-Eating Bacteria 'Mine' Valuable Metals." *National Geographic,* 5 November 2008. Available online. URL: http://news.nationalgeographic.com/news/2008/11/081105-bacteria-mining.html. Accessed September 21, 2009.

Kim, Byung Hong, and Geoffrey M. Gadd. *Bacterial Physiology and Metabolism.* Cambridge: Cambridge University Press, 2008.

Winogradsky, Sergei. "Recherches Physiologique sur les Sulfobactéries" (Physiological Studies on the Sulfur Bacteria). *Annales des l'Institut Pasteur* 3 (1889): 49–60. In *Milestones in Microbiology,* translated by Thomas Brock. Washington, D.C.: American Society for Microbiology Press, 1961.

methanogen Methanogens are archaea or bacteria that produce methane gas (CH_4) as part of their normal metabolism. Methanogenic prokaryotes use a type of anaerobic respiration in which a limited number of compounds accept electrons produced by the cell during energy metabolism and release methane as the end product. The main compounds that serve as the electron acceptors are carbon dioxide (CO_2), formate (HCOOH), methanol (CH_3OH), and acetate (CH_3CO_2). For example, methanogenesis from carbon dioxide produces one methane molecule per carbon dioxide molecule.

$$CO_2 + 4\ H_2 \rightarrow CH_4 + 2\ H_2O$$

Methanogens are strict anaerobes that cannot survive, for more than a few minutes, in the presence of oxygen. Archaea make up most of the known methanogens, comprising five orders (Methanobacteriales, Methanococcales, Methanomicrobiales, Methanosarcinales, and Methanopyrales) and almost 30 genera, all in phylum Euryarchaeota. These five orders are not made exclusively of methanogens, however, and contain some species that consume methane (methanotrophs) instead of emitting it.

Methanogenic archaea differ from bacteria in their cell wall composition and metabolism associated with extreme environments. Methanogen cell walls substitute the compound pseudomurein or a protein for the bacterial polymer peptidoglycan, and the methanogen's cell wall membrane consists of lipids, polysaccharides, and proteins that differ in structure from bacterial compounds. Methanogens also use various, unique auxiliary molecules, called cofactors, for enzyme activity, which differ from cofactors in nonmethanogens. Methanogens consist of both gram-positive and gram-negative species, shown in the table below.

All methanogens can be considered extremophiles, organisms that require conditions where other organisms cannot live, because they live in anaerobic habitats. The order Methanopyrales also possesses the ability to live at extremely high temperatures, with an optimal temperature of about 208°F (98°C). The single genus in this order, *Methanopyrus,* is thought to be the most ancient methanogen, on the basis of analysis of ribosomal ribonucleic acid called 16S rRNA. Methanogenic archaea are believed to be some of the oldest prokaryotes; *Methanopyrus* exists in harsh conditions that mimic those of early Earth.

Methanogens live in a diverse range of environments, alike only in how difficult they are to reach or sample. The main methanogenic environments are:

- deep sediments under water, usually in swamps, lagoons, bogs, and so on
- waterlogged soils, dead tree trunks or roots
- rice paddies
- sewage and manure
- wastewater treatment plants (anaerobic digester tanks)
- beneath landfills
- natural gas leaks
- animal and insect digestive tracts

Ruminant animals (cattle, sheep, goats, etc.) and insects possess diverse populations of methanogens that help in digestion. In the large four-compartment stomach of ruminant animals, methanogens play an important role in energy metabolism of the organ's microbial population. Methanogens that share habitat in the double compartment called the reticulorumen remove carbon dioxide, formate, and acetate from the contents. These compounds are the end products made by other reticulorumen bacteria in the digestion of dietary fibers and polysaccharides. Methanogens play a critical role by removing these end products, so that the overall metabolism can continue. Without methanogens, the ruminant animal would not be able to digest feeds. The methane produced in the reticulorumen is emitted by the animal by belching. A cow can belch 50–110 gallons (189–416 l) of methane a day.

The relationship between methanogens and the energy metabolism of other bacteria mimics the ruminant system. In all cases, the methanogen makes metabolism energetically favorable in environments rich in organic matter by constantly preventing the buildup of metabolic end products.

Together, Earth's methanogens produce a large volume of methane that drifts into the atmosphere. This is an ecological problem, because methane is a greenhouse gas. As a greenhouse gas, methane pro-

Methanogens

Order	Morphology	Cell Wall	Electron Acceptors
Methanobacteriales	long rods	pseudomurein	$H_2 + CO_2$ or formate
Methanococcales	irregular cocci (round)	protein	$H_2 + CO_2$ or formate
Methanomicrobiales	rods, cocci, and spirilla (helical)	protein	$H_2 + CO_2$ or formate
Methanosarcinales	irregular cocci	protein or polysaccharide	$H_2 + CO_2$, methanol, or acetate
Methanopyrales	cocci in chains	protein	$H_2 + CO_2$

vides a benefit by absorbing infrared light and warming the planet with other greenhouse gases. But since the Industrial Revolution, the composition of the atmosphere and greenhouse gases has been changing. This problem, known as *global warming,* has been due, in part, to methane emissions. The table on page 510 lists the major greenhouse gases in approximate order of concentration in the atmosphere. (Water vapor is sometimes also considered a greenhouse gas.)

The University of British Columbia physicist Michael Pidwirny has stated in his Web site PhysicalGeography.net, "Methane is a very strong greenhouse gas. Since 1750, methane concentrations in the atmosphere have increased by more than 150 percent. The primary sources for the additional methane added to the atmosphere (in order of importance) are: rice cultivation; domestic grazing animals; termites; landfills; coal mining; and, oil and gas extraction." Rice cultivation, grazing animals, termites, and landfills emit microbially produced methane. The preceding list does not include all of the ruminant wildlife that also depend on active microbial populations for digestion.

Some industries have made attempts to harness the excess methane produced in their operations. An increasing number of wastewater treatment plants collect methane from anaerobic digestion tanks, where organic matter decomposes, and use the methane in the same way buildings use natural gas for heating the plant's offices. Landfills have begun exploring similar technology. Microbially produced methane that can be converted to energy is called *biogas.* A small number of communities also have pursued the collection of methane that rises from manure produced at livestock farms and feedlots. These communities plan to convert the methane to electricity.

Microbial methanogenesis relates to Earth's energetics in ways that have been studied in detail, such as rumen microbiology, and in ways that remain largely unknown, such as the methanogens living in deep sediments. Regardless of the number of methanogens that remain undiscovered, they all use the same chemistry in their metabolism.

See also ANAEROBE; EXTREMOPHILE; RUMEN MICROBIOLOGY; WASTEWATER TREATMENT.

Further Reading

Miller, G. Tyler. *Environmental Science: Working with the Earth.* Belmont, Calif.: Thomson Learning, 2006.

Mullin, Rick. "Microbial Methane Farming." *Chemical and Engineering News,* 20 November 2006. Available online. URL: http://pubs.acs.org/cen/coverstory/84/8447cover4a.html. Accessed January 24, 2010.

Pidwirny, Michael. "Introduction to the Atmosphere. In *Fundamentals of Physical Geography,* 2nd ed. Kelowna, Canada: University of British Columbia Okanagan, 2009. Available online. URL: www.physicalgeography.net/fundamentals/7a.html. Accessed January 28, 2010.

Wall, Judy, Caroline S. Harwood, and Arnold L. Demain. *Bioenergy.* Washington, D.C.: American Society for Microbiology Press, 2008.

microbial community A microbial community is a defined population of different types of microorganisms living in association with one another. The association may take many forms, such as nutrient sharing, protection from the environment, or establishing a habitat. Communal associations among microorganisms usually provide benefits that the cells would not receive if living as individuals, called *planktonic* cells.

Microbial communities create conditions by which microorganisms can withstand stresses in the environment. Some species in the community perform activities that benefit the entire community, as well as themselves. Other species appear to benefit from the community but do not contribute to it, at least in ways that microbiologists have so far discovered. In the broadest sense, a microbial community helps all the members exploit the environment for all the members' survival.

In biology, a community is one component of the overall ecological organization. Ecologists define different levels of organization within Earth's biota to describe the roles various species have in Earth ecology. The main levels, from the least complex to the most complex, of biological organization are as follows:

- individual cells—the basic unit of all living things

Greenhouse Gases

Gas	Average Time in Atmosphere	Global Warming Potential Relative to CO_2
carbon dioxide (CO_2)	100–120 years	1
methane	12–18 years	23
nitrous oxide	115–120 years	296
chlorofluorocarbons	10–20 years	900–8,300
hydrochlorofluorocarbons	10–390 years	470–3,000
halons (fire extinguishers)	65 years	5,500
carbon tetrachloride (cleaning solvent)	42 years	1,400

- population—group of interacting individuals of the same species occupying a specific area
- community—all the populations of different species living in and interacting in an area
- ecosystem—one or more communities interacting with the physical features in an area
- biosphere—all Earth ecosystems

In microbiology, few species live by themselves in nature, and the terms *population, community,* and *ecosystem* have been used interchangeably. For example, the microorganisms that live in soil are often described as soil populations, soil communities, or the soil ecosystem. By the stricter definitions described here, soil microorganisms can be viewed in the following context:

- soil population—*Nitrobacter* species that participate in one of the major steps in nitrogen cycling, called nitrification
- soil community—nitrogen-fixing, nitrifying, nitrate-reducing, and denitrifying genera that operate the nitrogen cycle reactions in soil
- soil ecosystem—nitrogen cycle microorganisms; plants that absorb nitrogen; soil, water, and organic and inorganic nutrients; and soil inanimate matter that affects the chemical and physical conditions in the habitat (the physical place where an organism lives)

Microbial ecologists study all of these aspects to gain information on the roles that bacteria, fungi, algae, and protozoa play in the health of the biosphere. Examples of communities studied in microbial ecology are the following:

- photosynthetic bacteria and algae in a lake
- anaerobic bacteria and protozoa that digest fiber in the rumen (the stomach in ruminant animals)
- sulfur-metabolizing bacteria in microbial mats
- bacteria, yeasts, and fungi of human skin
- marine microorganisms living in the uppermost layer (photic zone) of the ocean
- bacteria that make up dental plaque
- biofilms that coat the interior of water distribution pipes

Because of the diversity of the world's microbial communities, it is not surprising that communities develop in their own distinctive ways. For example, a soil community contends with conditions of temperature extremes, changing water availability, and similar environmental factors. A community living in an animal's digestive tract enjoys constant environmental conditions but must react to variable influxes of food, potential changes in the animal's health, possible antibiotics or other drugs, and an escape route that takes a microorganism to an entirely different environment.

Populations within a community can also interact in distinct ways. Two main manners of interaction are synecology and autecology. Synecology is a situation in which populations in a community interact with each other but usually do not interact with other communities, such as the anaerobic bacteria and protozoa that inhabit a cow's rumen. Autecology is a situation in which the community has been defined more by the microorganisms' interactions with outside factors such as physical features in the environment (temperature, moisture, pH, oxygen levels, etc.).

Populations in a community can also be defined in terms of their history in the community and usually belong to the three following groups:

- autochthonous populations—native or indigenous to the community
- allochthonous—alien to the community
- zymogenous—originally alien, but adapted to the community's conditions

Unnatural ecosystems do not develop on Earth; nor do unnatural communities. Microorganisms or higher organisms derive no benefit from living in a place that offers no advantages at all to their survival. Species must vacate inhospitable places; otherwise, they will become extinct. Therefore, an organism benefits by having a job within a community. This serves the following two purposes: (1) having a job in a community allows a population to mold the community to its advantage, and (2) possessing a valued skill makes a population less likely to be destroyed by other members of the community. For example, species of the bacterial genus *Pseudomonas* have an important role in biofilm communities because this genus excretes large polymer compounds that form a glycocalyx. This material is sticky and binds the community together and to an underlying surface, while collecting nutrients from the water that flows over and through the biofilm. All members of a biofilm community would find survival much more difficult without *Pseudomonas*'s contribution.

The role of biofilm *Pseudomonas* provides an example of a term that is often misused in the biology: *niche*. A niche is the way of life of an organism in regard to how it acquires resources. To *occupy a niche* is to play a role in a community. (Students often incorrectly define a niche as a place where an organism lives. The habitat is the place where an organism lives.) Biofilm *Pseudomonas*'s niche is as a producer of the biofilm's physical matrix, that is, the glycocalyx.

Resource use is as crucial in microbial communities as it is in the communities of higher plants and animals. Other than sunlight, Earth's resources are finite. The manner in which communities share resources determines the success of species, populations, and the community itself. From a microorganism's point of view, resources fall into different categories as described in the table below.

Communities contain diverse species and might receive a rich resource supply or a scarce resource supply that community members must share. Important resources in microbial growth are nutrients, water, space, light, and air. Within communities, individual species with similar metabolism might enter into an antagonistic relationships in their fight to survive. Common antagonistic, called negative, relationships are:

- amensalism—one microorganism possesses an activity that inhibits others, such as antibiotic secretion
- competition—two microorganisms struggle for advantage for a resource
- parasitism—one microorganism benefits by harming a second organism
- predation—one microorganism eats another, thereby getting all the resources that the prey had acquired for its own

These relationships are also forms of symbiosis, in which organisms live in close association with each other. Microorganisms can adapt positive relationships with each other for survival, rather than the negative relationships described. Positive symbiosis benefits one or both microorganisms. In a community, a preponderance of positive relationships over negative ones gives the entire community a better chance for long-term survival.

Communities operate at their best when members have a communication system. In microbiology, cells use different forms of taxis, or "feeling" the conditions around them. For example, chemotaxis is a process by which microorganisms detect and react to compounds in their environment. Chemotaxis helps microorganisms in a community move toward stimuli that provide benefits, such as nutrients, and avoid stimuli that are harmful, such as antibiotics.

Microbial communities also rely on quorum sensing to monitor the conditions around them. Quorum sensing is used, for example, for detecting the community's density. If the cell density in a community rises to a certain threshold, a microorganism can slow its met-

Resource Use in Microbial Communities

Type of Resource	Description
antagonistic	two resources that, if consumed together, have a toxic effect, so for survival, a microorganism consumes only one
complementary	two resources contain different proportions of two nutrients, so a microorganism grows faster by consuming both resources rather than taking in large amounts of just one
essential	absolutely necessary for survival
hemiessential	one resource (resource A) supplies all essential nutrients, and a second (resource B) supplies all nutrients except one; a microorganism can use both as long as resource A is always present in levels sufficient to supply the essential nutrient
switching	microorganism selects the one of two resources that yields the highest growth rate; when this resource is used up, the microorganism can switch to the second resource and grow at a slower rate
substitutable	two resources can substitute for each other with equal effect

abolic rate, produce antimicrobial substances, move, or take other similar actions to assure its survival.

Microbial communities create complex study models that can require years of study to be fully understood. All communities have an overriding theme: survival. By studying communities, microbiologists learn about natural ecosystems, the activity of communities most important to human well-being, and microbial associations that contribute to infection and disease.

See also BIOFILM; CHEMOTAXIS; MICROBIAL ECOLOGY; RUMEN MICROBIOLOGY; SYMBIOSIS.

Further Reading

Jjemba, Patrick K. *Environmental Microbiology: Principles and Applications.* Enfield, N.J.: Science Publishers, 2004.

Miller, G. Tyler. *Environmental Science: Working with the Earth.* Belmont, Calif.: Thomson Learning, 2006.

Pacific Northwest National Laboratory. "Microbial Communities Initiative." Available online. URL: www.pnl.gov/biology/research/mci. Accessed January 27, 2010

microbial ecology Microbial ecology is the study of interactions of microorganisms with their environment and the manner in which these interactions define conditions on Earth. This discipline is related to the newer science of environmental microbiology, which studies the specifics of microbial diversity in the environment. As biologists learn more about the connections among plants, animals, microorganisms, and nonliving things, microbial ecology and environmental microbiology may become indistinguishable.

Microbial ecology began in earnest with the research of two biologists working independently to explain the role of soil bacteria in plant health. The Dutch botanist Martinus Beijerinck (1851–1931) investigated the activities of nitrogen-fixing bacteria in root nodules of legume plants. Beijerinck's studies, in the late 1800s, led to understanding of nitrogen fixation, the process of absorbing nitrogen from the atmosphere. Inside the nodules, the bacteria begin a series of steps that convert nitrogen gas to a form that the legume (beans, peas, alfalfa) can use. To recover the soil bacteria, Beijerinck formulated a medium that favored the growth of certain environmental bacteria over others. In 1901, he wrote, "The enrichment culture experiment makes it possible to isolate a large variety of microorganisms which are adapted to particular environmental conditions, and to bring them into development side by side in liquid culture media." Beijerinck saw the value of enrichment for studying medically important pathogens, in addition to soil bacteria. Martinus Beijerinck had invented enrichment medium, which is an important technique in today's microbial ecology studies.

At the same time as Beijerinck's work, the Russian microbiologist Sergei Winogradsky (1856–1953) examined processes by which soil bacteria derived energy from inorganic compounds. Winogradsky proposed that bacteria contribute to nitrogen, sulfur, and iron chemistry in soil. He studied many of the same biochemical reactions as Beirjerinck and developed methods for recovering difficult-to-grow anaerobes from soil habitats.

Beijerinck and Winogradsky had invented a new field in microbiology concerned with microbial actions on Earth's elements. Scientists now refer to the movement of elements through plants, animals, and nonliving things as biogeochemical cycles. Research that followed the work of Beijerinck and Winogradsky led to current understanding of carbon, nitrogen, sulfur, phosphorus, iron, and oxygen cycling. Microorganisms are also now known to participate in mineral cycles, in addition to their actions on iron, which Winogradsky first described. Soil bacteria contribute to the following mineral cycles: antimony, arsenic, copper, germanium, lead, mercury, nickel, and selenium.

COMPONENTS OF MICROBIAL ECOLOGY

Microbial ecology begins with an understanding of the ecology inside a microorganism. Microorganisms contain the same elements as the cells of higher organisms. To make cellular constituents, microorganisms use carbon, nitrogen, sulfur, phosphorus, oxygen or carbon dioxide, water, nonmetals such as calcium and sodium, and metals. While bacteria incorporate these elements into cellular substances, they also draw the elements into a food chain, an essential part of nutrient cycling (see the color insert, page C-5, for a community involved in nutrient cycling). A food chain is a series of organisms that transfer nutrients and energy from prey to predators. The organisms, at each successive step up in a food chain, become more complex and usually bigger than those in the level below. Microorganisms represent the smallest organisms in food chains. Biogeochemical cycles, thus, support food chains by drawing in new nutrients—the microorganisms' role—and food chains support cycles by converting the nutrients' chemical form.

Microbial ecology also includes biological microenvironments and macroenvironments. A microenvironment is a physical location defined by specialized conditions. For example, deep sediments that contain few organic nutrients and receive little oxygen and no light represent a microenvironment. The macroenvironment is a defined area that provides certain constant conditions to organisms living in it. The ocean might be considered a macroenvironment that supplies a fairly constant range of temperatures, sunlight, nutrients, and salt. Microbial ecologists study microenvironments and macroenvironments to learn the factors that contribute to nutrient and energy transfer to and from microorganisms.

Another important component of microbial ecology relates to populations and communities. Because most microorganisms exist in nature in a community, only by studying communities can microbiologists discover how bigger ecosystems work. Microorganisms in a community work together in a way that gives the community abilities that individual species would not possess on their own. For example, a biofilm is a community of mixed bacteria attached to a submerged surface, usually below a flowing liquid. The community members become embedded in polysaccharides excreted by some of the bacteria, plus other materials that stick to the film. The conglomeration of materials protects all the community members from toxic chemicals, temperature changes, and predator organisms.

Biofilms are important in microbial ecology because they change the nutrient levels in water and may also affect pollutants, wastes, or pathogens that approach the film. Other communities degrade decaying organic matter (saprophytic organisms that live on dead matter), transfer gases into and out of the atmosphere (microbial mats), or control the chemical features of the surroundings (soil communities).

Finally, microbial ecology mimics Earth ecology in terms of biodiversity. Microbial ecologists evaluate the diversity of microbial species in the environment to learn about the prevalent type of metabolism in a habitat. For example, a microbial ecologist takes a sample of pond water and analyzes the types and groups of bacteria in the sample. A predominance of aerobic microorganisms suggests that the pond contains photosynthetic organisms that probably feed food chains—a healthy aquatic ecosystem. A predominance of anaerobic microorganisms suggests that the pond has decaying materials that provide nutrients, but that metabolism is limited and probably lacks photosynthesis and aerobic respiration—an unhealthy ecosystem.

Genetic diversity refers to the number and variety of certain genes in a microbial population instead of the diversity species. Before the introduction of genetic analysis methods, microbiologists studied genotype by observing phenotype. Genotype is the genetic makeup that characterizes an organism; phenotype is the outward expression of the genotype. The esteemed microbial ecologist Thomas Brock (1926–) wrote in *Microbial Ecology,* in 1966, "It is not possible to examine an organism directly for the presence of a given gene and all that can be observed is the phenotype; the genotype can be inferred only from the results of breeding studies." In 2009, a brochure from the J. Craig Venter Institute (JCVI), a research organization that studies microbial genes, "Until the advent of genomics tools we were unaware of just how prevalent and important these organisms were to life on our planet. . . . Since our sequencing of the first free-living bacterium, *Haemophilus influenzae,* [institute] scientists have been leading the field of microbial genomics. Today, we've sequenced more than 100 organisms." Microbial ecologists now plan projects for analyzing all the genes in microbial ecosystems, even an ecosystem as vast as the ocean.

Genomics, the analysis of an organism's entire genetic makeup, helps answer three main questions regarding Earth's microorganisms: (1) ubiquity, (2) abundance, and (3) metabolic activity. To answer these questions, microbial ecologists take samples from a diversity of environments. No single microbiologist can determine all of the jobs that a microorganism performs in an ecosystem, but years of accumulated research begins to draw a picture of how ecosystems work.

STUDYING MICROBIAL ECOSYSTEMS

Laboratory studies in microbial ecology begin with enrichment cultures that selectively favor certain species. Enrichment helps separate species from others in the sample. Microbiologists also enrich media to study how certain isolated microorganisms react to substances from an environment. If a pure bacterium grows well in medium enriched with an extract from decaying trees, for instance, then the microbiologist concludes that this bacterium plays a role in organic fiber digestion.

Microbiologists conduct in vitro testing of whole environmental samples in specialized apparatuses for measuring metabolism in microbial communities. Measurements such as cell mass, carbon dioxide production, and oxygen absorption provide clues to microbial communities' metabolic activity. One test, called biological oxygen demand (BOD), equates metabolic activity with the amount of oxygen a microbial culture needs to sustain growth. BOD provides a rough measurement of activity; in concert

Microbial ecology studies the interactions of microorganisms in distinct environments, such as this hot spring in Yellowstone National Park, with overall Earth ecology. ***(Mila Zinkova)***

The Dutch microbiologist Martinus Beijerinck increased microbiology's knowledge of soil communities by studying nitrogen and sulfur metabolism. ***(Delft School of Microbiology Archives)***

with other studies, this test helps determine respiration activity in a given habitat.

Genetic studies of the environment have superseded many whole-cell tests such as BOD. By determining the presence of specific known genes, a microbiologist learns about exact biochemical reactions taking place in an ecosystem. Molecular biologists have developed gene probes, or biosensors, that hold a piece of genetic material containing a specific set of genes. The probes react only with complementary nucleic acids in the environment. This technology takes advantage of the fact that deoxyribonucleic acid (DNA) separates into single strands from its normal double-stranded structure when heated. The strands recombine on their own by finding complementary strands. Biosensors use single strands to detect the presence of complementary strands in a water or soil sample. As a result, the biosensor signals the presence of specific microorganisms or specific genes. Microbial ecologists can detect the presence of genes unique to nitrogen cycling, sulfur cycling, or other actions. Biosensors, usually, contain a fluorescent dye that gives a visible signal when a reaction between probe and sample has occurred. Extensive analysis of genes present in the environment, such as the studies at JCVI, makes biosensors possible.

The microbiologist Moselio Schaechter once described the influence of microorganisms on ecosystems: "Microbes may be small, but they do things on a grand scale. They have a major impact on forming and setting the concentrations of the major gases in the atmosphere—nitrogen, oxygen, and carbon dioxide." That and the numerous other contributions in nutrient cycling, waste degradation, enzymatic reactions, nutrient sequestration, weathering of rocks, and facilitating food digestion make microorganisms part of Earth sciences. Microbial ecology condenses a tremendous amount of scientific data to support the principles of modern biology.

See also BIOFILM; BIOGEOCHEMICAL CYCLES; BIOLOGICAL OXYGEN DEMAND; BIOSENSOR; ENVIRONMENTAL MICROBIOLOGY; GENOMICS; MICROBIAL COMMUNITY; MICROENVIRONMENT.

Further Reading

Atlas, Ronald M., and Richard Bartha. *Microbial Ecology: Fundamentals and Applications,* 4th ed. San Francisco: Benjamin Cummings, 1997.

Beijerinck, Martinus W. "Anhäufungsversuche mit Ureumbakterien" (Enrichment Culture Studies with Urea Bacteria). *Centralblatt f. Bakteriologie* 7 (1901): 33–61. In *Milestones in Microbiology,* translated by Thomas Brock. Washington, D.C.: American Society for Microbiology Press, 1961.

Brock, Thomas D. *Principles of Microbial Ecology.* Englewood Cliffs, N.J.: Prentice-Hall, 1966.

International Society for Microbial Ecology. Available online. URL: www.isme-microbes.org. Accessed January 10, 2010.

J. Craig Venter Institute. "Microbial and Environmental Genomics: Overview." Available online. URL: www.jcvi.org/cms/research/groups/microbial-environmental-genomics. January 10, 2010.

Maier, Raina M., Ian L. Pepper, and Charles P. Gerba, eds. *Environmental Microbiology,* 2nd ed. San Diego: Elsevier, 2009.

Schaechter, Moselio, John L. Ingraham, and Frederick C. Neidhardt. *Microbe.* Washington, D.C.: American Society for Microbiology Press, 2006.

microbiology Microbiology is the study of microscopic organisms, or microorganisms, and their interactions with each other and with their environment. Microorganisms, also called *microbes,* are any living things too small to be seen with the unaided eye. For that reason, microbiology could not have developed without the invention of the microscope, and today's microbiology owes its existence to technologically advanced microscopes.

Microbiology involves studies of bacteria, fungi and yeasts, protozoa, viruses, and microscopic parasites. Most microorganisms have the capability to live independently in nature. Microorganisms that cannot live outside another living organism, for all or part of their life cycle, are called *parasites*. Some bacteria, fungi, and protozoa survive only as parasites; these are called *obligate parasites*. Viruses are obligate parasites that have, during their evolution, shed all mechanisms that would allow them to live as independent entities. Viruses reproduce only when they infect living cells, and, for this reason, many microbiologists consider viruses to be unique in microbiology because they are nonliving things.

Modern microbiology relates to various disciplines in medicine, industry, and academic research. The following list describes the relationships between major disciplines in science and microbiology:

- biochemistry—microbial metabolism for energy generation and synthesis of cellular constituents
- ecology—interrelationships of microorganisms with the living and nonliving things in the environment
- epidemiology—the study of the origin and transmission of an infectious disease through a population
- genetics—mechanisms for adaptation in microorganisms and passage of traits from one generation to the next
- immunology—the body's mechanisms for defending against infection from microorganisms
- molecular biology—the processes by which genetic material replicates and genes control life in microorganisms

Microbiology also contributes to a variety of industries. Some industrial activities that use microorganisms date to antiquity, while other industries have developed only in the past few decades. The roles of microbiology in industry are the following:

- bioremediation—the use of microorganisms to clean up pollution
- biotechnology—development of new products from bioengineered microorganisms
- consumer products—preservation of products from microbial contamination
- food—the use of microorganisms to produce foods, beverages, or food additives and to preservation of foods from spoilage by microorganisms
- medicine—development of protocols for stopping and preventing infectious disease
- pathology—the study of the progression and consequences of an infectious disease on the body's tissues
- pharmaceuticals—development of new drugs for killing infectious agents and new therapies and vaccines produced by microorganisms
- wastewater treatment—processes for removing harmful microorganisms from wastewaters
- water quality—processes for removing harmful microorganisms from drinking water and recreational waters, and drinking water disinfection

Academic research in microbiology contains specialties that deal with many of the topics in the preceding list. Microbiology can also be divided into specialized focus areas that concentrate on one type of microorganism, such as the following:

- algology—the study of algae
- bacteriology—the study of bacteria
- microscopy—the discipline of producing the best possible image of a magnified microorganism
- mycology—the study of fungi
- protozoology—the study of protozoa
- virology—the study of viruses
- parasitology—the study of the life cycles of microscopic parasites, ranging from microorganisms to helminths (worms)
- chemotherapy—the study and development of chemicals that inhibit infectious agents

Each specialized discipline allows microbiologists to pursue subspecialties, such as enzymology, genomics, identification, structure and function, systematics, and taxonomy. Systematics is the theoretical and practical classification of organisms. Taxonomy is a similar discipline, in which microbiologists assign biological organisms to groups based on similarities and differences in the organisms. Until the emergence of molecular biology in the 1980s, microbiologists classified microorganisms on the basis of cell and colony morphology, structures on the microbial cell surface, and the nutrients microorganisms used and the end products they excreted. Microbiology still uses these tools, but it has expanded into studies of the entire genetic makeup of microorganisms, a field called *genomics*.

An emerging area in microbiology relates to the development of renewable energies from the activities of microorganisms. Microbiologists, now, grow algae and bacteria to produce gases, such as methane and hydrogen that can be used in energy production. Microbiology has also investigated the utilization of photosynthetic microorganisms to help convert the sun's energy to usable energy for humans, such as electricity. Microbiology, therefore, plays a role in the young biofuel industry.

MICROBIOLOGY'S HISTORY

Microbiology dates only to the 17th century. In 1665, the English scholar Robert Hooke (1635–1703) used a rudimentary microscope assembled from lenses arranged in sequence to observe the microscopic world. Hooke became the first person to report that living things were made of small, basic structural units he called "cells." This finding marked the beginning of cell theory, which provides a foundation for all of the biological sciences. Hooke's contemporary, the Dutch cloth merchant Antoni van Leeuwenhoek (1632–1723), had also built a microscope for viewing flaws in threads. He soon looked at other objects from nature and has been credited as the first person to study microbial cells in detail. This finding of "little animalcules," as van Leeuwenhoek called them, began the science of microbiology.

With van Leeuwenhoek's discovery of a microscopic world that had been previously unknown, scientists and philosophers began to debate the theory of spontaneous generation. This prevailing theory of the time stated that life arose spontaneously from nonliving matter. The Italian physician Francisco Redi (1626–97) strongly opposed the idea that life arose spontaneously from inanimate matter. In 1668, Redi performed a study that showed that maggots did not rise spontaneously from decaying matter (decaying meat) and that only flies laying eggs on the meat could give rise to maggots.

Scientists took opposing sides on the origin of new life on Earth. In 1745, the English clergyman John Needham (1713–81) conducted an experiment to prove the veracity of spontaneous generation. Needham cooked and then cooled chicken and vegetable broths to rid them of life, what he called a "vital force," then left them undisturbed, until they were dense with microorganisms. Needham concluded that the microorganisms had arisen out of the sterile broth. The Italian scientist Lazzaro Spallanzini (1729–99) refuted this proposal, by repeating Needham's experiment with a major difference: He closed some of the broth vessels with a lid. Closed containers contained no microbial growth, and open containers similar to Needham's quickly became contaminated.

The scientific community would remain split for the next century, either in support of or in opposition to, spontaneous generation. In 1858, the German pathologist Rudolf Virchow (1821–1902) proposed that living cells could only be produced by other living cells, the concept called *biogenesis*. Biogenesis remained only a theory, until the mid- to late 1800s, when the French biologist Louis Pasteur (1822–95) conducted a series of experiments that persuasively showed that although microorganisms could be present on nonliving matter, they did not spontaneously grow out of nonliving matter. Pasteur had already gained a high level of esteem in science, and his experiments were so simple and clear that he convinced most scientists that spontaneous generation was not worthy of further argument.

Pasteur's work and that of several contemporaries launched a period called the *Golden Age of Microbiology*. This period, from about 1865 to the early 1900s, witnessed advances in the development of the microscope, which greatly aided microbiology experiments as well as studies on the relationships between microorganisms and disease. Several scientists contributed to the development of a new concept, the germ theory of disease, in which microorganisms were associated with many different infectious diseases. Pasteur and the Italian scientist Agostino Bassi (1773–1856), independently, showed that microorganisms could be linked to disease. In the 1860s, the English surgeon Joseph Lister (1827–1912) compiled evidence showing that preventing contact of microorganisms with surgery patients dramatically improved the patients' chance of surviving.

In 1876, the German physician Robert Koch (1843–1910) designed a study to show that a specific microorganism was the cause of a specific disease. Koch based his experiment on a list of criteria for proving a disease's cause and effect developed by his mentor, Jacob Henle (1809–85). Koch built upon Henle's criteria for proving the cause of an infectious disease. The criteria required for linking a microorganism with a disease would become known as Koch's postulates. Shortly after Koch demonstrated his postulates, using anthrax spores on mice, Pasteur developed studies to confirm Koch's results. Studies of individual infectious diseases and their causes would progress thereafter. The work of Koch, Lister, and Pasteur, collectively, led to the germ theory of disease, the idea that microorganisms could cause disease.

Laboratory studies in microbiology took a critical step forward with three developments: (1) the Gram stain technique, (2) the use of agar in nutrient mixtures for growing microorganisms, and (3) the petri dish. In 1884, the Danish microbiologist Hans Christian Gram (1853–1938) developed a technique of staining bacteria that divided bacteria into two major groups, now called gram-positive bacteria and gram-negative bacteria, based on the ability of the cells to retain crystal violet dye inside their cell walls. At about the same time, Walther Hesse (1846–1911) and Fannie Hesse (1850–1934) adapted agar, a gelatinlike material used in kitchen recipes, to microbiological media to create a solid surface for growing and observing bacterial colonies. Meanwhile, one of Koch's assistants, R. J. Petri (1852–1921), designed a shallow stackable dish for holding agar media. Each of these three developments is used today in all aspects of bacteriology.

Microbiology grew in various directions toward the end of its golden age. In 1884, Pasteur's associate Charles Chamberland (1851–1908) discovered viruses. Pasteur himself expanded on the introduction of a vaccine against smallpox, in 1798, by the English physician Edward Jenner (1749–1823) by developing vaccines against the anthrax bacterium and the rabies virus. Vaccine development and immunology continued, from this period on. In the 1940s, the Scottish biologist Alexander Fleming (1881–1955) led microbiology into a new age of antibiotics that stopped bacterial infections when he published his findings on penicillin, a substance he extracted from molds, which killed staphylococci.

Modern microbiology may be characterized by three main areas of interest: (1) molecular studies of the microbial genome, (2) the development of antibiotic-resistant microorganisms, and (3) the relationship between microorganisms and the environment. Molecular studies consists of the identification of deoxyribonucleic acid (DNA) structure, the sequencing of gene base units, and the transfer of genes from one microorganism into another, which is called recombinant DNA technology. The new science of genomics incorporates all of these aspects. The problem of new antibiotic-resistant microorganisms also involves studies of genes. This includes the genes that control antibiotic resistance and the methods by which resistance transfers from one microorganism to another. Genomics also contributes to microbial ecology, by investigating the traits of certain microorganisms that carry out specific actions in the environment, including the degradation of pollutants.

Microbiology will, soon, develop the means to manipulate genes and the potential benefits of microorganisms in the following areas: environmental rehabilitation, renewable energy, and new disease therapies. Appendix I lists the important advances in the history of microbiology that have led to today's environmental and medical breakthroughs using microorganisms. Microbiologists will, additionally, be involved in the emergence of new infectious diseases and reemergence of diseases thought to be under control. Changes in the mobility and habits of society, combined with environmental factors, will impact the study of microorganisms in the near-future.

WORKING WITH MICROORGANISMS

Most microorganisms have no effect on humans or benefit humans. Pathogens, the microorganisms that cause infection and disease, make up a very small proportion of all microorganisms. Some normally harmless microorganisms can, however, infect living things if the opportunity arises. Opportunity for infection may increase as a result of weakened resistance against infection. Microbiologists must, therefore, treat every microorganism with care to prevent being infected by pathogens or by the benign microorganisms called *opportunists* that become dangerous under certain conditions.

Several activities within microbiology have developed for the purpose of protecting against contamination, the presence of unwanted microorganisms in any material, and infection. Microbiologists categorize their activities in terms of the potential danger of the microorganism being studied. These categories are called *biosafety levels*. In microbiology, all microorganisms must be treated as potential hazards, that is, biological materials that can harm the body. Microbiologists work with biohazards, hazards that are of biological origin, by using three complementary processes:

- aseptic techniques—activities that reduce the chance of contamination by unwanted microorganisms
- sterilization—the elimination of all microbial life in a material
- containment—processes for preventing the escape of microorganisms into the surroundings or the environment

Microbiologists follow these three procedures for two purposes: to protect themselves from infection and to protect others and the environment from dangerous microorganisms. Simple techniques such as donning protective clothing, disinfecting surfaces after working with microbial cultures, and decontaminating discarded cultures by sterilization reduce the risks to microbiologists and others. A special technology within microbiology, containment, also plays a critical role in protecting people from infection by pathogens.

BIOSAFETY

Biosafety involves all procedures that prevent people from being harmed by a biohazard. One of the most effective ways of ensuring biosafety, other than killing microorganisms, is containment. Containment encompasses all physical barriers that prevent microorganisms from escaping into the environment. Containment begins with one or more laboratories that a facility dedicates exclusively to work with microorganisms and other unrelated activities. Such microbiology laboratories must keep doors closed and restrict entry to only individuals who work there.

Most microbiology laboratories manage airflow to reduce the chance of microorganisms' escaping. Airflow may be adjusted, so that the microbiology laboratory interior has a slightly negative air pressure relative to the exterior. Air flows into the laboratory and not out, thus keeping airborne microorganisms inside the laboratory. Microbiology laboratories also equip airflow units with filters that remove particles from the air. Microbiologists select these filters on the basis of the type of microorganisms that they work with in the laboratory. The types of microorganisms, which are classified by their risk of harming people, and the types of laboratories constructed for working with these microorganisms belong to categories called *biosafety levels* (BSLs).

All of the four levels recognized in microbiology have specific requirements, based on the health risks of the microorganisms being handled, from biosafety level 1 (BSL-1), for the least dangerous microorganisms, to BSL-4, for extremely hazardous microorganisms. BSL-1 includes school and college microbiology teaching laboratories. BSL-2 and -3 facilities are usually in research laboratories at universities, in industry, or in the government, and only a limited number of government or industry research facilities own BSL-4 laboratories. The types of microorganisms recommended for each biosafety level are the following:

- BSL-1—microorganisms not known to cause disease consistently in healthy adults, such as common water and soil microorganisms
- BSL-2—microorganisms present in the community and known to cause diseases of moderate severity, such as hepatitis B virus and *Salmonella*
- BSL-3—microorganisms that can transmit through the air and cause severe, life-threatening disease if inhaled, such as *Mycobacterium tuberculosis*
- BSL-4—microorganisms that are lethal in small doses and for which no known vaccine or treatment exists, such as Ebola virus

Each biosafety level consists of rules for the practices that microbiologists must follow and for the types of barriers the facility must contain to prevent escape of microorganisms, described in the table on page 519. Two types of barriers contain microorganisms: primary and secondary. Primary barriers consist of equipment that confines microorganisms; secondary barriers consist of features of a laboratory facility for containing microorganisms.

The table indicates that various biosafety levels require different types of dedicated work spaces called biosafety cabinets. The two main biosafety cabinets belong to either class II or class III levels of protection for the microbiologist and the laboratory's surroundings; class III provides the most protection. (Class I cabinets provide minimal protection because of uneven airflow and are no longer manufactured.)

In 2002, NSF International (formerly the National Sanitation Foundation) reclassified biosafety cabinets into more detailed categories of protection for workers. Either the conventional or the new terminology accurately describes the protection levels for laboratory workers and the surroundings, described in the table on page 520. In this scheme, all class II cabinets have the following three features: (1) an opening in front with inward airflow, (2) HEPA-filtered unidirectional airflow in the work area, and (3) HEPA-filtered exhaust. All class III cabinets have an airtight design in which a microbiologist handles items inside the cabinet by using gloves attached to the cabinet's clear front wall and extending into the cabinet interior.

The stringent requirements of various biosafety cabinets imply that microorganisms are very dangerous. On the contrary, most microorganisms are not dangerous and may be cultured on an open laboratory bench. Microbiologists, on occasion, still call the microorganisms that present low health risks class I microorganisms. Common class I microorganisms:

bacteria—*Aeromonas, Bacillus, Citrobacter, Pseudomonas*

fungi—*Aspergillus, Mucor*

protozoa—*Entamoeba* other than *E. histolytica*

viruses—bacteriophages, Newcastle virus.

Microbiologists have categorized other microorganisms into higher-risk biosafety levels in order to increase their containment. Microorganisms may, thus, be referred to as class I, II, III, or IV to correspond with BSL-1, -2, -3, and -4. The table at the bottom of page 520 provides examples of microorganism classifications according to their health risk.

People trained in aseptic techniques, sterilization, biohazard handling, and decontamination qualify for working with class I–II microorganisms. Working in a BSL-4 laboratory, by contrast, requires specialized expertise from added training.

BIOSAFETY LEVELS 1–3 LABORATORIES

Microbiology laboratories that contain microorganisms that pose low to moderately high danger use an array of equipment for the handling, culturing, containment, and disposal of bacteria, yeasts, fungi, viruses, protozoa, and algae. A typical microbiology laboratory has the following basic equipment:

- autoclave for sterilizing media and equipment and decontaminating waste
- water bath for holding molten agar media
- heat source (Bunsen burner or portable incinerator) for sterilizing inoculating loops
- natural gas source for Bunsen burners
- refrigerators and freezers for culture storage
- centrifuge for separating DNA or products from cells
- microscope for checking purity and identification

Biosafety Levels

BSL	Practices	Primary Barriers	Secondary Barriers
1	• microbiology training • limited access • hand washing after all work • no eating, drinking, contact lens application • no mouth pipetting • work to reduce aerosols • decontamination of surfaces • separate containers for biohazardous waste	none required	sink for hand washing
2	BSL-1 practices plus: • limited access for people with increased infection risk • personal protective equipment • immunization for appropriate microorganisms • special decontamination/accident procedures	• properly maintained class II biosafety cabinet • waste container for disposal clothing, gloves • goggles	• autoclave • lockable doors • sinks equipped with foot pedal or automatic operation • eyewash station
3	BSL-2 practices plus: • controlled access • mandatory protective clothing • disinfection and/or sterilization of all wastes • compliance with local, state, and federal regulations	• class II or enclosed, airtight class III biosafety cabinet • waste container for clothing disposal • face shield	BSL-2 barriers plus: • emergency shower • laboratory located away from the building's normal traffic
4	BSL-3 practices plus: • specialized training in containment methods for lethal microorganisms • laboratory certification before taking in these microorganisms • biohazard warning signs on all doors	• enclosed, airtight class III biosafety cabinet, or • full-body, air-supplied, positive-pressure protective suits	• laboratory in separate building from the main facility • interlocked locked-double-door entries • dedicated air supply, exhaust, vacuum, and decontamination systems • HEPA filtration systems for exhaust

Note: HEPA = high-efficiency particulate air

Biosafety Cabinets

Conventional System	New System	Microbial Risk Level	Features
class II, type A	A1	low to moderate	70 percent air recirculation; 75 feet per minute (FPM) intake; exhausts to the room; positive or negative pressure
class II, type A/B3	A2	low to moderate	70 percent air recirculation; 100 FPM intake; exhausts to the room; negative pressure
class II, type B3	A2	moderate	70 percent air recirculation; 100 FPM intake; exhausts to facility exhaust system; negative pressure
class II, type B1	B1	moderate to high	40 percent air recirculation; 100 FPM intake; dedicated exhaust to facility exhaust system; negative pressure
class II, type B2	B2	high	0 percent air recirculation; 100 FPM intake; dedicated exhaust to facility exhaust system; negative pressure
class III	class III	lethal	0 percent air recirculation; all inflow HEPA-filtered; all outflow filtered by two HEPA filters in sequence; exhaust receives additional chemical treatment; cabinet connects directly to other laboratory equipment with airtight seals

- mixers and shakers for putting cells into suspension
- incubators for growing microorganisms
- biosafety cabinet for routine procedures

Microbiology laboratories have additional codes of behavior for workers for the purpose of minimizing the chance of infection. These basic requirements are summarized in the following list.

- Only people trained in microbiology should handle live microbial cultures.
- Laboratories should have limited access when experiments are in progress.
- Microbiologists should wash hands before handling cultures and before exiting the laboratory.
- Eating, drinking, and storing food in the laboratory are prohibited.

Microorganism Risk Classifications

Class	Bacteria	Fungi	Protozoa	Viruses
II	*Clostridium* *Escherichia coli* *Listeria* *Shigella*	*Cladosporium* *Penicillium* *Cryptococcus*	*Cryptosporidium* *Giardia* *E. histolytica*	coronavirus hepatitis influenza rhinovirus
III	*Brucella* *Mycobacterium tuberculosis* *Yersinia pestis*	*Coccidioides immitis* *Histoplasma capsulatum*	none	dengue yellow fever
IV	none	none	none	Ebola Lassa Marburg some encephalitis

- Smoking, applying contact lenses, and applying cosmetics in the laboratory are prohibited.
- Mouth pipetting is prohibited.
- Sharps (needles, razors, scalpels, etc.) must have an acceptable method for safe disposal.
- Work surfaces must be decontaminated after using live cultures.
- All spills must be immediately cleaned up and decontaminated.
- Biohazardous wastes must be decontaminated and disposed of regularly.

Microbiology laboratories also require that tiny droplets called *aerosols* be kept to a minimum, because they can carry microorganisms through the air. Certain activities, such as using mixers, shakers, and centrifuges, release large amounts of aerosols, so the following precaution should be used in these circumstances: waiting one minute before removing tubes from a centrifuge or opening a culture that has just been mixed or shaken.

Biosafety suits protect microbiologists from lethal microorganisms in BSL-4 laboratories. Each suit provides an air supply and an impermeable barrier to the smallest microscopic particles. ***(Hampton Fire and Rescue)***

BIOSAFETY LEVEL 4

Because BSL-4 laboratories contain the world's deadliest known microorganisms, these places employ strict precautions against escape of any microorganism, even a single cell or virus particle, into the environment. Workers also take added precautions against being contaminated with these microorganisms. The National Institute of Allergy and Infectious Diseases (NIAID) outlines on its Web site the elements of operating a BSL-4 laboratory for maximal safety. The eight main features of BSL-4 safety and security are the following:

- interior-to-exterior containment—airtight facility with sealed windows and doors; airtight ventilation with a filtration system; bomb-proofing; and air, surface, and water decontamination systems
- exterior security—fencing, guards, closed-circuit television surveillance, nighttime lighting, and alarms
- interior security—employee identification, visitor credentials, locked laboratories, biometric security systems
- air decontamination—HEPA filtration of all exhausts, periodic sterilization of filters, and double-door airlocks to laboratories
- surface decontamination—work surfaces regularly disinfected, and all solid and liquid wastes decontaminated by sterilization
- changing room—facility for changing from street clothes to protective clothing before entering the laboratory
- biocontainment suits and biosafety cabinets—for worker protection and containment of the microorganisms within the laboratory
- decontamination shower—equipment for rinsing the biocontainment suit after leaving the laboratory and before returning to the changing room

Microbiologists learn the proper techniques for working in a biocontainment suit, decontaminating the suit after finishing work with class IV microorganisms, and exiting the BSL-4 laboratory without accidentally releasing a pathogen.

Each of the skills in microbiology, whether working with harmless water microorganisms or with deadly pathogens, requires practice and experience. Microbiologists know that biological systems do not behave exactly the same way each time they grow. The microbial world complicates this challenge by being the most numerous and diverse of all living things on Earth. Microbiology calls for a keen ability to visualize invisible organisms, understand the risks presented by some of these microorganisms, and protect the safety of the individual, other laboratory workers, and the environment.

See also ASEPTIC TECHNIQUE; CULTURE; DISINFECTION; ENVIRONMENTAL MICROBIOLOGY; FOOD MICROBIOLOGY; INDUSTRIAL MICROBIOLOGY; MICROSCOPY; PASTEUR, LOUIS; STERILIZATION.

Further Reading

American Society for Microbiology. Available online. URL: www.asm.org. Accessed September 18, 2009.

Microbes Meeting the Need for New Energy Sources

by Anne Maczulak, Ph.D.

The global energy crisis has created a new industry dedicated to meeting the increasing energy needs of the world's population in a sustainable manner. Wind, wave power, solar energy, geothermal energy, and fossil-fuel alternatives called *biofuels* have been on the forefront of political and community agendas for new energies. Many businesses and communities have made excellent progress in using these energy sources. As consumers become familiar with alternative energies, they see benefits, but they are also now learning of some disadvantages of each type of alternative energy. Solar panels are expensive, windmills use up land, biofuels from grains upset the normal supply and demand of agriculture markets, and so on. Before the potential of these first-generation alternative energies has been fully realized, critics already are calling for a new generation of clean energy sources.

The newest alternative energies have already emerged. Algae and bacteria produce energy, do not demand much energy input to convert chemical or solar energy to fuel energy, and are self-propagating. Microbial energy is hardly new. Wastewater treatment plants capture the methane gas from anaerobic digestion of organic matter called *sludge* and use this natural product for powering plant operations. Wastewater treatment is an energy-demanding process for running pumps that move the water through the treatment facility. By capturing at least enough energy from methane to allow the treatment plant to break even, wastewater treatment avoids being an energy drain on its community. Many community landfills have also built collectors to capture methane produced deep within the landfill wastes. This methane can be converted to electrical energy and sent by power lines to homes and businesses.

Several companies have developed methods for increasing the energy produced by algae for consumer use. These enterprises, if successful, aim not merely to break even, but to serve as an important energy provider. Algae offer more than one energy-producing route: carbohydrates to make ethanol, algal oils to make biodiesel, and the entire algae harvest to make biomass to be burned in place of coal. In addition, algal proteins would be recovered as a feed supplement for livestock. As do other alternative energies, algae present challenges before entrepreneurs can develop them into a viable energy source. For example, algae are photosynthetic, so they must be grown in large shallow tanks exposed to the sun. Algae in volumes large enough to produce energy will also have high demands for water, nutrients, and carbon dioxide. The technologies that turn algae into a meaningful energy source may be more complex than many proponents originally thought.

Biofuels from algae offer tremendous promise, and scientists should continue pursuing these fuels and overcoming the challenges. The energy consultant Stefan Unnasch remarked, in 2009, on algal biofuels being produced by Solazyme, Inc., "[The company's] advanced biofuels substantially reduce greenhouse gas emissions per mile driven over petroleum-based fuels and result in much lower carbon emissions than currently available first generation biofuels." First-generation biofuels, such as ethanol, have demanded large areas of land to grow the crops. The energy inputs for making and transporting the grains and final products have also been daunting. Algae cultivation is compact by comparison, a distinct advantage of this type of microbial energy over fields of biofuel-bound corn.

Microbiologists have also begun to explore the energy-production potential of bacteria. Photosynthetic bacteria such as cyanobacteria could be grown in a process similar to algal growth. But nonphotosynthetic bacteria also carry out diverse metabolic pathways that can be harnessed for energy. Many scientists believe that the ease with which bacteria can be bioengineered leads to new opportunities in energy that algae do not offer. The chemical engineer Thomas Wood of Texas A. & M. University has been researching the possibil-

Fleming, Diane O., and Debra L. Hunt, eds. *Biological Safety: Principles and Practices,* 4th ed. Washington, D.C.: American Society for Microbiology Press, 2006.

Kluyver, Albert J. "Unity and Diversity in the Metabolism of Micro-Organisms." In *Albert Jan Kluyver, His Life and Work,* translated by Thomas Brock. Amsterdam, The Netherlands: North-Holland, 1959.

Lawrence Berkeley National Laboratory. Environment, Health and Safety Division. "Biological Safety Program Manual." Available online. URL: www.lbl.gov/ehs/biosafety/Biosafety_Manual/biosafety_manual.shtml. Accessed September 21, 2009.

National Institute of Allergy and Infectious Diseases. "Biodefense: Biosafety Labs." Available online. URL: www3.niaid.nih.gov/topics/BiodefenseRelated/Biodefense/PublicMedia/BioLabs.htm. Accessed September 18, 2009.

Needham, Cynthia, Mahlon Hoagland, Kenneth McPherson, and Bert Dodson. *Intimate Strangers: Unseen Life on Earth.* Washington, D.C.: American Society for Microbiology Press, 2000.

Prescott, Lansing M., John P. Harley, and Donald A. Klein. *Microbiology,* 6th ed. New York: McGraw-Hill, 2005.

ity of bioengineering *Escherichia coli* to produce large amounts of hydrogen. The current chemical method of producing hydrogen uses a process called *cracking water* to separate hydrogen and oxygen. This process uses large amounts of energy. Wood has made *E. coli* into a super-hydrogen-producing cell by removing six of the bacterium's genes. He describes the advantages of *E. coli*'s biological production of hydrogen compared with chemical production as follows: "One of the most difficult things about chemical engineering is how you get the product. In this case, it's very easy because hydrogen is a gas, and it just bubbles out of the solution. You just catch the gas as it comes out of the glass. That's it. You have pure hydrogen." Bacterial hydrogen biofuel cells make hydrogen where they grow and require little in terms of energy input or space.

Bacteria also have a future in innovative battery power. Scientists have been constructing batteries called *microbial fuel cells* out of soil bacteria for creating an electrical current. Harvard scientists have tried microbial fuel cells in regions of Africa that do not have a source of electricity. The fuel cells contain five components: a graphite anode, chicken wire cathode, manure-mud mixture that contains a high concentration of actively metabolizing bacteria, sand and salt water as an electrolyte placed between the two poles, and an electronic power board. This simple battery produces enough electricity to light lamps and charge cell phones. Bacteria fuel cells have the added benefit of consuming organic wastes.

The major concern about microbial energy sources should be obvious: Can microscopic organisms ever be grown in amounts large enough to produce usable quantities of energy in an energy-demanding world? Algae and bacteria should not be asked to carry 100 percent of the energy-demand load in the near-future. Microorganisms might not ever contribute energy in sufficient quantities to take pressure off the first-generation alternative energies. On a global scale, the energy problem seems unsolvable. But what if bacterial batteries begin to serve as power sources for millions of small electronic devices? What if algal hydrogen fills fuel pumps that commuters can tap into after work? The lesson that everyone must rapidly learn is that the energy crisis cannot be solved by one technology.

Microorganisms possess a feature that few other materials have: versatility. With clever planning, scientists can exploit the best features of microorganisms, that is, rapid growth, self-replication, low energy demands, small size, and environmental safety. Microorganisms, furthermore, have a history on Earth longer than that of any other being. They contribute to every ecosystem and the nutrient cycles that regenerate the matter humans use for energy. Put another way, microorganisms are already managing Earth's energy. Humans may soon have no choice but to pursue the untapped energy of the microbial world.

See also ALGAE; BACTERIA; METABOLIC PATHWAYS; METABOLISM.

Further Reading

Gies, Erica. "New Wave in Energy: Turning Algae into Oil." *New York Times,* 29 June 2008. Available online. URL: www.nytimes.com/2008/06/30/business/worldbusiness/30iht-renalg.1.14044230.html. Accessed November 10, 2009.

Science*Daily*. "*E. coli* Bacteria: A Future Source of Energy?" January 31, 2008. Available online. URL: www.sciencedaily.com/releases/2008/01/080129170709.htm. Accessed November 10, 2009.

Sharp, William J. "Waste-Eating Bacteria Explored as Power Source." November 7, 2009. Available online. URL: www.af.mil/news/story.asp?storyID=123031108. Accessed November 10, 2009.

Solazyme, Inc. "Additional Emission Testing Demonstrates Solazyme's Algal-Biofuels Shown to Significantly Lower Tailpipe Emissions When Compared to Ultra-Low Sulfur Diesel." April 21, 2009. Available online. URL: www.solazyme.com/media/2009-06-26-3. Accessed November 10, 2009.

Ward, Logan. "10 Most Brilliant Innovators of 2009: Bacteria-Powered Battery." *Popular Mechanics,* 2009. Available online. URL: www.popularmechanics.com/technology/industry/4332914.html. Accessed November 10, 2009.

Richmond, Jonathan, Y. "The 1, 2, 3's of Biosafety Levels." Available online. URL: www.cdc.gov/OD/ohs/symp5/jyrtext.htm. Accessed September 21, 2009.

Schaechter, Moselio, John L. Ingraham, and Frederick C. Neidhardt. *Microbe.* Washington, D.C.: American Society for Microbiology Press, 2006.

Todar, Kenneth. "Todar's Online Textbook of Bacteriology." Available online. URL: www.textbookofbacteriology.net. Accessed September 18, 2009.

Tortora, Gerard J., Berdell R. Funke, and Christine Case. *Microbiology: An Introduction,* 10th ed. San Francisco: Benjamin Cummings, 2009.

U.S. Dept. of Health and Human Services, Centers for Disease Control and Prevention, and the National Institutes of Health. "Biosafety in Microbiological and Biomedical Laboratories." Available online. URL: www.cdc.gov/od/ohs/pdffiles/4th%20BMBL.pdf. Accessed September 21, 2009.

microenvironment A microenvironment is the physical location and immediate surroundings of a microbial cell's habitat. Certain factors define microenvironments and determine the role that microbial inhabitants play in these locations. Common microenvironment-defining factors are oxygen availability, nutrients, chemical oxidation and reduction conditions, temperature, acidity, barometric pressure, and osmotic pressure. These factors can create conditions that are found in few other places, so that a microenvironment often excludes many microorganisms and allows the growth of a select group of normal inhabitants.

Microenvironments can be fairly large or very small and unique. For example, soil microorganisms live in various microenvironments defined by soil composition, water content, pH, and particle size. By comparison, some microorganisms are intracellular; that is, they live inside a host cell that defines the microenvironment. The table below describes examples of well-known microenvironments.

Many microenvironments contain smaller regions of increasingly unique physical conditions. For example, the oral cavity is a microenvironment that also contains the following smaller or submicroenvironments: periodontal space between tooth and gum, plaque, epithelia, tongue, and oral fluids.

Advances in instrumentation have allowed scientists to examine the area directly surrounding individual microbial cells. These nanoscale microenvironments, measured in nanometers (nm), can be characterized by the molecules in direct contact with or within several nanometers of the cell surface. Electron microscopy offers a method to observe microenvironments within 40 nm, and X-ray spectromicroscopy helps determine chemical constitu-

Microenvironments

Microenvironment	Description
soil	various types of soil create microenvironments for aerobes, anaerobes, nitrogen-fixing bacteria, or chemoautotrophs (microorganisms that use inorganic compounds for energy and carbon dioxide for carbon)
layers of freshwater lakes	various depths create differing conditions of light, temperature, and nutrients
deep ocean	the ocean contains extreme conditions of barometric pressure, cold, darkness, and limited nutrients
hydrothermal vents	deep-ocean structures emit hot gases and acids
hot springs	natural sources of boiling water contain high concentrations of minerals
skin	regions of the human body differ in moisture, oils, aeration, and temperature, creating specialized microbial communities
invertebrate digestive tract	specialized microbial communities live in anaerobic systems in termites, cockroaches, earthworms, other invertebrates, and in mammals
plant root systems	root nodules on legume plant roots contain specialized nitrogen-fixing bacteria that participate in the nitrogen cycle
host cells	the composition of the particular eukaryotic host cell, such as cytoplasm and enzymes, defines the microenvironment

ents in a microenvironment by measuring emissions of certain wavelengths from molecules.

Microenvironments influence ecosystems, nutrient recycling (called *biogeochemical cycles*), and the metabolism of plants and animals. Microenvironments, together, can also create a natural community, such as the microorganisms of soil, the digestive tract, or lakes.

See also BIOFILM; ELECTRON MICROSCOPY; MICROBIAL COMMUNITY; ORAL FLORA; RUMEN MICROBIOLOGY; SOIL MICROBIOLOGY.

Further Reading

Jjemba, Patrick, K. *Environmental Microbiology: Principles and Applications*. Enfield, N.H.: Science Publishers, 2004.

Maier, Raina M., Ian L. Pepper, and Charles P. Gerba, eds. *Environmental Microbiology*, 2nd ed. San Diego: Elsevier, 2009.

Sigee, David C. *Freshwater Microbiology*. Chichester, England: John Wiley & Sons, 2005.

Yeung, Tony, Nicolas Touret, and Sergio Grinstein. "Quantitative Fluorescence Microscopy to Probe Intracellular Microenvironments." *Current Opinion in Microbiology* 8 (2005): 350–358.

microscopy Microscopy is the use of an instrument for viewing objects too small to be seen with the unaided eye. The instrument, called a *microscope*, forms an enlarged image of a specimen, which can range from 20 times (20 ×) to several thousand times the object's normal size. A microscope has a twofold purpose: magnification of the specimen and resolution between features of the specimen. Magnification is the increase in the specimen's size when viewed in a microscope. Increases in magnification enable microbiologists to study the shape and some of the structure of microbial cells. Resolution is the ability of a microscope's lenses to distinguish among different structures in the specimen so that the microbiologist sees details in structure.

The microscope forms the basis of all microbiological studies, and, without it, scientists would not have made the advances in microbiology in the past two centuries. The earliest microscopes began by placing two or more convex lenses in sequence to enlarge a specimen's image. This arrangement, eventually, led to the compound light microscope, "compound" because it uses a series of lenses and "light" because it uses visible light to illuminate the specimen. Derivations of the light microscope developed, in time, in which light hit the specimen from various orientations to give different attributes to the image. Fluorescent microscopes offered an altogether different approach by using fluorescent light of a specific energy rather than visible light. Electron microscopy became the next major step forward in the development of microscopes. Rather than use light, electron microscopes use an electron beam to create the image. This technique has enabled microbiologists to view objects of much smaller size than the single cells usually studied in light microscopy. The images produced by electron microscopes showed cell organelles in detail, the inner membrane, and macromolecules. Recent advances in microscopy now allow scientists to see molecules. The table on page 526 traces the milestones in the development of microscopes.

The inventions described in the table would not have been meaningful if they did not produce a clear image, in addition to a magnified image. Microscopy always strives for optimizing both features. The production of clear images allows a microbiologist to differentiate between two objects in the microscope's viewing area, called the *field*. This ability to distinguish between objects in order to make fine details visible is called *resolution*.

MAGNIFICATION AND RESOLUTION

Understanding magnification begins with knowledge of the size of various objects. Microscopy uses the following units:

- millimeter (mm) = one thousandth of a meter, or 0.0394 inch
- micrometer (μm) = one millionth of a meter
- nanometer (nm) = one billionth of a meter
- angstrom (A) = one tenth of a nanometer

Nanotechnology has developed ways to study objects on a scale much smaller than the nanometer. Angstroms are, therefore, being replaced by nanoscale units, such as the picometer (pm), which is one trillionth of a meter.

The human eye resolves about 150 μm between two points. This allows a person to see various protozoa, but not many other eukaryotic cells, bacteria, yeast cells, fungal spores, or viruses. (Only electron microscopy provides images of viruses.) Light and fluorescent microscopes have a maximal magnification of about 2,000 X and the smallest resolution to about 0.2 μm, which is the size of the smallest bacteria. Exceptions always exist in biology. The giant bacterium *Thiomagarita namibiensis* measures a rare 0.75 mm and is visible without a microscope. Scientists have recently proposed the existence of bacteria measured in nanometers, called *nanobacteria*. In 2002, the German microbiologist Karl Stetter (1941–) discovered archaea of only 400 nm in diameter. This microorganism—archaea are distinct from bacteria—has been named *Nanoarchaeum equitans* and lives near extremely hot hydrothermal vents at the ocean floor. Microscopy has now entered a new phase of technology called nanomicroscopy to explore improvements in visualizing nanoscale objects.

Light microscopy is confined to more traditional specimens in microbiology, such as bacteria and fungi. Magnification for this purpose is obtained from the combination of lenses used in a compound microscope. Magnification develops when light from the light source, called the *illuminator* and situated under the specimen, passes through a condenser, which contains lenses that direct the light through the specimen. After passing through the specimen, the light enters an objective lens, then the ocular lens, which is also called the *eyepiece*. The objective and the ocular lenses provide a microscope's magnification. For example, an objective lens that magnifies 60 × and an ocular lens that magnifies 10 × gives total magnification of 600 ×. Most ocular lenses magnify 10 ×, but objectives typically range from 10 × to 100 ×.

Resolution improves in relation to the wavelength of light. In general, the shorter wavelengths of light increase a microscope's resolution. The power of resolution represents the actual distance between two distinguishable points. For example, a resolving power of 0.3 μm enables a microbiologist to tell the difference between two objects that are 0.3 μm apart. Light and fluorescent microscopes usually cannot resolve objects less than 0.2 μm apart. Two types of electron microscopes greatly improve resolving power: Scanning electron microscopes resolve to about 10 nm, and transmission electron microscopes resolve to about 0.5 nm.

High-magnification objectives, such as 100 ×, have very small apertures to collect light. Some light is always lost when a specimen deflects it, as it passes through. To preserve light rays at high magnifications, microbiologists use a technique called oil immersion. Immersion oil is formulated to have the same refrac-

History of Microscopy

Person	Approximate Date	Contribution
Seneca (ca. 4 B.C.E.–65 C.E.), Pliny (ca. 23–79)	first century C.E.	described magnifying properties of glass
Roger Bacon (ca. 1214–92)	1262	investigated magnifying properties of glass
Salvino D'Armate (1258–1312)	1284	credited with inventing first eyeglasses
Zacharias Janssen (1580–1638)	1590s	first use of lenses in sequence (compound microscope)
Galileo Gallilei (1564–1642)	1608	experiments on light and focus
Johannes Kepler (1571–1630)	1604	described how magnification works
Marcello Malpighi (1628–94)	1661	first scientific experiments in microscopy
Robert Hooke (1635–1703)	1665	published drawing of cells in *Micrographia*
Antoni van Leeuwenhoek (1632–1723)	1677–84	simple compound microscope for examining microorganisms
Carl Zeiss (1816–88)	1823	first achromatic lens
Ernst Abbe (1840–1905) and Carl Zeiss	1869	first microscope lamps
Ernst Abbe	1873	published the theory of the microscope
Ernst Abbe, Carl Zeiss, and Otto Schott (1851–1935)	1886	first apochromatic lens
August Köhler (1866–1948)	1904	first ultraviolet, dark-field microscopes
Max Knoll (1897–1969), Ernst Ruska (1906–88)	1931	first electron microscope
Albert Coons (1912–78), Melvin Kaplan (1922–)	1940s	fluorescent dye-labeled antibodies
Frits Zernike (1888–1966)	1950s	developed phase contrast microscopes
Georges Nomarski (1919–97)	1950s	developed differential interference contrast microscopy

tive index as glass—refraction is the altering of the paths of light rays—so immersing a specimen in oil and then lowering the 100 × objective directly into the oil drop allow more light to enter the objective to produce a bright image. Using a 100 × objective without immersion oil results in a very fuzzy image.

The table on page 528 provides a general idea of the range of various microscopic specimens. Microbiologists use oil immersion for viewing bacteria, including mycoplasmas.

Scanning tunneling electron microscopes use an electron beam, as do other electron microscopes, but use electrons emitted from the specimen rather than a beam that passes through the specimen.

MICROSCOPY BASICS

Microscopy basics constitute one of the first skills learned in microbiology. In addition to operating the parts of a light microscope, microbiologists must understand the power and uses of objectives, staining techniques to enhance images, and methods for measuring the true size of specimens. Fluorescence microscopy requires additional skills in the use of fluorescent dyes, antibodies, and light wavelength.

Eyepiece objectives usually contain ocular lenses of 10 × or 20 ×. Most microscopes offer a selection of objectives on a turret of various magnifications, often 20 ×, 40 ×, 60 ×, and 100 × for oil immersion. A microbiologist rotates the turret, which swings a new objective into position without disturbing the specimen. Objectives contain the following pieces of information:

- a color band to help the microbiologist quickly find an objective on the turret
- the type and magnification, such as "Plan Apo 20"
- the objective's tube length and numerical aperture (N.A.), written "160/0.65"
- the thickness of the cover glass that must be used with the objective

Three types of objectives used in light microscopy, in increasing order of image quality, are (1) achromatic, (2) semiapochromatic, and (3) apochromatic. Each objective contains lenses that correct for different colored light, due to different wavelengths, and give all colors common focus. Lenses also have been designed to correct for curvature in images, producing a flat image viewed in the eyepiece. These lenses are called plan-achromats, plan-semiapochromats, and plan-apochromats. The objective shown in the example above is a plan-apochromatic 20 magnification type. Tube length is the length of the objective in millimeters, designed to give the clearest image. N.A. describes an inherent feature of an objective's design that allows it to take in light from an angle. Most microscopes use a tube length of 160 mm and range in N.A. from 0.04 mm for low-powered objectives (5 × to 10 ×) to 1.4 for high-powered objectives (80 ×–100 ×). Finally, a microbiologist should select a coverslip, the glass square that covers the specimen, of a thickness that matches the objective's requirement. An objective that has "0.17" etched beneath the tube length/N.A. requires the use of coverslips of 0.17-mm thickness.

Microscopes also have a micrometer ruler that is visible through the glass microscope slide. Microbiologists use this ruler to determine the actual size of specimens in the viewing field.

SPECIMENS

Microbiology uses three main techniques for preparing specimens for microscopic study: (1) dry mounts, (2) wet mounts (see the color insert, page C-6 [bottom]), and (3) hanging drop. Dry mounts, or fixed specimens, have been treated to adhere to the glass slide. The fixed specimen is called a *smear,* prepared by gently heating the slide over a flame from a Bunsen burner. After fixation, a specimen can be stained by any of a large variety of dyes intended for different purposes, such as staining whole cells, organelles, proteins, flagella, and the background. Fixed specimens serve as an early step in bacterial identification through use of the Gram stain. This type of preparation is used almost exclusively in clinical and food microbiology.

Wet mounts offer the advantage of viewing living cells. Cells suspended in water or saline solution retain their normal shape and structure, and motile cells have the opportunity to swim through the liquid. Liquid specimens can be studied under a coverslip in the same way as fixed specimens, or a wet mount can be prepared by the hanging drop method. In this technique, a microbiologist places a drop of specimen on a coverslip, then covers the drop with a special glass slide containing a concave depression, or well. When both are inverted, the drop hangs from the coverslip into the well. Hanging drop slides have been useful in detailed studies of cell motility.

Fluorescence microscopy uses special preparations to make specimens fluoresce when exposed to ultraviolet light. The wavelengths of light in the visible spectrum range from 750 nm to 400 nm; ultraviolet light wavelengths are 300–400 nm. Fluorescence microscopy involves treating a specimen with a reagent (often an antibody) that links to a molecule (or organelle) of interest and attaches to a chemical (a fluorochrome) that emits light, when exposed to

Microscopic Sizes and Resolution

Specimen	Size Range	Microscope
epithelial cells (skin cells)	25–35 μm	light
protozoa	20–150 μm	light
red blood cell	7.5 μm	light
most bacteria	1–8 μm	light
mycoplasmas	200–300 nm	light
viruses	20–100 nm	electron
proteins	5–10 nm	electron
small molecules	0.6–0.9 nm	electron
atoms	0.1–0.3 nm	electron (scanning tunneling)

fluorescent light of specific wavelength. Specimens appear brightly colored—color depends on the dye used—against a black background. A common technique involves antibody-dye complexes that react only with specific antigens. Fluorescence microscopy has been valuable for detecting specific pathogens, toxins, or proteins in the environment or in clinical samples.

LIGHT MICROSCOPY

Light microscopes are the most common microscopes used in microbiology, because of their low cost and ease of use. Light microscopes may be bright-field, dark-field, phase contrast, or confocal; some microscopes are designed to switch among different types, such as bright-field and phase contrast.

In bright-field microscopes, a lens called a *condenser* directs a beam of visible light, just after it leaves the light source, directly onto the specimen. As a result, the condenser produces bright-field illumination of the specimen. Because unstained microbial cells offer little contrast with the surroundings on a glass slide, microbiologists often stain specimens with a dye before viewing them.

Dark-field microscopes (see the color insert, page C-6 [top]) contain a dark-field condenser that blocks all light that would enter the objective directly and allows only light deflected from the specimen to be seen. Because dark-field images have no background light, specimens appear bright against a black background. Dark-field microscopy is useful for examining the motility of unstained live cells and for inspecting some internal structures of large cells such as protozoa.

Phase contrast microscopy makes use of light that has been refracted and diffracted by the specimen. Diffraction is the altering of a light ray's path by touching the edge of an object. A phase contrast microscope uses unaltered light rays directly from the specimen, plus rays that have been refracted or diffracted by the specimen. The phase contrast condenser contains a thin transparent ring that produces a hollow cone of light that strikes the specimen, a process called Köhler illumination. Above the specimen, a light-transmitting phase plate provides a ring-shaped structure (phase ring) that alters light's wavelength. After the hollow light cone passes the specimen, most diffracted rays pass through the phase plate unchanged, because they miss the phase ring. Light waves altered by a specimen are out of phase with unaltered light; the peaks and troughs of out-of-phase light waves no longer synchronize with regular light. The phase plate's ring readjusts all light so that some altered and unaltered waves cancel each other out.

Phase contrast images develop because different cell components have various densities and deflect light differently. By separating and then recombining altered and unaltered light rays, phase contrast microscopy produces a sharp, detailed image of an unstained specimen against a gray background. Phase contrast microscopy helps in viewing some internal structures, as well as surface appendages, such as cilia (small hairlike structures) and flagella (long tails).

Differential interference contrast (DIC) microscopy (also called Nomarski interference contrast microscopy) uses two beams of light. Prisms redirect one beam to a right angle relative to the unaltered beam. One beam passes through the specimen, and the other bypasses it; then, the two beams recombined interfere with each other. The differences in refractive indexes between beams produce the DIC image (see the color insert, page C-6 [center]). Because the prisms also split light beams, the image contains contrasting colors. DIC microscopy results in sharp three-dimensional images with high contrast between cell components.

The table (top, left) summarizes the features of different types of light microscopy in addition to fluorescent microscopy, which uses light outside the visible spectrum. As a result of the path of light traveled through the lenses, all of these microscopes create an upside-down backward image of the specimen, which usually does not create a problem when viewing microorganisms.

Microscopes can be equipped with digital cameras for capturing specimen images or connected to a computer for saving images directly as files. Either process is called photomicrography. The best images captured by camera can only be as good as the microscope's capability.

Most current microscopes contain two eyepieces (binocular) with adjustable eyepieces to accommodate distance between the eyes; they have replaced older

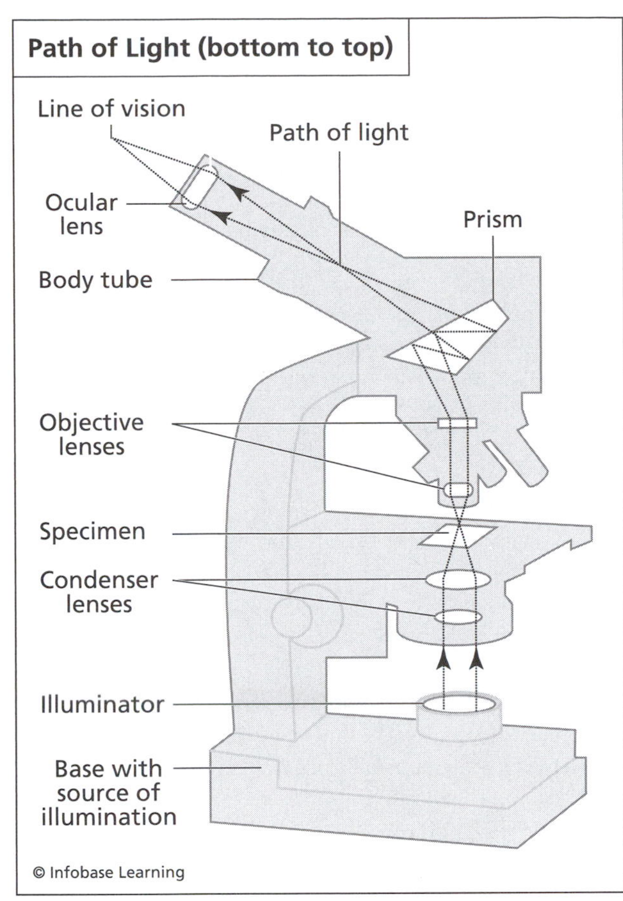

The light microscope is the oldest instrument used in microbiology and remains the main piece of equipment microbiologists use in pathogen identification, physiology studies, environmental monitoring, and morphology studies.

one-eyepiece (monocular) models. Although use of two eyepieces helps relieve eyestrain, long hours of work at a microscope remains a tedious task. Many laboratories also use teaching microscopes that hold two eyepiece heads, which allow two people to view the same image simultaneously.

ADVANCES IN MICROSCOPY

A stereo microscope, also called a dissecting microscope, contains low-powered lenses that produce magnifications of 7 × to 100 ×. Stereo microscopes aid in counting fungal spores and pollen, as well as examining fungal mycelia, the long stringy filaments that cover large surface areas when fungi grow. Stereo microscopes focus on the same point on a specimen from slightly different angles, resulting in a correctly oriented specimen (not upside down or backward).

Microscope technology has made recent advances in increased resolution of small specimens. Confocal scanning laser microscopes (CSLMs) use fluorochromes for highlighting specific regions of the specimen. The CSLM illuminates individual planes of the specimen with a laser beam. The laser scans over the plane by moving either the laser beam (beam scanning) or the stage holding the specimen (stage scanning). A computer combines the multitude of images to form a three-dimensional image. CSLM provides superior contrast and resolution through its ability to block all stray light and light scattered from the specimen. The electrical engineer Marvin Minsky (1927–), of the Massachusetts Institute of Technology, invented the confocal scanning microscope, in 1955. In his 1988 memoir published in *Scanning,* he said, "An ideal microscope would examine each point of the specimen and measure the amount of light scattered or absorbed by that point." Minsky elaborated on the trials of scientific discovery: "I demonstrated the confocal microscope to many visitors, but they never seemed very much impressed by what they saw on that radar screen. Only later did I realize that it is not enough for an instrument merely to have a high resolving power; one must also make the image *look* sharp." Minsky solved the dilemma by inventing stage scanning to replace the less precise beam scanning of his earliest CSLM models.

Scanning probe microscopy differs from CSLM by moving a sharp probe over a specimen to create a detailed image of surface features. In 1980, the German Gerd Binnig (1947–) and the Swiss Heinrich Rohrer (1933–), both physicists, developed scanning tunneling microscopy (STM). STM uses an extremely fine probe, called the *stylus,* that slowly scans across the specimen surface from a distance of only an atom's diameter. At this short distance, electrons tunnel between the surface and the stylus. A computer measures this electron flow to create an image in fine detail. In their published acceptance speech for the Nobel Prize in physics in 1986, Binnig and Rohrer recalled, "Perhaps we were fortunate in having training in superconductivity, a field which radiates beauty and elegance. For scanning tunneling microscopy, we brought along some experience in tunneling and angstroms, but none in microscopy or surface science. This probably gave us the courage and lightheartedness to start something which should 'not have worked in principle' as we were so often told." STM now achieves magnifications of 100 million times and allows scientists to observe individual atoms.

In 1986, Binnig invented atomic force microscopy (AFM) to overcome a drawback of STM: STM only creates images from materials that can produce a tunneling current, a movement of electrons between the specimen and the microscope's stylus. AFM produces three-dimensional images of nonconducting materials, such as polymers and biological samples. AFM uses the weak bonds that develop between substances, called van der Waals forces, to generate an image, also using a scanning stylus that never touches the surface.

Light microscopy has advanced as well, particularly a form of light microscopy called photo-

Types of Microscopy

Type	Main Feature	Main Uses
bright-field	visible light illuminates the specimen in a bright field	stained (dead) specimens
dark-field	condenser blocks direct light and permits only light reflected by the specimen against a black background	live specimens that do not stain well, motility
phase contrast	refracted and diffracted light intersect to form the image against a gray background	internal structures of live specimens, motility
DIC	prism effect produces colored image and high contrast	three-dimensional images of live or dead specimens
fluorescent	ultraviolet or near-ultraviolet light causes treated specimens to emit light	rapid differentiation of cells in mixtures or tissues, mainly in clinical and environmental samples

activated localization microscopy (PALM). Normal light microscopes cannot resolve objects below 0.2 μm because they are constrained by the wavelength of visible light. PALM achieves greater resolving power by activating fluorescent proteins with light photons. PALM resolves images of objects within living cells at about 60 nm apart.

The development of microscopy has enabled microbiologists to gain knowledge of the structure and function of microorganisms in ever finer detail. Microscopy has reached the level of the atom and, in the near-future, will probably approach subatomic particles. With these advances, microbiologists may soon answer the last unresolved questions in microbial reproduction and metabolism.

See also ELECTRON MICROSCOPY; FLUORESCENCE MICROSCOPY; LEEUWENHOEK, ANTONI VAN; NANOBIOLOGY; ORGANELLE; STAIN.

Further Reading

Binnig, Gerd, and Heinrich Rohrer. "Scanning Tunneling Microscopy—from Birth to Adolescence." Nobel lecture, Stockholm, Sweden, December 8, 1986. Available online. URL: http://nobelprize.org/nobel_prizes/physics/laureates/1986/rohrer-lecture.pdf. Accessed June 6, 2010.

Claxton, Nathan S., Thomas J. Fellers, and Michael W. Davidson. "Laser Scanning Confocal Microscopy." Available online. URL: www.olympusconfocal.com/theory/LSCMIntro.pdf. Accessed September 24, 2009.

Microscopy Society of America. Available online. URL: www.microscopy.org/index.cfm. Accessed September 23, 2009.

Minsky, Marvin. "Memoir on Inventing the Confocal Scanning Microscope." *Scanning* 10 (1988): 128–138. Available online. URL: http://web.media.mit.edu/~minsky/papers/ConfocalMemoir.html. Accessed September 24, 2009.

Nikon MicroscopyU. Available online. URL: www.microscopyu.com. Accessed September 23, 2009.

Olympus Microscopy Resource Center. Available online. URL: www.olympusmicro.com. Accessed September 23, 2009.

WWW Virtual Library—Microscopy. Available online. URL: www.ou.edu/research/electron/www-vl. Accessed September 24, 2009.

minimum inhibitory concentration (MIC) A minimum inhibitory concentration (MIC) is the lowest concentration of an antimicrobial agent that prevents the growth of a specific pathogen. MIC tests determine the relative strengths of antibiotics against pathogens and can indicate whether a microorganism is building resistance to an antibiotic, over time. Strong antimicrobial agents produce low MIC values, and weak antimicrobial agents result in higher MIC values. MIC is usually measured in milligrams per milliliter (mg/ml), meaning that a certain amount in milligrams of antibiotic dissolved in broth medium inhibits microbial growth. The increase in resistance (decrease in susceptibility) can be evaluated by performing the same MIC test repeatedly over time.

MIC serves an important role in clinical microbiology for selecting an antibiotic to treat an ongoing bacterial or fungal infection. In this use, MIC relates to susceptibility testing, which is the determination of a pathogen's susceptibility or resistance to an antibiotic.

Two values related to MIC are the minimum lethal concentration (MLC) and the minimum bactericidal concentration (MBC). Although an antibiotic at an MIC inhibits a microorganism's growth, it might not necessarily kill the pathogen. Inhibited bacteria could be temporarily damaged by an antibiotic that slows or stops reproduction. Some cells can repair the damage and continue multiplying after a period of exposure to the antibiotic. Microbiologists sometimes prefer the MLC because it is the lowest drug concentration that kills the pathogen rather than inhibiting its growth.

In microbiology, antimicrobial compounds can be either cidal or static. Cidal drugs, such as penicillin, which kills susceptible *Streptococcus,* are called bactericidal agents. The MLC test determines this cidal activity. Static drugs merely inhibit the growth of a microorganism, for a short period (days to weeks) or a longer period (months). Because static compounds do not kill all of a microbial population, some cells eventually start multiplying again, and large numbers of microorganisms override the compound's effectiveness.

Some antimicrobial compounds can be both microbicidal and microbistatic. By raising the concentration above the static MIC level, the drug might begin killing a pathogen. Clinical microbiologists tend not to use microbistats in this manner, because the drug's concentration must be increased many times over its MIC in order to kill the pathogens, if it can. Instead, certain drugs have been classified as microbistatic, and other drugs are categorized as microcidal. The MLC of microbicidal drugs usually occurs at only two to four times the MIC of a microbistatic drug for the same pathogen.

Static drugs have been implicated as a cause of antibiotic resistance in microorganisms even more than cidal agents, because static drugs promote the growth of the strongest cells and eliminate only the weakest. The succeeding generations of progeny develop stronger resistance to the antibiotic, until the drug is completely ineffective, even at cidal concentrations. In fact, microbicidal drugs have been used inappropriately for so many years that most human pathogens and many veterinary pathogens have developed resistance to both types of drugs.

Today, MIC testing provides a way to determine the resistance of a pathogen against a variety of drugs at the same time. The same test usually also identifies the most effective one or two drugs against the pathogen.

MIC BROTH TESTS

MIC values vary from pathogen to pathogen and may also vary within the same pathogen, depending on growth characteristics. MIC tests should, therefore, be performed under standard conditions to eliminate variability among microbial cultures of the same pathogen and between different laboratories.

Three different tests have been devised for determining the MIC of given antibiotics against pathogens. Clinical microbiologists test drugs directly against isolates from patients who have infectious disease. Drug manufacturers test a variety of antibiotics against type strains of common pathogens. A type strain is a species possessing growth characteristics and metabolism that best represent all members of the species. The tests used in both these instances are (1) dilution MIC tests, (2) disk diffusion, and (3) the E test (see the color insert, page C-7 [top]).

The dilution MIC test involves the following five steps:

1. Dissolve antibiotic in replicate broth tubes, each set of replicates containing a different concentration of antibiotic.

2. Set aside a set of tubes to serve as uninoculated controls.

MIC Test Results

Antibiotic Level (mg/ml)	Results in Five Replicates	Interpretation
0	+ + + + +	control indicating that the test pathogen grows in the test conditions
0.0001	+ + + + +	resistant to this level of antibiotic
0.005	+ + - + +	mostly resistant
0.01	- + - - -	intermediate effects caused by this level of antibiotic
0.05	- - - - -	susceptible
0.1	- - - - -	susceptible

Note: + = presence of growth; - = absence of growth

3. Inoculate all other tubes with a pathogen isolated from a patient specimen, called a *clinical isolate.*

4. Incubate all tubes.

5. Observe all tubes for the presence or absence of growth.

The table on page 531 presents hypothetical MIC results after completing the five steps described here. Microbiologists always run at least three replicates at each antibiotic concentration. In this case, the test uses five replicates per concentration.

The results in the table indicate that the antibiotic at an in vitro concentration of 0.01 mg/ml may be effective against the pathogen, but the MIC is reported as 0.05 mg/ml, because this is the lowest drug concentration that inhibited all growth. Because the MIC test can give an indication of a pathogen's response to an antibiotic, these tests are sometimes called *sensitivity testing.*

Microbiologists and physicians understand that MIC results do not translate directly to conditions in a patient's body. The test may be most valuable for comparing different antibiotics. For example, a clinical microbiologist would set up MIC tests for antibiotics A through E. After incubating the tubes and observing the results, the microbiologist might record the following MIC values for the five antibiotics:

- A = 0.001 mg/ml
- B = growth at all concentrations tested
- C = 0.005 mg/ml
- D = 0.0001 mg/ml
- E = 0.1 mg/ml

The MIC results given here indicate that antibiotics B and E are ineffective against the pathogen. Although antibiotic E has an MIC of 0.1 mg/ml, this concentration is very high for a typical MIC test, and a microbiologist would infer that this drug would not be effective in the body. Antibiotics A and C possess some activity against the pathogen, but antibiotic D gives a result superior to those of all the other antibiotics. From this MIC comparison, a clinical microbiologist would report to a physician that antibiotic D has the best chance of treating the patient.

The simple examples given here require a large number of tubes and the labor required to prepare the assays. Automated MIC tests, simultaneously, evaluate many different antibiotics against more than one clinical isolate. Automated MIC testing has the following four advantages over manual tests: (1) speed to results, (2) ease of use, (3) reproducible results, and (4) large numbers of replicates. Automated methods require, however, expensive equipment and routine equipment maintenance.

The MBC test expands on the information microbiologists receive from MIC testing. The MIC test does not indicate whether an antibiotic kills pathogens or merely inhibits their growth without killing them. MBC testing determines whether the test organism has been killed or inhibited.

In the table on page 533, the MIC equals 0.05 mg/ml. To determine MBC, a microbiologist inoculates the five 0.05 mg/ml replicate tubes with 10–100 pathogen cells. The microbiologist reincubates the tubes and interprets the possible results, as described in the table.

The MBC is always equal to or greater than the MIC. To narrow down the MBC from the preceding tests, a microbiologist reinoculates not only the 0.05-mg/ml tubes but also the 0.1-mg/ml tubes and additional replicates at higher concentrations.

Microbiology laboratories use Mueller Hinton broth for MIC or MBC testing. The Becton, Dickinson and Company's Web site has explained how this growth medium developed in microbiology: "Because clinical microbiology laboratories in the early 1960s were using a wide variety of procedures for determining the susceptibility of bacteria to antibiotic and chemotherapeutic agents, Bauer, Kirby, and others developed a standardized procedure in which Mueller Hinton Agar was selected as the test medium." This medium is also available in a broth formula.

MIC AGAR TESTS

In the 1950s, the medical microbiologist William Kirby and his colleague, Alfred Bauer, at the University of Washington School of Medicine began reviewing all of the scientific articles that had been published on susceptibility testing. The World Health Organization (WHO) noted Kirby and Bauer's literature review shortly after its publication and, in 1961, used the extensive research to recommend a standard procedure for all clinical antibiotic testing. Kirby, Bauer, and their colleagues then began developing a standard procedure for testing antibiotic activity on agar medium.

The Kirby-Bauer method built upon previous methods, in which small paper disks saturated with different antibiotics were placed on agar already inoculated with a sheet of bacterial culture. After incubation, a microbiologist could quickly assess the strength of the antibiotics by measuring the zone of clearing, due to cell lysis, surrounding the disk. In their 1966 article, "Antibiotic Susceptibility Testing by a Standardized Single Disk Method," the authors explained, "The importance of standardizing test conditions cannot be overstressed." As a result, the Kirby-Bauer procedure of disk diffusion on Muel-

MBC Test Results

Results in Replicate Tubes	Interpretation Regarding Antibiotic in the MIC Test
+++++	inhibitory but not bactericidal at 0.05 mg/ml
+---+	MBC is greater than 0.05 mg/ml
-----	bactericidal at 0.05 mg/ml

ler Hinton agar became the accepted method for all clinical laboratory antibiotic testing.

Clinical microbiologists use two main agar tests for determining an antibiotic's MIC against a specific pathogen: the Kirby-Bauer disk diffusion test and the E test. Both methods use Mueller Hinton agar. In disk diffusion, antibiotic spreads outward into the inoculated agar from the circular disk. The E test replaces the disk with a strip that also contains a scale that estimates MIC on the basis of the size of the clear zone.

Susceptibility testing is one of the most important procedures in the care of patients who have infectious disease. MIC testing enables clinical microbiologists to assess the relative strengths of antibiotics against clinical isolates. This assessment helps physicians choose the most effective antibiotics for arresting infections.

See also DIFFUSION; SUSCEPTIBILITY TESTING.

Further Reading

Bauer, Alfred W., William M. M. Kirby, John C. Sherris, and Marvin Turck. "Antibiotic Susceptibility Testing by a Standardized Single Disk Method." *American Journal of Clinical Pathology* 45 (1966): 493–496.

Becton, Dickinson Technical Center. "Mueller Hinton Agars." Available online. URL: www.bd.com/ds/technicalCenter/inserts/Mueller_Hinton_Agars.pdf. Accessed January 16, 2010.

Hudzicki, Jan. "Kirby-Bauer Disk Diffusion Susceptibility Test Protocol." December 8, 2009. Available online. URL: http://www.microbelibrary.org/asmonly/details.asp?id=2999. Accessed January 2, 2010.

RapidMicrobiology.com. "Antibiotic Sensitivity Testing—Establishing the Minimum Inhibitory Concentration." Available online. URL: www.rapidmicrobiology.com/test-methods/Antibiotic-Sensitivity-Testing.php. Accessed December 20, 2009.

mode of action The term *mode of action,* or *mechanism of action,* refers to the manner in which an antibiotic or a chemical called a biocide acts on a microbial cell. Antimicrobial compounds injure microorganisms in either of two main ways: by attacking the microbial cell's structure or by interfering with normal cell physiology. Mode of action may, thus, be considered as either physical or metabolic effects, respectively, on microorganisms.

ANTIBIOTIC MODE OF ACTION

Antibiotics can injure microbial cells by either physical or metabolic routes, but the interference of metabolism is the more common mode of action. One of various ways in which antibiotics are classified relates to the compound's main mode of action against bacteria or fungi. Against bacteria, antibiotics fall into five different groups, according to mode of action, as follows:

1. inhibitors of cell wall synthesis
2. inhibitors of protein synthesis
3. inhibitors of nucleic acid synthesis
4. antibiotics that injure membranes
5. inhibitors of the synthesis of essential metabolites

Antibiotics directed against fungi can be categorized into the following three groups:

1. disruptors of membrane sterols
2. disruptors of cell walls
3. inhibitors of nucleic acid synthesis

A small group of antibiotics fit into none of the main categories listed here. For example, the antifungal antibiotic griseofulvin interferes with mitosis, the process in which eukaryotic cells duplicate their chromosome and then divide the cell cytoplasm, just prior to replicating.

Classification schemes also use the antibiotic's molecular structure, and since structure very often relates to mode of action, this method of categorizing antibiotics gives microbiologists added information about the appropriate antibiotic to use against certain microorganisms. For example, beta-lactam (β-lactam) antibiotics, such as penicillin, disrupt the peptidoglycan synthesis bacteria need for building new cell wall. The relationships between mode of action and antibiotic structure are shown in the table on page 534.

Not all of the steps in antibiotic mode of action are known for each antimicrobial compound, but microbiologists and chemists have studied antibiotics, since the 1940s, and have deciphered the main points of how antibiotics work. Many aspects of these mechanisms involve complex biochemistry involving macromolecule synthesis or nucleic acid metabolism. The table on page 535 describes the essential features of antibiotic actions against bacteria.

Antibiotic Groups by Mode of Action

Mode of Action	Structure Type
ANTIBACTERIAL ANTIBIOTICS (EXAMPLES ARE IN PARENTHESES)	
inhibitors of cell wall synthesis	β-lactam; polypeptide; antimycobacterium
inhibitors of protein synthesis	aminoglycosides; tetracyclines; macrolides
antibiotics that injure membranes	polymyxins
inhibitors of nucleic acid synthesis	rifamycins (rifampin); quinolones
inhibitors of synthesis of metabolites	sulfonamides
ANTIFUNGAL ANTIBIOTICS	
disruptors of membrane sterols	polyenes (amphotericin B); azoles (clotrimazole, miconazole, ketoconazole); allylamines (naftifine)
disruptors of cell wall structure	echinocandins (caspofungin)
inhibitors of nucleic acid synthesis	flucytocine

The action of an antimicrobial compound can be helped by a second compound in an interaction called *synergism.* In antibiotics, synergism allows each antibiotic to work at a lower dose than if each were given alone. Synergism does not work, however, with any random antibiotic pairing. Antagonism, in which one antibiotic blocks the action of the other, can occur. For example, tetracycline acts as an antagonist to penicillin's action on cell wall peptidoglycan synthesis.

A general rule exists in antibiotic mode of action: that static, or inhibitory, antibiotics and cidal, or lethal, antibiotics are antagonistic to each other's mode of action. This depends on whether the antibiotic targets resting cells or multiplying cells. In general, static antibiotics target resting cells, and cidal antibiotics target cells in active multiplication. The clinical microbiologist L. P. Garrod noted, in 1973, that these general rules have many exceptions. "It is well recognized, for instance," he stated in a letter to the *British Medical Journal,* "that sulphonamides, which are purely static, do not interfere with the cidal action of penicillin." In other instances, Garrod found antagonism to hold to general patterns. Thus, the following antibiotics have been found to be antagonistic.

- streptomycin and chloramphenicol
- rifampin and penicillin
- methicillin and chloramphenicol

The effects of synergism and antagonism have been complicated with the emergence of bacteria resistant to most of these antibiotics. Doctors can no longer prescribe two antibiotics they believe to be synergistic without first testing the reaction between the antibiotics against bacteria in a clinical microbiology laboratory.

BIOCIDE MODE OF ACTION

Biocides usually disrupt the physical structure of microorganisms as their main effect, rather than interfering with metabolism. Strong biocides, such as chlorine, exert such fast and damaging effects on microbial cells that microbiologists have yet to define chlorine's mode of action fully, even though it has been used for centuries.

Chlorine's two main effects on microbial cells are probably a generalized damage to the membrane and the liberation of hypochlorite ions, which rapidly damage enzymatic functions inside and outside the membrane. In the first mode of action, chlorine combines with proteins of the cell membranes to form N-chloro compounds, which further act on cell metabolism. In the second mode of action, hypochlorite ions may shut down enzyme function, electron equilibrium across membranes, or both.

The reactions of chlorine in aqueous solutions, shown in the equations, are responsible for liberating the cidal activities of sodium hypochlorite (NaOCl), better known as household bleach. In water, NaOCl dissociates and forms hypochlorous acid (HOCl),

Key Features of Antibiotic Mode of Action against Bacteria

Mode of Action Key Feature	Example Antibiotic
INHIBITORS OF CELL WALL SYNTHESIS	
inhibit transpeptidation enzymes that build cross-linking in cell wall peptidoglycan	penicillin
activate cell wall lytic enzymes	penicillin
bind to select amino acids to prevent transpeptidation	vancomycin
interfere with cross-membrane transport system that supplies cell wall precursor compounds	bacitracin
INHIBITORS OF PROTEIN SYNTHESIS	
bind 30S ribosome subunit to inhibit synthesis of proteins with the correct amino acid sequence	streptomycin
bind 50S ribosome subunit to block peptide bond formation in proteins	chloramphenicol
bind to 30S ribosome subunit to block binding with transfer RNA (tRNA)	tetracycline
bind 50S ribosome subunit to stop peptide chain elongation	erythromycin
ANTIBIOTICS THAT INJURE MEMBRANES	
bind to membrane to disrupt structure and permeability	polymyxin B
INHIBITORS OF NUCLEIC ACID SYNTHESIS	
inhibit deoxyribonucleic acid (DNA) gyrase enzyme that forms the DNA supercoil	ciprofloxacin
block ribonucleic acid (RNA) synthesis by binding to DNA-dependent RNA polymerase enzyme	rifampin
INHIBITORS OF SYNTHESIS OF ESSENTIAL METABOLITES	
block folic acid synthesis	sulfonamide
interfere with glycine and serine metabolism	trimethoprim
inhibit mycolic acid (long lipid compound) required for cell walls	isoniazid

which is thought to be the principal biocidal form of bleach.

$$Cl_2 + H_2O \rightarrow HOCl + H^+ + Cl^-$$

$$HOCl \leftarrow\rightarrow H^+ + OCl^-$$

Normal tap water provides calcium that participates in intermediate stages of these reactions. Hard water, that is, water high in minerals and calcium, decreases the effectiveness of bleach for killing microorganisms. The solution's pH also affects chlorine's efficacy. A relationship exists among HOCl, OCl^-, and pH (the concentration of H^+ ions) in which the cidal activity of bleach can be optimized.

Chlorine- and iodine-containing biocides produce very reactive molecules in aqueous solution. These molecules, as shown in chlorine's case, destroy many components of the cell, almost on contact. As with chlorine, the mode of action of iodine has been only partially determined. Iodine probably kills microbial cells by the following four actions: (1) disrupting the amino (NH) portion of amino acids to halt protein function (example, lysine), (2) oxidizing the sulfhydryl (SH) group on sulfur amino acids to prevent normal protein folding (cysteine), (3) interfering with the normal bonds within the phenol ring structure of some amino acids (tyrosine), or (4) reacting with the carbon-carbon double bond of membrane fatty acids.

Peroxygen compounds are another group of highly reactive biocides. Hydrogen peroxide (H_2O_2) and peracetic acid, also called peroxyacetic acid (CH_3CO_3H), serve as the two most important compounds in this group. Peroxygen compounds kill cells by liberating oxygen from its bound form inside cells to its reactive form, which is toxic to all cells. In the reactions of H_2O_2 in cell membranes and cytoplasm, two reactive forms of oxygen appear: the superoxide ion (O_2^-) and the hydroxyl radical (OH·). The hydroxyl radical is a neutral form of the hydroxide ion (OH^-) common in many biological compounds. By losing the extra electron on the ion, the radical becomes chemically unstable and reacts with other cellular molecules, making them unstable. The hydroxyl radical is the strongest known oxidizing compound in nature.

Alcohols are biocides or antiseptics that contain a hydroxide ion or hydroxyl group on the molecule. Alcohols work mainly by denaturing proteins, meaning they cause the unfolding of proteins to take away their normal function. This is true for both structural proteins and enzymes. Because alcohols damage enzymes, they also interfere with almost every metabolic reaction inside the cell.

Detergents, or surface-active agents, work similarly to alcohols by denaturing the structure of cell components. Their three main modes of action are (1) disorganization of the cell membrane, (2) inhibition of enzymes, and (3) interference with or denaturation of transport systems and transport proteins.

Quaternary ammonium biocides (quats) have been synthesized into a wide range of structures. All of the compounds in this group have in common at least one quaternary ammonium ion, which is a nitrogen ion (N^+) bound to four other constituents of the molecule. The simplest quat is betaine, containing five carbons. One of the largest commonly used quats in disinfectants is methylbenzethonium chloride, which contains 28 carbon atoms.

Quats differ in their main modes of action, and the compound's concentration relative to other molecules in a solution may change their action to a small degree. Quats may affect microorganisms by adsorbing to the cell surface, diffusing through the cell wall, binding to membranes, disrupting membrane structure and function, or releasing ions from inside the cell, thus changing internal pH. Companies that produce new quats for disinfection or sanitization alter quat structures to achieve the best effects against pathogens. In general, quats have been designed to carry out one or more of the following specific modes of action:

- denature proteins
- separate the main active group from enzyme molecules
- interfere with enzymes in energy-generation systems in membranes
- damage membrane permeability
- interfere with metabolic pathways such as glycolysis

In addition to the common biocide groups described here, other compounds have been shown to act on microbial cells by unique processes. Dyes known as *acridines* can kill microorganisms by inserting into the DNA molecule between base subunits. This process, called *intercalation,* distorts the normal DNA structure. When DNA replication begins, the enzyme DNA polymerase cannot accurately read the sequence of DNA bases and, thus, cannot interpret information carried in the cell's genes. The cell cannot reproduce properly because of this action.

Ethylenediaminetetraacetic acid (EDTA) is another compound with a principal function other than as a biocide. EDTA is used in cosmetics and foods to stabilize formulas. EDTA also forms strong bonds around metal ions and makes important metals unavailable for microorganisms. For example, bacteria need cobalt to make vitamin B_{12} and magnesium to stabilize numerous enzymatic steps in metabolism. Through its activity, EDTA serves as a growth inhibitor, and it is now considered a preservative in many products.

A final major mode of action is produced by enzymes such as lysozyme that destroy cellular constituents. Lysozyme acts on the β, 1–4 linkages of peptidoglycan, causing bacteria to lyse. Lysozyme and any other compounds that have a major action against peptidoglycan (β-lactam antibiotics such as penicillin) are most effective against gram-positive bacteria that depend on a thick peptidoglycan layer for protection.

CONTACT TIME

No static or cidal action works instantaneously. Even the rapid action of bleach on germs needs a minimal time to work, called *contact time.* Contact time is the interval between exposure of a microbe to an antimicrobial agent and the time in which the action is complete, usually determined by cell death.

Contact time varies by the type of microorganism and its concentration and the antimicrobial agent and its concentration. Other factors, then, play secondary roles in lengthening contact time, because they interfere with the compound's ability to kill microorganisms. In general, concentration has the greatest effect on contact time, particularly when testing the action of biocides:

increased cell concentration → increased concentration of biocide needed

or increased contact time

decreased biocide concentration → increased contact time

Most biocides are formulated to work within 10 minutes or less of contact time to kill about one million bacterial cells. Quats usually require a contact time in a range of five to 10 minutes. Chlorine can work in less than five minutes and works as quickly as 30 seconds contact time if used as a sanitizer. A sanitizer lowers bacterial numbers to safe levels, while a disinfectant kills all bacteria other than bacterial spores. The following data give an example of how a disinfectant's use affects the contact time (data applicable to Clorox bleach):

- disinfecting water—eight drops bleach to one gallon (3.78 l) water; contact time is 30 minutes
- disinfecting hard surfaces of bacteria, fungi, and viruses—three-quarter cup (0.18 l) bleach per one gallon (3.78 l) water; contact time is five minutes
- sanitizing hard surfaces of bacteria—spray product onto contaminated hard, nonporous surface; contact time is two minutes

Mode of action is a critical aspect of understanding antibiotics and biocides, especially when designing new compounds to replace older, less effective versions. Many details of mode of action remain unknown. Bacterial resistance to antibiotics, and possibly to some biocides, complicates the knowledge microbiologists have of mode of action. This area of study continues to be crucial for discovering new antimicrobial agents.

See also ANTIBIOTIC; BIOCIDE; DISINFECTION; PEPTIDOGLYCAN.

Further Reading

Block, Seymour S., ed. *Disinfection, Sterilization, and Preservation,* 5th ed. Philadelphia: Lippincott Williams & Wilkins, 2000.

Clorox Company. Available online. URL: www.clorox.com. Accessed November 6, 2009.

Garrod, L. P. "Antagonism between Antibiotics." Letter to *British Medical Journal* 1 (1973): 110. Available online. URL: www.ncbi.nlm.nih.gov/pmc/articles/PMC1588805/?page=1. Accessed November 6, 2009.

Mayer, Gene. "Antibiotics—Protein Synthesis, Nucleic Acid Synthesis and Metabolism." In *Microbiology and Immunology Online.* Columbia: University of South Carolina Press, 2009. Available online. URL: http://pathmicro.med.sc.edu/mayer/antibiot.htm. Accessed November 6, 2009.

morphology Morphology is the study of all physical features of microorganisms. Along with cell metabolism and genetics, morphology is one of the three main methods for identifying microorganisms. In most microorganisms, morphology, alone, cannot be used for identifying a microorganism to species or genus. Morphology studies on unknown microorganisms do, however, help microbiologists correctly identify species.

Morphology makes up phenotype, which comprises the physical characteristics that reflect a microorganism's genetic makeup. This genetic makeup is called the *genotype.* Genotype cannot be seen in a microscope, while phenotype can be seen by viewing a microorganism microscopically and studying its metabolism by running biochemical tests.

Different types of microorganisms have their own morphological features that help in identifying them. Bacteria and yeast morphology encompasses cell shape and size, presence of appendages, presence of endospores, staining behavior, the manner in which cells group together, and colony morphology. Some microorganisms also have distinct color or cellular inclusions, special organelles that most other microorganisms lack. Fungi can also be identified by colony morphology, in addition to the microscopic features of reproductive structures, such as fungal spores. Light microscopes are adequate for studying the morphology of bacteria, yeasts, and fungi.

Virus morphology requires the use of electron microscopy, which enables microbiologists to view objects that are smaller than bacteria. Viruses belong to various classification schemes, among which is morphology of the virus particle called the virion. Virus studies cannot use colony morphology because viruses grow only inside other cells.

Because genera and some species have distinctive morphological structures, morphology has been used as an important aid to identification since the beginning of microbiology. Morphology now supplements biochemical tests and genetic analysis. (See the color insert, page C-3 [top] for a picture of a colony morphology.)

Morphological characteristics have aided taxonomists in placing microorganisms into the appropriate phylum, class, order, family, and genus. In most genera, all member species have the same cell morphology, so physical features alone cannot determine species. A morphotype is a microorganism that possesses the characteristics distinctive of its genus or species.

MORPHOLOGY STUDIES

The examination of microbial morphology entails three main aspects: (1) growth medium, (2) specimen preparation, and (3) microscopy.

The physical characteristics of colonies on agar media vary according to the medium's ingredients. Hundreds of different media have been formulated to differentiate microorganisms, mainly bacteria, from each other. The two main medium types used

Distinguishing Morphological Features of Bacteria

Feature	Description	Example
buds	produces a new cell as a bud that breaks off from the parent cell	*Hyphomicrobium*
capsules	outer layer of polysaccharides that hold water to protect against drying	*Bacillus*
fruiting bodies	a ball of resting cells held by a stalk	*Myxococcus* (resting cells called myxospores)
glycocalyx	network of polysaccharides that bind cells together in a protective matrix	*Pseudomonas*
magnetosomes	long chain of magnetic particles in the cytoplasm, used for chemotaxis	*Aquaspirillum magnetotacticum*
parallel membranes	layers of folded membrane traversing the cell, possibly for greater surface area for metabolic activities	*Nitrocystis*
shape	square, possibly for buoyancy	Walsby's square bacteria
size, large	750 µm diameter	*Thiomargarita namibiensis*
size, small	0.02–0.03 µm diameter	nanobacteria
S-layer	protective surface layer in a pattern resembling floor tiles	*Deinococcus* (common in archaea)
stalks	anchor cells to submerged surfaces	*Caulobacter*

in morphology are selective and differential media. Selective media contain ingredients that suppress the growth of some microorganisms, while promoting the growth of other microorganisms. A microorganism's ability to grow on selective medium and colony appearance help identify genera or larger groups. Differential media distinguish one type of microorganism from another on the basis of colony morphologies. For example, the differential medium SS agar distinguishes among various enteric bacteria by the following colony colors:

- *Salmonella* species—colorless colonies with black centers
- *Shigella* species—colorless colonies with good growth
- *Enterococcus* species—colorless colonies with weak growth
- *Escherichia coli*—pink to red colonies

Differential media have sped the process of identifying unknown bacteria.

Specimen preparation involves growing a pure culture of a microorganism on agar to study its colony characteristics. After noting the size, shape, and other physical features of colonies, a microbiologist can carry out fixation for use in light microscopy. In this process, a microbiologist transfers a small amount of a single colony to a glass microscope slide and heats the specimen by passing the slide briefly over a Bunsen burner flame. The specimen, thus fixed to the glass slide, may be stained with any variety of biological stains that make cells easier to view in a microscope. The Gram stain is the main staining technique for bacteria.

The advent of compound microscopes enabled early microbiologists to view the microbial world in greater detail than they could with a single lens. In the 1980s, electron microscopy became an additional tool for studying the ultrastructure of bacteria and the shape and size of viruses. Ultrastructure consists of the subcellular features of a microorganism, such as organelles, small surface appendages, and cell wall structure.

Electron microscopy requires different fixation chemicals and staining procedures than light microscopy. To observe specimens in electron microscopy, they must be sliced to a thickness measured in micrometers (µm). Electron microscopy also uses metals as part of its stain so that the specimen will diffract an electron beam to produce an image on a screen.

Specimens for light microscopy that do not require fixation may be viewed as a wet mount, in which liv-

Bacterial Cell Arrangements

Name	Description	Example
single	single cells not attached to others	*Bacillus*
pairs	diplococcus or diplobacillus	*Neisseria*
short chains	streptococci or streptobacillus, fewer than five cells	*Streptococcus*
long chains	streptococci, more than five cells	*Streptococcus*
tetrads	four cells	*Staphylococcus*
octets	sarcina, eight cells	*Sarcina*
clumps	grape clusters	*Staphylococcus*
filaments	long, thin arrangements	*Actinomyces*

ing cells are suspended in a small amount of liquid. Wet mounts work well for observing the presence of motility and the source of the motility.

Light microscopy with fixed specimens or wet mounts and electron microscopy serve as important tools in microbial identification. Regardless of the type of microorganism, microbiologists use some or all of the following features for classification and identification:

- cell shape
- cell size
- colony morphology
- color
- distinguishing organelles
- endospore production and morphology
- flagella and cilia
- organelles
- spore morphology
- type of motility

Microbiologists key on certain of these features that correlate with specific genera, then supplement the identification with biochemical and genetic testing.

INTERNAL CELL MORPHOLOGY

The main differentiating factor between prokaryotes and eukaryotes relates to internal morphology of these respective cells. Prokaryotes, the bacteria and archaea, have simpler morphology and lack membrane-bound internal organelles. Eukaryotes contain membrane-bound organelles and have a more complex internal structure, in general, than prokaryotes. Eukaryotes consist of fungi and yeasts, protozoa, plant cells, and animal cells.

All cells have four structures in common, although these organelles often function and are constructed differently in prokaryotes and eukaryotes: cell wall, plasma membrane, ribosomes, and vacuoles. Some species of each type of organism also have flagella.

Internal cell morphology rarely can be used as an identifying feature of bacteria. Some bacteria, however, have unusual features that distinguish them from most other species. The table on page 538 describes the major special features in bacteria.

As the technology of microscopes advances, additional new features will probably be discovered. Square bacteria, for instance, were discovered, in 1980, by the British microbiologist Anthony Walsby, but microbiologists did not culture the new microorganism—proposed name is *Haloquadratum walsbyi*—in a laboratory until 2004.

CELL MORPHOLOGY

Size differentiates most bacteria from yeasts. The majority of bacteria measure 0.2–0.5 μm wide and 0.5–2.0 μm long, and the majority of yeasts measure 3–5 μm diameter, or end to end. But some species grow to sizes far outside these ranges; the absolute size range of bacteria is 0.2–750 μm, and certain yeast species grow to 40 μm.

After size, microbiologists note the shape of the cells. Bacteria have the following shapes, which are distinctive for each species:

- cocci
- tapered at both ends (fusiform)
- coccobacillus—oval rods
- straight rods, including square-ended or round-ended
- rectangular
- curved rods
- spirals
- branched

- helices—twisted (helix) rod
- spirochetes—long corkscrews
- star-shaped

Some genera, such as *Corynebacterium,* display cells of various shapes. These are called *pleomorphic* shapes.

Yeast cells tend to be rounder than bacteria, with species having more oval or elongated shapes. Yeast colonies also grow slightly larger than bacteria on agar media. The use of yeast morphology agar with various nitrogen and carbon sources helps to differentiate several yeast species. A variety of growth responses include poor to no growth, cells with hyphae, colony color, and colony size.

Yeast cells usually remain single-celled without grouping together. Bacteria, however, often grow into characteristic arrangements as cells multiply. The table on page 539 describes multicell bacterial arrangements.

Bacterial cell arrangements develop out of the type of cell division, called *binary fission,* that bacteria undergo. For example, streptococci divide in a manner that forms an ever-lengthening chain of cells. Some species of *Streptococcus* develop branch chains off the main chain when they grow on specific sugars. *Staphylococcus* bacteria divide into diplococcic, which grow into tetrads, then octets, then form less symmetric clumps.

Various bacterial species have appendages on the cell's outer surface that aid motility and exchange of genetic material. The three main appendages are flagella for movement, fimbrae for attachment to surfaces, and pili for exchange of genetic material.

Flagella (singular; flagellum) are appendages 15–20 μm long that move in a whiplike fashion to propel the cell. The microbiologist Lori Snyder of the University of Birningham, United Kingdom, wrote in 2008, "The bacterial flagellum, the chief organelle of motility among bacteria, is a molecular motor and self-assembling nano-machine, built from approximately two dozen proteins." Bacterial flagella have four main arrangements, based on their orientation on the cell or their number:

- monotrichous—a single polar (at one end) flagellum
- amphitrichous—a group of flagella at each end
- lophotrichous—two or more flagella at one pole
- peritrichous—several flagella distributed over the entire cell

A bacterial flagellum consists of three parts. The filament comprises the long tail made of the different types of the protein flagellin, depending on species. The filament attaches at the cell membrane to the second piece, called the *hook.* Made of protein distinct from flagellin, the hook is constructed as a curved tube into which the flagellum docks. The hook and flagellum, then, attach to a complex structure called the basal body. The basal body straddles the outer and inner membranes in gram-negative bacteria and the peptidoglycan membrane in gram-positive bacteria.

The basal body that powers flagellar movement represents one of nature's cleverest engineering feats. The basal body in gram-negative bacteria is more complex than in gram-positive structures. The basal body contains five components, four disks called rings and a center rod that connects two rings in the outer membrane (the L ring and P ring) to the rings of the inner plasma membrane (the S ring and the M ring). The M ring, the largest of the four rings, measures about 20 nm in diameter.

The bacterial basal body has attracted attention from researchers because of its efficient conversion of electrochemical energy, a proton flow, into kinetic energy of movement. In monotrichous and lophotrichous bacteria, when flagella rotate counterclockwise (viewed from the tip), the cell swims forward. Clockwise rotation causes the cell to execute a random movement called *tumbling.* Many of the basal body's energy mechanisms and its function as a motor remain unknown.

Fimbrae are short, hairlike appendages that occur more often on gram-negative than gram-positive bacteria. These appendages number between 100 and 1,000, covering the entire cell surface or a region. Fimbrae measure 2–8 nm in width and about 100 nm in length and are too small to be seen by light microscopy. Electron microscopy has revealed that fimbrae lie on the surface as a tangled web at times and at other times extend.

Fimbrae act as an attachment device for adhering to surfaces. Pathogenic bacteria use fimbrae to attach to the mucosal membranes in the body, before initiating infection. In this role, fimbrae are considered one of many virulence factors that enable pathogens to invade the body. Other speculated uses for fimbrae are in cell-to-cell adhesion, motility, and cell protection from other cells in the environment or physical factors.

Pili (or sex pili) are slightly longer than fimbrae and number only a few to 10 per cell. Pili are not well understood but are believed to function in the direct transfer of genetic material between two cells. This process is called *conjugation,* a type of gene transfer.

Spirochetes, which are spiral-shaped cells, have unique outer structures called *axial filaments.* Axial filaments form ridges that spiral around the elongated

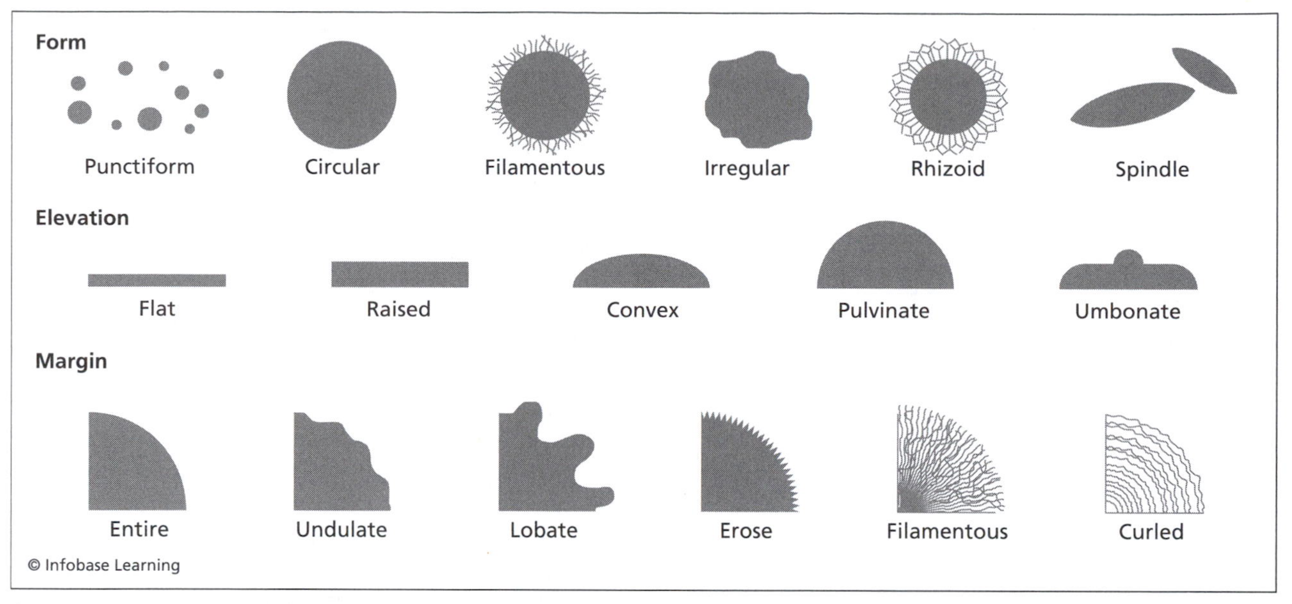

Observations of colony morphology provide an easy first step in the identification of unknown bacteria. Although colony morphology alone does not identify a microorganism, many species develop distinctive colonies that give a clue to identity.

cell in a candy cane fashion. As spirochetes move through a liquid, the axial filaments make the cells move forward in a corkscrew motion.

FUNGAL MORPHOLOGY

Fungal morphology differs from bacterial morphology, in that microbiologists rarely study a whole fungal cell. Identification of fungi by morphology focuses on the following six features:

- arthrospores—rod-shaped fragments of hyphae
- chlamydospores—swellings inside the hyphae
- conidia—thin-walled packets called a spore held by a hypha
- hyphae—long filaments that extend the growth of fungi outward into the environment
- sclerotic bodies—a mass of mycelia (a large tangle of hyphae)
- sporangia—spore-containing sacs

Mycologists observe the size and shape of arthrospores, chlamydospores, conidia, and sporangia to identify fungal species. Hyphae can be of two types: (1) Septate hyphae, which have septa, or walls that divide the hyphae into segments, and (2) coenocytic hyphae, which lack septa. Finally, the presence, size, and shape of sclerotic bodies provide information on fungal identity.

PROTOZOAL MORPHOLOGY

Protozoologists study protozoal morphology of single live cells in wet mounts. The main features of protozoal morphology are size, shape, number and positioning of appendages, presence of an oral groove called a cytosome, size and number of nuclei, characteristics of the mitochondria, and presence or size of food vacuoles. Some species, such as *Trichomonas vaginalis,* have a distinguishing undulating membrane, which is a membrane flap extending from the cell and bordered by a flagellum. Other species, such as *Giardia lamblia* and *Cryptosporidium parvum,* form a protective cyst that allows the microorganism to live outside a host.

Pliable protozoa cells may be either structured, so that a misshapen cell returns to its original shape, or amoeboid, in which the cell constantly changes shape. Amoeboid cells develop pseudopodia by extending part of their outer membrane and allowing cytoplasm to rush into the extension. By using pseudopodia, amoeba can "walk" through the environment.

Two main morphological groups in protozoa are flagellated and ciliated cells. Flagellated protozoa, or flagellates, contain one or more flagella. Different species vary in the number of flagella they have, from a few to hundreds that cover the cell like hairs. Eukaryotic flagella's composition and mechanism of action differ from those of bacterial flagella, but they function similarly in motility. Ciliated protozoa, or ciliates, have threadlike appendages that coordinate waves of motion along the cell surface for moving the cell. These cilia, at 7–10 µm in length, are shorter than flagella but enable ciliates to move much faster than flagellates.

COLONY MORPHOLOGY

Colony morphology is the appearance of bacteria, yeast, or fungal colonies grown on agar media. Colony morphology is a major component of identification and has been enhanced, in the past century, with the development of differential media. These medium formulas result in colonies of various size, color, shape, and presence of extracellular secretions. For example, beta-hemolytic staphylococci grown on blood agar lyse red blood cells near them to form a clear zone around the colony.

The main components of colony morphology are (1) form, (2) elevation, (3) margin, (4) color, and (5) texture. Form refers to the shape of the entire colony. Elevation is the height of the colony from the agar surface, and margin is the shape of the colony's outermost edge.

Microbiologists often inspect the underside of colonies grown on agar. In bacteria, microbiologists observe the translucence of colonies, and, in fungi, they note the colony underside's color.

Morphology is one of the oldest aspects of microbiology. Although morphology served as the main identification aid for almost 200 years, it has been supplanted, in part, by biochemical testing of microbial metabolism and genetics. Morphology will, nevertheless, remain the first characteristic a microbiologist interprets when examining unknown microorganisms in laboratory conditions.

See also BACTERIA; EUKARYOTE; GENE TRANSFER; GRAM STAIN; IDENTIFICATION; ORGANELLE; PROKARYOTE; PROTOZOA; STAIN.

Further Reading

Calkins, Gary N. *Protozoa Morphology and Physiology.* Palm Springs, Calif.: Wexford College Press, 2007.

Larone, Davise H. *Medically Important Fungi: A Guide to Identification.* Washington, D.C.: American Society for Microbiology Press, 2002.

Real World: Microbes in Action. "Introduction to Bacteria: Morphology and Classification." Available online. URL: www.umsl.edu/~microbes/pdf/introductiontobacteria.pdf. Accessed October 11, 2009.

Snyder, Lori A. S., Nicholas J. Loman, Klaus Fütterer, and Mark J. Pallen. "Bacterial Flagellar Diversity and Evolution: Seek Simplicity and Distrust It?" *Trends in Microbiology* 17 (2008): 1–5. Available online. URL: www.sciencedirect.com/science?_ob=ArticleURL&_udi=B6TD0-4V4130J-1&_user=10&_coverDate=01%2F31%2F2009&_rdoc=1&_fmt=high&_orig=browse&_sort=d&view=c&_acct=C000050221&_version=1&_urlVersion=0&_userid=10&md5=a57b19d6ed511 18de760fa72ab414c60. Accessed October 11, 2009.

Walsby, Anthony E. "A Square Bacterium." *Nature* 283 (1980): 69–71.

most probable number The most probable number (MPN) method is a technique for estimating the number of bacteria in a sample without inoculating agar medium. MPN is also called the *multiple-tube method* for enumerating bacteria, because it relies on dilution of the sample in five replicate test tubes, incubation, and then observation of the dilutions for the presence or absence of growth.

Environmental testing laboratories use MPN for determining the bacterial count (number of bacteria) in natural and treated water samples. Each tube in the MPN method contains broth medium that can be formulated in two ways. In the first method, the medium contains a variety of nutrients to support the growth of many different types of bacteria. In water microbiology, these diverse bacteria are called *heterotrophic bacteria.* MPN testing that enumerates heterotrophic bacteria gives an indication of the overall bacteria, called *bacterial load,* in the sample. In the second method, MPN tubes contain a selective medium, which promotes the growth of only a particular type of microorganism. This application of MPN can estimate the numbers of coliform bacteria in water. Coliform bacteria are a general group of gram-negative species common in natural waters, which may indicate fecal contamination. For this reason, coliform bacteria are called *indicator organisms:* They indicate the possible presence of other microorganisms.

Microbiologists usually use five replicate tubes at each dilution in the MPN method, but the technique also works with three replicates or 10 replicates. The greater the number of replicate tubes, the more accurate will be the estimate of bacterial numbers in the

MPN Test Preparation

Dilution	Undiluted (10^0)	1:10 (10^{-1})	1:100 (10^{-2})	1:1,000 (10^{-3})	1:10,000 (10^{-4})	1:100,000 (10^{-5})
aliquot to broth replicates	1 ml	1 ml	1 ml	1 ml	1 ml	1 ml
number of replicates	5	5	5	5	5	5

MPN Results in a Five-Tube Test

Dilution	Undiluted (10^0)	1:10 (10^{-1})	1:100 (10^{-2})	1:1,000 (10^{-3})	1:10,000 (10^{-4})	1:100,000 (10^{-5})
replicate 1	+	+	+	0	0	0
replicate 2	+	+	0	0	0	0
replicate 3	+	+	+	0	0	0
replicate 4	+	0	+	+	0	0
replicate 5	+	+	0	0	0	0
results		3	1	0		

Note: + = growth; 0 = no growth

original sample. Five-replicate tests offer accuracy and ease of use.

Microbiologists can enumerate bacteria in four major ways—(1) agar plating, (2) filtration counting, (3) direct counting—in addition to MPN. Agar plating involves two steps, beginning with serial dilution of the sample in sterile water, saline solution, or buffer (called the *diluent*). The microbiologist, then, inoculates a small volume from each of the most dilute subsamples to agar medium. After, incubation, the second step involves counting the number of colonies on the agar and correcting for the dilution to determine the number of bacteria in the original sample. This procedure is labor-intensive, but accurate and counts only living bacteria. Filtration counting also employs agar but differs from plating by passing a known volume of sample through a filter before laying the filter on the agar medium. After incubation, the number of colonies on the filter indicates the bacterial load of the filtered sample.

Five-Tube MPN Table

Number of Tubes Positive for Growth in: FIRST SET, 10^{-2}	MIDDLE SET, 10^{-3}	LAST SET, 10^{-4}	MPN in the Middle Set of Tubes
2	3	0	0.12
3	0	0	0.08
3	0	1	0.11
3	**1**	**0**	**0.11**
3	1	1	0.14
3	2	0	0.14
3	2	1	0.17
4	0	0	0.13
4	0	1	0.17

The direct counting method involves enumerating the bacteria in a very small sample size in a microscope. This method is tedious and may be inaccurate because of the small sample volume of less than one microliter (μl). Direct counting determines the number of both living and dead bacteria.

The MPN method gives an estimate of probable bacterial numbers rather than an exact bacterial concentration. It has the advantage, however, of allowing microbiologists to estimate the numbers of specific types of bacteria, such as coliforms, nitrogen utilizers, or acid producers. In addition to water testing, MPN may be applied to soil and food microbiology.

PRINCIPLE OF MPN

Mac H. McCrady developed the MPN method, in 1915, and the public health statistician William Cochran (1909–80) adapted the method for routine laboratory work. Cochran explained, in a 1950 article, "Estimation of Bacterial Densities by Means of the 'Most Probable Number,'" "The method consists in taking samples from the liquid [water or milk], incubating each sample in a suitable culture medium, and observing whether any growth of the organism has taken place. The estimation of density is based on an ingenious application of the theory of probability to certain assumptions." The first assumption

Cochran refers to is that all bacteria are distributed randomly through the sample as single cells rather than clumps. The second assumption is that diluted inocula containing as few as one bacterial cell will grow when incubated in the broth medium.

As dilution of the original sample increases, the inocula of replicate broth tubes contain fewer bacteria per millimeter. A microbiologist inoculates several sets of replicate tubes at increasing dilutions with the intent of achieving a dilution at which no bacteria grow. This is called *diluting to extinction*. The estimation of bacterial numbers depends on attaining this dilution, in which no growth occurs in any replicate tube regardless of whether the test is a three-, five-, or 10-tube MPN.

The MPN method depends on an algorithm for calculating probability that the sample contains no cells as well as the probability that the sample holds one or more cells that will grow when incubated. The microbiologist uses the growth results in the highest three dilutions to determine bacterial numbers, providing that the highest dilution contains no growth. After incubation, microbiologists observe the tubes for one of two results: Cloudy broth indicates growth and clear broth indicates no growth.

MPN can be run manually or with some of the steps automated. Automation has helped speed the dilution of samples and inoculation of tubes. Scanning instruments determine the number of growth tubes and nongrowth tubes. These advances enable microbiologists to test much larger replicate sets than in the past, thereby increasing the estimate's accuracy. A small-replicate manual test is, nevertheless, the best way to understand the principles of MPN.

MPN METHOD

Any microbial enumeration method begins with a representative sample in which a small sample volume accurately represents a larger system. For example, a water microbiologist who samples a lake for water quality takes 100 milliliters (ml) of water that seems to have the same cloudiness, color, silt levels, and other visible attributes characteristic of the entire lake. The microbiologist, then, dilutes the sample in 10-fold dilutions, by transferring 1 ml of sample to 9 ml of sterile diluent, and continuing this same process for several dilutions: The 1:10 dilution makes a 1:100 dilution, which makes a 1:1,000 dilution, and so forth.

The microbiologist transfers a 1-ml volume, called an *aliquot,* from each dilution to each of the replicate broth tubes for that dilution. The table at the bottom of page 542 provides an example of how to set up a five-replicate MPN test.

In the example shown, the microbiologist inoculated each of five replicate tubes with 1 ml of dilution, but the method works with any size aliquot and any size volume of broth medium. Microbiologists select volumes that save on the amount of medium to be prepared and the space the tubes occupy in an incubator.

Soil and food microbiologists adapt the MPN method to solid samples, by adding progressively smaller amounts of sample to the diluent tubes. For example, instead of the dilution series shown in the table, the microbiologist might add the following grams of sample to the replicate sets: 1.0, 0.1, 0.001, and 0.0001 gram.

Following the setup shown in the table, a microbiologist records the growth/no growth results in each tube after incubation. The table at the top of page 543 gives an example of a possible MPN outcome.

From the five-tube test results, a microbiologist selects the data from the 10^{-2}, 10^{-3}, and 10^{-4} dilutions. The results of 3, 1, and 0 derive from the number of tubes containing growth at each of those dilutions of the original sample. These three dilutions, in sequence, have been selected because they show the point at which the bacteria were diluted to extinction. The bacteria have been diluted to extinction at the 10^{-4} dilution, so the 10^{-3}dilution will be used for estimating the MPN of bacteria in the original sample. Statistical tables have been published that interpret the results from the three-, five-, or 10-tube method. A microbiologist uses the appropriate table for the type of MPN method performed. A portion of a five-tube MPN table is shown at the bottom page 543 .

For the MPN determination in this situation, the microbiologist uses the fourth row, containing the results 3, 1, and 0. For this row, the MPN value of 0.11 is used to determine the number of bacteria in the original sample. The MPN for this sample is 0.11×10^3 bacteria per milliliter of original sample or 1.1×10^4 bacteria per milliliter.

In summary, the MPN method of enumerating bacteria involves the following steps:

1. Dilute the original sample.
2. Add aliquots of each dilution to an individual set of replicate broth tubes.
3. Incubate the broth tubes.
4. Record the growth/no growth results for all tubes in all replicate sets.
5. Total the number of tubes positive for growth in each replicate set.

6. Select the three sets that demonstrate dilution to extinction.

7. Look up the results in the appropriate table to find the MPN of bacteria in the sample.

The time required to receive results after completing these steps ranges from 24 to 48 hours, with most of the time consumed by the incubation period.

MPN FOR COLIFORMS, FECAL COLIFORMS, AND *ESCHERICHIA COLI*

Coliforms, fecal coliforms and *E. coli* serve as indicators of possible contamination of water and food. Because the coliform group of bacteria can be present in many natural waters, fecal coliforms are now considered a better indicator of fecal contamination, mainly for testing shellfish or shellfish-harvesting waters. Microbiologists currently use coliforms as a general indicator of unsanitary conditions in drinking water and food production. The presence of *E. coli* indicates recent contamination with fecal matter.

The MPN method can be adjusted to determine the numbers of these three major indicator organism types by the following procedures:

- coliforms—incubate in lauryl tryptose broth to detect gas production from lactose. Transfer an aliquot from tubes positive for gas production to brilliant green lactose bile (BGLB) broth to detect the ability to grow in the presence of bile and produce gas at 95°F (35°C) for 48 hours. Use the MPN method on tubes positive for both gas and growth in the presence of bile

- fecal coliforms—transfer small aliquot from each gas-producing tube to EC broth. Check for the presence of growth after incubation at 113.9°F (45.5°C), for 24 hours, and determine MPN for these tubes

- *E. coli*—transfer a small aliquot from fecal coliform–positive tubes to agar medium selective for *E. coli;* perform Gram stain and additional biochemical tests

The MPN remains a standard method in many laboratories that determine bacterial load in natural waters, drinking water, treated wastewater, soil, and some foods. Its main disadvantage is the use of statistical probability to estimate bacterial numbers, rather than a determination of actual bacterial numbers. An advantage resides in the ability of this method to give simultaneous results on certain types of bacteria, such as coliforms.

See also COLIFORM; CULTURE; INDICATOR ORGANISM; SERIAL DILUTION; WATER QUALITY.

Further Reading

Cochran, William G. "Estimation of Bacterial Densities by Means of the 'Most Probable Number.'" *Biometrics* 6 (1950): 105–116. Available online. URL: www.unc.edu/depts/case/BMElab/MPNcalculator/Cochran_biometrics1950.pdf. Accessed October 11, 2009.

Feng, Peter, Stephen D. Weagant, and Michael A. Grant. "Enumeration of *Escherichia coli* and the Coliform Bacteria." In *Bacteriological Analytical Manual,* 8th ed. Silver Spring, Md.: Food and Drug Administration, 1988. Available online. URL: www.fda.gov/Food/ScienceResearch/LaboratoryMethods/BacteriologicalAnalyticalManualBAM/ucm064948.htm#authorswww.fda.gov/Food/ScienceResearch/LaboratoryMethods/BacteriologicalAnalyticalManualBAM/ucm064948.htm#authors. Accessed October 11, 2009.

Oblinger, J. L., and J. A. Koburger. "Understanding and Teaching the Most Probable Number Technique." *Journal of Milk and Food Technology* 38 (1975): 540–545.

motility Motility is the ability and mechanism of a microbial cell to move under its own power. Motility in microorganisms serves two main purposes: to enable cells to go toward a stimulus, such as food or light, or to enable cells to escape harmful conditions, such as chemicals or heat. Not all microorganisms are motile, and, among motile species, the manner in which they move varies by the type of microorganism.

Motility is an energy-demanding process in all species. In order to move across a surface or through liquid, the microorganism must generate enough force to overcome the force exerted on the cell by the environment. The expenditure of energy gives motile cells an advantage over nonmotile species because motility allows cells to react to stressors. Stressors are any changes in the environment. Important biochemical stressors are food, by-products, antimicrobial substances, toxic chemicals, acids and bases, salt, and oxygen level. Physical stressors are temperature, light, osmotic pressure, and barometric pressure. Motile cells do not gain an advantage over nonmotile species that can adapt to the changes instead of moving.

Motile microbial cells respond to biochemical and chemical stressors in the environment by sensing a change and signaling the cell to turn on motility mechanisms, an overall process called *chemotaxis.* Microorganisms do not use chemotaxis to recognize the presence of a stressor and react to it. Rather, chemotaxis works by the movement of a cell up or down a continuous concentration gradient. For example, a river may receive an influx of nitrogen compounds at a particular site. Bacteria that require more nitrogen respond to gradually increasing nitrogen levels, while

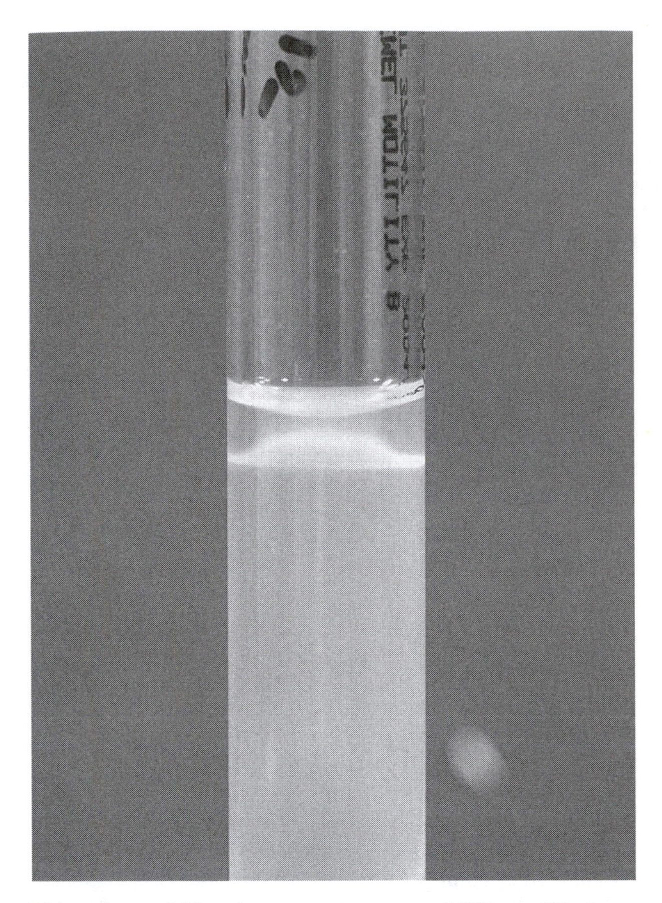

This culture of *Listeria monocytogenes* exhibits motility by growing upward into semisolid medium. ***(Clinical Laboratory Technology Department, St. Louis Community College)***

the compound disperses. The bacteria migrate until they reach an area of suitable nitrogen concentration; then, they stop. As nitrogen concentration falls, the bacteria again follow the gradient. In this case, the cells travel up a concentration gradient, from lower to higher concentration.

Two other mechanisms that are comparable to chemotaxis are phototaxis, movement in response to light, and aerotaxis, movement in response to oxygen levels. Microorganisms displaying phototaxis usually move toward light. Chemotaxis and aerotaxis both display movement toward or away from a stimulus.

TYPES OF MOTILITY

Types of motility are diverse in the microbial world. The table below describes the seven main types of motility.

Gliding motility takes place with the possible aid of two factors: pili and polysaccharides. Type IV pili are very short appendages, of no more than 10 micrometers (μm), that may help advance a gliding cell. As the microbiology authors Timothy Paustian and Gary Roberts explained the process in *The Microbial World,* the type IV pili "are extended away from the cell and stick to the surrounding surface. The microbe then pulls itself toward the tethered end by retracting the pilius back inside the cell. By repeating this process the cell drags itself along a surface." The bacteria *Myxococcus xanthus, Neisseria gonorrhoeae,* and *Pseudomonas aeruginosa* use this mechanism.

Some bacteria, such as motile cyanobacteria, extrude a polysaccharide from the cell onto the surface. The action of extruding the material may push the cell along. *M. xanthus* uses polysaccharide secretion, as well as pili, for motility. Paustian and Roberts explain: "*M. xanthus* is a social predator. It glides

Types of Microbial Motility

Type	How It Works	Microorganisms That Use It
amoeboid	cell fills an extension, called a pseudopodium, with cytoplasm; then the rest of the cell follows	amoebae
cilium-mediated	coordinated waves or beating of hairlike projections	protozoa
euglenoid	crawling along a surface by progressive waves of swelling and constriction	*Euglena*
flagellum-mediated	rotation of one or more whiplike tails,	bacteria, protozoa,
gliding	translocation of membrane proteins on a surface, then the rest of the cell follows the advancing edge	algae, bacteria, protozoa
gregarine	translocation of membrane proteins on a surface in a longitudinal orientation	*Gregarina* protozoa
spiraling	specialized structure called an axial filament, spirals around the corkscrew-shaped cell	*Rhodospirillum* bacteria

around in large groups of cells secreting toxins and degradative enzymes that kill other microbes. The leftovers of these dead microbes then provide nutrients for the marauding *Myxococcus*." Predatory bacteria of the *Bdellovibrio* genus use flagella for motility, rather than gliding.

The speed of motile bacteria ranges from a few micrometers per minute to more than 100 µm per second. For example, flagellated *Salmonella* moves at 20 µm/second, *Spirillum* travels at 50 µm/second, and *Vibrio* reaches 200 µm/second. Put another way, the *Spirillum* moves 100 times its body length in a second. Ciliated protozoa exceed absolute bacterial speeds, with smaller protozoa tending to be faster than larger cells. For example, the marine ciliate *Uronema,* which measures 28 µm long, swims faster than 450 µm/second, or 16 times its body length.

NONMOTILE MOVEMENT

Nonmotile microorganisms can move from one environment to a different one by adhering to a surface, such as an animal's fur, or by following the flow of water. Nonmotile bacteria also exhibit movements that have been confused with true motility. These actions are called *Brownian movement* and *twitching.*

Brownian movement is the random movement of particles suspended in a liquid. It is due to the bombardment of small entities of about 1 µm, such as bacteria, by molecules in the liquid. Brownian movement can be distinguished from motility by viewing bacteria in a wet mount in a microscope. A wet mount is a small liquid specimen on a glass microscope slide covered with glass coverslip. Brownian movement exhibits oscillating motions at a fairly fixed position. Motility, by contrast, has directional movement between two points.

Twitching occurs on surfaces rather than in liquid suspensions. The Danish bacteriologist Jørgen Henrichsen described twitching, in 1972: "The motility appears as small, intermittent jerks covering only short distances and often changing the direction of movement, which is not regularly related to the long axis of the cell." Later studies on twitching have provided evidence that this activity is similar, if not identical, to gliding. Twitching, as does gliding, involves more than one cell moving in the same direction with the help of type IV pili.

OBSERVING MOTILITY

Microbiologists observe motility in laboratory conditions using agar plates or agar tubes. Surface motility, such as gliding and euglenoid movement, can be demonstrated on a moist agar surface. After incubating the cultures at the species's optimal temperature, the evidence of motility appears as either concentric rings of growth from a starting colony at the axis or colonies followed by slime trails on the agar.

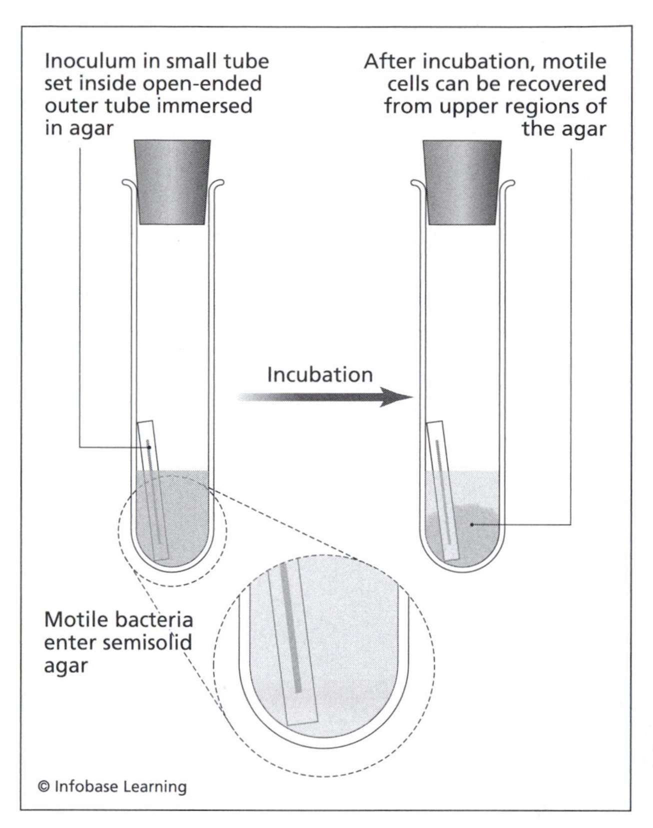

The Craigie tube method determines whether a microorganism is motile. This method also isolates motile microorganisms from nonmotile cells.

Three main methods demonstrate motility in liquids: the hanging drop, the semisolid agar tube, and the Craigie tube. The hanging drop method enables a microbiologist to observe the motility of living cells in a liquid. The microbiologist fills the shallow well of a specialized glass slide with the specimen in a suspension. The inverted slide, then, creates a drop of suspension that gives microorganisms free movement and can be viewed in a light microscope.

The hanging drop method has two disadvantages that prevent its widespread use. First, the glass slide dries out quickly and reduces the ability of microorganisms to swim in liquid. Second, the method involves handling cultures in an open system that is not confined in a more secure tube with a closure. Such an open system presents higher risks of infection to the microbiologist when studying pathogens.

One tube method for testing motility uses semisolid or semisoft agar medium formulations. Agar at a concentration of 0.4 percent creates a soft medium through which motile microorganisms can swim. A microbiologist takes an inoculum of bacterial culture onto a straight inoculating needle, then pushes the needle straight down into the tube's semisolid agar. The resulting culture is called a *stab culture.* After incubation, motile bacteria will have migrated into the agar away from the original inoculation. Nonmotile bacteria also grow, but they grow only in the immediate area of the stab.

The Craigie tube method also uses semisolid agar in a tube or wider bottle. A microbiologist inserts a short, narrow open-ended tube into the agar and, then, inoculates the microorganism into the opening of the tube that protrudes from the agar. After incubation, aerobic motile cells will have migrated from the tube and moved toward the agar's surface. The Craigie tube method offers the advantage of being able to separate motile bacteria from nonmotile bacteria in a mixed-species inoculum.

Motility is a key adaptation in microorganisms for responding to changes in the environment. The presence or absence of motility in bacteria also gives microbiologists evidence as to the identity of unknown species.

See also IDENTIFICATION; MICROSCOPY; MORPHOLOGY.

Further Reading

Aygan, Ashabil, and Burhan Arikan. "An Overview of Bacterial Motility Detection." *International Journal of Agriculture and Biology* 9 (2007): 193–196. Available online. URL: www.fspublishers.org/ijab/past-issues/IJABVOL_9_NO_1/44.pdf. Accessed October 13, 2009.

Henrichsen, Jørgen. "Bacterial Surface Translocation: A Survey and a Classification." *Bacteriological Reviews* 36 (1972): 478–503. Available online. URL: http://mmbr.asm.org/cgi/reprint/36/4/478?ijkey=558a8ca1d274d635d1cbda1109c9f51a5a9f3d85&keytype2=tf_ipsecsha. Accessed October 12, 2009.

Paustian, Timothy, and Gary Roberts. *Through the Microscope: A Look at All Things Small.* London: Lulu, 2006. Available online. URL: www.microbiologytext.com/index.php?module=Book&func=toc&book_id=4. Accessed October 12, 2009.

Semmler, Annalese B. T., Cynthia B. Whitchurch, and John S. Mattick. "A Re-Examination of Twitching Motility in *Pseudomonas aeruginosa.*" *Microbiology* 145 (1999): 2,863–2,873. Available online. URL: http://mic.sgmjournals.org/cgi/content/full/145/10/2863#R20. Accessed October 12, 2009.

Wang, Wei, Leslie M. Shor, Eugene J. LeBoeuf, John P. Wikswo, Gary L. Taghon, and David S. Kosson. "Protozoan Migration in Bent Microfluidic Channels." *Applied and Environmental Microbiology* 74 (2008): 1,945–1,949. Available online. URL: www.pubmedcentral.nih.gov/articlerender.fcgi?artid=2268297. Accessed October 12, 2009.

mutation rate A mutation is any permanent, heritable change in the genetic makeup of an organism that changes its phenotype, which is the outward expression of genes. The expression of genes can be either physical or metabolic. Mutation rate is the number of mutations formed per generation of a given population. Microbiologists also describe mutation rate as the probability that a single gene will mutate when a microbial cell divides.

Mutation rates have been studied mainly in bacteria, which require little time for a population to double in size, called its *generation time.* In some bacteria, under laboratory conditions, generation time may be as low as 20 minutes. Mutation rate is also important in pathogenic viruses and fungi.

The study of mutation rate in bacteria provides valuable insight into the manner in which species or strains adapt to adverse conditions. This has been particularly true in the worldwide problem of microbial resistance to antibiotics. Antibiotic resistance has developed since the first commercial use of antibiotics, in the 1940s, and this resistance was a direct result of mutation rates. A microorganism with a high mutation rate can adapt to changes in its environment faster than species with slower rates.

The molecular biologist Trudy Wassenaar described, in 2009, the relationships among mutation, adaptation, and evolution in microorganisms:

- As organisms reproduce, more offspring will be produced than can survive. This causes competition for sources of energy and habitat. Only the fittest organisms will survive.
- During cell division, minute changes will be introduced in the offspring. Such mutations will generate genetic diversity among the offspring.
- The result of these two processes, competition and diversity, is a selection pressure that will favour *[sic]* certain populations that are best qualified to survive. Over time, this will result in notable changes in a species.

The study of mutation and mutation rates enables microbiologists to understand the ancestors of today's bacteria and the changes in microbial populations currently occurring.

MUTATION

Mutations usually occur randomly anywhere along a microbial cell's chromosome, which is the entire amount of genetic material inside a cell. These random mutations, normally, occur at low frequency and can be unfavorable to the cell, favorable, or neither (neutral). Unfavorable mutations disappear from a population because the cell containing the mutation has a survival disadvantage compared with others in the population. Neutral mutations have no effect. Favorable mutations, however, confer an advantage on a mutated cell that allows it to survive conditions that kill others in the population. As this cell multiplies, the

Mutagens

Type of Mutagen	Mode of Action
	PHYSICAL
UV light	forms harmful covalent bonds between bases in DNA
X-rays and gamma rays	produce compounds that disrupt DNA replication and repair (an example of gene mutation)
	CHEMICAL
mutagenic chemicals (example: nitrous acid)	alters the structure of certain bases so they bond with a base other than the normal base pair (an example of point mutation)
nucleoside or base analogs (example: 5-bromouracil)	substitute for normal DNA bases during DNA synthesis
intercalating agents (example: acridine dyes)	insert between normal DNA bases
mispairing agents (example: methylnitrosoguanidine)	same as mutagenic chemicals

new trait goes to successive generations, until a new, improved population has replaced the previous one, which lacked this survival mechanism.

Mutations belong to one of two major categories: point and gene. Point mutations affect only one base pair in deoxyribonucleic acid (DNA). Gene mutations compose permanent changes to the DNA base sequences that define genes.

The random mutations normal in every population of living things represent spontaneous mutations, because they occur without any outside instigator. Induced mutations, by contrast, occur because a specific physical or chemical agent, called a *mutagen,* has affected the organism's chromosome. Induced mutations take place almost anytime a mutagen damages a cell's DNA molecule that makes up the chromosome, disrupts DNA repair mechanisms, or interferes with DNA's normal functioning.

Three types of radiation are common physical mutagens: X-rays, gamma rays, and ultraviolet (UV) radiation. Each of these forms of light causes reactions inside the cell that ultimately damage the DNA molecule. Bacteria have repair systems for minor damage from radiation but cannot withstand high doses or prolonged exposures. *Deinococcus* bacteria offer an exception, because this genus contains highly efficient repair processes that allow the bacteria to survive doses of radiation that would kill a human within hours.

The table above summarizes the main physical and chemical mutagens in bacteria.

Mutations that take place only under certain environment conditions are called *conditional mutations.* Conditions known to cause mutations in bacteria are temperatures outside the optimal range and lack of an essential nutrient.

Escherichia coli produces a mutant strain that does not use the sugar lactose as a carbon and energy source as do all other *E. coli* strains. But, in conditions when lactose is the only carbon and energy source available, the *E. coli* induces its own mutations so that it can use lactose and, thus, survive. This type of mutation is called *directed* or *adaptive mutation.* The biologists John Cairns and his colleagues proposed the theory of directed mutations, in 1988, and immediately created a firestorm of controversy. The idea that a bacterium could select when it would mutate went against all previous evidence of evolution as a random event, including random exposures to mutagens.

In 2004, the molecular geneticists Susan Rosenberg and P. J. Hastings discussed the implications of directed mutation in the *Journal of Bacteriology.* "Directed mutation (DM) models suggested the provocative possibility that mutations might be targeted specifically to those that relieve the stress, an idea tinged with Lamarckism [discredited theory that parents pass on characteristics acquired during their lifetime to their offspring]." Rosenberg and Hastings proposed that the directed mutation seen by Cairns was due to a phenomenon called *hypermutation.* In hypermutation, a special mutator gene activates multiple mutations in other genes with the purpose of affecting mutation rate. This, in turn, maximizes the chance of creating a population adapted to the new stress. In the case of *E. coli,* the presence of lactose and the absence of alternate carbon-energy sources may activate the mutator gene.

Since Cairns's study, more than 30 mutator genes have been discovered in *E. coli.* In 2008, a Canadian team of microbiologists, led by Irith Wiegand, found a mutator gene in *Pseudomonas aeruginosa* that caused an increase in resistance to several antibiotics. Directed mutations continue at the center of debate on how and why they work, but accumulating evidence points to a role of directed mutation and hypermutation in increasing mutation rates.

CALCULATION OF MUTATION RATE

Mutation rate is usually stated as a power of 10, and because mutations are very rare, the exponent is a negative number. Mutation rates in microorganisms normally range from 10^{-5} to 10^{-9}. A mutation rate of 10^{-5} means that when 10^5 cells (100,000 cells) divide to make 2×10^5 cells (200,000), an average of one mutant cell will be formed. The 10^{-9} mutation rate is much slower; a mutant will form on average only once per billion divisions. These exact same mutations rates apply to the genes of the microorganism. A mutant formed once every one billion cell divisions corresponds to one mutated gene per one billion gene replications.

The probability that two independent genes will mutate at the same time becomes astronomical. If one gene has a mutation rate of 10^{-8} and a second gene mutates at the rate of 10^{-9}, the chance that both will mutate in the same cell equals 10^{-17}.

Mutation rate calculation uses a Poisson equation, which is a mathematical description of the probability that any random event will occur. Because the equation for calculating mutation rate is complex, microbiologists use a shortcut estimate of mutation rate by calculating mutation frequency.

mutation frequency = ratio of mutants ÷ total number of bacteria in the population

Mutation frequency can estimate mutation rate because of two assumptions. First, the mutation rate equals the number of mutations per cell division. Second, because bacteria populations are very large, the number of cell divisions approximates the number of cells in the population. For example, a single culture tube contains 10^7 cells. The total number of mutations, at any one point in time, might be 0.6 mutation per culture. The mutation rate of that culture can be estimated using the mutation frequency equation.

$$\text{mutation rate} = 0.6 \text{ mutation} \div 10^7 \text{ cell divisions} = 6 \times 10^{-8} \text{ mutation per cell division}$$

Most bacteria have a mutation rate of about 10^{-8} per generation. DNA virus mutation rate ranges from 10^{-6} to 10^{-8}, and ribonucleic acid (RNA) viruses have a range of 10^{-3} to 10^{-5}. DNA viruses carry their genetic material in a DNA molecule, and RNA viruses carry theirs in the form of RNA. In eukaryotes, mutation rates are generally between 10^{-4} to 10^{-6}. Although bacteria mutate at a slower rate than complex organisms made of eukaryotic cells, bacteria replicate much faster than any other organism. This fast reproduction enables bacterial species to make rapid adaptations to changes in the environment.

THE IMPLICATIONS OF MUTATION RATE

Organisms with high mutation rates adapt to changes in the environment faster than organisms with a slow mutation rate, but there are limits to the advantages of very high mutation rates. High mutation rates lead to increased variability in a population and, thus, help the entire population weather harmful events. A too-high mutation rate may, however, have two consequences that hurt the population. First, high mutation rates and subsequent increased variability may create subpopulations so different from each other that they can no longer exchange genetic material. This is particularly true for eukaryotic cells. Second, the rate of mutation may exceed a cell's ability to maintain the integrity of the rest of its genome (the total genetic material of the cell).

Molecular biologists have developed drugs that kill pathogens by affecting their mutation rate. The antiviral drug ribavirin, which combats the poliovirus, may have as one of its modes of action an ability to speed the virus's mutagen rate until errors in gene replication reach a level that causes a population to begin to die out. This concept, called *lethal mutagenesis,* requires more study of its mechanism at molecular levels. The molecular biologists James Bull and Claus Wilke of the University of Texas studied lethal mutagenesis in bacteria, DNA-genome viruses, and RNA-genome viruses; they wrote, in 2008, in their article "Lethal Mutagenesis of Bacteria," "Lethal mutagenesis has been proposed as a mechanism to cure viral infections but has not been proposed for treating bacterial infections. The reason for this taxonomic bias may be simply that some viruses have RNA genomes, whose intrinsic mutation rates are already so high that it seems feasible to boost their mutation rates 'over the top.'" Poliovirus contains RNA as its genetic material.

If the theory of lethal mutagenesis is correct, certain microorganisms become extinct because their mutation rates are too slow to allow the organisms to adapt to environmental changes, and others' rates are so fast that the microorganisms cannot overcome the number of unfavorable mutations.

Mutation rates have been additionally used as a tool in studying evolution, called a *molecular clock.* In 1965, the Austrian molecular biologist Emile Zuckerkandl (1922–) and the American biochemist Linus Pauling (1901–94) introduced the concept of the *molecular evolutionary clock* for relating gene mutation rates to the phylogeny of species. Phylogeny is the history of an organism's lineage over its entire evolution. If genes mutate at a constant and predictable rate, the degree of difference between the same gene in two different species can be used to estimate how long ago the species diverged in their evolution. The biologist S. Blair Hedges of Pennsylvania State University simplified the premise, in 2008, explaining, "Unlike a wristwatch, which measures time from regular changes (ticks), a molecular clock measures time from random changes (mutations) in DNA." Mutation rates have, thus, given scientists insight into adaptations that organisms develop in the present and the adaptations of the past that led to today's biological diversity.

See also CHROMOSOME; GENERATION TIME; RESISTANCE.

Further Reading

Barazesh, Solmaz. "Probing Question: What Is a Molecular Clock?" Physorg.com News. November 20, 2008. Available online. URL: www.physorg.com/news146418967.html. Accessed October 14, 2009.

Birge, Edward A. *Bacterial and Bacteriophage Genetics,* 5th ed. New York: Springer, 2006.

Bull, James J., and Claus O. Wilke. "Lethal Mutagenesis of Bacteria." *Genetics* 180 (2008): 1,061–1,070. Available online. URL: www.genetics.org/cgi/reprint/180/2/1061. Accessed October 14, 2009.

Cairns, John, Julie Overbaugh, and Stephan Miller. "The Origin of Mutants." *Nature* 335 (1988): 142–145.

Foster, Patricia L. "Adaptive Mutation in *Escherichia coli.*" *Journal of Bacteriology* 186 (2004): 4,846–4,852. Available online. URL: http://jb.asm.org/cgi/content/full/186/15/4846#R11. Accessed October 14, 2009.

Rosenberg, Susan M., and P. J. Hastings. "Adaptive Point Mutation and Adaptive Amplification Pathways in the *Escherichia coli Lac* System: Stress Responses Producing Genetic Change." *Journal of Bacteriology* 186 (2004): 4,838–4,843. Available online. URL: http://jb.asm.org/cgi/reprint/186/15/4838. Accessed October 13, 2009.

Wassenaar, Trudy M. "Evolution in Bacteria." The Virtual Museum of Bacteria. January 6, 2009. Available online. URL: www.bacteriamuseum.org/cms/Evolution/evolution-in-bacteria.html. Accessed October 13, 2009.

Wiegand, Irith, Alexandra K. Marr, Elena B. M. Breidenstein, Kristen N. Schurek, Patrick Taylor, and Robert E. W. Hancock. "Mutator Genes Giving Rise to Decreased Antibiotic Susceptibility in *Pseudomonas aeruginosa.*" *Antimicrobial Agents and Chemotherapy* 52 (2008): 3,810–3,813. Available online. URL: http://aac.asm.org/cgi/content/full/52/10/3810. Accessed October 13, 2009.

Zuckerkandl, Emile, and Linus Pauling. "Molecules as Documents of Evolutionary History." *Journal of Theoretical Biology* 8 (1965): 357–366. Available online. URL: http://lectures.molgen.mpg.de/phylogeny_ws05/papers/zuckerkandl_pauling.pdf. Accessed October 14, 2009.

mycobacteria *Mycobacteria* is a general term for bacteria of the genus *Mycobacterium.* These gram-positive species are nonmotile, do not form endospores, and live in environments ranging from aerobic to microaerophilic (oxygen levels less than 10 percent). *Mycobacterium* cells are straight to slightly curved rods measuring 0.2–0.6 micrometer (μm) in diameter and 1.0–10.0 μm in length. The cells occasionally develop branches and can form long extensions called *filaments.* The prefix *myco-* refers to the funguslike appearance of these microorganisms.

Mycobacterial diseases are among the oldest known in civilization. Mycobacteria live as parasites and pathogens in humans, animals, and fish. Mycobacterial disease in humans is a zoonotic disease: a disease transmitted to humans from animals. *Mycobacterium* lives as a free-living microorganism, rather than a community, in soil and natural waters. These habitats are presumably the source of animal infections.

Mycobacteria contain two features of their cell wall that contribute to the survivability of the microorganism, as well as its ability to infect hosts. The first component comprises a group of compounds called *mycolic acids,* which are large fatty acids of 60–90 carbon molecules in length. The genera *Corynebacterium, Nocardia,* and *Rhodococcus* also contain mycolic acids in their cell wall. All belong to the same suborder, Corynebacterineae, of order Actinomycetales, class Actinobacteria, and phylum Actinobacteria. The second component, wax D, is unique to mycobacteria. Wax D is a large matrix of compounds containing mycolic acids linked to the bacterial cell wall polymer, peptidoglycan. These two cell wall constituents make mycobacteria more structurally like gram-negative species than gram-positive species.

The outer waxy layer of mycobacteria, sometimes called a *capsule,* enables these bacteria to withstand dramatic changes in the environment. The layer resists water yet holds in moisture, so that mycobacteria can survive in aerated, dry environments. (The layer also reduces the effectiveness of Gram staining.) Nutrients move very slowly through the thick waxy layer, making mycobacteria time-consuming to grow in laboratory conditions. Mycobacteria have a generation time of 12–18 hours and require weeks to develop visible colonies on growth media. Colonies display a characteristic waxy yellow appearance on agar media. The protective layer also repels damaging compounds such as antibiotics and makes antibiotic treatment of mycobacterial infections difficult and lengthier than the treatment period for other bacterial infections.

MYCOBACTERIA PATHOLOGY

Mycobacterium contains about 50 known species, of which more than 30 species participate in or cause disease. Several of these pathogenic mycobacteria belong to one of three complexes: (1) the *M. tuberculosis*-complex, (2) the *M. leprae*-complex, (3) and the *M. avium*-complex. The species within these three groupings cause tuberculosis (TB) in humans and animals, leprosy, and disease in birds, respectively. Important species of the three main mycobacteria complexes are

- *M. tuberculosis*-complex—*M. africanum, M. bovis, M. microti, M. tuberculosis*
- *M. leprae*-complex—*M. leprae, M. lepromatosis*
- *M. avium*-complex—*M. avium, M. intracellulare, M. scrofulaceum*

The *M. tuberculosis*-complex infects humans and other animals, particularly cattle and primates. Other mycobacteria not identified as part of the two human disease complexes can also cause minor to serious infections in humans. The main noncomplex human pathogens are:

- *M. branderi*—respiratory tract infection
- *M. celatum*—range of infections, especially in immunocompromised patients
- *M. chelonei*—isolated from infected wounds
- *M. gordonae*—pulmonary disease
- *M. haemophilum*—infections in immunocompromised adults and healthy children
- *M. interjectum*—associated with lymph disorders in children
- *M. kansasii*—similar to pulmonary tuberculosis
- *M. thermoresistibile*—infections of lungs, skin, lymph nodes, and breast tissue
- *M. triplex*—possible pathogen in immunocompromised adults
- *M. ulcerans*—cause of Buruli ulcer, a progressive skin lesion

Atypical or nontuberculosis mycobacterial infection can be caused by many of the species listed here but differs by infecting areas than the respiratory tract. Atypical mycobacterial infections may appear as chronic skin abscesses, infections of joints, or osteomyelitis (bone infection). These infections also spread to the lymph nodes in many cases. The main atypical mycobacteria are *M. avium-intracellulare* (a strain of *M. avium*), *M. chelonei, M. kansasii, M. marinum,* and *M. ulcerans*. Treatment for atypical mycobacterial infections requires long-term drug administration from six months to two years, but the severity of the infection determines whether a patient will fully recover.

TUBERCULOSIS

TB may be the earliest disease known to infect human civilization and probably existed before recorded human history. It remains a significant health threat worldwide, and mycobacteria are the number-one bacterial killer. In addition to humans, TB affects primates and other mammals, birds, amphibians, and fish. Humans can be infected by any of the three types of TB bacteria: human, bovine, or avian.

The French military physician Jean-Antoine Villemin (1827–92) discovered the TB pathogen, in 1865, by taking the exudate from the lung of a TB patient and injecting it into healthy rabbits, which then developed TB in their lungs. Villemin's critics thought that the disease they called *consumption* resulted from poverty, with its characteristic malnutrition, overwork, and unsanitary housing. In 1882, the renowned German physician Robert Koch (1843–1910) presented his evidence of the TB microorganism. Koch's more methodological approach than Villemin's convinced the medical community.

TB, most commonly, affects the respiratory system, causing inflammation, abscesses, necrosis (tissue death), and characteristic tubercles, or swelled lesions, in lung tissue after a four- to 12-week incubation period. In severe cases, TB causes fibrosis (formation of scar tissue on healed lesions) and calcification (accumulation of calcium deposits) of the lungs. This pulmonary TB causes the following symptoms: chronic cough, sputum production, fever, and weight loss.

TB can also spread to other organs, where the immune system arrests most, but not all, of the infection, leading to a chronic and progressively worse form of the disease and possible death. The overall mortality rate, deaths per 100,000 people, of TB varies by regions of the world, depending on level of nutrition and medical care. The global average TB mortality rate is 27 per 100,000.

Pulmonary TB begins when *M. tuberculosis* infects lung tissue and forms a tubercle. Upon detecting the bacteria's presence, the body sends macrophages and other cells of the immune system to the infection site. The immune cells build a mass called a *granuloma* around the lesion. In many cases, the granuloma confines the infection, until the body's defenses can kill the bacteria. In other instances, the bacteria recruit more macrophages to the infection in a defensive maneuver that may help protect them from further immune system activity. Science*Daily* explained in 2004, "The end game of the chess match remains unclear. While granulomas are required for protection against mycobacteria, they are not completely effective. Thus, these bacteria have developed a strategy to recruit the normally defensive cells of the host to their advantage." The New Jersey public health researcher Barry Kreiswirth added in *Microbe*, in 2009, "It's a brilliant mechanism for a pathogen.

Infect large numbers of people, not kill most of them, and then let those who eventually become so immunocompromised that they come down with active disease, spread you around." TB researchers continue to study the unique give-and-take relationship between the immune system and the TB bacteria.

In 2009, the New Jersey public health researchers Patricia Fontan and Issar Smith suggested a potential advantage of mycobacterial behavior in the lungs: "Because the *M. tuberculosis* [mutant strain] stimulates the host immune system during macrophage infections, that mutant might be evaluated for use as a vaccine strain." Since 1921, the main TB vaccine has been the bacille Calmette-Guérin (BCG) vaccine, made of a live weakened strain of *M. bovis,* the main cause of bovine TB. "But because it is a live vaccine, . . ." the *New York Times* reported, in 2009, "it can cause its own problem—'disseminated BCG disease,' a type of bacterial infection that can rage through the body. It is fatal in more than 70 percent of cases." TB vaccines require improved effectiveness and ease of administration.

The tuberculin skin test serves as a standard assay for determining whether a person is infected or ever has been infected with TB. The procedure involves an intradermal (into the skin) injection of a purified protein derived from TB bacteria. For this reason, the tuberculin test is also called the purified protein derivative (PPD) test. A small red bump at the injection site from the immune system's reaction to the protein indicates that the person has been exposed to *M. tuberculosis.*

TB declined steadily, from the 1950s, when antibiotic treatments became available, to the 1980s, when the acquired immunodeficiency syndrome (AIDS) epidemic began. Today, about one fourth of all AIDS patients are infected with TB, and one fourth of TB patients have AIDS. Other factors, in the 1980s, that contributed to TB's reemergence were an increase in the homeless population, an increase in international immigration from high-TB areas to areas with low incidence, and a decrease in surveillance by public health officials, who assumed the disease had been arrested.

The World Health Organization has estimated that the frequency of new TB cases, called *incidence,* exceeds nine million each year. The countries that currently have the highest incidence are, in order; South Africa (about 950 per 100,000), Zimbabwe (785), Cambodia (495), Mozambique (430), Democratic Republic of Congo (390), Ethiopia (380), Kenya (355), Uganda (330), Nigeria (310), and the United Republic of Tanzania (300). India and China have more than one million new cases annually, making these regions the highest centers of TB growth.

Current TB treatment involves four different antibiotics—isoniazid, rifampin, ethambutol, and pyrazinamide—given for several months. Because the TB bacteria have developed resistance to many current TB therapies, doctors alter the timing and dosage of these four antibiotics to get the best effects. Even so, antibiotic-resistant mycobacteria have become a global problem. Multidrug-resistant TB (MDR TB) now resists the activity of isoniazid and rifampin. Extensive drug-resistant TB (XDR TB) often resists all four of the conventional TB antibiotics. The Mayo Clinic has explained, "Treatment of drug-resistant TB requires taking a 'cocktail' of at least four drugs, including the first line medications that are still effective and several second line medications, for 18 months to two years or longer. Even with treatment, many people with these types of TB may not survive." Barry Kreiswirth summarized the main challenge in today's TB therapies, "Keeping patients on a complicated regimen of four toxic drugs isn't easy, even in the developed world." But the alternatives are not desirable; drug-resistant TB may, in some cases, be cured only with surgery to remove affected areas of the lungs.

LEPROSY

Leprosy is an ancient and chronic disease of the skin and peripheral nervous system caused by *M. leprae.* Leprosy is endemic (continuous in a population) in India and other tropical regions and infects about 12 million people worldwide.

The Norwegian physician Gerhard Armauer Hansen (1841–1912) was, in 1873, the first to identify leprosy as a bacterial disease, caused by *M. leprae.* Today, *Hansen's disease* is a synonym for *leprosy.* The disease manifests in two forms: borderline (lepromatous) and indeterminate (tuberculoid). Borderline leprosy involves both the skin and the nerves; indeterminate leprosy involves fewer bacteria-infected skin lesions and often resembles TB. As is TB, leprosy is spread via coughs, sneezes, and breathing. Left untreated, *M. leprae* can spread to the muscles, bones, and testicles and cause severe disfigurement in the extremities.

Effective drug treatments reduced the worldwide incidence of leprosy, by the 1980s, to less than one case in 10,000 people, but the disease remains endemic in the following countries: Angola, Brazil, Central African Republic, Democratic Republic of Congo, India, Madagascar, Mozambique, Nepal, and the United Republic of Tanzania.

The WHO has established a program called "Final Push" to rid the world of leprosy by focusing on the areas where the disease is endemic. The program includes the following action points:

- expand availability of multidrug therapy (MDT) where needed
- encourage patients to take MDT regularly and completely

Veterinary Mycobacterial Infections

Animal	Common Mycobacteria Susceptibility
birds	*M. avium*
cats	*M. avium*-complex, *M. bovis,* and *M. lepraemurium* (causes a leprosylike disease)
cattle	*M. avium, M. bovis, M. kansasii,* and *M. paratuberculosis* (Johne's disease)
deer and elk	*M. avium, M. bovis,* and *M. paratuberculosis*
dogs	*M. bovis, M. fortuitum,* and *M. tuberculosis*
elephants	*M. tuberculosis*
horses	generally resistant to TB
monkeys and apes	*M. avium, M. bovis, M. kansasii,* and *M. tuberculosis*
pigs	*M. avium, M. bovis, M. kansasii,* and *M. tuberculosis*
sheep and goats	*M. bovis*

- promote awareness in communities for halting the leprosy's spread and getting immediate treatment
- set targets and monitor progress toward elimination

Current leprosy therapy is similar to TB therapies, in that the risk of development of drug-resistant bacteria increases if only one drug is administered. The main drugs in leprosy MDT are rifampicin and clofazimine. The progress in ridding the world of leprosy has been promising for three main reasons: (1) MDT remains an effective treatment against the pathogen with no resistance yet detected, (2) leprosy is not highly infectious, and (3) leprosy has been confined mainly to endemic areas, making eradication easier to plan and complete.

MYCOBACTERIA IN VETERINARY HEALTH

Several different *Mycobacterium* species impact animal health, particularly among birds, cattle, primates, and fish. Each of the three TB-complexes that can infect species outside their main host is susceptible to a variety of animal-specific pathogens, in addition to the *M. tuberculosis* that infects humans.

The introduction of pasteurization to all milk and milk products was an important step in reducing TB in humans and remains a method of TB control. The Food and Drug Administration now regulates the pasteurization procedures used by milk producers, under the authority of the Pasteurized Milk Ordinance, which intends to eliminate bacteria in addition to TB bacteria in milk. Cattle become infected mainly by *M. bovis,* but this species also infects humans.

The table above lists the main mammal species or groups that are affected by mycobacteria.

M. paratuberculosis infections in domesticated and wild ruminant animals cause a slowly progressive nonpulmonary disease leading to diarrhea, weight loss, and emaciation. No treatment currently exists for this disease.

Cold-blooded animals and fish also contract TB infections through species other than those that infect warm-blooded species. *M. marinum,* which infects fish, has been shown, however, to cause disease in humans.

Mycobacteria have remained an insidious enemy to human health since, perhaps, the first humans walked the earth. Modern medicine has made progress in eradicating mycobacterial diseases and may successfully complete this challenge in the near-future. In the meanwhile, mycobacteria must be treated with vigilance and the most effective drug therapies.

See also PASTEURIZATION; PATHOGEN; PATHOGENESIS.

Further Reading

Blanc, Léopold, and Katherine Floyd. *Global Tuberculosis Control 2009.* Geneva, Switzerland: World Health Organization, 2009. Available online. URL: www.who.int/tb/publications/global_report/2009/pdf/full_report.pdf. Accessed October 19, 2009.

Centers for Disease Control and Prevention. "Tuberculosis (TB)." Available online. URL: www.cdc.gov/tb. Accessed October 14, 2009.

Fontán, Patricia A., and Issar Smith. "*M. tuberculosis* σΣ Protects against Environmental Stress, Immune Responses." *Microbe,* March 2009.

Han, Xiang Y., Kurt C. Sizer, Erika J. Thompson, Juma Kabanja, Jun Li, Peter Hu, Laura Gómez-Valero, and Francisco J. Silva. "Comparative Sequence Analysis of

Mycobacterium leprae and the New Leprosy-Causing *Mycobacterium lepromatosis.*" *Journal of Bacteriology* 191 (2009): 6,067–6,074.

Kahn, Cynthia M., ed. *The Merck Veterinary Manual,* 9th ed. Whitehouse Station, N.J.: Merck, 2008. Available online. URL: www.merckvetmanual.com/mvm/index.jsp. Accessed October 21, 2009.

Mayo Clinic. "Tuberculosis." Available online. URL: www.mayoclinic.com/health/tuberculosis/DS00372/DSECTION=treatments-and-drugs. Accessed October 21, 2009.

McNeil, Donald G. "Tuberculosis: TB Vaccine Too Dangerous for Babies with AIDS Virus, Study Says." *New York Times,* 2 July 2009. Available online. URL: www.nytimes.com/2009/07/07/health/07glob.html. Accessed October 19, 2009.

Ngan, Vanessa. "Atypical Mycobacterial Infection." June 15, 2009. Available online. URL: http://dermnetnz.org/bacterial/atypical-mycobacteria.html. Accessed October 21, 2009.

Science*Daily*. "A Clear View of Mycobacterial Infection." November 5, 2004. Available online. URL: www.sciencedaily.com/releases/2004/10/041030154143.htm. Accessed October 19, 2009.

Sherman, Irwin W. *Twelve Diseases That Changed Our World.* Washington, D.C.: American Society for Microbiology Press, 2007.

Stone, Marcia. "Determined Progress in War against Malaria, HIV-AIDS, and TB." *Microbe,* March 2009. Available online. URL: www.microbemagazine.org/index.php?option=com_content&view=article&id =91:determined-progress-in-war-against-malaria-hiv-aids-and-tb&catid=48:featured&Itemid=82. Accessed June 6, 2010.

World Health Organization. Available online. URL: www.who.int/en. Accessed October 19, 2009.

Mycoplasma *Mycoplasma* is a bacterial genus of about 60 species that lack a cell wall and require large sterol compounds for cellular structure. *Mycoplasma* and the closely related genus *Ureaplasma* are sometimes grouped generically as mycoplasmas. Both genera cause human disease, mainly in the respiratory and urogental tracts. Because of the association of mycoplasmas with various respiratory tract diseases, this group of microorganisms has also been referred to as pleuropneumonialike organisms (PPLOs).

Mycoplasma and *Ureaplasma* belong with other cell wall–less species in family Mycoplasmataceae, order Mycoplasmatales, and class Mollicutes of phylum XIII Firmicutes. Microorganisms in order Mycoplasmatales also cause disease in domesticated animals, birds, and plants. In plant diseases, the mycoplasma pathogen is often referred to as mycoplasma-like organism (MLO).

Mycoplasma cells are nonmotile, range from spherical to pear-shaped, and measure 0.3–0.8 micrometer (µm) in diameter. These microorganisms are among the smallest bacteria, but they can also grow long filaments, of up to 150 µm. Because of the diversity of forms *Mycoplasma* can produce, its morphology is said to be *pleomorphic.* Mycoplasmas contain sterols in their cell membranes, a feature common in eukaryotic cells but unusual in bacteria. Scientists have also detected unusual tip structures on *Mycoplasma* cells that may play a role as an anchor for attachment to surfaces. The tip always points in the direction of cell movement, suggesting it also has a role in cell motility. Most *Mycoplasma* are facultative anaerobes (can live with or without oxygen), and a smaller percentage lives as obligate anaerobes that require oxygen-free environments.

Mycoplasma colonies on agar medium grow to a small size, compared with other bacteria, and tend to be difficult to see because they are transparent. Many species form fried-egg colonies containing a dark center and translucent outer area. Common laboratory stains that react with most bacteria, such as the Gram stain, do not work well on mycoplasmas because mycoplasmas' unusual cell wall does not retain the dyes. Microbiologists apply a different stain—the Dienes stain containing methylene blue is common—directly to colonies growing on agar. The microbiologist, then, removes an entire block of agar holding the stained colony and embeds it in melted paraffin wax, which provides support when it solidifies. This specimen can, then, be examined in a microscope. Despite *Mycoplasma*'s distinctive structure, deoxyribonucleic acid (DNA) analysis indicates the microorganism has a genetic relationship to gram-positive *Bacillus, Streptococcus,* and *Lactobacillus.*

Mycoplasma's small size and lack of a rigid cell wall allow cells to pass through filters with pore size of 0.45 µm and sometimes 0.2 µm. For this reason, this genus is a frequent contaminant of tissue cultures that rely on filter-sterilized growth media. The cell biologists Maik Jornitz and Theodore Meltzer said in *Genetic Engineering and Biotechnology News,* in 2009, "Mycoplasmas, Accoleplasmas, and Ureaplasmas all belong to the Mollicutes class—small prokaryotes that do not have a rigid cell wall. Contamination of cell culture medium by these organisms has become a recurring problem for the biopharmaceutical industry." Tissue culture scientists have adapted specialized filters of 0.1 µm pore size or less and made of adsorptive materials that retain most mycoplasmas by sticking to them.

IMPORTANT MYCOPLASMAS

The most-studied *Mycoplasma* species are those that cause disease in humans or animals. Of these, the obligate anaerobe *M. pneumoniae* creates health concerns because it causes pneumonia, bronchitis, and a syndrome called *primary atypical pneumonia.*

Major Mycoplasma Diseases in Humans

Species	Disease	Description
M. genitalium	nongonococcal urethritis	sexually transmitted disease caused by pathogens other than *Neisseria gonorrhoeae*
M. hominis	pyelonephritis pelvic inflammatory disease postpartum fever	bladder-originated infection causing inflammation in kidney and renal pelvis inflammation of uterus, fallopian tubes, and pelvic structures infection after childbirth
M. pneumoniae	upper respiratory tract disease bronchitis (tracheobronchitis) mycoplasmal (atypical) pneumonia	any infection of nasal passages, pharynx, or bronchi inflammation of trachea and the two main passageways (bronchi) leading into the lungs also called walking pneumonia, minor infection of the lung
U. urealyticum	nongonococcal urethritis	sexually transmitted disease caused by pathogens other than *Neisseria gonorrhoeae*

M. pneumonia takes up to three weeks of incubation in a laboratory to form colonies of only 50–100 μm diameter on agar medium. Because of the difficulty of growing mycoplasmas in a laboratory, many mycoplasma illnesses may be undetected each year.

M. mycoides produces distinctive branching filaments when viewed in a microscope. This species causes a type of pneumonia in cattle called *contagious bovine pleuropneumonia*. Bovine pleuropneumonia spreads through the air in exhaled aerosols. Acute infections cause death within days or weeks; the disease has a 50 percent mortality rate. *M. bovis* and several other *Mycoplasma* species are responsible for massive outbreaks of mastitis in dairy herds and may create secondary infection after respiratory infection with *Mycoplasma*. *M. mycoides* and *M. putrefaciens* have also caused severe mastitis outbreaks in goats.

Ureaplasma species are called *microaerophiles* because they require oxygen to live but only at very low levels; oxygen levels less than 5–10 percent are typical conditions for microaerophiles. *Ureaplasma* species occur in humans and other animals in the mouth and the respiratory and urogenital tracts. *Ureaplasma* forms very small colonies on agar, earning it the nickname *T-strains,* for "tiny strains." Although this genus causes some diseases, *Ureaplasma* may be a less serious pathogen than *Mycoplasma*.

MYCOPLASMA DISEASES IN HUMANS

Mycoplasmas live inside host cells as parasites. This quality contributes to the difficulty of diagnosing infection and ridding the body of the pathogen. Three *Mycoplasma* species and one *Ureaplasma* species cause the most severe mycoplasma diseases in humans. The table above summarizes these diseases.

Respiratory infections by mycoplasmas usually begin with sore throat before turning into bronchitis or pneumonia. The pathogen uses either airborne or direct-contact transmission.

Mycoplasmas possess three main factors that contribute to pathogenesis, which is the development of disease in the body after the initial infection. The first factor, called a *virulence factor,* because it contributes to the severity of disease, involves adherence proteins on the outside of the mycoplasma cell. Mycoplasmas adhere mainly to epithelial cells, the proteins binding specifically with the cement between epithelial cells called sialic acid. For example, *M. pneumoniae* has a protein called P1 that is known to attach to sialic acid. In vitro experiments have demonstrated that P1 protein adheres to a variety of eukaryotic cells and red blood cells.

Mycoplasmas produce hydrogen peroxide, which poisons most other microorganisms. Many bacteria, as well as human cells, produce the enzyme catalase to destroy hydrogen peroxide. Mycoplasmas disrupt this safety mechanism in the body by secreting the poison while inhibiting production of catalase.

MYCOPLASMAS IN VETERINARY MEDICINE

Mycoplasmas cause several serious conditions in veterinary medicine. *Mycoplasma* causes mycoplasmal pneumonia in pigs, which is often complicated by other bacterial and viral infections. The disease's incidence can reach 80 percent in operations that house pigs in close quarters, that is, factory farming. Swine mycoplasmal pneumonia resembles the human form, in that it is generally mild and nonlethal.

Cattle and calves also contract a form of mycoplasmal pneumonia that might be caused by either *Mycoplasma* or *Ureaplasma*. The disease results in mild signs of respiratory infection and can be cured with antibiotics. The pathogen causing bovine mycoplasmal pneumonia had originally been called PPLO, a term also used in past years for the human pathogen.

Birds are susceptible to several different contagious mycoplasmal diseases. *M. gallisepticum, M. iowae,* and *M. meleagridis* cause different respiratory illnesses in chickens and turkeys. Chronic illness from each pathogen leads to reduced hatchability of eggs, transmission of the pathogen to eggs, and, sometimes, infections in reproductive organs. *M. gallisepticum* and *M. iowae* appear to be the most virulent; *M. meleagridis* occurs worldwide in turkeys.

Mycoplasmosis in goats caused by *M. mycoides* results in severe lameness, polyarthritis, and mastitis and can spread to the respiratory tract. Mortality rates can reach 90 percent in kids that have contracted the disease.

Most mycoplasma infections spread rapidly through herds or farm operations, where animals are confined close together for their entire lives. Antibiotic treatments have been effective in curing most outbreaks, but mycoplasma epidemics remain a health threat in commercial meat- and egg-producing businesses.

Microbiology has not spent as much time on studies of mycoplasmas as it has for other pathogens. The microorganism's morphology, life cycle, and growth characteristics in laboratories have slowed the rate of new discoveries. Analysis of the mycoplasma genome, which has already made progress, holds promise for a new phase of study of these unique bacteria.

See also CELL WALL; PATHOGEN; SEROLOGY; STAIN.

Further Reading

Drasbek, Mette, Gunna Christiansen, Kim R. Drasbek, Arne Holm, and Svend Birkelund. "Interaction between the P1 Protein of *Mycoplasma pneumoniae* and Receptors on HEp-2 Cells." *Microbiology* 153 (2007): 3,791–3,799. Available online. URL: http://mic.sgmjournals.org/cgi/reprint/153/11/3791. Accessed January 7, 2010.

Jornitz, Maik W., and Theodore H. Meltzer. "Mycoplasma Contamination in Culture Media." *Genetic Engineering and Biotechnology News,* 15 March 2009.

Mayer, Gene. "Mycoplasma and Ureaplasma." In Microbiology and Immunology Online. Columbia: University of South Carolina, 2007. Available online. URL: http://pathmicro.med.sc.edu/mayer/myco.htm. Accessed January 7, 2010.

mycoses (singular: mycosis) Mycoses are diseases caused by a fungus or resembling a fungal disease. Mycoses are either on the skin or in the body as systemic diseases and can begin as opportunistic infections. Opportunistic pathogens are microorganisms that do not normally cause disease unless conditions change to favor infection.

The study of mycoses calls upon two different disciplines in science: mycology, which is the study of fungi, and dermatology, which is the study of medically important skin conditions. Experts in both fields categorize mycoses into five general groups:

- superficial—localized on outer epidermis (skin) and in hair shafts
- cutaneous (dermatomycosis)—limited to epidermis, hair, and nails
- subcutaneous—infection spreads to beneath the skin
- systemic—in body fluids or organs
- yeast—caused by a yeast, often opportunistic

A mucocutaneous mycosis, in both mucus and skin, occurs in infrequent instances, when fungus infects both skin and mucous membranes.

Superficial and cutaneous mycoses occur in most people, sometime in their lives but tend to occur more frequently in the Tropics, where constant moist conditions and warm temperatures promote fungal growth. The following superficial mycoses can progress into cutaneous forms:

- tinea barbae—"barber's itch"
- tinea capitis—scalp
- tinea corporis—entire body
- tinea cruris—anal and genital areas
- tinea pedia—"athlete's foot"

Tinea is a term for any fungal infection that occurs on different parts of the body.

Dermatophytes are fungi that infect only skin, hair, or nails. These organisms cause cutaneous mycosis, which begins when a person receives the fungus by person-to-person contact or via an intermediate inanimate object. Dermatophytes secrete the enzyme keratinase, which breaks down keratin, a protein of hair and nails that few other organisms degrade. Keratinase enables the fungus to burrow into skin or nails as the infection progresses.

Subcutaneous mycosis results from saprophytic fungi native to soil, organisms that get their nutrition by degrading dead organic matter. In nature, saprophytes return carbon and other elements from dead plant and animal matter to the earth and, thus, play a vital role in cycling nutrients through the biosphere, a process called *biogeochemical cycles*. Subcutaneous infections often occur when soil contaminates a puncture wound.

Systemic mycosis occurs when a fungus enters the body, usually by inhaling of airborne fungal spores. Infection begins in the lungs and enters the bloodstream in capillaries serving the alveoli, small sacs that provide air-blood interfaces.

Fungal infections challenge dermatologists for two reasons. First, fungi grow more slowly than most other microorganisms and do not respond quickly to drug treatment. Second, fungi are eukaryotic cells that have similarities with human cells. The drugs effective for killing fungal cells can also injure human cells. For these reasons, mycoses other than systemic mycoses tend to be chronic diseases.

Systemic mycoses are life-threatening diseases that progress on the basis of the health status of a person's immune system. Weakened immunity increases the risks of infection by fungi, as it does for other pathogens. The New Zealand Dermatological Society has identified the following systemic mycoses as the highest threats to people who have healthy immune systems:

- *Blastomyces dermatitidis*—blastomycosis
- *Coccidioides immitis*—coccidioidomycosis
- *Histoplasma capsulatum*—histoplasmosis
- *Paracoccidioides brasiliensis*—paracoccidioidomycosis
- *Penicillium marneffei*—peniciliosis

Immunocompromised people and others with higher-than-normal risk for infection are subject to added threats from the following infections and mycoses:

- *Aspergillus*—aspergillosis
- *Candida* (yeast)—candidiasis
- *Cryptococcus*—cryptococcosis
- *Zygomycetes*—zygomycosis

The immune system of people in high-risk health groups does not defend against fungal infections, which are difficult to cure even in people with healthy immune systems. Mycoses cause the greatest health threats to the following groups:

- acquired immunodeficiency syndrome (AIDS) patients
- cancer or chemotherapy patients
- diabetes mellitus cases
- elderly adults and youngsters
- organ transplant recipients
- postsurgery patients
- pregnant women

In 2007, the microbiologist Anupriya Wadhwa wrote in the *Journal of Medical Microbiology*, "Though HIV is the causative agent of AIDS, most morbidity and mortality in AIDS patients result from opportunistic infections; approximately 80 percent of these patients are seen to die as a result of such an infection rather than from HIV. Mostly the infections seen in AIDS patients are endemic to the geographical region, and involve many organ and organ systems simultaneously with a tendency to disseminate." The most common opportunistic mycoses in AIDS patient are, in approximate order; candidiasis, cryptococcosis, pneumocystinosis (from *Pneumocystis*), aspergillosis, and histoplasmosis. Mycoses and the pathogens most common in immunocompromised and healthy people are listed in the table on page 559.

FUNGI THAT CAUSE MYCOSIS

Mycoses often result from fungi common in soil, where they are normal inhabitants of the environment. Inhalation of the fungal spores and infected puncture wounds are the two most common modes of transmission. Most mycoses do not transmit by person-to-person direct contact except for people in one of the high-risk health groups. If a person has an immunocompromised health condition, *Aspergillus*, *Cryptococcus*, or *Penicillium* can transmit from a healthy person to the immunocompromised person. Opportunistic mycoses occur in imunocompromised people more often than in immune-healthy people. The table lists the main causes of common mycoses that often begin as opportunistic infections.

Superficial and cutaneous mycoses such as the examples in the table occur more often than subcutaneous or systemic mycoses. These mycoses require swift treatment to prevent the patient from entering a life-threatening situation. Clinical microbiologists look for the fungi associated with systemic mycoses when examining a specimen from inside a patient's body. Fluids, exudates, and biopsy samples should be checked for the following fungi: *Acremonium*, *Aspergillus*, *Blastomyces*, *Candida*, *Chrysosporium*, *Cladosporium*, *Coccidioides*, *Cryptococcus*, *Histoplasma*, *Paracoccidioides*, *Rhizopus*, *Sporothrix*, and *Torulopsis*. Bacteria in genus *Actinomycetes* look like fungi because the cells produce filaments similar to fungal hyphae. Clinical microbiologists, therefore, also check the specimens for *Actinomycetes*.

A small number of soil fungi become lethal pathogens if they enter the body and cause a systemic mycosis. The plant pathogen *Fusarium* provides an important example of one such fungus, which causes a mortality rate of more than 50 percent in systemic infections in humans and animals. Infection results from contaminated grains in which *Fusarium* has secreted its mycotoxin, which is any poison produced by a fungus.

Fusarium belongs to a group of fungi that produce a health threat because of its mycotoxin and because of its contamination of indoor environments. A phenomenon called *sick building syndrome* is any allergy or mycosis attributed to high levels of mold spores or mycotoxins inside a building, and certain fungi have been implicated in this general type of mycosis: *Aspergillus, Cladosporium, Fusarium, Penicillium,* and *Stachybotrys.* People refer to these fungi, collectively as *black mold* because dense concentrations of *Aspergillus* and *Stachybotrys* in moist indoor areas produce a thick, black growth. Molds may be referred to as either molds or fungi; *mold* is a general term for any fungus that grows spreading colonies that look fluffy or fuzzy.

CLINICAL TESTING FOR MYCOSES

Dermatologists use a Wood's light to detect mycosis on patients' skin. The handheld lamp emits ultraviolet light that causes some fungal species to fluoresce. The following conditions or infections can be detected by using a Wood's light:

- *Malassezia* tinea versicolor—pale yellow
- *Microsporum* tinea capitis—bright, light green
- *Trichophyton schoenleini* tinea capitis—green
- erythrasma (bacterial infection)—bright coral red

Clinical microbiologists familiar with dermatological infections often identify mycoses by observing the morphology of the fungus in a patient's specimen. Mycology uses four main features to identify fungi in a microscope: (1) shape and size of spores, (2) features of the sporangia and conidia (structures that hold fungal spores), (3) the supporting filaments called the *hyphae,* and (4) morphology of a colony grown on agar. Genetic testing supports morphological methods, and, for many fungal species, genetic testing has become the standard method of identification.

Staining techniques that work well on bacteria do not give the same results with most fungi. The Gram stain can be used for the yeasts *Candida* and *Malassezia,* but few other fungi react with the Gram stain dyes. The following methods make fungal features easier to see and, thus, make it easier to identify a mycosis:

- potassium hydroxide, used with phase contrast microscopy
- Calcofluor white, used with fluorescent microscopy
- India ink, used with light microscopy

Sites of Superficial and Cutaneous Fungal Mycoses

Fungus	Mycosis	Skin	Hair	Nails
Aspergillus	aspergillosis	yes	no	yes
Candida (yeast)	candidiasis	yes	no	yes
Epidermophyton	epidermophytosis	yes	no	yes
Malassezia (yeast)	pityriasis, tinea versicolor	yes	yes	no
Microsporum	tinea capitis and ringworm	yes	yes	yes
Phaeoannellomyces	mycosis	yes	no	no
Piedraia	piedra	no	yes	no
Scopulariopsis	mycosis	no	no	yes
Scytalidium	onychomycosis	yes	no	yes
Sporothrix	sporotrichosis (often subcutaneous)	yes	no	no
Trichophyton	various tineas	yes	yes	yes
Trichosporon	piedra	no	yes	no

In all cases of fungus identification, experience in mycology helps in differentiating organisms that look almost alike.

In 2008, the British mycologist Andrew Borman said, "Systemic fungal infections represent a major cause of morbidity and mortality in immunocompromised patients. The ever-increasing number of yeast species associated with human infections that are not covered by conventional identification kits, and the fact that moulds isolated from deep infections are frequently impossible to identify using classical methods due to a lack of sporulation, has driven the need for rapid, robust molecular identification techniques." Microbiologists use polymerase chain reaction (PCR) to make large amounts of fungal nucleic acids, then analyze the nucleic acids by sequencing the molecule's base units, called *nucleotides*. Molecular methods in mycology involve sequencing a fungus's ribosomal ribonucleic acid (rRNA) and provide more accurate identification than was ever achieved with morphological methods.

VETERINARY MYCOSES

Domesticated animals contract many of the same systemic and opportunistic mycoses as humans. Veterinarians often see the following mycoses in their practices:

- aspergillosis—birds, cattle, dogs, horses
- blastomycosis—cats, dogs, variety of wildlife
- candidasis—birds, cattle and calves, foals, pigs, and sheep
- coccidioidomycosis—cats, dogs, primates
- cryptococcosis—birds, cattle, dogs
- epizootic lymphangitis *(Histoplasma)*—horses
- geotrichosis *(Geotrichum)*—cattle, dogs, fowl
- histoplasmosis—cats, dogs

Many animal mycoses occur worldwide, and the most contagious cases spread in regions with poor veterinary care. Veterinarians use many of the same antifungal drugs for their patients used for humans, as well as the same methods for genetic testing.

All mycoses present a health concern because of their chronic nature and limited treatment choices. Most are not contagious, a characteristic that helps reduce transmission in dense populations. Mycology continues to seek more effective treatments for mycoses, treatments that selectively kill persistent pathogens but do not harm the host.

See also *ASPERGILLUS;* BIOGEOCHEMICAL CYCLES; FUNGUS; IDENTIFICATION; INFECTION; TOXIN; YEAST.

Further Reading

Borman, Andrew M., Christopher J. Linton, Sarah-Jane Miles, and Elizabeth M. Johnson. "Molecular Identification of Pathogenic Fungi." *Journal of Antimicrobial Chemotherapy* 61, Suppl. 1 (2008): i7–i12. Available online. URL: http://jac.oxfordjournals.org/cgi/reprint/61/suppl_1/i7.pdf. Accessed January 11, 2010.

Infectious Diseases Society of America. "Infections by Organism: Fungi." Available online. URL: www.idsociety.org/content.aspx?id=920. Accessed January 10, 2010.

Kahn, Cynthia M., ed. *Merck Veterinary Manual.* Whitehouse Station, N.J.: Merck, 2005.

New Zealand Dermatological Society. "Skin Manifestations of Systemic Mycoses." Available online. URL: http://dermnetnz.org/fungal/systemic-mycoses.html. Accessed January 10, 2010.

Richardson, Malcolm D., and Elizabeth M. Johnson. *Pocket Guide to Fungal Infection,* 2nd ed. Malden, Mass.: Blackwell, 2005.

Toxic Black Mold Information Center. "Common Species of Mold." Available online. URL: www.toxic-black-mold-info.com/moldtypes.htm. Accessed January 11, 2010.

Wadhwa, Anupriya, Ravinder Kauer, Satish Kumar Agarwal, Shyama Jain, and Preena Bhalla. "AIDS-Related Opportunistic Mycoses Seen in a Tertiary Care Hospital in North India." *Journal of Medical Microbiology* 56 (2007): 1,101–1,106. Available online. URL: http://jmm.sgmjournals.org/cgi/content/full/56/8/1101. Accessed January 10, 2010.

nanobiology Nanobiology is the science of developing biological systems with activity on a size scale (nanoscale) measured in nanometers (nm), or one billionth of a meter. Nanotechnology is the study of how materials behave at the nanoscale level. The value of nanotechnology derives from the fact that inanimate materials exhibit chemical behavior different from the behavior of the same chemicals in larger quantities. Microbiology, before the advent of electron microscopy, studied systems no smaller than a micrometer:

- 1 meter (m) = 3.28084 feet
- 1 centimeter (cm) = 0.01 m
- 1 millimeter (mm) = 0.001 m
- 1 micrometer (μm) = 0.000001 m = 1×10^{-6} m
- 1 nanometer (nm) = 0.000000001 m = 1×10^{-9} m = 1×10^{-3} μm

The National Institute of Science and Technology explained on its Web site how the meter originated in 1791: "The meter was intended to equal 10^{-7} or one ten-millionth of the length of the meridian through Paris from pole to the equator. However, the first prototype was short by 0.2 millimeters because researchers miscalculated the flattening of the earth due to its rotation. Still this length became the standard." Modern sciences in electronics, materials, and biology operate mainly at the micrometer (μm) and nanometer (nm) levels. Advances in electron microscopy now enable microbiologists to study microbial cell structures to about 0.05 nm, or the approximate size of a hydrogen atom.

Nanotechnology is an emerging science; most materials that chemists have studied, for centuries, have never been studied in nanoscale amounts. The *National Geographic* reporter Jennifer Kahn explained, in 2006, "Substances behave magically at the nanoscale because that's where the essential properties of matter are determined. Arrange calcium carbonate molecules in a sawtooth pattern, for instance, and you get fragile, crumbly chalk. Stack the same molecules like bricks, and they help form the layers of the tough, iridescent shell of an abalone." Nanobiology is the discipline that seeks knowledge of processes of molecules inside the cell; the manner in which some materials undergo self-assembly, such as the self-assembly that occurs in microbial membranes; and the compartmentalization of synthesis reactions in the cell or in organelles. Nanobiology may also help microbiologists detect the chemical gradients that microorganisms use as their immediate stimulus for sensing the environment.

The table on page 562 provides a scale of biological objects measured in nanometers.

APPLICATIONS IN NANOBIOLOGY

Nanobiology research institutes have begun to investigate applications for microbially based tools in medicine, environmental science, and materials science, the study of how matter behaves under various physical conditions.

In medicine, numerous projects in nanobiology have been planned for studying disease and for

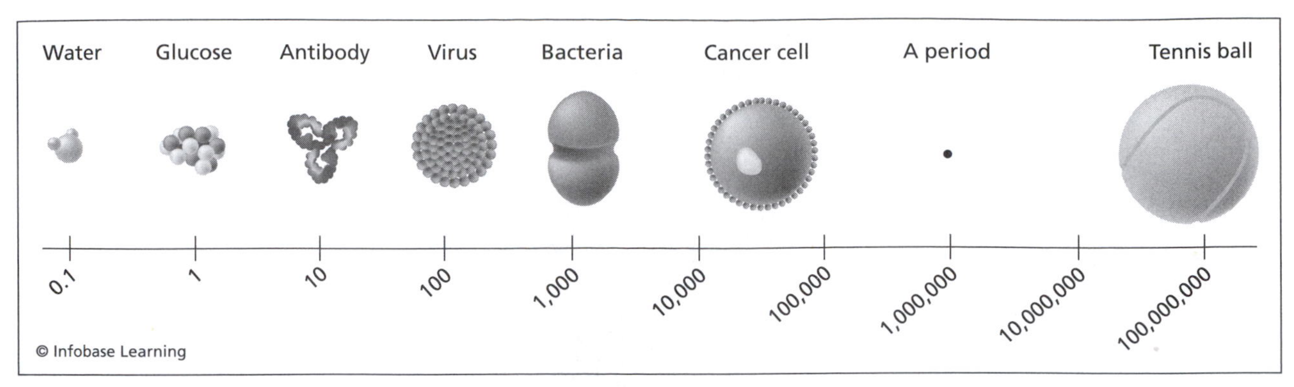

Nanobiology is the study of materials or objects on a nanoscale. Microscopy has only recently developed methods for measuring and studying objects measured in nanometers (nm).

detecting disease-related processes in the cell. The main areas of interest in medical nanobiology are described in the table.

Microbiologists employ biosensors in environmental science and food microbiology for detecting low levels of pathogens. Many biosensors contain an enzyme that is from a microorganism or a piece of DNA that detects specific microorganisms in heterogeneous mixtures such as soil, surface waters, or food. Despite the sensitivity of these biosensors, they have not detected many substances in nanoscale amounts. New nanosensors have the potential to detect substances in the body's tissue or in a single cell. The most promising approach to developing nanosensors resides in a structure called a *nanowire.* A nanowire is a cylindrical particle that can access structures inside a living cell; most important, the wire conducts a current so that it can produce a signal when it detects a target molecule or reaction within the cell.

Nanotechnology scientists began developing wirelike structures called nanotubes constructed of carbon atoms, many years ago, without a specific purpose for the structures. In 1991, the Japanese chemist Sumio Iijima (1939–) reinvestigated carbon nanotubes for their electrical conductivity. Carbon nanotubes offer promise in biology because, as Iijima put it, "the atoms of metal move about like amoeba" and "the carbon itself has a high affinity with the human body." Microbiological subjects have the same compatibility with the organic constituents of cells and offer the advantage of assembling on their own.

Virus nanowires have equal or greater promise than carbon nanowires for inspecting life at the nanoscale. The *Scientific American* reporter Philip Ross explained, in 2006, when virus nanowire technology was just emerging, "Put a phage [virus] with a highly specific taste for gold into a solution containing gold ions, and it will gild itself into a wirelet a micron long, suitable for connecting adjacent elements in a microcircuit. A variant of this phage will even link up with its fellows to form a gold wire many centimeters long, which can be spun like thread and woven into cloth fabric. Such a wire, bonded to chemically sensitive receptors, might detect toxic or biologically threatening agents." While virus nanowire research has continued, other laboratories have investigated the protein-synthesis capabilities of viruses as mini–production plants for nanoscale materials. Viruses contain genetic material wrapped in a protein coat. Scientists may, soon, put viruses to work for self-assembling tiny protein-based batteries. The Massachusetts Institute of Technology's materials chemist Angela Belcher (1958–) built such a battery, in 2006, by engineering viruses to coat themselves with cobalt oxide and gold and, then, self-assemble into nanowires. Now

Nanometer Measurements in Biology

Object	Approximate Nanometer Diameter
orange	100,000,000
cancer cell	10,000–100,000
normal human cell	5,000–100,000
red blood cell	6,000–8,000
bacteria	1,000 (range 300–5,000)
Mycoplasma bacteria	200
nanobacteria	50
rhinovirus (cold virus)	25
bacterial flagellum	20
ribosome	20
prion	13
antibody	10
glucose molecule	1
water molecule	0.1

Nanobiology Applications in Medicine

Application	Description
biochips	particles containing substances that bind with cellular molecules and track their movements in the body
magnetic nanoparticles	particles injected into cell or nucleus for manipulating structures inside living cells
nanobiocides	nanosilver particles added to bandages to speed wound healing
nanopores	voltage applied to nanoscale pores to help draw in cellular enzymes for studying enzyme-substrate interactions, such as deoxyribonucleic acid (DNA) and ribonucleic acid (RNA) processes
nanosensors	particles with microbial enzymes that detect toxins, pollutants, or disease markers in cells or in the nucleus
nanowires	virus assemblies for conducting a current
quantum dots	protein-coated particles that penetrate cells when the cells mistake them for nutrients; particles, then, track the movements of proteins in the live cell

that researchers have shown that viruses can make conducting materials, scientists are pursuing virus batteries with increasing voltage and capacitance.

Microbiologists have also studied a converse approach to nanodevices by developing nanowires to detect dangerous pathogens at low levels. Such sensitive detection of a few pathogen cells could give early warning of an imminent outbreak. New nanowires contain a piece of synthetic antibody on one end and a connection to an electrical base at the other end. The *DotMed News* writer Lynn Shapiro explained, in 2009, "If the protein the antibody binds to is present in the liquid, it will bind to these antibodies, immediately creating a sharply measurable jump in current through the nanowire." The authors of the study said, "We believe that nanowire biosensor devices functionalized with engineered proteins can have important applications ranging from disease diagnosis to homeland security." A prototype nanowire has, already been made to detect the virus that causes severe acute respiratory syndrome (SARS), which has been responsible for deadly outbreaks, since 2002.

NANOBACTERIA

Nanobacteria are bacteria smaller than any previously discovered bacteria. Nanobacterial cells measure less than 0.2 μm, or about one tenth the size of a typical bacterium, and smaller than *Mycoplasma* bacteria, previously thought to be the smallest independently living things. The size of the smallest nanobacteria has not been determined, but these cells are known to pass through filters with pore sizes of 0.1 μm.

In 1993, the geologist Robert Folk, at the University of Texas, first discovered nanobacteria in rock specimens, but many in microbiology doubted the tiny cells were living things. Critics thought that cells measured in nanometers could contain all the proteins and other large molecules needed by a cell to live independently. In 1996, new evidence for nanobacteria arose when scientists found what looked like fossils of nanobacteria in a meteorite that had crashed on Earth and dated to 4.5 billion years old. The meteor opened two lines of debate: one on the possibility of bacteria on other planets and another over the existence of nanobacteria. In 2004, the excitement rekindled, when researchers at the Mayo Clinic found what they thought were nanobacteria in calcified blood vessels of patients.

By 2009, studies on the proteins that researchers had identified as nanobacterial proteins gave evidence of contamination, meaning that the nanobacteria may never have been alive, after all. The nanobacteria researchers John Young and Jan Martel wrote in *Scientific American,* in 2010, "Although nanobacteria have been conclusively shown to be nonliving nanoparticles crystallized from common minerals and other materials in their surroundings, these nanoentities may still play an important role in human health. We believe that nanobacterialike particles are generated through a natural process that normally protects the body against unwanted crystallization but that can also promote nanoparticle formation under certain conditions." The study of nanobiology may require advances in technology comparable to improvements in microscopy in the 19th century. Nanobacteria symbolize the vast unknown world existing at the nanometer level.

See also BIOSENSOR.

Further Reading

Foster, Lynn E. *Nanotechnology: Science, Innovation, and Opportunity.* Upper Saddle River, N.J.: Pearson Education, 2006.

Gruner, George. "Carbon Nanonets Spark New Electronics." *Scientific American,* May 2007.

Heath, James R., Mark E. Davis, and Leroy Hood. "Nanomedicine Targets Cancer." *Scientific American,* February 2009.

Hochheiser, Sheldon. "Sumio Iijima." IEEE Global History Network. September 3, 2008. Available online. URL: www.ieeeghn.org/wiki/index.php/Sumio_Iijima. Accessed January 2, 2010.

Jain, Kewal K. "Nanobiotech in Genomics and Proteomics." *Genetic Engineering and Biotechnology News,* 1 April 2006.

Kahn, Jennifer. "Nano's Big Future." *National Geographic,* June 2006. Available online. URL: http://ngm.nationalgeographic.com/2006/06/nanotechnology/kahn-text. Accessed January 2, 2010.

Nam, Ki Tae, Dong-Wan Kim, Pil J. Woo, Chung-Yi Chiang, Nonglak Meethong, Paula T. Hammond, Yet-Ming Chiang, and Angela M. Belcher. "Virus-Enabled Synthesis and Assembly of Nanowires for Lithium Ion Battery Electrodes." *Sciencexpress,* 6 April 2006. Available online. URL: www.rifters.com/real/articles/Science_Viral_Nanowires.pdf. Accessed January 2, 2010.

National Institute of Science and Technology. "Historical Context of the International System of Units." Available online. URL: http://physics.nist.gov/cuu/Units/meter.html. Accessed January 2, 2010.

Patolsky, Fernando, and Charles M. Lieber. "Nanowire Nanosensors." *Materials Today,* April 2005. Available online. URL: http://nano.nd.edu/ESTS40403/Materials Today_8_20.pdf. Accessed January 25, 2010.

Ross, Philip E. "Viral Nano Electronics." *Scientific American,* October 2006.

Shapiro, Lynn. "Nanowire Biosensor Device Detects SARS Virus." DotMed News, 31 May 2009. Available online. URL: www.dotmed.com/news/story/9218. Accessed January 25, 2010.

Society for Industrial Microbiology News. "Engineering Biological Systems for Nanotechnology." March–April 2008.

Trafton, Anne. "New Virus-Built Battery Could Power Cars, Electronic Devices." *Massachusetts Institute of Technology News,* 2 April 2009. Available online. URL: http://web.mit.edu/newsoffice/2009/virus-battery-0402.html. Accessed January 2, 2010.

Young, John D., and Jan Martel. "The Truth about Nanobacteria." *Scientific American,* January 2010.

nitrogen fixation Nitrogen fixation is the conversion of nitrogen gas (N_2) into ammonia (NH_3) and a phase in the nitrogen cycle. The nitrogen cycle represents one of many processes, called biogeochemical cycles, that make nutrients available for use and reuse by the chemical conversions of nutrients as they move from the atmosphere to the earth or water, into living organisms, and again to the atmosphere.

All living things require nitrogen. Plants, animals, and microorganisms need nitrogen for synthesizing amino acids and proteins, enzymes, nucleic acid, and other specialized substances, such as the peptidoglycan that makes up bacterial cell walls. Nitrogen makes up about 14 percent of a bacterial cell and 3 percent of a human body. Despite the abundance of nitrogen in the atmosphere, 78 percent of the air's volume, plants and animals cannot access this nitrogen reservoir. Nitrogen fixation belongs exclusively to prokaryotes.

The nitrogen cycle helps put nitrogen into forms that higher organisms can use in their metabolism. This cycle, also, recovers nitrogen from decaying organic matter, which usually contains nitrogen compounds, and recycles it to soil or the air. Nitrogen bound in nonliving matter also reenters the environment, but this occurs so slowly that living things cannot depend on it. For example, rocks contain some nitrogen, but the weathering of rock lasts decades or longer. For all practical purposes, nitrogen fixation by prokaryotes provides plants and animals with the only way they can obtain nitrogen.

The nitrogen cycle can be thought of as beginning with bacteria. Many bacteria get nitrogen by breaking down amino acids in plants or animals. Eventually, however, amino acid nitrogen cannot sustain all of the organisms that need the element. Nitrogen fixation taps the enormous reserves of nitrogen in the atmosphere.

Microorganisms that fix gaseous nitrogen, called diazotrophs, consist of three main groups: (1) soil and water microorganisms, (2) microorganisms that live in symbiotic relationship with plants, and (3) some cyanobacteria. Soil and water microorganisms that do not require a symbiotic association with plants are called *free-living* organisms. No one group is more important than the other two in the nitrogen cycle, although bacteria that live with plants are critical to the plants' health.

Details of how bacteria operate in the nitrogen cycle were, first, elucidated by two microbiologists trained in botany. In the late 1800s, Martinus Beijerinck (1851–1931) of the Netherlands investigated the bacteria that live in root nodules of certain plants called *legumes.* Beijerinck's background in plant disease and the discovery of small nodules on plant roots probably led him to question the nodules' role in plant health. Beijerinck became the first scientist to isolate the bacteria from root nodules and grow them in a laboratory. The complexity of nitrogen fixation in bacterium-plant symbiosis (unrelated organisms living in close association) would be beyond Beijerinck's technology, but he, nonetheless, understood the generalities of nitrogen fixation. In 1888, Beijerinck acknowledged "the extremely complicated situation we have in the symbiotic relationship which exists in the nodule. When the living plant cell must live with another organism which is actually a part of its protoplasm, it is then necessary that a subtle balance must exist between the growth of the plant and the growth of the bacterium." Beijerinck's discovery influenced various disciplines, including agriculture, environmental microbiology, and nutrient cycling.

The Russian bacteriologist Sergei Winogradsky (1856–1953) worked independently of Beijerinck but, almost simultaneously, studied nitrogen use by bacteria, mainly free-living soil and water bacteria. Winogradsky made the first investigations into nitrification, which is the conversion of the ammonia to nitrite (NO_2^-), and then to nitrate (NO_3^-).

Both scientists laid the foundation for studying the microbial biochemistry of nitrogen. Neither would elucidate all of the processes in nitrogen cycling; generations of microbiologists to follow would shed light on the complicated relationship of nitrogen fixers with plants and with other nitrogen-metabolizing bacteria.

NITROGEN-FIXING BACTERIA

Nitrogen-fixing bacteria belong to a diverse group that use a variety of metabolisms and include gram-positive and gram-negative species. These bacteria use photosynthesis, aerobic respiration, or anaerobic pathways. Although most live in soils and water, a few species inhabit the digestive tract of humans and animals. The species that form symbiotic relationships with plants do so with land plants and some marine plants.

Free-living diazotrophs inhabit soil, freshwaters, and marine waters. The main microorganisms in this group are *Azospirillum, Azotobacter, Beijerinckia, Clostridium, Desulfovibrio,* and *Klebsiella*. Bacteria that belong to the following general groups also fix nitrogen: purple sulfur, purple nonsulfur, and green sulfur bacteria. Not all species of *Klebsiella* fix nitrogen; this genus lives in natural freshwaters and the digestive tract. *Azospirillum* is unique because it fixes nitrogen in a region of the soil called the *rhizosphere* but also infects some grass and tuber roots in symbiosis with the plant. The rhizosphere is an area surrounding plant roots and high in compounds excreted by the roots. The root substances serve as nutrients for microorganisms; thus, rhizospheres usually contain higher microbial numbers than other areas in the soil. Some microbiologists define *Azospirillum* as a symbiotic bacterium with plants because of the close connection between the rhizosphere and the plant.

The bacteria that live in symbiosis with plants are called *plant symbionts*. Nitrogen-fixing plant symbionts participate in a type of symbiosis called *mutualism,* in which both organisms benefit from the association. By putting nitrogen into a usable form, the bacteria benefit the plant, and the plant benefits the diazotrophs by providing them with a habitat. The main symbionts in this relationship are *Rhizobium, Bradyrhizobium,* and *Frankia. Rhizobium* and *Bradyrhizobium* infect the roots of legumes, which are plants that form a pod fruit, such as peas and beans.

Frankia associates with nonleguminous plants in an association called *actinorhiza.* These plants (example: alder trees) form root nodules that *Frankia* infects and where nitrogen fixation occurs. *Actinorhiza* symbionts typically act as the first plants to colonize barren, nutrient-poor land. Their symbiotic relationship with *Frankia* represents a vital aid in moving nitrogen into the plant's tissue.

Cyanobacteria of the genera *Anabaena, Nostoc, Plectonema,* and *Trichodesmium* fix nitrogen in water. Three examples of cyanobacterium-plant associations that use nitrogen fixation are (1) *Anabaena* and the floating fern *Azolla,* (2) *Nostoc* and *Anthoceris* (aquatic hornworts), and (3) *Nostoc* and cycads (woody plants that produce cones). A large population of cyanobacteria also lives in association with a fungus to forms lichens. Lichens serve a major role in terrestrial nitrogen fixation.

Of all the earth's diazotrophs, the symbionts of legume plants fix the most nitrogen, followed in order by cyanobacteria, bacteria living in the rhizosphere, and free-living soil and water bacteria. Although diazotrophs principally live in association with legumes, the diazotroph *Frankia* is an exception because it lives in symbiosis with nonlegume plants. When free-living in soil without associating with a plant, *Frankia* does not fix nitrogen. The following examples of diazotrophs indicate the diversity of this group:

- free-living, aerobic—*Azotobacter*
- free-living, anaerobic—*Clostridium*
- legume plant-associated—*Rhizobium*
- nonlegume plant-associated—*Frankia*

All bacteria classified as diazotrophs play a vital role in Earth physiology by running the nitrogen cycle.

MECHANISMS OF NITROGEN FIXATION

The enzyme nitrogenase controls the reaction by which bacteria add hydrogen atoms to a nitrogen atom to form ammonia. Nitrogenase consists of two proteins; one is an iron-containing protein, and the other is an iron-molybdenum-protein complex. Both proteins must combine in a specific way for the enzyme to work. Oxygen at low levels inactivates nitrogenase, so aerobic diazotrophs must develop mechanisms for protecting the enzyme from oxygen. The diverse population of diazotrophs have developed varied means of keeping oxygen away from nitrogenase. Some of the known protective methods are the following:

- water-saturated extracellular polysaccharide that slows the diffusion of oxygen into the cell
- respiratory systems that rapidly use up oxygen in the cell
- use of oxygen-scavenging molecules that bind oxygen and make it metabolically unavailable
- compartmentalization of nitrogenase reactions in the cell to prevent exposure to oxygen-utilizing systems

Cyanobacteria use the last approach for keeping nitrogenase distant from photosynthesis, which produces oxygen. Anaerobic bacteria, such as *Clostridium,* do not confront these problems, because they live in oxygen-free environments and use metabolism that does not require oxygen.

The nitrogenase reaction requires the bacterial cell to expend energy, which is bound in the cell's adenosine triphosphate (ATP), the main energy-storage and transfer compound in biology. As the reaction progresses, ATP gives up energy held in bonds connecting phosphate (PO_4^{-3}) to the ATP molecule. As a result, the expenditure of energy produces adenosine disphosphate (ADP):

$$N_2 + H^+ + ATP \rightarrow NH_3 + H_2 + ADP$$

The overall equation shows how hydrogen ions (H^+) bind with nitrogen to form ammonia and hydrogen gas (H_2):

$$N_2 + 8\ H^+ + 8\ \text{electrons} + 16\ ATP + 16\ H_2O \rightarrow 2\ NH_3 + H_2 + 16\ ADP + 16\ \text{phosphates}$$

In addition to oxygen sensitivity, nitrogenase stops working when ammonia builds up. This process is called *feedback inhibition* or *end product inhibition.* In the simplest form of feedback inhibition, the buildup of end product suppresses the gene that encodes for nitrogenase. When the end product levels subside, the gene becomes active again, and the cell starts making more nitrogenase.

Microbiologists have made extensive studies into nitrogenase's mechanism and the manner by which the cell regulates the enzyme's synthesis and activity. Nitrogenase provides a helpful study model for feedback mechanisms. The enzyme's role in making nitrogen available for biota attests to its importance in biology. In 2009, University of California–Irvine biochemists discovered a new form of nitrogenase in *Azotobacter vinelandii* that requires the metal vanadium. "The vanadium (V)-nitrogenase is very similar to the 'conventional' molybdenum (Mo)-nitrogenase," the biochemists Chi Chung Lee, Yilin Hu, and Markus Ribbe said; "yet it holds unique properties of its own that may provide useful insights into the general mechanism of nitrogenase catalysis." Microbiologists continue to conduct studies on the relationship between nitrogenase, its metal cofactors, and cell energetics.

ROLE OF NITROGEN FIXATION IN THE NITROGEN CYCLE

Nitrogen fixation occurs in the soil, in root nodules, or in water to begin the nitrogen cycle. Plant roots either absorb ammonia from the soil or receive it from the bacteria in their roots. The plant, then, assimilates the ammonia, meaning it builds the nitrogen compounds it needs by using the ammonia nitrogen. When the plant dies, decomposer bacteria degrade large compounds such as proteins to their base units. In the case of proteins, the units are amino acids. As decayed matter sinks into the soil, bacteria and fungi carry out ammonification, the process of releasing ammonia from nitrogen-containing molecules. Similarly to nitrogen fixation, aerobes or anaerobes perform ammonification.

Ammonia can either cycle back into new plant life or return to the atmosphere by two processes in sequence: nitrification and denitrification. Nitrification is an aerobic process to convert ammonia to nitrites and nitrates. *Nitrosomonas* is a genus that performs the first step, producing nitrite, and *Nitrobacter* carries out the second step in nitrification to make nitrates. Several different anaerobic denitrifying bacteria convert these forms of nitrogen to the gases nitrous oxide and nitrogen. Most of the nitrogen eventually ends up as N_2 in the atmosphere, from which nitrogen fixation again captures it.

BACTERIA–ROOT NODULE ASSOCIATIONS

Plant symbionts in nitrogen fixation begin by infecting the legume plant's roots just as a pathogen infects mammalian cells. The rhizosphere attracts nitrogen-fixing species in the same way as it attracts other soil microbiota. The infection of the roots starts with attachment of the bacteria to the root surface due to the bacteria's recognition of certain structures on the root hair. The bacteria cluster at microscopic thread-like projections called *root hairs* and secrete substances that weaken the root hairs' defenses. The bacteria enter the root hairs and travel to the main root, where they begin to accumulate and form densely packed bunches of cells called bacteroids. A root nodule forms on the infected root, partially by the accumulation of bacteroids and partially by the plant's response to the infection. The plant builds an ever-thickening wall around the bacteroids, which turns into the root nodule.

Only specific strains of *Rhizobium* and *Bradyrhizobium* infect certain legumes and cannot infect others. Genes in the diazotroph's deoxyribonucleic acid (DNA) called *specificity genes* determine the legumes that a particular bacterial strain can infect. Not all strains that infect the target plant can produce nitrogen-fixing root nodules. Strains that possess the capacity to form functioning root nodules are called *effective strains,* and strains that cannot induce nodule formation after infecting are called *ineffective strains.*

The following plants are legumes that use symbiosis with nitrogen-fixing bacteria:

- alfalfa
- beans
- clover
- lentils
- mesquite
- peanuts
- peas
- soybeans

The legumes listed here each contain numerous species. Agriculture depends on nitrogen fixation for a significant part of its yearly crop worldwide. Nitrogen fixation enables legumes to grow in nitrogen-poor soils that would not support other crops. Because legumes have good nitrogen content, when they decay, they return nitrogen to the soil, a type of fertilization. Farmers take advantage of this by alternating legume crops with nonlegumes, each growing season. Nitrogen fixation, thus, is as important in agricultural economy as it is in ecology.

See also BIOGEOCHEMICAL CYCLES; GREEN BACTERIA; MICROBIAL ECOLOGY; PURPLE BACTERIA; SYMBIOSIS.

Further Reading

Beijerinck, Martinus W. "Die Bacterien der Papilionaceenknöllchen" (The Root-Nodule Bacteria). *Botanische Zeitung* 46 (1888): 725–804. In *Milestones in Microbiology,* translated by Thomas Brock. Washington, D.C.: American Society for Microbiology Press, 1961.

Deacon, Jim. "The Microbial World: The Nitrogen Cycle and Nitrogen Fixation." Available online. URL: www.biology.ed.ac.uk/research/groups/jdeacon/microbes/nitrogen.htm. Accessed January 16, 2010.

Haselkorn, Robert, and William J. Buikema. "Nitrogen Fixation in Cyanobacteria." In *Biological Nitrogen Fixation,* edited by Gary Stacey, Robert H. Burris, and Harold J. Evans. New York: Chapman & Hall, 1992.

Lee, Chi Chung, Yilin Hu, and Markus W. Ribbe. "Unique Features of the Nitrogenase VFe Protein from *Azotobacter vinelandii.*" *Proceedings of the National Academy of Sciences* 106 (2009): 9,209–9,214. Available online. URL: www.pnas.org/content/106/23/9209.full. Accessed January 16, 2010.

Society for General Microbiology. "Rhizobium, Root Nodules and Nitrogen Fixation." 2002. Available online. URL: www.microbiologyonline.org.uk/forms/rhizobium.pdf. Accessed January 16, 2010.

University of Connecticut. "*Frankia* and Actinorhizal Plants." Available online. URL: http://web.uconn.edu/mcbstaff/benson/Frankia/FrankiaHome.htm. Accessed January 16, 2010.

normal flora Normal flora consist of bacteria, fungi, and yeasts that live naturally on the body. Normal flora consist of the microorganisms that live on the skin, in the mouth, in the gastrointestinal tract, and in the respiratory tract. These organisms are also called *native,* indigenous, or resident flora. Transient flora, by contrast, are microorganisms that stay on the body for a short period and fail to establish a permanent population there.

The numbers of normal flora on the body exceed the number of human cells in the body. Normal bacterial numbers vary from person to person, but, on average, people have the following amounts of bacteria on the body: 10^{10} in the mouth, 10^{12} on the skin, and 10^{14} in the gastrointestinal tract. These numbers do not account for the yeasts and fungi that also inhabit the body. If all of the microbial chromosomes on the body were collected and mixed together, the result would show that, in chromosomal content, a human body is more microbial than it is human.

PROTECTIVE ROLE OF NORMAL FLORA

Normal flora add to the body's nonspecific defenses against infection by pathogens through a process called *antagonism.* Antagonism is a relationship between organisms in which they compete for resources. On humans, animals, or plants, the resources over which microorganisms compete are nutrients, water, space, light, and protective factors. Examples of protective factors on human skin are specialized locations that favor some microorganisms over others, such as the nasal passages, underarms, ear canal, or mouth.

The activities of normal flora, such as the following, often enhance antagonism in a way that allows them to outcompete nonnative microorganisms:

- fatty acid production that lowers pH
- antimicrobial secretions such as antibiotics and bacteriocins (antibiotics that inhibit members of the same genus or species)

- oxygen depletion of same locations to inhibit aerobes
- overgrowth of locations to prevent nonnative organisms from attaching to host receptor sites (proteins on cell surfaces recognized by certain microorganisms)

These methods of outcompeting nonnative flora become especially important in repelling any pathogens that have contact with the body. Secretions of fatty acids and antimicrobial substances serve as important defenses produced by skin flora. Oxygen depletion occurs in numerous places, including the mouth. In this situation, facultative anaerobes that can live with or without oxygen use up all the oxygen in a certain location, thus reducing the variety of transient organisms that can colonize the site. The final mechanism through which normal flora grow over an entire body surface to disallow attachment by nonnative organisms helps protect the digestive tract from pathogens that have been ingested.

The process whereby normal flora prevent potentially dangerous nonnative organisms from colonizing the body is called *competitive exclusion.* All of the tactics known to be used by normal flora—and probably additional unknown factors—defend the body against all microorganisms but those recognized by the body as native. The immune system participates in this process of differentiating microorganisms, called self versus non-self matter. Immune system cells are programmed to recognize surface structures of normal flora that differ or are lacking on nonnative flora. Normal flora possess their own recognition system, which allows native microorganisms to stay on the body but excludes nonnative microorganisms. The biologist Noreen Murray of Scotland's Institute of Cell and Molecular Biology explained, in 2001, "Bacteria commonly endow their DNA with an identity mark. When DNA is transferred from one bacterium to another strain of the same species, DNA that lacks the identification mark of the recipient strain is recognized as 'foreign' rather than 'self.' Foreign DNA is commonly degraded." Enzymes called *restriction endonucleases* perform this job of cleaving foreign DNA molecules, thus preventing other bacteria or infective viruses, called *bacteriophages,* from changing the composition of the native population.

People become exposed to a diversity of microorganisms every day and often in very high numbers. A very small proportion of microorganisms in the environment cause illness in humans, but the body's natural defenses and hygiene practiced by a person are directed mainly against those pathogens. Good hygiene involves proper hand washing, avoiding touching hands to the face, covering the mouth when sneezing or coughing, and maintaining sound skin care and oral care. General overall health should be maintained to strengthen the immune system, which can fight off infections from two types of pathogens: frank pathogens and opportunistic pathogens. Frank pathogens are microorganisms whose major relationship with humans is in disease. Opportunistic pathogens, by comparison, are part of the body's normal flora and do not infect the body unless conditions change to favor infection. One such change would be the decline of the host's health, in which case, this person is referred to as a *compromised host.*

In the right conditions, almost any member of the normal flora can be an opportunistic pathogen, so it can be difficult to distinguish opportunistic from nonopportunistic species on the body. *Staphylococcus aureus* and *Candida* represent the most common of the normal flora that cause opportunistic infections. As a rule, microorganisms that stay in their normal habitats on a healthy body present very low risk of opportunistic infection. The risks increase if the host's defenses have been weakened by damage to the immune system, injury or trauma, or a debilitating chronic disease. Otherwise, the normal flora shown in the table on page 569 do not behave as health threats. The table also shows the principal habitats for normal flora on the human body.

Diphtheroids are a general group of nonpathogenic bacteria represented mainly by species of *Corynebacterium.* Diphtheroid cells are pleomorphic (many shapes occur in one species). Many species of diphtheroids are lipophilic, meaning they require fats, and, therefore, inhabit places on the body where sebum levels are high. *Propionibacterim acnes,* the cause of skin acne, represents the only harmful diphtheroid that is a member of the normal flora.

Everyone carries a normal population of flora that contains most or all of the microorganisms in the table. Some people have additional species that make up their normal flora. People in certain occupations or living conditions have normal flora that reflect their surroundings. Pet owners versus non–pet owners, urban versus rural living, farming versus nonfarming occupations, and other criteria cause groups of people to differ from each other in the general types and proportions of microorganisms in normal flora.

The following genera (including those listed in the table) have been identified as the bacteria, fungi, yeasts, and protozoa that can be part of a person's normal flora: *Abiotrophia, Acoleplasma, Acidaminococcus, Acinetobacter, Actinobacillus, Actinomyces, Aerococcus, Aeromonas, Anaerorhabdus, Arcanobacterium, Arthrobacter, Bacillus, Bacteroides, Bifidobacterium, Bilophila, Blastocystis, Blastoschizomyces, Brachyspira, Brevibacterium,*

Burkholderia, Butyrivibrio, Campylobacter, Candida, Capnocytophaga, Cardiobacterium, Chilomastix, Chryseobacterium, Citrobacter, Clostridium, Corynebacterium, Cryptococcus, Dermabacter, Dermacoccus, Desulfomonas, Desulfovibrio, Eikenella, Endolimax, Entamoeba, Enterobacter, Enterococcus, Enteromonas, Epidermophyton, Escherichia, Eubacterium, Fusobacterium, Gardnerella, Gemella, Haemophilus, Hafnia, Helcococcus, Helicobacter, Iodamoeba, Kingella, Klebsiella, Kocuria, Kytococcus, Lactobacillus, Lactococcus, Leptotrichia, Listeria, Malassezia, Megasphaera, Micrococcus, Microsporum, Miksuokella, Mibiluncus, Moraxella, Morganella, Mycoplasma, Neisseria, Oligela, Pasteurella, Peptostreptococcus, Porphyromonas, Prevotella, Propionibacterium, Propioniferax, Proteus, Providencia, Pseudomonas, Psychrobacter, Retortamonas, Rothia, Ruminococcus, Selenomonas, Serpulina, Staphylococcus, Stomatococcus, Streptococcus, Succinivibrio, Tissierella, Treponema, Trichomonas, Trichophyton, Turicella, Ureaplasma, Veillonella, Weeksella, and *Yersinia.* Depending on

Normal Flora of the Human Body

Habitat	Predominant Conditions	Predominant Microorganisms
eyes (conjunctiva)	moist and containing lysozyme	• staphylococci • streptococci
nose	drying conditions, warmer than outer skin but cooler than mouth, interior tracts, and armpits	• *Staphylococcus aureus* • *Streptococcus pneumoniae* • *Neisseria* • *Haemophilus*
mouth and upper respiratory tract (lower respiratory tract has few to none)	moist and anaerobic in mouth, drying conditions in pharynx	• streptococci • various anaerobes • *Candida* (yeast) • *Actinomyces* • diphtheroids
outer ear	thick secretions	• staphylococci • diphtheroids • *Pseudomonas*
stomach	acidic (pH 2) gastric juices	• *Streptococcus* • *Lactobacillus* • *Helicobacter pylori* • *Peptostreptococcus*
intestines	high-nutrient, anaerobic	• enteric bacteria • *Lactobacillus* • *Clostridium*
urethra	moisture, salts	• staphylococi • streptococci • diphtheroids • *Bacteroides* • *Fusobacterium*
vagina	pH maintained at 4.4–4.6	• *Lactobacillus* • *Candida* • diphtheroids • *Bacteroides* • *Gardnerella vaginalis*
skin	dry in some places and moist at armpits, feet, and urogenital tract	• staphylococci • diphtheroids • streptococci • *Malassezia furfur* (yeast) • *Candida* • *Trichophyton* (fungus)

the specific microorganism, these genera have been isolated from the skin, ear, eyes, respiratory tract, digestive tract, or urogenital tract.

NORMAL FLORA AND SYMBIOSIS

A person exists in a symbiotic relationship with the normal flora, meaning two different types of organisms live in close association. Symbiosis can be either a benefit to one or both organisms or a detriment to one of the organisms. (If a relationship became a detriment to both organisms, it probably would not last.) Normal flora tend to occupy three types of symbiotic roles on the body: (1) commensalism, (2) mutualism, or (3) parasitism. In commensalism, one organism receives a benefit, and the other is unaffected. For example, many of the skin species may not be particularly important for a person's well-being. The microorganism may benefit from the body's supply of nutrients and habitat, but the host has no perceived benefit. Many normal flora, in fact, give the person a benefit by excluding pathogens from attaching to the skin. In this case, the symbiosis is called mutualism, in which both organisms benefit from the relationship. A small portion of normal flora act as parasites by gaining a benefit at the expense of the host's well-being. For example, the fungus *Trichophyton mentagrophytes,* which lives on or in the skin covering the feet, causes athlete's foot. Despite this ailment it causes its host, *T. mentagrophytes* often makes up part of the normal flora.

As microbiologists have learned more about the interactions among skin flora and between the flora and the host, they have likened the normal flora to an ecosystem. The molecular biologist Julie Segre of the National Human Genome Research Institute remarked to Live*Science,* in 2009, "We use the analogy that the skin is like a desert with large dry areas, but then there are these streams, or creases of your body. There's much richer bacterial life in the streams." The deserts-versus-jungles analogy accurately describes the varied habitats on the skin that affect the types and numbers of microorganisms living there.

Research on the activities of normal flora often employs two types of in vivo test models: gnotobiotic animals and germ-free animals. Gnotobiotic animals are animals that are delivered by cesarean section, so that they cannot receive microorganisms from the mother, and then inoculated with a mixture of known microorganisms. Researchers compare the growth, nutrient requirements, and general health of gnotobiotic animals with those of animals of the same age for the purpose of learning more about the role of normal flora. Testing a group of known microorganisms against a complete population of normal flora can make the activities of the known species more obvious. Later experiments might use additions or subtractions of known species.

Germ-free animals lack all flora because they do not receive an inoculation. Germ-free conditions are difficult to establish and maintain in animals; usually, small animals, such as hamsters, guinea pigs, or chickens, lead the way in this area of research. Chickens work best as germ-free models because a scientist need only take a fertile egg to raise the germ-free animal. The scientist sterilizes the egg by injecting a biocide into it that kills bacteria but does not harm the chick. When the chick hatches, a scientist transfers it to a specially constructed pen that supplies filtered air, removes waste aseptically, and uses a double-door port for moving equipment in and out of the pen. Germ-free animals can move about and access sterilized feed and water. Most germ-free chambers include a tube of sterile microbiological medium to assure the conditions inside the pen remain germ-free. Any growth on the medium indicates a contaminant has entered and possibly the animal.

Experiments with germ-free animals have resulted in the following observations:

- germ-free animals returned to a normal environment are very susceptible to infections
- infectious dose of pathogens is smaller in germ-free animals than in normal animals
- return to the environment causes rapid establishment of a population on the body
- no dental plaque or caries develops on germ-free animals
- germ-free animals resist contracting amoebic dysentery when inoculated with the pathogen *Entamoeba histolytica*
- germ-free animals grow more slowly to smaller body weight than normal animals
- skin and hair quality are poor in germ-free animals

(*E. histolytica* cannot establish an infection in germ-free animals because the animals lack bacteria that the protozoan uses for food.)

Germ-free experiments have helped confirm that normal flora serves a generally beneficial role in animal health. Poor weight gains and growth rates and outward signs of nutrient deficiency indicate that animals need their intestinal flora for food digestion, amino acid production, and synthesis of certain vitamins such as vitamin K and B-complex.

The poor condition of animals that lack their normal complement of flora has prompted scientists to investigate the value of giving microorganisms to animals and people for better health. *Probiotics* refers to the administration of living microorganisms to a person as a dietary supplement. Probiotics are usually taken for the purpose of reestablishing a normal population of microorganisms in the intestines after abrupt changes in diet, health stresses, or antibiotic therapy. Each of these events may cause a condition called *microbial imbalance*. In imbalance, a person does not usually lose all of the intestinal flora, but certain species may disappear, allowing the growth of others that usually exist in low numbers. Imbalanced gut flora may reduce the efficiency of nutrient digestion and cause general discomfort in the gastrointestinal tract.

Probiotic formulas have been developed for dogs, cats, cattle, swine, horses, sheep, goats, and poultry. Probiotic formulas vary for these animals because of different digestive systems; however, most of the formulas developed for humans contain the following bacteria: *Bifidobacterium bifidus*, *Lactobacillus acidophilus*, and additional species from each of these genera. Formulas can also include yeasts such as *Saccharomyces*.

Probiotic microorganisms do not repopulate the digestive tract with their own numbers, but, rather, they make conditions suitable for the normal flora to begin growing again. The National Institutes of Health's National Center for Complementary and Alternative Medicine recommends probiotic treatment for the following ailments:

- severe diarrhea
- *Clostridium difficile* infection of the gastrointestinal tract
- urogenital tract infections
- irritable bowel syndrome
- side effects of surgeries on the colon
- atopic dermatitis (eczema) in children

Probiotics must meet certain requirements to work as they are intended. First, the microorganisms in the formula must be able to tolerate the acid conditions of the stomach and bile salts in the intestines, both of which inhibit nonenteric species. Second, the organisms must be nontoxic and have minimal side effects. Third, probiotic microorganisms should be able to adhere to the intestinal lining. Finally, microorganisms that are easy to grow, such as *Bifidobacterium* and *Lactobacillus*, help probiotic manufacturing.

Normal flora probably acts on and in the body in many ways that microbiology has yet to discover. There can be no question that the overall benefits of normal flora outweigh the health risks of these organisms.

See also ENTERIC FLORA; HYGIENE; IMMUNITY; MICROBIAL COMMUNITY; OPPORTUNISTIC PATHOGEN; ORAL FLORA; SYMBIOSIS.

Further Reading

Davis, Charles P. "Normal Flora of the Skin." In *Medical Microbiology*, 4th ed., edited by Samuel Baron. Galveston: University of Texas Press, 1996. Available online: URL: www.ncbi.nlm.nih.gov/bookshelf/br.fcgi?book=mmed&part=A512. Accessed January 1, 2010.

Hsu, Jeremy. "Rich Streams of Bacterial Life Found on Skin." May 28, 2009. Available online. URL: www.livescience.com/health/090528-skin-bacteria.html. Accessed January 1, 2010.

Marples, Mary J. *The Ecology of the Human Skin*. Springfield, Ill.: Thomas, 1965.

Murray, Noreen E. "Immigration Control of DNA in Bacteria: Self versus Non-Self." *Microbiology* 148 (2002): 3–20. Available online. URL: http://mic.sgmjournals.org/cgi/reprint/148/1/3. Accessed January 2, 2010.

Murray, Patrick R., and Yvonne R. Shea. *Pocket Guide to Clinical Microbiology*, 3rd. ed. Washington, D.C.: American Society for Microbiology Press, 2004.

National Center for Complementary and Alternative Medicine. "An Introduction to Probiotics." August 2008. Available online. URL: http://nccam.nih.gov/health/probiotics. Accessed January 1, 2010.

Noble William C., ed. *The Skin Microflora and Microbial Skin Disease*. Cambridge: Cambridge University Press, 2004.

Rosebury, Theodor. *Microorganisms Indigenous to Man*. New York: McGraw-Hill, 1962.

nosocomial infection A nosocomial infection is an infection that starts during the course of a hospital stay and was not present when the patient was admitted. Sometimes, this definition is expanded to include infections that result from a visit to an outpatient clinic, a doctor's office, or a dentist's office.

Nosocomial infections are a major concern in health care. In 2009, the Centers for Disease Control and Prevention (CDC) attributed 250 deaths daily to infections acquired in a hospital. Two million people became sick each year from microorganisms picked up in health care settings from staff or from other patients.

Hospital environments increase the risk of infection for three reasons: (1) a population of patients who have weakened immunity due to disease or injury, (2) easy transmission between staff and patients, and (3) a high concentration of microorganisms in the hospital concurrently with infectious diseases. Hospital patients whose immune systems are already compromised have an increased risk of infection from pathogens and from opportunistic pathogens, which are normally harmless but can cause illness if the conditions favor infection. Hospital patients in the following categories often have a higher-than-normal risk of infection due to a weak or overtaxed immune system:

- acquired immunodeficiency syndrome (AIDS)
- cancer chemotherapy
- organ transplant recipients
- chronic, debilitating disease
- surgery or trauma
- elderly adults
- late-term pregnancy and childbirth
- newborns
- addictions

These high-risk patients, called *compromised hosts,* have a greater chance of infection inside the hospital than they do away from the hospital environment.

A hospital also provides an easy chain of transmission for pathogens. Staff-to-patient contact and patient-to-patient contact serve as efficient and common transmission routes in hospitals. Hospitals also use a large number of fomites (inanimate objects that carry germs) where microorganisms can remain infective from a few hours to several days. Typical hospital fomites for disease transmission are:

- surgical instruments
- needles
- intravenous systems
- respiratory aids
- gowns and bedding
- trays and bedside tables
- bedside handrails
- bathrooms

Perhaps the factor most unique to hospitals is the population of microorganisms confined there. Many hospitals have developed bacterial populations that differ from anything seen outside the hospital. The two main differences between nosocomial settings and places outside the hospital are the relative proportions of species and the frequency of antibiotic-resistant bacteria.

NOSOCOMIAL MICROORGANISMS

Nosocomial infections may soon become synonymous with antibiotic-resistant infections, and the makeup of these resistant populations has been changing since antibiotics first had commercial use, in the 1940s. In the 1940s and 1950s, *Staphylococcus aureus* caused most nosocomial infections. Penicillin decreased *S. aureus* incidence, for the next few decades, and *Escherichia coli* and *Pseudomonas aeruginosa* emerged as the main nosocomial bacteria. Antibiotic-resistant *S. aureus* reemerged, in the 1980s, along with other resistant bacteria, such as streptococci and enterococci. By the 1990s, multiple drug-resistant (MDR) species dominated hospitals more than any other environment.

In 2008, the National Safety Healthcare Network produced its most recent report on the microorganisms that cause most nosocomial infections. Ten microorganisms accounted for 84 percent of all infections, which are shown in the table on page 573 along with the predominant MDR bacteria.

Other bacteria that also cause a small percentage of MDR infections are *A. baumannii, K. pneumoniae, K. oxytoca,* and *E. coli.*

Clostridium difficile has been increasing rapidly as a major threat of infection in hospital patients. *Clostridium* species resist many physical hardships in the environment, as well as antibiotics, because of the strong endospore they form. An endospore is a thick-walled form of the cell that protects the dormant cell against extremes in temperature, drying, and chemicals. Infected patients shed *C. difficile* in feces, which may cause oral-fecal routes of infection in hospital patients. The microbial pathologist Brendan Wren of the London School of Hygiene and Tropical Medicine explained, in 2009, "*C. difficile* is a high-profile and rapidly emerging pathogen and is responsible for the death of a patient every hour in our hospitals—but its biology and transmission are so far poorly understood." Researchers must also determine why some strains of *C. difficile* in hospitals are more virulent and transmissible than others. The rugged endospores probably help this organism evade antibiotics and disinfectants in hospitals.

All pathogens have a preferred portal of entry, which is the main site by which they enter the body and cause infection. Nurses care for hospital patients by assuring that portals of entry are protected from contamination by microorganisms. These protections involve the use of antiseptics on skin, antimicrobial treatments or dressing for open wounds, sterilized medical equipment, and aseptic technique, which are all the processes used to prevent contamination. Aseptic techniques in hospitals encompass proper disinfection of inanimate surfaces, sterilization of devices such as needles and syringes, and discarding of contaminated items (biohazards). Aseptic techniques provide a valuable barrier against germ transmission in hospitals, because they do not depend on drugs but rather on actions that reduce the chance of contamination.

Two main portals of entry for nosocomial infection are surgical wounds and the urinary tract. Most of the main nosocomial pathogens can use either route, except *Candida,* which uses only the urinary tract, and *C. difficile,* which enters the digestive tract. The table on page 574 describes how nosocomial microorganisms use portals of entry.

Staphylococcus, Pseudomonas, and *E. coli* have been called the *nosocomial troika* because of their prevalence in hospitals and hospital infections. Nosocomial infection is not a static situation, however, because resistant microorganisms change, over time, as do medical procedures. Microorganisms responsible for nosocomial infections on the rise are *Candida* species, *Klebsiella, Enterobacter,* and *C. difficile.* Some emergences occur in a cyclical pattern, as new antibiotics arrest a pathogen for a time, then antibiotic-resistant strains arise, followed by another antibiotic regimen. For example, enterococci did not emerge as a nosocomial problem until cephalosporin antibiotics increased in use to kill organisms that had become resistant to penicillin. Antibiotic-resistant enterococci emerged as a nosocomial population, in the 1980s, and is now as critical as MRSA. Likewise, MRSA that seemed to succumb only to vancomycin is now developing resistance to this antibiotic, too. The new microorganism is called *vancomycin-resistant MRSA,* or VRSA.

Advances in medical techniques also change the patterns of hospital infections over time. The following medical activities contribute to the increase or decrease in specific nosocomial pathogens:

Main Nosocomial Infection Microorganisms

Microorganism	Percentage of Total Nosocomial Infections	Antibiotic-Resistant Microorganism	Percentage of Total Infections from MDR Bacteria
staphylococci	15	MRSA	8
Staphylococcus aureus	15	VRE	4
Enterococcus species	12	carbapenem-resistant *P. aeruginosa*	2
Candida yeast species	11	XCR *K. pneumoniae*	1
Escherichia coli	10	XCR *E. coli*	0.5
Pseudomonas aeruginosa	8	others	0.5
Klebsiella pneumoniae	6		
Enterobacter species	5		
Acinetobacter baumannii	3		
Klebsiella oxytoca	2		

Note: MRSA = methicillin-resistant *S. aureus;* VRE = vancomycin-resistant *Enterococcus faecium;* XCR = extended-spectrum cephalosporin-resistant

Portals of Entry for Nosocomial Infections

Portal	Percentage of Infections	Main Causes
urinary tract	35–40	*Candida, Enterobacter,* enterococci, *E. coli, P. aeruginosa*
lower respiratory tract	13–18	*Acinetobacter, Enterobacter, K. pneumoniae, P. aeruginosa, S. aureus*
surgical wounds	17	*Enterobacter,* enterococci, *E. coli, P. aeruginosa*
skin	8	*E. coli, P. aeruginosa,* staphylococci, *S. aureus*
septicemia (infection in bloodstream)	6	*Candida, Enterobacter,* enterococci, *P. aeruginosa,* staphylococci, *S. aureus*

- aging population
- increasingly invasive surgeries
- implantation of medical devices
- organ transplantation

Changing trends in health care demand extra vigilance in the methods for controlling the spread of infections in hospitals.

CONTROLLING NOSOCOMIAL INFECTIONS

A large contributor to the spread of nosocomial infection relates to poor hand washing in hospitals. The CDC has concluded that health care workers wash their hands only 27–31 percent of the time between patient visits. Many hospitals use people as inconspicuous monitors to watch whether hospital staff wash their hands properly before entering patient rooms. The University of Florida has developed an automatic system to do the same. The university newspaper reported, in 2009, "After washing their hands, health care workers, who wear small badges, run their hands underneath the sensors. When they have scrubbed their hands thoroughly, the badge will give off a green light that tells the worker that he or she may proceed to work with a patient. Another monitor is set up near a patient's bed. When the health care worker approaches, the bedside monitor sends out infrared and acoustic signals to the badges. If the monitor reads clean, it will give off a green light. If health care workers need to go back and wash their hands more thoroughly, the monitor will vibrate." A growing number of hospitals are evaluating this new hygiene-monitoring system, called *HyGreen.*

In addition to proper hand washing, nosocomial infections can be prevented by the following actions:

- disinfection of surfaces and tubs that patients contact
- disinfection and/or sterilization of instruments used on multiple patients, such as stethoscopes or thermometers
- replacement of reusable instruments with sterile disposable instruments whenever possible
- aseptic handling of bandages, tubing, or invasive devices
- avoidance of unnecessary invasive medical procedures

Physicians can help reduce the problem of antibiotic resistance in hospitals by prescribing antibiotics only when necessary. Hygiene and infection control committees operate in almost every U.S. hospital to develop programs for reducing nosocomial infections and train staff in better methods for stopping germ transmission. Conscientious infection committees run monitoring programs for sampling equipment, tubing, catheters, respiratory devices, and other items that patients contact for the presence of microorganisms. These tasks require continued vigilance as medical technology changes and microorganisms find new ways to cause infection.

See also ASEPTIC TECHNIQUE; HYGIENE; OPPORTUNISTIC PATHOGEN; PORTALS; RESISTANCE; SPORE.

Further Reading

Centers for Disease Control and Prevention. "Guideline for Hand Hygiene for Healthcare Settings—2002." Available online. URL: www.cdc.gov/Handhygiene. Accessed January 2, 2010.

Hidron, Alicia I., Jonathon R. Edwards, Jean Patel, Teresa C. Horan, Dawn M. Sievert, Daniel A. Pollock, and Scott K. Fridkin. "Antimicrobial-Resistant Pathogens Associated with Healthcare-Associated Infections: Annual Summary of Data Reported to the National Healthcare Safety Network at the Centers for Disease Control and Prevention, 2006–2007." *Infection Control and Hospital Epidemiology* 29 (2008): 996–1,011. Available online. URL: www.cdc.gov/nhsn/PDFs/AR_report2008.pdf. Accessed January 2, 2010.

National Nosocomial Infections Surveillance System. Centers for Disease Control and Prevention. Available online. URL: www.cdc.gov/ncidod/dhqp/nnis.html. Accessed January 2, 2010.

National Safety Healthcare Network. Centers for Disease Control and Prevention. Available online. URL: www.cdc.gov/nhsn. Accessed January 2, 2010.

Pruner, C. J. "UF Hygreen System Recognized by Popular Science." *Independent Florida Alligator,* 20 November 2009. Available online. URL: www.alligator.org/news/campus/article_750d5ff6-d594-11de-a95a-001cc4c002e0.html. Accessed January 2, 2010.

Science*Daily*. "C. difficile Spores Spread Superbug." July 21, 2009. Available online. URL: www.sciencedaily.com/releases/2009/07/090720134522.htm. Accessed January 2, 2010.

Weinstein, Robert A. "Nosocomial Infection Update." *Emerging Infectious Diseases* 4 (1998): 416–420. Available online. URL: www.cdc.gov/ncidod/eid/vol4no3/weinstein.htm. Accessed January 2, 2010.

Yale University. *Yale New Haven Hospital Infection Control Manual.* New Haven, Conn.: Yale University Press, 2001. Available online. URL: www.med.yale.edu/ynhh/infection/welcome.html. Accessed January 2, 2010.

opportunistic pathogen An opportunistic pathogen is a microorganism that does not ordinarily cause disease but can produce an infection under certain circumstances. Opportunistic species (bacteria, fungi, viruses, or protozoa) are normally harmless in their natural habitat but can infect a host whose natural defenses have been weakened. A person of this higher-than-normal risk of infection is called a *compromised host.* The major reasons for compromised health condition are chronic debilitating disease, immune system diseases, trauma, poor nutrition, and treatment with broad-spectrum antibiotics. Broad-spectrum antibiotics kill a wide variety of microorganisms on and in the body, many of which are helpful in preventing infection. Many opportunistic pathogens, in fact, belong to the population known as *normal flora,* which are the microorganisms that live on the body and usually cause no harm.

An infection caused by a normally harmless microorganism is called an *opportunistic infection.* Such infections result from two characteristics of a microorganism: invasiveness and pathogenicity. Invasiveness is the ability of a microorganism to spread from one type of tissue to another, for example, spreading from a skin infection to internal organs. Pathogenicity is the ability of a microorganism to cause disease. Features called *virulence factors* help microorganisms become pathogenic, and many normal flora have one or more of these virulence factors. The infectious dose of a microorganism that a host receives also determines whether infection will occur. Small numbers of cells of normal flora infecting a cut, for instance, may not be adequate to sustain an infection in the face of the body's immune defenses. Very large numbers of a microorganism increase the chance of infection. Overall, the likelihood of opportunistic infection depends on the relationship between the host and the microorganism, depicted by the physician Art DeSalvo of the University of South Carolina School of Medicine as the disease equation:

$$(\text{number of microorganisms} \times \text{virulence}) \div \text{host resistance} = \text{disease}$$

DeSalvo emphasized the special risks opportunistic infections pose for people who have a weakened immune system: "With opportunistic infections, the equation is tilted in favor of 'disease' because resistance is lowered when the host is immunocompromised. In fact, for the immunocompromised host, there is no such thing as a nonpathogenic fungus."

The table on page 577 describes common opportunistic pathogens in humans and the virulence factors or other characteristics they use to invade and survive in the body.

Normal flora also cause secondary infections, which occur when a microorganism colonizes an area already fighting an infection by an unrelated microorganism. On the skin, secondary infections at the site of a cut or wound are often caused by *S. aureus* and streptococci. Both microorganisms are common on the surface of the hands and easily transferred to other parts of the body.

C. albicans also causes a common secondary infection when a previous bacterial infection has been treated with broad-spectrum antibiotics. Broad-spec-

Normal Flora Opportunistic Infections

Microorganism	Normal Habitat on Body	Disease
Candida albicans	urogenital area	candidiasis (thrush in the mouth)
Corynebacterium acnes	sebum-rich areas of the face	acne
Haemophilus species	oropharynx	otitis media
Malassezia furfur	scalp	dandruff
Staphylococcus aureus	nose	skin infections, necrotizing fasciitis, secondary infections
streptococci	skin	necrotizing fasciitis, sepsis, secondary infections
Streptococcus pneumoniae	mouth and oropharynx	bronchitis, otitis media, pneumonia, sinusitis

trum treatment can eliminate members of the normal flora that limit the relative numbers of *Candida* yeasts. For example, vaginal yeast infections often result from treatment for bacterial infection.

S. pneumoniae can cause secondary infection in cases of upper respiratory tract illnesses treated with antibiotics. Secondary infections from streptococci are prone to spreading in lymph and the bloodstream to internal organs. The infection of internal fluids and organs is called sepsis, a serious and potentially fatal condition.

OPPORTUNISTIC INFECTIONS AND AIDS

Health conditions that weaken the immune system cause the highest risk for opportunistic infection. Chemotherapy for cancer patients and antirejection drugs for organ transplant recipients increase the opportunities for infection because of their damaging effect on immune defenses. The opportunistic pathogens enter from the environment, as well as normal flora. Acquired immunodeficiency syndrome (AIDS) patients suffer similar risks, because AIDS attacks cells in the immune system specifically made by the body to remove infectious agents. When the AIDS epidemic accelerated, in the 1980s, AIDS patients did not die of the human immunodeficiency virus (HIV), which causes the disease, but of a variety of opportunistic infections. Bacteria, fungi, viruses, and protozoa cause equally dangerous diseases in patients who have compromised defenses against infection.

The opportunistic infections that have the highest incidence in AIDS patients are the following:

- bacterial pneumonia from various species, especially (in order) *S. pneumoniae, Haemophilus influenzae, Pseudomonas aeruginosa,* and *S. aureus*
- candidiasis of the upper respiratory system or esophagus
- cryptococcosis of the central nervous system caused by the fungus *Cryptococcus neoformans*
- chronic cryptosporidiosis in the intestines caused by *Cryptosporidium*
- CMV infection of the eyes, causing retinitis, blurred vision, and blindness, due to cytomegalovirus
- fungal pneumonia from *Pneumocystis carinii* (also called *P. jirovecii*)
- herpes simplex virus of the mouth and genitals
- herpes zoster infection or shingles
- histoplasmosis fungal infection of the lungs
- isosporiasis caused by the protozoan *Isospora belli,* resulting in severe intestinal damage and nutrient malabsorption
- leishmaniasis of the skin or internal organs caused by *Leishmania* protozoa

- toxoplasmosis affecting the brain from *Toxoplasma* protozoal infection
- tuberculosis from *Mycobacterium avium* complex (MAC) and *M. tuberculosis* bacteria

HIV-positive people and AIDS patients must be careful to avoid exposure to these infections by taking the following actions: avoiding touching raw meat, avoiding contact with domestic animals, eliminating contamination from human excrement, and refraining from entering lake and river waters.

Pneumonias may also develop in rare occasions from *Legionella pneumophila, Mycoplasma pneumoniae,* and *Chlamydia pneumoniae.* The AidsMeds Web site has emphasized, "Not only are HIV-positive people more likely to develop bacterial pneumonia as a result of one of these infections, they are also more likely to experience recurrent pneumonia. People with CD4 [the immune cell attacked by HIV] counts below 100, and those whose bacterial infection has spread beyond the lungs, are at increased risk of death from bacterial pneumonia." Despite advances in anti-HIV treatments, opportunistic infections remain a significant health concern for HIV-positive people and AIDS patients.

OPPORTUNISTIC MYCOSES

Opportunistic mycoses are fungal infections produced by normally harmless organisms. As Art DeSalvo's disease equation suggests, people who have weakened immune systems are at particular risk from mycoses for two reasons: (1) Immunocompromised conditions increase the risk of infection, and (2) fungal diseases are harder to cure than bacterial diseases because drugs targeted at the eukaryotic fungal cell can also harm the body's cells.

Except *Candida* infections, most opportunistic mycoses are caused by environmental fungi. The major environmental fungus involved in opportunistic infections is *Aspergillus,* which enters the body when a person inhales the *Aspergillus* spores. Infection can grow in the upper and lower respiratory tracts. Allergies and asthmas are common. In the lower respiratory tract, *Aspergillus* infection leads to a variety of ailments of increasing severity called aspergillosis: bronchitis, bronchopulmonary aspergillosis, and invasive aspergillosis. In the invasive infection, the mold forms colonies called *aspergillomas,* or *fungus balls.* As the colony grows, it spreads out a tangle of mycelial filaments that can fill the entire lung.

The table above describes the common opportunistic mycoses that affect immunocompromised people, elderly adults, chemotherapy patients, people who have genetic deficiencies, and those who have chronic, debilitating disease.

Opportunistic Mycoses

Fungus	Description
Aspergillus	upper and lower respiratory tract infections, systemic infection
Candida	skin and mucous membrane infections of mouth, moist surfaces (groin, skin folds, underarms) of skin, subcutaneous infections, systemic infection
Cryptococcus	meningitis
Torulopsis	form of candidiasis caused by *C. glabrata* affecting lungs, kidneys, heart, and central nervous system
Zygometes	zygomycosis (subcutaneous masses of fungal growth), systemic infection

Opportunistic mycoses present a serious health threat especially when the organism enters the bloodstream. Antifungal therapies can reduce mortality rates to 20–70 percent, but systemic mycoses left untreated approach a 100 percent mortality rate.

Candida produces a range of opportunistic illnesses, most of them attributed to *C. albicans.* Skin and mouth infections account for most *Candida* infections. The *Candida* diseases paronychia and onychomycosis are subcutaneous infections in the fingers and of the nails, respectively. *Candida* overgrowth in moist regions of the skin that receive little aeration is called *intertriginous candidiasis.* Opportunistic candidiasis also involves a variety of illness that affects women mainly. For example, candidal vaginitis can strike females who are taking antibiotics or oral contraceptives, who have diabetes, or who are pregnant. *Candida* yeast infections can be transmitted to males as a sexually transmitted disease (STD).

Candidiasis has been a difficult mycosis to diagnose quickly for three reasons: (1) *Candida* causes many secondary infections, (2) *Candida* is often present in mixed-organism infections even when not causing disease itself, and (3) rapid and accurate tests for *Candida* have not yet been developed.

Various antifungal drugs help treat opportunistic mycoses, and the chance of a full recovery increases when the infection is treated early. Opportunistic infections are, nevertheless, difficult health challenges because they accompany other underlying health issues in the host.

See also HUMAN IMMUNODEFICIENCY VIRUS; INFECTIOUS DOSE; MYCOSES; NORMAL FLORA; PATHOGEN; PATHOGENESIS; SEXUALLY TRANSMITTED DISEASE.

Further Reading

AIDSmeds.com. "Bacterial Pneumonia." Available online. URL: www.aidsmeds.com/articles/BacterialPneumonia_6703.shtml. Accessed December 20, 2009.

Avert. "What Are Opportunistic Infections?" Available online. URL: www.avert.org/hiv-opportunistic-infections.htm. Accessed December 20, 2009.

DeSalvo, Art. "Opportunistic Mycoses." In Microbiology and Immunology On-Line. Columbia: University of South Carolina, 2008. Available online. URL: http://pathmicro.med.sc.edu/mycology/opportunistic.htm. Accessed January 1, 2010.

optimal growth conditions Optimal growth conditions are the physical and biochemical conditions that give the highest yield of microbial in a laboratory or in a fermentation facility.

The physical conditions that enhance a microorganism's growth relate to the temperature, water, and mixing that a culture requires for best growth. These physical factors, then, affect the microorganism's metabolism or biochemical conditions. For example, a bacterial species may grow better with constant mixing to increase aeration. The aeration makes more oxygen available to be part of the bacterium's metabolic pathways. Because physical and biochemical factors are related, such as in this example, some microbiologists call these related growth requirements the *biophysical growth conditions.*

Microbiologists test new microorganisms under a variety of circumstances to determine the optimal growth conditions. Some testing relies on trial and error until the microbiologist finds the best combination of biophysical conditions to make the microorganism grow fastest and to a high cell density. To expedite the process of finding these optimal conditions, experiments can be set up to study a range of conditions using the same culture for the inoculum. The table below gives an example of a typical range-finding experiment to narrow down an unknown microorganism's optimal growth conditions.

The microorganisms most familiar to people, because they are pathogens, food spoilage species, or important environmental microorganisms, have been well defined in respect to their optimal growth conditions. The majority of these microorganisms live in the same conditions humans require. These microorganisms tend to be mesophiles; their optimal growth temperature range mimics that of warm-blooded animals. Mesophiles, often, grow best at neutral pH, use simple carbon sources such as sugars, but, unlike humans, can grow either aerobically or anaerobically.

Microorganisms with optimal growth requirements far outside these ranges that warm-blooded animals find comfortable are called *extremophiles.* Extremophiles live in environments that few other life-forms can survive, because of high temperature, intense pressure, high salt, high acid or alkali, or other toxic conditions, such as high radioactivity. Microbiology laboratories that grow extremophiles require specialized incubation equipment to culture these microorganisms.

Douglas Bartlett of Scripps Institution of Oceanography said of ocean extremophiles, "These organisms live in a world that is very different from the skin of the planet in which we humans reside." Microbiologists who study extremophiles usually become experts in their specific focus area, that is, culturing anaerobes, halophiles (high-salt-requiring), barophiles (high-pressure-tolerant), hyperthermophiles (very-high-temperature-tolerant), and other microorganisms.

BIOPHYSICAL GROWTH FACTORS

Physical growth requirements must be met by the medium in which the microorganisms grow and by external factors. Medium conditions consist of pH, water activity (the amount of available water), and osmotic pressure (the pressure outside the cell relative to the inside). The main external factors affecting growth are temperature, oxygen level, and pressure. The reduction-oxidation potential, or

Testing for Optimal Growth Conditions

Study	Range				
			PARAMETER		
Temperature	22	32	37	45	55
pH	4.0	5.0	6.0	7.0	8.0
Carbon Source	glucose	maltose	carbon dioxide	fatty acid	cellulose
Oxygen Level	0.5 percent	1 percent	5 percent	10 percent	21 percent

redox potential (Eh), develops as a result of factors in the medium and external factors. For instance, dissolved oxygen in the growth medium in combination with the culture's external atmosphere determines whether a fastidious anaerobe will grow or die.

The medium's pH is a measure of the hydrogen ion (H^+) activity. In very dilute solutions, the activity almost equals H^+ concentration, so pH is often referred to as a measure of H^+ concentration. The pH scale ranges from 0, wherein H^+ concentration (designated $[H^+]$) equals 1 molar, to 14 in which $[H^+]$ equals 10^{-14} molar. (A *molar concentration* refers to a number equal to the number of atoms in exactly 12.0 grams of carbon 12; this number equals one mole.)

Microbiologists measure pH in a freshly prepared batch of medium using a pH meter. Because bacteria can shift the medium's pH as they grow, by producing acids, the medium often includes a buffer. Buffers are mixtures of weak acids and bases that enable a solution to stay in a narrow pH range even when a small amount of strong acid or strong base is added.

Current medium formulations for growing microorganisms meet the requirements for water activity and osmotic pressure. The formulation must have an amount of free or available water that the microorganism can absorb for taking in nutrients and that will dilute microbial wastes. Osmotic pressure relates to water activity due to the amount of salt in the medium. Salts such as sodium chloride (NaCl), potassium chloride (KCl), or sodium sulfate (Na_2SO_4) have been used for maintaining the osmotic pressure of the medium in most laboratory formulas. Presence of too much salt, however, unbalances the outside-inside salt concentrations and affects the amount of free water available for normal metabolism. Together, water activity and osmotic pressure play critical roles in controlling the cell membrane's permeability to nutrients and its overall function in protecting the cell interior.

Incubators can control the other biophysical growth factors, of temperature, oxygen level, and pressure. Standard laboratory incubators provide a range of temperatures from room temperature to 150°F (65°C). For temperature extremophiles, specialized incubation chambers can provide temperatures from below freezing to 200°F (93°C) or higher.

Anaerobic conditions are maintained by using special techniques adapted from rumen microbiology. In the culturing of anaerobes, oxygen may be purged out of the medium during its preparation by boiling and the cooled medium, then, sealed in vessels filled with nitrogen or a nitrogen gas blend to keep out oxygen.

Most microorganisms survive in barometric pressures of one to 100 atmospheres (atm) and require no specialized culture techniques. For barophiles, microbiologists use an apparatus that maintains a high hydrostatic pressure in the culture. Most studies of barophiles have been on deep-sea bacteria and archaea, which survive not only hydrostatic pressures of up to 700 atm, but also cold temperatures of 34–36°F (1–2°C) or extremely hot conditions (390–750°F [200–400°C]) near hydrothermal vents on the sea bottom. Special pressure vessels that can be exposed to a wide temperature range provide the optimal conditions for these microorganisms.

BIOCHEMICAL GROWTH FACTORS

Biochemical growth factors encompass all of a microorganism's nutrient requirements. Optimal growth conditions differ between heterotrophic species, which obtain energy from a variety of organic compounds, and autotrophic species, which have simple nutrient requirements of water, carbon dioxide, inorganic salts, and trace elements. The salts required by most microorganisms are sodium, potassium, calcium, iron, and magnesium salts. Common trace elements needed by various species are cobalt, copper, manganese, molybdenum, and zinc.

Microbiologists who grow heterotrophs design media that fulfill the microbial requirements for carbon, nitrogen, sulfur, inorganic ions, vitamins, amino acids, peptides or proteins, purines and pyrimidines, fatty acids and other fat compounds, plus miscellaneous factors. Miscellaneous factors that contribute to optimal growth can often be obtained by adding a small amount of any of the following substances to the medium: serum, yeast extract, beef extract, or specific polyamines (putrescine, spermine, or spermidine).

The development of media for microorganisms with very specific growth requirements can be a tedious process of trial and error. Many hundreds of different types of media have already been formulated in the past century for culturing microbiology's most familiar species. Microbiologists rely on two primary sources for information on media that will provide the biochemical factors for optimal growth: the *Difco and BBL Manual* and *Bergey's Manual of Determinative Bacteriology,* originally published in 1957 and updated yearly. Microbial depositories such as the American Type Culture Collection (ATCC) also provide information on the media that elicit the best growth from known microorganisms.

Despite the vast amount of information that microbiology has available regarding optimal growth conditions, most microorganisms on Earth remain a mystery. Microbiologists have estimated that they have cultured no more than 10 percent, perhaps much less, of all microorganisms. This is mainly due to the absence of knowledge of optimal growth conditions and/or the inability to mimic these conditions in a laboratory.

The majority of Earth's microorganisms may be in a state called *viable but nonculturable* (VBNC).

VNBC microorganisms are known or suspected to exist, but microbiologists have not yet found a way to grow them in a laboratory for the purpose of identifying these genera or species. The microbial ecologist Rita Colwell explained, in the 2000 journal article "Viable but Nonculturable Bacteria: A Survival Strategy," "The viable but nonculturable phenomenon is not new for microbial ecologists. It has long been recognized that one of the major limitations to research in microbial ecology is the inability to isolate and grow in culture the vast majority of bacteria which occur in nature." If bacteria have the genetic ability to enter a VBNC form when taken out of their normal environment, microbiologists may never be able to duplicate the optimal growth conditions that nature provides these organisms. The question of VBNC microorganisms and the reasons for this condition continue to be important research areas in microbiology.

GROWTH CONDITIONS IN THE ENVIRONMENT

Microorganisms grow much differently in their normal habitats in nature than they do in laboratory conditions. Optimal growth conditions relate exclusively to the laboratory, when microorganisms are grown in pure culture with no competition from other microorganisms. Optimal growth conditions, furthermore, provide a surplus of nutrients so that the growing cells can use the least energy-demanding metabolic pathways. This situation is rarely found in nature.

The environment often presents microorganisms with challenges to their survival. The main obstacles microorganisms must overcome are the following:

- antimicrobial substances (antibiotics and bacteriocins) from other microorganisms
- limited nutrients, oxygen, sunlight, or other growth factors
- competition among species for nutrients, water, space, and light
- temperature, pH, and other physical factors outside the optimal ranges
- predation by other organisms

In the environment, most microorganisms attach to surfaces as a survival aid. By attaching to a surface, a cell has a better chance to avoid being washed away from an area of high nutrients and, furthermore, saves energy that would be needed to swim after nutrients. Attachment to communities of organisms called *biofilms* offers some protection for all of the community's species and helps cells pull scarce nutrients out of the milieu around them. This situation differs from laboratory conditions, in which cells are bathed in a nutrient-rich broth medium at perfect temperature with no interference from other microorganisms.

Food microbiologists have studied the phenomenon of injured bacteria. These are bacterial cells that have been damaged to prevent optimal growth but are not dead. With adequate time and specific growth factors, injured bacteria can sometimes repair damage and again begin reproducing. In food microbiology, the damage done to contaminants by preservatives may injure cells but not kill them. Similar circumstances may exist in the environment, where any of the suboptimal conditions found there might injure a microorganism temporarily.

Microbiology students can take advantage of decades of studies on newly discovered microorganisms and how best to grow and maintain these species in a laboratory. The work involved in determining a new microorganism's optimal growth conditions is often underappreciated in microbiology, but it represents the starting point for all microbial studies to follow.

See also EXTREMOPHILE; MESOPHILE; METABOLIC PATHWAYS; OSMOTIC PRESSURE; RUMEN MICROBIOLOGY; WATER ACTIVITY.

Further Reading

American Type Culture Collection. Available online. URL: www.atcc.org. Accessed November 12, 2009.

Becton-Dickinson. "Difco and BBL Manual." 2009. Available online.URL:www.bd.com/ds/technicalCenter/inserts/difcoBblManual.asp. Accessed November 12, 2009.

Bergey, David H., and Robert S. Breed. *Bergey's Manual of Determinative Bacteriology.* Baltimore: Williams & Wilkins, 1957. Available online. URL: www.archive.org/details/bergeysmanualofd1957amer. Accessed November 12, 2009.

Colwell, Rita R. "Viable but Nonculturable Bacteria: A Survival Strategy." *Journal of Infection and Chemotherapy* 6 (2000): 121–125.

———, and D. Jay Grimes. *Nonculturable Microorganisms in the Environment.* Washington, D.C.: American Society for Microbiology Press, 2000.

Hoyle, Brian. "Bacteria from Ocean Depths Respond to Pressure, Thrive on Unusual Nutrients." *Microbe,* July 2005.

Kato, Chiaki, Takako Sato, and Koki Horikoshi. "Isolation and Properties of Barophilic and Barotolerant Bacteria from Deep-Sea Mud Samples." *Biodiversity and Conservation* 4 (1995): 1–9. Available online. URL: www.springerlink.com/content/r371252452603u55/fulltext.pdf. Accessed November 12, 2009.

Oliver, James D. "The Viable but Nonculturable State in Bacteria." *Journal of Microbiology* 43 (2005): 93–100. Available online. URL: www.ncbi.nlm.nih.gov/pubmed/15765062. Accessed November 12, 2009.

Todar, Kenneth. "Nutrition and Growth of Bacteria." "Todar's Online Textbook of Bacteriology." 2008. Available online. URL: www.textbookofbacteriology.net/nutgro.html. Accessed November 12, 2009.

oral flora The oral flora of mammals comprise the microorganisms that normally live in the mouth and soft part of the throat before the epiglottis, called the oropharynx. These flora are part of a body's normal or native flora.

In the mouth and throat, the native microorganisms are almost exclusively bacteria. Only the yeast *Candida* and the protozoa *Entamoeba* and *Trichomonas* are part of the normal flora in some people.

Within the oral cavity, several microenvironments exist, mainly the teeth, periodontal spaces, tongue, and palate. The areas share several of the same bacterial species, but certain bacteria also dominate in these regions and no other. The tonsils, also, have bacteria similar to the oral flora, but with slightly different proportions. The major oral bacteria of humans, listed in approximate order of prevalence, are:

- *Streptococcus mitis, S. salivarius, S. mutans*
- *Lactobacillus*
- *Neisseria*
- *Actinomycetes*
- *Staphylococcus epidermidis*
- spirochetes (mainly *Treponema*)
- *Streptococcus pneumoniae, S. pyogenes*
- *Proteus*
- *Haemophilus*
- *Enterococcus faecalis*

Various other normal flora of the skin, such as *Staphylococcus aureus* and the group of bacteria known as diphtheroids, are also common in the mouth in many people. The proportions of these species vary slightly between the mouth and the oropharynx, streptococci, staphylococci, spirochetes, and lactobacilli more common in the mouth, and *Neisseria* species, *S. salivarius,* and *S. epidermidis* are more prevalent in the oropharynx.

Of the oral flora listed here, some may be opportunistic pathogens that cause infections only if favorable conditions for infecting arise. The potential pathogens are *S. mutans, S. pyogenes, Haemophilus influenzae, Neisseria meningitidis, S. aureus,* and *E. faecalis. S. mutans* serves a pivotal role in causing dental caries. All the other potential pathogens are more likely to infect the bloodstream during incidences of oral disease.

The oral flora also includes a mixture of anaerobic bacteria that appear in varying concentrations and relative proportions in different people. *Veillonella, Fusobacterium,* and *Escherichia coli,* a facultative anaerobe that can live with or without oxygen, are the main anaerobes of the mouth. The important oral flora are described in the table (left).

Important Oral Flora in Humans

Microorganism	Significance
Streptococcus mutans	primary microorganism of plaque formation and initiation of dental caries (cavities)
Streptococcus pneumoniae	responsible for almost 95 percent of bacterial pneumonia cases
Lactobacillus species	acid formation from this microorganism leads to dental caries
Neisseria meningitidis	colonizer of the pharynx; frequently infects the upper respiratory tract
Staphylococcus epidermidis	one of the most adaptable microorganisms on the human body
Streptococcus pyogenes	cause of strep throat; can lead to rheumatic fever or nephritis

The table illustrates that many of the major oral bacteria cause more damage than benefits to human health. Oral bacteria contribute to the digestion of food to a small degree through the activity of proteases (break down proteins), lipases (lipids), and amylases (starch). The overall oral digestion of nutrients by either host-excreted enzymes or bacteria contributes little, however, to host nutrition.

Almost half of all people harbor either of two protozoa: *Entamoeba gingivalis* and *Trichomonas tenax*. Both microorganisms ingest oral flora as their nutrient and energy sources; *E. gingivalis* also feeds on *T. tenax* cells. Neither species is pathogenic in healthy individuals who have good dental hygiene habits.

FLORA OF THE TEETH

Studies on the flora that attach to and live on the tooth surface focus mainly on two features of oral care: dental plaque and dental caries.

Dental plaque is a thin film of up to 0.5 millimeter (mm) in thickness that contains a mixture of bacterial and nonbacterial species held to the surface and to each other by a matrix of bacterial polysaccharides and salivary polymers. The initial formation of plaque occurs when streptococci and *Neisseria* species attach to the clean tooth enamel. High levels of dietary sucrose have been associated with higher proportions of *S. mutans* and faster development of plaque. As the plaque builds up, it is sometimes defined as a biofilm because it is a mixed microbial community attached to a hard surface with periodic rinsing with liquid. As the biofilm increases in complexity, anaerobic bacteria, mainly *Bacteroides* and *Veillonella,* start to inhabit the layers nearest the tooth, and filamentous bacteria, such as *Actinomycetes,* also join.

The composition of plaque varies from person to person. People have slightly different relative amounts of plaque bacteria. Plaques also differ by their amount and proportions of two bacterial polymers: mutan excreted by *S. mutans* and levan produced by species of *Streptococcus* and *Bacillus.*

Jose Alvarez and Juan Navia of the University of Alabama–Birmingham describe dental caries "as a plaque-dependent, chronic disease that affects humans in all parts of the world." After the development of a diverse biofilm community on the teeth, bacteria called *cariogenic bacteria* begin to break down the enamel in various locations. This breakdown, or removal of minerals such as calcium, is called *demineralization* and is enhanced by acid-producing bacteria. The effectiveness of caries formation relates to water-insoluble polymers called *glucans* that hold the demineralizing bacteria in place and allow them to eat through the enamel and expose the tooth dentin, the bonelike material making up the major part of teeth.

The details of dental caries formation may differ in different people, mainly because individuals differ in plaque composition, diet, and immunity, all of which interact in caries formation. A small percentage of people never experience dental caries.

The American Dental Association (ADA) recommends the addition of fluoride to toothpaste or drinking water because fluoride acts as a potent anticaries agent. Fluoride converts a compound in tooth enamel, called *apatite,* to a more acid-resistant fluorapatite, which becomes part of the tooth structure. The ADA has stated on its Web site, "For over five decades, the American Dental Association has continuously endorsed the fluoridation of community water supplies and the use of fluoride-containing products as safe and effective measures for preventing tooth decay." Some members of the public and entire communities have opposed the fluoridation of their water supply, however, as a potential health threat, damaging the kidneys and possibly causing harm to growing children. Since fluoride has begun being added to community water systems, researchers have found that fluoride works best locally, at the teeth, and systemic intake of fluoride may not be the best way of delivering this element.

While the Centers for Disease Control and Prevention (CDC) continues to support water fluoridation as an important step in oral health, fluoridation opponents such as the Fluoride Action Network (FAN) vehemently denounce the practice. Paul Connett, executive director of FAN, has stated that fluoridation "is simply not ethical; we simply shouldn't be forcing medication on people without their 'informed consent.'" The Nobel Prize in medicine recipient (2000) Arvid Carlsson may be the most prominent scientist to oppose fluoridation publicly, saying, "Fluoridation is against all principles of modern pharmacology. It's really obsolete." Local water districts offer the public information on the fluoride levels present in community water systems.

DISORDERS IN NORMAL ORAL FLORA

Dental caries are the most prevalent disorder involving oral flora, but additional severe disease states can occur, caused by specific microorganisms. The main oral disorders are thrush, periodontitis, and stomatitis.

Thrush, or oral candidiasis, is an infection of the mouth caused by the yeast *Candida.* Thrush occurs when the host immune defenses become weakened or when the relative proportions of organisms in normal oral flora become unbalanced. Antibiotic treatment, which kills oral bacteria, and systemic corticosteroid therapy, which decreases white blood cells, both increase the risk for thrush. Chronic immunocompromised condition, such as that in human immunodeficiency virus/acquired immunodeficiency syndrome (HIV/AIDS) patients, also leads to this disease.

The oral lesions of thrush appear as raised, white patches on the mucosa and tongue. Much of the patches can be scraped off, but full treatment requires antifungal drugs such as fluconazole, clotrimazole, or nystatin. Full recovery usually occurs in 14 days.

Periodontitis is an inflammation of tooth-supporting tissues, resulting in a loosening of the tooth and recession of the gingivae (gums). This condition often leads to gingivitis, or inflammation of the gums. The anaerobes *Bacteroides, Eubacterium,* and *Fusobacterium* have been implicated as the main bacteria responsible for infection, but heavy plaque buildup also can cause periodontitis. Plague developing in or near the area called the gingival sulcus, a shallow groove (about 1 mm deep) at the tooth-gum junction, probably initiates the process leading to periodontitis. Many of the oral flora that participate in this infection have probably not been identified. Moderate to serious cases of periodontitis must be treated by a dentist for root planing (physical removal of plague from the tooth surfaces) and possible gum surgery.

Stomatitis is an inflammation of tissues in the mouth. It can be caused by an infection by oral flora or by virus infection, often initiated by ill-fitting dentures or vitamin deficiency. Most cases of stomatitis can be corrected by better dental hygiene.

ORAL HYGIENE

Hygiene is the observance of rules regarding cleanliness for the purpose of preventing infection and maintaining health. Oral hygiene pertains to the practices that should be followed to prevent infection of the mouth by microorganisms; *dental hygiene* refers to the care of the teeth.

Oral hygiene includes behavioral practices that reduce the risk of oral infection. The following practices have been recommended by health care providers for oral health:

- discontinuing the use of tobacco products
- regular brushing of the teeth with fluoride toothpaste
- daily use of dental floss
- removal of impacted debris between teeth and between teeth and gums
- removal of plaque at least twice a year by a dental hygienist to prevent periodontal disease

People who use orthodontic devices (braces) or false teeth require extra measures for assuring oral hygiene is sound and prevents infection.

Oral flora make up a specialized community within the normal flora of the human body. For this reason, microbiologists studying these microorganisms require extra training in anaerobic culture techniques and oral diseases.

See also CANDIDA ALBICANS; MICROENVIRONMENT; NORMAL FLORA; OPPORTUNISTIC PATHOGEN.

Further Reading

Alvarez, Jose O., and Juan M. Navia. "Nutritional Status, Tooth Eruption, and Dental Caries: A Review." *American Journal of Clinical Nutrition* 49 (1989): 417–426. Available online. URL: www.ajcn.org/cgi/reprint/49/3/417. Accessed November 13, 2009.

American Dental Association. Available online. URL: www.ada.org. Accessed November 14, 2009.

Connett, Paul. "The Absurdities of Water Fluoridation." Fluoride Action Network, 28 November 2002. Available online. URL: www.fluoridealert.org/absurdity.htm. Accessed November 14, 2009.

Medical News Today. "End Fluoridation, Say 500 Physicians, Dentists, Scientists and Environmentalists." August 10, 2007. Available online. URL: www.medicalnewstoday.com/articles/79326.php. Accessed November 14, 2009.

Todar, Kenneth. "Normal Bacterial Flora of Humans." Todar's Online Textbook of Bacteriology. 2008. Available online. URL: www.textbookofbacteriology.net/normalflora.html. Accessed June 6, 2010.

Wantland, Wayne W., and Edna M. Wantland. "Incidence, Ecology, and Reproduction of Oral Protozoa." *Journal of Dental Research* 39 (1960): 863. Available online. URL: http://jdr.sagepub.com/cgi/content/citation/39/4/863. Accessed November 14, 2009.

organelle An organelle is any structure that forms part of a microbial cell and performs a special function.

Prokaryotes and eukaryotes differ mainly in the collection of organelles they possess. In general, all archaea and bacteria share the same organelles, and all eukaryotes also have a majority of organelles in common. Eukaryotes have been differentiated in biology from prokaryotes specifically, because eukaryotic organelles are membrane-bound inside the cell and prokaryotes lack these internal membranes. All of the cell's organelles, other than the outer cell wall and the plasma membrane that lies inside the cell wall, reside in the cytoplasm. Also called the *cytoplasmic matrix* or *cytosol,* the cytoplasm is the watery material that fills the majority of the cell's interior.

A small number of organelles are unique to certain species or groups of microorganisms, however. For example, *Magnetospirillum magnetotacticum* is an unusual bacterium that contains iron crystals called magnetite; the organelles are called *magnetosomes*. Photosynthetic cells contain chloroplasts that all nonphotosynthetic cells lack. The presence or absence of distinctive organelles makes up a species's morphology, and microbiologists use these structures to help identify unknown microorganisms.

Organelles are responsible for the intracellular organization of microorganisms. These structures play a part in energy metabolism, nutrient transport and use, cell component synthesis, motility, and management of the cell's genome. The genome is one complete copy of a cell's genetic material. No single organelle is more important than another. Each organelle has evolved a structure and function that contribute directly to cell survival. Microbiology has, nevertheless, focused on the structure and function of certain organelles because of their critical roles in genetic engineering, nutrient cycling, and disease. The table on page 585 describes the functions of prokaryotic and eukaryotic structures.

Advances in electron microscopy (EM) have enabled scientists to study organelles in much greater detail than allowed by light microscopy. EM has revealed the internal structure of organelles, many of which contain additional internal components. For example, the eukaryotic chloroplasts, mitochondria, and nucleus contain internal structures that have specific functions within the organelle's overall function.

Other than the cell wall and the cell membrane, most organelles can occur in microorganisms as multiple copies. For example, some eukaryotic cells have more than one nucleus. In most cases, microorganisms have more than one of the following organelles: ribosomes, endoplasmic reticulum, mitochondria, Golgi apparatus, chloroplasts, vacuoles, lysosomes, flagella, cilia, fimbriae, and pili.

PROKARYOTIC ORGANELLE STRUCTURE AND FUNCTION

Prokaryotic organelles reside in the cytoplasm without enclosure in a membrane, but the plasma membrane sometimes contains folds that break away from the main membrane and form a vesicle called a *mesosome* or lamella within the cytoplasm. Mesosomes occur slightly more often in gram-negative species than gram-positive and often attach to the cell's chromosome, the collection of genetic material. Although mesosomes were once thought to carry out a function for the cell, they are now believed to be artifacts

Functions of Prokaryotic and Eukaryotic Structures

Structure	Present in Prokaryotes	Present in Eukaryotes	Function
cell wall	+	+	shape, rigidity, and protection
plasma	+	+	selectively permeable barrier for nutrient
membrane			uptake and waste excretion
periplasmic	+	-	nutrient processing and uptake
space nucleus	-	+	repository for genetic information and the cell's control center
nucleolus	-	+	ribosome construction
nucleoid	+	-	localized area of genetic material
microfilaments	-	+	cell structure and movement
ribosomes	+	+	protein synthesis
vacuoles	+	+	depending on microorganism, serves for buoyancy, water balance, or temporary storage
Golgi apparatus	-	+	packaging and secretion site and lysosome formation
lysosomes	-	+	intercellular digestion
mitochondria	-	+	energy production
chloroplasts	-	+	photosynthesis
inclusion bodies	+	-	nutrient storage
endoplasmic reticulum	-	+	transport and protein and lipid synthesis; site of the ribosomes
fimbriae and pili	+	-	attachment to surfaces
flagella	+	+	cell movement (motility)
cilia	-	+	cell movement (motility)

Note: (+) denotes the organelle is present; (-) denotes the organelle is absent

that develop during preparation for EM study. Similar invaginations called *thylakoids* that are found only in photosynthetic bacteria (cyanobacteria and purple bacteria) can extend so deep into the cell that they wrap multiple times around the interior. Because photosynthesis requires a large membrane surface to run efficiently, thylakoids probably play the role of increasing membrane surface in photosynthetic bacteria.

The nucleoid is a general area containing the cell's deoxyribonucleic acid (DNA) arranged in a dense mass. Various microbiology texts also call this organelle the *nuclear body*, nuclear region, or chromatin. Although the nucleoid appears microscopically to be a disorganized tangle of DNA, the DNA molecule actually contains folds at specific points that serve two purposes: The folds make the DNA more compact, and they facilitate repair of damaged DNA. During cell replication, the nucleoid becomes less compact and forms branches that extend into the cytoplasm. The nucleoid contains all of the cell's genome unless the prokaryote also possesses plasmids, small circular pieces of DNA in the cytoplasm.

Bacteria contain various inclusion bodies that hold either organic or inorganic material and can often be seen with a light microscope. Some inclusion bodies have a thin (2–4 nanometers thick) membrane cover, but this membrane does not have the typical bilayer structure of most membranes, that is, proteins on the outer surfaces and lipids on the membranes' inner region. Inclusion bodies vary in composition; some are predominantly protein, and others are lipid. In addition, inclusion bodies that store nutrients vary in size according to the availability of nutrients.

Organic inclusion bodies mainly contain glycogen, a polymer of glucose units, or poly-beta-hydroxybutyrate as energy storage compounds. Inorganic inclusion bodies typically contain phosphate or sulfur compounds.

Gas vacuoles in bacteria are a type of inclusion body made of several small cylindrical gas vesicles. Bacteria use gas vacuoles to regulate their buoyancy in aqueous environments by taking up or letting out air as if the vacuole were a balloon.

Prokaryotic ribosomes have the same function they have in eukaryotes, but the ribosomal structure differs between prokaryotes and eukaryotes. In a microscope, ribosomes appear as small (about 14–15 by 20 nm) featureless particles, numbering in the thousands. The ribosomes contribute crucial activities in the replication of the cell's genetic matter and protein synthesis. In detail, ribosomes are complex structures made mainly of protein and ribonucleic acid (RNA). Bacterial ribosomes are 70S ribosomes; the designation *S* for "Svedberg unit" results from the sedimentation characteristics of the particle when centrifuged. Each 70S ribosome consists of two subunits: the 50S subunit and the 30S subunit. The Svedberg units of these structures result from their molecular weight, volume, and shape.

The ribosomes are the site of conversion of the genetic code that RNA receives from DNA and subsequent synthesis of protein from the RNA code or sequence of nucleotide base units. The activity of ribosomes makes up a complex subject in cell genetics. The following summary provides the key points of prokaryotic and eukaryotic ribosomes:

- main site of protein synthesis
- actively growing cells have more ribosomes than resting cells
- each ribosome is constructed of two subunits
- prokaryotic ribosomes are smaller and less dense than eukaryotic
- several antibiotics work specifically to stop ribosome function

Because of discrete differences between prokaryotic and eukaryotic ribosomes, prokaryotes are more susceptible to ribosome-targeting antibiotics than eukaryotes.

EUKARYOTIC ORGANELLE STRUCTURE AND FUNCTION

Eukaryotic cells have a more complex cellular structure than prokaryotes. Eukaryotic cells are generally larger than prokaryotic cells, typically 10–100 micrometers (μm) in diameter, compared with 0.2–5.0 μm in prokaryotes. Membrane-bound organelles make the eukaryotic cell appear more organized than the prokaryotes cell, especially during cell division. Prokaryotes divide by simple binary fission. Eukaryotes reproduce by mitosis, the process of cell division, and meiosis, a method of sexual recombination of genetic material. Eukaryotic cells also have flagella of a complex structure consisting of many microtubules. Prokaryotic flagella consist mainly of two proteins.

Only in cell wall composition do prokaryotes have a more complex structure than eukaryotes. The bacterial cell wall contains a netlike matrix dominated by the polymer peptidoglycan. The eukaryotic cell wall lacks peptidoglycan and is chemically simple. Some eukaryotic cells, such as amoeba, lack a cell wall.

The nucleus is spherical or oval and is frequently the largest organelle in eukaryotes. This organelle contains most of the cell's DNA, other than a small amount residing in the mitochondria and some in the chloroplasts of photosynthetic organisms. A membrane called the *nuclear envelope* encloses the nucleus and contains tiny pores called *nuclear pores* that allow the nucleus to transfer material into and out of the cytoplasm. Within the nucleus, reside one or more nucleolus (plural: nucleoli): the region of condensed genetic material, where a type of RNA called ribosomal RNA (rRNA) is synthesized.

The DNA of eukaryotes is organized around proteins called *histones,* which prokaryotes lack. Every 165 base pairs of DNA and nine histones make up a unit called a *nucleosome.* When eukaryotic cells are not reproducing, the DNA is arranged around the histones in loose threads. The entire mass is called *chromatin.* During division of the nucleus in reproduction, the chromatin coils into thicker, shorter bodies called the *chromosomes.*

Eukaryotic cells contain two types of endoplasmic reticulum (ER): rough ER and smooth ER. Rough ER connects with the nuclear membrane and contains thousands of ribsomes attached to its many folds. Rough ER functions as the main site for protein synthesis and membrane components. Smooth ER extends off some of the rough ER folds and does not have ribosomes. Smooth ER contains a unique set of enzymes devoted to synthesizing lipids, including fats, sterols, and phospholipids.

The ribosomes attached to the rough ER (called *membrane-bound ribosomes*) and occasionally free in the cytoplasm (free ribosomes) synthesize protein. In general, free ribosomes produce proteins used inside the cell, and membrane-bound ribosomes produce proteins intended for the plasma membrane or excretion from the cell. The 80S ribosomes in eukaryotes consist of two subunits: 60S and 40S. The nucleus makes each subunit separately, then excretes them into the cytoplasm, where they, then, connect. The 60S subunit contains three molecules of rRNA, and the 40S subunit contains one molecule.

Mitochrondia play the role of energy generators for the eukaryotic cell. These rod-shaped organelles occur throughout the cytoplasm. Depending on the type of eukaryotic cell, there may be a few mitochondria to thousands. The inner membrane of mitochondria folds into multiple layers called *cristae* that extend toward the mitochondrial interior, called the *matrix.* The many folds of cristae greatly increase the surface area where enzymes work in cellular metabo-

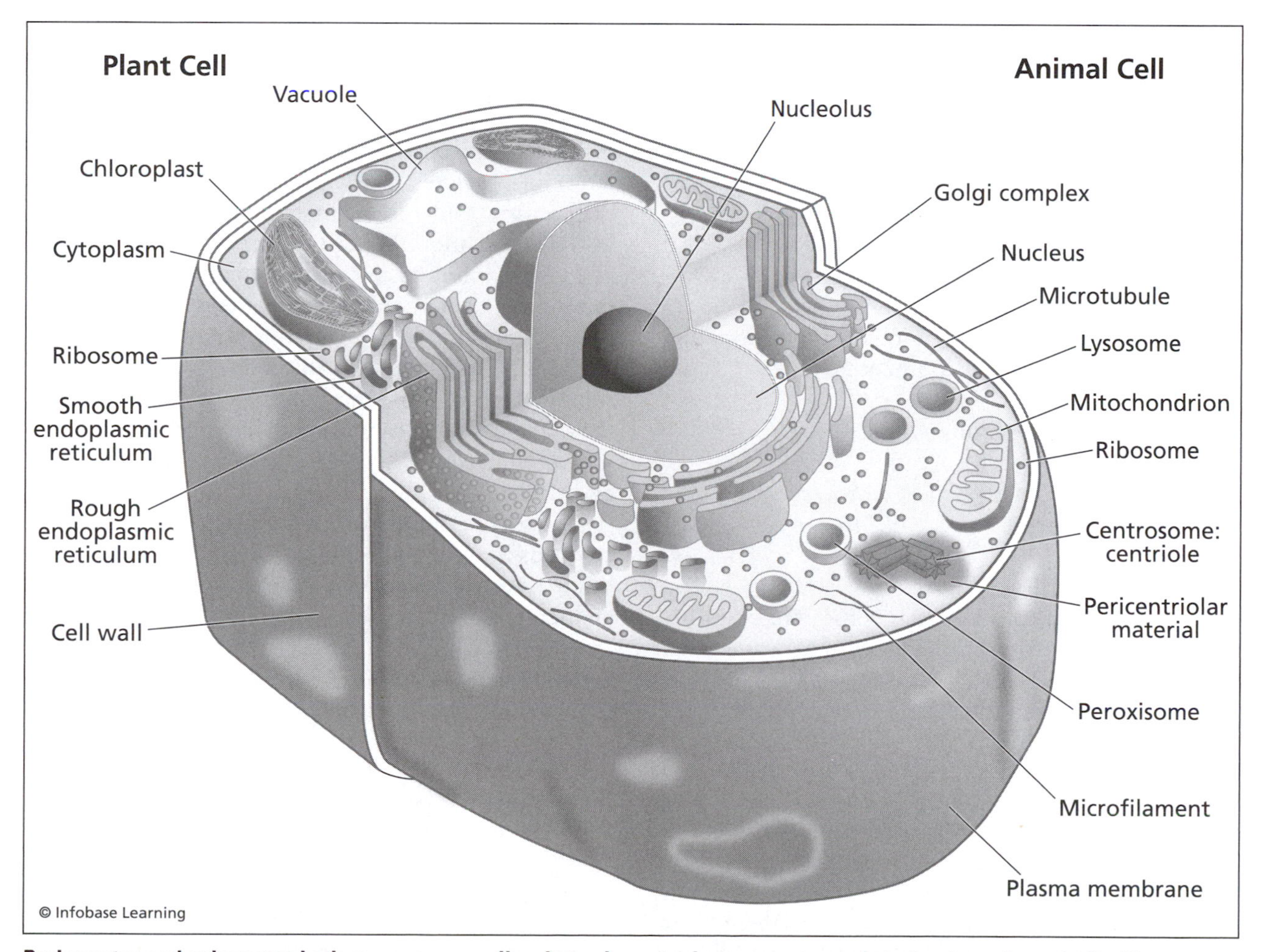

Prokaryotes and eukaryotes both possess organelles, but eukaryotes have a more complex structure characterized by membrane-bound organelles. The eukaryotic cytoskeleton consists of microfilaments and microtubules that contribute to cytoplasmic streaming, the movement of cytoplasm inside the eukaryotic cell.

lism. Mitochondria also contain some ribosomes (70S) and a small amount of DNA. These features allow mitochondria to replicate on their own inside the cell.

Eukaryotes contain various organelles not found in prokaryotes:

- lysosomes—produce strong digestive enzymes that destroy foreign molecules and participate in cell death (called *apoptosis*)
- peroxisomes—produce enzymes that metabolize fatty acids and amino acids
- centrosome—an organizing structure made of small protein fibers, which facilitates mitosis

The peroxisomes play an additional protective role for the cell. During metabolism of organic compounds, inside the peroxisomes, the toxic chemical hydrogen peroxide accumulates. The peroxisomes produce the enzyme catalase in addition to the other enzymes involved in nutrient breakdown. Catalase destroys hydrogen peroxide before it can damage other parts of the cell.

MEMBRANES

The plasma membrane, also called the *cytoplasmic membrane,* has been studied extensively in biology because of its multiple functions for the cell. The main functions of this membrane are:

- structure—holds in the cytoplasm
- selective permeability for exchanging ions and water with the environment
- nutrient uptake and waste excretion
- production of the energy storage compound adenosine triphosphate (ATP)

Selective permeability is the process, within the membrane, through which some molecules are allowed to move through the membrane while other molecules are restricted.

The plasma membrane assembles into a structure called the *phospholipid bilayer.* In this arrangement, hydrophilic (water-attracting) or polar ends of long fatty acids point toward the aqueous material that occurs outside the cell or in the cytoplasm. The fats are called *phospholipids* because each polar end contains a phosphate group (PO_4^-). The hydrophobic (water-repelling) hydrocarbon chains of these same fatty acids point into the center of the membrane, away from water.

Plasma membranes contain proteins embedded in the bilayer. Some of the proteins, called *integral proteins,* that extend from the cytoplasm across to the outer surface of the membrane form pores for transport of large molecules. Small molecules (salts, oxygen, carbon dioxide, and simple sugars) pass through the membrane without using pores. Other membrane proteins, called *peripheral proteins,* do not extend all the way across the membrane but stay mainly on the cytoplasmic side. Peripheral proteins may serve mainly for support.

The plasma membrane transports materials in one of two ways: simple diffusion and facilitated diffusion. Simple diffusion moves compounds from a region of high concentration to a region of lower concentration. Cells use simple diffusion for transporting small molecules until the concentrations inside and outside are no longer different, called a *state of equilibrium.* Facilitated diffusion, unlike simple diffusion, requires energy. In this type of diffusion, a protein called a transporter carries a large molecule into the cell in three steps: (1) attaches to the molecule on the outside of the cell, (2) changes shape to change orientation in the membrane, and (3) releases the molecule inside the cell.

Membranes also play an important role in maintaining cellular osmotic pressure, the pressure required to prevent water from moving into or out of the cell. Osmosis is the net movement of dissolved molecules (solutes) through a selectively permeable membrane, from a site of high concentration to one of low concentration. Water uses osmosis as its method of moving into and out of biological cells. By maintaining the osmotic pressure, the membrane prevents a cell from swelling and bursting when its inner concentration of solutes exceeds the outside concentration. If the inner concentration of a solute is less than the outside concentration, the cell shrivels.

In eukaryotes, the plasma membrane serves an additional function called *endocytosis.* In endocytosis, a segment of the membrane surrounds a particle, envelops it, and then breaks off from the main membrane to draw the particle into the cell. Two types of endocytosis used by microbial cells are:

- phagocytosis—extensions from the membrane reach out to enclose a particle and take it into the cell
- pinocytosis—the membrane folds inward to take a substance dissolved in fluid into the cell

In microbiology, the plasma membrane remains a captivating aspect of cell physiology, structure, and transport.

UNIQUE ORGANELLES

Some microorganisms can be distinguished from others by unusual organelles visible by light microscopy or EM. An important advance in microbiology occurred with the discovery of plasmids, which are short circular pieces of DNA separate from bacteria's nucleoid. Plasmids carry fewer genes than the cell's larger DNA molecule, but they have the ability to replicate on their own and move easily among cells of the same species or a different species. Plasmids have gained importance in microbiology because they often carry genes that confer antibiotic resistance on a species.

Bacteria also form different structures lying outside the cell wall. A capsule is a layer of polysaccharides that cannot be washed off bacterial cells. Capsules make cells more resistant to phagocytosis by predators or to the host immune cells during an infection. A slime layer, by contrast, contains polysaccharides that can be washed off cells. Capsules and slime layers can be referred to as the cell's glycocalyx, a mesh of polysaccharides and other substances that surround the cell and can be stained and viewed in a light microscope. Glycocalyx also protects cells from drying out when water is not available and aids in the attachment of cells to surfaces. Glycocalyx is important, in this latter role, for biofilm formation.

Some bacteria form a structure that is more protective than a glycocalyx layer. These bacteria, notably *Bacillus* and *Clostridium*, form endospores. An endospore is a thick outer layer that enables the dormant cell inside to resist intense heat, drying, and harsh chemicals. Endospores, furthermore, preserve cells for centuries, so that once the cell leaves the spore form, a process called *germination*, it multiplies normally.

Bacteria of the genus *Aquaspirillum* develop chains of magnetite (Fe_3O_4) particles called *magnetosomes*. Less common are magnetosomes made of greigite (Fe_3S_4) or pyrite (FeS_2). These particles of 40–100 nm in diameter act as tiny magnets, so that the bacteria can orient in northward or southward directions. Magnetosomes are thought to help bacteria locate nutrients in sediments by a process called *chemotaxis*, which is the movement of a cell toward (or away from) a stimulus. The microbiologist Sandi Clement on the Microbe Zoo Web site states that "magnetotactic bacteria were discovered in 1975 by Richard P. Blakemore. Blakemore noticed that some of the bacteria that he observed under a microscope always moved to the same side of the slide. If he held a magnet near the slide, the bacteria would move towards the north end of the magnet." In addition to finding nutrients, the directional compass provided by the magnetosomes allows the anaerobic *Aquaspirillum* to find deep sediments, which are low in oxygen.

Cyanobacteria contain two distinctive inclusion bodies that play a role in nitrogen and carbon metabolism: cyanophycin granules and carboxysomes. Cyanophycin granules store extra nitrogen for the bacteria, mainly in the amino acids arginine

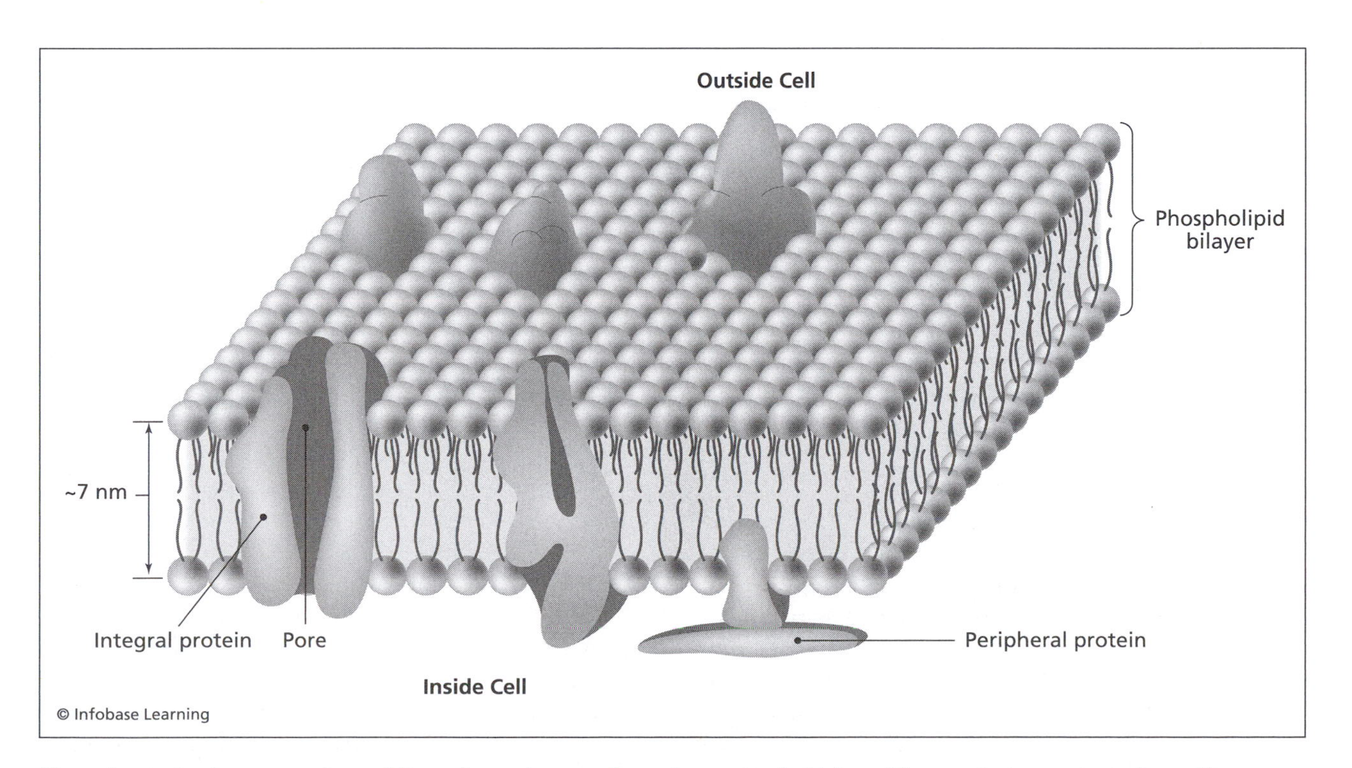

The eukaryotic plasma membrane bilayer is an almost universal structure in biology. The membrane protects the cell's internal environment, provides the location for energy metabolism, transports materials, and participates in endocytosis.

and aspartic acid. Carboxysomes also occur in nitrifying bacteria (bacteria that convert ammonia to usable forms of nitrogen) and *Thiobacillus* (sulfur-using soil bacteria). These organelles may serve as a site of carbon dioxide capture.

Some bacteria and algae construct a holdfast that attaches the cell to a solid surface. Protein- and polysaccharide-composed holdfasts are helpful in aqueous habitats, in which the cells must stay in place against strong currents. The microbiologist Pamela Brown wrote in *Microbial Physiology,* in 2009, "Recent biophysical analysis of the holdfast has shown that it is an elastic, gel-like substance, with impressive adhesive properties." Holdfasts benefit microorganisms because the cells can remain in place and let nutrients move to them, rather than expend energy to swim after their nutrients. When Brown sprayed holdfast-producing *Caulobacter crescentus* cells attached to a hard surface with a jet of water, she could not dislodge the cells.

The study of organelles is essential to understanding the physiology of microorganisms. Microbiologists, today, have a variety of microscopic techniques that aid in uncovering new details about these structures.

See also BIOFILM; CELL WALL; CENTRIFUGATION; DIFFUSION; ELECTRON MICROSCOPY; FRACTIONATION; MORPHOLOGY; OSMOTIC PRESSURE; PHOTOSYNTHETIC BACTERIA; PLASMID; SPORE.

Further Reading

Brown, Pamela J. B., Gail G. Hardy, Michael J. Trimble, and Yves V. Brun. "Complex Regulatory Pathways Coordinate Cell-Cycle Progression and Development in *Caulobacter crescentus.*" In *Advances in Microbial Physiology,* edited by Robert K. Poole. London: Academic Press, 2009.

Clement, Sandi. "Magnetic Microbes." Digital Learning Center for Microbial Ecology. Available online. URL: http://microbezoo.commtechlab.msu.edu/curious/caOc96SC.html. Accessed November 14, 2009.

Tortora, Gerard J., Berdell R. Funke, and Christine Case. *Microbiology: An Introduction,* 10th ed. San Francisco: Benjamin Cummings, 2009.

Willey, Joanne, Linda Sherwood, and Chris Woolverton. *Prescott, Harley, Klein's Microbiology,* 7th ed. New York: McGraw-Hill, 2007.

osmotic pressure Osmotic pressure is the pressure needed to prevent water from moving into or out of a cell. Because the membrane of microorganisms is a selectively permeable membrane (also called a *semipermeable membrane*), meaning it lets some substances pass across unimpeded but restricts others, microorganisms can be affected by the concentration of substances in their environment.

Water naturally moves across membranes, from an area where the water contains no solutes (dissolved compounds) to an area where the solute concentration is higher. For example, the cell cytoplasm contains a number of substances in solution or in suspension. If a cell protected only by a membrane were to be dropped into distilled water containing no salts, water would rush into the cell, and it would swell and then burst. The cell must constantly work to maintain equilibrium of concentrations inside the cell compared with the outside.

Water moves freely through biological membranes. Rather than control the movement of water into or out of cells, microorganisms can more easily control the movement of solutes to reach equilibrium with the environment. *Osmosis* is a term for the net movement of a given solute across the semipermeable membrane, from an area of high concentration to the low concentration of the solute.

Bacteria and most algae and fungi have a rigid cell wall that holds the cell's shape regardless of the environmental conditions. But the plasma membrane lying under the cell wall remains semipermeable, making the cell as vulnerable to osmotic pressure as if no cell wall existed. When cells are placed in a solution of high solute concentration, water rushes out of the cell and the membrane shrivels, even though the cell wall remains intact, a state called *plasmolysis.* This condition injures the cell or can cause cell death. The solutions that present microorganisms with a range of solute concentrations are called *osmotic solutions.*

OSMOTIC SOLUTIONS

Three types of osmotic solutions control the behavior of microbial cells: isotonic (isoosmotic), hypotonic (hypoosmotic), and hypertonic (hyperosmotic). These three solutions are distinguished by the following characteristics:

- isotonic solution—solute concentration in the solution equals the concentration inside the cell
- hypotonic solution—solute concentration in solution is lower than inside the cell
- hypertonic solution—solute concentration in solution is higher than inside the cell

No one can prepare a solution in a laboratory and call it iso-, hypo-, or hypertonic. These are relative terms; they having meaning solely when cells possessing a semipermeable membrane also occupy the solution.

The French physicist Jean-Antoine Nollet (1700–70) first studied the movement of water and substances through membranes. In a simple experiment, Nollet showed how water rushed across a piece of paper covering a tube partially filled with a concen-

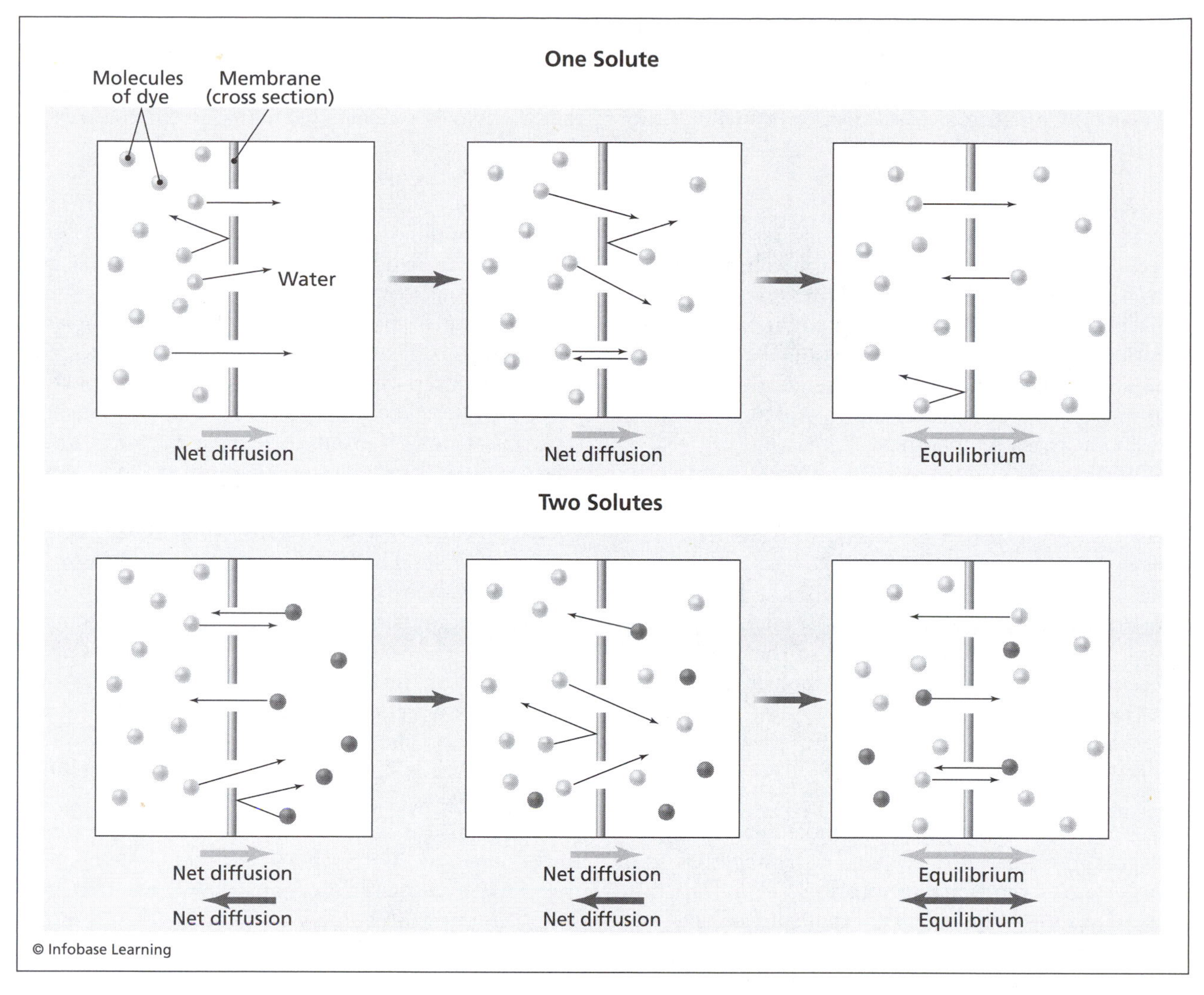

Osmosis occurs in biology as both a protective mechanism for cells and a means of taking nutrients and water into the cell without expending energy. Microorganisms must counter the effects of osmosis to maintain a favorable osmotic pressure.

trated sugar solution. The Scottish chemist Thomas Graham (1805–69) built on the theory of osmosis demonstrated by Nollet, by showing that some materials always crossed semipermeable membranes, and others could not cross. Using a membrane from an animal, Graham set up a more sophisticated version of Nollet's model to study the behavior of various solutes.

In 1861, Graham presented his findings on osmosis to London's Royal Society. He explained that certain compounds he called *crystalloids* had a greater affinity for crossing semipermeable membranes than other substances he named *colloids*. Inorganic salts, sugars, and alcohols belonged to the crystalloid group and starches, gums, and large proteins were colloids. Most importantly, Graham made the connection between osmosis in beaker experiments and the same osmosis taking place in biological cells. He reasoned, "Chemical osmose *[sic]* appears to be an agency particularly adapted to take part in the animal economy." Graham proposed that the term *dialysis* be applied to the selective diffusion of substances across membranes. Medical history has bestowed on Thomas Graham the moniker The Father of Dialysis for his concepts, which are still in use in medical treatment, today, in artificial kidney devices.

CALCULATING OSMOTIC PRESSURE

Sugar or salt solutions have been used to demonstrate the effects of osmotic pressure on biological cells. In this case, a semipermeable material formed as a sac acts as a cell. Scientific suppliers offer several different types of synthetic semipermeable membranes.

The German botanist Wilhelm Pfeffer (1845–1920) developed mathematical equations to describe osmosis. In 1899, Pfeffer published "Osmotic Investigations" in *The Modern Theory of Solution*, in which he described an apparatus for measuring osmotic pressure. The device later known as an *osmometer* or Pfeffer cell, advanced the physical study of osmosis.

Pfeffer expressed the relationship between a cell's inside pressure, called *turgor pressure,* and osmosis:

$$S_Z = O_Z - W,$$

where S_Z is the suction force of the cell, O_Z is the osmotic value, and W is the wall pressure.

Physicists later developed an equation to calculate the osmotic pressure that would be needed to keep a material from flowing through a semipermeable membrane. In this equation, π equals osmotic pressure at room temperature (RT):

$$\pi = (n_2 \div V) \times RT$$

where the expression n_2 refers to the number of moles of solute and V equals the volume of the solution surrounding the membrane. A mole is the number of grams of a molecule equal to that molecule's molecular mass. For example, the molecular mass of H_2O is:

$$(2 \times 1) + 16 = 18$$

where two hydrogen atoms of molecular mass, or molecular weight, of one added to a single oxygen atom of molecular mass 16 totals 18. Therefore, one mole of water is equivalent to 18 grams of water at room temperature.

Osmotic pressure in nature relates to the water activity inside the cell, a measure of the free water available to take part in biochemical reactions. The cytoplasm of microorganisms has a high solute concentration; thus, cytoplasm's water activity, A_W, is low. Increasing the concentration measure of a solute, called osmolality, decreases the A_W.

OSMOTIC PRESSURE CONTROL

Microorganisms—including osmophiles, which are extremophiles that thrive in environments with high concentrations of solutes—must constantly maintain osmotic pressure. Simple diffusion and facilitated diffusion, which uses a membrane protein to help move substances into the cytoplasm, work well when the concentration of a substance, such as a nutrient, is higher outside than inside a cell. Both processes use little energy. Microorganisms must use energy-demanding active transport to absorb nutrients that are in the environment at a low concentration.

Bacteria use two main active transport processes: active transport and group translocation. In active transport, the cell moves substances across the membrane with a concurrent use of the energy storage compound adenosine triphosphate (ATP). Active transport usually depends on a transporter protein that binds to a nutrient outside the cell, changes its conformation inside the membrane, and deposits the nutrient into the cytoplasm. Transporters are specific for a single compound or a group of structurally similar compounds.

Active transport occurs in prokaryotes and eukaryotes and often does not change the chemical structure of the compound being transported. Group translocation occurs only in prokaryotes and chemically alters the compounds being transported. Once the altered compound is inside the cell, its new structure blocks it from recrossing the membrane. Sugar transport provides an example of group translocation. For bacteria to absorb glucose, a bacterial enzyme must add a phosphate group (PO_4^-) to the sugar molecule. The sugar is, then, said to be in its *phosphorylated form.*

Microorganisms do not live in optimal conditions in the environment. Many environments challenge microorganisms to maintain osmotic pressure. Soil, marine waters, the digestive tract, and food are some of the environments that exert osmotic pressure on the cell. The capacity to adjust intracellular concentration is called *osmoregulation.*

Most microorganisms tolerate salt levels of up to 0.2 mole per liter (0.2 M). Extremophile archaea called *halophiles* withstand conditions saturated with salt, such as the 30 percent or higher salt levels of the Great Salt Lake in Utah. Halophiles have been, further grouped according to the conditions they tolerate:

- obligate halophiles—require high salt concentration to grow; also called *extreme halophiles*
- facultative halophiles—do not require high salt but can grow at salt concentrations up to 2 percent
- halotolerants—withstand moderately high salt concentrations no higher than 2 percent for short periods

Hypersaline (high-salt) environments are about 3.5 moles per liter. Halophilic archaea use a sodium pump to shunt sodium ions (Na^+) continuously out of the cell and back into an environment high in sodium. Meanwhile, the cells accumulate potassium ions (K^+) to balance the ion concentrations. Halophiles balance osmotic pressure by taking in organic molecules such as glycerol.

"Although halophilic Archaea survive in the hypersaline environments through the control of inorganic ion concentrations," wrote the microbial ecologist Carol Litchfield, in 1998, "the halotolerant Bacteria and algae have developed an alternative osmotic pressure balancing mechanism: the synthesis of specific organic molecules as their compatible solutes." Compatible solutes are the internal molecules that help maintain osmotic pressure. Some bacteria take up the solute from the environment; others synthesize the needed compounds. The common compatible solutes in microorganisms are sugars, amino acids, glycerol, and ectoine. Ectoine ($C_6H_{10}N_2O_2$) is a compound made by some microorganisms for cellular protection.

Halophiles tolerate high-salt conditions. Osmophiles are similar to halophiles but tolerate high osmotic pressure from substances in addition to salts. Most studies on osmoregulation in microorganisms have been conducted using high-salt solutions, so the two names are used almost interchangeably. Halophiles have now been discovered in salt lakes, evaporation salt ponds, and salt mines.

All living cells require a method to maintain osmotic pressure. Osmosis is a simple concept of dissolved compounds flowing through a semipermeable membrane from places of high concentration to places of low concentration.

See also EXTREMOPHILE; WATER ACTIVITY.

Further Reading

Allred, Ashlee, and Bonnie K. Baxter. "Microbial Life in Hypersaline Environments." Available online. URL: http://serc.carleton.edu/microbelife/extreme/hypersaline/index.html. Accessed November 16, 2009.

Eknoyan, Garabed, Spyros G. Marketos, Natale G. De Santo, and Shaul G. Massry. *History of Nephrology 2.* Basel, Switzerland: S. Karger, 1997.

Litchfield, Carol D. "Survival Strategies for Microorganisms in Hypersaline Environments and Their Relevance to Life on Early Mars." *Meteoritics and Planetary Science* 33 (1998): 813–819. Available online. URL: http://adsabs.harvard.edu/full/1998M&PS...33..813L. Accessed November 14, 2009.

Moat, Albert G. *Microbial Physiology.* New York: Wiley-Liss, 2002.

Pfeffer, Wilhelm. "Osmotic Investigations." In *The Modern Theory of Solution,* translated and edited by Harry C. Jones. New York: Harper, 1899. Available online. URL: http://books.google.com/books?id=n_tLAAAAIAAJ&dq=%22Wilhelm+pfeffer&printsec=frontcover&source=bl&ots=JNtsDJCdkf&sig=GGx77a HoKCX-HmoJIYP4oLvRDfk&hl=en&ei=H-YBS4-fIdD8nAf5n-WWCw&sa=X&oi=book_result&ct=result&resnum=10&ved=0 CCIQ6AEwCQ#v=onepage&q=&f=false. Accessed November 16, 2009.

Paramecium *Paramecium* is a genus of freshwater protozoa. These single cells typically have an elongate shape resembling a slipper, with the total length varying by species. *Paramecium* has been used as an aid in studying eukaryotic cell morphology, motility, and sexual reproduction by a process called *conjugation.*

Several thousand species of *Paramecium* have been identified, the most-studied of which are the following:

- *P. multimicronucleatum,* 200–300 micrometers (µm) in length
- *P. caudatum,* 180–300 µm in length
- *P. aurelia,* 120–200 µm in length
- *P. bursaria,* 90–150 µm in length
- *P. trichium,* 50–120 µm in length

Paramecium lives in aqueous environments, and it is easy to recover in samples from rivers, lakes, ponds, and moist soils. The cells' large size makes the organism just visible without a microscope. As do other freshwater protozoa, *Paramecium* constantly takes in water by osmosis, which is the movement of water or dissolved materials across a semipermeable membrane.

Paramecium feeds on bacteria by employing a coat of small hairlike appendages called *cilia* that grasp and move bacterial cells toward an invagination called the *oral groove.* The protozoan engulfs the bacteria within the oral groove and, then, gathers the bacteria in food vacuoles that release digestive enzymes. After digesting the bacteria, the vacuoles dispense nutrients throughout the cell. Other vacuoles absorb the wastes and carry them to the outer plasma membrane for excretion.

Paramecium contributes to the diversity of freshwater ecosystems and probably is a member of many different food chains. For example, an aqueous food chain begins with bacteria, which the *Paramecium* eats. Tiny invertebrates feed on the protozoa, and these organisms are, in turn, eaten by larger organisms, such as insects, insect larvae, and small fish. Each animal becomes larger with each step up a food chain.

Because *Paramecium* lacks the rigid cell wall that bacteria and many plants have for cell protection, the organism must remain in aqueous environments. The water in which many *Paramecium* species live is usually hypotonic in relation to the inside of the cell, meaning the concentration of dissolved materials is higher inside the cell than outside. This condition requires the protozoan constantly to adjust its inner pressure by a process called osmoregulation. *Paramecium* possesses an unusual attribute to aid osmoregulation: Its outer plasma membrane is stronger and less water-permeable than other biological membranes. For this reason, the organism can slow the constant diffusion of water into the cell. Vacuoles inside the cell, further, help in osmoregulation by taking in excess water and, quickly, forcing it back out into the surroundings. This membrane-vacuole system evolved in *Paramecium* as a survival mechanism and distinguishes the genus from most other protozoa.

Within the domain Eukarya, which contains other protozoa along with all other eukaryotes, *Paramecium* belongs to phylum Protozoa, class Ciliata, order Hymenostomatida, and family Parameciidae.

PARAMECIUM MORPHOLOGY

Paramecium belongs to a general group of protozoa called *ciliates* because of the thousands of hairlike cilia that cover the entire cell surface. *Paramecium* uses the cilia for movement, in addition to capturing food. The organism swims by slowly rotating with waves of cilia movement and can also make rapid changes in direction for chasing a meal or evading predators.

The plasma membrane in *Paramecium* works the same way as in other microorganisms by holding in the aqueous interior, called *cytoplasm,* and regulating the inflow and outflow of water. Just beneath the plasma membrane lies a more rigid layer called the *pellicle,* which provides strength to the cell and maintains its shape.

Protozoologists sometimes refer to *Paramecium*'s inner and outer regions as the *endoplasm* and the *ectoplasm,* respectively. The endoplasm consists of the cell's cytoplasm. The ectoplasm usually is defined as containing the pellicle, membrane, and ciliated surface.

The oral groove, also called the buccal cavity, measures one quarter to one third the cell's length and usually leads directly to a balloonlike food vacuole. Several additional food vacuoles are distributed throughout the cell's endoplasm. The food vacuoles secrete the digestive enzyme lysosome into their interior, when stimulated by the presence of food particles. *Paramecium* also contains two or more contractile vacuoles that serve in water regulation. Contractile vacuoles develop a star shape when they contract and expel water from the cell. Water-filled contractile vacuoles are evident in a microscope by their swollen, round appearance.

Most *Paramecium* species possess a macronucleus and a micronucleus. *P. aurelia* differs by carrying a second micronucleus, and *P. multimicronucleatum* has three to four total micronuclei. The larger macronucleus contains dozens to hundreds of copies of the cell's genetic material that makes up its genome. Genes inside the macronucleus control the cell's maintenance functions, such as feeding, excreting wastes, and osmoregulation. The micronucleus plays a central role in reproduction. Sexual reproduction in *Paramecium* involves splitting diploid (carrying an entire complement of genes) micronuclei into haploid (carrying one half the genome) micronuclei. Conjugation serves as the process by which each newly formed cell acquires a full complement of *Paramecium* genome.

CONJUGATION AND REPRODUCTION

Depending on species, *Paramecium* divides every two to 10 hours. Such rapid growth is somewhat unusual for a larger microorganism, such as *Paramecium,* because it must begin duplicating its internal organelles almost as soon as young cells form and power up their metabolism. Rapid cell growth also requires repair mechanisms for the inevitable mistakes that occur when duplicating organelles and synthesizing proteins, lipids, carbohydrates, and plasma membrane components.

Paramecium uses a well-coordinated system of asexual reproduction by binary fission, the splitting of two cells into two identical offspring, and sexual reproduction highlighted by conjugation to exchange genetic traits. Binary fission in paramecia proceeds as does that in prokaryotes, with the exception that organelles must divide and distribute into new cells.

Paramecium reproduces using specific mechanisms for replicating the nuclei and combining genetic material at the start of each new generation. Reproduction begins with a pairing of two complementary cells, analogous to male and female, called *conjugants.* To begin conjugation, these two cells contact each other and fuse their pellicles at the contact site. At the same time, their macronuclei disintegrate from organized organelles to general areas containing deoxyribonucleic acid (DNA). Micronuclei divide twice by meiosis to form four haploid nuclei, three of which disintegrate. The fourth micronucleus, called the pronucleus, then divides by using mitosis. In meiosis, a diploid cell containing a full complement of chromosome divides to form two haploid cells that contain only half of the chromosome. In mitosis, which takes place in the nucleus, two new nuclei, after division, receive the same number of chromosomes as the parent.

Mitosis in *Paramecium* micronuclei forms two distinct nuclei: stationary and migratory. During conjugation, the migratory nuclei pass between the conjugants and fuse with the haploid nucleus in the conjugant cell. A diploid zygote nucleus results from this step in conjugation, followed by separation of the two cells. The zygote undergoes replication, going from one to two nuclei, two to four, and finally eight nuclei. In most *Paramecium* species, these nuclei have the following fates in a process called *nuclear modification*:

- One nucleus becomes a new micronucleus.
- Four nuclei develop into macronuclei.
- The three remaining nuclei disintegrate.

Once a cell completes nuclear modification, it may divide asexually to produce two new *Parame-*

cium cells with half the number of micronuclei and macronuclei of the parent. Shortly thereafter, the cells are again ready for another conjugation process.

In *Paramecium* reproduction, division of the macronucleus does not require much precision because this organelle usually contains many copies of the same set of genes. The micronucleus, by contrast, must replicate with exact precision by mitosis so that the microorganism's genome has been passed on to the next generation without error.

Despite *Paramecium*'s active reproductive life, cells do not reproduce forever. Because of a process called *senescence, Paramecium* dies after a certain number of cell divisions. Many of the control mechanisms of protozoan senescence have not yet been identified. For example, no one knows what acts as the determining factor for initiating the process that leads to cell death. For *Paramecium* to avoid becoming extinct, the organism must occasionally complete conjugation to make a new generation. The factors that control the frequency of conjugation also remain unknown.

PARAMECIUM MOTILITY

Paramecium serves as a visual study of movement in ciliated organisms. The organism's cilia beat in a coordinated fashion to produce waves. These waves make the cell rotate, change direction, and swim forward or backward.

Most microbial cilia beat in two distinct stages. In the first type of cilia movement, called the *effective stroke,* the cilia pulls through the water as oars propel a rowboat. The *Paramecium* cell bends slightly to aid forward movement during the effective stroke. In the second movement, called the *recovery stroke,* the cell lets some of the cilia rest while other cilia continue working. By alternating these two types of strokes, a *Paramecium* maintains smooth motility through water.

Paramecium regulates the style and speed of its swimming with three factors: (1) cyclic adenosine monophosphate (cAMP), (2) cyclic guanosine monophosphate (cGMP), and calcium ions (Ca^{2+}). By controlling the calcium transport through the plasma membrane, *Paramecium* can adjust the cilia beat frequency, swim speed, and backward-forward motion. The calcium's antagonistic effect on the two energy compounds cAMP and cGMP determines the cell's overall direction and speed. In general, *Paramecium* movement is influenced in the following ways by its three motility factors, elucidated by the neuroscience researchers Nancy Bonini and David Nelson, in 1988:

- cAMP increases forward swimming speed
- addition of calcium to cAMP reverses the motion to backward
- cGMP antagonizes the calcium effect to slow backward movement and sometimes change direction to slow forward movement
- cAMP alone causes forward swimming in tight helices
- cGMP alone causes forward swimming in wide helices
- cGMP with calcium changes the forward motion to tight helices

The effects of cAMP, cGMP, and calcium relate to the *Paramecium*'s membrane potential, the difference in electrical charge from one side of the membrane to the other side.

PARAMECIUM SURVIVAL

Paramecium uses the ingestion of other microorganisms as a defensive tactic by eliminating competition for food. In addition to bacteria, *Paramecium* can ingest other protozoa, including other *Paramecium* species.

Some cells, called *killer paramecia,* possess a defensive ploy to avoid destruction if engulfed by another *Paramecium.* Some *Paramecium* cells carry bacteria inside them that they do not digest as food. When such a protozoan is engulfed by another, the bacteria, called *endosymbionts,* release a toxin that kills the predator. The *Paramecium* that acts as a host to endosymbionts possesses a type of immunity to the toxin so it can survive attacks. Killer *Paramecium* offers an example of symbiosis, a cooperative and often beneficial relationship between two different organisms.

Paramecium undoubtedly possesses additional survival tools that remain mysteries in microbiology. This microorganism continues to be a model for studying cell physiology, motility, and symbiosis.

See also EUKARYOTE; OSMOTIC PRESSURE; PROTOZOA; SYMBIOSIS.

Further Reading

Bonini, Nancy M., and David L. Nelson. "Differential Regulation of Paramecium Ciliary Motility by cAMP and cGMP." *Journal of Cell Biology* 106 (1988): 1,615–1,623. Available online. URL: http://jcb.rupress.org/cgi/reprint/106/5/1615. Accessed November 24, 2009.

Caprette, David R. "Studies on Paramecium." August 10, 2007. Available online. URL: www.ruf.rice.edu/~bioslabs/studies/invertebrates/paramecium.html. Accessed November 24, 2009.

Samworth, Mike, and Mike Morgan. "*Paramecium.*" *Micscape,* September 1999. Available online. URL: www.microscopy-uk.org.uk/mag/indexmag.html?http://www.microscopy-uk.org.uk/mag/article s/param1.html. Accessed November 16, 2009.

Pasteur, Louis (1822–95) ***French Microbiologist*** Louis Pasteur was the single most influential scientist in microbiology. His contributions ranged from theoretical experiments, such as his disproving the theory of spontaneous generation, to practical work in industrial microbiology. The knowledge that microbiologists have today of alcohol and lactic acid fermentations, vaccines, pasteurization, and disease diagnosis was gained, in large part, through Louis Pasteur's brilliant scientific career.

Louis Pasteur was born on December 27, 1822, amid vineyards in Dôle, France. Generations of the Pasteur family had made a living tanning hides; that background may have inspired Louis to study the anthrax bacillus, a soil bacterium that often contaminates hides and pelts, years later. Pasteur had a second family characteristic that would shape his life. He once said of his father, who had served in the French army, "In teaching me to read, you made sure that I learned about the greatness of France." His fierce loyalty to France impacted Pasteur's later career.

Almost all biographies describe the young Pasteur as an average student, but he applied himself, in his teens, with an eye toward attending school in Paris. After a few months in Paris, only 16 years old, Pasteur became homesick and returned to Dôle. At the college of Besançon, Pasteur pursued his love of art and received a bachelor of letters degree, in 1840. To improve his career prospects, he took an additional two years at the school studying science, especially chemistry, which piqued his interest more than any other subject. Pasteur returned to Paris for additional courses, at the prestigious school Normale, and began to excel at chemistry and physics, winning a student physics competition in 1843.

After graduating with a degree in chemistry, Pasteur returned to Paris. The chemist Auguste Laurent (1808–53), a leader in the young science of crystallography, took Pasteur under his wing. Pasteur began his career studying the behavior of crystals. In his first research project, he worked with a class of compounds, called tartrates, that were known to rotate light to the right. Pasteur discovered that tartrates dissolved in water produced similar compounds he called *paratartrates,* which rotated light to the left. By producing mirror images of the same compound, Pasteur opened up the new field of stereochemistry, the science of examining the arrangement of atoms in a compound.

Pasteur, next, studied the effect of tartrates and paratartrates on microorganisms that carried out alcohol-producing fermentations. The bacteria fermented the right-handed (rotating light to the right) version but did not use the left-handed version. Biochemists would, later, show that the bacteria's use or nonuse of mirror compounds, called *stereoisomers,* was due to very specific requirements of the microorganism's enzymes.

In 1849, Pasteur became chemistry professor at the University of Strasbourg, France, and the same year married Marie Laurent, whose father was head of the university. In 1854, Pasteur moved with Marie and their two young children to the university at Lille, where Pasteur became dean of the faculty of science, as well as chemistry professor. During Pasteur's period at Lille, the philosophy of science had changed in France, to an emphasis on practical studies for industry or agriculture. Either by choice or by directives from the university, Pasteur began studies in biology, particularly of the fermentation microorganisms that he had investigated years before.

Pasteur understood that microorganisms in the fermentation mixture were carrying out the reactions that transformed sugars to other compounds. But others in the scientific community doubted that tiny cells could exert such a powerful effect. The German chemist Justus Liebig (1803–73) thought that fermentation and putrefaction were chemical reactions that proceeded without the need for microorganisms. Many in science also believed, at the time, that the chemical alterations of organic matter arose spontaneously and had no biological origin. Others who thought microorganisms might be playing a role in the reactions speculated that these processes occurred because the microorganisms in the organic material had died. Putrefaction, for instance, was the result of thousands of dead bacterial cells that spoiled meat.

Perhaps frustrated by those who did not see the importance of microorganisms in fermentation, Pasteur took samples of spoiled meat, as well as samples from beer and wine fermentations, and in a microscope demonstrated the presence of live microorganisms. The presence of live microorganisms did not, however, satisfy the proponents of spontaneous generation. Supporters of this theory believed they had also run convincing experiments to prove spontaneous generation as the origin of life.

In 1857, Pasteur moved his studies back to Paris to teach at École Normale Superiéure. Pasteur began in earnest to put the theory of spontaneous generation to rest. In 1861, he would accomplish this by publishing an article, "On the Organized Bodies Which Exist in the Atmosphere; Examination of the Doctrine of Spontaneous Generation," which convinced the majority of the scientific community that microorganisms were the cause of many of the biochemical reactions involving organic compounds in the environment.

Pasteur set up flasks, each filled with broth boiled to sterilize it and, then, allowed to cool. He heated and reshaped the necks of the flasks by pulling them to a thin tube and adding one or more twists and turns. The broth remained open to air because the tip of the tube remained open, and he did not occlude the tube itself. No growth ever occurred in those flasks, while vessels left open to the air above soon filled with living microorganisms. Pasteur repeated the experiment with milk and got the same result. The experiment showed that airborne particles, many of which were microorganisms, could contaminate open flasks but could not reach the liquid in flasks that prevented the entry of airborne particles. "I do not know any more convincing experiments than these," a confident Pasteur said, "which can be easily repeated and varied in a thousand ways." The proponents of spontaneous generation would fade away, and most of the scientific community began looking to Pasteur as a leader.

Louis Pasteur, the Father of Microbiology, contributed to the science in the areas of germ theory, fermentation reactions, pathogen identification, vaccines, immune system function, food microbiology, and food preservation. Pasteur advanced microbiology's studies of bacteria, yeasts, and viruses. ***(Dibner Library of the History of Science and Technology)***

SPONTANEOUS GENERATION

Spontaneous generation is a theory, held since antiquity, stating that living organisms can develop from nonliving matter. Proponents of this theory believed that a power called a *vital force* existed in every inanimate object and could become a living thing on its own. This view predominated for centuries, until the Italian physician Francesco Redi (1626–97) conducted several experiments on decaying meat.

Redi set up three flasks, each containing a piece of raw meat: One flask he left uncovered and open to the air, the second flask he covered with paper, and the third flask received a gauze covering that would exclude flies but allow free movement of air into and out of the flask. Redi's experiment contained all the important elements of a scientific experiment: a control group (the open flask), test groups (the paper- and gauze-covered flasks), and replicates (several flasks in each group). The experiment gave the following results:

- open flask was contaminated by flies that laid eggs and produced maggots
- paper-covered flask excluded flies
- gauze-covered flask drew flies, which laid their eggs on the gauze, not the meat

Redi's experiment showed that life had not arisen spontaneously from the meat; rather, a biological presence (flies) had propagated new life.

Redi won over some scientists, but not all, especially with the discovery of microorganisms, in 1677. The discovery of bacteria by the Dutch haberdasher Antoni van Leeuwenhoek (1632–1723) suggested to some that tiny organisms arose spontaneously even if larger beings did not. The theory of spontaneous generation continued to hold sway over scientists, the clergy, and the public.

In 1748, the English priest John Needham (1713–81) boiled meat extract to drive out any possible life, then stoppered the flasks. After cooling, the liquids became cloudy with microbial growth, thus proving, he believed, that life had arisen from the lifeless broth. Microbiologists today know that Needham had not sterilized the liquid; he merely had not heated it for a long enough time to kill all microorganisms.

The Italian naturalist and priest Lazzaro Spallanzani (1729–99) expanded on how spontaneous generation worked by, first, sealing flasks of water containing seeds, then boiling them for three quarters of an hour. As long as the flasks were stoppered, no growth appeared, but opened flasks soon grew cloudy with microorganisms. Many who heard of the experiment reasoned that spontaneous generation does occur, but nonliving matter needs air to produce life. Spallanzani finally fueled both sides of the debate with a follow-up experiment, in which he sealed the boiled flasks with wooden stoppers or with cotton or left them open. The open flasks contained numerous microorganisms in a few days, but the cotton-sealed flasks held microorganisms at a lower concentration. The wood-stoppered flasks held no growth. "The number of animalcula developed, is proportioned with the communication with the external air. The air either conveys the germs to the infusions, or assists the expansion of those already there." Proponents of spontaneous generation took the second supposition to mean that life arose on its own, provided it had air.

Scientists would invent variations of the experiments, with flasks sealed versus unsealed, boiled versus not boiled, and air-exposed versus air-excluded. Each result seemed to prove what the scientist wanted it to prove. Pasteur's experiment dealt a blow to the theorists supporting spontaneous generations.

After Pasteur's flask studies, the English scientist John Tyndall (1820–93) demonstrated, in 1877, that dust could carry microorganisms, thus supporting Pasteur's idea. The German botanist Ferdinand Cohn (1828–98) would, soon, find evidence for the existence of bacterial endospores that survived heating that killed other bacteria. Pasteur, Tyndall, and Cohn had finally proven that spontaneous generation was a false hypothesis.

PASTEUR'S CAREER

Pasteur had also been commissioned by the brewing industry to explain why their product often became sour, instead of fermenting properly to produce alcohol. He discovered that in the case of spoiled beer, bacterial populations had overtaken the yeasts needed to ferment the grains properly in beer. In addition to giving the brewing industry a practical answer, he uncovered more details of alcohol and lactic acid fermentations, mainly the presence of anaerobic versus aerobic microorganisms.

The microbiologist also suggested to beer brewers and winemakers that heating the liquid before starting yeast fermentations could eliminate spoilage bacteria. Pasteur himself called the technique *pasteurization*. Microbiologists in the dairy industry now know that pasteurization does not eliminate all microorganisms from foods but reduces their numbers to safe levels. For this reason, pasteurization remains in use to aid food safety and preservation.

Pasteur had discovered the distinction between aerobes and anaerobes when he put a glass coverslip on top of a glass slide holding a drop of fermentation liquid. Some microorganisms swam to the outer edges of the drop, which were exposed to the air; other microorganisms stayed in the center of the drop. He, thus, became the first microbiologist to propose the existence of anaerobes, as opposed to aerobes. This discovery led him to revisit fermentation with the intent of finding the role played by oxygen.

In 1861, Pasteur demonstrated the fermentation of grains that takes place when oxygen has been excluded from the reaction vessel. The experiment demonstrated yeast fermentations that in anaerobic conditions yielded butyric acid as an end product. This and subsequent studies showed three characteristics of fermentation: (1) Microorganisms can live under conditions of no available oxygen, (2) aerobic metabolism and anaerobic metabolism yield different end products, and (3) by changing the conditions of fermentation, a microbiologist can alter the end products. In the process, Pasteur had also forwarded the germ theory, which posited that microorganisms might be responsible for many, if not all, of the biochemical activities in nature.

In the 1860s, the silk industry had called on Louis Pasteur to find the cause of a silkworm disease that affected worldwide silk production. Pasteur identified disease-causing parasites in the silkworms and recommended that the industry convert to a disease-free strain of worms instead of propagating disease with the current worms.

Pasteur would connect the fields of fermentation and disease. If the germ theory showed that microorganisms controlled biochemistry, such as the reactions of fermentation or the putrefaction of meat, perhaps they also affected disease. By the middle of the 1800s, science would be entering the Golden Age of Microbiology, when many biologists contributed to better understanding of microorganisms, their transmission, and their role in infection and disease. The English surgeon Joseph Lister (1827–1912) convinced many of his peers in medicine that surgical patients must be protected from infection by using antiseptics on surgical incision wounds and disinfectants in the surgery room to prevent the spread of germs. The German physician Robert Koch (1843–1910) developed a set of criteria, now called Koch's postulates, for proving that a given microorganism is the cause of a specific disease.

Pasteur had been drawn to the relationship between the body's ability to fight disease and disease-causing bacteria and viruses. In the 1880s, Pasteur investigated ways to develop vaccines. He experimented with the attenuation, or weakening, of live rabies, anthrax, and cholera microorganisms for the purpose of inoculating people against these diseases. (A virus such as rabies is not a true living thing, because it does not live independently, but viruses are usually thought of as microorganisms). Although Pasteur's first attempts at attenuation were crude because science did not yet have a full grasp of immunity and the influence of heredity, he developed a successful rabies vaccine. Pasteur's attenuation method involved passing a virulent microorganism through animals and recovering the pathogen from any individuals that survived the disease. Pasteur assumed that the pathogen had sufficiently weakened in the animal to make it safe for human or veterinary use. In 1881, the anthrax vaccine developed in Pasteur's laboratory helped end an epidemic that had devastated Europe's sheep industry.

Pasteur would continue work on vaccines, for the rest of his career, almost to his death. In his career, Pasteur received many awards but refused more than one that had been offered by German universities or science organizations. The French patriot Pasteur held Germany responsible for a conflict that had arisen over the Alsace and Lorraine regions, which lie between the two countries. A rivalry between Pasteur and the German Robert Koch prevented any communication on microbiology, at a time when both men were developing theories on disease that are still accepted today.

Many giants in science history have drawn controversy along with their triumphs. Pasteur's colleagues had accepted the scientist's penchant for drawing important scientific conclusions with limited data to support them. But Pasteur differed from most other scientists in history in having keen intuition about the essence of biology. Many of his theories proved to be true through the work of later scientists, even though Pasteur himself may have produced few data on his own.

Not everyone would be satisfied with Pasteur's approach to science, however. Koch made several public comments deriding Pasteur's studies. More recently, Gerald Geison of Princeton University, author of *The Private Science of Louis Pasteur,* expressed his doubts that Pasteur had performed all of the experiments for which he was credited. In 1995, the University of California–San Francisco virologist Jay Levy remarked on the confirmation of past experiments, "It is when you cannot repeat what they have published that you had better have good notebooks so you can go back and find out a reason for the discrepancy." Pasteur had, in 1886, commented on the vagaries of scientific thought and experimentation, saying, "In the sciences, certain persons have convictions, others have only opinions. Conviction supposes proof; opinions generally rest upon hypotheses." These words were ironic, considering that a later generation of scientists would question Pasteur's method of drawing conclusions from conjecture rather than data. The Pasteur biographer Geison inspected Pasteur's notebooks and concluded that Pasteur had left out important information that would have made those experiments repeatable by others. But, despite the doubts that Pasteur raised in others, no one has doubted the contribution of Louis Pasteur in advancing microbiology further than any other scientist had.

See also ANAEROBE; FERMENTATION; GERM THEORY; KOCH'S POSTULATES; LISTER, JOSEPH; PASTEURIZATION.

Further Reading

Altman, Lawrence K. "The Doctor's World; Revisionist History Sees Pasteur as a Liar Who Stole Rival's Ideas." *New York Times,* 16 May 1995. Available online. URL: www.nytimes.com/1995/05/16/science/doctor-s-world-revisionist-history-sees-pasteur-liar-who-stole-rival-s-ideas.html?scp=1&sq=Pasteur's%20Deception&st=nyt&pagewanted=1. Accessed December 18, 2009.

Debré, Patrice. *Louis Pasteur,* translated by Elborg Forster. Baltimore: Johns Hopkins University Press, 1998.

Fankhauser, D. B., and J. Stein Carter. "Spontaneous Generation: Protocol, Background, Procedure." 2004. Available online. URL: http://biology.clc.uc.edu/courses/Bio114/spontgen.htm. Accessed December 18, 2009.

Geison, Gerald L. *The Private Science of Louis Pasteur.* Princeton, N.J.: Princeton University Press, 1995.

Pasteur, Louis. "Influence de l'Oxygène sur le Développement de la Levure et la Fermentation Alcoolique" (Influence of Oxygen on the Development of Yeast and on the Alcoholic Fermentation). *Bulletin de la Sociétéchimique de Paris,* 28 June 1861, 79–80. In *Milestones in Microbiology,* translated by Thomas D. Brock. Washington, D.C.: American Society for Microbiology Press, 1961.

———. "Sur les Virus-vaccins du Choléra des Poules et du Charbon" (On a Vaccine for Fowl Cholera and Anthrax). *Comptes rendus des Travaux du Congrès International des Directeurs des Stations Agronomiques, Session de Versailles,* June 1881, 151–162. In *Milestones in Microbiology,* translated by Thomas D. Brock. Washington, D.C.: American Society for Microbiology Press, 1961.

Porter, J. R. "Louis Pasteur. Achievements and Disappointments, 1861." *Bacteriological Reviews* 25 (1961): 389–403. Available online. URL: www.ncbi.nlm.nih.gov/pmc/articles/PMC441122/#fn1. Accessed December 19, 2009.

U.S. National Library of Medicine, History of Medicine Division. "Louis Pasteur, 1822–1895." Available online. URL: www.nlm.nih.gov/hmd/resources/index.html. Accessed December 19, 2009.

pasteurization Pasteurization is a heat treatment of certain beverages and foods that kills pathogens and/or spoilage microorganisms.

The dairy industry has been the primary beneficiary of pasteurization technology. This industry has, in turn, developed improvements in pasteurization since its introduction. Early forms of pasteurization held milk at a minimum of 145°F (62.8°C), for at least 30 minutes, to reduce the number of microorganisms in it. This method is now referred to as *low-temperature long-time* (LTLT) pasteurization.

More recent methods use a minimal temperature of 161°F (71.7°C), for at least 15 seconds, a process called *flash pasteurization*, or the high-temperature short-time (HTST) method.

The U.S. Food and Drug Administration (FDA) publishes the current allowances, according to the Grade A Pasteurized Milk Ordinance of 2007, for pasteurizing milk. The ordinance expands the options that milk producers have for treating their product. According to this regulation, milk may be heated and then held in the following conditions:

- 145°F (63°C), 30 minutes
- 161°F (72°C), 15 seconds
- 191°F (89°C), 1.0 second
- 194°F (90°C), 0.5 second
- 201°F (94°C), 0.1 second
- 204°F (96°C), 0.05 second
- 212°F (100°C), 0.01 second

Some of these parameters might require adjustment if the fat content of the milk is 10 percent or greater or the milk contains at least 18 percent total solids or contains sweeteners. In these cases, the specified temperatures should be increased by 5°F (3°C).

MILK SAFETY

The French microbiologist Louis Pasteur (1822–95) developed the heat treatment that takes his name, when he was hired by the French wine industry, in 1864, to improve preservation methods without harming the wine's quality. By heating the liquid to below the boiling point, Pasteur found he could kill most harmful bacteria but preserve the wine's flavor.

The Industrial Revolution led to an increase in milk production and distribution to growing cities. Milk-borne diseases, such as typhoid fever, scarlet fever, diphtheria, and strep throat and diarrhea, increased as milk became more available to a larger population. Public health officials in Europe began calling for the pasteurization of milk, in the 1800s, to reduce the chance of such infections. Pasteurization was introduced in the United States, in 1889, and, in the early 1900s, the United States began requiring pasteurization of all commercially sold milk. In 1924, the U.S. Public Health Service instituted the Standard Milk Ordinance, to help states comply with new requirements for pasteurizing milk produced and sold in their state.

States initially had a difficult time setting up inspection programs to assure that dairies followed the new ordinance. The Missouri health commissioner James Stewart wrote, in 1927, that he supported the ordinance "because of the increasing evidence of the need of safer municipal milk supplies. . . . To date the ordinance has been passed in 26 cities in Missouri, representing a total population of 324,500, whose milk supply is produced under standard ordinance requirements." The ordinance led to a steady introduction of pasteurization programs in large U.S. cities, with fewer in rural areas. In 1947, Michigan became the first state to require that all milk sold in the state be pasteurized.

The original pasteurization ordinance is now the Grade A Pasteurized Milk Ordinance (PMO), administered by the U.S. Departments of Health and Human Services and Public Health. The U.S. Food and Drug Administration (FDA) takes responsibility for defining the procedures in dairy sanitation, milking methods, milk handling, and milk products standards that states can use to comply with the ordinance. In the United States, milk pasteurization continues to be regulated by state public health departments.

The PMO has undergone several revisions, since its introduction. The major areas related to milk production now covered by the ordinance are:

- milk quality examination methods
- dairy farm and milk plant inspections
- barn, milking house, and equipment sanitation and personnel hygiene
- aseptic techniques for milk product manufacture
- bottling and packaging
- animal health
- milk transportation

Milk quality examination methods include bacteriological tests for identifying potential pathogens in milk. Today, dairy and milk product microbiology is a specialized area of food microbiology. In all of food microbiology, a microbiologist's goal is to test an adequate amount of samples to ensure that a product contains no pathogens that could endanger consumers.

Dairies use the Hazard Analysis and Critical Control Program (HACCP) to identify the critical points in food production for sanitation and contamination monitoring. HACCP has helped the food industry reduce the packaging of contaminated foods and the subsequent spread of food-borne illnesses.

HACCP is particularly important in milk production because milk is a good source of nutrients

Potential Pathogens in Raw Milk

Microorganism	Health Threat
BACTERIA	
Bacillus cereus	diarrhea, abdominal cramping
Brucella species	brucellosis, undulating (intermittent) fever
Campylobacter jejuni	gastroenteritis
Escherichia coli O157:H7	hemorrhagic colitis
Listeria monocytogenes	listeriosis
Mycobacterium bovis or *M. tuberculosis*	tuberculosis
Salmonella species	salmonellosis, typhoid fever
Yersinia enterolitica	gastroenteritis, vomiting, joint infection
FUNGI (MOLDS)	
Aspergillus, Fusarium, and *Penicillium*	mycotoxin poisoning

for mammals and for microorganisms. Peter Wareing wrote in *International Dairy Topics,* in 2005, "Milk is essentially sterile at the point of production. Problems occur if the animal is diseased, or poor hygienic practices are employed on the farm. . . . Milk is also a very nutritious medium for microbial growth, and spoils very quickly if not handled properly after collection. In addition, many pathogenic bacteria can also grow in milk." For these reasons, milk requires strict HACCP and other sanitation procedures that many other foods do not. Part of the efforts to ensure milk quality relate to the monitoring of microorganisms to assess the effectiveness of a pasteurization program.

DAIRY PRODUCT MICROBIOLOGY

Milk is sterile in the udder but becomes contaminated with microorganisms as it leaves the udder. Most microorganisms in healthy animals present no health hazards to the suckling animals or to humans. During milking, handling, storage, or other prepasteurization steps, however, harmful microorganisms can enter the milk. These contaminants belong to three main groups of microorganisms: (1) bacteria, (2) yeasts, and (3) molds.

Bacteria far outnumber yeasts and molds in milk. The dairy industry divides the bacteria common in unprocessed milk into two groups: pathogenic and spoilage bacteria. Pasteurization does not sterilize milk by removing these microorganisms; pasteurization's goal is to kill disease-causing microorganisms and reduce the levels of spoilage microorganisms.

Pathogenic bacteria, although rare in most properly processed milk products, are important to monitor because they can cause typhoid fever, brucellosis, or tuberculosis, plus a variety of digestive upsets. The table (left) lists the main pathogenic microorganisms, including molds, of greatest health concern in raw (unpasteurized) milk.

Milk is rarely sterile when a consumer buys it. Pasteurization reduces microbial numbers to make milk safe to drink and prolong its shelf life, but almost all milk products eventually spoil, because of a mixture of bacteria that alter its taste, color, and odor. The main spoilage bacteria are:

- lactic acid–producing bacteria
- *Pseudomonas*
- *Microbacterium*
- *Streptococcus*
- *Bacillus*
- *Clostridium*
- *Micrococcus*
- *Arthrobacter*
- *Corynebacterium*

Lactic acid–producing bacteria (LAB) and *Streptococcus* spoil milk, but these bacteria also are used in the production of foods such as cheeses and yogurt. Both groups of bacteria form lactic acid from the sugar lactose, especially when the milk's temperature rises above 75°F (24°C). When lactic acid reaches a level of about 0.2 percent, milk is considered spoiled. At 0.55 percent acid, the milk protein casein coagulates and forms a smooth solid in the milk. The main LAB in milk are:

- *Lactococcus lactis, L. dulbrueckii*
- *Lactobacillus casei*
- *Leuconostoc*

Spoilage bacteria find milk a good substrate for growth for two reasons. First, milk products are stored refrigerated after pasteurization, during transportation, and in distribution. Most of the spoilage bacteria listed here can live as psychrophiles: That is, they grow in a temperature range of about 23–65°F (-5–18°C). Second, milk spoilage bacteria produce the extracellular enzymes lipase and protease, which break down milk fats and proteins, respectively. Since milk is a good source of fat and protein, the population of spoilage bacteria grows to high concentration in a short time.

After pasteurization, dairy products should follow aseptic processing and packaging procedures to reduce the chance of microbial contamination of the milk. These processes include, but are not limited to, presterilized containers, sterilized filling equipment, aseptic closures, and container atmosphere free of microorganisms.

Microbiologists determine the amount of microorganisms in milk by either of two methods: standard plate counts and direct microscopic counts. To prepare a standard plate count, the microbiologist dilutes a milliliter (ml) of milk 10-fold and repeats this 10-fold dilution from five to seven times. This procedure is called *serial dilution*. A small volume (an aliquot) of several of the highest dilutions can be added to molten agar in a petri dish. After the agar solidifies, the microbiologist incubates the cultures. After incubation, the agar contains visible colonies that can be counted in a low-powered microscope. Direct counts require no plating in agar or incubation. In this method, the microbiologist uses a microscope to count the number of cells in a drop of diluted milk placed on a glass microscope slide.

Raw milk contains 1,000 microorganisms or fewer per milliliter. In cows that have an inflammation of the udder, called *mastitis,* the counts can exceed 1,000/ml. Properly pasteurized milk contains fewer microorganisms than raw milk, from a few cells to several dozen cells per milliliter.

The dairy industry has accumulated data on the relationship between bacteria in milk and its shelf life. The Moseley Keeping Quality Test helps predict the shelf life of refrigerated (40°F [4.4°C]) pasteurized milk, which is based on the number of bacteria detected in the milk. As a generalization, milk lasts at least 10 days refrigerated before spoiling if it starts out with 6,650 bacteria/ml or fewer. Shelf life extends to 15 days or longer if the starting counts are 355 bacteria/ml or fewer. Only milk with no detectable bacteria on the first day of refrigeration does not spoil for more than 20 days. The food science expert J. Russell Bishop has concluded, "Application of the Moseley [Keeping Quality Test] has improved shelf life of fluid pasteurized milk and has represented an economic advantage to the [dairy] plants adopting the program." Several commercial laboratories in the United States conduct the Moseley test and other monitoring methods for assuring milk quality.

APPERTIZATION

The French confectioner Nicholas Appert (1752–1841) invented the heat treatment named for him, as part of his development of the canning process followed today. Appertization is a food preservation technique similar to pasteurization. Appertization requires heat treatment of certain foods, followed by sealing in a clean container at a temperature not to exceed 104°F (40°C).

Modern appertization of fluid milk is also defined as ultra-high temperature (UHT) pasteurization. In UHT, a milk producer heats milk to at least 284°F (140°) for at least two seconds. Nutritionists have questioned the effect of such high temperatures on the nutrients in milk, particularly milk proteins, which may lose their structure, called *denaturing,* when heated. UHT treatment has been shown to affect the digestion of milk proteins in appertized milk compared with conventionally pasteurized milk. Many of these heat treatments have caused consumers to question the quality of their milk, and some people have decided to purchase raw milk products rather than pasteurized. Pasteurization does not seem to cause significant damage to milk nutrients, however.

THE PASTEURIZATION DEBATE

Many consumers have decided against pasteurized food products in favor of raw or otherwise unprocessed products. A return to more natural foods without extensive processing has driven this trend. Nina Planck, author of *Real Food: What to Eat and Why,* wrote, in 2008, that pasteurization is "a form of sterilization that kills the living organisms in milk, both beneficial and pathogenic." In fact, pasteurization has never been intended as a sterilization method, which is the complete elimination of all life in a substance.

The nutrition professor B. M. Pickard of Britain's Leeds University supported Planck's assertion that pasteurization harms the nutritional value of milk. In a 2008 issue of *E Magazine,* Pickard said, "Evidence shows that untreated milk has a higher nutritional value, providing more available vitamins and minerals than pasteurized milk." The medical community disagrees with the raw milk movement. The Centers for Disease Control and Prevention (CDC), the American Medical Association (AMA), and other medical associations strongly endorse pasteurization as a preventative measure against milk-borne diseases. Cornell University's chair of food science, Kathryn Boor, has correctly pointed out, "You can't always tell when a cow is sick. And cows can sometimes kick the milking machine off [increasing opportunity for contamination]. Generally, what's on the barn floor is not something I want in a glass." The debate over the nutritional value of pasteurized versus raw milk continues, but as Boor implies, microorganisms are invisible, and a glass of raw milk with thousands of pathogens looks the same as a glass of safer pasteurized milk.

See also ASEPTIC TECHNIQUE; FOOD MICROBIOLOGY; HACCP; HYGIENE; LACTIC ACID BACTERIA; PASTEUR, LOUIS; SERIAL DILUTION.

Further Reading

Bishop, J. Russell. "Establish a Dairy Product Quality Assessment Program." Available online. URL: www.arrowscientific.com.au/petrifilmdairyprogram.html. Accessed November 25, 2009.

Cornell University. "Heat Treatments and Pasteurization." Available online. URL: http://milkfacts.info/Milk%20Processing/Heat%20Treatments%20and%20Pasteurization.htm. Accessed November 24, 2009.

Fast, Yvona. "Spilt Milk: From Fat Free to Flavored to Fresh." *E Magazine,* May/June 2008.

Lacroix, Magali, Cyriaque Bon, Cécile Bos, Joëlle Léonil, Robert Benamouzig, Catherine Luengo, Jacques Fauquant, Daniel Tomé, and Claire Gaudichon. "Ultra High Temperature Treatment, but Not Pasteurization, Affects the Postprandial Kinetics of Milk Proteins in Humans." *Journal of Nutrition* 138 (2008): 2,342–2,347. Available online. URL: http://jn.nutrition.org/cgi/content/abstract/138/12/2342. Accessed November 25, 2009.

Park, Alice. "The Raw Deal." *Time,* 12 May 2008.

Planck, Nina. *Real Food: What to Eat and Why.* London: Bloomsbury, 2007.

Stewart, James. "Discussion." *American Journal of Public Health* 21 (1931): 806–808. Available online. URL: http://ajph.aphapublications.org/cgi/reprint/21/7/806.pdf. Accessed November 24, 2009.

University of Guelph. "Dairy Microbiology." Available online. URL: www.foodsci.uoguelph.ca/dairyedu/micro.html. Accessed November 25, 2009.

U.S. Food and Drug Administration. "Grade A Pasteurized Milk Ordinance (2007 Revision)." Available online. URL: www.fda.gov/Food/FoodSafety/Product-SpecificInformation/MilkSafety/NationalConferenceonInterstateMilkShipmentsNCIMSModelDocuments/PasteurizedMilkOrdinance2007/default.htm. Accessed November 25, 2009.

Wareing, Peter. "On Farm HACCP for Milk Production." *International Dairy Topics* 4 (2005). Available online. URL: www.milkproduction.com/Library/Articles/On_farm_HACCP_for_milk_production.htm. Accessed November 25, 2009.

pathogen A pathogen is any microorganism that causes disease in another organism through the direct action of the microbial cells on the infected host. Pathogens, therefore, exclude microorganisms that are normally safe and participate in disease only in the presence of other microorganisms. Some textbooks also exclude toxin-producing microorganisms if the toxin but not the microorganism causes illness.

The term *pathogen* has generally been reserved for prokaryotes, viruses, and some fungi. Higher eukaryotic organisms that cause disease, such as worms and insects, are referred to as *parasites.* Disease-causing protozoa are referred to as either pathogens or parasites.

Pathogens can be specific for humans, various animal species, plant species, and other microorganisms. Some pathogens are able to infect more than one species. For example, the same strain of a seasonal influenza virus can infect humans, birds, swine, and possibly other animals.

Microbial pathogens cause *infectious disease,* a term that distinguishes the illness from genetic, autoimmune, environmental, and other diseases not caused by a microorganism. An infectious disease commences when a pathogen invades a susceptible host and, then, spends part of its life cycle in that host. Pathology is the study of all diseases, so it includes the study of infectious diseases. Pathogenesis is the manner in which a disease develops in the susceptible host. Clinical microbiology supports these sciences by developing methods for identifying pathogens that are recovered from someone who has an undiagnosed disease.

The terms *infection* and *disease* have been used almost interchangeably in microbiology, even though they have different meanings. Infection is the invasion of the body by pathogens. Disease occurs when an infection progresses to a state in which the body's overall health is changed.

The ability of pathogens to initiate infection or disease relates to factors that make the pathogenic cells more or less able to overcome the body's defense mechanisms. The host's three main defenses against pathogens are (1) the skin, (2) external factors such as lysozyme in tears, and (3) the immune system. Pathogens depend on two properties, pathogenicity and virulence, which determine whether they will successfully invade a host and remain there. Pathogenicity is the ability of a pathogen to overcome the host's defenses to cause infection or disease. The food-borne bacterium *Salmonella* possesses pathogenicity because it can set up infection in the digestive tract. Virulence is the degree of pathogenicity possessed by a pathogen. To continue the example, *Escherichia coli* O157:H7 is more virulent than *Salmonella* because O157 can cause irreparable harm to organs such as the kidneys and can lead to death.

Although *Salmonella* causes some deaths each year of people who have higher-than-normal health risk, this microorganism usually does not progress beyond the typical symptoms of food-borne illness: diarrhea, headache, nausea, fever, and abdominal cramping.

DEVELOPMENT OF THE STUDY OF PATHOGENS

Most of science did not immediately understand the role of microorganisms in disease. The Roman philosopher Lucretius (ca. 98–55 B.C.E.) may have been the first to allude to unseen forces that carried disease, and, in 1546, the Italian physician Girolamo Fracastoro (1478–1553) discussed the nature of a condition called *contagion,* or "an infection that passes from one thing to another. The term is more correctly used when infection originates in very small imperceptible particles." In 1676, Antoni van Leeuwenhoek (1632–1723) observed such particles, which he called "animalcules" and provided the world's first study of bacteria. Later scientists did not make the same connection Fracastoro did between disease and microscopic living things. For 200 years after van Leeuwenhoek's observations, most believed that disease was an ethereal entity in the air. People, including scientists, believed that a thick, poisonous atmosphere called *miasma* caused disease.

Scientists could not investigate the role of microorganisms and disease until microscopes had advanced beyond the crude lenses used by van Leeuwenhoek. The 1800s saw the rapid growth in microbiological discovery, mainly due to better microscopes. From the mid-1800s to the early 20th century, science entered the Golden Age of Microbiology. During this period, scientists made critical advances, upon which today's microbiology builds.

The Italian scientist Agostino Bassi (1773–1856) was the first to observe that a microorganism, in this case, a fungus, caused a disease, when, in 1835, he investigated the cause of a silkworm disease. Soon afterward, others linked fungi with potato blight, which had caused devastating famine in Ireland, in the 19th century, and the crop diseases called *smut* and *rust.*

The British surgeon Joseph Lister (1827–1912) drew a definitive connection between microorganisms and disease in a career devoted to improving the aseptic (containing no unwanted microorganisms) handling of patients and surgical wounds. Only the use of antiseptics on surgical incision wounds, he argued, would prevent microorganisms from entering the body and causing deadly results. At the same time, Lister was trying to convince his peers of the need for aseptic conditions, the German microbiologist Robert Koch (1843–1910) developed a system, in 1884, for proving that a given microorganism was the cause of a specific disease. The criteria he established for proving this relationship became known as Koch's postulates. These postulates provide the foundation for diagnosis of infectious diseases today.

Lister's and Koch's work complemented that of the French microbiologist Louis Pasteur (1822–95) in developing the germ theory of disease. This theory put forth the idea that the activities of microorganisms in the body changed tissues and organs in a way that created disease. Before the work of these three men, scientists had made only tentative connections between microorganisms and infection. Some persisted in believing that disease was carried in the air with bad odors.

The discovery of pathogens would not have taken place without the contributions of microbiologists who improved on the techniques for studying bacteria and fungi. Robert Koch developed methods for growing pure cultures of bacteria. Koch's colleague Walther Hesse (1846–1911) devised a new type of medium for growing bacteria containing a solidifying material called agar. The Danish physician Hans Christian Gram (1853–1938) developed a biological stain that greatly improved on the differentiation of cells in a microscope, and the German bacteriologist Richard J. Petri invented a small flat dish for holding agar cultures and increasing the number of cultures a microbiologist could incubate at one time.

As the Golden Age of Microbiology began to end, microbiologists increasingly studied individual species to learn about their role in the environment or in human health. Microbiologists began to gain knowledge of individual pathogens such as smallpox, bubonic plague, rabies, and diphtheria.

A large part of the study of pathogens related to the ways in which the body's immune system tries to fight off infections. The Russian biologist Elie Metchnikoff (1845–1916) advanced the science of immunology by discovering that some white blood cells had the ability to engulf infection-causing bacteria and destroy them. He had discovered phagocytes and the process called *phagocytosis.*

The modern study of pathogens delves into the genetic makeup of hosts that make some more susceptible to infection than others. At the same time, molecular biologists are identifying the genes that contribute to pathogen virulence to find ways of blocking some of the virulence factors. New drug development focuses on the actions of pathogens in tissue or in individual cells. By studying pathogens at the molecular level, microbiologists try to find the most specific treatments for specific infectious disease. Developing drugs that are very specific to a pathogen's metabolism makes them safer for humans because the drugs affect only the pathogen and do not cause side effects in the body. A completely safe drug remains a long way off, however. Microbiologists still have many questions to answer regarding infection, the spread of disease in the body, transmission of pathogens between people or between animals and people, and the defensive tactics used by pathogens to avoid being destroyed.

TYPES OF PATHOGENS

Pathogens can be bacteria, fungi, yeasts, protozoa, and viruses. Only algae appear to cause harm to humans by the secretion of toxins rather than direct invasion of the body.

Different types of pathogens and their various modes of invading a host determine the type of infection that develops. Infectious diseases can, therefore, be classified on the basis of how they and how the pathogen behave in a population. Communicable diseases spread from one person to another, by either direct contact or indirect contact. Indirect contact occurs when a pathogen transmits between two hosts by way of an intermediate, such as a doorknob or an insect. Contagious diseases are communicable diseases that transmit easily between people. For example, acquired immunodeficiency syndrome (AIDS) is a communicable disease, and influenza is a contagious disease. Pathogens also cause noncommunicable diseases that a host contracts from conditions other than person-to-person direct or indirect transmission. *Clostridium tetani*, for example, can cause tetanus if the bacteria infect an untreated wound. A host normally becomes exposed to the *C. tetani* in soil or objects contaminated with the bacteria (fences, farm equipment, gardening tools, livestock-handling equipment, etc.).

In addition to the type of microorganism, pathogens can be classified into two broad categories by their relation to infection. Obligate pathogens require part of their life cycle to cause infection or disease in order for the pathogenic species to survive. *Mycobacterium tuberculosis* is an obligate pathogen. Opportunistic pathogens do not normally cause infection or disease when a person is exposed to them, but these microorganisms can cause infection if host conditions change. For example, the normal skin bacterium *Staphylococcus aureus* does not cause infection in the body unless the host receives an injury that gives *S. aureus* an opportunity to infect.

The Centers for Disease Control and Prevention (CDC) and other health agencies classify pathogens by the mode of transmission, as well as the system within the host where infection occurs. All pathogens, whether bacterial, fungal, or viral, can, thus, belong to these general categories. Microbiologists specialize in studying the pathogens specific to these categories, summarized in the table on page 607.

Pathogens in the categories defined here can also be classified in terms of their entry into a population, either emerging or reemerging. An emerging pathogen is a pathogen new to a population or newly discovered. A reemerging pathogen is a microorganism that was thought to be eradicated or under control but whose incidence of infection has again increased in a population. Appendix VII lists the main human pathogens and the diseases they cause.

Many pathogens require a certain mode of transmission and a defined portal of entry into a susceptible host. A pathogen may use more than one mode of transmission to spread through a population, but most pathogens have a main mode of transmission associated with most of their infections. For example, the influenza virus often uses airborne transmission, while *Vibrio cholerae* is almost exclusively a waterborne pathogen, causing cholera. The portal of entry represents the main route into the body that a pathogen uses, called the *preferred portal of entry*. Common portals of entry are the mucous membranes (a barrier composed of mucus, for example, parts of the respiratory tract), skin, gastrointestinal tract, genitourinary tract, or a parenteral route (directly into the blood).

PATHOLOGY, INFECTION, AND DISEASE

Pathology of infectious disease encompasses many interrelated factors that determine the success or failure of a host in resisting invasion by a pathogen. As single entities, pathogens are seldom dangerous. Pathogens require help to set up an infection in an individual or an outbreak in a population successfully. The factors that contribute to this ever-changing balance between the host and the pathogen are:

- pathogenicity—capacity to infect
- transmission—method of having contact with a potential host
- virulence factors—attributes that help the pathogen overcome the body's defenses
- host susceptibility—preexisting health conditions that increase opportunity for infection, called *predisposing factors*
- host defenses—strength of the host immune system and other defenses (skin, normal population of flora, antimicrobial factors such as lysozyme in tears and fatty acids in sweat)

People have contact with pathogens every day, and most pathogens fail to establish a connection on the body and disappear without the host's knowledge. Normal flora that live on the skin, the digestive tract, and the respiratory tract repel most pathogens. The body's physical barrier of healthy, unbroken skin

Types of Pathogens

Category	Subcategory
mode of transmission	• airborne • blood-borne • food-borne • nosocomial (hospital-associated) • sexually transmitted • vector-borne (insect-transmitted) • waterborne • zoonotic (from animals)
target tissue, organ, or metabolic system	• blood • cardiovascular • eye or ear • gastrointestinal • immune system • lymphatic system • neural (nervous system) • oral • reproductive system • respiratory • urinary or renal

offers a very strong defense against invasion of the bloodstream by pathogens. Finally, the host's immune system creates a multilayered array of defenses that not only detect foreign matter in the body, but destroy the particles and then develop antibodies to protect the host from future infections by the same pathogen. Many functions in the immune system have alternate systems that take over if a pathogen defeats the first line of defense. This characteristic is called the *duality of the immune system* and provides yet more insurance against invasion by pathogens. Pathogens have, nonetheless, evolved with human and animal populations and have adapted virulence factors designed specifically to evade many of the host's intricate defenses.

STUDYING PATHOGENS

Early microbiologists focused on human pathogens, in the 1800s, as the methods for studying and characterizing microorganisms advanced. The development of the microscope became the single most important tool for identifying pathogens in the specimens taken from sick or deceased patients. Louis Pasteur explored the various metabolic features of bacteria and fungi. By understanding the varied metabolisms in the microbial world, microbiologists began to see that each pathogen attacks a host and lives in the body in its own unique way.

One microbiologist in particular, Robert Koch, sought a way to link specific pathogens with the specific diseases they caused in the body. In 1884, Koch developed a set of criteria (Koch's postulates) for proving the connection between Pathogen A and Disease A1, Pathogen B and Disease B1, and so on. In brief, Koch's postulates state that (1) the pathogen must be present in the disease but absent in healthy organisms, (2) a microbiologist must isolate the pathogen from a diseased individual, (3) the same disease must ensue when the pathogen is put into another healthy host, and (4) the same pathogen must be isolated again from this second individual. These postulates remain the principles upon which disease diagnosis takes place today.

Studying pathogens in a population also requires the field of epidemiology. Epidemiology is the study of a pathogen's entry, growth, and transmission in a population, and the incidence and frequency of disease. Epidemiology allows the medical community to predict the movement of a disease (or a pathogen) through a population. With this information, public health officials can establish warnings and recommendations for a community for the purpose of ending the pathogen's transmission. Epidemiologists also try to identify a pathogen's reservoir, the organism that maintains a population of pathogens capable of entering a human population. For example, rodents such as ground squirrels and mice act as the reservoir for the hantavirus pathogen. Various pathogens use animal, human, or nonliving reservoirs, such as soil.

Pathology also requires expertise in two areas of medicine: etiology of disease and immunology. Etiology is the general study of the cause of disease, including the pathogen's identity, its modes of transmission and infection, and its activity in the body to cause harm. Immunology covers the broad field focused on the host's defenses against infection and the pathogen's establishment in the body.

Despite enormous attention medicine and microbiology devote to pathogens, these microorganisms make up a small percentage of all known microorganisms. The vast majority of microorganisms on Earth cause humans, animals, or plants no harm. Human pathogens can, however, appear in almost any place used frequently by many people. Public places such as offices, public restrooms, and mass transit vehicles hold pathogens on almost every surface. Frequently touched items, such as doorknobs, refrigerator handles, keyboards, and automated teller machines (ATM) screens, also harbor pathogens transferred to these sites by others.

In a study at the University of Arizona, the microbiologist Charles Gerba collected nearly 2,600 samples from various surfaces touched repeatedly by many different people. The science writer Brian Hoyle described the study results in a 2007 issue of *Microbe,* writing, "Those swab samples often contained MRSA [methicillin-resistant *Staphylococcus aureus*], which appeared to be widely but unevenly distributed along sampled surfaces. For instance, MRSA was recovered from grab bars and door exit handles of buses almost 70% of the time and, even more frequently, from airplane tray tables. Meanwhile, floors near toilets in public restrooms are a particular hot spot for recovering viable MRSA." Washing hands is an excellent way to stop the transmission of pathogens picked up in public places, but even soap has been found to present a hazard. Gerba's colleague Marisa Chattman studied 150 soap dispensers in public restrooms and restaurants and found 23 percent contaminated with the pathogens *Serratia marcescens, Enterobacter aerogenes,* and *Klebsiella pneumoniae,* all of which can cause gastrointestinal illness. Chattman said, "A person usually washes with at least a couple of milliliters of soap, so they are getting a big load (of bacteria)." Despite the presence of microorganisms in soap, hand washing remains important to good hygiene for avoiding pathogens.

Pathogens will always be a source of surprising discoveries and a critical research topic in microbiology. Humans have failed to defeat most of the pathogens that have plagued society. At best, perhaps, people can only hope to learn enough about the most deadly pathogens to create the strongest defenses possible against those microorganisms. In the long term, pathogens will probably never disappear from the human population, so humans must learn how to manage lifestyles knowing pathogens will always be in their midst.

See also EMERGING DISEASE; EPIDEMIOLOGY; GERM THEORY; KOCH'S POSTULATES; LISTER, JOSEPH; PATHOGENESIS; PHAGOCYTOSIS; PORTALS; RESERVOIR; TRANSMISSION; VIRULENCE.

Further Reading

Baquero, Fernando, Cesar Nombela, Gail H. Cassell, and Jose A. Gutierrez-Fuentes. *Evolutionary Biology of Bacterial and Fungal Pathogens.* Washington, D.C.: American Society for Microbiology Press, 2008.

Black, Jacquelyn G. *Microbiology: Principles and Explorations,* 7th ed. Hoboken, N.J.: John Wiley & Sons, 2008.

The Day Care Dilemma

Anne Maczulak, Ph.D.

Economic circumstances, in the past few decades, have forced increasing numbers of parents to use day care. One of the many questions related to this decision involves the potential exposure of children in the day care setting to infectious microorganisms. Recently, parents have begun to consider the true effect of germ exposure: Is it a hazard or a benefit for their children?

Germs, a generic term for infectious bacteria, fungi, and viruses that cause illness, congregate where people crowd together for long periods and constantly touch shared surfaces. These are certainly features of day care centers, where many young children share the facility and toys, and the adult staff provide constant supervision. Infectious microorganisms transmit easily in these circumstances. Very young children, further, help spread germs by the following behavior: playing on the floor, putting shared objects into the mouth, touching hands to the face, washing hands improperly, and using poor hygiene methods when eating or using the restroom. Young children, furthermore, do not yet have fully developed immune systems to fight infections.

The day care staff is often overlooked as potential germ transmitters. Poor hygiene of teachers, aides, and cooks can easily transmit infections. Day care, therefore, gives parents plenty of cause for worry about their child's health. Good hygiene does not occur if only one person practices it. Adults at the day care facility and parents at home have a responsibility to teach good hygiene as early in children's upbringing as possible. The main rules of good hygiene are (1) washing hands for 20 seconds with warm, soapy water before eating and after using restrooms; (2) avoiding touching the face with the hands; (3) covering the mouth and nose when sneezing and coughing; and (4) staying home when sick. These rules apply to adults as well as children.

Day care centers have been sources of infections, particularly rotavirus and other enteric viruses, ear infections, and respiratory infections. Enteric infections and diarrhea are common ailments among children in day care. In 1999, Wenche Nystad concluded from a study of almost 1,500 children that day care increased the risk of respiratory disease. Many researchers have joined the day care debate.

Some parents have concluded that children would be healthier at home than in the day care facility to avoid these common infections. But increasing support of the hygiene hypothesis has led others to view group settings such as day care as a way to decrease the chance of infection. How can this be? The hygiene hypothesis proposes that constant exposure to the normal microorganisms we confront every day in the environment strengthens the immune system. But this hygiene hypothesis is still highly controversial. Applying it to the well-being of children in day care may take a risky leap of faith.

When the German pediatrician Erika Von Mutius developed the hygiene hypothesis, in the 1990s, she drew her conclusions from children rather than adults. Von Mutius found a perfect experimental situation in segregated East Berlin and West Berlin. She compared children from poorer circumstances and a more polluted environment in the eastern part of the city with more

Centers for Disease Control and Prevention. Available online. URL: www.cdc.gov. Accessed November 26, 2009.

De la Maza, Luis M., Marie T. Pezzlo, Janet T. Shigei, and Ellena M. Peterson. *Color Atlas of Medical Bacteriology.* Washington, D.C.: American Society for Microbiology Press, 2004.

Hoyle, Brian. "Pathogens Seem to Be Everywhere—Even in Soap Dispensers." *Microbe,* August 2007.

National Institute of Allergy and Infectious Diseases. Available online. URL: www3.niaid.nih.gov. Accessed November 26, 2009.

Tortora, Gerard J., Berdell R. Funke, and Christine Case. *Microbiology: An Introduction,* 10th ed. San Francisco: Benjamin Cummings, 2009.

Wassenaar, Trudy M. "Pathogenic Bacteria." Virtual Museum of Bacteria. January 6, 2009. Available online. URL: www.bacteriamuseum.org/cms/Pathogenic-Bacteria/pathogenic-bacteria.html. Accessed November 26, 2009.

Willey, Joanne, Linda Sherwood, and Chris Woolverton. *Prescott, Harley, Klein's Microbiology,* 7th ed. New York: McGraw-Hill, 2007.

pathogenesis In microbiology, pathogenesis is the manner in which an infectious disease develops. Pathogenesis makes up one third of pathology, the study of diseases, which also covers etiology (the study of a disease's cause) and disease progression through the body and its effects. A pathogen is any microorganism that causes disease.

Pathogenesis encompasses two aspects of infectious disease: infection and pathogenicity. Infection is the establishment of a pathogen in the body (the host). A pathogen's ability to develop disease after infecting a host is called *pathogenicity.*

The host and the specific pathogen have a bearing on the course of pathogenesis. For instance, the host's general health and immune system strength contribute in part to the ability of a microorganism to sustain an infection and cause disease. The physical, chemical, and biological features of the host that combat infectious disease are called *host factors.* For example, the skin barrier is a physical host factor, fatty acids in sebum make up a chemical factor that

well-to-do children in the cleaner, less decrepit buildings of the western sector. Though Von Muitius had expected to find poorer health among East Berlin children, she was surprised to see that the West Berlin children suffered more respiratory ailments. From this initial study, Von Mutius began developing the hygiene hypothesis.

Von Mutius and other researchers have continued to explore the relationship between germ exposure and health. Some studies have suggested that children raised on farms with livestock may have fewer respiratory infections than village children who are not exposed to large amounts of environmental microorganisms.

Skeptics correctly point out that an enormous number of factors may be contributing to these findings, including household habitats, heredity, general environmental influences, and nutrition. The hygiene hypothesis has not garnered unequivocal support among scientists, but many parents have heard enough to believe that it has some validity. For this reason, they encourage the use of day care for all children so that all will develop stronger immunity over time.

Parents facing the day care dilemma have challenges related to income levels, long work hours, long commutes, and other factors that affect the decision to use day care. They hardly need the additional worry of infections and whether infections occur more frequently at home or in day care.

No one has solved the questions of how day care impacts children's health and the incidence of childhood infections. Common sense helps allay some fears, regardless of which side a parent takes in the hygiene hypothesis debate. Teaching young children good hygiene, as early as possible, to stop the spread of germs goes furthest in preventing infection. Allowing healthy children to participate in day care cannot be thought of as a health threat. But when a child is sick, the child must be sent home or kept home. Day care was never intended to substitute for good decisions in raising healthy children. Forcing a child into day care for the purpose of germ exposure seems as nonsensical as holding a child at home to keep him or her out of the path of transmitted germs.

The best choice in dealing with infectious pathogens is to use commonsense measures to avoid infections. Too little diligence or too much diligence in respect to pathogens does not work well for children or for adults.

See also IMMUNITY; TRANSMISSION.

Further Reading

Carr, Coeli. "Child-Care Centers and Parents Brace for Flu Season." *Time,* 16 October 2009. Available online. URL: www.time.com/time/health/article/0,8599,1930877,00.html. Accessed November 8, 2009.

National Library of Medicine. National Institutes of Health. "Day Care Health Risks." Available online. URL: www.nlm.nih.gov/medlineplus/ency/article/001974.htm. Accessed November 8, 2009.

Nystad, Wenche, Anders Skrondal, and Per Magnus. "Day Care Attendance, Recurrent Respiratory Tract Infections and Asthma." *International Journal of Epidemiology* 28 (1999): 882–887. Available online. URL: http://ije.oxfordjournals.org/cgi/reprint/28/5/882. Accessed November 8, 2009.

WGBH Educational Foundation. "Hygiene Hypothesis." Available online. URL: www.pbs.org/wgbh/evolution/library/10/4/l_104_07.html. Accessed November 8, 2009.

inhibits bacteria, and immunity is a biological host factor.

The strength of the pathogen, likewise, contributes to pathogenicity. A combination of characteristics called *virulence factors* determines the severity of infectious disease. Virulence factors enable a pathogen to avoid, neutralize, or overcome the host's defenses. For example, some bacteria form a coating on the cell called a *capsule*. Bacteria do not need capsules to live, but the formation of a capsule helps a pathogen's virulence and pathogenicity. Capsules decrease the ability of the body's phagocytes to engulf and digest bacteria (called *phagocytosis*).

Decades of study by microbiologists and pathologists have uncovered the phases of pathogenesis. The details of pathogenesis differ among bacteria, viruses, fungi, and protozoa, but all use the following general steps:

- attachment to a host
- entry into the host
- establishment of an initial population
- invasion of target tissue
- growth and multiplication in the body
- damage to host cells, tissues, or organs
- exit from the host

Scientists study each of these aspects of pathogenesis for the purpose of developing treatments and vaccines to combat the disease. Avoidance of the body's defenses by the pathogen is of paramount importance, in most of these events. Not all phases of pathogenesis have been delineated, however. Certain diseases need more study by medical researchers in some or all of the phases listed here. Each year, medical research worldwide adds to the store of knowledge on numerous diseases.

STUDIES ON PATHOGENESIS

Society has always noticed signs of its most prevalent diseases even before people understood what caused infection. The Greek historian Thucydides (ca. 460–395 B.C.E.). recorded clinical signs of the diseased during a smallpox epidemic in Athens, from 430 to 427 B.C.E. Various descriptions of illness during later epidemics of smallpox, plague, and syphilis indicate that scientists (usually philosophers) watched the progression of disease and attempted to draw conclusions about the disease signs they observed. The Roman philosopher Lucretius (98–55 B.C.E.) surmised that invisible things in the air infected the body and caused disease, but little progress occurred along this path of medicine, until the 16th century.

In 1546, the Italian physician Girolamo Fracastoro (1478–1553) wrote an essay on the possible origins of *contagion,* the term used for centuries for disease. Fracastoro constructed a rational discussion of disease, even noting the differences in transmission known at the time: direct contact with a sick person, transmission by touching a contaminated inanimate object, and transmission through the air. Fracastoro proposed diseases originated in "small imperceptible particles." With his essay, Fracastoro introduced the germ theory of disease.

The germ theory proposed that infectious diseases were caused by microorganisms. Although this concept seems simple to comprehend today, scientists spent the three centuries after Fracastoro's death arguing the relationship between disease and invisible cells. The sciences of microbiology and pathology could not have progressed without the introduction of the microscope by the Dutch merchant Antoni van Leeuwenhoek (1632–1723).

The accumulated work of several microbiologists convinced skeptics that microorganisms could cause disease. Into the 19th century, many believed that disease was caused by God as a punishment for evil acts. In 1835, the Italian scientist Agostino Bassi (1773–1856) connected an outbreak of disease in silkworms with a fungus, and, in 1865, the esteemed French microbiologist Louis Pasteur (1822–95) made a similar discovery and began a career-long study of germs and disease. The German physician Jacob Henle (1809–85) gave an insightful description of a pathogen, in 1840, observing that "the contagium is a material endowed with an independent living existence, which reproduces in the manner of plants or animals, and can multiply by the assimilation of organic substances, and lives as a parasite on the sick body." At about the same time, the British surgeon Joseph Lister (1827–1912) provided compelling evidence to the medical community that sterility maintained during surgery prevented germs from infecting patients. In the 1880s, the German bacteriologist Robert Koch (1843–1910) expanded on Henle's work and developed a list of criteria for linking a given disease to a specific pathogen. The criteria, to be known as *Koch's postulates,* opened the door to more refined studies on pathogenesis.

PATHOGENESIS PROCESS

The first step in pathogenesis involves adherence of a pathogen to the host. Bacteria display various methods of adherence, called adhesion factors, such as the following:

- capsule—heterogeneous polysaccharide layer that promotes adherence and protects against phagocytosis

- fimbriae—filamentous appendages that enable bacteria to grasp a surface

- glycocalyx—exterior layer of polysaccharides that help the pathogen stick to host tissue, such as skin and mucous membranes

- lectins—proteins on the pathogen that bind to carbohydrates on host tissue

- ligand—surface molecule on the pathogen that forms a specific connection with a surface molecule on a host cell

- S-layer—glycoprotein (protein with sugars attached) layer that promotes adherence

- slime layer—similar to a capsule but with a weaker connection to the pathogen; can be washed off

- teichoic and lipoteichoic acids—components in gram-positive cell wall that aid adherence

Viruses and certain bacteria possess specific adherence factors called receptors. By this mechanism, adherence to a host cell involves a specific structure on the pathogen's surface that recognizes and binds to a protein or glycoprotein on the target cell's surface. The glycoprotein gp 120 on the surface of human immunodeficiency virus (HIV), for example, forms a bond with a receptor structure (CD4) on immune system cells (T cells) in the initial step of infection. HIV particles able to carry out the rest of the steps in pathogenesis cause acquired immunodeficiency syndrome (AIDS).

Adherence relates to the portal of entry a pathogen usually uses to enter a host. Pathogens that infect humans and other animals use three main tissue types for entering a host: (1) skin, (2) mucous membranes, and (3) blood. Each pathogen has what researchers call a preferred portal of entry, but some pathogens also use secondary portals to cause infection, as an alternative to the preferred portal. For example, the anthrax pathogen *(Bacillus anthracis)* is a soil bacterium that enters a host through the mucous membranes of the respiratory tract but can also enter through cuts in the skin or by ingestion.

Pathogens use tools that help them get through a portal of entry. A group of enzymes called *spreading factors* help pathogens overcome some of the body's physical barriers and enter the bloodstream, for example:

- collagenase—breaks down collagen, which provides structure to muscles

- hyaluronidase—breaks down hyaluronic acid that forms connective tissue

- neuraminidase—degrades sialic acid, which holds together the epithelial cells lining the intestines

- kinase—converts plasminogen to plasmin, which digests the blood-clotting factor fibrin

The portal of entry and the target tissue are often the same. *Mycobacterium tuberculosis* uses the mucous membrane of the respiratory tract to invade a host and sets up its infection in the lungs.

After adherence and entry, a pathogen employs a variety of virulence factors to avoid destruction by the host's immune response. Pathogens evade the body's phagocytes, cells that engulf and destroy foreign cells, by producing enzymes or other compounds to kill phagocytes or surface structures, such as capsules, that make phagocytosis more difficult to perform. Some pathogens enter host cells to hide from phagocytes or enter the phagocytes themselves with defenses that prevent digestion. Survival inside a phagocyte usually involves reactions that stop the secretion or activity of the enzyme lysozyme, which phagocytes would normally use to degrade the pathogen. Other tactics used by pathogens to sustain pathogenesis by neutralizing the immune response are:

- initiating a hyperinflammatory response to weaken host metabolism

- rapid evolution to change surface structures, thus weakening the capacity of the immune system to attack the pathogen, called *antigenic variation*

- disguise by coating the pathogen cell with host proteins

- destruction of antibodies (immune system proteins designed to inactivate specific pathogens)

The pathogen must call upon these evasive actions to establish itself in the body, either at the site of infection or somewhere else in the body. This establishment of a stable population of pathogens in the host is called *colonization*. Many virulence factors are the same as adherence factors because these qualities help pathogens resist the immune response while trying to colonize the host. Some virulence factors protect the pathogen at the same site of infection. For example, influenza virus changes a variety of surface structures called *antigenic factors* to make itself difficult for the immune system to recognize. By contrast, HIV uses latency (lying dormant in a host cell for long periods) to hide inside T cells after invading the body via the blood. HIV ultimately damages nerve tissue during the progression of AIDS. The capacity of a pathogen to adhere, infect, and colonize a host constitutes the invasiveness of a pathogen.

Every tissue type in the body is vulnerable to some pathogen: blood, lymph, bone, nerve cells, liver, lungs, and so forth. Once a pathogen is inside a target tissue, pathogenesis develops by three main mechanisms executed by the pathogen: (1) using the host's nutrients, (2) causing direct damage to host tissue, or (3) producing toxins.

Different pathogens employ a variety of techniques for robbing a host of nutrients. Iron acquisition by pathogens provides a well-studied example in which a pathogen secretes a substance to gather iron from blood and tissue, thus preventing the host tissue from using this essential nutrient. Most higher organisms have 100 or more enzymes and proteins that require iron to function, such as the protein hemoglobin. The molecular biologists Marcus Miethke and Mohamed Marahiel explained, in their 2007 review article "Siderophore-Based Iron Acquisition and Pathogen Control," "The indirect strategies [of pathogens] of iron acquisition are quite diverse. One of them, which is found in gram-negative bacteria, employs specialized secreted proteins called hemophores to acquire heme [molecular structures that carry an iron molecule] from different sources. . . . In contrast, another indirect strategy is capable of exploiting all available iron sources independent of their nature, thus making it the most widespread and most successful mechanism of high-affinity iron acquisition in the microbial world." Miethke and Marahiel had described compounds called *siderophores,* which many prokaryotes and eukaryotes use to grasp iron molecules. When a pathogen needs iron, it secretes siderophore proteins that take iron away from host cell iron-transport systems; siderophores bind more tightly to iron than the transport proteins.

Pathogens can also cause direct damage to host tissue by a variety of methods, such as using up nutrients inside host cells, excreting toxic waste products, damaging host cell membranes during entry into or exit from the cell, or multiplying inside the host cell until the cell ruptures.

Toxin production is another common mechanism by which a pathogen advances disease inside a host animal. A toxin is a poison produced by a pathogen, which can be either an endotoxin or an exotoxin. Endotoxins constitute part of the pathogen cell and are released when the pathogen dies and lyses. Exotoxins can be excreted by the pathogen during its growth or released upon lyses. Exotoxins, therefore, cause damage to host tissue even after the pathogen has left the body naturally or as a result of drug therapy.

Pathogenesis will be more likely to progress if a pathogen has invasiveness and/or toxigenicity. By employing all of the factors that allow a pathogen to adhere, enter, and spread, the pathogen gives the immune system a very difficult task in protecting the host. Toxigenicity, the capacity to produce a toxin, creates additional challenges for the body because it must expend energy and nutrients to repair tissue damage even while the toxin remains in the body.

PATHOGENESIS AND MEDICAL TREATMENT

Molecular pathology is the study of molecules that affect pathogenesis. Within this discipline, molecular biologists determine the genes that control substances secreted by pathogens for sustaining disease. These scientists have made extraordinary advances in uncovering the details of pathogenesis in hundreds of pathogen species or individual strains within a species. Molecular biologists currently study the genes that control factors in adherence, spread, and virulence. Additional genes in pathogens determine the pathogen's "decision" to turn on or turn off certain activities in pathogenicity. The following list indicates general areas of study known to influence pathogenesis in addition to adherence, spreading, and virulence factors:

- body temperature
- pH of body tissues or fluids
- osmotic pressure (pressure inside a cell relative to the outside)
- available nutrients, especially iron, on or in the host
- infection of a bacterial pathogen by a bacteriophage (virus that attacks bacteria)

Gene analysis laboratories determine the entire base sequences of the genomes of pathogens for the purpose of identifying genes that play a part in pathogenesis. The University of Texas microbiologist and molecular geneticist George Weinstock explained the advantage of genome (a species's entire genetic material) sequencing in a 2000 issue of *Emerging Infectious Diseases:* "If the ultimate aim of pathogen genome sequencing is the development of vaccines, therapeutics, and diagnostics, candidate genes may be identified before the mechanism of infection is understood." Such gene-targeted therapies represent a major step forward in interfering with the course of pathogenesis.

Molecular biologists have identified large segments of deoxyribonucleic acid (DNA) from pathogens containing a high proportion of virulence genes, called *pathogenicity islands*. A microorganism with one or more pathogenicity islands has a greater capacity for virulence than a pathogen without these islands. For example, pathogenicity islands have been detected in several pathogenic *Escherichia coli* strains but are absent from nonpathogenic *E. coli* strains.

See also GERM THEORY; INFECTIOUS DISEASE; KOCH'S POSTULATES; PORTALS; TOXIN; VIRULENCE.

Further Reading

Casadevall, Arturo, and Liise-Anne Pirofski. "The Damage-Response Framework of Microbial Pathogenesis." *Nature Reviews Microbiology* 1 (2003): 17–24.

Henle, Jacob. "Concerning Miasmatic, Contagious, and Miasmatic-Contagious Diseases." In *Milestones in Microbiology*, translated by Thomas Brock. Washington, D.C.: American Society for Microbiology Press, 1961.

Miethke, Marcus, and Mohamed A Marahiel. "Siderophore-Based Iron Acquisition and Pathogen Control." *Microbiology and Molecular Biology Reviews* 71 (2007): 413–451. Available online. URL: http://mmbr.asm.org/cgi/reprint/71/3/413. Accessed January 1, 2010.

Todar, Kenneth. "Mechanisms of Bacterial Pathogenicity." In Todar's Online Textbook of Bacteriology. Madison: University of Wisconsin Press, 2008. Available online. URL: www.textbookofbacteriology.net/pathogenesis.html. Accessed January 7, 2010.

Weinstock, George M. "Genomics and Bacterial Pathogenesis." *Emerging Infectious Diseases* 6 (2000): 496–504. Available online. URL: www.cdc.gov/ncidod/EID/vol6no5/weinstock.htm. Accessed January 7, 2010.

penicillin Penicillin is a natural antibiotic against bacteria produced by species of the mold fungus *Penicillium* and one of the first antibiotics to be used in medicine. Penicillin represents a class of antibiotics called beta-lactam (β-lactam) antibiotics because the molecule's activity depends on an intact nitrogen-containing β-lactam ring structure, believed to give the molecule its antibiotic activity.

Penicillin works by interfering with cell wall synthesis in actively growing bacterial cells. The antibiotic targets multiple sites in the cell wall synthesis process, particularly proteins on the outer layers of gram-positive and gram-negative cells that bind to the antibiotic. These proteins, called penicillin-binding proteins, make the bacteria susceptible to the antibiotic. In general, penicillin is more effective in killing gram-positive bacteria than gram-negative species because of the difference in cell wall composition between these two groups. One of penicillin's activities is to interfere with cell synthesis of peptidoglycan, a large molecule that provides the strength and structure to the cell wall. Because gram-positive bacteria have a peptidoglycan layer many times thicker than that in gram-negative species, penicillin appears to be more effective against gram-positive species.

DISCOVERY OF PENICILLIN

The discovery of antibiotics did not occur all at once. One of the most important medical discoveries of the modern era resulted from a series of revelations by several scientists. In 1871, the British surgeon Joseph Lister (1827–1912), who had proposed a few years earlier the use of antiseptics on surgical patients, noticed that a urine sample that had been contaminated with mold did not grow bacteria. Three years later, another British physician, William Roberts (1830–99), noted that *Penicillium glaucum* mold cultures grown in his laboratory seldom became contaminated with bacteria (see color insert, page C-4 [bottom]). Similar findings occurred in other laboratories: Either a mold contaminant stopped bacterial growth, or a prepared mold culture repelled contamination from bacteria.

The French doctoral student Ernest Duchesne (1874–1912) may have been the first to study penicillin in a laboratory. During the completion of his dissertation, Duchesne noticed that *P. glaucum* inhibited the growth of the bacterium *Escherichia coli*. Testing the idea that the mold could kill pathogens, Duchesne, then, inoculated animals with a virulent strain of the typhoid bacterium *Salmonella typhi*, followed by an inoculation with *P. glaucum*. None of the animals contracted typhoid, and Duchesne suggested in his notes that molds might provide a good treatment for bacterial infections. Others were not as excited about Duchesne's idea; Duchesne enlisted in the French army but died young of tuberculosis, and his laboratory findings were never published.

In the 1920s, the microbiologists André Gratia (1893–1950) and Sara Dath also noticed that mold contamination in *Staphylococcus aureus* cultures caused the gram-positive bacteria to stop growing. Gratia's son, Jean-Pierre Gratia, recounted, in 2000, in a retrospective on microbiology, "At the time of their first studies on antibiosis, Gratia and Dath (1925) mentioned that a *Penicillium* strain exerted a highly bacteriolytic activity against anthrax-causing bacteria *[Bacillus anthracis]*. Gratia's attention was then diverted to the study of 'coli. V.' Unfortunately, a serious illness prevented him from studying this antibiosis due to *Penicillium*. Once back in his laboratory in 1929, he found that his strain had died. The substance produced was thus never identified." Jean-Pierre Gratia lamented that his father had never pursued what might have been the discovery of penicillin.

In 1928, at St. Mary's Hospital in London, the staff scientist Alexander Fleming made a casual discovery in his laboratory. Historians have described Fleming's laboratory as a site of permanent disarray. One of the many petri dishes containing bacteria scattered throughout the laboratory developed mold contamination. During a break, Fleming took from the messy laboratory, mold spores had probably drifted into the laboratory from a mycology laboratory a floor below. Upon returning, Fleming noticed mold contamination in one of the petri dishes containing *Staphylococcus* and observed that bacteria did not grow near the mold colony. Fleming spent the remainder of 1928 studying the mold-bacteria connection and identified the mold as a *Penicillium.* He coined the word *penicillin* for the substance he extracted from mold cultures and showed that it also killed streptococci, pneumococci, gonococci, meningococci, and diphtheria bacilli. Fleming's crude extract could be diluted 1,000-fold and still inhibit susceptible species. Fleming also discovered some of the microorganisms that were unaffected by penicillin: viruses such as influenza and gram-negative bacteria. Fleming published all of these results in a scientific journal, in 1929, but his findings raised little interest.

Alexander Fleming has been described as a dull speaker and uninspiring writer, perhaps explaining why his discovery was unnoticed for a time. Soon after the publication of Fleming's article, Britain entered World War II against Germany, and British doctors knew they needed a drug to stop infections sustained on the battlefield. Even minor injuries received by soldiers in all previous wars had accounted for 50 percent of all deaths. Without a drug to halt the spread of infection into the body, a condition called *sepsis,* field medics would be helpless in saving many of the injured. Few of the doctors realized, however, the potential of penicillin.

Shortly after Fleming's publication, the British surgeon Cecil Paine (1904–94) experimented with applying *Penicillium* extract to patients' eye infections. He had been one of the few scientists who had been intrigued by Fleming's study. The extract cured the infections from pneumococci and gonococci. Paine, however, did not pursue work on the promising potential drug.

Fortunately, Alexander Fleming did pursue unusual experimental outcomes. Fleming tried in vain to purify the extract—the first step in making any drug. By the end of the 1930s, the scientific community had increased their efforts to find better drugs for troops facing imminent war with Nazi Germany. Two chemists working in Oxford, England, the Briton Howard Florey (1898–1968) and the German refugee Ernst Chain (1906–79), rifled through biochemistry journals seeking a compound that offered promising results. When they found Fleming's 1929 article, they added penicillin to their list of compounds to study. By luck, Florey and Chain found in their building a pure culture of *Penicillium notatum* that Fleming had sent to another researcher years before, for an entirely different path of study. Florey and Chain set about growing large amounts of *P. notatum* and extracting the penicillin. The two scientists also had more success in purifying penicillin, and, by 1940, they published results on how penicillin had cured laboratory animals of disease. The chemist Edward Abraham (1913–99) joined Florey and Chain in purifying penicillin, by 1943.

With war under way, Great Britain and the United States needed large-scale production of penicillin for their troops. With considerable coaching from Florey and Chain, who made dangerous crossings between England and the United States, university laboratories and drug companies began to turn out large amounts of penicillin. Fleming offered his insight into the microbial actions of penicillin; Florey and Chain developed the purification and spearheaded large-scale production of the drug. The medical historian Ronald Rubin wrote, in 2007, "from D-Day (June 1944) onward, the rate of Allied soldiers dying from infected war wounds inexorably declined." The material that, 50 years earlier, Ernest Duchesne thought would make a good drug finally became what people would call a *miracle drug.*

In his acceptance speech for the 1945 Nobel Prize in chemistry, Fleming said, "I had the opportunity this summer of seeing in America some of the large penicillin factories which have been erected at enormous cost and in which the mould *[sic]* was growing in large tanks aerated and violently agitated. To me it was of especial interest to see how a simple observation made in a hospital bacteriology laboratory in London had eventually developed into a large industry and how what everyone at one time thought was merely one of my toys had by purification become the nearest approach to the ideal substance for curing many of our common infections." Penicillin's discovery involved luck and perseverance, as well as the intuition of many scientists, in addition to Fleming.

B-LACTAM ANTIBIOTICS

Penicillin was the first β-lactam antibiotic discovered. Other antibiotics containing a β-lactam structure in the molecule have since been found. Penicillin and cephalosporin (spelled with either a *ph* or an *f*) are produced by fungi, and bacteria produce β-lactam antibiotics that gave rise to the synthetic or semisynthetic carbapenems, clavams, monobactams, and nocardicins. Synthetic antibiotics consist of structures made entirely by chemistry; semisynthetic antibiotics are constructed partly from a natural compound and partly from a synthetic structure.

Most new generations of antibiotics used today are either synthetic or semisynthetic. In antibiotics, a generation is a version of the original molecule that has been changed to improve its bacteria-killing performance. The original molecule, such as that made by fungi in Fleming's laboratory, represents the natural form. Changes made to this molecule by a chemist represent a first-generation antibiotic. When the first-generation version receives more modification, the new drug becomes a second-generation form, and so on. The table (bottom, right) describes the main next-generation forms of penicillin.

Chemists began developing next-generation antibiotics for three main purposes: (1) to broaden the drug's spectrum of activity, meaning the breadth of organisms the antibiotic kills; (2) to improve uptake by the body in oral or intravenous form; and (3) to kill bacteria that developed resistance to penicillin.

Cephalosporin produced by the fungus *Cephalosporium acremonium* has a structure similar to penicillin's, but the antibiotic was found to possess less antibacterial activity than penicillin. But because cephalosporin worked better in killing gram-negative bacteria than penicillin, chemists began developing new versions of the antibiotic. As with the penicillins, second, third, and fourth generations of the original structure have been synthesized for the purposes of working better in killing bacteria and attacking bacteria that have developed resistance to previous versions. Although natural cephalosporin C produced by the fungus had less overall antibacterial activity than penicillin, chemists improved its spectrum of activity by making semisynthetic versions, such as the following cephalosporins:

- first-generation—cephalexin, cephalothin, cephazolin
- second-generation—cefonicid, ceforamide, cefuroxime
- third-generation—cefixime, ceftazidime, ceftazoxime
- fourth-generation—cefepime, cefpirome

Bacteria have developed a mechanism for resisting cephalosporins similar to the one they use for neutralizing penicillin's activity. Bacterial resistance to all β-lactam antibiotics results from the ability to destroy the antibiotic before it can move into the cell by crossing the cell wall and cell membrane.

PENICILLIN-RESISTANT BACTERIA

Bacteria resistant to penicillin and other β-lactam antibiotics avoid being killed by these drugs by producing the enzyme β-lactamase. β-lactamase acts by cleaving the carbon-nitrogen bond in the antibiotic's β-lactam ring.

Both gram-positive and gram-negative bacteria of various species produce β-lactamase. Depending on species, bacteria may be induced to produce it (called an *inducible enzyme*) when they detect the presence of an antibiotic, then excrete the enzyme into the surroundings. In other cases, the enzyme is called a *constitutive enzyme,* which is present in the cell at all times but becomes active only when needed. Constitutive β-lactamase works at or near the cell membrane and does not leave the cell.

In *Staphylococcus,* a protein that is a normal constituent of the cell membrane (called an *integral*

Penicillin Antibiotics

Antibiotic	Description
penicillin G	natural, activity confined mostly to gram-positive cocci
procaine penicillin	first-generation combination of penicillin G and procaine, increases retention time in the body
benzathine penicillin	first-generation combination of penicillin G and benzathine further increases retention time in the body
ampicillin	second-generation more effective against gram-negative bacteria than natural forms
amoxycillin	third-generation derivative of ampicillin
carbenicillin	second-generation more effective against gram-negative bacteria than natural forms
carfecillin	third-generation derivative of carbenicillin
methicillin	semisynthetic derivative of penicillin to kill penicillin-resistant bacteria
oxacillin, nafcillin	first-generation derivatives of methicillin

protein) makes up part of the bacterium's control of inducible β-lactamase production. This integral protein is called the *signaler* for the presence of penicillin. When the signaler becomes activated by binding to penicillin, it turns on gene expression of β-lactamase production. Gene expression is the process of turning the information carried in genes into a functioning protein. A second protein, called a *repressor*, normally blocks the gene expression. At the same time that the signaler-penicillin reaction notifies the cell that penicillin is in the vicinity, the repressor also becomes inactivated. The signaler and the repressor work in tandem as a control mechanism, for turning on β-lactamase production, when the cell needs protection from penicillin, and turning it off, when the danger has passed.

All β-lactamases have the same goal of deactivating a β-lactam antibiotic before the substance can kill the cell. The enzymes differ slightly among species in the base sequences in their genes and in the antibiotics against which they work best. Microbiologists have categorized β-lactamases according to the main antibiotics they inactivate. More recent studies of the genes that control β-lactamase and of the enzyme's structure have provided more detail on the enzyme at a molecular level. The table below describes β-lactamase categories based on enzyme structure and, thus, activity against specific antibiotics.

The table indicates that plasmids play an important role in antibiotic resistance. These small circular pieces of deoxyribonucleic acid (DNA), in fact, have been linked with various types of antibiotic resistance, in addition to β-lactam resistance. Different strains and species can transfer resistance among them by sharing plasmids.

Biochemists have invented countermeasures to bacteria's use of β-lactamase. Compounds called β-*lactamase inhibitors* inactivate β-lactamase and, thus, allow the antibiotic to regain activity. Physicians prescribe β-lactamase inhibitors to be given with β-lactam antibiotics to boost the antibiotics' effectiveness.

Bacteria have developed resistance to third- and even fourth-generation penicillins; therefore, microbiologists now seek entirely different antibiotics, rather than planning for additional generations of the drug. The *Staphylococcus* known as methicillin-resistant *S. aureus* (MRSA) has developed resistance against all penicillin generations, so physicians must now treat MRSA infections with the antibiotic vancomycin, which is still effective against this bacterium. Hints of MRSA resistance to vancomycin are beginning to appear in microbiology. If the microorganism develops strong resistance to this antibiotic, doctors will have few or no weapons against MRSA infections.

PENICILLIUM

Penicillium is a fungus found throughout nature and most familiar as the mold that spoils bread and cheese. Most *Penicillium* species cause no direct harm to health, except *P. marneffei,* which causes penicilliosis, a fatal disease in humans and animals, when the organism infects the bloodstream and affects organs. Several *Penicillium* species produce mycotoxins, which are poisons secreted by fungi.

Penicillium also serves useful roles in making cheese and producing antibiotics. In cheese production, *P. roqueforti* is mixed with milk curd and grows, during ripening, to form the typical blue veins of Roquefort cheese.

In addition to penicillin, the mold produces the antibiotics griseofulvin and mycophenolic acid. Griseofulvin, produced by *P. griseofulvum,* inhibits the growth of fungi that cause skin, hair, and nail infections. This antibiotic acts in two ways. First, the antibiotic inhibits the formation of eukaryotic cell structures called *microtubules.* Because microtubules are a component of mitosis, the antibiotic's action stops cell division. Griseofulvin's second mode of action involves

Classes of β-Lactamases

Class	Description	Main Producers
A	characterized by containing the amino active serine at the enzyme's active site (where it binds with the antibiotic); genes carried mainly on a plasmid	gram-positive and many gram-negative
B	requires zinc for activity, mainly against cephalosporins and mainly on plasmids	*Bacillus*
C	similar to type A but with genes held in the chromosome	gram-negative bacteria
D	similar to type A but active against more next-generation penicillins	staphylococci, *Pseudomonas*

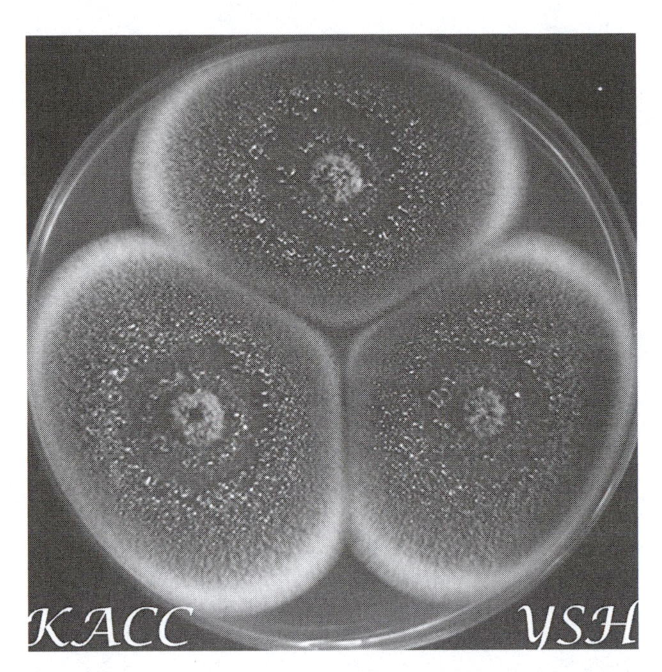

Alexander Fleming has been credited with discovering the antibiotic penicillin when the mold *Penicillium notatum* contaminated bacterial cultures in his laboratory. ***(Michael Gregory)***

interfering with the cell's synthesis of its cell wall. To do this, the antibiotic disrupts the normal deposition of two cell wall constituents: glucan and glycoprotein.

Penicillium species such as *P. brevicompactum* produce mycophenolic acid, which has two purposes in medicine. In 1946, Howard Florey discovered the compound's first use, as an antibiotic that kills *Candida* yeasts and some viruses. The second, more prevalent use of mycophenolic acid is as part of mycophenylate mofetil, a drug presently used in medicine to fight tumor growth and reduce organ rejection in organ transplantation patients.

During *Penicillium*'s growth, the mold produces a network of filaments called *mycelia* that spread over and through substances as the mold reproduces. Mycelia that grow upward remain colorless, but, when growing along a surface or in substratum, they develop muted reds, oranges, yellows, and purples. *Penicillium* releases spores called *conidiospores* from a conidium, a collection of numerous spores at the end of a single mycelium filament, or hypha.

The antibiotic penicillin occupies a significant place in microbiology history. Despite being the first mass-produced commercial antibiotic, the original natural form of penicillin has been largely defeated by penicillin-resistant bacteria. New antibiotics have taken over many of the jobs once performed by penicillin. This antibiotic, nevertheless, has helped scientists learn about antibiotic mode of action and the molecular basis of resistance.

See also ANTIBIOTIC; FLEMING, ALEXANDER; MODE OF ACTION; PEPTIDOGLYCAN; PLASMID; RESISTANCE; SPECTRUM OF ACTIVITY; *STAPHYLOCOCCUS*; SUSCEPTIBILITY TESTING.

Further Reading

Fleming, Alexander. "On the Antibacterial Action of Cultures of a *Penicillium* with Special Reference to Their Use in the Isolation of *B. influenzae.*" *British Journal of Experimental Pathology* 10 (1929): 226–236. Available online. URL: http://202.114.65.51/fzjx/wsw/newindex/wswfzjs/pdf/1929p185.pdf. Accessed December 20, 2009.

———. "Penicillin." Acceptance lecture, Nobel Prize ceremony, Stockholm, Sweden, December 11, 1945. Available online. URL: http://nobelprize.org/nobel_prizes/medicine/laureates/1945/fleming-lecture.pdf. Accessed December 20, 2009.

Gratia, Jean-Pierre. "André Gratia: A Forerunner in Microbial and Viral Genetics." *Genetics* 156 (2000): 471–476. Available online. URL: www.genetics.org/cgi/reprint/156/2/471.pdf. Accessed December 20, 2009.

Perez-Llarena, Francisco J., and German Bou. "β-Lactamase Inhibitors: The Story So Far." *Current Medicinal Chemistry* 16 (2009): 3,740–3,765.

Zhang, H. Z., C. J. Hackbarth, C. M. Chansky, and H. F. Chambers. "A Proteolytic Transmembrane Signaling Pathway and Resistance to β-Lactams in Staphylococci." *Science* 291 (2001): 1,962–1,965. Available online. URL: www.sciencemag.org/cgi/content/abstract/291/5510/1962. Accessed December 20, 2009.

peptidoglycan Peptidoglycan, also called *murein*, is a large molecule that acts as the main structural component of bacterial cell walls. Only members of domain Bacteria synthesize peptidoglycan and contain this substance in the cell. Genera such as *Mycoplasma* and *Ureaplasma* are exceptions and do not use have peptidoglycan because these bacteria lack a cell wall. The archaea contain a variety of complex polymers in the cell walls, but none contains peptidoglycan. The cell walls of algae and plants mainly contain cellulose.

The peptidoglycan polymer is composed of many repeating subunits consisting of the glucose-like compounds N-acetylmuramic acid (NAM) and N-acetylglucosamine (NAG) and several different amino acids. Alternating NAM and NAG units create peptidoglycan's backbone and provide a large portion of the polymer's strength. Three amino acids in peptidoglycan do not occur in proteins: D-glutamic acid, D-alanine, and meso-diaminopimelic acid. The *D* in *D-amino acids* refers to the molecule's structural orientation. The D-amino acids in bacterial cell walls help protect the cell because they resist digestion by peptide-degrading enzymes in nature.

Chains of alternating NAM and NAG connect by peptide (small strands of amino acids) bridges of alternating amino acids. Extensive cross-linking also occurs to make peptidoglycan more like a mesh than

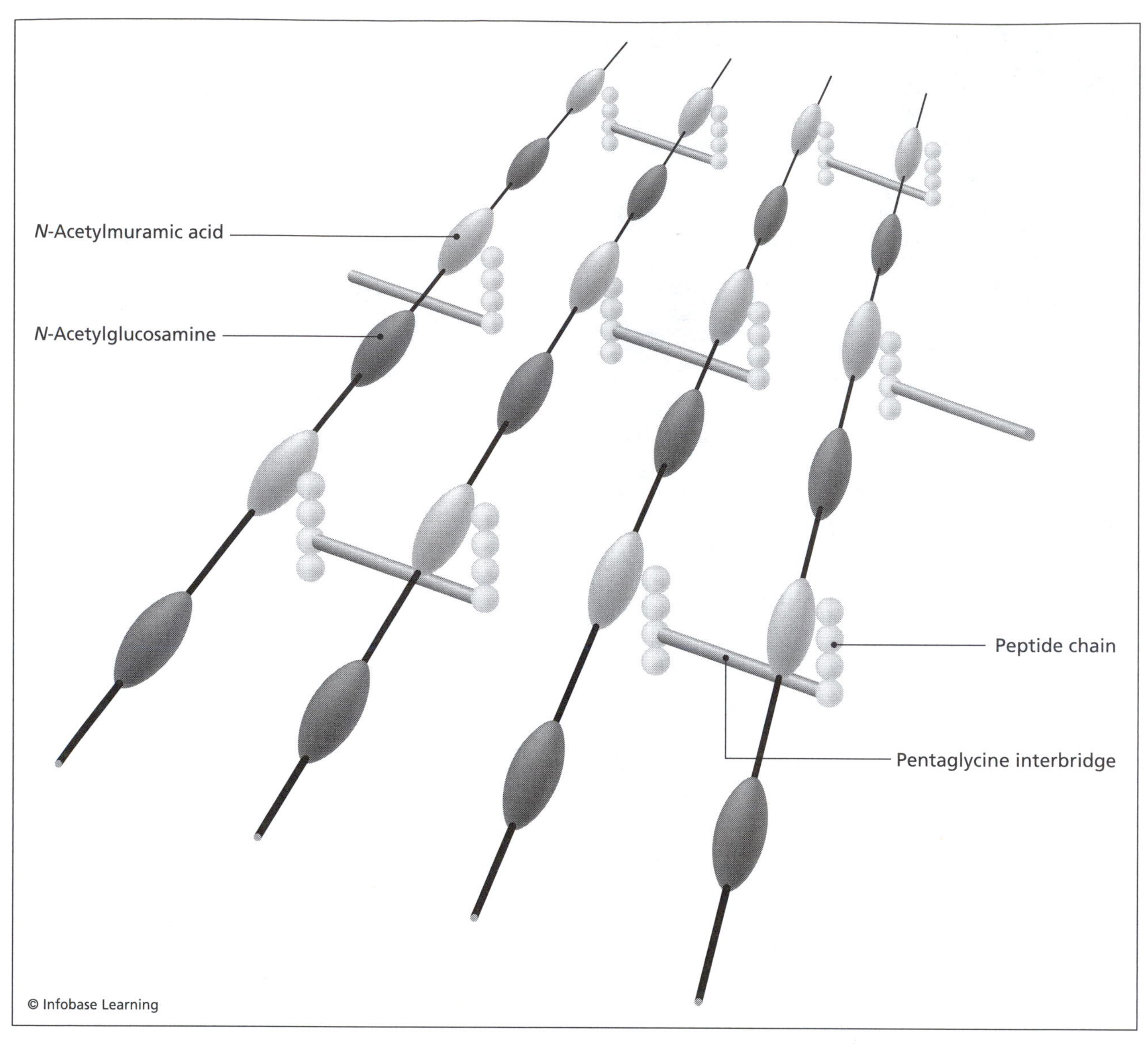

The peptidoglycan molecule contains peptide cross-connections that create a pleated sheet arrangement in the polymer. This structure stiffens the bacterial cell wall, making bacteria very resistant to physical damage.

like an arrangement of orderly chains. The peptide-based linkages are called *peptide interbridges.* When fully composed, the peptidoglycan layer resembles a complex woven cover for the cell.

Genera may differ slightly in their final peptidoglycan structure, but all peptide linkages that connect adjacent chains have tetrapeptide side chains of four amino acids that connect with the NAM units in the backbone.

The peptidoglycan mesh manages to provide bacteria with tremendous strength and yet is porous, allowing materials to flow through to the fragile cell membrane underneath. Peptidoglycan possesses some elastic qualities; it can stretch and bend, but it always returns to give the cell a shape that is distinctive for its genus.

Gram-positive and gram-negative bacteria differ in the way peptidoglycan is oriented in the cell wall. The peptidoglycan layer in gram-positive bacteria measures 20 nanometers (nm) to 80 nm thick and essentially makes up the gram-positive cell wall. Gram-negative cells contain a peptidoglycan less than one fifth that thickness and situated between an outer membrane and the inner cytoplasmic membrane.

PEPTIDOGLYCAN SYNTHESIS

Peptidoglycan synthesis is one of the most intricate processes performed by bacteria. Some of the eight major steps in peptidoglycan synthesis take place inside the cytoplasmic membrane. After making a section of the polymer, the cell transfers it outside the membrane, as a prefabricated house is constructed in pieces, then shipped to where the complete house will be built.

The gram-positive species *Staphylococcus aureus* has been the main model for studying peptidoglycan synthesis. The synthesis begins with two compounds needed for the assembly steps in the cytoplasm: bactoprenol and uridine diphosphate (UDP). Bactoprenol is a large alcohol that carries newly made polymer pieces through the membrane to the cell surface. UDP acts as a base for the assembly steps before handing off each new peptidoglycan segment to bactoprenol. The entire peptidoglycan synthesis takes place in the following eight steps:

1. NAM and NAG are made in the cytoplasm and each attached to a UDP molecule.

2. Enzymes add amino acids in sequence to the UDP-NAM complex to build a five-amino-acid pentapeptide

3. The NAM-pentapeptide is transferred from UDP to bactoprenol at the membrane's inner surface.

4. UDP-NAG adds NAG to NAM-pentapeptide to form the peptidoglycan repeatable unit. Peptide interbridges are added at intervals during this step.

5. Bactoprenol transports the NAM-NAG unit across the membrane to the outer surface.

6. Enzymes attach the new NAM-NAG unit to the preexisting polymer.

7. Bactoprenol returns to the cytoplasm to pick another NAM-pentapeptide complex.

8. Outside the cell, enzymes build cross-links between the peptidoglycan chains in a step called *transpeptidation*.

Transpeptidation adds strength and durability to peptidoglycan but also gives the polymer a modest amount of flexibility.

The cell reuses bactoprenol by adding or subtracting a phosphate (PO_4^{-3}) group to the molecule. Before bactoprenol carries a polymer segment through the membrane to the outer cell surface, an enzyme attaches a phosphate structure called *pyrophosphate* to bactoprenol. When bactoprenol returns to the cytoplasm, it releases the phosphate group.

Gram-positive and gram-negative species use very similar synthesis processes. The two groups differ in the way peptide interbridges are added to the growing structure. In *Mycobacterium*, peptidoglycan uses N-glycollylmuramic acid in place of NAM; the two compounds are structurally similar.

PEPTIDOGLYCAN AND ANTIBIOTICS

Bacteria are susceptible to a group of antibiotics called *beta-lactam* (β-lactam) antibiotics, including the penicillins and the cephalosporins. These antibiotics kill bacteria by interfering with multiple sites on the cell. A well-studied mode of action involves β-lactam interference with the building of the cross-linkages in peptidoglycan, during the polymer's synthesis. The lack of cross-links makes the bacterial cell vulnerable to physical forces and leads to much faster death and lysis than in cells with fully formed peptidoglycan. For this reason, β-lactam antibiotics always work better against actively growing bacteria than cells nearing slower growth phases or death. β-Lactam antibiotics also exert greater activity against gram-positive than gram-negative cells because of the thick peptidoglycan layer that dominates the gram-positive cell wall.

Bacterial resistance to β-lactam antibiotics results either from physical factors or enzymatic activity. Cells with complex outer cell membrane constituents, such as found on gram-negative species, make penetration of the antibiotic to the peptidoglycan layer more difficult than in gram-positive cells. Another large group of antibiotic-resistant gram-positive and gram-negative bacteria contain an enzyme called β-lactamase. Bacteria secrete this enzyme for the purpose of cleaving the active structure in the antibiotic's molecule, called the β-lactam ring, thus inactivating the antibiotic.

Bacteria have developed methods to protect against peptidoglycan destruction, which attest to the critical role of this substance in almost all bacteria.

See also CELL WALL; GRAM STAIN; RESISTANCE.

Further Reading

Todar, Kenneth. "Structure and Function of Bacterial Cells." In Todar's Online Textbook of Bacteriology. Madison: University of Wisconsin, 2008.

Ullrich, Matthias. *Bacterial Polysaccharides: Current Innovations and Future Trends.* Norwich, England: Caister Academic Press, 2009.

photosynthetic bacteria Photosynthetic bacteria conduct the process of photosynthesis, the important metabolic pathway that converts the Sun's energy into chemical energy, which can, then, be used by all living things to sustain life. The main photosynthetic bacteria are the purple photosynthetic and green photosynthetic bacteria and the cyanobacteria.

Bacteria were the first organisms to develop the ability to turn solar energy into chemical energy. Algae and green plants evolved from these ancient photosynthetic precursors. The combined action of photosynthetic bacteria and photosynthetic green plants and algae serves three purposes on a global scale: (1) produces the oxygen in the atmosphere, (2) produces the carbon food source for Earth's microbial and animal life, and (3) helps balance the atmosphere's composition by absorbing carbon dioxide, a process called *carbon fixation*. By converting carbon into different chemical forms and making the carbon compounds available as food for herbivores, omnivores, and carnivores, photosynthetic bacteria contribute to the carbon cycle. The carbon cycle is one of several biogeochemical cycles on Earth that recycle Earth's nutrients through living and nonliving things.

Bacteria do not store large amounts of energy in their compact cells. For this reason, metabolic pathways such as photosynthesis run almost constantly.

PHOTOSYNTHESIS

Photosynthesis is life's first step in synthesizing complex organic compounds from simple inorganic substances. Photosynthesis consists of several steps to achieve the overall conversion of carbon dioxide and water to sugar (glucose). In cyanobacteria, the reaction is the same as it is in green plants and algae:

$$6\,CO_2 + 12\,H_2O + \text{light energy} \rightarrow C_6H_{12}O_6 + 6\,O_2 + 6\,H_2O$$

Two phases make up the process described by the equation: light-dependent reactions and light-independent reactions, or light and dark reactions, respectively.

Photosynthesis's light reactions begin when chlorophyll pigments in bacteria absorb light energy. Cyanobacteria use chlorophyll *a*, as do green plants and algae, for this step. Other photosynthetic bacteria use bacteriochlorophylls.

Energy from light excites electrons in the chlorophylls, beginning a series of steps by which electrons move from compound to compound in a process called *electron transport*, or the *electron transport chain* (ETC). As electrons hop from one pigment in the ETC to the next (sometimes called a *downhill* electron flow), protons move through the cell membrane. A difference in total protons on one side of the membrane versus the other side, which is known as a *proton gradient*, develops. The proton gradient serves as the source of energy generation. The proton movement releases a spurt of energy, which the energy-storage compound adenosine triphosphate (ATP) captures in its chemical bonds. The process by which this energy transfers from the ETC to ATP is called *chemiosmosis*.

Bacteria deal with the electrons that reach the end of the ETC, in either of two ways, by processes called *cyclic phosphorylation* or *noncyclic phosphorylation*. Most photosynthetic use the noncyclic process, which results in the following three end products:

- ATP
- oxygen (O_2)
- nicotimamide adenine dinucleotide phosphate + hydrogen ion (NADP + H^+, or NADPH)

The compound NADPH collects the extra electrons emerging from the ETC.

The dark reactions of photosynthesis involve a metabolic pathway called the *Calvin-Benson cycle*, which begins with carbon fixation. As the five-step Calvin-Benson cycle operates, carbon dioxide (CO_2) molecules are converted into a glucose molecule.

PHOTOSYNTHETIC BACTERIAL GROUPS

Microorganisms that derive energy from sunlight are called *phototrophs*. Sometimes, they are called *photoautotrophs*, because the microorganisms seem to live practically on their own, meaning they exist on sunlight for energy and carbon dioxide for carbon. The cyanobacteria and the purple and green bacteria accomplish this metabolism by different mechanisms.

Two light-capturing systems operate in cyanobacteria: photosystem I and photosystem II. By having both systems, cyanobacteria maximize their efficiency at capturing light of varying wavelengths. Photosystem I contains pigments that absorb long-wavelength light of greater than 680 nm. Photosystem II captures light at wavelengths shorter than 680 nm. Visible light serves as the main energy source for the cell.

Cyanobacteria are aerobic and run noncyclic phosphorylation. The energy-producing reactions generated in electron flow take place inside chloroplasts. Each chloroplast contains rows of membrane folds called *thylakoid membranes*, which act as the site of proton gradients and ATP production.

Cyanobacteria inhabit freshwater and marine environments but have also been isolated from deserts, from glaciers, and on plants. In addition to their crucial contribution to Earth's energy and carbon use, the biologist and author Betsey Dexter Dyer pointed out, "the great evolutionary innovation of cyanobacteria was oxygen-generating photosynthesis. It probably evolved about 3 ½ billion years ago and, once established, became the dominant metabolism for producing fixed carbon (in the form of sugars) from carbon dioxide." Biologists have theorized, as well, that cyanobacteria that infected larger bacterial cells may have been the ancestors of the chloroplasts inside modern green plants and algae.

Purple and green photosynthetic groups, so named for the range of colors produced by their pigments, employ cyclic phosphorylation (although green bacteria can also carry out the noncyclic version). In cyclic phosphorylation, a mechanism exists for taking electrons from the end of the ETC and returning them to the chlorophylls. Green photosynthetic bacteria appear to be slightly more versatile than purple photosynthesizers; green bacteria can use either the Calvin-Benson cycle or the tricarboxylic acid cycle for carbon fixation.

Other major differences between the purple and green bacteria compared with cyanobacteria reside in the role of oxygen in the pathways. Purple and green bacteria lack photosystem II, which is the system needed for producing oxygen from water. Because these bacteria lack this capability, their metabolism is called *anoxygenic photosynthesis*. Rather than use water as a source of electrons, the purple and green bacteria use compounds such as elemental sulfur (S), hydrogen (H_2), and hydrogen sulfide (H_2S).

Purple bacteria inhabit aqueous environments where some sunlight reaches (photic zone), and these microorganisms depend mainly on infrared light. These bacteria also live in some sediments and extreme conditions such as salt lakes and alkaline environments. Green bacteria live in alkaline to neutral hot springs, geysers, and microbial mats consisting of various species with complementary metabolisms. In mat communities, the end products of one type of metabolism feed other species. For example, in microbial mats that lie above hot springs, green sulfur-utilizing bacteria use sulfur compounds produced by other species as part of their energy generation. They, simultaneously, inhabit a layer of the mat that gives them access to sunlight for photosynthesis. In this community, green bacteria participate in the sulfur cycle and have a greater role in Earth's sulfur cycling than in carbon cycling.

The major characteristics of the photosynthetic bacteria are summarized in the table below.

The American microbiologist Howard Gest considered, in his 2003 book *Microbes: An Invisible Universe* the ways in which photosynthesis connects oxygenic versus anoxygenic photosynthesis. "The similarities and differences between the cyanobacteria and the purple bacteria evoke many questions, particularly about biochemical and cellular evolution. It seems very plausible that these two groups of bacteria are related from an evolutionary standpoint. . . . Many considerations indicate that the first photosynthetic organisms on Earth were nonoxygenic bacteria that resembled some of the purple bacteria species now extant. Evolutionary changes in such ancient purple bacteria presumably led to the appearance of oxygenic cyanobacteria, which are regarded as the precursors of green plants." The relationship between purple bacteria and cyanobacteria holds vast information on the evolution of Earth's first respiratory organisms.

Photosynthetic Bacteria

Characteristic	Cyanobacteria	Purple Bacteria	Green Bacteria
general metabolism	aerobic	anaerobic	anaerobic (or aerobic in the dark)
photosynthetic pigments	chlorophyll *a* and phycobiliproteins	bacteriochlorophylls	bacteriochlorophylls
photosystem	I and II	I	I
type of photosynthesis	oxygenic	anoxygenic	anoxygenic
electron donors	H_2O	S, H_2, and H_2S	S, H_2, and H_2S
primary energy conversion compounds	ATP and NADPH	ATP	ATP
carbon source	CO_2	CO_2 or organic compounds	CO_2 or organic compounds

Scientists investigating fossil fuel alternatives have looked at photosynthetic algae and bacteria as possibilities for a new generation of solar energy technologies. In 2009, the biotechnology professor Donald Bryant at Penn State University told *Thaindian News,* "We found that the orientation of the chlorophyll molecules make green bacteria extremely efficient at harvesting light. The ability to capture light energy and rapidly deliver it to where it needs to go is essential to these bacteria, some of which see only a few photons of light per chlorophyll per day."

Energy for human use from photosynthetic bacteria and algae represents an enormous untapped area in ecology and energy sustainability. But many aspects of the energy mechanisms in these bacteria, especially purple and green bacteria, have remained undiscovered. Research on photosynthetic microorganisms may, soon, become the next important breakthrough in microbiology since genetic engineering.

See also BIOGEOCHEMICAL CYCLES; CYANOBACTERIA; GREEN BACTERIA; METABOLIC PATHWAYS; METABOLISM; MICROBIAL COMMUNITY; PURPLE BACTERIA.

Further Reading

Dyer, Betsey D. *A Field Guide to Bacteria.* Ithaca, N.Y.: Cornell University Press, 2003.

Gest, Howard. *Microbes: An Invisible Universe.* Washington, D.C.: American Society for Microbiology Press, 2003.

Hu, Xiche, Ana Damjanović, Thorstem Ritz, and Klaus Schulten. "Architecture and Mechanism of the Light-Harvesting Mechanism of Purple Bacteria." *Proceedings of the National Academy of Sciences* 95 (1998): 5,935–5,941. Available online. URL: www.ks.uiuc.edu/Publications/Papers/PDF/HU98/HU98.pdf. Accessed January 28, 2010.

Thaindian News. "'Green Bacteria' May Be Used to Build Artificial Photosynthetic Systems." Available online. URL: www.thaindian.com/newsportal/health/green-bacteria-may-be-used-to-build-artificial-photo synt hetic-systems_100188456.html. Accessed January 28, 2010.

Ting, Clare S., Gabrielle Rocap, Jonathan King, and Sallie W. Chisholm. "Cyanobacterial Photosynthesis in the Oceans: The Origins and Significance of Divergent Light-Harvesting Strategies." *Trends in Microbiology* 10 (2002): 134–142. Accessed January 28, 2010.

University of California. "Introduction to the Cyanobacteria." Available online. URL: www.ucmp.berkeley.edu/bacteria/cyanointro.html. Accessed January 28, 2010.

University of Hamburg Faculty of Biology. "Bacterial Photosynthesis." Available online. URL: www.biologie.uni-hamburg.de/lehre/bza/photo/ebacphot.htm. Accessed January 28, 2010.

Vermass, Wim F. J. "An Introduction to Photosynthesis and Its Applications." June 12, 2007. Available online. URL: http://bioenergy.asu.edu/photosyn/education/photointro.html. Accessed January 28, 2010.

———. "Photosynthesis and Respiration in Cyanobacteria." Available online. URL: http://dels.nas.edu/banr/gates1/docs/mtg5docs/bgdocs/Photosynthesis_Respiration.pdf. Accessed January 28, 2010.

plankton (singular: plankter) *Plankton* is a term for microscopic organisms, small invertebrates, and fish larvae that float passively near the surface in freshwaters and marine waters. Although some plankton contain appendages, such as flagella, to give them motility, the water's currents provide the main force behind the movement of these organisms.

Plankton forms the foundation of almost all aqueous food chains, meaning they provide food for tiny organisms that are eaten by larger organisms, eaten by still larger organisms, and so on, passing nutrients and energy upward in the food chain.

Various schemes for classifying plankton have been used in biology. The following main groups of plankton are based on the types of organisms:

- bacterioplankton—nonphotosynthetic, heterotrophic bacteria
- phytoplankton—algae, cyanobacteria, and plants or plant debris
- zooplankton—protozoa, larvae, and invertebrates

Heterotrophic metabolism involves the use of a variety of organic compounds for energy and carbon. Bacterioplankton and zooplankton use this metabolism, called *heterotrophy.* Phytoplankton, by contrast, consist of nonanimal organisms that carry out photosynthesis. Phytoplankton play a key role on Earth by capturing the Sun's energy and converting it through photosynthesis to the energy held in the chemical bonds of sugars, mainly glucose. Because of this activity, microbial ecologists define phytoplankton as part of a larger group of organisms, called *producers.* Producers consist of phytoplankton and green terrestrial and aquatic plants. On Earth, producers convert the Sun's energy into a usable form for all life.

Another means of classifying plankton involves the main habitats in which they live: freshwater and marine water. Thousands of species make up these groups. The table on page 624 lists the microorganisms that make up the largest proportion of plankton.

Plankton represent an ancient life-form that contributed to the evolution of aquatic life and probably helped compose Earth's atmosphere. Ocean samples from material known as *radiolarian ooze* have provided fossil evidence of radiolaria dating to

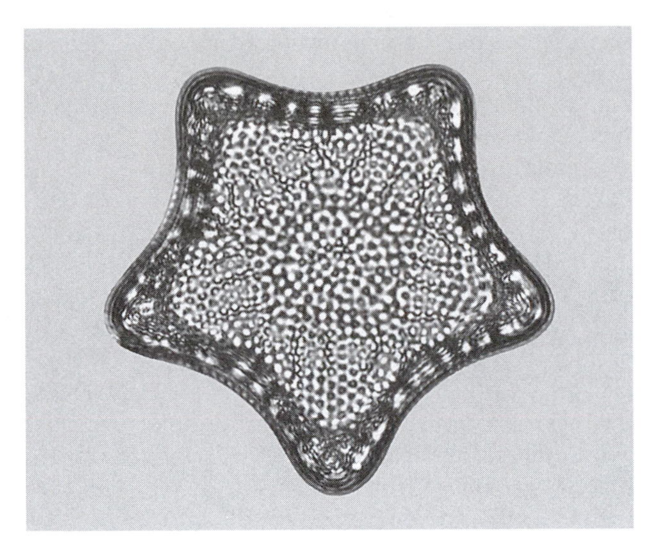

The marine alga *Triceratium* is one of many species that make up plankton. Plankton comprises one of the most numerous populations of living things on Earth, and it forms the foundation of thousands of food chains. ***(Alain Coute)***

the Mesozoic (250 million–65 million years ago) or early Cenozoic (65 million years ago–present) eras.

Marine biologists also classify plankton on the basis of size as a fast way to distinguish organisms in water samples. The main size categories of phytoplankton, for example, are as follows:

- macroplankton = more than 200 micrometers (µm) in diameter
- microplankton = 20–200 µm in diameter
- nanoplankton = 2–20 µm in diameter
- picoplankton = 0.2–2 µm in diameter

The smaller phytoplankton have faster growth rates, in general, than larger cells.

PLANKTON ECOLOGY

Microbial oceanography is the science of collecting, enumerating, and studying microorganisms of the ocean. Scientists have made forays onto the oceans for studying microbial life, since the beginning of the 20th century. In 1931, the marine biologist Henry Bigelow (1879–1967) inspired detailed studies on marine habitats with his book *Oceanography: Its Scope, Problems, and Economic Importance.* Bigelow proposed a new way of studying the ocean by moving beyond simple observations to more in-depth experiments on how ocean organisms live as a community. Bigelow became the first director of the Woods Hole Oceanographic Institution (WHOI), which would become and remains a leader in marine ecology. One of Bigelow's first acts was to hire Selman Waksman (1888–1978), a renowned soil microbiologist from Russia, who had conducted research on antibiotic production by soil species. Waksman set up laboratories at WHOI to approach marine microbial ecology in the manner already established for studying the soil ecosystem.

In the 1940s, the medical microbiologist Claude ZoBell (1904–89) transferred his expertise in medical microbiology to the study of marine microorganisms. His 1946 book, *Marine Microbiology: A Monograph on Hydrobacteriology,* began the modern era of microbial oceanography. This discipline devotes a large quantity of resources to plankton ecology, the activities of plankton communities and the relationship of those activities to the larger freshwater or marine ecosystem.

Marine ecology includes research on the central and unique activities of plankton in carbon and nitrogen cycling. The WHOI graduate student Annette Hynes explained, in 2007, "The ocean is filled with thousands of species of phytoplankton that coexist though they all seem to fill the same niche—a phenomenon known as the 'paradox of the plankton.' Among these paradoxical plankton is *Trichodesmium.* Although they seem very similar, the six species of *Trichodesmium* live together, making the plankton paradox a tantalizing question to apply to these phytoplankton." Upon deeper study, Hynes uncovered small differences among the *Trichodesmium* species, such as that some lived in waters at slightly higher or lower temperatures than the others; some played greater or lesser roles in nitrogen cycling. Earth's phytoplankton, in fact, contribute significantly to these and other biogeochemical cycles by which Earth's biota acquire and reuse nutrients and energy.

Microbial oceanography may, soon, become a critical area of work in marshaling Earth's energy reserves, optimizing nutrient supply, and affecting the status of the atmosphere.

See also ALGAE; BIOGEOCHEMICAL CYCLES; CYANOBACTERIA; HETEROTROPHIC ACTIVITY; MARINE MICROBIOLOGY.

Further Reading

Bigelow, Henry B. *Oceanography: Its Scope, Problems, and Economic Importance.* New York: Houghton Mifflin, 1931.

Hynes, Annette. "Most Ingenious Paraxodical Plankton." *Oceanus Magazine,* 25 November 2008. Available online. URL: www.whoi.edu/page.do?pid=7501&tid=282&cid=53146. Accessed January 30, 2010.

International Nannoplankton Association. Available online. URL: www.nhm.ac.uk/hosted_sites/ina. Accessed January 30, 2010.

Predominant Types of Unicellular Plankton

Plankton	Description
	FRESHWATER
cyanobacteria	large, varied group of photosynthetic bacteria (used to be called *blue-green algae*)
desmids	green algae in division Chlorphyta
diatoms	large, diverse group of brown algae
heliozoea	free-floating protozoa that extend their membranes and cytoplasm to form starlike shapes (sometimes called sun animalcules)
	MARINE
diatoms	same as freshwater but adapted to salt water
dinoflagellates	photosynthetic or heterotrophic microorganisms that can belong to either the algae or protozoa
Globigerina	amoeboid protozoa that can live in subpolar and cold temperate oceans
Globigerinoides	amoeboid protozoa common in tropical and subtropical oceans
Prymnesiophyceae	flagellated protozoa (sometimes classified as algae) with chloroplasts for photosynthesis
Radiolaria	spherical protozoa with a fragile internal shell containing silica

Monterrey Bay Aquarium Research Institute. Available online. URL: www.mbari.org/microbial. Accessed January 30, 2010.

New Hampshire Public Television. "Plankton." NatureWorks, 2010. Available online. URL: www.nhptv.org/NATUREWORKS/nwep6d.htm. Accessed January 30, 2010.

ZoBell, Claude E. *Marine Microbiology: A Monograph on Hydrobacteriology.* Waltham, Mass.: Chronica Botanica, 1946.

plant pathogen A plant pathogen is any bacterium, fungus, or virus that infects plants or trees. Plant species probably have many more infectious agents that cause serious disease than the number of animal pathogens. Various fungi cause devastating diseases in food crops and forests. Blights, rusts, smuts, and rots are major groups of plant diseases caused by fungi. Bacteria cause various leaf spots, wilting, and rotting, and viruses also are associated with dwarfed plants, mottled or diseased leaves, and plant tumors. A group of pathogens called *phytoplasmas* resemble mycoplasmas that cause disease in humans, because both organisms lack a cell wall. The flexible cell structure enables phytoplasmas to infiltrate plant cells, where the phytoplasma lives as an intracellular parasite. Most other plant pathogens are extracellular parasites that cause damage from the outside of the plant cell. Pathogens that live on the surface of the plant body or leaves are called *epiphytes*. Pathogens that, in addition to infecting the plant, infiltrate plant debris and decaying matter in the soil are called *saprophytes*.

Plant, animal, and human health all relate on a global scale. Major outbreaks of plant disease can affect the feeding habits of wildlife and human agriculture. Stresses to one sector of crop production or natural forests and grasslands cause a domino effect on animal life, particularly on animals that are near extinction and human societies living in poverty. The European Commission (EC) has elevated the importance of maintaining plant health worldwide in the face of climate change and its expected effects on plant populations. The EC has stated on its Web site, "Climate change also has effects on the health of the plants and on the world which surrounds them. Considering that climate is the prime determinant of their geographical distribution, various landscapes are likely to change if no measures are implemented. The threats differ according to the regions concerned; threats can result from more frequent droughts and/or floods, hotter summers leading to an increase in fungi, and warmer winters causing a proliferation in the population of insects." After insect infestation, fungi present the greatest threat to plant health and crop yields, especially because many fungi are carried from plant to plant on insects.

Plant diseases range from parasites that infect algae, to grasses, simple plants, complex plants (bushes, shrubs), garden and ornamental varieties, agricultural food and feed crops, and natural grasslands, wetlands, and forests. Each of these areas represents a specialty in the broad science of plant pathology.

PLANT DISEASES CAUSED BY FUNGI

Fungi make up the largest group of plant pathogens. Fungi infiltrate plant tissue by developing a network of threadlike filaments, called *hyphae*, that lengthen as the fungus grows. A mycelium results when the hyphae grow into an extensive tangled mass of branched hyphae. Most pathogen fungi are from the soil, and disease transmits through roots, in soil particles carried in the air, in splashing water, or on the bodies of insects, wildlife, domesticated animals, or people.

As do plants, proliferation, of fungi depends on climate. Fungi can exist in moist or dry conditions in temperate climates and are especially prevalent in places containing large amounts of decaying organic matter. These conditions characterize most gardens, meadows, and forests, so it is not surprising to learn that fungi are called ubiquitous organisms in nature. Among microorganisms, only fungi live in a wider range of habitats on Earth.

Most fungal pathogens can infect more than one plant species or type of plants; other pathogens are specific to a certain host plant. For example, many powdery mildews exist, but each has a particular plant species that the mildew infects. With few exceptions, fungal plant pathogens belong to differ genera than the fungi that infect animal life. *Fusarium* is an important exception, because this genus can cause a fatal sepsis (infection in the blood or tissues) in humans.

The table on below describes some of the important plant diseases caused by fungi and general groups of host plants. Because plant diseases are extensive, the table lists diseases by type rather than by individual disease names.

Fungal Plant Pathogens

Pathogen	Plant Affected	Disease
Alternaria	vegetables	leaf blight
Armillaria	fruit trees	root rot
Fusarium	vegetables	leaf wilting
Plasmopara	grapes	downy mildew
Phythium	most plants	seed rot, root rot
Phytophthora	vegetables, fruit trees	root rot
Puccinia	grains	leaf rust, stem rust
Rhizoctonia	herbs	root rot, stem rot
Rhizopus	fruits, vegetables	soft rot
Ustilago	grains	corn smut

Gardeners, agriculture specialists, and forest managers learn to recognize the major signs of many fungal diseases. Plant pathology lacks the sensitive diagnostic tests that human and veterinary medicine use in identifying a pathogen. Most plant disease diagnosis requires some experience with healthy and sick plants and a knowledge of the disease's outward signs. Diagnosis of fungal diseases thus relies on skill in evaluating the following for signs of disease:

- general view of a stand of trees or plot of crops to recognize differences between healthy and potentially sick plants
- close-up views of leafs, stems, roots, flowers, and fruits for spots, wilting, color, and health in the vessels of cut-open stems
- progression in a population over time to note differences in general health and growth rate and yield
- sending of samples laboratories to find parasites in the plant tissue

Samples should be wrapped in dry paper towels, not put into plastic bags, and individual samples should be kept separate. Collecting an entire plant (if feasible), including roots, is best for helping an agricultural expert make an accurate diagnosis.

The following common disease groups have recognizable hallmarks that help in diagnosing a disease:

- cankers—sores on stem or twigs
- powdery mildews—white-gray powder on leaf and stem surfaces; weakened plants
- rots in root or crown—stunted growth, yellowing
- rots, soft—softening of fruit or vegetable from within
- rusts—orange or red spores on leaves; usually develop in cool weather
- spots and blights—various-colored spotting on leaves
- smuts—dark brown or black blotches on leaves, stems, or fruit

These general signs are common for fungi but may also apply to bacterial diseases or nonpathogenic mold growth

PLANT DISEASES CAUSED BY BACTERIA

Most bacterial plant pathogens are gram-negative rods with the major exception *Streptomyces*, which is a filament-producing genus that resembles mold. Bacteria enter plants through wounds or through natural openings, such as stomata, and can multiply very quickly in plant tissue and can produce toxins that poison the plant independently of bacterial damage. (Stomata are small openings in the under part of leaves used by the plant for gas exchange.) The culturing of resistant plant varieties and use of antibiotics have been useful for reducing the incidence of bacterial plant diseases.

The table below presents common general types of bacterial plant diseases.

Pierce's disease results from an infection of the vessels called *xylem* that carry fluids in plants. Leaves turn brown from the outer edge inward and eventually drop off. The pathogen *X. fastidosa* is transmitted by the insects sharpshooters and leafhoppers.

Phytoplasmas are bacterialike microorganisms that cause a group of diseases called *yellows*. The main pathology in yellows relates to a gradual disintegration of the internal vasculature of the plant.

PLANT DISEASES CAUSED BY VIRUSES

Viruses usually enter plant tissue through wounds to the plant body and survive only by infecting the plant cells. Virus transmission between plants can be by insects, worms, or fungi or inside bacteria that infect alone or infect by transmission in an insect.

As do the viruses that infect humans and other animals, plant viruses contain genetic material in the form of single- or double-stranded ribonucleic acid (RNA) or deoxyribonucleic acid (DNA). After a virus has penetrated a plant cell, it takes over the cell's DNA replication machinery for the purpose of making new virus particles. Viruses move among plant cells by using natural channels in the cell wall called *plasmodesmata*. Plasmodesmata allow plant cells to communicate by exchanging molecules, but even small viruses cannot fit through these channels. Plant viruses overcome this hindrance by producing a protein that modifies the plasmodesmata and allows viral genetic material to migrate to the adjacent plant cell.

Viruses cannot be seen with light microscopy because of their small size (25–300 nanometers); only electron microscopes provide views of viruses. Botanists and agricultural scientists usually diagnose viral diseases by noting the plant's outward signs of disease. The signs of disease result from three main effects of the virus: (1) cell death caused by virus multiplication; (2) hypoplasia, or retarded growth; and (3) hyperplasia, or excessive cell growth in an area, resulting in tumors. Viral diseases can sometimes be controlled by removing the infected part of the plant.

Viruses produce a variety of signs that resemble bacterial infections. Viruses are very host-specific; the viruses that infect plants do not infect humans or animals. The table on page 627 lists important viral plant pathogens.

Because of easy methods for removing infected areas from the plant body, many viral diseases can be controlled. Where surgery is not practical, vital diseases have cost millions of dollars in crop losses. For viruses carried by insects, insect control programs, simultaneously, solve virus infections.

Bacterial Plant Pathogens

Pathogen	Plant Affected	Disease
Agrobacterium tumefaciens	fruit trees	crown gall
Erwinia carotovora	fruits, vegetables	soft rot
E. tracheiphila	cucumbers, melons	vascular wilting
Pseudomonas fluorescens	potatoes	soft rot
P. syringae	vegetables, tobacco	leaf spots
Streptomyces scabies	potatoes	potato scab
Xanthomonas campestris	cereals, fruits, cabbages	leaf spots, black rot
Xylella fastidosa	grapes	Pierce's disease

Viral Plant Pathogens

Pathogen	Plant Affected	Disease
Caulimovirus	cauliflower	stunted growth
Phytoreovirus	rice	tumors
Potexvirus	potato	stunted growth
Potyvirus	beans	leaf chlorosis (yellowing)
tobacco mosaic virus (tobamovirus)	tobacco	leaf chlorosis
wheat mosaic virus (furovirus)	wheat	dwarfed plants, mottled leaves

Plant pathology gives agriculture, botany, and microbiology daunting challenges in identifying the many undiscovered plant diseases, while controlling the existing known diseases. Plant pathogens belong to a separate population of organisms that deserve continued research, diagnostics, and treatment because of the vital role played by plants in the biosphere.

See also FUNGUS; PATHOGEN; PATHOGENESIS.

Further Reading

European Commission. Available online. URL: http://ec.europa.eu/index_en.htm. Accessed January 28, 2010.

Ohio State University Extension. "Plant Pathology." In Master Gardener Training Manual and Online Resource Center. Available online. URL: www.hcs.ohio-state.edu/mg/manual/path2.htm. Accessed January 1, 2010.

Oklahoma State University. Department of Entomology and Plant Pathology. Available online. URL: http://entoplp.okstate.edu. Accessed January 28, 2010.

U.S. Department of Agriculture. Available online. URL: www.usda.gov/wps/portal/usdahome. Accessed January 1, 2010.

plaque A plaque is a small area of clearing in a dense bacterial culture grown on agar caused by the lytic activity of a virus on specific host cells. The formation of plaques in agar cultures of bacteria has been used for determining the presence and concentration of bacteriophages. Plaque formation has also been adapted to biotechnology for tracking the presence or absence of specific genes in deoxyribonucleic acid.

The term *plaque* also refers to a bacterial community attached as a thin film to the surface of teeth. This community, referred to as dental plaque to prevent confusion with viral plaques in agar, consists of a heterogeneous mixture of bacteria embedded in polysaccharides and constituents from saliva. Dental plaque is also a type of biofilm, which is any microbial film that attaches to a surface, is intermittently or continually submerged in liquid, and contains organisms that interact in a community.

VIRUS PLAQUE ASSAYS

Microbiologists use three types of plaque assays, each of which serves a different purpose. First, plaques can be used as an aid in virus cultivation and recovery of a desired virus particle. Second, plaque formation serves as part of a procedure called *titer determination*. The titer procedure is used for two main reasons: (1) to determine the concentration of viruses in a suspension and (2) to determine the strength or lethal dose (LD_{50}) of a virus. Finally, plaque formation serves in bacteriophage typing, the use of a known bacteriophage to identify an unknown bacterium.

All plaque assays consist of the same basic elements. A plaque develops when a virus infects a cell, replicates, and lyses the cell. If the host cell is a bacterium, the virus specific only for bacteria is called a *bacteriophage*. A single lysed bacterial cell can liberate hundreds of new virus particles, called *virions*, that infect other bacteria near the original host cell. Microbiologists use a technique called an *agar overlay* to make plaques easier to see after the virus has been incubated with host cells. An agar overlay consists of a petri dish half-filled with sterile solidified agar medium. A microbiologist, then, mixes bacteria and the diluted virus in medium containing a reduced amount of agar. This agar forms the overlay, when poured on top of the solidified layer. The soft overlaid agar restricts the movement of virus particles and, thus, produces distinct, clear plaques after incubation. Microbiologists count the number of plaques, called *plaque-forming units* (PFUs), in the same way they count bacterial colonies on agar.

The virologist Teri Shors wrote, in 2009, "Plaques are visualized by staining cells with dyes such as neutral red or crystal violet. Theoretically, each virus in the original inoculation gives rise to a clearing (plaque) or plaque forming unit (PFU). . . . In other words, a plaque assay measures infectivity. . . . It is possible that viral particles are present in a given sample but are not infectious. More specifically, then, a plaque assay measures only the number of infectious particles in a sample, not the total number of particles." Viruses only affect living organisms when they infect the host, so microbiologists generally do not care about noninfectious viruses.

The plaque assay for determining the concentration of virus particles in suspension begins with 10-fold serial dilutions of the suspension. The microbiologist inoculates a constant volume of each dilution to overlay agar plates. After the overlay agar solidifies, the plates incubate for several hours at about 99°F (37°C). After incubation, the microbiologist counts the number of PFUs at each dilution to determine the number of infectious viruses in the original suspension. For example, duplicate plates that each received 0.1 milliliter (ml) of the 1×10^{-5} dilution have 43 and 40 plaques. The average number of plaques equals 41.5, so the number of virus particles in the original suspension is:

$$(41.5 \times 10^5) \times 10 = 4.15 \times 10^7 \text{ per ml}$$

The plaque assay can be used for bacteria infected with bacteriophages and animal cells infected with animal viruses.

Virologists determine the strength or LD_{50} of a virus by a variety of similar plaque assays. LD_{50} is the dose of pathogen that kills 50 percent of a test population. The Reed-Muench assay determines virus strength by calculating the proportion of positive results (virulent virus) to the total of positive and negative (avirulent virus). This assay can use plaque formation as a method for determining the amount of viruses in the test's dilution series.

Bacteriophage typing refers to the identification of an unknown bacterial species by reacting it with bacteriophages of known specificity. For example, a microbiologist prepares a test plate of the unknown bacterium by spreading the bacterial suspension in a thin film over an agar surface in a petri plate. The plate can, then, be marked off in sections, and each section receives a small drop of different bacteriophages. The table on page 629 describes a hypothetical bacteriophage typing assay for four gram-negative rod-shaped enteric bacteria. In this example, a microbiologist divided the agar plate into quadrants for testing the unknown bacterium against four different bacteriophages.

In bacteriophage typing, a bacterial strain can be susceptible to more than one bacteriophage. For example, a bacterium's susceptibility may be designated as 10/4210/106, meaning the particular bacterial strain is susceptible to bacteriophages 10, 4210, and 106. Any collection of strains in the same species that has this same susceptibility pattern is termed a phagovar, or phagotype.

Medical microbiology and food microbiology use phagovars to study clinical isolates and food-borne pathogens, respectively. Clinical microbiologists might determine that an unknown bacterium isolated from a patient belongs to a phagovar known to be susceptible to a certain antibiotic. The antibiotic becomes the main treatment method for stopping the patient's infection. Food microbiologists can use phagovars to track specific strains of bacteria from a food-borne outbreak to a particular food processing plant.

DENTAL PLAQUE

Dental plaque contains as many as 400 different species of bacteria of such a high cell density that it may hold up to 10^{10} bacteria in 1 milligram (mg). Plaque, in fact, is probably the densest collection of bacteria on the body.

Dentistry colleges study the development, species, and persistence of plaque because it is antagonistic to good oral health. Dental plaque leads to two health conditions when a person fails to follow oral care recommendations: dental caries and periodontal disease. Dental caries, or cavities, begin with lesions in the tooth outer surface called enamel with progressively worse decay of the tooth. Periodontal disease encompasses various diseases of the gum-tooth interface, the gingival pockets beneath the gums at the tooth's neck, and the jawbone in serious cases.

Oral care specialists recommend that everyone use the actions listed to reduce plaque buildup and subsequent oral health problems:

- brushing at least twice daily
- flossing at least once daily
- use of mouthwashes
- limited intake of high-sugar foods and drinks
- no smoking
- plaque removal by a dental care professional twice yearly

Despite the preventive measures listed here, dental plaque can be a persistent problem. The main reason for plaque's tenacity relates to its structure. As do most biofilms, dental plaque develops features that make it difficult to dislodge from a surface once a community of bacteria has attached to the surface.

Dental plaque formation begins with the deposit of pellicle to a clean tooth enamel surface. Pellicle constituents arise mainly from saliva and include the enzymes amylase and lysozyme, the large proteins albumin and immunoglobulin, and mucin, which is a protein with numerous sugars attached. The pellicle forms a sticky surface on which bacteria called *early colonizers* attach, including *Streptococcus gor-*

Bacteriophage Typing

Bacteriophage	Specificity	Reaction	Result
A	*Citrobacter freundii*	no plaques	negative
B	*Escherichia coli*	no plaques	negative
C	*E. coli* O157:H7	plaques present	identified as *E. coli* O157:H7
D	*Shigella flexneri*	no plaques	negative

donii, S. mitis, S. mutans, S. oralis, S. sanguis, Actinomyces naeslundii, and *A. viscosis.* The *Actinomyces* and *S. gordonii* adhere to the pellicle more strongly than other oral bacteria, and these bacteria serve as the foundation layer of the biofilm.

Additional bacteria attach to the foundation bacteria. As the layer thickens, bacteria called *late colonizers* attach. The main late colonizers of dental plaque are the anaerobes *Bacteroides, Fusobacterium, Haemophilus, Propionibacterium,* and *Veillonella.* Late colonizers might have little or no contact with the bacteria at the underlying layers. In thick plaque, the anaerobes seek microscopic microenvironments where oxygen is at low levels. The entire plaque community and smaller spaces within the plaque represent the phenomenon of microenvironments. These places offer narrowly defined conditions for growth, where only a limited number of microorganisms can thrive.

The members of an established plaque community secrete substances that help hold the community together and to the surface, especially to resist being washed away by beverages or saliva. The streptococci produce enzymes that connect glucose molecules from the diet into polymers that support the biofilm. For example, these bacteria produce various glucans, or branched polysaccharides, that serve more than one purpose. Glucans store food in the matrix, protect the plaque from washing away, and protect the plaque members from harmful chemicals, such as toothpaste. From the perspective of plaque bacteria, toothpaste is a harmful chemical.

When sugars are trapped in plaque, the bacteria metabolize it and produce lactic acid and smaller amounts of other organic acids as wastes. Saliva would normally neutralize the acids, but saliva does not penetrate thick plaques. The acid demineralizes the tooth enamel by drawing out calcium, and caries can soon develop in the weakened tooth.

Severe cases of plaque lead to various periodontal ailments, such as the following:

- gingivitis—inflammation of the gums
- periodontitis—inflammation of the gums and subgingival areas with tissue breakdown
- periodontosis—bone destruction

Relationships between plaque and periodontal diseases have been difficult to define because plaque is an ever-changing community that differs in composition from person to person. Most plaque communities contain the main species mentioned here, but hundreds of other bacteria can make up a greater or lesser proportion in different people. Plaque bacteria live under the influence of factors from the host and from other bacteria. For example, a person might make less lysozyme than someone else, thus differing in the buildup and persistence of plaque. Certain bacteria in plaque might inhibit other bacteria, and this, too, contributes to a person's unique plaque composition.

Carol Potera wrote in *Microbe,* in 2009, "More than 700 types of microorganisms grow in the mouth, many of them helping to form plaque along tooth surfaces. Within minutes after a professional hygienist removes plaque, however, it begins reforming." Some people wage a losing battle against plaque because of their unique oral chemistry and because of plaque's resiliency. Few chemicals have been found to remove plaque while being safe for oral use. The meat tenderizer ficin has been shown to repel the early colonizers in plaque formation. For reasons unknown, ficin's effect fades over time, and the plaque community becomes resistant to it.

Dental plaque ranges from an annoyance to a more serious health threat, but dentistry has made only modest progress in fighting this form of biofilm.

See also AGAR; BACTERIOPHAGE; BIOFILM; IDENTIFICATION; MICROENVIRONMENT; VIRULENCE; VIRUS.

Further Reading

Haake, Susan K. "Microbiology of Dental Plaque." Available online. URL: www.dent.ucla.edu/pic/members/microbio/mdphome.html. Accessed January 4, 2010.

Potera, Carol. "A Potpourri of Probing and Treating Biofilms of the Oral Cavity." *Microbe,* October 2009. Available online. URL: www.microbemagazine.org/index.php?option=com_content&view=article&id =974:antifungal-compounds-from-seaweed-show-antimalarial-potential&catid=296:current-topics&Itemid=378. Accessed January 4, 2010.

Reed, L. J., and H. Muench. "A Simple Method of Estimation of Fifty Per Cent Endpoints." *American Journal of Hygiene* 27 (1938): 493–497.

Shors, Teri. *Understanding Viruses.* Sudbury, Mass.: Jones & Bartlett, 2009.

plasmid A plasmid is a linear or circular piece of double-stranded deoxyribonucleic acid (DNA) in the cell cytoplasm and distinct from the chromosomal DNA in the nucleoid of gram-positive or gram-negative bacteria or the nucleus of eukaryotes.

The plasmid reproduces independently of chromosomal DNA, and while chromosomal DNA is indispensable for species survival, many species can survive without plasmids. The daughter cells can, however, inherit plasmids from the parent. Plasmids can insert their genes into the host cell's DNA, or, in some cases, plasmids called *episomes* exist in the host cell with or without the need to become part of the chromosome.

Bacterial plasmids have received the most study in microbiology, particularly because many bacterial plasmids confer antibiotic resistance and may also carry virulence factors in pathogens.

Plasmid naming uses several conventions. The most common designations for plasmids relate to one of the following four characteristics:

- colicin plasmids are made by certain enteric bacteria such as *Escherichia coli* and *Shigella* (a colicin is a type of bacteriocin, a substance made by a bacterium to kill other closely related bacteria)
- antibiotic resistance plasmids carry genes that enable bacteria to survive the activity of antibiotics
- recombinant plasmids are made by combining genes from two unrelated organisms
- miscellaneous named plasmids receive arbitrary names not related to the other three naming systems

The table on page 631 gives examples of plasmid names.

Copy number refers to the number of plasmids typically present in a particular bacterial strain. Some plasmids, such as R plasmids, number only one to three copies in most cells that carry them. Col plasmids, by contrast, can reach copy numbers as high as 40 per cell. Host cells, furthermore, can carry more than one type of plasmid.

Plasmids are much smaller than chromosomal DNA. Molecular biologists describe the size of plasmids in terms of kilobase pairs (kbp), which equals 1,000 basic units of DNA called *nucleotides* (paired because DNA is a double-stranded molecule). DNA or plasmid size units common in molecular biology are:

- bp = base pair = one pair of complementary bases (adenine paired with thymine or cytosine paired with guanine)
- kpb = kilobase pairs = 1,000 bp
- Mb = megabase pairs = 1,000,000 bp
- Gb = gigabase pairs = 1,000,000,000 bp

Some well-studied plasmids have the following kilobase pair (kbp) sizes:

- ColV-K30 = 2 kbp
- ColE1 = 9 kbp
- SAL = 56 kbp
- R100 = 90 kbp
- Ti = 200 kbp
- CAM = 230 kbp

Plasmids contain few genes compared to chromosomal DNA; plasmids usually contain no more than 30 genes. Every plasmid contains a replicon, which is a section of the DNA that serves as the origin of replication. Chromosomal DNA has its own unique replicon, explaining why chromosomal and plasmid DNA can replicate independently of one another.

Plasmids' importance in biology relates to their ability to move from the native cell into another cell, called the *host.* Plasmids can, then, give the host characteristics that the host species would not normally possess, such as resistance to an antibiotic or ability to degrade a pesticide.

In nature, a host cell might receive an advantageous attribute from a plasmid, such as the examples shown in the table, or no perceivable advantage. A host cell that derives no benefit from a plasmid will probably turn on systems that rid it of the invasive

Plasmid Naming

Naming Convention	Plasmid Type	Example
colicins	Col plasmids	ColE1, ColB-K98, Collb-P9
antibiotic resistance	R plasmids	RP4, R1, R100
F factor	fertility factor	F factor (for DNA transfer by cell conjugation)
none	virulence plasmids	K88, pZA10, Ti, ColV-K30
none	metabolic plasmids	CAM (camphor degradation) SAL (salicylate degradation) TOL (toluene degradation) sym (symbiosis with plants in nitrogen fixation)

DNA. Bacteria possess enzymes called restriction endonucleases, which act in replicating cellular DNA but also cleave any foreign DNA that infects the cell. Since restriction endonucleases are as efficient at cleaving host DNA (a necessary step in DNA replication) as they are at destroying virus or plasmid DNA that has entered the cell, bacteria must protect their native DNA from harm. Host cells do this by putting a methyl group (CH_3) onto adenine or cytosine bases in the DNA. These groups help the restriction endonuclease distinguish between native DNA, which must not be destroyed, and foreign DNA, which is potentially dangerous and should be destroyed. This method of distinguishing self from non-self DNA is called *restriction modification*.

Bacteria use other techniques for removing plasmids, such as curing. In curing, the host cell turns on mechanisms that interfere with plasmid replication. Even though this plasmid will pass to successive generations each time cells divide, the plasmid concentration in the total population of cells becomes increasingly dilute. Only by blocking the plasmid's replication can curing rid a species of an unwanted plasmid.

RESISTANCE FACTORS

Bacterial R plasmids (or R factors) have become important in medicine because they contain genes that encode for enzymes capable of destroying or neutralizing antibiotics. R plasmids have two features that aid in developing antibiotic resistance: (1) location in a transposon and (2) combination with conjugative factors. Transposons are sections of DNA that can move from one site to another in a DNA molecule. When transposons also carry a resistance gene, the transposon can carry the gene from plasmid to chromosome or chromosome to plasmid, increasing the likelihood of resistance building within a species. Transposons also interfere with drug treatments that target a pathogen's DNA.

Conjugative factors, or fertility factors, contain genes that give cells the ability to develop sex pili and perform conjugation. The pili are thin, short tubes that connect two bacterial cells, to allow the exchange of genetic material. This form of gene transfer helps bacteria of the same species or different species pass R factors among them. The genes associated with conjugation are collectively called the *resistance transfer factor* (RTF).

The rise of antibiotic resistance in bacteria has prompted research on the types, numbers, and mechanisms of R factors. Researchers have, thus far, revealed the capabilities of R factors to move through the environment. In 2000, the microbiologists Malik Ajamaluddin, Mohn Khan, and Asad Khan wrote in the *Indian Journal of Clinical Biochemistry,* "R-plasmids that transfer antibiotic resistance are common in the non-pathogenic *Escherichia coli* of the gastro-intestinal tract of human beings and domestic animals, which in turn may enter into sewage. Therefore we have isolated 30 *Escherichia coli* isolates from hospital sewage in Aligarh city. These isolates were tested for their resistance and sensitivity against 10 antibiotics. Ninety percent isolates showed resistance against ampicillin and sulphamethizole. Of the total 30 *E. coli* isolates 86.6 percent were resistant to erythromycin and rifampicin but none of them was resistant to kanamycin and streptomycin." Many similar studies have demonstrated how R factors can access environmental populations of microorganisms.

The study of R factors has been advancing quickly, with new plasmids being discovered every year. The table on page 632 contains some well-known R factors that have helped increase antibiotic resistance to the crisis levels it has reached today. In most cases, an R factor gives an organism resistance to multiple

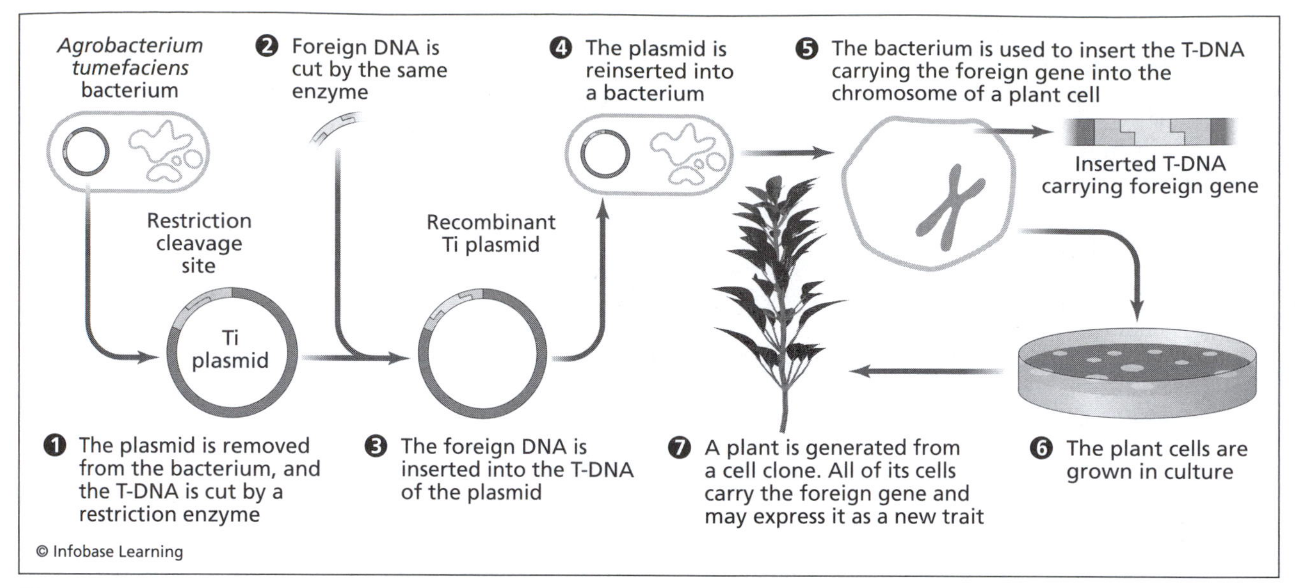

The Ti plasmid of *Agrobacterium tumefaciens* serves as a vector for delivering genes into plant cells. Similar techniques in genetic engineering have been applied to agriculture and marine biology. ***(RE is restriction endonuclease)***

antibiotics, called multidrug resistance (MDR), rather than resistance to a single antibiotic. Resistance to ampicillin implies that the host organism is already resistant to ampicillin's precursor, penicillin.

The RP4 plasmid also carries a fertility factor. The R6 plasmid offers hosts the unusual ability to resist damage from mercury in addition to multiple antibiotics.

METABOLIC PLASMIDS

Metabolic plasmids carry genes for functional enzymes that help the host use certain compounds in nutrition. These plasmids offer an advantage in allowing a microorganism to obtain carbon and energy from some substances that are toxic in the environment, such as the organic solvent toluene and other aromatic hydrocarbon solvents. In addition to toluene, camphor, and salicylate, metabolic plasmids have been discovered that enable bacteria to degrade oil hydrocarbons, octane, naphthalene, nicotinic acid, sugars, and the herbicide 2,4-dichlorophenoxyacetic acid (2,4-D).

Metabolic plasmids have an obvious value for bioremediation of polluted environments. Recombinant DNA technology involves the insertion of genes from an organism into a different, unrelated organism. By inserting into a cell a plasmid that confers the ability to degrade a pollutant, bacterial populations can be put to work in pollution cleanup. These bioengineered microorganisms would be expected to contain safety mechanisms to prevent the uncontrolled release of genetically modified organisms (GMOs) into the environment. One safety tactic is suicide gene, which automatically destroys a GMO after a period in the environment.

Plasmids have additional applications in agriculture. Bacterial plasmids have been used for putting new genes into plants that improve the yield, environmental tolerance, or disease resistance of crops.

R Factors

Type	Common Hosts	Antibiotics
R1	gram-negative bacteria	ampicillin, chloramphenicol, kanamycin, streptomycin, sulfonamides
R6	*E. coli, Proteus mirabilis*	chloramphenicol, kanamycin, neomycin, streptomycin, sulfonamides, tetracyclines
R100	*E. coli, Proteus, Salmonella, Shigella*	chloramphenicol, kanamycin, streptomycin, sulfonamides, tetracyclines
RP4	*Pseudomonas* and other gram-negative bacteria	ampicillin, kanamycin, neomycin, tetracyclines

The discovery of plasmids and techniques to manipulate them into and out of organisms will continue to expand the capabilities of agriculture, industry, and medicine.

See also BACTERIOCIN; BIOREMEDIATION; GENE TRANSFER; MORPHOLOGY; RECOMBINANT DNA TECHNOLOGY; RESISTANCE.

Further Reading

Ajamaluddin, Malik, Mohn A. Khan, and Asad U. Khan. "Prevalence of Multiple Antibiotic Resistance and R-Plasmid in *Escherichia coli* Isolates of Hospital Sewage of Aligarh City in India." *Indian Journal of Clinical Biochemistry* 15 (2000): 104–109.

Funnell, Barbara E., and Gregory J. Phillips, eds. *Plasmid Biology.* Washington, D.C.: American Society for Microbiology Press, 2004.

Higgins, N. Patrick, ed. *The Bacterial Chromosome.* Washington, D.C.: American Society for Microbiology Press, 2004.

Pandey, Gunjan, Debarati Paul, and Rakesh K. Jian. "Conceptualizing 'Suicide Genetically Engineered Microorganisms' for Bioremediation Applications." *Biochemical and Biophysical Research Communications* 237 (2005): 637–639.

Shukla, O. P., U. N. Rai, Smita Dubey, and Kumkum Mishra. "Bacterial Resistance: A Tool for Remediation of Toxic Metal Pollutants." International Society of Environmental Botanists. *EnviroNews,* April 2006. Available online. URL: http://isebindia.com/05_08/06-04-2.html. Accessed January 28, 2010.

Tortora, Gerard J., Berdell R. Funke, and Christine Case. *Microbiology: An Introduction,* 10th ed. San Francisco: Benjamin Cummings, 2009.

Willey, Joanne, Linda Sherwood, and Chris Woolverton. *Prescott, Harley, Klein's Microbiology,* 7th ed. New York: McGraw-Hill, 2007.

Yoon, Kyung Pyo. "Stabilities of Artificially Transconjugated Plasmids for the Bioremediation of Cocontaminated Sites." *Journal of Microbiology* 43 (2005): 196–203. Available online. URL: www.msk.or.kr/jsp/downloadPDF1.jsp?fileName=p.196-203.pdf. Accessed January 28, 2010.

pneumonia Pneumonia is an acute lower respiratory infection of the lungs (pulmonary disease) caused by a variety of bacteria, viruses, and fungi (rare). Pneumonia involves inflammation of the alveoli of the lungs and may infiltrate interstitial spaces, which are areas between tissues or organs. The inflammation can spread to both lobes of the lung; *lobar pneumonia* is the term often used to describe inflammation in one lobe, and *double pneumonia* involves both lobes. Pneumonitis is an inflammation of the lung from irritants other than infectious agents.

Pneumonia is the leading cause of death in children worldwide, killing about 1.8 million children every year. In the United States, about 4.5 million people contract pneumonia each year, and the disease is the sixth-most-common cause of death and the most common cause of death due to infectious disease.

The disease in children can be caused by microorganisms that are part of a child's normal flora in the nose. Transmission occurs by the transfer of nasal discharge to others by sneezing or by way of an inanimate object, called a *fomite,* such as a kitchen counter. Inhalation to the lungs is the pathogen's main portal of entry into the body, but the blood and lymph also serve as secondary routes. Pneumonia's main symptoms are fever, chills, aches, malaise, difficulty in breathing, and chest pain. The disease often starts with a dry cough that progresses to a cough containing bloody sputum.

Debilitating diseases increase the risks of catching pneumonia, similarly to other infectious diseases. Chronic lung disease or other chronic diseases and immunocompromised conditions predispose adults and children to pneumonia. The immobile and the elderly have a higher risk of infection, which is due to fluid buildup in the lungs.

Diagnosis of pneumonia involves noting the classic signs of disease, signs of blood in the sputum, and microbiological examination of a tracheal specimen. Clinical microbiologists use culture and/or serology methods for identifying the pathogen. X-rays detect the infiltration of the infection in the lung tissue.

Pneumonia caused by bacteria can be successfully treated by giving antibiotics. Treatment for viral and fungal pneumonias is more difficult. Many patients are hospitalized and receive supplemental oxygen with antibiotic treatment for potential secondary infections, that is, infections that start after the primary pneumonia has begun. Vaccines are available for some forms of pneumonia.

Pneumonias belong to categories based on either pathology or the causative organism. The table on page 634 describes common types of pneumonias according to pathogenesis, which is the manner in which a disease progresses in the body.

Classifying pneumonias by the pathogen that causes the primary infection helps physicians determine the best treatments for the disease. Clinical microbiologists conduct tests on the microorganisms they have isolated from sputum samples. Once the causative pathogen has been identified, an appropriate course of treatment can begin. Any microorganism recovered from the lower respiratory tract is a potential pathogen, because this organ should not have microbial populations (although the lungs are not necessarily sterile). The main tests performed

Types of Pneumonia

Type	Description of the Pneumonia
abortive	mild with brief course
aspiration	caused by inhalation of gastric contents or other fluids
community-acquired	various infections in outpatients
embolic	after embolization (intentional obstruction of a blood vessel to remove plaque or perform other repairs) of a pulmonary blood vessel
eosinophilic	infiltration of the lungs with eosinophil white blood cells
fibrous	accompanied by formation of scar tissue
giant cell	infiltration of lungs with immune system giant cells, often associated with cases of childhood measles
hypostatic	in elderly or bedridden patients confined to the same position for long periods
interstitial	infiltration of the areas between tissues (interstitium)
lipoid	caused by aspiration of oils
nosocomial	occurs after 48 hours of hospitalization
secondary	occurs in connection with another disease, such as diphtheria, plague, or typhoid
walking	mild condition attributed to an unidentified bacterium or virus
woolsorter's	pulmonary anthrax disease

on sputum or tracheal samples for determining the possibility of pneumonia are Gram stain, type of streptococcus hemolysis on blood agar, yeast and mold tests, and specific tests for *Staphylococcus, Haemophilus, Neisseria, Moraxella,* and *Mycobacterium*. The table on page 635 (top) describes pneumonia categories based on the main infectious agents that cause this disease.

The Centers for Disease Control and Prevention (CDC) offers information on how people can reduce their chance of contracting pneumonia. Pneumonias associated with health care settings have become so common that the CDC places them into one of three groups: (1) health care–associated pneumonia (HCAP), (2) hospital-associated pneumonia (HAP), and (3) ventilator-associated pneumonia (VAP). HCAPs can occur when outpatients visit doctors' offices, outpatient clinics, long-term care facilities, nursing homes, and dialysis centers. HAPs make up a large proportion nosocomial infections

VAP is a serious condition caused by the use by a hospital patient of an endotracheal tube and artificial ventilator. *S. aureus* and *P. aeruginosa* can develop biofilms on the inner surfaces of invasive equipment and heighten the risk of infection in the lower respiratory tract. A biofilm is a layer of microorganisms acting as a community, which forms a strong attachment to a submerged surface. VAPs can lead to acute respiratory distress syndrome (ARDS) or acute lung injury (ALI), either of which can be fatal.

International agencies and individual nations have made concerted efforts to control all types of pneumonia. A major part of this initiative involves pneumonia vaccine development to address the various forms of this disease. The CDC has explained, "Access to vaccines and treatment (like antibiotics and antivirals) can help prevent many pneumonia-related deaths. Pneumonia experts are also working to prevent pneumonia in developing countries by reducing indoor air pollution and encouraging good hygiene practices." In the United States, several vaccines are available for diseases that often involve pneumonia, such as the following:

- pneumococcus
- *H. influenzae* type b (Hib)
- pertussis (whooping cough)
- Varicella (chicken pox)
- measles
- seasonal influenza
- H1N1 influenza (2009 strain)

Pneumonia preventive measures resemble preventatives for most other upper respiratory tract infections. Tips for preventing the transmission of pneumonia are:

- regular, proper hand washing
- cleaning hard surfaces that are touched repeatedly by many people (doorknobs, faucets, refrigerator handles, countertops, etc.)
- coughing or sneezing into a tissue, elbow, or sleeve

Causes of Pneumonia

Type of Pneumonia	Type	Description
aspergillosis	fungus	lower respiratory infection of *Aspergillus fumigatus* and other species
atypical, Friedländer's	bacterium (*Klebsiella pneumoniae*)	usually a complication of chronic lung disease
Haemophilus influenzae	bacterium	common cause of primary pneumonia in children and secondary in adults who have preexisting viral lung infection
Legionella	bacterium	legionnaires' disease
pneumococcal	bacterium	most common form of bacterial pneumonia; *Streptococcus pneumoniae* carried in the upper respiratory tract; incubation time one to two days
Pneumocystis carinii (also called *P. jirovecii*)	fungus	common secondary infections in AIDS patients
Staphylococcus aureus	bacterium	often interstitial; causes yellow and sometimes bloody sputum
Streptococcus pyogenes	bacterium	primary or secondary and often leads to empyema (fluid accumulation)
tuberculosis	bacterium	*Mycobacterium* infection
viral	virus (adenoviruses, coxsackieviruses, parainfluenza, respiratory syncytial virus)	common form in small children

- remaining home from school or work when sick
- avoiding others who have obvious signs of illness

Additional important considerations in pneumonia control involve programs to quit smoking, control of diabetes, and control and treatment of human immunodeficiency virus (HIV) infections.

VETERINARY PNEUMONIAS

Several contagious lower respiratory infections affect cattle and other livestock, horses, and pets. Bacterial, viral, and fungal forms of pneumonia in veterinary medicine have similar causative agents to those in humans, but fungal pneumonias may be more prevalent in animals than in humans. The table (right) describes important pneumonias seen in veterinary medicine.

Veterinary pneumonias receive treatments similar to those for the human forms of the disease.

Pneumonia may be overlooked by the public as a global health problem, especially because HIV/AIDS, tuberculosis, diabetes, heart disease, and cancer consume research dollars. Pneumonia is, nonetheless, a critical health concern that demands serious research for preventions and treatment.

Veterinary Pneumonias

Name	Description
aspiration	any animal receiving contamination from faulty dosing via esophagus
chlamydial	*Chlamydia* infection in cats, calves, sheep, piglets, foals, goats
hypostatic	fluid accumulation due to immobilization in any animal
mycotic	fungal infection mainly from *Cryptococcus neoformans, Histoplasma capsulatum, Coccidioides immitis, Blastomyces dermatiditis, Pneumocystis carinii, Aspergillus,* or *Candida;* occurs mostly in small animals and pets

See also *ASPERGILLUS*; BIOFILM; IDENTIFICATION; NORMAL FLORA; SEROLOGY; *STREPTOCOCCUS*.

Further Reading

British Society for Antimicrobial Chemotherapy. "Nosocomial Pneumonia—Introduction." Available online. URL: www.bsac.org.uk/pyxis/RTI/Nosocomial%20pneumonia/Nosocomial%20pneumonia.htm. Accessed January 28, 2010.

Centers for Disease Control and Prevention. National Center for Infectious Diseases. "Pneumonia." Available online. URL: www.cdc.gov/ncidod/Diseases/submenus/sub_pneumonia.htm. Accessed January 28, 2010.

———. "Pneumonia Can Be Prevented—Vaccines Can Help." *CDC Features,* 27 October 2009. Available online. URL: hwww.cdc.gov/Features/Pneumonia. Accessed January 28, 2010.

Global Pneumonia Summit. Available online. URL: http://worldpneumoniaday.org. Accessed January 30, 2010.

World Health Organization. "Pneumonia." Available online. URL: www.who.int/mediacentre/factsheets/fs331/en. Accessed January 27, 2010.

polymerase chain reaction Polymerase chain reaction (PCR) is an in vitro procedure that makes unlimited copies of a gene from a small starting quantity of the gene. PCR begins with a small sample of deoxyribonucleic acid (DNA), which can be as low as microgram (µm) amounts, and makes more copies, a process called *amplification.* Biologists use amplification to produce DNA quantities large enough for further analysis. Prior to PCR technology, DNA extracted from small amounts of microorganisms, the environment, or crime scenes could not be analyzed because of insufficient quantities.

Currently, PCR is used in medicine to detect pathogens, genetic diseases, and abnormalities of the blood or muscle. Food microbiologists and environmental scientists also employ PCR to detect foodborne pathogens and fecal contaminants in soil and water, respectively.

In the mid-1980s, the Cetus Corporation scientist Kary Mullis (1944–) developed the PCR technique by exploiting three features in biology: (1) DNA is a double-stranded molecule, in which the different base units (nucleotides) in each ladder bind in complementary fashion; (2) DNA unwinds when warmed up (called *denaturation*) and spontaneously reconnects (called *renaturation* or *annealing*) when cooled; and (3) extremophile microorganisms called *thermophiles* produce enzymes that work at the high temperatures needed for denaturing DNA.

The PCR technique consists of the following basic steps for making large quantities of a gene:

- synthesize single-strand DNA fragments with base sequences identical to the sequences of the target gene in natural DNA
- add the fragments, called *primers,* in excess amounts to the small piece of DNA to be amplified, plus the enzyme DNA polymerase and excess base units, to the mixture
- heat the mixture to denature the DNA and separate the two individual strands
- cool the mixture to allow primers to attach to complementary sections of the DNA strands
- allow time for DNA polymerase to extend the primers and synthesize copies of the target DNA containing the desired gene
- repeat the cycle

Because each heating-cooling cycle in PCR doubles the amount of the desired gene, exact copies of the gene increase in number exponentially. Each PCR cycle, furthermore, occurs in a few minutes, so that millions of gene copies can be made in an hour. Web sites such as the Dolan DNA Learning Center's provide interactive tools for plotting the increase in the desired gene by PCR cycling over 30 cycles.

Most PCR techniques require no more than 25 cycles. From a single piece of target DNA, 25 cycles produces an estimated 33 million copies of the gene in about one hour. Thirty cycles yield about one billion copies. Automated PCR cyclers now reduce the workload of technicians, who need only prepare the starting mixture of DNA, enzyme, primer, and bases to begin the procedure.

PCR PROCEDURE

DNA of all living things contains four nitrogen-containing bases: adenine, guanine, cytosine, and thymine. Whether in nature or when using a PCR machine called a *thermocycler,* DNA assembles in the same way: The sugar ribose is added to each base, then three phosphate groups (PO_4^{-3}) to the sugar base.

DNA polymerase synthesizes new DNA but can do this only from an existing piece. For this reason, the starting mixture must include the primer. To amplify DNA using PCR, a scientist must know at least part of the base sequence in the fragment. Knowing a small part of the target DNA allows an appropriate primer to be assembled. In addition to DNA polymerase and primer, a technician adds four nucleotides that construct the backbone of all DNA molecules. The table on page 638 describes synthesis of DNA's nucleotides

Technicians add the primer and the nucleotides in quantities much larger than the target DNA. By adding extra amounts of these components, they improve the efficiency of DNA synthesis during the cooling phase for two reasons: (1) Extra primer raises the chance

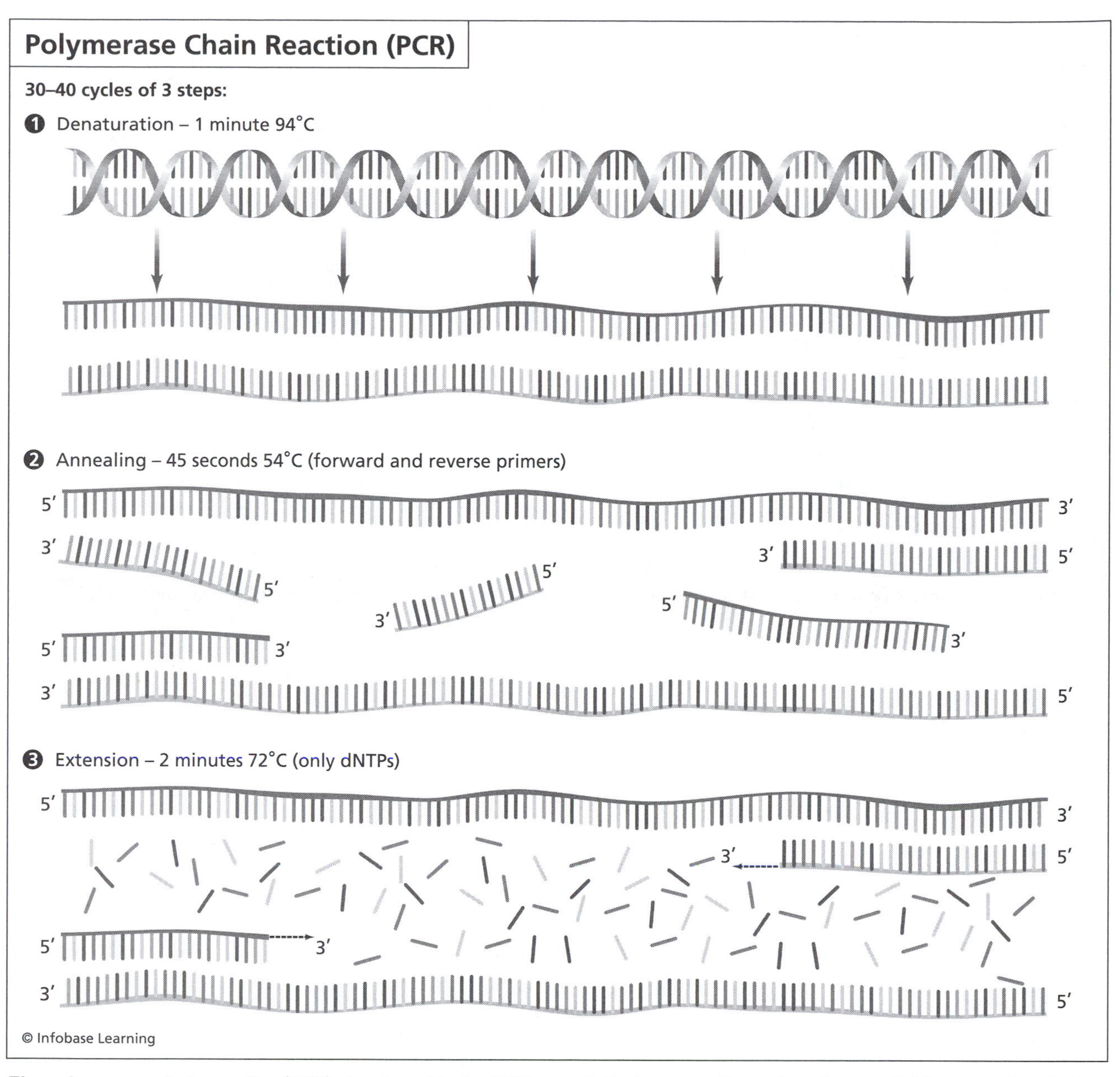

The polymerase chain reaction (PCR), developed in the 1980s, revolutionized genetic engineering and DNA sequencing. PCR remains a fast and easy way to make large amounts of DNA copies from a tiny starting amount of DNA.

that DNA single strands will bind with primer during the cooling step rather than with each other, and (2) excess nucleotides assure that the reaction does not slow down. For example, a mason can build a brick wall faster if supplied with a big stack of bricks than if each new brick must be located from somewhere else.

The starting mixture, thus, contains the following components:

- small amount of target DNA
- excess quantities of a DNA primer of known sequence
- excess quantities of dATP, dCTP, dGTP, and dTTP
- DNA polymerase.

Biologists use the following standard cycling conditions for the PCR method:

- denaturation step—201°F (94°C) for 15 seconds
- renaturation step—154°F (68°C) for 60 seconds

Modifications to the standard PCR method have been developed so that ribonucleic acid (RNA) can be transcribed into DNA and, then, the DNA can be amplified. A technique called *quantitative PCR* estimates the amount of original DNA.

Real-time PCR (RT-PCR) is the newest major advance in PCR technology. RT-PCR enables a scien-

DNA Components

Base	Step 1	Nucleoside	Steps 2, 3, and 4	Nucleotide
adenine	+ ribose	deoxyadenosine	+ 3 PO_4^{-3}	deoxyadenosine triphosphate (dATP)
cytosine	+ ribose	deoxycytidine	+ 3 PO_4^{-3}	deoxycytidine triphosphate (dCTP)
guanine	+ ribose	deoxyguanosine	+ 3 PO_4^{-3}	deoxyguanosine triphosphate (dGTP)
thymine	+ ribose	deoxythymidine	+ 3 PO_4^{-3}	deoxythymidine triphosphate (dTTP)

tist to monitor the amount of DNA being produced as it happens. The main RT-PCR method uses a tool called a *Taq*Man probe. The probe consists of a short strand of nucleotides with the same purpose as the primer in conventional PCR. The *Taq*Man probe also contains a visible signal-producing reporter dye, on one end, and a quenching dye, on the opposite end of the strand. The two dyes are located close to each other when the strand is short. In this situation, the quenching dye blocks the ability of the reporter dye to give off a signal. As the strand lengthens during amplification, the quenching dye gradually unblocks the reporter dye. As a result, the reporter dye's signal increases in amplitude as the number of DNA copies increases. An electronic monitor shows the DNA amount increasing as the PCR cycling progresses.

PCR BACTERIA

The *Taq* portion of *Taq*Man originates in either of two thermophiles that serve as the sources of DNA polymerase that works at the elevated temperatures of the PCT reactions: *Thermus aquaticus* and *Thermococcus litoralis*. DNA polymerase from mesophile organisms, which grow only at temperatures between 40°F and 120°F (4.4–50°C), cannot withstand PCR temperatures. The discovery of the PCR techniques, therefore, relied on these enzymes, now referred to as the Taq polymerases (for *T. aquaticus,* the first organism used for PCR technology).

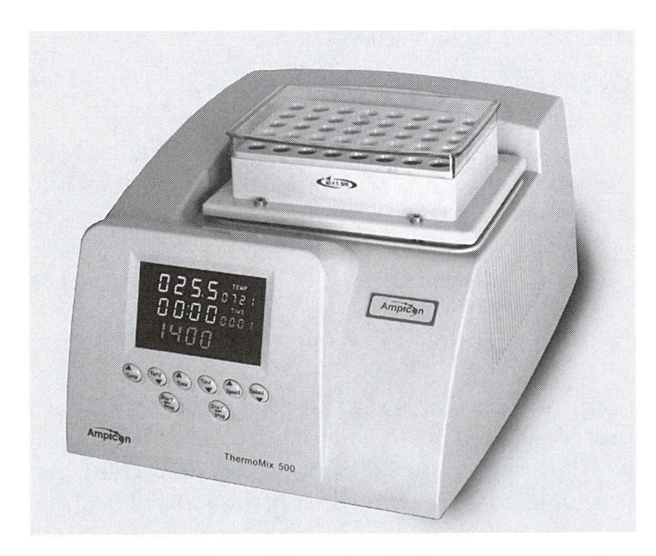

Thermocyclers raise and lower incubation temperatures to automate the polymerase chain reaction (PCR). ***(PCRStation)***

Thermophiles can be isolated from hot springs and hydrothermal vents located on the ocean floor. Most thermophiles contain cell walls and cell membranes of special composition to maintain fluidity among the membrane fats and protect the cell contents.

Kary Mullis recalled in his lecture when receiving the 1993 Nobel Prize in chemistry, "I had solved the most annoying problems in DNA chemistry in a single lightning bolt. Abundance and distinction. With two oligonucleotides, DNA polymerase, and the four nucleosidetriphosphates I could make as much of a DNA sequence as I wanted and I could make it on a fragment of a specific size that I could distinguish easily. Somehow, I thought, it had to be an illusion. Otherwise it would change DNA chemistry forever." The PCR technique is in fact simple and uses straightforward logic that surely has made scientists wonder why they had not thought of it first.

See also DNA FINGERPRINTING; EXTREMOPHILE.

Further Reading

Cold Spring Harbor Laboratory's Dolan DNA Learning Center. Available online. URL: www.dnalc.org/resources/animations/pcr.html. Accessed January 24, 2010.

Mullis, Kary B. "The Polymerase Chain Reaction." Nobel lecture presented in Stockholm, Sweden, December 8, 1993. Available online. URL: http://nobelprize.org/nobel_prizes/chemistry/laureates/1993/mullis-lecture.html.

———, ed. *The Polymerase Chain Reaction.* Boston: Birkhäuser, 1994.

Polymerase Chain Reaction (PCR). Pasadena, Calif.: Sumanas, 2006. Available online. URL: www.sumanasinc.com/webcontent/animations/content/pcr.html. Accessed January 28, 2010.

portals The term *portals* refers to the routes taken by pathogens to enter a host and to leave a host. Portals of entry are the sites pathogens use to enter the body; portals of exit are the ways in which a pathogen leaves an infected person's body.

Portals belong to the broader subject in infectious disease of *pathogenicity,* the ability of a microorganism to cause disease. Without a preferred portal of entry, a pathogen cannot infect and establish a stronghold in the body. In this situation, the pathogen does not cause a high risk for disease. Portals of entry are important in pathogenicity, also, because they determine how easily a pathogen can move from an infected person to a healthy person, the process of transmission.

After a pathogen accesses the body through a portal of entry, it begins the pathogenesis process, which has the following general steps: (1) adherence to host tissue; (2) invasion of body fluids or tissue; (3) development of mechanisms for evading the host's defenses, called *virulence factors;* (4) replication; and (5) development of disease symptoms.

PORTALS OF ENTRY

A pathogen must enter the body before it can cause disease. The portals through which pathogens enter can be categorized by tissue type or by system in the body. Tissue types identified as main portals of entry are mucous membranes, skin, and internal tissue, including blood. Mucous membranes are layers of specialized tissue that cover epithelial cells exposed to the environment. Pathogens reach mucous membranes and internal tissue by entering the respiratory system, digestive tract, or urogenital system. In most infections, pathogens, first, contact mucous membranes at the following sites in each of the five systemic portals of entry, also called *routes of infection:*

- respiratory system—nasal passages, trachea, alveoli, and bronchi in the lungs
- digestive tract—mouth, stomach, small and large intestines
- urinary tract—urethra, bladder, kidney
- genital tract—vagina, cervix, fallopian tubes, ovaries, urethra
- skin—skin, scalp, outer ear canal, lips, nails

Mucous membranes protect the eyes, respiratory tract, digestive tract, and urogenital surfaces. Mucus is a thick substance consisting of water, salts, and glycoproteins, which give mucus its slippery, viscous feel. The glycoproteins, called mucin, contain two factors that combat pathogens that try to enter: immune cells called *macrophages* that digest microorganisms and nutrient-absorbing compounds. Nutrient-absorbing lactoferrin collects iron molecules so that the pathogens cannot access this nutrient.

Mucous membranes and skin also contain similar defensive cell layers; in mucous membranes, the cell layer is called mucosal-associated lymphoid tissue (MALT). MALT possesses mechanisms for engulfing and digesting foreign microorganisms, a process called *phagocytosis.* Other immune system cells and antibodies also migrate to the site and support the MALT in defending against invasion by a pathogen.

The skin creates a physical barrier to pathogen entry. The skin is one of the largest organs in the body, and, when healthy and unbroken, it prevents entry of most microorganisms. Some fungi (dermatophytes) have enzymes that allow them to enter the skin barrier and grow in the epithelial layers. Other than this exception, the skin is the body's strongest external barrier.

Many pathogens gain entry into the body by going directly through the skin into underlying layers or the blood. Pathogens use this portal of entry, called the *parenteral route,* in the following events: bites, burns, cuts, punctures, surgery, or split skin due to dryness.

Most infectious diseases begin when a pathogen traverses the mucous and skin barriers. Dermatitis and conjunctivitis offer two common exceptions, because the diseases occur in the skin or mucous membrane, respectively. Various dermatitis infections can occur when a healthy person touches affected skin of an infected person. Fungi often cause this form of dermatitis. Conjunctivitis is a highly contagious viral or bacterial infection of the eye's mucous membranes.

Pathogens that enter the respiratory or digestive tract or the urogenital system often cause disease at those sites because the pathogen first encounters a mucous membrane at those sites. For example, influenza enters the body by infecting the respiratory tract, and this is where the disease develops. *Salmonella* is a food-borne pathogen that has contact with the body inside the digestive tract and causes disease in digestive organs. Sexually transmitted diseases (STDs) behave in similar fashion by causing infection at the site where they first have contact with mucosal cells (individual cells of the mucous membrane).

Some diseases follow a second pattern of pathogenesis by causing disease at a place unrelated to the route of infection. For example, human immunodeficiency virus (HIV) uses the mucous membrane of the urogenital system but causes disease in the nervous and immune systems. HIV and most other pathogens have a preferred portal of entry regardless of the organ where the infectious disease will develop.

PREFERRED PORTAL OF ENTRY

Many pathogens must use their preferred portal of entry, or they will not cause disease. For example, a cold virus on the skin does not cause a cold, but the same virus causes health problems when it enters the nasal passages. By contrast, the anthrax bacterium *Bacillus anthracis* can cause disease when using the respiratory tract, the digestive tract, or the parenteral route. The table below gives examples of familiar diseases and the pathogens' portal of entry.

Portals of entry relate to the infectious dose of a pathogen, which is the lowest number of pathogen cells or virus particles that usually cause disease in a healthy person. For a pathogen to cause disease, it must overcome the host's skin, mucous membranes, and immunity defenses and must infect in numbers large enough to sustain an infection.

PORTALS OF EXIT

Pathogens leave the body by specific routes called *portals of exit*. Portals of exit play a role in disseminating disease because by releasing pathogens into the environment, the host spreads disease. Epidemiologists use their knowledge of common portals of exit to determine how a disease spreads through a population, called *transmission*. Information about a pathogen's portal of exit helps public health officials make recommendations on how to prevent infection.

In 2009, the H1N1 influenza virus reached pandemic proportions. A pandemic is an epidemic that has spread worldwide. Epidemiologists began studying the characteristics of the virus's portal of exit and its relation to rate of transmission. Medical care providers and the public already knew that influenza's portal of exit is the respiratory tract. In late 2009, Seth Segel, an executive in the medical supply industry, offered advice on the use of face masks to stop transmission. Segel had explained the importance of portals of exit when he said, "We know that masks provide one of the simplest, most cost-effective infection prevention tools in the fight against spreading the flu. . . . The more people who wear masks, the lower the risk of transmitting disease, especially in densely populated areas." Managing portals of exit is almost as important to disease management as disease prevention and treatment.

Pathogens sustain their numbers in a host population by using portals of exit that usually relate to the portal of entry. Thus, the respiratory and digestive tracts and the urogenital system act as common portals of exit. The pathogen exits the body in any of the following substances: blood, feces, semen, sputum, tissue, vaginal secretions, and wound discharges such as pus.

The table on page 641 lists types of diseases and the pathogens' portal of exit.

Bloodborne diseases use portals of exit that involve medical devices such as dental instruments, colonoscopes, and needles when these items have

Portals of Entry

Mucous Membranes	Pathogen	Disease
respiratory tract	*Aspergillus* (F)	aspergillosis
	Bordetella pertussis (B)	whooping cough
	influenza (V)	influenza
	Mycobacterium tuberculosis (B)	tuberculosis
digestive tract	*Escherichia coli* (B)	gastroenteritis
	Clostridium botulinum (B)	botulism
	hepatitis A (V)	hepatitis A
	Salmonella typhi (B)	typhoid fever
	Vibrio cholerae (B)	cholera
urogenital system	*Candida albicans* (Y)	candidiasis
	Chlamydia trachomatis (B)	urethritis
	herpes simplex (V)	herpes
	Neisseria gonorrhoeae (B)	gonorrhea
Skin	*Clostridium tetani* (B)	tetanus
	hepatitis B (V)	hepatitis B
	Malassezia (Y)	dermatitis
	Plasmodium (P)	malaria
	rabies (V)	rabies

Note: B = bacterium; F = fungus; P = protozoan; V= virus; Y = yeast

Portals of Exit

Portal	Mode of Exit	Diseases
respiratory tract	coughing, sneezing, talking	chicken pox, influenza, measles, pneumonia, tuberculosis, whooping cough
digestive tract	fecal contamination of food and water	amoebic dysentery, cryptosporidiosis, hepatitis A, salmonellosis, typhoid fever
urogenital system	sexual contact	AIDS, genital herpes, gonorrhea, syphilis

not been sterilized. Blood-borne pathogens can also use blood as a natural portal when an injury causes bleeding, or by way of animal bites and insect bites.

Self-limiting diseases often do not continue because the pathogen causes high mortality rates and thus runs out of hosts. For example, bubonic plague caused by the bacterium *Yersinia pestis* has such a high level of virulence that it can eliminate a large percentage of a human population if left untreated. *Y. pestis*, then, disappears from the population and grows in its rodent reservoir before emerging in a new population. Bubonic plague is self-limiting because it is not contagious and it causes high mortality rates.

Portals of entry occupy an important place in pathogenesis because a host's defenses often relate to the health of the body's portals of entry. Portals of exit play an equally important role in disease transmission because they influence the manner and ease with which pathogens spread from person to person.

See also INFECTIOUS DOSE; PATHOGENESIS; SEXUALLY TRANSMITTED DISEASE; VIRULENCE.

Further Reading

Baron, Samuel, ed. *Medical Microbiology*, 4th ed. Galveston: University of Texas Medical Branch, 1996. Available online. URL: www.ncbi.nlm.nih.gov/bookshelf/br.fcgi?book=mmed&part=A2644. Accessed January 17, 2010.

Faegermann, Jan. "Atopic Dermatitis and Fungi." *Clinical Microbiology Reviews* 15 (2002): 545–563. Available online. URL: www.ncbi.nlm.nih.gov/pmc/articles/PMC126862. Accessed January 17, 2010.

Medical News Today. "Face Masks Offer Protection against Spreading H1N1." October 3, 2009. Available online. URL: www.medicalnewstoday.com/articles/166104.php. Accessed January 17, 2010.

preservation Preservation is the use of physical or chemical means to prevent the growth of spoilage microorganisms in a material. Preservation is an ancient technique. The earliest civilizations required methods for preserving foods for later use, presumably by burying foods underground or in ice in cold climates or drying them in warm climates. These methods and others that emerged early in history, such as smoking, salting, or adding sugar, served the same purpose: to reduce the water activity of the food. Water activity is a measure of how much water in a material is chemically available for biological activities. Modern food production now uses chemical additives as preservatives, but the ancient methods still play a part in food production and preservation.

Preservation also refers to the act of keeping stored populations of microorganisms alive for extended periods. Bacterial cultures can be preserved by refrigerating, freezing, or freeze-drying (desiccation). In this application, the purpose of preservation is to keep the microorganisms alive and uncontaminated.

Food microbiology is most closely associated with preservation, but other industries have as great a need for preservatives for the purpose of preventing spoilage of their products by bacteria, algae, or fungi. The following industries rely on preservatives for maintaining the quality of their products:

- adhesives
- cooling water
- cosmetics
- inks
- food and food additives
- paints
- paper
- pharmaceuticals
- plastics
- textiles
- wood, fabricated woods, and veneers

Other industries such as metals, automobile manufacturing, and shipbuilding use preservative coatings to prevent corrosion by microbial activities.

Preservatives do not necessarily need to kill microorganisms. In fact, preservatives usually do not kill all the microorganisms in a material, but they inhibit the growth of microorganisms for a period called the product's shelf life. In this manner, preservatives mainly target spoilage organisms; only the food, drug, and cosmetic industries use preservation to prevent both spoilage organisms and pathogens.

Two main approaches to preservation reduce spoilage: removal or inhibition. By removing all or most microorganisms from a substance, microbiologists ensure that no spoilage will occur, unless the substance later receives a contaminant. The main removal methods are sterilization, irradiation, filtration, or methods that dramatically lower the number of microorganisms, such as pasteurization and centrifugation.

Inhibiting the growth of microorganisms has been the traditional approach to preservation for centuries, and long before companies began producing chemical biocides. Physical or chemical methods can reduce water activity and, thus, stop or greatly slow microbial growth. Chemical methods of synthesized biocides act directly on microbial cells and do not target water availability.

Most preservatives use one of the following modes of action for stopping microbial growth:

- block transport of nutrients into the cell
- inhibit enzymes that protect the cell against toxic substances
- lyse the cell membrane
- bind to nucleic acids
- react with amino acid groups needed for enzyme activity
- disrupt membrane permeability
- inhibit metabolic enzymes

In some cases, a preservative's exact mode of action is not known, but it, nonetheless, works well at killing microorganisms.

FOOD PRESERVATION

Physical methods for removing water from food, or at least immobilizing water molecules, have always been successful means of preservation. Meats can be smoked, salted, cured (a combination of smoke and salt), treated with sugar, dried, or wrapped to prevent exposure to oxygen. Vegetables and fruits tend to have very high water content, so the tactics for assuring long-term storage have been limited in number. Storing fruits in sugar and air- or sun-drying vegetables and fruits have been useful. Ancient societies also discovered the benefits of fermentations for preserving fruits, vegetables, and some meats.

Fermentation preserves food in two ways, both of which chemically alter the food. First, alcohol fermentations produce ethanol, which inhibits the growth of many microorganisms. Second, other types of fermentations produce lactic acid and other organic acids. Acid-producing fermentations inhibit the growth of all microorganisms except acidophiles, acid-loving organisms. While acid fermentations eliminate most spoilage organisms, they can also favor the growth of certain microorganisms that prefer pH at 4 or lower.

Food producers have at their disposal additional preservative methods that do not require chemical biocides. Low-temperature production, packaging, and storage help deter the growth of microorganisms in dairy products. Refrigeration (41°F [5°C]) slows microbial metabolism, so that even if microorganisms are present in milk, they take many days to grow to harmful numbers. Colder production and storage further reduce the chance of microbial growth.

Heating can be more effective than cooling because heating often kills microorganisms, whereas cooling only slows them down. Cooking and pasteurization are two important preservative methods in the food industry. Canning involves putting foods into large containers called *retorts* that heat the food to about 240°F (115°C), for up to 100 minutes. The food producer, then, cools the canned product quickly to keep the food out of a temperature range called the *danger zone*. In a temperature range from about 104°F (40°C) to 140°F (60°C), microorganisms multiply quickly in foods. The principle behind cooking and refrigeration of foods is to keep them in the danger zone for as short a time as possible. Pasteurization also kills many microorganisms by heat treatment but does not kill all of them, as becomes very apparent when milk spoils. Even refrigeration cannot forestall, indefinitely, the spoiling of milk.

Some foods can be treated by irradiation. Gamma radiation and electromagnetic radiation damage cellular constituents of microorganisms and, thus, reduce or eliminate their numbers.

Every type of food preservation method seems to have at least one microorganism that can resist the preservative's action. The following examples hint at the challenges of keeping unwanted microorganisms out of foods:

- *Penicillium* mold grows on foods with low water activity, such as cured meats, cheese, and bread.

- *Lactobacillus* bacteria survive in acid foods such as tomato products and salad dressings.
- *Clostridium* produces an endospore that protects the cell from high temperatures, drying, and chemicals.
- *Listeria* grows well at low temperatures of refrigerated foods.
- *Deinococcus* possesses cell repair mechanisms that allow it to withstand radiation.

The microorganisms that grow in foods despite physical preservation methods have led the food industry to add chemical preservatives to food as insurance against spoilage or illnesses.

CHEMICAL PRESERVATIVES

Food preservatives can be static or cidal, meaning they inhibit or kill microorganisms, respectively. Most preservatives have static action against bacteria (bacteriostatic) and molds (fungistatic) because the chemical has been designed to inhibit microorganisms but remain safe when ingested.

Safety in food, drug, and cosmetic preservatives must take precedence over all other attributes a chemical preservative might possess. Chemical preservatives must be tested to show that they inhibit microorganisms but will not harm consumers or the environment. One list of chemicals does not require stringent testing, however, because they have been defined by the U.S. Food and Drug Administration (FDA) as Generally Recognized As Safe, or GRAS. The GRAS list is published by the U.S. Department of Health and Human Services to let consumers know which preservatives are considered safe because of their decades of use without affecting health. For example, societies have used salt (sodium chloride), for centuries, to preserve food. There would seem to be no point in testing salt as a preservative for a new food, when centuries of practical evidence shows sodium chloride is safe.

The GRAS list changes often, as new compounds are added to it, but the following list shows commonly used chemical preservatives on the current GRAS list:

- ascorbic acid
- ascorbyl palmitate
- benzoic acid
- butylated hydroxyanisole (BHA)
- butylated hydroxytoluene (BHT)
- calcium ascorbate, sodium ascorbate
- calcium propionate, dilauryl thiodipropionate, sodium propionate
- calcium sorbate, potassium sorbate, sodium sorbate
- caprylic acid
- erythorbic acid
- gum guaiac
- methyparaben, propylparaben
- potassium bisulfite, potassium metabisulfite, sodium bisulfite, sodium methisulfite, sodium sulfite
- propionic acid, thiodipropionic acid
- propyl gallate
- sodium benzoate
- sorbic acid
- stannous chloride
- sulfur dioxide
- tocopherols

Of the GRAS list preservatives used in foods, most combat molds. For example, propionic acid/propionates, sorbic acid/sorbates, benzoic acid/benzoates, parabens, and caprylic acid are all effective against molds. Parabens and benzoic acid/benzoates also inhibit yeasts. The sulfites, by contrast, work best in inhibiting bacteria (and insects).

Other common food preservatives are ethylene and propylene oxides, sodium diacetate, dehydroacetic acid, sodium nitrite, and ethyl formate. Absence from the GRAS list does not mean a chemical is unsafe, because to be approved for use in foods, it has passed strict safety testing.

Food preservatives can be effective at very low concentrations. Most preservatives do not exceed 0.3 percent in food. Many chemicals preserve foods at levels from 0.1 percent to parts per million (ppm) amounts. A ppm equals the number of milligrams of a substance per liter or kilogram of food.

Food producers try to use chemical preservatives in the lowest amount that will extend shelf life. Food companies do this to contain costs, but they also try to minimize the chemicals in food to prevent flavors

Hurdle Effect in Preservation

Ingredients	Formula 1	Formula 2	Formula 3
preservative 1	+	+	-
preservative 2	+	+	-
acid	+	-	-
Growth Response	none	poor growth	good growth

that might result from high levels of preservative. Producers can keep the quantity of any one preservative in food low by using more than one chemical. Chemicals often work better in very low doses if they are used with another chemical. The reason for this is called the *hurdle effect*. As a result of the hurdle effect, microorganisms are forced to evade more than one type of preservative, and the strength of the microorganism weakens with each additional preservative (hurdle) it must evade. This method of preservation is analogous to a very long high-hurdle race. The first few hurdles are easy to clear, but as the runner tires, each succeeding hurdle becomes harder to jump. The table above gives an example of the hurdle effect in a product formulated with two chemical preservatives in an acidic food.

The hypothetical formula in the table above shows how microorganisms in food might respond as the formula's ingredient list changes. Food manufacturers test different formulas varied by the amount, type, and number of preservatives to find a food that appeals to consumers while maintaining good preservation against microorganisms.

PRESERVATIVE TESTING

The *challenge test* is the name for a procedure whereby a microbiologist tests the strength of a food preservative by purposely adding microorganisms to the food. The addition of microorganisms, the challenge, represents a worst-case scenario should the food become contaminated by a consumer. After adding the microorganisms, the microbiologist monitors microbial numbers in the food, over time (days to weeks), to determine how well the preservative works.

An example preservative challenge test involves the following steps:

- Weigh a fixed amount of food product.
- Add a known amount of microorganisms to the food.
- Take a sample (day 0) and determine the microbial numbers by standard culturing methods.
- Incubate the food.
- Repeat sampling and microbial load determination at intervals.

A typical pattern of intervals for sampling the food is day 0 (time of inoculation), days 1, 3, 7, 14, 21, 28, and 56. This example tells the microbiologist what might happen in the two months following contamination of the food. Another aspect of the challenge derives from the very large inoculum added to the food on day 0. By giving the food an unrealistically large dose of contaminants, a microbiologist pushes the limits of the preservative's capabilities.

The table on page 645 gives five hypothetical results that can occur in a challenge test.

Challenge tests can be set up to use a single microorganism or a mixture of different microorganisms such as several types of bacteria or diverse bacteria plus fungi.

PRESERVATIVE SELECTION

Food, drugs, and cosmetics contain many ingredients that have the potential to interfere with a preservative's effectiveness. Some preservatives work better in acidic foods, others work better in dry foods, and so forth. Food companies and their microbiologists have extensive experience in selecting preservatives that will be compatible with a food and will inhibit the type of microorganisms likely to contaminate that food. In addition to these considerations, the preservative must be used only within safe levels. Preservatives should also meet the additional requirements:

- cause no irritation or other side effects in consumers
- inhibit or kill the microorganisms most likely to contaminate the formula

Preservative Challenge Test

Day	Result	Interpretation
1	no growth	food might be overpreserved
1 and 7	growth on day 1; no growth on day 7	preservative eliminated the challenge within a week
28	same microbial numbers as day 0	preservative inhibits growth but does not kill the contaminant
1, 7, and 56	gradual decrease in microbial numbers	preservative is working but too slowly
28 and 56	day 28 numbers equal day 0, but numbers are higher on day 56	preservative inhibited growth for a month until it lost strength and could no longer preserve the food

- be miscible or soluble in different phases of multiple-phase foods, such as oil and vinegar
- remain stable with no chemical breakdown over the product's shelf life while withstanding elevated temperatures
- be compatible with all other ingredients in the formula
- cause no damage to the container
- possess acceptable odor and color at a practical cost

Most microbiologists build in extra safety to their products by testing a preservative's stability far longer than the product would be expected to last in a household.

No preservative used today meets all of these criteria, so manufacturers try to fulfill as many of the criteria as possible.

See also CULTURE; FERMENTATION; FOOD-BORNE ILLNESS; FOOD MICROBIOLOGY; PASTEURIZATION; STERILIZATION; WATER ACTIVITY.

Further Reading

Block, Seymour S., ed. *Disinfection, Sterilization, and Preservation,* 5th ed. Philadelphia: Lippincott Williams & Wilkins, 2000.

U.S. Food and Drug Administration. GRAS Substances (SCOGS) Database. Available online. URL: www.fda.gov/Food/FoodIngredientsPackaging/GenerallyRecognizedasSafeGRAS/GRASSubstancesSCOGSDatabase/default.htm. Accessed January 20, 2010.

prokaryote A prokaryote is one of two major types of cells in biology, characterized by the lack of membrane-enclosed inner structures, a dense area of storage for deoxyribonucleic acid (DNA) called the *nucleoid,* and asexual replication by binary fission. These characteristics help distinguish prokaryotic cells from the other main group, eukaryotes, which possess a more complex internal structure and often use sexual reproduction. Prokaryotes are the simplest independently living cells in biology.

The two main divisions among prokaryotes are bacteria and archaea. Until the introduction of molecular studies on microbial deoxyribonucleic acid (DNA) and ribonucleic acid (RNA), bacteria and archaea were not distinguished by many features. In 1990, the University of Illinois microbiologist Carl Woese (1928–) used analyses of RNA associated with cell ribosomes, called 16S rRNA, to classify all organisms into three domains: Archaea, Bacteria, and Eukarya. The analysis showed that archaea and bacteria were more dissimilar than originally believed and that their evolution separated early in the evolutionary history of all living things. The term *bacteria* replaced *eubacteria,* for "true bacteria," and the term *archaea* replaced *archaebacteria.* Domain Eukarya contained fungi and yeasts, protozoa, algae, and higher plants and animals.

PROKARYOTIC CELL STRUCTURE

Prokaryotes contain more species and diversity than all other living things. All prokaryotes are unicellular (made of a single cell) and tiny. Although cell size among prokaryotes ranges from 0.2 micrometers (μm) to 750 μm, big enough to be seen without a microscope, most species have cells that measure 0.5–3.0 μm in length. With a few exceptions, prokaryotic cells are round (coccus), oval, rod-shaped (bacillus); or curved (vibrio), spiral (spirillum), or corkscrew (spirochete) rods. Microbiologists use cell morphology of stained specimens viewed through a microscope to gain clues on a microorganism's identity. The genus *Corynebacterium* offers an exception to this rule: Corynebacteria are pleomorphic, meaning a single pure culture contains many different cell shapes.

The arrangement of cells as they multiply in liquid culture broth also indicates certain types of prokaryotes. For example, different species of cocci divide in different planes as they multiply. Chains of cocci begin as a pair, called a *diplococcus,* then extend to longer streptococci. Other species divide into two planes early in cell replication to form a tetrad of four cells in a single plane, which divides to form a block-like packet of eight cells, called *sarcinae.* Continued replication forms irregular bunches of cells called *straphylococci.* The names of many prokaryotes give clues as to their cell morphologies: *Bacillus subtilis, Streptococcus mutans, Staphylococcus aureus, Vibrio cholerae, Rhodospirillum rubrum,* and so on.

The internal structure of prokaryotes is simpler than that of any other living organism. Almost all species have an outer cell wall that provides rigidity and physical protection to the cell contents. In bacteria, a large polymer called *peptidoglycan* makes up a layer in the cell wall. Bacteria are further divided into two main groups based on their reaction in a technique called the *Gram stain.* Gram-positive bacteria contain a thick peptidoglycan layer that makes up most of the cell wall. Gram-negative bacteria have a thinner peptidoglycan layer that lies between an outer membrane and the inner cytoplasmic membrane. The cytoplasmic membrane of bacteria resembles the membrane of eukaryotes; it contains a double layer (bilayer) of phospholipids with proteins interspersed throughout.

The prokaryotic cytoplasm differs markedly from eukaryotic cells. As do other cells, prokaryotic cytoplasm consists of about 80 percent water, but it lacks clearly demarcated substructures called *organelles.* The cytoplasm contains three main areas that carry out specific functions: (1) The nucleoid, which holds most or the cell's entire DNA; (2) ribosomes by the thousands, which dot the cytoplasm and provide a site for protein synthesis; and (3) inclusions, a variety of structures in different species usually involved in nutrient storage. Many cells also contain circular strands of DNA, called plasmids, in the cytoplasm. Plasmids carry some of the microorganism's genes, especially genes that confer antibiotic resistance on the cell. Bacteria can share plasmids and in this way exchange genes between strains or between species.

The following main inclusion bodies exist principally in bacteria:

- carboxysomes—in photosynthetic species, contain an enzyme for capturing carbon dioxide
- gas vacuoles—contain cylinders of gas vesicles that take in or let out air to help aquatic species stay at certain depths
- lipid inclusions—store fatty compounds
- magnetosomes—stored iron oxide that acts as a magnet and may contribute to cell orientation
- metachromatic granules—store inorganic phosphorus in a form that can be used for making the energy metabolism compound adenosine triphosphate (ATP)
- polysaccharide granules—store sugar polymers
- sulfur granules—store sulfur as deposits of insoluble elemental sulfur

Archaea and bacteria of the genus *Mycoplasma* have different cell wall structure from the rest of the prokaryotes. Archaea do not contain peptidoglycan but use a similar polymer called pseudomurein that supports the cell wall. *Mycoplasma* lack a cell wall, and only the membrane provides containment for the cytoplasm. This makes *Mycoplasma* vulnerable to the environment, so these species must live as parasites inside eukaryotic cells to survive.

Bacteria develop additional unique structures that make certain species motile, able to stick to surfaces, or pathogenic. Various outer structures protect bacteria from harsh conditions in the environment or help the cells avoid destruction from a host's immune system. Species of *Bacillus* and *Clostridium* form tough coats called *endospores* that protect the dormant cell inside, when nutrients run low or when the environment becomes unfavorable, such as very hot, cold, dry, or high in toxic chemicals. Other outer protective coatings are capsules or slime layers. These similar structures are composed of a network of sugar polymers collectively called extracellular polysaccharide (EPS). The EPS makes up the main component of capsules and slime layers, called the *glycocalyx.* Glycocalyx protects cells from certain immune system activities and toxic chemicals in the environment and helps cells stick to surfaces. Bacteria that form biofilms on living or nonliving surfaces rely on the glycocalyx matrix for these benefits.

Motile bacteria contain appendages on the outside of the cell for swimming. Whip- or taillike flagella or flaplike axial filaments serve this purpose. Other appendages, the *fimbriae* and *pili,* aid adherence to surfaces and exchange of genetic material, respectively.

The various morphological features on prokaryotes enable these microorganisms to live in every environment on Earth. Enormous numbers of bacteria reside in the upper layers of soil and the photic zone (where light penetrates) of freshwater and marine water. The upper layer of soil contains at least 100 million cells per dry gram. Many archaea and bacteria are equipped to survive in much more extreme environments, defined by heat, cold, salts, pressure, acids, bases, radioactivity, or lack of oxygen. These extreme-loving prokaryotes are collectively called *extremophiles*.

The diversity of cell morphologies and habitats implies that prokaryotes can carry out a wide variety of metabolisms. Prokaryotes, indeed, represent every type of metabolism in biology. The following general types of metabolism can be performed by at least one group of prokaryotes:

- aerobic and anaerobic
- fermentation and respiration
- photosynthesis and heterotrophy (use of organic compounds for energy)
- organic and inorganic energy sources
- organic compounds and carbon dioxide as sole carbon sources

Prokaryotes also contain members that are pathogenic and others that are nonpathogenic and ecologically important. (No known archaea cause disease in humans.) Other prokaryotes such as *Bacillus* are ubiquitous in the environment, while others, such as *Rhizobium,* reside in unique microenvironments. In this example, *Rhizobium* inhabits root nodules in leguminous plants. *Rhizobium* is one of hundreds of bacterial genera that operate nutrient cycles, also called *biogeochemical cycles,* in which various nutrients are converted into different chemical forms as they move through the atmosphere, soil or water, and living things. Prokaryotes decompose organic matter, thereby releasing these nutrients into the soil or the atmosphere. Different bacteria absorb elements from the atmosphere and incorporate them into the cell. The microbial compounds, then, become available for other plant and animal life. When their life ends, the decomposer bacteria resume the cycle. Many bacteria have been identified as participants in biogeochemical cycles, but there undoubtedly exist many more unidentified microorganisms that carry out steps in cycling.

Ecologists estimate that about 1 percent of all microorganisms have been grown and studied in laboratories. Earth contains a diversity of unknown microorganisms that far outnumber the already diverse collection of known species. Microbiology depends on culture techniques and genetic analysis such as Carl Woese's method to learn all microbiologists know of prokaryotes. But culture techniques have not been figured out for many microorganisms living in the environment; therefore, microbiologists have learned very little about these nonculturable species. Many of the unidentified prokaryotes in nature may never be cultured because they depend on communities made up of other unknown species. The American microbial ecologists Edward DeLong and Norman Pace pointed out, in 2001, "To a large extent, the lack of alternatives to cultivation has severely limited the abilities of microbial biologists to characterize naturally occurring microbes. The unexpected . . . breadth and diversity of many of the newly discovered microbial lineages imply novel evolutionary innovations, new phenotypic properties, and unanticipated ecological roles." Despite sophisticated technologies for studying the relatedness of prokaryotes by analyzing their genes, the world of prokaryotes presents microbiology with a vast unknown.

EARTH'S PROKARYOTES

Prokaryotes became the first life on Earth 3.8 billion–3.5 billion years ago. Earth and the other planets of the solar system developed about 4.6 billion years ago; the oldest rocks on Earth have been dated to 3.8 billion years.

Conditions on the planet 3.5 billion years ago looked nothing like they do today. Volcanic eruptions produced an atmospheric stew of nitrogen oxides, carbon dioxide, methane, ammonia, hydrogen, and hydrogen sulfide. Earth would start to cool as volcanic activity slowed. During this phase, water condensed, and much of the atmosphere's hydrogen disappeared into space. In this milieu, the first organic compounds formed, perhaps using this atmosphere at Earth's surface or perhaps at subterranean volcanic structures. At breaks in Earth's crust under the oceans, deep-sea vents (hydrothermal vents) released inorganic sulfur and iron compounds, as they still do today. These vents may have served as sources for minerals for early Earth.

With the building-block elements of life in place, simple molecules formed, possibly by reactions energized by lightning strikes or volcanic activity. Scientists have devoted careers to unraveling the possible steps that produced increasingly complex molecules such as amino acids and nucleotides, the basic units of nucleic acids. Regardless of the route by which large molecules developed, fatty compounds and peptides are likely to have combined spontaneously in aqueous environments to form rudimentary membranes. Cell precursors of a simple membrane enclosing an RNA molecule developed in anaerobic conditions. Photosynthesis arose as an early form of energy generation, but the ancient photosynthetic organisms did not produce oxygen. Cyanobacteria would slowly develop between 3.0 billion and 2.5

billion years ago, and, only then, did oxygen begin to build up in the atmosphere. Oxygen levels may not have reached levels to support aerobic respiration similar to today's until 2.2 billion years ago. Carl Woese's ribosomal RNA (Rrna) studies indicated that bacteria and archaea separated evolutionary paths early on. Eukaryotes began developing no earlier than 2.2 billion years ago.

This thumbnail version of evolution cannot convey the various points at which different paths diverged and sometimes converged. As a consequence, biologists continue to confront questions of how the incredible diversity of prokaryotes developed and the meaning of *species* among organisms that share genes.

Moselio Schaechter, John Ingraham, and Frederick Neidhardt discussed, in 2006, in their book *Microbe,* the estimates of prokaryotic numbers based on phosphorus and carbon content of biological mass. They reported the "total number of their cells to be a mind-boggling 4×10^{30} to 6×10^{30}. Altogether they weigh more than 50 quadrillion metric tons! This huge population, coupled with a capacity for rapid multiplication and their ubiquity, constitutes a reservoir for immense genetic variation among prokaryotes. That means that a vast number of prokaryotic species must exist." The job of estimating all these species is daunting, but, in microbial ecology, the activity of microbial communities means more than the identity of individual species.

Future studies of prokaryotic roles in nature will probably consider whole communities and the ecosystems to which prokaryotes contribute. In some instances, communities may be critical because many prokaryotes are known to need the metabolic end products of other species for their own metabolism. Of the following metabolic activities that only prokaryotes perform, each makes up part of a microbial community or microorganism-eukaryote relationship:

- nitrogen fixation—conversion of nitrogen gas (N_2) to ammonia (NH_3)
- denitrification—conversion of nitrate (NO_3^-) to nitrogen gas
- nitrification—ammonia to nitrite (NO_2^-) and nitrite to nitrate
- sulfate reduction—sulfate (SO_4^{-2}) to sulfide (S^{-2})

Prokaryotes are part of every activity in biology. The study of prokaryotes encompasses evolution, ecology, biochemistry, medicine, and industrial pursuits that use prokaryotic cells or products. Prokaryotes are the most numerous organisms on Earth, and their metabolism makes possible the activities in all of Earth's ecology.

See also ARCHAEA; BACTERIA; BINARY FISSION; BIOGEOCHEMICAL CYCLES; CELL WALL; EUKARYOTE; EXTREMOPHILE; GRAM STAIN; METABOLISM; MORPHOLOGY; ORGANELLE; PEPTIDOGLYCAN; SPORE.

Further Reading

Brun, Yves V., and Lawrence K. Shimkets. *Prokaryotic Development.* Washington, D.C.: American Society for Microbiology Press, 2000.

DeLong, Edward F., and Norman R. Pace. "Environmental Diversity of Bacteria and Archaea." *Systematic Biology* 50 (2005): 470–478. Available online. URL: http://sysbio.oxfordjournals.org/cgi/reprint/50/4/470.pdf. Accessed January 24, 2010.

Gest, Howard. *Microbes: An Invisible Universe.* Washington, D.C.: American Society for Microbiology Press, 2003.

Schaechter, Moselio, John L. Ingraham, and Frederick C. Neidhardt. *Microbe.* Washington, D.C.: American Society for Microbiology Press, 2006.

Woese, Carl R., Otto Kandler, and Mark L. Wheelis. "Towards a Natural System of Organisms: Proposal for the Domains Archaea, Bacteria, and Eucarya." *Proceedings of the National Academy of Sciences* 87 (1990): 4,576–4,579. Available online. URL: www.pnas.org/content/87/12/4576.full.pdf+html. Accessed January 24, 2010.

proteobacteria Proteobacteria make up the largest group of bacteria characterized in bacteriology. Most of the known gram-negative bacteria with chemoheterotrophic metabolism belong to phylum Proteobacteria. *Chemoheterotrophy* refers to metabolism that uses a wide variety of organic compounds for carbon and energy.

The term *proteo-* is from the name of the mythological Greek god Proteus, who had the ability to change into different shapes. The proteobacteria also encompass a broad range of cell types and morphologies. Proteobacteria have been divided into five subgroups, named with a letter from the Greek alphabet: alpha, beta, delta, epsilon, and gamma.

All proteobacteria subgroups bear similarities in the composition of their deoxyribonucleic acid (DNA) and ribosomal ribonucleic acid (rRNA), a component of protein synthesis from the information carried in DNA. Physiology and morphology are so varied, however, that no single metabolism or cell feature represents all species of proteobacteria, other than being gram-negative.

All five subgroups contain pathogens to animals or plants or are parasites in other bacteria. Proteobacteria are ubiquitous in nature and occur in high numbers in soil, sediments, freshwater, and marine water.

PROTEOBACTERIA SUBGROUPS

The alpha proteobacteria (α-proteobacteria) use various metabolisms. Members of this group perform nitrogen fixation *(Rhizobium)*, use methane as a carbon-energy source *(Methylobacterium)*, or possess chemolithotrophic metabolism *(Nitrobacter)*, which is the use of inorganic compounds for energy and carbon dioxide as a carbon source. Some alpha proteobacteria live on very low levels of nutrients and contribute to Earth's cycling of nitrogen from the atmosphere to the land and biota. For example, the genus *Rhizobium* lives in nodules on the roots of legume plants and pulls nitrogen gas directly from the air (nitrogen fixation). The partnership between the microorganism and the plant serves the plant by giving it nitrogen in a form it can use, and the root nodule provides a habitat for the bacteria. This mutually beneficial partnership is called *mutualism*, a type of symbiosis. Other alpha proteobacteria called *purple nonsulfur bacteria* live in freshwater and marine water, with oxygen needs varying among species. These bacteria gather energy from sunlight and use organic compounds for carbon and energy metabolism.

Alpha proteobacteria include human and plant pathogens and nonpathogenic species important to agriculture. The following list of alpha proteobacteria illustrates the subgroup's diversity in morphology, ecological role, or pathogenicity (the ability to cause disease):

- *Acetobacter*—acetic acid producer
- *Agrobacterium*—plant pathogen (crown gall and cane gall)
- *Beijerinckia*—free-living nitrogen fixation in soil
- *Brucella*—human and animal pathogen (brucellosis)
- *Caulobacter*—aquatic species use stalks grown out of the cell membrane and wall to hold onto submerged surfaces
- *Nitrobacter*—conversion of ammonia form of nitrogen to nitrate, called *nitrification*, a step in nitrogen cycling
- *Rickettsia*—human pathogen that must live intracellularly (Rocky Mountain spotted fever)
- *Rhizobium*—nitrogen fixation in symbiotic relationship with plants
- *Rhodospirillum*—photosynthetic without using oxygen
- *Wolbachia*—parasite in insects

As does the alpha proteobacterium *Nitrobacter*, many beta proteobacteria perform nitrification. The soil bacteria convert ammonia (NH_3) or nitrites (NO_2^-) to nitrates (NO_3^-) in a chemical step called *oxidation*. Beta proteobacteria receive the reduced forms of nitrogen mainly from other anaerobic soil bacteria that excrete them as wastes. In addition to ammonia, members of the subgroup use hydrogen gas or methane in energy metabolism. Many beta proteobacteria also live in water environments or sewage, where they carry out the same oxidation reactions. The following list shows some of the most important beta proteobacteria in ecology and health:

- *Bordetella*—human pathogen (whooping cough)
- *Burkholderia*—potential pathogen called *opportunistic pathogen* (secondary infection in cystic fibrosis)
- *Neisseria*—human pathogen (gonorrhea)
- *Nitrosomonas*—nitrification
- *Thiobacillus*—sulfur compound oxidation

The delta proteobacteria include species that prey on other bacteria *(Bdellovibrio* and *Myxococcus)* and members that play an important role in Earth's sulfur cycle *(Desulfovibrio* and *Desulfuromonas)*. The anaerobic sulfate- and sulfur-reducing bacteria live in watery habitats, especially mud, and often grow well in polluted streams and lakes.

Myxobacteria is a general name for gliding bacteria that produce stalklike fruiting bodies when nutrients become scarce. Many are soil inhabitants, but others live in moist environments containing smooth surfaces the bacteria glide over in search of food. The presence of these myxobacteria becomes evident with the slimy trail the cells leave behind. Other myxobacteria form broad sheets rather than a single trail. As the cells in the sheet move forward, they coordinate their motility, so that they move in an organized group of bacteria. For this reason, myxobacteria have also been called *swarming bacteria*.

In conditions of low nutrient supply, myxobacteria such as *Myxococcus*, *Stigmatella*, and *Chondromyces* build fruiting bodies, each of which contains thousands of dormant (live but nonmetabolizing) cells. Some species also form a protective outer shell for the dormant cells inside the stalk, called *myxospores*. Myxospores wait inside the structure until nutrients, again, become available. They, then, burst free to resume their normal lifestyle. *Stigmatella* cells are unique in another way; they have a characteristic long and tapered rod shape.

The genus *Bdellovibrio* is a delta proteobacterium and a predator of the microbial world. Motile *Bdellovibrio* cells stalk gram-negative bacteria, including *Escherichia coli*. (The enteric species *E. coli* does not normally inhabit soil unless the soil has fecal contamination.) When *Bdellovibrio* finds its prey, it penetrates the gram-negative cell wall and burrows into a region called the *periplasm*, a layer between the gram-negative outer membrane and inner membrane. Inside the periplasm, the parasite elongates into a spiral, which soon breaks apart into smaller cells. After this replication step, the parasite releases enzymes that lyse the host cell and, thus, frees new predators into the environment. The lifestyle of this organism mimics that of viruses, which also infect bacteria and eukaryotic cells, then often end the cycle by lysing the host cell. In 2003, Megan Núñez and her research team at Occidental College in Los Angeles wrote, "Depending on the prey and the environment, this life cycle takes roughly 3–4 h. On the course of the invasion and digestion cycle, the bdellovibrio makes significant changes in the structure of the prey cell, not all of which are fully understood. The predator modifies the prey's exterior membrane and peptidoglycan layer [part of cell wall] without destroying them." Núñez reported four different broad classes of enzymes that *Bdellovibrio* uses during its occupation of the prey. These delta proteobacteria occur in all soils, sewage, freshwater, and marine water.

Epsilon proteobacteria make up a small group containing two important pathogens of humans and animals: *Campylobacter* and *Helicobacter*. Both gen-

Characteristics of Gamma Proteobacteria Genera

Genus	Characteristics
Azotobacter and *Azomonas*	nitrogen fixation, soil inhabitants (A)
Beggiatoa	microaerophilic, gliding motility, uses hydrogen sulfide and forms intracellular sulfur granules (F)
Chromatium	photolithoautotroph (N)
Coxiella	parasitic pathogen that causes Q fever (A)
enteric bacteria	includes *Escherichia coli* and several food-borne and waterborne pathogens (F)
Haemophilus	pathogen that attacks mucous membranes; requires blood for nutrients (A) or (F)
Legionella	pathogen carried in warm-water systems in buildings, lives inside aquatic amoebae (A)
Leucothrix	forms long filaments with a holdfast to attach it to surfaces (A)
Methylococcus	methane, methanol, or formaldehyde is sole carbon and energy source (A)
Photobacterium	two species emit blue-green light called bioluminescence (F)
Pseudomonas	prevalent water inhabitant and opportunistic pathogen, yellow-green pigment fluoresces in ultraviolet light (A)
Vibrio	curved rod, inhabitant of coastal waters, *V. cholerae* and *V. parahaemolyticus* are waterborne pathogens (F)
Xanthomonas	used industrially to make xanthan polysaccharides (A)

Note: (A) = aerobe; (F) = facultative anaerobe; (N) = anaerobe

era contain microaerophilic (require very low levels of oxygen) curved rods that move by using a single flagellum. *C. jejuni* is a dangerous food-borne pathogen, often associated with contaminated meat products. *C. fetus* causes spontaneous abortion in domestic animals. *H. pylori* has gained notoriety as a cause of stomach ulcers and possibly stomach cancer in humans.

Epsilon proteobacteria in nature form large mats on water surfaces rich in hydrogen sulfide. These mat communities have been discovered living in harsh oil- or sulfur-polluted places, so they are thought of as extremophiles. Microbiologists who seek new bacteria for cleaning up pollution (a process called *bioremediation*) have made epsilon proteobacteria one of their research priorities because of the bacteria's ability to degrade toxic chemicals.

Extremophile epsilon proteobacteria have recently been found near thermal vents that exist on parts of the ocean floor and emit superheated steam and gases (up to 750°F [400°C]) from Earth's core. Genetic analysis of the epsilon proteobacteria from hydrothermal vents has progressed more slowly than for other bacteria because of the difficulties involved in reaching and recovering these organisms. A few of the strains that have been recovered can grow in the laboratory at temperatures of 105.8–113°F (41–45°C). Current genetic studies on this subsubgroup have identified unique genes that presumably relate to its lifestyle at extreme temperatures and intense pressures of the ocean floor.

Gamma proteobacteria represent the largest and most diverse class in the phylum Proteobacteria. The subgroup contains 14 orders and 25 families. Many members are facultative anaerobes, meaning they can live with or without oxygen; others are aerobes. Gamma proteobacteria also contain a diversity of metabolisms for energy generation: chemoheterotrophy, chemolithotrophy, photolithotrophy, or methylotrophy. The features of each type of metabolism are described in the following list:

- chemoheterotrophy—also called *chemoheterotropy,* use of organic molecules as carbon and energy sources
- chemolithotrophy—also called *chemoautotrophy,* use of inorganic chemicals as an energy source and carbon dioxide as a carbon source
- photolithotrophy—also called *photoautotrophy,* use of light as an energy source and carbon dioxide as a carbon source
- methylotrophy—use of one-carbon compound methane or methanol as carbon and energy sources

The diversity in metabolism enables gamma proteobacteria, as a group, to live in very different environmental conditions. Many members of the subgroup run part of the sulfur cycle. Other beneficial genera inhabit the inside of insect digestive tracts, where they aid food digestion, or live on human skin as part of the body's normal flora. The table on page 650 describes bacteriology's familiar gamma proteobacteria.

Gamma proteobacteria contain three subgroups of photosynthetic microorganisms. These are collectively called the *purple sulfur bacteria.* The purple sulfur bacteria require strict anaerobic conditions, use hydrogen sulfide or hydrogen gas in energy metabolism, and use carbon dioxide as a carbon source. Purple sulfur bacteria live in sulfide-rich oxygen-free zones in still lakes, swamps, bogs, and lagoons.

Gamma proteobacteria also include the large order Enterobacteriales, which contains the enteric bacteria. Many of the enteric species are normal, harmless inhabitants of the human intestines but help in food digestion to a small degree. These same species can cause illness if fecal matter contaminates food or water or if a person takes in an infective dose on the hands. Familiar enteric gamma proteobacteria are *E. coli, Klebsiella, Salmonella, Serratia, Shigella,* and *Yersinia.*

Proteobacteria are the most diverse bacteria studied. Microbiologists have become familiar with a large proportion of species because these bacteria play critical roles in the following areas: environmental microbiology and nutrient cycling; symbiotic relationships with humans, animals, and plants; pathogens; photosynthesis; and representatives of distinctive morphology or life cycle. The alpha and gamma proteobacteria subgroups are, in themselves, more diverse than many other families or orders in bacterial taxonomy.

See also ENTERIC FLORA; HETEROTROPHIC ACTIVITY; METABOLISM; NITROGEN FIXATION; SOIL MICROBIOLOGY; SYMBIOSIS.

Further Reading

Campbell, Barbara J., Christian Jeanthon, Joel E. Kostka, George W. Luther, and S. Craig Cary. "Growth and Phylogenetic Properties of Novel Bacteria Belonging to the Epsilon Subdivision of the *Proteobacteria* Enriched from *Alvinella pompejana* and Deep-Sea Hydrothermal Vents." *Applied and Environmental Microbiology* 67 (2001): 4,566–4,572. Available online. URL: http://aem.asm.org/cgi/reprint/67/10/4566. Accessed January 1, 2010.

Dyer, Betsey D. *A Field Guide to Bacteria.* Ithaca, N.Y.: Cornell University Press, 2003.

Nakagawa, Satoshi, Ken Takai, Fumio Inagaki, Hisako Hirayama, Takuro Nunoura, Koki Horikoshi, and Yoshihiko Sako. "Distribution, Phylogenetic Diversity and Physiological Characteristics of Epsilon-Proteobacteria in a Deep-Sea Hydrothermal Field." *Environmental Microbiology* 7 (2005): 1,619–1,632.

Núñez, Megan E., Mark O. Martin, Lin K. Duong, Elaine Ly, and Eileen M. Spain. "Investigations into the Life Cycle of the Bacterial Predator *Bdellovibrio bacteriovorus* 109J at an Interface by Atomic Force Microscopy." *Biophysical Journal* 84 (2003): 3,379–3,388. Available online. URL: www.sciencedirect.com/science?_ob=MImg&_imagekey=B94RW-4V411DM-1X-1&_cdi=56421&_user=10&_coverDate=05%2F31%2F2003&_sk=%23TOC%2356421%232003%23999159994%23757131%23FLA%23display%23Volume_84,_Issue_5,_Pages_2793-3490_(May_2003)%23tagged%23Volume%23first%3D84%23Issue%23first%3D5%23date%23(May_2003)%23&view=c&_gw=y&wchp=dGLbVzb-zSkzS&md5=8e34030446d3c68a703a05608a3d52dd&ie=/sdarticle.pdf. Accessed January 1, 2010.

Prescott, Lansing, John Harley, and Donald Klein. "Bacteria: The Proteobacteria." In *Microbiology,* 4th ed. Boston: McGraw-Hill, 1999. Available online. URL: www.mhhe.com/biosci/cellmicro/prescott/outlines/ch22.mhtml. Accessed January 1, 2010.

protoplast A protoplast is a gram-positive bacterial, fungal, algal, or plant cell that lacks a cell wall. In intact cells, the term *protoplast* refers to the cytoplasmic membrane and all cell components inside the membrane. Preparations of membranes without the internal constituents are called *whole cell ghosts.*

Without a cell wall, protoplasts exhibit characteristics not seen in intact cells (see the color insert, page C-7 [bottom]). Without the rigid cell wall, protoplasts swell into a sphere that is sensitive to changes in the osmotic pressure of the environment. Osmotic pressure relates to the difference in pressure between the outside of the cell and the cell contents. Protoplasts shrivel in aqueous solutions of a higher solute (dissolved materials) concentration than the cell interior; protoplasts burst in aqueous solutions of lower solute concentration than the cell contents.

Protoplasts live without a cell wall provided they remain in isotonic solution, which is a liquid with the same solute concentration as in the protoplast interior. With the proper nutrient supply, protoplasts carry out the same metabolism as intact cells, but protoplasts differ because they cannot reproduce. Protoplasts resist infection by bacteriophages (bacteria-specific viruses) that attack whole cells.

Microbiologists use protoplasts for two main purposes: nutrient transport studies on membranes and studies of deoxyribonucleic acid (DNA) uptake from the environment by bacterial cells.

PROTOPLAST PREPARATION

Laboratory preparations of protoplasts depend on selecting the correct treatment for degrading the cell walls of bacteria, fungi, or plants. In bacteria, gram-positive species serve as a better source of protoplasts because the cell wall, although thick, consists mainly of peptidoglycan. Lysozyme is a natural enzyme that degrades bacteria, including gram-positive species. Gram-negative bacteria contain a complex outer wall composed of an outer membrane, a peptidoglycan layer thinner than that in gram-positives, and an inner cytoplasmic membrane. The configuration of the gram-negative cell wall makes its removal difficult and usually results in a partially degraded cell wall. This partially prepared cell, called a *spheroplast,* usually cannot be treated further to produce a protoplast. Gram-positive species also produce sphaeroplasts when enzyme treatments have not been run properly.

Protoplasts from fungi can be made by treating the fungal cells with the enzyme chitinase. The polymer chitin that makes up fungal cell walls resists most degradative enzymes in nature. Enzyme supply companies offer chitinases made by the bacteria *Streptomyces* and *Bacillus.* Plant cell walls also have distinctive fibrous materials that require digestion. The enzymes cellulose, pectinase, and xylanase degrade cellulose, pectin, and hemicelluloses (xylans), respectively.

Protoplast preparation from gram-positive bacteria begins by growing the cells in a nutrient-rich broth medium. The microbiologist harvests the cells by centrifugation and washes them, at least once, in isotonic buffer to excess medium. The intact cells can be suspended in an isotonic solution containing lysozyme and ethylenediaminetetraacetic acid (EDTA). EDTA is a compound that binds to metals or component salts such as calcium. (A compound that captures elements in a clamplike grip is called a *chelator.*) In many species, lysozyme treatment alone can degrade the cell wall; in other species, lysozyme-EDTA helps protoplast recovery. Additional steps may include specific substances known to weaken cell walls. For example, the antimicrobial compound lysostaphin helps break the cell wall of *Staphylococcus,* prior to the lysozyme-EDTA step. Protoplasts can be recovered from cell debris by centrifuging the mixture in a solution of sucrose and membrane-stabilizing salts such as magnesium chloride.

When cells produce lysins such as *Staphylococcus*'s lysostaphin, the lysin sometimes acts on the same cell that produced it. A spheroplast or protoplast produced in this manner is called an *autoplast,* and the lysine is an *autolysin.*

Microbiologists have, over time, modified the methods that give the best yield of healthy protoplasts from bacteria. Although the modifications vary by gram-negative and gram-positive species, general aids to the protoplast procedure are:

- osmotic conditions of the growth medium of starting cells
- exposure of anaerobic cultures to the air, for 30 minutes, before beginning the protoplast preparation
- harvesting the starting cells at a specific point in the growth curve—early (lag phase), exponential growth (logarithmic phase), or slowed growth (stationary phase)

Within a species, different strains can be more or less difficult for making protoplasts. In general, gram-positive cocci require more enzymatic treatment to cause lysis than rod-shaped cells (bacilli).

PROTOPLASTS IN GENETIC ENGINEERING

Protoplasts help scientists move deoxyribonucleic acid (DNA) from one cell type to another. This process, called *gene transfer,* is a critical part of recombinant DNA technology, which has the goal of growing organisms with genes from another unrelated organism in their chromosome. Several in vitro methods exist for transferring genes from one microorganism to another. This science, called *combinational biology,* uses the following methods:

- transformation—cells take up DNA molecules directly from the medium
- conjugation—one bacterium passes a plasmid (short circular DNA) to another bacterium
- transduction—use of bacteriophage virus to carry DNA into a bacterial cell
- microinjection—use of a device to inject or propel DNA into a cell (used mainly in algae and plants)
- protoplast fusion—protoplasts from two different organisms merge and combine their DNA

In protoplast fusion, a microbiologist combines two suspensions, both containing protoplasts from unrelated bacteria. The microbiologist, then, adds polyethylene glycol, which has detergent properties and, thus, predisposes the protoplast membranes to fuse. When the protoplasts incubate together, they slowly undergo spontaneous fusion and create hybrid protoplasts. Hybrid protoplasts that contain part of the genetic material from each starting protoplast are called *biparentals.* Because the hybrid protoplast contains all the information needed to make an intact cell, by returning the protoplasts to growth medium, a microbiologist can induce the hybrids to build new cell walls. The entire protoplast fusion process involves the following events:

1. protoplast preparation from intact cells
2. combination of protoplast preparations
3. treatment with polypropylene glycol
4. incubation to allow time for fusion
5. fused protoplast forms
6. segments of the two chromosomes combine to form single hybrid chromosome
7. protoplasts transferred to growth medium to induce cell wall synthesis
8. new intact hybrid cells resume replication

Protoplast fusion has become widely used in gene transfer studies on bacteria, plants, yeasts, and molds. Organisms vary in their efficiency in protoplast formation. Protoplast yields—proportion of intact cells that become usable protoplasts—from intact starting cells range from about 30 percent to 70 percent.

L-FORMS

A bacterial cell that in nature loses part or all of its cell wall has, in the past, been called an L-form. The terminology for L-forms has been gradually replaced by the term *spheroplast,* to indicate a partially formed protoplast. But in medical microbiology, L-forms are any bacteria that because of a compromised cell wall become flexible and can pass through openings they would not normally cross. Filtration sterilization methods that remove intact bacteria might not remove L-forms. L-forms cause a second problem in medicine because they are unaffected by antibiotics that act on bacterial cell walls. These bacteria have been called *stealth pathogens* or *cell wall deficient* (CWD) bacteria.

Microbiologists have suspected that certain bacteria develop an L-form as a temporary tactic against antibiotics. The microbiologist Josep Casadesús of the University of Seville in Spain commented, in 2007, "Treatment with penicillin does not merely select for L-forms (which are penicillin resistant) but actually induces L-form growth." When penicillin is applied to a plate containing gram-positive bacteria (penicillin's main target), L-forms of the bacteria begin to form at the edge of the agar surface. L-form bacteria are now believed to be another survival mechanism of bacteria because, in addition to resisting antibiotics, L-forms can infiltrate host cells and live there safe from the actions of the immune system.

L-forms cause two additional problems for clinical microbiologists. First, because L-forms are small (0.01 micrometer [µm]) and flexible, they pass through filters used for sterilizing laboratory solutions. Second, their tiny size and ever-changing shapes make L-forms impossible to recognize through a microscope.

The following genera or species have been implicated in chronic diseases because of their ability to develop an L-form: *Bacillus anthracis, Borrelia burgdorferi, Helicobacter pylori, Mycobacterium tuberculosis, Rickettsia prowazekii, Treponema pallidum,* and species of *Proteus, Streptococcus,* and *Vibrio.*

Mycoplasma is a rare bacterium that lives naturally without a cell wall. This genus displays many of the same characteristics as L-forms: small size, pleomorphic (inconstant) shapes, and resistance to conventional antibiotics.

Protoplasts were once believed to be an aberrant form of cells that increased the microorganism's vulnerability to the environment. Although protoplasts are more sensitive to osmotic changes than intact cells, the protoplast form may give cells temporary advantages for survival.

See also GENE TRANSFER; *MYCOPLASMA;* OSMOTIC PRESSURE; PEPTIDOGLYCAN.

Further Reading

Casadesús, Josep. "Bacterial L-Forms Require Pepitidoglycan Synthesis for Cell Division." *Bioessays* 29 (2007): 1,189–1,191.

Proal, Amy. "Understanding L-Form Bacteria." *Bacteriality.* August 15, 2007. Available online. URL: http://bacteriality.com/2007/08/15/l-forms/#footnote_2_35. Accessed January 25, 2010.

Temeyer, Kevin B. "Comparison of Methods for Protoplast Formation in *Bacillus thuringiensis.*" *Journal of General Microbiology* 133 (1987): 503–506. Available online. URL: http://mic.sgmjournals.org/cgi/reprint/133/3/503.pdf. Accessed January 24, 2010.

Weiss, Richard L. "Protoplast Formation in *Escherichia coli.*" *Journal of Bacteriology* 128 (1979): 668–670. Available online. URL: http://jb.asm.org/cgi/reprint/128/2/668.pdf. Accessed January 24, 2010.

protozoa (singular: protozoan) Protozoa are single-celled eukaryotes of diverse cell structure. Protozoa live only in moist habitats, such as wet soils, water, decaying matter, and the digestive tract. In these habitats, protozoa use organic compounds for carbon and energy, and some have developed the ability to prey on other microorganisms for food. Protozoa possess cilia that cover all or part of the cell or flagella that usually connect at one end of the cell. Each type of appendage contributes to cell motility. The threadlike cilia extend from the cell surface and beat in a coordinated fashion to make the cell swim and grasp food. Flagella are longer, whiplike tails mainly for swimming. Some protozoa use a third type of motility by extending part of the cell as a pseudopod, a footlike appendage. As cell contents flow into the pseudopod, the entire cell migrates in that direction.

Protozoa range in size from 1 micrometer (µm) to 4 millimeters (mm), which can be seen with the naked eye. Light microscopy using phase contrast works best for viewing protozoa. No single cell shape is characteristic of all protozoa, but unique cellular features aid in identification of several genera. In addition to the organelles typical of other eukaryotic cells, protozoa contain more distinctive structures, such as the following:

- pellicle—plasma membrane and related layers directly beneath it
- ectoplasm—moderately rigid layer just beneath the pellicle for strength and shape
- endoplasm—aqueous interior substance of the cell
- macronucleus—large nucleus, not always present, that controls metabolic activities
- micronucleus—small nucleus that controls reproduction and production of the macronucleus
- cytostome—oral groove, or "mouth"
- hydrogenosome—small organelle used in anaerobic metabolism; produces hydrogen
- contractile vacuole—pressure and solute regulation by collecting excess water and expelling it
- secretory vacuoles—store specific enzymes
- phagocytic vacuoles—sites of food digestion

Protozoa belong to four morphological groups, the ciliates and flagellates, amoebae, and sporozoa. Amoeba consist of cells with constantly changing cell shapes that move via pseudopods. Sporozoa produce a strong outer spore coat for part of their life cycle.

Taxonomy places protozoa in domain Eukarya, which contains all organisms made of cells that have membrane-bound organelles and a distinct site for genetic material called the *nucleus.* Protozoa do not belong, however, to either the plant or the animal kingdom. Some microbiologists classify protozoa as protists, a general term for single-celled or simple multicellular organisms, consisting mainly of protozoa and algae.

Protozoology, the study of protozoa, categorizes the organisms into seven phyla (see the table below), although some texts report more than seven phyla. The number of recognized species also ranges, from 20,000 to 120,000. The number of protozoal species may be less important to ecology than the role the organisms play in nature. Protozoa make up part of a diverse population of tiny organisms called *plankton*. These free-floating aquatic organisms have a critical role in Earth ecology because they create the foundation of almost all of Earth's food chains and food webs. Protozoa also live inside the digestive organs of various animals, where they contribute to food digestion.

A small portion of all protozoa cause disease. Pathogenic protozoa live in the environment but also spend some part of their life cycle in a host. During the time the organism lives inside a host, the protozoan is a parasite, an organism that harms its host while receiving a benefit for itself. For this reason, the science of parasitology also includes protozoa.

PROTOZOA LIFE CYCLES

Protozoal cells have probably been observed since scientists first trained lenses on water samples from the environment. The Dutch merchant Antoni van Leeuwenhoek (1632–1723) began the science of microbiology by studying protozoa and bacteria in water, using a rudimentary microscope. The German biologist Ernst Haeckel (1834–1919) has been credited with initiating the study of protozoa, in 1878, when he proposed the term *protist* for the organisms he observed in nature. Subsequent attempts to classify protozoa and study their physiology have trailed the same areas in bacteriology. Protozoa do not adapt to laboratory studies using solid agar media like those used for bacteria, and the nutrient and environmental needs of a protozoan are often unknown.

Investigations into protozoal life cycles have been easier to conduct than physiological studies. Many protozoa change cell structure in different phases of the life cycle, and these changes can usually be seen using a light microscope. Phase contrast microscopy, which uses deflected light to illuminate specimens, works well for studying protozoa behavior in wet mounts, which are specimens suspended in water rather than dried to the microscope slide.

Protozoa reproduce asexually by three methods: (1) fission of a parent cell into two identical daughter cells, (2) budding off a progeny cell from a mother cell, or (3) multiple nuclear division, called *schizogony*. In schizogony, the cell replicates its nuclei multiple times, before dividing into two daughter cells. In any of these asexual forms of reproduction, the progeny receive a complete genome from the parent.

Sexual reproduction takes place mainly by conjugation, in which two cells fuse temporarily and exchange micronuclei. The micronuclei contain half of the chromosome (haploid). Each haploid nucleus fuses with the opposite cell's macronucleus to produce a new diploid nucleus. When the two conjugating cells separate, each goes on its way with a diploid chromosome containing genes from the mate.

Each phylum contains members that have modified the standard reproductive methods described. Protozoologists have learned the purpose of most but not all of these modifications. The following list describes some of the special structures that participate in reproduction or other cell functions:

- gametes—haploid cells that fuse to form a diploid zygote in sexual reproduction
- cyst—protective capsule formed for withstanding harsh environmental conditions
- oocyst—cyst that also takes part in asexual reproduction
- trophozoite—cell released from a cyst and ready to reproduce

Protozoal Phyla

Phylum	Description
Apicomplexa	parasitic; single nucleus; no cilia and usually no flagella
Ascetospora	parasitic in invertebrates; develops a spore
Ciliophora	ciliated; two nuclei; widespread in soil, waters, and rumen
Labyrinthomorpha	lives on plants and algae in marine waters
Microspora	obligate parasite of vertebrates and invertebrates; develops a spore
Myxozoa	intracellular parasite of cold-blooded vertebrates (fish) and invertebrates; develops a spore
Sarcomastigophora	flagella or amoeboid movement; single nucleus

- undulating membrane—flap of membrane along one length of the cell exterior controlled by a flagellum inside it

- pseudopods—extensions of the cell that fill with endoplasm and thus "walk" the cell forward

- skeletal plate—rigid wall along all or most of the cell length, presumably for strength

Electron microscopes have enabled protozoologists to study the structures listed here, as well as the manner in which protozoa feed.

Protozoa get nutrients by two types of feeding: holozoic nutrition and saprozoic nutrition. Holozoic nutrition depends on engulfment of a food particle by the protozoal cell, a process called *phagocytosis*. After the protozoan has engulfed the food, it pushes it into one of its phagocytic vacuoles. The presence of food prompts the vacuole to release digestive enzymes into the interior and, thus, digest the meal. Protozoa eat bacteria by using phagocytosis.

Saprozoic nutrition involves the crossing of soluble nutrients, such as sugars, through the plasma membrane. Protozoa perform this mode of nutrition by pinocytosis, diffusion, or active or passive transport. Pinocytosis involves engulfing a portion of the aqueous environment containing nutrients, carrying the drop across the membrane, then pinching off that section of the membrane to deliver the nutrients inside. In diffusion, nutrients move across the membrane without any energy expenditure by the cell. Active or passive transport uses another molecule, which helps the nutrient cross the membrane barrier. These two methods differ in the energy demands they put on the cell. Active transport demands energy when the cell takes in a nutrient; passive transport resembles diffusion because it works without energy input by the cell.

NONPATHOGENIC PROTOZOA

Protozoa retrieve nutrients directly from the environment or prey on smaller protozoa and bacteria. The organisms, then, pass the nutrients to other life-forms that prey on protozoa. This process helps recycle the earth's nutrients and energy by making them available to other species. Nutrient cycling represents one of the most important roles of nonpathogenic protozoa.

Protozoa help ecosystems by serving as a food source for tiny invertebrates and by ingesting microscopic particles from the water. Protozoa might keep bacterial populations in check, in certain environments, by preying on bacteria. For example, in the rumen, protozoa might be playing a crucial role in controlling the high numbers of bacteria living there. The interactions of rumen protozoa and bacteria have not been fully explored, but researchers already know that cattle, sheep, goats, and other ruminants are less healthy when they lose their protozoa population.

Termites and other insects that digest woody fibers contain a population of mixed protozoa (more than one species) in their gut. Termites do not possess fiber-digesting enzymes such as cellulase that break down plant material. The insects depend on their anaerobic protozoa to break down the wood, but the gut protozoa also depend on a different population of microorganisms living inside their cell. Bacteria inside termite protozoa secrete cellulase to break down woody fibers and liberate nutrients for the bacteria, the protozoa, and the termite. Protozoa inside insects are called *endosymbionts* because they have a symbiotic relationship with a host by living inside the host. The bacteria are endosymbionts of the protozoa.

Ruminant animals possess a large compartmentalized stomach for digesting fibrous foods. The largest compartment, called the reticulorumen (made up of the reticulum and the rumen), contains a dense and diverse population of strictly anaerobic bacteria and protozoa that digest cellulose and similar fibers. The partially digested foods, then, move to the omasum and abomasum, which are the distant two compartments of the stomach, before flowing into the duodenum section of the small intestine. Microbiologists had learned many of the activities of rumen bacteria, but protozoa's role has remained a mystery. For decades, microbiologists believed the protozoa had a minor part in ruminant physiology. In the 1960s, the American microbiologist Robert Hungate (1906–2004) expanded the science of rumen microbiology. Hungate's students created a new generation of protozoologists, who focused entirely on rumen protozoa.

Studies by Hungate and others on rumen protozoa produced evidence of a wider role for the eukaryotes. Hungate wrote, in 1966, "Whereas protozoa in the rumen and reticulum were viable and active, those in the omasum and abomasum were immobile and disintegrating, and the duodenum showed no traces of the bodies of the protozoa. It was concluded that the digested protozoa composed part of the food of the host. Since at that time the relative advantages of animal versus plant food was a topic of lively speculation, the protozoa were greeted as an unexpected source of animal food in the supposedly herbivorous ruminant." In tracing the history of rumen protozoology, Hungate continued, "for a period of almost 90 years after the discovery of the rumen pro-

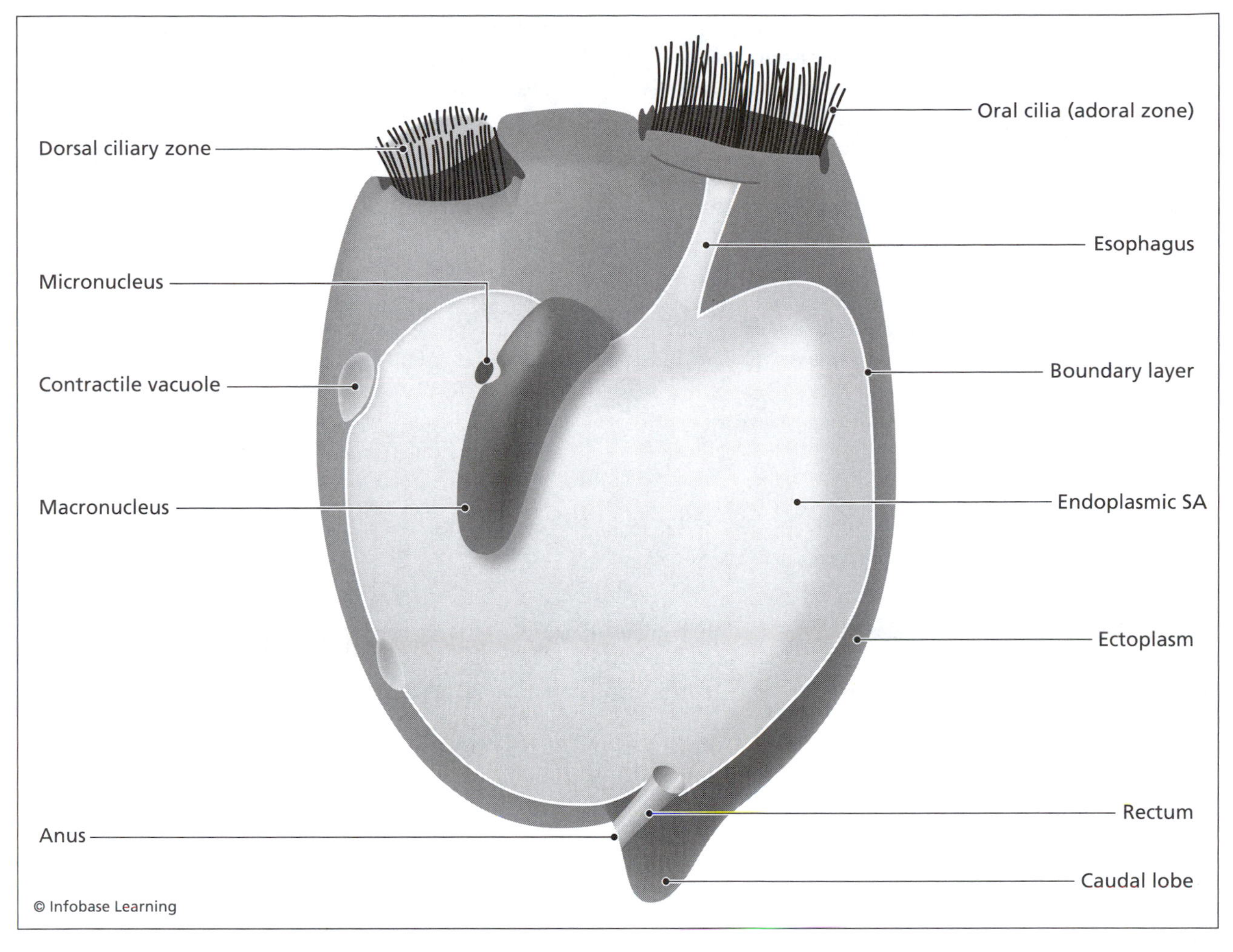

Protozoa play a vital role in the rumen and reticulum, two of the four sections of the ruminant animal's stomach. The protozoa in this habitat, such as the *Diplodinium* genus, include a diversity of ciliated species that aid in fiber digestion.

tozoa, reports and investigations were concerned with their specificity for this habitat. The question was answered conclusively when means were devised for artificially freeing the ruminant of its protozoa." Ruminant animals that have lost their protozoal populations show signs of nutrient deficiency. Rumen protozoa serve the animal in at least two important ways: (1) supplying highly digestible nutrients to the intestines for absorption and (2) controlling the bacterial population in the rumen by ingesting bacteria and, thus, contributing to the ecosystem's efficiency. By ingesting certain bacteria, protozoa probably help maintain relative proportions of rumen bacteria, a situation that creates the best rumen environment for feed digestion.

PATHOGENIC PROTOZOA

Some of the protozoa in the environment or living inside a host create severe health concerns. Protozoa that build a cyst form can withstand changes in the environment and stay viable until they have a chance to infect a host. For example, the flagellate *Giardia lamblia* forms cysts that contaminate natural waters. People who drink natural surface waters without first filtering them, risk ingesting *G. lamblia* cysts, which cause the digestive illness giardiasis.

G. lamblia spends part of its life cycle in an animal's digestive tract. Parasitic protozoa rely on two different types of hosts, intermediate or definitive. An intermediate host is an organism in which a parasite undergoes asexual reproduction. A definitive host, by contrast, provides a site for a parasite's sexual reproduction. In malaria, for example, the protozoan *Plasmodium vivax* uses a mosquito as a definitive host for sexual reproduction; then, after being transmitted to a human in a mosquito bite, *P. vivax* uses the person as an intermediate host to replicate asexually.

The table on page 658 presents protozoal parasites important in human and veterinary medicine.

Protozoology provides opportunities for learning about parasite life cycles, symbiosis, environmental pathogens, and anaerobic microbiology. Most important, protozoa are excellent study models for eukaryotic cell structure and life cycles.

See also EUKARYOTE; MORPHOLOGY; ORGANELLE; PHAGOCYTOSIS; PLANKTON; RUMEN MICROBIOLOGY; SYMBIOSIS.

Further Reading

Dennis Kunkel Microscopy. "Protozoa." Available online. URL: www.denniskunkel.com/index.php?cPath=12. Accessed January 9, 2010.

Hungate, Robert E. *The Rumen and Its Microbes.* New York: Academic Press, 1966.

International Society of Protistologists. Available online. URL: www.uga.edu/protozoa. Accessed January 1, 2010.

McAllister, Milton M. "A Decade of Discoveries in Veterinary Protozoology Changes Our Concept of 'Clinical' Toxoplasmosis." *Veterinary Parasitology* 132 (2005): 241–247. Available online. URL: http://vetmed.illinois.edu/faculty/path/documents/02.pdf. Accessed January 10, 2010.

Patterson, David J. *Free-Living Freshwater Protozoa: A Color Guide.* Washington, D.C.: American Society for Microbiology Press, 1996.

Van Egmond, Wim. "Flagellated Protozoa." Available online. URL: www.microscopy-uk.org.uk/mag/indexmag.html?http://www.microscopy-uk.org.uk/mag/wimsmall/flagdr.html. Accessed January 1, 2010.

Important Pathogenic Protozoa

Pathogen	Disease	Known Hosts	Known Sources
Babesia species	babesiosis	humans, cattle, horses, sheep, goats, pigs, dogs, cats	wild and domestic animals
Acanthamoeba species	meningoencephalitis	humans	fecally contaminated water
Cryptosporidium parvum and others	cryptosporidiosis (gastrointestinal)	humans, calves, horses, reptiles, caged birds	wild animals and cattle
Cyclospora species	diarrhea, gastroenteritis	humans	unknown
Entamoeba histolytica	amebic dysentery	humans, primates	humans, pigs, primates
Giardia lamblia	giardiasis (gastrointestinal)	humans, horses, caged birds	fecally contaminated water
Isospora belli	coccidiosis (diarrhea)	humans	domestic animals
Leishmania braziliensis	leishmaniasis (skin, mucous membranes)	humans, primates	sand flies
Naegleria fowleri	amebic meningoencephalitis	humans	free-living in environment
Neospora caninum	encephalitis	dogs	contaminated food, water
Plasmodium vivax	malaria	humans, primates	*Anopheles* mosquito
Sarcocystis neuroma	encephalitis	horses, opossums	contaminated feed, water
Toxoplasma gondii	toxoplasmosis (various organs)	humans, primates, dogs, cats	cats, beef, congenital in humans
Trichomonas vaginalis and others	Trichomoniasis (urethritis, vaginitis)	humans, birds	humans, birds
Trypanosoma species	sleeping sickness (Chagas' disease)	humans, domestic animals, primates	*Triatoma* (kissing bug) or tsetse flies

Wiser, Mark F. "Intestinal Protozoa." 2000. Available online. URL: www.tulane.edu/~wiser/protozoology/notes/intes.html. Accessed January 10, 2010.

purple bacteria *Purple bacteria* is a general term for two types of gram-negative bacteria involved in the earth's sulfur cycling. Almost all use photosynthesis in energy generation. Purple nonsulfur bacteria comprise 10 species of alpha proteobacteria (except one species classified with the beta proteobacteria) of order Rhodospirillales and family Rhodospirillaceae. Purple sulfur bacteria belong to the gamma proteobacteria in order Chromatiales and family Chromatiaceae and total about 25 species.

Purple bacteria morphology is so diverse that no single identifying feature connects them. The color of the bacteria in the environment results from mixtures of bacteriochlorophylls and carotenoid pigments that take part in photosynthesis. All purple bacteria carry either bacteriochlorophyll *a* or *b* located in a membrane that merges with the cell's plasma membrane. Many of the species also have flagella and are motile.

Until the 1930s, biologists believed that photosynthesis exclusively characterized green plants containing plant chlorophyll pigments. But, in 1937, the American biochemist C. Stacey French (1907–95) found a new type of photosynthesis in *Spirillum rubrum* (now called *Rhodospirillum rubrum*). "The group known as 'purple bacteria' which use light for the synthesis of organic matter, as do green plants, are generally purple, brown, or red in color," he said. "Oxygen is not produced by photosynthesis in [the] bacteria, but instead a hydrogen donator is used up in the process of photochemical CO_2 [carbon dioxide] reduction." French's work hinted at the unusual metabolism of purple bacteria. The early discoveries by French and others opened the door to a new branch of microbial ecology.

Today, purple bacteria are usually classified with another group, called green bacteria, that use similar metabolisms. Thus, purple nonsulfur bacteria relate to green nonsulfur bacteria because they use anoxygenic (no oxygen produced) photosynthesis and use organic compounds as a hydrogen source. Purple sulfur bacteria act as green sulfur bacteria do in also depending on anoxygenic photosynthesis but requiring inorganic compounds, usually hydrogen sulfide (H_2S), in metabolism. All types use carbon dioxide as a carbon source.

PURPLE NONSULFUR BACTERIA

Purple nonsulfur bacteria grow in anaerobic aqueous environments that receive some sunlight. These bacteria have no preferences among freshwater, marine water, or high-salt water, such as salt lakes. These bacteria grow best in still waters that have stratified according to sunlight penetration and oxygen availability. In a swamp or stagnant pond, the purple nonsulfur bacteria occupy a layer just above a layer of purple sulfur and green sulfur bacteria.

Although this group is called *nonsulfur,* its species use sulfur in their metabolism but can only tolerate sulfur at low levels. The amounts of sulfur used by the purple sulfur bacteria are toxic to the nonsulfur group.

Purple nonsulfur bacteria obtain adequate nutrients from most of their habitats because of their versatility in using various organic compounds. The photosynthesis in these bacteria differs from other microbial photosynthesis, such as in algae or cyanobacteria, because lower levels of filtered light are adequate. While algae and cyanobacteria cluster at the top of the water's photic zone (the layer that receives sunlight), purple nonsulfur bacteria exist at the deep portion of the photic zone. Oxygen inhibits the bacteria's pigments, making a deeper layer more suitable than growth at the surface.

Purple nonsulfur bacteria can live in deep sediments underneath bodies of water that receive neither light nor oxygen. In those situations, the bacteria switch their metabolism to respiration rather than photosynthesis. Species such as the versatile *Rhodospirillum rubrum* use a variety of metabolisms depending on the conditions in the environment. The spiral-shaped cells of *R. rubrum* are motile through several polar flagella (all based at one end). Depending on conditions in the environment, *R. rubrum* uses aerobic or anaerobic metabolism, such as fermentation, photosynthesis, photoautotrophic metabolism (inorganic compounds and carbon dioxide only), or respiration.

PURPLE SULFUR BACTERIA

The purple sulfur bacteria are strict anaerobes and usually depend on photoautotrophic metabolism. In this energy-generating process, the bacteria convert sunlight energy to chemical energy and absorb carbon dioxide from the atmosphere as the carbon source. The pigments used for photosynthesis reside in stacked membranes that connect with the cell's plasma membrane.

Purple sulfur photosynthesis differs from that in higher plants, which convert water and carbon dioxide to glucose and liberate a molecule of water and oxygen. Purple bacteria substitute hydrogen sulfide for water:

$$2\ H_2S + CO_2 \rightarrow CH_2O + H_2O + 2\ S^0$$

The S^0 in the preceding reaction refers to elemental sulfur, or the pure insoluble form of the element.

Sunlight provides the energy to power this reaction. As metabolism progresses, purple sulfur bacteria start accumulating elemental sulfur inside the cell. Soon, the cell cannot hold any more, and the bacteria deposit surplus sulfur on the outside of the cell. When viewed in a microscope, the elemental sulfur is visible as yellowish granules attached to the cells.

Purple sulfur bacteria live in environments that receive low light, such as a pond that filters sunlight as it penetrates the water. Some purple species move into and out of light and dark areas, especially when they are part of a microbial community. Other common habitats include slow-flowing banks along estuaries and mudflats (see the color insert, page C-8 [top]).

Microbial mats represent an aquatic microbial community characterized by the interaction of several species with different types of metabolism. In stagnant bodies of water, mats can form if the water's surface receives sunlight and the sediments at the bottom exclude oxygen but provide sulfur compounds. A typical mat consists of the following types of microorganisms, in order from top to bottom and from high oxygen and light conditions to anaerobic and dark conditions. This list includes the colors likely to be visible:

- aerobic photosynthetic algae and cyanobacteria—green
- oxidized iron—a thin layer that sometimes forms at the interface between oxidation and reduction reactions, reddish brown
- purple nonsulfur bacteria—orange-brown
- purple sulfur bacteria—purple
- diverse anaerobes—dark brown-black

The microorganisms that form mats withstand hot temperatures and high-salt environments, although they can also be found in temperate waters. The following habitats can develop conditions that promote the formation of mats: freshwater lakes, bogs, lagoons, shallow marine waters near beaches, and hot springs.

Purple sulfur bacteria play a role in the earth's sulfur cycle by taking in the hydrogen sulfide that rises from anaerobic metabolism in aqueous environments. The process by which the bacteria chemically convert hydrogen sulfide to elemental sulfur and elemental sulfur to sulfate is called *phototrophic oxidation:*

$$H_2S \rightarrow S^0 \rightarrow SO_4^-$$

Other bacteria continue the cycle by putting the sulfur from sulfate into a form that higher organisms can use. Bacteria can do this by making the sulfur-containing amino acids cysteine and methionine.

The purple sulfur genus *Chromatium* contains species that vary mainly in cell size and sulfide requirements. Cell size ranges from 2–6 micrometers (μm) to about 15 μm in length. *Chromatium* inhabits microbial mats in a region where it can scavenge light that passes through the aerobic, photosynthetic layer above and receive hydrogen sulfide rising from the deeper sediments. *Chromatium* prefers to migrate downward in the direction of the sediments but will swim upward and lie just below the aerobic organism in the surface layer if light is limited. The microbiologist and author Betsey Dexter Dyer explained, in 2003, "*Chromatium* bacteria are highly motile and are known to swarm up from the sediment in a pink cloud when oxygen-generating photosynthesis has shut down for the day. A stronger sulfur smell at night may therefore indicate the activities of *Chromatium*."

As microbial ecologists build a store of information on purple sulfur bacteria, they have expanded the list of places where these bacteria live. In addition to the high-sulfide waters discussed here, purple sulfur bacteria grow in hot springs reaching 104°F (40°C), halophilic environments such as salt lakes and salt evaporation pools, alkaline lakes (soda lakes), and slightly acidic waters such as sulfur springs. Examples of purple sulfur currently studied in microbial ecology are:

- *Allochromatium*
- *Chromatium*
- *Halorhodospira*
- *Thermochromatium*
- *Thiocapsa*
- *Thiocystis*
- *Thiopedia*
- *Thiorhodospira*
- *Thiospirillum*

Purple sulfur bacteria grow well in laboratories at their optimal conditions. *Chromatium* grows at incubation temperatures of 77–86°F (25–30°C) in medium supplied with macrominerals (manganese, magnesium, calcium, and iron) and trace minerals plus elevated levels of salt (NaCl).

PURPLE BLOOMS

Purple blooms can develop in stratified lakes containing higher-than-normal anaerobic activity in the sediments. Sulfide compounds rise and feed a rapid growth of purple sulfur bacteria. The purple bacteria numbers can increase to concentrations that turn the lake water pink, reddish, or purplish. Purple blooms occur with a mixture of species or single species. The following purple sulfur bacteria have caused mixed species blooms: *Chromatium, Thiocystis, Thiopedia,* and *Thiospirillum.*

In 1996, the microbiologists Jörg Overmann, J. Thomas Beatty, and Ken J. Hall detected a single-species purple bloom in Mahoney Lake, British Columbia, Canada. The sulfur bacterium *Amoebobacter purpureus* rose to levels that dominated, in both cell numbers and cell mass, a layer deep in the photic zone. The subsequent death of the purple bacteria served as a carbon and phosphorus source for the heterotrophic population of bacteria in adjacent layers.

Purple blooms do not destroy the water's ecosystem as cyanobacterial and algae blooms do. The latter blooms cause devastating damage to water ecosystems because they consume oxygen. All aerobic prokaryotes and eukaryotes, including higher organisms, can suffocate in these blooms. Purple sulfur blooms, by contrast, do not consume oxygen but produce it.

Purple bacteria have recently made news as a model for new alternative energy sources. William Matthews reported in *Defense News,* in 2009, "In purple bacteria, which thrive in shallow lakes, the pigment uses the energy from photons to convert carbon dioxide into carbohydrates, which the bacteria consumes as food. In the Air Force, scientists are using a synthetic version of the dye to convert sunlight into electrons—electricity that it hopes can power an unmanned aerial vehicle on extended-duration flights." The prototype bacteria-energy cells are almost as efficient as conventional solar cells and will be far less expensive to make in thin-film versions.

In 2009, the microbiologists Michael Madigan and Deborah Jung summarized the value of purple bacteria in Earth ecology: "Purple bacteria have emerged as ideal model systems for dissecting the physiology, biochemistry and molecular biology of photosynthesis. Moreover, anoxygenic photosynthesis preceded oxygenic photosynthesis on Earth by billions of years. Thus, studies of purple and other anoxygenic phototrophs have contributed in major ways to our understanding of the evolution of photosynthesis." Purple nonsulfur and sulfur bacteria occupy an important niche in microbial ecology and serve as an excellent example of the interrelatedness of species in microbial ecosystems.

See also GREEN BACTERIA; MICROBIAL COMMUNITY; PROTEOBACTERIA.

Further Reading

French, C. Stacey. "The Rate of CO_2 Assimilation by Purple Bacteria at Various Wave Lengths of Light." *Journal of General Physiology* 21 (1937): 71–87. Available online. URL: http://jgp.rupress.org/cgi/reprint/21/1/71.pdf. Accessed January 1, 2010.

Madigan, Michael T., and Deborah O. Jung. "An Overview of Purple Bacteria: Systematics, Physiology, and Habitats." In *Advances in Photosynthesis and Respiration,* vol. 28, *The Purple Phototrophic Bacteria,* edited by C. Neil Hunter, Fevzi Daldal, Marion C. Thurnauer, and J. Thomas Beatty. Dordrecht, The Netherlands: Springer Science, 2009.

Matthews, William. "Flying on Sunlight." Defense News, July 6, 2009. Available online. URL: http://defensenews.com/story.php?i=4172521&c=FEA&s=TEC. Accessed January 1, 2010.

MicrobeWiki. "Rhodospirillum." Available online. URL: http://microbewiki.kenyon.edu/index.php/Rhodospirillum#Cell_Structure_and_Metabolism. Accessed January 1, 2010.

Overmann, Jörg, J. Thomas Beatty, and Ken J. Hall. "Purple Sulfur Bacteria Control the Growth of Aerobic Heterotrophic Bacterioplankton in a Meromictic Salt Lake." *Applied and Environmental Microbiology* 62 (1996): 3,251–3,258. Available online. URL: http://aem.asm.org/cgi/reprint/62/9/3251. Accessed January 1, 2010.

Uffen, Robert L., and Ralph S. Wolfe. "Anaerobic Growth of Purple Nonsulfur Bacteria under Dark Conditions." *Journal of Bacteriology* 104 (1970): 462–472. Available online. URL: www.ncbi.nlm.nih.gov/pmc/articles/PMC248231/pdf/jbacter00376-0492.pdf. Accessed January 1, 2010.

recombinant DNA technology Recombinant deoxyribonucleic acid (DNA) technology comprises the methods scientists use to make new DNA from two different DNA molecules. This technology is part of a broad science called genetic engineering, in which genetic material in cells is changed into new forms or made to carry out new processes.

Recombinant DNA technology developed, in the 1970s and 1980s, when scientists learned methods for studying individual genes in an organism's main mass of genetic matter called the *chromosome*. Geneticists had studied the principles of heredity since the Austrian priest and naturalist Gregor Mendel (1822–84) first demonstrated, using pea plants, that traits of parents could be inherited by offspring. Within the next century, scientists would discover that DNA was the genetic material that held those traits in the form of genes. In the 1950s and 1960s, the structure of DNA was elucidated, and scientists delved into the mechanisms by which the large molecule replicated and passed to successive generations.

In 1970, the American biologists Hamilton Smith (1931–) and Kent Wilcox discovered an enzyme that bacteria used as a defensive mechanism for destroying foreign DNA. The enzyme, called *restriction endonuclease,* cleaved DNA molecules. With a tool in hand to manipulate DNA artificially, biologists began a new era that Smith called modern genetics. In his acceptance speech when he was awarded the Nobel Prize in physiology, in 1978, Smith said, "The basic ingredients of this new technology are the cleavage-site-specific restriction enzymes: a special class of bacterial endonucleases that can recognize specific nucleotide sequences in duplex DNA and produce double-stranded cleavages. Using a collection of these enzymes, each with its own particular sequence specificity, DNA molecules may be cleaved into unique sets of fragments useful for DNA sequencing, chromosome analysis, gene isolation, and construction of recombinant DNA." The biotechnology industry arose from the first experiments on recombinant DNA—in those early years people called the method *gene splicing.*

Scientists quickly saw the value of an enzyme that could open the DNA at a specific site. In 1971, Kathleen Danna and Daniel Nathans at Johns Hopkins University used the restriction endonuclease that Smith and Wilcox had discovered to cut a simian virus and showed the specific fragments that it produced.

The manner in which DNA is treated, cleaved with enzyme, and used for accepting a new DNA insertion has been refined, but its foundation began with the experiment by Danna and Nathans.

RECOMBINANT DNA PROCEDURES

The basic procedure of obtaining DNA for use in recombinant technology covers four steps: (1) Extract DNA from an organism, (2) cleave the DNA into fragments, (3) isolate the fragment of interest, and (4) clone the fragment. In cloning, a biologist inserts the fragments into *Escherichia coli* bacterial cells and grows the cells, through several generations, by incubating them. Each identical daughter cell (a clone) of an ancestor cell makes an identical copy of DNA. The microbiologist, then, tests a number of clones to find the clone carrying the gene of interest.

Recombinant DNA technology has become almost routine in microbiology laboratories. The following procedure outlines the main steps and their purposes.

1. Preparatory steps—The biologist prepares solutions, buffers, microbial cultures, and DNA.

2. Recovery of the desired DNA fragment—A biologist cleaves the isolated DNA with restriction enzymes to break the molecule into many smaller fragments. The scientist, then, separates the fragments from each other in a procedure called *agarose gel electrophoresis*. The fragment of interest can be recovered after this electrophoresis step.

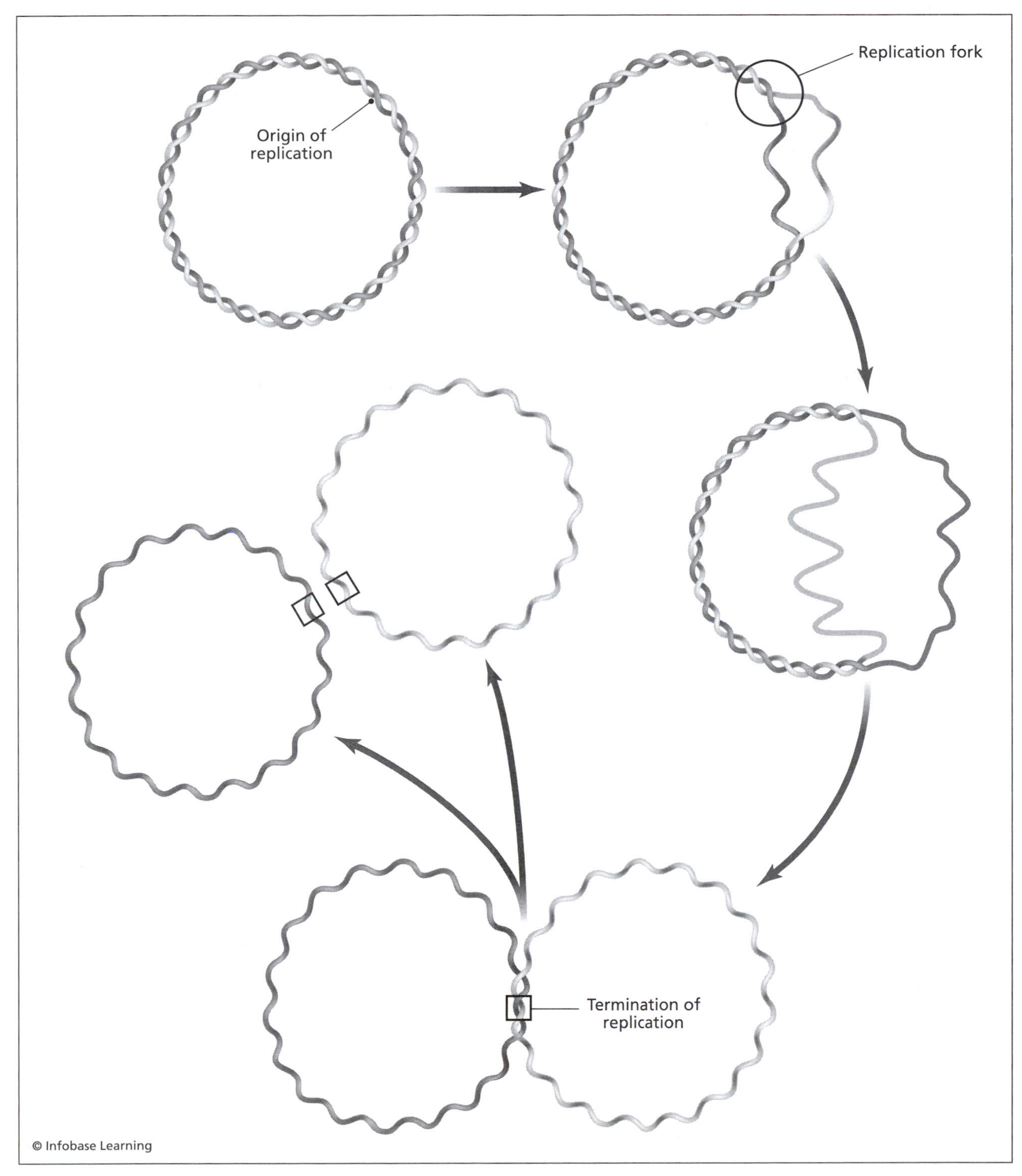

Recombinant DNA technology begins with a thorough understanding of DNA replication in prokaryotes and eukaryotes. Bacterial DNA replication is bidirectional. Two replication forks progress along the double-stranded molecule until two new copies have been made from the original DNA molecule.

3. Insertion into a vector—The biologist inserts the DNA fragment of interest into vector DNA, that is, a carrier that will convey the fragment into a host cell. A plasmid is a short piece of DNA often found in the cytoplasm of bacteria, which serves as a common vector in recombinant DNA technology. Vectors must be self-replicating so that they will reproduce as the host cell reproduces. Either plasmids or viruses used as vectors have this quality.

4. Gene transfer—Bacteria take the vector into their cells. During incubation, many clones containing the desired fragment multiply to a high concentration.

5. Recovery—The biologist recovers either the cloned DNA containing the useful gene or an end product made by the clones.

Biologists need restriction endonucleases for steps 2 and 3. In step 2, the enzyme breaks the original DNA into pieces, so that a target gene becomes isolated on a DNA fragment. In step 3, the endonuclease cleaves the plasmid so that the fragments can insert into the vector's DNA. In this scheme, the plasmid is the recombinant DNA.

Molecular biologists have refined recombinant DNA technology for the purpose of assuring the gene(s) of interest are cloned and recovered. For example, in step 2, the entire DNA of an organism can be cleaved into fragments, only one of which carries the desirable gene. All the fragments represent a gene library. Molecular biology laboratories, today, retain gene libraries from various organisms because the DNA holds potentially valuable genes. The biologists transfer all the fragments into a bacterial culture that will incubate and produce clones. By using the entire gene library, the biologist can be sure that the gene of interest will be included in one of the clones.

After the incubation is complete, a culture of millions of bacterial cells must be tested. The biologist must find the clones containing the important gene amid a sea of other cells carrying the other DNA fragments. This separation of one useful gene from millions of others can be accomplished by a step called *screening*. Screening relies on the inclusion of a second gene on the DNA fragment holding the target gene. The second gene helps the scientist distinguish the desirable clone from all the others. A gene carrying resistance for the antibiotic tetracycline has been very useful as the second gene used for screening. Screening works as follows:

1. Select a vector plasmid that contains a tetracycline-resistance gene, called the *Tet*R gene, and a second gene with a specific enzymatic activity.

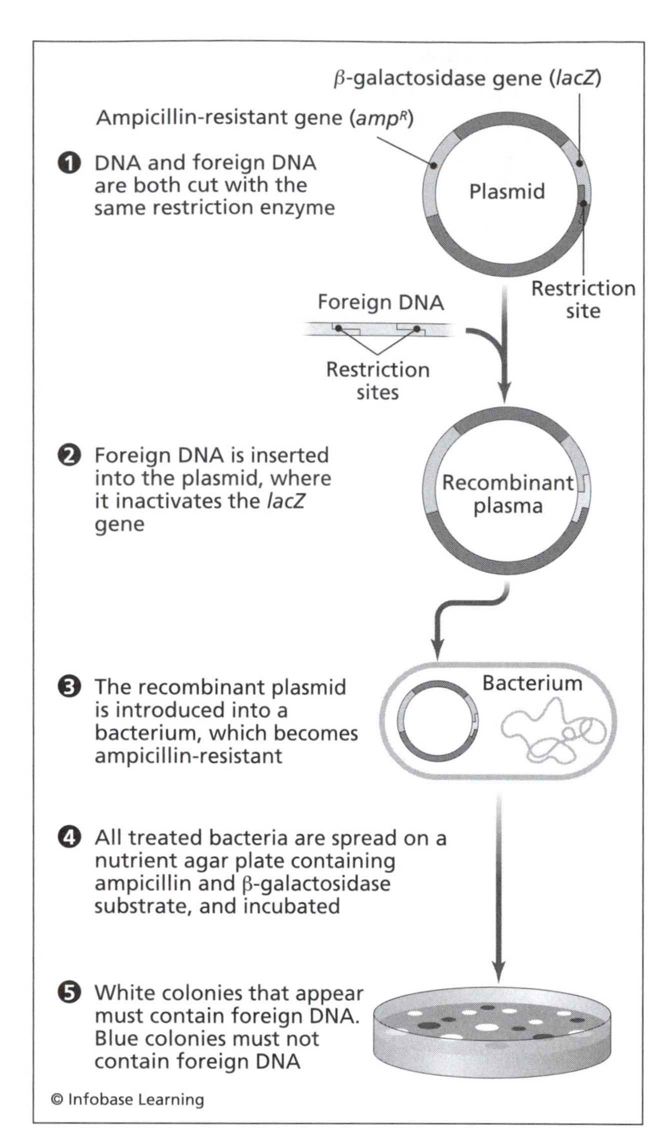

The blue-white screening technique provides a fast way to determine whether recombinant bacteria have integrated a piece of foreign DNA into their genome. A color reaction linked with gene activity enables a scientist to find bacteria containing the recombinant DNA.

2. Insert the DNA fragments into a population of plasmids in a location on the plasmid that does not disrupt the *tet*R gene but does disrupt the enzyme gene.

3. Transfer the plasmid into a host.

4. Incubate the host cells.

5. Dilute the host cells and inoculate them to agar containing tetracycline, then incubate these cultures.

6. After incubation, observe the agar to find any colonies that have grown. Only cells that have successfully taken in the plasmid will be tetracycline-resistant.

Bioengineered Microbes in the Environment

by Anne Maczulak, Ph.D.

A bioengineered microbe is any microorganism that contains DNA from another organism in its genome. The new DNA is called *recombinant DNA*. Bioengineering, which grew into an established industry in the 1980s, can now be performed in almost any biology laboratory. Companies supply microorganisms, growth media, enzymes, and genes so that professionals and students can insert genes into DNA by an easy process that has become routine in biology.

Bioengineering has become a major part of academic research on the function of particular genes and in industry, mainly in medicine and agriculture. In medicine, bioengineered bacteria and yeasts produce substances in large quantities that would otherwise be prohibitively expensive to obtain from nature. For example, hormones (insulin), enzymes (streptokinase that dissolves blood clots), therapeutic proteins, and the substances used in gene therapy all rely on the genetic engineering of microbial DNA. In agriculture, bioengineered bacteria have been valuable as pesticides and as means for protecting plants from freeze damage. In the former role, *Bacillus thuringiensis* (Bt) contains a toxin that kills more than 150 different species of insects. Scientists now put the Bt toxin gene into various other microorganisms and onto plants themselves. In the latter role, a group of bacteria known as ice-minus bacteria *(Erwinia, Pseudomonas,* and *Xanthobacter)* have the ability to make ice crystals less damaging to the leaves of plants.

Bioremediation scientists who focus on cleaning up pollution with microorganisms have been developing bioengineered varieties that degrade toxic chemicals more completely and more rapidly than nature does alone. The genes from bacteria that degrade organic solvents, chemical pesticides, radioactive chemicals, and neutralize toxic metals will be valuable additions to natural microorganisms.

The benefits of bioengineering have been questioned, since the first foreign gene was put into an organism's DNA. A large number of scientists and non-scientists vehemently oppose genetic engineering despite its benefits. Many opponents have convinced lawmakers to put strict limits on bioengineering or to outlaw it altogether. A new science worthy of more than one Nobel Prize has become a lightning rod for anger and fear.

The U.S. economist Jeremy Rifkin has opposed genetic engineering, since its emergence in the mid-1970s. Rifkin's Foundation on Economic Trends (FET) has given credit to biotechnology for developing new drugs, therapies, food sources, increased life span. Genetically engineered microorganisms for energy production also have been accepted by some critics for their potential for replacing fossil fuels. The FET, nonetheless, has posed the questions "Will the artificial creation of cloned, chimeric and transgenic animals mean the end of nature and the substitution of a 'bio-industrial' world? Will the mass release of thousands of genetically engineered life forms into the environment cause catastrophic genetic pollution and irreversible damage to the biosphere?" In this context, bioengineering experts certainly must address serious concerns of the public before their science progresses further.

The main criticism of bioengineering relates to the accidental release of genetically modified organisms (GMOs) (or genetically engineered microorganisms [GEMs]) into the environment, where they will permanently change natural systems. In the early days of field testing of bioengineered bacteria, scientists assured the public that safety procedures would be strictly enforced to prevent accidents. Even if a GMO escaped into the environment, it would not be able to compete with the natural microbial communities in soil, in water, or on other plants.

The promises of safety seemed well founded, especially as the use of genetically modified crops began to increase in agriculture. Genetic engineering has improved the hardiness and yield of corn, rice, soybeans, tomatoes, and various other important crops. These improvements have been critical for subsistence farmers in impoverished regions. But mistakes have occurred, and genetically modified crops have entered foods where they do not belong. These publicized events raise awareness of other types of genetic engineering.

No one knows the fate of a bioengineered microorganism let loose in nature. Scientists have conducted controlled experiments to build safety mechanisms into current bioengineered microorganisms. These features ensure a microorganism cannot survive in the environment for long. The main safety features are weakened strains that do not survive in natural conditions and microorganisms equipped with suicide genes that cause the cells to die when their desired actions have been completed. But these approaches have also worried GMO opponents, because of the risks of unusual gene's entering vital microbial populations in nature.

Technologies have almost no limits given enough time and effort. Scientists will undoubtedly create newer and more failsafe ways to confine GMOs. But bioengineered organisms in the environment cause another problem that technology cannot fix: the ethical issues of bioengineering. Bioengineered microor-

(continues)

Bioengineered Microbes in the Environment

(continued)

ganisms differ from other microorganisms that have evolved naturally in the environment. GMO opponents argue that humans have no right to rearrange genes, whether the genes belong to a sheep, a salmon, a tomato plant, or bacteria, all organisms that are now bioengineered. The effect of bioengineered microbes on ecosystems may be unknown for decades. Some of the effects may remain hidden much longer, because ecologists have yet to define all the interactions of hundreds of species that make up some ecosystems, such as soil and ocean ecosystems.

The production of large quantities of bioengineered microbes raises additional concerns. Biotechnology companies use fermenters of 3,000-gallon (11,355 l) capacity or more. An accident causing a spill has the potential to release a huge amount of bioengineered cells into the environment. Other critics cite the potential for misuse of GMOs designed by bioterrorists to be extremely virulent. These scenarios have not become reality, but the concerns remain very real.

Bioengineering's future seems to hinge on the promise that is tantalizingly close versus the variety of perceived dangers. In 2009, British researchers developed a genetically engineered enteric bacterium, *Bacteroides ovatus,* to deliver a therapeutic protein to the digestive tract to treat inflammatory bowel disease. So far, the bioengineered *Bacteroides* has been tested only in animals, but a safety mechanism already looks promising: The bioengineered bacteria will only produce protein in the presence of the sugar xylan. Simon Carding of the University of East Anglia Medical School said of the therapy, "This is the first time that anyone has been able to control a therapeutic protein in a living system using something that can be eaten. The beneficial [bacteria] could be activated when they are needed." This may represent one of many bioengineering ideas in which safety mechanisms automatically become part of the scientific planning.

The use of recombinant DNA has been likened to the invention of the microscope as a pivotal event in microbiology's history. Bioengineering has already changed biology, medicine, and food production. Stanford University's bioengineering department has stated, "Bioengineers are focused on advancing human health and promoting environmental sustainability, two of the greatest challenges for our world." But bioengineering is a young science that has not explored all of its potential benefits or risks. Members of the public are right to question revolutionary technologies that emerge with little oversight from academic groups or the government. Of all new sciences, bioengineering faces perhaps the greatest potential for enhancing health and saving the environment while creating dangerous health and environmental hazards.

See also GENE THERAPY; GENETIC ENGINEERING.

Further Reading

Foundation on Economic Trends. Available online. URL: www.foet.org. Accessed November 8, 2009.

Science*Daily*. "Genetically Engineered Bacteria Are Sweet Success against Inflammatory Bowel Disease." August 21, 2009. Available online. URL: www.sciencedaily.com/releases/2009/08/090820204456.htm. Accessed November 8, 2009.

Stanford University. Bioengineering. Available online. URL: http://bioengineering.stanford.edu. Accessed November 8, 2009.

7. Test these cells for enzymatic activity. The absence of activity by the specific enzyme indicates that the DNA fragment has inserted properly into the plasmid.

These basic steps have been used in diverse biotechnology applications. Molecular biology companies provide enzymes, cloning kits, screening tests, and equipment that make today's recombinant DNA technology similar to following a recipe in a cookbook.

See also CHROMOSOME; GENETIC ENGINEERING; GENE TRANSFER; PLASMID; POLYMERASE CHAIN REACTION.

Further Reading

Brownlee, Christine. "Danna and Nathans: Restriction Enzymes and the Boon to Modern Molecular Biology." *Proceedings of the National Academy of Sciences* 102 (2005): 5,909. Available online. URL: www.ncbi.nlm.nih.gov/pmc/articles/PMC1087965. Accessed January 28, 2010.

Roberts, Richard J. "How Restriction Enzymes Became the Workhorses of Molecular Biology." *Proceedings of the National Academy of Sciences* 102 (2005): 5,905–5,908. Available online. URL: http://www.pnas.org/content/102/17/5905.full.pdf+html. Accessed January 28, 2010.

Smith, Hamilton O. "Nucleotide Sequence Specificity of Restriction Endonucleases." Nobel lecture presented in Stockholm, Sweden, December 8, 1978. Available online. URL: http://nobelprize.org/nobel_prizes/medicine/laureates/1978/smith-lecture.pdf. Accessed January 28 2010.

Tortora, Gerard J., Berdell R. Funke, and Christine Case. *Microbiology: An Introduction,* 10th ed. San Francisco: Benjamin Cummings, 2009.

reservoir A reservoir of infection is a living or nonliving site that normally harbors a pathogen and acts as a continuous source of that pathogen for infecting living things. Reservoirs may be human, animal, or a nonliving thing.

Humans can act as a reservoir for human diseases transmitted to others, and, in fact, the human body acts as the principal reservoir for human disease. Humans and animals can be reservoirs for each other. For example, skunks are a common reservoir for rabies in humans, and humans can be a reservoir for tuberculosis transmission to primates.

Two main nonliving reservoirs for human disease are soil and water. Soil contains fungi and bacteria that can infect humans when ingested, inhaled, or taken into the body through a cut or other injury to the skin. For example, the soil bacterium *Bacillus anthracis,* which causes anthrax, uses all three routes to infect humans. Water contains pathogens that live naturally in water, such as *Entamoeba histolytica,* the protozoan that causes amebic dysentery, or *Cryptosporidium,* a protozoan that originates in fecal contamination of surface waters.

Reservoirs are important in enabling a disease to perpetuate itself, especially virulent pathogens that cause a high mortality rate in a host population. Pathogens that eliminate a large portion of a population in a single outbreak receive an advantage from a reservoir because the pathogens can persist in the reservoir until the host population recovers to normal numbers. A highly virulent pathogen with no known reservoir would be at a disadvantage for survival compared with pathogens that have a stable reservoir.

Epidemiologists studying a new outbreak try to find the source and the reservoir of the pathogen as quickly as possible for the purpose of preventing further transmission into a human population. A source of infectious disease differs slightly from a reservoir. For instance, a source is a location from which a pathogen is directly and immediately transmitted to a host. For *Cryptosporidium,* water can be considered a source of infection. The reservoir is the storage site of the pathogen in nature. For *Cryptosporidium,* cattle and various wildlife serve as the reservoir. Some pathogens, such as the rabies virus, do not use a source because they infect humans directly from the reservoir when an infected animal bites a person. In this case, the source and the reservoir become synonymous. The *period of infectivity* refers to the time in which the source can disseminate a living, active pathogen.

HUMAN RESERVOIRS

Some pathogens live as obligate parasites, meaning they cannot survive outside a host. The human immunodeficiency virus (HIV) is such a pathogen, which can be viewed in two ways in respect to its reservoir. First, an obligate parasite does not have any reservoir because it cannot exist outside the target host. By contrast, a second viewpoint would classify HIV as a pathogen that depends on a human reservoir as its source for infecting other humans.

Human reservoirs require a carrier, which is a person who has the pathogen in the body and infects healthy people. Carriers may be symptomatic, asymptomatic, or recovering from the disease but still able to transmit pathogens. Human reservoirs are necessary for the following diseases or types of diseases: diphtheria, hepatitis, sexually transmitted diseases, *Streptococcus* infections, *Staphylococcus* infections, tuberculosis, and typhoid fever.

Humans also act as reservoirs for some animals, a reversal of more typical zoonotic diseases, in which animals transmit disease to humans. In 2010, Melissa Thaxton wrote for the Population Reference Bureau, "Although tracing a disease's origin back to a specific host can be extremely difficult, scientists working in African national parks such as Gombe, Kibali (in Uganda), Parc de Volcans (Rwanda), and Bwindi Impenetrable Forest (Uganda) believe that humans are most likely the transmitters of diseases such as polio, pneumonia, measles, and scabies, which have affected the parks' respective ape populations over the past four decades." An increasing number of tourists who desire to see primates in their natural habitats, such as Gombe's chimpanzees, may be carrying diseases into the animal communities.

In veterinary hospitals, zoos, and wildlife rehabilitation centers, the rules for preventing human-to-animal disease transmission resemble the rules for preventing human-human or animal-human transmission: sanitation to reduce pathogens in animal housing, hand washing before and after contact with animals, and avoidance of animals by people who are sick.

Details of human-to-animal transmission might become increasingly important in the study of reservoirs because the route can mimic the animal-to-human transmission route. This is especially true for direct-contact transmission other than bites. Animals probably receive pathogens from humans mainly by skin-to-skin (or coat) contact and respiratory transmission.

ANIMAL RESERVOIRS

Both wild and domestic animals serve as important reservoirs for human disease. Diseases that humans contract from animals are called *zoonotic diseases,* or zoonoses. The medical community has identified about 150 zoonoses, which use a variety of transmission routes:

- direct contact with infected animals
- direct contact with animal waste
- food and water contaminated with animal waste or by direct contact with the animal
- inhaling or otherwise being infected by contaminated hides, fur, or feathers
- ingesting infected animal meat
- insect vectors carrying pathogen from an animal to a human
- bites

The American pathologist Theobald Smith (1859–1934) pioneered the study of animal-human

Nonhuman Reservoirs for Human Diseases

Disease	Pathogen	Main Reservoirs
brucellosis	*Brucella* species (B)	livestock, dogs, cats, rabbits
bubonic plague	*Yersinia pestis* (B)	rodents
cat-scratch disease	*Bartonella henselae* (B)	domestic cats
cowpox	cowpox virus	cattle, horses
cryptosporidiosis	*Cryptosporidium parvum* (P)	cattle (mainly calves)
encephalitis (St. Louis)	arbovirus	birds
encephalitis (eastern equine)	arbovirus	horses, fowl
encephalitis (Venezuelan equine)	arbovirus	horses, rodents
encephalitis (Western equine)	arbovirus	horses, birds, snakes, squirrels
glanders	*Pseudomonas mallei* (B)	horses
hantavirus pulmonary syndrome	hantavirus	rodents (mainly deer mice)
herpes B viral encephalitis	*Herpesvirus simiae*	monkeys
influenza	influenza virus A and B	swine, fowl
leptospirosis (Weil's disease)	*Leptospira interrogans* (B)	dogs, wildlife
Lyme disease	*Borrelia burgdorferi* (B)	field mice
malaria	*Plasmodium* species (P)	monkeys
psittacosis	*Chlamydia psittaci* (B)	birds (mainly parrots)
rabies	rabies virus	dogs, cats, bats, opossums, skunks, foxes, raccoons
ringworm	*Trichophyton, Microsporum,* or *Epidermophyton* (F)	domestic mammals
Rocky Mountain spotted fever	*Rickettsia rickettsii* (B)	rabbits, groundhogs, rodents
toxoplasmosis	*Toxoplasma gondii* (P)	mammals (mainly cats)
tuberculosis	*Mycobacterium bovis* (B)	cattle
tularemia	*Francisella tularensis* (B)	wild rabbits
typhus fever	*Rickettsia* species (B)	rats

disease transmission and consideration of the human-to-animal route, as well as the reverse. In 1928, Smith wrote, "Underlying this phenomenon of the passing of infectious agents from animals to man and the reverse is the far more important and fundamental one as to the fate of such aberrant parasites. Do they become modified in the human or other aberrant host in one passage? Is the process irreversible? Bearing on this question, it is of interest to note that the guinea-pig, the most useful laboratory animal, is victimized by a number of diseases which may have been derived at least in part from close association with man." Smith also speculated on the importance of pathogens' losing or gaining virulence while in either the human or the animal host.

Smith studied about a dozen zoonotic diseases that affected human health and agriculture, in the early 20th century, during a time in the United States when agriculture was the country's main industry. The threat of zoonoses was greater to a larger proportion of the population, in that period, than in the decade that immediately followed, when U.S. manufacturing began a steady growth.

Today, zoonoses cause the greatest risks in human health to animal handlers, veterinarians, and communities that still depend mainly on livestock for their livelihood. Some of the important zoonoses from non-human hosts are described in the table on page 668.

Health agencies such as the World Health Organization (WHO) conduct ongoing studies to find animal reservoirs of certain diseases suspected of being zoonoses. The WHO's purpose is to improve control of the emergence and transmission of global diseases, and controlling reservoirs is a major part of this mandate. Some of the diseases that the WHO currently is studying for identifying all possible reservoirs are Ebola virus hemorrhagic fever, leishmaniasis, Marburg virus, and sudden acute respiratory syndrome (SARS).

In some influenza outbreaks, the reservoir has not been confirmed by the WHO or national health agencies before the disease moves into the human population. Countries, sometimes, take dramatic action to remove a suspected reservoir to head off an epidemic. In late 2003, a highly pathogenic avian influenza (bird flu) called *H5N1* emerged in Asia, and health agencies believed that the virus had already become established in poultry populations throughout the Far East. The WHO recommended large-scale culling (removal of infected birds by killing and disposing of them in a safe manner) of commercial farms as well as "backyard" farms. The agency's recommendation resulted from years of successes in which countries acted quickly to remove a suspected reservoir before a human epidemic developed. In January 2004, the WHO pointed to an instance of successful reservoir control, stating in a health update, "In the Netherlands in 2003, an outbreak of highly pathogenic H7N7 avian influenza caused infection, with mild illness, in 89 persons, and fatal illness in a veterinarian. An estimated 30 million poultry were culled within a week." Similar reaction to H5N1 by removing poultry and wild bird populations may have averted a pandemic.

The outbreak of the H1N1 virus that became pandemic (a worldwide epidemic) was first identified, in 2008, when it emerged as swine flu, but questions soon arose as to whether swine were the reservoir for humans. Genetic testing on the new H1N1 strain led epidemiologists to propose a more complicated route from reservoir to humans for the pathogen that eventually became pandemic by 2009. The public health physicians Shanta Zimmer and Donald Burke of the University of Pittsburgh concluded, in 2009, "The emergence of influenza A (H1N1) 91 years ago led to a disastrous global pandemic. That virus is thought to have emerged almost simultaneously from birds into humans and swine. . . . Although the immediate genetic event that led to the emergence of the new [2009] pandemic threat was a reassortment between two influenza A (H1N1) swine viruses, these two viruses were actually the products of at least four independent avian-to-mammalian cross-species transmissions, with at least four previous reassortments of gene segments among avian, human, and swine-adapted viruses." Zimmer and Burke's results demonstrate the difficulty that can arise when trying to identify a disease reservoir, even using sophisticated genetic analysis on reservoirs that have received more study than almost all others.

LATENT RESERVOIRS

A latent reservoir is a site where an infectious agent remains dormant for a period before becoming active. The concept of latent reservoirs has been studied in the human immunodeficiency virus (HIV), which causes acquired immunodeficiency syndrome (AIDS), because controlling this pathogen's latency may be key to controlling AIDS.

HIV's latent reservoir is an immune system cell called the $CD4^+$ T cell. The following characteristics have been established for this reservoir in the human body:

- small; frequency of infected $CD4^+$ T cells is one per million
- stable with little to no deterioration of the reservoir for as long as seven years
- establishes early in the infection

The ability of HIV to enter the latent reservoir, early in the infection, assures that the pathogen can hide from destruction by the body's immune system—T cells are themselves members of the immune system. Stability also helps maintain the pathogen population for a long latency period. HIV's latent population stability results from one of two potential causes that AIDS researchers have proposed: (1) $CD4^+$ T cells are stable, do not decompose readily, and so provide the virus with a long-term reservoir; and (2) a low level of virus replication replenishes any low numbers of HIV destroyed over time.

The AIDS researchers Rachel Simmons and Robert Siliciano of Massachusetts General Hospital wrote, in 2004, in *AIDS Reader:* "Eradication of virus is not a reasonable goal in anti-HIV therapy because of the latent reservoir of HIV-1 in resting memory $CD4^+$ T cells, which guarantees lifetime persistence of virus. Any drug-resistant viruses that arise and circulate for significant periods can be stored in the reservoir, limiting future treatment options, and mistakes in treatment can be 'remembered' by the virus. Wild-type virus [the most prevalent genetic form] is also preserved in the reservoir and can reemerge if therapy is stopped." As of 2009, AIDS researchers had not uncovered a way to affect HIV latency in the T cell reservoir.

Infectious disease control depends on management and control of animal reservoirs, especially since a large number of infectious diseases in humans reside in this type of reservoir, from months to decades. Good reservoir control will probably call for continued efforts in four areas proposed by the public health experts Zimmer and Burke in 2009: (1) complete understanding of zoonotic pathogens, (2) in vivo studies of zoonotic pathogenesis, (3) epidemiology studies of the diseases in populations, and (4) surveillance of animal population health.

Reservoirs are not always easy to identify. Even after a reservoir has been identified, epidemiologists and pathologists must learn the source of infection and the modes of transmission to humans. Reservoir control continues to be of significant importance in global infectious disease control.

See also LATENCY; PATHOGEN; PATHOGENESIS; TRANSMISSION.

Further Reading

Simmons, Rachel, and Robert F. Siliciano. "Can Antiretroviral Therapy Ever Be Stopped? An Update." *AIDS Reader* 14 (2004): 441–442.

Smith, Theobald. "Animal Reservoirs of Human Disease with Special Reference to Microbic Variability." *Bulletin of the New York Academy of Medicine* 4 (1928): 476–496. Available online. URL: www.ncbi.nlm.nih.gov/pmc/articles/PMC2393908. Accessed January 22, 2010.

Thaxton, Melissa. "What Can Be Done to Protect the Chimpanzees and Other Great Apes of Africa?" Population Reference Bureau, August 2006. Available online. URL: www.prb.org/Articles/2006/WhatCanBeDonetoProtecttheChimpanzeesandOtherGreatApesofAfrica.aspx?p=1. Accessed January 22, 2010.

World Health Organization. "Avian Influenza H5N1 Infection in Humans: Urgent Need to Eliminate the Animal Reservoir—Update 5." January 22, 2004. Available online. URL: www.who.int/csr/don/2004_01_22/en. Accessed January 20, 2010.

Zimmer, Shanta M., and Donald S. Burke. "Historical Perspective—Emergence of Influenza A (H1N1) Viruses." *New England Journal of Medicine* 361 (2009): 279–285. Available online. URL: http://content.nejm.org/cgi/reprint/NEJMra0904322.pdf. Accessed January 20, 2010.

resistance Resistance is the ability of a microorganism to avoid harm from antibiotics or antimicrobial chemicals, or the ability of a host to repel infection by a pathogen. *Susceptibility* is the opposite; susceptible microorganisms remain vulnerable to antibiotics, and susceptible hosts are vulnerable to infection.

Drug companies introduced antibiotics, in the 1940s, to treat major and minor infections that had previously led to more serious conditions, even death. The new "wonder drugs" substantially reduced the threat of infectious diseases. Since their introduction, antibiotics have saved millions of lives, controlled infectious diseases, and increased life expectancy. The widespread use of antibiotics—and often their overuse or incorrect use—has caused the development of antibiotic-resistant microorganisms. In many cases, antibiotic resistance has become a graver medical concern than the infectious disease the antibiotic is intended to treat.

Host resistance involves all of the nonspecific and specific actions the body takes to prevent infection and disease. Nonspecific resistance results from the physical, chemical, and biological barriers the body uses to stop infection or remove an infectious agent as soon as it contacts the body. Intact skin offers an example of a nonspecific physical form of host resistance. Specific host resistance develops as a means of killing pathogens that the body recognizes. Antibodies provide the host with specific resistance to disease.

Factors associated with a pathogen and other host factors interact to determine whether an infection will take hold in a person. In general, factors that increase the pathogenicity (ability to cause disease) of a microorganism help it overcome the host's defenses. These factors, called *virulence factors,* contribute a pathogen's ability to evade host defenses. The more virulent a pathogen, the greater capacity it has to cause serious disease. At the same time, in hosts with weakened defenses, the likelihood of infection from pathogens of both high virulence and low virulence increases. Microorganisms that do not normally cause disease but can do so if a host's defenses are weak are called *opportunistic pathogens.* To understand how resistance works, scientists always consider resistance capabilities of both the pathogen and the host.

MICROBIAL RESISTANCE

Some microorganisms possess a natural resistance to antibiotics. For example, gram-negative bacteria can better resist antibiotics called beta-lactam antibiotics than gram-positive species, because these drugs interfere with peptidoglycan synthesis in cell wall construction. Gram-positive species depend on a thick peptidoglycan layer to compose their protective cell walls, whereas gram-negatives contain a much thinner peptidoglycan layer and gain most of their protection from their outer membrane.

Microorganisms also develop resistance by acquiring genes that encode for resistance factors, meaning any features that enable the cell to repel an antibiotic's action. Microorganisms acquire resistance genes by two main processes: mutation and transfer from another microorganism.

Mutations occur in two major ways: (1) spontaneous natural mutations and (2) induced mutations due to an outside factor such as a chemical, virus, or ultraviolet light. Each of these outside factors causes small alterations in a cell's deoxyribonucleic acid (DNA) that can change the correct sequence of base units, called *nucleotides*, that make up genes. If the cell keeps the altered gene during DNA replication and cell reproduction, the new cells are called *mutants*. Organisms survive because of a phenomenon called *selective pressure*. Individuals, in any population, must respond to changes in their environment by adapting, or they will die. Mutations that harm a microorganism disappear with the next generation because they do not help an organism adapt, but mutations that give a cell a benefit or allow it to survive in changed conditions will remain in the DNA through successive generations. Either spontaneous or induced mutations satisfy this need to respond to the environment's selective pressure. In induced mutation, the substance that causes the mutation to happen is called a *mutagen*.

The rate of spontaneous mutations varies by organisms. In general, in archaea and bacteria, mutations occur about once in every 300 replications of the chromosome (the cell's entire amount of DNA). This translates to from 10^{-6} to 10^{-7} mutation per gene per generation. Most organisms also have places in the chromosome that exceed this rate and other spots that mutate more slowly than this rate. In other words, microorganisms do not mutate often. Microorganisms do, however, reproduce often and can make 50–75 new generations in 24 hours. An advantageous mutation such as one that makes a microorganism resistant to an antibiotic can develop quickly in a population of the microorganisms. These defined populations can occur in a school, a dormitory, a hospital, or a person's body.

A gene that gives a cell resistance against one or more antibiotics represents an obvious advantage. Bacteria, fungi, and plants, in nature, make antibiotics that microorganisms in soil and water must resist to survive. In the same way, a population of bacteria in a hospital where antibiotics are given to most patients holds a vital survival advantage.

Microorganisms pass around genes by a process called gene transfer. Cells that die and then lyse release their genes into the environment, where other microorganisms can take them in by using a process called *transformation*. Bacteria also use conjugation to transfer genetic material. The material that bacteria often exchange is in the form of a plasmid, which is a circular strand of DNA that resides in the cell cytoplasm separate from the main chromosome (the mass of DNA in a general area called the *nucleoid*). Many plasmids carry resistance genes and are, thus, sometimes called R factors, for resistance factors. In conjugation, two cells connect via a short appendage called a sex pilus (plural: pili) and exchange plasmids (or R factors). As R factors spread through a bacterial population, antibiotic resistance also spreads.

R factors confer a variety of mechanisms by which microorganisms can resist antibiotics or antimicrobial chemicals. The four main mechanisms in drug resistance are:

- destruction or inactivation of the antibiotic
- prevention of antibiotic entry into the cell
- rapid ejection of an antibiotic that has entered the cell
- alteration of the antibiotic's main target sites in the cell

Microorganisms have developed variations on all of these mechanisms. Some microorganisms excrete large amounts of the antibiotic's target substance so that the drug attaches to the excretion and does not attack as many microbial cells. Other microorganisms do the converse, by making less of an antibiotic's target substance, so that the antibiotic has less opportunity to attach to the cell.

Bacteria inactivate some antibiotics by secreting an enzyme that cleaves the antibiotic. The best-known example of this mechanism belongs to bacteria that produce the enzyme β-lactamase, which breaks the active portion of antibiotics such as penicillin, called the *β-lactam ring*. Other microorganisms chemically modify the antibiotic's molecule by adding a piece or removing a piece from its structure to take away the antibiotic's activity.

Mechanisms of Antibiotic Resistance

Antibiotic Type	Drug's Mode of Action	Main Mechanism of Resistance
beta-lactam (β-lactam)	inhibits cross-linking of amino acids during peptidoglycan synthesis	β-lactamase, a penicillin- destroying enzyme
tetracycline	binds to ribosome subunit to prevent protein synthesis	efflux pump
aminoglycoside	binds to ribosome subunit to prevent protein synthesis	aminoglycoside-modifying enzymes

Prevention of the antibiotic's entry also effectively blocks antibiotic action. Many gram-negative bacteria resist antibiotics that cannot penetrate the gram-negative outer membrane. Other microorganisms change the structure of proteins on the cell surface when they detect an antibiotic, thereby changing the antibiotic's normal docking site on the cell.

The ejection method uses mechanisms called *efflux pumps*. Proteins in the cell membrane react to specific antibiotics and turn on a transport system that reverses inflow and ejects any substance that tries to penetrate the membrane, including antibiotics. Efflux pumps are usually nonspecific, provided the membrane protein, first, recognizes the presence of an antibiotic or other harmful chemical.

A more specific mechanism involves changing a target in the cell so the antibiotic cannot connect to it. Removing, adding, or replacing amino acids on a cell surface protein can produce this result. For antibiotics that disrupt metabolic enzymes in the cell, some resistant cells switch metabolic pathways in the presence of the antibiotic and, in this way, evade the antibiotic's mode of action.

The diverse pathogens that infect humans and animals have developed additional modifications of the four main mechanisms of resistance. Some classes of antibiotics have been rendered almost useless in fighting disease because mechanisms for resisting these drugs have spread in microbial populations worldwide. Three groups of antibiotics have reached the point where they are no longer foolproof treatments for stopping infection: (1) beta-lactam antibiotics (penicillin and its derivatives), (2) tetracyclines, and (3) aminoglycosides (streptomycin and related antibiotics). The table above describes the main mechanisms of resistance against these drugs.

More than one antibiotic from each of the three antibiotic classes discussed here have now caused the evolution of resistant populations. The first, original form of the antibiotic is called the natural form. When resistance develops in a microbial population against this form, drug companies develop a new version of the molecule, called a *first-generation* drug. Some antibiotics have second-, third-, or even fourth-generation versions because pathogens have grown resistant to the previous versions. The table below illustrates why beta-lactam, tetracyclines, and aminoglycosides have decreasing effectiveness against infections: Each class has required several new generations to replace previous ineffective versions. This table presents some of the many derivatives in each of these three antibiotic classes.

The United States produces millions of pounds of antibiotics a year, of which about 70 percent are added to livestock feeds. The massive amounts of antibiotics used in meat production and in medical care have changed microbial populations, since the 1940s, when sulfa drugs and penicillin first became available. When doctors first began prescribing antibiotics, *Shigella*, *Staphylococcus aureus*, *Streptococcus pneumoniae*,

Next-Generation Antibiotics

Antibiotic Type	Original Antibiotic	Next-Generation Antibiotics
beta-lactam (β-lactam)	penicillin	amoxicillin, carbenicillin, cloxacillin, methicillin, oxacillin
tetracyclines	aureomycin	doxycycline, methacycline, minocycline, oxytetracycline
aminoglycosides	streptomycin	gentamicin, kanamycin, neomycin, tobramycin

Multiple Antibiotic– (Multidrug-) Resistant Microorganisms

Name	Microorganism	Antibiotics Resisted	Disease
VRE	vancomycin-resistant *Enterococcus*	penicillins, cephalosporins, vancomycin	gastroenteritis
MDR *Klebsiella*	*Klebsiella pneumoniae*	penicillins, cephalosporins, chloramphenicol, ciprofloxacin, trimethoprim-sulfamethoxazole	pneumonia
MDR TB	*Mycobacterium tuberculosis*	isoniazid, rifampicin	tuberculosis
XDR TB	extensively drug-resistant TB *(M. tuberculosis)*	MDR TB plus fluoroquinolones and at least one of amikacin, kanamycin, or capreomycin	tuberculosis
MDR gonococci or antibiotic-resistant gonorrhea (ARG)	*Neisseria gonorrhoeae*	quinolones, cephalosporins, chloromphenicol, tetracycline, erythromycin	gonorrhea
MDR *Salmonella*	*Salmonella* species	cephalosporins, fluoroquinolones	salmonellosis
MRSA	methicillin-resistant *S. aureus*	penicillins, cephalosporins, gentamicin	skin and urinary tract infections, septicemia (systemic infection)
VRSA	vancomycin-resistant MRSA	MRSA antibiotics plus vancomycin	same as MRSA
MDR *S. pneumoniae*	*Streptococcus pneumoniae*	penicillins, cefotaxime, trimethoprim-sulfamethoxazole, erythromycin	pneumonia

Mycobacterium tuberculosis, and enteric cocci were sensitive to antibiotic treatment. All of these pathogens now resist the antibiotics first prescribed to fight them.

Concern has grown among microbiologists about potential connections between the use of antimicrobial products in the home and antibiotic resistance. Microbiologists long believed that disinfectant chemicals did not lead to chemical resistance in bacteria, but evidence has accumulated to show that bacteria may use the same mechanisms for resisting chemicals that they use against antibiotics. The National University of Ireland microbiologist Gerard Fleming studied *Pseudomonas aeruginosa*'s response to levels of a common disinfectant called benzalkonium chloride. *P. aeruginosa* is a common contaminant of aqueous solutions used in hospitals for intravenous systems and rinsing of burn wounds. Fleming said of *P. aeruginosa*'s reaction to the chemical, "They're taking in the disinfectant and they're kicking it out as fast as they're taking it in." Bacteria may, in fact, use the same or a similar efflux pump for ejecting chemicals that they use for antibiotics.

HEALTH CONCERNS DUE TO ANTIBIOTIC RESISTANCE

Antibiotic-resistant microorganisms first emerged about two decades after the use of antibiotics became widespread. The first antibiotic-resistant organisms evaded the actions of the single antibiotic prescribed for them. Today, the same pathogens have resistance to multiple antibiotics. For example, *S. aureus* that was once susceptible to penicillin now resists penicillin, methicillin, and gentamicin and may be developing resistance to the only antibiotic that had been working against it, vancomycin. This strain of *S. aureus,* called methicillin-resistant *S. aureus* (MRSA), emerged, in the late 1970s, and scientists have since traced how this microorganism spread. Using a technology called *high-throughput gene sequencing,* scientists analyzed MRSA strains collected from patients around the world, between 1982 and 2003. The project member Simon Harris of the United Kingdom's Wellcome Trust Sanger Institute explained, in 2010, "We wanted to test whether our method could successfully zoom in and out to allow us to track infection on a global scale—from continent to continent, and also on the smallest scale—from person to person." By tracking minor DNA variations in the strains, the researchers traced MRSA's emergence to strain ST239, which originated in Europe, in the 1960s, when antibiotic use boomed; then spread to South America and then Asia. By 2005, a new MRSA emerged: community-acquired MRSA (ca-MRSA). The Los Angeles County epidemiologist Elizabeth Bancroft was quoted in *Time,* in 2006, saying, "This bug has gone from 0 to 60, not in five seconds but in about five years. It spreads by contact, so if it gets into any community that's fairly close-knit, that's all it needs to be passed." Close-knit communities such as college dormitories, sports teams, and hospitals are at the greatest risk for this contact-transmitted MRSA. Resistant pathogens have become a troubling aspect of patient care in hospitals. Many hospital-associated, or *nosocomial,* infections involve multidrug-resistant (MDR) pathogens.

The table on page 673 lists important multidrug-resistant microorganisms that are creating the high health risks. These microorganisms are often called MDR organisms, or MDROs.

Resistance develops quickly in many microorganisms. Some of the pathogens listed in the table are beginning to show low to moderate resistance to antibiotics in addition to the drugs listed here.

Antibiotic resistance began growing, in about the 1960s, for the following five main reasons: (1) inappropriate use of antibiotics such as for pathogens unaffected by antibiotics, (2) incorrect use by patients by taking too low a dose or discontinuing the drug before the prescription was used up, (3) self-medication, (4) growth in nosocomial resistance, and (5) growth in antibiotic use in food animals. Of these factors, inappropriate prescribing, overuse, and incorrect compliance with the drug's directions have been blamed for today's resistance crisis.

Other unexpected factors related to societal changes and the economy have indirectly contributed to increasing risks from antibiotic-resistant pathogens:

- increased urbanization and overcrowding—increases risk of disease transmission
- pollution—debilitates health and weakens immunity
- changing weather patterns—increase disease vectors (insects)
- poverty—decreases nutrition and proper health care
- demographic changes toward an older population—increase the proportion of high health risks
- acquired immunodeficiency syndrome (AIDS) epidemic—increases immunocompromised population
- globalization—increased global trade and travel accelerate the spread of disease

With the potential increase in infections that accompany the factors listed here more antibiotic use is likely. As a result, the chances for developing new antibiotic-resistant microbial populations also increase.

RESISTANCE TO ANIMAL ANTIBIOTICS

Beginning in the 1960s, antibiotic use in food-producing animals increased. Low levels of antibiotics had been instituted to reduce infections in animals living in close quarters, where infections could quickly transmit through the population. Farm managers also noticed that the growth rate of animals bound for slaughterhouses increased when feeds included antibiotics. Research on the effects of antibiotics on meat-producing animals suggests that antibiotics alter the populations of digestive tract bacteria and presumably increase the efficiency of nutrient digestion. This type of antibiotic administration for animals with no obvious illness is called *subtherapeutic use.* About 70 percent of U.S. antibiotic use is in agriculture for subtherapeutic use in animals; humans use 9 percent of antibiotics, 6 percent is used for infection control in agricultural animals, and the remainder of the antibiotics are used in small animal veterinary medicine, research, and emergencies.

Groups such as the Chicago-based organization Keep Antibiotics Working and the Boston-based Alliance for the Prudent Use of Antibiotics (APUA) warn that agricultural antibiotics can enter the human population in three ways: (1) retail meats, (2) close contact with livestock, and (3) contaminated groundwater, surface water, and soil. These and likeminded organizations have put pressure on agriculture to reduce subtherapeutic antibiotic use. Research on antibiotic resistance in meat-producing animals shows trends toward increased resistance to most antibiotics.

In 2005, a Johns Hopkins University study compared fluoroquinolone-resistant *Campylobacter* in meat from antibiotic-fed chickens versus non-antibiotic-fed chickens. The authors reported, "Overall, *Campylobacter* was detected on 84 percent of the chickens tested, and FQ [fluoroquinolone]-resistant strains were detected on 17 percent." These results are surprising when considering that large U.S. poultry producers discontinued fluoroquinolone use, in 2002. The study authors concluded,

Factors in Host Resistance

Nonspecific Resistance	Resistance Factors
first line	• intact skin • mucous membranes • skin, sweat gland secretions • normal flora
second line	• phagocytosis • inflammation • fever • nonspecific antibodies
Specific Resistance	**Immune System**
third line	• specialized lymphocytes • white blood cells • antibodies

"Accepting the veracity of these announcements, our data suggest that past FQ use may have persistent effects on *Campylobacter* populations in poultry houses. This is consistent with reports from Denmark indicating that vancomycin-resistant enterococci could be isolated from broiler flocks five years after [the antibiotic] avoparcin was banned for use in broilers in that country." Similar studies have indicated that resistant microorganisms occur in places no longer using antibiotics, meaning that resistant populations have remained on the farm or people are introducing contamination.

Antibiotic-resistant bacteria spread from animal populations to humans. In 1983, people in the Midwest fell ill from eating *Salmonella*-contaminated hamburger. Clinical microbiologists isolated MDR *Salmonella newport* from 20 patients, all of whom had eaten meat from animals that had been raised with subtherapeutic doses of chlortetracycline. The rise of MDR organisms in animals and its relation to human health involve many subtle factors that make studies very difficult. Evidence points, nevertheless, to an antibiotic-resistance problem that needs solving.

HOST RESISTANCE

The body has three lines of defenses that resist infectious agents, the first two of which are nonspecific defenses. If pathogens get past the nonspecific defenses, the immune system provides the final defense for helping a host resist infection. The table above describes the main factors in host resistance.

Nonspecific resistance in the host has physical, chemical, and biological sources. Specific resistance, by contrast, is contained in biological factors produced by the body's immune system.

The skin and mucous membranes supply the strongest physical defenses to resist infection. Saliva, tears, and sweat wash some microorganisms from the skin and dilute the numbers of any pathogens that happen to be on the skin. The skin and mucous membranes also provide chemical and biological resistance factors, such as the following:

- secretions containing enzymes, such as lysozyme, that degrade bacteria
- fatty acids in sebum that lower pH and inhibit some microbial species
- sebum oils to prevent drying of the skin and, thus, reduce breaks in the outer skin layer
- gastric juice acids that kill many infested microorganisms

The normal flora on the skin and in the mouth and digestive tract add another line of resistance by crowding out transient pathogens, outcompeting nonnative microorganisms for food, and producing secretions to inhibit pathogens.

Resistance, whether in microorganisms or a host, ensures the survival of organisms. Life would be perilous for either host or infectious agent without defenses that have evolved over generations. In this way, resistance plays a pivotal part in infectious disease and the persistence of these diseases in populations of humans or animals.

See also ANTIBIOTIC; GENE TRANSFER; IMMUNITY; NORMAL FLORA; NOSOCOMIAL INFECTION; OPPORTUNISTIC PATHOGEN; PLASMID; VIRULENCE.

Further Reading

Alliance for the Prudent Use of Antibiotics. Available online. URL: www.tufts.edu/med/apua/index.html. Accessed January 2, 2010.

Centers for Disease Control and Prevention. Available online. URL: www.cdc.gov. Accessed January 2, 2010.

Keep Antibiotics Working. "Factsheet: Antibiotic Resistance and Animal Agriculture." Available online. URL: www.keepantibioticsworking.com/new/resources_library.cfm?refID=69872. Accessed January 2, 2010.

Kelland, Kate. "High-Resolution Gene Technique Zooms in on Superbug." January 21, 2010. Available online. URL: http://abcnews.go.com/Technology/wirestory?id=9625552&page=1. Accessed January 2, 2010.

Khan, Amina. "Cleansers versus Superbugs." *Los Angeles Times*, 2 January 2010.

Price, Lance B., Elizabeth Johnson, Rocio Vailes, and Ellen Silbergeld. "Fluoroquinolone-Resistant *Campylobacter* Isolates from Conventional and Antibiotic-Free Chicken Products." Available online. URL: www.ncbi.nlm.nih.gov/pmc/articles/PMC1257547. Accessed January 2, 2010.

Do Disinfectants Cause Antibiotic Resistance?

by Nokhbeh M. Reza, Susan Springthorpe, and Syed A. Sattar, Ph.D.
Centre for Research on Environmental Microbiology, Faculty of Medicine, University of Ottawa, Ottawa, Ontario, Canada

The past century is often called the "era of antimicrobials" for the unprecedented successes in introducing naturally occurring and synthesized chemicals to counter disease-causing and other undesirable microbes. There are three broad categories of such chemicals: (a) preservatives—added to foods and other perishable items to extend their shelf life or usability by preventing or slowing the growth of microbes that can cause spoilage or decay; (b) anti-infectives (e.g., antibiotics)—drugs to prevent or treat infections in humans, animals, and plants; and (c) microbicides (disinfectants and antiseptics)—to kill undesirable microbes in the environment, as well as on living skin/mucous membranes. Here, we will, first, focus on how human pathogenic bacteria can develop resistance to antibiotics and, then, discuss the potential of preservatives and microbicides to contribute to such resistance.

We now know that the rampant and essentially unbridled use of antibiotics has caused many harmful microbes to become resistant to those once-beneficial chemicals, and this continues to happen at an alarming rate. For example, the bacterium that causes tuberculosis *(Mycobacterium tuberculosis)* is now resistant to virtually every available antibiotic. Multidrug-resistant tuberculosis (MDR-TB), extensively drug-resistant tuberculosis (XDR-TB), as well as methicillin-resistant *Staphylococcus aureus* (MRSA), and vancomycin-resistant *Enterococcus* (VRE) are just some examples of antibiotic-resistant pathogens of increasing concern. At the same time, the rate of introduction of new antibiotics has also slowed considerably, as a result of shrinking profit margins and increasingly complex government regulations. Therefore, many believe that we have already entered the postantibiotic era. The questions of immediate relevance here are:

1. How do bacteria develop resistance to antibiotics?
2. Can bacteria also become resistant to chemicals such as microbicides and preservatives?
3. Can resistance to microbicides and preservatives also make bacteria cross-resistant to antibiotics?
4. If such cross-resistance does occur, can the indiscriminate use of microbicidal and preservative chemicals exacerbate the already-serious problem of antibiotic resistance?

To consider this, we must, first, understand how microbes can develop resistance to chemicals in general.

In ancient times, salt, spices, sugar, and wood smoke were often used to preserve many types of foods; when and where possible, storage at lower temperatures was also common. Such practices not only preserved perishable foods, but also reduced the risk of food-borne illnesses by suppressing the growth of disease-causing microbes; we continue to use those time-honored practices even today. The past 150 years, however, has witnessed rapid-fire discoveries of microbes and their association with food spoilage and infectious diseases, thus providing science-based approaches to using chemicals to preserve perishables and prevent/treat a wide variety of infections. While the invention of canning has become an enormously convenient and safe means of preserving perishable foods, many chemicals are still relied upon as food preservatives.

Ignaz Semmelweis (1818–65) and Joseph Lister (1827–1912) were among the first to introduce the use of chemicals as antiseptics to reduce the risk from hospital-associated infections. Lister also pioneered the use of carbolic acid for disinfecting surgical instruments and for general decontamination of surgical suites.

The work of Semmelweis and Lister not only catalyzed the wider acceptance of chemicals as antiseptics and disinfectants, but also encouraged others to find chemicals for the treatment of infections. Paul Ehrlich (1854–1915) was among the first to venture into this new and potentially highly impactful area of scientific endeavor. He coined the term *chemotherapy* to denote the administration of chemicals into the body to prevent the growth of disease-causing microbes. This was truly a bold and revolutionary concept for its time. Not surprisingly, the initial attempts at chemotherapy were fraught with enormous problems of toxicity of the administered chemicals, many of which contained harmful substances such as arsenic and mercury. The early attempts at chemotherapy also often failed because they were no more than a shot in the dark without the crucial knowledge of the target pathogen.

Even though the subsequent work of Ehrlich and others led to the marketing of certain synthetic drugs (sulfonamides, or sulfa drugs), the next major breakthrough in this area occurred in the chance observation of antibiosis by Alexander Fleming (1881–1955) and the successful mass production of the first antibiotic (penicillin) by Howard Florey and Ernst Chain, in the early 1940s. The unprecedented success of penicillin as an antibiotic, especially in the period of World War II, led to the rapid discovery and marketing of many other similar chemicals.

Unlike earlier types of drugs, antibiotics proved much less toxic and highly selective in their action against many common bacterial pathogens. These "magic bullets," together, generated a strong sense of euphoria and encouraged even some experts prematurely to declare the demise of infectious diseases as problems of human health. Antibiotics, were, thus regarded as a panacea to all our ills and were used with gay abandon. Antibiotic resistance, a phenomenon that had indeed been noted soon after the discovery of penicillin, was either forgotten or ignored to our own peril. As a result, we have rendered many of those magic bullets virtually useless against some infectious agents through misuse and overuse. But how could this happen in a matter of a few years?

To understand this, we must remember that bacteria are the most abundant, prolific, and adaptable of all life-forms. These innate abilities allow them to respond to and survive exposure to many environmental stresses and toxins. Bacteria, in general, can become resistant to such stresses mainly through the following interconnected factors:

- Population doubling times of as short as 20–30 minutes under favorable conditions can generate millions of bacterial cells in just a few hours.
- Their relatively small and highly pliable genomes allow bacteria to adapt rapidly to selective pressures. Thus, a single resistant mutant can not only survive, but also give rise to a large progeny of such mutants relatively fast.

Resistance of bacteria to potentially damaging chemicals may be triggered by two related means. The chemical itself may be mutagenic. Alternatively, or in addition, the presence of the chemical may act as a powerful selective force to eliminate the unfit and enrich for subpopulations of resistant mutants.

Mutations can arise spontaneously or be induced by physical and/or chemical agents that cause genetic change (mutagens). In the case of induced mutations, the ultimate effect of these agents is exerted on bacterial deoxyribonucleic acid (DNA) by formation of modified nucleotide(s). During replication, the modification causes the nucleotide to pair with a nucleotide other than its normal complement, resulting in a mismatched base pair. During a second round of replication, the newly synthesized strand of DNA, with the incorrect base included, will act as a template and a normal base pairing will occur, but the change already made in its DNA will persist. The outcome is alteration of a single nucleotide base (base substitution mutation). Many such changes occur continually, but only a handful may confer on the bacterium an advantage that allows it to become dominant over its neighbors. Some mutations require additional energy expenditure by bacteria, and these are only likely to be maintained in a population while it is advantageous to do so, for example, under a selective pressure such as might be exerted by sublethal levels of an antimicrobial. On the other hand, maintaining a selective pressure on the population for prolonged periods might allow stepwise adaptation through successive mutations.

Though mutations can occur randomly in virtually any gene, there are some "hot spots" with a higher tendency to mutate. Mutations can also occur spontaneously through errors in the genome replication process itself. Microbes can also simply assimilate fragments of DNA from other microbes in their surroundings (transformation). The process of mating (conjugation) can transfer DNA from a donor to a recipient cell. Viruses that can infect bacteria (bacteriophages) can also mediate the transfer of either genomic or plasmid DNA between bacterial cells. All of these processes confer on bacteria a higher resistance to a variety of chemicals.

Bacteria can, routinely, correct errors in DNA replication to maintain the integrity of their genetic traits from one generation to the next. But, from time to time, escaped errors can lead to altered and fixed genetic traits. Such surviving errors may be in genes directly related to antibiotic sensitivity or in those involved in housekeeping functions such as DNA repair; bacteria with damaged DNA repair capacity are nearly 1,000-fold more prone to accumulating mutations.

Antibiotics such as quinolones can also be bacterial mutagens, in themselves, adding another layer of complexity to this already-complex issue. In addition, antibiotics can weaken the immune system, compromising the body's ability to deal with harmful bacteria. Thus, antibiotics, themselves, may play multiple roles in inducing antibiotic resistance in bacteria and in the host's ability to deal with the infectious agent.

Antibiotics are also used routinely in veterinary medicine, animal feed, and sprays against plant pathogens. Such uses, which account for hundreds of metric tons annually, can lead to widespread exposure of bacteria to sublethal levels of the chemicals and probably

(continues)

Do Disinfectants Cause Antibiotic Resistance?
(continued)

contribute much to the global development of antibiotic resistance.

What about bacterial resistance to microbicides and preservatives? To answer this question properly, one must understand the basic differences and similarities in the nature and usage of antibiotics and other antimicrobials. At least in human and veterinary medicine, antibiotics are often administered in dosages that have been determined to be as safe as possible to the host, while being inhibitory to the target pathogen. In other words, critical factors such as host toxicity, the rate of clearance of the drug from the body, and market competitiveness must be carefully balanced to determine the dosage to be applied. If the correct type of antibiotic is selected and the prescribed therapeutic regimen is followed, the chance of success is high in countering the infection and quite low in inducing strains resistant to the administered drug. However, quite frequently, the prescribed course is aborted if the patient begins to feel better, or the drug itself may be adulterated, expired, or substandard, thus enhancing the risk of selecting for resistant mutants. These factors may also contribute to the development of antibiotic resistance.

Antibiotics, in general, produce their antibacterial action by working on specific metabolic sites on dividing bacterial cells. For example, penicillin can interfere with the process of cell wall synthesis, and resistance to penicillin can develop if mutants that have an altered target site are generated.

With some notable exceptions (see later discussion), microbicides and preservatives are quite different from antibiotics with regard to the two factors described, as well as their chemistry. However, with the exception of some of the chemicals used for food preservation, the chemicals used in biocides and preservatives have significant overlap, though some chemicals are specific to each class of use. Since microbicides are not meant to be either injected or ingested, their in-use concentrations can be much higher than those of antibiotics; even antiseptics are often used at dosages substantially greater than those for antibiotics. The concentration of preservatives varies widely with intended use. Moreover, microbicides and preservatives are much less discriminating in the way they inactivate bacteria. For example, chlorine-based formulations could quickly and irreversibly damage many essential components of the target bacteria cell rather than selectively affect a specific site in/on it. This means that microbicides, when used correctly, are far less likely to give rise to mutant strains. Why, then, the concern with cross-resistance?

Certain types of microbicides can either break down into innocuous by-products (hydrogen peroxide) or evaporate (ethanol) at the site of application. Many others can be a lot more stable and eventually end up in soil or water, possibly at sublethal levels. Exposure of bacteria to such sublethal concentrations could select for resistant mutants.

Triclosan is a common microbicidal ingredient in many disinfectants, antiseptics, and personal care products. Much attention focused on this chemical when laboratory-based studies showed that when bacteria develop resistance to it, they also show higher levels of resistance to certain types of drugs—a classic case of cross-resistance. Subsequent work revealed that Triclosan behaves much as an antibiotic does, in being site-specific to produce its antibacterial activity, and it does so by inhibiting bacterial fatty acid synthesis at the enoyl-acyl carrier protein reductase (FabI) step. There are also increasing numbers

San Diego State University. "Mutation Rates." Available online. URL: www.sci.sdsu.edu/~smaloy/MicrobialGenetics/topics/mutations/fluctuation.html. Accessed January 24, 2010.

Tapsall, John. "Multidrug-Resistant *Neisseria gonorrheae.*" *Canadian Medical Association Journal* 180 (2009): 268–269. Available online. URL: www.cmaj.ca/cgi/content/full/180/3/268. Accessed January 2, 2010.

Time. "Surviving the New Killer Bug." June 18, 2006. Available online. URL: www.time.com/time/magazine/article/0,9171,1205364,00.html. Accessed January 2, 2010.

World Health Organization. "Antimicrobial Resistance." January 2002. Available online. URL: www.who.int/mediacentre/factsheets/fs194/en. Accessed January 20, 2010.

Rickettsia *Rickettsia* is a genus of gram-negative, aerobic, nonmotile bacteria that can grow only inside other living cells; thus, they are called *obligate parasites.* Cells are short and small, measuring 0.3 micrometer (μm) in width and no more than 2 μm in length. Most cells are rod-shaped, but the genus also produces oval-shaped and pleomorphic cultures, which are cultures of the same species containing various cell shapes.

Rickettsia belongs to family Rickettsiaceae, order Rickettsiales, and class Alpha-proteobacteria of phylum Protobacteria. Two important genera that also belong to family Rickettsiaceae are *Ehrlichia,* which is also an obligate parasite in humans, and *Wolbachia,* a microorganism that lives in symbiosis

of reports from laboratories as well as field-based studies on the higher levels of resistance of a variety of common human pathogenic bacteria *(Pseudomonas aeruginosa* and *Staphylococcus aureus)* to many microbicidal chemicals.

The application of properly formulated microbicides at the right concentration is unlikely to give rise to resistant mutants. Inevitably, though, such chemicals are discarded and, thereby, diluted to sublethal levels in the environment. What happens, then, when microbes are exposed to them at these reduced levels? The answer to this is by no means clear for either microbicides or preservatives, but one can look, perhaps, at the microbial responses for clues.

Bacteria have a limited repertoire for resisting exposure to toxic chemicals. Perhaps, the most obvious means of protection is keeping the toxic out through changes to cell permeability. While this may happen more frequently than is recognized, the capability may also put the cell at a selective disadvantage for acquiring nutrients and other beneficial chemicals and may, thus be restricted to certain types of organisms or situations when growth is, in any case, limited. The other case, which is more frequently seen in resistant mutants, is the "pump it out" scenario. Many microbes possess pumps to rid themselves of toxic chemicals, and frequently these are induced by the presence of the chemical(s). Often, these pumps are relatively nonspecific, in terms of the type of chemical that can be removed from the cell. Changes to such pumps in relation to range of chemicals or efficiency may permit the organism to survive more or less well in the presence of toxins, but obviously those mutations that confer more sensitivity are out-selected when the chemical is present in the environment of the organism.

These two classes of modification/mutation might suggest that resistance to one chemical might also mediate resistance to another related or unrelated chemical, including perhaps antibiotics. On the other hand, more specific mutations that might detoxify the chemical by enzymatic activity, sequester it with a specific metabolite, or alter the specificity or number of targets in the cell might not be so readily transferable across chemical classes. Thus, one could argue that whether or not there is cross-resistance occurring among miscellaneous antimicrobial chemicals and antibiotics might depend on the exact mechanism by which resistance is achieved. In spite of logic, this issue remains unresolved. There are, however, additional arguments that can be made to suggest that antimicrobial chemicals other than antibiotics are also being greatly overused and misused. In most cases, demand for such products is driven more by market forces than by outright demonstration of their value and effectiveness; they may also give a false sense of security to the user.

See also ANTIBIOTIC; ANTISEPTIC; BIOCIDE; DISINFECTION; RESISTANCE.

Further Reading

Levy, Stuart B. "Antibacterial Household Products: Cause for Concern." Presented at the 2000 Emerging Infectious Diseases Conference, Atlanta, July 16–19, 2000. Available online. URL: www.cdc.gov/ncidod/eid/vol7no3_supp/pdf/levy.pdf. Accessed November 8, 2009.

Mascaretti, Oreste A. *Bacteria versus Antibacterial Agents: An Integrated Approach.* Washington, D.C.: American Society for Microbiology Press, 2003.

Price, Christopher T. D., Vineet K. Singh, Radheshyam K. Jayaswal, Brian J. Wilkinson, and Johm E. Gustafson. "Pine Oil Cleaner–Resistant *Staphylococcus aureus:* Reduced Susceptibility to Vancomycin and Oxacillin and Involvement of SigB." *Applied and Environmental Microbiology* 68 (2002): 5,417–5,421. Available online. URL: http://aem.asm.org/cgi/content/full/68/11/5417. Accessed November 8, 2009.

with insects. *Rickettsia* that infect humans transmit by way of ticks, fleas, mites, flies, or lice and enter the bloodstream when the insect bites.

Inside the bloodstream, *Rickettsia* induces phagocytosis by immune system cells. In this process, a variety of human cells called phagocytes envelop and engulf particles. By inducing this action, *Rickettsia* hides from other destructive reactions of the immune system to infectious agents. The phagocyte holds the bacterial cell in a sac called the *phagosome,* which is normally the site where digestive enzymes break down foreign matter. But *Rickettsia,* unlike other types of bacteria captured in phagocytosis, is adept at escaping the phagosome and entering the phagocyte's cytoplasm, where it, then, reproduces by binary fission. This reproductive method is used by almost all bacteria. Although *Rickettsia* reproduce in the host cell, they do not immediately cause the cell to burst, as other intracellular parasites do. The microorganisms slowly produce about 10 daughter cells, before lysing the host cell. The rickettsial membrane is more permeable than membranes of most other bacteria, meaning it allows more substances to pass in and out of the cell. By staying in a host cell's cytoplasm, *Rickettsia* can reduce its vulnerability to substances toxic to the bacteria.

Rickettsia requires special culture techniques for growing them in a laboratory. Rather than use traditional growth medium, microbiologists grow *Rickettsia* by inoculating mammalian cell cultures or chicken embryos. *Rickettsia* growth appears as a purple film over the host tissue when colored by a technique called the Giemsa stain. This method uses Giemsa powder in a mixture of glycerol and methanol, which must remain on a specimen for one hour before it is studied microscopically.

Rickettsia's small cell size makes it very difficult to study with light microscopy. The Gram stain makes cells almost indistinguishable from background matter. The Giemsa stain or fluorescent staining helps solve this problem. The following two stains have also been helpful for study *Rickettsia* by light microscopy:

- Giménez stain—basic fuchsin, ethanol, and phenol in water; cells appear reddish against a green background
- Macchiavello stain—basic fuchsin, ethanol, citric acid, and methylene blue in water; cells appear bright red against a blue background

Despite the choices in staining methods, many of the cellular features of *Rickettsia* have been discovered by using electron microscopy, which produces images of objects measured in nanometers (nm) rather than micrometers (μm). A nanometer is one billionth of a meter; a micrometer is one millionth of a meter.

This genus has a metabolism and physiology that differ from those of other bacteria. *Rickettsia* is one of the rare organisms in nature that do not have glycolysis, the metabolic pathway by which most animals and prokaryotes derive energy from glucose. As a consequence, *Rickettsia* cannot grow on glucose. The microorganism depends on the host to supply potassium, albumin, and the sugar sucrose. The cells also use the host cell's adenosine triphosphate as an energy source.

RICKETTSIA PATHOGENESIS

Because *Rickettsia* species are obligate parasites, this microorganism, unsurprisingly, causes disease in humans. *Rickettsia* multiplies in the digestive tract of insect carriers, called *vectors*. Cells migrate to the insect's saliva and remain there until being transmitted to a human, when the insect bites.

Rickettsia can live anywhere in the host but tend to target the skin, brain, and heart. *Rickettsia*, additionally, nestles in blood vessel walls, such as capillaries, and, eventually, causes leaks in those vessels. As a result, a condition called spotted rash indicates a possible *Rickettsia* infection.

A group of diseases called the *rickettsial diseases* includes illnesses caused by *Ehrlichia*, *Coxiella*, and *Bartonella* through similar modes of infection and pathogenesis. Several species now part of *Bartonella* were recently grouped with *Rickettsia*. The table on page 681 lists the major rickettsial diseases seen worldwide.

Ehrichliosis is a sometimes-fatal disease in humans caused by *Ehrlichia* bacteria that attack white blood cells in a similar manner to *Rickettsia*. Pathologists often group viral and rickettsial diseases together because of their similar need to infect animal cells to propagate.

Veterinary rickettsial diseases are not major health threats. Dogs are susceptible to two severe diseases: ehrlichiosis and tick fever. Ehrlichiosis results from a bite of the brown dog tick that infects the animal with *E. canis* and transmits easily to other dogs when ticks are passed on to them. Ehrlichiosis often develops into a chronic disease marked by loss of appetite and stamina, depression, and stiffness.

Tick fever is the canine version of Rocky Mountain spotted fever, caused by *R. ricketsii*. This disease is endemic in the Americas and carried by the American dog tick. Early signs of infection include fever, loss of appetite, and coughing. Tick fever progresses to the nervous system and infects the conjunctiva. Both tick fever and ehrlichiosis respond to antibiotic treatment.

Rickettsia and related bacteria occupy a special place in pathology because they behave more as viruses do than bacteria do by infecting host cells. *Rickettsia* are not thought to infect other bacteria, but the study of these organisms lags behind study of many other pathogens because of the difficulty of working with these species in vitro.

See also BINARY FISSION; PATHOGENESIS; PHAGOCYTOSIS.

Further Reading

Centers for Disease Control and Prevention. Division of Viral and Rickettsial Diseases. Available online. URL: www.cdc.gov/ncidod/dvrd/index.htm. Accessed December 22, 2009.

Merck Manual. Whitehouse Station, N.J.: Merck, Sharp and Dohme, 2009.

Raoult, Didier, and Philippe Parola, eds. *Rickettsial Diseases*. London: Informa Healthcare, 2007.

rumen microbiology Rumen microbiology is a specialization in the study of enteric microorganisms living in the digestive tract of some herbivorous animals. Ruminant animals possess a compartmentalized stomach that evolved for the digestion of fibrous diets (grasses, hay, woody plants), through fermentations carried out by a dense and diverse population of bacteria and protozoa. Because no oxygen accumulates in the ruminant digestive organs, this microbial population contains only strict anaerobes. Rumen microbiology studies mainly the populations of the rumen, the largest part of the animal's four-part stomach and the site of the most active fermentations of feedstuffs.

Rickettsial Diseases

Disease	Pathogen	Known Vectors
cat scratch fever	*Bartonella henselae*	fleas
ehrlichiosis	*Ehrlichia* species	ticks
murine typhus	*Rickettsia typhi*	fleas
Oroyo fever	*Bartonella bacilliformis*	flies
Rocky Mountain spotted fever	*Rickettsia rickettsii*	ticks
scrub typhus	*Rickettsia tsutsugamushi*	mites
trench fever	*Bartonella quintana*	lice
typhus epidemics	*Rickettsia prowazekii*	lice

RUMINANT ANATOMY AND FUNCTION

Mammals belong to one of two major groups based on the anatomy and function of their digestive tract: ruminants and nonruminants. Ruminant animals are herbivores that rely on a large fermentation organ to digest fibrous materials. The main fibers in the ruminant's diet are cellulose, hemicellulose (xylans), and lignin. Digestible fibers such as cellulose are often interwoven with indigestible lignin and are thus referred to as *lignocellulosic fibers*. The ruminant diet also contains lesser amounts of the following polymers: dextrins, pectins, xylans, amylose (starch), and polygalacturonic acid. Each polymer differs from the others in its composition of sugars and the bonds that link the sugars in a chain.

Nonruminant animals can be subdivided into true monogastrics (single-stomach) and monogastrics adapted for hindgut fermentation. True monogastrics are carnivores or omnivores, such as humans, possessing a single stomach of highly acidic excretions that performs some food digestion before partially digested matter moves to the intestines. The second group of monogastric animals are herbivores that also use a stomach for digestion but also possess a fermentation organ called the *cecum*. The distal part of the small intestine (the ileum) opens into the cecum, where fibrous matter can be digested by a process, similar, to ruminant fermentations. After the fibers have been degraded, the matter flows to the large intestine, as it would in true monogastrics. Bacteria and protozoa in the cecum contain many similarities to the rumen microorganisms, and these animals are sometimes called *pseudoruminants*. Many of the same species occur in both, but in different proportions between the two groups. All animals, regardless of the three groups listed here, contain dense populations of microorganisms in their intestines, especially in the colon (large intestine), for final digestion of matter before they excrete the wastes as feces. The colon populations carry out both fermentative digestion and enzymatic digestion. The table on page 682 lists animals by their type of digestive tract.

The type of diet given to ruminant animals affects the types and proportions of microorganisms in the rumen. Diets high in fiber (grasses, hay) increase the proportion of cellulose-degrading, or *cellulolytic,* bacteria. Diets high in starches (grains) increase the proportion of species that degrade these polysaccharides.

The ruminant digestive system begins in the mouth, leads to the esophagus, and, then, expands to a large complex of digestive organs, before continuing to the small intestine, large intestine, and rectum. The large complex consists of the reticulum, rumen, omasum, and abomasum. The reticulum and rumen have the same function and are separated by a partial divider called the reticulorumen fold, which allows the liquid contents to flow freely back and forth between these compartments. For this reason, the two compartments are often referred to as the reticulorumen. The omasum is the third compartment, and contents flow from the reticulorumen through it to the abomasum. The abomasum has been referred to as the *true stomach,* because the enzymatic activities occurring here resemble digestion in monogastric animals. The reticulum, rumen, and omasum, together, make up the forestomach. The table on page 683 describes the features of the four compartments that the ruminant stomach comprises.

Ruminant animals carry out a digestive process called rumination, which consists of four stages: (1) regurgitation of partially digested fibers from the reticulorumen, (2) mastication (chewing the cud), (3) forming a bolus of masticated matter, and (4) reswallowing. The animal spends the largest portion of rumination time chewing the cud.

Rumination helps to break down woody fibers physically and separate difficult-to-digest lignin from more digestible cellulose and hemicelluloses, which serve as the main substrates for the rumen microorganisms.

RUMEN FERMENTATION

Rumen microorganisms have been studied in detail, since the 1950s, but rumen microbiology did not take shape until research by the American microbiologist Robert Hungate (1906–2004) at the University of California–Davis. In a 2005 memoriam to Hungate,

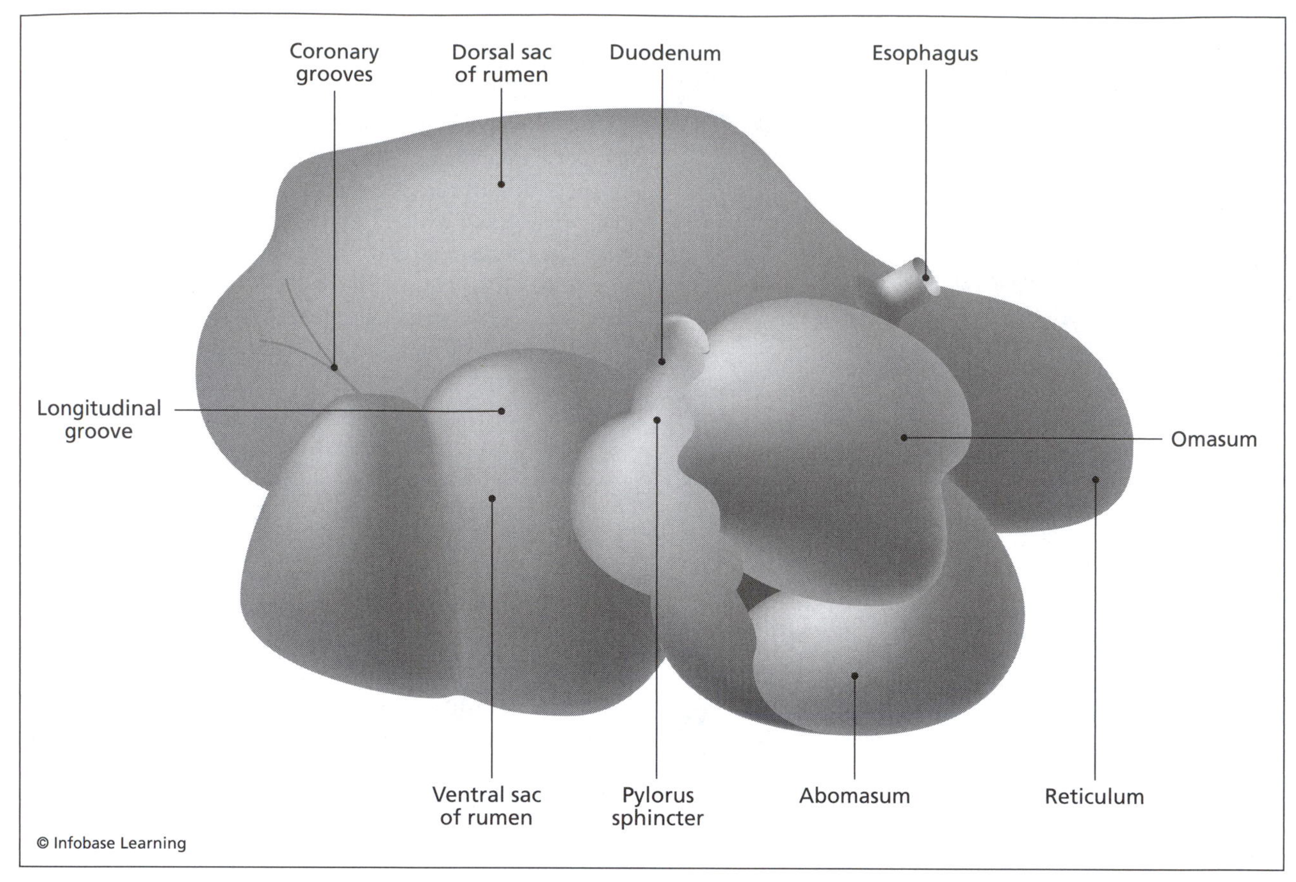

The ruminant stomach is a four-part organ. Fiber digestion and fermentations by the anaerobic microbial population occur in the rumen and reticulum. The omasum moves fluid to the abomasum, which has been called the "true stomach" because it functions similarly to the human stomach.

the authors wrote, "Hungate's success as a microbial ecologist was largely based on a simple but ingenious technique of maintaining anaerobiosis. Anaerobiosis was achieved by flushing the culture tubes with O_2 [oxygen]-free CO_2 [carbon dioxide]. The CO_2 was initially purified . . . by passing it through a hot copper column previously reduced with hydrogen. After the tubes were filled, they were closed with butyl rubber stoppers. The tubes containing molten agar were 'rolled' in an ice-bath to create a film of agar on the inside surface." The agar inoculated with a small volume of rumen contents would contain numerous bacterial colonies after incubation. The microbiologist removed desired colonies with a CO_2-flushed pipette, quickly transferred them to similar tubes of sterile broth medium instead of agar, and then incubated the tubes. This method for culturing fastidious (strict or obligate) anaerobic bacteria is called the *Hungate technique*.

Microbiologists have since modified the Hungate technique by using rubber-stoppered serum vials instead of test tubes and transfer inocula with CO_2-flushed syringes. These steps help reduce the introduction of oxygen to the medium.

Animal Digestive Tracts

Type of Digestive Tract	Type of Digestion in Order of Importance	Animals
ruminant	microbial fermentations, digestive enzymes	antelope, camel, cattle, deer, giraffe, goat, sheep
monogastric	digestive enzymes, microbial fermentations	bear, cat, dog, human, pig, rat
monogastric with active cecum	microbial fermentations, digestive enzymes	elephant, horse, guinea pig, hamster, koala, rabbit

Compartments of the Ruminant Stomach

Compartment	Anatomy	Digestion	Percentage of Total Volume
reticulum	inner surface lined with folds resembling a honeycomb to increase area for absorbing nutrients	fermentations	85 in reticulorumen
rumen	muscular wall for mixing contents; inner surface lined with papillae protuberances (3–10 millimeters long) for increased surface area	fermentations	
omasum (manyplies)	lined with numerous flat sheets (leaves) of tissue	screening of large particles and water absorption	12
abomasum	mucus-lined inner surface	enzymatic digestion	4

The reticulorumen provides a habitat of constant temperature and rich nutrient supply. Because ruminant animals spend about one third of their time eating, one third ruminating, and one third resting, the influx of fresh nutrients is intermittent. The overall fermentation remains fairly constant, however, with masticated and fresh food entering and partially digested matter exiting to the intestines. For this reason, the ruminant stomach has been likened to a continuous-flow culture with the reticulorumen acting as a chemostat, a vessel that supports inflow of nutrients and outflow of wastes over an extended period.

The continuous nature of nutrient inflow and waste outflow helps maintain a stable population of bacteria and protozoa that fill niches in feed digestion. For example, some species specialize in breaking down cellulose to smaller chains of sugars, other bacteria liberate individual monosaccharides (single sugars) or disaccharides (two sugars linked together), and additional bacteria use the end products of sugar fermentation, such as carbon dioxide and hydrogen. The latter group of bacteria, called methanogens, make methane gas and, thus, reduce the buildup of fermentation end products in the organ. By removing end products such as hydrogen (H_2), the methanogens make the fermentation reactions energetically favorable. The various rumen species have, in fact, been studied as an example of symbiosis, in which the relationships favor the microorganisms as well as the host. The animal prevents methane from building up in the rumen by belching.

Rumen (or reticulorumen) bacteria exist in total concentrations of 10^{10}–10^{11} per milliliter (ml); protozoa numbers range from 10^4 to 10^6 per milliliter. These enormous numbers and their fermentation activities help heat the animal and provide the host with nutrients and energy. The main fermentation end products absorbed by ruminants and used in metabolism are formate, acetate, propionate, butyrate, various branched-chain acids, ethanol, lactate, succinate, vitamin K, B vitamins, and amino acids. Acetate, propionate, butyrate, and the branched-chain acids are called volatile fatty acids (VFAs); they serve as the animal's main energy sources. The animal also absorbs vitamins and amino acids. Because rumen microorganisms convert poor-quality dietary proteins into a wide variety of microbial amino acids, ruminant animals do not have essential dietary amino acid requirements. The host absorbs the microbial amino acids in the intestines after the microorganisms have lysed.

RUMEN BACTERIA

Hundreds of bacterial species exist in the ruminant forestomach, representing more than 50 genera. Many receive nutrients from the metabolism of other species, but several species have requirements for vitamins and minerals that result directly from the animal's diet. Some species are, thus, limited in their carbon and energy needs, and others are more versatile. The Cornell University rumen ecologist Peter van Soest estimated, in 1982, "Microbes that utilize products of others as an energy source may represent 60 percent or more of the total numbers and constitute a very important group in terms of net rumen output." Rumen microbiology cannot be studied one species at a time and give an accurate picture of rumen microbial ecology. Many species of bacteria have, nevertheless, been investigated initially as pure cultures in a laboratory to gain an idea of their probable role in the rumen. The table on page 684 describes the main rumen bacteria.

The table illustrates the variety of interactions that rumen bacteria have in their metabolism. The methanogens and *V. succinogenes* owe their survival to the metabolic reactions of other bacteria that digest dietary fibers.

The ruminant diet contains low levels of protein, and various species secrete adequate amounts of the enzyme protease to digest it. Few rumen bacteria can degrade lignin. Mastication serves as the main way ruminants break up lignocellulosic fibers and make the digestible fibers available to the bacteria.

RUMEN PROTOZOA

Microbiologists have characterized about 25 different genera of protozoa in the rumen. Rumen protozoa cell volume can be several thousand to one million times the volume of a bacterial cell. For this reason, protozoa mass in the rumen can exceed bacterial mass, even though the total numbers of protozoa rarely exceed 10^5 per milliliter.

The Indian rumen microbiologist Devki N. Kamra wrote in a 2005 review of rumen ecology, "The rumen protozoa were detected in domestic animals as early as the nineteenth century by Gruby and Delafond [in 1843]. Not much work was done for several decades after their first report. . . . It was only after 1920 that the researchers paid any significant attention towards the identification, morphology and biochemical functions of protozoa in the rumen." The gap in knowledge resulted from the difficulty of growing the strict anaer-

Rumen Bacteria

Bacteria	Main Role	Main End Products
Anaerovibrio lipolytica	fat digestion	acetate, propionate, butyrate, succinate
Bacteroides amylophilus	starch digestion	formate, acetate, succinate
Bacteroides succinogenes	cellulose digestion	acetate, succinate
Bifidobacterium species	starch, dextrin digestion	
Butyrivibrio fibrisolvens	starch, cellulose, xylan digestion	butyrate, lactate, H_2, CO_2
Clostridium lockheadii	starch and cellulose digestion	H_2, CO_2
Eubacterium ruminantium	sugar fermentation	formate, acetate, butyrate, lactate, CO_2
Lactobacillus species	sugar fermentation	lactate
Lachnospira multiparus	pectin digestion	acetate, propionate
*Methanobacterium*formate, H_2 utilization *ruminantium* and other methanogens	methane	
Peptostreptococcus elsdenii	lactate utilization	propionate
Prevetella ruminicola	starch digestion	acetate, succinate
Ruminicoccus albus	cellulose, xylan digestion	H_2, ethanol, lactate
Ruminococcus flavefaciens	cellulose, xylan digestion	formate, acetate, propionate, lactate, succinate
Selenomonas ruminatium	starch digestion, lactate/succinate utilization	H_2, CO_2
Streptococcus bovis	starch digestion	lactate
Succinimonas amylolytica	starch digestion	acetate, propionate, succinate
Succinovibrio dextrinosolvens	dextrin, pectin digestion	formate, acetate, succinate, lactate
Veillonella alcalescens	lactate utilization, fat digestion	acetate, propionate, H_2, CO_2
Vibrio succinogenes	H_2, formate utilization	succinate, ammonia

obes in laboratory cultures. In recent decades, more time has been spent on maintaining protozoa cultures and studying them under laboratory conditions.

Until the advent of genetic analysis for determining microbial species of protozoa, rumen microbiologists used cell morphology to identify the protozoa. Rumen protozoa cell morphology is dramatic compared with that of many of the protozoa found elsewhere in nature; thus, morphology became an accurate way to identify genera.

Rumen protozoa belong to either of two groups divided by cell morphology: holotrichs and entodiniomorphs. The table below summarizes the main characteristics of these two groups.

The contractile vacuole in protozoa is a bladder-like sac that regulates water intake, expulsion, and storage. Cilia play an important part in motility, feeding, and capture of bacterial cells upon which the protozoa prey. In genera that possess cilia only at the oral opening, called an *oral groove,* the protozoa feed by using the cilia to sweep particles from rumen fluid into the oral groove.

For many years, rumen protozoa were thought to obtain nutrition by predation on bacteria. Subsequent studies have shown that the protozoa secrete a variety of digestive enzymes and play an important part in feedstuff digestion. Kamra said, in 2005, "The enzymatic profile of holotrich protozoa indicates that they have amylase, invertase, pectin esterase and polygalacturonase in sufficiently large quantities for using starch, pectin and soluble sugars as energy source." Entodiniomorphs contribute the major part of cellulose digestion carried out by rumen protozoa.

Prominent genera of holotrichs that have been isolated from the rumens of domesticated and wild animals are *Buetschlia, Charonina, Dasytricha, Isotricha,* and *Oligoisotricha.* The most-studied entodiniomorphs are of the following genera: *Caloscolex, Diplodinium, Entodinium, Epidinium, Eremoplastron, Eudiplodinium, Metadinium, Ophryoscolex, Ostracodinium,* and *Polyplastron.*

The protozoa listed here are almost exclusive to the rumen. Hungate explained in his classic text *The Rumen and Its Microbes,* "In ages past, when progenitors of modern ruminants consumed grass and water, ciliates similarly appeared in stomachs containing this natural infusion. Through evolution, during millions of years they became specifically adapted to their habitat." Newborn animals become inoculated, within the first hours after birth, by the mother's nurturing (licking, cleaning).

Rumen Protozoa

morphology	main role
HOLOTRICHS	
cilia cover the cell in most species	sugars, starch, pectin digestion
ENTODINIOMORPHS	
single oral opening surrounded by cilia; large contractile vacuole; some have long, rigid skeletal plates	cellulose, hemicellulose digestion

RUMEN FUNGI AND BACTERIOPHAGES

Single-celled eukaryotes propelled by flagella have been seen in the rumen, since the early 1900s, and categorized as protozoa. In the mid-1970s, the British microbiologist Colin Orpin discovered that the flagellate *Neocallimastix* produced mycelia and reproductive structures characteristic of fungi. Rumen fungi have now been estimated to number 10^3–10^5 zoospores (motile reproductive structures) per milliliter, representing five to 10 genera.

Rumen fungi take part in fiber digestion. Because certain fungal species have the rare ability to degrade lignin, these organisms might be important in releasing digestible fibers from lignocellulose. Electron microscopy studies of natural rumen fluid show that fungi prefer to attach to the fibers containing the highest lignin content.

In addition to fungi, bacteriophages live in the rumen in numbers approaching 10^9 per milliliter. Bacteriophages propagate by infecting bacteria, using the bacterial genome replication system to produce more viruses, then lysing the cell to escape back into the environment. The lysis processes may serve to make bacterial constituents such as amino acids more available for absorption by the host animal.

FISTULATION

Microbiologists gained a valuable way to study rumen microorganisms when, in the 1950s and 1960s, anatomists developed the method called *fistulation.* Fistulation is a surgical method whereby a permanent opening, or fistula, connects the rumen interior to the outside of the animal. The large rumen, which can hold up to 40 gallons (151 l), lies close to the left side of the abdominal wall. A veterinary surgeon makes an initial opening to connect the body wall to the peritoneum, the thin tissue that lines the abdominal cavity. After the tissues grow together, the surgeon continues the procedure by entering the rumen and stitching the rumen wall to the peritoneum-body wall.

A completed fistula creates a painless opening through which microbiologists take rumen fluid samples. A rubber ring secures the entire opening, which the microbiologist closes with a rubber plug after retrieving a sample. Microbiologists sample small fistulated animals, such as sheep, with a tube called a *cannula*. Larger animals, such as cattle, can be sampled by using a cannula or by reaching directly into the rumen, wearing rubber gloves, and scooping or grabbing a sample of rumen contents.

Fistulation has become a valuable aid in studying bacteria, protozoa, and fungi in symbiotic relationship with the host ruminant animal. In vivo and in vitro techniques have provided a window on ruminant digestion, host metabolism, and the interrelated microbial activities in the rumen. Rumen microbiology serves as an excellent study model for symbiosis and microbial ecology, as well as a training system for learning techniques in growing anaerobic microorganisms.

See also ANAEROBE; CONTINUOUS CULTURE; ENTERIC FLORA; FERMENTATION; METHANOGEN; MORPHOLOGY; SYMBIOSIS.

Further Reading

Chung, King-Thom, Bruce Hungate, and James Russell. "Robert E. Hungate." *American Society for Microbiology Newsletter,* 31 January 2005. Available online. URL: www.ars.usda.gov/research/publications/publications.htm?seq_no_115=176367. Accessed January 5, 2010.

Duncan, Alan J., and Dennis P. Poppi. *The Ecology of Grazing and Browsing Ruminants.* Berlin: Springer, 2008.

Hungate, Robert E. *The Rumen and Its Microbes.* New York: Academic Press, 1966.

Kamra, Devki N. "Rumen Microbial Ecosystem." *Current Science* 89 (2005): 124–135. Available online. URL: www.ias.ac.in/currsci/jul102005/124.pdf. Accessed January 5, 2010.

Karnati, S. K. R., Z. Yu, J. T. Sylvester, B. A. Dehority, M. Morrison, and J. L. Firkins. "Specific PCR Amplification of Protozoal 18S rDNA Sequences from DNA Extracted from Ruminal Samples of Cows." *Journal of Animal Science* 81 (2003): 812–825. Available online. URL: http://jas.fass.org/cgi/reprint/81/3/812.pdf. Accessed January 5, 2010.

Orpin, Colin G., and Andrew J. Letcher. "Utilization of Cellulose, Starch, Xylan, and Other Hemicelluloses for Growth by the Rumen Phycomycete *Neocallimastix frontalis.*" *Current Microbiology* 3 (1979): 121–124.

Van Soest, Peter J. *Nutritional Ecology of the Ruminant,* 2nd ed. Sacramento, Calif.: Comstock, 1994.

Saccharomyces *Saccharomyces* is a genus of heterogeneous yeasts, in the family Saccharomycetaceae and an important organism in food production and biotechnology. The brewing, wine making, and baking industries use *Saccharomyces* to carry out sugar fermentations. Industry also uses this eukaryote for producing vitamins or bioengineers the yeast to produce other commercial products. Bioengineered yeast carries genes from another organism in the yeast's deoxyribonucleic acid (DNA) and for the purpose of producing large amounts of the gene's product in industrial fermentations of several thousand liters.

Saccharomyces cells are usually ellipsoidal (oblong) but also form spherical and cylindrical forms. The ellipsoidal cells range in size from 2.5–6.5 micrometers (μm) in width and 3.5–8.5 μm in length.

Saccharomyces reproduces asexually by budding, a process typical of yeasts, in which a daughter cell with a full chromosome breaks away from the parent cell. Young budded cells, then, grow to normal size. Some *Saccharomyces* also grow filaments called *pseudomycelia* that make them resemble molds, but the yeast lacks true mold filaments called *hyphae.* Hypha-like structures, if present, are called *pseudohyphae.*

Mycologists refer to *Saccharomyces* as a *perfect yeast* because in sexual reproduction the genus produces spherical ascospores inside a saclike structure called an *ascus.* Yeasts lacking the ability to produce asci or similar reproductive structures called *basidia* are termed *imperfect yeasts.*

Saccharomyces lives in a diversity of habitats using different types of metabolism, versatility that may explain why this organism has been useful for various purposes since antiquity. Most species live in soil, on trees and plants, and on fruits, as well as inside the fruit fly *Drosophila.* Various species metabolize sugary exudates from plants, and in anaerobic conditions they ferment the sugars glucose, galactose, sucrose, and maltose. None of the *Saccharomyces* can ferment the milk sugar lactose, so this has been used as an identifying feature of the genus. The organism also uses aerobic respiration when oxygen is available, metabolizing glycerol, ethanol, and lactate.

In the environment, some *Saccharomyces* withstand extreme environments; thus, they belong to the group of organisms known as *extremophiles. Saccharomyces* represents a psychrophile because it grows at temperatures down to 32°F (0°C), yet it also thrives in a range up to 104°F (40°C). Some acidophile species withstand acidic conditions of pH 3 or lower.

On humans, *Saccharomyces* and *Candida* make up the main yeasts that normally live on the body. *Saccharomyces* lives in the external ear canal and on rare occasions inhabits the upper respiratory tract.

Industry and research use *S. cerevisiae,* also called *brewer's yeast* or *baker's yeast,* more than other species, and this yeast is among the most-studied organisms in microbiology. Additional *Saccharomyces* species have been fully or partially characterized in their life cycle and metabolism. The table on page 688 describes important *Saccharomyces* species and includes tools for differentiating them. All species have been isolated from soil, in addition to the sites listed in the table.

SACCHAROMYCES CEREVISIAE

Studies on *S. cerevisiae* occurred centuries before the genus was named, in 1837. The Dutch merchant Antoni van Leeuwenhoek (1632–1723), who, in the

Saccharomyces Species

Species	Isolation	Ferments	Does Not Ferment	Limit of Sensitivity to Cycloheximide (ppm)
S. cerevisiae	alcohols, fruits, human skin, pseudohyphae when grown on cornmeal	glucose, some also galactose, sucrose, maltose	lactose	100
S. dairiensis	buttermilk, cucumber brines	galactose, glucose	lactose, maltose, sucrose	1,000
S. exiguus	fruit, cucumber brines	galactose, glucose, sucrose	lactose, maltose	1,000
S. kluyveri	pseudohyphae when grown on cornmeal	galactose, glucose, sucrose	lactose, maltose	100
S. servazii	mainly in soil	galactose, glucose	lactose, maltose, glucose sucrose	>1,000
S. telluris	poultry, rodents	glucose	galactose, lactose, maltose, sucrose	100
S. unisporis	cheese, kefir	galactose, glucose	lactose, maltose, sucrose	1,000

Note: ppm = parts per million or milligrams per liter (mg/l)

17th century, developed a skill for making microscope lenses, had been the first to study beer yeast microscopically. Not until 1860, would studies on the metabolism of beer yeasts begin. In that year, the French microbiologist Louis Pasteur (1822–95) observed the transformations in a sugar solution inoculated with a small amount of yeast he isolated from beer. Pasteur described, in one of his memoirs, an experiment he had conducted on nutrient use by the yeast: "In a solution of pure candy sugar containing 10 grams, I put the ash of 1 gram of yeast [mineral source], 0.100 grams of ammonium tartrate and the amount of fresh beer yeast that would fit on the head of a pin. . . . Very remarkably, the globules which were added under these conditions developed, multiplied and the sugar fermented, at the same time that the minerals slowly dissolved and the ammonia disappeared." Pasteur's experiment launched subsequent studies, which continue to the present, on nutrient requirements and metabolism in microorganisms.

Other research on microbial physiology followed, usually using the *Saccharomyces* species because of its ubiquitous nature and ability to grow well in a variety of laboratory conditions. When genetic engineering first began, in the 1970s, *S. cerevisiae* formed the foundation of almost as many studies as the bacterium *Escherichia coli.*

In 1978, the Cornell University geneticists Albert Hinnen, James Hicks, and Gerald Fink showed that *S. cerevisiae* could pick up a small piece of bacterial deoxyribonucleic acid (DNA) called a *plasmid* directly from its environment, a process called *transformation*, and incorporate the bacterial genes into its DNA. They wrote, "Transformation of yeast makes possible the cloning of eukaryotic genes in a eukaryotic host with a sophisticated genetic system. Baker's yeast has several advantages over enteric bacteria [*E. coli*] as a host for pharmacologically important genes such as insulin. It is not a pathogen under any known circumstances and, because it is a eukaryote, it probably will allow more efficient expression of such eukaryotic genes." The authors' prediction proved true. Eukaryotic organisms usually worked better for eukaryotic gene transfers than prokaryotes.

S. cerevisiae has been called a model organism for genetic studies for the following five reasons: It is (1) easy to culture because it is single-celled and has a short generation time (population doubling time of 1.5–2 hours at 86°F [30°C]) and has simple nutrient requirements; (2) undergoes transformation, an easier method of gene transfer than most others; (3) grows either as haploid (half the chromosome) or diploid (full chromosome), expanding its usefulness as a genetics model; (4) contains few noncoding segments in its DNA, a problem that complicates gene transfer on higher eukaryotes; and (5) as a eukaryote shares the basic internal structure of plant and animal cells.

In 1996, the Belgian geneticist André Goffeau and his research team published the complete gene sequence of *S. cerevisiae*'s 16 chromosomes. The authors wrote in *Science,* "The genome of the yeast *Saccharomyces cerevisiae* has been completely sequenced through a worldwide collaboration. The sequence of 12,068 kilobases defines 5885 potential protein-encoding genes." With considerable understatement they added, "One of the major problems to be tackled during the next stage of the yeast genome project is to elucidate the biological functions of all of these genes." At present, more than 73 percent of the *S. cerevisiae* genes have been characterized as to their function. Researchers have refined the 1996 data to show that the yeast contains almost 12,156 kilobases and 6,275 genes.

BREWER'S AND BAKER'S YEAST

S. cerevisiae is a facultative anaerobe, meaning it can use oxygen when available but continue growing in the absence of oxygen. This adaptability has made the organism suitable for a variety of industrial uses.

Brewing is the production of beer or lager by yeast fermentation of extracts from malted barley, which is barley grain that has been partially degraded by starch- and protein-digesting enzymes. The table (right) describes the general steps in brewing. As brewer's yeast, *S. cerevisiae* carries out the fermentation of the sugar glucose to the alcohol ethanol:

$$C_6H_{12}O_6 \rightarrow 2\ CH_3CH_2OH$$

Different *S. cerevisiae* strains can be either top-fermenting yeasts used for beer brewing or bottom-fermenting yeasts used for making lager.

After fermentation, brewers remove the yeasts. Top-fermenting strains float to the top of the liquid and create a foam; bottom-fermenting yeasts settle to the bottom. Brewers, then, mature the liquid at about 30°F (-1°C). A small amount of remaining yeasts, meanwhile, degrade the residual compounds maltotriose and diacetyl. Carbon dioxide purges push out any residual gases. The brewer completes the production by filtering, bottling, and pasteurizing the beer.

The beverage industry uses *S. cerevisiae* also for fermentations for making natural wines, sparkling wines (champagne), sake, brandy, and whiskey. In wine making, producers crush grapes for the juice, to which they add sulfur dioxide to kill wild yeast contaminants, and then inoculate the liquid with *S. cerevisiae.* Red wines incubate at a warm 77°F (25°C), age in oak casks for three to five years, then age in the bottle five to 15 years. White wines incubate at cooler temperatures (50–59°F [10–15°C]) and age two to three years in bottles. Sparkling wines undergo a second fermentation in the bottle, after adding 2.5 percent sugar and inoculating with the yeast. Incubation at 59°F (15°C) finishes the process.

Baker's yeast is another specialized *S. cerevisiae* strain, which ferments dough under conditions of low oxygen, low water activity, and high osmotic pressure. Baker's yeast is produced by growing the strain in an aqueous medium containing molasses, vitamins, minerals, and a nitrogen source. After aerobic incubation, centrifugation of the culture removes the liquids from the yeast. The yeast can be pressed to exclude moisture, then packaged. In baking, the yeast ferments sugars from starch, producing carbon dioxide. Dough yields to the gas buildup a little but prevents it from escaping. As a result, the bread rises.

INDUSTRIAL USES FOR *SACCHAROMYCES*

Saccharomyces produces natural and genetically engineered substances for a variety of industries.

Beer Brewing

Step	Activity
barley malting	germinated grain releases starches and proteins for degradation by enzymes
kilning	halting germination by drying, then heating the malt
milling	grinding the malt exposes the grain's starches or grist
mashing	grist mixed with water and other optional carbohydrates called *adjuncts* enters temperature-time cycles to degrade starches to fermentable sugars and nonfermentable materials to amino acids and peptides; addition of yeast growth factor inositol
wort separation	separation of the liquor (wort) from the grain
wort boiling	wort boiled with hops to inactivate enzymes and kill spoilage organisms
wort cooling	clarification of the cooled wort
inoculation	addition of *S. cerevisiae* and beginning of lag period (about 12 hours), in which organism prepares for rapid growth
fermentation	aerobic respiration uses up the oxygen, then yeast switches to anaerobic fermentation, converting wort sugars to ethanol and carbon dioxide

Enzyme manufacturers use *Saccharomyces* as a source of natural amylase. *S. cerevisiae* produces the enzyme invertase, which breaks apart the sugar sucrose to its component sugars glucose and fructose, producing sugar syrup used in candy making.

Saccharomyces has contributed to some of the first products developed from genetic engineering. Some of the main uses of genetically engineered *S. cerevisiae* are:

- alpha-interferon—treatment for hepatitis and some cancers
- colony-stimulating factor—drug for counteracting chemotherapy side effects
- hepatitis B vaccine
- influenza vaccine—trial vaccines
- lactic acid—industrial chemical
- superoxide dismutase—helps repair damaged tissue
- xylitol—sweetener

S. cerevisiae and other *Saccharomyces* also cause spoilage in foods, when microorganisms use up oxygen in a food container. *S. aceti* that contaminates wine produces acetic acid (vinegar) instead of ethanol. The species *S. bailii, S. bisporus,* and *S. rouxii* also cause various forms of food spoilage, mainly in fruits such as dates, honey, syrups, mayonnaise, and salad dressings. *S. bisporus* and *S. rouxii* withstand very high sugar concentrations that inhibit most other microorganisms. Both of these osmophiles tolerate sugar concentrations up to 60 percent.

Saccharomyces, the "sugar fungus," has been used by humans, since the earliest civilizations, for making beer, wine, and bread. From antiquity to today, this yeast has been an important part of the most advanced technologies in genetics and bioengineering.

See also FOOD MICROBIOLOGY; OSMOTIC PRESSURE; PASTEURIZATION; WATER ACTIVITY; YEAST.

Further Reading

Dickinson, J. Richard, and Michael Schweizer, eds. *The Metabolism and Molecular Physiology of* Saccharomyces cerevisiae, 2nd ed. Washington, D.C.: American Society for Microbiology Press, 2004.

Feldmann, Horst. "Yeast Molecular Biology." 2005. Available online. URL: http://biochemie.web.med.uni-muenchen.de/Yeast_Biol. Accessed January 1, 2010.

Goffeau, André et al. "Life with 6000 Genes." *Science* 247 (1996): 546–567. Available online. URL: www.sciencemag.org/cgi/content/abstract/274/5287/546. Accessed January 10, 2010.

Goldammer, Ted. *The Brewer's Handbook*. Clifton, Va.: Apex, 2009. Available online. URL: www.beer-brewing.com/default.htm. Accessed January 1, 2010.

Hinnen, Albert, James B. Hicks, and Gerald R. Fink. "Transformation of Yeast." *Proceedings of the National Academy of Sciences* 74 (1978): 1,929–1,933. Available online. URL: www.pnas.org/content/75/4/1929.full.pdf+html. Accessed January 1, 2010.

Pasteur, Louis. "Mémoire sur la Fermentation Alcoölique" (Memoir on the Alcoholic Fermentation). *Annales de Chimie et de Physique* 58 (1860): 323–426. In *Milestones in Microbiology,* translated by Thomas Brock. Washington, D.C.: American Society for Microbiology Press, 1961.

Salmonella *Salmonella* is a gram-negative, nonendospore-forming food-borne pathogen of the general group called *enteric bacteria. Salmonella* species cause typhoid fever and illnesses of varying severity, collectively called *salmonellosis* or *Salmonella enterocolitis.*

The straight rod-shaped *Salmonella* cells measure 0.5–1.0 micrometer (μm) wide and 1–5 μm long. Most species are motile, using a single polar (at one end of the rod) flagellum. *Salmonella* grows on a variety of nutrients as a facultative anaerobe. The genus belongs to family Enterobacteriaceae, order Enterobacteriales, and class Gammaproteobacteria of the phylum Proteobacteria.

In 1885, the veterinary assistant Theobald Smith (1859–1934) named a newly discovered swine bacterium *Salmonella* (pronounced with a silent *l*) after his mentor, Daniel E. Salmon (1850–1914). *Salmonella* naming conventions differ from those of most other genera because the names refer to serovar rather than species. A serovar (or serotype) is a group of bacteria with surface structures in common that elicit a response from the human immune system. By this style of organization, *Salmonella* contains more than 2,400 serovars. For ease, many microbiologists use a common name for a species in place of the serovar designation, for example:

- *S. typhimurium* for *S. enterica* serovar *typhimurium*
- *S. typhi* for *S. enterica* serovar *typhi*

Salmonella occurs in the digestive tract of a variety of food-producing animals, especially poultry and swine. Fecal contamination puts *Salmonella* into the environment; the microorganism has been isolated from soil, surface freshwaters, and insects. Microbiologists studying the incidence of *Salmonella* have recovered this pathogen from raw meat, raw chicken, and a variety of inanimate surfaces in homes, schools, day care facilities, hospitals, and workplaces, presumably caused by fecal contamination carried on the hands.

SALMONELLA FOOD-BORNE ILLNESS

An estimated 1.2 million people are sickened each year by *Salmonella*. *Salmonella*-contaminated foods that have been linked with illness in humans are undercooked eggs and poultry, raw meats, dairy products, sauces and salad dressings, cake mixes, cream-filled desserts, peanut butter, and chocolate.

In 2009, the U.S. Food and Drug Administration (FDA) issued a massive recall of peanut products nationwide that had been produced by the Peanut Corporation of America (PCA). FDA inspectors cited as evidence "test results from consumer samples that match the outbreak strain of *Salmonella* Typhimurium, and FDA positive samples of finished product (post-processed peanut meal collected at the PCA Texas facility) that match the outbreak strain." The consumer samples had been identified by clinical microbiology laboratories as people visited their doctors at the height of the outbreak. FDA microbiologists, then, used a technology called *DNA fingerprinting.* Sensitive techniques for matching the base sequences of deoxyribonucleic acid (DNA) of a clinical isolate from a patient to bacteria from a suspected source can identify the original source of contamination. This technology works for food-borne pathogens in addition to *Salmonella.*

The 2009 outbreak resembled the classic signs of *Salmonella* outbreak. Illness symptoms began six hours to four days after ingesting contaminated food, and most people had symptoms nine to 48 hours after ingestion. As few as 15 *Salmonella* cells in food can cause illness, but some people withstand doses up to 1,000 cells before becoming sick. Illness includes nausea, vomiting, abdominal cramps, fever, headache, and diarrhea, which last for one to four days. Because these symptoms relate to many food-borne illnesses, public health officials have a difficult time confirming a *Salmonella* outbreak without serology testing on patient stool samples by a clinical microbiology laboratory. Laboratory culture methods take four to five days, and this delay presents a problem when an outbreak is under way and health agencies need to give the public advice. Many small outbreaks of *Salmonella* occur and disappear before ever being diagnosed.

The amount of pathogens needed to make someone sick depends on the health condition of a person and susceptibility to infection. All adults and children are susceptible to *Salmonella,* but people who have weakened immune systems have the highest risk of infection. *Salmonella* causes inflammation of the small intestine when the bacteria penetrate the epithelial lining. Some strains produce a toxin called an *endotoxin,* which stays attached to the bacterial outer membrane and releases into the epithelium only after the bacterium dies and lyses. Most laboratory studies that have been done on bacterial endotoxins have used *Escherichia coli* and *Salmonella.* The *Salmonella* endotoxin is a lipopolysaccharide (LPS), which is a sugar polymer with several fatty acids attached. The LPS contains three portions: (1) R polysaccharide, (2) O polysaccharide, and (3) lipid A, which is the toxic portion of the LPS.

During infection, the O polysaccharide adheres to the epithelium and introduces lipid A into the host cell. The O polysaccharide, also called the O-antigen, repels various actions by the immune system by resisting phagocytosis and the action of antibodies and complement. Phagocytosis is a nonspecific defense mechanism of hosts, by which an immune cell engulfs and digests the pathogen. Antibodies and complement carry out more pathogen-specific actions in immunity.

If *Salmonella* cells enter the bloodstream, they trigger a dramatic immune response, resulting in release of cytokines and nitric acid by the body. Both substances lead to a condition called endotoxic shock or septic shock. Sepsis is the contamination of the bloodstream or internal organs with pathogens. This severe condition causes lowered blood pressure, rapid heart rate, shortness of breath, and possible disorientation.

The *Los Angeles Times* reporter Joel Rubin contracted salmonellosis, in 2007, at the same time a major *Salmonella* outbreak had hit Los Angeles. Rubin said, "My eyes popped open sometime after midnight and I knew I was in trouble. This was not a typical bellyache. It radiated from my gut. . . . Beads of sweat rose suddenly on my forehead. A sharp chill hit me. My teeth chattered, my body shuddered. . . . I couldn't have known it at the time, but in those early Monday morning hours, dozens of other people across Los Angeles were suffering just like I was. By sunrise, some of us would wind up in hospital emergency rooms. We were men and women, old and young, linked only by our unfortunate decision to eat a certain meal at a certain place at a certain time. An outbreak had begun." Los Angeles County health officials spent considerable energy tracing the outbreak to eggs that had been used in traditional hollandaise sauce, that is, sauce made by using raw eggs.

Although most cases of salmonellosis cause some uncomfortable days, the illness is usually not fatal.

TYPHOID FEVER

Typhoid fever is an acute disease caused by *S. enterica* serovar *typhi* contamination in food or water. A similar disease, called parathyphoid fever; is caused by *S. paratyphi* type A, B, or C. Humans are the only known reservoir for these pathogens.

S. typhi bacteria travel from the intestines to the gallbladder, during infection, and can maintain a stable population there for many years. During this time, the carrier sheds live *S. typhi* cells in the stool. Many carriers of this disease are called *asymptomatic carriers* because they never experience symptoms but transmit the diseases to others, possibly to hundreds of other people, in their lifetime. The most famous asymptomatic carrier in history was Mary Mallon, known better as "Typhoid Mary" (1869–1938), who cooked for families in the New York City area, between 1884 and 1910, and again in 1915, when she was a cook at Sloane Maternity Hospital. During her career, Mallon had transmitted typhoid fever to at least 50 people and had caused three deaths. Doctors discovered that Mallon had an unusually high concentration of *Salmonella* in her stool but had never been sick with typhoid fever or salmonellosis. When the New York sanitation engineer George Soper (1870–1948) traced the outbreaks to Mallon, the newspapers gave her the nickname that today refers to any person who spreads disease or other tragedies.

Typhoid fever has an incubation period in the body of five to seven days, after which symptoms appear: headache, fever, malaise, muscle aches, loss of appetite, and possible constipation or diarrhea. Fevers can reach 104°F (40°C), characterized by lowest fever in the morning and reaching a peak by evening. The disease causes characteristic inflammation in the lymphoid tissues of the small intestine called Peyer's patches. Severe inflammation at these areas can kill epithelial tissue, perforate the intestinal wall, and cause death.

Antibiotics and a vaccine are both available for *S. typhi,* and today the disease is much less prevalent than salmonellosis. Some patients require surgery to remove the gallbladder for the purpose of eliminating the *Salmonella* from the body.

Some typhoid and paratyphoid fever pathogens have developed resistance to standard antibiotic treatments of the past. Microbiologists have isolated nalidixic acid–resistant strains and multidrug-resistant (MDR) strains that cannot be killed by ampicillin, chloramphenicol, or trimethoprim-sulfamethoxazole.

Salmonella outbreaks are preventable by conscientious personal hygiene to reduce germ transmission. Outbreaks require swift diagnosis and intervention, but, as with most food-borne illnesses, medical providers often recognize an outbreak only after it has already spread throughout a community.

See also DNA FINGERPRINTING; ENTERIC FLORA; FOOD-BORNE ILLNESS; IMMUNITY; SEPSIS; SEROLOGY.

Further Reading

Mintz, Eric. "Typhoid and Paratyphoid Fever." In *Travelers' Health Yellow Book*. Atlanta: Centers for Disease Control and Prevention, 2010. Available online. URL: wwwnc.cdc.gov/travel/yellowbook/2010/chapter-v/typhoid-paratyphoid-fever.aspx. Accessed January 2, 2010.

Public Broadcasting System. "The Most Dangerous Woman in America." Available online. URL: www.pbs.org/wgbh/nova/typhoid. Accessed January 2, 2010.

Rubin, Joel. "Making the Right Sick Call." *Los Angeles Times,* 3 November 2007. Available online. URL: http://articles.latimes.com/2007/nov/03/local/me-foodpoison3?pg=5. Accessed January 2, 2010.

University of Maryland Medical Center. "*Salmonella* Enterocolitis—All Information." Available online. URL: www.umm.edu/ency/article/000294all.htm. Accessed December 20, 2010.

U.S. Food and Drug Administration. "Bad Bug Book: *Salmonella* spp." Available online. URL: www.fda.gov/Food/FoodSafety/FoodborneIllness/FoodborneIllnessFoodbornePathogensNaturalToxins/BadBugBook/ucm069966.htm. Accessed January 20, 2010.

———. "FDA's Investigation." Available online. URL: www.fda.gov/Safety/Recalls/MajorProductRecalls/Peanut/default.htm. Accessed June 6, 2010.

sanitization Sanitization is the reduction of bacterial numbers on inanimate surfaces to safe levels. Sanitization makes up one of three methods in microbiology for eliminating unwanted microorganisms from materials. Sterilization is the elimination of all life from a material, and disinfection is the elimination of all life except bacterial spores from a material. The objective of sanitization is only the reduction of bacterial numbers, rather than the complete elimination of bacteria.

Disinfection and sanitization can be distinguished from sterilization in the types of materials they treat. Disinfection and sanitization apply only to hard, inanimate surfaces, with a few exceptions; sterilization applies to both solid and liquid substances. The exceptions to the sanitization of hard, inanimate surfaces are sanitizers of fruits and vegetables, hands, carpets, and air. A disinfectant is any product formulated to eliminate all organisms except bacterial spores from applicable surfaces, and a sanitizer is any product formulated to reduce the number of bacteria to a predetermined and required level on applicable surfaces.

Microbiologists in the United States follow regulations set and enforced by government agencies for sanitizers. The following agencies regulate the uses for sanitizers:

- U.S. Environmental Protection Agency (EPA) regulates sanitizer requirements for food- and non-food-contact surfaces
- U.S. Food and Drug Administration (FDA) regulates the use of sanitizers for commercial

food preparation equipment, fruit and vegetable sanitizers, and hand sanitizers

Both agencies use a parameter called *log reduction* to assess sanitization. Log reduction is the decrease in the number of bacteria caused by a sanitizer, expressed as the logarithm of the bacterial number. A logarithm is the power to which a number is raised. For example, for the number 7.1×10^9, 10 has been raised to the power of 9. The number 7.1×10^9 can be converted to a logarithm (on a base 10 scale because the result is a multiple of 10) by using a calculator or looking up the value in a published table of logarithms. The base 10 logarithm ($\log_{10}$) of 7.1×10^9 equals 9.8513.

LOG REDUCTION IN SANITIZATION

The numbers of bacteria on a surface can reach several thousand or millions over a given area. Microbiologists use either log reduction data or percentage reduction to assess sanitization. For a surface containing 10,000 bacteria per square inch (6.45 square centimeters), reducing the number to 10 bacteria in the same-size area represents a 3 log, or 99.9 percent, reduction:

$$(10 \div 10{,}000) \times 100 = 0.1 \text{ percent of the original bacteria remaining, and}$$

0.1 percent remaining is equivalent to 99.9 percent of the bacteria eliminated.

The same calculation performed with logarithms is:

$$\log_{10} \text{ of } 10 = 1.00$$

$$\log_{10} \text{ of } 10{,}000 = 4.00$$

$$\text{log reduction} = 4.00 - 1.00 = 3.00$$

The following percentage reductions in bacterial numbers always equal the following log reductions:

99.9 percent = 3 logs

99.99 percent = 4 logs

99.999 percent = 5 logs

99.9999 percent = 6 logs

Most sanitizers must meet 3 log, 4 log, or 5 log reduction of bacteria requirements in laboratory tests.

People who use sanitizers to remove bacteria from a hard surface should understand that if the number of bacteria starts out at a high level, sanitization may leave behind a large number of live bacteria. For example, a sanitizer that eliminates 99.9 percent of 1,000 bacteria leaves only one bacterial cell:

$$1{,}000 - 999 = 1$$

The same sanitizer applied to a surface containing 1,000,000 bacteria leaves 1,000 cells! (In $\log_{10}$, 6.00 - 3.00 = 3 log reduction.)

Restaurants and food manufacturers avoid many instances of contamination by maintaining clean conditions in their facilities. By reducing dust, dirt, hair, and other particle contamination, a daily sanitization program keeps pathogens out of most operations. Workers also must follow strict rules regarding personal hygiene to prevent inoculating a surface or a food with germs.

The food and other industries maintain sanitation programs, as well as sanitization programs. *Sanitation* refers to a scheduled cleaning program designed to remove gross dirt. Sanitation cleans surfaces but does not necessarily reduce the number of bacteria to safe levels, as sanitization does.

SANITIZER TESTING

All bacterial species do not behave in the same way in the presence of a sanitizer. Microbiologists made significant strides, in the 19th century, in elucidating the role of bacteria in disease and food production and spoilage. By the early 1900s, many had turned their attention to finding new chemicals to kill harmful and nuisance microorganisms. Various scientists soon learned that microorganisms follow a spectrum of resistance and susceptibility to chemical biocides. The microbiologists Robert Berube and Gordon Oxborrow explained, in 1991, that scientists designing the first sanitizer tests realized "that a sanitizing rinse should give an antimicrobial result significantly greater than that obtained by the use of cold water. Unfortunately . . . the 'significantly greater' stipulation was a bit vague."

Today's sanitizer tests overcome most of the early worries on finding a standard way to compare different chemicals. Laboratory tests, nevertheless, cannot replicate what happens in the environment. A 99.9 percent reduction of bacteria in a laboratory test may not always translate to the same effect in nature. With these constraints in mind, Berube and Oxborrow summarized the scope of today's sanitizers in reducing potentially harmful bacteria:

- mitigate the spread of nonpathogenic microorganisms
- prevent development of microbial malodors
- prevent development of dangerous microbial end products

Types of Sanitizers

Product	Main Use	Reduction Requirements (%)
nonfood-contact	floors, walls, furniture, bathroom surfaces, etc.	99.9
food-contact	eating/drinking utensils, food-processing equipment, food-contact hard surfaces	99.999
halide- (chlorine-)containing air	food or nonfood hard surfaces reduction of bacteria and odors	see note 99.9
residual, nonfood-contact	floors, walls, furniture, bathroom surfaces, etc.	99.9 for an extended period
laundry additive, nonresidual	wash loads	99.9
laundry additive, residual	wash loads	99.99 for an extended period
carpet	carpets	99.9
porous surfaces	wood, some plastics, concrete	99.9

Note: chlorine-containing sanitizers do not need to meet percentage reduction requirements, but they must show that the chlorine stays active at a concentration of 200 parts per million (ppm)

- control potential pathogens in their reservoirs of infection
- control contaminants of economic significance, such as in foods and manufacturing plants
- prevent spoilage of foods and raw materials
- control airborne microorganisms

Biocide chemicals such as quaternary ammonium compounds (quats), chlorine compounds, and surfactants (detergentlike chemicals) make up most commercial sanitizers. Test methods differ according to two features of the sanitizer: (1) the biocide in the formula and (2) the sanitizer's intended use. The table above presents the different categories of sanitizers listed in the approximate order of their prevalence.

In the table above, the percentage reduction requirements are determined by comparing the product to a control substance such as water or the product minus the active ingredient. Each product must also meet the reduction requirements within a specified period. Depending on the type of sanitizer, this contact time ranges from 30 seconds to 10 minutes.

FOOD INDUSTRY SANITIZATION METHODS

The food industry must pass requirements set by the FDA and by other national, state, and local agencies. Because the public's health can be directly affected by poor sanitation and sanitization practices, food producers must pass government inspections, but they must also self-regulate by running cleaning programs and the HACCP method for monitoring food facilities. *HACCP* refers to the Hazard Analysis and Critical Control Point system. Each facility develops an HACCP system for its unique circumstances to accomplish three main tasks: (1) identify locations and activities at the highest risk of contamination, (2) institute a monitoring system targeted mainly at the high-risk points, and (3) develop corrective actions and intensified monitoring for points with the highest levels of contamination.

The food industry is one of the most difficult industries to monitor for safety because of the enormous amounts of food produced, shipped, and imported in the United States each year. Self-regulating by companies is the only way to ensure safety because the FDA does not have the staff, money, or time to inspect every food item. The food safety consultant Tom Weschler wrote in *FoodSafety Magazine,* in 2008, "HACCP has been a cornerstone of regulatory and industry practices since the 1990s. In fact, companies like Pillsbury have been using HACCP-based plans for over 40 years." Weschler acknowledged that microbiology has a hard time keeping up with the speed required in food production. Microbiological techniques can be lengthy, and microorganisms need minimal periods to incubate. For this reason, new rapid tests that give results in hours or minutes rather than one to two days have increased in food laboratories more than 40 percent, since 2005.

Sanitization programs in manufacturing consist of some or all of the following components:

- setting up a sanitation and worker personnel hygiene program
- selecting one or more sanitizer products
- establishing an HACCP program
- determining the items to be sampled and test; how many and when
- assigning microbial analyses
- recording and reviewing data
- completing corrective actions for problems

The main microbiological problem in manufacturing occurs when the number of microorganisms exceeds acceptable limits. This situation is called *out-of-spec* for "outside specifications." Out-of-spec conditions indicate that a sanitization step was missing or performed incorrectly. The effect of a sanitizer on a surface does not last forever. Within a period of hours or minutes, new microorganisms can be transferred to a sanitized surface. Sanitization programs, therefore, require that sanitization take place repeatedly and always to the same level of competency.

FOOD AND HAND SANITIZERS

Diluted solutions of chlorine bleach—about 200 ppm in contact for at least one minute—have been tried as sanitizers for fresh fruits and vegetables. The FDA permits the sale and use of chlorine-based food sanitizers as long as the products meet two criteria:

- sanitizer concentration in the wash water cannot exceed 2,000 ppm
- sanitizer must be rinsed off with potable (drinkable) water after sanitization

In food sanitization, no requirements for log reduction of microorganisms exist.

The FDA monitors the use of alcohol-based hand sanitizers, which usually include a polymer to help spread the alcohol over the skin. These products have been recommended by microbiologists for preventing cold and flu transmission, but questions exist regarding their effectiveness. In 2006, researchers at East Tennessee State University reported to the Centers for Disease Control and Prevention (CDC) of a hand sanitizer they had tested, "Despite a label claim of reducing 'germs and harmful bacteria' by 99.9%, we observed an apparent increase in the concentration of bacteria in handprints impressed on agar plates after cleansing." The researchers had demonstrated the phenomenon in which a brief hand washing removes some dirt from the skin surface but also exposes many bacteria that hide in the skin's crevices.

A combination of frequent hand washing and the use of hand sanitizers may be more effective in reducing bacteria and viruses on the skin than use of sanitizers alone. In 2009, the microbiologist Charles Gerba at the University of Arizona advised that for cold and flu season, "Good hygiene practices that emphasize handwashing or hand sanitizers, along with proper maintenance, can help reduce the number of infections in schools each year."

Questions regarding the usefulness of hand sanitizers remain, and neither the FDA nor the CDC has stated unequivocally that these products reduce the spread of germs.

Sanitization is an oft-misunderstood aspect of germ control. Sanitizers have been confused with disinfectants even by some microbiologists. Sanitizers have strict testing regulations they must pass to be registered with the EPA, as have disinfectants and sterilizers. When sanitization is done properly, it is an effective way to reduce the number of bacteria on surfaces.

See also BIOCIDE; DISINFECTION; FOOD MICROBIOLOGY; HACCP; HYGIENE; SERIAL DILUTION; STERILIZATION.

Further Reading

Berube, Robert, and Gordon S. Oxborrow. "Methods of Testing Sanitizers and Bacteriostatic Substances." In *Disinfection, Sterilization, and Preservation,* 4th ed. Philadelphia: Lea & Febiger, 1991.

Gapud, Veny. "Food Safety Trends in Retail and Foodservice." *FoodSafety Magazine,* December 2009–January 2010. Available online. URL: www.foodsafetymagazine.com/article.asp?id=3479&sub=sub1. Accessed January 4, 2010.

Gerba, Charles P. "The Importance of Infection Control in Education Institutions." *American School and University Magazine,* 1 June 2009. Available online. URL: http://asumag.com/Maintenance/infection-control-tips-200906. Accessed January 4, 2010.

McGlynn, William. "Guidelines for the Use of Chlorine Bleach as a Sanitizer in Food Processing Operations." Oklahoma State University. Food Technology Factsheet. Available online. URL: http://pods.dasnr.okstate.edu/docushare/dsweb/Get/Document-963/FAPC-116web.pdf. Accessed January 4, 2010.

Reynolds, Scott A., Foster Levy, and Elaine S. Walker. "Hand Sanitizer Alert." Letter to editor in *Emerging*

Infectious Diseases 12 (2006): 527–529. Available online. URL: www.cdc.gov/NCIDOD/eid/vol12no03/pdfs/05-0955.pdf. Accessed January 4, 2010.

U.S. Environmental Protection Agency. Available online. URL: www.epa.gov. Accessed January 4, 2010.

U.S. Food and Drug Administration. Available online. URL: www.fda.gov. Accessed January 4, 2010.

Weschler, Tom. "Testing Food Micro—We Need to 'Bridge the Gap.'" FoodSafety Magazine, June–July 2008. Available online. URL: www.foodsafetymagazine.com/article.asp?id=2483&sub=sub1. Accessed January 4, 2010.

sepsis Sepsis is an infection inside the body accompanied by inflammation. Sepsis (also called *septicemia* or systemic inflammatory response syndrome) can occur in any body fluid or organ interior to the body's skin or mucous membrane barriers. For example, sepsis occurs when blood becomes infected with bacteria, whereas a localized urinary tract infection is not sepsis. Localized infections can, however, enter the bloodstream to begin sepsis. Sepsis begins with fever, shortly after the body's immune system detects microorganisms, then, if untreated, progresses to hypothermia (lowered body temperature), tachycardia (rapid heart rate), tachypnea (rapid breathing), and reduced blood flow to organs, generalized in the following:

- body temperature—above 100.4°F (38°C) or below 96.8°F (36°C)
- tachycardia—more than 90 beats per minute
- tachypnea—more than 20 breaths per minute
- white blood cell count—more than 12,000 cells per microliter

Numerous bacteria, fungi, protozoa, and viruses are capable of causing this serious health condition. Definitions for various types of sepsis are:

- bacteremia—bacteria in the blood
- fungemia—fungi, usually *Candida* (yeast) or *Aspergillus*, in the blood
- viremia—viruses in the blood

Terminology may also be applied to specific causes of sepsis, such as meningococcal septicemia caused by members of the bacterial genus *Neisseria*.

Pathogenesis refers to the events in the body related to disease development. The damage caused by a systemic infection occurs either through toxin release from a microorganism or direct damage of tissue by the microorganism.

The blood is normally sterile in the body. When microorganisms enter through a cut or wound, the immune system should eliminate the invader within a few hours of infection. More serious infections may last longer, but a healthy immune system will eventually rid the blood of potential pathogens. If the body's defenses do not control microorganisms in the blood, sepsis develops with dangerous consequences.

A person can assume that an infection has entered the bloodstream if fever, chills, and accelerated heart rate and breathing develop. This simple sepsis can lead to two additional, more serious, stages: severe sepsis and septic shock. Severe sepsis correlates with the spread of microorganisms to at least one organ whereby organ function is impaired. Severe sepsis results in a drop in blood pressure. Septic shock develops when blood pressure continues to plunge and cannot be reversed. Single or multiple organ failure and altered mental acuity often accompany advanced stages of sepsis.

SEPSIS IN MEDICINE

The general term *sepsis* means any presence of microorganisms in places where they do not belong. Aseptic techniques—*aseptic* means the absence of microorganisms—involve activities intended to keep unwanted organisms out of sterile media or sterilized equipment. Medical care providers use aseptic techniques when treating medical or dental patients. Physicians also use antiseptics to remove microorganisms from the skin before giving an injection or making a surgical incision. An antiseptic is any chemical that removes microorganism from the skin.

The British surgeon Joseph Lister (1827–1912) first addressed the problem of sepsis in surgical patients and developed methods that surgeons still use for keeping germs out of open wounds. In the mid-1860s, Lister had grown discouraged by the high mortality rates of surgical patients in hospitals. Even minor surgeries resulted in infections, and about half of all surgical patients died of sepsis, although most physicians at the time did not attribute the problem to microorganisms. Many doctors assumed that postsurgery deaths were a natural reaction to the trauma of surgery and inevitable for certain unlucky patients. Lister refused to believe that infections were inevitable. He believed patient deaths had resulted from poor sanitation in the hospital and a lack of safeguards against airborne pathogens in surgery rooms.

In 1867, Lister wrote in the *British Medical Journal*, "In the course of an extended investigation into the nature of inflammation, and the healthy and morbid conditions of the blood in relation to it, I arrived, several years ago, at the conclusion that the essential cause of suppuration [formation of pus] in wounds is decomposition, brought about by the influence of the atmosphere upon blood or serum retained within them, and, in the case of contused wounds, upon portions of tissue destroyed by the violence of the injury." Lister continued by explaining his use of carbolic acid as an antiseptic on surgery patients before starting an operation.

Lister's contemporaries doubted the connection between infection and germs in the air. They, furthermore, hesitated to put an irritating chemical on their patients' skin. But Lister persevered and continued using aseptic techniques during surgeries. His colleagues noticed, over time, that far fewer of Lister's patients than of their patients died after surgery; medicine began to adopt methods for sterilizing equipment and keeping germs out of the body. Today's sterilization practices and aseptic techniques are conducted for the purpose of decreasing the risk of sepsis from surgery.

Although state-of-the-art technologies in patient care exist, sepsis remains a serious medical concern in the United States. Sepsis causes about 6 percent of all deaths in the United States and is the 10th leading cause of death in the country.

The physician Burke Cunha of the State University of New York School of Medicine–Stony Brook has identified the following main sources for bacterial sepsis cases:

- contaminated central intravenous lines (not peripheral lines to arms or legs)
- intraabdominal or pelvic infections
- abscesses related to appendicitis, diverticulitis, colecystitis, or Crohn's disease
- previous abdominal surgery
- urinary tract with previous patient history of kidney inflammations, kidney stones, and prostate or renal surgery
- patients who have diabetes, systemic lupus erythematosus, or alcoholism who also take steroids

Sepsis that has resulted from a surgical procedure is called *postoperative sepsis.*

Postoperative sepsis presents a severe health threat because a surgery patient's immune system and tissue repair processes may already be stressed. Surgery-related infections result from errors in aseptic technique or breaches in the peritoneum, the thin membrane that lines the abdominal cavity. Decades of improvements in patient care have decreased the number of deaths due to postoperative sepsis, but a study of more than two million elective-surgery patients, between 1997 and 2006, indicated that the overall cases of sepsis increased. Sepsis cases increased in incidence from 0.7 percent to 1.3 percent in the study period, an 85 percent increase; severe sepsis (inflammation with drop in blood pressure) tripled from 0.3 percent to 0.9 percent. The anesthesiologist Mervyn Maze of the University of California–San Francisco School of Medicine said about the trend, "Hospitals are still dangerous places to get ill." Patients should not assume that sepsis will be a forgone conclusion with hospital stays; the overall incidence rates in the United States for sepsis are low. Joseph Lister probably would warn, however, that aseptic techniques in medicine have no shortcuts.

GRAM-NEGATIVE SEPSIS

Severe sepsis and septic shock have been correlated with sepsis caused by gram-negative bacteria more than gram-positive. The dramatic drop in blood pressure associated with serious cases of sepsis has been called gram-negative sepsis, or endotoxic sepsis, due to endotoxins (toxin retained inside the bacterial cell) produced by gram-negative species.

Gram-negative sepsis can result from a variety of gram-negative bacterial infections. The main groups usually implicated are enterococci, *Klebsiella,* and *Bacteroides.* The term *enterococci* refers to species with spherical cells that live in the human or animal digestive tract. The genus *Klebsiella* inhabits the digestive tract. An infection in the blood with either enterococci or *Klebsiella* suggests the infection resulted from a gallbladder or urinary tract infection. *Bacteroides*'s main entry to the blood, by contrast, tends to be through colon or pelvic injury.

Pseudomonas aeruginosa inhabits soil, water, and plant surfaces. This bacterium causes most infections of burn wounds, but it uses numerous portals of entry for developing sepsis: skin, respiratory tract, gastrointestinal tract, urinary tract, eye, and ear. *P. aeruginosa* has also caused rare instances of intravenous contamination.

GRAM-POSITIVE SEPSIS

The general groups staphylococci, streptococci, and enterococci may cause sepsis more than other gram-positive bacteria. Staphylococci, such as *Staphylococcus aureus,* form clusters of spherical cells when viewed through a microscope, rather than the long chains of streptococci. Both cause toxic shock syndrome, which is an acute disruption to the skin accompanied by fever and caused by the bacteria's toxin.

Enterococci belong to the normal flora of the intestines and were long thought to be harmless

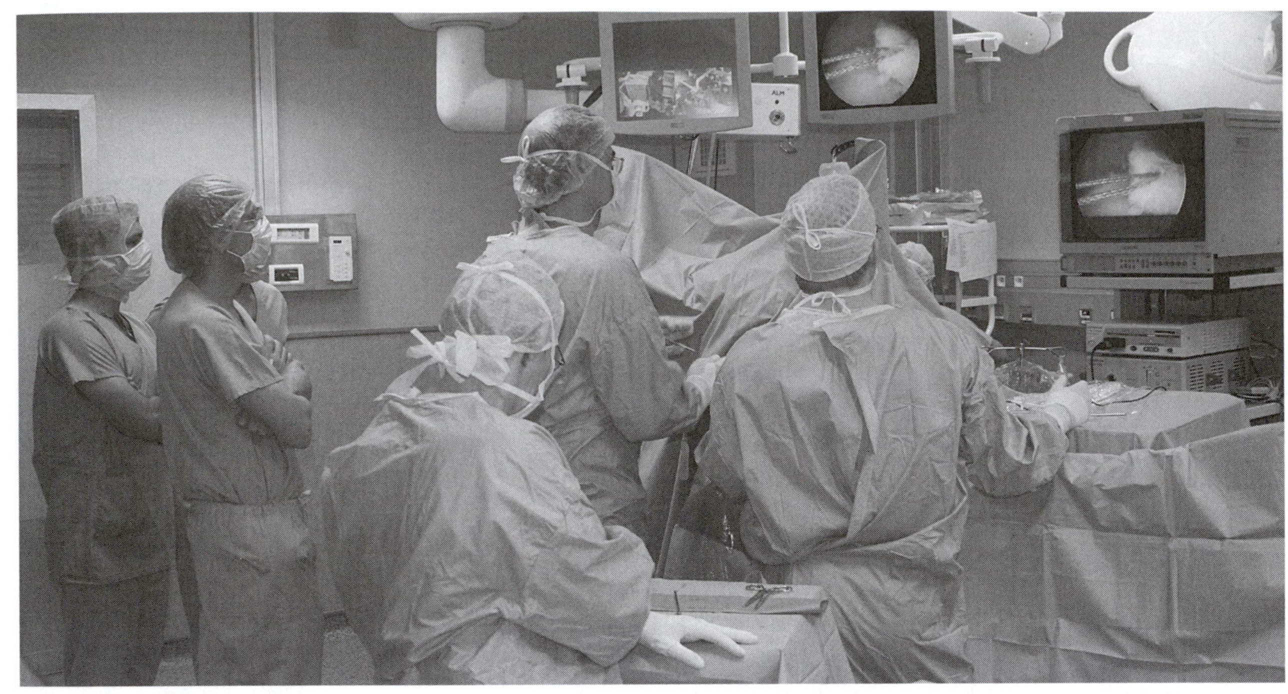

Surgical techniques that require preoperative cleaning and sterile equipment greatly reduce the risk of sepsis. ***(Alps Surgery Institute)***

until doctors noticed increasing incidences of nosocomial (hospital-associated) infections associated with these bacteria. Enterococci are opportunistic pathogens, meaning they live with humans without causing harm but are capable of causing disease if an opportunity arises.

MANAGING SEPSIS

The following actions help stop or reverse the series of events associated with sepsis:

- intravenous fluids
- oxygen supply
- vasopressor drugs
- broad-spectrum or narrow-spectrum antibiotics

Physicians order broad-spectrum antibiotics when patients are infected with more than one microorganism or when an infectious agent is unknown. If the infectious microorganism has been identified, the physician treats it with a narrow-spectrum antibiotic specific for the microorganism. Effective drug selection, therefore, calls for swift and accurate identification of microorganisms in a patient's blood specimen by a clinical microbiologist.

Postoperative sepsis has always been a serious health threat to surgery patients. Despite the outstanding contributions made by Joseph Lister in reducing the incidence of surgery-associated infections, sepsis continues to be a serious concern in hospitals worldwide.

See also ANTISEPTIC; ASEPTIC TECHNIQUE; LISTER, JOSEPH.

Further Reading

Bone, Roger C. "Gram-Negative Sepsis: A Dilemma of Modern Medicine." *Clinical Microbiology Reviews* 6 (1993): 57–68. Available online. URL: http://cmr.asm.org/cgi/reprint/6/1/57.pdf. Accessed January 1, 2010.

Cunha, Burke A. "Sepsis, Bacterial." EMedicine from WebMD. October 19, 2009. Available online. URL: http://emedicine.medscape.com/article/234587-overview. Accessed January 1, 2010.

Helwick, Caroline. "Postoperative Sepsis Rates Are Up but Mortality Is Down." *Medscape Medical News,* 21 October 2009. Available online. URL: www.medscape.com/viewarticle/711071. Accessed January 1, 2010.

Lister, Joseph. "On the Antiseptic Principle in the Practice of Surgery." *British Medical Journal* 2 (1867): 246–248. Available online. URL: www.ncbi.nlm.nih.gov/pmc/articles/PMC2310614/pdf/brmedj05631-0002.pdf. Accessed January 1, 2010.

Melamed, Alexander, and Frank J. Sorvillo. "The Burden of Sepsis-Associated Mortality in the United States from 1999 to 2005: An Analysis of Multiple-Cause-of-Death Data." *Critical Care* 13 (2009): 138–146. Available online. URL: http://ccforum.com/content/13/1/R28. Accessed January 1, 2010.

Surviving Sepsis Campaign. Available online. URL: www.survivingsepsis.org/Pages/default.aspx. Accessed January 1, 2010.

University of Dayton. "Sir Joseph Lister." Available online. URL: http://campus.udayton.edu/~hume/Lister/lister.htm. Accessed January 1, 2010.

serial dilution Serial dilution is a procedure used in microbiology for diluting concentrated suspensions several times to produce a lower concentration.

Cultures or samples of bacteria, yeasts, fungi, and viruses often reach very high concentrations of particles per milliliter (ml). High concentrations from 1,000 to 10 million microorganisms per milliliter are too dense for most calculations needed in microbiology. A serial dilution produces concentrations of 1,000 or fewer microorganisms per milliliter. This lower concentration range allows microbiologists to do simple calculations and enables them to count the number of cells from a diluted sample on an agar surface. For example, a volume of 0.1 ml of a diluted sample containing 1,150 bacterial cells per milliliter and inoculated to an agar plate produces 115 colonies after incubation. In this example, the 0.1 ml inoculums, called an *aliquot,* represent only one tenth the volume of a milliliter. Since 1,150 cells are present in 1 ml, 115 cells can be expected in 0.1 ml. Bacteria and yeast that grow into isolated discrete colonies on agar are usually reported as *colony-forming units* (CFUs) in microbiology.

Two types of cell counting are used for determining the concentration of microorganisms in a growing culture, a clinical sample, or a sample taken from the environment: direct counting and indirect counting. Direct counting involves a single step, in which a microbiologist determines the number of microorganisms per volume by counting them in the suspension using a microscope or an electronic counter. Specialized glass slides, such as the Petroff-Hausser counting chamber, contain a tiny well for the sample and a grid etched into the bottom of the well to help convert number of cells per grid into number of cells per milliliter. Electronic counting machines automatically register the number of cells per volume of liquid that flows past the machine's detector.

Direct counting offers the advantage of indicating results quickly, within minutes to hours of receiving a sample in the laboratory. The method has disadvantages, however, because it does not distinguish between live and dead cells. On occasion, a microbiologist at a microscope or an electronic counting device will also accidentally count cell debris or other nonmicrobial particles, which cause the final results to overestimate true cell numbers.

Indirect techniques involve an incubation step to determine the number of CFUs per volume. Altogether, indirect counting involves the following steps from original sample to the final result as number of microbial cells per milliliter:

1. Prepare serial dilution of the sample.
2. Inoculate several of the highest dilutions to agar media plates.
3. Incubate for 18–48 hours, depending on the microorganisms expected.
4. Count the number of CFUs per plate, called *plate counts,* on plates containing 30–300 total CFUs.
5. Convert plate counts to number of microorganisms per milliliter.

Each of the steps described here includes detailed substeps that increase the accuracy of the overall method. For example, a microbiologist uses aseptic techniques throughout the dilution process to ensure

Serial Dilution for a High-Concentration Sample

Starting Concentration	Dilution Step	Dilution Ratio	Final Concentration (also the dilution)
10^9	1.0–9.0 ml	1:10	10^{-1}
10^8	1.0–9.0 ml	1:100	10^{-2}
10^7	1.0–9.0 ml	1:1,000	10^{-3}
10^6	1.0–9.0 ml	1:10,000	10^{-4}
10^5	1.0–9.0 ml	1:100,000	10^{-5}
10^4	1.0–9.0 ml	1:1,000,000	10^{-6}
10^3	1.0–9.0 ml	1:10,000,000	10^{-7}
10^2	1.0 ml to 9.0 ml	1:100,000,000	10^{-8}

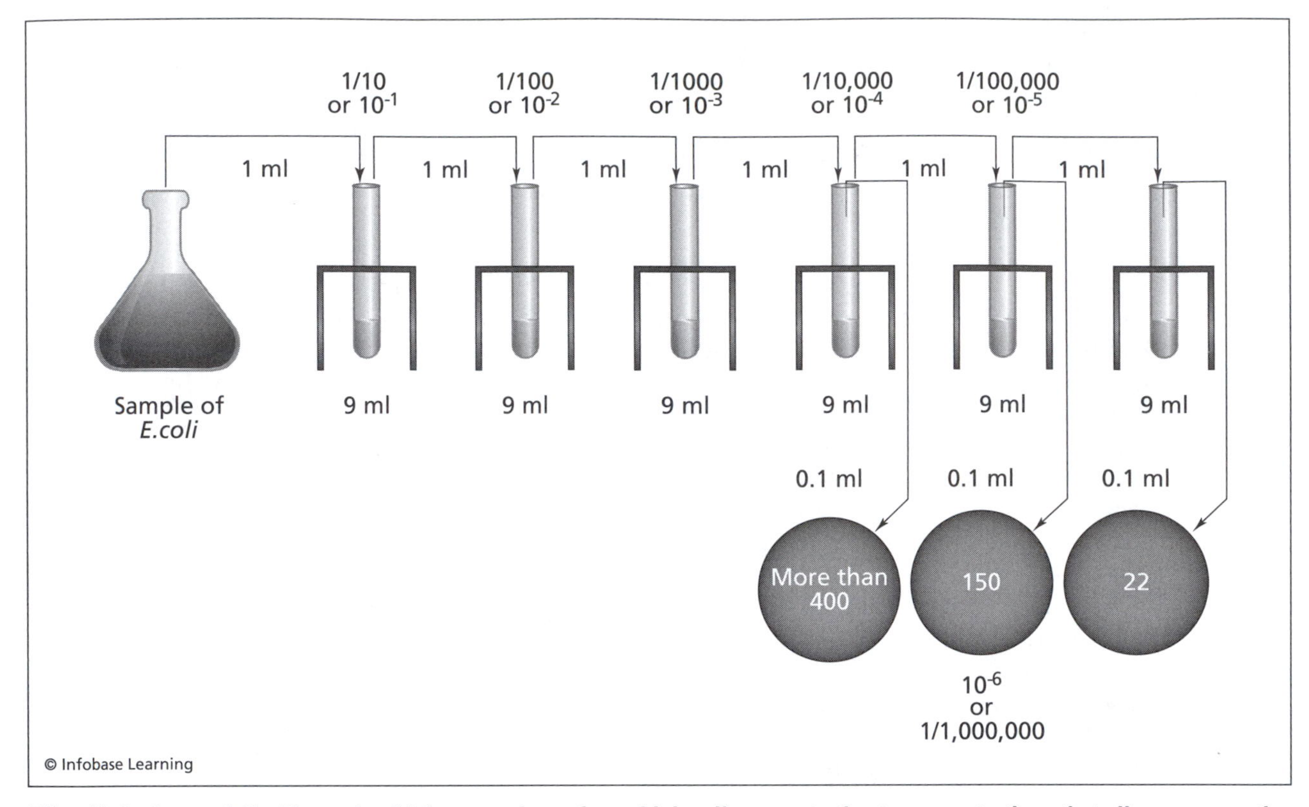

Microbiologists serially dilute microbial suspensions of very high cell concentration to concentrations that allow enumeration and further studies. Few sciences other than microbiology involve working with concentrations as high as 10^6–10^8 per milliliter, common for microbial samples.

that no contaminant enters. Any contamination could cause the procedure to overestimate the concentration of microbial cells in the original undiluted sample.

A serial dilution consists of a series of tubes containing 9.0 milliliters each of sterile buffer or water, called a dilution blank. A microbiologist makes a one-to-ten dilution, written as 1:10, of the sample by transferring 1 ml of sample to 9.0 ml of dilution blank. A similar transfer of 1 ml of the 1:10 dilution produces a 1:100 dilution, and so forth, until a very concentrated initial sample has been diluted five to seven times.

Serial dilutions provide two advantages. First, the dilutions produce CFU numbers that can be easily counted. Nondiluted samples inoculated directly to an agar plate would contain millions of CFUs impossible to distinguish from each other. Plates from serial dilutions contain low numbers of single, pure and isolated CFUs that are visible to the unaided eye. The second advantage of serial dilutions relates to the wide range of concentrations that microorganisms may be found in a sample of unknown composition. When microbiologists receive a sample from a patient, food, or the environment, they have no idea whether the sample contains millions of bacteria or few. Serial dilution helps span the range of possible concentrations.

Bacteria from places in nature such as the soil, surface waters, or the inside of cattle and other ruminant animals harbor very high numbers of bacteria. A microbiologist wishing to study a sample of rumen fluid, in the following example, requires serial dilu-

Hypothetical Plate Counts from Serial Dilution

Dilution	Number of CFUs on Duplicate Plates	Average CFUs per Plate
10^{-1} to 10^{-5}	TNTC (for Too Numerous to Count)	—
10^{-6}	more than 300 on each duplicate	—
10^{-7}	73, 69	71.0
10^{-8}	2, 8	—

Monitoring a Bacterial Growth Curve

Time, hours	Aliquot, ml	Average Plate Count	Log_{10}
0	0.1	0.5	less than 0
4	0.1	9.7×10^2	2.9868
8	0.1	1.6×10^6	6.2041
12	0.1	8.0×10^6	6.9031
16	0.1	4.3×10^7	7.6335
18	0.1	4.9×10^7	7.6902

tions. This is because the number of bacteria in this habitat can reach 10^9 per milliliter or higher.

A sample suspected of containing a microbial density in the order of 10^9 per milliliter requires a dilution series, as described in the table on page 699.

A serial dilution like that shown in the table would be adequate to dilute the 10^9 cells per milliliter sample to a more dilute version. A high dilution means the serial dilution contains many transfers of 1 ml to 9.0 ml. For example, 10^{-7} is a high dilution. A low dilution means that the serial dilution used few transfers. For example, 10^{-2} is a low dilution.

The set of tubes containing the desired dilutions is called the *dilution series,* which is ready for inoculation onto agar plates. Inoculation can be either of two types: spread plates and pour plates. A spread plate receives the inoculum after the agar has already solidified in a petri dish. The microbiologist spreads 0.1 ml of the dilutions over the agar surface using a sterile glass or plastic rod bent at one end about an inch from the end of the rod. Because of their shape, microbiologists call these rods *hockey sticks.* Pour plates are prepared by transferring the aliquots from the dilutions into plates in which heated agar is still molten. After the inoculums are gently mixed throughout the agar, the agar is allowed to cool and solidify. Petri dish spread plates and pour plates must be covered before incubating.

Many bacteria sampled in temperate environments and human pathogens grow at body temperature, so incubators can be set to about 98.6°F (37°C) for growing these bacteria. Bacteria native to water prefer a slightly cooler incubation at 95°F (35°C). Fungi require even cooler temperatures at 77–80.6°F (25–27°C).

After the plates have incubated for a sufficient period to produce visible CFUs, the microbiologist counts the number of CFUs on each plate. This counting step can be done manually, by viewing each plate under a magnifying glass, or electronically, by using a counter that scans the agar surface with a laser beam.

The undiluted original sample and low dilutions often contain plates covered with a contiguous sheet of colonies. High dilutions produce plates that have CFU counts of 30–300. Microbiologists prefer to count plates having CFUs numbers in this range. Plates containing fewer 20–30 colonies are too dilute to give consistently accurate results, and plates containing 300 or more colonies have too many colonies to count manually. These densely populated plates also cause the colonies to start inhibiting nearby colonies by using up nutrients or excreting antimicrobial substances.

Returning to the sample from a cattle rumen, consider the example counts in the table on page 700. In this example, duplicate plates had been prepared for each dilution.

The average plate counts in the table indicate that 71×10^7 bacteria were present per volume of inoculum. The inoculum volume in this example was 0.1 ml. Because the plate counts represent the number of bacteria in only one tenth of a milliliter, the microbiologist multiplies the count by 10 to convert the plate count to a per milliliter value.

$$(71 \times 10^7) \times 10 = 71 \times 10^8 \text{ or } 7.1 \times 10^9 \text{ CFU/ml}$$

The sample taken from the cow rumen in this example had 7.1 billion bacteria in 1 ml of rumen fluid. Numbers of this magnitude are not unusual in microbiology. Soil, marine waters, freshwaters, fecal samples, and untreated wastewater can all attain very high bacterial concentrations. Since studying samples with such enormous numbers would be impractical, microbiologists depend on serial dilutions to make microbial numbers easier to use. Logarithms provide another tool used in conjunction with the serial dilution results like these.

A logarithm is the power to which a number is raised. For the number 7.1×10^9, 10 has been raised to the power of 9. The entire number 7.1×10^9 can be converted to a logarithm (on a base 10 scale because the result is a multiple of 10) by using a calculator or looking up the value in a published table of logarithms. The base 10 logarithm, also written $\log_{10}$, of 7.1×10^9 equals 9.8513. Using 9.85 in calculations is much easier than using 7.1×10^9.

The same principles help microbiologists track the growth of a culture of microorganisms in a laboratory. Within a few hours, a culture of bacteria such as *Escherichia coli* can increase from about 10 cells per milliliter to millions per milliliter. Microbiologists can track the changes in a culture's population, called a *growth curve* when plotted on a graph, by making serial dilutions of small aliquots from the culture at specific time points. The table on page 701 provides a hypothetical growth curve of a 10-ml culture monitored over time. The microbiologist arrives at the results by taking aliquots from the culture at the time points, serially diluting them, inoculating the dilutions to agar plates, and counting CFUs after incubation.

The $\log_{10}$ values in the preceding table allow a microbiologist to plot a growth curve as integers rather than using a logarithmic scale and graph. Serial dilutions facilitate the handling of the very large numbers that are common in microbiology and few other sciences.

See also ASEPTIC TECHNIQUE; CULTURE; GROWTH CURVE; LOGARITHMIC GROWTH.

Further Reading

Maier, Raina M., Ian L. Pepper, and Charles P. Gerba, eds. *Environmental Microbiology*, 2nd ed. San Diego: Elsevier, 2009.

Tortora, Gerard J., Berdell R. Funke, and Christine Case. *Microbiology: An Introduction*, 10th ed. San Francisco: Benjamin Cummings, 2009.

Willey, Joanne, Linda Sherwood, and Chris Woolverton. *Prescott, Harley, Klein's Microbiology*, 7th ed. New York: McGraw-Hill, 2007.

serology Serology is a discipline within immunology that studies reactions between antigens and antibodies. An antigen, also called an *immunogen*, is any substance that the body identifies as foreign matter and responds to by producing an antibody. The antibody consists of protein that binds to the antigen in a specific manner, such as a lock and key connection. Because of the specific connection between an antigen and its corresponding antibody, microbiologists can use serology tests to identify microorganisms.

Serology takes advantage of the antigen-antibody complex that plays a pivotal role in immunity. When microorganisms invade the body, the body recognizes them as antigens, that is, matter foreign to the body. The immune system launches an immediate defensive response to destroy any antigen it finds, called *nonspecific defense*. A coordinated set of actions ensues, involving immune cells, messenger molecules, and nonspecific antibodies. When the infection has passed and the body has healed, the host will depend on specific defenses should the same microorganism infect in the future. Antibodies specially made by the body to recognize the microorganism (the antigen) make up specific defense.

The immune system has evolved to identify any biological particle or cell as "self" or "non-self." The host's cells and substances represent self entities. Bacteria, viruses, pollen, nonnative proteins, and large polysaccharides represent nonself entities that are seen by the immune system as antigens. Specific defense works because most antigens have distinctive features on their surface that signal the presence of foreign matter. If the microorganism that had infected a person in the past appears again in a later infection, the immune system now recognizes the microorganism and mounts a speedy response. Serology uses that specific recognition between antibody and antigen in an in vitro test to identify unknown pathogens.

DEVELOPMENT OF SEROTYPING

In the 1930s, the American microbiologist Rebecca Lancefield (1895–1981) decided to organize streptococci on the basis of the antigenic features on their cell

Slide Agglutination Testing

Unknown Bacterium	Antiserum Added	Reaction	Result
1 drop	A	no clumping	negative
1 drop	B	clumps	positive
1 drop	C	no clumping	negative
1 drop	D	no clumping	negative
1 drop	E	no clumping	negative

Lancefield Streptococcal Groups

Group	Source	Species
A	human pathogens	*S. pyogenes*
B	cows with mastitis; human throat and vagina	*S. agalactiae*
C	cattle and other animals; human throat	*S. equi, S. equisimilis, S. dysgalactiae*
D	cheese	*S. bovis*
E	milk products	*S. porcinus*
F	human throat in tonsillitis	*S. mutans, S. milleri*
G	humans, monkeys, dogs	*S. canis*
H, K, O	human respiratory tract but nonpathogenic	H—*S. sanguis* K—*S. salivarius* O—*S. mitis*

surface. Part of Lancefield's motivation may have been the prevalence of rheumatic fever and scarlet fever, during those years. Doctors knew a streptococcus like that of strep throat caused the diseases, but they lacked a reliable test to narrow down the species.

The renowned geneticist Maclyn McCarty (1911–2005) recalled, in 1973, "Dr. Lancefield had her first introduction to the streptococcus when, as a graduate student, she joined O. T. Avery and A. R. Dochez at the Rockefeller Institute in studies of strains isolated from a severe epidemic in military camps during World War I. This study resulted in the demonstration of serological differences between several of the epidemic strains and was the initial step in a series of investigations that culminated some years later in her discovery of the type-specific M antigen of group A streptococci." The painstaking differentiation of streptococci, in fact, dominated the rest of Lancefield's career, and her methods became the serology tests used today.

The different groups would be known as *serotypes*—sometimes also called *Lancefield groups*—and the in vitro testing of serotypes would be called *serotyping,* or typing. Lancefield differentiated serotypes as A through T, based on the reaction between bacterial cells and known antibodies, a test called *antibody agglutination.* The test requires antiserum, which is blood minus red blood cells but containing a specific known antibody. By putting a drop of each different antiserum on replicate drops of bacteria, a microbiologist looks for the formation of the antigen-antibody complex. The complex, if present, develops within minutes and appears as clumps in the mixture. Clear liquid indicates no reaction occurred between antigen and antibody. Because microbiologists usually perform the test on a glass microscope slide, it is also called the *slide agglutination test.* The table on page 702 summarizes a hypothetical agglutination test.

The table indicates that the unknown bacterium reacted with antiserum B but no other antiserum. If the antiserum contains antibodies for *Streptococcus pyogenes,* for example, the test has identified the unknown bacterium as *S. pyogenes.* Companies sell to laboratories antisera for various microorganisms. Before these commercial preparations were developed, scientists injected rabbits with pathogens to induce antibody production in the animals. A scientist, then, drew some blood and removed the red blood cells to make antiserum.

Rebecca Lancefield's system targeted cell wall carbohydrates (C polysaccharides) as the antigen. She had found that all the streptococci tested contained a C antigen but they differed in the other antigens they held. Lancefield also discovered that some serotypes, such as group A, could be divided further, by the presence of proteins called *M antigens.* Microbiologists have, since, further refined the test by including additional bacteria and finding more unique antigens such as the following:

- H antigen—flagella
- K antigen—capsule
- antigen—cell wall

Almost 200 serotypes for *Escherichia coli* have been found, the most notorious of which is the lethal food-borne pathogen O157:H7. *E. coli* serotypes O55, O111, and O127 have been associated with infant diarrhea. The table above summarizes important streptococcus serotypes in the Lancefield classification system.

SEROTYPING PROCEDURES

Microbiologists use serotyping today for identification of specific strains and substrains of bacteria, called either serotypes or *serovars*. A *strain* is a microorganism of a certain species that possesses a unique trait. Serotyping serves as a useful tool in food microbiology, as well as medicine, for the purpose of identifying food-borne microorganisms.

Serotyping tests are types of immunoassays, which are in vitro tests based on the antigen-antibody complex. Serotyping techniques commonly used in microbiology laboratories are the antigen agglutination test, the precipitin test, the enzyme-linked immunosorbent assay (ELISA), and Western blotting.

The precipitin test differs from agglutination by the form of the antigen to be tested. Agglutination uses whole bacterial cells for creating antigen-antibody complexes, while the precipitin test uses cell wall fragments separated from the rest of the cell. The test also includes a step that separates soluble cell wall antigens from other parts of the cell wall. The separation of cells into their components is a technique called *fractionation*. A microbiologist records a positive test result when the cell extract–antiserum mixture turns hazy white. A precipitate forms when the mixtures incubate at 98.6°F (37°C).

Many immunoassays depend on fluorescent dyes to make the results easier to interpret. The ELISA test is fast and has been automated so that hundreds of tests run simultaneously. In ELISA testing, an antibody used in the reaction has been tagged with a fluorescent dye. Positive tests results correlate with specific color reactions.

The Western blot test also uses an enzyme to produce a color reaction, rather than a fluorescent tag, described in the following steps:

1. Proteins in a patient's serum sample separate in a procedure called electrophoresis.

2. The electrophoresis produces a flat, thin gel that holds the separated proteins stationary.

3. A technician transfers the entire set of proteins to a large sheet of filter paper; the paper receives the proteins in the same orientation as they had been in the gel.

4. The technician washes the entire filter with a solution of a specific antibody tagged with a color-producing enzyme.

5. A positive reaction between a protein and antibody appears as a color change.

The Western blot test produces very specific results because it analyzes the protein composition of the test species. These unique compositions serve as a type of fingerprint that identifies strains and individuals within a strain. Western blot tests take longer to reach a result than the ELISA test. Microbiologists select these and other methods in serology according to the priorities of their work. A doctor who needs a fast identification of a pathogen may turn to automated serotyping or ELISA testing. The need to be very specific when identifying a microorganism calls for Western blotting or derivatives of this test.

Serotyping can play a lifesaving role when a doctor needs to know the potential virulence of a microorganism causing infection in a patient. For example, food-borne *E. coli* O157:H7 makes up a small percentage of all *E. coli* strains, but it also is more virulent than the others. The strain can cause kidney failure and death in a small percentage of people who become infected. Serotyping, thus, remains an important method in clinical microbiology and medical care.

See also FRACTIONATION; IDENTIFICATION; IMMUNITY; IMMUNOASSAY.

Further Reading

Lancefield, Rebecca C. "A Serological Differentiation of Human and Other Groups of Hemolytic Streptococci." *Journal of Experimental Medicine* 57 (1933): 571–595. Available online. URL: http://jem.rupress.org/cgi/reprint/57/4/571. Accessed January 9, 2010.

McCarty, Maclyn. "Presentation of the Academy Medal to Rebecca C. Lancefield, Ph.D." *Bulletin of the New York Academy of Medicine* 49 (1973): 949–953. Available online. URL: www.ncbi.nlm.nih.gov/pmc/articles/PMC1807098/pdf/bullnyacadmed00189-0011.pdf. Accessed December 5, 2009.

O'Hern, Elizabeth M. "Rebecca Craighill Lancefield, Pioneer Microbiologist." *American Society for Microbiology News*, December 1975. Available online. URL: www.asm.org/ccLibraryFiles/FILENAME/0000000266/411275p805.pdf. Accessed December 5, 2009.

sexually transmitted disease (STD) A sexually transmitted disease (STD) (formerly called *venereal disease*) is any disease that can be transmitted by direct contact with the mouth, genitals, or rectum. Most pathogens that cause STDs do not withstand environmental conditions well, so transmission must be by intimate contact. Most sexually transmitted pathogens infect the body through the mucosal membranes. More than 30 different bacteria,

Sexually Transmitted Diseases

Microorganism	Type	Disease
Chlamydia trachomatis	bacterium	urethritis (inflammation of urethra), pelvic inflammatory disease (PID)
Neisseria gonorrhoeae	bacterium	gonorrhea, PID
Treponema pallidum	bacterium	syphilis
Trichomonas vaginalis	protozoan	trichomoniasis (vaginitis)
herpes simplex (HSV-2)	virus	genital herpes
human papillomavirus type 6 (HPV)	virus	genital warts
hepatitis B (HBV)	virus	serum hepatitis
human immunodeficiency virus (HIV)	virus	acquired immunodeficiency syndrome (AIDS)
Mycoplasma hominis	bacterium	PID
Candida albicans	yeast (fungus)	candidiasis
Ureaplasma urealyticum	bacterium	urethritis
Haemophilus ducreyi	bacterium	chancroid genital sores (soft chancre)
cytomegalovirus (CMV)	virus	congenital cytomegalic inclusion disease (white blood cell infection transmitted from mother to fetus)

viruses, fungi, and protozoa cause sexually transmitted infections (STIs), most from bacteria and viruses.

STDs are a worldwide health problem, made worse by the fact that these diseases often cause hidden epidemics, meaning that people are reluctant to report them for societal and cultural reasons. STDs are closely linked with human behavior of the most personal nature, and discussing these infections often makes people feel uncomfortable. Two consequences of these societal factors are that many STDs are underreported each year, and many young people and adults do not receive the education on STDs they need to prevent infection.

In 2008, the most recent year the Centers for Disease Control and Prevention (CDC) tabulated statistics on STDs in the United States, the infection rate exceeded 213 cases per 100,000 people. Cases of the main three STDs, *Chlamydia* infection, gonorrhea, and syphilis, totaled 1.5 million. Worldwide, between 300 million and 400 million new STD cases occur every year. As a result of untreated STDs, the following problems have put strain on health agencies and economies: rise in acute illnesses, infertility, psychological problems, long-term disability, and death.

MAJOR SEXUALLY TRANSMITTED DISEASES

STDs occur in every part of the world, often at epidemic proportions, even though most of the diseases have a treatment or a cure. The table above describes the world's major STDs in general order of their prevalence worldwide.

Many bacterial, yeast, and protozoal STDs have effective drug treatments. Tetracyclines, erythromycin, and metronidazole antibiotics cure most bacterial STDs, with the exception of gonorrhea. *Candida* infections usually can be eliminated with either nystatin or terconazole, and *Trichomonas* is susceptible to oral metronidazole. *N. gonorrhoeae* has developed resistance to multiple antibiotics, in the past few decades, making gonorrhea an increasingly serious health problem worldwide. The CDC has said of the antibiotic-resistant gonorrhea (ARG) crisis, "The development of antibiotic resistance in *Neisseria gonorrhoeae* is a growing public health concern, particularly because only one remaining class of antibiotics is recommended for its treatment. Currently, the CDC STD guidelines recommend that cephalosporin antibiotics be used to treat all gonococcal infections in the United States." *N. gonorrhoeae* is resistant to penicillins, tetracyclines, spectinomycin (an aminoglycoside antibiotic), and ciprofloxacin (a fluoroquinolone antibiotic).

Various female STDs are manifested as vaginitis (inflammation of the vagina), and the correct treatment depends on an accurate diagnosis. STDs can be diagnosed by culturing a vaginal specimen from a patient by using methods in clinical microbiology. Other characteristics of infection, such as the following, can help the patient and her physician guess the cause of vaginitis before laboratory tests have been completed:

- bacterial vaginitis—copious amounts of gray-white, thin discharge with fishy odor and pink mucosa (lining of the vagina)
- *Candida*—varying amounts of white, thick discharge with slight yeast odor (sweet) and dry, red mucosa
- *Trichomonas*—copious amounts of green-yellow, frothy discharge with sting odor and tender, red mucosa

All of the causative organisms of vaginitis also establish infection at different pH ranges: bacteria grow above pH 4.5, *Candida* grows below pH 4, and *Trichomonas* grows in a pH range of 5–6. Vaginitis can be cured with antibiotics. Infections by these pathogens without signs of inflammation (redness, tenderness, swelling) are collectively called *vaginosis*.

The incidence of gonorrhea in the United States has remained steady, since the 1970s, at about 120 cases per 100,000 people. Gonorrhea and *Chlamydia* infections have been concentrated in the age group of 15- to 24-year-olds—this age group acquires about half of all STDs in the United States. Since 2005, gonorrhea rates have increased more than 2 percent in the 15- to 19-year-old age group and 0.7 percent for 20- to 24-year-olds, according to the Resource Center for Adolescent Pregnancy Prevention.

Viral STDs have been much more difficult to control with treatments for infections that have already become established in a host. The following treatments or preventions for viral STDs have been used in the past with varying degrees of success:

- AIDS—drug "cocktails" of reverse transcriptase enzymes to prevent viral replication inside cells and protease inhibitors to prevent the virus infiltration of cellular processes; vaccines are currently in testing
- genital herpes—acyclovir and similar drugs that reduce the symptoms but may not eliminate the virus
- genital warts—surgery
- hepatitis B—hepatitis B virus (HBV) vaccine
- congenital cytomegalic inclusion disease—limited treatments (ganciclovir and cidofovir) for high-risk patients

Vaccines against STIs have raised concerns unlike those provoked by other communicable diseases because many people think vaccinations would promote high-health-risk behavior. In 1997, the STD researchers Cibele Barbosa-Cesnik, Antonio Gerbase, and David Heymann emphasized, "If the aim is eradication or elimination, universal vaccination is required (with a vaccine that provides lifelong immunity), which is usually accomplished during infancy or childhood. However, this is an expensive strategy and might not be widely accepted." At present, arguments among parents and medical care providers continue on the controversial decision to immunize a newborn against an STD that might occur in the child's teens. Only one STD vaccine intended for teenaged girls has emerged at present: a vaccine for the human papillomavirus (HPV). The CDC's justification for this vaccine is that of the approximately 40 types of HPV, "some types can cause cervical cancer in women and other less common genital cancers—like cancers of the anus, vagina, and vulva (area around the opening of the vagina)." The CDC recommends the vaccine for girls 11 to 12 years old, but females younger (to age nine) or older can also consider vaccination.

STDs complicate health care in young and mature adults when the infection occurs with other underlying health problems, such as malnutrition, or another STD. For example, the CDC has reported that people infected with an STD are two to five times more likely than uninfected people to acquire an HIV infection. STDs increase a person's susceptibility by two mechanisms: (1) Open sores or ulcers caused by STDs give HIV a portal of entry into the body, and (2) the inflammation caused by STDs at certain sites on the body draws immune system cells ($CD4^+$ T cells) that HIV specifically targets for infection.

STDs involve numerous other complicating health factors that make these diseases difficult to treat. Epidemiologists study disease demographics to determine possible preventatives. Epidemiologists have found some associations between three major STDs and U.S. population demographics.

- *Chlamydia* infection—rate of infection in women is almost three times that of men, but infections have been steadily increasing in both sexes; prevalence is greater among poor women
- gonorrhea—rates have plateaued, for more than a decade, but incidence is rising in the

southern states while declining in the Northeast, Midwest, and West

- syphilis—incidence decreased almost 90 percent, in the decade between 1990 and 2000, but has increased slowly since then; primary and secondary syphilis infections are eight times more prevalent in blacks than in whites

Of the three STDs described here, syphilis may be nearest to elimination in the United States. The success in eliminating any infectious disease, not just STDs, depends on coordinated efforts that usually involve national programs and local resources for information, vaccine administration, or treatments.

CONTROL PROGRAMS FOR SEXUALLY TRANSMITTED DISEASES

The Syphilis Elimination Effort (SEE) is a program sponsored by the U.S. government that may become the first to succeed in eliminating an STD from the U.S. population. In 2006, midway through the SEE's history, the CDC explained the need for this initiative: "Elimination of syphilis would have far-reaching public health benefits because it would remove two serious consequences of the disease—increased likelihood of HIV transmission, and serious complications in pregnancy and childbirth, such as spontaneous abortions, stillbirths, and congenital syphilis (syphilis among newborns who acquired it from their mothers)." The program has, to date, made significant progress in lowering the incidence of syphilis cases, since its launch in 1999. (Syphilis rates were already declining in the United States when the SEE began.)

Abstinence from sexual contact is the only 100 percent effective way to avoid STI. Medical care providers understand that recommending abstinence is not realistic and that it would not be a practical way to control the spread of STDs. Some STDs spread by nonsexual means, such as the following transmission methods: contaminated hypodermic needles and syringes, contaminated blood used for transfusions, and infected mother–to–infant transmission, called vertical transmission. Controlling STDs, therefore, requires more than altering sexual behavior.

Behavioral changes work better than drugs or vaccines for controlling STDs. The main recommendations of numerous health agencies for controlling STD transmission are:

- limiting the number of sexual partners or forming a monogamous relationship
- using condoms
- avoiding direct skin contact with anyone with outward signs of herpes or papillomavirus
- checking partners for obvious signs of infection
- questioning partners about sexual history
- visiting a doctor after having unprotected (no condom) sex with a casual partner
- undergoing immediate treatment when symptoms of an STD appear and completing the treatment according to doctor instructions

STDs have been an insidious illness in society, since antiquity. Accurate means of diagnosing STDs have arrived in medicine, but reliable preventive measures have been difficult to achieve because STDs relate to personal human behavior.

See also ANTIBIOTIC; *CANDIDA ALBICANS;* HUMAN IMMUNODEFICIENCY VIRUS; TRANSMISSION.

Further Reading

Advocates for Youth. Available online. URL: www.advocatesforyouth.org/index.php. Accessed January 28, 2010.

Avert. "STD Reporting in the USA." January 25, 2010. Available online. URL: www.avert.org/stdstatisticusa.htm. Accessed January 28, 2010.

Barbosa-Cesnik, Cibele T., Antonio Gerbase, and David Heymann. "STD Vaccines—an Overview." *Genitourinary Medicine* 73 (1997): 336–342. Available online. URL: www.ncbi.nlm.nih.gov/pmc/articles/PMC1195888/pdf/genitmed00005-0006.pdf. Accessed January 28, 2010.

Centers for Disease Control and Prevention. Division of STD Prevention. "Sexually Transmitted Diseases." Available online. URL: www.cdc.gov/std/dstdp. Accessed January 28, 2010.

———. "HPV Vaccination Information for Young Women." Available online. URL: www.cdc.gov/std/hpv/STDFact-HPV-vaccine-young-women.htm.

———. "The Role of STD Prevention and Treatment in HIV Prevention." CDC Fact Sheet. April 10, 2008. Available online. URL: www.cdc.gov/std/hiv/stdfact-std&hiv.htm#WhatIs. Accessed January 28, 2010.

———. "Syphilis Elimination Effort (SEE)." Available online. URL: www.cdc.gov/StopSyphilis. Accessed January 28, 2010.

Resource Center for Adolescent Pregnancy Prevention. Available online. URL: www.etr.org/recapp/index.cfm?fuseaction=pages.home. Accessed January 28, 2010.

Royal Adelaide Hospital. Sexually Transmitted Diseases Services. Available online. URL: www.stdservices.on.net. Accessed January 28, 2010.

U.S. Department of Health and Human Services. "Sexually Transmitted Disease Surveillance 2007 Supplement."

March 2009. Available online. URL: www.cdc.gov/STD/gisp2007/GISPSurvSupp2007Short.pdf. Accessed January 2, 2010.

soil microbiology Soil microbiology is a specialty within environmental science covering the bacteria, archaea, fungi, algae, and protozoa native to soils. Soil microbiologists study microbial metabolism as it relates to living and nonliving soil constituents. This broad subject concerns additional focus areas, such as biogeochemical cycles, plant-microorganism relationships, decomposition and bioremediation, subsurface environments, soil microbial communities, and environmental pathogens in soil. Biogeochemical cycles, or nutrient cycles, are the pathways by which nutrients pass through different chemical forms as they circulate from soil to water, living organisms, the atmosphere, and back to soil. The biosphere depends on biogeochemical cycles to control the movement of Earth's elements between living and nonliving things, which, in turn, determines the health of ecosystems. Soil microbiology and chemistry have vital roles in operating these cycles for the benefit of all biota. Because of the importance of nutrient cycling, all of the specialty areas described in the table on page 709 relate directly or indirectly to these cycles.

More microorganisms exist on Earth than all plant and animal life combined, and most of this population resides in the upper few meters of surface soils. The Soil and Crop Science Department at Texas A. & M. University estimates the following amounts of microorganisms in a gram of soil:

- 100 million–1 billion bacteria
- 100,000–1 million fungi
- 1,000–1 million algae and cyanobacteria
- 1,000–100,000 protozoa

By their relative sizes—a fungus can be several thousand times bigger than a bacterium—fungi make up the greatest mass of microorganisms in soil, followed by actinomycetes (filamentous bacteria that resemble fungi), bacteria, protozoa, and algae. Although the total mass of soil remains largely inorganic, organic living matter makes up a tremendous portion of soil.

The physical and chemical characteristics of soil affect the type of microorganisms found there and the activities they perform. Soil characteristics, moreover, change in horizontal and vertical directions. In other words, moving a few feet in any direction from a given point will probably lead to different soil characteristics. Microbial populations in soil follow a similar continuum. Because it would be impossible to study the effect of thousands of different soil characteristics on microorganisms, soil microbiologists learn the basic chemical and physical properties of major soil types.

SOIL CHARACTERISTICS

Soil is not an inert lump meant only to support the life walking on it, rooted to it, or living within it. Soil has specific defining characteristics and influences the plant, animal, and microbial life that interacts with it.

Soil science covers the details of soil's chemistry and physical features. Although all of the features probably affect the microorganisms in soil, microbiologists focus on a group of characteristics known to impact microbial growth. The principal features of soil that affect microorganisms are the following:

- amount of organic carbon
- moisture content
- aeration
- pH
- temperature
- amount and variety of inorganic nutrients

Microorganisms display a remarkable range of environments in which they live on Earth, from frozen tundra to boiling hot springs. Considering only the microorganisms of temperate climates, where most humans live, soil microorganisms prefer the following ranges: moisture content of 50–60 percent of total water-holding capacity, pH of 6.0–8.0, and temperature of 50–104°F (10–40°C). Other populations of extremophiles inhabit extreme environments in soil, that is, places having conditions that most organisms cannot tolerate. Soil extremophiles can exist at cold temperatures, at high salt concentration, in toxic chemicals, and at intense pressure.

Most soil inhabitants require small amounts of the following inorganic nutrients: nitrogen, sulfur, phosphorus, potassium (K), iron, magnesium (Mg), manganese; trace minerals (minerals needed only in extremely low levels, such as boron, cobalt, copper, and molybdenum); plus calcium (Ca), chlorine C1 and sodium (Na). Carbon compounds (organic chemicals) usually supply part or all of the oxygen and hydrogen needed by a cell. Some microorganisms have additional and unusual requirements, such as certain algae called *diatoms,* which require silicon.

Many of these inorganic nutrients reside in soil in a positively or negatively charged form called an *ion.* The charge results from chemical reactions that either

Disciplines in Soil Microbiology

Specialty Area	Subspecialties
biogeochemical cycles	• carbon • nitrogen • sulfur • phosphorus • iron and other metals
plant-microorganism relationships	• symbiosis • tri- and tetrapartite associations • rhizosphere environments
decomposition and bioremediation	• plant decomposition • composting • biodegradation of pollutants
subsurface environments	• microfossils • subsurface gas flux • aquifers
microbial communities	• biofilms • nitrogen metabolism • sulfur metabolism
environmental pathogens	• human and animal pathogens • plant pathogens

add or take away an element's electrons. Soil scientists use the term *cation exchange capacity* (CEC) to describe this chemical characteristic. A cation is a positively charged ion. CEC is a measure of the amount of cations a soil can hold. Agriculturalists use CEC to estimate the value of soil for supplying plants with nutrients. Soil microbiologists similarly use CEC to gauge the availability of inorganic nutrients for soil microorganisms. In general, aluminum (Al) binds more tightly to soils than many other elements:

$$Al^{+3} \rightarrow Ca^{+2} = Mg^{+2} \rightarrow K^{+} \rightarrow Na^{+}$$

The CEC of a soil affects the ability of a microorganism to obtain the essential metals it needs for metabolism. CEC also influences whether soil can bind and immobilize a toxic metal, for example, mercury, or allow the metal to spread in the environment. Bioremediation scientists study soil chemistry and CEC to determine whether microorganisms will be likely to remove a toxic metal pollutant from the soil.

TYPES OF SOIL MICROORGANISMS

Microbiologists can define the microorganisms they wish to study in terms of the type of organism, type of soil, or type of interaction between the two. The table on page 710 describes some of the broad classifications in soil microbiology.

Microbiologists who use ecological roles to group soil microorganisms are able to demonstrate the relationship of these organisms to biogeochemical cycling and human health. In general, the main role of each of these groups is:

- decomposers—release nutrients from decaying organic matter for reuse
- nitrogen fixers—absorb atmospheric nitrogen, thus making this limited nutrient available to biota
- pathogens—cause disease in plants and animals
- chemoautotrophs—make inorganic chemicals available for biological use

Of the groups listed here, decomposers, nitrogen fixers, and chemoautotrophs function in biogeochemical cycles, and pathogens play an obvious role in health. But pathogens also contribute to nutrient cycling because infectious diseases are a major cause of plant and animal deaths. Thus, dead plant and animal tissues release their elements back to the biosphere with the help of decomposers.

DECOMPOSERS

Decomposers are mainly bacteria and fungi that break down the organic compounds of dead plant and animal matter into simple inorganic compounds. These microorganisms carry out their activities solely for the purpose of sustaining their own populations. All other life receives a benefit, however, because decomposers return nutrients to the soil, where other organisms can absorb them. These organisms begin a first step in distributing nutrients and energy to higher organisms by way of food chains.

If biochemistry were to be divided into compound-building activities or compound-degrading activities, decomposers carry out degradations. Other than making new cell material, decomposers carry out dissimilatory processes, the dispersal of elements from organized large molecules to various small molecules. Assimilatory processes do the opposite, by assembling elements into organized compounds.

Decomposers accomplish their task by using the same enzymes that people use for digesting a meal. Leaves, branches, fallen trees, dead vertebrates and invertebrates, and macromolecules degrade under the action of amylases (digest polysaccharides), cellulases (cellulose), proteases (proteins), lipases (lipids), and other specialized enzymes that can decompose heterogeneous fibers or compounds containing a carbon ring.

In decomposition, macromolecules break down to their constituent base units: polysaccharides to sugars, proteins to amino acids, and so on. Other soil microorganisms use the subunits in their metabolisms until all the matter in a fallen leaf, for example, ends as carbon dioxide and other gases, water, and minerals.

Decomposers as a group contain a broad variety of bacteria; the soil genera *Bacillus, Pseudomonas, Arthrobacter, Clostridium,* and *Streptomyces* dominate in many areas with high decomposition activity. The fungi belong to a general group called *saprophytes,* which are organisms that live on dead organic matter. *Penicillium* and *Aspergillus* join many other saprophytes in the decomposition process.

The table on page 711 presents various bacteria that act as either primary decomposers or secondary decomposers. Primary decomposers are the first organisms to begin degradation of large, complex molecules. Secondary decomposers aid the process further by degrading intermediate molecules to end products or carrying out functions that assist the decomposition process. For example, anaerobic bacteria decomposing a log from the inside might slow their metabolism as end products accumulated. A bacterial group called the methanogens helps the reactions proceed by taking away the end products carbon dioxide and formic acid (HCOOH).

Types of Soil Microorganisms

Method of Categorizing	Categories
type of organism	bacteria, fungi, protozoa, algae
domain	archaea, bacteria, eukaryotes
soil type	sandy, silt, clay, loam
soil depth	surface, leaching zone, mineral zone, sediment
energy metabolism	aerobic, anaerobic
nutrient metabolism	chemoheterotrophs (organic compounds for carbon energy), chemoautotrophs (carbon dioxide for carbon, chemical compounds for energy)
water saturation	unsaturated, immobile water, nearly saturated, water movement, saturated
ecological role	decomposers, nitrogen fixers, pathogens, chemoautotrophs

SUBSURFACE ENVIRONMENTS

Deep subsurface environments have not received the same intensive study as surface soils, partly because of the difficulty in reaching subsurface soils. Geomicrobiologists study the interactions between microorganisms and geological structures. Most of the studies these scientists conduct have been possible by using gold mines to access sampling sites or deep wells drilled by certain industries to retrieve samples. In these ways, geomicrobiologists have reached places more than 1.8 miles (3 km) deep.

The microorganisms of deep sediments are extremophiles that live in complete darkness, with no oxygen, scarce nutrients, slow water flow, and very intense pressure. The proportion of chemoautotrophs to heterotrophs increases with increasing depth. Chemoautotrophy suits an environment where only inorganic nutrients exist. Chemoautotrophs live on these nutrients and carbon dioxide. Heterotrophs use a much wider variety of organic compounds than are available in deep sediments.

Despite chemoautotrophs' ability to live in spartan conditions, even these microorganisms must make adjustments to their habitat by growing very slowly and reducing their cell size. Both tactics decrease the amount of nutrients needed to keep their populations alive. The microorganisms, then, subsist on iron, manganese, and sulfur, which they use in reduction-oxidation (redox) reactions.

At the Homestake mine in Lead, South Dakota, the U.S. government is planning the Deep Underground Science and Engineering Laboratory (DUSEL). The deepest mine in North America will provide space at an onsite laboratory for biology, physics, and geology studies. Large underground laboratories already exist in Canada, Italy, and Japan, and each has a waiting list of scientists eager to use the space for experiments. DUSEL will increase the availability worldwide of deep subsurface environments with the goal of advancing the science of subsurface microbiology.

PATHOGENS IN SOIL

Many of the pathogens in soil result from fecal contamination in water or direct entry to the soil, but two pathogens are normal inhabitants of soil: *Bacillus anthracis* and *Naegleria fowleri. B. anthracis* is the bacterium that causes anthrax, and people who have close contact with soils have a higher risk of infection. The amoeba *N. fowleri* lives in water and soil. Infection occurs by an unusual route in which a motile reproductive stage called a trophozoite swims into the nasal canal.

The protozoan *Toxoplasma gondii* lives in soil, where it turns into a spore form called an *oocyst.*

Common Soil Microorganisms

Name	Type	Metabolism	Soil Type	Cycling
Actinomycetes	filamentous bacteria	aerobic	dry, high pH	N
Amoeba	protozoa	aerobe	surface, moist	C
Arthrobacter	bacteria	aerobic	nonspecific	C
Bacillus	spore-forming bacteria	aerobic	nonspecific	C
Chlorophyta	algae	aerobe	surface, moist, low pH	C, O
Chrysphycophyta	algae (diatoms)	aerobe	neutral to high pH	silica, Ca
Clostridium	spore-forming bacteria	anaerobe	nonspecific	C
methanotrophs	bacteria	aerobic	oxygen-limited	C
Nitrobacter	nitrification	aerobe	nonspecific	N
Nitrosomonas	nitrification	aerobe	nonspecific	N
Pseudomonas	bacteria	aerobic	nonspecific	C
Rhizobium	nitrogen fixation	aerobic	rhizosphere	N
Thiobacillus	bacteria	anaerobic	oxygen-limited	S

Note: C = carbon; Ca = calcium; N = nitrogen; O = oxygen; S = sulfur

People become infected with *T. gondii* oocysts by putting contaminated items in their mouth.

Many fungi also use soil as a natural habitat. Ubiquitous organisms such as *Aspergillus* and *Penicillium* present the highest risk of infection. *A. fumigatus* spores, when inhaled into the lungs, are the major cause of aspergillosis in humans.

The rest of the pathogens likely to be found in soil belong to the general group called *enteric microorganisms*. The following bacteria contaminate soils and from there can adhere to fruits and vegetables: *Campylobacter, Clostridium, Escherichia coli, Salmonella, Shigella, Vibrio,* and *Yersinia*. Viruses such as hepatitis A and E, echovirus, coxsackievirus, and poliovirus can enter soils from sewage. Soil is not a high-risk source of infection. Good hygiene and hand washing after handling soil, farm animals, or crops are usually adequate to reduce the chance of infection.

Soil microbiology is a complex science that requires some knowledge of earth science and chemistry. But soil microbiology also builds a foundation for understanding Earth's ecology.

See also ANTHRAX; BIOGEOCHEMICAL CYCLES; BIOREMEDIATION; ENVIRONMENTAL MICROBIOLOGY; FUNGUS; MICROBIAL COMMUNITY; NITROGEN FIXATION; SPORE.

Further Reading

Lesko, Kevin T. "The Deep Underground Science and Engineering Laboratory at Homestake." Available online. URL: www.slac.stanford.edu/econf/C0805263/ProcContrib/lesko_k.pdf. Accessed January 2, 2010.

MicrobiologyProcedure.com. "The Rhizosphere." Available online.URL:www.microbiologyprocedure.com/rhizosphere-phyllosphere/rhizosphere-introduction.html. Accessed January 2, 2010.

Oak Ridge National Laboratory. "Amazing Microbes." Available online. URL: www.ornl.gov/info/ornlreview/rev32_3/amazing.htm. January 1, 2010.

Onstott, Tullis C., Frederick S. Colwell, Thomas L. Kieft, Lawrence Murdoch, and Thomas J. Phelps. "New Horizons for Deep Subsurface Microbiology." *Microbe,* November 2009. Available online. URL: www.microbemagazine.org/index.php?option=com_content&view=article&id=1045:new-horizons-for-deep-subsurface-microbiology&catid=310:featured&Itemid=394. Accessed January 1, 2010.

Orfanoudakis, Michail, Athanasios Papaioannou, and Evangelos Barbas. "Preliminary Studies in *Alnus glutinosa* Root Symbiosis in the Field." Presented at Eurosoil 2004. Available online. URL: www.bodenkunde2.uni-freiburg.de/eurosoil/abstracts/id262_Orfanoudakis_full.pdf. Accessed June 6, 2010.

Santamaria, Johanna, and Gary A. Toranzos. "Enteric Pathogens and Soil: A Short Review." *International Microbiology* 6 (2003): 5–9. Available online. URL: www.springerlink.com/content/mxhhnhjpqm6mhmgd/fulltext.pdf. Accessed January 2, 2010.

United States Department of Agriculture. "The Soil Biology Primer." Available online. URL: http://soils.usda.gov/sqi/concepts/soil_biology/biology.html. Accessed January 2, 2010.

Yamanaka, Takashi, Akio Akama, and Ching-Yan Li. "Growth, Nitrogen Fixation and Acquisition of *Alnus sieboldiana* after Inoculation of *Frankia* Together with *Gigapora margarita* and *Pseudomonas putida*." *Journal of Forest Research* 10 (2005): 21–26. Available online. URL: www.fs.fed.us/pnw/pubs/journals/pnw_2005_yamanaka001.pdf. Accessed January 2, 2010.

Zak, Donald R., William E. Holmes, David C. White, Aaron D. Peacock, and David Tilman. "Plant Diversity, Soil Microbial Communities, and Ecosystem Function: Are There Any Links?" *Ecology* 84 (2003): 2,042–2,050. Available online. URL: www.cedarcreek.umn.edu/biblio/fulltext/t1908.pdf. Accessed January 11, 2010.

species (singular and plural) In microbiology, a species is successive generations of like microorganisms with stable properties in common, different from all other species. In higher organisms, *species* refers to interbreeding natural populations of like members. Each species represents the lowest, most specific level of a taxonomy hierarchy, that is, the organization of all living things according to their degree of relatedness.

Microbiologists may also define a species as a collection of strains, each a subpopulation of identical cells descended from a single ancestor cell. To taxonomists, a species represents the least inclusive grouping of individuals. By contrast, a domain represents the most inclusive grouping of living things. Biologists currently recognize three domains:

- bacteria, containing only bacteria
- archaea, containing microorganisms that look like bacteria but have characteristics of both bacteria and eukaryotic cells, plus unique characteristics
- eukarya, containing organisms made of cells with membrane-bound organelles (algae, protozoa, fungi, higher plants and animals)

In the hierarchy of biota, every organism belongs to the following categories, from the most inclusive to the narrowest: domain, phylum, class, order, family, genus, and species.

Speciation is the procedure of assigning an organism to an existing species or a new species. In prokaryotes, bacteria and archaea have presented biologists with problems in arriving at accurate species identifications. Much of the speciation conducted prior to the development of molecular analysis techniques of nucleic acids was based on cell morphology and biochemical tests that give information on cell metabolism. Although these methods served well for placing microorganisms into logical genera and species, some mistakes occurred. Recent analysis of a structure called 16S rRNA has clarified the true species of many misidentified microorganisms.

16S rRNA

Prokaryotic and eukaryotic cells contain thousands of small structures, called *ribosomes,* that act as the site of protein synthesis. Ribosomes measure 20–30 nanometers (nm) in diameter and in bacteria are distributed in the cell cytoplasm. Each prokaryotic ribosome is classified as a 70S ribosome, on the basis of the manner in which it migrates through a suspension when centrifuged. The 70S ribosome contains two subunits, one 30S and one 50S.

The ribonucleic acid (RNA) associated with ribosomes, or rRNA, participates in the conversion of a cell's genetic code held in its deoxyribonucleic acid (DNA) to proteins. The ribosome is said to have been strongly conserved during evolution, meaning that its structure has changed very little from the earliest ancestors. This feature of conservation makes ribosomes valuable as aids in determining the evolution of various species and the relationships among them. The 30S subunit in bacteria contains one molecule of 5S rRNA and one molecule of 16S rRNA.

In 1977, the University of Illinois microbiologists Carl R. Woese (1928–) and Ralph S. Wolfe (1921–) reported on studies of 16S rRNA composition suggesting that the world's biota should be classified into three domains (listed earlier) instead of the traditional divisions of prokaryotes (bacteria and archaea) and eukaryotes (all other life). Woese said, "The central task of biology in the new century will be to elaborate and lay out this overarching framework of relationships among living organisms." In fact, despite Woese and Wolfe's important breakthrough, the relationships of species remain a difficult question to answer.

The German evolutionary biologist Ernst Mayr (1904–2005) once described species as "the basic unit of ecology . . . no ecosystem can be fully understood until it has been dissected into its component species and until the mutual interactions of these species are understood." Microbiologists grapple with this problem, and today's microbial taxonomy consists of a combination of morphology, physiology, and rRNA analysis.

The Montana State University environmental scientist David Ward pointed out, in 2006, "Debat-

Strain Terminology

Term	Definition
strain	population of microorganisms distinguishable from other members of the species
type strain	the most-studied member of a species, which, therefore, defines many characteristics typical of the species
biovar	variant strain with biochemical or physiological differences from the rest of the species
serovar	variant strain with antigenic (surface structures that react with antibodies) properties distinctive from the rest of the species
morphovar	variant strain with morphological differences from the rest of the species
ecotype	group of similar microorganisms that occupy a certain ecological niche
geotype	group of similar microorganisms with a certain geographical distribution

ing species concepts is not merely an esoteric exercise. On the contrary, the species concept is central to achieving a predictive understanding of the composition, structure, and function of microbial communities, the population biology of disease outbreaks, and the emergence of new diseases." Ward's point is correct, but a complication exists in microbiology that does not occur in higher organisms: horizontal gene transfer.

SPECIES ORGANIZATION

Microbial species have been difficult to define because of a phenomenon called *horizontal gene transfer.* Microorganisms share genes almost at will compared with higher organisms. For this reason, thousands of microbial species can be considered to be related, so much so that bacteria sometimes seem to be members of one giant species.

Horizontal gene transfer has made the task of determining the lineage of bacterial species daunting. Ward has reported the "'gold standards' of > 30 percent variation in DNA-DNA hybridization (and > 2 to 3 percent variation in 16S rRNA sequence) have been proposed as genetic distances needed to ensure that two strains belong to different species, and these standards have been used to conservatively estimate the number of microbial species in nature." No one has determined the number of bacterial or archaeal species that exist. About 5,000 bacterial species have been characterized and another 10,000 partially characterized. The evolutionist Edward O. Wilson estimates there are 1.8 million species of living things on Earth. Insects dominate, but prokaryotes also make up a large percentage of total species. The Harvard University correspondent wrote, in 2006, "Scoop up a handful of soil in a place like Franklin Park in Boston and, Wilson assures us, you could be holding 10 billion bacteria, representing 5,000–6,000 different species. Scoop up a ton of soil and the number of varieties of bacteria could jump to 4 million, considerably more than the number of animals and plants now known." The number of archaea may be more difficult to estimate than the number of bacteria because many archaea live in extreme environments that are inaccessible to humans. Most scientists believe humanity will never attain the technology to determine the actual number of species.

A hierarchy of microorganisms can be developed without knowing how many species exist. Current taxonomy accommodates the discovery of new species in the future. Species receive a two-part Latin name that derives from their place in the hierarchy of biota. This binomial system of naming organisms means that each species is referred to by two names as follows: *Genus species. Streptococcus pyogenes,* for example, belongs to the genus *Streptococcus* and the term *pyogenes* identifies the species. The genus name alone cannot distinguish species, because *Streptococcus* contains several species: *Streptococcus pyogenes, Streptococcus mutans, Streptococcus faecalis, Streptococcus bovis,* and so on. If the meaning is obvious, a species may be abbreviated as follows: *S. pyogenes.*

A common misconception in biology relates to the nature of species as static versus dynamic. All of Earth's current species are changing and not static. Species have changed and continue to change through evolution. Microorganisms naturally become extinct, and others emerge in evolution. The theory of speciation summarizes this concept by stating that all species are changing irreversibly toward greater diversity.

In 2006, the American Academy of Microbiology published a report entitled "Reconciling Microbial Systematics and Genomics" to summarize the state of the art in speciating microorganisms by traditional phenotypic (the outward expression of genes) methods and nucleic acid or gene analysis. The report stated, "Microorganisms are fantastically diverse, both genetically and phenotypically. The amount and depth of genetic diversity represented among the named microbial species is greater than that represented among the animals and plants combined. If the distinctions and testing in use today could be applied to all microorganisms (even those that cannot be grown in the laboratory), the number of microbial species easily could reach into the billions. Moreover, the current statistical approaches all tend to underestimate microbial diversity." A definition for *species* that serves all of microbiology remains elusive. On an everyday basis, however, the current methods for classifying species have been adequate for clinical, food, industrial, and other areas of microbiology.

SUBSPECIES CLASSIFICATIONS

Different strains within a species can be as important as the species itself. For example, the food-borne pathogen *Escherichia coli* O157:H7 is an especially virulent strain of *E. coli*. The O157 strain can cause more severe food-borne illnesses than most other *E. coli* strains. For this reason, special microbiological media and tests have been developed for O157, as if it were a separate species. The table on page 713 describes other uses for the strain and corresponding terminology.

In the table's definitions, ecotype and geotype do not describe traditional species but, rather, different microorganisms that behave in the same way in the environment. Microbial ecologists tend to study such groups of microorganisms rather than individual species. In the environment, the activity of a single species becomes almost meaningless because microorganisms in nature live in communities of diverse organisms. As communities, microorganisms affect the biosphere.

The exact identification of a species becomes more important in medicine and in food microbiology. In these areas, species have different effects on health. Speciation in clinical and food microbiology laboratories remains a central task of the microbiologist.

Natural classification systems depend more on the activities of microorganisms in nature than on genotype (genetic makeup). By this system, a microbiologist might be more interested in the activity of methanogens as a group than that of *Methanococcus, Methanosarcina,* and other species that emit methane. Other groups of bacteria that contain more than one species but are usually studied as a single physiological group are:

- photosynthetic—use sunlight, water, and carbon dioxide for energy and carbon metabolism
- green sulfur—use sunlight, hydrogen sulfide (H_2S), and carbon dioxide for energy and carbon metabolism
- spirochetes—spiral-shaped cells
- proteobacteria—diverse group of important environmental and pathogenic bacteria

These classifications and many additional groups provide microbiologists with more information about natural microbial activities than pure cultures of single species can in test tubes. Microbiology reserves a role for studies of pure and defined species as well as diverse groups. Studies on single species have given microbiologists critical information on optimal growth, toxin production, end products, and gene expression. Despite the drawbacks of current methods for categorizing microorganisms, the species concept still lays the foundation for all of microbiology.

See also DOMAIN; PROKARYOTE; PROTEOBACTERIA; SYSTEMATICS; TAXONOMY.

Further Reading

American Academy of Microbiology. "Reconciling Microbial Systematics and Genomics." 2006. Available online. URL: http://academy.asm.org/images/stories/documents/reconcilingmicrobialsystematicsandgenomicsfull.pdf. Accessed December 4, 2009.

Barlow, Jim. "Microbiologist Carl R. Woese Named Winner of National Medal of Science." University of Illinois. November 13, 2000. Available online. URL: http://news.illinois.edu/news/00/1113woese.html. Accessed December 4, 2009.

Cromie, William J. "Naturalist E. O. Wilson Is Optimistic." *Harvard University Gazette,* 15 June 2006. Available online. URL: www.news.harvard.edu/gazette/2006/06.15/03-biodiversity.html. Accessed December 4, 2009.

Ward, David M. "A Macrobiological Perspective on Microbial Species." *Microbe,* June 2006.

specimen collection Specimen collection is the process of obtaining tissue or fluids from a diseased patient for the purpose of diagnosis. Clinical microbiologists analyze specimens for the presence of pathogens that could be the cause of the disease. The combination of information gleaned from the specimen analysis and the patient's signs and symptoms enables physicians to diagnose infectious disease accurately.

Each type of specimen that is studied in a clinical microbiology laboratory has specific requirements on the best methods for collecting, handling, and transporting the material to prevent contamination and deterioration. Technicians who collect specimens use aseptic techniques, procedures that reduce the chance of contamination, and only sterilized containers are used for transporting specimens. For example, a technician collecting a urine sample collects urine midstream to ensure that the fluid contains microorganisms from the urinary tract and not from the skin outside the body. This technique ensures that the specimen reflects only the state of health inside the urinary tract.

TYPES OF SPECIMENS

Patient specimens collected in hospitals or in doctors' offices can be any of the following: throat or

Clinical Specimens

Specimen	Main Collection Methods	Significance for Diagnosis
blood	needle aspiration	sepsis (microbial presence in the bloodstream)
cerebrospinal fluid	needle aspiration	nervous system infection such as meningitis
pus	needle aspiration	wound infection
semen	semen cup	sexually transmitted disease
skin	swab	bacterial or fungal skin infection or rash
sputum, saliva	sputum cup	upper respiratory tract infection
stomach contents	intubation	infection of upper gastrointestinal tract
stool	stool cup	food-borne or waterborne infection
urine	catheter	urinary tract infection

nostril swab, wound swab, urine, stool, pus, sputum, blood, semen, cerebrospinal fluid, or stomach contents. Two rules apply to the collection of specimens. First, the specimen must be representative, meaning it must be from the diseased area of the patient's body, rather than a healthy part of the body. By taking specimens from diseased tissue or organs, technicians increase the chance of capturing the pathogen responsible for the disease. Second, specimen size must be large enough to enable a microbiologist to carry out all necessary identification activities. Fortunately, most specimen sizes do not have to be large for microbiological analysis: A few cubic centimeters (cc) of liquid or a gram or two of solid specimen is usually adequate for clinical microbiology.

Each type of specimen requires its own type of collection. Technicians or nurses can collect specimens from the outside of the body by swab (skin, wound, throat, nostrils, or ears), but specimens from the inside of the body require invasive sampling. Sampling for retrieving a specimen from inside the body involves using receptacles for excreted specimens (urine, stool, sputum, pus, or semen), but this method is not invasive. Invasive collection involves using a device that enters the body to withdraw the specimen. The most common invasive sampling methods are needle aspiration (cerebrospinal fluid, pus, blood, or urine in veterinary patients), intubation (tube sampling, usually of hollow organs such as the stomach), and catherization (use of catheter tube for collecting urine).

The table above describes the various specimens studied in clinical microbiology. The collection methods listed provide the best way of preventing contamination.

Health care workers have the responsibility not only to protect the specimen from damage, but also to protect themselves from infection. Infections in health care settings, called *nosocomial infections*, threaten the health of employees and their patients. The main precautions against infection taken by health care workers are the following:

- use of barriers to pathogen transmission—gloves, masks, protective eyewear, gowns, lab coats
- thorough hand washing after removing gloves or after touching patient skin surfaces
- proper handling of needles and scalpels (called *sharps*)
- use of mouthpieces for mouth-to-mouth resuscitation
- refraining from working with patients if the health care worker has open wounds or broken skin
- proper spill cleanup

Handling of needles and scalpels has become a major concern in health care. Needlestick injuries increase the risk of contracting human immunodeficiency virus/acquired immunodeficiency syndrome (HIV/AIDS) or hepatitis virus. The international AIDS education organization AVERT has stated on its Web site, "Since the beginning of the HIV/AIDS epidemic, health care workers across the world have become infected with HIV as a result of their work. The main cause of infection in occupational settings is exposure to HIV-infected blood via a percutaneous injury (i.e. from needles, instruments, bites which break the skin, etc.)." The risk of this type of injury is low, about three per 1,000 injuries of every type in health care. The severity of the diseases that can be transmitted by percutaneous injury makes this a major concern among health care workers, however. In 2008, the American Nurses Association said in a press release, "According to the latest research, nearly two-thirds (64 percent) of U.S. nurses say needlestick injuries and blood borne infections remain major concerns." Handling used sharps in hospital wastes may be as risky as collecting patient specimens. The following procedure gives an example of proper spill cleanup methods for body fluids:

1. Put on gloves.
2. Wipe up excess material with paper towels and discard for sterilization.
3. Disinfect the area with an Environmental Agency Protection (EPAs)-registered biocide.

Hospitals have on hand various disinfectants for cleaning up specimen spills.

Most specimen sampling equipment used today is disposable, helping to reduce the chance of the spread of infection by contaminated instruments. After used items have been discarded in a biohazardous waste bag, a technician will decontaminate the bag contents in an autoclave, equipment that kills all microorganisms by treating them with steam under high pressure.

SPECIMEN HANDLING

Specimen handling consists of two tasks: labeling and transport. Proper labeling ensures that the clinical laboratory receives all of the patient information it needs to process the specimen correctly. Specimen labels should contain most or all of the following information:

- hospital name and registration number
- patient's location in the hospital
- admission date
- attending physician
- diagnosis
- antibiotic therapy, if applicable
- type of specimen

Microbiologists recommend that any special instructions for transport (on-ice, in anaerobic bag, etc.) be included on either the label or the form that accompanies the specimen. On the form, physicians request the tests they want the clinical microbiology to run on the specimen.

Transport under the proper conditions is essential for obtaining the most accurate results from specimen analysis. Companies provide a selection of transport containers that help this aspect of specimen handling. The following common transport methods aid pathogen identification:

- in growth medium to keep fragile microorganisms alive
- agar medium to maintain the same ratio of species as in the patient
- sealed anaerobic transport vials
- buffered preservative solutions
- refrigeration for specimens in transport longer than one hour

Although clinical microbiologists try to process all specimens within one hour of their sampling, some specimens, such as anaerobic samples or cerebral spinal fluid, should be processed within 15 minutes, if possible. Any specimen suspected of containing a virus is packed on ice before transport to a laboratory and held refrigerated for up to 72 hours. Beyond 72 hours, the specimens should be frozen.

LABORATORY METHODS FOR SPECIMENS

Inside the microbiology laboratory, speed is at a premium for processing the specimens. Processing entails all of the preparation and testing done on a specimen to determine whether it contains pathogens, and what kinds. Many processing steps take place, simultaneously, to expedite processing. The principal specimen analysis methods are described in the table on page 717 in the general order in which microbiologists perform them.

Molecular methods have moved to the fore in clinical microbiology as the most accurate methods for determining pathogen identification. But many of these techniques require more time to complete than the time a microbiologist has to identify a pathogen. Molecular probes have helped speed this analysis. A probe is a short segment of a deoxyribonucleic acid (DNA) strand from a known pathogen. When a microbiologist exposes the specimen to this probe, also called a *biosensor,* the probe's DNA strand binds with any complementary DNA in the speci-

Specimen Processing Methods

Method	Description	Uses
microscopy	examination of wet samples or stained specimens in a microscope	primary test for determining whether the pathogen is bacterial, fungal, or protozoal
growth requirements	incubating part of the specimen on various media and at various temperatures	determine growth patterns that suggest certain groups of microorganisms
biochemical testing	tests for the presence/absence of specific enzymes; oxygen requirements	differentiates various groups and genera of bacteria
serology (immunology)	antibody reactions to determine presence of antigenic structures on the cell surface	identifies species and strains of certain bacteria
bacteriophage typing	exposing bacteria to a phage (virus that attacks only bacteria) known to be specific for species	identifies genus, species, or strains of bacteria
fatty acid analysis	analysis of unique lipid composition of bacteria	identifies bacterial species
molecular methods	analysis of nucleic acid base unit sequences	identifies bacterial species and strains

men. Binding, indicated by using a fluorescent compound attached to the probe, indicates the presence of the pathogen in the specimen. No signal probe indicates that the specimen lacks the probe's specific microorganism.

Molecular test kits cost more than other tests, and they often require the microbiologist to increase the DNA level by a method called *polymerase chain reaction* (PCR). These drawbacks have limited the use of molecular technology in many clinical microbiology laboratories. The molecular epidemiologist Michael Pfaller wrote, in 2001, "Despite the probability that improved patient outcome and reduced cost of antimicrobial agents and length of hospital stay will outweigh the increased laboratory costs incurred through the use of molecular testing, such savings are difficult to document." In larger studies of disease outbreaks in a population (the field of epidemiology), molecular methods have been powerful aids in finding the outbreak's source. "Newer DNA-based typing methods," Pfaller reported, "have eliminated most of these limitations and are now the preferred techniques for epidemiologic typing." Molecular methods will undoubtedly overcome some of their disadvantages—time, cost, variable results, and labor-intensive steps—and become the main means for detecting pathogens in clinical specimens.

Clinical microbiology has also increased its use of computers to keep patient records and track incidences of infectious disease. Computer databases also manage the progress of specimen processing, report overdue tests, store test results, and keep data for epidemiologic uses.

See also BIOSENSOR; IDENTIFICATION; MINIMUM INHIBITORY CONCENTRATION; POLYMERASE CHAIN REACTION; SEROLOGY.

Further Reading

American Nurses Association. "Workplace Safety and Needlestick Injuries Are Top Concerns for Nurses." June 24, 2008. Available online. URL: www.nursingworld.org/MainMenuCategories/OccupationalandEnvironmental/occupationalhealth/SafeNeedles/WorkplaceSafetyTopConcerns.aspx. Accessed December 4, 2009.

AVERT. "Occupational Exposure to HIV." September 10, 2009. Available online. URL: www.avert.org/needlestick.htm. Accessed December 4, 2009.

Central Dupage Hospital. "HealthLab Microbiology Specimen Collection Guidelines." Available online. URL: www.cdh.org/uploadedFile/59%20Microbiology%20Guidelines.pdf. Accessed December 4, 2009.

Pfaller, Michael A. "Molecular Approaches to Diagnosing and Managing Infectious Diseases: Practicality and Costs." *Emerging Infectious Diseases,* 22 December 2001. Available online. URL: www.cdc.gov/ncidod/eid/vol7no2/pfaller.htm. Accessed December 4, 2009.

spectrum of activity Spectrum of activity is the range of diverse microorganisms affected by an antimicrobial substance. An antibiotic or a chemical biocide that kills or inhibit diverse microorganisms is called a broad-spectrum antimicrobial agent. By contrast, a substance that kills or inhibits a limited

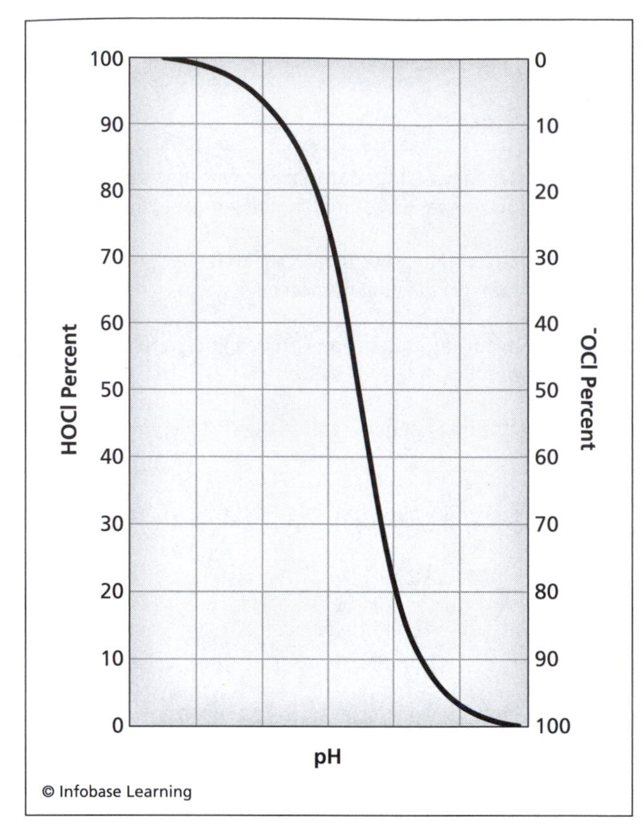

External conditions affect the spectrum of activity of many biocides. The pH of the environment, for example, influences the efficacy of chlorine disinfectants in killing microorganisms.

range of microorganisms, such as affecting only gram-negative bacteria, is called a narrow-spectrum or limited spectrum antimicrobial agent.

SPECTRUM OF ACTIVITY OF ANTIBIOTICS

Drug manufacturers planning to develop a new antibiotic make an initial decision on the drug's untended targets. The initial decision involves prokaryotes versus eukaryotes. The structure and reproduction of these two groups differ markedly, so drugs developed for one group will probably not affect the other group. Among eukaryotes, scientists develop drugs that affect fungi or protozoa. Likewise, for bacteria, antibiotics can affect both gram-negative and gram-positive species, either group, or an exact genus.

Antibiotics are substances made by a microorganism to kill other unrelated microorganisms. For example, *Penicillium* mold produces penicillin to inhibit the growth of gram-positive bacteria. Antibiotics serve the purpose of decreasing competition from other organisms for nutrients and habitat. Very few antibiotics target a single species within a genus. Bacteriocins behave similarly to antibiotics; microorganisms secrete bacteriocins to inhibit other microorganisms. But bacteriocins inhibit specific microorganisms at the species level or sometimes a strain within a species.

Broad-spectrum and narrow-spectrum antibiotics each have appropriate uses, advantages, and disadvantages. Broad-spectrum antibiotics treat patients in two main health situations: (1) when an unknown microorganism is suspected of causing an infection and (2) when a patient has infection from more than one type of microorganism. Broad-spectrum antibiotics have disadvantages, however, that can affect long-term health. First, broad-spectrum drugs kill members of the body's normal flora that usually benefit a person by outcompeting pathogens. Second, microorganisms that survive exposure to broad-spectrum therapy may have survived because of the presence of resistance genes. The therapy, thus, creates a new population of microorganisms in the body more antibiotic-resistant than before.

Narrow-spectrum antibiotics have the disadvantage of attacking a limited number of microorganisms. Physicians must be certain of the pathogen causing a patient's illness before prescribing a narrow-spectrum drug, or the drug will be ineffective. The advantage of narrow-spectrum antibiotics relates to their specificity: These drugs kill fewer normal flora and may have less influence on creating resistant populations.

Any antibiotic treatment can cause a condition called *overgrowth,* although broad-spectrum therapy is more likely to cause this situation than narrow-spectrum therapy. Overgrowth is a rapid increase in one type of normal flora when other members of the body's flora have been eliminated by an antibiotic. For example, *Candida albicans* yeast infections occur with greater frequency in people who receive broad-spectrum antibacterial therapy because yeasts resist these drugs.

C. albicans causes the following three types of candidiasis in overgrowth conditions:

- oropharyngeal—candidiasis of the mouth and throat, also called *thrush*
- esophageal—candidiasis in the esophagus, which can lead to inflammation (esophagitis)
- genital—rash, itching, or discharge in male or female genitals, also called *yeast infection*

Overgrowth of *Candida* or other antibiotic-resistant flora can lead to a condition called *superinfection.* A superinfection can be an infection that results from overgrowth of one type of microorganism or the overgrowth of an antibiotic-resistant target organism. In medical care related to acquired immunodeficiency syndrome (AIDS), a superinfection results from infection by two or more different strains of the human immunodeficiency virus (HIV).

Antibiotic Spectrum of Activity

Antibiotic Class	Antibiotics	Activity
aminoglycosides	• streptomycin • kanamycin • gentamicin	many gram-negative, some gram-positive, ineffective against anaerobes
beta-lactam antibiotics	• penicillins • cephalosporins	many gram-positive, some gram-negative
macrolides	• erythromycin	gram-positive, *Mycoplasma, Legionella*
membrane disruptors	• polymyxin B	gram-negative
quinolones	• ciprofloxacin • nalidixic acid	gram-positive cocci
sulfonamides	• sulfisoxazole • sulfamethizole • trimethoprim • sulfacetamide	many gram-negative and gram-positive
tetracyclines	• tetracycline • doxycycline • minocycline	many gram-negative, gram-positive, and intracellular *(Rickettsia)*

The dramatic rise in antibiotic-resistant microorganisms has narrowed the spectrum of activity for many antibiotics. For example, the original natural form of penicillin, first used on patients in the 1940s, killed many gram-positive bacteria and several types of gram-negative bacteria. Current resistance to penicillin is pervasive, and the spectrum of activity of this antibiotic has diminished. The table above provides information on commonly used antibiotics.

The table shows that aminoglycosides, sulfonamides, and tetracyclines are broad-spectrum antibiotics. Quinolones possess a much more limited spectrum. Other narrow-spectrum antibiotics in use today are:

- chloroquine—exclusively for *Plasmodium vivax* (malaria)
- clindamycin—gram-positive except *Enterococcus* and methicillin-resistant *Staphylococcus aureus* (MRSA)
- isoniazid—exclusively for *Mycobacterium* (tuberculosis)
- ketoconazole—fungi

The medical community must consider the ramifications of overgrowth, superinfection, and antibiotic resistance when selecting a treatment with an appropriate spectrum of activity. The *Merck Manual,* published by the pharmaceutical company Merck, Sharp and Dohme, advises the following regarding drug selection: "Culture and antibiotic sensitivity testing are essential for selecting a drug for serious infections. However, treatment must often begin before culture results are available, necessitating selection according to the most likely pathogens (empiric antibiotic selection). Whether chosen according to culture results or not, drugs with the narrowest spectrum of activity that can control the infection should be used." The authors recommend broad-spectrum antibiotics for multiple-pathogen infections but also point out, "The most likely pathogens and their susceptibility to antibiotics vary according to geographic location (within cities or even within a hospital) and can change from month to month." Antibiotic spectra of activity can be expected to shift in the future, as they have for the past several decades. Selection of antibiotics for treating infection requires, perhaps, a measure of intuition in addition to science.

SPECTRUM OF ACTIVITY OF CHEMICAL BIOCIDES

The spectrum of activity of biocides relates to the range of microorganisms a chemical kills, from the easiest-to-kill microorganisms to bacterial endospores, which are the most difficult microorganisms to kill. Biocides formulated to kill microorganisms belong to two groups: sterilants and disinfectants.

Sterilants have the broadest spectrum of activity of all chemical biocides: Sterilants kill all microorganisms. Disinfectants also provide broad-spectrum activity one level below sterilants: Disinfectants kill all microorganisms except bacterial endospores.

Disinfectants have been formulated to contain either broad-spectrum or narrow-spectrum activity and follow one of two specific classifications:

- broad spectrum—kills all gram-positive and gram-negative bacteria (except endospores) and usually can also kill certain fungi and viruses

- limited spectrum—kills gram-positive or gram-negative bacteria, not both, and not fungi or viruses

Unlike antibiotics, biocides list the microorganisms they kill on the product label. Disinfectants that claim they kill gram-positive *Staphylococcus aureus* and gram-negative *Salmonella choleraesuis* are broad-spectrum products. Disinfectants that include one or the other of these bacteria are limited-spectrum products.

Consumers usually do not care whether a disinfectant kills a broad range or a narrower range of microorganisms. An exception occurs in hospitals, where only broad-spectrum disinfectants may be used. Biocide selection based on spectrum of activity, thus, differs from antibiotic selection in which the health of a patient might be at stake.

See also ANTIBIOTIC; BACTERIOCIN; BIOCIDE; *CANDIDA ALBICANS;* DISINFECTION; NORMAL FLORA; RESISTANCE.

Further Reading

Mayer, Gene. "Antibiotics—Protein Synthesis, Nucleic Acid Synthesis and Metabolism." In *Microbiology and Immunology On-Line.* Columbia: University of South Carolina, 2009. Available online. URL: http://pathmicro.med.sc.edu/mayer/antibiot.htm. Accessed January 2, 2010.

Merck Manual. Whitehouse Station, N.J.: Merck, Sharp and Dohme, 2009. Available online. URL: www.merck.com/mmpe/sec14/ch170/ch170a.html#sec14-ch170-ch170a-231a. Accessed January 2, 2010.

spore A spore is a specialized form of a microbial cell developed for the purpose of reproduction, dissemination in the environment, or survival of adverse conditions. In general, a spore is a thick-walled, dormant form of a microorganism. Not all microorganisms form spores, but for those that produce this structure, the spore provides a survival advantage over other non-spore-forming species. Fungi, algae, and some funguslike bacteria form spores that serve mainly as a reproductive form with secondary functions in physical protection and cell dissemination. The term *endospore,* often used interchangeably with *spore,* refers to specialized spores made by certain bacterial genera, notably *Bacillus* and *Clostridium.*

Spores and endospores have in common a strong resistance to damaging conditions in the environment, including heating, drying, freezing, nutrient scarcity, and exposure to chemicals. Endospores, in particular, have been known to protect live bacteria for centuries. Micropaleontologists study the artifacts of ancient microorganisms to determine the history of living things. Spores have been discovered by micropaleontologists in sediments 1,850 feet (564 m) deep in New Mexico and dating to 250 million years old. Assuming these spores have not been contaminated by bacteria of more recent origin, compelling evidence exists for spores surviving for millennia. Microbiologists have yet to explain all of the mechanisms that contribute to spores' incredible longevity. "Despite significant advances in our understanding of the process of spore formation," the molecular biologist Sigal Ben-Yehuda has stated, "little is known about the nature of the mature spore. It is unrevealed how dormancy is maintained within the spore and how it is ceased, as the organization and dynamics of the spore macromolecules remain obscure." Microbiologists have delineated the phases of endospore formation, but the reactions inside the durable structure have been difficult to decipher.

The term *sporulation,* or sporogenesis, refers to the formation of a spore from a normal, growing form of a cell, called the *vegetative form.* When spores detect favorable conditions in the environment, they revert to the vegetative form, in a process called *germination.* Sporulation and germination, thus, make up two major phases in the life cycle of spore-forming microorganisms. Ben-Yehuda added, "The unusual biochemical and biophysical characteristics of the dormant spore make it a challenging biological system to investigate using conventional methods." Although many aspects of bacterial sporulation remain a mystery, biologists know that sporulation can result from exogenous (external) or endogenous (internal) factors. Exogenous factors that lead to spore formation arise from the environment, while endogenous factors are part of a cell's regulation of its metabolism, for instance, as a response to lowered nutrient supply.

SPORE FORMATION IN REPRODUCTION

Fungi reproduce by either asexual or sexual processes, both of which use a spore in part of the reproductive cycle. Several types of spores contribute to asexual reproduction. In all cases, the spores develop through mitosis, the process by which cellular genetic material replicates before cell division.

The five major types of fungal spores of asexual reproduction with example genera are:

- sporangiospores—develop within a saclike structure called a *sporangium* situated at the tip of fungal hyphae (branched, tubular filaments that act as the support structure of a fungus); *Mucor*
- conidiospores—produced at the tip of hyphae but not within a sac; *Aspergillus*
- arthrospores—created from pieces that break off from hyphae; *Trichophyton*
- blastospores—developed by budding off a vegetative mother cell; *Cryptococcus*
- chlamydospores—cells surrounded by a thick wall within a hypha before breaking free; *Microsporum*

The types of asexual spores and their morphology help mycologists (microbiologists who study fungi) identify fungal genera. Morphology serves as a tool for fungal identification more than any other feature and more than in most other microorganisms. Mycologists use three primary morphological features: (1) distinctive structure of hyphae, (2) features of the structure that holds spores or spore sacs, and (3) spore morphology. Toward the tip of hyphae in some species, the filament widens into a section that holds or supports spores. In sporangiospore-producing fungi, the supportive structure is a *sporangiophore,* and the spore-containing sac is a *sporangium*. In conidiospore-producing species, the structure that holds bunches of conidiospores is called the *conidiophore*.

Sexual reproduction in fungi requires the union of two nuclei called *haploid* because they contain half the chromosome after meiosis. Compatible nuclei may be from the same organism or from another organism. For a short period, the fungal cell, thus, contains two separate haploid nuclei, before they fuse. This period of two nuclei in one cell is called the *dikaryotic stage*. After fusion of the nuclei, the fungus produces one of three types of spores, each type serving as a way to group fungi morphologically because the spores have distinctive shapes, sizes, and pigmentation. The three spore types of sexual reproduction with example genera are:

- ascospores made by ascomycetes fungi and formed within a sac called an ascus; *Saccharomyces*
- basidiospores from basidiomycetes fungi and contained in a structure called a basidium; *Puccinia*
- zygospores from zygomycetes fungi and contained in a thick-walled structure called a zygosporangium; *Basidiobolus*

Ascomycetes fungi possess a characteristic method of dispersing ascospores by forcibly ejecting the spores from the ascus. Species vary by the mechanisms used for this spore dispersal, and many of the details have yet to be identified. Ejection probably occurs at least in part by the buildup of pressure, most likely by water absorption, inside the enclosed ascus before the sac bursts.

Environmental factors, such as light and humidity, can also affect ascospore dispersal. Many ascomycete species disperse spores on a diurnal (day-night) schedule. For example, *Daldinia concentrica* ejects ascospores almost exclusively at night (or in the dark).

Fungal Spores Associated with Airborne Disease

Organism	Disease
HUMAN PATHOGENS	
Aspergillus fumigatus	aspergillosis
Blastomyces dermatiridi	blastomycosis
Coccidioides immitis	coccidioidomycosis
Histoplasma capsulatum	histoplasmosis
Paracoccidioides brasiliensis	paracoccidioido-mycosis
Stachybotrys alternans	pulmonary infection
ANIMAL PATHOGENS	
Aspergillus species	aspergillosis
Chrysosporium species	adiaspiromycosis
Coccidioides immitis	coccidioidomycosis
Cryptococcus species	cryptococcosis
PLANT PATHOGENS	
Ceratocystis species	Dutch elm disease
Cronartium species	pine rusts
Gymnosporangium species	apple rust
Phytophthora infestans	potato blight
Puccinia species	various rusts
Ustilago species	wheat smut

Basidiospores are also ejected, but, unlike in ascospores, the ejection occurs as a rapid-fire action, one spore at a time. The spores travel a few millimeters (mm) or less before falling to the earth by gravity. Basidiospore shape affects the distance and the pattern of dispersal after ejection from the basidium.

Outdoor wind and humidity impact the dispersal of fungal spores to a great degree. The science of aeromicrobiology involves the study of how microorganisms, including fungal spores, travel through the air before settling. Dispersal can vary from a few millimeters, as in basidiospores, to distances measured in feet to miles.

The rugged nature of fungal spores and their ability to travel long distances, blown by winds, increase the risk of fungal disease transmission in humans, domesticated animals, and plants and trees. The table on page 721 lists important fungal pathogens in which airborne spores are transmitted through the environment.

Some algae have three different types of asexual reproduction: binary fission, fragmentation, and spore formation. Binary fission resembles that in bacteria, in which a cell splits into two identical daughter cells. In fragmentation, the main body, called the *thallus,* breaks into pieces, and each piece develops into a new cell. Spore formation develops by either of two routes: (1) conversion of a vegetative (nonreproducing) cell into a spore or (2) production of spore-producing sporangia. Depending on species, algal spores can be either motile zoospores, powered by flagella, or nonmotile aplanospores. These two terms are used in general in microbiology to describe motile and nonmotile spores, respectively.

Several bacterial genera of the order Actinomycetales produce branched filaments, called *mycelia,* that make these microorganisms resemble fungi. Two notable spore-forming actinomycetes (a general name for this group of bacteria) belong to the genera *Streptomyces* and *Nocardia.* Actinomycetes produce spores by either breaking off a piece of a hypha in the process called *fragmentation* or developing a sporangium.

Spore formation from actinomycete hyphae requires that a hypha be divided into segments by a cross wall called a *septum.* As septa divide up the hypha, each segment develops a thick wall and, thus, a spore. Chains of spores that originate in hyphae use the following four different types of development (example genera in parentheses):

- acropetal—spores begin forming in the hypha closest to the hyphal base *(Pseudonocardia)*
- basipetal—spores begin forming at the tip of the hypha and progress toward the hyphal base *(Micropolyspora)*
- random—spores develop anywhere in the hypha *(Nocardiopsis)*
- simultaneous—spores develop in generally simultaneous manner throughout the hypha *(Streptomyces)*

Microbiologists use spore formation to help identify actinomycetes genera by staining the spores by either the Gram stain or the acid-fast stain. Positive or negative staining is indicative of genera. More specialized stains have been developed for bacterial endospores, which possess much stronger and thicker cell walls and resist chemical and physical damage more than any other type of spore.

ENDOSPORES AND SPORULATION

The bacterial endospore is a specialized protective structure that is also more resistant to chemical and physical damage than any other structure in microbiology. Because of the resistance of endospores to damage, microbiologists use endospores as a criterion for sterilization, the removal of all life-forms from a material. Sterilization is the elimination of all life, including endospores; disinfection is the elimination of all life except endospores.

The German botanist Ferdinand Cohn (1828–98) discovered endospores made by *Bacillus* in experiments designed to study the theory of spontaneous generation, the premise held for centuries that life arose from inanimate matter. Cohn investigated the process of heating liquids and foods as a sterilization method. Although heating at 212°F (100°C) killed many microorganisms, Cohn observed that not all succumbed to the same temperature and heating regimen. Some scientists took the result as proof that life arose spontaneously in "sterilized" material, but Cohn believed that different microorganisms simply had varying resistance to heating. "Why is it," he wrote in 1877, "that 100° is not sufficient to kill bacteria, when much lower temperatures are sufficient for killing other living organisms?" Cohn experimented on increasing sterilization temperatures and sterilization times, until be noticed that *Bacillus* cells formed "strongly refracting [light-deflecting] bodies." He added, "From each of these bodies develops an oblong or shortly cylindrical, strongly, refracting, dark-rimmed spores." The spores developed in place of *Bacillus* rod-shaped cells normally arranged in long chains called filaments. Instead of chains of cells placed end to end, Cohn observed chains of spores. He, additionally, noted that *Bacillus* spores remain viable and that they come to life in a process called *germination.*

Microbiologists after Cohn gathered more information on the processes of sporulation in *Bacillus,* which remains the most studied endospore-forming

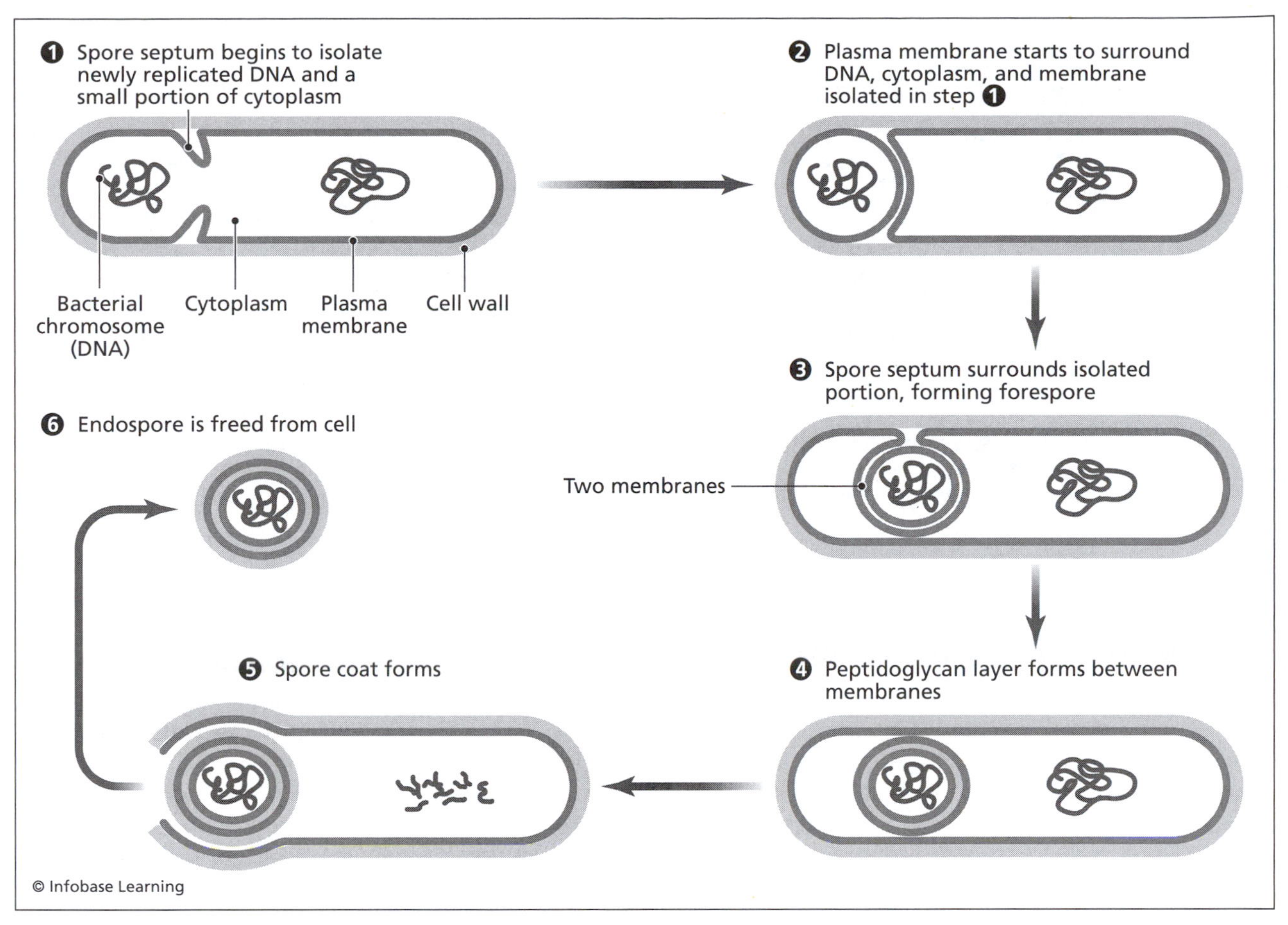

Bacteria such as *Bacillus* develop an endospore when conditions in the environment threaten cell survival. The endospores can protect the cell in a dormant state for centuries. When conditions are favorable, the endospore germinates, meaning it transforms into an active, replicating cell.

genus. In addition to *Bacillus*, the following bacteria have been identified as endospore formers: *Acyclobacillus, Brevibacillus, Clostridium, Coxiella, Desulfotomaculum, Geobacillus, Paenibacillus, Sporolactobacillus, Sporomusa, Sporosarcina,* and *Sporospirillum.*

Endospores develop in cells stressed by conditions in the environment, such as low nutrient levels. When a spore develops inside a vegetative cell, the cell becomes known as a *sporangium*. The position of the spore inside the thick protective coating of the sporangium is characteristic of species. Some endospores develop at one end of the cell (terminal spores), others develop near one end (subterminal spores), and others develop in the center of the cell (central spores). Some bacteria make distinctive endospore shapes such as those produced by *Clostridium*. This genus develops spores on one swollen end of a cell to create a tennis racket or bowling pin shape.

Sporulation proceeds in the following general stages in most endospore-forming species:

1. invagination of the plasma membrane that underlies the cell wall

2. creation of a plasma septum across the cell to differentiate the cell's vegetative portion from the spore portion

3. engulfment of the spore portion by the vegetative portion to create a double-membrane-enclosed spore

4. cortex formation, a heat-resistant peptidoglycan layer with less than normal amount of crosslinking between peptidoglycan chains

5. exosporium formation, a thin and delicate covering

6. thickening of spore coat that lies between the cortex and the exosporium

After the endospore formation steps, the final structure contains the following components, from the inside to the outside: core containing ribosomes and a nucleoid, core wall, cortex, spore coat, and exosporium. Although the structure contains ribosomes (used for protein synthesis) and a nucleoid (gen-

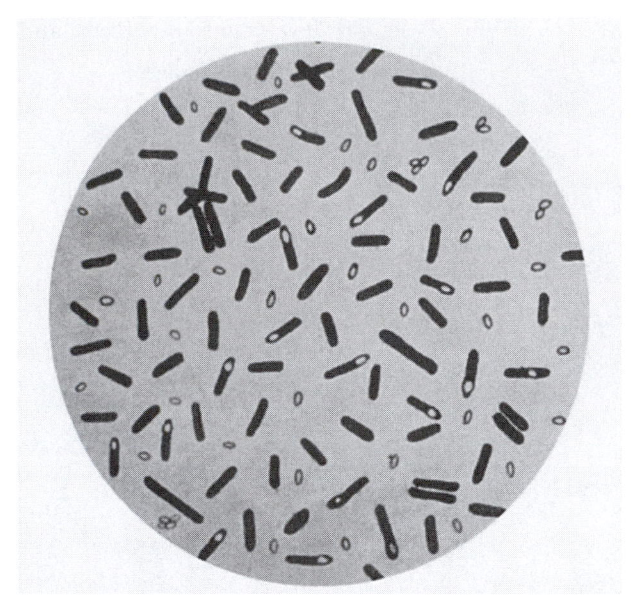

Endospores of *Clostridium botulinum* have a distinctive tennis racket or bowling pin shape. ***(CDC Image Library)***

eralized area of high deoxyribonucleic acid [DNA] density), neither of these components is active in the dormant endospore. Cells such as spores with no detectable metabolism are referred to as *cryptobiotic.*

Microbiology has not uncovered all of the mechanisms by which endospores resist almost all types of physical and chemical damage, including heat, drying, freezing, radiation, radioactivity, lysozyme, and strong chemicals. The spore coat can become very thick compared with the other endospore layers and probably contributes to the structure's hardiness. In addition, the entire endospore consists of 15 percent dipicolinic acid combined with calcium. Dipicolinic acid is a simple nitrogen-containing molecule arranged in a ring shape that may aid in stabilizing the spore's DNA when exposed to heat.

Microbiologists have proposed a number of mechanisms that enhance the durability of endospores in addition to the dipicolinic acid–calcium complex: DNA repair enzymes active during germination, proteins that stabilize DNA and other components, dehydration of the core, and the thick spore coat. The spore coat contains an amino composition unique in biology, which is thought to contribute to the endospore's resistance to chemicals that dissolve most other proteins in nature.

After sporulation, an endospore can remain in its cryptobiotic state for several to thousands or even millions of years. Germination occurs when factors in the environment trigger the endospore to revert to a vegetative form. In the vegetative form, the cell metabolizes and reproduces.

The formation of a vegetative cell from an endospore proceeds in three phases: activation, germination, and outgrowth. In activation, molecules inside the endospore begin to reconfigure in anticipation of starting up metabolic processes. The exact mechanisms that initiate this phase are poorly understood, but it is known that factors in addition to increased nutrient levels must be present for activation to take place. Sublethal heating for short periods (minutes), changes in pH or nutrient levels, exposure to certain chemicals, and possible unidentified aging factors lead to activation.

In germination, the endospore swells, and the spore coat ruptures. The presence of certain substances, called *germinants,* promotes germination. Some of the known germinants are glucose, amino acids, surfactants (detergentlike compounds), and ions of calcium, manganese, and potassium. Very specific relationships often exist between certain bacteria and germinants. For example, L-alanine acts as a germinant for some species, and D-alanine represses germination in the same species. (The prefixes *L*- and *D*- refer to the orientation of atoms around the amino acid molecule's plane; L, or levorotary, molecules rotate light to the left, and D, or dextrorotary, molecules rotate light to the right.)

The final phase is outgrowth, in which the endospore has reverted completely to a vegetative cell capable of metabolizing nutrients and replicating. Outgrowth is a reversible process. If nutrient levels fall or other factors threaten the vegetative cells, they sporulate again and remain spores until conditions improve.

ENDOSPORE STAINS

Endospores are refractive in a light microscope compared with vegetative cells, which are nonrefractive. This quality makes endospores easy to differentiate microscopically from cells, but endospores resist many of the dyes used in routine staining of vegetative bacteria for light microscopy. Special staining methods have, therefore, been developed for endospores.

Endospores that resist the Gram or acid-fast stains, commonly used for other bacteria, usually can be colored by using the Schaeffer-Fulton endospore stain. In this method, heat applied to a dried endospore specimen on a glass microscope slide makes the spores more receptive to taking up a dye. Malachite green acts as the primary stain that penetrates the endospore's outer coat. After water rinsing to wash the dye away from all of the cell parts other than the endospore, the method uses a counterstain, which colors all other constituents in the specimen. The counterstain safranin (the same counterstain used in the Gram stain) turns all nonendospore matter pink to red. Staining, thus, provides a more detailed view of endospores than is revealed by unstained specimens observed in a light microscope.

DORMANCY

Dormancy, also called *hypobiosis,* is a state of minimal metabolic and physical activity in a spore or endospore over time. Exogenous dormancy occurs under specific external factors in the environment. Endogenous, or constitutive, dormancy comes about by an activation and repression of genes to cause a change in cellular metabolism. Three main changes that lead to dormancy are the following: (1) development of a block against membrane permeability to nutrients, (2) existence of reversible blockers to metabolic pathways, or (3) germination inhibitors. Some fungi and bacteria react to specific compounds in the environment that repress the steps leading to germination. For example, the microbiologists Harold Curran and Georges Knaysi showed, in 1961, that ethanol served as a germination inhibitor in the common soil bacterium *Bacillus subtilis,* the spore former first discovered by Ferdinand Cohn, in 1872.

Dormancy can be ended by chemical factors, called *activators,* or physical factors, such as damage to the spore coat. *Bacillus* endospore dormancy, for instance, ceases by heating a suspension of endospores at 122–140°F (50–60°C) for about 10 minutes. This sublethal heating may weaken the spore coat and, thereby, cause activation genes to turn on.

ENDOSPORES AND FOOD SAFETY

Bacillus and *Clostridium* endospores can survive in the environment for many years before germinating, but when they enter the body, the bacteria readily germinate into the vegetative form. This becomes a health threat when endospores are present as contaminants in food.

Bacillus, Clostridium, and *Desulfotomaculum* represent the main food spoilage spore-forming bacteria. Various *Bacillus* species, such as *B. cereus,* and *Clostridium botulinum* must be regarded as serious food-borne pathogens.

Bacillus endospores withstand temperatures ranging from 23°F to 113°F (-5°C to 45°C) and a pH range from 2.0 to 8.0. The endospores have been detected in soil, water, fecal material, and plant materials and contaminate foods from any of these sources. Food microbiologists pay special heed in monitoring the following foods known to carry *Bacillus* endospores: flour, spices, starch, and sugar. Foods made from these ingredients, such as fermented sausages, breads, and canned foods, also become contaminated. Because the endospores resist acidic conditions, they can contaminate canned tomato products, which are naturally acidic.

C. botulinum can be a deadly food-borne pathogen when the endospores germinate in the body after ingestion and the growing cells begin to produce the botulism toxin. Endospores survive anaerobic conditions, heat treatment, salt levels up to 6 percent, and the food preservative sodium nitrate up to about 1 percent. Endospores do not tolerate acidic conditions, so *C. botulinum*–associated illnesses often occur in neutral foods that have been canned under anaerobic conditions.

Desulfotomaculum species have not been implicated in serious food-borne illness but can spoil canned foods if the spores germinate in the food and begin producing sulfide compounds from sulfur-containing food constituents. The gas hydrogen sulfide, detected by its rotten egg odor, is indicative of possible *Desulfotomaculum* spoilage.

The food industry uses a combination of treatments to ensure that bacterial endospores have been killed during the food production process. Moist heat using steam or dry heating serves as the best method for killing the bacteria. In addition to heating, additional food constituents provide secondary barriers against contamination. For example, adjusting pH and adding chemical preservatives can supplement the heating step in reducing the risk from endospores. A food microbiology professor at the Ohio State University advised, in 1989, "The temperatures used for heat processing or canning vary from 100°C [212°F] for high-acid foods to 121°C [250°F] for low-acid foods. For HTST [high-temperature, short-time] processes, temperatures in the range of 120° to 150°C [248–302°F] or higher may be used." Food microbiologists have devised the best methods for eliminating endospores from the particular types of foods they monitor. One constant remains: Bacterial endospores are the most difficult microorganisms to kill in food products and in all of microbiology.

See also AEROMICROBIOLOGY; *BACILLUS; CLOSTRIDIUM;* FOOD MICROBIOLOGY; FUNGUS; STERILIZATION.

Further Reading

Banwart, George J. *Basic Food Microbiology,* 2nd ed. New York: Chapman & Hall, 1989.

Ben-Yehuda, Sigal. "Bacterial Spores." European Research Council Legal Notice. June 26, 2009. Available online. URL: http://erc.europa.eu/index.cfm?fuseaction=page.display&topicID=195. Accessed December 29, 2009.

Cano, Raul J., and Monica K. Borucki. "Revival and Identification of Bacterial Spores in 25- to 40-Million-Year-Old Dominican Amber." *Science* 268 (1995): 106–164.

Cohn, Ferdinand. "Untersuchungen über Bacterien. IV. Beiträge zur Biologie der Bacillen" (Studies on the Biology of the Bacilli). *Beiträge zur Biologie der Pflanzen* 2 (1877): 249–276. In *Milestones in Microbiology,* translated by Thomas Brock. Washington, D.C.: American Society for Microbiology Press, 1961.

Curran, Harold R., and Georges Knaysi. "Survey of Fourteen Metabolic Inhibitors for Their Effect on Endospore Germination in *Bacillus subtilis.*" *Journal of Bacteriol-*

ogy 82 (1961): 793–797. Available online. URL: www.ncbi.nlm.nih.gov/pmc/articles/PMC279260. Accessed January 1, 2010.

Todar, Kenneth. "Structure and Function of Bacterial Cells." In Todar's Online Textbook of Bacteriology. Madison: University of Wisconsin, 2009. Available online. URL: http://textbookofbacteriology.net/structure_10.html. Accessed December 31, 2009.

Vreeland, Russell H., William D. Rosenzweig, and Dennis W. Powers. "Isolation of a 250-Million-Year-Old Halotolerant Bacterium from a Primary Salt Crystal." *Nature* 407 (2000): 897–900.

Weiner, Matthew A., and Philip C. Hanna. "Macrophage Mediated Germination of *Bacillus anthracis* Endospores Requires the *GerH* Operon." *Infection and Immunity* 71 (2003): 3,954–3,959. Available online. URL: www.ncbi.nlm.nih.gov/pmc/articles/PMC161980. Accessed December 31, 2009.

stain The term *stain* refers to the act of coloring a specimen containing microorganisms or tissue with a dye that makes features easier to see through a microscope. The dyes that are used in staining techniques are often themselves called *stains*.

The staining procedure provides microbiologists with the valuable ability to differentiate different types of microorganisms and visualize subcellular structures called *organelles*. Some stains have been formulated for more specific purposes, such as highlighting or showing the presence of capsules or endospores, which are both protective coats that some bacteria produce. The development of new stains allowed the science of microscopy to develop. Without stains, many microorganisms appear almost colorless or transparent through a microscope. Even greater magnifications cannot give an image that differentiates parts of a cell. From the lowest powers of light microscopes to the most advanced electron microscopes used today, staining has become a crucial step for studying objects that are too small to be seen without a microscope.

Staining makes microorganisms easier to study through a microscope. Negative staining can further enhance images because negative stains color the background without staining the specimen. Negative staining is particularly useful in electron microscopy of virus particles because viruses are much smaller than bacteria.

Most staining techniques require that a microbiologist, first, dry a microbial specimen, either from broth or from a colony on agar, onto a glass slide. This step, called *fixing* the specimen, produces a dried specimen called a *smear*. The specimen can, then, be introduced into any of hundreds of different staining methods now available in microbiology. After a specimen has been stained, the smear can be preserved almost indefinitely in the dry form stored at room temperature. Microbiologists also have an option of coating the smear with specially formulated substances that prolong the quality of the stained specimen for many years.

TYPES OF STAINS

Stains can be divided into two main categories of simple stains and differential stains. Simple stains contain a single dye dissolved in water or alcohol, which highlights an entire microbial cell. In some instances, microbiologists add a second factor, called a *mordant*, that holds the dye to the cell or makes the final image brighter. Differential stains react differently with various microorganisms so that microbiologists can distinguish between two or more groups of species. For example, the Gram stain, developed by the Danish microbiologist Hans Christian Gram (1853–1938), in 1884, groups bacteria in two main categories: gram-positive and gram-negative cells. Microbiologists have, since, depended on this differential stain in the first step in species identification, and the stain has further been used in describing the ability of certain species to infect a host and the probable reaction to antibiotics. For example, the antibiotic penicillin tends to kill more susceptible gram-positive species than gram-negative species. The Gram stain remains the most useful procedure in microbiology, and it is one of the first methods microbiology students learn.

Simple stains use either of two types of dyes: basic and acid. Basic dyes have positively charged chemical groups that react with the surface of microbial cells, which are negatively charged as a result of the presence of proteins. Nucleic acids inside the cell, such as deoxyribonucleic acid (DNA) and ribonucleic acid (RNA), also have a negative charge, which aids in basic staining. Basic dyes have been most useful for highlighting the size, shape, and arrangement of bacterial cells. Acid dyes contain negatively charged chemical groups (mainly carboxyl [-COOH] and hydroxyl [-OH]). Acid dyes bind with the positively charged cell structures, mainly bacterial capsules, endospores, glycogen, and the reproductive and nonproductive stages of yeasts. Acid dyes offer an advantage in binding to their target without nonspecific binding to cell debris. The table on page 727 lists examples of basic and acid dyes.

In the table, safranin is an example of a counterstain employed in differential staining. A counterstain colors the cells or parts of cells that do not react with the primary stain. For example, in Gram staining, crystal violet is a primary stain that colors bacteria dark blue to purple. Safranin colors the rest of the bacteria that did not react with the crystal violet; these cells turn dark pink. In the Gram stain, gram-positive bacteria react with crystal violet, and gram-negative bacteria react with safranin, after exposure (and no reaction) to crystal violet.

Basic and Acid Dyes Used in Simple and Differential Staining

Basic Dyes	Acid Dyes
• basic fuchsin—differential for bacteria; protozoan nuclei	• acid fuchsin—differential for bacteria
• crystal violet—differential in Gram staining	• aniline blue—fluorescence microscopy
• malachite green—parasites; endospores	• light green SF—eukaryotic organelles
• methylene blue—nucleus stain	• erythrosin B—bacteria in plants
• safranin—counterstain in Gram staining	• rose bengal—eukaryotic tissue

Neutral dyes contain mixtures of basic and acid dyes and work best in staining complex eukaryotic cells that contain a mixture of positively and negatively charged structures. Each category of dyes has been formulated in numerous ways by combining different dyes or altering the solvent (the liquid that dissolves the dye). These derivations make up differential and specialty stains, which offer to the microbiologist more information on cell structure. Simple staining retains the advantage of being simple to prepare and easy to use on specimens.

Differential stains have gained acceptance in microbiology because they have the power to give more information about cell structure than simple stains. Differential stains have been additionally important in identifying new species or in identifying pathogens recovered from a person who has a disease. The two major differential stains used in microbiology are the Gram stain and the acid-fast stain. By using these two methods, microbiologists can differentiate almost all of the main bacteria studied today in environmental and medical microbiology.

ACID-FAST BACTERIA

The acid-fast stain has become second only to the Gram method as an important differential stain. In the late 1800s, the German biologist and immunologist Paul Ehrlich (1854–1915) sought a method for finding tubercle bacilli in respiratory tract specimens from tuberculosis patients. A stain that could quickly and accurately differentiate the pathogen from all other matter in the specimen would give doctors an advantage in diagnosing and treating the disease. Ehrlich explained, in 1882 (during the same period Gram developed his staining method), "I have worked almost exclusively with dried preparations from sputum, but I have also made control experiments which show that the method is also useful for tissue sections." The German bacteriologist Franz Ziehl (1857–1926) and the physician Friedrich Neelson (1854–98) later modified Ehrlich's method to the version used today; the principles of the method have not changed since Ehrlich's work.

The Ziehl-Neelson acid-fast stain is used today for differentiating *Mycobacterium* bacteria from other genera, especially because these hardy organisms do not stain well by using the Gram method. *Mycobacterium tuberculosis* and *M. leprae* cause tuberculosis and leprosy, respectively, in humans. Basic fuchsin dissolved in phenol penetrates the high-lipid exterior of these species. The fuchsin, then, binds with mycolic acid, a major constituent of the fatty mycobacterial cell wall.

The acid-fast stain has also become useful for identifying bacteria of the genus *Nocardia* that can infect the lungs, kidney, and nervous system of humans and other animals.

The general steps of the Ziehl-Neelson acid-fast stain are the following:

1. Prepare a thin smear of specimen with as little physical damage as possible.
2. Gently heat-fix over a flame or on a hotplate set at about 150°F (65°C) for two hours.
3. Cover smear with piece of filter paper and flood it with Ziehl's carbolfuchsin (basic fuchsin dissolved in dilute phenol).
4. Heat the smear with additions of fuchsin mixture to prevent drying.
5. Rinse with tap water and wash off excess stain with a solution of hydrochloric acid in ethanol; this step is called *decolorizing*.
6. Rinse again and flood the slide with alkaline blue or methylene blue solution as a counterstain.
7. Rinse and pat dry.

Acid-fast microorganisms appear bright red through a microscope, while all other organisms and the background material appear blue.

Tuberculosis (TB) is a severe disease worldwide that causes death in regions where drug treatments are unavailable or the lengthy therapy is not properly

Assessment of Acid-Fast Specimens

Result	Grading	Meaning
numerous acid-fast-positive cells observed within one minute	+++	TB pathogen present
10 or more positive cells observed in several minutes	++	TB pathogens "few"
three to nine positive cells observed in several minutes	+	TB pathogens "rare"
one or two positive cells observed in 20 minutes	report 1 or 2	pathogen present at low number
no positive cells observed in 20 minutes	negative	negative for TB

Source: Clark, George, ed., *Staining Procedures,* 4th ed. (Baltimore: Williams & Wilkins, 1981), 321.

followed. Clinical microbiologists, therefore, take care in examining specimens suspected of harboring the TB organism. A general scheme for examining acid-fast specimens appears in the table above.

Ehrlich made an additional and crucial discovery, when examining stained specimens in his laboratory, by relating staining reaction to the bacteria's physiology. No one had made this connection before. "A further point which I would like to mention," he wrote, "is the question of what information on the nature of the bacillus is revealed by the staining characteristics." Noticing that the outer surface of *Mycobacterium* resisted damage or penetration by strong acids, he suggested "that this condition would have a practical interest . . . on the question of sterilization or disinfection." Microbiologists now know that microorganisms differ in their susceptibility to chemical disinfectants and sterilization. *Mycobacterium* is second only to bacterial endospores in resistance to disinfection and sterilization.

Ehrlich's observation on the staining characteristics opened new avenues in studying pathogenesis, the processes used by pathogens to cause disease. Ehrlich would build on this idea to develop the new science of chemotherapy, or the treatment of pathogenic cells or microorganisms while they are in the body.

SPECIAL STAINS IN MICROBIOLOGY

Microbiologists have developed new stains as needed for locating specific cell organelles or extracellular structures. Special stains can also color certain compounds and leave all other cellular constituents unstained.

Stains formulated specifically to highlight endospores help microbiologists distinguish between true endospores and other cells that resemble spores. The Schaeffer-Fulton method for endospore staining uses the dye malachite green at 0.5 percent in water to differentiate endospores of *Bacillus* and *Clostridium* bacteria from other microorganisms. After successive heating and rinsing steps, endospores appear bright green, and other materials are brownish red. The color of the background is due to the same counterstain used in the Gram stain: safranin.

The endospore stain reported by the American microbiologists Alice Schaeffer and MacDonald Fulton, in 1933, is superior to other stains because malachite green can penetrate the spore wall, whereas most other dyes cannot. If properly done, the technique produces green spores inside the rest of the cellular components, which become brown-red.

Cyst production in bacteria occurs less frequently than endospore formation. Cysts differ from spores mainly in being cells enclosed inside a thick wall, whereas endospores occur when vegetative cells (reproducing and not in spore form) enter a dormant phase and transform into the endospore. The best-known bacterium that forms cysts is the soil genus *Azotobacter.* A commonly used cyst stain uses the dye light green SF for staining the cyst green and neutral red for staining all other parts of the smear brownish red. The various reactions that can result from the cyst stain are:

- vegetative cells—yellowish green
- early encystment stage—green, darker than vegetative cells
- completely formed cyst—dark green
- unstained materials and background—brownish red

Capsules made of a gelatinous material also require special stains. Capsule-forming bacteria make this protective coating out of polysaccharides or polypeptides. Both of these substances dissolve in water, so microbiologists avoid rinsing the smear with water after staining. Capsule material also damages easily if heated or dried, so these steps must be avoided, too.

Three capsule stains can also be used for staining the slime layer produced by some microorganisms. (Slime is a disorganized coating of glycocalyx, or gelatinlike material, loosely attached to the cell wall.) The simplest method for observing capsules is to immerse bacteria in an India ink suspension. The capsule does not stain, but the background turns black and thus helps highlight capsule (or slime) species.

Microbiologists also use special methods for staining the flagella of motile microorganisms. Because flagella can be as small as 10–30 nanometers (nm) in diameter, light microscopy cannot produce a good image of these organelles. Without a staining method suitable for light microscopy, only electron microscopes have the power to make images of flagella. Flagella stains coat the appendage with tannic acid and potassium alum, two chemicals that act as mordants. Stains such as basic fuchsin or paraosaniline, then, attach to the mordant to make flagella visible by light microscopy.

The following specialty stains have been used for certain substances or genera in microbial studies:

- Sudan black—colors poly-beta-hydroxybutyrate black against a pink cytoplasm background
- toluidine blue—stains polyphosphate granules red against the blue cytoplasm
- alcian blue—turns polysaccharides blue against red cytoplasm
- silver nitrate in basic solution—spirochete bacteria turn brownish black
- Giemsa stain—turns rickettsia bacteria purple

Each of the staining methods listed here has additional steps that help the dye attach to the targeted substance or enhance the overall color.

Fluorescent microscopy employs different dyes called *fluorochromes* that emit light when exposed to light of specific wavelengths. This type of imaging relies on the energy emitted by the fluorochrome in light rather than a simple color stain. This field of microscopy produces images of cellular features, differentiates microorganisms, differentiates living from dead cells, and serves in antibody-antigen reactions.

Staining techniques range from simple methods for observing microorganisms through a light microscope to sophisticated techniques needed in fluorescence and electron microscopy. Simple and differential staining methods remain the foundation of microbiology for helping in identification, determining the presence or absence of contaminants, and characterizing unique features in certain species.

See also ELECTRON MICROSCOPY; FLUORESCENCE MICROSCOPY; GRAM STAIN; MICROSCOPY; ORGANELLE.

Further Reading

Biological Stain Commission. Available online. URL: www.biologicalstaincommission.org. Accessed December 19, 2009.

Clark, George, ed. *Staining Procedures,* 4th ed. Baltimore: Williams & Wilkins, 1981.

Ehrlich, Paul. "A Method for Staining the Tubercle Bacillus." *Deutsche Medizinische Wochenschrift* 8 (1882): 269–270. In *Milestones in Microbiology,* translated by Thomas Brock. Washington, D.C.: American Society for Microbiology Press, 1961.

Schaeffer, Alice B., and MacDonald Fulton. "A Simplified Method of Staining Endospores." *Science* 77 (1933): 165–194.

Staphylococcus

Staphylococcus is a genus of gram-positive, nonmotile, spherical (coccus) bacteria native to the skin on many animals, including humans, where it mainly inhabitats the external nasal passages. The microorganism belongs to family Staphylococcaceae, order Bacillales, class Bacilli, in phylum Firmicutes.

Staphylococcus cells measure about 1 micrometer (µm) in diameter—their range is 0.9–1.3 µm—and, when multiplying, they form characteristic irregular clumps resembling grape bunches. The generic term *staphylococci* can be used for any cocci that grow in clumps, in contrast to streptococci, which grow in chains.

Staphylococcus species produce the enzyme catalase to decompose the toxic chemical hydrogen peroxide (H_2O_2) to oxygen and water. *Staphylococcus* is halotolerant, meaning it withstands elevated salt levels, and has chemoheterotophic metabolism, in which the cells use a variety of organic compounds as carbon and energy sources. The bacterium is a facultative anaerobe; it can grow well with or without oxygen.

The *Staphylococcus* cell wall consists of a thick peptidoglycan layer, typical of gram-positive genera, but also contains high levels of teichoic acids, which are polymers characterized by the amount of phosphate groups (PO_4^{-2}) attached to the base units glycerol and ribitol. *Staphylococcus* exhibits sensitivity to the bacteriocin (a type of antibiotic) lysostaphin, which cleaves the linkages in the peptidoglycan polymer. Some species produce a protective outer layer called a *capsule,* and some species make orange/yellow carotenoid pigments.

This genus contains human pathogens, commensals, and opportunistic pathogens. Commensal organisms live in association with a host and derive a benefit but have no effect on the host's well-being. Opportunistic pathogens live with a host without causing harm but are capable of infecting and causing disease if opportunities arise. As a commensal organism that is part of the skin's normal flora, *Staphylococcus* is harmless. It has become one of the most prevalent opportunistic pathogens in part because of its habitat on the skin, which is vulnerable to injury, and in part because of the genus's development of resistance to many different antibiotics.

The most prevalent *Staphylococcus* species on humans are *S. aureus* (often called *Staph aureus*) and *S. epidermidis*. Several other species can be found occasionally on humans or are native to other animals. *S. aureus* is the most medically important species because of its activity as an opportunistic pathogen and its resistance to multiple antibiotics. A resistant *S. aureus* called methicillin-resistant *Staphylococcus arueus* (MRSA) originally created infections only in hospitals, when discovered in the 1960s. Microbiologists have now found MRSA on numerous surfaces in public places such as schools, dormitories, gyms, locker rooms, workplaces, prisons, and public transportation. Although MRSA can be confined to mild skin infections, it has grown into a public health worry because of the limited number of antibiotics that kill it.

STAPHYLOCOCCUS AUREUS

S. aureus grows well in nasal secretions and the nasal passages because it withstands both high osmotic pressure and low moisture. Nasal secretions provide a material in which the concentration of substances outside the cell exceeds that inside the cell. This creates an osmotic pressure difference from outside to inside that could kill other types of bacteria. The inflow and outflow of air in the nasal passages also inhibit most bacteria other than *Staphylococcus*.

Multiple traits make *S. aureus* pathogenic. *S. aureus* can use more than one portal of entry, which is the route taken to get inside a host. Although the microorganism typically invades broken skin, it can also cause infection by way of the digestive tract or the urinary tract. At an injury site, this species produces the enzyme coagulase, which clots blood plasma. Since *Staphylococcus* often infects skin wounds, a protective clot surrounding the infected site gives the pathogen time to multiply outside the reach of immune system defenses. Some *S. aureus* strains can, then, excrete a protein called fibrinolysin that breaks down the blood clot. Other strains use a similar tactic by producing hemolysin, which lyses red blood cells. Inside the body, non-MRSA strains possess natural defenses against antibiotics. Beta-lactamase enzymes break apart the penicillin molecule, plus other antibiotics structurally similar to penicillin. Finally, *S. aureus* produces toxins that damage a variety of tissues in the body.

Some strains of *Staphylococcus* also excrete slime that helps cells adhere to smooth surfaces and stay viable for extended periods by preventing them from drying out. This attribute increases the probability that a person will pick up the pathogen, perhaps on the hands, and transfer it to the body. Children touch their hands to their face and nose hundreds of times a day—adults do this almost as much as children—thus giving *S. aureus* a convenient way to move from person to person.

In *Staphylococcus,* species other than *S. aureus* can cause infections in humans and animals, but *S. aureus* is the most medically important because it can infect almost every organ and tissue in the body. The table on page 731 describes the human diseases associated with *S. aureus*.

S. aureus has also been implicated in playing a part in the pathogenesis of osteomyelitis (bone) and endocarditis (heart).

SSSS is caused by strains of *S. aureus* that contain a gene for exfoliative toxin. The toxin causes a peeling of the skin that exposes a red area underneath. The disease can spread rapidly, especially in neonatal facilities and day care centers that include infants.

TSS is a serious syndrome experienced most often by women who use contraceptive sponges or menstruating women who use superabsorbent tampons, although it is not confined to this group. TSS causes irritation of the skin, followed by skin shedding, rash, fever, and low blood pressure. Occurrences of TSS have been associated with skin infections and infections at surgical incision sites. The disease spreads quickly through the body and can be fatal.

S. aureus also causes an acute (rapid onset and short duration) food-borne illness when food preparers contaminate the food by handling or by sneezing or coughing. *Staphylococcus* food poisoning develops from a toxin called an enterotoxin that acts on the cells lining the intestines. The toxin's activity results in most or all of the following symptoms: nausea, vomiting, abdominal cramping, and weakness. The person must receive a large infective dose of *S. aureus* to produce enough toxin to develop the illness. For most people with no preexisting illness, about 1 microgram or less of toxin causes the illness. A person must ingest more than 100,000 *S. aureus* cells per gram of food to get this much toxin.

ANTIBIOTIC-RESISTANT *STAPHYLOCOCCUS AUREUS*

S. aureus has caused considerable concern in medical care because it resists multiple antibiotics. Hospital-associated infections are called *nosocomial infections,* which would be expected in any confined

Diseases Associated with *S. aureus*

Location of the Disease	Diseases
skin	• boils, carbuncles, impetigo, and pimples • abscesses at wound infections • tissue necrosis • staphylococcal scalded skin syndrome (SSSS)
urogenital area	• toxic shock syndrome (TSS) • nephritis (inflammation in the kidney)
nervous system	• meningitis
respiratory system	• bronchitis • laryngitis • pneumonia
digestive tract	• gastroenteritis

community of sick people and a high concentration of pathogens. Antibiotic use increased steadily from the time penicillin was introduced, in 1945. Doctors took advantage of penicillin and other new antibiotics for battling localized infections to life-threatening diseases. Hospitals became hot spots of antibiotic use. By the 1960s, doctors began noticing that penicillin and derivatives of this drug, such as ampicillin and methicillin, had lost effectiveness against particular pathogens. MRSA became the first important discovery of a pathogen that resisted not only its natural antagonist, penicillin, but also a second generation of penicillinlike antibiotics. MRSA and an increasing number of other antibiotic-resistant species have been called *superbugs*.

MRSA is no longer confined to hospitals. In addition to schools and gyms, where the microorganism was first detected outside the hospital, microbiologists find MRSA in a widening array of places. A 2008 study of 10 West Coast beaches revealed *Staphylococcus* in the sand at nine of the beaches and MRSA on five of the beaches. A University of Washington microbiologist commented, in 2009, "The fact that we found these organisms suggests that the level is much higher than we had thought." Microbiologists advised that people wash off sand after visiting beaches with particular care to any open cuts or wounds. These actions would be prudent for stopping infection regardless of the potential pathogens in the environment.

Genetic analysis of MRSA has not answered the question of whether MRSA developed in different places or from a single ancestor cell. By the 1980s, MRSA had spread internationally, and the pattern of spread suggests the organism developed in more than one place. Each year, more than 90,000 people in the United States have an MRSA infection; MRSA now kills more people even than acquired immunodeficiency syndrome (AIDS).

When MRSA first emerged as a health threat, doctors turned to the alternative antibiotic vancomycin, which is not as effective against the pathogen as the penicillin family of drugs had once been. Vancomycin has continued to be the primary treatment for MRSA infections, but rare strains with resistance against this antibiotic have been detected. Doctors have also tried the drug linezolid against MRSA to prevent the emergence of vancomycin resistance. Linezolid is expensive and has side effects, so it does not appear to be the ultimate drug against MRSA if vancomycin resistance grows.

Two derivations of MRSA have entered medical research: community-associated MRSA (CA-MRSA) and epidemic MRSA (EMRSA). CA-MRSA infections are MRSA infections in people who did not visit a hospital or have an invasive medical procedure within the past year. CA-MRSA begin as skin infections characterized by redness and small pimples. The infection can progress to skin abscesses, fever, and pain. Physicians avoid using antibiotics to treat CA-MRSA but use vancomycin in serious cases.

EMRSA is any strain of MRSA that moves swiftly through a population and builds an epidemic. EMRSA may have first emerged, in the 1980s, in the United Kingdom, but today multiple antibiotic-resistant EMRSA strains have been isolated in many countries. A 2002 study in the United Kingdom discovered an EMRSA strain carrying genes for resistance to 10 different antibiotics.

Two new antibiotic-resistant *S. aureus* strains that resist vancomycin have been differentiated by the strength of their resistance. Microbiologists use a laboratory test called *minimum inhibitory concentration* (MIC) to determine the relative susceptibilities of bacteria to an array of antibiotics. Low concentrations of antibiotic needed to inhibit a pathogen indicate the pathogen susceptibility. The two vancomycin-resistant strains are, thus, identified with the following MIC results:

- vancomycin-resistant *S. aureus* (VRSA)—susceptible to vancomycin at 16 micrograms per milliliter (µg/ml) or higher doses
- vancomycin-intermediate *S. aureus* (VISA)—susceptible to vancomycin at 4 to 8 µg/ml

VRSA is considered resistant to vancomycin because it withstands high concentrations of the antibiotic. (MIC tests usually study antibiotics at microgram per milliliter [µg/ml] levels.) VISA appears to possess an intermediate level of resistance because it withstands very low vancomycin levels (less than 4 μg/ml) but can be inhibited with slightly raised levels of the antibiotic.

CATALASE TEST

The catalase test serves as a simple screening method for differentiating *Staphylococcus* from other gram-positive cocci. The test also differentiates bacilli such as endospore-forming *Bacillus* and *Clostridium*.

The catalase test detects the presence of this enzyme in broth cultures or in colonies grown on agar. A microbiologist puts a drop of 3 percent hydrogen peroxide on an isolated colony or into one-half milliliter of broth. The presence of catalase becomes evident with the visible formation of oxygen bubbles. *Staphylococcus* gives a positive result (oxygen produced), and *Streptococcus* gives a negative result (no oxygen bubbles).

For differentiating gram-positive bacilli, *Bacillus* is positive for the presence of catalase, and *Clostridium* is negative.

Staphylococcus is ubiquitous in the human environment because it makes up part of the body's normal flora. *Staphylococcus* has grown into an increasingly important pathogen as a result of the development of strains resistant to more than one antibiotic at the same time. Because of its prevalence as a human health concern, *S. aureus* will receive continued attention in microbiology research and pathology.

See also MINIMUM INHIBITORY CONCENTRATION; NORMAL FLORA; NOSOCOMIAL INFECTION; PATHOGENESIS; RESISTANCE.

Further Reading

Aucken, Hazel M., Mark Ganner, Stephen Murchan, Barry D. Cookson, and Alan P. Johnson. "A New UK Strain of Methicillin-Resistant *Staphylococcus aureus* (EMRSA-17) Resistant to Multiple Antibiotics." *Journal of Antimicrobial Chemotherapy* 50 (2002): 171–175. Available online. URL: http://jac.oxfordjournals.org/cgi/reprint/50/2/171. Accessed January 7, 2010.

Centers for Disease Control and Prevention. "*Staphylococcus aureus* Infections." Available online. URL: www.cdc.gov/ncidod/diseases/submenus/sub_staphylococcus.htm. Accessed January 7, 2010.

———. "VISA/VRSA: Vancomycin Intermediate-Resistant *Staphylococcus aureus*." Available online. URL: http://www.cdc.gov/ncidod/dhqp/ar_visavrsa.html. Accessed January 7, 2010.

Enright, Mark C., D. Ashley Robinson, Gaynor Randle, Edward J. Feil, Hajo Grundmann, and Brian G. Spratt. "The Evolutionary History of Methicillin-Resistant *Staphylococcus aureus*." *Proceedings of the National Academy of Sciences* 99 (2002): 7,687–7,692. Available online. URL: www.pnas.org/content/99/11/7687.full. Accessed January 7, 2010.

Marchione, Marilynn. "MRSA Discovered at Five Public Beaches." *Seattle Times*, 13 September 2009. Available online. URL: http://seattletimes.nwsource.com/html/localnews/2009859381_webmrsabeach13m.html?syndic ation=rss. Accessed January 7, 2010.

New York State Department of Health. "Community-Associated Methicillin-Resistant *Staphylococcus aureus* (CA-MRSA) Fact Sheet." Available online. URL: www.nyhealth.gov/diseases/communicable/staphylococcus_aureus/methicillin_resistant/co mmu nity_associated/fact_sheet.htm. Accessed January 7, 2010.

Wegner, Dennis L. "No Mercy for MRSA: Treatment Alternatives to Vancomycin and Linezolid." *Medical Laboratory Observer*, January 2005. Available online. URL: http://findarticles.com/p/articles/mi_m3230/is_1_37/ai_n9770627/?tag=content;col1. Accessed January 7, 2010.

sterilization Sterilization is the chemical or physical process of eliminating all life from a substance. In microbiology, this process requires the removal of all microorganisms, including bacterial endospores, from a material. Endospores are singled out in this definition because they are the biological forms most resistant to damage by chemicals, heat, pressure, drying, or irradiation. The endospores made by the genera *Bacillus* and *Clostridium* are usually the target organisms in sterilization. The active agent, such as a gas or a chemical that sterilizes a material, is called the *sterilant*.

Sterilization makes up part of a hierarchy of actions to control the growth of unwanted microorganisms in inanimate materials. For specific materials, microbiologists adhere to the following hierarchy:

- sterilization—elimination of all microorganisms
- disinfection—elimination of all microorganisms except bacterial endospores
- sanitization—reduction of unwanted microorganisms to safe levels

The following materials require sterilization before they can be used for their specific purposes: surgical and dental instruments, surgical coverings, medical devices that enter the body or penetrate the skin, microbiological growth media, and vessels, containers, inoculating instruments, and any other object that has contact with a sterile surface, a clinical sample from a patient, or a pure culture of a microorganism.

Disinfection has been confused with sterilization for good reason. Some microbiologists use a different terminology than the hierarchy explained here. For example, in some texts, the term *high-level disinfection* is used to describe sterilization as defined here. In this alternative hierarchy, *medium-level disinfection* resembles disinfection, and *low-level disinfection* is similar to sanitization.

In microbiology and medicine, sterilization serves to prevent the following events:

- microbial contamination of sterilized growth media, diluents, and nutrient solutions
- entry of unwanted microorganisms into a pure culture, that is, a microbial culture containing only one species
- transfer of environmental microorganisms to instruments used for medical or dental patients
- contamination by microorganisms of surgical incision sites, surgical devices, and dental devices

The food industry also relies on sterilization during food canning processes. Packaged foods must be sterilized to kill endospores of *Clostridium botulinum,* a deadly food-borne pathogen. Food producers use physical sterilization methods, such as heating or irradiation, to ensure the safety of their products.

To achieve these objectives, microbiologists use sterilization in conjunction with aseptic techniques, which are all the activities performed to eliminate contamination. In other words, sterilization removes all life from materials, and aseptic techniques ensure that the materials remain free of contamination as they are being used.

Sterilization also serves the important purpose of decontaminating materials that contain pathogens—these materials are called *biohazards*—so that the materials can be disposed of without creating a health threat in the environment.

PHYSICAL STERILIZATION METHODS

Four main means of physical sterilization are heat, filtration, ultraviolet irradiation, and ionizing irradiation. Chemical sterilization mainly includes exposure to gas or chemical biocides.

Heat sterilization represents the most common method used by microbiology laboratories and hospitals. Also called *thermal destruction,* heat sterilization can be accomplished by two routes: moist heat and dry heat. A piece of equipment called an *autoclave* produces the conditions of pressurized steam to sterilize by moist heat. An autoclave is a large pressure cooker–like instrument. Ovens that can reach high temperatures for a minimal required period of time perform dry heat sterilization.

Sterilization using an autoclave requires training in the physics of heat exchange in materials. This method also involves biological factors that determine the autoclave settings needed to kill all life. In general, sterilization using an autoclave requires that a microbiologist consider the following factors:

- time-temperature relationship—microbial killing capacity of a given temperature on a specific microorganism in a minimal period
- temperature-pressure relationship—microbial killing capacity of moist heat under pressure
- innate heat (or thermal) resistance—capacity of a microbial strain to resist killing by heat
- protective and thermodynamic effects—factors such as the composition of materials containing microorganisms that protect against heat transfer and, thus, cell death

Most microbiology laboratories have determined the conditions required for meeting these conditions. As a typical example, autoclaves kill all microorganisms in small loads of materials when achieving 250°F (121°C) at 15 pounds pressure per square inch (psi), for at least 15 minutes. The pressure value defined as psi refers to the amount of pressure above atmospheric pressure at sea level.

Dry heat sterilization may be accomplished by heating a metal inoculating loop in the flame of a Bunsen burner to kill any microorganisms. For glass or metal flasks, tubes, instruments such as forceps, or other nonliquids, hot-air sterilization in an oven involves a set temperature for a given minimal period. For example, a temperature of 338°F (170°C) maintained for two hours ensures the sterilization of most dry materials. Finally, combustion of biohazardous materials in an incinerator also serves as a means of both destruction and sterilization.

Filtration involves passing a liquid through a small-pore filter to remove microorganisms. After passing through the filter, the liquid can be considered sterile as long as the pore size is small enough to remove the smallest microorganisms. Membrane filters are sheets of about 0.1 millimeter (mm) in thickness made of cellulose esters or synthetic plastics and with pore sizes of a predetermined size. Filter manufacturers make filters with pore sizes of 0.45 micrometer (μm) in diameter, 0.22 μm, and a variety of smaller pore sizes. These pore sizes retain different types of microorganisms on the filter and prevent these particles from contaminating the filtered liquid, called the *filtrate.* The following examples give the general filtration capacities of different pore sizes:

- 0.45 μm—removes most bacteria and all yeasts and fungal spores
- 0.22 μm—removes all but the smallest bacteria

- 0.01 μm—removes viruses and some large molecules

Filtration in devices that apply pressure to speed the passage of liquid through the filter can also expand the pore size. Therefore, pressure filtration is not recommended for removing particles of a certain size. Filtration devices preclude this problem by using suction to pull the liquid through the filter's pores.

Sterilization by irradiation uses light of certain wavelength and energy to damage microorganisms irreparably. Microbiology includes two types of radiation: nonionizing and ionizing. Nonionizing radiation has a long wavelength relative to ionizing radiation, measuring 1 nm or longer. Ultraviolet (UV) light is a type of nonionizing radiation that damages deoxyribonucleic acid (DNA) in cells and prevents further reproduction. UV light in a wavelength range of 220–300 nm has been called the *abiotic range* because it is most effective at killing microorganisms. (*Abiotic* means "without life.") DNA absorbs light of about 260 nm wavelength. For this reason, laboratory sterilizing lamps for the air or for inanimate surfaces usually produce UV light at this wavelength. UV light, even in the most effective wavelength range, does not penetrate deep into organic matter, so cells must be at a material's surface for UV to work best.

Ionizing radiation consists of gamma rays, X-rays, and electron beams or beta particles. Each type of radiation kills microorganisms by penetrating organic matter and decomposing water molecules. As a by-product of the reaction, ionizing radiation creates chemically unstable molecules called *hydroxyl radicals,* which disrupt the cell's normal mechanisms.

Gamma rays are energy beams that are emitted by radioactive elements and penetrate organic matter. Given an adequate amount of time in hours, gamma rays and X-rays kill all microorganisms in a mixture. High-energy electron beams kill the same number of microorganisms but do so in seconds rather than hours. But the disadvantage of high-energy electron beams lies in their poor ability to penetrate materials, so that they work best on microorganisms on the surface of a material.

Microbiologists study the effect of irradiation on microorganisms through the use of survival curves. Survival curves result from plotting the fraction of cells surviving out of the original number on the y-axis and the radiation dose in rads on the x-axis. The survival fraction equals the number of live cells divided by the number of cells in the original sample before irradiation. A rad is a measure of exposure to radioactivity. Survival curves for heterogeneous mixtures of microorganisms indicate that increasing the radiation dose causes a rapid decline in the number of surviving cells. As dose increases further, radiation has less effect on killing the remaining survivors. Irradiation of pure cultures creates more of a straight-line or proportional relationship between survival fraction and dose.

AUTOCLAVES

An autoclave is a piece of equipment that contains in inner chamber surrounded by an outer chamber, called a *steam jacket*. The jacket receives steam from a facility's steam supply and controls its entry into the inner chamber, where the material to be sterilized has been placed. Autoclaves can hold the pressurized steam due by means of stout, airtight door. After the steam circulates around the materials, it exits the chamber through a trap and condenses to liquid in the equipment's pipes.

Autoclaves offer an effective and fast way to kill microorganisms, including bacterial endospores. The main mode of action is denaturing cellular proteins, meaning that the heat-pressure exposure unravels the protein from the configuration it requires for activity. The principle behind autoclave sterilization resides in the interaction between temperature and pressure; the higher the temperature inside the autoclave's chamber, the higher the pressure that will be produced, summarized in the table below.

Microbiologists who study the optimal conditions for autoclave sterilization have developed relationships between heat, pressure, and time of exposure and the resistance of different microbial species to heat. Usually, the microorganisms that have been evaluated for these relationships have been bacteria. Heat resistance of various bacteria can be expressed as a thermal death point (TDP), which is the lowest temperature that will kill all the bacteria inside the autoclave in 10 minutes. A related measurement, the thermal death time (TDT), is the minimal number of minutes required to kill all the bacteria at a given temperature. As temperature increases, the TDT decreases.

Decimal reduction time (DRT), also known as the *D value,* has superseded TDP and TDT as measures of sterilization efficacy in many laboratories. The *D* value is the number of minutes required for killing 90 percent of all bacteria in the autoclave at a given temperature. *D* values can be used to describe the relative resistance of different microorganisms to the moist heat generated by an autoclave. A plot of *D* value on the y-axis and temperature on the x-axis shows that *D* value decreases in proportion to increased temperature. Two values related to the *D* value provide additional details on sterilization conditions for certain microorganisms:

- *z* value = increase in temperature needed to reduce the *D* value to 10 percent of its original value

- *F* value = time in minutes at a given temperature to kill bacteria

To kill bacterial endospores—the definition of sterilization—an autoclave must produce an *F* value of 15 minutes when the sterilization temperature is 250°F (121°C).

CHEMICAL STERILIZATION METHODS

The vast majority of gaseous sterilizations use the gas ethylene oxide (ETO). Other gases, such as chlorine dioxide, ozone, and vaporized hydrogen peroxide, have been investigated as gases for sterilization, but they have not replaced ETO.

ETO reacts with various chemical groups attached to compounds that make up living cells: amino, carboxyl, hydroxyl, phenolic, or sulfhydryl groups. The American biologist Charles R. Phillips developed most of the theories on how ETO kills vegetative (nonspore, reproductive form) cells and bacterial endospores. In 1959, Phillips and colleagues at the U.S. Army's research facility at Fort Detrick, Maryland, proposed a design for a chamber in which instruments could be safely exposed to ETO for sterilization.

ETO's main use today is as a low-cost sterilization method for hospital instruments and plastics that cannot withstand high temperatures. ETO cannot be used for sterilizing liquids or powders. Hospitals use ETO-sterilized bedding, laboratory plastic instruments, bandages, and surgical instruments.

A technology invented in 1987 by the Johnson & Johnson Company combines hydrogen peroxide vapor and radio wave energy to create sterilization conditions at a low temperature (104°F [40°C]). The frequency of the radio waves causes breakup of the hydrogen peroxide molecules into reactive chemicals called *free radicals*. These chemicals, then, create a gas plasma, which is a more reactive form of the vaporized hydrogen peroxide. The plasma penetrates microbial cells and interferes with various cellular constituents. In 2001, Cynthia Spry wrote in *Infection Control Today*, "The sterilization process is designed for safe use. There are no toxic fumes, by-products, or residues." By contrast, ETO sterilization requires aeration to clear out fumes that can be harmful to workers.

Temperature-Pressure Relationships in Autoclave Sterilization

Temperature, °F (°C)	Pressure (psi)
212 (100)	0
230 (110)	5
241 (116)	10
250 (121)	15
259 (126)	20
275 (135)	30

An autoclave sterilizes large amounts of liquids or solid items with pressurized steam. ***(Department of Health and Environmental Safety, Princeton University)***

As in other sterilization methods, gases can be made more effective or less effective by other factors within the sterilization process. The main factors affecting the efficacy of gas sterilization are:

- gas concentration
- temperature
- relative humidity
- material containing the microorganisms
- microorganism's resistance to gas sterilization

Survivor curves similar to irradiation survivor curves can be developed for individual species. Increases in gas concentration, temperature, or relative humidity increase the killing effectiveness of sterilizing gases. Certain absorbent materials as well as highly resistant microorganisms can decrease gases' effectiveness.

Few chemical sterilants are safe and effective enough to replace ETO or physical sterilization methods. Sodium hypochlorite (household bleach) can sterilize solid materials under certain conditions, but the exposure time required to kill endospores is prohibitively long (more than 30 minutes). Glutaraldehyde replaced the effective sterilant formaldehyde, in the 1960s, because of health hazards associated with formaldehyde. Glutaraldehyde kills cells by reacting with individual amino acids in cellular proteins and nucleotide units in DNA. As does bleach, glutaradehyde requires several minutes to a few hours to kill bacterial endospores. The chemical kills most other microorganisms within 10 minutes; viruses usually require several minutes, while nonendospore bacteria can be killed in less than a minute's exposure.

Workers exposed to glutaraldehyde by inhalation or skin contact are susceptible to a number of irritations of the respiratory tract, eye irritation, and conjunctivitis. For these reasons, few establishments use this sterilant.

STERILIZATION CONFIRMATION

To sterilize an object, steam must penetrate all parts of the item. A straight glass rod is easier to sterilize than an endoscope that has several small spaces. Microbiologists ensure that a physical or chemical sterilization process has operated completely and properly by including a sterilization indicator with each load.

Sterilization indicators change from one form to another during a sterilization process. Indicators commonly use the appearance of words or symbols, color change, or test organisms to signal the effectiveness of the sterilization run. The microbiologist checks the indicator, after the sterilization, to confirm that the process ran as expected. Special paper indicator strips produce words, lines, or symbols or change color when required sterilization conditions have been met.

Indicator test organisms provide stronger proof of sterilization than paper strips. In this method, a microbiologist includes an ampule containing endospores or a test strip impregnated with endospores with each load of material to be sterilized. For heat sterilization, such as in autoclaves or ovens, the test endospores are often of the species *Bacillus stearothermophilus,* which tolerates temperatures reaching 150°F (65°C). After the sterilization run, the microbiologist retrieves the ampule or strip and releases the contents into sterile growth medium. After incubation of the inoculated medium at no less than 131°F (55°C), no growth in the tube indicates the sterilization conditions were met. Growth indicates that the process did not kill all microorganisms, and the materials were not sterilized.

Various sterilization methods are available to meet the specific needs of laboratories or other facilities that must eliminate unwanted microorganisms from their environment. The conditions for sterilization cover a range of treatments, doses, or exposures to the sterilant. Any method is acceptable as long as it can be confirmed that all bacterial endospores have been killed.

See also ASEPTIC TECHNIQUE; BIOCIDE; DISINFECTION; FILTRATION; SANITIZATION.

Further Reading

Block, Seymour S., ed. *Disinfection, Sterilization, and Preservation,* 5th ed. Philadelphia: Lippincott Williams & Wilkins, 2000.

Fraise, Adam P., Peter A. Lambert, and Jean-Yves Maillard, eds. *Russell, Hugo and Ayliffe's Principles and Practices of Disinfection, Preservation and Disinfection,* 4th ed. Malden, Mass.: Blackwell Science, 2004.

Schley, Donald G., Robert K. Hoffman, and Charles R. Phillips. "Simple Improvised Chambers for Gas Sterilization with Ethylene Oxide." *Applied Microbiology* 8 (1960): 15–19. Available online. URL: http://aem.asm.org/cgi/reprint/8/1/15.pdf. Accessed December 20, 2009.

Spry, Cynthia. "Low-Temperature Sterilization." *Infection Control Today,* 1 May 2001. Available online. URL: www.infectioncontroltoday.com/articles/412/412_151steriliz.html. Accessed December 20, 2009.

Streptococcus *Streptococcus* is a genus of gram-positive bacteria involved in a variety of human diseases. *Streptococcus* cells are round or ovoid and measure about 1 micrometer (μm) in diameter. The general term *streptococcus* (plural: streptococci) describes any bacterial cocci that form unbranched or branched chains of cells. *Streptococcus* usually produces short chains or pairs, called diplococci. *Streptococcus* is a facultative anaerobe that does not form endospores, is nonmotile, and has chemoheterotrophic metabolism, meaning it uses a variety of organic compounds for carbon and energy.

Streptococcus belongs to family Streptococcaceae, order Lactobacillales, and class Bacilli of phylum Firmicutes. The genus contains at least 40 species that have been well characterized and includes pathogens, commensals, and species used in food production. Commensal relationships occur when one organism receives a benefit from a close association with another organism, which is unaffected. Of the pathogens, the main groups are oral pathogens associated with dental caries and a pyogenic group that causes pus formation when infecting the skin.

Streptococcus contains at least 10 different groups based on serology, a study of the structures on the outer surface of bacterial cells. Groups A, B, and C are the most medically important *Streptococcus* groups in humans, having the following characteristics:

- group A—endometritis, glomerulonephritis, impetigo, necrotizing fasciitis, otitis media, pharyngitis, pneumonia, rheumatic fever, scarlet fever, sinusitis, strep throat, and tonsillitis

- group B—neonatal and postpartal infections and urinary tract infections
- group C—pharyngitis and sore throat

S. pyogenes belongs to group A and acts as one of the most prevalent pathogens in human health. *S. pyogenes* lives on 5–15 percent of all people as a normal inhabitant of skin. Skin infections from this organism occur most often when cells invade a previously infected wound. In this circumstance, the streptococcus causes a suppurative infection, an infection that develops soon after another pathogen has invaded a host.

Clinical microbiologists use both serology assays and the type of hemolysis (destruction of red blood cells) streptococci display on blood agar to differentiate pathogenic from nonpathogenic species. The two main reactions associated with streptococcus colonies grown on blood agar are described in the table (right) with an example species. The reactions refer to a visible zone of clearing that surrounds isolated colonies on the agar after incubation.

The hemolysis reactions result from substances called hemolysins that damage red blood cells. Beta-hemolysin lyses red blood cells completely; alpha-hemolysin lyses the cells and reduces hemoglobin (red) to methemoglobin (green), which gives the clearing zone its typical color.

PATHOGENIC *STREPTOCOCCUS*

Oral pathogens cause infections and caries in the mouth but also are related to disease in the heart, joints, muscle, nervous system, and skin. The main oral streptococcal pathogens are *S. acidominimus*, *S. mutans*, *S. oralis*, *S. pneumoniae*, *S. salivarius*, *S. sanguis*, and *S. thermophilus*.

S. mutans has been implicated as a major cause of dental caries in humans. Most other oral streptococci produce complex polysaccharides that coat surfaces in the mouth and may contribute to biofilm formation. Biofilm formation on teeth causes the initial deposition and subsequent thickening of dental plaque. *S. pneumoniae* has been implicated in the development of periodontitis, an inflammation of the area between the tooth surface and gum.

S. pneumoniae, a major cause of pneumonia in humans, produces a virulence factor called a *capsule*. Capsules cover various pathogenic bacteria for the purpose of evading destruction from the immune system's phagocytes. *S. pneumoniae* includes more than 80 strains, which differ in the composition of the capsular material. This species also causes forms of arthritis, osteomyelitis, otitis media, and peritonitis. As can many pathogenic streptococci, *S. pneumoniae* can invade the bloodstream to cause often-fatal sepsis.

Among the pyogenic streptococci, *S. pyogenes* is a prevalent pathogen, causing the following diseases:

- erysipelas—skin infection and rash
- scarlet fever—contagious pharyngitis with pimply red rash
- necrotizing fasciitis—infection of fascia (layer of tissue covering muscle)
- puerperal sepsis—genital tract infection in women
- strep throat—sudden, severe, inflamed sore throat with difficulty in swallowing

In its role as the cause of necrotizing fasciitis, *S. pyogenes* has been nicknamed "flesh-eating bacteria" because it reproduces rapidly while excreting enzymes and toxins that kill the surrounding tissue. Recent research on *S. pyogenes* strains has focused on the genetic control of the bacterium. In early 2010, James Musser of Houston's Methodist Hospital said, "When we identify a gene mutation that has a direct effect on a disease—like we have done for the flesh-eating bacteria—this opens up doors to designing drugs that provide treatments and cures." A single gene mutation in *S. pyogenes* can turn off the processes that lead to fasciitis and, thus, make certain mutant strains less pathogenic.

Some species formerly identified as *Streptococcus* have been classified in a general group called enterococci that cause gastrointestinal illnesses as food-borne pathogens. The two most important members are group D streptococci: *S. faecium* and *S. faecalis*. Both species cause fecal contamination of foods.

The viridans group of pathogenic *Streptococcus*, sometimes given the name *S. viridans*, normally lives in the upper respiratory tract of humans

Streptococcus Hemolysis

Type	Description	Species
alpha (α)	greenish zone of partial hemolysis	*S. pneumoniae*
beta (β)	completely clear zone of hemolysis	*S. pyogenes*
gamma or nonhemolytic	no reaction	*S. bovis*

and animals and has been connected with cases of endocarditis, an inflammation of the heart valves. In endocarditis, the bacteria enter the bloodstream and produce a layer of plaque, or biofilm, on the heart valves. In 2005, the pathologist Elizabeth Presterl at the Medical University of Vienna (Austria) reported, "Endocarditis is a serious disease associated with a mortality of up to 50 percent despite antibiotic treatment. A main feature of endocarditis is the endocarditic plaque on the leaflets of one or more heart valves. The formation of the endocarditic plaque is unique, involving bacteria, platelets, coagulation factors and leucocytes, and is considered a rather special kind of biofilm." Bacteria growing in biofilms are more resistant to antibiotics than free-living (planktonic) bacteria outside biofilms.

The viridans group mainly contains alpha-hemolytic species: *S. acidominimus*, some strains of *S. mutans, S. oralis, S. pneumoniae, S. salivarius*, and *S. sanguis*.

STREP THROAT

Pathogenic *Streptococcus* can access and damage heart tissue, as evidenced by the endocarditis caused by the viridans group. Strep throat caused by group A *Streptococcus* provides another example of how *Streptococcus* infections can lead to more serious diseases involving the heart.

Rheumatic fever is an inflammatory disease of the heart valves that can lead to permanent heart damage. This disease, which is most common in children five to 15 years old, begins with strep throat that has been neglected or inadequately treated. *S. pyogenes* infiltrates the blood and confuses immune system defenses through surface proteins that mimic host cell proteins. The body's immune reactions attack the tissue of the body at the same time it seeks to destroy the bacteria, mainly damaging the heart, joints, skin, and central nervous system. Symptoms of rheumatic fever consist of fever, joint pain, chest pain, heart palpitations, fatigue, and shortness of breath. The long-term consequences of rheumatic fever are heart murmurs and arthritis.

Clinical microbiologists test throat specimens (called a *throat culture*) from patients who have sore throat to distinguish possible strep throat from more common virus-caused sore throats. Specimens can be tested by either a rapid test kit for strep or bacterial culture methods. Rapid strep tests give results in about 15 minutes, an advantage in making a quick diagnosis, but may not always be as accurate as bacterial culture methods.

The traditional test for strep throat, the throat culture, involves taking a swab sample from the back of the patient's throat and applying the specimen on the swab to agar media. Completing this procedure takes a total of two to three days. Since *S. pyogenes* is known as a group A beta-hemolytic strep, clinical microbiologists run a program of tests to identify possible strep throat as accurately and as quickly as possible. General steps for diagnosing strep throat are the following:

1. Gram stain procedure to differentiate gram-positive streptococci from gram-negative bacteria
2. serology testing to identify group A
3. specimen grown on blood agar incubated at 98.6°F (37°C) for 24 hours to identify type of hemolysis
4. beta-hemolytic colonies tested for catalase enzyme activity (*Streptococcus* lacks catalase activity) to differentiate from other streptococci
5. confirmation of beta-hemolysis on blood agar

Strep throat is confirmed by the following result: group A, beta-hemolytic, catalase-negative, gram-positive streptococcus

NONPATHOGENIC *STREPTOCOCCUS*

Nonpathogenic *Streptococcus* has been used in making a variety of food products, particularly in the conversion of fresh milk to fermented dairy products. Some examples of foods made by Streptococcus species are

- buttermilk, cottage cheese, and sour cream—*S. cremoris* or *S. lactis*
- cheddar cheese—mixture of *S. cremoris* and *S. lactis*
- sour dough bread—*S. lactis* and *Lactobacillus bulgaricus* with yeast
- Swiss cheese—*S. thermophilus* with *L. bulgaricus*
- yogurt—*S. thermophilus*

In addition to these foods, several streptococci that once belonged to the *Streptococcus* genus have been reclassified as *Lactococcus*, which is also used in making dairy products.

Streptococcus species contribute to the ripening stage of a variety of cheeses, and *S. thermophilus* is used for making yogurt. *Streptococcus* species that have been classified as *Lactococcus* and involve milk fermentations are:

- *S. cremoris* = *L. cremoris*
- *S. lactis* = *L. lactis*
- *S. plantarum* = *L. plantarum*
- *S. raffinolactis* = *L. raffinolactis*

Lactococcus belong to a group of bacteria called *lactic acid bacteria* because they produce lactic acid from fermentations of the sugar glucose.

Additional *Streptococcus* species that are either nonpathogenic to humans or related only to veterinary care are the following isolates:

- *S. agalactiae*—bovine mastitis
- *S. avium*—fowl
- *S. bovis*—domesticated animals
- *S. canis*—bovine mastitis and dog skin infections
- *S. cricetus*—hamsters
- *S. dysgalactiae*—cattle and pigs
- *S. equi*—horses
- *S. parauberis*—cattle
- *S. porcinus*—pneumonia in pigs
- *S. rattis*—rats
- *S. saccharolyticus*—cattle
- *S. suis*—various diseases in pigs

Streptococcus is a homogeneous genus compared with many other bacteria, but it causes a diversity of diseases in humans and can be found in a variety of mammals and birds. The genus is medically important and has been valuable since ancient times as an organism used in food preservation.

See also LACTIC ACID BACTERIA; PLAQUE; SEROLOGY.

Further Reading

Mayo Clinic. "Rheumatic Fever." Available online. URL: www.mayoclinic.com/health/rheumatic-fever/DS00250. Accessed January 23, 2010.

Presterl, Elizabeth, Andrea J. Grisold, Sonja Reichmann, Alexander M. Hirschl, Apostolos Georgopoulos, and Winfried Graninger. "Viridans Streptococci in Endocarditis and Neutropenic Sepsis: Biofilm Formation and Effects of Antibiotics." *Journal of Antimicrobial Chemotherapy* 55 (2005): 45–50. Available online. URL: http://jac.oxfordjournals.org/cgi/reprint/55/1/45. Accessed January 23, 2010.

Science*Daily*. "Mutant Gene Lessens Devastation of Flesh-Eating Bacteria." January 3, 2010. Available online. URL: www.sciencedaily.com/releases/2009/12/091231111725.htm. Accessed January 23, 2010.

Todar, Kenneth. "*Streptococcus pyogenes* and Streptococcal Disease." In *Todar's Online Textbook of Bacteriology.* Madison: University of Wisconsin, 2008. Available online. URL: www.textbookofbacteriology.net/streptococcus.html. Accessed January 24, 2010.

Streptomyces

Streptomyces is a bacterial genus native to soil belonging to a general group known as the *actinobacteria,* characterized by colonies that develop branching filaments, or hyphae, that make these bacteria resemble molds. *Streptomyces* forms an extensively branched network of hyphae called mycelia that give colonies a lichenlike or leathery appearance. Cells grow from 0.5 to 2.0 micrometers (μm) in length and line up in chains to form the hyphae. Hyphae that extend upward from a surface begin to divide in a single plane to form chains of three to 50 nonmotile conidiospores. When a conidiospore breaks off from the main structure, it germinates to form a new colony. Because the bacteria are strict aerobes, they must live at or near the surface of soils or exposed to air when grown on laboratory medium.

Streptomyces contains about 500 species and belongs to family Streptomycetaceae, suborder Streptomycineae, order Actinomycetales, class Actinobacteria, and phylum Actinobacteria. *Streptomyces* may be the best-known genus in this category of bacteria, also called the actinomycetes, because of a variety of compounds *Streptomyces* produces that are useful in industrial microbiology.

Streptomyces uses chemoheterotrophy as its form of metabolism; it uses a variety of organic compounds for carbon and energy. The bacteria use a wider variety of compounds for nutrition than many other bacteria in nature. Species can grow on glucose, lactate, starch, or chitin, a difficult-to-digest polysaccharide in the shells of insects and crustaceans.

STREPTOMYCES IN THE ENVIRONMENT

Streptomyces produces numerous extracellular enzymes that help degrade organic matter in soil. The genus's main enzyme groups are proteases for digesting proteins, amylases for starch, cellulases for cellulose, and chitinases for chitin. Some species also digest fibers such as pectin, lignin, latex, and keratin, in addition to aromatic compounds. In some soils, *Streptomyces* makes up 20 percent of the culturable microorganisms in the environment; thus, it is one of the most active saprophytes in soil. Saprophytes are bacteria or fungi that serve ecology by decomposing dead organic matter. The bacterium can degrade many of these compounds completely to carbon dioxide, water, and salts. Decomposition of complex molecules to the simplest possible end products is called *mineralization*. Because of the microorganism's capacity to degrade a variety of chemical structures, *Streptomyces* may also make important contributions to removal of pollutant from contaminated soils.

The overall significance of *Streptomyces* in ecology relates to its role as a decomposer, which is a microorganism that decomposes organic matter to simple compounds that allow nutrients to recycle into the biosphere. The bacterium, thus, helps carry out nonspecific steps in various biogeochemical cycles, which recycle the earth's nutrients through different chemical forms in atmosphere, soils and waters, and living things. Upon decomposition, the nutrients return to the soil or to the atmosphere and enable the cycle to continue.

Streptomyces is so dominant in some soils that its presence can be detected by the odor of a volatile nontoxic compound called *geosmin*. Geosmin gives soil its characteristic earthy odor. Water treatment plants consider this compound to be a nuisance, however, because it gives drinking water an earthy or musty odor that ruins its flavor.

STREPTOMYCES IN INDUSTRY

Streptomyces produces about two thirds of the world's antibiotics, plus a spectrum of other commercially important products. In 2001, microbiologists at Abasaheb Garware College in India developed a mathematical model to estimate the number of antibiotics produced by *Streptomyces* based on the number of antibiotics already discovered from this genus. The lead scientist, Milind Watve, wrote in *Archives of Microbiology*, "The number of antimicrobial compounds reported from the species of this genus per year increased almost exponentially for about two decades, followed by a steady rise to reach a peak in the 1970s. . . . the model estimated the total number of antimicrobial compounds this genus is capable of producing to be of the order of 100,000—a tiny fraction of which has been unearthed so far." This estimate, at first glance, may seem improbable, but the number of *Streptomyces* in existence probably exceeds the number already discovered 100- to 1,000-fold, if not more. More bacteria live in soil than in any other habitat in Earth.

The microbiologist Flavia Marinelli of the University of Insubria in Italy recounted the rise of *Streptomyces* antibiotics, in 2009: "The majority of antibacterial and antifungal agents in clinical use nowadays were discovered during the 'Golden Age' of antibiotics in the 1940s–1960s through massive isolation and screening of soil actinomycetes and fungi. It has been estimated that almost 12,000 bioactive secondary metabolites were discovered during that time." *Streptomyces* accounted for about 55 percent of the antibiotics discovered in that period.

The table below highlights the main *Streptomyces* antibiotics used commercially today in human and veterinary medicine.

Because *Streptomyces* is easy to grow in laboratories and is a prolific producer of biologically active compounds, this genus holds the potential for future compounds for use in medicine and industry. The bacteria of this genus currently make more chemical

Streptomyces Antibiotics

Antibiotic	Species	Activity
amphotericin B	*S. nodosus*	for mycoses and pathogenic protozoa
chloramphenicol clavulanic acid	*S. venezuelae* *S. clavuligerus*	bacteriostatic and inhibits some fungi active against penicillin-resistant bacteria
erythromycin	*S. erythreus*	kills susceptible group A *Streptococcus*
neomycin	*S. fradiae*	gram-positive and gram-negative bacteria, including streptomycin-resistant *Mycobacterium tuberculosis*
nystatin	*S. noursei*	for topical treatment of superficial (skin surface) mycoses
streptomycin	*S. griseus*	active against *M. tuberculosis*, some staphylococci, *Brucella*, and some gram-negative bacteria
tetracyclines	*S. aureofaciens*, *S. rimosus*	*Brucella*, *Chlamydia*, *Mycoplasma*, *Rickettsia*, and *Vibrio* bacteria and fungal plant pathogens

products than any other commercially viable microorganism. The potential of this bacterium for making medical and industrial products explains why it is one of the most-studied organisms in microbiology.

See also ANTIBIOTIC; BIOGEOCHEMICAL CYCLES; FUNGUS; HYBRIDIZATION; SOIL MICROBIOLOGY; SYMBIOSIS.

Further Reading

Angier, Natalie. "Ant and Its Fungus Are Ancient Cohabitants." *New York Times,* 13 December 1994. Available online. URL: www.nytimes.com/1994/12/13/science/ant-and-its-fungus-are-ancient-cohabitants.html?scp=1&sq=ants&st=nyt. Accessed January 24, 2010.

Chu, Wai Lang. "Researchers Use Bacteria to Make Anti-Cancer Drugs." November 1, 2006. Available online. URL: www.drugresearcher.com/Emerging-targets/Researchers-use-bacteria-to-make-anti-cancer-drugs. Accessed January 24, 2010.

Lin, Xiuping, Ying Wen, Meng Li, Zhi Chen, Jia Guo, Yuan Song, and Jilun Yi. "A New Strain of *Streptomyces avermitilis* Produces High Yield of Oligomycin A with Potent Anti-Tumor Activity on Human Cancer Cell Lines In Vitro." *Applied Microbiology and Biotechnology* 81 (2009): 839–845. Available online. URL: www.springerlink.com/content/d055078p76h5n87q/fulltext.pdf. Accessed January 24, 2010.

Marinelli, Flavia. "Antibiotics and *Streptomyces:* The Future of Antibiotic Discovery." *Microbiology Today,* February 2009. Available online. URL: www.sgm.ac.uk/pubs/micro_today/pdf/020903.pdf. Accessed January 2, 2010.

Swansea University. "*Streptomyces* Genetics." Available online. URL: www.swan.ac.uk/ils/Research/BioMed/MicrobiologyandInfection/StreptomycesGenetics. Accessed January 24, 2010.

Watve, Milind, R. Tickoo, Maithili Jog, and B. D. Bhole. "How Many Antibiotics Are Produced by the Genus *Streptomyces?*" *Archives of Microbiology* 176 (2001): 386–390. Available online. URL: www.ncbi.nlm.nih.gov/pubmed/11702082. Accessed January 24, 2010.

susceptibility testing Susceptibility testing is the determination of whether an antimicrobial substance, usually an antibiotic, kills bacterial, fungal, or viral pathogens. Testing an antibiotic's effectiveness against a pathogen is crucial in medical care. By determining whether a pathogen is susceptible or resistant to an antibiotic, a clinical microbiologist can advise a physician on the right choices in prescribing drugs to fight infection or disease.

Susceptibility testing begins with proper specimen collection from a patient so that the specimen contains the pathogen but does not contain contaminants. Types of specimens collected in medical care are blood, cerebrospinal fluid, fecal, oral-tracheal, pus, sputum, tissue, and urine. A clinical microbiologist uses specific culture techniques to isolate a suspected bacterial or fungal pathogen from specimens. A clinical microbiologist, first, selects growth medium that will aid in recovering the pathogen from the specimen. This process becomes especially important when testing samples high in a heterogeneous population of microorganisms, such as a fecal sample. Enrichment medium is used to induce the growth of one type of microorganism over others. Selective medium differentiate one group of microorganisms from other groups. After the clinical microbiologist recovers the presumed pathogen in pure form, the organism is known as a *clinical isolate*.

Clinical microbiologists endeavor to identify the pathogen by using rapid biochemical tests and genetic testing. Some pathogens need no further testing, once they have been identified. For example, salmonellosis is a food-borne illness known to be caused by *Salmonella* bacteria. Standard treatments exist for this disease. In other cases, a clinical microbiologist cannot predict the behavior of a clinical isolate in relation to various drugs without performing susceptibility testing.

A physician often orders a clinical microbiology laboratory to do susceptibility testing for either of the two following reasons: (1) The pathogen has not responded to other treatments, or (2) more than one pathogen has been isolated.

SUSCEPTIBILITY TESTS

Regardless of the reason a physician requires a susceptibility test, speed and accuracy are of equal importance. Clinical microbiologists must produce fast test results because the patient's health depends on rapid treatment. But physicians also need accurate results in order to prescribe an effective treatment.

Clinical microbiologists use three types of bacterial and fungal susceptibility testing: (1) qualitative, (2) quantitative (sometimes called *semiquantitative*), and (3) nucleic acid analysis. These tests are listed in order of increasing complexity and costs.

Qualitative testing indicates antibiotics likely to work against pathogens and others unlikely to work. Most qualitative testing involves agar diffusion tests, in which an antibiotic diffuses from a point into the surrounding agar. When the antibiotic contacts bacterial growth in the agar, its efficacy becomes visible by a zone of clearing in the cloudy agar. The clearing results from lysed bacterial cells.

Qualitative testing cannot indicate the relative strength of antibiotics. While the zone of clearing indicates antibiotic activity, it might also mean that the antibiotic diffuses well through agar. Therefore, the size of a zone of clearing does not necessarily correspond to the antibiotic's strength.

Quantitative tests such as minimum inhibitory concentration (MIC) estimate the relative strength of antibiotics in inhibiting the growth of a common pathogen. A related test, called the *minimum bactericidal test* (MBC), also estimates the relative strength. The two tests provide the following information:

- MIC—minimal concentration of antibiotic needed to inhibit or kill a pathogen in broth or agar medium
- MBC—minimal concentration of antibiotic needed to kill a pathogen in broth or agar medium

MIC and MBC results equate antibiotic strength to a value in milligrams per milliliter (mg/ml). Agar-based MIC tests such as the Kirby-Bauer test also give an MIC value read off a calibrated paper strip.

Nucleic acid analysis involves probes to detect resistance genes in the pathogen's deoxyribonucleic acid (DNA) or ribonucleic acid (RNA). The probe carries a single strand of DNA or RNA containing a base sequence known to correlate with the resistance gene's sequence. When the probe finds a resistance gene, it gives a visible signal.

Microbiologists use nucleic acid testing to determine the presence of antibiotic resistance without interference from factors in growth medium. The presence of resistance genes is proof enough that resistance exists, but the results are not quantitative. Nucleic acid testing does not estimate the relative strength of resistance. The testing also does not account for other factors in the pathogen's physiology that build susceptibility or resistance.

SUSCEPTIBILITY RESULTS

The goal of susceptibility testing is to find the most effective antibiotic for killing a pathogen that has infected a patient. Qualitative and quantitative tests produce one of the three following results:

- susceptible—the antibiotic has a high probability of killing the pathogen in the body
- intermediate—the antibiotic has some activity against the pathogen but may require higher or more frequent doses
- resistant—the antibiotic does not kill the pathogen in in vitro testing

A pathogen can be susceptible in qualitative or quantitative tests, but this finding provides little guarantee that the antibiotic will kill the pathogen in the body. Most pathogens express one or more virulence factors inside the body. Virulence factors make the pathogen better at neutralizing the body's defenses so it can further a disease. Protective capsules or blood-clotting substances, or the ability to invade host cells, helps pathogens evade the body's immune system but also provides security from antibiotics. The table below describes how susceptibility relates to an antibiotic's activity in the body.

Clinical microbiologists have defined break points for interpreting the susceptibility results shown in the table. A break point is a set of conditions that determines whether a result should be interpreted as susceptible, intermediate, or resistant. The British medical microbiologists Alasdair MacGowan and Richard Wise have cautioned, "The setting of clinical breakpoints is a controversial subject and the focus of much debate among infection specialists, [government] regulators and industry. In the past this process was seen as arbitarary and lacking in consistency. In the last decade, an increasing knowledge of pharmacodynamics has allowed more rationality to be introduced into discussion about breakpoints." Pharmacodynamics is the study of how drugs work in the body or against microbial pathogens. Only by studying the effects of antibiotics on people with an infectious disease can scientists determine the in vivo susceptibility of pathogens to an antibiotic.

VIRUS SUSCEPTIBILITY TESTING

Clinical microbiologists use different methods for testing susceptibility in viruses than for bacteria and fungi. The infectious disease specialist Pablo Tebas

Susceptibility Test Interpretations

Result	Factors Affecting Susceptibility
susceptible	host factors, virulence factors, site of infection, presence of antibiotic-repelling biofilms, toxin production by the pathogen (not affected by antibiotics)
intermediate	higher antibiotic doses or frequency may be effective at certain infection sites or tissues; virulence factors, biofilms, or toxin production may reduce effectiveness
resistant	qualitative and quantitative testing can be imprecise and underestimate a pathogen's susceptibility to antibiotic

explained the drawbacks of viral testing as late as the mid-1990s, saying, "Antiviral susceptibility testing is not routinely performed by most clinical virology laboratories. This omission is in large part because the most widely accepted method, the plaque reduction assay (PRA), is cumbersome to perform and results are rarely available in time to influence treatment." Virus testing also requires that a microbiologist be trained in the special techniques required for keeping viruses active in cell tissue cultures.

Newer tests use antigen-antibody reactions linked to a compound that produces a color reaction based on the presence or absence of active virus. A microbiologist exposes the virus to different levels of antiviral drug and, then, inoculates the treated virus suspensions to human cell tissue. Microbiologists select specific human cell types that lyse (break apart) if infected with the active virus. The virologist Rožena Stránská described such an assay, developed in 2004: "A yield reduction assay is set up in which virus is inoculated on human fibroblasts in the presence of antiviral drug. Subsequently, reporter ELVIRA cells . . . are added. The β-galactosidase [enzyme] activity in the cell lysates [lysed cell suspension] reflects the number of infected reporter cells and, thereby, the yield of infectious virus after drug action." In this study, *ELVIRA* stands for *enzyme-linked virus inhibitor reporter assay*. Although tests similar to this ELVIRA method indicate a virus's susceptibility to a drug, they are more complicated and labor-intensive to run than bacteria tests.

Susceptibility testing will remain a crucial aspect of medical care when patients need prompt treatment for an infectious disease. For this reason, microbiologists continue devising faster methods to determine the drug susceptibility of pathogens.

See also ANTIBIOTIC; CLINICAL ISOLATE; DIFFUSION; MEDIA; MINIMUM INHIBITORY CONCENTRATION; PLAQUE; SPECIMEN COLLECTION; VIRULENCE.

Further Reading

British Society for Antimicrobial Chemotherapy. "Susceptibility Testing Guide." Available online. URL: www.bsac.org.uk/susceptibility_testing/guide_to_antimicrobial_susceptibility_testing.cfm. Accessed December 20, 2009.

MacGowan, Alasdair P., and Richard Wise. "Establishing MIC Breakpoints and the Interpretation of in Vitro Susceptibility Tests." *Journal of Antimicrobial Chemotherapy* 48, Suppl. S1 (2001): 17–28. Available online. URL: http://jac.oxfordjournals.org/cgi/reprint/48/suppl_1/17. Accessed January 16, 2010.

Merck Manuals Online Medical Library. "Susceptibility Testing." Available online. URL: www.merck.com/mmpe/sec14/ch168/ch168d.html. Accessed December 20, 2009.

Stránská, Rožena, Rob Schuurman, David R. Scholl, Joseph A Jollick, Carl J. Shaw, Caroline Loef, Merjo Polman, and Anton M. van Loon. *Antimicrobial Agents and Chemotherapy* 49 (2004): 2,331–2,333. Available online. URL: http://aac.asm.org/cgi/reprint/48/6/2331. Accessed January 16, 2010.

Tebas, Pablo, Erik C. Stabell, and Paul D. Olivo. "Antiviral Susceptibility Testing with a Cell Line Which Expresses β-Galactosidase after Infection with Herpes Simplex Virus." *Antimicrobial Agents and Chemotherapy* 39 (1995): 1,287–1,291. Available online. URL: www.ncbi.nlm.nih.gov/pmc/articles/PMC162728/pdf/391287.pdf. Accessed January 16, 2010.

University of Pennsylvania Medical Center. "Antimicrobial Susceptibility Testing: What Does It Mean?" Available online. URL: www.uphs.upenn.edu/bugdrug/antibiotic_manual/amt.html. Accessed December 20, 2009.

symbiosis Symbiosis is a stable condition in which two or more organisms live in close physical association. The individual organisms in the association are called *symbionts*.

In recent years, a symbiotic relationship is usually thought of as an association beneficial to each symbiont. Symbiosis actually has a broader definition, which covers relationships that range from harmful to beneficial for the symbionts.

The principal symbiotic relationships seen in microbiology are the following:

- commensalism—one organism benefits and the other is unaffected
- mutualism—both organisms benefit
- neutralism—neither organism is affected
- opportunism—one organism changes from a commensal, mutual, or neutral organism into a harmful organism
- amensalism—one organism produces a substance that has a harmful effect on the other
- parasitism—one organism benefits at the expense of the other

Additional relationships that derive from these forms of symbiosis undoubtedly occur in nature. In many known cases of symbiosis, microbiologists have not discovered all of the interactions among organisms.

Antagonism (also called *competition*) represents the opposite of symbiosis, specifically mutualism. In antagonism, the symbionts compete with each other and do not benefit by living in close association.

In each form of symbiosis, one member fulfills either an optional or an obligatory role. An optional symbiont can live outside the symbiotic relationship; an obligatory symbiont must be in a symbiotic relationship to survive.

Some symbiotic relationships persist, from generation to generation, without dissolving. Symbioses can persist in at least two ways. In the first method, both members synchronize reproduction, while they are together, so that new generations are about the same age. In the second method, the symbionts form an intermediate structure called a propagule. The propagule establishes in a new location and the symbionts resume their coexistence.

Many unknowns exist regarding the degree to which symbionts share metabolism, nutrients, end products, protective mechanisms, or genes. The only requirements for symbiosis are that the association is stable and that it persists over generations. Stable associations continue even when stressed by factors in the environment such as limited nutrients or water. Symbiosis in microbiology does not require that one or both organisms belong to the same or different domains. In other words, symbiotic relationships exist in the following configurations: prokaryote-prokaryote, prokaryote-eukaryote, eukaryote-eukaryote, archaean-prokaryote, and archaean-eukaryote.

Symbiosis between different archaea species is thought to be very rare. In 2008, the molecular biologist Mircea Podar reported in *Genome Biology,* "The relationship between the hyperthermophiles *Ignicoccus hospitalis* and *Nanoarchaeum equitans* is the only known example of a specific association between two species of Archaea. Little is known about the mechanisms that enable this relationship." Podar's research team showed that the two archaea pass genes between them, a process known as lateral gene transfer.

COMMENSALISM, MUTUALISM, AND NEUTRALISM

In commensalism, one symbiont, often called the *commensal,* derives a benefit from the other symbiont, often called the *host.* Using these terms, the host is the symbiont unaffected by the relationship. Because one symbiont does not receive a benefit or become harmed, commensalism is called a *unidirectional* process.

Commensalism exists in a diversity of forms. The following list describes some of the benefits that commensals might receive from a host or another symbiont:

- commensal gets nutrients captured by the host
- commensal uses the waste products of the other organism

Root nodules of legume plants contain *Rhizobium* and *Bradyrhizobium* in a symbiotic relationship with the plant. These bacteria play a pivotal part in the nitrogen cycle. ***(R. Ford Dennison)***

- host provides shelter for the commensal
- host makes the habitat favorable for the commensal's growth
- products from one symbiont protect the other
- one symbiont colonizes a surface, making it suitable for the other symbiont's growth

Usually, a commensal organism derives no more than one or two of the benefits listed here.

In mutualism, both organisms benefit, often by deriving commensalism benefits listed. Mutualism is a bidirectional process. A well-known example of mutualism occurs between legume plants (beans, peas, peanuts, alfalfa, etc.) and the nitrogen-metabolizing bacteria that infect nodules on the plants' roots. Bacteria of the genus *Rhizobium* benefit the plant by converting nitrogen gas to ammonia. Plants do not have the ability to carry out this process, called *nitrogen fixation,* so they depend on the bacteria to provide the nitrogen in a usable chemical form. *Rhizobium* benefits by occupying a habitat with little competition from other microorganisms.

Mutualism plays an important role in ecology by giving each symbiont a capability or benefit that it would not enjoy if living alone. In this perspective, mutualism may play a central role in the evolution of organisms that enter such a symbiotic relationship, at some point in their lives. The bacteria living in the digestive tract of ruminant animals, some insects, and some worms digest dietary materials for the host animal as the host provides a nutrient-rich habitat for the bacteria. Ruminant animals need a fibrous diet to survive, but the animal digests none of the fiber. A large population of bacteria and protozoa in the rumen carries out the fiber digestion and releases fatty acids that the ruminant depends on for energy. A ruminant's metabolism has evolved, in fact, to use the end products of metabolism that microorganisms produce in the rumen. The bacteria also supply amino acids and vitamins to the host animal.

Insect-microorganism relationships are among the earliest known symbioses. The American entomologist Augustus Imms (1880–1949) first discovered that termites possessed a system similar to that in ruminants by depending on protozoa for part of their nutrition. Termites and wood-eating cockroaches cannot digest woody fibers, but the protozoa inside the insect gut can. In termites, the protozoa *Trichomonas, Trichonympha,* and *Streblomastix* decompose wood's cellulose fibers to sugars the insect can use.

What makes the termite system unusual is the dependence by the protozoa on smaller symbionts to break down cellulose. Bacteria inside the protozoa help digest the wood that termites ingest. The protozoa harbor a second population of bacteria that fit into grooves running along the outside of the eukaryote's cell. The rod-shaped bacteria synchronize the beating of their flagella to create waves that enable the protozoa to swim in search of the next meal. Without bacterial symbionts, the protozoa would be of little service to the termite host.

The termite microenvironment provides examples of endo- and ectosymbionts. The bacteria living inside the protozoa and the protozoa living inside the termite are called endosymbionts because they live inside the host. The bacteria that live on the protozoa's outer surface are ectosymbionts. The protozoa-ectosymbiont also gives an example of phoresis, which is a situation in which one organism uses another for moving.

Other insect-microorganism relationships exist in nature. For example, aphids depend on the bacterium *Buchnera aphidicola* to supply amino acids that plants lack. Termites depend on bacteria aside from the species that participate in wood digestion. The termite hosts nitrogen-fixing bacteria that absorb nitrogen gas from the air and supply this nutrient to the insect in a usable form.

Neutralism is an association that is not positive (a symbiont gains a benefit) or negative (a symbiont is injured). Two symbionts live in close proximity but have very little or no effect on each other. Bacteria that make up biofilm build neutralism. Thick biofilm is a community of diverse microorganisms attached to a submerged surface. Mature biofilm grows so thick that bacteria in the inner layers never have contact with the bacteria at the surface layer, yet they are connected by the biofilm matrix of polysaccharides. Each symbiont participates in building the biofilm, but it would be difficult to describe a direct benefit received by either.

The more microbiologists learn about neutralism, the more likely it is that they will reclassify the association to another category of symbiosis. The biofilm symbionts, for instance, receive an indirect benefit from the protective, nutrient-storing biofilm community. This viewpoint suggests that biofilms are mutualistic communities.

OPPORTUNISM, AMENSALISM, AND PARASITISM

Opportunism, amensalism, and parasitism are all unidirectional negative processes. In opportunism and parasitism, a host can be identified: It is the organism in which the other symbiont lives. Amensalism lacks a true host.

Opportunism is a relationship that begins as positive and, then, becomes negative when circumstances change. The microorganism that makes the transition from harmless to harmful is called an *opportunistic pathogen. Staphylococcus aureus* lives as a normal inhabitant on human skin. As part of a person's normal flora, *S. aureus* is harmless and may even provide a benefit by preventing pathogens from inhabiting the skin. *S. aureus* can cause infection, however, if the person sustains an injury to the skin. In this situation, the bacteria can infect the wound and, if left unattended, cause a more serious systemic infection. By taking the opportunity to infect, *S. aureus* becomes an opportunistic pathogen.

Amensalism occurs when one symbiont produces a substance that injures the other symbiont. The following activities are examples of amensalism:

- antibiotic secretion
- bacteriocin secretion (bacteriocins are antibiotics that inhibit closely related microorganisms)
- harmful waste products such as acids

Milk souring involves amensalism when various types of bacteria succeed each other. The bacteria that survive milk pasteurization begin to grow on milk's nutrients, and their population size increases quickly—the population expansion takes days in refrigerated milk but occurs overnight in unrefrigerated milk. The first population of bacteria curdles proteins and breaks down lactose to the sugars glucose and galactose. A second wave of bacteria metabolizes the sugars and produces lactic acid as an end product of this metabolism. The accumulation of lactic acid causes the pH in the milk to drop (causing more curdling), and that, in turn, inhibits the growth of other bacteria, including the first-generation species that first acted on the proteins.

Escherichia coli that resides in the intestines creates a reverse type of amensalism by consuming a molecule instead of producing a harmful substance. *E. coli* is a facultative anaerobe, meaning it uses oxygen, but it can switch metabolism and survive in anaerobic conditions. *E. coli* is one of the first bacteria to colonize a newborn's digestive tract. After *E. coli* uses up all the oxygen that might be present in the intestines, the anaerobic conditions favor the growth of strict anaerobic bacteria. Eventually, the anaerobes outnumber *E. coli* by at least 1,000 to one. Meanwhile, *E. coli*'s ability to consume oxygen makes the newborn's digestive tract inhospitable to bacteria that the baby takes in by putting toys, hands, and various other objects in the mouth. By removing oxygen, *E. coli* inhibits bacteria living in close proximity.

Parasitism receives considerable attention and research because it is a component of infectious disease. In parasitism, a symbiont acts as a pathogen and thrives in or on the host to the detriment of the host.

Parasitism is a complex process that involves the parasite's ability to infect (infectivity), capacity to cause disease (pathogenicity), and virulence (capacity to cause severe disease). The health condition of the host also determines whether a parasite can establish a population in the host. Host factors that repel parasitism consist of physical barriers and the actions of the immune system.

Bacteria, fungi, yeasts, protozoa, and viruses have the ability to develop parasitism in human hosts. Bacteria, fungi, and viruses infect plants. Parasitism also reigns among microorganisms. Bacteria can parasitize other bacteria or protozoa, and viruses called bacteriophages infect bacteria.

Parasitologists have discovered specialized types of parasitism that distinguish some parasites from others. The following types of parasitism can be found in host-microorganism unions:

- facultative—prefers to infect a host but can live independently, too
- permanent—lives with host for a long period, perhaps throughout the host's life
- temporary—lives with host for a short period
- periodic—infects a host on occasion with no set pattern
- latent—while living in the host, does not cause any detectable signs

Predation in which one microorganism attacks and ingests another can be viewed as a derivative of symbiosis. The predator receives an obvious advantage, and the prey loses its life. Predation should not be included with classic examples of symbiosis because it fails to meet one of the criteria for a symbiotic relationship: Predation cannot persist. A predator must continually find new prey, so it never establishes an association with prey as a symbiont. Predation by protozoa on bacteria is common in aqueous environments.

Although many details of symbiosis have not yet been uncovered, the study of all types of symbiosis promises to help science build a picture of how organisms evolved together. Symbiosis studies may soon provide information on the evolution of metabolic pathways, as well.

See also BIOFILM; LICHEN; NORMAL FLORA; OPPORTUNISTIC PATHOGEN; RUMEN MICROBIOLOGY.

Further Reading

Podar, Mircea et al. "A Genomic Analysis of the Archaeal System *Ignicoccus hospitalis–Nanoarchaeum equitans.*" *Genome Biology* 9 (2008): R158.1–R158.18. Available online. URL: http://genomebiology.com/content/pdf/gb-2008-9-11-r158.pdf. Accessed January 4, 2010.

Science*Daily*. "Evolution of Symbiosis." April 11, 2007. Available online. URL: www.sciencedaily.com/releases/2007/04/070410084136.htm. Accessed January 4, 2010.

———. "Symbiosis: Bacterial Gut Symbionts Are Tightly Linked with the Evolution of Herbivory in Ants." December 2, 2009. Available online. URL: www.sciencedaily.com/releases/2009/12/091201132431.htm. Accessed January 4, 2010.

systematics Systematics is the study of the evolutionary history of organisms. Systematics relies on a related science called *taxonomy,* which uses scientific tests to group organisms into categories called *taxa.* Taxa show the similarities among organisms, and systematics uses these similarities to build a hierarchy of biota that shows organisms' evolutionary paths. Such a hierarchy is called *phylogeny,* or a phylogenetic hierarchy.

Seventh-century botanists first attempted to classify plants in a logical manner based on morphology. The English naturalist John Ray (1627–1705) created, perhaps, the most detailed categorization of plants and animals. In 1686, Ray published *Historia Plantarum,* a massive set of volumes describing his plant classifications. *Historia Plantarum* served as a blueprint, several years later, for the Swedish botanist Carl Linnaeus (1707–78). Although Linnaeus also developed a classification system, he invented an equally important aid to classification: a standard naming system for every plant in every category. His systematics book, *Species Plantarum,* organized plants from general groups based on reproductive features all the way to detailed genus and species. Systematics continues to use the genus and species system of giving each organism a unique name, a system called *binomial nomenclature.* Thus, a species of bacteria is called *Staphylococcus aureus,* a species of protozoa is *Paramecium caudatum,* and so on. The binomial genus-species name is always italicized.

Microbial classifications followed the same general scheme that had been devised for plants by including physical and biochemical features. In 1872, the German botanist and bacteriologist Ferdinand Cohn (1828–98) collected the notes he had amassed on studies of bacteria and set about categorizing them into genera and species. His book *Untersuchungen über Bacterien* organized bacteria as previous scientists had plants and animals. But microorganisms contain many submicroscopic features, which make them a difficult study. Scientists struggled with sorting out the groups to which species would be assigned. Over two centuries, the phylogenetic hierarchy changed from two kingdoms consisting of plants and animals to three domains, which also received continual rearrangement. In 1977, the American biologist Carl R. Woese (1928–) revolutionized the traditional classification methods by using analysis of ribosomal ribonucleic acid (rRNA). For the first time, microorganisms would be categorized on the basis of genetic makeup.

A phylogenetic hierarchy assumes that all of biota began with a common ancestor. Woese's studies of rRNA have thrown some doubt on that long-held theory, however. His hierarchy suggested that bacteria and archaea had evolved along different paths early in the evolution of higher organisms. Prior to Woese's 1977 study, dividing bacteria, archaea, and eukaryotes into three domains, archaea were thought to be ancestors of bacteria. Biology still refers to a common ancestor without definitive proof that some organisms evolved by entirely different paths. Phylogenetic trees, thus, remain useful for studying the relationships among species and diversity. At present, biology operates on the assumption that bacteria and archaea split early in evolution, and eukaryotes developed later. For this reason, diagrams of phylogenetic hierarchies are often called evolutionary trees.

Before genetic testing, systematics used fossil records of extinct plants and animals and compared them to known living organisms. A strong resemblance to a fossil indicated that a modern organism had a long evolution. Fossil records are incomplete, however, and can be used for only a fraction of the species that have ever existed. Fossils embedded in rock also may become distorted over the long period between their origin and the discovery. Systematics, today, uses many of the same technologies taxonomists use to classify organisms: DNA and RNA analysis, morphology, and physiology.

Modern systematics combines information from fossil records with methods adapted from taxonomy to accomplish the following five main tasks:

- confirm the accuracy of fossil records
- learn about the evolution of complex features, such as flagella
- gain insight into poorly understood species
- determine the order of species in evolution
- study the evolution of diversity

For these tasks, systematic relies on the theory of parsimony, which states that scientists should choose the simplest scientific explanation to fit the evidence. In constructing an evolutionary tree, those with fewer branches needed to represent species are more likely to be correct compared with complex trees with excess branches.

PROKARYOTIC PHYLOGENY

The group of organisms known as prokaryotes consists of bacteria and archaea. These microorganisms differ from eukaryotes in several morphological features, the main being that prokaryotes lack membrane-enclosed organelles. Prokaryotic genetic material is loose in the cell cytoplasm and not confined in a distinct nucleus, as is the eukaryotic genome. (A genome is the entire amount of genetic material in a cell.)

Prokaryotic diversity derives from the differences between bacteria and archaea. Although these two types of prokaryotes can behave alike in nature, they differ in several morphological and physiological features, such as the following:

- Archaea have pseudomurein cell walls; bacteria have peptidoglycan.
- Archaea contain longer lipid chains in their membranes than bacteria.
- The archaean genome is often smaller than the bacterial genome.
- Archaean protein synthesis uses elements from both bacteria and eukaryotes.
- Archaea often live in extreme environments of temperature, salt concentration, or acid conditions.
- Archaea contain ribosomes that are unlike bacterial or eukaryotic ribosomes.

In addition to the differences between bacteria and archaea listed here, molecular studies are revealing that transfer RNA (tRNA) composition and proteins contain sequences more like those of eukaryote than of bacteria. Transfer RNA is an RNA molecule that delivers the correct amino acid during protein synthesis, according to the cell's genetic code.

BERGEY'S MANUAL OF SYSTEMATIC BACTERIOLOGY

In 1923, five members of the Society of American Bacteriologists (now the American Society for Microbiology) published a manual that explained the classification and names of all known bacteria. The manuals' editorial board consisted of David H. Bergey (1860–1937), Francis C. Harrison, Robert S. Breed, Bernard W. Hammer, and Frank M. Huntoon. Bergey took lead editorial control and published updates, in 1925 and 1930. Now known as *Bergey's Manual of Systematic Bacteriology,* this five-volume book provides the most comprehensive classification of bacteria in the world.

In 1934, Bergey instituted a trust to which he transferred ownership of the manual. Upon turning over control of the manual to the trust, Bergey wrote in the *Journal of the American Medical Association,* "Whether bacteriologists agree or not with the classification and nomenclature of organisms as published in this manual, they accept the work as an effort in the right direction." *Bergey's Manual* continues to serve as a primary resource in laboratories that identify and characterize new bacteria or archaea.

Bergey's Manual currently has five volumes by various editors. The table below provides a brief overview of the contents of each.

The manual describes the known morphological features of species, as well as their metabolism, end products, and nutrient requirements. Genetic analysis data have been included since the earliest editions, but the manual remains primarily a compendium of species phenotypes, the outward display of information carried by genes. Most entries also recommend growth media for the microorganism and any special growth requirements; carbon dioxide gas, light, and so on.

The trust representative Noel Krieg wrote in the newsletter, in 2009, "[James Staley] and other editors decided to ask the most knowledgeable authorities on the various taxa to prepare the individual chapters, thus allowing the *Manual* to become a truly international project."

MICROORGANISM NAMES

Microbiologists and other biologists continue to use the binomial nomenclature devised by Carl Linnaeus. The names of microorganisms appear in Latin, unlike higher organisms that scientists and nonscientists identify by common names as well as the Latin name. Naming of prokaryotes follows a number of conventions regarding the derivation of a species name. The basic requirement is that no other organism in biology has the same name. The table on page 749 gives examples of common and acceptable naming conventions.

The Latin names from the class level to genus level also follow certain conventions. Most but not all microorganisms follow this scheme of suffixes (examples for phylum Chloroflexi in parentheses):

- class Chloroflexi
- order suffix: *-ales* (Chloroflexales)
- family suffix: *-aceae* (Chloroflexaceae)
- genus *Chloroflexus*

Additional descriptive terms as part of the name give information on many microorganisms, such as the following:

- suffix *-coccus,* round cells
- suffix *-bacillus,* rod-shaped cells

Bergey's Manual of Systematic Bacteriology

Volume	Main Features	Examples
1	complete classification of archaea; 11 phyla of bacteria, including photosynthetic	methanogens, halophiles, thermophiles, cyanobacteria
2	covers the five classes of proteobacteria bacteria	enteric bacteria, nitrogen-fixing
3	low G + C gram-positive bacteria (low guanine and cytosine content)	anaerobes, endospore-forming bacteria
4	high G + C gram-positive bacteria (high guanine and cytosine content)	filamentous bacteria, mycobacteria (no cell walls)
5	nine phyla of bacteria	spirochetes, anaerobes

Bacterial Names

Source of the Name	Example	Description
a historic event	*Legionella pneumophila*	cause of a new disease that occurred at an American Legion convention, in 1976
color	*Cyanobacterium*	blue-green color
cell shape and arrangement	*Streptococcus pyogenes*	long, twisting chains (strepto-) of spherical (-coccus) cells
place of discovery	*Thiomargarita namibiensis*	found off the coast of Namibia
discoverer	*Escherichia coli*	discovered by Theodor Escherich, in 1885
in honor of a famous microbiologist	*Pasteurella multocida*	genus named for Louis Pasteur
unique feature	*Magnetospirillum magnetotacticum*	spiral-shaped bacteria with magnet-containing magnetosomes inside their cells
extreme growing conditions	*Thermus aquaticus*	grows in very hot waters such as hot springs

- suffix *-vibrio*, curved shape
- prefix *spiro-*, spiral-shaped
- *thio* or *sulfo*, sulfur-metabolizing
- *ferro* or *ferrus*, iron-metabolizing
- prefix *rhodo-* or *rubri-*, red
- prefix *cyano-*, blue-green
- prefix *chloro-*, green
- prefix *halo-*, salt-loving
- prefix *thermo-*, heat-loving
- prefix *methyl-*, methane-metabolizing
- prefix *methano-*, methane-producing
- prefix *aceto-*, acetate-producing
- prefix *acido-*, acid-producing

Microbiologists and all of biology depend on systematics to provide an orderly and logical classification to microorganisms. Systematics becomes particularly important when discovering new organisms that have never before been studied.

See also ARCHAEA; BACTERIA; DOMAIN; EUKARYOTE; MORPHOLOGY; PROKARYOTE; PROTEOBACTERIA; TAXONOMY.

Further Reading

Bergey, David H. "Bergey's Manual of Determinative Bacteriology: A Key for the Identification of Organisms of the Class Schizomycetes." *Journal of the American Medical Association* 103 (1934): 437.

Bergey's Manual Trust. Available online. URL: www.bergeys.org/index.html. Accessed June 6, 2010.

Krieg, Noel. "James T. Staley—an Inquiring Mind." *Microbial Taxonomist* 3 (2009): 2–3. Available online. URL: www.bergeys.org/newsletter/microbialtaxonomist_3_1.pdf. Accessed November 30, 2009.

Society of Systematic Biologists. Available online. URL: http://systbio.org. Accessed November 30, 2009.

University of California–Berkeley. "Phylogenetic Systematics, a.k.a. Evolutionary Trees." Available online. URL: http://evolution.berkeley.edu/evolibrary/article/phylogenetics_01. Accessed November 30, 2009.

T

taxonomy Taxonomy is the science of the classification of organisms. By arranging every form of life into taxonomic categories called *taxa,* biologists help prevent confusion when communicating about specific organisms. This is because taxonomy provides a naming system that gives each organism a unique name.

Taxonomy also facilitates the study of biota because taxa are organized according to similarities among organisms. These similarities have historically been based on phenotype, or the physical characteristics of an organism determined by its genes. Newer technologies in studying genes of organisms have already begun to rearrange some taxa as gene analyses reveal heretofore-unknown relationships among organisms.

Taxonomy is often confused with a related field called *systematics.* Systematics groups organisms according to their relationship in evolution. Taxonomy is a classification based on relatedness. This discrete difference makes taxonomy and systematics complementary sciences. Systematics includes the naming of new species according to a scientifically based scheme called a phylogenetic hierarchy. This hierarchy groups organisms by common properties, which imply that related organisms have common evolutionary paths that all originate with a single ancestor cell. A taxomonic hierarchy provides more detail to phylogenetic hierarchy by assigning every organism from large generalized domains to specific groups containing only one member. The following classification scheme for humans provides an example:

- domain—Eukarya
- kingdom—Animalia
- phylum—Chordata
- class—Mammal
- order—Primate
- family—Homoidae
- genus—*Homo*
- species—*sapiens*

New technologies in analyzing the nucleic acids (deoxyribonucleic acid [DNA] and ribonucleic acid [RNA]) or amino acids are replacing the classic methods that were used when microbiology was a young science. Previous to today's molecular analysis methods, microorganisms were assigned to taxa based on the following features:

- cell morphology and staining characteristics
- colony morphology
- enzyme systems
- nutrient requirements
- oxygen requirements

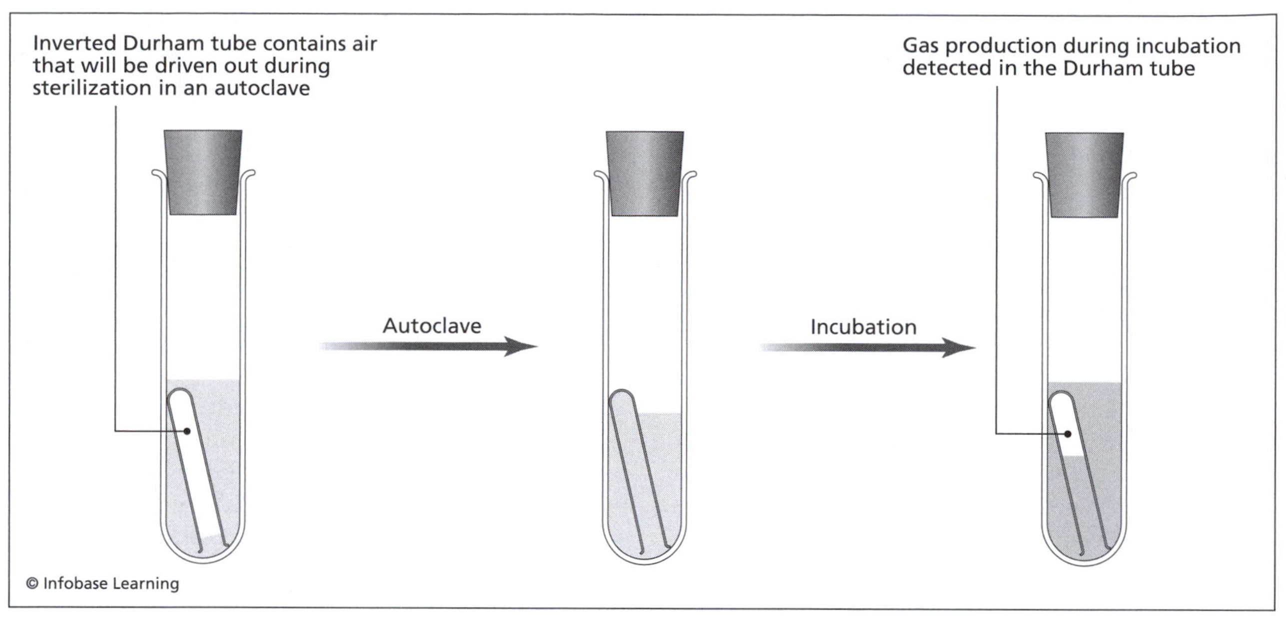

Taxonomy traditionally has used simple tests such as this Durham tube for detecting gas production. Newer taxonomic methods increasingly depend on DNA and rRNA analyses.

- physical structures (endospores, flagella, capsules, etc.)
- fatty acid composition
- serotypes determined by surface molecules

Microbiologists continue to use these tests in identifying new microorganisms, but molecular techniques are believed to provide more accurate classifications based on relatedness of species.

DEVELOPMENT OF ORGANISM CLASSIFICATION

The English naturalist Charles Darwin (1809–82) proposed, in 1859, that natural selection resulted in the similarities and differences among species. In evolution, species kept adaptations that helped them survive and dispensed with characteristics that made individuals vulnerable to their environment. But gross features made the task of differentiating very similar organisms difficult. Is a finch with a thick bill a different species from a finch with a sharp bill, or are these two members of the same species, that is, finch?

Biologists had endeavored to classify organisms since antiquity but discovered the inconsistencies that could arise when unrelated organisms looked similar. In 1735, the Swedish botanist Carl Linnaeus (1707–78) developed a formal system of classifying organisms and naming them. The naming system, now called *binomial nomenclature,* gives every organism a unique two-part name. In other words, no species shares its name with any other species. Using Latin as the language for his naming system, Linnaeus divided biota into two kingdoms: Plantae and Animalia.

A century after Linnaeus's proposed taxonomic hierarchy, the Golden Age of Microbiology commenced, in which biologists delved deeper into the physiology of microorganisms than ever before. The Swiss botanist Carl von Nägeli (1817–91) proposed that bacteria and fungi belonged in the kingdom Plantae with higher plants. With improvement in the sophistication of microscopes, biologists continued examining the intracellular structures of eukaryotic cells. The German biologist and philosopher Ernst Haeckel (1834–1919) considered the associations among organisms laid out by Linnaeus and von Nägeli. He proposed that a new kingdom, Protista, suited bacteria, protozoa, algae, and fungi rather than either kingdom of plants or animals.

Protists would be a difficult group for most biologists to define, particularly since the eukaryotes in this new kingdom (protozoa, algae, and fungi) differed in organization, lifestyle, and size from bacteria. Within the next 100 years, fungi were placed in their own kingdom and returned to the protists, repeatedly. While biology puzzled over the best way to develop a new hierarchy based on additional information microbiology had gathered on these organisms, von Nägeli's scheme remained the classification method preferred by most.

Electron microscopy was introduced in 1937. This dramatic improvement in the power of microscopy enabled microbiologists to see in unprecedented detail the structures that Haeckel had studied by light microscopy. Electron microscopes showed that bacteria had no distinct nucleus, whereas eukaryotes possessed one or more distinct nuclei and other organelles bound by membranes. Such an obvious distinction led the Canadian biologist Roger Stanier (1916–82) to suggest that the simplest cells without membrane-bound nuclear material should belong to the prokaryotes, distinct from the eukaryotic world. In 1968, another Canadian, Robert G. E. Murray (1968–), proposed the new kingdom Prokaryotae. This kingdom would capture the bacteria and bacterialike archaea (yet to be studied). The term *protists* remains in use today as an informal term for unicellular eukaryotes. The American Herbert F. Copeland (1902–68) reclassified members of the protists to provide more accuracy to the hierarchy.

In 1969, the American biologist Robert H. Whitaker (1924–80) expanded the three kingdoms to five, based on cell structure: Plantae, Fungi, Animalia, Protista, and Monera. As a result, biology began reconsidering how the diversity of life should be classified. Scientists would soon develop six-kingdom and eight-kingdom systems. Taxonomy seemed to be becoming more complicated, rather than simplifying the classification of biota, as Linnaeus had, than two centuries earlier.

After electron microscopy, the next great technological step in microbiology belonged to the study of microbial genomes. Methods for extracting DNA from cells improved. Biologists refined their ability to analyze the base composition of DNA and RNA. In 1978, the American biologist Carl R. Woese (1928–) revolutionized taxonomy by using the base sequences from microbial ribosomal RNA (rRNA) as the basis for classification. Up to that point, science had relied on phenotype, the outward expression of an organism's genes, as the main tool for classifications. Woese used the actual genes, called the *genotype,* to discern the relatedness among microorganisms. The molecular biologist Norman R. Pace called Woese's 1977 publication on the new rRNA classification method the "foundation of the modern era of microbiology." The article "Phylogenetic Structure of the Prokaryotic Domain: The Primary Kingdoms" by Carl Woese and George E. Fox described the domains containing eukaryotes, prokaryotes, and archaea.

Woese and Fox wrote, in their 1977 article, "With the identification and characterization of the [prokaryotes] we are for the first time beginning to see the overall phylogenetic structure of the living world." The three-kingdom system remains the basis of taxonomy and uses Linnaeus's original binomial naming system. Taxonomy has retained the hallmarks from most of the earlier hierarchies, however, so that in addition to genetic analysis, microorganisms fit into classes, orders, and so forth, according to morphology and biochemical activity.

Woese's taxonomy method was revolutionary because it changed the way many biologists had viewed the evolution of higher organisms. Robert Whitaker's five-kingdom system assumed that the single-celled Monera contained the ancestor for all other life. Some of the Monera evolved into more structurally complex Protista. Plant, fungi, and animals then descended from the protists. Genetic analysis, by contrast, revealed more divergent evolutionary paths, with archaea splitting from bacteria early and evolving along a separate path from bacteria. When archaea were first discovered, microbiologists assumed these cells were more closely related to bacteria than any other cell type. Analysis of rRNA indicates that archaea possess some characteristics of bacteria and other characteristics similar to those of eukaryotes.

Advances in nucleic acid and amino acid analyses have further refined the following three-kingdom system in use today:

- domain Bacteria—all prokaryotes excluding archaea, that is, bacteria
- domain Archaea—prokaryotes without peptidoglycan in their cell wall
- domain Eukarya—all organisms with complex cell structure and membrane-bound organelles

An organelle is a structure within a microorganism, such as the nucleus or the cell membrane.

PROKARYOTE CLASSIFICATION

A prokaryotic species is any population of cells with similar characteristics. These characteristics used for assigning bacteria to a species are DNA and RNA composition (called the *base sequence*), cell morphology, staining (gram-negative versus gram-positive), biochemical tests, serology, phage typing, and fatty acid profiles. *Serology* refers to the analysis of structures on the surface of bacterial cells that identify the cells as specific antigens to a host's immune system. Phage typing involves exposing bacteria to certain viruses known to infect only specific bacteria. An unknown gram-positive coccus that can be killed by a phage specific for *Staphylococcus aureus* indicates that the unknown microorganism is probably *S. aureus*.

Microbiologists also study the relatedness of bacteria, that is, how close or distant different species are from each other. With this information, the microbiologist gets an idea of the evolution of the two bacteria. Closely related bacteria probably followed similar evolutionary roads before splitting onto their unique paths. Distant relations indicate the two species developed independently of each other after splitting from a common ancestor.

Microbiology uses four main methods to study the relatedness among bacteria: DNA fingerprinting, rRNA sequencing, hybridization, all of which can be aided by the method called polymerase chain reaction (PCR). PCR is a method invented, in the 1980s, for rapidly producing many copies of DNA (or RNA) from a very small starting amount of the nucleic acid. By this method, called *amplification* of nucleic acid, microbiologists can make large amounts of material for the other three methods.

DNA fingerprinting is the analysis of the exact sequence of DNA's base units adenine, thymine, cytosine, and guanine. Because the sequence of bases is unique for each species, it acts as an identification fingerprint. The DNA fingerprint of a newly discovered bacterium can be compared to a database of other fingerprints from known species. Matching fingerprints indicate related bacteria; mismatched fingerprints indicate different bacteria.

Woese's rRNA sequencing technique remains an important way to classify microorganisms, and molecular biologists continue to refine this method. Sequencing begins by recovering the DNA from a microorganism. PCR amplifies the DNA, which, then, is analyzed in an instrument called a *sequencer* to determine the DNA molecule's base sequence. From DNA's base sequence, microbiologists can determine rRNA's complementary sequence; RNA substitutes the base uracil for cytosine. This method provides a more detailed assessment of a microorganism's genetic makeup than use of DNA because the proteins that run the cell are made from the information in rRNA. This information must be specific and error-free for a cell to function properly or simply survive.

Nucleic acid hybridization determines the extent of similarity between a single DNA strand from one microorganism and a strand from a second microorganism. If the strands in the assay mixture reattach to each other along long stretches of complementary bases, the two microorganisms are closely related. Failure of the two strands to develop a strong attachment indicates fewer complementary base pairings and, thus, less related species.

EUKARYOTE CLASSIFICATION

Organisms composed of eukaryotic cells belong to the fungus, plant, or animal kingdom. The rRNA sequencing technique has been especially useful for differentiating protists in this domain, but higher organisms continue to be classified by a combination of genetic and morphological features.

The kingdom containing the fungi includes unicellular yeasts; multicellular molds, which are both microscopic at a point in their life cycle; and mushrooms, which are not microorganisms. Yeasts and fungi have been classified principally by morphology of the reproductive and vegetative (nonreproductive) cells, spores, sporangia that hold a cluster of spores, and the structure of hyphae, the thin filaments that enable molds to spread over a surface.

The plant kingdom consists of microscopic algae, macroscopic algae (seaweeds), mosses, ferns, conifers, and flowering plants. Only the unicellular microscopic algae are microorganisms. The rest of the members of this kingdom are multicellular higher organisms.

The animal kingdom consists of multicellular organisms: the invertebrates sponges, worms, insects, and crustaceans, and the vertebrates (animals with backbones).

CLASSIFICATION EXCEPTIONS

Viruses do not belong to any of the three domains. A virus is not a true cell because it lacks the structures and enzymes to reproduce on its own. Viruses replicate only by taking over the reproduction machinery of a prokaryotic or eukaryotic cell.

Virus classification includes groups that contain families, subfamilies, and genera. *Species* is synonymous with *genus,* a population of viruses with similar characteristics and occupying a particular ecological niche. A genus, therefore, has a distinctive morphology and a unique action inside the host.

The following main characteristics classify viruses:

- type of genetic material, DNA (double-stranded or single-stranded) or RNA
- type of host
- specific mode of replicating its genetic material
- shape

The taxonomy of viruses makes up a science separate from taxonomy of the three domains because of the unique features of these cellular parasites.

Microbiology continues to rely on traditional classification tools, such as cell morphology, while using nucleic acid analysis. This practice to occurrences in which a bacterial species remains in a genus even though rRNA or DNA analysis suggests it is not closely related to other members of the genus. Evolving technologies mean that taxonomy is not static. Taxonomists move species from one genus to another. Sometimes, taxonomists eliminate a species because analysis has shown two different species with two different names are actually the same microorganism.

Two techniques enable taxonomists to classify new organisms that do not yet belong to any hierarchy. The first technique, called *dichotomous keys,* uses a series of tests and morphological features laid out as a scheme. A microbiologist might begin by answering the question regarding the microorganism, "gram-positive or gram-negative." Additional tests are performed in an orderly fashion and answered "yes" for a positive response and "no" for a negative response. Each test result narrows down the possible identification of a microorganism, from a very broad group to a genus and then a species.

Cladograms are diagrams of an evolutionary tree starting from a common ancestor and branching upward into orders, families, genera, and species. General cladograms of the three domains usually show only groups of organisms for clarity. For example, a cladogram starting with a universal ancestor branches into the three domains, then, within each domain contains subbranches. In archaea, these subbranches might illustrate the groups methanogens, halophiles, and thermophiles.

Taxonomy is an important learning tool for studying evolution and the relationship among organisms. Taxonomy is also a dynamic science that uses sophisticated technology for analyzing genetic makeup of cells. This science also incorporates its roots by making use of traditional techniques in classifying microorganisms, such as the Gram stain, gross morphology, and nutrient use. Taxonomy helps students understand the relationships microorganisms have with all other biota.

See also APPENDIX I; ARCHAEA; DNA FINGERPRINTING; DOMAIN; GRAM STAIN; POLYMERASE CHAIN REACTION; SEROLOGY; SYSTEMATICS.

Further Reading

Michigan State University. "Taxonomy: Scientific Names." Available online. URL: www.msu.edu/~nixonjos/armadillo/taxonomy.html. Accessed November 30, 2009.

National Center for Biotechnology Information. "The NCBI Taxonomy Homepage." Available online. URL: www.ncbi.nlm.nih.gov/Taxonomy/taxonomyhome.html. Accessed November 30, 2009.

Rosselló-Mora, Ramon. "Updating Prokaryotic Taxonomy." *Journal of Bacteriology* 137 (2005): 6,255–6,257. Available online. URL: http://jb.asm.org/cgi/reprint/187/18/6255.pdf. Accessed November 29, 2009.

Stanier Institute. "The Genius of Roger Stanier." Available online. URL: www.stanier.ca/geniusEN.htm. Accessed November 30, 2009.

University of Colorado. "Norman R. Pace—Biographical Sketch." Available online. URL: http://mcdb.colorado.edu/spotlights/faculty/norm-pace. Accessed November 30, 2009.

Woese, Carl R., and George E. Fox. "Phylogenetic Structure of the Prokaryotic Domain: The Primary Kingdoms." *Proceedings of the National Academy of Sciences* 74 (1977): 5,088–5,090.

toxin A toxin is a poison of animal, plant, or microbial origin. In microbiology, the ability of a microorganism to produce a toxin, called *toxigenicity,* contributes to its virulence, or capacity to cause serious disease. In pathology of infectious diseases, toxins are classified as virulence factors.

Bacteria, fungi, and algae produce the major toxins studied in microbiology. Bacterial toxins number in the hundreds and have received, perhaps, the most research on toxin production and secretion, mode of action, and genes. Many algal and fungal toxins, however, also cause serious illnesses or death. Exposure to algal toxins usually occurs when a person has contact with natural fresh and marine waters. Fungal toxins, called *mycotoxins,* enter the body when fungal spores are inhaled or the toxin has been produced in a contaminated food and then eaten.

Microorganisms produce toxins as part of their normal metabolism. The benefit of toxin production resides in the damage done by the toxin to host tissue, thereby allowing the pathogen to establish a population. *Intoxication* refers to the effects on the body caused by the toxin. Intoxication does not require that the toxin-producing microorganism be present in the body. For example, several types of food-borne illnesses occur because a toxin remains in a food even though cooking killed the pathogen that made the toxin. *Toxemia* refers to the entrance of toxin into the host's bloodstream.

TYPES OF BACTERIAL TOXINS

Exotoxins

Toxins belong to various categories based on their location in the pathogen's cell, the type of organism producing them, target tissue, or mode of action, which is the way a poison or a drug normally works in the body. The following examples illustrate the different ways of classifying toxins:

- location in cell—exotoxins versus endotoxins
- organism producing the toxin—mycotoxins made by fungi
- target tissue—neurotoxins that affect the nervous system
- mode of action—membrane-disrupting toxins

Bacteria produce exotoxins and endotoxins in the cell but excrete exotoxins into the cell's surroundings. Endotoxins, by contrast, remain part of

Exotoxins and Endotoxins

Feature	Exotoxins	Endotoxins
main source	gram-positive or gram-negative bacteria	gram-negative bacteria
structure	mainly protein	mainly lipid
secretion	excreted into environment	released when cell dies and lyses
heat stability	destroyed at high temperatures reaching 176°F (80°C)	stable at 250°F (121°C) for one hour
toxicity	high	low
lethal dose	small (microgram per kilogram)	large (milligram per kilogram)
symptoms	varies by toxin	fever, chills, weakness, general aches
mode of action	interfere with specific metabolic processes in the host	stimulate immune system macrophages to produce levels of cytokines, which are toxic to the host's cells

the cell, located in the cell wall. The table above summarizes the main features and differences between exo- and endotoxins.

Exotoxins and endotoxins can both be fatal in high doses. Examples of exotoxin-related diseases from bacteria are gas gangrene, tetanus, and botulism (all from *Clostridium*); cholera; diphtheria; and scarlet fever. Diseases associated with endotoxins are typhoid fever and meningococcal meningitis and miscarriages.

Some exotoxins are among the deadliest substances in biology and lethal in microgram (μg) amounts. Exotoxins are dangerous, also, because they can be active even if the bacteria producing them have been killed. The chemicals formaldehyde and iodine inactivate exotoxins, however, into a harmless form called a toxoid. Toxoids of various pathogens can be injected into the body for immunizing against further infection. This is possible because toxoids stimulate the immune system to produce compounds called *antitoxins* that neutralize toxins.

Three main types of exotoxins exist:

- type I—superantigens
- type II—membrane-disrupting toxins
- type III—A-B toxins that act as enzymes or block cellular enzymes

Superantigens cause damage to the body by eliciting an intense immune response. Lymphocytes called *T cells* react dramatically to the presence of a superantigen and multiply to high numbers. The high concentration of T-cells releases huge amounts of chemical called *cytokines*. These small proteins travel the bloodstream and wreak havoc on cellular functions, shutting down some enzymes and overstimulating others. Cytokine activity results in nausea, vomiting, diarrhea, and possible fever.

Membrane-disrupting toxins enter the plasma membrane of host cells and destroy membrane function. These toxins work mainly by ruining the channels through which nutrients and secretions pass through the membrane. When membrane-disrupting toxins contact immune cells, such as white blood cells and phagocytes, they weaken the host's immunity. The toxins, similarly, attack and lyse red blood cells. Toxins that attack white blood cells are called *leukocidins*, and those that lyse red blood cells are *hemolysins*.

A-B toxins consist of two polypeptides, A and B. In most toxins, the A component is an enzyme that gives the toxin its activity but is incapable of binding to a host cell and penetrating it. The B component binds to the target tissue at a specific site called the *receptor* and helps the entire molecule enter the cell. Glycoproteins (G-proteins) usually compose the receptor site. Once inside, the A component becomes an active enzyme that interferes with cellular processes. The B polypeptide does not appear to possess any toxic activity, but the A portion cannot harm the host without it.

A-B toxins use two similar methods of entering host cells. The first method involves direct entry, by which the B component opens a pore through the membrane that allows entry of the whole toxin molecule into the cell. In a second approach, the toxin binds to the cell receptor to cause a change in the configuration of membrane proteins. The cell

Types of Microbial Toxins

Toxin	Mode of Action	Type	Microorganism
botulism	A-B toxin	neurotoxin	*Clostridium botulinum*
cholera	A-B toxin	enterotoxin	*Vibrio cholerae*
streptolysins	membrane-disrupting	cytotoxin	*Streptococcus pyogenes*
toxic shock	superantigen	cytotoxin	*Staphylococcus aureus*

membrane envelops the toxin, then incorporates it into the cytoplasm; the entire process is called receptor-mediated endocytosis (RME). The following list contains other pore-forming bacterial toxins (toxin-producing bacteria in parentheses):

- anthrax toxin *(Bacillus anthracis)*
- alpha toxin and leukocidin *(S. aureus)*
- hemolysin *(Escherichia coli)*
- listeriolysin *(Listeria monocytogenes)*
- perfringiolysin O *(Clostridium perfringens)*
- streptolysin O *(Streptococcus pyogenes)*

Type I, II, and II toxins act on various host tissues, which can be used as another way to group the toxins. The three main categories according to the toxin's target tissue are:

- cytotoxins—general tissues but each toxin binds with specific organs; cardiotoxin binds to heart tissue, hepatotoxin binds to liver, and nephrotoxin binds to kidney
- enterotoxins—mucous membrane of the gastrointestinal tract
- neurotoxins—nerve cells, ganglia, brain

The table above provides examples of microbial toxin categories.

Several bacterial toxins have received extensive research because of their importance in human health. Most of these toxins relate to food-borne or waterborne illnesses.

Streptolysins consist of the O type and an S type, both produced by *Streptococcus pyogenes.* Oxygen-sensitive streptolysin O (SLO) is a hemolysin that binds to cholesterol-containing sites on red blood cells and creates large pores in the membrane, which contribute to cell lysis. SLO, as can many other hemolysins, can kill cells other than red blood cells. Streptolysin L (SLS) tolerates exposure to oxygen and acts as a leukocidin.

Staphylococcus aureus enterotoxin and the cholera enterotoxin made by *Vibrio cholerae* cause the epithelial cells lining the digestive tract to release large amounts of fluids and electrolytes (ion form of salts). The toxins slow the intestines' normal motility by affecting muscle contractions. As a consequence, a person experiences severe diarrhea and vomiting.

Corynebacterium diphtheriae, the only exotoxin-producer discussed here that is not a food-borne pathogen, makes diphtheria toxin only when the bacterium has been infected by a bacteriophage (bacteria-specific virus) that carries a *tox* gene. Diphtheria cytotoxin uses either direct entry or RME to penetrate a host cell. Once inside the cell, the A-B toxin inhibits protein synthesis in a variety of host cells.

Tetanus toxin, also called *tetanospasmin,* produced by *Clostridium tetani,* is an A-B type neurotoxin. This toxin binds to nerve cells that control muscle contraction, specifically blocking the relaxation phase of the muscle. As a result, muscles remain in a perpetual state of contraction or spasm, a condition called *lockjaw.*

Botulism toxin (or botulinum), made by *Clostridium botulinum,* is one of the most lethal toxins known. *C. botulinum* produces an endospore form to protect the cell during periods of harsh environmental conditions. The bacterium does not produce toxin until it returns to a growing vegetative form from the endospore. Cells produce the toxin late in their growth cycle, shortly before cell death and lysis.

The botulism toxin is a zinc-dependent protease, an enzyme that cleaves protein molecules that are active at the site between nerve cells called the *synapse.* At the active site, the enzyme blocks steps required for the release of the compound acetylcholine, the nervous system's main neurotransmitter. By blocking nerve transmission, the toxin causes flaccid paralysis, in which no muscle impulses can occur, the opposite of spasm paralysis caused by the tetanus toxin.

The food-borne pathogen *Shigella dysenteriae* produces A-B type shiga toxin containing five B subunits and specific for epithelial cells of the gastrointestinal tract. Shiga toxin binds to very specific receptors on the epithelial cell surface, before entry. Unlike with most other A-B toxins, part of the A component is cleaved off, but the component retains activity inside the cell. Shiga toxin binds to the host cell's ribosomes to inhibit protein synthesis and is distinctive because it acts as a cytotoxin, neurotoxin, and enterotoxin.

Enterohemorrhagic *E. coli,* including the virulent O157:H7 strain, produces shigalike toxin. The shigalike toxins share the same host cell receptor site with *Shigella*'s toxin and use a very similar mode of action inside the cell. One of shigalike toxins' main modes of action is as a vasculotoxin: The toxin causes lesions in blood vessels and possibly stimulates high amounts of cytokine release that further damage vessels.

The bacteria toxins discussed here have a very high degree of toxicity at very low doses, the main reason why microbial toxins have been studied as potential bioweapons. These substances, furthermore, retain their activity at high dilutions. The following bacterial toxins have high toxicity at low doses relative to other known toxins. For example, diphtheria toxin is 2,000 times more lethal than strychnine and 200 times more lethal than snake venom. The other three toxins listed here range from 10^5 to 10^6 times more lethal than these poisons. The toxic dose in milligrams (mg) is shown:

- diphtheria toxin, 6×10^{-5}
- shiga toxin, 2.3×10^{-6}
- tetanus toxin, 4×10^{-8}
- botulinum toxin, 0.8×10^{-8}

These extremely low doses needed to achieve lethality in laboratory test animals attest to the perceived usefulness of microbial toxins as weapons of mass destruction. Toxin placed into food or water would be a health threat, but not as grave threat as many people believe. Large quantities of food or water cause a circumstance called the *dilution effect.* Because of the dilution effect, very large quantities of toxin would be needed to contaminate food or water sufficiently to cause massive sickness or deaths. These large quantities are impractical, if not impossible, for a bioterrorist to make.

Endotoxins

Gram-negative bacteria contain a lipopolysaccharide (LPS) (a polysaccharide and fat complex) component in their cell walls that acts as a toxin. These endotoxins stay bound to the bacterial cell and are liberated at cell death and lysis; some cells excrete endotoxins into the environment, but only during cell replication.

Endotoxins exert less toxicity and a lower immune response in host tissue than most exotoxins. Endotoxins also have a more general and indirect effect on the host then exotoxins. Most endotoxins influence host cells to turn on responses that become the ultimate cause of damage to the body. For example, endotoxins indirectly induce fever by stimulating the immune system's macrophage cells to secrete compounds called *pyrogens.* Pyrogens reset the body's thermostat in the hypothalamus, thus leading to fever. The LPS constituent of endotoxins has a second effect in the host by binding to specific host proteins called *LPS-binding proteins,* which are specifically called CD14 receptors for the type of protein. Immune system cells, such as macrophages and monocytes, form this LPS-receptor complex, and the complex triggers a cascade of steps leading to elevated levels of cytokines, interleukins, and tumor necrosis factor (TNF), all fever-inducing compounds. (One of the interleukins, IL-1, was formerly called endogenous pyrogen.)

The complex of endotoxin with CD14 receptors has been implicated as a contributor to miscarriages associated with endotoxin poisoning. Endotoxin-producing bacteria can infiltrate the endometrial lining of the uterus, in infections during pregnancy, and cytokines are known to play several roles in maintaining pregnancies. Findings of genetic and biochemical studies intended to determine the relationship between LPS endotoxins and miscarriages have been inconclusive. The pediatrician Aaron Hirschfeld at the British Columbia Children's Hospital in Canada explained, in 2007, "Toll-like receptors (TLR) are a recently identified group of vertebrate receptors that play a central role in determining the [cytokine] balance of immune responses. The human TLR family consists of 10 receptors that orchestrate the innate immune response by linking pathogen recognition with immune cell activation. Individual TLRs recognize a distinct, but limited, repertoire of conserved [genetically related] microbial products, and the best-characterized receptor-ligand pair is THR4 and lipopolysaccharide (LPS or endotoxin)." Similar relationships have been proposed by other researchers, but the relationship between endotoxins and frequency of miscarriage has not yet been fully explained.

In addition to causing fever, endotoxins activate blood-clotting proteins that begin blocking blood flow in small vessels, such as capillaries. In severe cases, the surrounding tissue dies of lack of blood supply, a condition called disseminated intravascular clotting (DIC). In DIC, blood-clotting factors of various types begin to accumulate in the blocked vessels, and general bleeding through the vessel walls fills the surrounding tissue. At this stage, which appears as bruising on the skin, the condition becomes known as *consumptive coagulopathy*. Treatments to alleviate the discomfort brought on by DIC include heparins, which reduce the clotting reaction in blood. Severe cases call for blood transfusion.

The main endotoxins causing health concerns in humans are from the following pathogens:

- *Bordetella pertussis*—whooping cough
- *E. coli*—gastroenteritis and urinary tract infections
- *Haemophilus influenzae*—respiratory infections
- *Neisseria*—sexually transmitted disease
- *Pseudomonas*—nosocomial infections
- *Salmonella typhi*—typhoid fever
- *Shigella*—some forms of shigellosis

Overall, endotoxins are less potent in their actions on the host than exotoxins and thus less life-threatening.

Endotoxins complicate the human trials that pharmaceutical companies run on new drugs and can also interfere with in vitro experiments on living tissue. Microbiologists, therefore, use sensitive tests to detect the presence of endotoxin. The *Limulus* amoebocyte lysate (LAL) assay detects the presence of nanogram levels of endotoxin due to the toxin's ability to form clots in preparations of lysed blood cells (amoebocytes) from the horseshoe crab *Limulus polyphemus*. A positive result in the LAL assay tells the microbiologist that a tissue culture cannot be used for further drug tests because of the presence of endotoxins. Hospitals also eliminate potential endotoxins from medical devices by heating the devices at 482°F (250°C) for 30 minutes.

ALGAL TOXINS AND MYCOTOXINS

The main health concerns from algal toxins, called *phycotoxins*, occur when large numbers of certain algae dominate freshwater or marine environments. These algal overgrowths are called harmful algal blooms (HABs). HABs contain enormous numbers of algae that grow rapidly in waters that have received a sudden influx of nutrients. Algal blooms occur worldwide and may be increasing in certain areas that experience blooms along shorelines every year. A famous type of bloom is called red tide for the algae's red pigments. The table below describes the prevalent algal blooms that are associated with toxic poisoning of fish and humans or the people or animals that eat infected fish. Each of the toxins listed in the table is a neurotoxin.

Algal toxins are directed to very specific metabolic pathways in fish or animals. Because of this specificity, biochemists have used algal toxins as blockers of certain pathways to gain knowledge of enzyme activities. For example, biochemists know that okadaic acid affects calcium currents in muscle and inhibits protein phosphatases, enzymes important in activating or deactivating enzymes.

Mycotoxicology is the study of the thousands of environmental and pathogenic fungi that produce mycotoxins. Mycotoxins can be a significant threat to food-producing animals, such as cattle, sheep, and poultry, and to horses, because these animals

Harmful Algal Blooms

Type of Bloom	Phycotoxin	Alga	Disease
red tide	saxitoxin	*Gymnodinium, Gonyaulax*	paralytic shellfish poisoning
red tide	brevetoxin	*Ptychodiscus brevis, Gymnodinium breve*	neurotoxic shellfish poisoning
golden-brown algae	domoic acid	diatom *Pseudonitschia*	amnesic shellfish poisoning
associated with coral reefs	ciguatoxin	*Gambierdiscus toxicus*	ciguatera fish poisoning
fish kills	unknown	*Pfiesteria piscicida*	intoxication including skin lesions, disorientation, and memory loss
unnamed	okadaic acid	*Prorocentrum, Dinophysis*	diarrhetic shellfish poisoning

Mycotoxins

Mycotoxin	Fungus	Disease
aflatoxin	*A. flavus, A. parasiticus*	aflatoxicosis
amanitin	*Amanita verna* (mushroom)	mushroom poisoning
citrinin	*Penicillium*	kidney pathology
ergot alkaloids	*Claviceps purpurea*	ergotism
ergot alkaloids	*Acremonium coenophialum*	tall fescue toxicosis
patulin	*Aspergillus, Byssochlamus, Penicillium*	poisoning
rubratoxin	*Penicillium rubrum*	hemorrhaging
slaframine	*Rhizoctonia leguminicola*	black patch disease
sterigmatocystin	*Aspergillus nidulans, A. versicolor*	liver cancer
trichothecenes	*Fusarium, Myrothecium, Trichothecium*	neurological and digestive symptoms

depend on grains in their diets; fungi (mostly molds) are a common contaminant in stored grains, even after the fungus has died. In stored grains, the fungus remains dormant if the air and moisture supply are limited but begins growing if the humidity rises and the grains absorb moisture. Diseases that result from mycotoxin poisoning are called *mycotoxicoses*.

Aflatoxins represent the largest group of mycotoxins and are made by several different species of *Aspergillus*. *Aspergillus flavus* produces aflatoxins that contaminate rice, cereal grains, corn, peanuts, sorghum, and soybeans, as well as meals (cornmeal, feathermeal, etc.) used as feedstuffs. Cattle, sheep, swine, and poultry are susceptible to aflatoxin poisoning, and dogs have also become sick from ingesting toxin-contaminated dog foods. In 2006, several dog food brands were recalled after aflatoxin contamination had caused dog deaths throughout the United States. The Cornell University veterinarian Sharon Center said at the height of the outbreak, "I've been working with liver disease in dogs for 30 years, and I've never seen such miserably ill dogs." Aflatoxins cause liver damage, cirrhosis, and possible cancer of the liver in affected animals.

The table above describes mycotoxins important in human and animal health.

Mycotoxins can cause serious neurological damage, particularly the trichothecenes, which have been tried in the past as a bioweapon called *yellow rain*. The Centers for Disease Control and Prevention (CDC) reported, in 2006, "Trichothecene mycotoxins might be weaponized and dispersed through the air or mixed in food or beverages. . . . Dermal exposure leads to burning pain, redness, and blisters, and oral exposure leads to vomiting and diarrhea. Ocular exposure might result in blurred vision, and nasal inhalation might cause nasal irritation and cough. Systemic symptoms can develop with all routes of exposure and might include weakness, ataxia [loss of muscle control], hypotension, coagulopathy, and death." Other general health conditions known to be associated with mycotoxin poisoning are:

- infantile pulmonary hemorrhage—in infants, bleeding in lungs
- encephalopathy—short-term memory loss, difficulty in concentrating, poor attention span
- fatigue
- gastrointestinal distress

Mycotoxin poisoning has been difficult to diagnose in animals and humans because of the broad range of symptoms caused by the toxins and a very large number of potential species that produce one or more toxins. Mycotoxicology is one of the most challenging areas in microbiology and in medicine.

ANTITOXINS

An antitoxin is an antibody that combines with and neutralizes a specific microbial toxin, usually a bacterial exotoxin. The process by which antibodies inactivate toxins is called *toxin neutralization*.

After an antibody has attached to a toxin, toxin neutralization works in any of the three following ways: (1) Antibody-toxin complex is unable to bind to the host cell receptor site, (2) antibody-toxin complex cannot enter the target cell, or (3) the antibody holds the toxin until macrophages arrive to engulf and digest it. Three important antitoxins used in medicine act against the diphtheria, gas gangrene (made by *Clostridium perfringens*), and tetanus toxins.

The concept that the body could develop antibodies against toxins in addition to whole pathogen cells, expanded with the work of the German medical researcher Emil von Behring (1854–1917). In 1890, von Behring published results of work that he and the Japanese microbiologist Shibasaburo Kitasato (1853–1931) had completed on diphtheria and tetanus bacteria. They had discovered that broth cultures of the pathogens, sterilized to kill all the bacteria and injected into laboratory animals, could induce the animals to develop immunity to the diphtheria and tetanus toxins. Von Behring and Kitasato had developed effective antitoxins. A year later, the French physician Émile Roux (1853–1933) conducted a real-life test of the diphtheria antitoxin in a procedure medical researchers now call clinical testing, or drug trials in humans. The author Alistair Lax wrote, in 2005, in *Toxin: The Cunning of Bacterial Poisons,* "This new preparation of anti-toxin was tested at two Paris hospitals. In the Hôpital des Enfant-Malade the death rate among treated children was 24 percent, whereas at the Hôpital Trousseau, where anti-toxin was not used, it was 60 percent." The Centers for Disease Control and Prevention (CDC) now recommends antitoxins for additional poisonings such as botulism and a variety of plant toxins.

Toxin production in pathogens is a crucial part of their virulence. Some toxins cause general illnesses; others are among the strongest poisons in biology. For this reason, the study of microbial toxins remains essential in pathology and veterinary medicine and must be included in programs focused on biological weapons.

See also ALGAE; BIOWEAPON; FUNGUS; MODE OF ACTION; VIRULENCE.

Further Reading

Balashova, Nataliya. "Bacterial Toxins: How They Cause and Sustain Disease." *University of Medicine and Dentistry of New Jersey Research,* Fall 2008. Available online. URL: www.umdnj.edu/research/publications/fall08/index.htm. Accessed January 22, 2010.

Centers for Disease Control and Prevention. "Case Definition: Trichothecene Mycotoxin." Available online. URL: www.bt.cdc.gov/agent/trichothecene/casedef.asp. Accessed January 23, 2010.

Eddleston, Michael, and Hans Persson. "Acute Plant Poisoning and Antitoxin Antibodies." *Journal of Toxicology and Clinical Toxicology* 41 (2003): 309–315. Available online. URL: www.ncbi.nlm.nih.gov/pmc/articles/PMC1950598. Accessed January 23, 2010.

Hirschfeld, Aaron F., Ruby Jiang, Wendy P. Robinson, Deborah E. McFadden, and Stuart E. Turvey. "Toll-Like Receptor 4 Polymorphisms and Idiopathic Chromosomally Normal Miscarriages." *Human Reproduction* 22 (2007): 440–443. Available online. URL: http://humrep.oxfordjournals.org/cgi/reprint/22/2/440. Accessed January 23, 2010.

Karthukorpi, Jari, Tarja Laitinen, and Riitta Karttunen. "Searching for Links between Endotoxin Exposure and Pregnancy Loss: CD14 Polymorphism in Idiopathic Recurrent Miscarriage." *American Journal of Reproductive Immunology* 50 (2003): 346–350. Available online. URL: www.ncbi.nlm.nih.gov/pubmed/14672339. Accessed January 23, 2010.

Lang, Susan S. "Dogs Keep Dying: Too Many Owners Remain Unaware of Toxic Dog Food." *Cornell University Chronicle,* 6 January 2006. Available online. URL: www.news.cornell.edu/stories/Jan06/dogs.dying.ssl.html. Accessed January 2, 2010.

Lax, Alistair. *Toxin: The Cunning of Bacterial Poisons.* Oxford: Oxford University Press, 2005.

Perez, Roberto, Li Liu, Jose Lopez, Tianying An, and Kathleen S. Rein. "Diverse Bacterial PKS Sequences Derived from Okadaic Acid–Producing Dinoflagellates." *Marine Drugs* 6 (2008): 164–179. Available online. URL: www.mdpi.org/marinedrugs/papers/md6020164.pdf. Accessed January 2, 2010.

Schmitt, Clare K., Karen C. Meysick, and Alison D. O'Brien. "Bacterial Toxins: Friends or Foes?" *Emerging Infectious Diseases* 5 (1999): 224–234. Available online. URL: www.cdc.gov/ncidod/eid/vol5no2/pdf/schmitt.pdf. Accessed January 23, 2010.

Todar, Kenneth. "Bacterial Protein Toxins." In *Todar's Online Textbook of Bacteriology.* Madison.: University of Wisconsin, 2008. Available online. URL: www.textbookofbacteriology.net/proteintoxins.html. Accessed January 22, 2010.

Van Dolah, Frances M. "Marine Algal Toxins: Origins, Health Effects, and Their Increased Occurrence." *Environmental Health Perspectives Supplements* 108 (2000): 133–142. Available online. URL: http://ehp.niehs.nih.gov/docs/2000/suppl-1/133-141vandolah/abstract.html. Accessed January 23, 2010.

Zaccaroni, Annalisa, and Dino Scaravelli. "Toxicity of Sea Algal Toxins to Humans and Animals." In *Algal Toxins: Nature, Occurrence, Effect and Detection,* edited by Valtere Evangelista, Laura Barsanti, Anna Maria Frassanito, Vincenzo Passarelli, and Paulo Gualtieri. Dordrecht, the Netherlands: Springer, 2008. Available online. URL: http://books.google.com/books?id=4Ub6-nLS_zYC&pg=PT96&lpg=PT96&dq=algal+toxins&source=bl&ots=GgZADdMM4m&sig=9DhPntAUlFf7OQbBeJ2RACoJ2XQ&hl=en&ei=RzBaS4X1N5O2swOnutzNBA&sa=X&oi=book_result&ct=result&resnum=5&ved=0CBYQ6AEwBA#v=onepage&q=&f=false. Accessed January 23, 2010.

transmission Transmission is the manner by which a pathogen moves from an infected person to a healthy person. Transmission contributes to the pathogenicity of a microorganism, which is the ability to cause disease. Without a reliable mode of transmission, a pathogen cannot spread among individuals in a population.

Transmission relates to two factors in the process of disease progression in the body, called *pathogenesis.* The two factors are the pathogen's portal of entry and portal of exit. A portal of entry is the main route used by a pathogen to invade the body. Transmission ends when a transmitted pathogen accesses its preferred portal of entry. For example, a sneeze transmits the influenza (flu) virus. Flu transmission culminates with the virus's entry into a new host's respiratory tract. The respiratory tract of a person suffering with the flu, in this case, also represents the portal of exit. Pathogens, thus, depend on these three factors to move from infected people or animals by a specific mode of travel to a healthy, susceptible host:

portal of exit → transmission →
portal of entry

The germ theory of disease, which proposed that microorganisms cause infectious disease, developed in the 19th century. Before then, people believed that disease was produced by bad odors, such as the odors that arise from sewage, animal waste, decaying matter, or carcasses. Microbiologists took advantage of improved microscopes to detect pathogens in specimens taken from sick patients, and many began to suspect that the tiny organisms had more to do with disease than bad smells. From the 1840s to about 1910, the period called the Golden Age of Microbiology, microbiologists made dramatic strides in relating microorganisms to specific diseases. The table (right) describes key events during this period that reinforced the germ theory in biology and provided the foundation for studying germ transmission.

Development of the Germ Theory

Date	Event
1835	Italian microscopist Agostino Bassi (1773–1856) shows that a fungus causes a certain silkworm disease
1840	German physician Jacob Henle (1809–85) describes the process of pathogenesis
1849	British physician John Snow (1813–58) shows that infectious diseases can be waterborne
1850	Hungarian physician Ignaz Semmelweis (1818–65) shows that washing hands reduces the spread of infection
1861–65	French microbiologist Louis Pasteur (1822–95) proves the existence of live airborne microorganisms; demonstrates that a disease in silkworms was caused by a fungus
1867	British surgeon Joseph Lister (1827–1912) advocates aseptic techniques in surgeries
1876	German microbiologist Robert Koch (1843–1910) develops criteria for proving a specific microorganism causes a specific disease
1892	Russian microbiologist Dmitri Ivanovski (1864–1920) discovers viruses and later shows they can cause plant disease
1897	Scottish physician Ronald Ross (1857–1932) discovers that malaria is transmitted by mosquitoes

Acceptance of the idea of germs transported in air and water or on the hands marked a critical point in microbiology's history. From that point, physicians would follow in Joseph Lister's footsteps and study the manner in which disease-causing organisms moved through populations.

TYPES OF TRANSMISSION

Four types of transmission carry pathogens in human, animal, and plant populations: (1) airborne, (2) contact, (3) vehicle, and (4) vector.

Airborne disease transmission is part of a broader discipline in microbiology called *aeromicrobiology,* the study of microorganisms in the air. Most airborne pathogens reside in moisture droplets in water or in mucus expelled by sneezing or coughing. For this reason, airborne transmission is sometimes called droplet transmission, and the tiny droplets are called bioaerosols, meaning aerosols containing a biological entity. The main microbiological items carried in bioaerosols are viruses, bacteria, fungal spores, and pollen.

Airborne pathogens can also travel in dry particles such as dust, soil particles, and soot. In dry and droplet transmission, the pathogen's main portal of entry is the mucous membrane of the respiratory tract.

Contact transmission occurs when a healthy person has physical contact with an object that holds pathogens. Two types of contact transmission exist: direct and indirect. Direct, also called person-to-person, transmission, takes place when a healthy person receives a pathogen through physical contact with an infected person. Common modes of direct contact transmission are touching (skin to skin), kissing, and sexual intercourse (sexually transmitted disease). Indirect contact transmission uses an intermediate object to carry pathogen from one person to another. These objects, called *fomites,* are any nonliving objects that can transmit an infectious pathogen from one person to another. The following household items serve as common fomites in disease transmission:

- clothes
- towels
- tissues and handkerchiefs
- bedding
- drinking cups
- kitchen utensils
- money
- toys
- diapers

Microbiologists also warn that other objects probably transmit pathogens: keyboards, remote controls, faucet handles, kitchen appliance handles, countertops, handrails, doorknobs, telephones, automated teller machine (ATM) buttons, and any other surface repeatedly touched by a large number of people.

Medical devices have been implicated as effective transmitters of pathogens because they often deliver the pathogen to a person who already is at a high risk of infection. Surgical patients and dental patients have body sites where the chance of a pathogen's entering the bloodstream has been increased, for example, surgical incisions. Many hospital patients have weakened immunity, which also increases the opportunity for infection. The following devices act as possible fomites in disease transmission:

- dental instruments
- scalpels
- syringes
- endoscopes
- thermometers
- blood pressure cuffs
- patient gowns
- hospital bed rails
- hospital bedside tables

The physician and specialist in infection control William Rutala stated, in 2002, in *Infection Control Today,* "The acquisition of nosocomial [hospital-associated] pathogens depends on a complex interplay of the host, pathogen and environment." In addition to the strength of the host's immune system, different pathogens stay active on fomites for different amounts of time. Temperature, relative humidity, and moisture on the fomite contribute to survival time, in addition to the type of pathogen. Microbiologists have made the following estimates of the survival time of various pathogens on fomites:

- enterococcus bacteria—five days on countertops, 24 hours on bed railings, 60 minutes on telephones, and 30 minutes on stethoscopes
- influenza virus A and B—24–72 hours on hard, nonporous surfaces such as stainless steel; eight to 12 hours on cloth or tissues
- avian influenza—six days on hard surfaces or cloth
- cold virus (rhinovirus)—more than 24 hours on hard, nonporous surfaces
- cold virus (coronavirus)—three to 12 hours on hard, nonporous surfaces

Since 2002, medical studies have been run to investigate the incidence of germ transmission on medical devices. The *Infection Control Today* reporter Kelly Pyrek wrote, in 2009, "A plethora of studies in the medical literature has demonstrated that nearly everything in the healthcare setting—from surfaces to healthcare workers' hands, to medical equipment and everything in between—can serve as a reservoir and vector for opportunistic pathogenic organisms." Reservoirs and vectors are living things that provide a continual source of a pathogen or a transmitter of the pathogen, respectively. Opportunistic pathogens are microorganisms not normally dangerous but able to cause disease if conditions in a host change to increase the likelihood of infection, such as a weakened immune system. The proper sterilization of invasive medical devices that enter the body (syringes, scalpels, dental instruments, etc.) provides assurance against transmission.

Vehicle transmission involves the use of a medium such as food or water—air can also be considered a medium—to carry pathogens. Food-borne transmission occurs when food has been undercooked, inadequately preserved or refrigerated, or prepared under unsanitary conditions. Common types of pathogens that contaminate foods belong to the enteric (fecal contamination) group of bacteria or viruses, respiratory viruses, or normal flora of skin.

Vector-Borne Diseases in Veterinary Medicine

Vector	Pathogen	Disease	Animals Affected
fly	*Streptococcus* and *Staphylococcus* (bacteria) *Moraxella* (bacteria)	mastitis conjunctivitis	dairy cattle calves
mosquito	arboviruses West Nile virus	equine encephalitis encephalomyelitis	horses horses
tick	*Anaplasma* and *Ehrlichia* (rickettsia bacteria) *Cytauxzoon felis* (protozoan)	anaplasmosis, ehrlichiosis, Rocky Mountain spotted fever cytauxzoonosis	dogs, horses cats

The following microorganisms represent these three main sources of food-borne contamination:

- *Escherichia coli, Shigella, Salmonella,* rotavirus—fecal contamination
- rhinovirus—contamination from sneezing or coughing on food
- *Staphylococcus aureus*—contamination of foods from the hands of food preparers

Water can contaminate foods, such as when contaminated water is used for rinsing raw vegetables, or it can transmit pathogens directly to a host. Waterborne transmission takes place when drinking water, bathing water, or recreational water becomes contaminated with pathogens. Common waterborne diseases are cholera *(Vibrio cholerae),* cryptosporidiosis *(Cryptosporidium),* giardiasis *(Giardia lamblia),* Legionnaires' disease *(Legionella pneumophila),* and leptospirosis *(Leptospira interrogans).*

Waterborne pathogens sometimes differ in the types of waters where they are most prevalent. For example, *V. cholerae* contaminates freshwater sources, and *L. pneumophila* resides in ventilation cooling waters, from which they spread in airborne moisture droplets.

VECTOR TRANSMISSION

Vector transmission uses arthropods (insects) as the means for carrying pathogens through a population. This type of transmission causes a significant proportion of infectious diseases in human and veterinary medicine. The World Health Organization (WHO) identifies vector management as one of the major needs in infectious disease control. In 2004, the WHO report "Global Strategic Framework for Integrated Vector Management" stated, "Vector-borne diseases are responsible for a significant fraction of the global disease burden and have profound effects not only on health but also on the socioeconomic development of affected nations." The WHO cited malaria as an example. Malaria is a protozoal disease transmitted by a mosquito, which, according to the WHO, "is responsible for more than one million deaths every year . . . countries with intensive malaria have income levels of only 33 percent of those without malaria." Vector control programs have become as important in veterinary medicine as in human medicine because vectors have economic impacts and because animals are often the reservoir for human diseases.

Common vectors that transmit human diseases are fleas, lice, mites, mosquitoes, and ticks. Many plant diseases also transmit by way of vectors, especially aphids, leafhoppers, mites, and whiteflies. Vector transmission can be either biological or mechanical. Biological transmission occurs mainly through insect bites. Various types of processes unfold in biological transmission, depending on the pathogen. Some pathogens reproduce in the insect gut and migrate to the saliva, which the insect injects into a host when it bites; this process is called circulative transmission. Other pathogens are transferred to wounds in the insect's feces or vomit and, then, enter the host's bloodstream.

In biological vector transmission, the life cycle of the pathogen in the host and insect takes many different forms. Some pathogens do not undergo any life cycle changes when inside the insect but go through different stages in the host. For example, *Yersinia pestis,* the cause of bubonic plague, does not reproduce in the insect (rat flea) but waits to reproduce until it enters the human host. This type of vector transmission is called *harborage.* Other pathogens spend part of their life cycle in the insect and part in the host, as in the case of malaria. In malaria, *Plasmodium* protozoa undergo three main stages inside the insect (*Anopheles* mosquitoes) and several more stages once inside the human host. Cyclical transmission is the process whereby more than one life cycle stage occurs in the vector.

Vector-Borne Zoonoses

Disease	Pathogen	Vector	Main Animal Reservoir
babesiosis	*Babesia* (P)	tick	rodents, cattle
bubonic plague	*Yersinia pestis* (B)	flea	rodents
Colorado tick fever	virus	tick	rodents
encephalitis, various	viruses	mosquito	birds, horses, primates, rodents, sheep, swine
eperythrozoonosis	*Mycoplasma* (B)	various biting insects	livestock
leishmaniasis	*Leishmania* (P)	sand fly	dogs
Lyme disease	*Borrelia burgdorferi* (B)	tick	deer
malaria	*Plasmodium* (P)	mosquito	primates
relapsing fever	*Borrelia hernsii* (B)	tick	rodents
rickettsial pox	*Rickettsia akari* (B)	mite	mice
Rocky Mountain spotted fever	*Rickettsia rickettsii*	tick	rabbits, field mice, dogs
Ross River fever	virus	mosquito	unknown
tick bite fever	*Rickettsia*	tick	rodents, dogs
trypanosomiasis	*Trypanosoma* (P)	tsetse fly	dogs, cattle
typhus (murine)	*Rickettsia typhi*	flea	rats, cats, opossums, skunks, raccoons
typhus (Queensland)	*Rickettsia siberica*	tick	rodents
typhus (scrub)	*Orientia tsutsugamushi* (B)	mite	rodents

Note: B = bacteria; P = protozoa

Mechanical transmission occurs when a pathogen attaches to the outside of an insect and uses it for transport to a new host. For example, household flies can carry pathogens for bacillary dysentery *(Shigella)* and typhoid fever *(Salmonella)* and deposit the microorganisms on food. Mechanical transmission of disease causes particular concern in veterinary medicine, because horses and domesticated food animals are exposed to flies every day. Two veterinary health problems are examples of vector-borne diseases that use mechanical transmission: conjunctivitis caused by flies carrying *Moraxella* bacteria and dairy cow mastitis caused by flies carrying *Streptococcus* and *Staphylococcus* bacteria.

Biting insects such as ticks, biting flies, and mosquitoes cause most veterinary vector-borne diseases. The table on page 763 provides examples of the many vector-borne diseases that afflict domesticated animals.

Vector-borne diseases in animals affect humans in two ways. First, diseases in food-producing animals harm the food supply and the economic strength of agriculture. Second, many vector-borne diseases are also zoonotic diseases, or zoonoses, which are diseases that transmit from animals to humans. The table above lists zoonoses transmitted from animals via an insect vector.

The vectors in many rare zoonoses have been easier to identify than the animal reservoir. In many cases, an insect gets a pathogen from an unknown wildlife reservoir.

CARRIERS

Transmission that depends on human carriers affects epidemiology, the study of how disease originates and spreads through populations. Public health specialists recognize the following four types of disease carriers:

- active—carrier displays outward signs of the infectious disease
- convalescent—person has recovered from the disease but still carries large numbers of the pathogen
- healthy—carrier harbors the pathogen but shows no signs of illness
- incubatory—carrier contains increasing numbers of pathogens but does not yet have signs of illness

People who carry a pathogen for long periods of time, from several months to a lifetime, are called *chronic carriers*. Transient carriers, also called *casual carriers*, contain the pathogen for hours, days, or weeks.

TRANSMISSION CONTROL

Transmission control is a multifaceted endeavor that interrupts the movement of pathogens from their source to susceptible hosts. Sanitation and good personal hygiene provide the most trustworthy measures of stopping disease transmission. Sanitation involves processes to keep a building or a community clean, which specifically include frequent cleaning, trash pickup, sewage collection and removal, and separation of sick and contagious people from the healthy population. The latter step can be accomplished by staying home from school or work when sick with an infectious disease.

Good personal hygiene also breaks the route of disease transmission. The following basic personal hygiene practices have proved to be valuable in preventing pathogen transmission: frequent and proper hand washing; covering the moth when sneezing or coughing; avoidance of sharing food, beverages, cooking utensils, and toothbrushes; and avoiding others when sick. The use of disinfectants also helps stop transmission in places where people have obvious signs of infectious disease. Disinfection is as important in the workplace and at home as it is in hospitals and doctors' offices.

Transmission is a crucial component of infectious disease pathogenesis. As a result, the medical profession makes special efforts to study the preferred modes of transmission of specific pathogens. Hygiene, sanitation, and disinfection and sterilization play key roles in stopping the transmission of pathogens in human, in animals, or from animals to humans.

See also AEROMICROBIOLOGY; DISINFECTION; GERM THEORY; HYGIENE; MALARIA; NOSOCOMIAL INFECTION; PATHOGENESIS; PORTALS; SEXUALLY TRANSMITTED DISEASE.

Further Reading

Bean, Bonnie, B. M. Moore, B. Sterner, Lance R. Peterson, Dale N. Gerding, and Henry H. Balfour. "Survival of Influenza Viruses on Environmental Surfaces." *Journal of Infectious Disease* 146 (1982): 47–51.

Boone, Stephanie A., and Charles P. Gerba. "Significance of Fomites in the Spread of Respiratory and Enteric Viral Disease." *Applied and Environmental Microbiology* 73 (2007): 1,687–1,696. Available online. URL: http://aem.asm.org/cgi/reprint/73/6/1687. Accessed January 21, 2010.

Dailey, F. Dean, Pauline M. Rakich, and Kenneth S. Latimer. "Cytauxzoonosis in Cats: An Overview." University of Georgia Clinical Pathology Clerkship Program. Available online. URL: www.vet.uga.edu/VPP/CLERK/Dailey/index.php. Accessed January 21, 2010.

Mount Sinai Hospital. "Methods of Disease Transmission." Available online. URL: http://microbiology.mtsinai.on.ca/faq/transmission.shtml. Accessed January 21, 2010.

Pyrek, Kelly M. "Breaking the Chain of Infection and Preventing Cross-Contamination." *Infection Control Today,* 29 April 2009. Available online. URL: www.infectioncontroltoday.com/articles/chain-of-infection-cross-contamination.html. Accessed January 21, 2010.

———. "Fomites' Role in Disease Transmission Is Still Up for Debate." *Infection Control Today,* 1 August 2002. Available online. URL: www.infectioncontroltoday.com/articles/281feat1.html. Accessed January 21, 2010.

Tellier, Raymond. "Aerosol Transmission of Influenza A Virus: A Review of New Studies." *Journal of the Royal Society Interface* 6 (209): S783–S790. Available online. URL: http://rsif.royalsocietypublishing.org/content/6/Suppl_6/S783.full. Accessed January 21, 2010.

Thomas, Yves, Guido Vogel, Werner Wunderli, Patricia Suter, Mark Witschi, Daniel Koch, Caroline Tapparel, and Laurent Kaiser. "Survival of Influenza Virus on Banknotes." *Applied and Environmental Microbiology* 74 (2008): 3,002–3,007. Available online. URL: http://aem.asm.org/cgi/reprint/74/10/3002. Accessed January 21, 2010.

World Health Organization. "Global Strategic Framework for Integrated Vector Management." 2004. Available online. URL: http://whqlibdoc.who.int/hq/2004/WHO_CDS_CPE_PVC_2004_10.pdf. Accessed January 21, 2010.

———. "Vector-Borne Disease." Available online. URL: www.who.int/heli/risks/vectors/vector/en. Accessed January 21, 2010.

How Safe Is Air Travel?

by Philip M. Tierno, Jr. Ph.D., Clinical Microbiology and Immunology, New York University Medical Center, New York City, New York

Certainly anyone who flies frequently and without mishap might conclude that air travel is a relatively safe phenomenon. Indeed, the Bureau of Transportation Statistics estimates that 700 million people fly annually in the United States, yet few significant events related to health have been reported. Nevertheless, many travelers experience bouts of a cold or flu, dizziness, sore throats, light-headedness, dry or watery eyes, fatigue, hearing loss, difficulty in breathing, or anxiety shortly after flying. Though these conditions occur at a relatively low frequency, concern has been growing regarding the increased risks air travel might cause for contracting a contagious disease.

The Centers for Disease Control and Prevention (CDC) has warned travelers that tuberculosis (TB) can spread during air travel. Such a spread occurred on a flight from Chicago to Honolulu, in 1995, when six passengers contracted TB; four of them had been sitting two rows away from a sick passenger who had boarded the plane infected with TB. Although the chance of contracting TB in this way is only about one in 1,000, the World Health Organization (WHO) says flights of eight hours or more are a concern when a person who has TB symptoms (persistent coughing) is on the plane. There are several other diseases that travelers might contract either from-person-to-person transmission or through the air while onboard an aircraft: severe acute respiratory syndrome (SARS), influenza, adenovirus, respiratory syncytial virus, pneumonia, legionella, meningitis, and measles. Of these infectious diseases, SARS is caused by a coronavirus and may be transmitted by airborne droplets or by direct contact. In 2003, on one three-hour flight from Hong Kong to Beijing, China, a single passenger sitting in the middle of the aircraft coughed continuously. Within eight days, 21 passengers and crew members contracted SARS. Most of the infected passengers had been sitting within five rows of the ill passenger.

Probably the best documented airborne infection is influenza, and health professionals, in fact, believe that air travel is the chief cause of the global spread of the flu. After the September 11, 2001, terrorist attacks in New York City and Washington, D.C., air travel dropped dramatically. A team of Harvard University epidemiologists led by John Brownstein showed a direct link between the numbers of people traveling by air and the rate of spread of the flu virus. By restricting air travel during a two-week period after September 11, the spread of flu was also delayed.

There are only a few reports of outbreaks of the common cold as a result of air travel. Colds are transmitted by direct and indirect contact with respiratory secretions rather than via airborne transmission. Direct contact is defined as person-to-person germ transmission, and indirect contact is transmission by touching contaminated inanimate surfaces, then touching a person's own eyes, nose, or mouth. The eyes, nose, and mouth are the body's portals of entry for pathogens. In addition, air travel can increase a person's susceptibility to colds because of low humidity aboard aircraft, as well as the tendency of people to become fatigued and have higher anxiety levels when traveling, or because they are more likely to become dehydrated. Perhaps most important, people in close quarters sharing common surfaces can more efficiently spread germs. Finally, aircraft may carry mosquitoes, cockroaches, rodents, or other pests and vermin that act as vectors in transmitting infectious microorganisms. For this reason, aircraft may occasionally have to be fumigated if rodents or other vermin or severe insect infestations are found. Cleaning and sanitation reduce food-borne and waterborne outbreaks, which have also been periodically reported. The common food-borne and waterborne microorganisms implicated in air travel–associated outbreaks are *Salmonella, Staphylococcus, Vibrio,* and a virus called norovirus (Norwalk-like viruses).

The table lists recommendations for maintaining good health during air travel. Almost all (80 percent) airplane air flows vertically through the cabin, a system that, to some extent, reduces the spread of germs from row to row. Air inside an airplane exchanges about 15–20 times per hour (compared to about 10–12 exchanges per hour in a typical office building), and half the air recirculates through special particulate filters. These high-efficiency particulate air filters (HEPA filters) take small infectious particles out of the recirculating air and, thus, limit passenger exposure to them. Unfortunately, no regulations currently require the use of HEPA filters on aircraft.

The risk of airborne infections is probably low, but not zero. Passengers can take precautions to help avoid contracting infections during air travel in addition to the recommendations described in the table. All air travelers should stay well hydrated, eat sensibly, and take an extra source of antioxidant vitamins like C, E, and A, as well as B complex vitamins. Numerous products on the market contain ingredients to boost immunity. Since fecal contamination is present on aircraft sinks, faucets, and door handles, passengers are wise to use an alcohol-based gel sanitizer or an antibacterial wipe to degerm hands after using the restroom and before eating, drinking, or touching the face. Finally, reducing

Healthy Air Travel Recommendations

Recommendation	Reason
drink at least eight ounces of fluid (preferably water) for every hour of flight; avoid alcohol and caffeine drink bottled water and avoid eating foods that appear improperly prepared	proper hydration helps the immune system stop germs from entering the common portals of entry: mouth, nose, and eyes bottled water often has fewer microorganisms than the water aboard airplanes; improperly cooked foods allowed to cool slowly before serving might carry dangerous levels of food-borne pathogens
change seats (if possible) if seated next to someone who has symptoms of illness (coughing, sneezing, runny nose, etc.) or create a barrier against transmission with a face mask or handkerchief avoid wearing contact lenses during a flight	the best defense against germ transmission is to block the transfer of germs between people when they touch, expel aerosols, or contaminate inanimate surfaces mucous membranes of the eyes are a portal of entry for infection; irritated and burning eyes offer less defense against infection
practice good personal hygiene	most surfaces on airplanes (seats, headrests, tray tables, armrests, etc.) are not disinfected, and surfaces inside airplane restrooms carry high numbers of fecal organisms; thorough hand washing before and after eating and using the restroom and avoiding touching hands to the face all reduce infection
assist the immune system by eating healthy foods, getting adequate sleep, and staying hydrated	after the skin, the immune system is a major defense against pathogens
avoid flying if already injured or sick, have had recent surgery or a heart attack; if flying is unavoidable, take along all medications to maintain good health	preexisting illnesses create higher risks for infection
take extra care with chronic respiratory conditions or immunocompromised conditions	chronic diseases (diabetes, acquired immunodeficiency syndrome [AIDS], heart disease) or people in high-risk health categories (elderly, very young, pregnant women, transplant patients) may have weak immune systems that create greater opportunity for infection

stress is important. Most travelers are under stress, which weakens immunity. Good planning, arriving early for flights, and keeping the challenges of travel in perspective all preserve good overall health.

Undoubtedly, travelers can minimize their risk of contracting an infection or other ill effects associated with air travel by using common sense and taking some simple precautions as outlined here. As Grandmother used to say, "An ounce of prevention is worth a pound of cure!"

See also AEROMICROBIOLOGY; IMMUNITY; TRANSMISSION.

Further Reading

Abraham, Thomas. *Twenty-First Century Plague: The Story of SARS.* Baltimore: Johns Hopkins University Press, 2007.

Brownstein, John S., Cecily J. Wolfe, and Kenneth D. Mandl. "Empirical Evidence for the Effect of Airline Travel on Inter-Regional Influenza Spread in the United States." *Public Library of Science Medicine* 3 (2006): 1,826–1,835. Available online. URL: www.plosmedicine.org/article/info%3Adoi%2F10.1371%2Fjournal.pmed.0030401. Accessed November 8, 2009.

Bureau of Transportation Statistics. Available online. URL: www.bts.gov. Accessed November 8, 2009.

Gilbert, Susan. "Travel Advisory: TB Spread on Airliner, but Risk Is Small." *New York Times,* 26 March 1995. Available online. URL: www.nytimes.com/1995/03/26/travel/travel-advisory-tb-spread-on-airliner-but-risk-called-s mall .html. Accessed November 8, 2009.

Mangili, Alexandra, and Mark A. Gendreau. "Transmission of Infectious Diseases during Commercial Air Travel." *Lan*

(continues)

How Safe Is Air Travel?

(continued)

cet 365 (2005): 989–996. Available online. URL: www.pall.com/pdf/Transmission_of_infectious_diseases_during_commercial_air_travel.pdf. Accessed November 8, 2009.

Ozonoff, David, and Lewis Pepper. "Ticket to Ride: Spreading Germs a Mile High." *Lancet* 365 (2005): 917–919.

Pavia, Andrew T. "Germs on a Plane: Aircraft, International Spread and the Global Spread of Disease." *Journal of Infectious Diseases* 195 (2007): 621–622.

Rainford, David J., and David P. Gradwell, eds. *Aviation Medicine,* 4th ed. Oxford: Oxford University Press, 2006.

Tierno, Philip M. *The Secret Life of Germs.* New York: Simon & Schuster, 2001.

Zitter, Jessica N., Peter D. Mazonson, Dave P. Miller, Stephen B. Hulley, and John L. Balmes. "Aircraft Cabin Air Recirculation and Symptoms of the Common Cold." *Journal of the American Medical Association* 288 (2002):483–486. Available online. URL: http://jama.ama-assn.org/cgi/reprint/288/4/483.pdf. Accessed November 8, 2009.

vaccine A vaccine is a substance administered to the body with the purpose of stimulating an immune response against a specific disease. For this reason, the act of giving a vaccine to a person or animal is called *immunization*.

The principle behind vaccines is to introduce an antigen into the body and, thus, initiate the production of antibodies. An antigen can be any biological particle identified by the body as foreign (non-self) matter. Bacteria, viruses, mold spores, proteins, and pollen represent the most common items that act as antigens in the body. Antibodies develop in response to a given antigen and are constructed to fit together with the antigen in a very specific way. This specificity gives antibodies a narrow role in defending the body from infection, but it also ensures that the next time the antigen appears, antibodies will be efficient at targeting the invader. Vaccines, thus, serve a twofold purpose: (1) to induce antibody production and (2) to ensure that the antibody is specific for the vaccine's particular antigen. By this process, vaccines provide the body with acquired immunity; that is, the body builds immunity through an outside stimulus.

Two major types of immunity defend against infection: acquired and innate. Innate immunity is supplied naturally by the body. Acquired immunity must be produced in the body in response to outside factors. A person develops natural acquired immunity when the body responds to foreign matter by releasing antibodies and specialized lymphocytes into the circulation system. By contrast, artificially acquired immunity forms by using the following processes:

- the body produces antibodies against a specific antigen (an entity foreign to the body) (active)
- vaccines put antigens into the body to elicit antibody production (active)
- preformed antibodies are put into the body (passive)

The first two methods of making acquired immunity are active methods, because an antigen induces the body to make antibodies. The third method is a passive way of forming immunity because an injection puts functioning antibodies into the body, sparing the body this task. In summary, a vaccine represents the passive form of artificially acquired immunity.

Oral and parenteral (by injection) vaccination makes up a major part of health care worldwide. Global vaccination programs have been successful at reducing the risk of certain diseases, notably smallpox, which was eradicated, in 1977. Most vaccination programs prevent viral and bacterial diseases, but fewer successful vaccines have been developed for protozoal or fungal diseases, because of the evasive mechanisms eukaryotes have against antibodies. Protozoa use various tactics that neutralize the action of the immune system:

- antigenic variation—frequent changes to surface proteins that antibodies cannot recognize
- enzyme systems that remove the digestive powers of the body's phagocytes
- proteases that degrade antibodies

The World Health Organization (WHO) states on its Web site, "Routine vaccination is now provided in all developing countries against measles, polio, diphtheria, tetanus, pertussis, and tuberculosis. To this basic package of vaccines, which served as the standard for years, have come new additions." The new vaccines that have become available, in almost 200 countries, are directed against the following pathogens or diseases: *Haemophilus influenzae*, hepatitis B, rubella, and yellow fever. Industrialized countries receive a broader array of available vaccines, including the vaccines available in developing countries, plus vaccines for influenza and strains that cause pneumococcal disease. Children in industrialized countries usually receive mumps vaccination in a single vaccine that also covers measles and rubella.

HISTORY OF VACCINATION

The ancient Greeks understood the principle of immunization through the observation that people who had recovered from yellow fever, plague, and other diseases never again contracted the same disease. More than one culture may have discovered the benefit of inoculating healthy people with exudates from a diseased person to protect against disease. In Turkey, doctors used secretions from smallpox sufferers to inoculate against this virus.

The first vaccine in Western medicine had an unlikely source, in the 18th century. Lady Mary Montagu (1689–1762), a member of the British upper class, had learned about vaccination during a period when she lived in Turkey. In her home country, Lady Montagu took the daring step of inoculating her son and the princess of Wales's children with secretions from the lesions of smallpox patients. Despite the success of these inoculations, doctors did not pursue the science behind vaccination until Edward Jenner (1749–1823), a village doctor in rural England, began studying smallpox.

In the English countryside, almost every family had a member who contracted cowpox (called *Variole vaccinae*) from dairy cows, and almost 40 percent of them died of the disease. But cowpox seemed to create an important health advantage: Many of the villagers believed that anyone who could recover from cowpox would never develop smallpox.

Jenner suspected the cowpox-smallpox connection to be a myth, until he noticed that young women hired to milk cows at local dairies seldom had the telltale pock lesions that he saw in smallpox patients. In 1796, Jenner discovered one girl who milked cows but, unlike the others, had pocks on her skin. He collected the contents of one of her active cowpox pustules and injected the preparation into the arm of an eight-year-old boy, who, a week later, developed only mild symptoms of cowpox. Jenner, then, inoculated the youngster with live smallpox virus, a process called *variolation*. As Jenner had suspected would happen, the boy never developed the smallpox disease.

Jenner began inoculating all his patients with exudates from cowpox pustules with the same promising results. Jenner had, in fact, collected the small quantity of cowpox exudates from the boy and used it in inoculation of others, repeating the same process five times in five groups of patients, each inoculated with material from the previous group. In 1798, Jenner wrote, "These experiments afforded me much satisfaction, they proved that the matter in passing from one human subject to another, through five gradations, lost none of its original properties." Jenner made key observations that now form the basis for vaccine development:

- different strains or mutants of the same pathogen can be used for immunization
- weakened strains of a pathogen retain the ability to induce antibody formation in the body

The French microbiologist Louis Pasteur (1822–95) carried out further work on immunization by growing pathogens in cultures until they lost virulence but still induced antibody formation when injected into a host. Pasteur had discovered attenuation, which is the process of weakening a pathogen so it can serve as a safe inoculation. Pasteur's method involved lengthening the incubation period for smallpox in laboratory cultures. After increasingly long periods, up to a month or more, Pasteur detected a decrease in the pathogen's virulence in chickens. He wrote in 1880, "When such intervals between culturing are used, one finds that instead of identical virulence, so that out of ten chickens inoculated, only nine, eight . . . three, two, or one out of ten die, and at times not a one dies, so that the sickness may develop in all of the inoculated chickens, but they all recover. . . . By merely lengthening the time between transfers, we have obtained a method for decreasing progressively the virulence of the virus, until we finally have a virus that is a true vaccine, in that it does not kill, but induces a benign illness which immunizes against a fatal illness." Attenuation remains an important method used by drug companies for vaccine preparation.

TYPES OF VACCINES

Drug companies make the following four different types of vaccines: (1) whole-cell, (2) macromolecule or subunit, (3) recombinant, and (4) nucleic acid. Whole-cell vaccines include bacterial or viral vaccines and contain either inactivated (killed) organisms or attenuated (live) organisms. Each type of whole-cell

Inactivated and Attenuated Whole-Cell Vaccines

Type	Advantages	Disadvantages
inactivated	• pathogen is killed • pathogen virulence cannot return • stable during storage	• may require booster shots • requires chemicals for inactivation
attenuated	• single shots, no boosters	• pathogen is live • may revert to virulent form • less stable during storage

vaccine has advantages and disadvantages, summarized in the table above.

Macromolecule vaccines consist of a subunit of a whole pathogen. Purified constituents of a pathogen do not have the drawbacks of whole-cell preparations. The main macromolecule vaccines are (1) polysaccharides from bacterial capsules, (2) surface antigens made by recombinant deoxyribonucleic acid (DNA) technology, and (3) inactivated toxins called *toxoids.* The *H. influenzae,* meningitis, and *Streptococcus pneumoniae* vaccines derive from polysaccharides; the hepatitis B vaccine is made from recombinant antigens; and the diphtheria and tetanus vaccines use toxoids made from inactivated toxins.

Vaccine makers often prepare macromolecule vaccines by using recombinant technology. In this type of production, a molecular biologist finds a gene that encodes for the macromolecule's synthesis in a cell. The biologist removes the gene from the cell and inserts it into a microorganism, that, can produce it rapidly in high amounts. Macromolecules from the surface of a pathogen cell can elicit the same immune reaction from the body that would occur if the whole pathogen were present.

Toxoid vaccines target poisons made by a pathogen rather than the pathogen itself. Since the toxin has been inactivated, it cannot cause harm to the body. Vaccines that deliver inactivated matter require the use of booster shots. Booster shots must be administered months to years after the initial vaccination. Vaccines require booster shots whenever the initial shot cannot stimulate an adequate level of immunity to protect against infection. The following viral diseases require booster shots to give a person complete immunity against the pathogen (approximate timing for the booster is in parentheses):

- hepatitis A—six to 12 months
- poliomyelitis—as needed
- influenza—yearly

Bacterial pathogens that require booster shots are:

- cholera—six months
- diphtheria—10 years
- pertussis—10 years
- tetanus—10 years
- tuberculosis—three to four years

Not all people need to receive all of these boosters if their risks of infection are low. For example, only people living in areas where cholera is endemic (always present in a population) or travelers to those areas need a cholera shot.

A newer type of vaccine, the DNA vaccine, activates both parts of the immune system: antibody production (the humoral system) and immune cells (the cellular system). DNA vaccines require a more complicated response from the body than the other vaccines. The U.S. Department of Agriculture explained, "Traditional vaccine development involves either passing a disease-producing virus through a different species or cell type until it no longer causes disease but does create immunity, or by killing the virus in such a manner that it allows it to produce immunity but no disease in the recipient. DNA vaccines, by contrast, use different fragments of a pathogen's unique genetic material to stimulate a target immune response from the host." The vaccine delivers a piece of DNA containing genes for an antigen into a muscle injection site. The muscle cells take up the DNA segment and incorporate the gene into the cells' DNA. As a result, the muscle starts producing antigens, which turns on both antibody production and immune system cells. In 2005, a vaccine for horses to protect against West Nile virus infection became the first DNA vaccine approved by the U.S. Food and Drug Administration (FDA). At the time of the approval, the acting head of the FDA Julie Gerberding said, "This is a truly exciting innovation

Does Vaccination Improve or Endanger Our Health?

by Anne Maczulak, Ph.D.

Vaccinations help the body defend against infection by boosting the immune system's antibody production against a specific pathogen. Vaccination, thus, plays a preemptive role in building the body's defenses against infection. Since the ancient Greeks, people have recognized that many diseases do not recur after a person, usually in childhood, first contracts them. A case of smallpox, mumps, measles, chicken pox, or rubella would protect a person for life, aided by the fact that people had much shorter life spans in the past than they do today.

Although a few people experimented with homemade vaccines by taking into their body the exudates from a sick person, the English doctor Edward Jenner (1749–1823) was the first to take a scientific approach to immunization. Since Jenner's first use of a smallpox vaccine in 1796, many millions of vaccinations have been given to adults and children the world over, for a variety of bacterial and viral pathogens. In science, knowledge accumulates, so people know more about the development and actions of vaccines than ever before. It seems surprising, then, that more than 200 years after Jenner convinced skeptical fellow physicians of the value of vaccination, the public has now turned against many vaccination programs that the medical community advocates.

Current vaccines are made in a variety of ways for the purpose of eliciting an antibody response from the immune system. These vaccines may have any of the following forms: attenuated (live but weakened) pathogen, killed pathogen, pieces of a pathogen, inactivated toxin, a toxin conjugated to a polysaccharide, or segments of a pathogen's nucleic acids. In other words, a person who receives a vaccine accepts that a pathogen or part of that pathogen is to be injected directly into an otherwise healthy body. Many people no longer view this process as an important step in maintaining good health and see vaccination as a risk too dangerous to take.

The public's resistance to vaccination has three main roots. The first is an understandable reluctance to introduce a pathogen into the body. Early smallpox vaccines would, on occasion, cause disease, rather than prevent it. Current oral polio vaccines have a very small risk of causing this disease. Sometimes, a vaccine's side effects worry people. Influenza vaccines cause scratchy throat, weakness, and fatigue in many vaccine recipients, one to three days after a flu shot. The second reason for avoiding vaccines relates to allergic reactions some people may have to the constituents of the vaccine. For example, proteins derived from the pathogen may lead to allergic reactions. Drug companies make flu vaccine by using eggs in which the influenza virus multiplies. People who have egg allergy have been advised to avoid the flu shot. Finally, many pregnant women and parents of small children think the benefits of vaccination do not outweigh the unknown risks to the child. This last reason for being antivaccination has emerged because of partial information or misinformation in complex areas of science.

The H1N1 influenza pandemic, of 2008–09, provided an example of how a lack of information can increase the health risks of an entire population. In 2009, four drug companies rushed to supply the world with vaccines for the H1N1 virus. These companies performed sound research on the vaccine's development and did their best to meet the demand for the 2009 flu season. Public health officials noted, however, that the public believed the drug did not have proper testing, had been rushed into production, and contained substances that caused more harm than flu would. These assumptions were wrong, but the per-

and an incredible scientific breakthrough that has potential far beyond preventing West Nile virus in horses. This science will allow for the development of safer or more effective human and animal vaccines more quickly." DNA vaccines for the human immunodeficiency virus (HIV), influenza, hepatitis C, tuberculosis, and possible bioweapons are in testing as the next generation in this new technology.

HOW VACCINES WORK

Vaccines force the body to detect foreign matter so that the immune system will develop memory of the antigen. Memory is a phenomenon is which the immune system response is faster and stronger the second time an antigen invades the body than in the first invasion, months or years earlier.

Nonspecific antibodies in the blood give people a primary response to pathogens. Primary responses do not target a particular pathogen and are not always adequate to stop infection without additional immune system help, but these antibodies are, at least, always at the ready. Memory cells (specialized lymphocytes), by contrast, provide targeted protection against a specific pathogen. The system gives the body a highly efficient defense, but the body needs time to manufacture

ception of an unsafe vaccine prompted many people to forgo a flu shot. Pregnant women had been particularly skeptical of vaccination even though they are known to belong to a high-risk group for infections of all kinds. In most communities, less than half of all pregnant women received the vaccine.

The World Health Organization (WHO) has warned the public that the H1N1 pandemic is unstoppable, and only vaccination can break its transmission. The organization has recommended that health care workers receive first priority for vaccination. Because health care workers take care of people who have chronic disease and other conditions that weaken the immune system, they must be extra careful to reduce the chance of transmission. Vaccinations for H1N1 were recommended for the next priority group, the young, between ages two and 24 years.

Decisions on whether to accept or reject vaccination are often based on the fear that some vaccines have been linked to an increased incidence of multiple sclerosis, diabetes, Alzheimer's disease, inflammatory bowel disease, or autism. Doctors usually can diagnose autism in young children, during the period when children are likely to be receiving vaccinations for the "childhood diseases" chicken pox, measles, mumps, and rubella. No definitive studies have proven a connection between autism and vaccination, but the fear among parents persists.

Herd immunity is a phenomenon in which a certain minimal percentage of immune individuals in a population confer immunity on the entire group. This can occur because the probability of a pathogen's finding a susceptible host has been greatly reduced in a community containing a high percentage of immune individuals. In many communities, people know of herd immunity and depend on it to protect them from infection. Of course, if every person relied on herd immunity in lieu of vaccination, then no one would be protected.

The reality of vaccination is that drug companies have decades of experience in perfecting the production of flu and other vaccines. These companies have, ironically, become a victim of the success of past vaccination programs. The smallpox virus was eradicated, in 1977, with a global vaccination effort. Measles and polio have been reduced through vaccination and might be eradicated as soon as 2015, and the malaria vaccine has had more modest success in eliminating that disease. Today, the public might assume that doctors are vaccinating for diseases that no longer exist; some doctors have never seen a polio patient.

Doctors must convince wary patients that the risk of becoming sick because of avoiding vaccination far outweighs the risk of side effects of vaccines. Current and a new generation of vaccines can only work if a majority of people believe in their value. New vaccines being developed for human immunodeficiency virus/acquired immunodeficiency syndrome (HIV/AIDS), addictions, unwanted pregnancy, and certain cancers will require the same diligence as childhood disease vaccines and the flu shot.

On the other hand, a vaccination can never absolve a person from practicing good hygiene to stop disease transmission. Vaccinations have always been a single weapon in an arsenal against disease. Too little faith or too much faith in a vaccine will never lead to success in eradicating disease.

See also IMMUNITY; JENNER, EDWARD; VACCINE.

Further Reading

Centers for Disease Control and Prevention. "Vaccines and Immunizations." Available online. URL: www.cdc.gov/vaccines. Accessed November 8, 2009.

World Health Organization. "WHO Recommendations on Pandemic (H1N1) 2009 Vaccines." Available online. URL: www.who.int/csr/disease/swineflu/notes/h1n1_vaccine_20090713/en/index.html Accessed June 6, 2010.

the correct antibodies. For most pathogens, memory cell immunity requires two to four weeks to reach its maximal strength. But the advantage of memory resides in the strength of the response; the secondary immune response (memory) can be 100 times stronger than the primary response to the same pathogen.

VACCINE CONTROVERSY

Many people harbor suspicions about vaccine safety and efficacy. The treatments for killing pathogens to be put into vaccines require chemicals or irradiation, which worries many consumers about the safety of the vaccine. Some people also worry about the possibility that live pathogens will accidentally become part of the vaccine. Vaccine opponents point to the following potential dangers of vaccines as a reason to avoid all vaccinations: (1) toxic chemicals in vaccine preparations, (2) constituents that accumulate in body tissues, (3) combination of the DNA from the vaccine with the body's DNA, (4) potential to cause the disease the vaccine is meant to prevent, (5) overstimulation of the immune response, (6) presence of genetically engineered components. None of these possibilities has been proved, but the fear of vaccination has prevailed in society, for decades.

In 1920, Charles M. Higgins wrote the treatise *The Horrors of Vaccination Exposed and Illustrated,* which made some of the earliest claims against the safety of vaccines. In his book, Higgins characterized smallpox vaccination "as being now actually more dangerous to public health and human life than natural smallpox" and charged doctors with "denying and concealing these facts from the people." Although today's consumers have more information about the science of vaccination, much of the criticism of vaccines reflects the same anger that Higgins displayed a century ago.

Many U.S. parents choose not to have vaccinations for their children, believing that avoiding the shots and boosters is safer than vaccination. As a result of increasing the percentage of unvaccinated children in schools, many doctors believe, the mechanism by which pathogens move through populations has changed to the detriment of community health. The Marin County, California, public health officer Anju Goel said, in 2009, "It's more than a decision about an individual child. Your decision not to vaccinate a child has a larger effect in the community. If you do make that decision, you need to be aware." Parents' reasons for refusing vaccinations for their children fall into three main groups: (1) doubt over the vaccine's effectiveness; (2) fear that the vaccine causes other medical conditions, such as autism; and (3) distrust of the government's vaccination programs.

Some of the fears of vaccine safety are caused by the ingredients used in vaccine preparations. The Centers for Disease Control and Prevention (CDC) publishes a list of ingredients approved for use in vaccines made or sold in the United States. Any person can access this list, the "Vaccine Excipient and Media Summary," on the CDC's Web site to check for ingredients that pose particular concerns.

Vaccinations protect community health through herd immunity. Herd immunity occurs when a certain minimal percentage of individual members of a population have immunity to a pathogen and, as a result, the probability of the pathogen's finding a susceptible individual becomes very low. Goel's colleague Fred Schwartz explained, "As the rates of unvaccinated children increase, the potential of [herd immunity] increases. In effect, it's contributing to the greater risk of the community." The same year, California chiropractor Donald Harte countered, "Autism was a rare thing 30 to 40 years ago. Now it's affecting one child in 170. It's an epidemic, physician-caused. And it's ruining a generation." Physicians concede that no drug is 100 percent safe. More than 20 scientific studies of vaccines and autism have shown no link between them. Parents understandably prefer to err on the side of safety when their children are involved.

The WHO has stated that immunization is a proven method for preventing infectious diseases and saves up to two million lives annually. But many people perceive the health risks of vaccines and the risks of vaccine side effects to be about equal. On the contrary, the majority of physicians agree with the WHO and emphasize that the danger of infectious disease far outweighs any harmful side effects of a vaccine.

Vaccines have been a crucial part of medical care since their discovery. Vaccination programs have been lifesaving measures in developing countries in regions where infectious disease is endemic. In industrialized countries, vaccination sometimes faces more scrutiny. Despite the perceived dangers, vaccines are considered by the medical community to be a safe and very effective way to prevent deaths of infectious disease.

See also IMMUNITY; JENNER, EDWARD; RECOMBINANT DNA TECHNOLOGY.

Further Reading

Bill and Melinda Gates Foundation. "Bill and Melinda Gates Pledge $10 Billion in Call for Decade of Vaccines." January 29, 2010. Available online. URL: www.gatesfoundation.org/press-releases/Pages/decade-of-vaccines-wec-announcement-100129 .aspx. Accessed January 30, 2010.

Centers for Disease Control and Prevention. "Vaccine Excipient and Media Summary." Available online. URL: www.cdc.gov/vaccines/pubs/pinkbook/downloads/appendices/B/excipient-table-1.pdf. Accessed January 19, 2010.

Center for Infectious Disease Research and Policy. "West Nile Shot for Horses Is First Licensed DNA Vaccine." July 21, 2005. Available online. URL: www.cidrapbusiness.us/cidrap/content/other/wnv/news/july2105wnv.html. Accessed January 2, 2010.

Higgins, Charles M. *The Horrors of Vaccination Exposed and Illustrated.* New York: Charles M. Higgins, 1920. Available online. URL: http://books.google.com/books?id=jKc3WdszeOYC&printsec=frontcover&dq=Charles+Michael+Higgins+Horrors+of+Vaccination&source=bl&ots=m_O4-_uIgB&sig=3z1aVcO74QwqlHMRUeG2nbzehro&hl=en&ei=hAlU S5eaH5DusQPUzPX9Bw&sa=X&oi=book_result&ct=result&resnum=8&ved=0CBkQ6AEwBw#. Accessed January 17, 2010.

Rogers, Bob. "Refusal to Vaccinate Puts Kids at Risk." *Marin Independent Journal,* 12 April 2009.

U.S. Department of Agriculture. Available online. URL: www.usda.gov/wps/portal/usdahome. Accessed January 2, 2010.

U.S. Food and Drug Administration. Available online. URL: www.fda.gov. Accessed January 2, 2010.

University of South Carolina. "Edward Jenner and the Discovery of Vaccination." 1999. Available online. URL: www.sc.edu/library/spcoll/nathist/jenner.html. Accessed January 17, 2010.

World Health Organization. "Immunization against Diseases of Public Health Importance." Fact sheet, March 2005. Available online. URL: www.who.int/mediacentre/factsheets/fs288/en/index.html. Accessed January 17, 2010.

virulence Virulence is the intensity of a microorganism's pathogenicity, which is the ability to cause disease. Virulence can refer to any pathogen, whether it is a bacterium, virus, fungus, yeast, or protozoan.

Virulence and pathogenicity describe pathogens in relative terms. For example, Lassa virus is more virulent than the cold virus; *Escherichia coli* O157:H7 is more virulent than other strains of *E. coli*. To determine whether one microorganism is more virulent than another, a microbiologist or physician assesses the severity of symptoms in an infected person. Severe symptoms correspond to a pathogen of high virulence, and mild to absent symptoms relate to low virulence.

The severity of any infectious disease that develops in a human or animal results from the interrelationship of three factors: virulence of the pathogen, infectivity of the pathogen, and immunity of the host. Infectivity is a pathogen's ability to establish a community on or in a host but not necessarily cause disease. Immunity is the collection of physical and biological factors that protect a host from infection and/or infectious disease.

The establishment of an infection or the development of a disease from an initial infection is due

Virulence Factors

Factor	Type	Mode of Action	Producer
α-1,3-glucan	carbohydrate	interferes with immune cell recognition and attachment	*Blastomyces* (F)
coagulase	enzyme	clots blood to protect the infecting pathogens from body's defenses	*Staphylococcus aureus* (B)
collagenase	enzyme	degrades connective tissue to allow pathogen to spread from an infection site	*Clostridium* (B)
elastase	enzyme	degrades tissue basement membranes to facilitate pathogen spread	*Pseudomonas aeruginosa* (B)
hyaluronidase	enzyme	degrades hyaluronic acid that holds host cells together, thus allowing penetration into tissue,	*Clostridium, Staphylococcus Streptococcus* (B)
hydrogen peroxide	chemical defenses	damages host cells and inhibits host	*Mycoplasma* (B)
lecithinase	enzyme	destroys a component of host cell membrane	*Clostridium* (B)
lysis	enzymatic action	destroys host cell to liberate a new generation of pathogens	viruses
O-antigens	carbohydrate	variability reduces the effectiveness of the host's antibodies	many gram-negative bacteria
porins	proteins	bacterial membrane channels inhibit attachment by host cells such as phagocytes	mainly gram-negative bacteria
protease	enzyme	degrades tissue basement membranes to facilitate pathogen spread	*Aspergillus* (F)
protein A	protein	on bacterial cell wall, interferes with host's normal antibody response to infectious agents	*Staphylococcus aureus* (B)
streptokinase	enzyme	digests blood clots to allow pathogen to spread from an infection site	*Staphylococcus, Streptococcus*

Note: (B) = bacterium; (F) = fungus

to the interactions between the pathogen and the host. Each of these aspects of infectious disease can be subdivided into specific virulence factors possessed by the pathogen and host factors that either strengthen or weaken a host's ability to fight infection.

VIRULENCE FACTORS

Pathogenicity factors are the mechanisms a pathogen calls upon to establish an infection and cause disease. For example, the methods various bacteria use for adhering to tissue constitute pathogenicity factors. Without adherence, a microorganism's ability to start an infection becomes reduced. Virulence factors, by comparison, encompass all of the mechanisms a pathogen uses to increase the severity and the persistence of the disease it has caused.

Virulence factors damage the host in either a direct or on indirect manner. Direct virulence factors are from the pathogen, such as a toxin or an enzyme, and disrupt normal functions in specific tissues in the host. Indirect virulence factors involve pathogen activities that eventually lead to harm in the host. For example, the use of host's nutrients, production of compounds that make nutrients unavailable to the host, damage to tissue surrounding an infection, and induction of a hyperimmune response represent indirect virulence factors.

Many bacteria, algae, and fungi use toxins as a virulence factor. Toxins are poisonous substances with effects ranging from mild cell damage to fatal damage of organs or metabolic systems. The capacity of a microorganism to produce toxins is called *toxigenicity.* A high degree of toxigenicity, therefore, contributes to a pathogen's virulence, because toxins act by either damaging parts of a host's cells or interfering with metabolic reactions.

Medical microbiology researchers regularly discover new virulence factors. Some are poorly understood and require more study. For example, pathogens might use chemotaxis to regulate the size of an infecting population. Chemotaxis is the response of a microorganism to a chemical stimulus. By monitoring nutrients or the density of an invading population, a pathogen can control its rate of growth until it develops a population density likely to overcome the body's defenses. Many other virulence factors have already been characterized in more detail, such as the well-known factors listed in the table on page 775.

The virulence factors listed in the table enable pathogens (1) to evade the immune system, (2) to create opportunities for spreading in the body, and (3) to penetrate tissue. Many pathogens are known to possess more than one virulence factor. Pathogens also usually possess a combination of pathogenicity and virulence factors working in tandem to foster infection and disease. For this reason, virulence and pathogenicity have become related in the study of infection and infectious disease.

The host cell's response to a pathogen impacts the pathogen's success or failure in persisting in the body. Many virulence factors evolved as defenses against these host immune responses, explaining why no discussion of virulence can ignore the corresponding host defenses against infection.

HOST FACTORS IN INFECTIOUS DISEASE

Host factors that affect a pathogen's virulence belong to two categories: nonspecific and specific defenses. Nonspecific defenses consist of physical barriers to infection. Specific defenses are the reactions to foreign substances in the body executed by the immune system.

Physical, chemical, and biological processes make up a host's nonspecific factors against infection. The following mechanical factors create physical barriers that prevent pathogens from entering the host:

- intact skin—epidermis in the upper layer contains a strong protein, called *keratin,* that resists penetration by pathogens
- mucous membranes—provide a weak physical barrier and secrete mucus, which can trap some pathogens
- tears and saliva—rinse surfaces to flush pathogens away from mucous membranes
- urine—flushes pathogens from mucous membranes of the genitourinary tract
- cilia—synchronized beating of small hairlike appendages lining the upper respiratory tract move foreign particle away from the lower respiratory tract

Chemical host factors on the human body consist of the following:

- sebum—prevents skin from drying out and, thus, prevents breaks in this barrier
- perspiration—flushes pathogens from the skin surface
- fatty acids—compounds in natural oils that inhibit bacteria
- lysozyme—enzyme in tears and perspiration that damages bacterial cells
- gastric juice—acidic fluid that kills most ingested microorganisms

Biological host defense factors occur outside and inside the body. On the outside of the body, native flora, which are the bacteria and fungi that are normal residents of the body, compete against pathogens by a variety of mechanisms. Native flora monopolize space and nutrients and, often, produce inhibitory substances such as antibiotics. Inside the body, nonspecific cells called phagocytes engulf and destroy any foreign particles they detect in the blood or in tissue. Bacteria have evolved virulence factors that seem to be designed to combat phagocytes. For example, some bacteria produce M protein on their surface or cover the entire cell in a capsule. These structures inhibit the attachment of phagocytes to the target bacteria. Pathogens that produce leukocidins can kill phagocytes before being destroyed by the host cells. Leukocidins cause the phagocyte to release its digestive enzymes into its own cytoplasm, leading to self-destruction.

The host's ability to set up an inflammation at an infection site also serves as an important protective nonspecific factor. Inflammation consists of four events: redness, pain, heat, and swelling. Each of these events is a sign that the inflammatory response is working and combating the infection.

In inflammation, the telltale redness indicates that vasodilation, expansion of blood vessels, has occurred. Vasodilation accelerates the body's ability to send defensive factors to the infection site. Pain results from the accumulation of compounds, fluids, and heat at the site. Heat is a function of an increased blood supply to the site and increased permeability of vessels, which permits fluids from the blood to enter intercellular spaces in host tissue. Finally, the increased fluid supply causes swelling, or edema. The increased blood supply and permeability, together, increase the level of the compound histamine at the infection site. Histamine signals other host factors of an inflammation and possible damaged tissue. Kinins, also supplied by permeable vessels, attract specialized phagocytes called *neutrophils* to the injured area. In the meantime, injured host cells release substances called *prostaglandins* that the body makes for the purposes of enhancing histamine and kinin activity and attracting more phagocytes.

The influx of fluids at an infection site helps the body override many of a pathogen's virulence factors. If the body's nonspecific factors work efficiently and quickly, blood clots will form around the infection to prevent pathogens from escaping into the bloodstream. This event can be detected by the accumulation of pus, which contains fluids with dead host and bacterial cells. When the body has successfully cordoned off the infection, additional phagocytic cells arrive to destroy the remaining pathogenic cells.

Fever accompanies many inflammations. When the brain's hypothalamus receives a message of ongoing inflammation, it raises the body's thermostat above its normal 98.6°F (37°C) to as high as 102.2°F (39°C). Short periods of high fever intensify the actions of several anti-infection processes and speed the body's tissue repair systems. But, as mentioned earlier, virulent pathogens capable of causing acute infections can elicit an exaggerated inflammation-fever response. The acute response weakens the host rather than protects it by damaging nerve tissue, speeding heart rate, unbalancing the body's chemicals called *electrolytes,* and causing dehydration. These actions contribute to a weakening of the immune system, which favors the virulent pathogen.

Some highly virulent pathogens kill so many hosts in such a short time that the disease they cause is called *self-limiting. Yersinia pestis,* the bacterium that causes bubonic plague, has caused some of history's most deadly epidemics. The black plague of the Middle Ages is estimated to have decreased Europe's population, in the mid-14th century, by one third. Doctors at the time of the black plague never found a cure for the disease; nor did others in subsequent epidemics, which recurred repeatedly into the 1800s. (Bubonic plague recurs sporadically today in isolated regions of the world.) Treatments did not defeat *Y. pestis.* Rather, the pathogen eliminated a large number of hosts quickly and disappeared from the population, thus ending each epidemic. A pathogen that possesses too much virulence seals its own fate by eliminating either its reservoir or its host. A reservoir is an organism that maintains a constant population of a pathogen and acts as a continual source of infection.

Microbiologists studying *Yersinia* have identified three main virulence factors, collectively called the Yops. The Yops consist of the following:

- YpkA, a kinase enzyme that breaks down blood clots
- YopE, a substance that breaks down muscle fiber
- YopH, the enzyme protein tyrosine phosphatase, which helps the *Yersinia* resist phagocytosis

The Yop factors confer dramatic virulence on *Y. pestis.* If untreated, the bubonic plague kills its victims in three to five days and the mortality rate is above 50 percent.

VIRULENCE IN MEDICAL MICROBIOLOGY

In 1928, the British geneticist Frederick Griffith (1879–1941) demonstrated the ability of pathogenic bacteria to transfer virulence genes to nonpathogenic bacteria. His procedure, now known as Griffith's transformation experiment, used two types of *Pneumococcus*: a

History of Virology

Person	Date	Event
Walter Reed (1851–1902), United States	1901	discovers yellow fever is caused by a virus
Karl Landsteiner (1868–1943), United States, and Erwin Popper (1879–1955), Austria	1908	shows that poliovirus causes poliomyelitis
Frederick W. Twort (1877–1950), Britain	1915	discovers that an extract from smallpox vaccine of a bacteriophage) inhibits bacterial cells growing in culture (discovery
Félix d'Herelle (1873–1949), Canada and France	1917	conducts detailed study of bacteriophages
Reinhold Rudenberg (1883–1961), Germany	1930	begins developing electron microscopy for the purpose of studying poliovirus
Wendell M. Stanley (1904–71), United States	1935	identifies part of tobacco mosaic virus (TMV) as protein
Frederick C. Bawden (1908–72) and Norman W. Pirie (1907–97), Britain	1938 1939	separates TMV into constituent proteins and nucleic acid
Emory L. Ellis (1906–?) and Max Delbrück (1906–81), United States	1939	delineates infective steps of viruses and invents plaque technique for studying bacteriophages
Alfred Hershey (1908–97) and Martha Chase (1927–2003) United States	1952	confirm with electron microscopy that DNA is the genetic material in the T2 bacteriophage

known virulent form possessing a protective capsule, which produced smooth colonies on agar, and an avirulent strain, which produced rough colonies. Griffith injected heat-killed smooth strains into mice and noted that the mice lived because the heating had destroyed the bacteria and their genes. He also mixed some of the killed smooth strains with live avirulent rough strains. When he inoculated mice with this mixture, the mice died. The only reason for this result could have been that the formerly avirulent rough strains had become deadly. Something in the mixture of dead virulent bacteria and live avirulent bacteria turned the innocuous rough strains into pathogens. Griffith rightly concluded that a component from the killed smooth strains had been transferred to the live rough strains, when he mixed both together. He called the unknown material a *transforming principle*. More than a decade later, other scientists would identify the transforming principle as deoxyribonucleic acid (DNA).

Griffith demonstrated that bacteria carry genes that confer virulence. Bacteria, furthermore, had the capability to pass the virulence genes to other bacteria by a process called *transformation*, the uptake of naked DNA from a solution. This revelation made the science of virulence all the more important in developing new disease treatments.

Researchers' main objective in characterizing virulence factors is developing effective treatments that neutralize or block pathogenic activities. This area of research has grown increasingly important, in light of the large number of bacteria now resistant to multiple antibiotics. A new generation of drugs will probably differ from an antibiotic mode of action and target the virulence factors of specific pathogens in order to control infectious disease. But despite microbiologists' best efforts, the most virulent pathogens have never disappeared, and the study of virulence will always be a vital part of medical microbiology.

See also GENE TRANSFER; IMMUNITY; INFECTION; PATHOGEN; PATHOGENESIS; TOXIN.

Further Reading

Hogan, Laura H., Bruce S. Klein, and Stuart M. Levitz. "Virulence Factors of Medically Important Fungi." *Clinical Microbiology Reviews* 9 (1996): 469–488. Available online. URL: http://cmr.asm.org/cgi/reprint/9/4/469. Accessed January 3, 2010.

Peterson, Johnny W. "Bacterial Pathogenesis." In *Medical Microbiology,* 4th ed. Galveston: University of Texas Medical Branch, 1996. Available online. URL: www.ncbi.nlm.nih.gov/bookshelf/br.fcgi?book=mmed&part=A560. Accessed January 3, 2010.

State Key Laboratory for Molecular Virology and Genetic Engineering (Beijing, China). "Virulence Factors of Pathogenic Bacteria." 2003. Available online. URL: www.mgc.ac.cn/VFs. Accessed January 3, 2010.

virus A virus is a submicroscopic particle consisting of a nucleic acid surrounded by a protein coat and capable of infecting animal, plant, or bacterial cells. Viruses do not contain all of the components needed to live as an independent organism so they must infect living cells to propagate new virus particles. For this reason, viruses are referred to as *obligate intracellular parasites.*

Viruses that infect animal cells often infect more than one species. For example, various seasonal influenza viruses have the ability to infect humans as well as swine, birds, and other animals. Animal viruses do not cross-infect plant or bacterial cells, however. Viruses specific for plants do not infect animals, and viruses that infect bacteria, called *bacteriophages,* do not infect either animals or plants.

The study of viruses, called *virology,* encompasses a broad range of subjects, including infectivity and infectious disease, immunity, pathogen transmission, and cell genetics. Most of these subjects related to viruses have aspects not seen in either prokaryotic or eukaryotic cell biology. Virologists specialize in these aspects of virus study, as well as the microscopic techniques needed to study biological objects much smaller than the smallest bacterial cell. Because viruses are measured in nanometers (nm; a scale of one billionth of a meter) and bacteria are measured in micrometers (μm; a scale of one millionth of a meter), virologists depend on electron microscopy, rather than light microscopy, to study the features of virus structure.

Virologists work in tandem with molecular geneticists to determine the origin of viruses and their relationship to cell evolution. These studies have raised interesting questions on whether viruses preceded bacteria or were derived from bacteria when life began forming on Earth.

ORIGIN OF VIRUSES

Today's known viruses depend entirely on living cells to host viral replication. This fact implies that viruses may have evolved very early in cell evolution, or cells and viruses coevolved. Modern advances in the analysis of deoxyribonucleic acid (DNA) and ribonucleic acid (RNA) have helped scientists propose theories on viral evolution. Studies of viral genes and their relationship to cellular genes have given rise to three main theories of the origin of viruses. In the first theory, viruses developed from primitive pieces of DNA or RNA, which became enveloped in protein during the evolution of true cells. Rather than developing the additional components that make up today's cells, viruses retained a primitive structure that could not live on its own. The second theory of virus origin proposes that after primitive cells developed on Earth, some lost certain features and regressed into a form that could not live on its own. These cells with limited capabilities to live independently became the obligate parasites today recognized as viruses. The final theory proposes a process by which DNA or RNA molecules developed rudimentary abilities to replicate, by using some of the enzymes present in cell replication. Each of these theories has compelling scientific evidence to support it.

The theory of coevolution between cells and viruses may incorporate features of each of the preceding scenarios. Genetic analysis of viruses and cells indicates the presence of shared genes. Gene transfers between viruses and cells take place by a mechanism known as *horizontal gene transfer,* meaning that genes move among different adult organisms, rather than from parent cells to progeny, the process of *vertical gene transfer.* The science of virogenomics has grown in recent years with the intent of comparing and interpreting the genetic makeup of viruses in relation to cells.

Viruses as parasites probably evolve inside the cells they infect. When viruses infect a cell, they take over the cell's DNA replication system. During this process, a virus has the opportunity to pick up genes or partial genes from the cell and incorporate the changes into viral nucleic acid. Such a change in the viral genome could contribute to the virus's evolution. A second occurrence that is known to take place in viral infections relates to a process called reassortment. In reassortment, two related viruses, such as two strains of the same influenza virus, infect the same animal. During infection, the two viruses mix and match genes, when they replicate inside a host cell. The *National Geographic* reporter Virginia Morell explained, in 2005, "Reassortment explains the two major flu pandemics [worldwide epidemics] of the 20th century, in 1957 and 1968. In each year a new flu subtype appeared, combining genes from the human virus that had been causing mild outbreaks in prior years with new genes from a bird virus." The predisposition of viruses to undergo rapid evolution explains why some pathogens, such as influenza, emerge in a new form every year.

Viruses have left no fossil records to help biologists trace their history. Even while virologists ponder the origin of virus particles, medical history shows that viruses may be humanity's oldest pathogens. Viral disease seems to have existed before recorded history. Evidence of smallpox has been uncovered from ancient human artifacts, and pustules on the 3,000-year-old mummy of the Egyptian Rameses V, suggest the ruler's death was due to the smallpox virus.

HISTORY OF VIROLOGY

Because viruses invade and disrupt cellular reproduction and because many diseases are caused by viruses, virology has developed simultaneously with other areas in microbiology. From the mid-1800s to the first decade of the 20th century, microbiology experienced a golden age of discovery, in which scientists increased their knowledge of bacteria, fungi, and protozoa.

In 1840, the German anatomist Friedrich Gustav Jacob Henle (1809–85) proposed the existence of additional microscopic organisms he called *contagia animata* in his essay "On Miasma and Contagia." Henle correctly suggested that these organisms were too small to be seen by using light microscopes. The microbiologists Louis Pasteur (1822–95), Henle's protégé Robert Koch (1843–1910), and the surgeon Joseph Lister (1827–1912) seized on Henle's idea to develop methods for monitoring microorganisms and preventing them from contaminating patients and foods. But in 1876, the German agricultural chemist Adolf Mayer (1843–1942) became the first scientist to study a virus when he inoculated healthy tobacco plants with juice extracted from the leaves of tobacco infected with mosaic disease (also called leaf spot disease). Although Mayer had discovered tobacco mosaic virus by showing that the inoculated plants developed the disease, within two to three weeks, he incorrectly attributed the occurrence to a pathogenic bacterium.

The Dutch soil microbiologist Martinus Beijerinck (1851–1931) would follow Mayer's studies with experiments of his own, in an attempt to identify the cause of mosaic disease. In 1899, he wrote, "In 1887 I attempted to discover if there was not a parasite which could be demonstrated to be the cause of the disease. Since microscopic studies were completely negative, the only type of bacteria that could be considered were those which could not be observed directly. But culture procedures showed that bacteria were completely absent, either from healthy or diseased plants." Beijerinck discovered that the pathogen existed in liquid squeezed out of tobacco leaves, which passed through a filter that collected bacteria and larger particles. The agricultural scientist Dimitri Ivanofsky (1864–1920) had shown the same filterable nature of mosaic disease in 1892, but Beijerinck was probably unaware of the Russian's work. Beijerinck, however, gave the first description in science of the possible mechanism of viral pathogens when he proposed, "A partial explanation would be the view that the *contagium* must be incorporated into the living protoplasm of the cell in order to reproduce, and its reproduction is so to speak passively brought about with the reproduction of the cell." Ivanofsky's and Beijerinck's work led to the present definition of a virus as a filterable parasite of living cells. The term *virus* also originated, during this time, to describe a virulent entity and was first applied to tobacco mosaic virus (TMV).

Filterable yet invisible viruses prompted a 25-year debate on whether these entities were solid or liquid. Many scientists in the debate may have overlooked the work of the German bacteriologists Friedrich Loeffler (1852–1915) and Paul Frosch (1860–1928), who theorized, in 1898, that the causative agent of foot-and-mouth disease could be a particle "only 1/10 or even 1/5 as large as [the smallest known bacterium], which really does not seem impossible." Loeffler and Frosch advanced the possibility of a pathogen existing in a realm beyond that which light microscopy could attain. The period between these early virus studies and the introduction of electron microscopy included additional milestones in virology, summarized in the table on page 778.

The advent of powerful microscopy methods enabled virologists to study the interrelationships of virus structure and infection. With this tool, virology could advance because the modes of action of viruses could be observed rather than postulated.

VIRUSES IN MICROBIOLOGY

Virus structure, classification, and mode of infection interconnect, so that studying one aspect automatically leads to studies of the others. These related specialties within virology have led to the following maxims of viral structure, metabolism, and infectivity:

- Viruses contain a single nucleic acid, either DNA or RNA, packaged in a protein coat.
- Viruses lack biochemical machinery to replicate themselves.
- Viral enzyme systems are limited to enzymes needed for nucleic acid replication and packaging of new virus particles.
- Viruses often destroy the host cell when they take over the reproductive processes for making new virus particles.
- A single virus invasion of a cell can result in the production and liberation of 100 or more new virus particles.

Although viruses are not microorganisms, in the sense that they cannot live on their own in any portion of their "life" cycle, microbiology includes viruses because they are observable only with a microscope, viruses and parasitic bacteria share many characteristics, and viruses infect higher organisms similarly to other pathogenic microorganisms.

VIRUS STRUCTURE AND CLASSIFICATION

Virus morphology has advanced with the development of technologies, in addition to electron microscopy, such as X-ray diffraction, which determines a mac-

Virus Classifications by Morphology

Feature	Categories of Viruses
nucleic acid	• DNA, double-stranded or single-stranded, linear or circular • RNA, double-stranded or single-stranded
lipid envelope	• enveloped • nonenveloped
capsid morphology	• helical—spiral surrounding a hollow core • icosahedral—made of 20 equilateral triangular surfaces and 12 vertices • mixed morphology—icosahedral capsid and helical core

romolecule or particle's shape on the basis of scattering of X-ray beams by the specimen. (The British physicist Rosalind Franklin [1920–58] used X-ray diffraction to help elucidate DNA's structure.)

Viruses differ from bacteria in their lack of a plasma membrane, cell wall, ribosomes, and other bacterial organelles. A virion is the fully developed virus particle capable of transmitting from host to host and infecting a new host. The virion consists of two components: nucleocapsid and capsid. The nucleocapsid is made up of either DNA or RNA (never both), and the capsid comprises the protein that enfolds these nucleic acids. Some viruses contain an additional layer outside the capsid, called the *envelope,* made of lipids in combination with proteins and/or carbohydrates. Envelopes sometimes support spikes made of proteins and carbohydrates. Spikes have become a useful feature for identifying viruses with electron microscopy.

Viruses have been classified in a variety of ways, mainly by the type of nucleic acid in the nucleocapsid, the presence or absence of an envelope, and capsid size and shape. The table above describes various classifications based on virus morphology.

Bacteriophages consist of mixed morphology of an icosahedral capsid and a helical tail that hold tail filaments for attaching to a bacterial cell.

Capsid morphology is distinctive of specific viruses, and electron microscopic studies reveal intricate features of both helical and icosahedral structures. A helical virus such as TMV is a rigid tube 15–18 nm in diameter and about 300 nm in length. TMV's RNA winds inside around the capsid in a spiral. By contrast, the helical influenza virus possesses a flexible capsid in which the RNA is folded. Icosahedral capsids vary by the subunits that make up the main structure. Small knobs called *capsomers* construct the capsid. The capsomer contains smaller subunits called *protomers.* Clusters of five protomers called *pentamers* (or pentons), usually, surround each of the vertices. Six-protomer capsomers called *hexamers* (or hexons) support the edges and faces of the icosahedron.

Morphology provides less information on how a virus behaves inside a host than nucleic acid content and presence/absence of an envelope. For this reason, virologists use these two means of classifying viruses more than morphology.

Enveloped viruses may belong to the DNA or RNA category. The main human enveloped viruses tend to carry only doubled-stranded DNA (dsDNA) or single-stranded RNA (ssRNA). Nonenveloped viruses, by contrast, may carry either DNA or RNA in double- or single-stranded configuration. Each type of virus also contains the enzymes needed for incorporating its DNA or RNA into a cell's DNA replication process.

When virologists classify viruses by capsid size and shape, they also group the viruses by nucleic acid to provide more information on the mechanisms of virus replication. The table on page 779 provides a common classification scheme for viruses using capsid features and nucleic acid content.

Some viruses within the classifications shown in the table possess distinctive features beyond size, shape, and nucleic acid content. For example, some DNA viruses, such as the hepatitis B virus, use an RNA intermediate to make more viral DNA inside a host cell. Some RNA viruses do the opposite, by using a DNA intermediate. This latter group, called the retroviruses, includes HIV and numerous tumor viruses (oncogenic viruses). Almost all of the information scientists have gathered on the genetic mechanisms of eukaryotic cells has been derived from studies of DNA-RNA systems in viruses, especially retroviruses.

In the 1960s and 1970s, virologists knew of animal retroviruses that carried RNA and built DNA by reversing the processing normally found in cells:

dsDNA → ssDNA → messenger RNA (mRNA)
→ ribosomal RNA (rRNA) and transfer RNA
(tRNA) → protein

The pathway has been described in biology by the mantra "DNA makes RNA makes proteins." But as virology developed, scientists discovered many variations on this theme, particularly the ability of viruses to run this pathway in reverse by retroviruses.

The first retrovirus studied caused leukemialike disease in fowl and leukemia and mammary tumors in mice. Little evidence existed for human retroviruses until, in the late 1970s, the American geneticist Robert Gallo (1937–) questioned the prevailing thought. He recalled, in 2005, that "pressure mounted against attempts to find human retroviruses. It was

Human Virus Classifications by Capsid Style and Nucleic Acids

Nucleic Acid Content	Family	Important Members	General Size (nm)
ssDNA, nonenveloped	Parvoviridae	human parvovirus	18–25
dsDNA, nonenveloped	Adenoviridae	mastadenovirus	70–90
	Papovaviridae	papillomoavirus	40–57
dsDNA, enveloped	Herpesviridae	herpes simplex	150–200
	Hepadnaviridae	hepatitis B	42
	Orthopoxvirus	smallpox	200–350
ssRNA, nonenveloped	Caliciviridae	hepatitis E	35–40
	Picornaviridae	rhinovirus (common cold)	28–30
ssRNA, enveloped	Bunyviridae	hantavirus	90–120
	Coronaviridae	coronavirus	80–160
	Deltaviridae	hepatitis D	32
	Flaviviridae	hepatitis C	40–50
	Orthomyxoviridae	influenza	80–200
	Retroviridae	human immunodeficiency virus (HIV)	100–120
	Rhabdoviridae	rabies	70–180
	Togaviridae	rubella	60–70
dsRNA, nonenveloped	Reoviridae	rotavirus	60–80

Note: ssDNA = single-stranded DNA; dsDNA = double-stranded DNA; ssRNA = singled-stranded RNA; dsRNA = double-stranded RNA

not only the history of failure but also scientific arguments such as: (1) little evidence for leukemia viruses in primates; (2) the knowledge that when retroviruses were found in animals they were not difficult to find. Extensive viremia preceded disease; therefore, if they infected humans, they would be easy to find and would havebeen discovered much earlier. (3) Human sera in the presence of complement [an immune system component] lysed animal retroviruses, thereby providing a rational mechanism for the conclusion that humans were protected against infections by retroviruses." Gallo would report the first retrovirus, human T cell lymphotropic virus type-1 (HTLV-1), in 1979. In the next decade, Gallo would join an international research effort in discovering and describing a new retrovirus, HIV, the cause of acquired immunodeficiency syndrome (AIDS).

INFECTION AND REPLICATION IN VIRUSES

Viruses have no energy-generating system of their own and no capability of reproducing sexually or asexually as other organisms do. The ability of a virus to infect a host and replicate inside the host's cells, thus, determines whether a virus will persist in a population or will disappear.

All viruses follow a general scheme described here for infecting a host cell:

- attachment—capsid attaches to the surface of a target cell
- penetration—virion crosses the cell membrane and sheds its protein coat, called *uncoating*
- early transcription and translation—host produces the enzyme RNA polymerase for the purpose of making viral nucleic acid replication enzymes
- late transcription and translation—host replicates viral genes carrying information for new capsid and viral proteins
- capsid synthesis—host synthesizes viral proteins in the cytoplasm
- packaging—the proteins migrate into the cell nucleus and envelop newly made viral nucleic acids
- release—virions emerge from the host cell

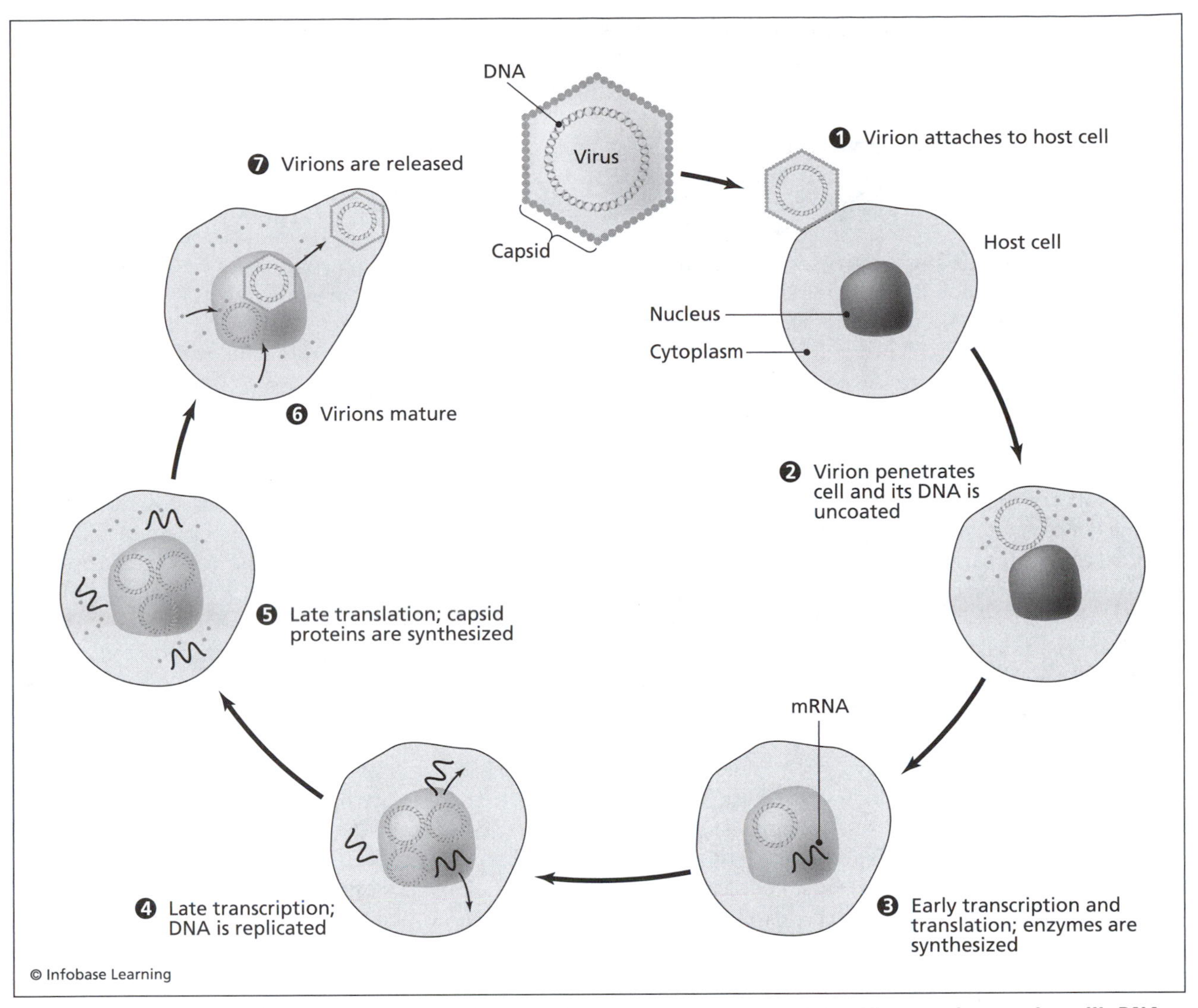

Viruses are obligate parasites that must invade a living cell to produce new virus particles. Viruses take over the cell's DNA replication and use it to make viral DNA or RNA.

The penetration step occurs in either of two ways: endocytosis or fusion. In endocytosis, the target cell engulfs the virus particle the same way it engulfs nutrients. In fusion, the viral envelope joins with the plasma membrane, and the viral contents release directly into the cell cytoplasm.

Outside these seven basic steps, DNA and RNA viruses differ in the manner in which they control the replication of viral genes by using the cell's reproduction mechanism. In general, different types of viruses use the following mechanisms:

- dsDNA and ssDNA viruses use the cell's enzymes to transcribe viral genes into new viral DNA
- dsRNA and ssRNA viruses use their RNA as a template for synthesizing the enzyme RNA polymerase, which controls the production of new viral RNA and subsequent viral proteins in the cell cytoplasm

Viral replication depends on the activity of the enzyme reverse transcriptase. Reverse transcriptase operates the normal DNA-to-RNA process in reverse, so that an RNA virus can replicate by using the host to make new viral genes carried in new viral RNA. After attachment, penetration, and uncoating, RNA viruses use alternative routes for replicating their genome. These processes are complicated inside the cell and summarized here in simplified language:

- Retroviruses use reverse transcriptase to make DNA from the viral RNA. The new DNA is called *proviral DNA* because it contains the virus's genes. Proviral DNA is transcribed into

(a) new viral RNA and (b) new viral proteins using the normal cellular protein synthesis system.

- Single-stranded RNA viruses produce proteins directly from the RNA and use a viral replicase enzyme to produce new viral RNA.

- Double-stranded RNA viruses use the enzyme RNA-dependent RNA polymerase as a transcriptase to make proteins and as a replicase to make new viral RNA.

The reactions summarized here have received intense study in virology to learn how the host and virus control certain genes, express some genes, and repress others. The retrovirus HIV uses the first scheme described; the cold virus, enteroviruses, hepatitis A, mumps, measles, and influenza use the second scheme described for ssRNA viruses; and dsRNA rotavirus (a common cause of infant diarrhea) uses the third scenario.

The virus's late genes direct the production of capsid proteins. Once the proteins are made and are available in the cytoplasm, the capsid assembles automatically. Virologists have yet to uncover the exact mechanism by which nucleic acids are included in the new virion during the self-assembly steps.

Enveloped and nonenveloped viruses use different mechanisms for exiting the host cell. Enveloped viruses leave the cell in a multistep process in which the virion takes up part of the cell membrane to form a portion of the envelope. Organelles such as endoplasmic reticulum, Golgi apparatus, and other internal membranes also contribute to building the virion's envelope. Nonenveloped viruses usually exit the cell when the cell lyses. In general terms, therefore, enveloped viruses may allow the cell to survive, but nonenveloped viruses require cell death for further propagation.

Virologists have mainly studied bacteriophages to learn the details of two different types of virus propagation: the lytic cycle and the lysogenic cycle. The lytic cycle releases new viruses only at the death and lysis of a host cell. The lysogenic cycle allows the host cell to remain alive even while producing hundreds of new viruses. In lysogeny, the bacteriophage inserts its genes into host cell DNA; the DNA of this new composition is called the *prophage*. Several viral genes are turned off in the prophage, so that the host cell can reproduce and, thus, reproduce the viral genes at the same time. In the presence of certain initiators, such as ultraviolet light or chemicals, lysogeny ends, and the lytic cycle proceeds. Within a defined time frame, the lysis of hundreds or thousands of host cells results in the release of new viruses.

The microbiologists and authors Moselio Schaechter, John Ingraham, and Frederick Neidhardt wrote, in 2006, "It is widely believed that [bacterio] phages strongly affected bacterial evolution. Acting as predators, virulent phages select for phage-resistant survivors, which favors the emergence of new bacterial types." Lysogeny would have favored such a circumstance more than the lytic cycle, which hardly allows for the evolution of new types of bacteria. Lysogeny also provides viruses with a favorable survival strategy by retaining a pool of host organisms in which to propagate.

VIRAL DISEASE

Viral diseases depend on the efficient replication of viral nucleic acids in the processes described here. Regardless of the nucleic acid replication mechanism or the type of release from the cell, viral diseases usually depend on high numbers of virions in the host animal or plant.

Viruses cause either acute or persistent infections. Acute infections have a sudden onset and last for a finite time, from days to a few months. Persistent infections last many years and take three main forms: chronic, latent, and slow. Chronic infections last a long time with mild or undetectable symptoms, even though the virus remains present in the host. Latent infections occur when a virus infects a host but, then, becomes dormant and nonreproducing for a long period, before becoming active and causing disease symptoms. The following viruses cause latent infections: herpes simplex, Epstein-Barr, varicella-zoster, and cytomegalovirus.

Slow viruses make up a poorly understood group that develops disease more slowly than latent viruses. Slow disease symptoms may develop decades after the initial infection. Lentiviruses, a group that includes HIV in the family Retroviridae, offer an example of slow viruses. Other human viruses identified as slow viruses are multifocal leukoencephalopathy, subacute sclerosing encephalitis caused by the rubella (measles) virus, and T cell leukemia virus.

Infectious proteins called *prions* have also been implicated as the cause of slow infections as part of virology. These infectious agents cause scrapie in sheep, bovine spongiform encephalopathy in cattle, and Creutzfeld-Jakob disease (CJD) in humans.

VIROIDS AND PRIONS

Viroids and prions do not conform to the definition of a virus in a strict sense, but they bear similarities to viruses, are submicroscopic, and can infect host cells. For these reasons, viroids and prions often make up a specialized subject within virology.

A viroid is a circular RNA molecule of 300–400 bases, without a protein coat, which infects plants. Viroid RNA contains a double-stranded middle section that connects two single-stranded loops. The RNA folds in a configuration believed to expose a pathogenic end and a variable, nonpathogenic end.

In 1971, the plant pathologist Theodor O. Diener, working for the U.S. Department of Agriculture, discovered an infectious agent of potato plants that appeared similar to a virus but also unlike a virus. Viroids lack genes that encode for structural proteins because they do not use proteins. As RNA viruses do, viroids depend on a host cell's RNA polymerase to assemble new RNA molecules from the viroid's RNA template. The RNA replication proceeds around the circular molecule in what scientists call rolling-circle replication, which is also used by DNA viruses containing circular DNA and plasmids, which are small, circular DNA molecules in bacteria.

Prions were discovered, in 1982, by the American neurobiologist Stanley Prusiner (1942–), who proposed a "naked" protein caused the disease scrapie in sheep. He coined the term *prion* as an abbreviation for *proteinaceous infectious* particle. In 1997, when Prusiner won the Nobel Prize in medicine for his discovery of prions, he recounted, "I had anticipated that the purified scrapie agent would turn out to be a small virus and was puzzled when the data kept telling me that our preparations contained protein but not nucleic acid." Prusiner continued research on the infectious agent with little support from a scientific community that doubted the existence of an active, infectious protein. When Prusiner published his results, in 1982, and the term *prion,* he set off a heated debate. "Virologists were generally incredulous," he said, "and some investigators working on scrapie and CJD were irate." Prusiner weathered considerable scientific and personal attacks on his credibility, for the next decade, until additional data from other laboratories began to prove that prions indeed existed.

Prions disrupt host cell function by converting a normal glycoprotein (protein connected to a carbohydrate) into an abnormal, infectious form. The abnormal glycoprotein can cause additional changes in healthy cells, and the disease slowly progresses, usually through nervous system tissue. The cause of cell damage is not fully known. Affected glycoproteins have been found in the brain tissue of diseased animals, but the exact mode of action of prions still holds many unknowns.

MEASURING VIRUSES

Virus studies can be accomplished in various ways. Bacteriophages offer the easiest avenue for studying viruses, because they can be propagated in bacterial cultures grown on agar media. Animal viruses are studied in living animals, tissue cultures derived from live animal cells, or embryonated eggs, that is, fertilized eggs containing an embryo. Plant viruses have, likewise, been cultivated in live plants and in tissue culture.

Microbiologists can study bacteriophage activity because of the ability of bacteriophages to lyse bacterial cells. This simple action is captured in a method called a plaque assay. In virology, a plaque is a clearing in a film or lawn of bacteria caused by the lysis of cells in the immediate area of an active bacteriophage. Plaque assays can be used for counting the number of phage particles in a suspension, with a procedure similar to the method microbiologists use to determine the number of bacteria in a volume of suspension. The plaque assay involves the following steps:

- mix a small volume of a bacteriophage-containing suspension with a dense concentration of susceptible bacteria
- pour the mixture into molten agar medium formulated to be soft when it cools, and pour the mixture onto petri dishes containing a layer of solidified agar
- incubate at the bacteria's optimal growth temperature
- after incubation, count the number of plaques (called plaque-forming units, or PFUs) per agar plate

In the plaque method, the soft agar called the overlay allows bacteriophages to diffuse through it during incubation. Each time a virus particle meets and infects a bacterium, the bacterium lyses, and new viruses diffuse outward. Uninfected bacteria continue to grow and turn the agar cloudy with billions of cells. After incubation, the lysed areas, or PFUs, stand out as clear spots in the cloudy agar.

In animal cell cultures, virologists measure the effect of pathogenic viruses by recording the cytopathic effect (CPE) of the virus on its host cell. Host cell CPEs that have been incorporated into laboratory screening methods for the presence or the activity of viruses are:

- altered cell shape
- changes in organelles or inclusion bodies, such as the development of multiple nuclei
- changes in membrane permeability

Titer Determination

Dilution	Logarithm of Dilution	Observed Units Positive	Observed Units Negative	Accumulated Units Positive (p)	Accumulated Units Negative (n)	Percentage Positive [p ÷ (n + p)] × 100
10^{-4}	-4.00	10	0	26	0	100.0
10^{-5}	-5.00	8	2	16	2	88.9
10^{-6}	-6.00	6	4	8	6	57.1
10^{-7}	-7.00	2	8	2	14	12.5
10^{-8}	-8.00	0	10	0	24	0.0

- cell lysis
- apoptosis, the programmed death of a cell

Virologists measure the presence and activity of viruses by a more detailed method called *titer determination*. Titer determination serves two purposes: (1) determining the infectious dose of a pathogenic virus and (2) measuring the effectiveness of disinfectants in killing specific viruses.

A titer is an estimate of the number of virus particles in a given volume of suspension, based on the principle that the strength (concentration) of a virus suspension relates to the extent the suspension must be diluted before the suspension can kill animals or bacteria growing in culture. Suspensions suspected of holding a high concentration of viruses require serial dilution, a sequence of 1:10 dilutions. The final estimate of virus concentration involves the use of the plaque assay, in which the number of PFUs correlates to the number of viruses in the diluted suspension.

More than one titer determination method has been developed in virology for the same purpose of estimating virus strength. Virologists have accepted the Reed and Muench method as the most accurate assessment of titer. This assay, developed in 1938, can be used in bacterial cultures, eggs, or test animals. In each case, a microbiologist calculates titer from the proportion of noninfected units (cultures, eggs, or animals) to infected units. The example in the table above shows a titer calculation when positive results equal the presence of viral activity and negative results equal the absence of viral activity.

The purpose of the Reed-Muench method is to estimate the virus dose at which 50 percent of the test units are killed. Since this 50 percent assay end point lies somewhere between the 10^{-6} and 10^{-7} dilutions, the lethal titer can be calculated by using the dilution next above 50 percent of the test units affected. In this example, the dilution is 10^{-7}. Using the logarithm of this value, the calculation is:

$$-7.00 + 0.5\,[(12.5 - 50.0) \div (12.5 - 57.1)] = -6.58$$

The effective dose against 50 percent of the test units, designated ED_{50}, is $10^{-6.58}$. This value expresses the following equivalent pieces of information:

- The original virus suspension contains $10^{6.58}$ ED_{50} doses of constant volume.
- The lethal dose, LD_{50}, of the virus against the test unit equals 2.63×10^{-7}, which is the antilogarithm of $10^{-6.58}$

Titer assays use biological systems to reach a final ED_{50} value for a given virus against a specific test organism. Because biological systems have a high amount of variability among organisms, virologists strive to reduce the errors in running assays such as the Reed-Muench method. Many virologists have added statistical tests to determine the probable accuracy of results from this and similar assays.

Virology encompasses one of the most complex disciplines in microbiology. Viruses require specialized culture techniques, technologies for studying structure and function, and methods for assessing their activity. Virology will remain a critical part of medical microbiology because of the historic connection between organism health and the presence and strength of various viruses.

See also BACTERIOPHAGE; ELECTRON MICROSCOPY; INFECTIOUS DISEASE; INFLUENZA; PLAQUE; PRION; SERIAL DILUTION.

Further Reading

Gallo, Robert C. "History of the Discoveries of the First Human Retroviruses: HTLV-1 and HTLV-2." *Oncogene* 24 (2005): 5,926–5,930. Available online. URL: www.nature.com/onc/journal/v24/n39/full/1208980a.html. Accessed January 2, 2010.

Knipe, David M., and Peter M. Howley, eds. *Field's Virology,* 5th ed. Baltimore: Lippincott, Williams & Wilkins, 2007.

Morell, Virginia. "Tracking the Next Killer Flu." *National Geographic,* October 2005.

Prusiner, Stanley B. "Autobiography." Nobel autobiography presented December 1997, Stockholm, Sweden. Available online. URL: http://nobelprize.org/nobel_prizes/medicine/laureates/1997/prusiner-autobio.html. Accessed January 2, 2010.

Schaechter, Moselio, John L. Ingraham, and Frederick C. Neidhardt. "Viruses, Viroids, and Prions." In *Microbe.* Washington, D.C.: American Society for Microbiology Press, 2006.

Shors, Teri. *Understanding Viruses.* Sudbury, Mass.: Jones & Bartlett, 2009.

Ter Muelen, Volker, and William W. Hall. "Slow Virus Infections of the Nervous System: Virological, Immunological and Pathogenetic Considerations." *Journal of General Virology* 41 (1978): 1–25. Available online. URL: http://vir.sgmjournals.org/cgi/reprint/41/1/1.pdf. Accessed January 2, 2010.

University of South Carolina School of Medicine. "Virology." In "Microbiology and Immunology On-Line." 2009. Available online. URL: http://pathmicro.med.sc.edu/book/virol-sta.htm. Accessed June 6, 2010.

U.S. Department of Agriculture. Agricultural Research Service. "Tracking the Elusive Viroid." *Agricultural Research,* May 1989. Available online. URL: www.ars.usda.gov/IS/timeline/viroid.htm. Accessed January 1, 2010.

Weiss, Robin A. "The Discovery of Endogenous Retroviruses." *Retrovirology* 3 (2006): 67–78. Available online. URL: www.retrovirology.com/content/3/1/67. Accessed January 2, 2010.

wastewater treatment Wastewater, or sewage, treatment comprises all of the processes for removing all materials from wastewater that could harm human or animal health or the environment. Wastewater consists of water that flows into drains in households and businesses, storm drains, and surface runoff. Each of these paths is called a *waste stream* because it represents a route taken by the wastewater from its source to its site of treatment. Because these waste streams collect a diversity of microorganisms during their flow, wastewater treatment involves significant efforts to remove pathogens from the water before the treated water can be released back into the environment.

Wastewater treatment also makes use of beneficial bacteria that help degrade the water's organic matter. Wastewater treatment plants rely on two main activities of beneficial bacteria: aerobic digestion of suspended organic matter and anaerobic digestion of sludge. Sludge, also called *biosolids,* is the material remaining after all of the treatment processes have taken place. Sludge requires a longer period to break down difficult-to-degrade organic matter, so it enters an additional process in a wastewater treatment plant. In this step, anaerobic bacteria degrade sludge organic matter in an anaerobic digester tank.

The U.S. Geological Survey (USGS) identifies four main reasons why wastewater treatment is important: (1) clean water for aqueous plants, animals, and fish; (2) clean riparian habitats (streamside) for migrating animals and birds; (3) recreational water quality at beaches, rivers, and streams; and (4) removal of disease-causing substances from water before it reenters the environment.

BIOLOGICAL HAZARDS IN WASTEWATER

Wastewaters contain high levels of organic matter from sewage, agricultural wastes (crop and animal wastes), food processing, industrial effluents, lawn runoff, and wastes from wildlife and forests. Sources such as sewage, agricultural wastes, and wildlife contribute a significant amount of pathogens to the water.

Wastewater's biological hazards fluctuate, depending on the source of the water and other factors such as proximity to densely populated areas, discharge from industries, and rainfall levels. In general, the following biological hazards should be expected in any wastewater: enteric bacteria and viruses, algae, protozoa, pathogenic cysts, environmental pathogens, worms, and crustaceans. Enteric microorganisms originate in the digestive tract of animals and may contain a large proportion of illness-causing microorganisms. Pathogenic cysts mainly include *Giardia* and *Cryptosporidium* shed by wildlife and livestock and present in most untreated surface waters. Environmental pathogens may include any microorganisms that are normally found in soils or waters, such as *Bacillus anthracis* (anthrax), *Clostridium tetani* (tetanus), and various coliforms. Coliforms are gram-negative bacteria normally found in water but also present in animal digestive tracts.

Even though many coliforms are not considered dangerous, in water and wastewater treatment, microbiologists use coliforms as indicator organisms that signal that more harmful microorganisms may be present. Fecal coliforms, which are coliforms known to be in animal wastes, and the enteric coliform *Escherichia coli* also serve as indicator organisms. Fecal coliforma and *E. coli* provide a more specific indication of the presence of fecal matter in water.

Wastewater Treatment

Treatment Type	Description
physical	solid particles removed by screens, traps, filters, and settling by gravity
chemical	disinfectants and flocculants that cause suspended solids to precipitate from the water as aggregates
biological	digestion of organic matter by aerobic and anaerobic bacteria
Treatment Phase	
primary	mainly a physical separation of large solids from the plant's incoming waste stream
secondary	biological degradation of suspended and dissolved solids by bacteria, followed by disinfection
tertiary	additional physical and chemical steps to clarify the water of nitrogen compounds, metals, and bacteria: coagulation, filtration, carbon adsorption of organic compounds, and additional disinfection

Wastewater contains a variety of organic compounds that affect treatment because these compounds either serve as nutrients for the bacteria, algae, and protozoa or interfere with the water cleanup process. In 2006, the Polish microbiologists Kangala Cehipasa and Krystyna Mędrzycka explained in the *Journal of Industrial Microbiology and Biotechnology,* "Lipids (characterized as oils, greases, fats and fatty acids) are one of the most important components of natural foods and many synthetic compounds and emulsions. The latter are mostly found in pharmaceutical and cosmetic industrial effluents. Further, lipids constitute one of the major types of organic matter found in municipal wastewater, which may find their way into surface waters." Lipids interfere with wastewater treatment in a number of ways: (1) clogging screens and filters, (2) coating beneficial bacteria to reduce their ability to metabolize wastes, (3) making other organic matter unavailable for decomposition, (4) causing foaming and aggregation of solids, and (5) making sludge more difficult to digest by anaerobes. Thus, the organic matter is not a hazard in itself, but in wastewater it affects the processes used to clean hazards out of the water.

Wastewater treatment microbiologists use a measurement called biological oxygen demand (BOD) to assess the amount of organic matter in wastewater, as assessed by the amount of oxygen the wastewater consumes over a set period. Water treatment's main goal is to reduce the BOD level in the water. By reducing BOD, treatment makes the water less able to nurture the growth of new microorganisms.

The BOD test is a five-day procedure—it is sometimes written BOD_5—that measures the amount of oxygen consumed by the bacteria in water when held in the dark at 68°F (20°C). By running the test in the dark, the BOD measures the activity of only nonphotosynthetic bacteria, a group likely to include pathogens. The following BOD results give a general idea of the water's organic load:

- BOD = less than 150 milligrams/liter (mg/l)—low level of organic matter
- BOD = 150–300 mg/l—moderate level of organic matter
- BOD = greater than 300 mg/l—high level of organic matter

Typical raw sewage has a BOD of 110–440 mg/l. Current U.S. wastewater treatment operations reduce these levels to 5.5–60 mg/l or a 95 percent reduction.

The length of time to run the BOD test and certain factors that interfere with its accuracy have limited the usefulness of this measurement. Wastewater treatment plants, nevertheless, continue to use BOD for the following four reasons: (1) It gives an approximate measure of the amount of oxygen that will be needed to stabilize the wastewater's organic matter, (2) it helps determine the size that a wastewater facility must attain to treat the local wastewater, (3) it helps assess the efficiency of current treatment processes, and (4) it gives an indication of whether the treatment process is meeting local requirements.

Wastewater treatment has adopted additional tests that are similar to BOD, in that they give a general indication of the relationship between the wastewater constituents and the microbial activity in the water. Typical supplementary tests are:

- nitrogenous biochemical oxygen demand (NBOD)—oxygen demand associated with the conversion of ammonia (NH_3) to nitrate (NO_3^-)
- carbonaceous biochemical oxygen demand (CBOD)—BOD when the NBOD reaction is suppressed to study only carbon compounds

Wastewater treatment uses settling tanks such as this one that give aerobic bacteria time to break down organic matter. ***(Epco Australia)***

- total organic carbon (TOC)—organic carbon present in the water that relates to BOD

- dissolved total organic carbon (DTOC)—dissolved carbon compounds that relate to BOD

The amount of organic compounds in water is important to assess for treatment plants that use ultraviolet (UV) radiation as part of the decontamination process. UV light kills microorganisms, but a high level of organic matter blocks the transmission of the light through the water. For this reason, wastewater of high TOC content is more difficult to treat and make safe than water low in TOC.

WASTEWATER TREATMENT

Wastewater treatment combines physical, chemical, and biological processes for making the water safe for return to the environment. These methods contribute to the three main phases of treatment: primary, secondary, and tertiary. The table on page 789 differentiates the various treatments that take place in a typical wastewater treatment plant.

The USGS estimates that primary treatment removes about 60 percent of wastewater's suspended solids, and secondary treatment removes 90 percent of solids exiting the primary phase.

Of the pathogens removed from wastewater during its treatment, primary treatment has been estimated to remove 50–98 percent of viral pathogens, the secondary phase to remove 53–99.9 percent, and tertiary to remove more than 99.99 percent. Primary, secondary, and tertiary treatments remove from 99 to more than 99.999 percent of fecal coliform bacteria.

Removal of *Giardia* and *Cryptosporidium* cysts can be more difficult than eliminating viruses and bacteria, because these cysts are very resistant to chlorine disinfectants. But because these organisms are large, filtration serves as an effective way of removing most of them. The common water contaminants, oval *Giardlia lamblia* and round *Cryptosporidium parvum,* are 8–18 micrometers (µm) long, 5–15 µm wide, 4–7 µm in diameter, respectively.

Wastewater treatment begins when raw wastewater enters a treatment plant by passing through chambers that remove large insoluble matter. The chambers usually consist of a bar screen that acts as a coarse filter, a grit chamber where sand and dirt sink to the bottom, and a settling tank that gives time for small particles to sink.

The water, then, flows to an aeration tank, where a heterogeneous mixture of suspended bacteria and biofilms digest much of the fine suspended and dissolved organic matter. Aeration tanks receive a constant supply of air bubbled through the contents to enhance the growth of the beneficial bacteria. After six to eight hours in the aeration tank, the partially clarified water flows to another settling tank, where chemicals may be added to encourage the formation of precipitates, which form over three to four days. The water may, additionally, flow to a trickling filter bed, which slowly allows the water to pass through a bed of sand and fine stones. The clarified water from these steps can be rerouted back to the aeration tank for more digestion or sent to a disinfection tank. Chlorine compounds added to the disinfection tank kill all remaining pathogens and bacteria from the aeration tank.

A thick sludge high in poorly digestible organic matter accumulates in most of these steps. The treatment plant pumps the sludge to a separate tank, where it is mixed with anaerobic bacteria to form a slurry called activated sludge. The activated sludge decomposes in the anaerobic digester tank over several days to a week. In anaerobic digestion, the activated sludge rises to 131°F (55°C), for about five days. The elevated temperature kills pathogenic bacteria and viruses that may have escaped the other treatment steps. During this high-temperature phase, the anaerobic population grows at its fastest and produces large amounts of methane and carbon dioxide. After the initial burst of microbial activity, the anaerobic process slows, and temperature falls to about 95°F (35°C), where it remains for up to 15 days. Additional organic matter decomposes during this longer phase.

The sludge collected from the anaerobic digester is referred to as biosolids. Biosolids have been used as ground covering, as a soil consistency additive, and for landfill cover. Biosolids have also been considered as a biofuel that can be burned to produce energy as an alternative to burning fossil fuels such as coal. At the same time as biosolids develop in the anaerobic digester, the anaerobic population of bacteria produces methane (CH_4), termed biogas, which also serves as an energy source. Biogas is about 60 percent methane and 40 percent carbon dioxide. Many wastewater treatment plants now recycle the methane from the digestion to provide energy for running the plant's pumps and providing heat for the facility's offices and laboratories.

Anaerobic digesters work optimally with a diverse mixture of anaerobes that carry out complementary reactions. For example, some of the bacteria degrade sludge organic matter in fermentations that produce carbon dioxide along with other end products. Methanogenic archaea keep these reactions moving in a forward direction by converting the carbon dioxide to methane (biogas), thereby preventing the carbon dioxide from building up in the tank. The table below lists the main anaerobic bacteria and archaea that carry out sludge digestion.

The U.S. Environmental Protection Agency (EPA) and local communities have requirements about where biosolids from wastewater treatment can be used. Because this material has a potential of containing high amounts of microorganisms and heavy metals, it can be used or dumped only at des-

Microorganisms of Anaerobic Sludge Digesters

Microorganisms	Type of Reactions	Starting Material	End Products
Bacteroides *Clostridium* *Eubacterium* *Peptococcus* *Peptostreptococcus*	fermentation	organic compounds	organic acids, ethanol, hydrogen, carbon dioxide
Acetobacterium *Syntrophobacter* *Syntrophomonas*	acetate production	organic acids (butyrate, propionate, lactate, succinate), ethanol	acetate, hydrogen, carbon dioxide
Methanosarcina *Methanothrix*	methane from acetate	acetate	methane, carbon dioxide
Methanobacterium *Methanobrevibacter* *Methanococcus* *Methanomicrobium* *Methanospirillum*	methane from carbon dioxide and hydrogen	hydrogen (H_2) and bicarbonate (HCO_3^-)	methane

Source: Prescott, Lansing M., John P. Harley, and Donald A. Klein, eds. *Microbiology,* 6th ed. (New York: McGraw-Hill, 2005), 640.

ignated places. Most states and cities have regulations on the distance from natural surface waters where sludge may be unloaded.

To ensure the safety of treated wastewater further, many municipal treatment plants include additional steps with the ones described here. These may involve any or all of the following:

- preliminary treatment—removal of large items such as rags, sticks, and floating grease before the water enters primary treatment
- advanced primary treatment—extra filtration or addition of a chemical flocculant to remove microscopic suspended particles before beginning secondary treatment
- nutrient removal—special chemical additions as part of secondary treatment that removes nitrogen and phosphorus compounds
- advanced tertiary treatment—removal of designated dissolved and suspended materials after normal treatment so that the water qualifies for certain reuses, such as irrigation

Several U.S. regulations have been established to guide wastewater treatment plants on required standards before the release of treated water into the environment, such as rivers or marine waters, or its reuse as irrigation or gray water. Gray water is treated water that may be used for purposes other than drinking, rinsing foods, or bathing.

WASTEWATER STANDARDS

In the United States, wastewater treatment plants follow requirements set by the Clean Water Act of 1972. This act and subsequent amendments, in 1987 and 2000, have established minimal national standards for treated wastewater. Because some communities have very clean wastewater entering their municipal treatment plant, tertiary treatment may be minimal in these places. For this reason, the standards in the table below are for secondary treatment of wastewater.

The EPA's Office of Wastewater Management (OWM) emphasizes, "Cleaning and protecting the nation's water is an enormous task. Under the Clean Water Act, OWM works in partnership with Environmental Protection Agency (EPA) regions, states and tribes to regulate discharges into surface waters such as wetlands, lakes, rivers, estuaries, bays and oceans. Specifically, OWM focuses on control of water that is collected in discrete conveyances (also called point sources), including pipes, ditches, and sanitary or storm sewers." Despite the challenges of a job this size, U.S. waters are among the world's cleanest.

SEPTIC SYSTEMS

A septic system is a device that treats wastewater from a single household. Wastewater from a home's drains and toilets goes to the underground septic tank in dedicated pipes. The tank operates in a manner similar to the anaerobic digester at municipal wastewater treatment plants. Wastewater inflow decomposes under the action of a diverse population of anaerobic bacteria. This digestion leads to three layers that are usually present in septic tanks: (1) sludge layer at the bottom, (2) liquid layer, and (3) scum layer of suspended materials (grease and oils) floating on the liquid.

As inflow occurs intermittently into the septic tank, the anaerobes digest organic matter and kill some of the pathogens that enter with human waste. Each influx of material pushes out an equal amount of treated water through an outlet pipe. A series of baffles in most modern septic tanks keep solid matter inside the tank for further digestion and allow only liquids to pass out of the tank.

Septic tank effluent flows in pipes to a leachfield or drainfield, which is an underground network of pipes perforated with tiny holes. Leachfields usually have been constructed to include a bed of small gravel surrounding the pipes. As treated water seeps from the pipes to the gravel, soil microorganisms continue breaking down any residual organic matter and pathogens. The leachfield spreads out a few to several inches below the soil surface so that the water continually evaporates. The combination of in-tank digestion and further treatment in the leachfield renders the wastewater safe for return to the environment.

As in sludge management, septic tank management must follow certain rules to assure that no pathogens are released. Leachfields must be a certain minimal distance from surface waters and should be inspected periodically, along with the tank. Septic tanks require periodic pumping to remove sludge buildup. With proper care, septic tanks can func-

Minimal U.S. Standards for Secondary Treatment

Effluent Constituent	Average 7-Day Level	Average 30-Day Level
BOD, mg/l	45	30
CBOD, mg/l	40	25
Total suspended solids, mg/l	45	30
pH	6.0–9.0	6.0–9.0

tion safely for many years and reduce the stress on municipal wastewater treatment systems.

See also BIOLOGICAL OXYGEN DEMAND; COLIFORM; *CRYPTOSPORIDIUM;* INDICATOR ORGANISM.

Further Reading

Chipasa, Kangala B., and Krystyna Mędrzycka. "Behavior of Lipids in Biological Wastewater Treatment Processes." *Journal of Industrial Microbiology and Biotechnology* 33 (2006): 635–645.

Gerba, Charles. "Domestic Wastes and Waste Treatment." In *Environmental Microbiology,* 2nd ed. San Diego: Elsevier, 2009.

Tchobanoglous, George, Franklin L. Burton, and H. David Stensel. *Wastewater Engineering: Treatment and Reuse,* 4th ed. New Delhi, India: Tata McGraw-Hill, 2003.

U.S. Environmental Protection Agency. Office of Wastewater Management. *Emerging Technologies for Biosolids Management.* Washington, D.C.: U.S. EPA, 2006. Available online. URL: www.epa.gov/OWM/mtb/epa-biosolids.pdf. Accessed November 28, 2009.

U.S. Geological Survey. "Wastewater Treatment, Water Use." Available online. URL: http://ga.water.usgs.gov/edu/wuww.html. Accessed November 28, 2009.

Western Lake Superior Sanitary District. "Wastewater Treatment Process." Available online. URL: www.wlssd.duluth.mn.us/wastewater_treatmentprocess.php. Accessed November 28, 2009.

water activity (a_W) Water activity is a measure of the water in a habitat available for an organism's use. The water activity of all living or nonliving things ranges from 0 to 1.0.

Water quality is an important component of a microorganism's environment because of the osmotic effect exerted by the surrounding solution on microbial cells. Cells must react to the concentration of solutes, or dissolved substances, in their environment to maintain osmotic pressure, which is the cell's internal pressure relative to the pressure exerted on the cell from the surroundings. Microorganisms maintain osmotic pressure to preclude either bursting or shriveling by a process called *osmoregulation.* The a_W of a solution in which the cells are suspended influences their osmoregulation activity.

A solution's a_W is one one hundreth of the solution's relative humidity. For liquid solutions, this is equivalent to the ratio of the solution's vapor pressure (P_{soln}) to the vapor pressure of pure water (P_{water}). The a_W of a solution can be determined by sealing an opened vessel of the solution in an airtight chamber. By measuring relative humidity (as a percentage) after the solution and the air in the chamber have reached to equilibrium, a_W can be calculated. If the air becomes 95 percent saturated with moisture compared with the saturation using pure water, the relative humidity is 95 percent. In this case, the solution's a_W is 0.95.

Microorganisms have a minimal, optimal, and maximal a_W that they tolerate for growth. Some groups of microorganisms can tolerate conditions of very low a_W. For example, halophilic bacteria are extremophiles (organisms that live in extreme conditions where other organisms do not survive) that survive in high salt conditions. The extremophile *Halobacterium* lives in salt lakes where a_W reaches 0.75. Such a microorganism is also called an *osmophile,* because it lives in conditions of high osmotic pressure from the environment. Some bacteria are osmotolerant, meaning they can survive at moderately low a_W, from about 0.85 to 0.90. Many fungi exist at a_W values lower than most bacteria. The genera *Xeromyces, Saccharomyces* (yeast), *Aspergillus,* and *Penicillium* can thrive at a_W values below 0.85.

Most microorganisms prefer a_W values of 0.98 or higher; 0.98 is the approximate a_W for seawater. A solution of pure water containing no solutes has a_W equal to 1.00. Microorganisms are not known to survive at values of 0.55 or lower, because in these conditions, the cell's deoxyribonucleic acid (DNA) becomes permanently damaged. At values of 0.70 and lower, enzyme activity begin to slow.

WATER ACTIVITY IN FOODS

Water activity is of critical importance to food microbiologists as a method for preserving foods. The food microbiologist George Banwart explained, in his classic 1989 book *Basic Food Microbiology,* "Some microorganisms can remain alive in a dried condition but cannot carry out their normal metabolic activities or multiply without water. Microorganisms can grow only in aqueous solutions. They cannot grow in pure water or in the absence of water." Water serves the microbial cell in several ways: (1) carrying nutrients to and into the cell; (2) dispersing wastes; (3) participating in chemical reactions, such as hydrolysis reactions; (4) donating hydrogen ions for regulating internal pH; and (5) serving in cell buoyancy through the use of contractile vacuoles that take in or expel water.

By adjusting the a_W of foods, microbiologists can create an efficient way to preserve them. The a_W can be lowered by adding salt or sugar or by freezing the food. Each of these methods stabilizes water in a form that makes it unavailable for microbial activities. Although a_W helps to preserve foods, other factors determine the shelf life of a food before it spoils. For example, high-sugar foods such as jams, jellies, or honey resist bacterial growth, but osmophilic yeasts grow well in them. High-salt foods such as cured meats can be spoiled by halotolerant or halophilic bacteria. Dried foods have a_W value of 0.75 and lower, which acts as a good preservative by itself because foods of a_W 0.70 or lower seldom support microbial growth.

For the best preservation by using a_W, a food microbiologist should understand the moisture levels of the food and the optimal a_W of common food contaminants. The table below presents approximate a_W values for foods and the optimal values for food spoilage microorganisms.

In addition to slowing the activity of enzymes, a low a_W affects other aspects that microorganisms use to survive in the environment or cause infection. In some of the instances summarized in the following list, a low a_W aids the survival of a microorganism even as it prevents growth:

- low a_W induces endospore formation in *Bacillus* and *Clostridium*
- high a_W induces *Bacillus* and *Clostridium* to germinate from endospores
- *S. aureus* requires a higher-than-optimal a_W for toxin production
- Molds *Aspergillus* (optimal a_W, 0.88) and *Penicillium* (optimal a_W, 0.90) require higher-than-optimal a_W for mycotoxin production
- osmotolerant microorganisms grow at lower a_W when the solute is sugar rather than salt
- halophiles grow at lower a_W when the solute is salt rather than sugar

Because bacteria grow faster than yeasts and molds, bacteria have been associated with most food spoilage and food-borne illness cases. Yeasts and molds receive an advantage, however, in foods that have been formulated to lower a_W to prevent bacterial growth. For example, the following foods have a_W values below 0.85 for the purpose of inhibiting bacterial growth: preserves, rice, flour, dried fruits, molasses, rolled oats, noodles, biscuits, and crackers. Molds are much more probable contaminants of these foods than bacteria.

ISOTHERMS

A sorption isotherm is a graphical illustration of the relationship between a food's moisture content and a_W. An isotherm consists of two related sigmoid curves: The adsorption curve depicts the moistening process of microorganisms in a food.

The desorption curve depicts the drying process in microorganisms.

At the lowest region of the isotherm, water binds to the outside of food surfaces and is unavailable for biological activity. At the middle section of the isotherm, additional water molecules layer on the layer

Water Activity in Foods and Major Spoilage Organisms

Food	Food a_W	Microorganism	Minimal a_W for Growth
fresh fruit, vegetables	0.97–1.00	*Salmonella*	0.93–0.96
fresh meat, poultry, fish	0.97–1.00	*Bacillus*	0.90–0.99
eggs	0.97	*Clostridium botulinum*	0.90–0.98
bread	0.94–0.97	most spoilage bacteria	0.90–0.91
brick and soft cheeses	0.91–1.00	*Lactobacillus*	0.90–0.96
cooked egg yolk	0.90	*Micrococcus*	0.90–0.95
cured meat	0.87–0.95	*Staphylococcus aureus*	0.88–0.92
dry, hard cheeses	0.68–0.76	most yeasts	0.87–0.94
jellies, jams	0.75–0.80	halophiles	0.75
cereal, candy bars	0.70	most molds	0.70–0.80
chocolate, honey	0.60	*Xeromyces bisporus*	0.60–0.61

of bound water molecules. The a_W increases rapidly in this region, and moisture content of the food also increases. At the isotherm's uppermost region, moisture content increases more than a_W increases. in this region, the food supplies ample free water for microbial metabolism.

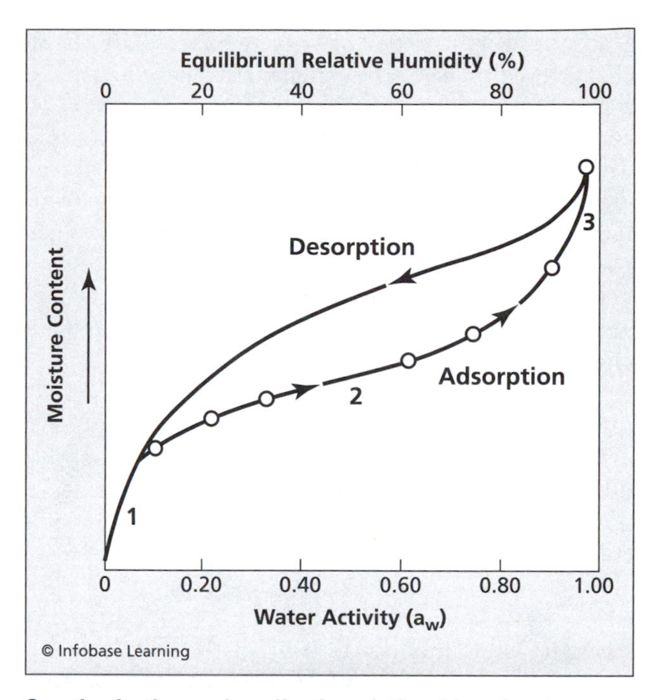

Sorption isotherms describe the relationships of water activity, moisture content, and relative humidity. Water availability for microbial metabolism occurs in the middle section of the isotherm.

INTERACTIONS OF WATER ACTIVITY

In addition to osmotic pressure, nutrient availability, temperature, and pH all influence a_W's effect on microbial growth. A microorganism's optimal a_W range shrinks or widens, depending on nutrient supply. The addition of growth factors such as amino acids, vitamins, or minerals can lower a microorganism's minimal a_W as much as 0.02 unit. Banwart said, "Since foods differ in their nutrient content, microorganisms may be able to grow in some foods at lower a_W than required in other foods." Food microbiologists consider these interactions when formulating processed and packaged foods.

At optimal temperature, microorganisms tolerate wider ranges in a_W than when temperature is a source of stress. Food storage temperatures must, therefore, remain outside the spoilage microorganisms' optimal range, usually by refrigerating or freezing, to make a_W most effective in preservation. Increases or decreases of pH from a species's optimal range have a similar effect. In general, decreasing the pH from the optimal range of a given microorganism increases the microorganism's minimal a_W that allows growth.

Microbiologists prepare growth media that are specially formulated to encourage the growth of halophiles or osmophiles. Elevated concentrations of any of the following ingredients have been used for lowering a_W:

- glucose, 50 percent
- glucose-fructose, 70 percent
- glycerol, 18 percent
- salt, 5–10 percent, plus glucose, 12 percent
- potatoes, crushed, 25 percent

Salts such as sodium chloride (NaCl), sodium sulfate (Na_2SO_4), and potassium chloride (KCl) have been used for adjusting a_W in growth media.

The ingredient used to adjust a_W also determines the response by individual microorganisms. For example, *S. aureus*'s minimal a_W increases, from 0.85 to 0.89, when glycerol replaces salt for adjusting a_W. The minimal a_W for *Pseudomonas aeruginosa,* by comparison, is 0.97 in sodium chloride and 0.95 using glycerol.

Knowledge of a food's a_W is a valuable tool in improving the preservation of the food, while maintaining the nutritional value and reducing the need for chemical preservatives. The drying, salting, or sugaring of foods is one of the most ancient forms of preservation. Today, microbiologists understand more about the principles of a_W and its interactions with other factors that promote microbial growth. Preservation methods employing a_W have, nonetheless, changed very little since their first uses centuries ago.

See also FOOD MICROBIOLOGY; OPTIMAL GROWTH CONDITIONS; OSMOTIC PRESSURE; PRESERVATION.

Further Reading

Banwart, George J. *Basic Food Microbiology,* 2nd ed. New York: Chapman & Hall, 1989.

Hocking, Ailsa D., and John I. Pitt. "Media and Methods for Detection and Enumeration of Microorganisms with Consideration of Water Activity Requirements." In *Water Activity: Theory and Applications to Food.* New York: Marcel Dekker, 1987.

U.S. Food and Drug Administration. "Water Activity (a_W) in Foods." Available online. URL: www.fda.gov/ICECI/Inspections/InspectionGuides/InspectionTechnicalGuides/ucm072916.htm. Accessed November 29, 2009.

water quality Water quality represents all of the characteristics and constituents in water that make it safe for drinking, bathing, or recreation. Physical, chemical, and biological factors contribute to safe drinking water, also called *potable water.* Part of the biological makeup of drinking water results from the variety of microorganisms in the water. Since bacteria are normal in water, water quality testing pertains to the determination of pathogenic microorganisms, and water quality results from the complete elimination of pathogens from water.

Three components combine to make up a drinking water system: (1) the source, (2) the treatment system, and (3) the distribution system. Microbiologists monitor each of these components to ensure high-quality water that contains no off-flavors, color, odor, or pathogens.

PHYSICAL, CHEMICAL, AND BIOLOGICAL CONSTITUENTS OF WATER

Physical factors in drinking water are turbidity (cloudiness), color, pH, and temperature. The main chemical constituents of water are dissolved oxygen, dissolved minerals, nitrogen and phosphorus nutrients, carbonate, and dissolved solids that may affect odor or flavor. The main minerals in drinking water are calcium and sodium, but local water systems might also contain small amounts of selenium, arsenic, copper, iron, lead, and magnesium. Some minerals (selenium, arsenic, and lead) are health hazards.

Turbidity is a measure of the cloudiness created by insoluble particles in water. Turbidity indicates a possible high concentration of microorganisms because the particles give the microorganisms a surface for attachment. Turbidity is reduced by a combination of physical and chemical methods, and water treatment plants must meet minimal allowable values for turbidity before releasing drinking water into the distribution system.

Water often contains compounds made by some soil bacteria and algae, which produce undesirable flavors. The main example of these compounds is geosmin, which is harmless to health but creates an earthy smell and flavor in water. David Cane, chemistry professor at Brown University, discovered the gene for a unique two-part enzyme system. One part of the enzyme produces a compound that combines with the opposite end of the enzyme to produce geosmin. Cane said, in 2007, "The two steps of the process that forms geosmin are metabolically related. This finding was a real surprise. This is the first bifunctional enzyme found for this type of terpene, the class of chemicals geosmin belongs to." Geosmin represents an aesthetic contaminant of water rather than a health concern.

Water contains biological factors such as bacteria and algae that may be normal for a municipality and do not affect the safety of the water. Source waters might also include enteric viruses and bacteria, protozoa, cysts, small crustaceans, insect larvae, and worms that a water treatment plant must remove. Physical and chemical treatment methods eliminate pathogenic viruses, bacteria, protozoa, crustaceans, larvae, and worms. Cysts such as *Giardia lamblia* and *Cryptosporidium parvum* resist chemical treatments and require filtration for their removal.

Water microbiology laboratories monitor bacteria and cysts as the main indicators of water quality. Coliform, fecal coliform, and *Escherichia coli* concentrations can all be determined using a technique called the most probable number (MPN) method. In the MPN method, a microbiologist inoculates diluted water samples to a series of tubes containing broth medium. After incubation, the microbiologist observes the number of tubes showing growth and no growth, then refers to statistical tables to estimate the number of bacteria in the original sample.

Generic water bacteria called heterotrophic plate count (HPC) bacteria can be determined by inoculating water samples to agar medium. An alternative method passes the water sample through a filter. Because the filter retains the bacteria, a microbiologist can determine the amount of bacteria per filtered volume of water by placing the filter on an agar plate and incubating. The standard sample size for either the MPN method or the HPC method is 100 milliliters (ml).

WATER QUALITY STANDARDS

The U.S. Environmental Protection Agency (EPA) publishes recommended minimal levels of water constituents that states must ensure for drinking water, recreational waters, and aquatic life. The EPA lists about 120 chemicals that it wants controlled in drinking water. The maximal allowable concentrations of some of these chemicals have already been set by the EPA and listed on the EPA Web site. The remaining chemicals on the list have yet to be assigned a maximal allowable level.

The EPA currently publishes methods for meeting the allowable levels for bacteria, viruses, molds, and cysts in drinking water. At minimum, water quality laboratories must monitor the following bacteria: *Aeromonas,* enterococci, *E. coli,* and total coliforms. Most water quality laboratories do additional testing for fecal coliforms. The EPA also sets requirements for individual viruses by using a coliphage as an indicator organism. An indicator organism is a microbe that microbiologists monitor as a signal that pathogens might be present. In this case, the coliphage is a natural virus that infects only coliform bacteria.

Chlorine has been used as the main water disinfectant in the United States for killing microorgan-

Drinking Water Standards for Microorganisms and Turbidity

Contaminant	Maximal Containment Level Goal	Significance
total coliforms (including fecal and *E. coli*)	5 percent	no more than 5 percent samples can be positive for coliforms in a month
heterotrophic plate count bacteria	no limit but should not exceed 500 colonies per milliliter (ml)	no health effects but a general indicator of microbial contamination
Legionella	zero	natural inhabitant of water
Cryptosporidium	zero	indicator of human and animal fecal waste
Giardia lamblia	zero	indicator of human and animal fecal waste
enteric viruses	zero	indicator of human and animal fecal waste
turbidity	1 nephelometric turbidity unit (NTU)	indicator of filtration effectiveness

isms other than cysts. This creates a problem for water treatment plants because chlorine can chemically combine with organic compounds in water to produce hazardous chemicals called *disinfection by-products.* The main chemical categories making up disinfection by-products are trihalomethanes and haloacetic acids. Trihalomethanes include chloroform, bromoform, bromodichloromethane, and dibromochloromethane. The maximal annual average the EPA allows for trihalomethanes is 80 parts per billion (ppb). Haloacetic acids consist of mono-, di-, and trichloroacetic acid and mono- and dibromoacetic acid. The EPA's limit for these compounds equals 60 ppb.

Treatment plants must balance the need for removing pathogens that can cause immediate illness with the potential long-term health effects of disinfection by-products. To avoid by-products formed by the chlorination of water with chlorine, municipalities have been increasingly changing to chloramines or chlorine dioxide as disinfectants. Some towns have installed ozone disinfection units, which avoid the chlorination by-products but may create by-products of their own.

The table above shows the maximal containment levels (MCLs) for microorganisms in drinking water published by the EPA.

Municipalities follow the guidelines in the table, but the size of the community determines how many samples must be tested per month. At minimum, a municipality should collect 40 water samples per month.

WATER TREATMENT

Source waters may be surface waters, such as reservoirs, lakes, and rivers, or groundwaters, also called *aquifers*. Groundwaters occur from a few feet to several feet beneath the earth surface. The quality of the water source from month to month determines the extent of treatments needed to meet the maximal allowable limits set by the EPA, state, and local municipality.

The simplest treatment is chlorination of source water before sending it into the distribution system. For water requiring more extensive treatment, the following activities can be added, listed in the order in which facilities usually add them:

- filtration—passing water through very small pores to retain particles on the filter
- coagulation—adding the chemicals alum, ferric sulfate, or ferric chloride to cause particles to aggregate by electrochemical forces, to be removed by filtration
- flocculation—stirring to cause particles to combine physically and then settle
- sedimentation—settling of suspended particles by gravity

Two or more of these processes used in sequence can reduce water microorganisms by more than 99 percent.

The United States enjoys a dependable supply of safe water from water utility companies and from privately owned wells. Water treatment in other places in the world often does not meet the same standards. As evidenced by the following statement by the World Health Organization (WHO), an accessible supply of safe water is diminishing in many parts of the world and is leading to a health crisis. "Human development and population growth exert many and diverse pressures on the quality and quantity of water resources and on access to them. Nowhere are the pressures felt so strongly as at the interface of water and human health. Infectious, water-related diseases are a major cause of morbidity and mortality worldwide." At least 35 new infectious agents have been identified, since 1972, in properly treated water. Of these emerging infectious agents, the WHO reported the following breakdown:

- viruses, 44 percent
- bacteria, 30 percent
- protozoa, 11 percent
- fungi, 9 percent
- helminths (worms), 6 percent

Of the bacteria and viruses, many waterborne pathogens are antibiotic-resistant.

After treatment, drinking water should contain four parts per million (ppm) chlorine before it enters the distribution system. Water distribution pipes contain biofilms that draw some chlorine from the water, and high levels of organic matter within the water also reduce chlorine levels. For this reason, buildings near water treatment plants usually receive more highly chlorinated tap water than buildings located farther from the plant.

WATERBORNE DISEASE

Waterborne illness has plagued civilization since the beginning. Not until the 1850s, when microbiologists had developed analytical tools, were water-associated infections connected with, specific pathogens. To that point, waterborne disease outbreaks occurred repeatedly. Outbreaks or a lack of any clean water can influence societies by causing mass migrations, conflicts over water, and privatization of water that may limit clean water to groups of people and exclude others. The former president of Russia Mikhail Gorbachev once remarked, "Water, not unlike religion and ideology, has the power to move millions of people." The need for clean water will never end. Several regions of the world have entered a situation known as *water stress,* in which the region's water demand exceeds its water supply.

Erik Peterson of the Center for Strategic and International Studies in Washington, D.C., said in 2005, "At any given time, close to half the population of the developing world is suffering from waterborne diseases associated with inadequate provision of water and sanitation services." Although hundreds of infectious agents can potentially contaminate water, some waterborne diseases have been particular problems in developing regions, as Peterson pointed out.

Vibrio cholerae has persisted for centuries and still contaminates inadequate water systems today. This bacterium causes cholera, an acute disease of profuse dehydrating diarrhea. Infection occurs when human fecal matter contaminates drinking water sources. In untreated cases, cholera can have a mortality rate of 50–75 percent, due mainly to dehydration. Two *V. cholerae* strains, O1 and O139, have been identified as the pathogens that cause cholera epidemics. In 2008, during political upheavals that shook the African nation of Zimbabwe, the *New York Times* reporter Celia Dugger wrote in December, "A ferocious cholera epidemic, spread by water contaminated with human excrement, has stricken more than 16,000

Major Waterborne Pathogens in Humans

Pathogen	Type of Microorganism	Disease
Cryptosporidium species	protozoal cyst	cryptosporidiosis
Cyanobacterium species	bacteria	toxin poisoning
Entamoeba histolytica	protozoa	dysentery
Giardia lamblia	protozoal cyst	giardiasis
hepatitis A, hepatitis E	virus	hepatitis
Salmonella typhi, S. paratyphoid	bacteria	typhoid fever
Vibrio cholerae	bacteria	cholera

people across Zimbabwe since August and killed more than 780. The outbreak is yet more evidence that Zimbabwe's most fundamental public services—including water and sanitation, public schools and hospitals—are shutting down, much like the organs of a severely dehydrated cholera victim." Although cholera can be prevented by water treatment, water quality is often a victim of political strife in many parts of the world.

The table on page 795 lists microbial diseases that can be transmitted in contaminated water.

Animals such as cattle, swine, and dogs are susceptible to leptospirosis, a waterborne disease caused by bacteria of the genus *Leptospira.* In humans, leptospirosis causes fever, headache, chills, muscle aches, and vomiting. As do all of the pathogens listed in the table, this microorganism causes health problems worldwide that can be avoided with adequate water treatment for a clean water supply.

Many waterborne diseases are due to contamination of water with human or animal fecal matter. But many waterborne pathogens also exist in the environment no need to infect a host for part of their life cycle. From the table, *V. cholerae* and *Cryptosporidium* occur in the environment even if surface waters have not been contaminated. Other microorganisms that also live in the environment and can cause a health threat in inadequately treated water are:

- bacteria—*Aeromonas hydrophila, Campylobacter* species, *Helicobacter pylori, Legionella pneumophila, Mycobacterium* species, *Pseudomonas aeruginosa, Vibrio parahaemolyticus,* and *Yersinia enterolitica*

- protozoa—*Acanthamoeba, Cyclospora cayetanensis,* and *Naegleria fowleri*

Most of these microorganisms cause diarrhea of varying severity and gastroenteritis.

Water quality is an important aspect of public health microbiology. Most techniques used for removing pathogens from water are effective, and, when combined with rigorous water testing, water quality should not present a health problem in any community. Water treatment and distribution infrastructure in many developing parts of the world are inadequate. As a result, waterborne disease is a major cause of deaths worldwide.

See also CRYPTOSPORIDIUM; DISINFECTION; MOST PROBABLE NUMBER.

Further Reading

Berman, Jessica. "WHO: Waterborne Disease Is World's Leading Killer." March 17, 2005. Available online. URL: www1.voanews.com/english/news/a-13-2005-03-17-voa34-67381152.html. Accessed November 28, 2009.

Dugger, Celia W. "Cholera Epidemic Sweeping across Crumbling Zimbabwe." *New York Times,* 11 December 2008. Available online. URL: www.nytimes.com/2008/12/12/world/africa/12cholera.html?pagewanted=1&sq=cholera&st=nyt&scp=4. Accessed November 28, 2009.

Lenntech. "Waterborne Diseases." Available online. URL: www.lenntech.com/library/diseases/diseases/waterborne-diseases.htm. Accessed November 28, 2009.

NSF International. "Disinfection Byproducts." Available online. URL: www.nsf.org/consumer/drinking_water/disinfection_byproducts.asp?program=WaterTre. Accessed November 28, 2009.

Science*Daily*. "Good Earth: Chemists Show Origin of Soil-Scented Geosmin." September 19, 2007. Available online. URL: www.sciencedaily.com/releases/2007/09/070916143521.htm. Accessed November 28, 2009.

U.S. Environmental Protection Agency. "Drinking Water Contaminants." Available online. URL: www.epa.gov/safewater/contaminants/index.html#3. Accessed November 28, 2009.

U.S. Geological Survey. "USGS Water-Quality Information Pages." Available online. URL: http://water.usgs.gov/owq. Accessed November 28, 2009.

Waterborne Disease Center. Available online. URL: www.waterbornediseases.org. Accessed November 28, 2009.

World Health Organization. *Emerging Issues in Water and Infectious Disease.* Geneva, Switzerland: WHO, 2003. Available online. URL: www.who.int/water_sanitation_health/emerging/emerging.pdf. Accessed November 28, 2009.

yeast Yeast is any fungus that spends all or most of its life cycle in a unicellular form or a fungus that has no known multicellular form. In the unicellular form, yeasts are called *nonfilamentous,* meaning they do not produce long thin filaments, or hyphae, which grow into a large mass of mycelia, the familiar colonies produced by molds.

Many studies of yeasts focus on two genera: *Candida* and *Saccharomyces. Candida* is a normal inhabitant of human skin. On the skin, native bacteria keep the yeast numbers in check, but when the normal flora become imbalanced, for example, as the result of antibiotic therapy, *Candida* can become an opportunistic pathogen. These microorganisms are not true pathogens, but they are capable of causing infection or disease if conditions in the host present an opportunity.

Saccharomyces (see the color insert, page C-6 [bottom]) includes many species that have been adapted to industrial microbiology. This genus includes yeasts commonly known as baker's yeast and brewer's yeast because they are used in breads and in wine brewing, respectively. The biotechnology industry also uses *Saccharomyces* as a bioengineered microorganism for producing a commercial product. A bioengineered microorganism is a strain that contains a piece of deoxyribonucleic acid (DNA) from an unrelated organism and makes proteins from the foreign DNA's genes.

Yeasts distribute throughout nature, and they can also act as human pathogens, such as *Candida albicans.* Depending on species, yeasts can grow aerobically or anaerobically. Most yeasts used today for commercial or teaching purposes have uncomplicated growth requirements and grow almost as fast as bacteria when incubated at an optimal growth temperature.

Environmental yeasts or potential pathogenic yeasts can be identified by gross morphology. To do this, a microbiologist mixes the cells with a drop of 10 percent of the fluorescent dye Calcofluor White stain. When observing the stained cells in a fluorescent microscope, the yeast appears light blue.

The table on page 801 describes some of the yeasts most familiar to microbiology students.

In 1680, Antoni van Leeuwenhoek (1632–1723) observed beer yeast with one of the first microscopes introduced to science. The French naturalist Cagniard de la Tour (1777–1859) made early studies on yeasts that had previously been thought to be plantlike organisms. De la Tour recognized yeasts as globules or microorganisms capable of fermenting sugar.

The French microbiologist Louis Pasteur (1822–95) advanced the knowledge of yeast metabolism during studies related to beer brewing and wine fermentations. In 1860, Pasteur explained the methods he had used for his studies: "On January 18, 1858, I placed 100 grams of sugar in a liter of water which contained in it the soluble substances from the beer yeast. To this I added a trace of the globules of fresh yeast. . . . On the 30th of April I repeated this experiment . . . but this time I used a very small amount of ordinary yeast, so that the fermentation could last for a longer time." Pasteur maintained this fermentation through November. In this study, Pasteur developed quantitative methods for monitoring fermentation products, such as alcohol, as well as cellulose digestion and nitrogen use.

Most modern investigations of yeast metabolism relate to yeast reproduction and metabolism that can be applied to industrial microbiology.

Major Yeast Genera

Genus	Significance
Candida	opportunistic pathogen in humans
Cryptococcus	cause of infection, especially in human immunodeficiency virus/acquired immunodeficiency syndrome (HIV/AIDS) patients
Hansenula	can form mycelia; infects trees
Malassezia	on the skin of humans and animals
Rhodotorula	may form mycelia; most species assimilate nitrates in soil
Saccharomyces	common microorganism in industrial microbiology
Trichosporon	yeastlike fungus that produces mycelia and budding cells
Torulopsis	sometimes classified as *Candida;* pathogenic in humans

REPRODUCTION IN YEASTS

Yeast reproduce by either sexual or asexual means, depending on nutrient supply. When food is plentiful, yeast cells reproduce asexually by producing bud cells. Bud cells contain the full complement of the parent yeast's chromosome, so they are called *diploid cells.* The smaller bud cell, then, grows to adult size before reproducing again. Each bud leaves a bud scar on the parent cell, and when a yeast cell becomes covered with bud scars, it dies.

If food it scarce, a yeast cell might divide by the process of meiosis to produce daughter haploid cells that contain one-half of the chromosome. The parent produces up to four daughter cells, which stay inside the parent cell until conditions in the environment improve. Once the haploid cells are released by the parent cell, they can fuse to again make a diploid cell, analogously to sexual reproduction.

In fungal reproduction, species can produce either of two structures: an ascus or a basidium. A saclike ascus releases conidiospores, small structures intended for asexual reproduction. Basidia differ from asci in two main ways: Basidia are clublike structures that extend from hyphae, and basidia contain basidiospores, which participate in sexual reproduction. Yeasts that form either asci or basidia have traditionally been called *perfect yeasts,* and those that do not produce either asci or basidia have been called *imperfect yeasts.* Of the genera in the table, *Candida, Cryptococcus, Malassezia, Rhodotorula,* and *Torulopsis* are imperfect yeasts.

Mycology, the study of fungi, has applied terms to the various forms of yeasts. For example, dimorphic yeasts are those known to have two different morphological forms in their life cycle. Other terms that apply to yeasts are:

- teleomorph—yeast in its perfect state by forming an ascus or a basidium
- anamorph—dimorphic yeast in its single-celled imperfect state
- holomorph—dimorphic yeast exhibiting both forms

Yeast terminology contains an additional idiosyncrasy in naming: Some yeasts have one name in the perfect state and another name in the imperfect state. For example, the unicellular *Candida variabilis* is called *Pichia burtonii* when producing ascopores.

GENETIC ENGINEERING IN YEAST

Yeasts work well in genetic engineering because they grow readily in a laboratory, mating is easy to induce, and yeasts have a medium to small genome (14 million bases) that receives foreign deoxyribonucleic acid (DNA) by a process called *transformation.* In transformation, microbial cells take up DNA directly from the environment.

Biotechnology takes advantage of characteristics of yeast cells that further make them good tools for genetic engineering. First, the small chromosome makes yeasts easy to study as a model for DNA replication in more complex or difficult-to-study eukaryotes. Second, yeasts have haploid and diploid phases that facilitate studies on mutation. Third, microbiologists have perfected the technique of performing transformation in yeasts. Because these cells have a strong cell wall, microbiologists enable transformation by either of two ways: electroporation and salt treatment. In electroporation, a microbiologist applies a sudden electric shock to the cells, which allows foreign DNA to cross the cell wall and membrane. Salt treatment has a similar effect by weakening the cell wall, usually by exposing cells to lithium acetate solution. Finally, yeast cells accommodate bacterial plasmids.

A plasmid is a circular strand of DNA in the bacterial cell cytoplasm separate from the dense mass of DNA in a region called the *nucleoid.* Yeasts replicate plasmids as if the plasmid DNA were their own. This feature has made yeasts an attractive model for studying genes inserted into bacterial plasmids and subsequent transformation of the yeast with the plasmid. A remarkable variety of genes can now be carried in plasmids, including human genes. When DNA is bioengineered specifically fort replication in yeast, the DNA is sometimes referred to as *yeast artificial chromosomes* (YACs).

Yeast mating processes have also aided in preparing recombinant DNA in yeasts. The yeast chromosome contains two sites that direct mating processes: the *a* gene and the alpha (α) gene. Both genes are inactive, called *silent genes,* unless they are transferred to a site on the DNA molecule to be transcribed (an action in which the enzyme transcriptase reads the genetic code). Only one mating gene can be expressed at a time. When either of the mating genes inserts into the DNA molecule, it initiates the steps required for genetic transfer in mating. Because the mating gene inserts into DNA as a cassette inserts into a cassette player, this method had been known as the *cassette model* (before cassette players became obsolete). The cassette model in yeasts enables microbiologists to have more control over the mating process in yeasts used for biotechnology.

YACs have enabled biologists to study specific regions of DNA from higher organisms. Yeasts replicate the YACs and, thereby, produce new copies of the original DNA fragment, a process called cloning. (When similar DNA amplification takes place in an instrument called a *thermocycler,* the process is known as *polymerase chain reaction* [PCR]). YACs put into yeasts can produce the same proteins that would be produced if the DNA had remained in the original cell. In some instances, molecular biologists can insert foreign genes in the same location of the yeast chromosome as in the DNA donor cell. In addition, because yeasts are eukaryotic cells, their cellular systems can fold newly synthesized proteins and carry out modifications to the proteins, just as in other eukaryotes. One common modification to proteins in eukaryotic cells involves glycosylation, which is the process of adding carbohydrates to a protein.

Biotechnology currently uses *Saccharomyces,* or has in the past, for making a diversity of commercial products. The following list provides important examples of products from this yeast:

- enzyme rennin—for curdling milk in cheese making
- enzyme invertase—hydrolyzes sucrose to glucose and fructose, used in candy making
- enzyme lactase—digestive aid that degrades lactose in milk
- lactic acid—by expression of cow's lactate dehydrogenase gene
- xylitol—food industry sweetener through the expression of xylose reductase from *Pichia*
- pediocin—wine preservative made by expressing the gene from *Pediococcus* bacteria
- hepatitis B virus vaccine—by expression of human gene for the virus antigen

The hepatitis B vaccine, developed in 1984, became the first vaccine produced by genetic engineering to defend against a human virus.

Genetically modifying yeasts that can be used directly in food production has not been as widely accepted as the products listed here. Many people have a strong aversion to genetically modified foods of any kind. The California winemaker Greg Fowler stated, in 1999, "No one wants to have anything to do with genetic engineering." At the time, yeasts were being considered for improving the wine fermentation process, flavor, and odor.

Molecular biologists continue to experiment with using bioengineered yeasts in wine making and in beer brewing. Microbiologists at Rice University reported, in 2008, on a project to engineer yeast to produce the purported anti-aging compound resveratrol, found in red wine in low levels. Thomas Segall-Shapiro, a member of the research team, commented on the value of using bioengineered yeast, "The amount in red wine's actually not that much compared to what might be possible with this process."

Yeast will continue to be an important vehicle for expressing genes from mammals, possibly from fish, and certainly from bacteria. Second only to *Escherichia coli* bacteria, yeast is biotechnology's critical microorganism for future technologies.

See also CANDIDA ALBICANS; FUNGUS; GENETIC ENGINEERING; PASTEUR, LOUIS; PLASMID; SACCHAROMYCES.

Further Reading

Davison, Anna. "Beer That's Good for You." Massachusetts Institute of Technology. *Technology Review,* 4 November 2008. Available URL: www.technologyreview.com/biomedicine/21628/page1. Accessed November 30, 2009.

Kurtzman, Cletus P., and Jack W. Fell. *The Yeasts: A Taxonomic Study.* Amsterdam, The Netherlands: Elsevier, 1998.

Larone, Davise H. *Medically Important Fungi: A Guide to Identification,* 4th ed. Washington, D.C.: American Society for Microbiology Press, 2002.

McAleer, William J., Eugene B. Buynak, Robert Z. Maigetter, D. Eugene Wampler, William J. Miller, and Maurice R. Hilleman. "Human Hepatitis B Vaccine from Recombinant Yeast." *Nature* 307 (1984): 178–180. Available online. URL: www.nature.com/nature/journal/v307/n5947/abs/307178a0.html. Accessed November 30, 2009.

Pasteur, Louis. 1860. "Mémoire sur la fermentation alcoölique" (Memoir on the Alcoholic Fermentation). *Annales de Chimie et de Physique* 58 (1860): 323–426. In *Milestones in Microbiology*, translated by Thomas Brock. Washington, D.C.: American Society for Microbiology Press, 1961.

Wines and Vines. "It's the Yeast We Can Do." June 1999. Available online. URL: http://findarticles.com/p/articles/mi_m3488/is_6_80/ai_54926734/?tag=content;col1. Accessed November 30, 2009.

APPENDIX I
Chronology

1595 The Dutch glassmaker Zacharias Janssen teams with his father, Hans, to assemble lenses in series, building the first compound microscope

1665 Robert Hooke, a brilliant inventor and scientist, first observes cells in nature and coins the term *cell*

1673 Antoni van Leeuwenhoek becomes the Father of Microbiology as the first person to observe microorganisms in a microscope

1735 Carl Linnaeus creates a system for naming organisms that lays the foundation for systematics and taxonomy

1798 British physician Edward Jenner develops the first vaccine (for smallpox)

1835 Italian Agostino Bassi proposes the theory of "contagion" to explain infectious diseases

1857–85 French microbiologist Louis Pasteur describes the steps of fermentation, disproves spontaneous generation, develops a new heating process for wine preservation (pasteurization), saves the silk industry by finding a parasite in silkworms, develops an anthrax vaccine, and develops and is the first to use a rabies vaccine

1867 British surgeon Joseph Lister develops aseptic techniques for surgery

1876–82 German microbiologist Robert Koch explains the germ theory of disease, invents a method for growing pure cultures, and discovers the cause of tuberculosis *(Mycobacterium tuberculosis)*

1881 Koch's colleague Walther Hesse first uses agar for solidifying growth media

1883 Immunologist Elie Metchnikoff first describes the role of phagocytosis in immunity

1884 Danish bacteriologist Hans Christian Gram develops a staining technique for differentiating bacteria, to become known as the Gram stain

1885 Theodor Escherich discovers a new bacterium associated with diarrhea *(Escherichia coli)*

1887 Bacteriologist Richard Petri designs a shallow dish for microbial cultures (petri dish)

1890 Paul Ehrlich proposes how the immune system recognizes antigens

1892 Russian scientist Sergei Winogradsky describes the steps of the sulfur cycle and introduces the science of microbial ecology. Russian microbiologist Dmitri Iwanowsk discovers viruses

1894 Public health microbiologists Alexandre Yersin and Shibasaburo Kitasato codiscover the plague bacterium *(Yersinia pestis)*

1915 Félix d'Herelle and Frederick Twort discover bacteriophages

1921–28 Alexander Fleming discovers lysozyme and penicillin

1928 Frederick Griffith demonstrates transformation of genetic material between bacteria

1933 German physicist Ernst Ruska develops the electron microscope

1934 American Rebecca Lancefield discovers streptococcus antigens

1937 French biologist Edward Chatton divides organisms into prokaryotes and eukaryotes

1938–40 Ernst Chain and Howard Florey purify and test penicillin

1941 George Beadle and Edward Tatum propose the one gene–one enzyme hypothesis

1944 Oswald Avery, Colin MacLeod, and Maclyn McCarty show that the DNA molecule is life's genetic material

1946 Joshua Lederberg and Edward Tatum discover conjugation in bacteria

1949 John Enders, Thomas Weller, and Frederick Robbins develop a technique for culturing viruses in laboratory conditions

1950 Robert Hungate develops a method for growing anaerobic bacteria in laboratory conditions

1952 Alfred Hershey and Martha Chase describe the consequences of viral protein and nucleic acid during viral infection of cells

1953 James Watson and Francis Crick propose double-helix structure for DNA

1957 François Jacob and Jacques Monod describe how cells regulate protein synthesis

1960 Leland Hartwell, Paul Nurse, and Timothy Hunt begin discovery of genes that encode for the proteins that control cell division

1971 Werner Arber, Daniel Nathans, and Hamilton Smith discover restriction enzymes, making genetic engineering possible

1977 American microbiologist Carl Woese defines archaea as a microbial group separate from bacteria

1983 Barbara McClintock discovers transposons. Kary Mullis develops the polymerase chain reaction method

1988 Kary Mullis develops the polymerase chain reaction

1990 W. French Anderson first uses gene therapy in medicine

1995 The first complete sequencing of microbial DNA (*Haemophilus influenzae*) is performed by a multidisciplinary team led by Robert Fleischmann and J. Craig Venter

1997 Stanley Prusiner discovers prions. Frederick Blattner sequences the entire genome of *E. coli*

2002 Eckard Wimmer assembles poliovirus from basic building block molecules

2006 International Census of Marine Microbes discovers that more than 20,000 kinds of microorganisms exist in a liter of seawater

2008 Molecular biologist Dan Gibson assembles entire bacterial genome *(Mycobacterium genitalium)* in a laboratory

2009 Josh Tickell develops an algae-powered car that crosses the United States

2010 Robert Gifford and Aris Katzourakis show the extent in which virus genomes have evolved as part of mammal, bird, and insect DNA. GlaxoSmithKline develops a boron-containing molecule with antimicrobial activity against antibiotic-resistant bacteria. National Aeronautics and Space Administration (NASA) scientists discover a microorganism in Mono Lake, California, that is the first known to use arsenic in its metabolism, expanding the possibilities for microbial life on other planets

APPENDIX II
Glossary

AIDS acquired immunodeficiency syndrome, an infectious disease caused by the human immunodeficiency virus (HIV)

amino acid organic molecule containing both a carboxyl group (-COOH) and an amino group (-NH2)

antitoxin antibody produced by the body in response to a specific biological toxin

biosynthesis production of any compound by a biological organism

biotechnology industry of producing microorganisms, higher eukaryotic cells, or cell components to make a useful product

catalyze action of a compound to alter the rate of a chemical reaction without being altered itself

clinical pertaining to the medical aspects of microbiology for the diagnosis of infectious disease and the selection of treatments for specific pathogens

compound chemical composed of two or more elements, for example, water (H_2O)

concentration amount of a material in solution or suspension and usually expressed as weight or mass per unit volume, for example, milligrams per liter (mg/l)

containment any physical or chemical means of preventing microorganisms from escaping into the environment, which could be a microbiology laboratory or the outdoor environment

cyst protective capsulelike structure of some protozoa, such as *Giardia;* in *Cryptosporidium,* the similar structure is an oocyst

daughter cell progeny of a microbial parent cell produced by either sexual or asexual reproduction

deoxyribonucleic acid (DNA) nucleic acid that holds the entire genetic history and makeup of cells and some viruses

dilution action of decreasing the concentration of a dissolved or suspended substance by transferring a small volume to a larger volume of water, buffer, or other pure material; a high dilution means the substance's concentration is low, and a low dilution means the substance's concentration is high

diphtheroids general group of nonpathogenic skin bacteria consisting mainly of the species *Corynebacterium*

disease an abnormal state of the body in which one or more functions are altered to the detriment of overall health

diversity amount of number and variety of different species in an environment

dormant condition of a microorganism that is alive but not actively metabolizing or reproducing, that is, a dormant state

electron acceptor also, electron sink, a molecule that takes one or more electrons from another molecule in a reduction-oxidation (redox) reaction

electron donor molecule that transfers one or more electrons to another molecule in a reduction-oxidation (redox) reaction

electrophoresis technique for separating molecules through differences in their migration rate in a medium (paper, slurry, or gel) influenced by an electric field

encode in genetics, the action of carrying information required for synthesizing a functioning protein

enumeration process of determining the amount of microorganisms in a unit of weight or volume, for example, 1,000 cells per milliliter

enzyme protein that catalyzes biochemical reactions inside a living organism or secreted by an organism

epithelium tissue of tightly packed cells that lines organs and body cavities

express in genetics, the action of converting the information in genes into a functioning protein

fatty acid hydrocarbon chain molecule made of repeating CH_2 units, containing a terminal methyl group (-CH_3) on the non-water-soluble end and a carboxyl group (-COOH) on the more water-soluble end

flora general term for any diverse group of microorganisms, usually used to describe microorganisms associated with the body: skin flora, oral flora, etc.

food chain flow of energy and matter in living organisms from producers, which convert the sun's energy to chemical energy, to more complex consumers

food web network of interlinked food chain containing producers, consumers, predators, and prey

gastrointestinal pertaining to the portion of the digestive tract in higher animals from the stomach to the anus

gene segment of deoxyribonucleic acid (DNA) that carries all the information for the cell's synthesis of a functional product

genus (plural: genera) the first name in a scientific binomial name—*Escherichia* is the genus name for *Escherichia coli*—and the taxonomic group between family and species

germ general term for a harmful or nuisance microorganism

global warming the gradual increase in Earth's average atmospheric and marine temperatures

glycogen large molecule or polymer made of glucose sugar units and acting as an energy storage compound

gradient gradual change in environmental conditions, usually to describe changes in temperature, pH, or chemical concentration

habitat the place where a microorganism normally lives, such as freshwater, marine water, soil, the digestive tract, skin, or the mouth.

host an organism that provides an environment for part or all of a microorganism's or parasite's life cycle

hydrocarbon molecule consisting of only carbon and hydrogen

hyphae (singular: hypha) long filament of cells in fungi and some bacteria

immunocompromised the condition of a human or animal in which the immune system is not fully functioning and cannot provide optimal defense against infection or disease by a pathogen

impedance resistance to movement or other change in conditions due to a physical barrier or an electrochemical force

incidence the frequency of new cases of a disease in a population

infection the invasion of the body or part of the body, or cells, by microorganisms

infectious agent general term for any microorganism that causes or participates in infection

inhibition any action of permanently or temporarily slowing or stopping a biological action such as enzyme activity, growth, or reproduction

inoculate to place a small number of microbial cells into broth medium or on agar medium for the purpose of growing large numbers of the same type of cells, species, or strain

ion positively or negatively charged atom (hydrogen ion, H^+) or group of atoms (phosphate, PO_4^{3-})

isolate verb: to separate one type of microorganism from all other different microorganisms; noun: a pure microorganism separated from a diverse population of microorganisms

kefir beverage of fermented cow's milk

kilobase piece of nucleic acid equal to 1,000 base units, called nucleotides

lipid non-water-soluble organic compound that includes fats (tri-, di-, and monoglycerides), phospholipids, and sterols

lymphocyte type of white blood cell involved in acquired immunity; lymphocytes from bone marrow are B-cells and lymphocytes that mature in the thymus are T-cells

lysis the breaking apart of a cell by either physical forces, chemical action, or enzyme activity that causes the cellular contents to escape, resulting in cell death

lysozyme antibacterial enzyme in tears, saliva, and sweat

macromolecule large organic molecule, usually including proteins, polysaccharides, fibers, and nucleic acids, as well as peptidoglycan, some antibiotics, and some vitamins

macronutrient an essential nutrient required in large amounts by cells, usually carbon, nitrogen, phosphorus, sulfur compounds, and water

matrix any complex, heterogeneous material that houses chemical or biochemical reactions or microbial activity

meiosis replication process in eukaryotic cells that results in cells with one-half the chromosome number of the original cell

micronutrient an essential nutrient required in very small amounts by cells, usually minerals and vitamins

mitosis replication process in eukaryotic cells that duplicates the chromosomes prior to cell division

membrane thin biological or physical barrier that may allow certain substances to pass through to the opposite side (a semipermeable membrane) or prevent all substances from passing through (an impermeable membrane)

metabolite the intermediary products and end products of a metabolic pathway; for example, pyruvate and ethanol are intermediate and end products of glucose fermentation, respectively

molar pertaining to the amount of moles in a solution; a mole is a gram amount of chemical equal to the atomic weights of all the atoms making up a molecule of the chemical; for example, a mole of water (H_2O contains two hydrogen atoms, total atomic weight of 2, and one oxygen atom, atomic weight of 16) is 18 grams

mold fungus that exists as multicellular colonies producing filaments, called mycelia, that give the colony a visible fuzzy or fluffy appearance

molecular pertaining to any process or study performed at the level of individual molecules, such as nucleic acids, rather than in a whole cell or multicellular organism

molecule combination of atoms that form a distinct chemical unit; hydrogen and oxygen atoms combine to form a molecule of water (H_2O)

mucous membrane the passages and cavities of the body that interface with the air, consisting of a base layer of connective tissue, a membrane, and a surface layer of epithelium that secretes mucus, a viscous fluid

neurotoxin poisonous substance of microbial, plant, or animal origin that injures an organism's nervous system

nonmotile microorganism's lack of the ability to move through the environment on its own power, perhaps due to the lack of flagella, cilia, or amoeboid movement

nucleic acid large molecule, either deoxyribonucleic acid (DNA) or ribonucleic acid (RNA), composed of nucleotide subunits

nucleotide basic unit of nucleic acids, composed of a purine or pyrimidine base, a five-carbon sugar, and a phosphate group

nutrient compound or molecule that a cell requires for energy metabolism, growth, or another specific function

nutrient cycle also biogeochemical cycle, a process conducted by microorganisms that recycles the earth's elements for reuse by microorganisms and higher organisms

osmosis net movement of solvent molecules, often water, across a semipermeable membrane from an area of low solute (dissolved substance) concentration to an area of high solute concentration

oxidation chemical reaction in which electrons are removed from a molecule

parasite an organism that must live all or part of its life inside another organism to survive and, in this process, causes harm to the host organism

peptide chain of amino acids like a protein but usually lacking the full function of proteins

phospholipid molecule containing a polar hydrophilic (water-attractant) head provided by a phosphate group (PO_4^{-3}) and a long hydrophobic (water-repellent) tail provided by a hydrocarbon chain

plate petri dish filled with solidified agar ready to be inoculated with microorganisms or already containing incubated microbial colonies

polymer any long, chainlike molecule made of repeating subunits called monomers, for example, starch (polysaccharide), cellulose, and deoxyribonucleic acid (nucleic acid DNA)

polysaccharide carbohydrate made of eight or more sugars, called monosaccharides, joined in a chainlike structure

population group of organisms of the same type either by species or by a common activity in the environment, such as a photosynthetic population of microorganisms

precipitate verb: to separate a material from a fluid by making it insoluble; noun: an insoluble aggregate separated from a solution or suspension, usually by gravity

prevalence the number of cases of a disease in a specific population at a given point in time

protocol a formal, standardized procedure

purines nitrogen-containing compounds that include adenine and guanine and make up part of a nucleotide

pyrimidines nitrogen-containing compounds that include uracil, thymine, and cytosine and make up part of a nucleotide

receptor proteinaceous structure on the outer surface of a cell that forms specific and complementary connections with other compounds in the environment

reduction chemical reaction in which electrons are added to a molecule

replicate verb: to duplicate or repeat a process; noun: one of several identical experiments, samples, or microorganisms

reservoir continual biological or nonbiological source of an infectious agent

respiration series of oxidation and reduction reactions that produce the energy-storage compound adenosine triphosphate (ATP) and usually use an inorganic compound in the final step of energy generation, for example, oxygen (O_2)

ruminant animal possessing a four-part stomach (abomasum, omasum, rumen, and reticulum) that houses dense microbial populations for digesting dietary fibers

ribonucleic acid (RNA) class of nucleic acids that participate in the conversion of genetic information in deoxyribonucleic acid (DNA) to functioning proteins; messenger RNA (mRNA), transfer RNA (tRNA), and ribosomal RNA (rRNA)

ribosome organelle that is the site of protein synthesis in the cell and composed of protein and ribonucleic acid (RNA)

sequence exact order of amino acids in a protein or of nucleotide bases in a nucleic acid

sign measurable change in the body due to disease, usually observed by the diseased person or animal, for example, elevated fever (compare to *symptom*)

solute substance dissolved in a liquid

solvent liquid that has one or more substances dissolved in it

sterile condition of being absent of all living things

sterol type of molecule containing a cyclic structure and at least one alcohol group (-OH), for example, cholesterol

strain population of microorganisms distinguishable from other members of its species, usually by the presence of unique structures on the cell surface

streaking action of distributing an inoculum of microorganisms over an agar surface for the purpose of growing distinct, isolated colonies during incubation

substrate starting compound of a metabolic pathway; for example, glucose is the substrate for glycolysis; the compound upon which an enzyme acts

susceptible condition of an organism lacking resistance to infection or a microorganism lacking resistance to an antimicrobial agent, such as an antibiotic

symptom change in body function due to a disease, which can be felt by a person or an animal, for example, lethargy (compare to *sign*)

systemic pertaining to a condition, disease, or infection affecting the whole body (opposite of *localized*)

technician scientist trained in specific procedures in industrial or medical microbiology

tissue community of cells of the same type with a cell-to-cell communication process; usually an individual cell cannot survive separately from the tissue

toxin poisonous substance of microbial, plant, or animal origin

transport conveyance of a substance across a matrix, such as cell cytoplasm, or across a membrane, with or without the consumption of energy

vegetative referring to cells that obtain and metabolize nutrients rather than reproduce

APPENDIX III
Further Resources

PRINT AND ONLINE RESOURCES

Black, Jacquelyn G. *Microbiology: Principles and Explorations,* 7th ed. Hoboken, N.J.: John Wiley & Sons, 2008. A well-illustrated resource on state-of-art microbiology.

Block, Seymour S., ed. *Disinfection, Sterilization, and Preservation,* 5th ed. Philadelphia: Lippincott Williams & Wilkins, 2000. The main reference book on antimicrobial compounds and sterilization techniques.

Dyer, Betsey D. *A Field Guide to Bacteria.* Ithaca, N.Y.: Cornell University, 2003. An enjoyable guide to environmental bacteria.

Jjemba, Patrick, K. *Environmental Microbiology: Principles and Applications.* Enfield, N.H.: Science Publishers, 2004. A useful resource on microbial ecology.

Maier, Raina M., Ian L. Pepper, and Charles P. Gerba, eds. *Environmental Microbiology,* 2nd ed. San Diego: Elsevier, 2009. An excellent overview of microorganisms of the air, water, and soil and new technologies.

Needham, Cynthia, Mahlon Hoagland, Kenneth McPherson, and Bert Dodson. *Intimate Strangers: Unseen Life on Earth.* Washington, D.C.: American Society for Microbiology Press, 2000. A collection of entertaining vignettes about today's microbiology.

Schaechter, Moselio, John L. Ingraham, and Frederick C. Neidhardt. *Microbe.* Washington, D.C.: American Society for Microbiology Press, 2006. This well-written book covers unique aspects of microbiology.

Tortora, Gerard J., Berdell R. Funke, and Christine Case. *Microbiology: An Introduction,* 10th ed. San Francisco: Benjamin Cummings, 2009. A standard resource in the microbiology profession.

Willey, Joanne, Linda Sherwood, and Chris Woolverton. *Prescott, Harley, Klein's Microbiology,* 7th ed. New York: McGraw-Hill, 2007. A standard resource in the microbiology profession.

INTERNET RESOURCES

American Society for Microbiology. Available online. URL: www.asm.org. Accessed October 3, 2009. The main international association for microbiologists.

Australian Society for Microbiology. Available online. URL: www.theasm.org.au. Accessed October 3, 2009. A good supplementary resource for international topics.

Centers for Disease Control and Prevention. Available online. URL: www.cdc.gov. Accessed October 3, 2009. The main source of updated information on diseases.

Doctor Fungus. Available online. URL: www.doctorfungus.org. Accessed October 3, 2009. An excellent resource on mycology.

International Society of Protistologists. Available online. URL: www.uga.edu/protozoa. Accessed October 3, 2009. A resource on protozoa.

J. Craig Venter Institute. Available online. URL: www.jcvi.org. Accessed October 3, 2009. An institute focused on molecular aspects of microorganisms.

Microbe World. Available online. URL: www.microbeworld.org. Accessed October 3, 2009. Contains useful educational materials in basic microbiology.

Microbiology Network. Available online. URL: www.microbiol.org. Accessed October 3, 2009. A clearinghouse for other online resources in microbiology plus microbiology news.

Microscopy Society of America. Available online. URL: www.microscopy.org. Accessed October 3, 2009. A good resource for the latest technologies.

National Institute of Allergy and Infectious Diseases. Available online. URL: www3.niaid.nih.gov. Accessed October 3, 2009. An essential resource for infectious diseases with some international content.

Pasteur Institute. Available online. URL: www.pasteur.fr/ip/easysite/go/03b-00002j-000/en. Accessed October 3, 2009. An interesting site that describes current research.

Society for Anaerobic Microbiology. Available online. URL: www.clostridia.net/SAM. Accessed October 3, 2009. This Web site covers a unique specialty in microbiology.

Society for General Microbiology. Available online. URL: www.sgm.ac.uk. Accessed October 3, 2009. Contains excellent references on a variety of subjects regarding microorganisms.

Todar, Kenneth. Todar's Online Textbook of Bacteriology. 2008. Available online. URL: www.textbookofbacteriology.net. Accessed October 3, 2009. An online resource that covers important topics about bacteria.

U.S. Food and Drug Administration. Bad Bug Book. Available online. URL: www.fda.gov/Food/FoodSafety/FoodborneIllness/FoodborneIllnessFoodbornePathogensNaturalToxins/BadBugBook/default.htm. Accessed October 3, 2009. Excellent resource on food-borne pathogens.

Virtual Museum of Bacteria. Available online.URL: http://bacteriamuseum.org. Accessed October 3, 2009. A good resource for the history of microbiology.

Woods Hole Marine Biological Laboratory. Available online. URL: www.mbl.edu. Accessed October 3, 2009. The primary research institute in marine microbiology.

World Health Organization. Available online. URL: www.who.int/en. Accessed October 3, 2009. The main international resource on infectious diseases.

APPENDIX IV

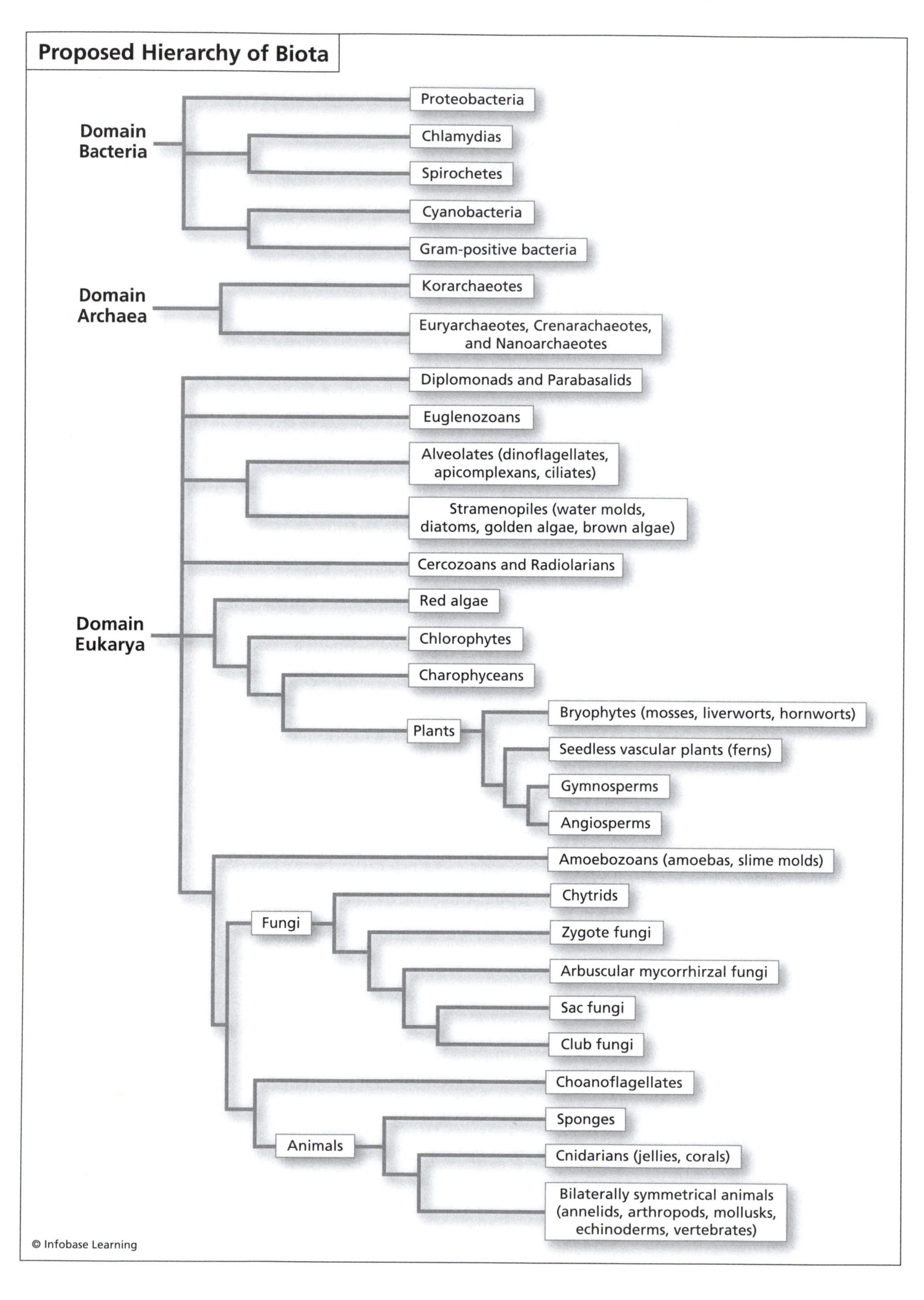
Proposed Hierarchy of Biota
Domain Bacteria
Proteobacteria
Chlamydias
Spirochetes
Cyanobacteria
Gram-positive bacteria
Domain Archaea
Korarchaeotes
Euryarchaeotes, Crenarachaeotes, and Nanoarchaeotes
Domain Eukarya
Diplomonads and Parabasalids
Euglenozoans
Alveolates (dinoflagellates, apicomplexans, ciliates)
Stramenopiles (water molds, diatoms, golden algae, brown algae)
Cercozoans and Radiolarians
Red algae
Chlorophytes
Charophyceans
Plants
Bryophytes (mosses, liverworts, hornworts)
Seedless vascular plants (ferns)
Gymnosperms
Angiosperms
Amoebozoans (amoebas, slime molds)
Fungi
Chytrids
Zygote fungi
Arbuscular mycorrhirzal fungi
Sac fungi
Club fungi
Choanoflagellates
Animals
Sponges
Cnidarians (jellies, corals)
Bilaterally symmetrical animals (annelids, arthropods, mollusks, echinoderms, vertebrates)
© Infobase Learning

APPENDIX V
Classification of Bacteria and Archaea

DOMAIN BACTERIA

Phylum I: Aquificae
- Class Aquificae
  - Order Aquificales
    - Family Aquificaceae
      - Genera: *Aquifex*, *Calderobacterium*, *Desulfurobacterium*, *Hydrogenobacter*, *Thermocrinis*

Phylum II: Thermotogae
- Class Thermotogae
  - Order Thermotogales
    - Family Thermotogaceae
      - Genera: *Fervidobacterium*, *Geotoga*, *Petrotoga*, *Thermosipho*, *Thermotoga*

Phylum III: Thermodesulfobacteria
- Order Thermodesulfobacteriales
  - Family Thermodesulfobacteriaceae
    - Genus: *Thermodesulfobacterium*

Phylum IV: Deinococcus-Thermus
- Class Deinococci
  - Order I: Deinococcales
    - Family Deinococcaceae
      - Genus: *Deinococcus*
  - Order II: Thermales
    - Family Thermaceae
      - Genera: *Meiothermus*, *Thermus*

Phylum V: Chrysiogenetes
- Class Chrysiogenetes
  - Order Chrysiogenales
    - Family Chrysiogenaceae
      - Genus: *Chrysiogenes*

Phylum VI: Chloroflexi
- Class Chloroflexi
  - Order I: Chloroflexales
    - Family I: Chloroflexaceae
      - Genera: *Chloroflexus*, *Chloronema*, *Heliothrix*
    - Family II: Oscillochloridaceae
      - Genus: *Oscillochloris*
  - Order II: Herpetosiphonales
    - Family Herpetosiphonaceae
      - Genus: *Herpetosiphon*

Phylum VII: Thermomicrobia
- Class Thermomicrobia
  - Order Thermomicrobiales
    - Family Thermomicrobiaceae
      - Genus: *Thermomicrobium*

Phylum VIII: Nitrospira
- Class Nitrospira
  - Order Nitrospirales
    - Family Nitrospiraceae
      - Genera: *Leptospirillum*, *Magnetobacterium*, *Nitrospira*, *Thermodesulfovibrio*

Phylum IX: Deferribacteres
- Class Deferribacteres
  - Order Deferribacterales
    - Family Deferribacter
      - Genera: *Deferribacter*, *Denitrovibrio*, *Flexistipes*, *Geovibrio*, *Synergistes*

Phylum X: Cyanobacteria
- Class Cyanobacteria
  - Subsection I
    - Genera: *Chamaesiphon*, *Chroococcus*, *Cyanobacterium*, *Cyanobium*, *Cyanothece*, *Dactylococcopsis*

Gloeobacter
Microcystis
Prochlorococcus
Prochloron
Synechococcus
Synechocystis

Subsection II

Genera: *Chroococcidiopsis*
Cyanocystis
Dermocarpella
Myxosarcina
Pleurocapsa
Stanieria
Xenococcus

Subsection III

Genera: *Arthrospira*
Borzia
Crinalium
Geitlerinema
Halospirulina
Leptolyngbia
Limnothrix
Lyngbya
Microcoleus
Oscillatoria
Planktothrix
Prochlorothrix
Pseudoanabaena
Spirulina
Starria
Symploca
Trichodesmium
Tychonema

Subsection IV

Genera: *Anabaena*
Anabaenopsis
Aphanizomenon
Calothrix
Cyanospira
Cylindrospermum
Nodularia
Nostoc
Rivularia
Scytonema
Tolypothrix

Subsection V

Genera: *Chlorogloeopsis*
Fischerella
Geitleria
Iyengariella
Nostochopsis
Stigonema

Phylum XI: Chlorobi

Class Chlorobi

Order Chlorobiales

Family Chlorobiaceae

Genera: *Ancalochloris*
Chlorobium
Chloroherpeton
Pelodictyon
Prosthecochloris

Phylum XII: Proteobacteria

Class I: Alpha-Proteobacteria

Order I: Rhodospirillales

Family I: Rhodospirillaceae

Genera: *Azospirillum*
Magnetospirillum
Phaeospirillum
Rhodocista
Rhodospira
Rhodospirillum
Rhodothalassium
Rhodovibrio
Roseospira
Shermanella

Family II: Acetobacteraceae

Genera: *Acetobacter*
Acidiphilium
Acidosphaera
Acidocella
Acidomonas
Asaia
Craurococcus
Gluconacetobacter
Gluconobacter
Paracraurococcus
Rhodopila
Roseococcus
Stella
Zavarzinia

Order II: Rickettsiales

Family I: Rickettsiaceae

Genera: *Orientia*
Rickettsia
Wolbachia

Family II: Ehrlichiaceae

Genera: *Aegyptianella*
Anaplasma
Cowdria
Ehrlichia
Neorickettsia
Xenohaliotis

Family III: Holosporaceae

Genera: *Caedibacter*
Holospora
Lyticum
Odyssella
Polynucleobacter
Pseudocaedibacter
Symbiotes
Tectibacter

Order III: Rhodobacterales

Family Rhodobacteraceae
Genera: *Ahrensia*
Amaricoccus
Antarctobacter
Gemmobacter
Hirschia
Hyphomonas
Maricaulus
Methylarcula
Octadecabacter
Paracoccus
Rhodobacter
Rhodovulum
Roseobacter
Roseibium
Roseinatronobacter
Roseovarius
Roseovivax
Rubrimonas
Ruegeria
Sagittula
Staleya
Stappia
Sulfitobacter
Order IV: Sphingomonadales
Family Sphingomonodaceae
Genera: *Blastomonas*
Erythrobacter
Erythromicrobium
Erythromonas
Porphyrobacter
Rhizomonas
Sandaracinobacter
Sphingomonas
Zymomonas
Order V: Caulobacterales
Family Caulobacteraceae
Genera: *Asticcacaulis*
Brevundimonas
Caulobacter
Phenylobacterium
Order VI: Rhizobiales
Family I: Rhizobiaceae
Genera: *Agrobacterium*
Carbophilus
Chelatobacter
Ensifer
Rhizobium
Sinorhizobium
Family II: Bartonellaceae
Genus: *Bartonella*
Family III: Brucellaceae
Genera: *Brucella*
Mycoplana
Ochrobactrum
Family IV: Phyllobacteriaceae
Genera: *Allorhizobium*
Aminobacter
Aquamicrobium
Defluvibacter
Mesorhizobium
Phyllobacteriaceae
Pseudaminobacter
Family V: Methylocystaceae
Genera: *Methylocystis*
Methylopila
Methylosinus
Family VI: Beijerinckiaceae
Genera: *Beijerinckia*
Chelatococcus
Derxia
Family VII: Bradyrhizobiaceae
Genera: *Afipia*
Agromonas
Blastobacter
Bosea
Bradyrhizobium
Nitrobacter
Oligotropha
Rhodopseudomonas
Family VIII: Hyphomicrobiaceae
Genera: *Ancalomicrobium*
Ancylobacter
Angulomicrobium
Aquabacter
Azorhizobium
Blastochloris
Devosia
Dichotomicrobium
Filomicrobium
Gemmiger
Hyphomicrobium
Labrys
Methylorhabdus
Pedomicrobium
Prosthecomicrobium
Rhodomicrobium
Rhodoplanes
Seliberia
Starkya
Xanthobacter
Family IX: Methylobacteriaceae
Genera: *Methylobacterium*
Protomonas
Roseomonas
Family X: Rhodobiaceae
Genus: *Rhodobium*
Class II: Beta-Proteobacteria
Order I: Burkholderiales
Family I: Burkholderiaceae
Genera: *Burkholderia*
Cupriavidus

Lautropia
Pandoraea
Thermothrix
Family II: Ralstoniaceae
Genus: *Ralstonia*
Family III: Oxalobacteraceae
Genera: *Duganella*
Herbaspirillum
Janthinobacterium
Massilia
Oxalobacter
Telluria
Family IV: Alcaligenaceae
Genera: *Achromobacter*
Alcaligenes
Bordetella
Pelistega
Sutterella
Taylorella
Family V: Comamonadaceae
Genera: *Acidovorax*
Aquabacterium
Brachymonas
Comamonas
Delftia
Hydrogenophaga
Ideonella
Leptothrix
Polaromonas
Rhodoferax
Rubrivivax
Sphaerotilus
Tepidimonas
Thiomonas
Variovorax
Order II: Hydrogenophilales
Family Hydrogenophilus
Genera: *Hydrogenophaga*
Thiobacillus
Order III: Methylophilales
Family Methylophilaceae
Genera: *Methylobacillus*
Methylophilus
Methylovorus
Order IV: Neisseriales
Family Neisseriaceae
Genera: *Alysiella*
Aquaspirillum
Catenococcus
Chromobacterium
Eikenella
Formivibrio
Iodobacter
Kingella
Microvirgula
Neisseria
Prolinoborus
Simonsiella
Vitreoscilla
Vogesella
Order V: Nitrosomonadales
Family I: Nitrosomonadaceae
Genera: *Nitrosomonas*
Nitrosospira
Family II: Spirillaceae
Genus: *Spirillum*
Family III: Gallionellaceae
Genus: *Gallionella*
Order VI: Rhodocyclales
Family Rhodocyclaceae
Genera: *Azoarcus*
Azonectus
Azospira
Azovibrio
Propionibacter
Propionivibrio
Rhodocyclus
Thauera
Zoogloea
Class III: Gamma-Proteobacteria
Order I: Chromatiales
Family I: Chromatiaceae
Genera: *Allochromatium*
Amoebobacter
Chromatium
Halochromatium
Halothiobacillus
Isochromatium
Lamprobacter
Lamprocystis
Marichromatium
Nitrosococcus
Pfennigia
Rhabdochromatium
Thermochromatium
Thioalkalicoccus
Thiocapsa
Thiococcus
Thiocystis
Thiodictyon
Thiohalocapsa
Thiolamprovum
Thiopedia
Thiorhodococcus
Thiorhodovibrio
Thiospirillum
Order II: Acidothiobacillales
Family Acidothiobacillaceae
Genus: *Acidothiobacillus*
Order III: Xanthomonadales
Family: Xanthomonadaceae
Genera: *Frateuria*

Lureimonas
Lyssobacter
Nevskia
Pseudoxanthomonas
Rhodanobacter
Stenothrophomonas
Xanthomonas
Xylella

Order IV: Cardiobacteriales
Family: Cardiobacteriaceae
Genera: *Cardiobacterium*
Dichelobacter
Suttonella

Order V: Thiotrichales
Family I: Thiotrichaceae
Genera: *Achromatium*
Beggiatoa
Leucothrix
Macromonas
Thiobacterium
Thiomargarita
Thioploca
Thiospira
Thiothrix
Family II: Piscirickettsiaceae
Genera: *Cycloclasticus*
Hydrogenovibrio
Methylophaga
Piscirickettsia
Thiomicrospira
Family III: Francisellaceae
Genus: *Francisella*

Order VI: Legionellales
Family I: Legionallaceae
Genus: *Legionella*
Family II: Coxiellaceae
Genera: *Coxiella*
Rickettsiella

Order VII: Methylococcales
Family: Methylococcaceae
Genera: *Methylobacter*
Methylocaldum
Methylococcus
Methylomicrobium
Methylomonas
Methylosphaera

Order VIII: Oceanospirillales
Family I: Oceanospirillaceae
Genera: *Balneatrix*
Fundibacter
Marinomonas
Marinospirillum
Neptunomonas
Oceanospirillum

Order IX: Pseudomonadales
Family I: Pseudomonadaceae
Genera: *Azomonas*
Azotobacter
Cellvibrio
Chryseomonas
Flavimonas
Lampropedia
Mesophilobacter
Morococcus
Oligella
Pseudomonas
Rhizobacter
Rugamonas
Serpens
Thermoleophilum
Xylophilus
Family II: Moraxellaceae
Genera: *Acinetobacter*
Moraxella
Psychrobacter

Order X: Alteromonadales
Family: Alteromonadaceae
Genera: *Allishewanella*
Alteromonas
Colwellia
Ferrimonas
Idiomarina
Marinobacter
Marinobacterium
Microbulbifer
Moritella
Pseudoalteromonas
Shewanella

Order XI: Vibrionales
Family: Vibrionaceae
Genera: *Allomonas*
Enhydrobacter
Listonella
Photobacterium
Salvinivibrio
Vibrio

Order XII: Aeromonadales
Family I: Aeromonadaceae
Genera: *Aeromonas*
Oceanomonas
Tolumonas
Family II: Succinivibrionaceae
Genera: *Anaerobiospirillum*
Ruminobacter
Succinomonas
Succinovibrio

Order XIII: Enterobacteriales
Family: Enterobacteriaceae
Genera: *Alterococcus*
Arsenophonus
Brenneria
Buchnera

Budvicia
Buttiauxella
Calymmatobacterium
Cedecea
Citrobacter
Edwardsiella
Enterobacter
Erwinia
Escherichia
Ewingella
Hafnia
Klebsiella
Kluyvera
Leclercia
Leminorella
Moellerella
Morganella
Obesumbacterium
Pantoea
Pectobacterium
Photorhabdus
Plesiomonas
Pragia
Proteus
Providencia
Rahnella
Saccharobacter
Salmonella
Serratia
Shigella
Sodalis
Tatumella
Trabulsiella
Wigglesworthia
Xenorhabdus
Yersinia
Yokenella

Order XIV: Pasteurellales
Family: Pasteurellaceae
Genera: *Actinobacillus*
Haemophilus
Lonepinella
Pasteurella
Mannheimia
Phocoenobacter

Class IV: Delta-proteobacteria
Order I: Desulfurales
Family: Desulfurellaceae
Genera: *Desulfurella*
Hippea

Order II: Desulfovibrionales
Family I: Desulfovibrionaceae
Genera: *Biophila*
Desulfovibrio
Lawsonia
Family II: Desulfomicrobiaceae
Genus: *Desulfomicrobium*
Family III: Desulfonalobiaceae
Genera: *Desulfohalobium*
Desulfomonas
Desulfonatronovibrio

Order III: Desulfobacterales
Family I: Desulfobacteraceae
Genera: *Desulfobacter*
Desulfobacterium
Desulfobacula
Desulfococcus
Desulfofaba
Desulfofrigus
Desulfonema
Desulfosarcina
Desulfospira
Desulfocella
Desulfotalea
Desulfotignum
Family II: Desulfobulbaceae
Genera: *Desulfobulbus*
Desulfocapsa
Desulfofustis
Desulforhopalus
Family III: Nitrospinaceae
Genera: *Nitrospina*
Desulfobacca
Desulfomonile

Order IV: Desulfuromonadales
Family I: Desulfomonadaceae
Genera: *Desulfuromonas*
Desulfuromusa
Family II: Geobacteraceae
Genus: *Geobacter*
Family III: Pelobacteriaceae
Genera: *Melonomonas*
Pelobacter
Trichlorobacter

Order V: Syntrophobacterales
Family I: Synthrophobacteraceae
Genera: *Desulfacinum*
Synthrophobacter
Desulforhabdus
Desulfovirga
Thermodesulforhabdus

Order VI: Bdellovibrionales
Family: Bdellovibrionaceae
Genera: *Bacteriovorax*
Bdellovibrio
Micavibrio
Vampirovibrio

Order VII: Myxococcales
Family I: Myxococcaceae
Genera: *Angiococcus*
Myxococcus
Family II: Archangiaceae

Genus: *Archangium*
Family III: Cystobacteraceae
Genera: *Cystobacter*
Melittangium
Stigmatella
Family IV: Polyangiaceae
Genera: *Chondromyces*
Nannocystis
Polyangium
Class V: Epsilon-proteobacteria
Order: Campylobacterales
Family I: Campylobacteraceae
Genera: *Arcobacter*
Campylobacter
Sulfurospirillum
Thiovulum
Family II: Helicobacteraceae
Genera: *Helicobacter*
Wolinella
Phylum XIII: Firmicutes
Class I: Clostridia
Order I: Clostridiales
Family I: Clostridiaceae
Genera: *Acetivibrio*
Acidaminobacter
Anaerobacter
Caloramator
Clostridium
Coprobacillus
Natronineola
Oxobacter
Sarcina
Sporobacter
Thermobrachium
Thermohalobacter
Tindallia
Family II: Lachnospiraceae
Genera: *Acetitomaculum*
Anaerofilum
Butyrivibrio
Catenibacterium
Catonella
Coprococcus
Johnsonella
Lachnospira
Pseudobutyrivibrio
Roseburia
Ruminococcus
Sporobacterium
Family III: Peptostreptococcaceae
Genera: *Filifacter*
Fusibacter
Helcococcus
Micromonas
Peptostreptococcus
Tissierella
Family IV: Eubacteriaceae
Genera: *Acetobacterium*
Anaerovorax
Eubacterium
Mogibacterium
Pseudoramibacter
Family V: Peptococcaceae
Genera: *Anaeroaarcus*
Anaerosinus
Anaerovibrio
Carboxydothermus
Centipeda
Dehalobacter
Dendrosporobacter
Desulfitobacterium
Desulfonispora
Desulfosporosinus
Desulfotomaculum
Mitsuokella
Peptococcus
Propionispira
Succinispira
Syntrophobotulus
Thermaterrobacterium
Family VI: Heliobacteriaceae
Genera: *Heliobacterium*
Heliobacillus
Heliophilum
Heliorestis
Family VII: Acidaminococcaceae
Genera: *Acetonema*
Anaeromonas
Acidaminococcus
Dialister
Megasphaera
Papillibacter
Pectinatus
Pascolarctobacterium
Quinella
Schwartzia
Selenomonas
Sporomusa
Succiniclasticum
Veillonella
Zymophilus
Family VII: Synthrophomonadaceae
Genera: *Acetogenium*
Aminobacterium
Aminomonas
Anaerobaculum
Anaerobranca
Caldicellulosiruptor
Dethiosulfovibrio
Pelospora
Synthrophomonas
Synthrophospora

Synthrophothermus
Thermoaerobacter
Thermoaerovibrio
Thermohydrogenium
Thermosynthropha

Order II: Thermoanaerobacteriales
Family: Thermoanaerobacteriaceae
Genera: *Ammonifex*
Carboxydobrachium
Coprothermobacter
Moorella
Sporotomaculum
Thermacetogenium
Thermoanaerobacter
Thermoanaerobacterium
Thermoanaerobium

Order III: Haloanaerobiales
Family I: Haloanaerobiaceae
Genera: *Haloanaerobium*
Halocella
Halothermothrix
Natroniella

Family II: Halobacteroidaceae
Genera: *Acetohalobium*
Haloanaerobacter
Halobacteroides
Orenia
Sporohalobacter

Class II: Mollicutes
Order I: Mycoplasmatales
Family: Mycoplasmataceae
Genera: *Mycoplasma*
Eperythrozoon
Haemobartonella
Ureaplasma

Order II: Entoplasmatales
Family I: Entoplasmatales
Genera: *Entomoplasma*
Mesoplasma

Family II: Spiroplasmataceae
Genus: *Spiroplasma*

Order III: Acholeplasmatales
Family: Acholeplasmataceae
Genus: *Acholeplasma*

Order IV: Anaeroplasmatales
Family: Anaeroplasmataceae
Genera: *Anaeroplasma*
Asteroleplasma

Order V: Incertae sedis (uncertain taxonomy)
Family: Erysipelothrichaceae
Genera: *Bulleidia*
Erysipelothrix
Holdemania
Solobacterium

Class III: Bacilli
Order I: Bacillales
Family I: Bacillaceae
Genera: *Amphibacillus*
Anoxybacillus
Bacillus
Exiguobacterium
Gracilibacillus
Halobacillus
Saccharococcus
Salibacillus
Virgibacillus

Family II: Planococcaceae
Genera: *Filibacter*
Kurthia
Planococcus
Sporosarcina

Family III: Caryophanaceae
Genus: *Caryophanon*

Family IV: Listeriaceae
Genera: *Brochothrix*
Listeria

Family V: Staphylococcaceae
Genera: *Gemella*
Macrococcus
Salinicoccus
Staphylococcus

Family VI: Sporolactobacillaceae
Genera: *Marinococcus*
Sporolactobacillus

Family VII: Paenibacillaceae
Genera: *Ammoniphilus*
Aneurinibacillus
Brevibacillus
Oxalophagus
Paenibacillus
Thermicanus
Thermobacillus

Family VIII: Alicyclobacillaceae
Genera: *Alicyclobacillus*
Pasteuria
Sulfobacillus

Family IX: Thermoactinomycetaceae
Genus: *Thermoactinomyces*

Order II: Lactobacillales
Family I: Lactobacillaceae
Genera: *Lactobacillus*
Paralactobacillus
Pediococcus

Family II: Aerococcaceae
Genera: *Abiotrophia*
Aerococcus
Dolosicoccus
Eremococcus
Facklamia
Globicatella
Tetragenococcus
Ignavigranum

Family III: Carnobacteriaceae
Genera: *Agitococcus*
Alloiococcus
Carnobacterium
Desemzia
Dolosigranulum
Granulicatella
Lactosphaera
Trichococcus
Family IV: Enterococcaceae
Genera: *Atopobacter*
Enterococcus
Melissococcus
Tetragenococcus
Vagococcus
Family V: Leuconostocaceae
Genera: *Leuconostoc*
Oenococcus
Weissella
Family VI: Streptococcaceae
Genera: *Lactococcus*
Streptococcus
Family VII: Incertae sedis
Genera: *Acetoanaerobium*
Oscillospira
Syntrophococcus
Phylum XIV: Actinobacteria
Class: Actinobacteria
Order I: Acidimicrobiales
Family: Acidimicrobiaceae
Genus: *Acidimicrobium*
Order II: Rubrobacterales
Family: Rubrobacteraceae
Genus: *Rubrobacter*
Order III: Coriobacteridae
Family: Coriobacteriaceae
Genera: *Atophobium*
Collinsella
Coriobacterium
Cryptobacterium
Denitrobacterium
Eggerthella
Slakia
Order IV: Sphaerobacterales
Family: Sphaerobacteraceae
Genus: *Sphaerobacter*
Order V: Actinomycetales
Suborder I: Actinomycineae
Family: Actinomycetaceae
Genera: *Actinobaculum*
Actinomyces
Arcanobacterium
Mobiluncus
Suborder II: Micrococcineae
Family I: Micrococcaceae
Genera: *Arthrobacter*
Kocuria
Micrococcus
Nesterenkonia
Renibacterium
Rothia
Stomatococcus
Family II: Bogoricellaceae
Genus: *Bogoriella*
Family III: Rarobacteriaceae
Genus: *Rarobacter*
Family IV: Sanguibacteraceae
Genus: *Sanguibacter*
Family V: Brevibacteriaceae
Genus: *Brevibacterium*
Family VI: Cellulomonadaceae
Genera: *Cellulomonas*
Oerskovia
Family VII: Dermabacteraceae
Genera: *Brachybacterium*
Dermabacter
Family VIII: Dermatophilaceae
Genus: *Dermatophilus*
Family IX: Dermacoccaceae
Genera: *Dermacoccus*
Demetria
Kytococcus
Family X: Intrasporangiaceae
Genera: *Intrasporangium*
Janibacter
Ornithinicoccus
Ornithinomicrobium
Terrabacter
Terracoccus
Tetrasphaera
Family IX: Jonesiaceae
Genus: *Jonesia*
Family XII: Microbacteriaceae
Genera: *Agrococcus*
Agromyces
Aureobacterium
Clavibacter
Cryobacterium
Curtobacterium
Frigoribacterium
Leifsonia
Leucobacter
Microbacterium
Rathayibacter
Subtercola
Family XIII: Beutenbergiaceae
Genus: *Beutenbergia*
Family XIV: Promicromonosporaceae
Genus: *Promicromonospora*
Suborder III: Corynebacterineae
Family I: Corynebacteriaceae
Genus: *Corynebacterium*

Family II: Dietziaceae
Genus: *Dietzia*
Family III: Gordoniaceae
Genera: *Gordonia*
Skermania
Family IV: Mycobacteriaceae
Genus: *Mycobacterium*
Family V: Nocardiaceae
Genera: *Nocardia*
Rhodococcus
Family VI: Tsukamurellaceae
Genus: *Tsukamurella*
Family VII: Williamsiaceae
Genus: *Williamsia*
Suborder IV: Micromonosporineae
Family: Micromonosporaceae
Genera: *Actinoplanes*
Catellatospora
Catenuloplanes
Couchioplanes
Dactylosporangium
Micromonospora
Pilimelia
Spirilliplanes
Verrucosispora
Suborder V: Propionibacterineae
Family I: Propionibacteriaceae
Genera: *Luteococcus*
Microlunatus
Propionibacterium
Propioniferax
Tessaracoccus
Family II: Nocardioidaceae
Genera: *Aeromicrobium*
Friedmaniella
Hongia
Kribella
Micropruina
Marmoricola
Nocardiodes
Suborder VI: Pseudonocardineae
Family I: Pseudonocardiaceae
Genera: *Actinoalloteichus*
Actinopolyspora
Amycolatopsis
Kibdelosporangium
Kutzneria
Prauserella
Pseudonocardia
Saccharomonospora
Saccharopolyspora
Streptoalloteichus
Thermobispora
Thermocrispum
Family II: Actinosynnemataceae
Genera: *Actinokineospora*
Actinosynnema
Lentzea
Saccharothrix
Suborder VII: Streptomycineae
Family: Streptomycetaceae
Genera: *Streptomyces*
Streptoverticillium
Suborder VIII: Streptosporangineae
Family I: Streptosporangiaceae
Genera: *Acrocarpospora*
Herbidospora
Microbispora
Microtetraspora
Nonomuraea
Planobispora
Planomonospora
Planopolyspora
Planotetraspora
Streptosporangium
Family II: Nocardiopsaceae
Genera: *Nocardiopsis*
Thermobifida
Family III: Thermomonosporaceae
Genera: *Actinomadura*
Spirillospora
Thermomonospora
Suborder IX: Frankineae
Family I: Frankiaceae
Genus: *Frankia*
Family II: Geodermatophilaceae
Genera: *Blastococcus*
Geodermatophilus
Modestobacter
Family III: Microsphaeraceae
Genus: *Microsphaera*
Family IV: Sporichthyaceae
Genus: *Sporichthya*
Family V: Acidothermaceae
Genus: *Acidothermus*
Family VI: Kineosporaceae
Genera: *Cryptosporangium*
Kineococcus
Kineosporia
Suborder X: Glycomycineae
Family: Glycomycetaceae
Genus: *Glycomyces*
Order VI: Bifidobacteriales
Family I: Bifidobacteriaceae
Genera: *Bifidobacterium*
Falcivibrio
Gardnerella
Family II: Unknown affiliation
Genera: *Actinobispora*
Actinocorallia
Excellospora
Pelczaria

Turicella
Phylum XV: Planctomycetes
Class: Planctomycetacia
Order: Planctomycetales
Family: Planctomycetaceae
Genera: *Gemmata*
Isosphaera
Pirellula
Planctomyces
Phylum XVI: Chlamydiae
Class: Chlamydiae
Order: Chlamydiales
Family I: Chlamydiaceae
Genera: *Chlamydia*
Chlamydophila
Family II: Parachlamydiaceae
Genus: *Parachlamydia*
Family III: Simkaniaceae
Genus: *Simkania*
Family IV: Waddliaceae
Genus: *Waddlia*
Phylum XVII: Spirochetes
Class: Spirochaetes
Order: Spirochaetales
Family I: Spirochaetaceae
Genera: *Borrelia*
Brevinema
Clevelandina
Cristispira
Diplocalyx
Hollandina
Pillotina
Spirochaeta
Treponema
Phylum XVIII: Fibrobacteres
Class: Fibrobacteres
Order: Fibrobacterales
Family: Fibrobacteraceae
Genus: *Fibrobacter*
Phylum XIX: Acidobacteria
Class: Acidobacteria
Order: Acidobacteriales
Family: Acidobacteriaceae
Genera: *Acidobacterium*
Geothrix
Holophaga
Phylum XX: Bacteroidetes
Class I: Bacteroidetes
Order: Bacteroidales
Family I: Bacteroidaceae
Genera: *Acetofilamentum*
Acetomicrobium
Acetothermus
Anaerorhabdus
Bacteroides
Megamonas
Family II: Rikenellaceae
Genera: *Marinilabilia*
Rikenella
Family III: Porphyromonadaceae
Genera: *Dysgonomonas*
Porphyromonas
Family IV: Prevotellaceae
Genus: *Prevotella*
Class II: Flavobacteria
Order: Flavobacteriales
Family I: Flavobacteriaceae
Genera: *Bergeyella*
Capnocytophaga
Cellulophaga
Chryseobacterium
Coenonia
Empedobacter
Flavobacterium
Gelidibacter
Ornithobacterium
Polaribacter
Psychroflexus
Riemerella
Saligentibacter
Weeksella
Family II: Myroideaceae
Genera: *Myroides*
Psychromonas
Family III: Blattabacteriaceae
Genus: *Blattabacterium*
Class III: Sphingobacteria
Order: Sphingobacteriales
Family I: Sphingobacteriaceae
Genera: *Pedobacter*
Sphingobacterium
Family II: Saprospiraceae
Genera: *Haliscomenobacter*
Lewinella
Saprospira
Family III: Flexibacteraceae
Genera: *Cyclobacterium*
Cytophaga
Dyadobacter
Flectobacillus
Flexibacter
Hymenobacter
Meniscus
Microscilla
Runella
Spirosoma
Sporocytophaga
Family IV: Flammeovirgaceae
Genera: *Flammeovirga*
Flexithrix
Persicobacter
Thermonema

Family V: Crenotrichaceae
Genera: *Chitinophaga*
Crenothrix
Rhodothermus
Toxothrix
Phylum XXI: Fusobacteria
Class: Fusobacteria
Order: Fusobacteriales
Family: Fusobacteriaceae
Genera: *Cetobacterium* (incertae sedis)
Fusobacterium
Ilyobacter
Leptothrichia
Propionigenium
Sebaldella
Streptobacillus
Phylum XXII: Verrucomicrobia
Class: Verrucomicrobiae
Order: Verrucomicrobiales
Family I: Verrucomicrobiaceae
Genera: *Prosthecobacter*
Verrucomicrobium
Family II: Xiphinematobacteriaceae
Genus: *Xiphinematobacter*
Phylum XXIII: Dictyoglomas
Class: Dictyoglomi
Order: Dictyoglomales
Family: Dictyoglomaceae
Genus: *Dictyoglomus*

DOMAIN ARCHAEA
Phylum I: Crenarchaeota
Class: Thermoprotei
Order I: Thermoproteales
Family I: Thermoproteaceae
Genera: *Caldivirga*
Pyrobaculum
Thermocladium
Thermoproteus
Family II: Thermofilaceae
Genus: *Thermofilum*
Order II: Desulfurococcales
Family I: Desulfurococcaceae
Genera: *Acidolobus*
Aeropyrum
Desulfurococcus
Ignicoccus
Staphylothermus
Stetteria
Sulfophobococcus
Thermodiscus
Thermosphaera
Family II: Pyrodictiaceae
Genera: *Hyperthermus*
Pyrodictium
Pyrolobus
Order III: Sulfolobales
Family: Sulfolobaceae
Genera: *Acidianus*
Metallosphaera
Stygiolobus
Sulfolobus
Sulfurisphaera
Sulfurococcus
Phylum II: Euryarchaeota
Class I: Methanobacteria
Order: Methanobacteriales
Family I: Methanobacteriaceae
Genera: *Methanobacterium*
Methanobrevibacter
Methanosphaera
Methanothermobacter
Family II: Methanothermaceae
Genus: *Methanothermus*
Class II: Methanococci
Order I: Methanococcales
Family I: Methanococcaceae
Genera: *Methanococcus*
Methanothermococcus
Family II: Methanocaldococcaceae
Genera: *Methanocaldococcus*
Methanotorris
Order II: Methanomicrobiales
Family I: Methanomicrobiaceae
Genera: *Methanoculleus*
Methanofollis
Methanogenium
Methanolacinia
Methanomicrobium
Methanoplanus
Family II: Methanocorpusculaceae
Genus: *Methanocorpusculum*
Family III: Methanospirillaceae
Genus: *Methanocalculus* (incertae sedis)
Methanospirillum
Order III: Methanosarcinales
Family I: Methanosarcinaceae
Genera: *Methanococcoides*
Methanohalobium
Methanolobus
Methanomicrococcus
Methanosalsum
Methanosarcina
Family II: Methanosaetaceae
Genus: *Methanosaeta*
Class III: Halobacteria
Order: Halobacteriales
Family: Halobacteriaceae
Genera: *Haloarcula*
Halobacterium
Halobaculum
Halococcus

Haloferax
Halogeometricum
Halorhabdus
Halorubrum
Haloterrigena
Natrialba
Natrinema
Natronobacterium
Natronococcus
Natronomonas
Natronorubrum

Class IV: Thermoplasmata
Order: Thermoplasmatales
Family I: Thermoplasmataceae
Genus: *Thermoplasma*
Family II: Picrophilaceae
Genus: *Picrophilus*
Family III: Ferroplasmataceae
Genus: *Ferroplasma*

Class V: Thermococci
Order: Thermococcales
Family: Thermococcaceae
Genera: *Paleococcus*
Pyrococcus
Thermococcus

Class VI: Archaeoglobi
Order: Archaeoglobales
Family: Archaeoglobaceae
Genera: *Archaeoglobus*
Ferroglobus

Class VII: Methanopyri
Order: Methanopyrales
Family: Methanopyraceae
Genus: *Methanopyrus*

APPENDIX VI

Viruses of Animals and Plants

(main hosts in parentheses)

VIRUSES OF VERTEBRATE AND INVERTEBRATE ANIMALS

1. Viruses with Double-Strand DNA (dsDNA)

Family: Adenoviridae		
Genera:	*Atendovirus*	(infects bovines)
	Aviadenovirus	(infects birds)
	Mastadenovirus	(infects mammals)
	Siadenovirus	(infects frogs, turkeys)
Family: Baculoviridae		
Genera:	*Granulovirus*	
	Nucleopolyhedrovirus	
Family: Herpesviridae		
Subfamily: Alphaherpesvirinae		
Genera:	*Simplexvirus*	(humans, primates)
	Varicellovirus	(humans, chickenpox)
Subfamily: Betaherpesvirinae		
Genera:	*Cytomegalovirus*	(humans)
	Muromegalovirus	(mice)
	Roseolavirus	(humans)
Subfamily: Gammaherpesvirinae		
Genera:	*Lymphocryptovirus*	(humans, Epstein-Barr)
	Rhadinovirus	(humans, primates)
Family: Iridoviridae		(insects, fish, frogs)
Family: Papilloviridae		(humans and numerous other species)
Family: Polyomaviridae		(humans, primates, birds)
Family: Poxviridae		
Subfamily: Chondropoxvirinae		
Genera:	*Avipoxvirus*	(fowl)
	Capripoxvirus	(sheep)
	Leporipoxvirus	(rabbits)
	Molluscipoxvirus	(mollusks)
	Orthopoxvirus	(humans)
	Parapoxvirus	(humans)
	Suipoxvirus	(swine)
	Yatapoxvirus	(monkeys)
Subfamily: Entomopoxvirinae		
Genera:	*Entomopoxvirus* A, B, and C	(insects)

2. Viruses with Single-Strand DNA (ssDNA)
    Family: Circoviridae
        Genus: *Circovirus* (chickens, humans)
    Family: Parvoviridae
        Subfamily: Parvovirinae
            Genera: *Erythrovirus* (humans)
            Parvovirus (canines, felines, mice)
        Subfamily Desnovirinae (insects)
3. Viruses with Double-Strand RNA (dsRNA)
    Family: Birnaviridae
        Genera: *Aquabirnavirus* (fish)
        Avibirnavirus (fowl)
        Entomobirnavirus (insects)
    Family: Reoviridae
        Genera: *Aquaaerovirus* (fish and crustaceans)
        Coltivirus (various vertebrates)
        Cypovirus (insects)
        Orbivirus (sheep)
        Orthoreovirus (humans, dogs, bovines, birds)
        Rotavirus (humans and other mammals)
4. Viruses with Single-Strand RNA (ssRNA)
    Family: Arenaviridae (humans)
    Family: Bornaviridae (horses)
    Family: Bunyaviridae (various vertebrates)
        Genera: *Bunyavirus* (humans, primates)
        Hantavirus (humans, rodents)
        Nairovirus (sheep)
        Phlebovirus (vertebrates)
    Family: Caliciviridae (humans, rabbits, swine)
    Family: Coronaviridae
        Genera: *Coronavirus* (birds, humans)
        Torovirus (horses, bovines)
    Family: Filoviridae
Genus: *Filovirus* (humans)
    Family: Flaviviridae
        Genera: *Flavivirus* (humans)
        Hepacivirus (humans)
        Pestivirus (sheep, bovines, swine)
    Family: Orthomyxoviridae
        Genera: *Influenzavirus* A (seashore birds, various vertebrates)
        Influenzavirus B (humans)
        Influenzavirus C (humans)
        Thogotovirus (salmon)
    Family: Paramyxoviridae
        Subfamily: Paramyxoviriniae
            Genera: *Morbillivirus* (humans, canines)
            Respirovirus (humans)
            Rubulavirus (humans, fowl)
        Subfamily: Pneumovirinae
            Genera: *Metapneumovirus* (turkeys)
            Pneumovirus (humans, bovines, mice)
    Family: Picornaviridae
        Genera: *Aphthovirus* (bovines)
        Cardiovirus (mice)
        Enterovirus (humans, primates)

Hepatovirus (humans, primates)
Parechovirus (humans)
Rhinovirus (humans, horses)

Family: Rhabdoviridae

Genera: *Ephemerovirus* (bovines)
Lyssavirus (mammals)
Norirhabdovirus (fish)
Vesiculovirus (vertebrates)

Family: Tetraviridae (unknown)

Family: Togaviridae

Genera: *Alphavirus* (vertebrates)
Rubivirus (humans)

5. RNA Viruses That Replicate Using a DNA Intermediate

Family: Retroviridae

Genera: *Alpharetrovirus* (birds)
Betaretrovirus (monkeys)
Deltaretrovirus (humans, bovines)
Epsilonretrovirus (fish)
Gammaretrovirus (mice)
Lentivirus (humans, primates, cats, ruminants)

Mammalian type C retrovirus (mammals)

Spumavirus (humans)

6. DNA Viruses That Replicate Using an RNA Intermediate

Family: Hepadnaviridae

Genera: *Avihepadnavirus* (ducks)
Orthohepadnavirus (humans, woodchucks)

VIRUSES OF PLANTS

1. Viruses with dsDNA

Family: Reoviridae

Genera: *Fijivirus*
Oryzavirus
Phytoreovirus

2. Viruses with ssDNA

Family: Geminiviridae

Genera: *Begomovirus*
Curtovirus
Mastrevirus

3. Viruses with ssRNA

Family: Bromoviridae

Genera: *Alfamovirus*
Bromovirus

Family: Bunyaviridae

Genus: *Tospovirus*

Family: Comoviridae

Genera: *Comovirus*
Fabavirus
Nepovirus

Family: Luteoviridae

Genera: *Enamovirus*
Luteovirus
Polerovirus

Family: Potyviridae

Genera: *Ipomovirus*
Macluravirus

Potyvirus
Rymovirus
Tritimovirus
Family: Rhabdoviridae
Genera: *Cytorhabdovirus*
Nucleorhabdovirus
Family: Tombusviridae
Genera: *Aureusvirus*
Avenavirus
Carmovirus
Dianthovirus
Machlomovirus
Necrovirus
Panicovirus
Tombusvirus
4. DNA Viruses That Replicate Using an RNA Intermediate
Family: Caulimoviridae
Genera: *Badnavirus*
Caulimovirus

APPENDIX VII

MAJOR HUMAN DISEASES CAUSED BY MICROORGANISMS

Bacterial Diseases	Microorganism
anthrax	*Bacillus anthracis*
botulism	*Clostridium botulinum*
cholera	*Vibrio cholerae*
dental caries	*Streptococcus mutans*
diphtheria	*Corynebacterium diphtheriae*
gangrene	*Clostridium perfringens*
gastritis	*Helicobacter pylori*
gastroenteritis	*Bacillus cereus, Campylobacter jejuni, Clostridium perfringens, Escherichia coli, Staphylococcus aureus, Vibrio* species, *Yersinia enterocolitica*
gonorrhea	*Neisseria gonorrhoeae*
legionellosis (Legionnaires' disease)	*Legionella pneumophila*
leprosy	*Mycrobacterium leprae*
listeriosis	*Listeria monocytogenes*
Lyme disease	*Borrelia burgdorferi*
meningitis	*Haemophilus influenzae, Neisseria meningitides, Streptococcus pneumoniae*
necrotizing fasciitis	*Streptococcus pyogenes*
otitis media	*Moraxella catarrhalis, Streptococcus pneumoniae*
pelvic inflammatory disease	*Neisseria gonorrhoeae*
peptic ulcers	*Helicobacter pylori*
plague	*Yersinia pestis*
pneumonia	*Haemophilus influenzae; Streptococcus pneumoniae*
pyelonephritis	*Escherichia coli*
Q fever	*Coxiella burnetti*
rheumatic fever	*Streptococcus pyogenes*
Rocky Mountain spotted fever	*Rickettsia rickettsii*
salmonellosis	*Salmonella enterica*
scarlet fever	*Streptococcus pyogenes*
strep throat	*Streptococcus pyogenes*
syphilis	*Treponema pallidum*
tetanus	*Clostridium tetani*

(continues)

Bacterial Diseases	Microorganism
tuberculosis	*Mycobacterium tuberculosis*
tularemia	*Francisella tularensis*
typhoid fever	*Salmonella enterica (S. typhi)*
urethritis	*Mycoplasma hominis, Ureaplasma ureolyticum*

Fungal Diseases	Microorganism
aspergillosis	*Aspergillus fumigates*
blastomycosis	*Blastomyces dermatidis*
candidiasis	*Candida albicans*
coccidioidomycosis	*Coccidioides immitis*
histoplasmosis	*Histoplasma capsulatum*
meningitis	*Cryptococcus neoformans*
pneumonia	*Pneumocystis jiroveci*
ringworm	*Microsporum* species, *Trichophyton* species
sporotrichosis	*Sporothrix schenckii*

Protozoal Diseases	Microorganism
African trypanosomiasis	*Trypanosoma brucei*
amoebic dysentery	*Entameoba histolytica*
babesiosis	*Babesia microti*
cryptosporidiosis	*Cryptosporidium parvum*
Cyclospora infection	*Cyclospora cayetanensis*
giardiasis	*Giardia lamblia*
leishmaniasis	*Leishmania* species
malaria	*Plasmodium* species
paralytic shellfish poisoning	*Alexandrium* species
toxoplasmosis	*Toxoplasma gondii*

Viral Diseases	Virus	Virus Group
AIDS	human immunodeficiency	retrovirus
chicken pox	varicella	herpesvirus
common cold	rhinovirus	picornavirus
	coronavirus	coronavirus
cold sores	herpes simplex	herpesvirus
dengue fever	dengue fever	togavirus
encephalitis	California, eastern equine, Japanese B, St. Louis, and western equine encephalitis	bunyavirus and togavirus
gastroenteritis	human enteric calicivirus	calicivirus
	Norwalk (norovirus)	calicivirus
	rotavirus	reovirus
genital herpes	herpes simplex type 2	herpesvirus
hantavirus pulmonary syndrome	hantavirus	bunyavirus
hemorrhagic fever	Ebola, Lassa, Marburg, Rift Valley	bunyavirus
hepatitis A	hepatitis A	picornavirus

Viral Diseases	Virus	Virus Group
hepatitis B	hepatitis B	hepadnavirus
hepatitis C	hepatitis C	flavivirus
hepatitis D	hepatitis D (infects only with hepatitis B)	deltavirus
hepatitis E	hepatitis E	calicivirus
influenza	influenza A, influenza B	orthomyxovirus
measles	measles	paramyxovirus
mononucleosis	Epstein-Barr	herpesvirus
mumps	mumps	paramyxovirus
poliomyelitis	polio	picornavirus
rabies	rabies	rhabdovirus
RSV infection	respiratory syncytial virus	paramyxovirus
rubella	rubella	togavirus
shingles	latent varicella-zoster	herpesvirus
smallpox	smallpox (eradicated)	poxvirus

INDEX

Note: Page numbers in **boldface** indicate main entries; *italic* page numbers denote illustrations. C refers to color insert pages.

C

E

F

Q

R

T

U

V